蜂胶研究

胡福良 等著

浙江大学出版社

PROPOLIS

Fu-Liang Hu et al.

Zhejiang University Press

《蜂胶研究》作者团队

胡福良

张翠平　朱　威　玄红专　李雅晶　王　凯

卢媛媛　黄　帅　张江临　曹雪萍　申小阁

张　挺　程晓雨　郑宇斐　李英华　郑火青

图书在版编目（CIP）数据

蜂胶研究 / 胡福良等著. —杭州：浙江大学出版社，2019.5(2019.7重印)
ISBN 978-7-308-18769-5

Ⅰ.①蜂… Ⅱ.①胡… Ⅲ.①蜂胶—研究 Ⅳ.①S896.6

中国版本图书馆CIP数据核字（2018）第283472号

蜂胶研究

胡福良　等著

责任编辑	杜玲玲　张　鸽
责任校对	梁容
封面设计	闳江文化
出版发行	浙江大学出版社
	（杭州天目山路148号　邮政编码310007）
	（网址：http://www.zjupress.com）
排　　版	杭州中大图文设计有限公司
印　　刷	浙江省邮电印刷股份有限公司
开　　本	787mm×1092mm　1/16
印　　张	48.5
字　　数	1242千
版 印 次	2019年5月第1版　2019年7月第2次印刷
书　　号	ISBN 978-7-308-18769-5
定　　价	398.00元

版权所有　翻印必究　　印装差错　负责调换

浙江大学出版社市场运营中心联系方式：0571-88925591；http://zjdxcbs.tmall.com

序 一

蜜蜂,大自然里的小精灵,在地球上已生活了数千万年。从远古时代的攀岩取蜜到如今的现代养蜂,蜜蜂与人类社会的关系源远流长,与人类文明的发展息息相关。作为大自然生态系统中的重要一员,蜜蜂的"月下老人"作用促进了植物物种多样性的形成和发展,提高了农作物的产量和品质。同时,蜜蜂也为人类提供了营养丰富、功能独特的蜂产品。

蜂胶作为一种古老的蜂产品,其应用历史十分悠久,古埃及、古希腊、古罗马人早就认识到蜂胶的药理特性并且将其作为药物广泛应用。作为一种新兴的蜂产品,从药学角度,它是一味动植物双性的中药,已收录于《中华人民共和国药典》;从化学成分角度,它含有20余类500多种天然成分,具有高度的复杂性和可变性;从生物学活性角度,它具有抗菌、抗炎、抗氧化、抗肿瘤、调节血脂血糖等广泛且显著的作用。近年来,蜂胶已成为蜂产品研究与开发的一大热点,如火如荼,方兴未艾。

我国是世界第一养蜂大国,也是世界第一蜂产品生产和出口大国。近几十年来,我国的蜂胶产业从无到有,从小到大,发展迅猛,蜂胶产量和出口量均跃居世界首位。但是,我国蜂胶领域"重应用、轻基础"的现象较为突出,蜂胶的质量控制、生物学活性及其作用机制等基础研究严重滞后,亟待科技工作者潜心研究,不断创新。

本书作者胡福良教授是我国著名的蜂学专家。20世纪末,因为蜂产品科研协作,我与他相识于西子湖畔。他的勤勉谦逊和对蜂业科研事业的热爱,给我留下了深刻的印象。他带领的浙江大学科研团队瞄准国内外学科前沿,在蜂胶的基础和应用研究领域取得了大量原创性的科研成果,先后发表了100余篇蜂胶方面的学术论文,获得了20多项授权发明专利,为我国蜂胶产业的科技创新做出了重要贡献。本专著就是胡福良教授团队在蜂胶的生物学活性、质量控制和产品综合开发利用方面取得的原创性研究成果的展示,内容丰富,创新性强,学术水平高,是一部不可多得的蜂产品学术专著。

特此作序,以致祝贺!

中国工程院院士
中国检验检疫科学研究院首席科学家

2018年夏,于北京亦庄

序 二

  蜂产品不仅营养十分丰富、全面,而且具有独特的生理、药理功能,是医食同一、食药同源、食药兼优的特殊物质,是人类永恒的保健食品,千百年来一直受到各国不同民族、性别、年龄消费者的喜爱。

  蜂胶是蜜蜂采集胶源植物树脂等分泌物与其上颚腺、蜡腺等分泌物混合形成的胶黏性物质。作为一种传统的蜂产品,蜂胶的应用历史十分悠久。早在3000多年前,古埃及人就已经认识蜂胶,并将蜂胶用作木乃伊的防腐材料。在2300多年前古希腊科学家亚里士多德的《动物志》、1900多年前的古罗马百科全书《自然史》和1000多年前的阿拉伯医学名著《医典》中都有关于蜂胶的特性及医疗用途的记载。20世纪50年代以来,国内外许多学者对蜂胶进行了较为全面的研究,主要包括蜂胶的植物来源、化学成分、理化特性、生物学活性、安全性及临床应用等。迄今为止,已从蜂胶中分离鉴定出20余类500多种天然成分。研究表明,蜂胶具有抗菌、消炎、抗病毒、抗氧化、抗肿瘤、降血脂、降血糖、增强免疫、保肝护肝、促进创伤修复等广泛的作用。蜂胶在现代医学中得到了越来越多的应用,已被《中华人民共和国药典》收录,因而是一味法定的中药。

  我国是世界第一养蜂大国,蜂群数量、蜂产品产量和出口量均位居世界首位。由于我国土生土长的中华蜜蜂不产蜂胶,所以我国研究、开发蜂胶的历史起步较晚。自20世纪90年代以来,我国对蜂胶的生产技术、化学成分、加工工艺、生物学活性和保健功能等方面的研究逐步展开,蜂胶产业得到迅速的发展。目前,我国蜂胶产量已位居世界首位。蜂胶已成为蜂产品研究和开发的热点,在医药、食品、日化、畜牧业等领域的应用越来越广泛。

  与迅猛发展的蜂胶开发形成鲜明对比的是,我国对蜂胶的研究还不够深入、广泛,特别是基础研究方面尤为薄弱,严重制约了我国蜂胶产业的发展。本书作者浙江大学胡福良教授,博士生导师,系中国养蜂学会副理事长、国家蜂产业技术体系产品加工研究室主任、营养品质评价岗位科学家,是我国蜂学界顶级的学者,长期从事蜂产品的科研与教学工作,主持了多项有关蜂胶基础研究和应用研究的科研项目,带领他的研究团队,瞄准学科前沿,取得了大量原创性的科研成果,解决了长期困扰蜂胶行业的许多技术难题,发表了100余篇蜂胶方面的学术论文,获得了20多项授权发明专利,为我国蜂胶的基础研究与应用研究做出了杰出的贡献,并为我国在世界蜂胶研究领域争得了一席之地。

  2005年8月,胡福良教授所著的国内第一部蜂胶学术专著《蜂胶药理作用研究》由浙江大学出版社出版,受到了专业人士的广泛好评。在此基础上,他们将团队10多年来在蜂胶领域的研究成果写成此鸿篇巨著,全面系统地介绍了国内外蜂胶的植物来源与化学成分、生物学活性、质量控制与标准化、有效成分的提取与产品加工等领域的研究进展,重点展示了该团队在蜂胶的生物学活性、质量控制及产品综合开发利用方面取得的原创性研究成果,包括:蜂胶抗氧化作用,蜂胶对炎症性疾病及炎症微环境的影响,蜂胶对控制糖尿病的作用,蜂胶对血管内皮细胞的作用，蜂胶促创伤修复等生物学活性及其作用机制；蜂胶中 β-葡萄糖苷酶的特性，蜂胶与杨树胶的鉴别方法，蜂胶质量控制与质量评价方法研究；蜂胶与蜂胶残渣中挥发性成分，纳米蜂胶及其降血脂降血糖作用等蜂胶综合开发利用研究。该专著内容丰富全面，研究方法先进，数据翔实可信，创新性强，是目前国内外学术水平最高的一部蜂胶学术专著，是蜂胶及相关领域的广大科研工作者、技术人员、管理者和大专院校师生很好的参考工具书。它的出版必将有助于我国蜂产品基础科学的深入研究，进一步推动我国蜂业科技创新，促进我国蜜蜂产业的可持续发展。

中国养蜂学会理事长
国家蜂产业技术体系首席科学家
吴杰
2018 年夏，于北京香山

目 录

第一章 绪 言 ··· 1
 第一节 蜂胶的应用历史 ··· 1
 第二节 蜂胶的研究历史 ··· 3
 第三节 中国的蜂胶研究与应用 ··· 6

第一篇 国内外蜂胶研究进展

第二章 蜂胶的植物来源与化学成分 ··· 19
 第一节 蜂胶的植物来源 ·· 19
 第二节 蜂胶的化学成分 ·· 23

第三章 蜂胶的生物学活性 ··· 65
 第一节 蜂胶的抗病原微生物作用 ·· 65
 第二节 蜂胶的抗氧化作用 ··· 86
 第三节 蜂胶的抗炎作用 ·· 99
 第四节 蜂胶的抗肿瘤作用 ·· 111
 第五节 蜂胶的免疫调节作用 ··· 127
 第六节 蜂胶的降血糖和降血脂作用 ·· 139
 第七节 蜂胶的保肝作用 ··· 157
 第八节 蜂胶的促创（烧）伤修复作用 ··· 171
 第九节 蜂胶的其他生物学活性 ··· 183

第四章 蜂胶的质量控制与标准化 ·· 192
 第一节 蜂胶化学成分的多样性 ··· 192
 第二节 蜂胶化学成分的分析方法 ·· 200
 第三节 蜂胶的质量控制与标准化 ·· 215

第五章 蜂胶有效成分的提取与产品加工 ·· 266
 第一节 蜂胶有效成分的提取工艺 ·· 266
 第二节 蜂胶产品的加工与应用 ··· 281

第二篇　蜂胶生物学活性研究

第六章　蜂胶抗氧化作用及其机制研究 ………………………………………………… 313
第一节　三种植物来源蜂胶和杨树胶乙醇提取物成分及其抗氧化活性 ……………… 313
第二节　不同植物来源蜂胶和杨树胶在 RAW264.7 中的抗氧化活性及其分子机制的比较 …………………………………………………………………………… 322

第七章　蜂胶对炎症性疾病及炎症微环境的影响及其作用机制研究 ………………… 346
第一节　中国蜂胶及其植物来源杨树芽提取物对细菌脂多糖诱导小鼠急性肺损伤的影响研究 ……………………………………………………………………… 346
第二节　中国蜂胶对硫酸葡聚糖诱导大鼠溃疡性结肠炎的影响研究 ………………… 353
第三节　中国蜂胶对肠道上皮细胞 Caco-2 屏障功能的影响及其机制研究 ………… 360
第四节　中国蜂胶和巴西蜂胶对巨噬细胞免疫调节作用及其机制研究 ……………… 371

第八章　蜂胶对糖尿病的作用及其机制研究 …………………………………………… 391
第一节　蜂胶改善 T1DM 大鼠效果的研究 …………………………………………… 391
第二节　蜂胶改善 T2DM 大鼠效果的研究 …………………………………………… 410
第三节　蜂胶对 T2DM 大鼠肾病的作用机制研究 …………………………………… 428

第九章　蜂胶对血管内皮细胞的作用及其机制研究 …………………………………… 451
第一节　去除血清和生长因子条件下蜂胶对血管内皮细胞的影响 …………………… 451
第二节　蜂胶对脂多糖诱导的血管内皮细胞的影响 …………………………………… 470
第三节　蜂胶对 ox-LDL 诱导的血管内皮细胞的影响 ………………………………… 476

第十章　蜂胶促创伤修复机制研究 ……………………………………………………… 488
第一节　两种植物来源蜂胶提取物的成分分析 ………………………………………… 488
第二节　蜂胶对 L929 细胞增殖、迁移及胶原等基因表达的影响 …………………… 492
第三节　蜂胶对 H_2O_2 致 L929 细胞损伤的保护作用及可能机制 …………………… 499

第三篇　蜂胶质量控制研究

第十一章　蜂胶中 β-葡萄糖苷酶的研究 ………………………………………………… 525
第一节　蜂胶中 β-葡萄糖苷酶的检测 ………………………………………………… 525
第二节　蜂胶中 β-葡萄糖苷酶的性质及其在成分转化中的作用 …………………… 530
第三节　储存条件对蜂胶中葡萄糖苷酶活力的影响 …………………………………… 538

第十二章　蜂胶与杨树胶鉴别方法研究 ………………………………………………… 542
第一节　HPLC 指纹图谱鉴别蜂胶与杨树胶 ………………………………………… 542
第二节　以水杨苷和邻苯二酚为参照的蜂胶与杨树胶鉴别方法 ……………………… 547

第十三章　蜂胶质量控制与质量评价方法研究 …………………………………… 567
　第一节　多指标成分定量结合指纹图谱定性的中国蜂胶质量评价方法研究 …… 567
　第二节　基于多指标指纹图谱及抗氧化活性的巴西绿蜂胶质量控制研究 ……… 577

第四篇　蜂胶综合开发利用研究

第十四章　蜂胶中挥发性成分的研究 …………………………………………… 605
　第一节　蜂胶中挥发性成分的提取工艺研究 …………………………………… 605
　第二节　巴西蜂胶与中国蜂胶中挥发性成分的分析与比较 …………………… 616
　第三节　蜂胶中挥发性成分的生物学活性 ……………………………………… 622

第十五章　蜂胶残渣中挥发性成分的研究 ……………………………………… 651
　第一节　响应面法优化微波辅助提取蜂胶残渣中的挥发性成分 ……………… 651
　第二节　巴西蜂胶残渣挥发性成分与巴西原胶挥发性成分的分析比较 ……… 663
　第三节　巴西蜂胶残渣挥发性成分的生物学活性 ……………………………… 671

第十六章　纳米蜂胶及其降血脂降血糖作用的研究 …………………………… 685
　第一节　纳米蜂胶的制备及体外释放实验 ……………………………………… 685
　第二节　纳米蜂胶对高脂血症 SD 大鼠的作用 ………………………………… 691
　第三节　纳米蜂胶对 2 型糖尿病大鼠代谢紊乱与胰岛素抵抗的影响 ………… 699

缩略词表 …………………………………………………………………………… 721
附录
　1　国家标准《蜂胶》(GB/T 24283) ……………………………………………… 727
　2　国家标准《蜂胶中杨树胶的检测方法　反相高效液相色谱法》(GH/T
　　　34782—2017) ……………………………………………………………… 737
　3　行业标准《蜂胶中阿替匹林 C 的测定方法　高效液相色谱法》(GH/T
　　　1114—2015) ………………………………………………………………… 747
　4　行业标准《蜂胶中咖啡酸、p-香豆酸、阿魏酸、短叶松素、松属素、短叶松素
　　　3-乙酸酯、白杨素和高良姜素含量的测定方法　反相高相液相色谱法》(报批稿) … 755

Contents

Chapter 1 Preface 1

Part 1 Recent Advances in Propolis

Chapter 2 Plant Sources and Chemical Composition of Propolis 19
  2.1 Plant Sources 19
  2.2 Chemical Composition 23
Chapter 3 Biological Activities of Propolis 65
  3.1 Antiviral and Antimicrobial Activity of Propolis 65
  3.2 Antioxidant Activity of Propolis 86
  3.3 Anti-inflammatory Activity of Propolis 99
  3.4 Antitumor Activity of Propolis 111
  3.5 Immunomodulatory Activity of Propolis 127
  3.6 Antidiabetic Activity of Propolis 139
  3.7 Hepatoprotective Activity of Propolis 157
  3.8 Wound Healing Activity of Propolis 171
  3.9 Other Biological Activities of Propolis 183
Chapter 4 Quality Control and Standardization of Propolis 192
  4.1 Diversity of Chemical Composition in Propolis 192
  4.2 Analytical Methods for Chemical Composition in Propolis 200
  4.3 Quality Control and Standardization of Propolis 215
Chapter 5 Extraction of Effective Composition in Propolis and Product Processing 266
  5.1 Extraction Process of Effective Composition in Propolis 266
  5.2 Processing and Application of Propolis Products 281

Part 2 Biological Activities of Propolis

Chapter 6 Antioxidant Activity of Propolis and Its Mechanism 313
  6.1 Chemical Composition and Anti-inflammatory Effects of Poplar Tree Gum and Ethanol Extracts of Propolis from Three Different Plant Sources 313

  6.2 Comparisons of Antioxidant Effects and Molecular Mechanisms of Poplar Tree Gum and Ethanol Extracts of Propolis from Different Plant Sources in RAW64.7 Cells ⋯ 322

Chapter 7 Effects and Mechanisms of Propolis on Inflammatory Diseases and Inflammatory Microenvironment ⋯ 346

  7.1 Anti-inflammatory Effects of Ethanol Extracts of Chinese Propolis and Poplar Buds in Mice with Lipopolysaccharide (LPS)—Induced Acute Lung Injury ⋯ 346

  7.2 Anti-inflammatory Effects of Chinese Propolis in Rats with Dextran Sulfate Sodium-Induced Colitis ⋯ 353

  7.3 Anti-inflammatory Effects of Chinese Propolis on Intestinal Barrier Function of Intestinal Epithelial Caco-2 Cells and Its Mechanism ⋯ 360

  7.4 Anti-inflammatory Effects of Chinese and Brazilian Propolis on Modulating Macrophage Activation and Its Mechanism ⋯ 371

Chapter 8 Antidiabetic Activity of Propolis and Its Mechanism ⋯ 391

  8.1 Antidiabetic Activity of Propolis in T1DM Rats ⋯ 391

  8.2 Antidiabetic Activity of Propolis in T2DM Rats ⋯ 410

  8.3 Effects of Propolis in T2DM Rats with Renal Disease and Its Mechanism ⋯ 428

Chapter 9 Effects of Propolis on Vascular Endothelial Cells and Its Mechanism ⋯ 451

  9.1 Effects of Propolis on Vascular Endothelial Cells Deprived of Basic Fibroblast Growth Factor (FGF-2) and Serum ⋯ 451

  9.2 Effects of Propolis on Lipopolysaccharide-Induced Vascular Endothelial Cells ⋯ 470

  9.3 Effects of Propolis on Oxidized-LDL-Stimulated Human Umbilical Vein Endothelial Cells ⋯ 476

Chapter 10 Wound Healing Activity of Propolis and Its Mechasim ⋯ 488

  10.1 Comparison of Chemical Composition of Propolis from Two Different Plant sources ⋯ 488

  10.2 Effects of Propolis on Genes Expression Associated with Proliferation, Migration and Collagen on L929 Cells ⋯ 492

  10.3 Protective Effects and The Underlying Mechanism of Propolis on L929 Cells Injured by Hydrogen Peroxide (H_2O_2) ⋯ 499

Part 3 Quality Control of Propolis

Chapter 11 Studies on β-glucosidase in Propolis ⋯ 525

  11.1 Detection of β-glucosidase in Propolis ⋯ 525

11.2 Properties of β-glucosidase in Propolis and Its Role in Composition Transformation 530

11.3 Effects of Storage Conditions on The Activity of Glucosidase in Propolis 538

Chapter 12 Methods for Distinguishing Poplar Tree Gum from Propolis 542

12.1 Distinguishing Poplar Tree Gum from Propolis by High-Performance Liquid Chromatography (HPLC) 542

12.2 Distinguishing Poplar Tree Gum from Propolis Based on the Markers Salicin and Catechol 547

Chapter 13 Quality Control and Quality Evaluation of Propolis 567

13.1 Quality Evaluation of Chinese Propolis by Combining Multi-Ingredient Quantitative Analysis and Fingerprint Analysis 567

13.2 Quality Control of Brazilian Green Propolis Based on Its Antioxidant Activity and Multi-Target Fingerprint 577

Part 4 Comprehensive Development and Utilization of Propolis

Chapter 14 Volatile Composition in Propolis 605

14.1 Extraction Process of Volatile Composition in Propolis 605

14.2 Analysis and Comparison of Volatile Composition in Chinese and Brazilian Propolis 616

14.3 Biological Activities of Volatile Composition in Propolis 622

Chapter 15 Volatile Composition in Propolis Extraction Residue 651

15.1 Optimization of Microwave-Assisted Extraction of Volatile Composition from Propolis Residue by Response Surface Methodology 651

15.2 Analysis and Comparison of Volatile Composition in Brazilian Propolis and Its Residue 663

15.3 Biological Activities of Volatile Composition in Brazilian Propolis Residue 671

Chapter 16 Encapsulated Propolis and its Effects of Decreasing Blood Glucose and Cholesterol 685

16.1 Preparation of Encapsulated Propolis and its In Vitro Release Test 685

16.2 Effects of Encapsulated Propolis in Hyperglycaemia SD Rat 691

16.3 Effects of Encapsulated Propolis on Metabolic Disorders and Insulin Resistance in Type 2 Diabetes Mellitus Rats 699

Abbreviations 721

Appendix
1. National Standard *Propolis*(GB/T 24283) ··· 727
2. National Standard *Determination of Poplar Tree Gum in Propolis by Reversed-Phase High-Performance Liquid Chromatography (RP-HPLC)* (GB/T 34378—2017) ··· 737
3. Industry Standard *Determination of Artepillin C in Propolis by High-Performance Liquid Chromatography (HPLC)*(GH/T 1114—2015) ·································· 747
4. Industry Standard *Determination of Caffeic Acid, p-Coumaric acid, Ferulic acid, Pinobanksin, Pinocembrin, 3-O-acetylpinobanksin, Chrysin and Galangin in Propolis by Reversed-Phase High-Performance Liquid Chromatography (RP-HPLC)* ············ 755

第一章 绪 言

蜂胶是一种古老而新兴的蜂产品。说它古老,是因为其应用历史十分悠久,古埃及、古希腊和古罗马人早就认识到蜂胶的药理特性,并且将其作为药物加以利用;说它新兴,是因为近20年来,蜂胶是蜂产品乃至天然产物研究开发的一大热点,新的胶源植物、蜂胶中新的化学成分、生物学活性和新的产品被不断地发现和创造出来。蜂胶研究与开发的历史,在某种程度上是人类探索和了解天然产物的一个缩影,令人回味,更激励着人们不断前行。

"蜂胶"一词源于古希腊语中的"προ-˘σσω",是指"预先软化,通过摩擦或揉捏使之软化"。在英文中,蜂胶(propolis)是由"pro"(在前)和"polis"(城市)组成,意思是用蜂胶缩小蜜蜂蜂巢这座"城市"的城门。人们普遍认为蜜蜂采集蜂胶是用来保护蜂巢的。因此,拉丁语词典中给"蜂胶"下的定义是:"蜜蜂用于修补巢房缝隙的黏性物质"(Lewis and Short,2012)。后来研究发现,除了填补孔洞、修补裂缝、润滑巢房内壁作用外,蜂胶似乎还作为一种抗菌剂来防止蜜蜂幼虫、储蜜和蜂巢的微生物感染(Seeley and Morse,1976)。蜜蜂作为一种社会性昆虫,生活在由数万只个体密集组成的蜂群内,这就增加了疾病暴发和寄生虫危害的风险(Schmid-Hempel,1998)。然而,相对于非社会性昆虫,虽然蜜蜂个体的免疫力较其他非社会性昆虫弱(Evans et al.,2006),但整个蜂群却能抵抗大部分病原菌的侵蚀。这就表明蜜蜂很可能在个体或群体水平潜在地存在着弥补其个体免疫力低下的一系列行为学机制(Evans and Spivak,2010)。蜜蜂采集有多种抗菌活性的复杂植物分泌物,并以蜂胶的形式融入蜂巢中,这就是一种蜂群水平的抗病原微生物方式。也就是说,蜂胶在蜂群社会性免疫力、蜂群自我治疗中发挥了重要作用(平舜等,2014)。同时,蜂胶涂抹于巢脾表面形成的薄层相当于一个不透水的内层,可以限制水分散失,有助于保持蜂巢内湿度的恒定(Visscher,1980)。此外,当一些比蜜蜂大许多的"庞然大物",如老鼠、壁虎和蜥蜴等闯入蜂箱内时,被蜜蜂群起而攻之,它们被蜜蜂蜇死后,蜜蜂是无法将其拖出蜂巢外的。此时,蜜蜂就用蜂胶将这些入侵者庞大的尸体包裹起来,形成一个"木乃伊",这些尸体就不会腐烂而影响蜂箱中的环境。因此,蜂胶在蜂巢中起到了神奇、独特且重要的作用。

第一节 蜂胶的应用历史

蜂胶的应用历史十分悠久,古埃及、波斯和古罗马都有应用蜂胶的记录(Houghton,1998)。早在3000多年前,古埃及人就已经认识了蜂胶,将蜂胶用作木乃伊的防腐材料,并记载在与木乃伊同期保存下来的有关医学、化学和艺术的草纸书中。古埃及人在花瓶和其他装饰品上描绘采集蜂胶的蜜蜂,并将蜂胶用作防治疾病的药物(Nicolas,1947)。

古犹太人认为 tzori(蜂胶,犹太语)是一种药物。《圣经》中也提到了蜂胶及其治疗功效(耶利米 8,第 22 节;耶利米 46,第 11 节;耶利米 51,第八节)。《圣经》中的乳香是示巴(Sheba)女王给所罗门王的礼物,而乳香和蜂胶几乎是相同的。乳香树 1500 年来生长在死海周围,因其独特的香味和药用价值闻名于世。乳香由香脂杨(*Populus balsamifera*)、黑杨(*P. nigra*)和另一种杨树(*P. gileadensis*)的树脂加工而成(Broadhurst,1996)。耶路撒冷圣殿每日 2 次使用的熏香中就含有乳香。

蜂胶最早记载于 2000 多年前的传世巨著《动物志》(*Historia Animalium*)第 9 卷第 14 章中。该书的第 1—8 卷由亚里士多德(Aristotle,公元前 384—前 322 年)撰写,第 9 卷的作者佚名。书中对蜂胶的特征是这样描述的:"蜜蜂最初建造的巢房是空的,之后它们修建蜡制的巢脾壁,采集多种花儿和柳树、榆树等植物分泌的树胶汁液。蜜蜂用之修补巢脾,抵御其他外来生物的入侵以及缩小巢门"(Aristotle,1910a)。该书还指出:"这种'黑蜡'具有一股刺鼻的气味,可用来治疗瘀伤和化脓性溃疡"(Aristotle,1910b)。蜂胶被古希腊人称为继蜂蜜和蜂蜡之后的"第三种天然蜂产品"。古希腊人用蜂胶与乳香、苏合香、芳香药草混合制成西洋樱草香水(Bogdanov,2012)。

西方医学奠基人希波克拉底(Hippocrates of Cos Ⅱ,约公元前 460—前 377 年)曾用蜂胶治疗内外伤口和溃疡(Dealey,2005)。古希腊医生狄欧斯考里德斯(Pedanios Dioscorides)在其《药物学》(*De materia Medica*)中详述了蜂胶的医疗用途:"黄蜂胶以其芬芳、柔软性以及与苏合香的相似性成为继乳香后得以广泛应用的物质;蜂胶相当温和,有吸引力,有助于拔出荆棘和碎片;用蜂胶烟熏可以缓解慢性咳嗽"(Fearnley,2001)。

古罗马人普遍敬畏蜜蜂和蜂胶,老普林尼(Pliny the Elder,23—79 年)在其名著《自然史》(*Natural History*)中写道:"蜂胶来自葡萄和白杨树分泌的芳香树脂,并混入了花蜜,密度较大。不过,蜂胶不能被称为蜂蜡,它用来修补巢房内的孔洞和缝隙以抵御寒流和其他有害因素的侵入。同时,因为蜂胶有一种强烈的气味,很多人用它代替白松香使用。"(Pliny the Elder,1855a)书中还描述了蜂胶的实用性,即蜂胶具有清除身体上的异物,转化肿瘤,消除硬块,缓解肌肉疼痛和治愈顽固性溃疡的疗效(Pliny the Elder,1855b)。

根据古罗马学者和作家马库斯·特伦提乌斯·瓦罗(Marcus Terentius Varro,公元前 116—前 27 年)的描述:"蜂胶是一种蜜蜂用于建造在蜂巢前方入口处的保护设施的物质。蜂胶因为被医生用作草药,所以比蜂蜜享有更高的价值。"(Varro and Rustica,1934)

在公元 1 世纪,罗马百科全书编纂者塞尔苏斯(Cornelius Celsus)在《医学论》(*De Medicina*)中写道:"蜂胶作为药物可以促进化脓,愈合伤口和治疗脓肿。"(Celsus,1971)

阿拉伯人也早就知道蜂胶。例如,穆斯林医生和哲学家阿维森纳(Avicenna,980—1037 年)描述了 2 种蜂蜡,即纯蜡和黑蜡,后者就是蜂胶。他写道:"黑蜡因其强烈的气味会让你打喷嚏,它可以起到净化、清洁以及吸收的作用。"在波斯手稿中,蜂胶是一种治疗湿疹、肌肉痛和风湿病的药物(Fearnley,2001)。

到了中世纪,蜂胶不再是一个热门的话题,也不再被用作主流的药物,只有少数记载蜂胶的手稿得以保存下来。12 世纪的一些文献记录了用于预防龋齿、治疗口咽部感染以及龋齿的含有蜂胶的药用制剂。幸运的是,蜂胶的药用价值在民间传统医学,尤其是在东欧地区的"草药"医学中得以流传下来(Kuropatnicki et al.,2013)。

随着文艺复兴的兴起,蜂胶的利用与"返璞归真"思潮一道重返欧洲。约翰·杰勒德

(John Gerard,1545—1612 年)在其著名的草药书《植物的历史》(*The History of Plants*)(1597)中提到了使用黑杨树叶芽所分泌的粘湿的树脂制作疗伤药膏。这种由白杨叶芽制作的药膏可以治疗几乎所有的炎症、挫伤、拉伤和侧索硬化症等。17 世纪的英国药典中记载,蜂胶是疗伤药膏的主要成分(Murray and Pizzorno,2005)。尼古拉斯·库尔佩珀(Nicholas Culpeper)在他的《草药大全》(*Complete Herbal*)中指出,由白杨树生产的"Populneon"药膏对身体任何部位的高温和炎症都有作用,可以降低伤口的温度(Culpeper,1995)。1824 年出版的《通用草药》(*The Universal Herbal*)中记载了"黑色的杨树、白杨树":"治疗硬而痛肿块的药膏中含有幼叶成分,这种成分春天时闻起来清香宜人,若用手将其压碎便会产生一种香树脂物质(蜂胶),用酒精提取的这种物质闻起来像安息香。肉汤中添加蜂胶酊剂可以清除顽固性肠道表皮脱落。"

自 19 世纪以来,随着蜂胶众多的生理药理活性逐渐被人们所认识,蜂胶的应用领域也越来越广泛。在欧洲、日本、美国等地,蜂胶越来越广泛地被应用于食品、医药、化工、农牧业等多个领域。研究表明,它能增进健康,预防与治疗外伤、皮肤病、炎症、心脏病、糖尿病、肿瘤等多种疾病(Marcucci,1995)。

第二节 蜂胶的研究历史

早在 19 世纪初,法国药剂师和化学家沃克兰(Nicolas Louis Vauquelin)就对蜂胶进行了研究。他在论文中写道:"蜂胶是蜜蜂采集的一种柔软、有味道、红棕色的树脂状物质。大块的蜂胶呈黑色,而在薄板中呈半透明状。蜂胶拥有类似于蜂蜡的柔软性,手的热量可使其软化,但韧性却比蜂蜡更强。它的气味芬芳,类似于草木樨、秘鲁香脂或香蕉杨树。"沃克兰将 100g 的蜂胶用乙醇提取 3 次,计算出的组分含量如下:纯蜂胶 57g,蜂蜡 14g,外来物质 14g,损失的芳香物 15g(Vauquelin,1803)。这是人类历史上对蜂胶成分进行的首次分析研究。

蜂胶的研究进展与化学分析技术,尤其是人们对黄酮类化合物研究的不断深入紧密相连(Kuropatnicki et al., 2013)。19 世纪初期,法国化学家谢弗勒尔(Michel Eugene Chevreul)提取到一些结晶态的黄酮,包括:从桑橙树中提取的桑色素,从黄犀草中提取的木樨草素,从漆树中提取的非瑟酮,从栎树皮中提取的槲皮素。19 世纪中后期,Piccard(1873)从常见的白杨树幼芽中分离出了第一个黄酮物质——白杨素;德国化学家 Carl Liebermann 从槲皮甙中提取了槲皮素;奥地利化学家 Herzig(1891)探明了栌木色素——非瑟酮和槲皮素的结构;瑞士伯尼尔大学的波兰籍教授 Stanislaw Kostanecki 发现苯并芘的苯基衍生物是白杨素的天然产物(Kabzi'nska,1993)。Kostanecki(1895)对白杨素结构提出了质疑,并在不久后证实了白杨素的结构;同时他给母环系统及其 3-羟基衍生物起名为黄酮和黄酮醇。在这前后,大量的天然黄色色素得以检测,其中很多属于黄酮和黄酮醇类化合物(Emilewicz et al., 1899;Piccard,1873,1874,1877)。1898 年,Emilewicz 等报道了白杨素的合成方法。次年,Kostanecki 等精制了黄酮,并在 1900 年合成了芹黄素和毛地黄黄酮。1904 年,Kostanecki 等合成了非瑟酮、槲皮素和山柰酚,后又以相似的方法合成了桑色素(Perkin and Everest,1918)。Kostanecki 还确定了姜黄素的结构,同时也研究了巴戊、苏木精和胭脂红。

之后，他又分析了将近2000种不同的物质，结果显示其中有200多种物质含有黄酮衍生物(Perkin and Everest，1918)。

蜂胶化学成分的现代研究始于20世纪初的德国。早期，人们试图用简单的分馏法确定蜂胶的组分。最早的报道是Dieterich和Helfenberg于1908年各自提出的蜂胶组分在乙醇、氯仿和乙醚中的分离方法(Dieterich，1908；Helfenberg，1908)。3年后，Dieterich和Küstenmacher又分别在蜂胶中分离鉴定出了香草醛、肉桂酸和肉桂醇(Dieterich，1911；Küstenmacher，1911)。1926年，Jaubert发现了蜂胶中的一种天然形成的黄酮——色素白杨素(Jaubert，1926)。次年，Rösch证实了Plinius的假说，即蜂胶来源于植物的叶芽(Rösch，1927)。随后，美国学者进行的一系列研究发现蜂胶含有少量的维生素B_1、B_2、B_6、C、E以及烟酸和泛酸(Haydak and Palmer，1940，1941，1942)。

20世纪50年代，各国学者对黄酮类化合物的生理作用及医药用途进行了一系列研究。Powers研究发现，黄酮类化合物对1个甚至多于10个的菌株有抑制作用(Proceedings of the 4th International Symposium on Food Microbiology，1964)。

60年代，对蜂胶化学成分的研究得以继续开展。起初，人们认为蜂胶像蜂蜡一样是一种成分复杂但相对稳定的物质(Lindenfelser，1967)。但后来，通过对不同地区大量蜂胶样本的分析研究发现，蜂胶的化学成分是高度变化的(Bankova，2000)。1969年，Popravko等分离鉴定了蜂胶中2种黄烷酮和异香兰素以及6种黄酮类色素(Popravko et al.，1969，1970)。继Lavie(1957)证实蜂胶对枯草杆菌、芽孢杆菌和普通变形杆菌有抑制作用后，法国科学家从蜂胶提取物中分离出了具有抑菌作用的高良姜素，之后又分离鉴定了松属素、杨芽黄素和伊砂黄素(Villanueva et al.，1964，1970)。1970年和1973年，Cizmarik和Matel先后从蜂胶中分离鉴定出了3,4-二羟基肉桂酸和4-羟基-3-甲氧基肉桂酸(Cizmarik and Matel，1970，1973)。Herold(1970)在蜂胶残渣中发现了铁、钙、铝、钒、锶、锰和硅等矿质元素。Nikiforov等(1971)检测发现蜂胶中含有铜和锰。

70年代，色谱分析方法(如柱层析法和薄层色谱法)的出现，使得对蜂胶更多组分的分离鉴定成为可能。Heinen和Linskens(1972)研究了蜂胶中的脂肪酸成分，结果表明，这些脂肪酸包括$C_7 \sim C_{18}$。Schneidweind等(1975)鉴定出了蜂胶中的17种成分，包括9种以前检测过的物质。同时，Metzner等(1975)用生物自显影法证明蜂胶提取液中只有少数化合物具有明显的抗真菌作用。Ghisalberti等(1977)澳大利亚学者分离鉴定了7种黄酮，即球松素、樱花亭、异樱花亭、紫檀芪、白杨素、3,5-二甲氧基苯乙腈醇和紫檀。Popravko(1978)分离鉴定出了蜂胶中的18种化合物，其中14种属于黄酮类化合物。

蜂胶复杂而丰富的化学成分决定了其独特而广泛的生物学活性。随着蜂胶化学成分研究的逐步深入，其生理药理活性的研究及应用也得到了快速发展。

早在40年代，Kivalkina就对蜂胶抗菌性能进行了系统研究，证明蜂胶对金黄色葡萄球菌(*Staphylococcus aureus*)、伤寒杆菌(*Salmonella enterica*)和其他细菌有明显的抑菌活性(Kuropatnicki et al.，2013)。Lindenfelser(1967)研究了美国不同地区和季节生产的蜂胶样本的抗菌活性，结果表明，在体外实验条件下，蜂胶对39种细菌中的25种表现出强烈的抑菌作用，而相同数量的真菌中有20种被抑制。同年，波兰学者研究了蜂胶对细菌的广谱抗菌活性，结果表明，球菌科对蜂胶敏感，而念珠菌属(*Nostoc*)和棒状杆菌属(*Corynebacterium*)的菌株对蜂胶部分敏感；$H_{37}Rv$以及从患者身上分离到的菌株对蜂胶醇提液表现出敏感性（Scheller et al.，1968）。

50 年代，动物实验证实，蜂胶提取物作为一种表面麻醉剂具有轻微的穿透能力，可以用于牙科临床（Prokopovich，1956，1957）。Todorov 等（1968）研究证明，蜂胶的渗透作用相当于普鲁卡因。后来，保加利亚学者研究表明，5%普鲁卡因蜂胶液比蜂胶水醇提取液具有更好更快的效果（Tsacov，1973）。蜂胶醇提液以及一些蜂胶分离组分，如 5,7-二羟黄烷酮（松属素）、5-羟基-7-甲氧基黄烷酮（球松素）和咖啡酸酯类的混合物，在兔和小鼠的角膜上进行的全麻实验结果表明，每种化合物的作用强度是总提取物的 3 倍（Paintz and Metzner，1979）。

蜂胶及其提取物对组织再生有着阳性效应。斯洛伐克学者研究结果证明蜂胶醇提液加速了组织再生进程（Sutta et al. 1975）。Scheller 等（1977）研究表明蜂胶醇提液以其酸性或中性 pH 来维持抗菌活性。后来波兰学者研究结果表明，蜂胶醇提液可以促进受损软骨的愈合过程，以及增强人为诱导骨质缺损的成骨作用（Stojko et al. 1978）。其他研究还发现蜂胶醇提液可促进牙髓再生，减少炎症和退化过程（Scheller，1978）。

在利用蜂胶治疗大鼠实验性胃溃疡取得理想效果（Aripov et al.，1968）的基础上，蜂胶醇提液被用于治疗人的胃溃疡（Gorbatenko，1971）和幽门十二指肠炎（Makarov，1972）。苏联学者利用水溶性蜂胶来治疗动脉硬化闭塞症引起的下肢营养性溃疡（Lutsenko and Pisarenko，1980）。此外，还有蜂胶对化脓性手术创伤（Damyanliev et al.，1982）和慢性胃十二指肠溃疡（Korochkin and Poslavskii，1986）治疗作用的报道。

蜂胶制剂还被应用于治疗骨头的化脓性炎症。蜂胶可抑制炎症进程，修复骨组织（Scheller et al.，1977）。蜂胶对牙髓再生的作用得以证实，蜂胶提取物在牙科上被广泛应用，包括用于治疗干槽症、黏膜和牙龈炎（Scheller et al.，1978）。此外，研究还证实了蜂胶对感冒和慢性扁桃体炎也有治疗效果（Doroshenko，1983；Szmeja et al.，1989）。

丹麦生物学家 Karl Lund Aagaard 博士自 20 世纪 60 年代以来一直致力于蜂胶生物学活性及其临床应用研究，历经 20 多年时间，观察了蜂胶对斯堪的纳维亚地区超过 5 万名患者的影响后，得出了如下结论：蜂胶的生物学活性范围极其广泛，可治疗的疾病包括尿道感染、喉咙肿胀、痛风、开放性创伤、感冒、流感、支气管炎、胃炎、耳朵疾病、牙周疾病、肠道感染、溃疡、急性湿疹、肺炎、关节炎、头痛、癌症、帕金森氏病、胆汁感染、硬化症、疣、结膜炎和声音嘶哑等（Elkins，1996）。

进入 20 世纪 90 年代以来，随着对蜂胶化学成分和生物学活性研究的不断深入，蜂胶越来越受到食品、医药、日化、农业等行业科研工作者的广泛关注，逐渐成为蜂产品乃至天然产物研究与开发的热点，获得了"天然免疫调节剂""血管清道夫"和"紫色黄金"等美誉。国内外每年发表的有关蜂胶的研究论文数量呈逐年递增趋势，研究领域广泛，研究成果丰硕（张翠平和胡福良，2010，2011，2012；王凯等，2013，2014，2015）。据 Pubmed、Elsevier、Springerlink 等数据库公布的数据显示，2000—2017 年的 18 年间共发表涉及蜂胶的英文研究性论文 2057 篇。各年度发表的论文数量见图 1.1。

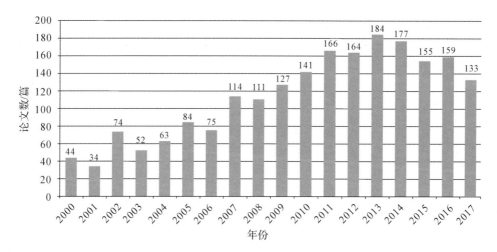

图 1.1 2000—2017 年发表的与蜂胶相关的英文论文数

Fig. 1.1　The number of English papers on propolis from 2000 to 2017

有关近 20 年来蜂胶的植物来源与化学成分、生物学活性、质量控制以及加工工艺等相关领域的研究进展将在本书的第一篇"国内外蜂胶研究进展"中作详细介绍。

第三节　中国的蜂胶研究与应用

由于中华蜜蜂（Apis cerana cerana）不产蜂胶，我国历史文献中对蜂胶的应用和生产的记载几乎是空白，仅对露蜂房、蜂房等含有蜂胶成分的中药有所记载。露蜂房作为中药，最早收载于公元前 1—前 2 世纪所著、我国现存最早的中药经典著作《神农本草经》中，其主要成分是蜂蜡、蜂胶（树脂）和蜂房油 3 种，具有"主治惊痫、寒热邪气、癫疾……肠痔"的作用。明朝李时珍的《本草纲目》记载，蜂房主治"臃肿不消，为末，醋调涂之，干更易之。不入服食，药性疗疗肿疮毒"。我国药典记载蜂房的功能与主治为："攻毒杀虫，祛风止痛。用于疮疡肿毒，乳痈，瘰疬，皮肤顽癣，鹅掌风，牙痛，风湿痹痛"（国家药典委员会，2010）。

我国对纯正蜂胶的认识、研究和开发起步较晚，20 世纪 50 年代，随着我国大量引入西方蜜蜂（Apis mellifera），商品化蜂胶开始生产以后，蜂胶的研究才得以展开。

房柱先生从 1956 年开始研究蜂胶的医药效用，并率先在《中华皮肤科杂志》1959 年第 4 期上发表了我国第一篇有关蜂胶研究的论文（房柱，1959）。70 年代，经试验证实，蜂胶对真菌具有抑菌作用，并用于体表癣病和深部皮肤霉菌病的治疗，取得了显著疗效（房柱和匡友成，1976，1978）；房柱先生主持研制的内服制剂蜂胶片具有降血脂、防治动脉粥样硬化效用（朱道程，1981；李景祥和房柱，1981），并于 1981 年 8 月获得江苏省药政准字批号，正式载入《江苏省药品标准》。

70—80 年代，中国科学院昆明动物研究所、中国林业科学研究院林产化学工业研究所、中国医学科学院药物研究所、中国农业科学院蜜蜂研究所等多家单位的科研人员先后开展了有关中国蜂胶的化学成分、生物学活性、质量标准等基础性研究工作，分离、鉴定了中国蜂胶的有效成分，从蜂胶挥发油中鉴定出桉叶油素、愈创木酚、桉叶醇等新组分（钱锐和陈志媛,1979;陈友地等,1983);证实了类黄酮化合物是中国蜂胶的主要功效成分,而且发现了新的类黄酮化合物(王秉极和张慧娟,1988);对蜂胶乙醇提取物、蜂蜡含量、杂质、酚酸含量、碘值、氧化指标、黄酮反应及锌、铅的含量进行了较全面的分析,为蜂胶质量标准的制定提供了基础性资料(尚天民和徐景耀,1983)。随着对蜂胶化学成分、生物学活性研究的不断深入,蜂胶在医药、保健、日化、畜牧、兽医、果蔬保鲜等领域的应用得以不断拓展。沈志强等将蜂胶应用于兽医免疫佐剂,从单纯的新城疫疫苗发展到各式各样的二联苗和三联苗,取得了多项原创性成果(沈志强和杨永福,1989;沈志强 等,1990a,1990b;沈志强和徐可利,1990,1991)。

90年代以来,不同领域的众多科研人员对蜂胶开展了全方位的研究,使我国蜂胶产品的研究与开发得到迅猛的发展。

中国知网(CNKI)数据库公布的数据显示,我国每年发表的有关蜂胶研究的中文论文数量从20世纪80年代初的几篇,到80年代中期的10多篇,到2002年超过100篇,此后逐年递增并维持在较高数量水平(图1.2)。

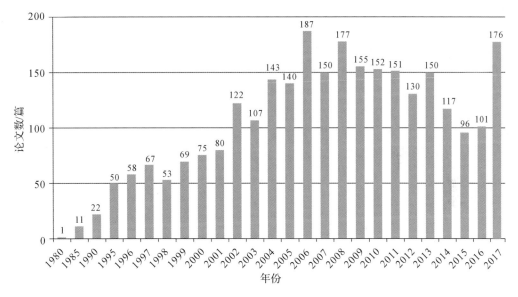

图 1.2 自 1980 年以来发表的中文蜂胶研究论文数量

Fig. 1.2 The number of Chinese research papers on propolis since 1980

根据"中国专利全文数据库"及"专利之星"检索结果,截至2017年年底,共检出1059条蜂胶专利记录(含中国台湾专利16件),如图1.3所示。在我国大陆1043件蜂胶专利中,共有724件为发明专利,占专利总量的69.4%,其数量远高于实用新型专利(70件)和外观设计专利(249件),表明我国对蜂胶知识产权保护的意识正在不断增强(王凯和胡福良,2014)。

蜂胶也已载于《中华本草》第九卷,并且已被收录到《中华人民共和国药典》(2005年版、2010年版、2015年版,一部)。记载的功能与主治为:"补虚弱,化浊脂,止消渴;外用解毒消肿,收敛生肌。用于体虚早衰,高脂血症,消渴;外治皮肤皲裂,烧烫伤"(国家药典委员会,2010)。

目前在国家食品药品监督管理总局登记备案的以蜂胶为原料的产品多达数百种,涵盖药品、保健食品、化妆品各种类(国家食品药品监督管理总局,2015),其中以保健食品为主,

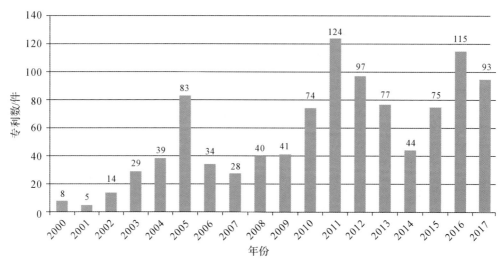

图 1.3 我国历年申请的蜂胶专利数量

Fig. 1.3 The number of propolis patent applications in China

药品较少。以蜂胶为主要成分的药品仅有6只，剂型包括酊剂、膜剂和片剂。以蜂胶为主要功效成分的化妆品共有68只，其中国产2只，进口66只。产品涉及润唇膏、洗发液、护发素、精华液、眼霜、营养霜、乳液、面膜、洗面奶、护手霜等多个种类。

据统计，从1996年我国开始实施《保健食品管理办法》以来，截至2016年年底，国家卫生部和食品药品监督管理局批准的以蜂胶为主要原料的保健食品共计421只（程晓雨和胡福良，2015）。这些保健食品的年度分布、保健功能、剂型以及地域分布情况分别见图1.4—1.7。

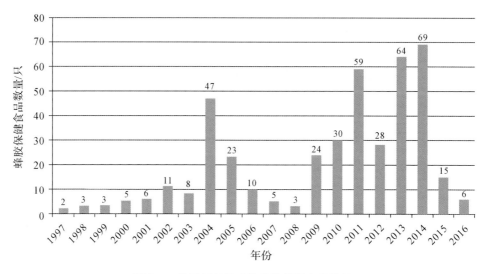

图 1.4 我国历年批准的蜂胶保健食品数量

Fig. 1.4 The number of approved propolis health food in China

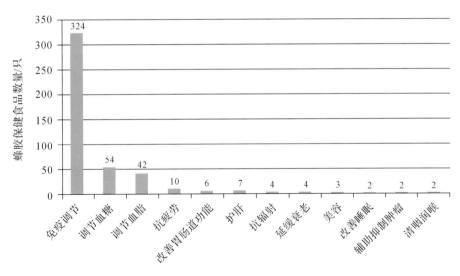

图 1.5 我国蜂胶保健食品的保健功能分布

Fig. 1.5 The distribution of the health functions of propolis health food in China

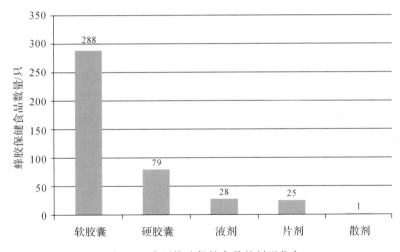

图 1.6 我国蜂胶保健食品的剂型分布

Fig. 1.6 The distribution of the formulations of propolis health food in China

与迅猛发展的蜂胶开发热形成鲜明对比的是,我国对蜂胶各个方面的研究还不够深入、广泛,基础研究方面尤为薄弱。我国的蜂胶科研水平跟不上产业发展的需求,蜂胶产品开发缺少突破与创新,产品同质化、低水平重复现象严重,市场监管不力,假冒伪劣产品横行,这些问题严重制约着我国蜂胶行业的健康发展。

笔者长期从事蜂产品的科研与教学工作,近年来得到了国家自然科学基金、国家蜂产业技术体系、浙江省杰出青年科学基金等项目的大力资助,取得了一批原创性的科研成果。研究团队中先后有 10 多位博士和硕士研究生从事与蜂胶相关的研究,发表了 100 多篇蜂胶方面的学术论文,其中 SCI 收录 30 余篇,获得授权国家发明专利 20 余项。2005 年 8 月,笔者所著的国内第一部蜂胶基础性研究专著《蜂胶药理作用研究》由浙江大学出版社出版(胡福良,2005),受到专业人士的好评与欢迎。在此基础上,现主要根据我们科研团队 10 多年来

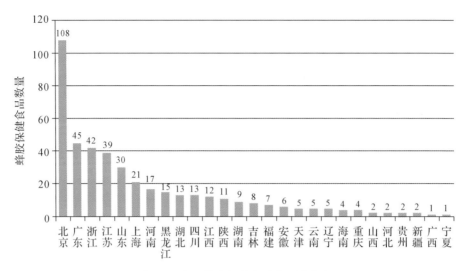

图 1.7 我国蜂胶保健食品的地域分布

Fig. 1.7 The geographical distribution of propolis health food in China

的研究成果，写成此专著，旨在抛砖引玉，推陈出新，为推动我国蜂产品尤其是蜂胶的科学研究和产业发展添砖加瓦，做出自己的一份贡献。限于作者的能力与水平，不足与谬误之处在所难免，敬请批评指正。

胡福良

2018 年春，于浙江大学紫金港校区

参考文献

Aripov KLA, Kamilov IK, Aliev KU (1968) Effect of propolis on experimental stomach ulcers in rats[J]. *Medskii Zh Uzbek*, 5:50-52. (Russian).

Aristotle (1910a) *The Works of Aristotle Translated into English under the Editorship of J. A. Smith and W. D. Ross, Volume IV by D'Arcy Wentworth Thompson* [M], sig. Ee6v, The Clarendon Press, Oxford, UK.

Aristotle (1910b) *The Works of Aristotle Translated into English under the Editorship of J. A. Smith and W. D. Ross, Volume IV by D'Arcy Wentworth Thompson* [M], sig. Ee7r, The Clarendon Press, Oxford, UK.

Bankova VS, de Castro SL, Marcucci MC (2000) Propolis: recent advances in chemistry and plant origin[J]. *Apidologie*, 31(1):3-15.

Bogdanov S (2012) Propolis: Composition, Health, Medicine: A Review[M]. http://www.bee-hexagon.net/files/file/fileE/Health/PropolisBookReview.pdf.

Broadhurst CL (1996) *Benefits of bee propolis* [M]. Health Supplement Retailer, 46-48.

Celsus A (1971) *De Medicina. Book 5* [M]. edited by W. G. Spencer, chapter 3, Harvard University Press, Cambridge, Mass, USA.

Cizmarik J, Matel I (1970) Examination of the chemical composition of propolis I. Isolation and identification of the 3,4-dihydroxycinnamic acid (caffeic acid) from propolis[J].

Experientia, 26(7):713.

Cizmarik J, Matel I (1973) Examination of the chemical composition of propolis 2. Isolation and identification of 4-hydroxy-3-methoxy-cinnamic acid (ferulic acid) frompropolis [J]. *Journal of Apicultural Research*, 12:52-54.

Culpeper N (1995) *Complete Herbal*[M]. Wordsworth Editions Ltd., Ware, UK.

Damyanliev R, Hekimov K, Savova E, Agopian R (1982) The treatment of suppurative surgical wounds with propolis[J]. *Folia Medica*, 24(2):24-27.

Dealey C (2005) *The Care of Wounds*[M]. Blackwell Publishing.

Dieterich K (1908) The analysis of beeswax in its several stages of formation and concerning the bee resin (propolis)[J]. *Pharmazeutische Post*, 40:369.

Dieterich K (1911) Further contributions to the knowledge of bee resin (propolis)[J]. *Pharmazeutische Zentralhalle für Deutschland*, 52:1019-1027.

Doroshenko PN (1983) Treatment of chronic tonsillitis patients with a propolis-wax paste [J]. *Meditsinskaia Sestra*, 42(11):36-37.

Elkins R (1996) *Bee pollen, royal jelly, propolis and honey: an extraordinary energy and health-promoting ensemble*[M]. Houghton Mifflin Harcourt, USA.

Emilewicz T, Kostanecki S, Tambor J (1899) Synthese des Chrysins[J]. *Berichte der Deutschen Chemischen Gesellschaft*, 32(2):2448-2450.

Evans J, Aronstein K, Chen Y, Hetru C, Imler JL, Jiang H, Kanost M, Thompson G, Zou Z, Hultmark D (2006) Immune pathways and defence mechanisms in honey bees *Apis mellifera*[J]. *Insect Molecular Biology*, 15(5):645-656.

Evans JD, Spivak M (2010) Socialized medicine: individual and communal disease barriers in honey bees[J]. *Journal of Invertebrate Pathology*, 103:S62-S72.

Fearnley J (2001) *Bee Propolis: Natural Healing from the Hive*[M]. Souvenir Press, London, UK.

Ghisalberti EL, Jefferies PR, Lanteri R (1977) Potential drugs from propolis. in *Mass Spectrometry in Drug Metabolism*[M]. 111-130, Plenum Press, New York, NY, USA.

Gorbatenko AG (1971) Treatment of ulcer patients with a 30 percent alcohol solution of propolis[J]. *Vrachebnoe Delo*, 3(1):22-24. (Russian).

Haydak MH, Palmer LS (1940) Vitamin content of bee foods. II Vitamin B1 content of royal jelly and beebread[J]. *Journal of Economic Entomology*, 33:396-397.

Haydak MH, Palmer LS (1941) Vitamin content of bee foods. III vitamin A and riboflavin content of beebread[J]. *Journal of Economic Entomology*, 34:37-38.

Haydak MH, Palmer LS (1942) Royal jelly and beebread as sources of vitamin B1, B2, B6, C, nicotinic and panthothenic acid[J]. *Journal of Economic Entomology*, 35:319.

Heinen W, Linskens HF (1972) Occurrence of fatty acids in propolis[J]. *Portugaliae Acta Biologica*, 12:56-76.

Helfenberg KD (1908) The analysis of beeswax and propolis[J]. *Chemiker Zeitung*, 31:987-988.

Herold E (1970) *Heilwerte aus dem Bienenvolk*[M]. Ehrenwirth,München,Germany.

Herzig J (1891) Studies on quercetin and its derivatives,treatise VII[J]. *Monatshefte für Chemie*,12(1):177-190.

Houghton PJ (1998) Propolis as a medicine. Are there scientific reasons for its reputation? in *Beeswax and Propolis for Pleasure and Profit*[M]. Munn P,Ed. ,p. 10,International Bee Research Association,Cardiff,UK.

Jaubert GE (1926) Origin of the colour of beeswax and the composition of propolis[J]. *Comptes Rendus Hebdomadaires des Seances de l' Académie des Sciences*, 184: 1134-1136.

Korochkin IM,Poslavskii MV (1986) Treatment of chronic gastroduodenal ulcers by local administration of propolis[J]. *Sovetskaya Meditsina*,10:105-107.

Kuropatnicki AK,Szliszka E,Krol W (2013) Historical aspects of propolis research in modern times[J]. *Evidence-Based Complementary and Alternative Medicine*,Volume 2013,Article ID 964149,11 pages.

Kustenmacher M (1911) Propolis[J]. *Berichte der Deutshen Pharmacologische Gesellschaft*,21: 65-92.

Langenheim JH (2003) *Plant Resins:Chemistry,Evolution,Ecology,Ethnobotany* [M]. Timber Press,Cambridge,UK.

Lavie P (1957) Étude de substance antibiotiques presentés chez *Apis mellifica* at chez quelques insects sociaux[J]. *Comptes Rendus Hebdomadaires des Seances de l'Academie des Sciences*,24:2653-2655.

Lewis ChT,Short C (2012) *A Latin Dictionary*[M]. http://www. perseus. tufts. edu/hopper/text? doc=Perseus:text:1999. 04. 0059:entry=propolis.

Lindenfelser LA (1967) Antimicrobial activity of propolis[J]. *American Bee Journal*,107: 90-92.

Lutsenko SM,Pisarenko AS (1980) Use of water-soluble propolis for trophic ulcers of the lower extremities in arteriosclerosis obliterans[J]. *Klinicheskaya Khirurgiya*,7:74.

Makarov FD (1972) Propolis treatment of ulcer disease and pyloroduodenitis[J]. *Vrachebnoe Delo*,4:93-96.

Marcucci MC (1995) Propolis:chemical composition,biological properties and therapeutic activity[J]. *Apidologie*,26(2):83-99.

Metzner J,Bekemeier H,Schneidewind E,Schwaiberger R (1975) Bioautographische Erfassung der antimikrobiell wirksammen Inhaltstoffe von propolis[J]. *Pharmazie*, 30:799-800.

Murray MT,Pizzorno JE. (2005) Bee products:pollen,propolis,and royal jelly,in *Textbook of Natural Medicine*[M]. Eds. ,chapter 70,Elsevier Health Sciences. http://books. google. pl/books? id = e9mcO1EhfokC&pg = PT2302&lpg = PT2302&dq ♯ v = onepage&q&f=false.

Nikiforov A,Kosin P,Alekseeva AL (1971) Spectral study of copper andmanganese levels in pollen, beebread and excrement of bees[J]. *Uchenye Zapiski Kazanskogo Gosudarstvennogo Veterinarnogo Instituta*, 108: 180-181.

Paintz M, Metzner J (1979) On the local anaesthetic action of propolis and some of its constituents[J]. *Pharmazie*, 34(12): 839-841.

Perkin AG, Everest AE (1918) *The Natural Organic Colouring Matters*[M]. Longmans, London, UK.

Piccard J (1873) On chrysin and its haloid derivatives[J]. *The Chemical News and Journal of Physical Science*, 27(717): 97-98.

Piccard J (1874) Ueber die Constitution des Chrysins und des Tectochrysins[J]. *Berichte der Deutschen Chemischen Gesellschaft*, 7(1): 888-892.

Piccard J (1877) Ueber Chrysin, Tectochrysin und hohere Homologe[J]. *Berichte der Deutschen Chemischen Gesellschaft*, 10(1): 176-180.

Pliny the Elder (1855a) The meaning of the terms commosis, pissoceros, and propolis, in Pliny the Elder, the Natural History, Book XI. *The Various Kinds of Insects*[M]. Bostock J and Riley HT, Eds., chapter 6(5), Taylor and Francis, London, UK. http://www.perseus.tufts.edu/hopper/text?doc=Perseus%3Atext%3A1999.02.0137%3Abook%3D11%3Achapter%3D6.

Pliny the Elder (1855b) Propolis: five remedies, in Pliny the Elder, the Natural History, Book XXII. *The Properties of Plants and Fruits*[M]. Bostock J and Riley HT, Eds., chapter 50 (24), Taylor and Francis, London, UK. http://www.perseus.tufts.edu/hopper/text?doc=Perseus%3Atext%3A1999.02.0137%3Abook%3D22%3Achapter%3D50.

Popravko SA, Gurevich AI, Kolosov MN (1969) Flavonoid components of propolis[J]. *Khimiya Prirodnykh Soedinenii*, 5: 476-482.

Popravko SA, Gurevich AI, Kolosov MN (1970) Isolation and identification of the main components of propolis [C]. In Proceedings of the 22nd International Beekeeping Congress: Summaries, 163-164.

Popravko SA (1978) Chemical composition of propolis, its origin and standardization. *A Remarkable Hive Product: Propolis, Apimondia Publ*[M]. House, Bucharest, 15-18.

Proceedings of the 4th International Symposium on Food Microbiology (1964) [C] Goteborg, Sweden, 59-75.

Prokopovich NN, Flis ZA, Frankovskaya ZI, Kopeeva EP (1956) An anaesthetizing substance for use in stomatology[J]. *Vrachebnoe Delo*, 1: 41-44.

Prokopovich NN (1957) Propolis a new anaesthetic[J]. *Vrachebnoe Delo*, 10: 1077-1080. (Russian).

Rösch GA (1927) Beobachtungen an Kittharz sammelnden Bienen (*Apis melliferca* L.) [J]. *Biologisches Zentralblatt*, 47: 113-121.

Scheller S, Rogala D, Stasiak E, Zurek H (1968) Antibacterial properties of propolis[J]. *Polskie Archiwum Weterynaryjne*, 11: 391-398.

Scheller S, Szaflarski J, Tustanowski J, Nolewajka E, and Stojko A (1977) Biological properties and clinical application of propolis. I. Some physico chemical properties of propolis[J]. *Arzneimittel-Forsch*, 27(4):889-890.

Scheller S, Stojko A, Szwarnowiecka I, Tustanowski J, Obuszko Z (1977) Biological properties and clinical application of propolis. VI. Investigation of the influence of ethanol extracts of propolis (EEP) on cartilaginous tissue regeneration[J]. *Arzneimittel-Forsch*, 27(11):2138-2140.

Scheller S, Ilewicz L, Luciak M, Skrobidurska D, Stojko A, Matuga W (1978) Biological properties and clinical application of propolis. IX. Experimental observation on the influence of ethanol extract of propolis (EEP) on dental pulp regeneration[J]. *Arzneimittel-Forsch*, 28(2):289-291.

Schmid-Hempel P (1998) *Parasites in Social Insects*[M]. Princeton University Press.

Schneidwein EM, Kala H, Linzer B, Metzner J (1975) Zur kenntnis der inhaltsstoffe von propolis[J]. *Pharmazie*, 30:803.

Seeley TD, Morse RA (1976) The nest of the honey bee (*Apis mellifera* L.)[J]. *Insectes Sociaux*, 23(4):495-512.

Sutta J, Hanko J, Janda J, Nolewajka E, and Stojko A (1975) Experimental and clinical experience of the treatment of wounds of domestic animals with a locally applied solution of propolis[J]. *Folia Veterinaria*, 18:143-147.

Stojko A, Scheller S, Szwarnowiecka I, Tustanowski J, Ostach H, and Obuszko Z (1978) Biological properties and clinical application of propolis. VIII. Experimental observation on the influence of ethanol extract of propolis (EEP) on the regeneration of bone tissue[J]. *Arzneimittel-Forsch*, 28(1):35-37.

Szmeja Z, Kulczynski B, Sosnowski Z, Konopacki K (1989) Therapeutic value of flavonoid in rhinovirus infections[J]. *Otolaryngologia Polska*, 43(3):180-184. (Polish).

Todorov V, Drenovski S, Vasilev V (1968) Pharmacodynamics of propolis[J]. *Farmatsiya*, 18:23-31. (Russian).

Tsacov T (1973) Investigations on anesthetic properties of propolis (bee's glue)[J]. *Farmatsiya*, 23(2):38-41. (Bulgarian).

Villanueva VR, Bogdanovsky D, Barbier M, Gonnet M, Lavie P (1964) Sur l'identification de la 3,5,7-trihydroxy flavones (galangine) á partir de la propolis[J]. *AnnaLes de L'Institut Pasteur*, 106:292-302.

Villanueva VR, Barbier M, Gonnet M, Lavie P (1970) Les flavonoides de la propolis. Isolement d'une nouvelle substance bacteriostatique: la pinocembrine[J]. *AnnaLes de L'Institut Pasteur*, 118:84-87.

Varro MT, Rustica D (1934) *On Agriculture. Book Three*[M]. Loeb Classical Library.

Vauquelin LN (1803) Analysis of the propolis or mastic of bees[J]. *Chemistry and the Arts*, 5:48-49.

Visscher P (1980) Adaptations of honey bees (*Apis mellifera*) to problems of nest hygiene[J]. *Sociobiology*, 5:249-260.

陈友地,姜紫荣,崔宁惠(1983)蜂胶挥发物成分的研究[J].中国养蜂,(5):10-12.
程晓雨,胡福良(2015)我国蜂胶保健食品概况[J].中国蜂业,66(4):45-47.
房柱(1959)蜂膠治疗鸡眼、胼胝、蹠疣和寻常疣的初步观察[J].中华皮肤科杂志,7(4):240-241.
房柱,匡友成(1976)蜂胶抗医学真菌的实验简报[J].中国养蜂,(3):42.
房柱,匡友成(1978)蜂胶对医学真菌抑菌试验[J].微生物学通报,(1):12-14.
国家药典委员会(2015)中华人民共和国药典(2015版,一部)[M].北京:中国医药科技出版社.
胡福良(2005)蜂胶药理作用研究[M].杭州:浙江大学出版社.
李景祥,房柱(1981)蜂胶治疗高脂血症167例疗效观察(摘要)[J].江苏医药,(9):40.
平舜,蔺哲广,胡福良(2014)蜂胶对蜂群病虫害的防治作用及机制研究[J].环境昆虫学报,36(3):427-432.
钱锐,陈志媛(1979)蜂胶的有效成分提取和鉴定[J].蜜蜂杂志,(2):32-33.
尚天民,徐景耀(1983)蜂胶化学成分的研究[J].中国养蜂,(4):20-22.
沈志强,杨永福(1989)禽霍乱蜂胶菌苗的研究:Ⅰ.制苗工艺,安全性与免疫原性试验[J].中国畜禽传染病,11(5):1-3.
沈志强,徐可利,杨永福(1990a)禽霍乱蜂胶菌苗的研究:Ⅱ.菌苗的免疫期、保存期及免疫强度试验[J].中国畜禽传染病,12(4):15-16.
沈志强,徐可利,陆峰,赵国平(1990b)禽霍乱蜂胶菌苗的研究:Ⅲ.禽霍乱蜂胶菌苗与氢氧化铝菌苗的免疫对比试验[J].中国畜禽传染病,12(5):18-19.
沈志强,徐可利(1990)禽霍乱蜂胶菌苗的研究:Ⅳ.菌苗的田间免疫试验[J].中国畜禽传染病,12(6):12-13.
沈志强,徐可利(1991)禽霍乱蜂胶菌苗与新城疫Ⅰ系苗联合免疫的探讨[J].中国畜禽传染病,13(6):11-12.
王秉极,张慧娟(1988)北京蜂胶的化学成分研究[J].中药通报,1988,13(10):37-38.
王凯,胡福良(2014)国内外蜂胶专利分析[J].中国蜂业,65(12):41-43.
王凯,张翠平,胡福良(2013)2012年国内外蜂胶研究概况[J].蜜蜂杂志,33(5):4-8;(6):3-7.
王凯,张翠平,胡福良(2014)2013年国内外蜂胶研究概况[J].蜜蜂杂志,34(3):1-5;(4):1-5.
王凯,张翠平,胡福良(2015)2014年国内外蜂胶研究概况[J].蜜蜂杂志,35(3):1-6;(4):1-4.
张翠平,胡福良(2010)2008—2009年国内外蜂胶研究概况[J].中国蜂业,61(4):35-38.
张翠平,胡福良(2011)2010年国内外蜂胶研究概况[J].蜜蜂杂志,31(6):1-4;(7):5-7.
张翠平,胡福良(2012)2011年国内外蜂胶研究概况[J].中国蜂业,63(1-3,中):64-73.
朱道程(1981)三种剂量蜂胶片治疗高脂血症319例的临床疗效[J].中国养蜂,(1):17-19.

第一篇

国内外蜂胶研究进展

第二章　蜂胶的植物来源与化学成分

第一节　蜂胶的植物来源

公元1世纪，古罗马学者普里尼乌斯（Puri Arrhenius）通过对蜜蜂采集行为的观察，认为蜂胶是蜜蜂从柳树、杨树、栗树以及其他植物幼芽上采集的树脂分泌物。然而，由于缺乏相应的科学依据，加之当时的技术水平较为有限，有关蜂胶来源的问题一直没有得到证实。直到1956年，Meyer运用照相法详细记录了蜜蜂采集制造蜂胶的全过程："蜂胶采集蜂用其上颚咬下一颗胶粒，用两前足把持住，再用一只中足伸向口器下的两前足，将胶粒送到同侧后足的花粉筐，当它将胶粒向花粉筐上填装的同时，又伸出前足去接应新的胶粒"；"蜜蜂反复剥离胶粒和向花粉筐中装填要花很长时间，最后才能满载蜂胶回到巢中，在内勤蜂的帮助下，将蜂胶从花粉筐中取出（可能需要等候一个小时或几小时才能卸完）"；"内勤蜂用上颚将蜂胶撕咬下来，并用上颚腺分泌物和蜂蜡调制蜂胶，用蜂胶加固蜂巢、填补缝隙或送至其他需要的地方"（Meyer，1956）。这项研究探明了蜂胶的来源及其形成机制，即蜂胶是蜜蜂采集植物组织及其分泌物，加入上颚腺、蜡腺等自身腺体分泌物混合加工而成的。这极大地推动了蜂胶植物来源的研究。

由于蜜蜂的采胶行为相对较为少见，并且经常发生在胶源植物的顶端位置，这就给研究者通过观察蜜蜂采胶行为来确定胶源植物带来了很大的困难。目前，研究者主要通过观察蜜蜂采胶行为，结合对比分析植物组织和蜂胶化学组成的方法来确定蜂胶的胶源植物。此外，由于蜜蜂在采集植物树脂分泌物时，会将黏在上面的植物组织碎片一同运输到蜂箱中制成蜂胶，因此对比蜂胶中所含植物组织碎片与胶源植物的组织形态学特征，也能够为蜂胶植物来源的确定提供一定的参考依据（Teixeira et al.，2005）。

温带、亚热带和热带地区气候条件的不同，造成植物分布的不同，因此不同地区蜂胶的植物来源也有很大的差异。温带地区主要的胶源植物有杨属（*Popolus*）、桦树（*Betula* spp.）、榆树（*Ulmus* spp.）、桤木（*Alnus* spp.）、豚草（*Ambrosia* spp.）、桉树（*Eucalyptus* spp.）、山毛榉（*Fagaceae* spp.）、橡树（*Quercus* spp.）、七叶树（*Aesculus hippocastanum*）以及柏属（*Cupressaceae*）、松属（*Pinus*）植物等（Ghisalberti，1979）；亚热带地中海地区蜂胶的化学成分与温带地区蜂胶的化学成分存在差异，但胶源植物仍然是以杨属植物为主，还有桉树、血桐属（*Macaranga*）、柏属以及松属等其他胶源植物；热带地区气候复杂，植物种类繁多，所以胶源植物的种类也存在多样性，主要包括杨属、克鲁西属（*Clusia*）、南洋杉属（*Araucuria*）、桉属、酒神菊属（*Baccharis*）等。目前，根据蜂胶植物来源的不同，全世界已发现的蜂胶大体可以分为5种类型：杨属型、酒神菊属型、克鲁西属型、血桐属型和地中海属型。

(1) 杨属型蜂胶

杨属及其杂交属植物作为蜂胶的主要胶源来源,广泛分布于欧洲、北美、亚洲、大洋洲以及非洲的温带地区。该植物来源的蜂胶被定义为"杨属型蜂胶",其特征性化学成分是 B 环无取代基的类黄酮以及苯丙酸及其酯类,如松属素、短叶松素、高良姜素、柯因和咖啡酸苯乙酯(Caffeic acid phenethyl ester, CAPE)。

(2) 酒神菊属型蜂胶

酒神菊属型蜂胶的胶源植物是酒神菊属植物,主要分布于巴西东南部(Matsuda et al., 2008),俗称"巴西绿蜂胶",主要成分为异戊烯苯丙类(如阿替匹林 C 和咖啡酰奎尼酸)以及二萜类化合物(dos Santos Pereiraa et al., 2003)。近年来,该类型的蜂胶也在南美的玻利维亚被发现(Nina et al., 2016)。

(3) 克鲁西属型蜂胶

克鲁西属型蜂胶的植物来源主要是克鲁西属植物花的分泌物,其特征性成分是异戊烯基苯甲酮,主要分布于委内瑞拉(Tomás-Barberán et al., 1993)、古巴(Hernández et al., 2005)、巴西亚马逊(de Castro Ishida et al., 2011)和巴西东北部(Trusheva et al., 2006)。

(4) 血桐属型蜂胶

血桐属型蜂胶的植物来源是血桐属植物的果皮分泌物,其特征性化学成分为带有香叶基的黄酮类化合物。该类型蜂胶主要分布于中国台湾(Huang et al., 2007)、日本冲绳岛(Kumazawa et al., 2008)等太平洋岛屿地区以及埃及(El-Bassuony, 2009)、肯尼亚(Petrova et al., 2010)等非洲西北部地区。

(5) 地中海属型蜂胶

根据所含特征性化合物推断,地中海属型蜂胶的主要胶源植物可能为柏属植物和松属植物。西西里岛(Trusheva et al., 2003)、克里特岛(Popova et al., 2009)以及希腊西北部地区(Melliou et al., 2004)的蜂胶,由于其含有多种柏属植物的特征性二萜类化合物如铁锈醇、桃柘酚等,因此推断该地区蜂胶的胶源植物为柏属植物。希腊其他地区的蜂胶因含有大量的丁二酰衍生物等多种松属植物的特征性化学成分,因此推断其胶源植物可能为松属植物。

此外,随着研究的不断深入以及现代分析技术的不断发展,蜂胶中的新成分不断被鉴定出来,同时越来越多的胶源植物被发现或证实。例如,黄檀属植物(*Dalbergia ecastophyllum*)为巴西红蜂胶的胶源植物(Alencar et al., 2007),金合欢属植物(*Acacia paradoxa*)、鳞子莎属植物(*Lepidosperma*)为产自澳大利亚袋鼠岛蜂胶的胶源植物(Tran et al., 2012; Duke et al., 2017),等等。表 2.1 详细地列出了不同地理来源蜂胶的主要胶源植物。

表 2.1 不同地理来源蜂胶的胶源植物

Table 2.1 The plant origin of propolis from different geographical locations

蜂胶类型	地理来源	胶源植物	参考文献
杨属型 *Populus*	阿尔巴尼亚	黑杨 *P. nigra*	(Bankova et al., 1994)
	保加利亚	黑杨 *P. nigra* 欧洲山杨 *P. tremula*	(Marcucci, 1995)
	比利时	欧洲山杨 *P. tremula*	(Marcucci, 1995)

续表

蜂胶类型	地理来源	胶源植物	参考文献
杨属型 Populus	蒙古	甜杨 P. suaveolens	(Marcucci, 1995)
	美国（大陆）	佛里蒙杨 P. fremontii	(Marcucci, 1995)
	英国	欧美杨 P. euramericana	(Marcucci, 1995)
	欧洲、北美、亚洲温带地区	黑杨 P. nigra	(Greenaway et al., 1990; Bankova et al., 2000)
	巴西南部	银白杨 P. alba	(Park et al., 2002)
	阿根廷	银白杨 P. alba	(Park et al., 2002)
	乌拉圭	银白杨 P. alba	(Kumazawa et al., 2004b)
	匈牙利	杨属 Populus	(Marcucci, 1995)
	新西兰	杨属 Populus	(Markham et al., 1996)
	土耳其	杨属 Populus	(Silici et al., 2005)
	墨西哥	杨属 Populus	(Li et al., 2010)
	伊朗	杨属 Populus	(Trusheva et al., 2010)
	中国	杨属 Populus	(Zhang et al., 2014)
酒神菊属型 Baccharis	巴西（绿蜂胶）	酒神菊 B. dracunculifolia	(Kumazawa et al., 2003)
克鲁西属型 Clusia	赤道地区	克鲁西属 Clusia	(Marcucci, 1995)
	委内瑞拉	大克鲁西 C. major 小克鲁西 C. minor	(Trusheva et al., 2004)
	巴西东北部	克鲁西属 Clusia	(Trusheva et al., 2006)
	巴西亚马逊	克鲁西属 Clusia	(de Castro Ishida et al., 2011)
	古巴	C. nemorosa	(Camargo et al., 2013)
血桐属型 Macaranga	中国台湾	血桐属 Macaranga	(Huang et al., 2007)
	日本冲绳岛	血桐 M. tanarius	(Kumazawa et al., 2008)
	埃及	血桐属 Macaranga	(El-Bassuony, 2009)
	肯尼亚	血桐属 Macaranga	(Petrova et al., 2010)
	印度尼西亚	M. tanarkus	(Trusheva et al., 2011)
	所罗门群岛	血桐属 Macaranga	(Inui et al., 2012a)
桦木属型 Betula	北温带地区	桦木属 Betula	(Ghisalberti, 1979)
	乌拉圭	桦木属 Betula	(Bonvehi et al., 1994)
	波兰、匈牙利	桦木属 Betula	(Marcucci, 1995)
	俄罗斯北部	疣皮桦 B. verrucosa	(Bankova et al., 2000)
	乌克兰	桦木属 Betula	(Kumazawa et al., 2004b)

续表

蜂胶类型	地理来源	胶源植物	参考文献
柏属型 Cupressaceae	西西里岛	柏属 Cupressaceae	(Trusheva et al.,2003)
	希腊西北部	柏属 Cupressaceae	(Melliou et al.,2004)
	克里特岛	柏属 Cupressaceae	(Popova et al.,2009)
松属型 Pinus	土耳其	南意松 P. brutia	(Kartal et al.,2002)
桉树属型 Eucalyptus	乌拉圭	蓝桉 E. globules	(Bonvehí et al.,1994)
	埃及	桉树属 Eucalyptus	(El Hady et al.,2000)
	巴西	桉树属 Eucalyptus	(Pereira et al.,2002)
	土耳其	桉树属 Eucalyptus	(Silici et al.,2007)
桤木属型 Alnus	波兰	桤木属 Alnus	(Marcucci,1995)
黄檀属型 Dalbergia	巴西东北部	D. ecastophyllum	(Alencar et al.,2007;Bueno-Silva et al.,2013)
	墨西哥	Dalbergia	(Lotti et al.,2010)
南洋杉属 Araucuria	巴西	A. angustifolia A. heterophylla	(Banskota et al.,2001a)
柳属 Salix	乌拉圭	柳属 Salix	(Bonvehí et al.,1994)
	土耳其	白柳 S. alba; S. fragillis	(Silici et al.,2005)
金合欢属 Acacia	澳大利亚（袋鼠岛）	A. paradoxa	(Tran et al.,2012)
檀栗属 Castanea	土耳其	C. sativa	(Silici et al.,2007)
岩蔷薇属 Cistus	突尼斯	岩蔷薇 Cistus	(Martos et al.,1997)
	葡萄牙	C. ladanifer	(Falcão et al.,2013)
阿魏属 Ferula	伊朗	F. gumosa; F. asafetida	(Trusheva et al.,2010)
	马耳他	Ferula	(Popova et al.,2011)
南美槐属 Myroxylon	萨尔瓦多	M. balsamum	(Popova et al.,2002)
杧果属 Mangifera	印度尼西亚	M. indica	(Trusheva et al.,2011)
其他类型	澳大利亚	Xanthorrhoea Lepidosperma	(Ghisalberti,1979) (Duke et al.,2017)
	赤道地区	Delchampia	(Marcucci,1995)
	索诺兰沙漠	Ambrosia deltoidea	(Silici et al.,2007)
	巴西东北部 墨西哥东南部	Hyptis divaricate Bursera simaruba Lysiloma lasisiliquum	(Castro et al.,2009) (Biosard et al.,2016)

第二节 蜂胶的化学成分

对蜂胶化学成分的研究始于德国,并经历了一个漫长的过程。1910年,Küstenmacher在蜂胶中发现了肉桂酸和肉桂醇(Küstenmacher,1911)。次年,Dietrich在蜂胶中发现了香草(Dietrich,1911)。但直到1927年,才确认蜂胶的类黄酮化合物来源于胶源植物——杨树。由于蜂胶的化学成分极其复杂,加之现代科学技术的发展还处于起步阶段,当时尚无适用于蜂胶化学成分分离与鉴定的技术,所以在很长一段时间内,对蜂胶化学成分的研究处于相对停滞状态。20世纪50年代以来,色谱技术的发展与质谱技术的出现,给蜂胶化学成分的研究注入了新的活力,极大地促进了蜂胶化学成分研究的发展。

蜂胶中既有胶源植物分泌物,又有蜜蜂腺体分泌物,集动植物精华于一身,是蜜蜂以一种人力不可及的配伍方式,经过复杂的生化过程加工转化而成。从基本组成成分来看,以中国杨属型蜂胶为例,蜂胶中主要包括大约50%的树脂和树香、30%的蜂蜡、10%的芳香挥发油、5%的花粉以及5%的杂质。从化学组成物质来看,蜂胶中主要包括类黄酮化合物、萜类化合物、酚酸类化合物、醛酮类化合物、烃类化合物、甾体化合物、糖类化合物、维生素、酶类、氨基酸、常量以及微量元素等。迄今为止,世界各国的蜂胶研究者运用不同的技术,已从蜂胶中分离鉴定出600多种化合物。其中,类黄酮化合物204种、萜类化合物231种、酚类化合物194种。

1 类黄酮化合物

类黄酮是一大类以 C_6-C_3-C_6 为基本骨架的低分子量多酚化合物,广泛存在于自然界中,大多数类黄酮化合物具有颜色。蜂胶中含有的类黄酮化合物种类之多和含量之丰富,是任何一种植物药物所不能比拟的。

作为蜂胶的主要功效成分,类黄酮化合物很大程度上决定了蜂胶的特殊药理活性(朱彤等,1997)。研究表明,类黄酮化合物具有抗氧化、抗肿瘤、抗衰老、抗炎症、调节免疫、促进组织再生等功效(唐传核和彭志英,2001,2002)。因此,类黄酮化合物的含量通常作为评价蜂胶质量的主要指标之一(GB/T 24283—2009)。迄今为止,从世界各地不同蜂胶中分离出的类黄酮化合物已达204种,主要包括黄酮、黄酮醇、二氢黄酮、二氢黄酮醇、异黄酮、二氢异黄酮、查耳酮、二氢查耳酮、黄烷、异黄烷和新类黄酮类化合物(主要为紫檀素和4-苯基香豆素类化合物)等(张翠平和胡福良,2009)。此外,研究者还在蜂胶中发现了鼠李糖苷(Popova et al.,2001)、槲皮素-3-氧芸香糖苷(Dondi et al.,1984;Bonvehi et al.,1994)、异鼠李素-3-氧芸香糖苷(Popova et al.,2009)和黄酮碳苷(Righi et al.,2011)等4种黄酮苷类化合物。

1.1 黄酮和黄酮醇类化合物

黄酮及黄酮醇类化合物广泛地分布于各种植物体中。常见的黄酮类化合物为芹菜素、木樨草素、茨菲醇、槲皮素和杨梅黄素(张翠平和胡福良,2009),这5种化合物及其衍生物在蜂胶中都广泛存在。由于黄酮醇类化合物所含的羟基可以提供氢离子,阻碍自由基链的氧化,从而表现出较强的抗氧化活性(唐传核,2005)。近年来,随着研究的不断深入,发现该类化合物除了具有优越的抗氧化功能之外还具有较好的抗菌活性(孟申,1992;王玮和王琳,

2002)。蜂胶中已发现的黄酮类化合物和黄酮醇类化合物见表 2.2 和表 2.3。

表 2.2 蜂胶中的黄酮类化合物
Table 2.2 Flavones in propolis

序号 No.	英文名称 English name	中文名称 Chinese name	参考文献 Ref.
1	Acacetin	刺槐素(金合欢素)	(Popravko et al.,1969)
2	5-Hydroxy-4′,7-dimethoxy flavone	5-羟基-4′,7-二氧甲氧基黄酮	(Popova et al.,2011)
3	Ermanin	5,7-二羟基-3,4′-二甲氧基黄酮	(Popova et al.,2011)
4	Apigenin	芹菜素(洋芹素)	(Greenaway et al.,1987)
5	Chrysin	柯因(白杨素)	(Greenaway et al.,1987)
6	4′,5,7-Trihydroxy-6-methoxyflavone	4′,5,7-三羟基-6-甲氧基黄酮	(Walker and Crane.,1987)
7	5,7-Dihydroxy-3,4′,6-trimethoxyflavone	5,7-二羟基-3,4′,6-三甲氧基黄酮	(Walker and Crane.,1987)
8	5,7-Dihydroxy-3-methoxyflavone	5,7-二羟基-3-甲氧基黄酮	(Walker and Crane.,1987)
9	5,7-Dihydroxy-2′-methoxyflavone	5,7-二羟基-2′-甲氧基黄酮	(Wang and Zhang.,1988)
10	Pectolinarigenin	柳穿鱼素	(Marcucci,1995)
11	Tectochrysin	柚木柯因(柚木杨素)	(Markham et al.,1996)
12	Genkwanin(Apigenin-7-methyl ether)	芫花(黄)素(芹菜素-7-甲醚)	(Chi et al.,1996)
13	5,7,4′-Trihydroxy-6,8-dimethoxy flavone	5,7,4′-三羟基-6,8-二甲基黄酮	(Wollenweber et al.,1997)
14	Siderite flavone	5,3′,4′-三羟基-6,7,8-三甲氧基黄酮	(Wollenweber et al.,1997)
15	5,6,7-Trihydroxy-3,4′-dihymethoxy flavone	5,6,7-三羟基-3,4′-二甲氧基黄酮	(Boundourova-Krasteva et al.,1997)
16	Pilloin	3′,5-二羟基-4′,7-二甲氧基黄酮	(Maciejewicz et al.,2001)
17	Hexamethoxy flavone	六甲氧基黄酮	(Faten,2002)
18	6-Cinnamylchrysin	6-苯丙烯柯因	(Usia et al.,2002)
19	Luteolin	木樨草素	(Cao et al.,2004)

表 2.3 蜂胶中的黄酮醇类化合物
Table 2.3 Flavonols in propolis

序号 No.	英文名称 English name	中文名称 Chinese name	参考文献 Ref.
1	Kaempferide	山柰素(莰非素)	(Popova et al.,2011)
2	Quercetin	槲皮素(栎精)	(Ghisalberti,1979)
3	Kaempferol	山柰酚(莰菲醇)	(Greenaway et al.,1987)
4	Kaempferol-3-methyl ether	山柰酚-3-甲醚	(Greenaway et al.,1987)

续表

序号 No.	英文名称 English name	中文名称 Chinese name	参考文献 Ref.
5	7-Methoxyquercetin	槲皮素-7-甲氧基	(Walker and Crane,1987)
6	3,7-Dimethoxyquercetin	槲皮素-3,7-二甲氧基	(Walker and Crane,1987)
7	Galangin	高良姜精	(Greenaway et al.,1987)
8	Galangin-3-methyl ether	高良姜精-3-甲醚	(Greenaway et al.,1987)
9	Izalpinin	良姜素(高良姜精-7-甲醚)	(Greenaway et al.,1987)
10	Alnusin	赤杨黄酮	(Walker and Crane,1987)
11	Rhamnetin	鼠李黄素(鼠李素)	(Greenaway et al.,1987)
12	Rhamnocitrin	鼠李柠檬素	(Greenaway et al.,1987)
13	3,5-Dihydroxy-4′,7-dimethoxyflavone	3,5-二羟基-4′,7-二甲氧基黄酮	(Walker and Crane,1987)
14	Isalpinin	3,5-二羟基-7-甲氧基黄酮	(Walker and Crane,1987)
15	3,4′,5,7-Tetrahydroxy-3′-methoxyflavone	3,4′,5,7-四羟基-3′-甲氧基黄酮	(Walker and Crane,1987)
16	3,4′,5-Trihydroxy-3′,7-dimethoxyflavone	3,4′,5-三羟基-3′,7-二甲氧基黄酮	(Walker and Crane,1987)
17	Kaempferol-7,4-′dimethyl ether	山奈酚-7,4′-二甲醚	(Greenaway et al.,1990)
18	Fisetin	非瑟酮	(Greenaway et al.,1990)
19	3-Methylkaempferol	山奈酚-3-甲基	(Johnson et al.,1994)
20	Quercetin-3,3′-dimethyl ether	槲皮素-3,3′-二甲醚	(Johnson et al.,1994)
21	Isorhamnetin	异鼠李素	(Marcucci,1995)
22	Quercetin-3,7,3′-trimethyl ether	槲皮素-3,7,3′-三甲醚	(Martos et al.,1997)
23	Myricetin 3,7,4′,5′-tetramethyl ether	杨梅素-3,7,4′,5′-四甲醚	(Martos et al.,1997)
24	Betuletol	3,5,7-三羟基-6,4′-二甲氧基黄酮	(Tazawa et al.,1999)
25	5-Methoxykaempferol	5-甲氧基山奈素	(郭伽和周立东,2000)
26	3,7-Methoxykaempferol	3,7-二甲氧基山奈素	(郭伽和周立东,2000)
27	Quercetin-3-methyl ether	槲皮素-3-甲醚	(郭伽和周立东,2000)
28	Quercetin-3′-methyl ether	槲皮素-3′-甲醚	(郭伽和周立东,2000)
29	Galangin-3,7-dimethyl ether	高良姜素-3,7-二甲醚	(郭伽和周立东,2000)
30	Quercetin-7,3′-dimethyl ether	槲皮素-7,3′-二甲醚	(郭伽和周立东,2000)
31	Galangin-5-methyl ether	高良姜素-5-甲醚	(郭伽和周立东,2000)
32	(7″R)-8-[1-(4′-Hydroxy-3′-methoxyphenyl)prop-2-en-1-yl]-galangin	(7″R)-8-[1-(4′-羟基-3′-甲氧苯)烯丙基]-高良姜素	(Li et al.,2010)
33	Macarangin	—	(Petrova et al.,2010)
34	2′-(8″-Hydroxy-3″,8″-dimethyl-oct-2″-enyl)-quercetin	2′-(8″-羟基-3″,8″-二甲基-辛-2″-烯基)-槲皮素	(Inui et al.,2012a)
35	8-(8″-Hydroxy-3″,8″-dimethyl-oct-2″-enyl)-quercetin	8-(8″-羟基-3″,8″-二甲基-辛-2″-烯基)-槲皮素	(Inui et al.,2012a)
36	2′-Geranylquercetin	2′-香叶基槲皮素	(Inui et al.,2012a)

1.2 二氢黄酮和二氢黄酮醇类化合物

二氢黄酮和二氢黄酮醇类化合物是自然界中广泛存在的一类具诸多生理活性的重要化合物。作为类黄酮化合物中的微量化合物,二氢黄酮和二氢黄酮醇具有抗氧化、抗癌、抗炎症等广泛的生物活性(孟申,1992;王玮和王琳,2002)。蜂胶中最常见的二氢黄酮为球松素和乔松素,常见的二氢黄酮醇类为短叶松素及其衍生物。截至2015年年底,世界各国共从蜂胶中分离鉴定出二氢黄酮类化合物43种、二氢黄酮醇类化合物18种(表2.4和表2.5)。

表2.4 蜂胶中的二氢黄酮类化合物
Table 2.4 Flavanones in propolis

序号 No.	英文名称 English name	中文名称 Chinese name	参考文献 Ref.
1	Pinostrobin	乔松酮(球松素)	(Popova et al.,2011)
2	5-Hydroxy-7,4′-dimethoxy flavanone	5-羟基-7,4′-二甲氧基二氢黄酮	(Popova et al.,2011)
3	Sak(a)uranetin	樱花亭(野樱素)	(Ghisalberti et al.,1978)
4	Isosakuranetin	异樱花素	(Ghisalberti et al.,1978)
5	2,5-Dihydroxy-7-methoxy flavanone	2,5-二羟基-7-甲氧基二氢黄酮	(Bankova et al.,1983)
6	Alpinetin	山姜素(良姜亭)	(Walker and Crane,1987)
7	Naringenin	柚皮素	(Greenaway et al.,1990)
8	Pinocembrin	乔松素(生松素,松属素)	(Greenaway et al.,1990)
9	3-O-[(S)-2-Methylbutyroyl] pinobanksin	3-O-[(S)-2-甲基丁酰]短叶松素	(Usia et al.,2002)
10	Nymphaeol-A (Propolin C)	5,7,3′,4′-四羟基-6-C-香叶基二氢黄酮	(Chen et al.,2003)
11	Nymphaeol-B (Propolin D)	5,7,3′,4′-四羟基-2′-C-香叶基二氢黄酮	(Chen et al.,2003)
12	Propolin A	5,7,3′,4′-四羟基-2′-C-(8′-羟基香叶基)二氢黄酮	(Chen et al.,2003)
13	Propolin B	5,7,3′,4′-四羟基-5′-C-(8′-羟基香叶基)二氢黄酮	(Chen et al.,2003)
14	Propolin E	5,7,4′-三羟基-3′-C-(8′-羟基香叶基)二氢黄酮	(Chen et al.,2003)
15	Sigmoidin B	5,7,3′,4′-四羟基-C-5′-异戊烯基二氢黄酮	(Chen et al.,2003)
16	Nymphaeol-C	5,7,3′,4′-四羟基-2′-C-香叶基-6-异戊烯基二氢黄酮	(Kumazawa et al.,2004)
17	7-O-Prenylstrobopinin	—	(Melliou and Chinou,2004)
18	Isonymphaeol-B(Propolin F)	5,7,3′,4′-四羟基-5′-C-香叶基二氢黄酮	(Kumazawa et al.,2004)

续表

序号 No.	英文名称 English name	中文名称 Chinese name	参考文献 Ref.
19	3′,4′,6-Trihydroxy-7-methoxy flavanone	3′,4′,6-三羟基-7-甲氧基二氢黄酮	(Shrestha et al.,2007a)
20	7,4′-Dihydroxy flavanone	7,4′-二羟基二氢黄酮	(Shrestha et al.,2007a)
21	3′,4′,7-Trihydroxy flavanone	3′,4′,7-三羟基二氢黄酮	(Shrestha et al.,2007a)
22	7-Hydroxy flavanone	7-羟基二氢黄酮	(Shrestha et al.,2007a)
23	7-Methoxy flavanone	7-甲氧基二氢黄酮	(Shrestha et al.,2007a)
24	(2S)-7-Hydroxyflavanone	(2S)-7-羟基二氢黄酮	(Li et al.,2008)
25	(2S)-Liquiritigenin	(2S)-甘草素	(Li et al.,2008)
26	(2S)-7-Hydroxy-6-methoxyflavanone	(2S)-7-羟基-6-甲氧基二氢黄酮	(Li et al.,2008)
27	(2S)-Naringenin	(2S)-4,5,7-三羟二氢黄酮	(Li et al.,2008)
28	(2S)-Dihydrobaicalein	(2S)-二羟基黄芩黄素	(Li et al.,2008)
29	(2S)-Dihydrooroxylin A	(2S)-二羟基木蝴蝶素 A	(Li et al.,2008)
30	(2R,3R)-3,7-Dihydroxyflavanone	(2R,3R)-3,7-二羟基二氢黄酮	(Li et al.,2008)
31	Garbanzol	3,7,4′-三羟基二氢黄酮	(Li et al.,2008)
32	(2R,3R)-3,7-Dihydroxy-6-methoxyflavanone	(2R,3R)-3,7-二羟基-6-甲氧基二氢黄酮	(Li et al.,2008)
33	Alnustinol	—	(Li et al.,2008)
34	Hesperitin-5,7-Dimethyl-ether	橙皮素-5,7-二甲基-酯	(Falcão et al.,2010)
35	pinobanksin-5-Methyl-ether-3-O-pentanoate	短叶松素-5-甲基-酯-3-氧-戊酸酯	(Falcão et al.,2010)
36	(2R,3R)-3,5-Dihydroxy-7-methoxyflavanone 3-(2-methyl)-butyrate	(2R,3R)-3,5-二羟基-7-甲氧基二氢黄酮 3-(2-甲基)-丁酸盐	(Li et al.,2010)
37	(7″R)-8-[1-(4′-Hydroxy-3′-methoxyphenyl)prop-2-en-1-yl]-chrysin	(7″R)-8-[1-(4′-羟基-3′-甲氧基苯基)烯丙基]-白杨素	(Li et al.,2010)
38	(2R,3R)-6[1-(4′-Hydroxy-3′-methoxyphenyl)prop-2en-1-yl] pinobanksin	(2R,3R)-6[1-(4′-羟基-3′-甲氧苯基)烯丙基]短叶松素	(Li et al.,2011)
39	(2R,3R)-6[1-(4′-Hydroxy-3′-methoxyphenyl)prop-2en-1-yl]-pinobanksin-3-acetate	(2R,3R)-6[1-(4′-羟基-3′-甲氧苯基)烯丙基]-短叶松素-3-乙酸酯	(Li et al.,2011)
40	Solophenol A	所罗门酚	(Inui et al.,2012a)
41	Sophoraflavanone A	槐属二氢黄酮	(Inui et al.,2012a)
42	Bonannione A	—	(Inui et al.,2012a)
43	(2S)-5,7-Dihydroxy-4′-methoxy-8-prenylflavanone	(2S)-5,7-二羟基-4′-甲氧基-8-异戊二烯基二氢黄酮	(Inui et al.,2012a)

表 2.5 蜂胶中的二氢黄酮醇类化合物

Table 2.5 Dihydroflavonols in propolis

序号 No.	英文名称 English name	中文名称 Chinese name	参考文献 Ref.
1	Pinobanksin	短叶松素	(Ghisalberti,1979)
2	3,7-Dihydroxy-5-methoxy flavanone	短叶松素-5-甲醚	(Bankova et al.,1983)
3	Alnusitol	3,5,7-三羟基-6-甲氧基二氢黄酮醇	(Walker and Crane,1987)
4	Pinobanksin-3-acetate	短叶松素-3-乙酸酯	(Greenaway et al.,1990)
5	Pinobanksin-3-butyrate	短叶松素-3-丁酸酯	(Greenaway et al.,1990)
6	Pinobanksin-3-hexanoate	短叶松素-3-已酸酯	(Greenaway et al.,1990)
7	Pinobanksin-3-methyl ether	短叶松素-3-甲醚	(Greenaway et al.,1990)
8	Pinobanksin-3-pentanoate	短叶松素-3-戊酸酯	(Greenaway et al.,1990)
9	Pinobanksin-3-pentenoate	短叶松素-3-戊烯酸酯	(Greenaway et al.,1990)
10	Pinobanksin-3-propanoate	短叶松素-3-丙酸酯	(Greenaway et al.,1990)
11	Pinobanksin-5-methylether-3-O-propanoate	短叶松素-5-甲醚-3-氧-丙酸酯	(Alday et al.,2015)
12	Pinobanksin-5-methylether-3-O-butyrate	短叶松素-5-甲醚-3-氧-丁酯酸	(Alday et al.,2015)
13	Dihydrokaempferol (Aromadendrin)	香树精(二氢山奈酚)	(Tazawa et al.,1999)
14	3-O-[2-Methylbutyroyl]pinobanksin	短叶松素-3-O-[2-甲基丁酰基]	(Usia et al.,2002)
15	Aromadendrine-4′-methyl Ether (dihydrokaempferide)	香树精-4′-甲醚(二氢山奈素)	(Kumazawa et al.,2004)
16	(2R,3R)-3,6,7-Trihydroxyflavanone	(2R,3R)-3,6,7-三羟基二氢黄酮醇	(Shrestha et al.,2007a)
17	5-Methoxy-3-hidroxyflavanone	5-甲氧基-3-羟基二氢黄酮醇	(Falcão et al.,2010)
18	5,7-Dihydroxy-6-methoxy-2,3-Dihydroflavonol-3-acetate	5,7-二羟基-6-甲氧基-2,3-二羟基二氢黄酮醇-3-乙酸酯	(Tran et al.,2012)

1.3 异黄酮和二氢异黄酮类化合物

异黄酮类化合物是广泛存在的黄酮类的异构体,主要分布于被子植物的豆科和蔷薇科中。与其他类黄酮化合物有所不同,它们的结构变化更大,而且异戊二烯取代的概率也更大,从而表现出较强的抗菌活性(Li et al.,2008;Alencar et al.,2007)。研究者运用气相色谱-质谱联用、高效液相色谱等技术从巴西、古巴等地的蜂胶中共分离鉴定出 20 种该类化合物(表 2.6)。

表 2.6　蜂胶中的异黄酮和二氢异黄酮类化合物

Table 2.6　Isoflavones and dihyroisoflavone in propolis

序号 No.	英文名 English name	中文名称 Chinese name	参考文献 Ref.
1	7-Hydroxy-4′-methoxyisoflavonoid	7-羟基-4′-甲氧基异黄酮	(Piccinelli et al.,2005)
2	Odoratin	7,3′-二羟基-6,4′-二甲氧基异黄酮	(Januario et al.,2005)
3	5,7 Dihydroxy-4′-methoxyisoflavonoids	5,7-二羟-4′-甲氧基异黄酮	(Piccinelli et al.,2005)
4	7,4′-Dimethoxy-2′-isoflavone	7,4′-二甲氧基-2′-异黄酮	(Alencar et al.,2007)
5	7,4′-Dihydroxyisoflavone	7,4′-二羟基异黄酮	(Alencar et al.,2007)
6	7,3′,4′-Trihydroxy-5′-methoxyisoflavonoids	7,3′,4′-三羟基-5′-甲氧基异黄酮	(Shrestha et al.,2007b)
7	6,7,3′-Trihydroxy-4′-methoxyisoflavonoids	6,7,3′-三羟基-4′-甲氧基异黄酮	(Shrestha et al.,2007b)
8	7,3′-Dihydroxy-6,5′-methoxyisoflavonoids	7,3′-二羟基-6,5′-甲氧基异黄酮	(Shrestha et al.,2007b)
9	Calycosin	毛蕊异黄酮	(Li et al.,2008)
10	Daidzein	大豆黄素	(Li et al.,2008)
11	Formononetin	芒柄花黄素	(Li et al.,2008)
12	Xenognosin B	—	(Li et al.,2008)
13	Biochanin A	鹰嘴豆芽素 A	(Li et al.,2008)
14	Pratensein	红车轴草素	(Li et al.,2008)
15	2′-Hydroxybiochanin A	2′-二氢鹰嘴豆芽素 A	(Li et al.,2008)
16	(3S)-Vestitone-	—	(Li et al.,2008)
17	(3S)-Violanone	—	(Li et al.,2008)
18	(3S)-Ferreirin	—	(Li et al.,2008)
19	(3R)-4′-Methoxy-2′,3,7-trihydroxyisoflavanone	(3R)-4′-甲氧基-2′,3,7-三羟基二氢异黄酮	(Li et al.,2008)
20	Biochanin	鹰嘴豆芽素	(Campo et al.,2008)

1.4　查耳酮和二氢查耳酮类化合物

查耳酮和二氢查耳酮类化合物是合成类黄酮化合物的主要中间体,由于自身的不稳定性,其在自然界中的存在受到了一定的限制。蜂胶中报道的查耳酮全部为 2′-羟基查耳酮,其衍生物为二氢黄酮的异构体。由于 2′-羟基查耳酮在酸作用下可转化为无色的二氢黄酮,碱化后又可转化为深黄色的 2′-羟基查耳酮(Trusheva et al.,2006),所以目前还不能完全确定这些查耳酮是蜂胶中天然存在的,还是在蜂胶提取分离中形成的。因其都具有碳氧双键共轭体系,所以该类物质表现出较强的细胞毒性。蜂胶中已鉴定出的查耳酮和二氢查耳酮类化合物见表 2.7 和表 2.8。

表 2.7 蜂胶中的查耳酮
Table 2.7 Chalcones in propolis

序号 No.	英文名称 English name	中文名称 Chinese name	参考文献 Ref.
1	Pinobanksin chalcone	短叶松素查耳酮	(Greenaway et al.,1987)
2	Pinobanksin-3-acetate-chalcone	短叶松素-3-乙酰-查耳酮	(Greenaway et al.,1987)
3	Pinocembrin chalcone	生松素查耳酮	(Greenaway et al.,1987)
4	PinoStrobin chalcone	乔松酮查耳酮	(Greenaway et al.,1987)
5	Alpinetin chalcone	山姜素查耳酮	(Greenaway et al.,1987)
6	2,6-Dihydroxy-4-methoxy chalcone	2,6-二羟基-4-甲氧基查耳酮	(Walker and Crane.,1987)
7	2,4′,6-Trihydroxy-4-methoxy chalcone	2,4′,6-三羟基-4-甲氧基查耳酮	(Walker and Crane.,1987)
8	2′,6′,α-trihydroxy-4′-methoxy-chalcone	2′,6′,α-三羟基-4′-甲氧基查耳酮	(Greenaway et al.,1987)
9	Sakauranetin chalcone	樱花亭查耳酮	(Greenaway et al.,1990)
10	Naringenin chalcone	柚配基(柚皮素,柑桔素)查耳酮	(Greenaway et al.,1990)
11	Isoliquiritigenin	异甘草素	(Li et al.,2008)
12	4,4′-Dihydroxy-2′-methoxychalcone	4,4′-二羟基-2′-甲氧基查耳酮	(Li et al.,2008)
13	2′,4′-Dihydroxychalcone	2′,4′-二羟基查耳酮	(Li et al.,2008)
14	3,4,2′,3′-Tetrahydroxychalcone	3,4,2′,3′-四羟基查耳酮	(Righi et al.,2011)

表 2.8 蜂胶中的二氢查耳酮类化合物
Table 2.8 Dihydrochalcones in propolis

序号 No.	英文名称 English name	中文名称 Chinese name	参考文献 Ref.
1	2′,6′-Dihydroxy-4′-methoxyDihydrochalcones	2′,6′-二羟-4′-甲氧基二氢查耳酮	(Greenaway et al.,1987)
2	2′,4′,6′-TrihydroxyDihydrochalcones	2′,4′,6′-三羟基二氢查耳酮	(Greenaway et al.,1990)
3	2′,4′,6′-Tihydroxy-4-methoxydihydrochalcone	2′,4′,6′-三羟基-4-甲氧基二氢查耳酮	(Awale et al.,2005)
4	2′,6′,4-Tryhydroxy-4′-methoxydihydrochalcone	2′,6′,4-三羟基-4′-甲氧基二氢查耳酮	(Awale et al.,2005)
5	2′,6′-Dihydroxy-4′,4-dimethoxydihydrochalcone	2′,6′-二羟基-4,4′-二甲氧基二氢查耳酮	(Awale et al.,2005)
6	(αS)-α,2′,4,4′-Tetrahydroxydihydrochalcone	(αS)-α,2′,4,4′-四羟基二氢查耳酮	(Li et al.,2008)

1.5 黄烷和异黄烷类化合物

黄烷类化合物是广泛存在于自然界中的一类黄酮类化合物,其分子含有3,4-二氢-2-

苯基-1-苯并吡喃环结构骨架。简单的黄烷类化合物结构类型主要有黄烷、黄烷-3-醇、黄烷-4-醇以及黄烷-3,4-二醇4种类型。蜂胶中存在的黄烷类化合物主要为黄烷-3-醇类化合物,如表2.9所示。除此之外,研究者从巴西蜂胶中分离鉴定出6种异黄烷类化合物(表2.10)。

表 2.9　蜂胶中的黄烷类化合物

Table 2.9　Flavans in propolis

序号 No.	英文名称 English name	参考文献 Ref.
1	8-[(E)-4-Phenylprop-2-en-1-one]-(2R,3S)-2-(3,5-dihydroxyphenyl)-3,4-dihydro-2H-2-benzopyran-5-methoxyl-3,7-diol	(Sha et al.,2009)
2	8-[(E)-4-Phenylprop-2-en-1-one]-(2S,3R)-2-(3,5-dihydroxyphenyl)-3,4-dihydro-2H-2-benzopyran-5-methoxyl-3,7-diol	(Sha et al.,2009)
3	8-[(E)-4-Phenylprop-2-en-1-one]-(2R,3S)-2-(3-methoxyl-4-hydroxyphenyl)-3,4-dihydro-2H-2-benzopyran-5-methoxyl-3,7-diol	(Sha et al.,2009)
4	3-Hydroxy-5,6-dimethoxyflavan	(Lotti et al.,2010)

表 2.10　蜂胶中的异黄烷类化合物

Table 2.10　Isoflavans in propolis

序号 No.	英文名称 English name	参考文献 Ref.
1	7,4′-Dihydroxy-2′-methoxyisoflavone	(Piccinelli et al.,2005)
2	(3S)-Vestitol	(Li et al.,2008)
3	(3S)-Isovestitol	(Li et al.,2008)
4	(3S)-7-O-Methylvestitol	(Li et al.,2008)
5	(3S)-Mucronulatol	(Li et al.,2008)
6	Neovestitol	(Campo et al.,2008)

1.6　新黄酮类化合物

近年来,研究者在古巴和巴西绿蜂胶中发现了大量紫檀素类化合物(表2.11)和4-苯基香豆素类化合物(图2.1),由于其具有C_6-C_3-C_6类黄酮化合物的基本结构骨架,因此该类合物被命名为新黄酮类化合物。

表 2.11　蜂胶中的紫檀素类化合物

Table 2.11　Pterocarpins in propolis

序号 No.	英文名称 English name	中文名称 Chinese name	参考文献 Ref.
1	4-Hydroxymedicarpin	4-羟基美迪紫檀素	(陈业高等,2004)
2	4′-Methoxy-5′-hydroxyvesticarpan	—	(陈业高等,2004)

续表

序号 No.	英文名称 English name	中文名称 Chinese name	参考文献 Ref.
3	Medicarpin	美迪紫檀素	(Piccinelli et al.,2005)
4	Homopterocarpin	后莫紫檀素	(Piccinelli et al.,2005)
5	Vesticarpan	—	(Piccinelli et al.,2005)
6	3,8-Dihydroxy-9-methoxypterocarpan	3,8-二羟基-9-甲氧基紫檀素	(Piccinelli et al.,2005)
7	3-Hydroxy-8,9-dimethoxypterocarpan	3-羟基-8,9-二甲氧基紫檀素	(Piccinelli et al.,2005)
8	3,4-Dihydroxy-9-methoxypterocarpan	3,4-二羟基-9-甲氧基紫檀素	(Piccinelli et al.,2005)
9	3,10-Dihydroxy-9-methoxypterocarpan	3,10-二羟基-9-甲氧基紫檀素	(Li et al.,2008)
10	6a-Ethoxymedicarpin	6a-乙氧基美迪紫檀素	(Li et al.,2008)
11	(6aR,11aR)-4-Methoxymedicarpin	(6aR,11aR)-4-甲氧基美迪紫檀素	(Li et al.,2008)

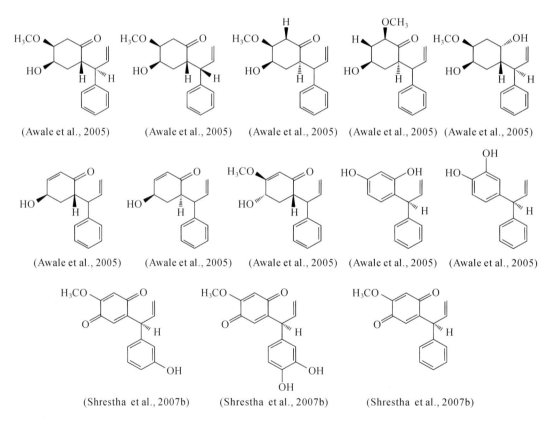

图 2.1 蜂胶中的新黄酮类似物
Fig. 2.1 Neoflavonoids in propolis

2 萜类化合物

萜类化合物是天然产物中一类非常重要的次生代谢产物(或称二级代谢产物),在自然界分布广泛,数量庞大,结构类型多。在天然产物化学研究中,萜类成分的研究是一个极其活跃的领域,也是寻找和发现天然药物先导性分子,以及其他功能分子的重要源泉。虽然萜类化合物数量多、类型多、结构复杂,但在众多复杂的结构中有基本的规律可循,正是这些规律的存在也为萜类化合物的划分和定义奠定了基础。目前,萜类化合物的定义主要有两种:从生物代谢的途径看,凡由甲戊二羟酸衍生,且分子式符合$(C_5H_8)_n$通式的衍生物称为萜类化合物;从化学结构特征看,萜类化合物是异戊二烯的聚合物及其衍生物,其基本骨架通常为具有5个碳的异戊二烯结构单元。根据其分子结构中异戊二烯结构的数目,可将萜类化合物分为:单萜、倍半萜、二萜、三萜、四萜以及多萜。萜类化合物是蜂胶挥发油中最主要的活性成分,蜂胶精油的多重生物学功能,如抗炎症、抗细菌、镇痛作用,都与其中含有大量萜类物质有关(鲁守平等,2006)。截至2017年2月,已从蜂胶中分离鉴定出231种萜类化合物。

2.1 单萜类化合物

单萜类化合物是由2个异戊二烯结构单元组成的具有10个碳原子的一类化合物。单萜类广泛分布于高等植物的分泌组织、昆虫激素、真菌及海洋生物中,同时也是植物精油的主要组成成分。单萜类化合物的含氧衍生物多具有较强的生物活性和香气,是医药、化妆品和食品工业的重要原料。单萜类化合物的研究进展很快,已经确定的单萜类化合物的基本骨架有30多种,而蜂胶中的单萜类化合物主要包括无环单萜、单环单萜和双环单萜及其含氧衍生物。蜂胶中无环单萜的基本骨架是月桂烯烷,主要成分为香叶醇和芳樟醇;单环单萜类化合物的基本碳架为薄荷烷和桉叶素类;双环单萜类化合物共有5种基本骨架,其中以蒎烷型和莰烷型最为稳定,在蜂胶中发现的相应化合物也最多。具体单萜类化合物如表2.12所示。

表 2.12 蜂胶中的单萜类化合物
Table 2.12 Monoterpene compounds in propolis

序号 No.	英文名称 English name	中文名称 Chinese name	参考文献 Ref.
1	γ-Terpinene	松油烯	(Borcic et al.,1996)
2	α-Pinene	α-蒎烯	(Borcic et al.,1996)
3	β-Pinene	β-蒎烯	(Borcic et al.,1996)
4	Geraniol	香叶醇	(Bankova et al.,1998)
5	Linalyl propionate	丙酸芳樟醇	(Bankova et al.,1998)
6	α-Terpineol	松油醇	(Kartal et al.,2002)
7	Linalool oxide	芳樟醇氧化物	(曾晞等,2004)
8	1,8-Cineole	1,8-桉叶素	(曾晞等,2004)
9		侧柏醛	(盛文胜等,2006)
10	α-Phellandrene	水芹烯	(Melliou et al.,2007)
11	p-Cymene	伞花烃	(Melliou et al.,2007)
12	p-Cymen-8-ol	—	(Melliou et al.,2007)

续表

序号 No.	英文名称 English name	中文名称 Chinese name	参考文献 Ref.
13	p-Cymenene	—	(Melliou et al.,2007)
14	α-Terpinene	松油烯	(Melliou et al.,2007)
15	Terpinolene	异松油烯	(Melliou et al.,2007)
16	Trans-β-Terpineol	松油醇	(Melliou et al.,2007)
17	α-Terpinyl acetate	松油基乙酯	(Melliou et al.,2007)
18	trans-Carveol	香芹醇	(Melliou et al.,2007)
19	Carvacrol methyl ether	香芹酚甲醚	(Melliou et al.,2007)
20	cis-m-Mentha-2,8-diene	—	(Melliou et al.,2007)
21	p-Mentha-1,5-dien-8-ol	—	(Melliou et al.,2007)
22	α-Thujene	侧柏烯	(Melliou et al.,2007)
23	cis-Sabinene hydrate	香桧烯	(Melliou et al.,2007)
24	Sabinene	香桧烯	(Melliou et al.,2007)
25	cis-Sabinol	松萜醇	(Melliou et al.,2007)
26	D-2-Carene	蒈烯	(Melliou et al.,2007)
27	trans-Pinocarveol	松香芹醇	(Melliou et al.,2007)
28	Pinocarvone	松油酮	(Melliou et al.,2007)
29	Verbenene	马鞭草烯	(Melliou et al.,2007)
30	trans-Verbenol	马鞭草烯醇	(Melliou et al.,2007)
31	cis-Verbenol	马鞭草烯酸	(Melliou et al.,2007)
32	α-Fenchol	小茴香醇	(Melliou et al.,2007)
33	α-Fenchene	茴香烯	(Melliou et al.,2007)
34	Myrtenal	桃金娘烯醛	(Melliou et al.,2007)
35	Camphene hydrate	莰烯水合物	(Melliou et al.,2007)
36	Borneol	冰片	(Melliou et al.,2007)
37	4-Terpineol	4-松油醇	(Junior et al.,2008)
38	β-Myrcene	月桂烯	(Junior et al.,2008)
39	β-Phellandrene	水芹烯	(Yang et al.,2010)
40	Linalool	芳樟醇	(Oliveira et al.,2010)
41	Limonene	柠檬烯	(Yang et al.,2010)
42	Eucalyptol	桉叶素	(Yang et al.,2010)
43	Camphene	莰烯	(Yang et al.,2010)

续表

序号 No.	英文名称 English name	中文名称 Chinese name	参考文献 Ref.
44	3-Carene	3-蒈烯	(Yang et al.,2010)
45	Tschimgin	冰片基羟苯酸盐	(Trusheva et al.,2010)
46	Tschimganin	冰片基香草酸盐	(Trusheva et al.,2010)
47	Camphor	樟脑	(Trusheva et al.,2010)
48	Bornyl p-Hydroxybenzoate	冰片基 p-羟基苯甲酸酯	(Trusheva et al.,2010)
49	Bornyl vanillate	冰片基香子兰酸酯	(Trusheva et al.,2010)

2.2 倍半萜类化合物

倍半萜类化合物是由3个异戊二烯结构单元组成的具有15个碳原子的一类化合物及其衍生物。倍半萜类化合物广泛分布于植物、海洋生物、微生物及昆虫组织中。多数倍半萜类化合物具有较强的生物活性和重要的生物功能,尤其是倍半萜内酯,具有抗菌、抗肿瘤、抗病毒和昆虫激素、昆虫拒食等作用。倍半萜类化合物是高沸点芳香精油的主要组成成分,也是芳香油香味差异的主要调节者。虽然倍半萜碳骨架仅有15个碳原子,但由于生物体内多种酶参与的一系列次生代谢过程产生了众多的基本碳骨架和复杂的衍生物,致使倍半萜无论在数量上,还是结构类型上,都居萜类之首。为此,倍半萜类化合物的研究一直是天然产物化学中非常重要和活跃的研究领域。目前,已在蜂胶中发现了82种倍半萜类化合物,其主要可分为无环倍半萜、单环倍半萜、双环倍半萜和三倍半萜类。

蜂胶中发现的无环倍半萜为金合欢烷的含氧衍生物;蜂胶中的单环倍半萜的基本骨架有4种,分别为吉马烷、没药烷、蛇麻烷和榄香烷;蜂胶中双环倍半萜的基本骨架有5种,分别为桉叶烷、杜松烷、艾里莫芬烷、愈创木烷和石竹烷。此外,还有 α-人参烯和 β-人参烯;蜂胶中三环倍半萜的基本骨架有10种,分别为柏木烷、广藿香烷、胡椒烷、依兰烷、香木兰烷、荜澄茄烷、环苜蓿烷、刺柏烷、波旁烷和长叶烷。蜂胶中已发现的倍半萜类化合物如表2.13所示。

表 2.13 蜂胶中的倍半萜类化合物

Table 2.13 Sesquiterpene compounds in propolis

序号 No.	英文名称 English name	中文名称 Chinese name	参考文献 Ref.
1	Nerolidol	橙花叔醇	(Junior et al.,2008)
2	Farnesol	金合欢醇	(Mohammadzadeh et al.,2007)
3	Germacrend	大根香叶烯	(Bankova et al.,1996)
4	Germacrenen	大根香叶酮	(李雅萍等,2007)
5	ar-Curcumene	芳姜黄烯	(曾晞等,2004)
6	γ-Curcumene	γ-姜黄烯	(Melliou et al.,2007)

续表

序号 No.	英文名称 English name	中文名称 Chinese name	参考文献 Ref.
7	α-Bisabolol	没药醇	(Silici et al.,2005)
8	α-Zingiberene	α-姜倍半萜	(Kartal et al.,2002)
9	ar-Turmerone	芳姜黄酮	(付宇新等,2009)
10	Curlone	姜黄新酮	(付宇新等,2009)
11	α-Humulene	α-蛇麻烯	(Junior et al.,2008)
12	δ-Humulene	δ-蛇麻烯	(刘波静,2009)
13	γ-Elemene	榄香烯	(Melliou et al.,2007)
14	α-Selinene	芹子烯	(李雅萍等,2007)
15	β-Selinene	β-芹子烯	(付宇新等,2009)
16	γ-Eudesmol	γ-桉叶醇	(盛文胜等,2006)
17	—	桉油-4-醇	(曾晞等,2004)
18	10-Epi-c-eudesmol	桉叶醇	(李雅萍等,2007)
19	α-Eudesmol	α-桉叶醇	(余兰平等,2006)
20	β-Eudesmol	桉叶醇	(余兰平等,2006)
21	4-βH,5α-Eremophil-1(10)-ene	—	(Mohammadzadeh et al.,2007)
22	γ-Cadinene	γ-杜松烯	(曾晞等,2004)
23	α-Cadinene	α-杜松烯	(Melliou et al.,2007)
24	δ-Cadinene	杜松烯	(Melliou et al.,2007)
25	α-Cadinol	杜松醇	(李雅萍等,2007)
26	Epi-alpha-cadinol	香榧醇	(Junior et al.,2008)
27	α-Calacorene	白菖考烯	(李雅萍等,2007)
28	Calamenene	去氢白菖烯	(赵强等,2007)
29	α-Muurolol	—	(Melliou et al.,2007)
30	epi-α-Muurolol	—	(Kartal et al.,2002)
31	α-Amorphene	α-紫穗槐烯	(赵强等,2007)
32	Valencene	瓦伦烯	(Oliveira et al.,2010)
33	Spatulenol	斯巴醇	(Bankova et al.,1996)
34	Isospatulenol	异斯巴醇	(Bankova et al.,1996)
35	Ledene	喇叭烯	(Bankova et al.,1996)
36	Ledol	喇叭茶醇	(Bankova et al.,1995)
37	Guaiol	愈创木醇	(崔庆新和刘国富,2001)

续表

序号 No.	英文名称 English name	中文名称 Chinese name	参考文献 Ref.
38	Bulnesol	异愈创木醇	（崔庆新和刘国富，2001）
39	α-Gurjunene	α-古芸烯	（李雅萍等，2007）
40	γ-Gurjunene	γ-古芸烯	（余兰平等，2006）
41	Ferutinin	—	(Trusheva et al.，2010)
42	Tefernin	—	(Trusheva et al.，2010)
43	Ferutinol p-hydroxybenzoate	—	(Trusheva et al.，2010)
44	Ferutinol vanillate	—	(Trusheva et al.，2010)
45	α-Caryophyllene	石竹烯	(Oliveira et al.，2010)
46	β-Caryophyllene	石竹烯	(Junior et al.，2008)
47	Caryophyllene oxide	石竹烯氧化物	（崔庆新和刘国富，2001）
48	E-caryophyllene	石竹烯	(Melliou et al.，2007)
49	Cedran-diol	雪松烷二醇	（李雅萍等，2007）
50	8-βH-Cedran-8-ol	—	(Kartal et al.，2002)
51	α-Cedrene	α-雪松烯	（李雅萍等，2007）
52	α-Cedrene oxide	α-雪松烯氧化物	（赵强等，2007）
53	α-Cedrol	α-雪松醇	(Mohammadzadeh et al.，2007)
54	Cedranoxide	香松烷氧化物	（余兰平等，2006）
55	Thujopsene	罗汉柏烯	（余兰平等，2006）
56	ά-Patchoulene	ά-绿叶烯	（李雅萍等，2007）
57	α-Copaene	胡椒烯	（付宇新等，2009）
58	—	α-胡椒烯-11-醇	（曾晞等，2004）
59	α-Ylangene	衣兰烯	（徐响等，2008）
60	α-Muurolene	α-衣兰油烯	（曾晞等，2004）
61	γ-Muurolene	γ-衣兰油烯	（徐响等，2008）
62	allo-Aromadendrene	异别香橙烯	（崔庆新和刘国富，2001）
63	Aromadendrene	香木兰烯	（李雅萍等，2007）
64	β-Maaliene	β-木兰烯	（崔庆新和刘国富，2001）
65	Aromadendrane	香木兰烷	（崔庆新和刘国富，2001）
66	Spathulenol	匙叶桉油烯醇	(Kusumoto et al.，2001)
67	Palustrol	喇叭茶碱	(Bankova et al.，1996)
68	Globulol	蓝桉醇	(Junior et al.，2008)

续表

序号 No.	英文名称 English name	中文名称 Chinese name	参考文献 Ref.
69	Veridiflorol	—	(李雅萍等,2007)
70	α-Cubebene	α-荜澄茄苦素	(Yang et al.,2010)
71	β-Cubebene	荜澄茄苦素	(Melliou et al.,2007)
72	—	α-荜澄茄油烯	(徐响等,2008)
73	Cubenol	荜澄茄油烯醇	(李雅萍等,2007)
74	Cyclosativene	环苜蓿烯	(Melliou et al.,2007)
75	Junipene	刺柏烯	(李雅萍等,2007)
76	β-Bourbonene	波旁烯	(Melliou et al.,2007)
77	α-Longipinene	长叶蒎烯	(Melliou et al.,2007)
78	Longifolenaldehyde	长叶醛	(余兰平等,2006)
79	Longifolene	长叶烯	(余兰平等,2006)
80	α-Panasinsene	α-人参萜烯	(徐响等,2008)
81	β-Panasinsene	β-人参萜烯	(徐响等,2008)
82	Patchouleneb	绿叶烯	(Syamsudin et al.,2009)

2.3 二萜类化合物

二萜类化合物作为萜类的 C20 代表化合物,是由 4 个异戊二烯前体物质"头-尾"相接合成而来的。它们的结构显示多样性,但来源上都是由前体物焦磷酸香叶基香叶酯(geranyl geranyl pyrophosphate,GGPP)衍生而成。在二萜类化合物中,链状二萜化合物在自然界存在相对较少,结构相对简单。蜂胶中的二萜类化合物主要为单环二萜、双环二萜、三环二萜和四环二萜。蜂胶中的单环二萜类化合物主要为西柏烷型,双环二萜类化合物主要为半日花烷型,三环二萜类化合物主要为松香烷、海松烷和桃柘烷型。仅从希腊蜂胶中鉴定出 1 个四环二萜:覆瓦南美杉醛酸(Imbricataloic acid)。具体二萜类化合物如表 2.14 所示。

表 2.14 蜂胶中二萜类化合物
Table 1.14 Diterpenoids in propolis

序号 No.	英文名称 English name	中文名称 Chinese name	参考文献 Ref.
1	ent-17-Hydroxy-3,13Z-clerodadien-15-oic acid	—	(Mastuno et al.,1995)
2	Communic acid	璎柏酸	(Bankova et al.,1996)
3	Isocupressic acid	异柏酸	(Bankova et al.,1996)
4	13-Symphyoreticulic acid	—	(Matsuno et al.,1997)
5	15-Oxo-3,13Z-kolavadiene-17-oic acid	—	(Matsuno et al.,1997)

续表

序号 No.	英文名称 English name	中文名称 Chinese name	参考文献 Ref.
6	Abietic acid	枞酸	(Kartal et al.,2002)
7	Dehydroabietic acid	脱氢枞酸	(Hegazi et al.,2002)
8	Manool	泪杉醇	(Melliou et al.,2007)
9	Isopimaric acid	异海松酸	(Mohammadzadeh et al.,2007)
10	15-Oxolabda 8(17),13Z-dien-19-oic acid	—	(Cvek et al.,2008)
11	Pimaric acid	海松酸	(Cvek et al.,2008)
12	Copalol	—	(Cvek et al.,2008)
13	Junicedric acid	—	(Popova et al.,2009)
14	Isoagatholal	—	(Meneses et al.,2009)
15	13-Epi-torulosal	—	(Meneses et al.,2009)
16	Epi-13-torulosol	—	(Meneses et al.,2009)
17	Palmitoyl isocupressic acid	—	(Popova et al.,2009)
18	Oleoyl isocupressic acid	—	(Popova et al.,2009)
19	13-Hydroxy-8(17), 14-labdadien-19-oic acid	—	(Popova et al.,2009)
20	15-Oxolabda-8(17),13E-dien-19-oic acid	—	(Popova et al.,2009)
21	8(17),13E-labdadien-15,19-dioic acid	—	(Popova et al.,2009)
22	8(17),13E-labdadien-15, 19-dioic acid 15-methyl ester	—	(Popova et al.,2009)
23	19-Oxo-8(17),13E-labdadien-15-oic acid	—	(Popova et al.,2009)
24	Totarolone	—	(Popova et al.,2009)
25	Manoyl oxide	迈诺氧化物	(Popova et al.,2010)
26	13-Epi-manool	13-表-泪杉醇	(Popova et al.,2010)
27	13-Epi-cupressic acid	13-表-柏酸	(Popova et al.,2010)
28	Agathadiol	—	(Popova et al.,2010)
29	14,15-Dinor-13-oxo-8(17)-labden-19-oic acid	—	(Popova et al.,2010)
30	18-Succinyloxyabieta-8,11,13-triene	—	(Popova et al.,2010)
31	18-Succinyloxyabietadiene	—	(Popova et al.,2010)
32	18-Succinyloxyhydroxyabietatriene	—	(Popova et al.,2010)
33	Acetylisocupressic acid	—	(Popova et al.,2010)
34	tran-Communal	—	(Popova et al.,2010)

续表

序号 No.	英文名称 English name	中文名称 Chinese name	参考文献 Ref.
35	18-Hydroxyabieta-8,11,13-triene	—	(Popova et al.,2010)
36	labda-8(17),12,13-Triene	—	(Popova et al.,2010)
37	Hydroxydehydroabietic acid	羟基脱氢枞酸	(Popova et al.,2010)
38	Neoabietic acid	新枞酸	(Popova et al.,2010)
39	Dihydroxyabieta-8,11,13-triene	二羟枞酸-8,11,13-三烯	(Popova et al.,2010)
40	13(14)-Dehydrojunicedric acid	13(14)-脱氢枞酸	(Popova et al.,2010)
41	Totarolon	—	(Popova et al.,2010)
42	trans-Totarol	陶塔酚	(Popova et al.,2010)
43	Ferruginol	菲洛醇	(Popova et al.,2010)
44	Ferruginolon	—	(Popova et al.,2010)
45	2-Hydroxyferruginol	—	(Popova et al.,2010)
46	6/7-Hydroxyferruginolq	—	(Popova et al.,2010)
47	Sempervirol	钩勿烷	(Popova et al.,2010)
48	Diterpenic acid	二萜稀酸	(Popova et al.,2010)
49	Imbricataloic acid	覆瓦南美杉醛酸	(Popova et al.,2010)
50	Imbricatoloic acid	—	(Popova et al.,2010)
51	Propsiadin	—	(Almutairi et al.,2014)
52	Rel-(5S,6S,8R,9R,10S,18R,19S)-18,19-epoxy-2-oxocleroda-3,12(E),14-triene-6,18,19-triol 18,19-diacetate 6-benzoate	—	(Tazawa et al.,2016)

2.4 三萜类化合物

三萜类化合物是一类基本母核由 30 个碳原子组成的萜类化合物，其结构根据异戊二烯规则可视为由 6 个异戊二烯结构单位聚合而成，是一类重要的天然产物化学成分。三萜及其苷类化合物在植物中分布广泛，菌类、蕨类、单子叶和双子叶植物、动物及海洋生物中均有分布。三萜类化合物结构复杂多样，这是由其生物合成途径的多样性决定的。目前已发现的三萜类化合物，多为四环三萜和五环三萜，少数为链状、单环、双环和三环三萜类化合物。蜂胶中发现的三萜类化合物多数也为四环三萜和五环三萜类化合物。蜂胶中四环三萜类化合物主要为羊毛脂烷型，五环三萜类化合物主要为齐墩果烷型、乌苏烷型以及羽扇豆醇型。蜂胶中已发现的三萜类化合物如表 2.15 所示。

表 2.15　蜂胶中的三萜类化合物
Table 2.15　Triterpenoids in propolis

序号 No.	英文名称 English name	中文名称 Chinese name	参考文献 Ref.
1	β-Amyrin	β-香树精	(Marcucci et al.,1998)
2	Cycloartenol	环阿乔醇	(Marcucci et al.,1998)
3	β-Amyrin alkanoates	β-香树精烷酸酯	(Pereira et al.,2000)
4	Lupenone	羽扇豆烯酮	(Pereira et al.,2002)
5	24-Methylene-9,19-ciclolanostan-3β-ol	—	(Pereira et al.,2002)
6	Lupeol alkanoates	羽扇醇烷酸酯	(Pereira et al.,2002)
7	Triterpenic acid methyl ester	三萜酸甲基酯	(Hegazi et al.,2002)
8	Cycloart-3,7-dihydroxy-24-en-28-oic acid	—	(Cvek et al.,2008)
9	3,4-Seco-cycloart-12-hydroxy-4(28),24-dien-3-oicacid	—	(Cvek et al.,2008)
10	(22Z,24E)-3-Oxocycloart-22,24-Dien-26-oic acid	—	(Li et al.,2009)
11	(24E)-3-Oxo-27,28-Dihydroxycycloart-24-en-26-oic acid	—	(Li et al.,2009)
12	Lupeol acetate	羽扇醇乙酯	(Marquez Hernandez et al.,2010)
13	Lupeol	羽扇醇	(Marquez Hernandez et al.,2010)
14	Lanosterol	羊毛甾醇	(Marquez Hernandez et al.,2010)
15	Lanosterol acetate	羊毛固醇乙酯	(Marquez Hernandez et al.,2010)
16	Germanicol acetate	日耳曼醇乙酯	(Marquez Hernandez et al.,2010)
17	Germanicol	计曼尼醇	(Marquez Hernandez et al.,2010)
18	β-Amyrin acetate	β-香树精乙酸酯	(Marquez Hernandez et al.,2010)
19	β-Amyrone	β-白檀酮	(Marquez Hernandez et al.,2010)
20	α-Amyrin acetate	α-香树精乙酸酯	(Marquez Hernandez et al.,2010)
21	α-Amyrin	α-香树精	(Marquez Hernandez et al.,2010)
22	α-Amyrone	α-白檀酮	(Marquez Hernandez et al.,2010)
23	α-Amyrenone	α-香树脂酮	(Boisard et al.,2016)
24	β-Amyrenone	β-香树脂酮	(Boisard et al.,2016)
25	Fucosterol	盐藻甾醇	(Boisard et al.,2016)
26	β-Sitosterol	β-谷甾醇	(Boisard et al.,2016)
27	γ-Sitosterol	γ-谷甾醇	(Boisard et al.,2016)
28	Cycloart-24-en-3β,26-diol	—	(Nina et al.,2016)
29	Cycloart-24-en-3-one	—	(Nina et al.,2016)
30	Cycloart-24-en-26-ol-3-one	—	(Nina et al.,2016)
31	Mangiferonic acid methyl ester	杧果酮酸甲酯	(Nina et al.,2016)
32	Lup-20(29)-en-3-one	—	(Nina et al.,2016)
33	methyl-3β,27-dihydroxycycloart-24-en-26-oate	—	(Talla et al.,2016)

3 酚类化合物

酚类化合物是一类携有一个或多个羟基的一个共同芳香环的植物化学物质,它们的范围非常广,既包括杨梅素、丁香酚等简单的物质,又包括灰黄霉素、鱼藤酮等结构复杂的物质。研究表明,大量酚类化合物对人类慢性疾病的预防起着非常重要的作用(郭新竹和宁正祥,2002)。酚类化合物主要以酸的形式存在于蜂胶中。此外,蜂胶中还含有二苯乙烯类、木酚素类及其他复杂的酚类化合物。目前,已从蜂胶中分离鉴定出种酚类化合物 194 种。

3.1 酚酸类化合物

酚酸类化合物是含有酚羟基和羧基的一类代谢产物,广泛分布在药用植物中。因为酚酸类化合物中的酚羟基是优良的氢或中子的给予体,对能引起生物组织膜因产生过氧化作用而导致结构和功能损伤的过氧自由基、羟自由基等自由基有明显的清除作用,所以酚酸类化合物是一种良好的抗氧化剂。迄今为止,已从蜂胶中鉴定出一百多种酚酸类化合物。这些酚酸类化合物被公认是蜂胶中的主要活性成分之一。蜂胶的抗氧化、抑菌、抗肿瘤等生物学活性均与其中所含的酚酸类物质有关(郭新竹和宁正祥,2002;Cos et al.,2002),尤其是咖啡酸苯乙酯(CAPE)的生物学活性是近年来研究的热点(Russo et al.,2002)。

蜂胶中发现的酚酸类化合物的主要骨架类型有以下两类:一是 C6-C1 型,基本骨架是苯甲酸,如没食子酸、原儿茶酸等;二是 C6-C3 型,基本骨架是苯丙酸,如咖啡酸、阿魏酸等。此外,从巴西蜂胶中还发现了许多绿原酸类化合物(张翠平和胡福良,2012)。

3.1.1 苯甲酸为母核的酚酸类成分

自然界中以苯甲酸为母核的酚酸类化合物主要是没食子酸及其衍生物。没食子酸广泛存在于葡萄、芍药、茶叶等植物中,具有抗炎、抗突变、抗氧化、抗肿瘤等多种生物活性,是生物体内参与生成酯键的最频繁的有机酸类。蜂胶中较为常见的该类酚酸化合物主要有没食子酸、原儿茶酸、香兰酸、龙胆酸等(表 2.16)。

表 2.16 蜂胶中以苯甲酸为母核的酚酸类成分
Table 2.16 Benzoic acid derivatives in propolis

序号 No.	英文名称 English name	中文名称 Chinese name	参考文献 Ref.
1	3,4-Dihydroxybenzoic acid	原儿茶酸	(Greenaway et al.,1987)
2	4-Methoxybenzoic acid	4-甲氧基苯甲酸	(Walker and Crane.,1987)
3	Benzyl salicylate	水杨酸苄酯	(Walker and Crane.,1987)
4	2,5-Dihydroxybenzoic acid	龙胆酸	(Walker and Crane.,1987)
5	3,4,5-Trihydroxybenzoic acid	没食子酸	(Bonvehí et al.,1994)
6	Methyl salicylate	邻羟基苯甲酸甲酯	(Marcucci,1995)
7	*p*-Hydroxy benzoic acid	*p*-苯甲酸	(Marcucci,1995)
8	2-Hydroxybenzoic acid	水杨酸	(Marcucci,1995)

续表

序号 No.	英文名称 English name	中文名称 Chinese name	参考文献 Ref.
9	Veratric acid	藜芦酸	(Marcucci,1995)
10	4-Hydroxy-3-methoxybenzoic	香兰酸	(Marcucci,1995)
11	Acetyl salicylic acid	乙酰水杨酸	(Krol et al.,1996)
12	4-Hydroxy-3-prenylbenzoic acid	4-羟-3-异戊二烯基苯甲酸	(Hayashi et al.,1999)

3.1.2 苯丙酸类化合物

苯丙酸类化合物广泛分布于中草药中,也是植物体内的芳香化合物生物合成过程中的一类重要中间体,其基本结构是由酚羟基取代的芳香环与丙烯酸构成的。蜂胶中最常见的苯丙酸类化合物是羟基肉桂酸类,如肉桂酸、p-香豆酸、咖啡酸、阿魏酸等结构单元及其衍生物。实验证明,阿魏酸对血小板凝集有抑制作用,能够抑制血小板的释放功能和黏附反应,具有抗凝血、抗血栓的作用(胡益勇和徐晓玉,2006)。阿魏酸在抗氧化、抗炎症、免疫调节方面也发挥着积极的作用(Graf,1992)。此外,咖啡酸以及衍生物已被证实具有抗癌的功效(Huang et al.,1988)。蜂胶中发现的苯丙酸类化合物主要可分为肉桂酸类(表2.17)、p-香豆酸类(表2.18)以及咖啡酸类化合物(表2.19)。

表2.17 蜂胶中的肉桂酸类化合物及其衍生物
Table 2.17 Cinnamic acid and derivatives in propolis

序号 No.	英文名称 English name	中文名称 Chinese name	参考文献 Ref.
1	Benzyl 3,4-dimethoxycinnamate	苯甲基3,4-二甲氧基肉桂酸盐	(Greenaway et al.,1987)
2	Cinnamic acid ethyl ester	肉桂酸乙酯	(Walker and Crane,1987)
3	Cinnamic acid methyl ester	肉桂酸甲酯	(Walker and Crane,1987)
4	Hexadecyl-4-methoxyhydrocinnamate	十六烷基-4-甲氧基羟基肉桂酸	(Bankova et al.,1992)
5	Hexadecenyl-4-methoxyhydrocinnamate	—	(Bankova et al.,1992)
6	Octadecyl-4-methoxyhydrocinnamate	十八烷基-4-甲氧基氢化肉桂酸	(Bankova et al.,1992)
7	Tetradecenyl-4-methoxyhydrocinnamate	—	(Bankova et al.,1992)
8	Tetradecyl-4-methoxyhydrocinnamate	—	(Bankova et al.,1992)
9	3,5-Dimethoxy-4-hydroxycinnamic acid	3,5-二甲氧基-4-羟基肉桂酸	(Bonvehí et al.,1994)
10	Dimethoxycinnamic acid	二甲氧基肉桂酸	(Bankova et al.,1998)
11	Capillartemisin A	茵陈香豆酸A	(Tazawa et al.,1999)
12	Allyl-3-prenyl cinnamate	烯丙基-3-异戊二烯基肉桂酸盐	(Marcucci et al.,2000)
13	cis-3-Methoxy-4-hydroxy-cinnamic acid	顺-3-甲氧基-4-羟基-肉桂酸	(Marcucci et al.,2000)
14	Hydrocinnamic acid	氢化肉桂酸	(Marcucci et al.,2000)
15	p-Hydroxycinnamic acid	p-羟基肉桂酸	(Marcucci et al.,2000)

续表

序号 No.	英文名称 English name	中文名称 Chinese name	参考文献 Ref.
16	p-Hydroxyhydrocinnamic acid	p-羟基氢化肉桂酸	(Marcucci et al.,2000)
17	trans-3,4-Dimethoxycinnamic acid	反-3,4-二甲氧基肉桂酸	(Marcucci et al.,2000)
18	trans-3-Methoxy-4-hydroxycinnamic acid	反-3-甲氧基-4-羟基肉桂酸	(Marcucci et al.,2000)
19	3,4-Dihydroxy-5-prenylcinnamic Acid	3,4-二羟基-5-异戊二烯基肉桂酸	(Hegazi and El Hady,2002)
20	3-Prenyl-4-(2,3-dihydrocinnamoyloxy) cinnamic acid	3-异戊二烯基-4-(2,3-二羟基桂皮酰氧基)肉桂酸	(Hegazi and El Hady,2002)
21	3-Prenyl-4-(2-methylpropionyloxy) cinnamic acid	3-异戊二烯基-4-(2-甲基丙酰氧基)肉桂酸	(Hegazi and El Hady,2002)
22	p-Methoxy-cinnamic acid cinnamyl ester	p-甲氧基-肉桂酸肉桂酯	(Hegazi and El Hady,2002)
23	Cinnamylidene acetic acid	亚桂皮基乙酸	(Hegazi and El Hady,2002)
24	Cinnamyl 3,4-dimethoxycinnamate	肉桂基3,4-二甲氧基肉桂酸盐	(EI Hady and Hegazi,2002)
25	Cinnamyl cinnamate	肉桂基肉桂酸盐	(EI Hady and Hegazi,2002)
26	3-Prenyl cinnamic acid allyl ester	3-异戊二烯基肉桂酸烯丙酯	(Teixeira et al.,2005)
27	p-Methoxycinnamic acid	p-甲氧基肉桂酸	(Salatino et al.,2005)
28	Dihydrocinnamic acid	二氢肉桂酸	(Salatino et al.,2005)
29	Cinnamic acid	肉桂酸	(Gardana et al.,2007)
30	Cinnamic acid-3,4 dimethoxy-tms-ester	—	(于世锋等,2007)
31	Cinnamic acid-trimethylester	肉桂酸-三甲基酯	(于世锋等,2007)
32	Trimethylsilyl 3-methoxy-4-cinnamate	三甲基硅烷基3-甲氧基-4-肉桂酸盐	(于世锋等,2007)
33	Artepillin C	阿替匹林C	(Silici et al.,2007)
34	Drupanin 3-prenyl-4-hydroxycinnamic acid	—	(Silici et al.,2007)

表2.18 蜂胶中的 p-香豆酸类化合物及其衍生物

Table 2.18 p-Coumaric acid and derivatives in propolis

序号 No.	英文名称 English name	中文名称 Chinese name	参考文献 Ref.
1	m-Coumaric acid	m-香豆酸	(Marcucci,1995)
2	o-Coumaric acid	o-香豆酸	(Marcucci,1995)
3	3-Prenyl-4-(2,3-dihydro-p-Coumaric acid)	3-异戊二烯-4-(2,3-二氢-p-香豆酸)	(Tazawa et al.,1998)
4	Baccharin	—	(Hegazi and El Hady,2000)
5	Coumaric acid methyl ester	香豆酸甲酯	(Hegazi and El Hady,2002)
6	p-Coumaric acid ester	p-香豆酸酯	(Hegazi and El Hady,2002)

续表

序号 No.	英文名称 English name	中文名称 Chinese name	参考文献 Ref.
7	p-Coumaric benzyl ester	p-香豆苄酯	(Hegazi et al.,2002)
8	p-Coumaric cinnamyl ester	p-香豆肉桂酯	(Hegazi et al.,2002)
9	p-Coumaric-methyl-butenyl ester	p-香豆-甲基-丁烯基酯	(Hegazi et al.,2002)
10	3,5-Diprenyl-p-Coumaric acid	3,5-二异戊二烯基-p-香豆酸	(Kumazawa et al.,2002)
11	3-Prenyl-p-Coumaric acid	3-异戊二烯-p-香豆酸	(Kumazawa et al.,2002)
12	p-Coumaric acid	p-香豆酸	(Ahn et al.,2007)

表 2.19 蜂胶中的咖啡酸类化合物及其衍生物

Table 2.19 Caffeic acid and derivatives in propolis

序号 No.	英文名称 English name	中文名称 Chinese name	参考文献 Ref.
1	Pentenyl ferulate	戊烯基阿魏酸	(Greenaway et al.,1987)
2	Cinnamyl isoferulate	肉桂基异阿魏酸	(Greenaway et al.,1987)
3	Pent-4-enyl isoferulate	4-戊烯基异阿魏酸	(Greenaway et al.,1987)
4	Benzyl ferulate	苯甲基阿魏酸	(Greenaway et al.,1987)
5	Isopentenyl-caffeate	异戊烯基咖啡酸	(Bankova et al.,1992)
6	Heindecenyl hydroferulate	—	(Bankova et al.,1992)
7	Octadecatrienyl ferulate	—	(Bankova et al.,1992)
8	Octenyl ferulate	—	(Bankova et al.,1992)
9	Pentadecenyl hydroferulate	十五烯基氢化阿魏酸	(Bankova et al.,1992)
10	Pentadecyl hydroferulate	十五烷基氢化阿魏酸	(Bankova et al.,1992)
11	Tridecatrienyl hydroferulate	—	(Bankova et al.,1992)
12	Tridecenyl hydroferulate	—	(Bankova et al.,1992)
13	Tridecyl hydroferulate	三癸基氢化阿魏酸	(Bankova et al.,1992)
14	Hydrocaffeic acid	氢化咖啡酸	(Chi et al.,1994)
15	Pentenyl isoferulate	戊烯基异阿魏酸	(Marcucci,1995)
16	Phenylethyl isoferulate	苯乙基异阿魏酸	(Marcucci,1995)
17	Prenyl isoferulate	异戊二烯基异阿魏酸	(Marcucci,1995)
18	2-Methyl-2-butenyl isoferulate	2-甲基-2-丁烯基异阿魏酸	(Marcucci,1995)
19	Benzyl isoferulate	苯甲基异阿魏酸	(Marcucci,1995)
20	Ethyl caffeate	乙烷基咖啡酸	(Marcucci,1995)
21	Prenyl ferulate	异戊二烯基阿魏酸	(Marcucci,1995)

续表

序号 No.	英文名称 English name	中文名称 Chinese name	参考文献 Ref.
22	3-Methyl-3-butenyl isoferulate	3-甲基-3-丁烯基异阿魏酸	(Marcucci,1995)
23	Cinnamyl caffeate	肉桂咖啡酸	(Marcucci,1995)
24	Butenyl caffeate	丁烯基咖啡酸	(Bankova et al.,1998)
25	β-Phenylethyl caffeate	β-苯乙基咖啡酸	(Bankova et al.,1998)
26	3-Methyl-2-butenyl ferulate	3-甲基-2-丁烯基阿魏酸	(Bankova et al.,1998)
27	3-Methyl-2-butenyl caffeate	3-甲基-2-丁烯基咖啡酸	(Bankova et al.,1998)
28	Butyl caffeate	丁基咖啡酸	(Bankova et al.,1998)
29	Pentyl caffeate	戊烷基咖啡酸	(Bankova et al.,1998)
30	Pentenyl caffeate	戊烯基咖啡酸	(Bankova et al.,1998)
31	Isoferulic acid	异阿魏酸	(Bankova et al.,1998)
32	Caffeic acid cinnamyl ester	咖啡酸肉桂酯	(Hegazi and El Hady,2002)
33	Caffeic acid isoprenyl ester	—	(Hegazi and El Hady,2002)
34	Caffeic acid phenethyl ester	咖啡酸苯乙酯	(Hegazi and El Hady,2002)
35	3-Methyl-3-butenyl ferulate	3-甲基-3-丁烯基阿魏酸	(EI Hady and Hegazi,2002)
36	2-Methyl-2-butenyl ferulate	2-甲基-2-丁烯基阿魏酸	(EI Hady and Hegazi,2002)
37	Ethyl caffeate	乙烷基咖啡酸	(EI Hady and Hegazi,2002)
38	Dimethylallyl caffeic acid	甲代烯丙基咖啡酸	(Usia et al.,2002)
39	Dimethylcaffeic acid	二甲基咖啡酸	(Park et al.,2002)
40	Phenylethyl caffeate	苯乙烯咖啡酸	(Uzel and Sorkun,2005)
41	3-Methyl-2-butenyl isoferulate	3-甲基-2-丁烯基异阿魏酸	(Salatino et al.,2005)
42	3-Methyl-3-butenyl caffeate	3-甲基-3-丁烯基咖啡酸	(Salatino et al.,2005)
43	Hexadecyl caffeate	十六烷基咖啡酸	(Salatino et al.,2005)
44	Benzyl caffeate	苯甲基咖啡酸	(Uzel and Sorkun,2005)
45	Caffeic acid	咖啡酸	(Souza et al.,2007)
46	Ferulic acid	阿魏酸	(Souza et al.,2007)

3.1.3 绿原酸类化合物

绿原酸类化合物一般是指一个或多个反式肉桂酸如咖啡酸、香豆酸、阿魏酸、芥子酸等和奎宁酸或其衍生物莽草酸、奎宁酸甲酯或丁酯、4-去氧奎宁酸等缩合形成的酯。目前,研究者已经从天然产物中发现大量该类化合物,并且证实大多数的绿原酸类化合物具有较强的生物学活性。蜂胶中的绿原酸类物质主要是由咖啡酸与奎尼酸组成的缩酚酸类物质,包括单咖啡酰奎尼酸(3-O-咖啡酰奎尼酸及4或5位取代的衍生物)、双咖啡酰奎尼酸(1,5-双咖啡酰奎尼酸、3,5-双咖啡酰奎尼酸、4,5-双咖啡酰奎尼酸和3,4-双咖啡酰奎尼酸)及三咖啡酰奎尼酸(如 3,4,5-三咖啡酰奎尼酸)等。蜂胶中发现的绿原酸类化合物均从酒神菊属型蜂胶中分离得到(表 2.20),并没有在其他类型的蜂胶中发现。

表 2.20　蜂胶中的绿原酸类化合物
Table 2.20　Chlorogenic acid compounds in propolis

序号 No.	英文名称 English name	中文名称 Chinese name	参考文献 Ref.
1	3,4-Dicaffeoylquinic acid	3,4-二咖啡酰奎尼酸	(dos Santos Pereira et al.,2003)
2	3,5-Dicaffeoylquinic acid	3,5-二咖啡酰奎尼酸	(dos Santos Pereira et al.,2003)
3	3-Caffeoylquinic acid	3-咖啡酰奎尼酸	(dos Santos Pereira et al.,2003)
4	4,5-Dicaffeoylquinic acid	4,5-二咖啡酰奎尼酸	(dos Santos Pereira et al.,2003)
5	4-Caffeoylquinic acid	4-咖啡酰奎尼酸	(dos Santos Pereira et al.,2003)
6	5-Caffeoylquinic acid	5-咖啡酰奎宁酸	(dos Santos Pereira et al.,2003)
7	4-Feruoylquinic acid	—	(Cao et al.,2004)
8	3,4,5-Tri-caffeoylquinic acid	3,4,5-三咖啡酰奎尼酸	(Matsui et al.,2004)
9	5-Ferruoyl quinic acids	—	(Cao et al.,2004)
10	Tricaffeoyl quinic acid	三咖啡酰奎尼酸	(Gardana et al.,2007)
11	Dicaffeoyl quinic acid	二咖啡酰奎尼酸	(Moura et al.,2011)
12	mono-Caffeoylquinic acid	单咖啡酰奎尼酸	(Moura et al.,2011)

3.2　二苯乙烯类化合物

二苯乙烯类化合物是以二苯乙烯为基本母核,苯环不同位置的氢原子被羟基取代而形成的多羟基酚酸类化合物,主要可分为 1,2-二苯乙烯类、菲类及其对应的二氢衍生物。此类化合物在自然界中广泛分布,存在于多种天然产物中,其代表性物质为来源于葡萄的白藜芦醇。研究表明,二苯乙烯类化合物具有抗炎、抗菌、抗病毒以及较强的抗癌功效,同时该类化合物在调节脂质代谢、保护心脏功能等方面发挥着重要作用(韩晶晶等,2008)。蜂胶中的二苯乙烯类化合物(表 2.21)也在蜂胶的生物活性方面发挥着积极的作用。

表 2.21　蜂胶中的二苯乙烯类化合物
Table1 2.21　Stibenoids in propolis

序号 No.	英文名 English name	中文名称 Chinese name	参考文献 Ref.
1	Schweinfurthin A	—	(Petrova et al.,2010)
2	Schweinfurthin B	—	(Petrova et al.,2010)
3	5′-Farnesyl-3′-hydroxyresveratrol	5′-法呢基-3′-羟基白藜芦醇	(Inui et al.,2012b)
4	5,4′-Dihydroxy-3′-methoxy-3-prenyloxy-E-stilbene.	5,4′-二羟基-3′-甲氧基-3-含氧异戊二烯-E-二苯乙烯.	(Abu-Mellal et al.,2012)

续表

序号 No.	英文名 English name	中文名称 Chinese name	参考文献 Ref.
5	3,5,3′,4′-Tetrahydroxy-2-prenyl-E-stilbene	3,5,3′,4′-四羟基-2-异戊二烯-E-二苯乙烯	(Abu-Mellal et al., 2012)
6	3,5,4′-Trihydroxy-3′-methoxy-2-Prenyl-E-stilbene	3,5,4′-三羟基-3′-甲氧基-2-异戊二烯-E-二苯乙烯	(Abu-Mellal et al., 2012)
7	5,3′,4′-Trihydroxy-3-methoxy-2-Prenyl-E-stilbene	5,3′,4′-三羟基-3-甲氧基-2-异戊二烯-E-二苯乙烯	(Abu-Mellal et al., 2012)
8	5,4′-Dihydroxy-3,3′-dimethoxy-2-Prenyl-E-stilbene	5,4′-二羟基-3,3′-二甲氧基-2-异戊二烯-E-二苯乙烯	(Abu-Mellal et al., 2012)
9	5,4′-Dihydroxy-3-prenyloxy-E-stilbene	5,4′-二羟基-3-含氧异戊二烯-E-二苯乙烯	(Abu-Mellal et al., 2012)
10	3′,4′-Dihydroxy-E-stilbene	3′,4′-二羟基-E-二苯乙烯	(Abu-Mellal et al., 2012)
11	3′,4′-Dihydroxy-3,5-dimethoxy-E-stilbene	3′,4′-二羟基-3,5-二甲氧基-E-二苯乙烯	(Abu-Mellal et al., 2012)
12	Di-prenylated dihydrostilbene	二异戊二烯基二羟基二苯乙烯	(Abu-Mellal et al., 2012)
13	3,5-Dihydroxy-2-prenyl-E-stilbene	3,5-二羟基-2-异戊二烯-E-二苯乙烯	(Abu-Mellal et al., 2012)
14	4-Prenyldihydroresveratrol	4-异戊二烯基二羟基白藜芦醇	(Abu-Mellal et al., 2012)
15	3-Prenylresveratrol	3-异戊二烯基白藜芦醇	(Abu-Mellal et al., 2012)
16	(E)-5-{2-[8-hydroxy-2-methyl-2-(4-methylpent-3-en-1-yl)-2H-chromen-6-yl]vinyl}-2-(3-methylbut-2-en-1-yl)ben-zene-1,3-diol	—	(Almutairi et al., 2014)
17	5-[(E)-3,5-dihydroxystyryl]-3-[(E)-3,7-dimethylocta-2,6-dien-1-yl]benzene-1,2-diol	—	(Almutairi et al., 2014)
18	Solomonin B	—	(Trusheva et al., 2016)
19	Solomonin C	—	(Trusheva et al., 2016)
20	(E)-4-(3-methyl-2-buten-1-yl)-3,4′,5-trihydroxy-3′-methoxystilbene	—	(Duke et al., 2017)
21	(E)-2,4-bis(3-methyl-2-buten-1-yl)-3,3′,4′,5-tetrahydroxystilbene	—	(Duke et al., 2017)
22	(E)-2-(3-methyl-2-buten-1-yl)-3-(3-methyl-2-butenyloxy)-3′,4′,5-trihydroxystilbene	—	(Duke et al., 2017)
23	(E)-2,6-bis(3-methyl-2-buten-1-yl)-3,3′,5,5′-tetrahydroxystilbene	—	(Duke et al., 2017)
24	(E)-2,6-bis(3-methyl-2-buten-1-yl)-3,4′,5-trihydroxy-3′-methoxystilbene	—	(Duke et al., 2017)

3.3 木脂素类化合物

木脂素是一类由苯丙素氧化聚合而成的天然产物,早期因其大量存在于植物树脂或木脂中,故称为木脂素。研究表明,该类物质具有很多医疗功效,特别是抗肿瘤及抗病毒活性,引起了研究者极大的关注(米靖宇和宋纯清,2002)。蜂胶中共发现了13种木脂素类化合物(表2.22)。

表 2.22 蜂胶中的木脂素类化合物
Table 2.22 Lignans in propolis

序号 No.	英文名称 English name	中文名称 Chinese name	参考文献 Ref.
1	1-(4-Hydroxy-3-methoxyphenyl)-1,2-bis{4-[(E)-3-acetoxypropen-1-yl]-2-methoxyphenoxy}propan-3-ol acetate	1-(4-羟基-3-甲氧基苯基)1,2-联[4-[(E)-3-乙酰氧基丙-1-烯基]-2-甲氧基苯甲氧基]-3-丙醇乙酸	(Valcic et al.,1998)
2	1-(4-Hydroxy-3-methoxyphenyl)-2-{4-[(E)-3-acetoxypropen-1-yl]-2-methoxyphenoxy}propan-1,3-diol 3-acetate	1-(4-羟基-3-甲氧基苯基-2-[4-[(E)-3-乙酰氧基丙-1-烯基]-2-甲氧基苯甲氧基]-1,3-二丙醇-3-乙酸	(Valcic et al.,1998)
3	3-Acetoxymethyl-5-[(E)-2-formylethen-1-yl]-2-(4-hydroxy-3-methoxyphenyl)-7-methoxy-2,3-dihydrobenzofuran.	3-乙酰氧基甲基-5-[(E)-2-甲酰基乙-1-烯基]-2-(4-羟基-3-甲氧基苯基)-7-甲氧基-2,3-二氢苯并呋喃	(Valcic et al.,1998)
4	Sesamin	芝麻明	(Christov et al.,1999)
5	Aschantin	—	(Christov et al.,1999)
6	Sesartenin	—	(Christov et al.,1999)
7	Yangambin	—	(Christov et al.,1999)
8	(+)-Tinoresinol dimethyl ether	(+)-松脂二甲基酯	(Li et al.,2008)
39	(+)-Tinoresinol	(+)-松脂醇	(Li et al.,2008)
10	(+)-Syringaresinol	(+)-丁香脂素	(Li et al.,2008)
11	Tetrahydrojusticidin B	四羟基爵床脂素	(Petrova et al.,2010)
12	6-Methoxydiphyllin	6-甲氧基山荷叶素	(Petrova et al.,2010)
13	Phyllam ricin C	—	(Petrova et al.,2010)

3.4 其他酚类化合物

蜂胶的化学成分极其复杂,除以上明确分类的酚类化合物之外,还含有一些其他种类或未能明确分类的化学组成。如具有芳香骨架结构的苯甲酮类、香豆素类、苯并呋喃类等。这些酚类化合物大多都具有抗肿瘤、抗炎症等生物活性(郭新竹和宁正祥,2002)。表2.23共列出蜂胶中的41种其他酚类化合物及其衍生物。

表 2.23 蜂胶中的其他酚类化合物
Table 2.23 Other phenols compounds and derivatives in propolis

序号 No.	英文名称 English name	化学分类 Chemical classify	参考文献 Ref.
1	1,3-Diferuloyl-2-acetylglycerol	—	(Maciejewicz,1985)
2	1-Feruloyl-3p-coumaroyl-2-acetylglycerol	—	(Maciejewicz,1985)
3	1′-O-eicosanyl glycerol	—	(Talla et al.,2016)
4	Methyl3-[4-hydroxy-3-(3-methyl-2-butenyl)phenyl]-2-(E)-propenoate	—	(黄卫平等,1998)
5	Tremetone	香豆素	(Banskota et al.,1998)
6	Viscidone	香豆素	(Banskota et al.,1998)
7	12-Acetoxyviscidone	香豆素	(Banskota et al.,1998)
8	6-(2-Carboxyethenyl)-2,2-dimethyl-2H-1-benzopyran	苯并呋喃类	(Hayashi et al.,1999)
9	8-(Methyl-butanechromane)-6-propenoic acid	—	(Marcucci et al.,2000)
10	3-Hydroxy-2,2-dimethyl-8-prenylchromane-6-propenoic acid	—	(Marcucci et al.,2000)
11	2,2-Dimethyl-8-prenylchromene-6-propenoic acid	—	(Marcucci et al.,2000)
12	2,2-Dimethylchromene-6-propenoic acid	—	(Marcucci et al.,2000)
13	2,2-Dimethyl-6-carboxyethenyl-8-prenyl-2H-1-benzopyran	苯并呋喃类	(Kumazawa et al.,2002)
14	2,2-Fimethyl-8-prenylchromene	—	(Teixeira et al.,2005)
15	2,2-Dimethyl-6-carboxyethenyl-2H-1-benzopyran	苯并呋喃类	(Silici et al.,2007)
16	3-(2-hydroxy-4-methoxybenzyl)-6-methoxy-2,3-dihydrobenzofuran	苯并呋喃类	(Omar et al.,2016)
17	1,3-bis(Trimethylsilylloxy)-5,5-proylbenzene	—	(Syamsudin et al.,2009)
18	3,4-Dimethylthioquinoline	—	(Syamsudin et al.,2009)
19	4-Oxo-2-thioxo-3-thiazolidinepropionic acid	—	(Syamsudin et al.,2009)
20	D-glucofuranuronic acid	—	(Syamsudin et al.,2009)
21	Dofuranuronic acid	—	(Syamsudin et al.,2009)
22	3-Quinolinecarboxamine	—	(Syamsudin et al.,2009)
23	Hyperibone A	苯甲酮类	(Castro et al.,2009)
24	Prenylated coumarin suberosin	香豆素	(Trusheva et al.,2010)
25	7-Epi-nemorosone	苯甲酮类	(de Castro Ishida et al.,2011)
26	7-Epi-clusianone	苯甲酮类	(de Castro Ishida et al.,2011)
27	Xanthochymol	苯甲酮类	(de Castro Ishida et al.,2011)
28	Gambogenone	苯甲酮类	(de Castro Ishida et al.,2011)

续表

序号 No.	英文名 English name	化学分类 Chemical classify	参考文献 Ref.
29	5-Pentadecylresorcinol	间苯二酚类	(Trusheva et al.,2011)
30	5-(8′Z,11′Z-Heptadecadienyl)-resorcinol	间苯二酚类	(Trusheva et al.,2011)
31	5-(11′Z-Heptadecenyl)-resorcinol	间苯二酚类	(Trusheva et al.,2011)
32	5-(14′Z-Heptadecenyl)-resorcinol	间苯二酚类	(Kardar et al.,2014)
33	5-Heptadecylresorcinol	间苯二酚类	(Trusheva et al.,2011)
34	2,2-Dimethyl-6-carboxyethnyl-2H-1-benzopyran	苯并呋喃类	(Abdallah et al.,2012)
35	6-Propenoic-2,2-dimethyl-8-prenyl-2H-1-benzopyran acid	苯并呋喃类	(Abdallah et al.,2012)
36	Methyl(E)-4-(4′-hydroxy-3′-methylbut-(E)-2′-enyloxy)cinnamate	苯丙烷类	(Abdallah et al.,2012)
37	3-(12′Z-Pentadecenyl)-phenol	烷基酚类	(Kardar et al.,2014)
38	3-(13′Z-Nonadecenyl)-phenol	烷基酚类	(Kardar et al.,2014)
39	3-(14′Z-Nonadecenyl)-phenol	烷基酚类	(Kardar et al.,2014)
40	Deperoxidised derivative of plukenetione C	—	(Almutairi et al.,2014)
41	Ethyl p-coumaryl ether	—	(Savka et al.,2015)

4 醛酮类化合物

蜂胶中的醛酮类化合物主要包括香草醛、异香草醛、苯甲醛、二羟基苯乙酮、二羟基甲氧基苯乙酮、羟基苯甲氧基苯乙酮、甲氧基苯乙酮、6,10,14-三甲氧基-2-十五酮、2-十七酮、4苯基-3-丁烯-2-酮、6-甲基-5 庚烯-2-酮（郭伽和周立东，2000）以及咕吨酮等（Sanpa et al.,2015）。

5 烃类化合物

从蜂胶中分离鉴定出来的烃类化合物主要包括戊烷、二十烷、二十一烷、二十二烷、二十三烷、二十四烷、二十五烷、二十六烷、二十七烷、二十八烷、二十九烷、三十烷、三十一烷、三十二烷、三十三烷、1-十八烯、9-二十三烯、9-二十五烯、9-二十七烯、9-二十九烯、8-二十九烯、10-三十烯、8-三十一烯、10-三十三烯、8,22-三十一碳二烯、9,23-三十三碳二烯等（Ahmed et al.,2002；Teixeira et al.,2005；Uzel and Sorkun,2005；郭伽和周立东,2000；Seifert and Haslinger,1989）。

6 甾体化合物

目前已从蜂胶中分离鉴定出羊毛甾醇、胆甾醇、岩藻醇、海绵甾醇、豆甾醇、β-二氢岩藻甾醇、β-二氢岩藻醇乙酸酯、豆甾醇乙酸酯、海绵甾醇乙酸酯等甾体化合物（Kholodova et al.,1981；Maciejew et al.,1983）。

7 糖类

蜂胶中的糖类有 D-葡萄糖、D-咩喃核糖、D-葡萄糖醇、D-果糖、塔罗糖、D-左乐糖、蔗糖、山梨糖醇、木糖醇、肌醇、D-古洛糖、D-呋喃果糖、甘露糖、茨鲜糖、半乳糖、半乳糖醇、乳糖、麦芽糖、葡萄糖酸、半乳糖醛酸等(Bankova et al., 1998; Mohammadzadeh et al., 2007; Silici et al., 2007; EI Hady and Hegazi, 2002)。

8 维生素

蜂胶中还含有维生素 A、维生素 B_1、维生素 B_2、维生素 B_6、维生素 C、维生素 E、烟酰胺、泛酸以及极微量的维生素 H 和叶酸等(黄卫平和刘放,1998;罗志刚和杨连生,2002)。

9 酶类

蜂胶中的活性蛋白质含量在 2.1%～3.8%,这些提取物中几乎都含有 α-淀粉酶和 β-淀粉酶,此外,蜂胶中还含有组织蛋白酶、胰蛋白酶、脂肪酶等(黄卫平和刘放,1998;罗志刚和杨连生,2002)。近年来,还在新鲜的杨树型蜂胶中检测到 β-葡萄糖苷酶(Zhang et al., 2012)。

10 氨基酸

蜂胶中的氨基酸包括天门冬氨酸、苏氨酸、半胱氨酸、谷氨酸、丙氨酸、异亮氨酸、丝氨酸、脯氨酸、亮氨酸、甘氨酸、酪氨酸、苯丙氨酸、组氨酸、缬氨酸、赖氨酸、精氨酸和蛋氨酸等多种氨基酸(董捷等,2007)。其中精氨酸和脯氨酸的含量最多,约占游离氨基酸总量的50%以上。精氨酸能够促进细胞分裂,在促进伤口复原、蛋白质生物合成过程中起着至关重要的作用(彭英和蔡力创,2011)。脯氨酸作为植物蛋白质的组分之一,是胶原蛋白的重要组成部分,在稳定生物大分子结构,降低细胞酸性以及调节细胞氧化还原势等方面起着重要作用(邵兴军等,2001)。

11 常量以及微量元素

蜂胶中的常量元素包括碳、氢、氧、氮、钙、磷。此外蜂胶中还含有丰富的微量元素,包括镁、铝、钾、铁、钠、磷、钛、钡、铅、锶、锂、铬、钒、镍、镧、钕、氯、硅、砷、硒、硫、钇、镓、钐、锆、钴、钍、钪、钼、镉、铋、铍、铋、钽、镝、铈、银、锑、汞、溴、硼等(郭伽和周立东,2000;邵兴军等,2001;胡福良和玄红专,2003;Cvek et al., 2008)。

小 结

蜂胶的化学组成极其复杂,并因受胶源植物、蜂种、采胶季节、气候条件以及储存时间等多种因素的影响而呈现出一定的差异性。蜂胶化学成分的研究,不仅有利于蜂胶胶源植物的判定,而且能够为蜂胶质量控制体系的建立、实现蜂胶标准化提供科学的理论依据。此外,蜂胶新成分的分离鉴定以及药理活性的研究,极大地促进了蜂胶新产品的研发。

目前,对蜂胶中化学成分的研究主要集中在其非挥发性成分方面,尤其是蜂胶中的黄酮类化合物,通常作为评价蜂胶品质及生物活性高低的重要指标,而对蜂胶中萜类和酚类化合物研究较少。因此，对蜂胶中不同类型化合物进行分离、鉴定以及药理活性的研究，对蜂胶的深入研究和开发具有重要的意义。

参考文献

Abu-Mellal A, Koolaji N, Duke RK, Tran VH, Duke CC (2012) Prenylated cinnamate and stilbenes from Kangaroo Island propolis and their antioxidant activity[J]. *Phytochemistry*, 77: 251-259.

Ahmed GH, Faten K, Abd EH (2002) Egyptian Propolis: 3. Antioxidant, antimicrobial activities and chemical composition of propolis from reclaimed lands[J]. *Zeitschrift für Naturforschung C*, 57: 395-402.

Ahn MR, Kumazawa S, Usui Y, Nakamura J (2007) Antioxidant activity and constituents of propolis collected in various areas of China[J]. *Food Chemistry*, 101(4): 1383-1392.

Alencar SM, Oldoni TLC, Castro ML, Cabral ISR, Costa-Netoc CM, Curyb JA, Rosalenb PL, Ikegakid M (2007) Chemical composition and biological activity of a new type of Brazilian propolis: red propolis[J]. *Journal of Ethnopharmacology*, 113(2): 278-283.

Alday E, Valencia D, Carreño AL, Picerno P, Piccinelli AL, Rastrelli L, Robles-zepeda Ramon, Hernandez J, Velazquez C (2015) Apoptotic induction by Pinobanksin and some of its easter derivatives from Sonoran propolis in a B-cell lymphoma cell line[J]. *Chemico-Biological Interactions* 242(2015): 35-44.

Almutairi S, Edrada-Ebel RA, Fearnley J, Igoli JO, Alotaibi W, Clements CJ, Gray AI, Watson DG (2014) Isolation of diterpenes and flavonoids from a new type of propolis from Saudi Arabia[J]. *Phytochemistry Letters* 10(2014): 160-163.

Almutairi S, Eapen B, Chundi SM, Akhalil A, Siheri W, Clements C, Fearnley J, Watson DG, Edrada-Ebel RA (2014) New anti-trypanosomal active prenylated compounds from African propolis[J]. *Phytochemistry Letters* 10: 35-39.

Awale S, Shrestha SP, Tezuka Y, Ueda J (2005) Neoflavonoids and related constituents from Nepalese propolis and their nitric oxide production inhibitory activity[J]. *Journal of Natural Products*, 68: 858-864.

Bankova V, Popov SS, Marekov NL (1983) A study on flavonoids of propolis[J]. *Journal of Natural Products*, 46(4): 471-474.

Bankova V, Dyulgerov A, Popov S, Evstatieva L, Kuleva L, Pureb O, Zamjansan Z (1992) Propolis produced in Bulgaria and Mongolia: phenolic compounds and plant origin[J]. *Apidologie*, 23(1): 79-85.

Bankova V, Christov R, Kujumgiev A, Marcucci MC, Popov S (1995) Chemical composition and antibacterial activity of Brazilian propolis[J]. *Zeitschrift für Naturforschung C*, 50: 167-172.

Bankova V, Marcucci MC, Simova S, Nikolova N, Kujumgiev A, Popov S (1996) Antibacterial diterpenic acids from Brazilian propolis[J]. *Zeitschrift für Naturforschung C*, 51: 277-280.

Bankova V, Christov R, Tejera AD (1998) Lignans and other constituents of propolis from the Canary Islands[J]. *Phytochemistry*, 49: 1411-1415.

Bankova V, Boudourova-Krasteva G, Popov S, Sforcin JM, Funari SC (1998) Seasonal variations of the chemical composition of Brazilian propolis[J]. *Apidologie*, 29(4): 361-367.

Bankova V, De Castro SL, Marcucci MC (2000) Propolis: recent advances in chemistry and plant origin[J]. *Apidologie*, 31(1): 3-15.

Bankova V (2005) Chemical diversity of propolis and the problem of standardization[J]. *Journal of Ethnopharmacology*, 100(1): 114-117.

Banskota AH, Tezuka Y, Prasain JK, Matsushige K, Saiki I, Kadota S (1998) Chemical constituents of Brazilian propolis and their cytotoxic activities[J]. *Journal of Natural Products*, 61: 896-900.

Banskota AH, Tezuka Y, Adnyana IK, Ishii E, Midorikawa K, Matsushige K, Kadota S (2001) Hepatoprotective and anti-helicobacter pylori activities of constituents from Brazilian propolis[J]. *Phytomedicine*, 8(1): 16-23.

Boisard S, Huynh THT, Escalante-Erosa F, Hernandez-Chavez LI, Pena-Rodriguez LM, Richomme P (2016). Unusual chemical composition of a Mexican propolis collected in quintana roo, Mexico[J]. *Journal of Apicultural Research*, 54(4): 350-357.

Bonvehí JS, Call FV (1994) Phenolic composition of propolis from Chania and from South America[J]. *Zeitschrift für Naturforschung C*, 49: 712-718.

Bonvehí JS, Coll FV, Jordà RE (1994) The composition, active components and bacteriostatic activity of propolis in dietetics[J]. *Journal of the American Oil Chemists Society*, 71 (5): 529-532.

Borcic I, Radonic A, Grzunov K (1996) Comparisom of the volatile constituents of propolis gathered in different regions of Croatia[J]. *Flavour and Fragrance Journal*, 11: 311-313.

Boudourova-Krasteva G, Bankova V, Sforcin JM, Nikolova N, Popov S (1997) Phenolics from Brazilian propolis[J]. *Zeitschrift für Naturforschung*, 52c: 676-679.

Bueno-Silva B, Alencar SM, Koo H, Ikegaki M, Silva GV, Napimoga MH, Rosalen PL (2013) Anti-inflammatory and antimicrobial evaluation of neovestitol and vestitol isolated from brazilian red propolis[J]. *Journal of Agricultural and Food Chemistry*, 61(19): 4546-4550.

Camargo MS, Prieto AM, Resende FA, Boldrin PK, Cardoso CR, Fernández MF, Molina-Molina JM, Olea N, Vilegas W, Cuesta-Rubio O, Varanda EA (2013) Evaluation of estrogenic, antiestrogenic and genotoxic activity of nemorosone, the major compound found in brown Cuban propolis[J]. *BMC Complementary and Alternative Medicine*, 13: 201.

Campo Fernández M, Cuesta-Rubio O, Rosado Perez A, Montes De Oca Porto R, Márquez Hernández I, Piccinelli AL, Rastrelli L (2008) GC-MS determination of isoflavonoids in seven red Cuban propolis samples[J]. *Journal of Agricultural and Food Chemistry*, 56

(21):9927-9932.

Cantarelli MA, Camina JM, Pettenati EM, Marchevsky EJ, Pellerano RG (2011) Trace mineral content of Argentinean raw propolis by nutron activation analysis (NNA): Assessment of geographical provenance by chemometrics[J]. *Food Science and Technology*, 44:256-260.

Cao YH, Wang Y, Yuan Q (2004) Analysis of flavonoids and phenolic acid in propolis by capillary electrophoresis[J]. *Chromatographia*, 59(1-2):135-140.

Castro ML, do Nascimento AM, Ikegaki M, Costa-Neto CM, Alencar SM, Rosalen PL (2009) Identification of a bioactive compound isolated from Brazilian propolis type 6 [J]. *Bioorganic & Medicinal Chemistry*, 17(14):5332-5335.

Chan SC, Chang YS, Luo SC (1997) Neoflavonoids from *Dalbergia odorifera*[J]. *Phytochemistry*, 46(5):947-949.

Chen CN, Wu CL, Shy HS, Lin JK (2003) Cytotoxic prenylflavanones from Taiwanese propolis[J]. *Journal of Natural Products*, 66(4):503-506.

Chi H, Hsieh AK, Ng CL, Lee HK, Li SFY (1994) Determination of components in propolis by capillary electrophoresis and photodiode array detection[J]. *Journal of Chromatography A*, 680(2):593-597.

Chi JP, Chen HS, Xue BW (1996) Isolation and identification of a New cinnamate ester from Liaoxi propolis[J]. *Acta Pharmaceutica Sinica*, 31(7):558-560.

Christov R, Bankova V, Hegazi A, Abd El, Hady F, Popov S (1998) Chemical composition of Egyptian propolis[J]. *Zeitschrift für Naturforschung C*, 53:197-200.

Christov R, Bankova V, Tsvetkovad I, Kujumgiev A, Delgado Tejera A (1999) Antibacterial furofuran lignans from Canary island propolis[J]. *Fitoterapia*, 70(1):89-92.

Christov R, Trusheva B, Popova M, Bankova V, Bertrand M (2006) Chemical composition of propolis from Canada, its antiradical activity and plant origin[J]. *Natural Product Research*, 19(7):673-678.

Cos P, Rajan P, Vedernikova I, Calomme M, Pieters L, Vlietinck AJ, Augustyns K, Haemers A, Vanden Berghe D (2002) In vitro antioxidant profile of phenolic acid derivatives [J]. *Free Radical Research*, 36(6):711-716.

Cvek J, Medic-Saric M, Vitali D, Vedrina-Dragojevic I, Šmit Z, Tomic S (2008) The content of essential and toxic elements in Croatian propolis samples and their tinctures[J]. *Journal of Apicultural Research*, 47(1):35-45.

de Castro Ishida VF, Negri G, Salatino A, Bandeira MF (2011) A new type of Brazilian propolis: Prenylated benzophenones in propolis from Amazon and effects against cariogenic bacteria[J]. *Food Chemistry*, 125(3):966-972.

Dietrich K (1911) Further contributions to the knowledge of bee resin (propolis)[J]. *Pharmazeutische Zentralhalle für Deutschland*, 52:10-19.

Dondi F, Blo G, Lodi G, Bighi C, Benfenati L; Moncalvo E (1984) Applications of high performance liquid chromatography in the analysis of cosmetics and toiletries[J].

Analytica Chimica acta,74:117-127.

dos Santos Pereira A,de Miranda Pereira A,Trugo LC,de Aquino Neto FR(2003)Distribution of quinic acid derivatives and other phenolic compounds in Brazilian propolis[J]. *Zeitschrift für Naturforschung C*,58(7-8):590-593.

Duke CC,Tran VH,Duke RK,Abu-Mellal A,Plunkett GT,King DI,Hamid K,Wilson KL,Barrett RL,Bruhl JJ(2017)A sedge plant as the source of Kangaroo Island propolis rich in prenylated p-coumarate ester and stilbenes[J]. *Phytochemistry*,134:87-97.

El Hady F,Hegazi A(2002)Egyptian propolis:2. Chemical composition,antiviral and antimicrobial activities of East Nile Delta propolis[J]. *Zeitschrift für Naturforschung C*,57:386-394.

El-Bassuony AA(2009)New prenilated compound from Egyptian propolis with antimicrobial activity[J]. *Revista Latinoamericana de Quimica*,México,37(1):85-90.

Falcão SI,Boas MV,Estevinho LM,Barros C,Domingues MR,Cardoso SM(2010)Phenolic characterization of Northeast Portuguese propolis:usual and unusual compounds[J]. *Analytical and Bioanalytical Chemistry*,396(2):887-897.

Falcão SI,Tomás A,Vale N,Gomes P,Freire C,Vilas-Boas M(2013)Phenolic quantification and botanical origin of Portuguese propolis[J]. *Industrial Crops and Products*,49:805-812.

Faten K(2002)Egyptian Propolis:2. chemical composition,antiviral and antimicrobial activities of east Nile delta propolis[J]. *Zeitschrift für Naturforschung C*,57:386-394.

Gardana C,Scaglianti M,Pietta P,Simonetti P(2007)Analysis of the polyphenolic fraction of propolis from different sources by liquid chromatography-tandem mass spectrometry[J]. *Journal of Pharmaceutical and Biomedical Analysis*,45(3):390-399.

GB/T 24283-2009 蜂胶[S].

Ghisalberti EL,Jefferies PR,Lanteri R,Matisons L(1978)Constituents of propolis[J]. *Experientia*,34(2):157-158.

Ghisalberti EL(1979)Propolis:A review[J]. *Bee World*,60(2):59-84.

Graf E(1992)Antioxidant potential of ferulic acid[J]. *Free Radical Biology and Medicine*,1992,13(4):435-448.

Greenaway W,Scaysbrok T,whatley FR(1987)The analysis of bud exudate of Populus euramericana, and of propolis, by Gas Chromatography-Mass Spectrometry[J]. *Proceeding of The royal Society*,232(1268):249-272.

Greenaway W,Scaysbrok T,whatley FR(1990)The composition and plant origins of propolis:A report of work at Oxford[J]. *Bee world*,71(3):107-118.

Hayashi K,Komura S,Isaji N,Ohishi N,Yagi K(1999)Isolation of antioxidative compounds from Brazilian propolis:3,4-dihydroxy-5-prenylcinnamic acid,a novel potent antioxidant[J]. *Chemical & Pharmaceutical Bulletin*,47(11):1521-1524.

Hegazi A,El Hady F,Abd Allah FA(2000)Chemical composition and antimicrobial activity of European propolis[J]. *Zeitschrift für Naturforschung C*,55(1):70-75.

Hegazi A,El Hady F(2002)Egyptian propolis:3. Antioxidant,antimicrobial activities and chemical composition of propolis from reclaimed lands[J]. *Zeitschrift für Naturforschung C*,2002,57(3-4):395-402.

Hernández IM,Fernandez MC,Cuesta-Rubio O,Piccinelli AL,Rastrelli L (2005) Polyprenylated benzophenone derivatives from Cuban propolis[J]. *Journal of Natural Products*,68(6):931-934.

Huang MT,Smart RC,Wong CQ,Conney AH (1988) Inhibitory effect of curcumin, chlorogenicanid,caffeic acid,and ferulic acid on tumor promotion in mouse skin by 12 o-tetradecanoylphorbol-13-acetate[J]. *Cancer Research*,48(21):5941-5946.

Huang WJ,Huang CH,Wu CL,Lin JK,Chen YW,Lin CL,Chuang SE (2007) Propolin G, a prenylflavanone, isolated from Taiwanese propolis, induces caspase-dependent apoptosis in brain cancer cells[J]. *Journal of Agricultural and Food Chemistry*,55(18):7366-7376.

Inui S,Hosoya T,Shimamura Y,Masuda S,Ogawa T,Kobayashi H,Shirafuji K,Moli RT,Kozone I,Shin-ya K,Kumazawa S (2012a) Solophenols B-D and solomonin: new prenylated polyphenols isolated from propolis collected from the Solomon Islands and their antibacterial activity[J]. *Journal of Agricultural and Food Chemistry*,60(47):11765-11770.

Inui S,Shimamura Y,Masuda S,Shirafuji K,Moli RT,Kumazawa S (2012b) A new prenylflavonoid isolated from propolis collected in the Solomon Islands[J]. *Bioscience Biotechnology and Biochemistry*,76(5):1038-1040.

Januario AH,Lourenco MV,Domzio LA (2005) Isolation and structure determination of bioactive isoflavones from callus culture of *Dipteryx odorata*[J]. *Chemical & Pharmaceutical Bulletin*,53(7):740-742.

Johnson KS,Eischen FA,Giannasi DE (1994) Chemical composition of North American bee propolis and biological activity towards larvae of greater wax moth[J]. *Journal of Chemical Ecology*,20(7):1783-1791.

Junior MRM,Daugsch A,Moraes CS,Queiroga CL,Pastore GM,Parki YK (2008) Comparison of volatile and polyphenolic compounds in Brazilian green propolis and its botanical origin *Baccharis dracunculifolia*[J]. *Food Science and Technology (Campinas)*,28(1):178-181.

Kartal M,Kaya S,Kurucu S (2002) GC-MS analysis of propolis samples from two different regions of Turkey[J]. *Zeitschrift für Naturforschung C*,57(9-10):905-909.

Kardar MN,Zhang T,Coxon GD,Watson DG,Fearnley J,Seidel V (2014) Characterisation of triterpenes and new phenolis lipids in Cameroonian propolis[J]. *Phytochemistry*,106(2004):156-163.

Krol W,Scheller S,Czuba Z,Matsuno T,Zydowicz G,Shani J,Mos M (1996) Inhibition of neutrophils' chemiluminescence by ethanol extract of propolis (EEP) and its phenolic components[J]. *Journal of Ethnopharmacology*,55(1):19-25.

Küstenmacher H (1911) Propolis[J]. *Berichte der Deutschen Pharmazeutischen Gesellschaft*,

21:65-92.

Kumazawa S, Hayashi K, Kajiya K, Ishii T, Hamasaka T, Nakayama T (2002) Studies of the constituents of Uruguayan propolis[J]. *Journal of Agricultural and Food Chemistry*, 50(17):4777-4782.

Kumazawa S, Yoneda M, Shibata I, Kanaeda J, Hamasaka T, Nakayama T (2003) Direct evidence for the plant origin of Brazilian propolis by the observation of honeybee behavior and phytochemical analysis[J]. *Chemical and Pharmaceutical Bulletin (Tokyo)*, 51(6):740-742.

Kumazawa S, Goto H, Hamasaka T, Fukumoto S, Fujimoto T, Nakayama T (2004) A new prenylated flavonoid from propolis collected in Okinawa, Japan[J]. *Bioscience Biotechnology and Biochemistry*, 68(1):260-262.

Kumazawa S, Nakamura J, Murase M, Miyagawa M, Ahn MR, Fukumoto S (2008) Plant origin of Okinawan propolis: honeybee behavior observation and phytochemical analysis[J]. *Naturwissenschaften*, 95(8):781-786.

Kusumoto T, Miyamoto T, Higuchi R, Doi S, Sugimoto H, Yamada H (2001) Isolation and structures of two new compounds from the essential oil of Brazilian propolis[J]. *Chemical and Pharmaceutical Bulletin (Tokyo)*, 49(9):1207-1209.

Lee MS, Lin YP, Hsu FL, Zhan GR, Yen KY (2006) Bioactive constituents of *Spatholobus suberectus* in regulating tyrosinase-related proteins and mRNA in HEMn cells[J]. *Phytochemistry*, 67(12):1262-1270.

Li F, Awale S, Tezuka Y, Kadota S (2008) Cytotoxic constituents from Brazilian red propolis and their structure-activity relationship[J]. *Bioorganic and Medicinal Chemistry*, 16(10):5434-5440.

Li F, Awale S, Zhang HY, Tezuka Y, Esumi H, Kadota S (2009) Chemical constituents of propolis from Myanmar and their preferential cytotoxicity against a human pancreatic cancer cell line[J]. *Journal of Natural Products*, 72(7):1283-1287.

Li F, Awale S, Tezuka Y, Esumi H, Kadota S (2010) Study on the constituents of Mexican propolis and their cytotoxic activity against PANC-1 human pancreatic cancer cells[J]. *Journal of Natural Products*, 73(4):623-627.

Li F, He YM, Awale S, Kadota S, Tezuka Y (2011) Two new cytotoxic phenylallylflavanones from Mexican propolis[J]. *Chemical and Pharmaceutical Bulletin (Tokyo)*, 59(9):1194-1196.

Lotti C, Campo Fernandez M, Piccinelli AL, Cuesta-Rubio O, Márquez Hernández I, Rastrelli L (2010) Chemical constituents of red Mexican propolis[J]. *Journal of Agricultural and Food Chemistry*, 58(4):2209-2213.

Maciejewicz W (2001) Isolation of flavonoid aglycones from propolis by a column chromatography method and their identification by GC-MS and TLC methods[J]. *Journal of Liquid Chromatography & Related Technologies*, 24(8):1171-1179.

Marcucci MC (1995) Propolis: chemical composition, biological properties therapeutic activity[J]. *Apidologie*, 26: 83-99.

Marcucci MC, Rodriguez J, Ferreres F, Bankova V, Groto R, Popov S (1998) Chemical composition of Brazilian propolis from São Paulo state[J]. *Zeitschrift für Naturforschung C*, 53: 117-119.

Marcucci MC, Ferreres F, Custódio AR, Ferreira MM, Bankova VS, García-Viguera C, Bretz WA (2000) Evaluation of phenolic compounds in Brazilian propolis from different geographic regions[J]. *Zeitschrift für Naturforschung C*, 55(1-2): 76-81.

Markham KR, Mitchell KA, Wilkins AL, Daldy JA, Lu Y (1996) HPLC and GC-MS identification of the major organic constituents in New Zealand propolis[J]. *Hytochemistry*, 42(1): 205-211.

Marquez Hernandez I, Cuesta-Rubio O, Campo Fernandez M, Rosado Pérez A, Montes de Oca Porto R, Piccinelli AL, Rastrelli L (2010) Studies on the constituents of yellow Cuban propolis: GC-MS determination of triterpenoids and flavonoids[J]. *Journal of Agricultural and Food Chemistry*, 58(8): 4725-4730.

Martos I, Cossentini M, Ferreres F, Tomás-Barberán FA (1997) Flavonoid composition of Tunisian honeys and propolis[J]. *Journal of Agricultural and Food Chemistry*, 45: 2824-2829.

Matsuda AH, de Almeida-Muradian LB (2008) Validated method for the quantification of artepillin C in Brazilian propolis[J]. *Phytochemical Analysis*, 19(2): 179-183.

Matsui T, Ebuchi S, Fujise T, Abesundara KJ, Doi S, Yamada H, Matsumoto K (2004) Strong antihyperglycemic effects of water-soluble fraction of Brazilian propolis and its bioactive constituent, 3,4,5-tri-O-caffeoylquinic acid[J]. *Biological and Pharmaceutical Bulletin*, 27(11): 1797-1803.

Matsuno T (1995) A new clerodane diterpenoid isolated from propolis[J]. *Zeitschrift für Naturforschung C*, 50: 93-97.

Matsuno T, Matsumoto Y, Saito M, Morikawa J (1997) Isolation and characterization of cytotoxic diterpenoid isomers from propolis[J]. *Zeitschrift für Naturforschung C*, 52 (9-10): 702-704.

Melliou E, Chinou I (2004) Chemical analysis and antimicrobial activity of Greek propolis [J]. *Planta Medica*, 70: 515-519.

Melliou E, Stratis E, Chinou I (2007) Volatile constituents of propolis from various regioun of Greece-Antimicrobial activity[J]. *Food Chemistry*, 103: 375-380.

Meneses E, Durango D, García C (2009) Antifungal activity against postharvest fungi by extracts from Colombian propolis[J]. *Química Nova*, 32: 2011-2017.

Meyer W (1956) Propolis bees and their activities[J]. *Bee world*, 37(2): 25-36.

Mohammadzadeh S, Shariatpanahi M, Hamedi M, Ahmadkhaniha R, Samadia N, Ostadb SN (2007) Chemical composition, oral toxicity and antimicrobial activity of Iranian propolis[J]. *Food Chemistry*, 103, (4): 1097-1103.

Moura SA, Negri G, Salatino A, Lima LD, Dourado LP, Mendes JB, Andrade SP, Ferreira

MA, Cara DC (2011) Aqueous extract of Brazilian green propolis: primary components, evaluation of inflammation and wound healing by using subcutaneous implanted sponges[J]. *Evidence-Based Complementary and Alternative Medicine*, 2011:748283.

Nina N, Quispe C, Jiménez-Aspee F, Theoduloz C, Gimenez A, Schmeda-Hirschmann G (2016) Chemical profiling and antioxidant activity of bolivian propolis[J]. *Journal of the Science of Food and Agriculture*, 96(6):2142-2153.

Oliveira AP, Franca HS, Kuster RM, Teixeira LA, Rocha LM (2010) Chemical composition and antibacterial activity of Brazilian propolis essential oil[J]. *Journal of Venomous Animals and Toxins including Tropical Diseases*, 16(1):121-130.

Omar RMK, Igoli J, Gray AI, Ebiloma GU, Clements C, Fearnley J, Ebel RAE, Zhang T, De Koning HP, Watson DG (2016) Chemical characterisation of Nigerian red propolis and its biological activity against trypanosoma brucei[J]. *Phytochemical Analysis*, 27:107-115.

Park YK, Alencar SM, Aguiar CL (2002) Botanical origin and chemical composition of Brazilian propolis[J]. *Journal of Agricultural and Food Chemistry*, 50(9):2502-2506.

Park YK, Alencar SM, Scamparini ARP, Aguiar CL (2002) Própolis produzida no Sul do Brasil, Argentina e Uruguai: evidências fitoquímicas de sua origem vegetal[J]. *Ciência Rural*, 32(6):997-1003.

Papotti G, Bertelli D, Plessi M, Cecilia Rossi MC (2010) Use of HR-NMR to classify propolis obtained using different harvesting methods[J]. *International Journal of Food Science & Technology*, 45(8):1610-1618.

Papotti G, Bertelli D, Bortolotti L, Plessi M (2012) Chemical and functional characterization of Italian propolis obtained by different harvesting methods[J]. *Journal of Agricultural and Food Chemistry*, 60(11):2852-2862.

Pereira AS, Norsell M, Cardoso J, Aquino Neto FR, Ramos MF (2000) Rapid screening of polar compounds in Brazilian propolis by high-temperature high-resolution gas chromatography-mass spectrometry[J]. *Journal of Agricultural and Food Chemistry*, 48(11):5226-5230.

Pereira AS, Nascimento EA, Aquino Neto F (2002) Lupeol alkanoates in Brazilian propolis [J]. *Zeitschrift für Naturforschung C*, 57(7-8):721-726.

Petrova A, Popova M, Kuzmanova C, Tsvetkova I, Naydenski H, Muli E, Bankova V (2010) New biologically active compounds from Kenyan propolis[J]. *Fitoterapia*, 81(6):509-514.

Piccinelli AL, Campo Fernandez M, Cuesta-Rubio O, Márquez Hernández I, De Simone F, Rastrelli L (2005) Isoflavonoids isolated from Cuban propolis[J]. *Journal of Agricultural and Food Chemistry*, 53(23):9010-9016.

Popova M, Bankova V, Tsvetkova I, Naydenski C, Silva MV (2001) The first glycosides isolated from propolis: diterpene rhamnosides[J]. *Zeitschrift für Naturforschung C*, 56(11-12):1108-1111.

Popova M, Bankova V, Chimov A, Silva MV (2002) A scientific note on the high toxicity of propolis that comes from *Myroxylon balsamum* trees[J]. *Apidologie*,33(1):87-88.

Popova M,Chinou I,Marekov I,Bankova V (2009) Terpenes with antimicrobial activity from Cretan propolis[J]. *Phytochemistry*,70(10):1262-1271.

Popova M,Graikou K,Chinou I,Bankova V (2010) GC-MS profiling of diterpene compounds in Mediterranean propolis from Greece[J]. *Journal of Agricultural and Food Chemistry*,58(5):3167-3176.

Popova M,Trusheva B,Antonova D,Cutajarb S,Mifsudc D,Farrugiab C,Tsvetkovad I,Najdenskid H,Bankova V (2011) The specific chemical profile of Mediterranean propolis from Malta[J]. *Food Chemistry*,126(3):1431-1435.

Popravko SA,Gurevich AI,Kolosov MN (1969) Flavonoid components of propolis[J]. *Khimiya Prirodnykh Soedinenii*,5(6):476-482.

Righi AA,Alves TR,Negri G,Marques LM,Breyer H,Salatino A (2011) Brazilian red propolis:unreported substances,antioxidant and antimicrobial activities[J]. *Journal of the Science of Food and Agriculture*,91(13):2363-2370.

Russo A,Longo R,Vanella A (2002) Antioxidant activity of propolis:role of caffeic acid phenethyl ester and galangin[J]. *Fitoterapia*,73:S21-S29.

Salatino A,Teixeira E,Negri G,Message D (2005) Origin and chemical variation of Brazilian propolis[J]. *Evidence-Based Complementary and Alternative Medicine*,2(1):33-38.

Savka MA,Dailey L,Popova M,Mihaylova R,Merritt B,Masek M,Le Phuong,Mat Nor SR,Ahmad M,Hudson AO,Bankova V (2015) Chemical composition and disruption of quorum sensing signaling in geographically diverse United States propolis[J]. *Evidence-Based Complementary and Alternative Medicine*,2015:472593.

Sanpa S,Popova M,Bankova V,Tunkasiri T,Eitssayeam S,Chantawannakul P (2015) Antibacterial compounds from propolis of *Tetragonula laeviceps* and *Tetrigona melanoleuca* (Hymenoptera:Apidae) from Thailang[J]. *Plos One*,DOI:10.1371.

Seifert M,Haslinger E (1989) Über die inhaltsstoffe der propolis,I[J]. *European Journal of Organic Chemistry*,1989(11):1123-1126.

Sha N,Guan SH,Lu ZQ,Chen GT,Huang HL,Xie FB,Yue QX,Liu X,Guo DA (2009) Cytotoxic constituents of Chinese propolis[J]. *Journal of Natural Products*,72(4):799-801.

Shrestha SP,Narukawa YJ,Takeda T (2007a) Chemical constituents of Nepalese propolis: isolation of new dalbergiones and related compounds[J]. *Natural Medicines*,61:73-76.

Shrestha SP,Narukawa YJ,Takeda T (2007b) Chemical constituents of Nepalese propolis (II)[J]. *Chemical & Pharmaceutical Bulletin*,55(6):926-929.

Silici S,Kutluca S (2005) Chemical composition and antibacterial activity of propolis collected by three different races of honeybees in the same region[J]. *Journal of Ethnopharmacology*,99:69-73.

Silici S, Ünlü M, Vardar-Ünlü G (2007) Antibacterial activity and phytochemical evidence for the plant origin of Turkish propolis from different regions[J]. *World Journal of Microbiology & Biotechnology*, 23 (12):1797-1803.

Simões LMC, Gregório LE (2004) Effect of Brazilian green propolis on the production of reactive oxygen species by stimulated neutrophils[J]. *Journal of Ethnopharmacology*, 94:59-65.

Souza JP, Tacon LA, Correia CC, Bastos JK, Freitas LA (2007) Spray-dried propolis extract, II: Prenylated components of green propolis[J]. *Pharmazie*, 62 (7):488-492.

Syamsudin, Wiryowidagdo S, Simanjuntak P, Heffen WL (2009) Chemical composition of propolis from different regions in Java and their cytotoxic activity[J]. *American Journal of Biochemistry and Biotechnology*, 5(4):180-183.

Talla E, Tamfu AN, Gade IS, Yanda L, Mbafor JT, Laurent S, Elst LV, Popova M, Bankova V (2016) New mono-ether of glycerol and triterpenes with DPPH radical scavenging activity from Cameroonian propolis[J]. *Natural Product Research*, 12:1-1.

Tazawa S, Warashina T, Noro T (1998) Studies on the constituents of Brazilian propolis[J]. *Chemical & Pharmaceutical Bulletin*, 46:1477-1479.

Tazawa S, Warashina T, Noro T (1999) Studies on the constituents of Brazilian propolis (II)[J]. *Chemical & Pharmaceutical Bulletin*, 47:1388-1392.

Tazawa S, Arai Y, Hotta S, Mitsui T, Nozaki H, Ichihara K. Discovery of a novel diterpene in brown propolis from the state of Parana, Brazil[J]. *Natural Product Communications*, 2016, 11(2):201-205.

Teixeira ÉW, Negri G, Meira RM, Message D, Salatino A (2005) Plant origin of green propolis: bee behavior, plant anatomy and chemistry[J]. *Evidence-Based Complementary and Alternative Medicine*, 2 (1):85-92.

Tomás-Barberán FA, GarcÍa-Viguera C, Vit-Olivier P, Ferreres F, Tomás-Lorente F (1993) Phytochemical evidence for the botanical origin of tropical propolis from Venezuela[J]. *Phytochemistry*, 34(1):191-196.

Toreti VC, Sato HH, Pastore GM, Park YK (2013) Recent progress of propolis for its biological and chemical compositions and its botanical origin[J]. *Evidence-Based Complementary and Alternative Medicine*, 2013:697390.

Tran VH, Duke RK, Mellal AA, Abu-Mellal A, Duke CC (2012) Propolis with high flavonoid content collected by honey bees from *Acacia paradoxa*[J]. *Phytochemistry*, 81:126-132.

Trusheva B, Popova M, Bankova V, Tsvetkova I, Naydenski C, Sabatini AG (2003) A new type of European propolis, containing bioactive labdanes[J]. *Rivista Italiana EPPOS*, 13:3-8.

Trusheva B, Popova M, Naydenski H, Tsvetkova I, Rodriguez JG, Bankova V (2004) New polyisoprenylated benzophenones from Venezuelan propolis[J]. *Fitoterapia*, 75(7/8):683-689.

Trusheva B, Popova M, Bankova V, Simova S, Marcucci MC, Miorin PL, Pasin FR, Tsvetkova I, Candida RM, Guilherme V (2006) Bioactive constituents of Brazilian red propolis[J]. *Evidence-Based Complementary and Alternative Medicine*, 3(2):249-254.

Trusheva B, Todorov I, Ninovs M, Najdenski H, Daneshmand A, Bankova V (2010) Antibacterial mono-and sesquiterpene esters of benzoic acids from Iranian propolis[J]. *Chemistry Central Journal*, 4(1):8.

Trusheva B, Popova M, Koendhori EB, Tsvetkova I, Naydenski C, Bankova V (2011) Indonesian propolis: chemical composition, biological activity and botanical origin[J]. *Natural Product Research*, 25(6):606-613.

Trusheva B, Stancheva K, Gajbhiye N, Dimitrova R, Popova M, Saraf R, Bankova V (2016) Two New Prenylated Stilbenes with an Irregular Sesquiterpenyl Side Chain from Propolis from Fiji Islands[J]. *Records of Natural Products*, 10(4):465.

Usia T, Banskota AH, Tezuka Y, Midorikawa K, Matsushige K, Kadota S (2002) Constituents of Chineses propolis and their antiproliferative activities[J]. *Journal of Natural Products*, 65(5):673-676.

Uzel A, Sorkun K (2005) Chemical compositions and antimicrobial activities of four different anatolian propolis samples[J]. *Microbiological Research*, 160(2):189-195.

Valcic S, Montenegro G, Timmermann BN (1998) Lignans from Chilean propolis[J]. *Journal of Natural Products*, 61:771-775.

Walker P, Crane E (1987) Constituents of propolis[J]. *Apidologie*, 18(4):327-334.

Wang BJ, Zhang HJ (1988) Studies on the chemical constituents of Beijing propolis[J]. *China Journal of Chinese Materia Medica*, 10:37-38.

Wollenweber E, Buchmann SL (1997) Feral honey bees in the Sonoran desert: propolis sources other than poplars (*Populus* spp.)[J]. *Zeitschrift für Naturforschung C*, 57:530-535.

Yang C, Luo L, Zhang H, Yang X, Lv Y, Song H (2010) Common aroma-active components of propolis from 23 regions of China[J]. *Journal of the Science of Food and Agriculture*, 90(7):1268-1282.

Zhang CP, Liu G, Hu FL (2012) Hydrolysis of flavonnoid glycosides by propolis β-glycosidase[J]. *Natural Product Research*, 26(3):270-273.

Zhang CP, Huang S, Wei WT, Ping S, Shen XG, Li YJ, Hu FL (2014) Development of high-performance liquid chromatographic for quality and authenticity control of Chinese propolis[J]. *Journal of Food Science*, 79(7):C1315-1322.

陈业高(2004)植物化学成分[M].北京:化学化工出版社.

曾晞,卢玉振,牟兰,张长庚(2004)GC-MS法分析比较贵州不同产地蜂胶挥发油化学成分[J].生命科学仪器,(2):28-29.

崔庆新,刘国富(2001)蜂胶乙醇提取物化学成分的GC/MS研究[J].天然产物研究与开发,13(6):36-38.

董捷,张红城,尹策,李春阳(2007)蜂胶研究的最新进展[J].食品科学,28(9):637-642.

付宇新,徐元君,陈滨,丽艳,罗丽萍(2009)气相色谱/质谱法分析内蒙古蜂胶挥发性成分[J].分析化学,37(5):745-748.

郭伽,周立东(2000)蜂胶的化学成分研究进展[J].中国养蜂,51(2):17-18.

郭伽,周立东(2001)北京蜂胶中的微量元素测定[J].中国养蜂,52(2):29.

郭新竹,宁正祥(2002)天然酚类化合物及其保健作用[J].食品工业,(3):28-29.

韩晶晶,刘炜,毕玉平(2008)白藜芦醇的研究进展[J].生物工程学报,24(11):1851-1859.

胡福良,玄红专(2003)蜂胶化学成分的最新研究进展[J].养蜂科技,(1):27-31.

胡益勇,徐晓玉(2006)阿魏酸的化学和药理研究进展[J].中成药,28(2):253-255.

黄卫平,刘放(1998)蜂胶的研究进展[J].养蜂科技,(5):6-11.

李雅萍,贺丽苹,陈玉芬,卢占列,郑尧隆,潘建国,林励(2007)SPME-GC/MS联用技术分析蜂胶中挥发性成分的研究[J].现代食品科技,23(7):78-80.

刘波静(2001)气相色谱/质谱法分析研究蜂胶中化学成分[J].分析化学,29(7):861.

鲁守平,隋新霞,孙群,孙宝启(2006)药用植物次生代谢的生物学作用及生态环境因子的影响[J].天然产物研究与开发,18(6):1027-1032.

罗志刚,杨连生(2002)蜂胶及其应用[J].广州食品工业科技,18(3):57-59.

孟申(1992)中药有效成分抗氧化作用的实验研究[J].中国药理学通报,8(5):326-330.

米靖宇,宋纯清(2002)牛蒡子中木脂素类化合物的抗肿瘤及免疫活性[J].时针国医国药,13(3):168-169.

彭瑛,蔡力创(2011)精氨酸的保健作用及其调控研究进展[J].湖南理工学院学报,(1):59-62.

邵兴军,陈黎红,丁德华编著(2001)蜂胶的力量[M].北京:中国医药科技出版社.

盛文胜,余兰平,李树岚,汪玲(2006)超临界CO_2萃取蜂胶气相色谱/质谱法分析研究[J].中国蜂业,57(8):5-7.

唐传核,彭志英(2001)类黄酮研究进展.抗氧化研究[J].中国食品添加剂,(5):12-16.

唐传核,彭志英(2002)类黄酮研究进展.生理功能[J].中国食品添加剂,(1):5-10.

唐传核(2005)植物生物活性物质[M].北京:化学工业出版社.

王玮,王琳(2002)黄酮类化合物的研究进展[J].沈阳医学院学报,4(2):115-119.

徐响,董捷,李洁(2008)固相微萃取与GC-MS法分析蜂胶中挥发性成分[J].食品工业科技,29(5):57-60.

余兰平,盛文胜,喻建辉(2006)气相色谱法-质谱法分析蜂胶中醇溶性物质的化学成分[J].江西食品工业,(2):28-30.

于世锋,符军放,陈卫军,王毕妮,曹炜(2007)高效液相色谱法测定蜂胶中4种酚酸的含量[J].中国蜂业,58(9):8-10.

赵强,张彬,周武,郭志芳(2007)微波辅助萃取-GC/MS联用分析蜂胶挥发油[J].精细化工,24(12):1192-1195,1203.

张翠平,胡福良(2009)蜂胶中的黄酮类化合物[J].天然产物研究与开发,21(6):1084-1090.

张翠平,胡福良(2012)蜂胶中的萜类化合物[J].天然产物研究与开发,24(7):976-984.

朱彤,吕效吾(1997)蜂胶在医学上的应用前景[J].中国养蜂,(5):35-36.

第三章 蜂胶的生物学活性

蜂胶具有十分广泛的生物学活性,这也正是蜂胶研究与开发的价值所在。虽然不同地区、不同来源蜂胶所含的化学成分差异较大,但各种蜂胶所具有的生物学活性基本类似。本章对国内外蜂胶、蜂胶提取物以及单体活性成分的生物学活性及其可能机制的研究进展进行综述,以期为蜂胶生物学活性的基础研究及产品开发提供参考。

第一节 蜂胶的抗病原微生物作用

蜂胶的抗病原微生物作用是蜂胶最重要的生物学活性之一。蜂胶抗病原微生物活性的应用最早可以追溯到古埃及时代。受蜜蜂使用蜂胶与蜂蜡包裹入侵者尸体的启发,古埃及人很早便学会将蜂胶作为一种防腐材料加以应用。而对于蜂胶抗菌活性的系统性研究,则最早开始于20世纪40年代(Kuropatnicki et al.,2013)。10多年来,随着人们对蜂胶关注度的逐渐升高,大量围绕蜂胶抗病原微生物活性的相关研究也随之展开,其中以蜂胶抗细菌活性的报道最多。

1 蜂胶提取物的抗病原微生物作用

1.1 蜂胶乙醇提取物(EEP)的抗病原微生物作用

乙醇是目前提取蜂胶最常用的溶剂,其成本低,提取率高,且提取物的抗菌效果好,因此蜂胶抗菌研究中以蜂胶乙醇提取物(ethanol extract of propolis,EEP)最为常见。国内外大量研究表明,尽管蜂胶的来源不同,但EEP的抗菌活性相似,且抗菌范围非常广泛,除了一些常见的革兰阳性及阴性菌,EEP对牙周致病菌甚至某些抗生素耐药菌都有良好的抗菌效果。

陈荷凤等(2001)比较研究了不同溶剂蜂胶提取物对金黄色葡萄球菌(*Staphylococcus aureus*)、大肠杆菌(*Escherichia coli*)及绿脓杆菌(*Pseudomonas aeruginosa*)等7种常见菌的抗菌效果,结果发现,95%乙醇提取物抗菌效果最好,其次是60%乙醇、乙醚、乙酸乙酯提取物,而30%乙醇、蒸馏水、1%氢氧化钠溶液和5%碳酸氢钠混合溶液提取的蜂胶提取物抗菌效果最差。这不仅表明EEP具有良好的抗菌效果,同时也表明EEP的抗菌效果与乙醇浓度有关。Koo等(2000)研究了巴西东北部一种新型意蜂蜂胶以及巴西东南部和南部的蜂胶对变形链球菌(*Streptococcus mutans*)生长、黏附及对其水不溶性葡聚糖合成的影响。结果表明,这几种EEP都表现出抗变形链球菌的活性,对变形链球菌的最小抑菌浓度(minimal inhibition concentration,MIC)为 $50\sim400\mu g/mL$,对远缘链球菌(*Streptococcus sobrinus*)及仓鼠链球菌(*S. cricetus*)的MIC为 $25\sim400\mu g/mL$,最小杀菌浓度(minimum bactericidal concentration,MBC)是最小抑菌浓度的 4~8 倍；变形链球菌、远缘链球菌细胞的黏附和水不溶性葡聚糖的合成被 EEP 显著抑制；东北部新型蜂胶对细胞黏附和水不溶葡聚糖合成的抑制达到 80% 时，浓度最低可分别至 12.5μg/mL 和 7.8μg/mL；高效液相色谱法(high-performance liquid chromatography,HPLC)分析结果显示，这种新型蜂胶的化学组分与巴西东南部和南部的蜂胶完全不同，既未检测到黄酮苷元也未检测到 p-香豆酸，说明蜂胶的抗菌活性并不仅仅取决于黄酮类和羟基肉桂酸类物质。Uzel 等(2005)评估了产自安纳托利亚(Anatolian)4 种不同的 EEP 对包括口腔病原菌在内的不同微生物组的抗菌活性,并比较了它们的化学成分。宏量稀释法测定 MIC 结果表明，效果最好的 EEP 对远缘链球菌和粪球球菌(*Enterococcus faecalis*)的 MIC 值为 2μg/mL，对滕黄微球菌(*Micrococcus luteus*)、白色念珠菌(*Candida albicans*)、克鲁斯假丝酵母(*C. krusei*)的 MIC 值为 4μg/mL，对变形链球菌、金黄色葡萄球菌、表皮葡萄球菌(*Staphylococcus epidermidis*)和产气肠杆菌(*Enterobacter aerogenes*)的 MIC 值为 8μg/mL，对大肠杆菌和热带念珠菌(*C. tropicalis*)的 MIC 值为 16μg/mL，对鼠伤寒沙门氏菌(*Salmonella typhimurium*)和绿脓杆菌的 MIC 值为 32μg/mL。HPLC 结合质谱分析蜂胶的化学成分结果发现，安纳托利亚蜂胶最主要的化学成分是黄酮，如松属素、槲皮素和短叶松素等。虽然这些蜂胶来自安纳托利亚的不同地方，但它们都对革兰阳性菌和酵母显示出比较强的抗菌活性。Najmadeen 和 Kakamand(2009)研究了产自伊拉克库德斯坦苏莱曼尼亚省几个不同地区的 EEP(28%)对金黄色葡萄球菌、表皮葡萄球菌、大肠杆菌、绿脓杆菌、克雷伯氏肺炎菌(*Klebsiella pneumonia*)、奇异变形杆菌(*Proteus mirabilis*)和白色念珠菌的抑制作用。抑菌圈结果显示，EEP 的抑菌效果甚至强于某些抗生素；对 EEP 最敏感的细菌是表皮葡萄球菌，接着是金黄色葡萄球菌和白色念珠菌，受试的革兰阴性菌最不敏感；MIC 和 MBC 测定结果表明，金黄色葡萄球菌最敏感(0.175~0.700mg/mL)，其次是表皮葡萄球菌和白色念珠菌(0.7~1.4mg/mL)，在受试的革兰阴性菌中，所有的 EEP 样品都显示出对奇异变形杆菌较强的抑制活性(14mg/mL)，而只有 2 个样品对绿脓杆菌和克雷伯氏肺炎菌有比较弱的活性(28mg/mL)。Zeighampour 等(2014)研究了伊斯法罕蜂胶 70% 乙醇提取物的抗菌活性，结果表明，该 EEP 对金黄色葡萄球菌的抗菌活性远高于绿脓杆菌，其 MIC 和 MBC 值分别为 0.0143mg/mL 和 0.0286mg/mL，对绿脓杆菌的 MIC 和 MBC 值分别为 0.75mg/mL 和 1.50mg/mL。Gomes 等(2016)研究了巴西棕蜂胶(Brown propolis)醇提物的体外抗菌活性，他们在 32 种革兰阳性菌、32 种革兰阴性菌上的抗菌实验结果表明，棕蜂胶对革兰阳性菌的最小抑菌浓度范围为 2.25~18.9mg/mL；对革兰阴性菌的最小抑菌浓度范围为 4.5~18.9mg/mL。

Gebara 等(2002)研究了蜂胶对牙周致病菌的抗菌活性，受试菌为中间普雷沃菌(*Prevotella intermedia*)、产黑色普雷沃菌(*P. melaninogenica*)、牙龈卟啉单胞菌(*Porphyromonas gingivalis*)、伴放线菌放线杆菌(*Actinobacillus actinomycetemcomitans*)、牙龈二氧化碳嗜纤维菌(*Capnocytophaga gingivalis*)和具核梭杆菌(*Fusobacterium nucleatum*)。肉汤稀释法结果表明，蜂胶对伴放线菌放线杆菌和牙龈二氧化碳嗜纤维菌的 MIC 为 1μg/mL，对中间普雷沃菌、产黑色普雷沃菌、牙龈卟啉单胞菌和具核梭杆菌的 MIC 为 0.25μg/mL，白色念珠菌的 MIC 为 12μg/mL，绿脓杆菌、大肠杆菌和金黄色葡萄球菌(野生型)的 MIC 为 14μg/mL。熊萍等(2009)研究了云南产 EEP 对牙周致病菌牙龈卟啉单胞菌(Pg)、具核梭杆菌(Fn)和伴放线菌杆菌(Aa)的抑制作用。结果表明，EEP 对 3 种细菌都有抑制作用，随着浓度增加，抑菌圈直径也随之增大。EEP 对 Pg、Fn 和 Aa 的 MIC 分别为 0.625、1.25 和 2.5g/100mL。

Kilic 等(2005)研究了土耳其蜂胶对一些常见的革兰阳性球菌的抗菌活性,将来自 Mamak(1 个样)和 Kemaliye(2 个样)地区的 3 个样品制成 EEP,研究其抗菌活性。受试菌为临床分离得到的 33 个耐甲氧西林的金黄色葡萄球菌(MRSA)和 33 个耐万古霉素的屎肠球菌(VREF)。宏量稀释法结果表明,这 3 个 EEP 对 MRSA 和 VREF 的 MIC 分别为 7.8~31.2μg/mL、70.3~281.2μg/mL 和 17.5~140.4μg/mL;所有的蜂胶对 MRSA 的活性都要强于对 VREF 的活性,其中来自 Mamak 的蜂胶活性最强。

Ozturk 等(2010)利用瘤胃模拟技术研究了不同浓度 EEP 对瘤胃微生物发酵的影响。结果表明,低浓度(20%)和高浓度(60%)的 EEP 都不影响瘤胃 pH、短链脂肪酸及乙酸盐的产生和合适的比例,也不影响总的原生动物数和干物质消化率;高浓度的 EEP 会显著降低丙酸的产生,而高低浓度的 EEP 均能显著降低丁酸生成量,显著降低瘤胃细菌总数。此外,瘤胃液中的 NH3-N 浓度也呈浓度依赖的方式在低浓度和高浓度 EEP 的作用下分别降低了 24% 和 39%。这表明蜂胶有可能成为一种有效的饲料添加剂,用于调节瘤胃微生物发酵,降低瘤胃氨的产生,提高氮的利用率。

此外,还有人研究了不同种类真菌对 EEP 的敏感性,并比较了细菌和真菌对 EEP 的敏感性。Ota 等(2001)研究了蜂胶对不同种假丝酵母的抗真菌活性,结果发现所有受试真菌对 EEP 都敏感,敏感顺序为:白色念珠菌(*Candida albicans*)>热带念珠菌(*C. tropicalis*)>克柔假丝酵母菌(*C. Krusei*)>吉利蒙念珠菌(*C. guilliermondii*)。同时,体内试验结果也表明蜂胶能显著降低假丝酵母的数量。杨鑫等(2013)对产自云南西双版纳的蜂胶原料进行 75%、85%、95% 乙醇常温浸提和热浸提,测定这些 EEP 的黄酮含量,并评估了 EEP 对杏鲍菇(*Pleurotus eryngii*)及污染菌的抑制作用。结果表明,75% 的乙醇热提法提取的 EEP 黄酮含量最高,为 10.51mg/mL;EEP 对杏鲍菇及污染菌均有一定的抑制作用,抑制强弱顺序为:细菌>杏鲍菇>霉菌;相关性分析表明,杏鲍菇、霉菌和细菌的抑制率与黄酮含量的相关系数均大于 0.8,有较强的相关性。

EEP 除了具有良好的抗细菌、真菌活性外,对其他病原微生物(寄生虫、病毒)也有一定的抑制作用。

de Carvalho Machado 等(2007)对产自巴西的 EEP 和产自保加利亚的 EEP 的成分及抗利什曼原虫(*Leishmania* spp)活性进行分析评估,结果发现 EEP 对 4 种不同的利什曼原虫的半抑制浓度(half-inhibitory concentration,IC_{50})在 2.8~229.3μg/mL,并且保加利亚 EEP 活性要强于巴西 EEP,这可能与保加利亚蜂胶中丰富的黄酮类化合物有关,但巴西 EEP 抗巴西利什曼原虫(*L. braziliensis*)的活性是保加利亚 EEP 的 3 倍,这表明除黄酮以外的其他酚类成分可能参与了蜂胶的抗利什曼原虫活性。此外,巴西 EEP 中的香树脂醇也可能是其中发挥作用的重要成分。Ayres 等(2007)评估了巴西 4 个不同地理来源的 EEP 的抗亚马逊利什曼原虫(*L. amazonensis*)活性。结果表明,所有蜂胶样本均能降低感染的巨噬细胞数量及胞内寄生虫载量,其中来自阿拉戈斯州的红蜂胶提取物活性最强,其中包含异戊烯基及苯丙酮物质。Pontin 等(2008)首次报道了巴西绿蜂胶的体内抗利什曼原虫活性,表明无论是局部应用还是口服治疗都能显著抑制病变的发展,而且口服加局部应用效果更好。da Silva 等(2016)采用利士曼原虫诱导小鼠足趾炎症模型,施用 5mg/kg 巴西蜂胶、10mg/kg 锑酸葡甲胺(一种抗血吸虫药)、巴西蜂胶联合锑酸葡甲胺。肝脏镜检结果表明,蜂胶处理能显著缓

解肝细胞的炎症过程,降低肝脏髓过氧化酶和 N-乙酰-β-葡糖胺酶水平,缓解肝脏胶原沉积,并提升抗炎症细胞因子的水平,对肝脾肿大也有一定的缓解作用。还有研究表明,产自土耳其亚纳达的 EEP 也具有抗利什曼原虫活性,当浓度在 250、500 及 750μg/mL 时能显著抑制热带利什曼原虫的增殖(Duran et al.,2008)。

Dantas 等(2006)研究发现保加利亚 EEP 及丙酮提取物具有抗克氏锥虫(*Trypanoma cruzi*)活性。保加利亚 EEP 作用于感染了克氏锥虫的骨骼肌细胞,使克氏锥虫无鞭毛体对细胞的感染及其在胞内的增殖减少。超微结构显示,提取物作用的靶点是前鞭体的线粒体及储备结合囊;此外,保加利亚 EEP 还能引起锥鞭体的动基体结构变化,表明线粒体动基体复合体也是 EEP 的潜在靶标。Gressler 等(2012)首次研究了巴西蜂胶抗伊氏锥虫的效果。体外实验结果表明,巴西蜂胶具有剂量依赖性杀锥虫活性,10μg/mL 的蜂胶提取物作用 1h 后就能将全部锥鞭体杀死;然而将蜂胶分别以 100、200、300、400mg/mL 的浓度连续 10d 经口给药予感染鼠,并没有显示出疗效,但是高剂量组能够延长感染鼠的寿命,表明蜂胶也许可以辅助治疗伊氏锥虫感染性疾病。

Freitasa 等(2006)研究了 EEP 在体外对十二指肠贾第虫滋养体生长和黏附的效果。结果表明,当蜂胶浓度为 125、250 和 500μg/mL 时,蜂胶对寄生虫生长的抑制最明显。当蜂胶的浓度为 125μg/mL 时,抑制率达到 50%,而且蜂胶还能抑制寄生虫的黏附;光学显微镜观察结果表明,蜂胶作用后,大多数的滋养体的形态有所变化,且鞭毛摆动频率降低。

关于 EEP 抗病毒活性相关的研究也早有报道。Harish 等(1997)研究表明蜂胶具有抗人类 1 型免疫缺陷病毒(HIV-1)的效果,能够显著抑制 HIV-1 在细胞内的复制。Gekker 等(2005)的研究进一步证实了这一结论。他们利用培养的 CD4+淋巴细胞和小胶质细胞研究了不同地区 EEP 抗 HIV-1 的活性。结果表明,蜂胶能有效抑制病毒在细胞内的表达,蜂胶作用浓度为 66.6μg/mL 时,对 HIV-1 在 CD4+淋巴细胞和小胶质细胞内的表达抑制率最大,可分别达到 85% 和 98%,并且这种作用具有浓度依赖性;不同来源的蜂胶表现出相似的抗病毒活性。此外,当把蜂胶和抗艾滋药物联合使用时,结果显示蜂胶与齐多夫定(zidovudine,反转录酶抑制剂)具有明显的协同作用,但对茚地那韦(indinavir,蛋白酶抑制剂)的协同作用表现不明显。

Shimizu 等(2008)首次研究发现蜂胶的体内抗流感活性。他们对产自巴西南部 13 个 EEP 的体内和体外抗流感活性进行了评估,经过体外的空白形成减少实验,筛选出了 4 种具有抗流感活性的提取物,并且通过鼠流感感染模型对这 4 种提取物的体内抗流感活性进行了进一步评估。结果表明,其中只有 1 种提取物在 10mg/kg 口服剂量时,能显著降低感染小鼠的体重损失;2mg/kg 和 10mg/kg 的剂量还能有效延长感染鼠的寿命;10mg/kg 剂量的 EEP 能显著减少感染鼠支气管肺泡灌洗液中的病毒量。然而,通过对感染鼠免疫相关的一些细胞因子水平检测发现,它们并没有发生变化,表明这种蜂胶可能本身就含有一些抗流感成分,并非是通过激活宿主免疫来发挥其抗流感作用。

Búrquez 等(2015)研究了墨西哥蜂胶乙醇提取物对感染了伪狂犬病病毒的马-达氏牛肾细胞(Madin Darby Bovine Kidney,MDBK)的影响,结果表明在感染之前用 0.5mg/mL 蜂胶提取物孵育 MDBK 细胞 2h 能显著降低空斑形成单位的数目;其机制可能与蜂胶损伤病毒包膜蛋白,影响病毒扩散和复制周期有关。

1.2 蜂胶水提物(WEP)的抗病原微生物作用

相对于蜂胶醇提物,蜂胶水提取物(water extract of propolis,WEP)的抗病原微生物作用报道较少。张红梅等(2008)采用琼脂稀释法测定 WEP 对中间普氏菌(*Prevotella intermedia*)、具核梭杆菌(*Fusobacterium nucleatum*)及二者混合菌的 MBC。结果表明,WEP 对中间普氏菌、具核梭杆菌及二者的混合菌的 MBC 分别为 0.025%、0.025% 和 0.05%,表明蜂胶对感染根管常见优势厌氧菌具有良好的抑菌活性。彭志庆等(2010)研究了中国产 WEP 对变形链球菌、远缘链球菌及其耐氟菌株的生长抑制作用以及对葡糖基转移酶(GTF)的影响。结果表明,WEP 对变形链球菌、变形链球菌耐氟菌株、远缘链球菌、远缘链球菌耐氟菌株的 MIC 分别为 0.39、0.78、0.20 和 0.39g/L,MBC 分别为 0.78、1.56、1.56 和 1.56g/L;随着蜂胶浓度增高,GTF 活性逐渐降低,表明中国产 WEP 能有效抑制变形链球菌、远缘链球菌及耐氟菌株的生长,对 GTF 具有较强的抑制作用。

此外,人们对 WEP 及其组分的抗病毒、抗寄生虫活性也进行了研究。Takemura 等(2012)研究发现,巴西 WEP 中的二咖啡酰奎宁酸能通过增强感染甲型流感病毒小鼠体内的肿瘤坏死因子相关的凋亡,诱导配体(TRAIL)的表达,促进被感染细胞的凋亡和病毒的清除,从而发挥体内抗流感病毒活性,并且这种作用比 WEP 本身的作用更强。Ferreira 等(2014)通过体内试验发现,巴西 WEP 能降低感染鼠肝脏中的寄生虫载量,并减轻感染鼠肝脾的病变程度。

1.3 其他类型蜂胶提取物的抗病原微生物作用

赵强等(2008)以 8 种细菌作为受试菌研究了蜂胶超临界 CO_2 萃取物的抑菌作用,结果表明,超临界 CO_2 蜂胶萃取物对 8 种受试菌种均有良好的抑制作用,MIC 为 78.1~2500μg/mL;超临界 CO_2 蜂胶萃取物的抑菌活性较 EEP(25mg/mL)更强,MIC 总体相当,但大大强于 WEP。王浩等(2010)也研究了蜂胶超临界 CO_2 萃取物的抗菌活性,并与其醇提物的抗菌活性作了对比。结果表明,蜂胶超临界 CO_2 萃取物对金黄色葡萄球菌和大肠杆菌的 MIC 分别为 14.5mg/mL 和 29mg/mL,并且抗菌活性强于其醇提物的抗菌活性。

此外,通过水蒸气蒸馏法、溶剂法、微波辅助提取等获得的蜂胶挥发性成分也有较好的抗菌活性。Melliou 等(2007)利用水蒸气蒸馏法将希腊 5 个不同地区蜂胶的挥发性成分进行提取,并对其抗菌活性进行了研究。结果表明,这几种挥发性成分具有良好的抗真菌效果,最强样品对白色念珠菌、热带假丝酵母、光滑念珠菌等 3 种真菌的 MIC 为 0.50~5.20mg/mL。此外,该样品对阴沟肠杆菌(*E. cloaceae*)和大肠杆菌的抗菌效果也最好,其 MIC 分别为 3.10mg/mL 和 3.40mg/mL。另一种样品对金黄色葡萄球菌、表皮葡萄球菌、绿脓假单胞菌和肺炎克雷伯氏菌这 4 种菌的抗菌效果最好,其 MIC 范围为 4.10~5.30mg/mL。

李雅晶(2011)对利用微波辅助石油醚提取法提取的巴西蜂胶和中国蜂胶挥发性成分的抗菌效果进行了研究,结果表明这两种蜂胶的挥发性成分对酵母菌、霉菌及革兰阳性菌有较强的抑菌效果,对革兰阴性菌的抑菌效果相对较弱,抑菌浓度范围在 39.06~625μg/mL;结果还发现,这两种蜂胶的醇提物对大肠杆菌均无抑制效果,而对绿脓杆菌只有中国蜂胶醇提物有较弱的作用;并且,这两种蜂胶挥发性组分的抗菌活性要强于其相应的乙醇提取物的抗菌活性。卢媛媛(2014)利用优化的微波辅助石油醚提取法对巴西蜂胶残渣中的挥发性组分进行提取,并对巴西蜂胶原胶挥发性组分、残渣挥发性组分和乙醇提取物三者的抗菌活性进行了比较研究。结果表明,三者对金黄色葡萄球菌有明显的抑制效果,而对大肠杆菌则无明显抑制效果,且这三者抑菌活性的顺序为依次减弱。由此可见,蜂胶中的挥发性成分具有良好的抗菌潜力,即便是工业生产中废弃的蜂胶残渣,仍有大量挥发性成分未被提取,其抗菌效果优越,这为蜂胶残渣的再利用提供了理论依据。

总之,不同提取方法得到的蜂胶提取物均有一定的抗菌效果,只是这些提取物的主要抗菌成分往往不同,因而在抗菌机制上也会存在一定的差异。例如,蜂胶挥发油水蒸气蒸馏提取物(HDE)中主要含萜烯类成分,黄酮含量极低,因此这种提取物可能主要依靠其中的萜烯类发挥作用;而蜂胶挥发油超临界CO_2提取物(SE)具有HDE中没有的黄酮类成分,同时比蜂胶醇提物含有更多的萜烯类成分,因此SE的抗菌机制中可能还包括黄酮及萜烯类物质的协同作用,所以抗菌效果更为优越(赵强,2007)。总的来说,蜂胶乙醇提取物和蜂胶挥发性组分的抗菌效果要远强于水提取物的抗菌效果。

除了对常见的病原微生物有抑制或杀灭效果,蜂胶的广谱抗菌活性还体现在对蜂群自身健康的维持上。蜜蜂一方面利用蜂胶加固巢房防止外界微生物的侵入,另一方面蜂胶可能通过抗菌及影响寄生虫代谢等方式阻止蜂群美洲幼虫腐臭病、瓦螨病等病虫害的发生(平舜等,2014)。

蜂胶抗病原微生物活性除了表现在其本身对病原微生物直接的抑制和杀灭作用外,还可以作为佐剂结合特定抗原注入体内,以增强机体特异性免疫应答的方式帮助宿主抵抗和清除病原微生物(沈志强,1995;沈志强等,2004;刘思伽等,2006;刘成芳等,2009;赵恒章等,2009;柴家前等,2010;Nassar et al.,2012;王凯等,2013)。

2 蜂胶抗病原微生物活性的影响因素

蜂胶作为一种天然产物,其来源丰富,成分复杂。活性成分的种类和含量是蜂胶发挥药理活性的物质基础。而蜂胶从采集到提取加工要经历多个步骤,其间会受到多种因素的影响,如植物源、采胶季节、蜂种、溶剂、提取温度、pH等。这些因素或多或少会影响到蜂胶活性成分的种类、提取率甚至结构的稳定性,从而最终对其生物学活性的发挥产生一定的影响。

2.1 乙醇浓度对蜂胶提取物抗病原微生物活性的影响

不同浓度的乙醇所提取的蜂胶醇提物的抗病原微生物活性具有显著差异。Sawaya等(2004)通过管中连续稀释法和琼脂扩散法确定巴西蜂胶不同浓度(30%、50%、70%、96%、100%)乙醇提取物对一些革兰阳性菌的抗菌活性,连续稀释法的结果比较一致,MIC在2.5~20.0mg/mL之间,用50%乙醇作为溶剂的蜂胶醇提物的杀菌活性最好。Muli(2007)研究了绿脓杆菌、伤寒氏沙门杆菌、大肠杆菌、金葡菌和枯草芽孢杆菌(*Bacillus subtilis*)对产自肯尼亚3个不同地区的EEP的敏感性。蜂胶分别用100%、70%、50%及30%的乙醇提取,使用滤纸片做琼脂扩散,通过抑菌圈大小来确定抗菌活性,结果发现3个地区EEP抗菌活性有明显差异,抗菌活性与受试菌种和乙醇浓度有关,枯草芽孢杆菌和金葡菌是最敏感的菌种,70%的EEP抗菌效果最好。Shahbaz等(2015)利用纸片扩散法评估了巴基斯坦蜂胶不同浓度乙醇及甲醇提取的蜂胶提取物的抗菌活性,受试菌为金黄色酿脓葡萄球菌、枯草杆菌和大肠杆菌,结果表明,65%乙醇制备的蜂胶乙醇提取物的抗菌效果最好。因此,EEP的抗菌效果受乙醇浓度影响,只是由于菌种及蜂胶来源不同,蜂胶醇提取获得最佳抗菌效果所要求的乙醇浓度也会有所不同。

2.2　pH 对蜂胶抗病原微生物活性的影响

艾予川等(2004)研究了河南蜂胶在不同 pH(5.0～8.5)条件下对金黄色葡萄球菌耐药株的抑菌作用,结果表明,河南蜂胶对金葡菌耐药株的抑菌效果随 pH 的增加,抑菌环直径变小;pH5.0 时抑菌环直径最大,抑菌活性最强,随 pH 增加,抑菌活性呈阶段性减弱。许兵红等(2004)研究了河南蜂胶对溶壁微球菌不同 pH(6.0～8.0)条件下的抑菌作用,结果表明,河南蜂胶在 pH6.5 时对溶壁微球菌的抑菌活性最强;pH7.0、7.5 和 8.0 各组间抑菌环直径无统计学差异,随 pH 增加,抑菌活性减弱。Yang 等(2006)研究了多种因素下 EEP 对单核细胞李斯特杆菌(*Listeria monocytogenes*)的作用和抗诱变作用,结果表明,蜂胶在 7.5μg/mL 甚至更高浓度时,对单核细胞李斯特杆菌具有杀菌效果;当温度为 37℃ 时,单核细胞李斯特杆菌对 EEP 的敏感性要比在 25℃ 或 4℃ 时更敏感;菌体细胞在酸性 pH 条件下要比在中性 pH 和碱性 pH 时对 EEP 更敏感;菌体细胞的生长阶段对 EEP 抗菌活性也有影响,指数生长中期的菌体细胞对 EEP 比指数生长后期和静止期对 EEP 更为敏感。曾莉萍等(2007a)研究了河南蜂胶对 3 种链球菌的抑菌作用,pH 设置为 5.5～8.0。结果表明,河南蜂胶对丙型链球菌无明显的抑菌活性,但对甲型链球菌和乙型链球菌有明显的抑菌环活性,而且抑菌活性随 pH 增加和浓度降低而减弱。曾莉萍等(2007b)还研究了河南蜂胶对金黄色葡萄球菌耐药株不同 pH(5.0～8.5)的抑菌作用,并在 pH6.5 下设置 10 个不同浓度组,将 31% 蜂胶溶液倍比稀释至 0.0621%。结果表明,随 pH 增加,河南蜂胶对金黄色葡萄球菌的抑菌活性呈阶段性减弱;随浓度降低,对金黄色葡萄球菌的抑菌活力减弱,在 pH6.5 下,产生抑菌环的最小浓度为 0.1242%(即实验剂量为 8.47μg)。许兵红和曾莉萍(2011)采用相同的方法研究了河南蜂胶对炭疽杆菌(*Bacillus anthraci*)的抑菌活性,结果表明河南蜂胶对炭疽杆菌有明显的抑菌活性,其抑菌活性随 pH 值增加或浓度降低而减小。许兵红等(2004)研究了河南蜂胶在不同 pH(5.5～8.5)和不同浓度(15.50%～0.12%)下对白色念珠菌的抑菌作用,结果表明,河南蜂胶对白色念珠菌的抑菌活性随蜂胶浓度降低而减弱,最低蜂胶浓度为 0.25%;同时,抑菌活性随着 pH 的降低而降低。孟良玉等(2010)利用金黄色葡萄球菌、大肠杆菌、沙门氏菌和蜡样芽孢杆菌(*Bacillus cereus*)对蜂胶提取物的抑菌性能进行了研究,同时以蜡样芽孢杆菌为指示菌研究蜂胶提取物抑菌成分的稳定性,结果表明蜂胶提取物具有较强的抑菌活性,对热处理、pH 值、紫外线、金属离子均具有较好的稳定性,仅在 121℃ 或 pH10 处理后其抑菌活性才明显下降。Ivančajić 等(2010)研究了 5 种不同溶剂提取的蜂胶在 3 种不同 pH(6、7、8)状况下对蜂胶抗菌作用的影响。通过对 12 种细菌抗菌活性的结果分析,发现除了沙门氏菌属(*Salmonella*),醚、丙酮、甲苯和氯仿蜂胶提取物无论哪种 pH 条件下,对所有受试微生物都表现出很强的抗菌效果;在大多数情况下,蜂胶在 pH6 时抗菌效果最佳。

总的来说,蜂胶在弱酸性条件下抗菌效果最佳,虽然 pH 对蜂胶抗菌活性的影响机制不是很清楚,但这可能是由于蜂胶中发挥抗菌作用的主要成分是黄酮和酚酸类物质,而这些成分在弱酸性条件下能最大限度地维持其结构稳定性,利于蜂胶抗菌活性的发挥(Friedman and Jürgens,2000)。

2.3 提取方式对蜂胶抗病原微生物活性的影响

Ugur 等(2004)对产自土耳其穆拉省 45 种不同的蜂胶样经二甲基亚砜(DMSO)和丙酮提取,测定其抗菌和抗真菌活性。结果表明,蜂胶抗微生物活性因蜂胶样、用药剂量及所用提取溶剂不同而不同,所有蜂胶样的抗微生物活性随着剂量的增加而加强,直到最高受试剂量也未到平台期。除了马耳他布鲁氏杆菌(*Brucella melitensis*),所有蜂胶样的 DMSO 提取物都比相同蜂胶样丙酮提取物的活性要强。阴性菌中对蜂胶最敏感的是索氏志贺氏菌(*Shigella sonnei*),阳性菌中最敏感的是变形链球菌,最不敏感的是白色念珠菌。与用标准抗生素的阳性对照组比较,蜂胶显示出相似甚至更强的抗变形链球菌、伤寒沙门氏菌、绿脓杆菌、索氏志贺氏菌和白色念珠菌能力。Garedew 等(2004)通过流动微量热法结合极谱法和培养皿生物测定方法确定 3 种蜂胶提取物的抗菌活性:水提取物、蜂胶挥发油和蜂胶乙醇提取物。与另外两种溶剂比较,蜂胶水提物的抗细菌和抗真菌活性最弱,而另外两种提取物的活性相似。丝状真菌不如细菌和酵母对蜂胶敏感,蜂胶既显示出抑菌活性又显示出杀菌活性,并与蜂胶类型、浓度及测试的菌种有关。大肠杆菌对大部分的蜂胶都不敏感,而且需要较高的蜂胶浓度以达到杀菌效果。

兰桃芳等(2006)采用水、95%乙醇及与超声波的综合作用从原蜂胶中提取蜂胶液,配制成不同浓度,并利用蜡样芽孢杆菌、大肠杆菌、金黄色葡萄球菌、沙门氏菌、黑曲霉(*Aspergillus*)等 5 种菌进行抗菌活性测定。结果表明,超声波作用于 95%的乙醇蜂胶液的方法提取效果最好,提取率为 27.3%;不同提取方法得到的蜂胶提取液的抑菌活性存在差异,以超声波和乙醇提取的蜂胶液对受试菌作用最强;蜂胶提取液对大肠杆菌的抑制作用最强,其次是沙门氏菌和蜡样芽孢杆菌,最差的是黑曲霉。杨明等(2006)以 EEP 和 CO_2 超临界萃取的蜂胶为材料,测定了两种蜂胶的抑菌效果和 MIC。结果表明,两种蜂胶对大肠埃希氏杆菌、枯草杆菌、金黄色葡萄球菌、白念珠菌和紫色癣菌都有一定抑菌效果,MIC 为 1.54~3.08mg/mL。但不同提取方法和浓度蜂胶的抑菌作用不同,EEP 抗菌活性强于 CO_2 超临界萃取的蜂胶。而赵强等(2008)的研究结果表明超临界 CO_2 蜂胶萃取物的抑菌活性较 EEP 更强。此外,赵强(2007)还曾将不同方法提取的蜂胶及蜂胶渣挥发油、蜂胶醇提物、蜂胶水提物及蜂胶超临界 CO_2 提取物的抗菌活性进行了比较研究。结果表明,蜂胶的 6 种挥发油中以蜂胶挥发油超临界 CO_2 提取物(SE)、水蒸气蒸馏提取物(HDE)、蜂胶渣石油醚提取物的抗菌效果较好,且这 3 种方法得到的挥发油提取物抗菌效果比较结果为:超临界＞水蒸气蒸馏＞溶剂法。蜂胶挥发油石油醚提取物与微波辅助提取物抗菌效果相当,略弱于超声辅助提取的挥发油,这一差异可能是由于提取率不同造成的。除 SE,其他几种挥发油的抗菌活性都低于 EEP 及超临界 CO_2 提取物。

陈炼等(2009)研究表明,江西蜂胶对幽门螺杆菌(*Helicobacter pylori*)具有明显的抑制作用,其中以 95%乙醇为溶剂的 EEP 抑菌效果最佳,远强于 WEP 的效果,且抑菌活性具有浓度依赖性。张秀喜(2009)比较了蜂胶各提取物的抑菌效果,结果表明,蜂胶对革兰阳性菌的作用强于对革兰阴性菌的作用;超声波辅助乙醇提取物的抑菌效果＞醇提物＞水提物;乳酸链球菌素和蜂胶的复合作用不能加强蜂胶的抑菌效果;蜂胶超声波辅助乙醇提取物对 5 种供试菌的 MIC 分别为:酵母菌 0.13mg/mL、枯草芽孢杆菌 0.25mg/mL、金黄色葡萄球菌 0.3mg/mL、大肠杆菌 0.8mg/mL、黑曲霉 10.5mg/mL;过高的温度会使蜂胶的抑菌效果变差;pH 值越低,蜂胶抑菌效果越好。张正强(2009)的研究结果也表明,蜂胶对革兰阳性菌的作用强于对革兰阴性菌的作用,且用超声波对蜂胶预处理,会增强蜂胶抗菌效果;EEP 和超临界萃取物溶液在浓度范围 0.1%~2.5% 之间对金黄色葡萄球菌、沙门氏菌、蜡样芽孢杆菌、产黄青霉(*Penicillium chrysogenum*)和牛肉表面混合菌均有显著的抑菌作用,且呈浓度依赖性;蜂胶超临界萃取物对蜡样芽孢杆菌和产黄青霉的抑菌作用优于 EEP,而对牛肉表面混合菌、金黄色葡萄球菌、沙门氏菌的抑菌作用略低于 EEP。Kashi 等(2011)评估了从德黑兰东北地区采集的 EEP 的抗菌活性,受试细菌有变形链球菌、唾液链球菌(*Streptococcus salivarius*)、金黄色葡萄球菌、粪肠球菌和干酪乳杆菌(*Lactobacillus casei*),通过琼脂扩散法检测这些细菌在蜂胶提取物为 20mg/mL 浓度时的敏感性。结果表明,EEP 的 MIC 和 MBC 在 250~500μg/mL;WEP 对变形链球菌和粪肠球菌的 MIC 为 500μg/mL,对变形链球菌的 MBC 则为 20mg/mL。这些结果表明 EEP 在对抗一些口腔病原菌上更有效。

总之,EEP 和蜂胶 CO_2 萃取物抗菌效果通常远强于 WEP,可能是前两种方法对蜂胶中抗菌物质的提取更为充分。蜂胶醇提过程中加入超声波辅助效果更为理想,可能是超声作用促进蜂胶活性成分的进一步提取。此外,超临界 CO_2 提取物和蜂胶醇提物的抗菌活性与菌种有关(张正强,2009),这可能是两种方法对各种抗菌成分提取程度不同,而不同的菌体对不同抗菌成分敏感性也不同。

2.4　蜂种对蜂胶抗病原微生物活性的影响

Silici 和 Kutluca(2005)对同一蜂场 3 种不同西方蜜蜂(*Apis mellifera*)采集的 3 种不同蜂胶的化学成分和抗菌活性进行了分析测定。蜂胶样品经 GC/MS 分析后,确定了 48 种成分。根据确定的成分,分析得出该蜂胶的主要植物源是银白杨(*Populus alba*)、颤杨(*P. tremuloides*)和白柳(*Salix alba*),采用的菌种有金黄色葡萄球菌、大肠杆菌、绿脓杆菌及白色念珠菌,蜂胶对革兰阳性菌金黄色葡萄球菌具有较高的抗菌活性,对革兰阴性菌大肠杆菌、绿脓杆菌和白色念珠菌的活性较差;高加索蜂(*A. m. caucasica*)采集的蜂胶比安纳托利亚蜂(*A. m. anatolica*)和卡尼鄂拉蜂(*A. m. carnica*)采集的蜂胶具有更强的抗菌活性。

Silici 等(2005)还对不同蜂种及不同地区蜂胶的抗真菌活性进行了比较研究,发现白色念珠菌、光滑念珠菌、丝孢酵母属(*Trichosporon*)以及红酵母菌属对低浓度的蜂胶就很敏感,后者显示出更强的敏感性。相对于其他蜂胶,高加索蜂采集的蜂胶显示出最强的抗真菌活性;相反,卡尼鄂拉蜂和安纳托利亚蜂采集的蜂胶抗真菌活性较差。

2.5　地理源对蜂胶抗病原微生物活性的影响

Hegazi 等(2000)对来自澳大利亚、德国及法国的 3 种蜂胶样的化学成分及抗菌活性进行了研究,发现苯乙基-反式-咖啡酸、苯甲基阿魏酸及高良姜素是德国蜂胶的主要成分,苯甲基咖啡酸脂及乔松素是法国蜂胶的主要成分,乔松素是澳大利亚蜂胶的主要成分,而 p-香豆酸是 3 种蜂胶中共有的主要成分。抗菌结果显示,德国蜂胶对金黄色葡萄球菌和大肠杆菌抗菌效果最好,澳大利亚蜂胶对白色念珠菌抗菌效果最好,法国蜂胶对这 3 种病原菌都有效果,但作用没有以上两种蜂胶强。Orsi 等(2005)对巴西两个不同地区(莫索罗 Mossoró 和乌鲁比西 Urubici)蜂胶的抗沙门氏菌活性进行了比较研究,发现莫索罗蜂胶的抗菌效果强于乌鲁比西蜂胶,表明蜂胶的抗菌活性与地理源有关。Silici 等(2005)的实验除了表明蜂胶的抗真菌活性与蜂种有关外,同时对土耳其几个不同地区蜂胶样的抗真菌活性比较发现,从亚达那地区收集的蜂胶的抗真菌活性要强于其他受试地区蜂胶的抗真菌活性,说明蜂胶抗真菌活性与地理源也有一定的关系。此外,Nina 等(2015)对智利不同地方收集来的 19 种蜂胶样品的抗菌活性进行了评估,结果发现来自中部山谷的智利蜂胶对大肠杆菌、假单胞菌、小肠结肠炎耶尔森菌(*Yersinia enterocolitica*)和肠炎沙门菌(*Salmonella Enteritidis*)的抗菌效果最强,其 MIC≤62.5μg/mL。

2.6　与抗生素配伍对蜂胶抗病原微生物活性的影响

蜂胶除了本身具有抗菌活性外,与某些抗生素配伍使用还能发挥协同作用,从而增强彼此的抗菌活性。Stepanović等(2003)对从塞尔维亚不同地区收集的 13 个蜂胶样对 39 种微生物(14 种对抗菌药物有耐药性或多重耐药性)的抗微生物活性进行了研究,并确定抗菌剂和蜂胶之间的协同作用。结果表明,不管是否耐药,EEP 对革兰阳性菌(0.078%～1.25%)和酵母菌(0.16%～1.25%)都表现出显著的抗微生物活性,而革兰阴性菌对蜂胶则没那么敏感(1.25%～5%或大于 5%)。粪球球菌是耐药性最强的阳性菌,沙门氏菌属是耐药性最强的阴性菌,白色念珠菌是耐药性最强的酵母菌。EEP 和选用的抗菌药之间显示出协同作用,并增强抗真菌药的作用。Scazzocchio 等(2006)研究了 EEP 亚抑菌浓度的抗菌活性及其对一些抗生素抗菌作用的影响。结果表明,EEP 对所有受试的临床菌株都有很好的抗微生物活性。此外,EEP 能显著增强一些抗菌药物如氨苄西林、万古霉素、头孢曲松等的作用,但不能增强红霉素的作用。此外,还观察到 EEP 能抑制一些葡萄球菌的脂肪酶活性和凝固酶活性和生物膜的形成。de Oliveira Orsi 等(2006)研究表明,产自巴西和保加利亚的 EEP 都有抗伤寒沙门氏菌的作用,保加利亚蜂胶作用效果更好,两种 EEP 都与一些抗生素存在协同作用,从而增强这些抗生素对伤寒氏沙门杆菌的作用。蒋琳等(2008)研究了蜂胶及蜂胶奥硝唑合剂对牙龈卟啉单胞菌生长的影响,结果表明蜂胶对牙龈卟啉单胞菌的 MBC 为 0.156%,相当于 0.125～0.25mg/L 的奥硝唑;单独采用蜂胶药敏片对牙龈卟啉单胞菌的抑菌效果不理想,但蜂胶和奥硝唑联合应用后,作用大于对应浓度的单独奥硝唑组,表明蜂胶与奥硝唑对牙龈卟啉单胞菌的生长抑制作用存在交互作用。Kalia 等(2016)利用蜂胶联合传统抗生素头孢克肟(Cefixime)治疗由沙门氏菌(*Salmonella*)导致的小鼠伤寒症。与单独使用头孢、蜂胶处理组相比,联合蜂胶能有效降低头孢的有效作用浓度,增强抗生素的有效性。

2.7　其他因素对蜂胶抗病原微生物活性的影响

周萍等(2006)对吐温-80、RH-40、辛癸酸甘油酯、聚甘油酯、蔗糖脂肪酸酯、倍他环糊精和食用乙醇等 7 种载体的蜂胶进行了抑菌试验,结果表明,以吐温-80 和 RH-40 为载体的蜂胶抑菌能力受到抑制,抗菌活性降低;以辛癸酸甘油酯、聚甘油酯、蔗糖脂肪酸酯为载体的蜂胶比食用乙醇为载体的蜂胶抑菌效果好,但抑菌能力在同一级别;以倍他环糊精为载体的蜂胶抑菌能力显著大于其他蜂胶,比食用乙醇为载体的蜂胶的抑菌能力高出近 500 倍,这可能与倍他环糊精具有纳米大小中空的特殊分子结构有关,溶于水时纳米大小的蜂胶便释放出来,显示出强大的抗菌活性。杨道锋等(2008)研究了普通蜂胶及纳米蜂胶对肺炎克雷伯菌、铜绿假单胞菌、伤寒沙门菌、金黄色葡萄球菌、表皮葡萄球菌、普通变形杆菌、大肠杆菌、肠球菌的抑菌作用,并测定了两者对苯唑西林敏感金葡菌(MSSA)和耐药金葡菌(MRSA)的 MIC,结果表明普通蜂胶及纳米蜂胶对上述细菌的生长均有一定的抑制作用,且随着浓度的升高抑制作用增强。除肠球菌和铜绿假单胞菌外,纳米蜂胶体外对上述细菌的抑制作用强于普通蜂胶;普通蜂胶对 MSSA 和 MRSA 的 MIC_{50} 和 MIC_{90} 均为 $128\mu g/mL$ 和 $256\mu g/mL$,纳米蜂胶对 MSSA 和 MRSA 的 MIC_{50} 和 MIC_{90} 均为 $32\mu g/mL$ 和 $64\mu g/mL$。Ma 等(2015)研究了纳米蜂胶黄酮(NPF)和蜂胶黄酮对猪细小病毒(PPV)在体外和体内的抑制效果。结果表明,NPF 可以显著抑制 PPV 感染猪肾 PK-15 细胞,并且纳米蜂胶黄酮比蜂胶黄酮效果更好。NPF 在高、中剂量时能够显著地抑制 PPV 在肺癌、性腺、血液、脾的复制,降低 PPV 对豚鼠体重的影响,并改善血清中 PPV 的血凝抑制作用。此外,它还可以增加血清中 IL-2 和 IL-6 含量,促进机体对病毒的抵抗。因此,总的来说,纳米蜂胶的抗病原微生物作用强于普通蜂胶。此外,还有研究表明蜂胶的储存时间也会对其抗菌活性产生影响,因为储存时间的长短影响了蜂胶内多酚成分的含量及组成,并进而影响了它抑制细菌生物膜形成的强度(Veloz etal. 2015)。

3 蜂胶抗病原微生物的机制

蜂胶的抗菌活性虽然很早就被人们发现,但在早期,其抗菌作用的确切机制一直不清楚,只是简单地将其归结为各种物质的协同作用。近年来,随着研究的不断深入,对蜂胶抗菌机制的探索有了一定的进展。

Koo 等(2002)研究发现,蜂胶中的黄酮和黄酮醇能有效抑制变形链球菌葡糖基转移酶活性,其中芹黄素最为有效。王晶等(2002)研究了 WEP 和 EEP 对变形链球菌生长和黏附的影响,结果表明蜂胶各组均可显著抑制变形链球菌的黏附,对变形链球菌和远缘链球菌的 MIC,WEP 为 0.032% 和 0.25%,EEP 均为 0.078%。

Tim 和 Lamb(2005)通过比较金黄色葡萄球菌细菌悬浊液中钾流失状况,以研究高良姜素对细菌细胞膜完整性的影响。结果表明,高良姜素、抑菌抗生素新生霉素和青霉素 G 对金黄色葡萄球菌(*Staphylococcus aureus* NCTC 6571)的 MIC 分别为 $50\mu g/mL$、$62.5 ng/mL$ 和 $31.3 ng/mL$,当 5×10^7 cfu/mL 的金黄色葡萄球菌在包含 $50\mu g/mL$ 的高良姜素的无钾培养介质中 12h 后,细菌活力下降了 60 倍。1×10^9 cfu/mL 的金黄色葡萄球菌在含有 $50\mu g/mL$ 的高良姜素条件下孵育 12h 后,比空白组多流失了 21% 的钾;新生霉素并不会造成钾流失,而细菌孵育在含有 $31.3 ng/mL$ 的青霉素 G 的培养介质中会有 6% 的钾流失。这些数据表明,高良姜素造成的钾流失可能是由于它对细菌细胞膜造成直接损伤或通过自溶作用造成间接损伤,即损伤细胞壁以及接下来的渗透性溶解。蔡爽等(2006)采用纸片琼脂扩散法观察了 10、25、50 和 100g/L 蜂胶防龋涂膜对变形链球菌 c 型和 d 型的抑菌作用,结果表明各浓度蜂胶涂膜及基质都能够抑制细菌生长和黏附,且抑菌作用呈明显的浓度依赖性,100g/L 涂膜组的抑菌效果与 1.6g/L 氯己定溶液无显著性差异。

Scazzocchio 等(2006)研究了 EEP 亚抑菌浓度的抗菌活性及其对一些抗生素抗菌作用的影响,还研究了 EEP 亚抑菌浓度对金黄色葡萄球菌一些重要的毒力因子,如脂肪酶、凝固酶的作用,以及对其生物膜形成的影响。结果表明,EEP 对所有受试的临床菌株都有很好的抗微生物活性。而且,EEP 能大大增强一些抗菌药物如氨苄西林、万古霉素、头孢曲松等的作用,但不能增强红霉素的作用,而同样的 EEP 浓度能抑制一些葡萄球菌的脂肪酶活性和凝固酶活性及生物膜的形成。Veloz 等(2016)对智利蜂胶的抗变异链球菌机制进行了研究,发现智利蜂胶多酚提取物抑制了糖基转移酶(GtfB、GtfC 和 GtfD)基因及

其下游调控基因(如 VicK、VicR 和 CcpA)的表达;且蜂胶参与调控了细菌毒力因子 SpaP 的表达。

张秀喜(2009)的研究表明,蜂胶可使金黄色葡萄球菌和大肠杆菌碱性磷酸酶和 β-半乳糖酸酶在胞外溶出,且对金黄色葡萄球菌的作用比大肠杆菌的要强,其抑菌机制表现为使细菌细胞壁和细胞膜的通透性增大。姜游帅(2010)研究了蜂胶对金黄色葡萄球菌毒力因子的抑制作用,结果表明蜂胶能抑制金黄色葡萄球菌 ATCC29213 生长,MIC 为 $512\mu g/mL$,并且亚抑菌浓度的蜂胶能降低金葡菌溶血活力和凝固酶效价,降低金葡菌肠毒素 A、肠毒素 B 及 α-溶血素表达量;在基因水平上,亚抑菌浓度蜂胶能减少 α-溶血素基因 hla 和 Agr 二元调控系统的 $agrA$ mRNA 表达量,而 Agr 二元调控系统对金葡菌对数生长后期分泌的许多毒力蛋白有调控作用,对 α-溶血素起到正向调控的作用。因此,预测蜂胶降低 α-溶血素的表达部分依赖于抑制 Agr 二元调控系统。祖力卡尔江·阿合买提等(2015)研究了伊犁黑蜂蜂胶对不同状态下变形链球菌乳酸脱氢酶及其相关基因的影响,结果表明,伊犁黑蜂蜂胶能够抑制浮游状态与生物膜状态下变形链球菌乳酸脱氢酶活性及其编码基因 ldh 表达,来抑制细菌产酸,表明伊犁黑蜂蜂胶可能是通过此途径发挥防龋效果。

朱明等(2013)研究了不同浓度新疆黑蜂蜂胶对大肠杆菌体外培养生物膜形成及清除的影响,结果表明,新疆黑蜂蜂胶醇提物对大肠杆菌的 MIC 为 12.50g/L,当浓度为 6.25g/L 时新疆黑蜂蜂胶就可在早期呈剂量依赖性干扰细菌生物膜的形成,当浓度达到 50.00g/L 时,其抑制生物膜的作用与庆大霉素无显著差异;当蜂胶浓度达 12.50g/L 时,能有效清除成熟生物膜,而当浓度达到 50.00g/L 时,其清除成熟 BF 的作用与庆大霉素无显著差异。Wojtyczka 等(2013)利用表皮葡萄球菌研究波兰蜂胶的抗菌活性,结果表明,波兰蜂胶乙醇提取物(EEPP)在浓度为 0.39~1.56mg/mL 时能抑制此菌的所有受试菌株的生物膜形成;EEPP 的 MIC 范围为 0.78~1.56mg/mL,平均 MIC 为 1.13 ± 0.39mg/mL。然而,EEPP 浓度为 0.025~0.39mg/mL 时,经过 12h 和 24h 的孵育,反而能促进此菌的增殖。总的结果分析表明,EEPP 的抗菌活性表现为细菌生长的抑制,生物膜形成能力的抑制,以及增殖的强度明显受到孵育时间、EEPP 浓度以及这些因素之间相互作用的影响。

综上所述,蜂胶抗菌机制主要有以下几种方式:①破坏菌体细胞膜及细胞壁的完整性,使细菌内容物外流;②抑制细菌生物膜形成,影响细菌的生长和黏附;③抑制细菌毒力因子活性,及其相关基因的表达,使其侵袭力和毒力下降。尽管蜂胶抗病原微生物机制上的研究已经取得一定进展,但研究还是偏少且不够深入。蜂胶化学成分复杂,其抗菌机制也必然多样,所以很值得进行更深入地研究。

蜂胶中丰富的黄酮及萜烯类物质对抗菌活性的发挥起到很大的作用,而黄酮及萜烯类物质的抗菌机制主要存在以下几种(Tim and Lamb,2011;赵强,2007):①损伤细菌细胞膜;②抑制细菌核酸合成;③干扰细菌的能量代谢;④抑制细菌细胞壁的形成;⑤抑制细菌细胞膜的形成。这些研究对蜂胶抗菌机制的深入研究有一定的参考意义。然而需要注意的是,由于一些实验设计不合理,或对实验数据、实验现象理解不当,所以实验结论可能需要进一步验证。例如,黄酮的抗菌机制到底是一种还是多种存在争议,因为机制研究实验中,存在将原因与结果混淆的可能。例如,抗菌成分损伤了菌体细胞膜,会扰乱质子动力势,而这会进一步影响 ATP 产生和物质转运。如果菌体细胞产生能力和摄取营养物质的能力受到影响,紧接着便会影响其核酸、肽聚糖等合成的能力,这种情况下,单一的作用机制有可能被误解为多种作用机制。蜂胶作为一种成分极为复杂的混合物,不同的成分其抗菌机制可能不同,有些成分由于结构的相似性,也可能具有相同的抗菌机制,所以对其抗菌机制的研究最好能建立在对其单体抗菌成分的研究之上,进一步研究蜂胶及其成分究竟是通过以上所提及的一种还是多种机制发挥抗菌作用的。

此外,关于蜂胶的抗菌研究,大多研究结果表明其对革兰阳性菌的作用强于对革兰阴性菌的作用,具体机制尚不是特别清楚。可能是革兰阴性菌复杂的外膜结构不利于蜂胶进入胞内发挥作用,也有可能是由于一些革兰阴性菌特殊的耐药机制,如主动外排系统(Savoia,2012;Tegos et al.,2002)。有研究表明,存在于革兰阴性菌细胞膜的外排泵系统 AcrAB-TolC(存在于肠科杆菌)及 MexAB-OprM(存在于绿脓杆菌)与革兰阴性菌对大多数植物成分耐受有一定关系(Kuete et al.,2011),而大肠杆菌及绿脓杆菌等阴性菌对蜂胶比较耐受的机制可能与此有很大的关系,有待进一步研究。

4 蜂胶中抗病原微生物的活性物质

蜂胶成分复杂,并且针对不同种类的微生物,其发挥活性的主要成分也有所差异。关于蜂胶中具有抗微生物作用的活性物质研究有很多,主要以抗菌活性成分的研究为主。

乔智胜和陈瑞华(1991)对河南蜂胶抗菌活性成分进行了研究,从蜂胶乙醇浸提液中分离到3个具有抗菌活性的结晶,经鉴定分别为芥子酸、异阿魏酸和咖啡酸,从醇提液中还分离得到白杨素。蜂胶中含有多种抑制金葡菌生长的成分,然而单体成分的 MIC 与蜂胶本身相近或略高,说明蜂胶对金葡菌的抗菌作用是多种成分协同作用的结果。

Amoros 等(1992)通过实验确定了蜂胶中抗1型单纯疱疹病毒的主要黄酮性成分,并发现黄酮醇的活性要强于黄酮的活性,活性顺序为高良姜素>山奈酚>槲皮素,且表明这些抗病毒成分之间还存在协同作用,后来 Lyu 等(2005)的实验也进一步证实了 Amoros 等的结论,并发现受试的黄酮类成分对1型单纯疱疹病毒的作用强于对2型单纯疱疹病毒的作用。

Velikovaa 等(2000)从巴西本地的一种无刺蜂(*Melipona quadrifasciata*)采集的蜂胶中分离出3种贝壳杉烯,而其中异贝壳杉烯酸与蜂胶总提取液具有相近的抗金葡菌的能力,说明这种物质在巴西无刺蜂蜂胶的抗菌活性中起到非常重要的作用。

Bosio 等(2000)从意大利西北部的不同地区采集到两种蜂胶样,用46种化脓性链球菌菌株检测其抗微生物活性。通过琼脂稀释法和琼脂扩散法,得到 MIC 和 MBC 不超过 $234\mu g/mL$,其中一种蜂胶抗菌作用更好,HPLC 分析发现这个样品提取物中含有更多的黄酮类成分松属素和高良姜素。Koo 等(2002)研究了蜂胶组分对变形链球菌生长和葡糖基转移酶(GTFs)活性的影响,通过评估蜂胶中已发现的几种不同化学组分对溶液中和唾液包被的羟基磷灰石珠子表面的葡糖基转移酶活性的影响。结果表明,黄酮和黄酮醇是溶液中葡糖基转移酶强有力的抑制物,而对不可溶性酶则效果较差。芹黄素(4,5,7-三羟基黄酮的一种)是最有效的 GTFs 抑制物,无论是在溶液中(浓度为 $135\mu g/mL$ 时达到 90.5%~95% 抑制)还是在 sHA 珠子表面($135\mu g/mL$ 时达到 30%~60%);通过 MIC、MBC 和 Time-kill 研究抗菌活性,结果表明,黄烷酮类和一些二氢黄酮醇以及倍半萜烯 *tt*-法尼醇能抑制变形链球菌和血链球菌的生长,*tt*-法尼醇是最有效的抗菌成分,其 MIC 和 MBC 分别为 $14\sim28\mu g/mL$ 和 $56\sim112\mu g/mL$。

Kartal 等(2003)对安纳托利亚喀山和马尔马里斯地区蜂胶样品的抗菌活性进行了评

估,研究表明其抗菌活性主要与蜂胶中的咖啡酸及其脂类有关,并从喀山蜂胶样中分离出了包含 3,3-二甲基咖啡酸烯丙酯(3,3-dimethylallyl caffeate)和 3-异戊烯咖啡酸酯(isopent-3-enyl caffeate)两个同分异构体混合物,这个混合物具有良好的抗菌活性。Popova 等(2005)的实验表明了酚类黄酮类物质对蜂胶发挥抗菌作用的重要性,因为总黄酮和总酚含量低的样品,其所需抗菌浓度会偏大,只有其中一种样品比较特殊,可能是其内含有的二萜酸发挥了抗菌活性。

de Paula 等(2006)研究了产自巴西的 EEP 和单体成分对 16 种口腔致病微生物的抗微生物活性,发现所有受试微生物都会受到 EEP 的影响,但没有哪一种单体成分的活性强于提取物,表明蜂胶抗微生物的作用是各种物质的协同作用。Gonsales 等(2006)研究表明,EEP 能有效抑制革兰阳性菌,而且抗菌能力与黄酮含量呈正相关。Melliou 等(2007)研究表明,蜂胶挥发油中含有的萜类物质具有很好的抗菌和抗真菌活性。Campana 等(2009)分析了两种 EEP 和几种黄酮成分对临床分离的 16 种人空肠弯曲杆菌(*Campylobacter jejuni*)和几种革兰阳性及阴性病原菌的抗菌活性。结果表明,EEP 能抑制空肠弯曲杆菌、粪肠杆菌和金黄色葡萄球菌的生长,活性最强的黄酮类成分是高良姜素;空肠弯曲杆菌对其敏感度达到 68.8%,槲皮素的活性稍差(50%);EEP 对人空肠弯曲杆菌的 MIC 为 0.3125~0.156mg/mL,高良姜素和槲皮素的 MIC 为 0.250~0.125mg/mL。

Trusheva 等(2010)从伊朗蜂胶中分离出 5 种单体成分,其中 4 种为萜脂:tschimgin(bornylp-hydroxybenzoate)、tschimganin(bornylvanillate)、ferutinin(ferutinolp-hydroxybenzoate)和 tefernin(ferutinol vanillate),它们均具有抗金黄色葡萄球菌的活性。这些抗菌成分属于苯甲酸类单萜和倍半萜烯酯类物质。

Ordóñez 等(2011)研究了从阿根廷北部采集的蜂胶对一些植物病原菌的抗菌活性。结果表明,活性最强的蜂胶来自图库曼省,并从中分离出抗菌活性成分 2,4-二羟基查耳酮,其抗菌作用比 EEP 还有效(MIC 分别为 0.5~1μg/mL 和 9.5~15μg/mL)。植物毒性试验结果表明,蜂胶提取物不会延迟莴苣种子的发芽和洋葱根的生长;蜂胶溶液喷洒在感染了丁香假单胞菌的西红柿上能降低病坏程度,表明蜂胶在防治水果病害上可能很有前景。

de Aguiar 等(2013)评估了 3 种巴西蜂胶提取物对瘤胃内一些细菌的作用,发现蜂胶提取物能抑制一些细菌的生长,但也有一些微生物对蜂胶提取物表现为耐受,被抑制的细菌对蜂胶提取物的敏感性不同,酚含量最低的提取物抗菌活性也最差。通过对蜂胶提取物中主要酚类化合物柚皮素、柯因、咖啡酸、p-香豆酸和阿替匹林 C(Artepillin C)对 4 个敏感菌株抗菌活性的评估,发现只有柚皮素对所有菌株表现出抑制活性,表明柚皮素是参与蜂胶抗菌作用的重要成分之一。Navarro-Navarro 等(2013)通过肉汤宏稀释法对墨西哥西北部索诺兰沙漠 3 个不同地区的蜂胶抗弧菌属抗菌活性的研究,发现产自乌雷斯地区蜂胶的抗菌活性最强,对大多数受试菌的 MIC_{50} 小于 50μg/mL,而且呈浓度依赖性。该蜂胶成分中的高良姜素和咖啡酸苯乙酯(CAPE)具有强烈抑制细菌(*V. cholerae* non-O1 和 *Serotype ogawa*)生长的活性,MIC_{50} 为 0.05~0.1mmol/L。

由此可见,蜂胶中含有多种已经证实的抗病原微生物成分,如松属素、高良姜素、山柰酚、槲皮素、芹黄素、咖啡酸苯乙酯、柚皮素、异阿魏酸、咖啡酸、芥子酸、苯甲酸类单萜和倍半萜烯酯类物质等。此外,还有文献报道了蜂胶中其他具有抗病原微生物作用的成分:球松素、杨芽黄素、对香豆苯酸酯、短叶松素、白杨素、靛红山姜素短叶松黄烷酮、乔松酮、柯因以及某些蜂胶中特有的活性成分,如巴西绿蜂胶中的 3-异戊二烯基-4-羟基肉桂酸、2,2-二甲基-6-羧乙基香豆素、3,5-二异戊二烯基-4-羟基肉桂酸(阿替匹林C)、2,2-二甲基-6-羧乙基-8-异戊二烯基香豆素等(郭芳彬,2004;杨书珍等,2009;黄文诚,2002)。

小　结

随着抗生素的广泛应用,耐药性问题也日益突出,因此寻找具有抗菌活性的天然药物成分也变得越来越重要和紧迫。蜂胶这一经过大量实验研究证实具有良好抗菌活性的天然产物,同时还具有抗氧化、抗炎、抗肿瘤、调节免疫等广泛的生物学活性,其应用前景广阔。然而尽管其抗菌作用已经被大量的体外实验证明,而相关的体内实验则非常少。由于体外实验只能证明蜂胶对一些微生物直接的抑制和杀灭作用,体外实验筛选的抗菌药物可能会存在有效作用浓度超过机体承受的毒性范围,或经机体代谢为抗菌能力大幅下降甚至消失的中间代谢产物等问题。因此,若要将蜂胶最终开发为临床用药还需要开展大量的体内试验,观察其在蜂胶、微生物及宿主三者相互作用动态条件下的实际效果,综合评价其抗病原微生物能力,为将来投入临床应用提供更确切的指导。此外,关于蜂胶抗病原微生物机制的研究还比较初步,这可能是由于蜂胶成分复杂,作用效果往往是各种活性成分通过各种作用机制交互协同作用的结果,从而加大了对其作用机制研究的难度。最后,关于蜂胶与其他植物性抗菌成分的配伍研究甚少,这方面的研究也有待深入。

参考文献

Amoros M, Simões CMO, Girre L, Sauvager F, Cormier M (1992) Synergistic effect of flavones and flavonols against herpes simplex virus type 1 in cell culture. Comparison with the antiviral activity of propolis [J]. *Journal of Natural Products*, 55(12): 1732-1740.

Ayres DC, Marcucci MC, Giorgio S (2007) Effects of Brazilian propolis on *Leishmania amazonensis* [J]. *Memorias do Instituto Oswaldo Cruz*, 102(2): 215-220.

Bosio K, Avanzin C, D'Avolio A, Ozino O, Savoia D (2000) In vitro activity of propolis against *Streptococcus pyogenes* [J]. *Letters in Applied Microbiology*, 31(2): 174-177.

Búrquez MJG, de Lourdes Juárez Mosqueda M, Mendoza HR, Soto Zárate CI, Miranda LC, Sánchez TAC (2015) Protective effect of a mexican propolis on MDBK cells exposed to aujeszky's disease virus (pseudorabies virus) [J]. *African Journal of Traditional, Complementary and Alternative Medicines*, 12(4): 106-111.

Campana R, Patrone V, Franzin ITM, Diamantini G, Vittoria E, Baffone W (2009) Antimicrobial activity of two propolis samples against human *Campylobacter jejuni* [J]. *Journal of Medicinal Food*, 12(5): 1050-1056

Dantas AP, Salomão K, Santos Barbosa H, De Castro SL (2006) The effect of Bulgarian propolis against *Trypanosoma cruzi* and during its interaction with host cells [J]. *Memorias do Instituto Oswaldo Cruz*, 101(2): 207-211 da Silva SS,Mizokami SS,Fanti JR,Miranda MM,Kawakami NY,Teixeira FH,Araújo EJA,Panis C,Watanabe MAE,Sforcin JM,Pavanelli WR,VerriJr WA,Felipe I,Conchon-Costa I (2016) Propolis reduces *Leishmania Amazonensis*-induced inflammation in the liver of Balb/c mice[J]. *Parasitology Research*,115:1557-1566.

de Aguiar SC,Zeoula LM,Franco SL,Peres LP,Arcuri PB,Forano E (2013) Antimicrobial activity of Brazilian propolis extracts against rumen bacteria in vitro[J]. *World Journal of Microbiology & Biotechnology*,29(10):1951-1959.

de Carvalho Machado GM,Leon LL,De Castro SL (2007) Activity of Brazilian and Bulgarian propolis against different species of *Leishmania* [J]. *Memorias do Instituto Oswaldo Cruz*,102(1):73-77

de Oliveira Orsi R,Sforcin JM,Funari SRC,Junior AF,Bankova V (2006) Synergistic effect of propolis and antibiotics on the *Salmonellatyphi*[J]. *Brazilian Journal of Microbiology*,37(2):108-112.

de Paula AMB,Gomes RT,Santiago WK,Dias RS,Cortés ME,Santos VR (2006) Susceptibility of oral pathogenic bacteria and fungi to brazilian green propolis extract[J]. *Pharmacologyonline*,3:467-473.

Duran G,Duran N,Culha G,Ozcan B,Oztas H,Ozer B (2008) In vitro antileishmanial activity of Adana propolis samples on *Leishmania tropica*:a preliminary study[J]. *Parasitology Research*,102:1217-1225.

FerreiraFM,CastroRAO,Batista MA,Rossi FMO,Silveira-Lemos D,Frézard F,Moura SAL,RezendeSA (2014) Association of water extract of green propolis and liposomal meglumine antimoniate in the treatment of experimental visceral leishmaniasis[J]. *Parasitology Research*,113(2):533-543.

FriedmanM,JürgensHS (2000) Effect of pH on the stability of plant phenolic compounds [J]. *Journal of Agricultural and Food Chemistry*,48(6):2101-2110.

Freitasa SF,Shinoharaa L,Sforcinb JM,Guimarães S (2006) In vitro effects of propolis on *Giardia duodenalis* trophozoites[J]. *Phytomedicine*,13:170-175.

Garedew A,Schmolz E,Lamprecht I (2004) Microbiological and calorimetric investigations on the antimicrobialactions of different propolis extracts:an in vitro approach. [J]. *Thermochimica Acta*,422(1-2):115-124.

Gebara ECE,Lima LA,Mayer1 MPA (2002) Propolis antimicrobial activity against periodontopathic bacteria[J]. *Brazilian Journal of Microbiology*,33(4):365-369.

Gekker G,Hu S,Spivak M,Lokensgard JR,Peterson PK (2005) Anti-HIV-1 activity of propolis in CD4+lymphocyte and microglial cell cultures[J]. *Journal of Ethnopharmacology*,102 (2):158-163.

Gomes MFF,Ítavo C,Leal CRB,Itavo LCV,Lunas RC (2016)*In vitro* biological activity of brown propolis[J]. *Pesquisa Veterinaria Brasileira*,36(4):279-282.

Gonsales GZ,Orsi RO,Fernandes Júnior A.,Rodrigues P,Funari SRC (2006) Antibacterial activity of propolis collected in different regions of Brazil[J]. *Journal of Venomous*

Animals and Toxins Including Tropical Diseases,12(2):276-284.

Gressler LT,Da Silva AS,Machado G,Rosa LD,Dorneles F,Gressler LT,Oliveira MS,Zanette RA,de Vargas ACP,Monteiro SG (2012) Susceptibility of *Trypanosoma evansi* to propolis extract in vitro and in experimentally infected rats[J]. *Research in Veterinary Science*,93:1314-1317.

Harish Z,Rubinstein A,Golodner M,Elmaliah M,Mizrachi Y (1997) Suppression of HIV-1 replication by propolis and its immunoregulatory effect[J]. *Drugs under Experimental and Clinical Research*,23(2):89-96.

Hegazi AG,Abd El Hady FK,Abd Allah FAM (2000) Chemical composition and antimicrobial activity of European propolis[J]. *Zeitschrift für Naturforschung-Section C Journal of Biosciences*,55(1-2):70-75.

Ivančajić S,Mileusnić I,Cenić-Milošević D (2010) In vitro antibacterial activity of propolis extracts on 12 different bacteria in conditions of 3 various pH values[J]. *Archives of Biological Sciences*,62(4):915-934.

Kalia P,Kumar NR,Harjai K (2016) Studies on the therapeutic effect of propolis along with standard antibacterial drug in salmonella enterica serovar typhimurium infected balb/c mice[J]. *BMC Complementary and Alternative Medicine*,16(1):485.

Kartal M,Yıldız S,Kaya S,Kurucu S,Topçu G (2003) Antimicrobial activity of propolis samples from two different regions of Anatolia[J]. *Journal of Ethnopharmacology*,86(1):69-73.

Kashi TSJ,Kermanshahi RK,Erfan M,Dastjerdi EV,Rezaei Y,Tabatabaei FS (2011) Evaluating the in-vitro antibacterial effect of Iranian propolis on oral microorganisms[J]. *Iranian Journal of Pharmaceutical Research*,10(2):363-368.

Kilic A,Baysallar M,Besirbellioglu B,Salih B,Sorkun K,Tanyukse M (2005) In vitro antimicrobiial activity of propolis against methicillin-resistant *Staphyloococcus aureus* and vancomycin-resistant *Enterococcus faecium*[J]. *Annals of Microbiology*,55(2):113-117.

Koo H,Rosalen PL,Cury JA,Ambrosano GMB,Murata RM,Yatsuda R,Ikegaki M,Alencar SM,Park YK (2000) Effect of a new variety of *Apis mellifera* propolis on *Mutans streptococci*[J]. *Current Microbiology*,41(3):192-196.

Koo H,Rosalen PL,Cury JA.,Park YK,Bowen WH (2002) Effects of compounds found in propolis on *Streptococcus mutans* growth and on glucosyltransferase activity [J]. *Antimicrobial Agents and Chemotherapy*,46(5):1302-1309.

Kuete V,Alibert-Franco S,Eyong KO,Ngameni B,Folefoc GN,Nguemeving JR,Tangmouo JG,Fotso GW,Komguem J,Ouahouo BM,Bolla JM,Chevalier J,Ngadjui BT,Nkengfack AE,Pagès JM (2011) Antibacterial activity of some natural products against bacteria expressing a ultidrug-resistant phenotype[J]. *International Journal of Antimicrobial Agents*,37(2):156-161.

Kuropatnicki AK,Szliszka E,Krol W (2013) Historical aspects of propolis research in

modern times[J]. *Evidence-Based Complementary and Alternative Medicine*, Volume 2013, Article ID 964149.

Lyu SY, Rhim JY, Park WB (2005) Antiherpetic activities of flavonoids against herpes simplex virus type 1 (HSV-1) and type 2 (HSV-2) in vitro[J]. *Archives of Pharmacal Research*, 28(11):1293-1301.

Ma X, Guo Z, Shen Z, Liu Y, Wang J, Fan Y (2015) The anti-porcine parvovirus activity of nanometer propolis flavone and propolis flavone in vitro and in vivo[J]. *Evidence-Based Complementary and Alternative Medicine*, 2015 Article ID:472876.

Melliou E, Stratis E, Chinou I (2007) Volatile constituents of propolis from various regions of Greece-Antimicrobial activity[J]. *Food Chemistry*, 103(2):375-380.

Muli EM, Maingi JM (2007) Antibacterial activity of *Apis mellifera* L. propolis collected in three regions of Kenya[J]. *Journal of Venomous Animals and Toxins Including Tropical Diseases*, 13(3):655-663.

Najmadeen HH, KH. Kakamand FA. (2009) Antimicrobial activity of propolis collected in different regions of sulaimani province-Kurdistan region/Iraq[J]. *Journal of Duhok University*, 12(1):233-239.

Nassar SA, Mohamed AH, Soufy H, Nasr SM, Mahran KM (2012) Immunostimulant effect of Egyptian propolis in rabbits[J]. *The Scientific World Journal*, Volume2012, Article ID 901516.

Navarro-Navarro M, Ruiz-Bustos P, Valencia D, Robles-Zepeda R, Ruiz-Bustos E, Virues C, Hernandez J, Domínguez Z, Velazquez C (2013) Antibacterial activity of Sonoran propolis and some of its constituents against clinically significant *Vibrio* species[J]. *Foodborne Pathogens and Disease*, 10(2):150-158.

Nina N, Quispe C, Jiménez-Aspee F, Theoduloz C, Feresín GE, Lima B, Leiva E, Schmeda-Hirschmann G (2015) Antibacterial activity, antioxidant effect and chemical composition of propolis from the Región del Maule, Central Chile[J]. *Molecules*, 20(10):18144-18167.

Ordóñez RM, Zampini IC, Nieva Moreno MI, Isla MI (2011) Potential application of Northern Argentine propolis to control some phytopathogenic bacteria[J]. *Microbiological Research*, 166(7):578-584.

Orsi RO, Sforcin JM, Rall VLM, FunariSRC, Barbosa L, Fernandes JRA (2005) Susceptibility profile of *Salmonella* against theantibacterial activity of propolis produced in two regions of Brazil[J]. *Journal of Venomous Animals and Toxins Including Tropical Diseases*, 11(2):109-116.

Ota C, Unterkircher C, Fantinato V, Shimizu MT (2001) Antifungal activity of propolis on different species of Candida[J]. *Mycoses*, 44 (9-10):375-378.

Ozturk H, Pekcan M, Sireli M, Fidanci UR (2010) Effects of propolis on in vitro rumen microbial fermentation[J]. *Ankara Universitesi Veteriner Fakultesi Dergisi*, 57(4): 217-221.

Pontin K, Da Silva Filho AA, Santos FF, e Silva MLA, Cunha WR, Nanayakkara PD, Bastos

JK,de Albuquerque S(2008) In vitro and in vivo antileishmanial activities of a Brazilian green propolis extract[J]. *Parasitology Research*,103(3):487-492.

Popova M,Silici S,Kaftanoglu O,Bankova V.(2005) Antibacterial activity of Turkish propolis and its qualitative and quantitative chemical composition[J]. *Phytomedicine*,12(3):221-228.

Savoia D(2012) Plant-derived antimicrobial compounds:alternatives to antibiotics[J]. *Future Microbiology*,7(8):979-990.

Sawaya ACHF,Souza KS,Marcucci MC,CunhaIBS,Shimizu MT(2004) Analysis of the composition of Brazilian propolis extracts by chromatography and evaluation of their In vitro activity against gram-positive bacteria[J]. *Brazilian Journal of Microbiology*,35(1-2):104-109.

Scazzocchio F,D'Auria FD,Alessandrini D,Pantanella F (2006) Multifactorial aspects of antimicrobial activity of propolis[J]. *Microbiological Research*,161(4):327-333.

Shahbaz M,Zahoor T,Randhawa A,Nawaz H (2015) In-vitro antibacterial activity of hydroalcoholic extract of propolisagainst pathogenic bacteria[J]. *Pakistan Journal of Life and Social Sciences*,13(3):132-136.

Silici S,Kutluca S (2005) Chemical composition and antibacterial activity of propolis collected by three different races of honeybees in the same region[J]. *Journal of Ethnopharmacology*,99(1):69-73.

Silici S,Koç NA,Ayangil D,Çankaya S (2005) Antifungal activities of propolis collected by different races of honeybees against yeasts isolated from patients with superficial mycoses[J]. *Journal of Pharmacological Sciences*,99(1):39-44.

Shimizu T,Hino A,Tsutsumi A,Park YY,Watanabe W,Kurokawa M (2008) Anti-influenza virus activity of propolis in virto and its efficacy against influenza infection in mice[J]. *Antiviral Chemistry and Chemotherapy*,19(1):7-13.

Stepanović S,Antić N,Dakić I,Švabić-Vlahović M (2003) In vitro antimicrobial activity of propolis and synergism between propolis and antimicrobial drugs[J]. *Microbiological Research*,158(4):353-357.

Takemura T,UrushisakiT,Fukuoka M,Hosokawa-MutoJ,HataT,Okuda Y,HoriS,TazawaS,ArakiY,Araki Y,Kuwata K (2012) 3,4-dicaffeoylquinic acid,amajor constituent of Brazilianpropolis,increases TRAIL expression and extends the lifetimesofmice infected with the influenza avirus[J]. *Evidence-Based Complementary and Alternative Medicine*,Volume 2012,Article ID 946867.

TegosG,Stermitz FR,Lomovskaya O,LewisK (2002) Multidrug pump inhibitors uncover remarkable activity of plant antimicrobials[J]. *Antimicrobial Agents and Chemotherapy*,46(10):3133-3141.

Tim Cushnie TP,Lamb AJ (2005) Detection of galangin-induced cytoplasmic membrane damage in *Staphylococcus aureus* by measuring potassium loss[J]. *Journal of Ethnopharmacology*,101(1-3):243-248.

Tim Cushniea TP, Lamb AJ (2011) Recent advances in understanding the antibacterial properties of flavonoids[J]. *International Journal of Antimicrobial Agents*, 38(2): 99-107.

Trusheva B, Todorov I, Ninova M, Najdenski H, Daneshmand A, Bankova V (2010) Antibacterial mono-and sesquiterpene esters of benzoic acids from Iranian propolis[J]. *Chemistry Central Journal*, 4:8.

Ugur A, Arslan T (2004) Anin vitro study on antimicrobial activity of propolis from Mugla Province of Turkey[J]. *Journal of Medicinal Food*, 7(1):90-94.

Uzel A, Sorkun K, Oncag O, Cogulu D, Gencay O, Salih B (2005) Chemical compositions and antimicrobial activities of four different Anatolian propolis samples[J]. *Microbiological Research*, 160(2):189-195.

Velikova M, Bankova V, Tsvetkova I, Kujumgiev A, Marcucci MC (2002) Antibacterial ent-kaurene from Brazilian propolis of native stingless bees[J]. *Fitoterapia*, 71(6): 693-696.

Veloz JJ, Saavedra N, Lillo A, Alvear M, Barrientos L, Salazar LA (2015) Antibiofilm activity of Chilean propolis on *Streptococcus mutans* isinfluenced by the year of collection [J]. *BioMed Research International*, 2015 Article ID:291351.

Veloz JJ, Saavedra N, Alvear M, Zambrano T, Barrientos L, Salazar LA (2016) Polyphenol-rich extract from propolis reduces the expression and activity of streptococcus mutans glucosyltransferases at subinhibitory concentrations[J]. *BioMed Research International*, Volume 2016, Article ID 4302706.

Wojtyczka RD, Kepa M, Idzik D, Kubina R, Kabala-Dzik A, Dziedzic A, Wdsik TJ (2013) In vitroantimicrobial activity of ethanolic extract of Polish propolis against biofilm forming *Staphylococcus epidermidis* Strains[J]. *Evidence-Based Complementary and Alternative Medicine*, Volume 2013, Article ID 590703.

Yang HY, Chang CM, Chen YW, Chou CC (2006) Inhibitory effect of propolis extract on the growth of *Listeria monocytogenes* and the mutagenicity of 4-nitroquinoline-N-oxide [J]. *Journal of the Science of Food and Agriculture*, 86(6):937-943.

Zeighampour F, Maryam MS, Shams E, Naghavi NS (2014) Antibacterial activity of propolis ethanol extract against antibiotic resistance bacteria isolated from burn wound infections[J]. *Zahedan Journal of Research in Medical Sciences*, 16(3):25-30.

艾予川,许兵红,曾莉萍,李进芬(2004)河南蜂胶对金黄色葡萄球菌耐药株不同pH抑菌作用研究[J].实用儿科临床杂志,19(8):710-711.

陈炼,廖旺娣,李国华,谢勇,吕农华,王崇文(2009)江西蜂胶对幽门螺杆菌的体外抑菌作用研究[J].江西医学院学报,49(12):27-32.

陈荷凤,韩文辉,李兵,崔海辉(2001)蜂胶各种溶剂提取物的抑菌效果比较[J].食品研究与开发,22:18-19.

柴家前,庞昕,刘金华,沈志强(2010)鸡新城疫纳米蜂胶灭活疫苗免疫作用机制[J].中国兽医学报,30(9):1151-1155.

蔡爽,时清,李玉晶,杨东梅,李金陆,郑焱(2006)蜂胶涂膜对变形链球菌生长和黏附的抑制作用[J].实用口腔医学杂志,22(2):171-174.

郭芳彬(2004)蜂胶的抗菌作用[J].蜜蜂杂志,(3):10-12.

黄文诚(2002)蜂胶中的抗菌和抗肿瘤成分—阿替匹林[J].蜜蜂杂志,(4):7-8.

蒋琳,林居红,刘明方,朱珠(2008)蜂胶及蜂胶奥硝唑合剂体外抑制牙龈卟啉单胞菌生长的实验研究[J].现代口腔医学杂志,22(5):497-500.

姜游帅(2010)蜂胶对金黄色葡萄球菌毒力因子抑制作用研究[D].大庆:黑龙江八一农垦大学硕士学位论文.

刘成芳,殷国荣,岳毅,赵云鹤,孔丽,刘红丽(2009)STAg联合蜂胶佐剂鼻内免疫小鼠抗弓形虫感染的保护作用[J].中国生物制品学杂志,22(9):887-890.

刘思伽,王凤阳,成子强,黄爱芳,李胜,邹永新,余双祥,罗映霞,李剑荣,沙才华(2006)番鸭呼肠孤病毒病蜂胶佐剂灭活疫苗的研制[J].中国预防兽医学报,28(2):225-227.

兰桃芳,孟良玉,白凤翎,王伟伟,马春颖(2006)蜂胶提取液的抑菌作用研究[J].食品科学,27(12):224-226.

李雅晶(2011)蜂胶中挥发性成分的提取方法、化学组成及生物学活性[D].杭州:浙江大学博士学位论文.

卢媛媛(2014)巴西蜂胶残渣中挥发性成分的提取、化学组成及生物学活性研究[D].杭州:浙江大学硕士学位论文.

孟良玉,兰桃芳,卢佳琨,张艺凡,初莹莹(2010)蜂胶提取物中抑菌成分稳定性研究[J].食品科学,31(21):98-100.

平舜,蔺哲广,胡福良(2014)蜂胶对蜂群病虫害的防治作用及机制研究[J].环境昆虫学报,36(3):427-432.

彭志庆,林居红,刘明方,邓一平(2010)国产水溶性蜂胶对主要致龋链球菌及其耐氟菌株致龋性的影响[J].第三军医大学学报,32(8):798-801.

王浩,李艳玲,高艳霞,辛晓明,刘宗昌,高允生(2010)蜂胶二氧化碳超临界萃取物体外抗菌作用研究[J].中国消毒学杂志,27(4):395-396.

王凯,何天骏,胡福良(2013)蜂胶佐剂在兽用疫苗中的应用[J].动物医学进展,34(10):111-115.

王晶,李玉晶,张春梅,赵静,王凤忠(2002)蜂胶抑制变形链球菌生长和黏附的体外研究[J].北京口腔医学,10(2):79-81.

乔智胜,陈瑞华(1991)河南蜂胶抗菌活性成分的研究[J].中国中药杂志,16(8):481-482.

沈志强(1995)禽霍乱蜂胶灭活疫苗研究与应用总结[J].中国畜禽传染病,(4):27-30.

沈志强,李强,庞万勇,管宇,贾杏林,庄金秋(2004)猪伪狂犬病蜂胶疫苗和油佐剂疫苗对小白鼠注射部位损伤的病理组织学比较[J].动物医学进展,25(1):91-93,106.

熊萍,张明珠,雷雅燕,税艳青,朱红,杨帆(2009)云南蜂胶对3种牙周致病菌的抑制作用[J].昆明医学院学报,(9):23-26.

许兵红,曾莉萍,李进芬,杨晓,艾予川(2004)河南蜂胶对溶壁微球菌不同pH抑菌作用研究[J].实用预防医学,11(3):423-424.

许兵红,曾莉萍,陈萍(2007)河南蜂胶对白色念珠菌的抑菌作用研究[J].中国病原生物学杂志,2(4):252-253.
许兵红,曾莉萍(2011)蜂胶对炭疽杆菌的抑菌活性[J].中国媒介生物学及控制杂志,22(3):245-247.
杨明,颜伟玉,曾志将(2006)蜂胶抗菌效果影响的研究[J].蜜蜂杂志,(5):5-7.
杨道锋,王静丽,朱慧芬,沈关心,龚非力(2008)普通蜂胶及纳米蜂胶对临床常见细菌的体外抑制效应研究[J].中国药师,11(10):1167-1169.
杨书珍,彭丽桃,姚晓琳,潘思轶(2009)蜂胶抗真菌作用研究进展[J].食品工业科技,30(11):349-352.
杨鑫,范家恒,段甯耀,赵风(2013)蜂胶乙醇提取液对杏鲍菇及污染菌的抑制作用[J].食品与发酵工业,39(7):128-131.
曾莉萍,许兵红,李进芬(2007a)蜂胶对链球菌的抑菌实验[J].现代预防医学,34(16):3042-3045.
曾莉萍,许兵红,牛杰,李进芬,毛泽善(2007b)蜂胶对金黄色葡萄球菌的抑菌作用[J].现代预防医学,34(13):2432-2434.
张红梅,林居红,蒋琳(2008)国产水溶性蜂胶对感染根管内厌氧菌的体外抑菌实验[J].重庆医科大学学报,33(12):1516-1519.
张秀喜(2009)蜂胶黄酮的提取及提取物的抑菌、抗氧化活性研究[D].合肥:合肥工业大学硕士学位论文.
张正强(2009)蜂胶提取物的抑菌作用及在冷却牛肉保鲜中的应用研究[D].上海:上海师范大学硕士学位论文.
赵恒章,赵坤,余燕,李国旺,马金友,张慧辉(2009)副猪嗜血杆菌蜂胶苗的制备及免疫效果试验[J].生物技术,19(6):72-74.
赵强(2007)蜂胶中挥发油有效成分的研究[D].南昌:南昌大学硕士学位论文.
赵强,刘文群,张彬,周率,黎新江(2008)蜂胶超临界CO_2萃取物抑菌作用[J].南昌大学学报,32(4):394-397.
周萍,胡福良,徐权华(2006)蜂胶在不同载体中对枯草芽孢杆菌的抑菌试验[J].中国蜂业,57(10):5-7.
朱明,徐琦,吴晔华,谢天宇,李芳芳,邵婷婷,崔雅静,赵姝月(2013)新疆黑蜂胶体外抑制大肠杆菌生物膜的研究[J].中国医药导报,10(36):18-21.
祖力卡尔江·阿合买提,林静,于倩,赵今(2015)伊犁黑蜂蜂胶对不同状态下变形链球菌乳酸脱氢酶及其相关基因影响的实验研究[J].中国微生态学杂志,27(5):543-547.

第二节 蜂胶的抗氧化作用

蜂胶的抗氧化活性一直被认为是蜂胶最重要的生物学活性之一。近年来,世界各国的研究者采用不同实验分析方法,针对蜂胶的抗氧化活性进行了广泛的研究,发现蜂胶具有良好的抗氧化效果,能有效清除自由基(Sulaiman et al.,2011;Thirugnanasampandan et al.,2012),缓解机体的氧化应激,改善组织细胞的生物功能(Capucho et al.,2012;Simões-

Ambrosio et al.,2010)。下面就国内外学者对不同地理来源、不同提取方法获得的蜂胶提取物以及蜂胶中几种重要的单体活性成分的抗氧化研究进展进行综述,并以此为基础,进一步探讨蜂胶及其单体活性成分的抗氧化分子机制,旨在加深对蜂胶抗氧化活性的认识,为更深入地研究与利用蜂胶的生物学活性提供参考。

1 不同地理来源蜂胶提取物的抗氧化活性

蜂胶的抗氧化活性与蜂胶中黄酮类、酚酸类及萜烯类物质的种类及含量密切相关。因不同地区蜜蜂采集的植物树脂不同,导致不同地理来源蜂胶的化学成分差异较大。此外,不同的提取方法也会对蜂胶提取物的化学成分造成一定影响。但总的来说,不同地理来源获得的蜂胶提取物都具有良好的抗氧化活性。有关蜂胶抗氧化的研究众多,世界各国学者对蜂胶抗氧化研究的有效部位、实验方法、检测指标和抗氧化作用评价情况见表3.1。

1.1 蜂胶乙醇提取物(EEP)的抗氧化活性

乙醇是蜂胶有效成分提取的最常用溶剂。Park 和 Ikegaki(1998)对多个浓度(10%~95%)乙醇提取的 EEP 进行了抗氧化作用研究,结果表明,70%与80%乙醇蜂胶提取物具有最强的抗氧化活性。Mavri 等(2012)用70%和96%的乙醇分别提取蜂胶,发现96%乙醇提取的 EEP 除了咖啡酸、阿魏酸、木樨草素的含量稍多于70%乙醇提取的 EEP 外,酚酸类及黄酮化合物的含量均明显少于前者。进一步研究显示,两种 EEP 均具有良好的抗氧化能力,但是70%乙醇提取的 EEP 具有更强的还原力、自由基及金属离子清除能力。

Cigut 等(2011)研究发现,96%乙醇提取的 EEP(酚酸质量浓度为0.05g/L)能有效缓解酵母菌胞内氧化,再用加压湿法萃取处理96%乙醇提取的 EEP,分离出强极性部分与温和极性部分,并且蜂胶提取物的温和极性部分能进入细胞并显著改善细胞内氧化-还原的平衡,据此推断蜂胶的抗氧化活性与此部分密切相关。进一步在线粒体蛋白质组水平的研究表明,EEP 以及蜂胶提取物的温和极性部分的酚类化合物可与线粒体 F0F1-ATP 合酶的 α 亚基(α subunit of mitochondrial F0F1-ATP synthase,Atp1)结合并修饰 Atp1 的转录调节以及 Atp1 的转录后修饰(如激酶、磷酸酶),进而调节抗氧化和 ATP 合成相关的蛋白质的含量。

Zhang 等(2016)比较了澳大利亚桉树型蜂胶与巴西酒神菊型蜂胶乙醇提取物的抗氧化活性差异,发现桉树型蜂胶体外自由基清除活性(DPPH 和 ABTS)较酒神菊型蜂胶更强,但酒神菊型蜂胶具有更好的细胞活性氧自由基(ROS)清除能力,并进一步证明这两种蜂胶在细胞层面抗氧化活性及抗氧化作用机制的差异。利用小鼠巨噬细胞(RAW264.7)研究发现,酒神菊树蜂胶能激活 p38 信号通路,同时诱导 Nrf2 蛋白的进核,进而激活细胞抗氧化系统,而桉树型蜂胶可以激活 ERK 信号通路来诱导 Nrf2 蛋白的激活。

1.2 蜂胶水提物(WEP)的抗氧化活性

Guo 等(2011)对中国不同地区蜂胶水提物(WEP)进行研究,发现 WEP 的主要有效成分是表儿茶素、香豆酸、二甲氧基肉桂酸、柚柑配基、阿魏酸、肉桂酸、松鼠素、柯因等酚类物质,并总结出中国热带、亚热带地区(云南、广西)的 WEP 的还原力以及 DPPH 自由基清除能力要低于中国温带地区(山东、河北等)的 WEP。Moura 等(2011)对巴西绿蜂胶 WEP 的研究

表 3.1 蜂胶抗氧化研究一览表

Table 3.1 Research list of antioxidant activity of propolis

地理来源	有效部位	实验方法	检测指标	抗氧化作用评价	参考文献
斯洛文尼亚	酿酒酵母胞液	DPPH 法、FRAP 法、SRSA 法、MCC 法、beta 亚油酸体系法	自由基清除活性、还原力、金属离子螯合力、总抗氧化力及细胞内抗氧化力	有效清除自由基、良好的还原力、降低细胞内氧化	(Mavri et al., 2012)
斯洛文尼亚	酚类 0.05 g/L	H_2DCFDA 法、BacTiter-Glo 细胞生存能力分析法、CFU、二维电泳等	ROS 含量、细胞生存能力、细胞抗氧化力、线粒体蛋白	降低细胞内氧化、改变线粒体水平抗氧化蛋白含量	(Cigut et al., 2011)
意大利	—	SDS 胶中的亚油酸过氧化方法、SPF 模拟检测方法	防晒系数、UVA/UVB 的比值、脂质过氧化抑制效果	抑制脂质过氧化、可作为药妆品的有效组分	(Gregoris et al., 2011)
罗马尼亚	—	DPPH 漂泊法、傅立叶变换红外光谱和紫外吸收光谱法	总抗氧化能力	良好的抗氧化效果	(Mot et al., 2011)
塞尔维亚	—	直流电极谱法、DPPH 法	总抗氧化能力、自由基清除能力	较好的自由基清除活性	(Potkonjak et al., 2012)
葡萄牙	红细胞、肾脏	TBARS 法、MTT 法等	检测脂质过氧化和红细胞溶血程度、蜂胶对正常细胞的毒性、癌细胞的生长抑制活性	抑制红细胞脂质过氧化和细胞溶解、对人肾癌细胞有毒性而对正常细胞无毒性	(Valente et al., 2011)
土耳其	血液、肾、肝、鳃	生化分析方法	血液学参数和主要的抗氧化酶	改善毒死蜱对鱼的毒性效果显著	(Enis Yonar et al., 2012)
土耳其	血液、肝脏等	生化分析方法、免疫学分析方法	脂质过氧化、抗氧化酶活性	减轻鱼的氧化应激和免疫抑制的效果	(Enis Yonar et al., 2011)
安纳托利亚	—	DPPH 法、ABTS 法	自由基清除活性	每克蜂胶提取物与 500mg trolox 自由基清除活性等效	(Erdogan et al., 2011)
中国	—	DPPH 法、FRAP 法	自由基清除活性、还原力	中国大部分水提蜂胶都表现出强的抗氧化活性	(Guo et al., 2011)

续表

地理来源	有效部位	实验方法	检测指标	抗氧化作用评价	参考文献
中国	—	DPPH法、ABTS法、FRAP法	自由基清除活性、还原力	天然抗氧化剂、营养添加剂,且乙酸乙酯蜂胶提取物自由基清除能力最强	(Yang et al.，2011)
日本	肾脏	MTT法、H₂DCFDA法、免疫印迹等	细胞内ROS含量,SOD酶、血红素氧化酶1的mRNA的表达量等	减轻细胞毒性和细胞内ROS的产生,提高血红素氧化酶的表达	(Kamiya et al.，2012)
印度	—	DPPH法	自由基清除活性	有效清除自由基	(Sulaimn et al.，2011)
乌拉圭	心血管潜在作用部位	ORAC法、生化分析方法	LDL、酪氨酸硝化作用、一氧化氮合成酶、NADPH氧化酶	抑制低密度脂蛋白与酪氨酸的硝化,内皮细胞的NO合成酶的表达和NAPDH氧化酶	(Silva et al.，2011)
巴西	睾丸	TBARS法、生化分析方法、组织总蛋白测定	脂质过氧化、抗氧化酶、组织总蛋白等指标	显著减低毒死蜱诱导小鼠睾丸的氧化应激	(Attia et al.，2012)
酒神菊属型蜂胶、棕蜂胶	皮肤	脂质过氧化分析、生化分析方法	皮肤的谷胱甘肽含量、皮肤蛋白酶	抗氧化效果良好,但不能抑制辐射引起的皮肤蛋白酶活性升高	(Fonseca et al.，2010)
巴西红蜂胶	—	β-胡萝卜素与亚油酸法	检测对亚油酸的抗氧化活性	可作为温和的抗氧化剂	(Oldoni et al.，2011)
巴西绿蜂胶	附睾	透射电子显微镜法、生化分析法、TBAR法	精子数、形态、寿命、抗氧化相关酶等	改善雄性鼠生殖功能	(Capucho et al.，2012)
澳大利亚	—	DPPH法	自由基清除活性	清除DPPH自由基力强于巴西绿蜂胶	(Abu-Mellal et al.，2012)

发现,咖啡酰奎宁酸以及苯丙酯类是其主要有效成分,二者都具有良好的抗氧化活性。Nakajima 等(2007)研究发现,10μg/mL 的巴西蜂胶 WEP 对视网膜神经有保护作用,并推断这可能是通过蜂胶的抗氧化活性发挥作用的。Saito 等(2015)利用紫外照射诱导细胞氧化应激模型,研究了巴西蜂胶 WEP 对皮肤纤维原细胞(NB1-RGB)内源抗氧化基因和抗氧化信号通路的作用。结果发现,WEP 能显著上调细胞抗氧化相关血红素加氧酶(HO-1)蛋白表达水平,并激活氧化和化学应激的防御性转导通路—Nrf2/ARE。

1.3 蜂胶水提物与醇提物的抗氧化活性比较

不同地理来源的蜂胶成分不同,能溶于溶剂的有效成分也不同,因而导致抗氧化活性有差异。Banskota 等(2000)分别对 9 个地区的蜂胶水提取物与甲醇提取物的抗氧化活性进行研究,发现巴西蜂胶和中国蜂胶的水提取物的 DPPH 自由基清除活性强于甲醇提取物,而秘鲁、荷兰蜂胶的甲醇提取物 DPPH 自由基清除活性强于水提取物。同时,由于蜂胶化学成分种类繁多,采用不同溶剂、不同提取方法得到的蜂胶提取物成分差别很大(表 3.2)。研究显示,中国蜂胶水提物还原力与蜂胶醇提物还原力大体相似,但其自由基清除能力要明显弱于蜂胶醇提取物(Guo et al.,2011)。而印度蜂胶水提物的总酚酸含量明显高于蜂胶乙醇提取物,此结果与使用欧洲和中国蜂胶提取物得到的数据并不一致,且印度蜂胶水提物表现出了比醇提物更强的抗氧化活性,这可能是由于蜂胶地理来源不同所致(Laskar et al.,2010)。总体来说,水对蜂胶有效成分的提取能力可能要弱于乙醇,同时蜂胶的地理来源也是影响蜂胶提取物抗氧化活性的一个重要因素。由于蜂胶水提取物也具有较好的抗氧化活性,且相较于乙醇提取的蜂胶来说,在外用以及口服时不会带来不适感。由此可知,水提蜂胶相关产品也具有良好的市场前景。

表 3.2 蜂胶水、醇提取物的成分及抗氧化活性比较

Table 3.2 Comparison of ingredients and antioxidant activities in WEP and EEP

蜂胶种类	总多元酚 (mg/g)	总黄酮 (mg/g)	黄烷酮 (mg/g)	黄酮-黄酮醇 (mg/g)	DPPH 自由基清除力 AAI	DPPH WEP:EEP AAI	参考文献
中国 EEP	260.59	—	52.54	79.68	—	0.41	(Guo et al.,2011)
中国 WEP	210.70	—	5.91	9.30	1.55		(Guo et al.,2011)
印度 EEP	159.10	57.25			0.28	1.39	(Laskar et al.,2010)
印度 WEP	269.10	25.50			0.39		(Laskar et al.,2010)
葡萄牙 EEP	62.7	—	17.8	13.0	0.80	—	(Miguel et al.,2010)
葡萄牙 WEP	17.5	—	5.9	0.23	—		(Miguel et al.,2010)

注:AAI=DPPH 终浓度(μg/mL)/IC_{50}(μg/mL);"—":未检测。

1.4 蜂胶其他溶剂提取物的抗氧化活性

澳大利亚蜂胶因含有丰富的烯化二苯乙烯类而具有很强的抗氧化活性,被认为是一种独特类型的蜂胶。Abu-Mellal 等(2012)用乙酸乙酯提取澳大利亚蜂胶,获得的乙酸乙酯蜂胶提取物对自由基的清除能力要比巴西蜂胶乙醇提取物更强,并从此类蜂胶中分离出了 6 种二苯乙烯类和 2 种黄酮类物质。Yang 等(2011)分别用氯仿、乙酸乙酯和正丁醇提取中国

蜂胶,研究结果表明,乙酸乙酯提取的蜂胶的自由基清除能力和还原力最强,正丁醇次之,而氯仿与乙醇相似。

2 几种单体成分的抗氧化活性

2.1 阿替匹林 C(Artepillin C)

巴西酒神菊属型蜂胶因富含酚酸类物质,使其生物学活性非常广泛,如抗炎、抗氧化、抗癌、抑制肝细胞 DNA 损伤等(AzevedoBentesMonteiroNeto et al.,2011)。阿替匹林 C(Artepillin C)作为巴西绿蜂胶特有的标志性成分,一直是科研人员研究的热点。

阿替匹林 C 含有两个异戊二烯基和简单的苯酚结构(图 3.1),异戊二烯基使其很难与别的物质共轭成对,但对细胞膜却有很强的吸引力,可以部分地穿过细胞膜,这加强了阿替匹林 C 抑制细胞膜脂质过氧化的能力。进一步的研究表明,阿替匹林 C(20nmol/mL)可穿过肝细胞膜并停留在细胞核附近,保护 DNA 免受氧化损伤(Shimizu et al.,2004)。Kawashima 等(2012)研究发现,阿替匹林 C 对 DPPH 自由基的清除可能是通过一步氢原子转移法将氢原子从酚羟基基团上转移到 DPPH 的途径,并不是通过伴随质子的转移而转移电子的途径。这些研究为进一步探究阿替匹林 C 和含阿替匹林 C 的巴西酒神菊属型蜂胶的抗氧化机制提供了重要参考。

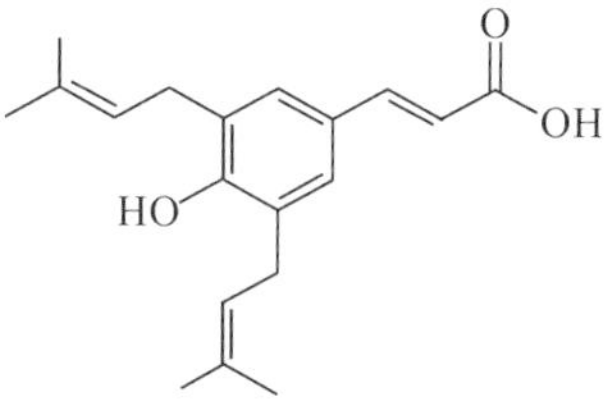

图 3.1 阿替匹林 C 的化学结构

Fig. 3.1 The chemical structure of Artepillin C

2.2 咖啡酸苯乙酯(CAPE)

咖啡酸苯乙酯(caffeic acid phenethyl ester,CAPE)是杨树型蜂胶中的代表性成分,因其具有良好的生物学活性而被广泛研究。Wang 等(2010)认为 CAPE 中邻苯二酚环上的 3,4-二羟基基团(图 3.2)是 CAPE 具有抗氧化活性的结构基础。然而,仅抗氧化活性不能保证 CAPE 对细胞的保护作用。构效关系研究发现,CAPE 的邻苯二酚环上的一个羟基甲基化或替换方能发挥细胞保护活性,此二者之间可能存在协同关系。

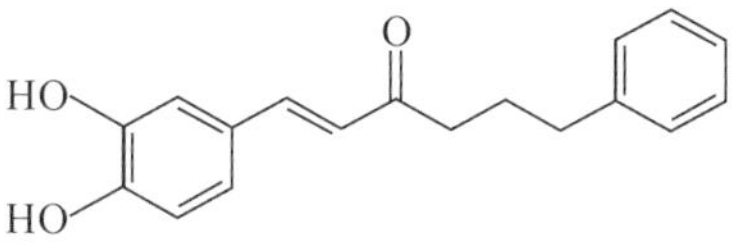

图 3.2 咖啡酸苯乙酯的化学结构式

Fig. 3.2 The chemical structure of caffeic acid phenethylester

Ozyurt 等(2007a)研究发现,利用 MK-801 诱导小鼠的精神分裂症会在附睾内产生明显的氧化应激,给药剂量为 10μmol/kg 的 CAPE 能通过降低总的超氧化物歧化酶(SOD)、丙

二醛(MDA)、一氧化氮(NO)、蛋白羰基等的含量显著降低氧化应激作用,减弱 MK-801 对附睾的伤害,使附睾维持正常的生理功能。CAPE 可通过影响相关酶的活性及含量,避免机体因局部缺血而带来损伤(Özeren et al.,2005;Ozyurt et al.,2007b)。Wang 等(2010)研究发现,血红素氧化酶的感应效应在 CAPE(20μM)对细胞的保护中起到关键作用,且效果要好于 CAPE 的抗氧化活性直接对细胞的保护效果。因此,蜂胶的抗氧化活性不仅仅是依靠蜂胶直接对 ROS 的调节而发挥作用的。

2.3 蜂胶中新发现化合物的抗氧化活性

蜂胶中新物质的发现与结构鉴定一直是蜂胶相关研究中的重要内容,近几年也有研究报道了蜂胶中新发现化合物的抗氧化活性。Righi 等(2011)研究表明,巴西红蜂胶具有很强的抗氧化活性,并鉴定发现了其中 2 种新化合物,包括含量较低的 narigenin-8-C-hexoside 和含量较高的 $3,4,2',3'$-tetrahydroxychalcone,并且后者因含有苯环间的双键而具有良好的抗氧化活性。Petrova 等(2010)从肯尼亚蜂胶中分离出了 2 种新的天然化合物(tetrahydrojusticidin 和 6-methoxydiphyllin)及 4 种酚醛树脂物质,其中 geranylflavonmacarangin 表现出较好的 DPPH 自由基清除能力。Sulaiman 等(2011)在研究伊拉克蜂胶的化学成分及抗氧化活性时,也发现了一种未知的化合物,其分子式为 $C_{18}H_{32}O_4$。

3 蜂胶抗氧化活性的作用机制

关于抗氧化作用的机制,总的来说,主要是通过直接或间接地调节机体活性氧(ROS)的途径来调节体内氧化-还原平衡,从而发挥抗氧化活性。蜂胶含有丰富的黄酮、酚酸及萜烯类物质,具有良好的抗氧化效果。另一方面,蜂胶的保护作用呈浓度依赖性,而浓度过高会有一定的毒性,且不同地理来源的蜂胶毒性也有差异。因此,蜂胶的给药浓度应选择在安全水平。有报道指出,蜂胶可调节 ROS 的含量。Xuan 等(2011)研究发现,低浓度($12.5\mu g/mL$)的巴西蜂胶可抑制 ROS 的水平,而高浓度($25\mu g/mL$、$50\mu g/mL$)的巴西蜂胶则会提升 ROS 水平。同时,蜂胶能通过不同途径有效调节多种酶的活性及其含量,体外化学分析实验也表明蜂胶具有良好的清除自由基的效果。其可能存在的作用机制如图 3.3 所示。

3.1 影响 ROS 自由基的通路

机体内的细胞呼吸会产生很多 ROS 自由基,如线粒体中的 ROS。ROS 能激活细胞内一系列的信号转导过程,蜂胶可能通过供氢或电子直接消除机体产生的 ROS。蜂胶中的生物活性成分大多都含有酚式羟基。研究表明,分子结构和反应介质是影响抗氧化机制的两个重要因素,含有酚式羟基的抗氧化剂可通过顺序质子损失电子转移(sequential proton loss electron transfer,SPLET)和氢原子转移(hydrogen atom transfer,HAT)两种途径发挥抗氧化作用(Qian et al.,2011)。阿替匹林 C 和 CAPE 具有独特的化学结构,并能通过供氢或者是传递电子抑制 ROS 自由基等的生成,这也与上述机制一致。因此,蜂胶可能是通过 SPLET 和 HAT 两条途径发挥抗氧化活性。

3.2 影响细胞内抗氧化酶系统相关信号通路转导

生物为保持机体内环境稳定,自身拥有完整的抗氧化酶体系(表 3.3),使机体保持氧化-还原系统的平衡。

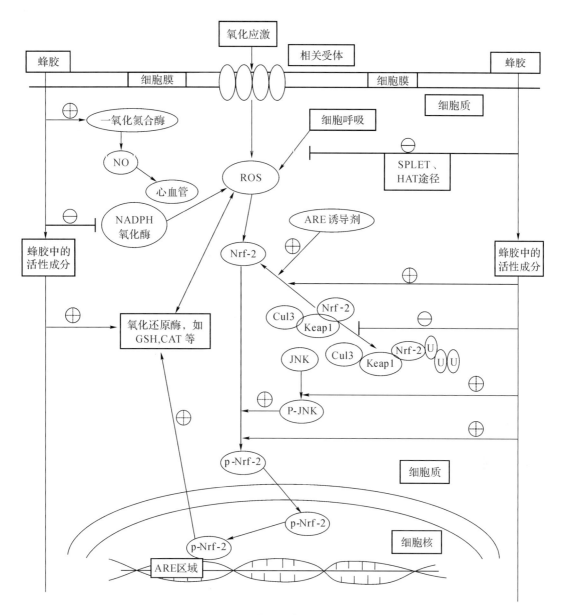

图 3.3 蜂胶抗氧化作用机制

Fig. 3.3 The mechanism of propolis antioxidant activity

表 3.3 主要的 ROS 分子及其代谢（Nordberg and Arnér，2001）

Table 3.3 The main ROS molecules and their metabolism

活性氧分子	主要来源	酶防御系统	产物
过氧化物（O_2^-）	电子从电子传递链泄露、激活的氧化酶等	超氧化物歧化酶、超氧化物还原酶	$H_2O_2 + O_2$ H_2O_2

续表

活性氧分子	主要来源	酶防御系统	产物
过氧化氢(H_2O_2)	来自于 O_2^- 通过超氧化物歧化酶 NADPH 氧化酶葡萄糖氧化酶等	谷胱甘肽过氧化物酶、过氧化氢酶、抗氧化蛋白	$H_2O+GSSG$ H_2O+O_2 H_2O
羟基(OH^-)	来自于 O_2^- 和 H_2O_2		
一氧化氮(NO)	一氧化氮合酶	谷胱甘肽/硫氧还蛋白还原酶	GSNO

Kamiya 等(2012)用氯化镉诱导 COS7 细胞,发现 $10\mu g/mL$ 的中国蜂胶、$5\mu g/mL$ 的巴西红蜂胶以及 $20\mu g/mL$ 的巴西绿蜂胶均能直接提高 SOD 的活性,还能通过信号转导途径提高细胞血红素氧化酶的表达而发挥抗氧化活性。Levites 等(2001)也研究发现,多酚类物质能提高抗氧化酶(如 SOD、CAT)的活性。Silva 等(2011)发现乌拉圭蜂胶提取物可以抑制内皮的 NADPH 氧化酶的活性,并降低 Nox4 基因的表达而发挥抗氧化活性,还可诱导心脏内皮细胞 NOS 的表达,提高 NO 在心脏内皮细胞的利用率而维持心血管的正常生理功能,而蜂胶亦可抑制炎症过程中 NO 的产生(Wang et al.,2013)。Zhu 等(2011)利用链佐星(STZ)诱导的小鼠肝肾损伤模型,对比研究了中国蜂胶和巴西蜂胶的生物学活性,发现在蜂胶给药剂量均为 $100mg/kg$ 条件下,中国蜂胶和巴西蜂胶均可提高血清中 SOD 的含量,而巴西蜂胶还可降低 MDA 和 NOS 的含量,进而抑制自由基对肝肾细胞的损伤。因此,蜂胶除了直接调节 ROS 的活性及含量外,还可能通过信号转导通路调节相关酶的活性及含量从而维持氧化-还原系统的平衡。

3.3 影响 Nrf-2 调节因子的信号转导过程

蜂胶还可能通过调节相关转录因子而控制氧化应激相关基因的表达,进而调控氧化-还原酶系统。核因子 E2 相关蛋白 2(Nrf-2)就是其中一种关键因子。Yeh 等(2009)研究发现,酚酸可调节心脏中抗氧化酶的表达,并伴随 Nrf-2 含量的升高。Nrf-2 是一种氧化还原敏感因子,在细胞质中,虽然 Nrf-2 不断生成,但是 Nrf-2 能与 Kelch-like 电解珩磨相关蛋白(Keap1)、E3 复合连接酶(Cul3)形成复合物,Keap1 和 Cul3 进一步诱导 Nrf-2 的泛素化,从而使 Nrf-2 一直保持低浓度。同时,Nrf-2 能感受氧化应激信号,激活后进入细胞核,与 DNA 上的抗氧化反应元件(DNA ARE-region)结合,诱导细胞保护酶(如 GST、SOD 等)相关基因的表达。Nrf-2 的信号转导通路受很多因子的调节,ARE 诱导剂(如槲皮黄酮)能提高 Nrf-2 在细胞质中的含量,并促使活化后的 Nrf-2 转运到细胞核与 ARE 区域结合,而 DNA 的 ARE 区域的基因可调控细胞保护酶,导致相关基因上调(Eggler et al.,2008)。López-Alarcón 等(2013)也研究发现,抗氧化剂可促使 Nrf-2/Keap1 分解出 Nrf-2,减少 Nrf-2 的泛素化,并加快 Nrf-2 从细胞质转入细胞核的过程。Vari 等(2011)研究发现,c-jun 氨基末端激酶(JNK)作为调节因子可促进 Nrf-2 的磷酸化,进一步促进 p-Nrf-2 转移进入细胞核。现已发现此通路可提高谷胱甘肽还原酶、谷胱甘肽过氧化物酶等的含量。有关研究还表明,NF-κB 和 AP-1 都是重要的调控氧化还原反应基因表达的重要介质,也均可提高 GPx 和 CAT 的含量(Zhou et al.,2001)。因此,蜂胶亦可能通过调节 NF-κB 和 AP-1 的相关信号通路,进而调节 GPx 和 CAT 等的含量。

小　结

蜂胶对机体的保护作用机制复杂,对整个机体起到协调的作用,而不是单纯的对某种物质或信号通路的抑制或促进作用。研究结果表明,不同地理来源、不同提取方法获得的蜂胶虽然成分有差异,但都表现出良好的抗氧化活性,且蜂胶对多种疾病有良好的治疗效果。这与蜂胶广泛的生物活性密切相关,而蜂胶其他生物学活性的发挥与蜂胶的抗氧化活性密不可分。对蜂胶中的主要单体成分的抗氧化活性及其作用机制的研究,在一定程度上也证实蜂胶的其他生物活性的发挥依赖于蜂胶的抗氧化活性。

参考文献

Abu-Mellal A, Koolaji N, Duke RK, Tran VH, Duke CC (2012) Prenylated cinnamate and stilbenes from Kangaroo Island propolis and their antioxidant activity[J]. *Phytochemistry*, 77:251-259.

Attia AA, ElMazoudy RH, El-Shenawy NS (2012) Antioxidant role of propolis extract against oxidative damage of testicular tissue induced by insecticide chlorpyrifos in rats [J]. *Pesticide Biochemistry and Physiology*, 103(2):87-93.

Azevedo Bentes Monteiro Neto M, Lima S, Mota I, Furtado RA, Bastos JK, Silva Filho AA, Tavares DC (2011) Antigenotoxicity of Artepillin C in vivo evaluated by the micronucleus and comet assays[J]. *Journal of Applied Toxicology*, 31(8):714-719.

Banskota AH, Tezuka Y, Adnyana IK, Midorikawa K, Matsushige K, Message D, Huertas AA, Kadota S (2000) Cytotoxic, hepatoprotective and free radical scavenging effects of propolis from Brazil, Peru, the Netherlands and China[J]. *Journal of Ethnopharmacology*, 72(1):239-246.

Capucho C, Sette R, de Souza Predes F, de Castro Monteiro J, Pigoso AA, Barbieri R, Dolder MAH, Severi-Aguiar GD(2012) Green Brazilian propolis effects on sperm count and epididymis morphology and oxidative stress[J]. *Food and Chemical Toxicology*, 50(11):3956-3962.

Cigut T, Polak T, Gasperlin L, Raspor P, Jamnik P(2011) Antioxidative activity of propolis extract in yeast cells[J]. *Journal of Agricultural and Food Chemistry*, 59(21):11449-11455.

Eggler AL, Gay KA, Mesecar AD (2008) Molecular mechanisms of natural products in chemoprevention: induction of cytoprotective enzymes by Nrf2[J]. *Molecular Nutrition & Food Research*, 52(S1):S84-S94.

Enis Yonar M, Mişe Yonar S, Silici S (2011) Protective effect of propolis against oxidative stress and immunosuppression induced by oxytetracycline in rainbow trout (*Oncorhynchus mykiss*, W.)[J]. *Fish & Shellfish Immunology*, 31(2):318-325.

Enis Yonar M, Yonar SM, Ural MŞ, Silici S, Düşükcan M (2012) Protective role of propolis in chlorpyrifos-induced changes in the haematological parameters and the oxidative/antioxidative status of *Cyprinus carpio carpio*[J]. *Food and Chemical Toxicology*, 50

(8):2703-2708.

Erdogan S,Ates B,Durmaz G,Yilmaz I,Seckin T (2011) Pressurized liquid extraction of phenolic compounds from Anatolia propolis and their radical scavenging capacities[J]. *Food and Chemical Toxicology*,49(7):1592-1597.

Fonseca YM,Marquele-Oliveira F,Vicentini FT,Furtado NAJ,Sousa JPB,Lucisano-Valim YM,Fonseca MJV (2010) Evaluation of the potential of Brazilian propolis against UV-induced oxidative stress[J]. *Evidence-Based Complementary and Alternative Medicine*, Volume 2011,Article ID 863917.

Gregoris E,Fabris S,Bertelle M,Grassato L,Stevanato R (2011) Propolis as potential cosmeceutical sunscreen agent for its combined photoprotective and antioxidant properties [J]. *International Journal of Pharmaceutics*,405(1):97-101.

Guo X,Chen B,Luo L,Zhang X,Dai X,Gong S (2011) Chemical compositions and antioxidant activities of water extracts of Chinese propolis[J]. *Journal of Agricultural and Food Chemistry*,59(23):12610-12616.

Kamiya T,Izumi M,Hara H,Adachi T (2012) Propolis suppresses CdCl 2-induced cytotoxicity of COS7 cells through the prevention of intracellular reactive oxygen species accumulation[J]. *Biological and Pharmaceutical Bulletin*,35(7):1126-1131.

Kawashima T,Manda S,Uto Y,Ohkubo K,Hori H,Matsumoto K-i,Fukuhara K,Ikota N, Fukuzumi S,Ozawa T (2012) Kinetics and mechanism for the scavenging reaction of the 2,2-diphenyl-1-picrylhydrazyl radical by synthetic artepillin canalogues [J]. *Bulletin of the Chemical Society of Japan*,85(8):877-883.

López-Alarcón C,Denicola A (2013) Evaluating the antioxidant capacity of natural products:a review on chemical and cellular-based assays[J]. *Analytica Chimica Acta*,763:1-10.

Laskar RA,Sk I,Roy N,Begum NA (2010) Antioxidant activity of Indian propolis and its chemical constituents[J]. *Food Chemistry*,122(1):233-237.

Levites Y,Weinreb O,Maor G,Youdim MB,Mandel S (2001) Green tea polyphenol (-)-epigallocatechin-3-gallate prevents N-methyl-4-phenyl-1,2,3,6-tetrahydropyridine-induced dopaminergic neurodegeneration[J]. *Journal of Neurochemistry*,78(5): 1073-1082.

Mavri A,Abramovic H,Polak T,Bertoncelj J,Jamnik P,Smole Možina S,Jeršek B (2012) Chemical properties and antioxidant and antimicrobial activities of Slovenian propolis [J]. *Chemistry & Biodiversity*,9(8):1545-1558.

Miguel MG,Nunes S,Dandlen SA,Cavaco AM,Antunes MD (2010) Phenols and antioxidant activity of hydro-alcoholic extracts of propolis from Algarve,South of Portugal[J]. *Food and Chemical Toxicology*,48(12):3418-3423.

Moţ AC,Silaghi-Dumitrescu R,Sârbu C (2011) Rapid and effective evaluation of the antioxidant capacity of propolis extracts using DPPH bleaching kinetic profiles,FT-IR and UV-vis spectroscopic data[J]. *Journal of Food Composition and Analysis*,24(4): 516-522.

Moura SALd, Negri G, Salatino A, Lima LDdC, Dourado LPA, Mendes JB, Andrade SP, Ferreira MAND, Cara DC (2011) Aqueous extract of Brazilian green propolis: primary components, evaluation of inflammation and wound healing by using subcutaneous implanted sponges[J]. *Evidence-Based Complementary and Alternative Medicine*, Volume 2011, Article ID 748283.

Nakajima Y, Shimazawa M, Mishima S, Hara H (2007) Water extract of propolis and its main constituents, caffeoylquinic acid derivatives, exert neuroprotective effects *via* antioxidant actions[J]. *Life Sciences*, 80(4): 370-377.

Nordberg J, Arnér ES (2001) Reactive oxygen species, antioxidants, and the mammalian thioredoxin system[J]. *Free Radical Biology and Medicine*, 31(11): 1287-1312.

Oldoni TLC, Cabral IS, d'Arce MA, Rosalen PL, Ikegaki M, Nascimento AM, Alencar SM (2011) Isolation and analysis of bioactive isoflavonoids and chalcone from a new type of Brazilian propolis[J]. *Separation and Purification Technology*, 77(2): 208-213.

Özeren M, Sucu N, Tamer L, Aytacoglu B, Bayrı Ö, Döndaş A, Ayaz L, Dikmengil M (2005) Caffeic acid phenethyl ester (CAPE) supplemented St. Thomas' hospital cardioplegic solution improves the antioxidant defense system of rat myocardium during ischemia-reperfusion injury[J]. *Pharmacological Research*, 52(3): 258-263.

Ozyurt B, Parlaktas BS, Ozyurt H, Aslan H, Ekici F, Atis Ö (2007a) A preliminary study of the levels of testis oxidative stress parameters after MK-801-induced experimental psychosis model: Protective effects of CAPE[J]. *Toxicology*, 230(1): 83-89.

Ozyurt H, Ozyurt B, Koca K, Ozgocmen S (2007b) Caffeic acid phenethyl ester (CAPE) protects rat skeletal muscle against ischemia-reperfusion-induced oxidative stress[J]. *Vascular Pharmacology*, 47(2): 108-112.

Park YK, Ikegaki M (1998) Preparation of water and ethanolic extracts of propolis and evaluation of the preparations[J]. *Bioscience, Biotechnology and Biochemistry*, 62(11): 2230-2232.

Petrova A, Popova M, Kuzmanova C, Tsvetkova I, Naydenski H, Muli E, Bankova V (2010) New biologically active compounds from Kenyan propolis[J]. *Fitoterapia*, 81(6): 509-514.

Potkonjak NI, Veselinovic DS, Novakovic MM, Gorjanovic SŽ, Pezo LL, Sužnjevic DŽ (2012) Antioxidant activity of propolis extracts from Serbia: A polarographic approach[J]. *Food and Chemical Toxicology*, 50(10): 3614-3618.

Qian YP, Shang YJ, Teng QF, Chang J, Fan GJ, Wei X, Li RR, Li HP, Yao XJ, Dai F (2011) Hydroxychalcones as potent antioxidants: Structure-activity relationship analysis and mechanism considerations[J]. *Food Chemistry*, 126(1): 241-248.

Righi AA, Alves TR, Negri G, Marques LM, Breyer H, Salatino A (2011) Brazilian red propolis: unreported substances, antioxidant and antimicrobial activities[J]. *Journal of the Science of Food and Agriculture*, 91(13): 2363-2370.

Saito Y, Tsuruma K, Ichihara K, Shimazawa M, Hara H (2015) Brazilian green propolis

water extract up-regulates the early expression level of HO-1 and accelerates Nrf2 after UVA irradiation[J]. *BMC Complementary and Alternative Medicine*, 15:421.

Shimizu K, Ashida H, Matsuura Y, Kanazawa K (2004) Antioxidative bioavailability of artepillin C in Brazilian propolis[J]. *Archives of Biochemistry and Biophysics*, 424(2):181-188.

Silva V, Genta G, Möller MaN, Masner M, Thomson L, Romero N, Radi R, Fernandes DC, Laurindo FR, Heinzen H (2011) Antioxidant activity of Uruguayan propolis. *In vitro* and cellular assays[J]. *Journal of Agricultural and Food Chemistry*, 59(12):6430-6437.

Simões-Ambrosio L, Gregório L, Sousa J, Figueiredo-Rinhel A, Azzolini A, Bastos J, Lucisano-Valim Y (2010) The role of seasonality on the inhibitory effect of Brazilian green propolis on the oxidative metabolism of neutrophils[J]. *Fitoterapia*, 81(8): 1102-1108.

Sulaiman GM, Sammarrae KWA, Ad'hiah AH, Zucchetti M, Frapolli R, Bello E, Erba E, D'Incalci M, Bagnati R (2011) Chemical characterization of Iraqi propolis samples and assessing their antioxidant potentials[J]. *Food and Chemical Toxicology*, 49(9):2415-2421.

Thirugnanasampandan R, Raveendran SB, Jayakumar R (2012) Analysis of chemical composition and bioactive property evaluation of Indian propolis[J]. *Asian Pacific Journal of Tropical Biomedicine*, 2(8):651-654.

Valente MJ, Baltazar AF, Henrique R, Estevinho L, Carvalho M (2011) Biological activities of Portuguese propolis: Protection against free radical-induced erythrocyte damage and inhibition of human renal cancer cell growth *in vitro*[J]. *Food and Chemical Toxicology*, 49(1):86-92.

Varì R, D'Archivio M, Filesi C, Carotenuto S, Scazzocchio B, Santangelo C, Giovannini C, Masella R (2011) Protocatechuic acid induces antioxidant/detoxifying enzyme expression through JNK-mediated Nrf2 activation in murine macrophages[J]. *The Journal of Nutritional Biochemistry*, 22(5):409-417.

Wang K, Ping S, Huang S, Hu L, Xuan H, Zhang C, Hu F (2013) Molecular mechanisms underlying the in vitro anti-inflammatory effects of a flavonoid-rich ethanol extract from Chinese propolis (poplar type)[J]. *Evidence-Based Complementary and Alternative Medicine*, Volume 2013, Article ID 127672.

Wang X, Stavchansky S, Kerwin SM, Bowman PD (2010) Structure-activity relationships in the cytoprotective effect of caffeic acid phenethyl ester (CAPE) and fluorinated derivatives: Effects on heme oxygenase-1 induction and antioxidant activities[J]. *European Journal of Pharmacology*, 635(1):16-22.

Xuan H, Zhao J, Miao J, Li Y, Chu Y, Hu F (2011) Effect of Brazilian propolis on human umbilical vein endothelial cell apoptosis[J]. *Food and Chemical Toxicology*, 49(1):78-85.

Yang H, Dong Y, Du H, Shi H, Peng Y, Li X (2011) Antioxidant compounds from propolis collected in Anhui, China[J]. *Molecules*, 16(4):3444-3455.

Yeh CT, Ching LC, Yen GC (2009) Inducing gene expression of cardiac antioxidant enzymes by dietary phenolic acids in rats[J]. *The Journal of Nutritional Biochemistry*, 20(3):163-171.

Zhang JL, Shen XG, Wang K, Cao XP, Zhang CP, Zheng HQ, Hu, FL (2016) Antioxidant activities and molecular mechanisms of the ethanol extracts of baccharis propolis and eucalyptus propolis in raw64.7 cells[J]. *Pharmaceutical Biology*, 54(10):2220-2235.

Zhou LZH, Johnson AP, Rando TA (2001) NFκB and AP-1 mediate transcriptional responses to oxidative stress in skeletal muscle cells[J]. *Free Radical Biology and Medicine*, 31(11):1405-1416.

Zhu W, Li YH, Chen ML, Hu FL (2011) Protective effects of Chinese and Brazilian propolis treatment against hepatorenal lesion in diabetic rats[J]. *Human & Experimental Toxicology*, 30(9):1246-1255.

第三节 蜂胶的抗炎作用

蜂胶具有良好的抗炎作用，这也是蜂胶最早受到关注的药理活性之一。早在公元前300多年古希腊科学家亚里士多德在其所著的《动物志》中就记载了蜂巢中一种具有刺激性气味的"黑蜡"（蜂胶）能治疗皮肤病、化脓和刀伤（胡福良，2005）。在19世纪末南非布尔战争中，蜂胶已被用于外伤治疗（Ghisalberti,1979；Sforcin,2007）。10多年来，针对蜂胶及其主要生物活性成分抗炎效果的研究，特别是对蜂胶抗炎作用机制方面的研究受到人们广泛的关注。下面针对不同种类蜂胶及其单体活性成分所具有的抗炎症活性及其可能的分子机制进行综述，为蜂胶产品的开发利用提供依据，同时也为蜂胶的标准化及质量控制提供参考。

1 蜂胶的抗炎症活性

炎症（Inflammation）是机体组织受外界有害刺激时（如病原体、受损细胞或其他刺激物等）所产生的一种保护性应答反应（Coussens and Werb,2002），同时局部组织会发生变质、渗出和增生等病理性变化。蜂胶是一种具有良好抗炎效果的天然产物，国内外的学者针对不同地理来源、采用不同提取方式对蜂胶及蜂胶中的主要单体成分的抗炎活性进行了广泛研究。

1.1 蜂胶提取物的抗炎症活性

研究人员利用相关急慢性炎症模型研究了蜂胶提取物的抗炎效果，所采用的炎症模型主要包括角叉菜胶致大鼠足肿胀模型、醋酸致小鼠腹腔毛细血管通透性增高模型、角叉菜胶致大鼠胸膜炎模型及油酸加LPS致大鼠急性肺损伤模型、小鼠棉球肉芽肿慢性炎症模型、完全弗式佐剂致大鼠免疫性炎症模型等。这些研究也都证实了蜂胶针对急性炎症（胡福良等，2003；胡福良 等,2007）、慢性炎症（Missima and Sforcin,2008；Orsatti and Sforcin,2012）都具有良好的抑制效果，推测了蜂胶中的一些活性成分能有效抑制环氧合酶的活性，从而减少了前列腺素E2的生成；同时也有研究发现蜂胶中的某些物质是脂氧合酶抑制剂，因而抑制了白细胞三烯的生成，缓解炎症的发生。蜂胶提取物抗炎相关研究见表3.4。

表 3.4 不同来源蜂胶提取物的抗炎症效果

Table 3.4 Anti-inflammatory effects of propolis extracts from different sources

研究材料	蜂胶来源	生物学活性	实验模型	抗炎效果	给药途径	参考文献
WEP	波兰	抗炎、抗微生物	大鼠关节炎,足跖肿胀模型;	抑制急性、慢性炎症	口服	(Dobrowolski et al.,1991)
WEP	不明	抗炎	大鼠足跖肿胀(角叉菜胶),佐剂性关节炎	减少给药组 PGE2、白三烯、组胺的产生	口服	(Khayyal et al.,1993)
EEP	英国	抗炎	鼠巨噬细胞;急性炎症	减少 PGE2 的生成;影响花生四烯酸的代谢	体外实验及离体实验	(Mirzoeva & Calder 1996)
EEP	韩国	抗炎	大鼠佐剂性关节炎模型;足跖肿胀(角叉菜胶)	止痛;减少肿胀程度;减少关节炎指数	口服	(Park & Kahng 1999)
HAEP	巴西	抗炎、免疫调节	干扰素激活巨噬细胞模型;	抑制 NO 的产生,减少过氧化氢的产生	体外实验	(Orsi et al.,2000)
EEP	韩国	抗炎	LPS 加 IFN-γ 诱导巨噬细胞模型;	抑制 iNOS 酶活及 mRNA 水平;抑制 NF-κB 活力	体外实验	(Song et al.,2002)
EEP 和 WEP	中国	抗炎	角叉菜胶诱导足跖肿胀、胸膜炎;弗氏佐剂关节炎;	降低 PGE2 及 NO 水平;增加 IL-6 水平;对 IL-2 及 IFN-γ 无影响	口服	(Hu et al.,2005)
EEP	巴西	抗炎	角叉菜胶诱导足跖肿胀	影响 L-精氨酸代谢;抑制 NO 生成	静脉注射	(Tan-No et al.,2006)
EEP	巴西	抗氧化、免疫调节	BALB/c 小鼠应激导致免疫抑制,组织病理学分析免疫器官,测定炎症因子生成	胸腺、骨髓、肾上腺无改变;但脾脏生发中心有增生	口服	(Missima & Sforcin 2008)
WEP	巴西	抗炎、抗血管新生	小鼠海绵植入模型;	VEGF 含量有所减少;调节细胞因子(TGF-β,TNF-α)产生量	口服	(de Moura et al.,2011)
WEP	巴西	抗炎、创伤修复	Swiss 小鼠皮下植入海绵诱导创伤模型;	减少胶原沉积;缓解炎症症状	口服	(Moura et al.,2011)
EEP	波兰	抑制胃炎	幽门螺旋杆菌致慢性胃炎	抑制幽门螺旋杆菌生长,减少 IL-8 生成	体外实验	(Skiba et al.,2011)
HAE MEP,WEP	葡萄牙	抗炎、抗微生物	体外抗菌实验;透明质酸酶活力	HAE 最有效,抑制透明质酸酶活力	体外实验	(Silva et al.,2012)
EEP	巴西	免疫调节,抗炎	胶原诱导关节炎小鼠	产生 IL-17 的细胞数量减少;抑制 Th17 细胞分化	体外实验	(Tanaka et al.,2012)

续表

研究材料	蜂胶来源	生物学活性	实验模型	抗炎效果	给药途径	参考文献
WEP, EEP	中国	抗炎	角叉菜胶以建立急性胸膜炎 Wistar 大鼠模型	对抗胸膜炎大鼠胸水增多,显著减少炎症症状和减少 NO 和 PGE2 的产生,	灌胃	(胡福良 等. 2007)
WEP, EEP	中国	抗炎、抗肿瘤	Wistar 大鼠急性关节炎模型和大鼠急性胸膜炎模型	抑制炎症病理过程中的渗出、肿胀,抑制白细胞聚集、增多;WEP/EEP 抗炎效果差别不大	灌胃	(胡福良 等. 2005a, 2005b)
WEP, EEP	中国	抗炎	ICR 小鼠毛细血管通透性、Wistar 大鼠急性关节炎、大鼠急性胸膜炎模型	减少 PGE2、NO 的生成,缓解炎症症状	灌胃	(胡福良 等. 2003)

缩略词(Abbreviations):WEP:Water extract of propolis,蜂胶水提物;EEP:Ethanol extract of propolis,蜂胶乙醇提取物,MEP:Methanol extracts of propolis,蜂胶甲醇提取物;HAEP:Hydroalcoholic extract of propolis,蜂胶水醇提取物。

1.1.1 蜂胶醇提物(EEP)的抗炎活性

相对于水提蜂胶而言,蜂胶中多数有效成分可以被醇提取,乙醇萃取法也是蜂胶各种提取方式中最传统、最经典的一种。Park 和 Kahng(1999)利用小鼠佐剂性关节炎模型、角叉菜胶诱导足跖肿胀模型研究 EEP 对急慢性炎症的影响,结果发现口服 EEP 剂量为 50mg/kg/d 和 100mg/kg/d 能显著降低关节炎指数。同时一次性口服剂量为 200mg/kg 小鼠能显著减轻由角叉菜胶诱导的足跖肿胀程度。他们推测蜂胶抗炎活性可能与抑制前列腺素的产生有关。

EEP 能对非特异性免疫反应产生影响,其主要的影响途径是通过激活巨噬细胞实现的。Orsi 等(2000)的研究发现 EEP 能减少炎症过程中的 H_2O_2、NO 的产生量,且存在剂量效应。Missima 和 Sforcin(2008)通过对应激状态下小鼠的巨噬细胞功能及相关免疫器官的组织病理学研究发现,巴西绿蜂胶 EEP 也能显著影响应激状态下小鼠巨噬细胞过氧化氢的产生量,并抑制 NO 的生成。这主要是因为蜂胶能通过影响 NF-κB 信号通路从而抑制 iNOS 基因的转录,进而减少 iNOS 产生及抑制 iNOS 催化合成 NO 的能力(Song et al.,2002;Tan-No et al.,2006)。

1.1.2 蜂胶水提物(WEP)的抗炎活性

Moura 等(2011)研究了巴西蜂胶水提物(WEP)对小鼠慢性炎症的影响,结果表明 WEP 抑制了细胞移行,同时胶原沉积并没有受到影响,这也说明 WEP 能在不影响组织器官修复的前提下抑制慢性炎症的发生。他们认为 WEP 中具有抗炎、促进伤口愈合作用的主要活性成分为咖啡奎宁酸及苯丙酯类物质。其他一些针对 WEP 的研究也发现其具有抗血小板凝集、抑制前列腺素合成、神经保护、免疫调节等作用,并能通过抑制 5-脂氧合酶(5-lipoxygenase,5-LOX)的活性发挥抗炎症效果(Khayyal et al.,1993;Nakajima et al.,

2007;Massaro et al. ,2011)。

1.1.3 蜂胶水提物、醇提物抗炎效果的比较研究

胡福良等通过研究 EEP 和 WEP 对小鼠急性炎症的影响,发现无论是 EEP 还是 WEP 对急性炎症均具有明显的抑制作用,特别是 EEP 和 WEP 均对急性炎症渗出液中 NO 及溶菌酶含量的升高表现出强烈地抑制作用(胡福良 等,2003;胡福良 等,2005a)。后续研究还证实蜂胶提取物对慢性炎症也具有显著的抑制作用。对相关细胞因子水平的变化研究发现,蜂胶提取物主要作用于 B 淋巴细胞系统。但 WEP 和 EEP 抗炎作用机制并不相同,EEP 发挥抗炎作用主要是因为醇提蜂胶中富含醇溶性的黄酮类化合物,而 WEP 中主要是一些水溶性酚酸类化合物的作用。但就抗炎效果而言,WEP 的抗炎效果与 EEP 是基本类似的(胡福良 等,2005b;胡福良 等,2007)。

1.1.4 蜂胶提取物对炎症因子、细胞因子产生的影响

在对蜂胶抗炎作用研究过程中,蜂胶提取物不仅能抑制 NO 的产生、减少 PGE2 的生成(Mirzoeva and Calder,1996;Hu et al. ,2005),同时也通过对相关细胞因子的影响发挥着免疫调节作用。Hu 等(2005)研究发现 EEP 和 WEP 能显著抑制小鼠炎症部位 IL-6 的上升,但对 IL-2 及 IFN-γ 的水平无明显影响。而 Sa-Nunes 等(2003)研究发现,蜂胶能减少由刀豆蛋白 A(concanavalin A,Con A)刺激造成的脾细胞增生,同时增加 IFN-γ 的生成。Sforcin(2007)认为蜂胶抑制淋巴细胞增生的主要原因是蜂胶影响了一些调节性细胞因子,(如 TGF-β、IL-10 等)的产生;而蜂胶的抗炎症、抗血管新生活性主要是通过影响 $TGF-\beta_1$ 而实现的(de Moura et al. ,2011)。研究还发现蜂胶能显著抑制病理状态下免疫细胞 IL-8、17 的产生(Skiba et al. ,2011;Tanaka et al. ,2012)。同时,Orsatti 等还发现,蜂胶对小鼠巨噬细胞的 Toll 样受体(Toll-like receptors,TLR)-2、4 的表达也有着一定的影响。特别是在应激状态下,蜂胶能通过下调小鼠 TLR2 和 TLR4 的 mRNA 表达水平从而发挥免疫调节作用(Orsatti et al. ,2010;Orsatti and Sforcin,2012)。而 TLR 在免疫系统,特别是天然免疫反应中发挥着重要作用,同时与细胞因子的产生、免疫细胞的激活也有着一定联系(Mills,2011;Steinhagen et al. ,2011)。Bueno-Silva 等(2016)研究了巴西红蜂胶对炎症反应中白细胞迁移的调控效果。体内试验结果表明,10mg/kg 巴西红蜂胶乙醇提取物能抑制中性白细胞向腹腔的迁移,同时降低其在肠系膜活体微循环中的水平,并抑制炎症细胞因子(TNF-alpha,IL-1 beta,CXCL1/KC 和 CXCL2/MIP-2)的释放;体外试验表明,巴西红蜂胶乙醇提取物能抑制 CXCL2/MIP-2 诱导的白细胞钙离子内流,但不影响 CXCR2 表达。

值得注意的是,由于蜂胶成分非常复杂,且不同种类、不同来源蜂胶中的有效活性成分差异很大,这些实验采用的炎症模型也不尽相同,这也导致这些研究的实验结果并不十分一致,特别是关于蜂胶抗炎症活性的分子机制方面目前为止国内外并没有一个统一的认识。未来的研究有必要在分析确定蜂胶生物活性成分的基础上,对蜂胶各种药理学活性作用成分单体进行进一步深入研究探讨,同时针对不同地区、不同植物来源的蜂胶作用机制进行研究,制定一个相对普遍的评价标准。

1.2 蜂胶中具有抗炎作用的有效生物活性成分

针对蜂胶中有效生物活性成分单体的报道近几年来呈上升趋势。越来越多的研究证实,蜂胶中丰富的黄酮类、酚酸类化合物是蜂胶具有多种生物学活性的主要原因。目

前相关研究报道蜂胶中具有良好抗炎效果的化合物主要包括:黄酮类化合物,如槲皮素(Quercetin)、柯因(Chrysin)、高良姜素(Galangin)、山奈酚(Kaempferol)等;酚酸类化合物,如咖啡酸(Caffeic acid)、阿魏酸(Ferulic acid)、肉桂酸(Cinnamic acid)等,同时还包括一些酚酸的衍生物如咖啡酸苯乙酯(CAPE)、阿替匹林C(Artepillin C)等。蜂胶醇提物和水提物中主要生物活性成分的化学结构分别如图3.4和3.5所示。

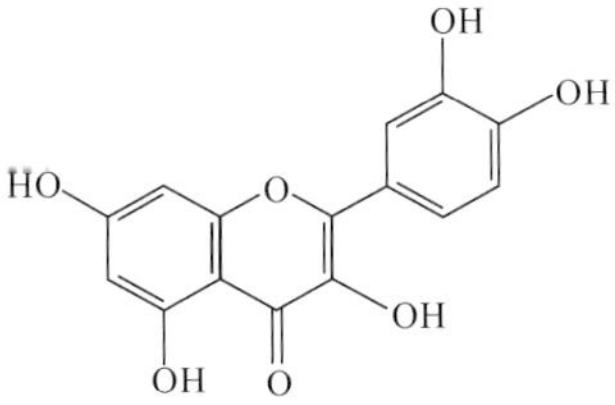

槲皮黄酮
Quercetin

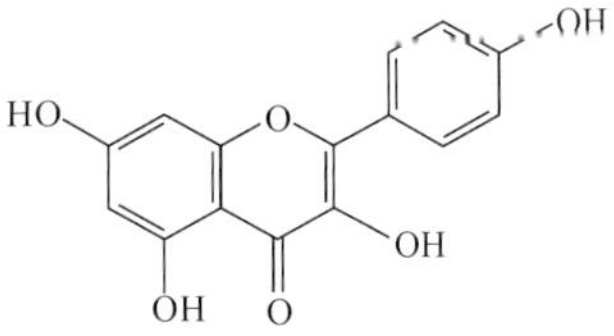

山奈酚
Kaempferol

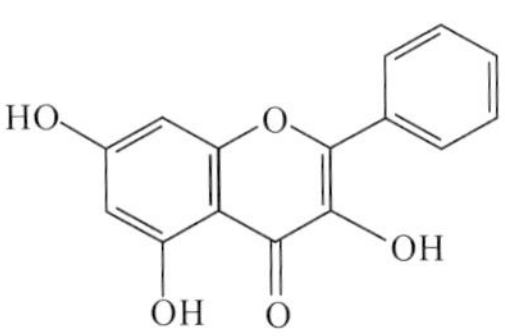

高良姜素
Galangin

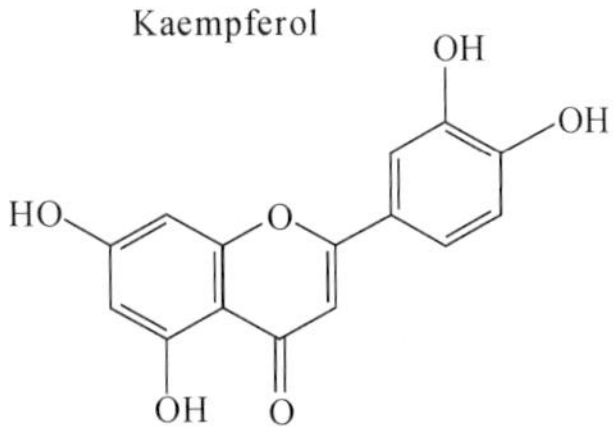

木犀草素
Luteolin

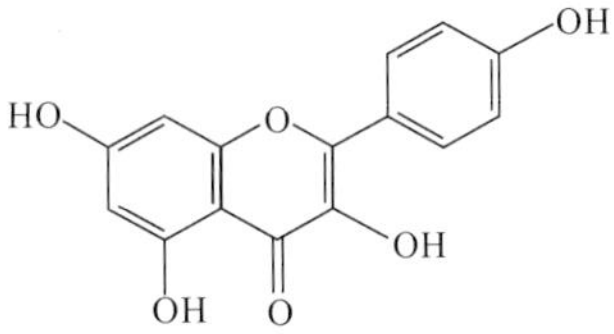

芹黄素
Apigenin

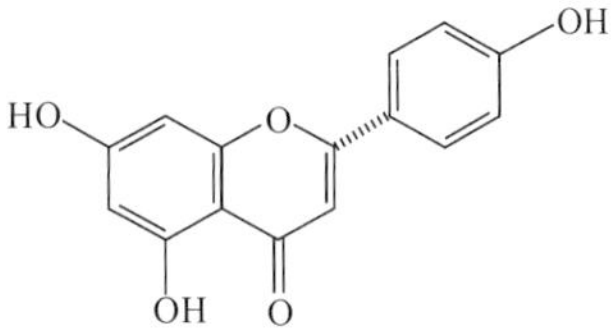

柚皮素
Naringenin

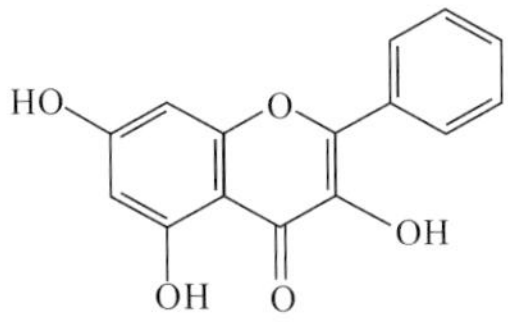

柯因
Chrysin

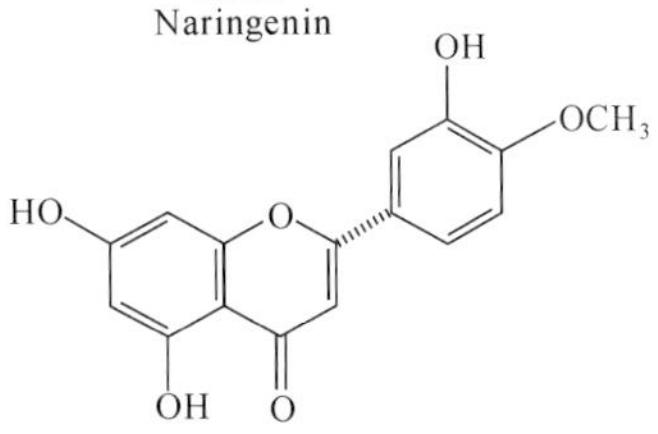

橙皮素
Hesperetin

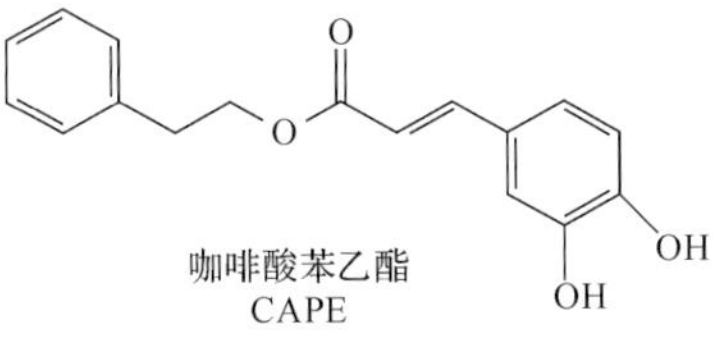

咖啡酸苯乙酯
CAPE

图3.4 蜂胶醇提物中主要生物活性成分的化学结构

Fig. 3.4 Main bioactive constituents in alcohol extract of propolis

	R1	R2	R3
Druoanin	prenyl	OH	prenyl
Baccharin	H	OH	prenyl
奎宁酸 Quinin acid	H	Hydrocinnamoyl	prenyl
咖啡酸 Caffeic acid	H	H	H
肉桂酸 Cinnamic acid	H	OH	OH
绿原酸 Chlorgenic acid	caffeoyl	H	H
3,4 二咖啡酰奎宁酸 3,4-dicaffeoyl quinic acid	caffeoyl	caffeoyl	H
3,5 二咖啡酰奎宁酸 3,5-dicaffeoyl quinic acid	caffeoyl	H	caffeoyl

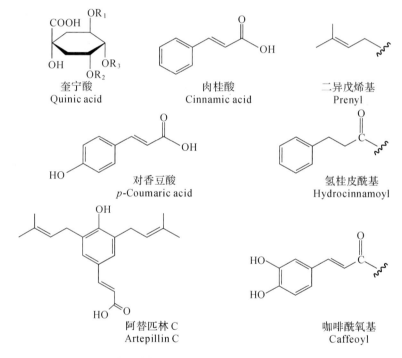

图 3.5　蜂胶水提物中主要生物活性成分的化学结构

Fig. 3.5　Main bioactive constituents in water extract of propolis

1.2.1　蜂胶中黄酮类化合物的抗炎作用

蜂胶中富含黄酮类化合物,其良好的抗炎症活性也与这些黄酮类化合物的种类、含量密切相关。Franchin 等(2016a,2016b)的研究结果表明,巴西红蜂胶中的类黄酮类化合物,如 Vestitol 和 Neovestitol 可能是红蜂胶中发挥抗炎效果的有效功能活性物质。Woo 等(2005)研究了柯因对脂多糖(LPS)刺激 RAW264.7 细胞株的影响,结果发现柯因能显著抑制 LPS 刺激下的 COX-2 的 mRNA 和蛋白表达水平,并通过突变体分析及凝胶电泳迁移率实验

(electrophoretic mobility shift assay,EMSA)结果表明柯因影响了 IL-6 核因子的转录。Raso 等(2001)在 J774A.1 细胞上的研究发现,槲皮素、高良姜素、芹黄素和柚皮素也能显著降低 COX-2 的表达与 PGE2 的生成。同时,蜂胶中的酚酸类化合物也有着较强的抗炎活性。动物实验证明咖啡酸能抑制花生四烯酸的生成,同时抑制 COX-2 的活力,且能抑制 COX-2 基因的表达(Borrelli et al.,2002)。阿魏酸对神经系统有着很好的保护作用,这主要是由于其具有良好的抗炎效果和自由基清除能力,同时也能抑制细胞间黏附因子(intercellular adhesion molecule-1,ICAM-1)的产生和 NF-κB 的转录(Cheng et al,,2008)。

其他一些研究针对蜂胶及其中的多酚类化合物(如黄酮类物质等)对 LPS 刺激下巨噬细胞 NO 和 iNOS 产生量的影响,发现多酚类化合物中起主要作用的主要是黄酮类物质,特别是黄酮和异黄酮。Hamalainen 等(2007)系统研究了 36 种黄酮类化合物的体外抗炎效果,发现其中的 8 种黄酮类化合物能有效抑制 LPS 刺激下细胞 NO 和 iNOS 的产生,且都能抑制 NF-κB 的活力。其中,染料木素、山奈酚、槲皮素、大豆苷元不仅能抑制 NF-κB,还能抑制 STAT-1(iNOS 另外一个重要的转录因子)的转录活性。Wang 等(2006)研究了黄酮醇类化合物抗氧化活性与抗炎症活性之间的构效关系,他们对几种黄酮醇类化合物的抗氧化及抗炎效果进行了研究,包括非瑟酮、山奈酚、桑色素、杨梅酮和槲皮黄酮,结果发现这些黄酮醇类化合物都有很好的抗氧化性能,对几种不同种类的自由基都有清除作用,且都有一定的抗炎功效,但它们的作用效果并非完全一致。Wang 和 Hamalainen 的研究都认为,尽管黄酮类化合物在结构上具有相似性,但由于取代基的种类和位置不同会导致其抗氧化和抗炎效果的显著差异。

1.2.2 咖啡酸苯乙酯(CAPE)的抗炎作用

咖啡酸苯乙酯(CAPE)是蜂胶中研究利用最早、最广泛的一种单体成分。它具有多种生物学活性,如抗肿瘤(Beltran-Ramirez et al.,2012;Lin,2012)、抗氧化(Aygun et al.,2012)、免疫调节(Yang,2011)、抗病毒(Shvarzbeyn and Huleihel,2011)、抗菌消炎(Gocer and Gulcin,2011;Saavedra et al.,2011)等。研究发现,CAPE 能刺激 T 细胞增殖,并抑制相关细胞因子的产生,从而发挥抗炎效果。相关研究也对 CAPE 发挥抗炎作用的分子机制进行了深入研究,发现 CAPE 本身就是一种强有力的 NF-κB 的抑制物,它能通过抑制 NF-κB 信号通路发挥抗炎效果,而 NF-κB 在转录水平调节着 COX-2 和 iNOS 基因的表达,并控制着许多细胞因子基因的转录,如 TNF、IL-1 等(Natarajan et al.,1996;Wang et al.,2009;Wang et al.,2010;Shvarzbeyn and Huleihel,2011)。

1.2.3 阿替匹林 C 的抗炎作用

与其他地区蜂胶富含黄酮类化合物不同,巴西绿蜂胶中主要活性成分主要是一些酚酸类化合物,特别是巴西绿蜂胶中富含阿替匹林 C(Artepillin C,3,5-二异戊烯基-4-羟基肉桂酸),它是巴西绿蜂胶中一种特有的活性成分,具有抗微生物、抗氧化、抗肿瘤、抗炎等多种药理活性(王凯 等,2013a)。Paulino 等(2008)采用体内试验(角叉菜胶致小鼠足跖肿胀模型、致小鼠腹膜炎模型)、体外试验(测定 RAW 264.7 细胞产生的 NO 水平、HEK 293 细胞中 NF-κB 活性),研究了阿替匹林 C 的抗炎效果,并研究了小鼠对阿替匹林 C 的吸收率及生物利用率。结果表明,阿替匹林 C 对小鼠足跖肿胀最多有 38% 的抑制效果,并显著减少患腹膜炎小鼠中性粒细胞的水平;细胞实验也证实了阿替匹林 C 可以减少 NO 的产生,同时能降低 NF-κB 的转录活性。

2 蜂胶抗炎作用的分子机制

尽管导致炎症发生原因多种多样,但一般认为炎症是由化学媒介物直接诱导产生的。目前已研究确认与炎症过程有关的化学媒介物主要包括血管活性胺(如组胺和 5-羟色胺等)、类花生酸物质(如前列腺素和白细胞三烯等)、血小板凝聚因子、细胞因子(白介素和肿瘤坏死因子等)、激肽(缓激肽)、氧自由基等。虽然不同地区、不同植物来源的蜂胶组成成分及有效活性成分差别很大,但令人惊奇的是,各种类型的蜂胶都具有相似的生物学功能,如抗氧化、抗癌、抗菌消炎等。早在 20 世纪 90 年代,人们就已经开始了蜂胶药理学活性作用机制方面的研究,特别是蜂胶抗炎和抗氧化作用机制一直受到研究人员的广泛关注。总的来说,蜂胶的抗炎机制主要是通过抑制炎症相关信号通路(NF-κB、MAPK 等)的转导,并通过影响这些化学媒介物的产生而发挥抗炎作用。

2.1 影响花生四烯酸代谢通路

炎症过程中花生四烯酸代谢通路起着非常重要的作用。花生四烯酸的代谢主要通过两种途径完成,包括环氧化酶途径(COX)和脂氧合酶途径(LO)。而机体受到的外界刺激会使 COX-2 表达量升高,从而催化合成大量的前列腺素类物质(如前列腺素 E2 等),引起和加重炎症反应。而蜂胶和蜂胶中的活性成分能通过抑制 COX-2 基因的表达,因而抑制了该酶的活力,从而影响了花生四烯酸的代谢,减少了前列腺素 E2 的生成。其可能存在的作用机制见图 3.6(王凯 等,2013b)。

2.2 影响 L-精氨酸代谢通路

除了上述提到的花生四烯酸代谢通路外,通过 L-精氨酸代谢通路产生的 NO 也对炎症的产生发展起着非常重要的作用。NO 是具有高度反应性的自由基,在体内作为一种细胞信号分子广泛存在,并在炎症反应中起着非常重要的作用。在炎症过程中,L-精氨酸在 iNOS 的作用下催化合成大量的 NO,造成细胞损伤,引起细胞相关病理反应。而蜂胶及蜂胶中的活性成分能通过抑制 iNOS 的表达而减少 NO 的生产量,发挥抗炎作用。

2.3 抑制 NF-κB 信号通路的激活

NF-κB 是一种重要的转录因子,因其能与免疫球蛋白 κ 链基因结合增强其转录而得名。NF-κB 信号通路与许多炎症、免疫相关基因的转录都有联系。在上游信号刺激下,与 NF-κB 蛋白结合的 IκBα 会发生磷酸化,磷酸化后的 IκBα 会与 NF-κB 分离开来,并在蛋白酶体的作用下发生降解。与 IκBα 分离开的 NF-κB 很快就会从细胞质转移进细胞核,进核后的 NF-κB 响应原件会结合在相关基因的 κB 位点上,从而起始下游基因的转录。与炎症过程相关的一些基因,如一些炎症因子(TNF-α,IL-12,IL-1β,IL-5 等)及一些生物酶类(sPLA2,COX2,LOX,NOS 等)都含有 NF-κB 结合位点,因此,NF-κB 的转录活性也是炎症发生发展过程中的关键因素。同时,由于蜂胶中含有大量的黄酮类、酚酸类物质,如咖啡酸、阿魏酸、咖啡酸苯乙酯、阿替匹林 C、槲皮素、柯因等,它们能有效地抑制炎症因子的生成,并能直接或间接地抑制 NF-κB 信号通路的激活,从而也抑制了相关基因的转录,有效缓解了炎症对细胞造成的伤害。

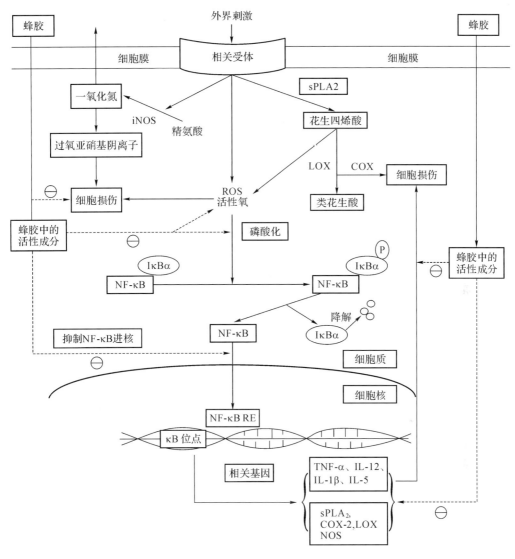

图 3.6 蜂胶抗炎作用的分子机制

Fig. 3.6　Molecular mechanism of anti-inflammatory effects of propolis

小　结

随着对蜂胶研究的不断深入,对其药理学活性的认识也一定会越来越全面、深入。同时,不同种类的蜂胶多种药理学活性已在临床上得到了证实,很多与蜂胶相关的保健品也出现在市场上。需要指出的是,蜂胶成分复杂,且不同生物活性成分的作用机制不尽相同,简单地把蜂胶的各种生物学活性看成是它们的协同作用的结果也并不是完全科学的。以蜂胶的抗炎活性为例,目前的研究都已经认识到这是蜂胶中某种或某几种主要成分作用的结果(如柯因、CAPE、阿替匹林 C 等),它们在蜂胶中含量的多少也会直接影响蜂胶抗炎作用的强弱。从这个角度来看,尽管不同地区、来源的蜂胶组成成分差异巨大,但从蜂胶中主要活性成分出发,联系其具有的生物活性,一定能更好地解释蜂胶的作用机制,同时也能促进蜂胶产品标准化的推进。

参考文献

Aygun FO, Akcam FZ, Kaya O, Ceyhan BM, Sutcu R (2012) Caffeic acid phenethyl ester modulates gentamicin-induced oxidative nephrotoxicity in kidney of rats[J]. *Biological Trace Element Research*, 145(2): 211-216.

Beltran-Ramirez O, Perez RM, Sierra-Santoyo A, Villa-Trevino S (2012) Cancer prevention mediated by caffeic Acid phenethyl ester involves cyp2b1/2 modulation in hepatocarcinogenesis [J]. *Toxicologic Pathology*, 40(3): 466-472.

Borrelli F, Maffia P, Pinto L, Ianaro A, Russo A, Capasso F, Ialenti A (2002) Phytochemical compounds involved in the anti-inflammatory effect of propolis extract[J]. *Fitoterapia*, 73 (Suppl. 1): S53-S63.

Bueno-Silva B, Franchin M, Alves CD, Denny C, Colon DF, Cunha TM, Alencar SM, Napimoga MH, Rosalen PL (2016) Main pathways of action of brazilian red propolis on the modulation of neutrophils migration in the inflammatory process [J]. *Phytomedicine*, 23(13): 1583-1590.

Cheng CY, Ho TY, Lee EJ, Su SY, Tang NY, Hsieh CL (2008) Ferulic acid reduces cerebral infarct through its antioxidative and anti-Inflammatory effects following transient focal cerebral ischemia in rats[J]. *American Journal of Chinese Medicine*, 36 (6): 1105-1119.

Coussens LM and Werb Z (2002) Inflammation and cancer[J]. *Nature*, 420 (6917): 860-867.

de Moura SAL, Ferreira M, Andrade SP, Reis MLC, Noviello MD, Cara DC (2011) Brazilian green propolis inhibits inflammatory angiogenesis in a murine sponge model[J]. *Evidence-Based Complementary and Alternative Medicine*, Volume 2011, Article ID 182703.

Dobrowolski JW, Vohora SB, Sharma K, Shah SA, Naqvi SAH, Dandiya PC (1991). Antibacterial, antifungal, antiamoebic, antiinflammatory and antipyretic studies on propolis bee products[J]. *Journal of Ethnopharmacology*, 35(1): 77-82.

Franchin M, Co'lon DF, Castanheira FVS, da Cunha MG, Bueno-Silva B, Alencar SM, Cunha TM, Rosalen PL (2016a) Vestitol isolated from brazilian red propolis inhibits neutrophils migration in the inflammatory process: Elucidation of the mechanism of action[J]. *Journal of Natural Products*, 79(4): 954-960.

Franchin M, Co'lon DF, da Cunha MG, Castanheira FVS, Saraiva ALL, Bueno-Silva B, Alencar SM, Cunha TM, Rosalen PL (2016b) Neovestitol, an isoflavonoid isolated from brazilian red propolis, reduces acute and chronic inflammation: Involvement of nitric oxide and il-6[J]. *Scientific Reports*, 79(4): 954-960.

Ghisalberti EL (1979) Propolis: a review[J]. *Bee World*, 60: 59-84.

Hamalainen M, Nieminen R, Vuorela P, Heinonen M, Moilanen E (2007) Anti-inflammatory effects of flavonoids: genistein, kaempferol, quercetin, and daidzein inhibit STAT-1 and NF-kappa B activations, whereas flavone, isorhamnetin, naringenin, and pelargonidin inhibit only NF-kappa B activation along with their inhibitory effect on iNOS expression and NO production in activated macrophages[J]. *Mediators of Inflammation*,2007:45673.

Hu F,Hepburn HR,Li Y,Chen M,Radloff SE,Daya S (2005) Effects of ethanol and water extracts of propolis (bee glue) on acute inflammatory animal models[J]. *Journal of Ethnopharmacology*,100(3):276-283.

Khayyal MT,Elghazaly MA,Elkhatib AS (1993) Mechanisms involved in the anti-inflammatory effect of propolis extract[J]. *Drugs under Experimental and Clinical Research*,19(5):197-203.

Moura SAL,Negri G,Salatino A,Lima LDC,Antunes DLP,Mendes JB,Andrade SP,Neves DFMA,Cara DC (2011) Aqueous extract of Brazilian green propolis:Primary components, evaluation of inflammation and wound healing by using subcutaneous implanted sponges[J]. *Evidence-Based Complementary and Alternative Medicine*, Volume 2011, Article ID 748283.

Lin HP,Jiang SS,Chuu CP (2012) Caffeic acid phenethyl ester causes p21 induction,Akt signaling reduction,and growth inhibition in PC-3 human prostate cancer cells[J]. *PLoS ONE* 7(2):e31286.

Massaro FC,Brooks PR,Wallace HM and Russell FD (2011) Cerumen of Australian stingless bees (*Tetragonula carbonaria*):gas chromatography-mass spectrometry fingerprints and potential anti-inflammatory properties[J]. *Naturwissenschaften*,98(4):329-337.

Mills KHG (2011) TLR-dependent T cell activation in autoimmunity[J]. *Nature Reviews Immunology*,11(12):807-822.

Mirzoeva OK and Calder PC (1996) The effect of propolis and its components on eicosanoid production during the inflammatory response[J]. *Prostaglandins Leukotrienes and Essential Fatty Acids*,55(6):441-449.

Missima F and Sforcin JM (2008). Green Brazilian propolis action on macrophages and lymphoid organs of chronically stressed mice[J]. *Evidence-Based Complementary and Alternative Medicine*,5(1):71-75.

Nakajima Y,Shimazawa M,Mishima S,Hara H (2007) Water extract of propolis and its main constituents,caffeoylquinic acid derivatives,exert neuroprotective effects via antioxidant actions[J]. *Life Sciences*,80(4):370-377.

Natarajan K,Singh S,Burke TR,Grunberger D,Aggarwal BB (1996) Caffeic acid phenethyl ester is a potent and specific inhibitor of activation of nuclear transcription factor NF-kappa B[J]. *Proceedings of the National Academy of Sciences*,93(17):9090.

Orsatti CL,Missima F,Pagliarone AC,Bachiega TF,Búfalo MC,Araújo JP,Sforcin JM (2010) Propolis immunomodulatory action in vivo on toll-like receptors 2 and 4 expression and on pro-inflammatory cytokines production in mice[J]. *Phytotherapy Research*,24(8):1141-1146.

Orsatti CL and Sforcin JM (2012) Propolis immunomodulatory activity on TLR-2 and

TLR-4 expression by chronically stressed mice[J]. *Natural Product Research*, 26(5): 446-453.

Orsi RO, Funari SRC, Soares AMVC, Calvi SA, Oliveira SL, Sforcin JM, Bankova V (2000) Immunomodulatory action of propolis on macrophage activation [J]. *Journal of Venomous Animals and Toxins*, 6: 1-17.

Park EH and Kahng JH (1999) Suppressive effects of propolis in rat adjuvant arthritis[J]. *Archives of Pharmacal Research*, 22(6): 554-558.

Paulino N, Abreu SRL, Uto Y, Koyama D, Nagasawa H, Hori H, Dirsch VM, Vollmar AM, Scremin A, Bretz WA (2008) Anti-inflammatory effects of a bioavailable compound, Artepillin C, in Brazilian propolis[J]. *European Journal of Pharmacology*, 587(1-3): 296-301.

Raso GM, Meli R, Di Carlo G, Pacilio M, Di Carlo R (2001) Inhibition of inducible nitric oxide synthase and cyclooxygenase-2 expression by flavonoids in macrophage J774A.1 [J]. *Life Sciences*, 68(8): 921-931.

Sa-Nunes A, Faccioli LH, Sforcin JM (2003) Propolis: lymphocyte proliferation and IFN-gamma production[J]. *Journal of Ethnopharmacology*, 87(1): 93-97.

Saavedra N, Leticia B, Herrera CL, Alvear M, Montenegro G, Salazar LA (2011) Effect of Chilean propolis on cariogenic bacteria *Lactobacillus fermentum*[J]. *Ciencia E Investigacion Agraria*, 38(1): 117-125.

Sforcin JM (2007) Propolis and the immune system: a review[J]. *Journal of Ethnopharmacology*, 113(1): 1-14.

Shvarzbeyn J and Huleihel M (2011) Effect of propolis and caffeic acid phenethyl ester (CAPE) on NF kappa B activation by HTLV-1 Tax[J]. *Antiviral Research*, 90(3): 108-115.

Silva JC, Rodrigues S, Feás X, Estevinho LM (2012) Antimicrobial activity, phenolic profile and role in the inflammation of propolis[J]. *Food and Chemical Toxicology*, 50(5): 1790-1795.

Skiba M, Szliszka E, Kunicka M, Krol W (2011) Effect of ethanol extract of propolis (EEP) on interleukin 8 release by human gastric adenocarcinoma cells (AGS) infected with *Helicobacter pylori* [J]. *Central European Journal of Immunology*, 36(2): 65-69.

Song YS, Park EH, Hur GM, Ryu YS, Kim YM, Jin C (2002) Ethanol extract of propolis inhibits nitric oxide synthase gene expression and enzyme activity[J]. *Journal of Ethnopharmacology*, 80(2-3): 155-161.

Steinhagen F, Kinjo T, Bode C, Klinman DM (2011) TLR-based immune adjuvants[J]. *Vaccine*, 29(17): 3341-3355.

Tan-No K, Nakajima T, Shoji T, Nakagawasai O, Niijima F, Ishikawa M, Endo Y, Sato T, Satoh S, Tadano T (2006) Anti-inflammatory effect of propolis through inhibition of nitric oxide production on carrageenin-induced mouse paw edema[J]. *Biological &*

Pharmaceutical Bulletin,29(1):96-99.

Tanaka M,Okamoto Y,Fukui T,Masuzawa T (2012) Suppression of interleukin 17 production by Brazilian propolis in mice with collagen-induced arthritis[J]. *Inflammopharmacology*,20(1):19-26.

Wang LC,Chu KH,Liang YC,Lin YL,Chiang BL (2010) Caffeic acid phenethyl ester inhibits nuclear factor-kappa B and protein kinase B signalling pathways and induces caspase-3 expression in primary human CD4+T cells[J]. *Clinical and Experimental Immunology*,160(2):223-232.

Wang LC,Lin YL,Liang YC,Yang YH,Lee JH,Yu HH,Wu WM,Chiang BL (2009) The effect of caffeic acid phenethyl ester on the functions of human monocyte-derived dendritic cells[J]. *BMC Immunology*,10:39.

Wang L,Tu YC,Lian TW,Hung JT,Yen JH,Wu MJ (2006) Distinctive antioxidant and antiinflammatory effects of flavonols[J]. *Journal of Agricultural and Food Chemistry*,54(26):9798-9804.

Woo KJ,Jeong YJ,Inoue H,Park JW,Kwon TK (2005) Chrysin suppresses lipopolysaccharide-induced cyclooxygenase-2 expression through the inhibition of nuclear factor for IL-6 (NF-IL6) DNA-binding activity[J]. *Febs Letters*,579(3):705-711.

Yang YH (2011) Caffeic acid phenethyl ester possessing various immunomodulatory effects is a potentially effective therapy for asthma[J]. *Pediatrics and Neonatology*,52(6):307-308.

胡福良（2005）蜂胶药理作用研究[M]. 杭州：浙江大学出版社.

胡福良,李英华,陈民利,应华忠,朱威（2003）蜂胶醇提液和水提液对急性炎症动物模型的作用[J]. 浙江大学学报（农业与生命科学版）,29(4):444-448.

胡福良,李英华,朱威,陈民利,应华忠（2005a）不同方法提取的蜂胶液中总黄酮含量的测定及抗肿瘤与抗炎作用研究[J]. 中国食品学报,5(3):11-15.

胡福良,李英华,朱威,陈民利,应华忠（2005b）蜂胶对大鼠佐剂性关节炎的作用及其机制的研究[J]. 中国药学杂志,40(15):1146-1148.

胡福良,李英华,朱威,陈民利,应华忠（2007）蜂胶提取液对大鼠急性胸膜炎的作用及其机制的研究[J]. 营养学报,29(2):189-191.

王凯,张翠平,胡福良（2013a）巴西绿蜂胶主要生物活性成分的研究进展[J]. 天然产物研究与开发,25(1):140-145.

王凯,张江临,胡福良（2013b）蜂胶抗炎活性及其分子机制研究进展[J]. 中草药,44(16):2321-2329.

第四节 蜂胶的抗肿瘤作用

癌症是全球范围内危害人类健康的"头号杀手"，而化疗是目前使用最为广泛的治疗癌症的方法之一。然而长期接受传统化疗药物会导致一系列副作用，如基因变异、DNA甲基

化、组蛋白修饰行为等,并产生抗药性。因此,寻找新的抗癌药物成了目前癌症治疗的新方向。而天然产物是一个很好的来源,其大光谱范围可以通过分子修饰来为临床治疗提供先导化合物。有研究表明,目前有超过70%的抗癌化合物来源于天然产物或天然产物的衍生物。此外,天然产物与单克隆抗体或者聚合载体的结合可以提供更有效、更有针对性的治疗(Karikas,2010)。

蜂胶作为一种生物活性成分极为丰富的天然产物,具有抗氧化、抗炎症、抗菌、抗病毒、免疫调节和抗寄生虫等广泛的生物学活性。近年来,有关蜂胶抗肿瘤的作用也有了一些报道(Sforcin and Bankova,2010)。最近几年,生物体内外的研究给蜂胶生物学活性的作用机制提供了新的信息,而蜂胶对肿瘤的抑制作用还是一个全新的领域,有很多值得研究的方向(Sforcin,2007;Bankova 2005)。

体外实验表明,蜂胶及其分离的化合物对于多种肿瘤细胞有细胞毒性作用,体内实验也表明蜂胶及其单体成分具有抗肿瘤活性。这一活性与蜂胶的免疫调节活性密切相关,主要表现在蜂胶能有效提高巨噬细胞的活化率,并产生可溶性因子直接干预肿瘤细胞或者其他免疫细胞功能,从而发挥非特异性抗肿瘤作用。

近年来,大量有关蜂胶及其单体化合物的抗肿瘤活性的报道证实了蜂胶作为新的抗肿瘤药物开发的潜力(Sforcin and Bankova,2010)。同时,由于蜂胶对人类没有明显的毒副作用,它在开发相对廉价的肿瘤治疗方案中也逐渐得到应用(Burdock,1998;Sforcin et al.,2002;Mani et al.,2006;Jasprica et al.,2007)。下面就国内外学者对蜂胶及其活性成分的抗肿瘤活性及可能的作用机制方面的研究进展进行综述,以期为蜂胶抗肿瘤活性的进一步研究以及在抗肿瘤药物中的开发利用提供参考。

1 不同地理来源蜂胶的抗肿瘤活性

蜂胶的化学成分十分复杂,不同的植物来源、地理分布、季节都会对蜂胶的化学组分造成影响。一般而言,蜂胶是由树脂、蜂蜡、花粉和挥发性物质等组成。蜂胶树脂中的主要化学成分为酚酸类及其酯、黄酮类化合物(黄酮,黄烷酮类,黄酮醇,黄烷酮醇,查耳酮)、萜烯类化合物、芳香醛和醇类化合物、脂肪酸类、对称二苯代乙烯和 β 类固醇等(Burdock,1998;Gardana et al.,2007)。生物实验中,可用不同溶剂例如甲醇、乙醇、水等来溶解和提取蜂胶中不同的组分(Cunha et al.,2004)。蜂胶乙醇提取物(EEP)是酚酸和黄酮类最丰富的提取物之一,具有多种生物学活性,包括抗氧化、抗菌、免疫调节、化学预防和抗肿瘤等作用(Oršolic et al.,2006)。

近二十年来,世界各国学者对不同地理来源蜂胶的生物学活性进行了广泛而深入的研究,然而由于地理来源不同的蜂胶其组分存在较大差异,这些实验结果无法平行比较(Sforcin and Bankova,2011),这使得蜂胶生物学活性评定的标准化过程变得相当艰难。人们普遍认为,由于蜂胶的这一特性,几乎不可能制定一个通用的标准。蜂胶的生物学活性研究应当与蜂胶的化学成分与地理来源相结合起来进行(Bankova,2005)。

为了检测不同地理来源的蜂胶活性,Orsolic研究团队曾在他们的肿瘤模型中评估过克罗地亚蜂胶和巴西蜂胶的生物调节活性。结果表明,经二者蜂胶水提物(WEP)处理后的小鼠巨噬细胞的杀瘤活性提高,淋巴激活因子产量增加,并都能有效抑制人宫颈癌细胞(HELA)和中国仓鼠肺成纤维细胞(V79)的增殖。同时,经蜂胶处理过的小鼠也对多克隆有丝分裂原表现出了更强的脾细胞反应。他们还发现蜂胶和其中一些单体化合物减少了肺中的肿瘤结节,而蜂胶所具有的抗转移效果也比其单体成分要强(Orsolic et al.,2005;Oršolic et al.,2004;Oršolicand Basic,2003)。

有研究发现,用巴西蜂胶 EEP 处理过的老鼠的自然杀伤细胞对抗细胞毒性的活性增加(Sforcin et al.,2002)。Missima 等(2009,2010)设计了实验用于评估 EEP 对应激状态下患有黑素瘤小鼠的作用,结果发现,在皮下接种 B16F10 细胞后,应激反应引发了更大的肿瘤面积,而喂食过蜂胶的小鼠,无论应激与否,其黑色素瘤的发展都与对照组相似。同时,研究还发现,无论是否发生应激反应,蜂胶都能在患有黑素瘤的小鼠上诱导产生更高浓度的白细胞介素 1-β 和白细胞介素-6。在实验过程中,蜂胶也刺激了 TH1 血细胞因子(白细胞介素-2 和干扰素 γ)的产生。因此,他们推测,干扰素 γ 和促炎症因子的协同作用可以通过诱发抗血管因子生成而抑制肿瘤在体内的生长。

Awale 等(2008)研究发现,巴西红蜂胶甲醇提取物对人体胰腺癌 PANC-1 细胞系表现出 100% 的细胞毒作用。在随后的植物化学分析中,他们分离鉴定出了 43 种化合物,其中 3 种是新的化合物。他们还分析了分离的单体化合物在营养缺失条件下对 PANC-1 细胞的影响。其中一种化合物 [(6aR,11aR)-3,8-dihydroxy-9-methoxypterocarpan] 表现出最强的(100%)有选择性的细胞毒作用,且具有时间、剂量依赖性。该细胞毒作用主要通过非凋亡通路介导,同时并不导致 DNA 碎片化,但会致使坏死细胞的形态发生变化。

研究表明,巴西蜂胶 EEP 能抑制人体前列腺癌细胞的增殖(Li et al.,2007)。日本蜂胶水提物可以在体外抑制 S-180 小鼠肉瘤生长,还能有效抑制小鼠体内移植瘤的生长(Inoue et al.,2008)。Khacha-ananda 等(2016)研究了泰国北部(帕尧、清迈和楠府)蜂胶乙醇提取物对不同癌细胞(A549 和 HeLa)生长的影响,发现采集于清迈的蜂胶样本具有最强的抗氧化活性和总酚含量,且该蜂胶处理过的癌细胞呈现出明显的 DNA 碎片化和明显的细胞凋亡,并激活内源性 Caspase 酶依赖的细胞凋亡信号通路。口服泰国蜂胶能有效延长荷瘤小鼠的生存时间。Taira 等(2016)研究发现,冲绳岛蜂胶对促癌相关激酶 PAK1 具有抑制效果,并能有效抑制人肺癌细胞 A549 生长,同时表现出抑制黑色素瘤形成作用,这些抑瘤效果均同 PAK1 激酶的失活密切相关。

有报道证实土耳其蜂胶具有抗肿瘤转移活性,其抗肿瘤效果可以归功于蜂胶中的黄酮类化合物抑制癌细胞对胸腺嘧啶脱氧核苷、尿嘧啶核苷和亮氨酸的摄入,从而阻碍了癌细胞的 DNA 复制(Ozkul et al.,2005,2006)。Eroglu 等(2008)认为膀胱癌细胞在组织培养中降低的有丝分裂指数率预示着蜂胶具有抗癌和抗有丝分裂的作用,并可用于提高人体健康水平。

中国和巴西蜂胶 EEP 在 4 种人体直肠癌细胞系(caco2、hct116、ht-29 和 sw480)上均表现出了一定的抗癌活性,且这两种蜂胶提取物都具有显著的剂量依赖的生长抑制现象。在 hect116 细胞系上,中国蜂胶提取物能诱导细胞凋亡,并导致剂量依赖性的肿瘤抑制蛋白(p21CIP1 和 p53)mRNA 基因表达量的增加(Ishihara et al.,2009)。泰国蜂胶对克隆癌细胞系 SW620 的抑制增殖作用和细胞毒性作用证实了蜂胶水提物比甲醇提取物具有更好地抑制增殖的活性效果(Umthong et al.,2009)。

Xuan 等(2014)利用两种不同来源的乳腺癌细胞系 MCF-7(原位 ER 阳性乳腺癌细胞系)和 MDA-MB-231(高转移性恶性乳腺癌细胞系)研究了中国蜂胶对乳腺癌细胞的细胞毒

作用。结果发现,中国蜂胶对这两种乳腺癌细胞系均呈剂量、时间依赖性的细胞毒性作用,同时蜂胶处理能显著增加膜联蛋白A7的表达量,并显著抑制ROS累积和线粒体膜电位水平、降低核因子kappaB p65的蛋白水平。此外,蜂胶对肿瘤抑制蛋白(p53蛋白)的调控作用在两种乳腺癌上呈现明显差异,但对正常细胞的细胞毒性作用非常小。

突尼斯蜂胶EEP对多种癌细胞亦表现出抗增殖活性。用丁醇从希腊蜂胶中提取和分离的成分主要是二萜类和黄酮醇,可用于抑制人类恶性和正常细胞株的细胞生长活性。蜂胶丁醇提取物对ht-1080人类纤维肉瘤和ht-克隆腺癌细胞有细胞毒作用,而对人类正常表皮成纤维细胞却没有此类作用(Kouidhi et al.,2010)。

葡萄牙蜂胶一直以其在人体红细胞和A-498细胞系上表现出的生物活性而被广泛关注。研究表明,低浓度的葡萄牙蜂胶提取物可以抑制和减少脂质过氧化作用和由过氧化自由基诱发的溶血现象(Valente et al.,2011)。另外,蜂胶能有效抑制人体肾癌细胞的增殖活性。更重要的是,体内针对正常细胞和癌细胞反应性的比较试验发现,蜂胶提取物只对癌细胞生长起抑制作用。总而言之,这些发现都揭示了葡萄牙蜂胶是预防由自由基引起的疾病的一种有开发利用前景的临床成分,尤其是在肾癌的化学预防中。

蜂胶在癌症及其他疾病的治疗上,它的低水溶性导致其系统生物利用率很低,限制了蜂胶在癌症及其他疾病上的临床应用。而基于纳米粒子的运动方式可以使蜂胶分散在水中,从而避免其疏水性所造成的影响。在小鼠异种移植物模型中,纳米蜂胶能有效抑制胰腺癌细胞的生长和具有抗血管生成活性,进一步拓展了蜂胶在临床上的应用。同时,即使采用最大蜂胶接种剂量也未对小鼠产生明显毒性,可推测蜂胶是一种对生物体毒性较低的物质。另一个体内细胞克隆形成实验也证实了蜂胶治疗人类胰腺癌细胞系的出色效果(Kim et al.,2008)。

2 蜂胶中活性单体成分的抗肿瘤活性

2.1 黄酮类化合物

黄酮类化合物是蜂胶的主要化学成分,具有良好的抗肿瘤活性。多酚/黄酮类化合物对于不同白血病细胞系的作用已有相关报道。通过分析5种化合物(懈皮素、咖啡酸、柯因、柚皮素、柚皮苷)作用在5种白血病细胞系上的实验结果,研究人员发现,懈皮素对5种细胞系表现出最强的细胞毒性作用,其次是柯因和咖啡酸(Josipovic and Orsolic,2008)。而从中国天然蜂胶中提取出来的新外消旋黄烷醇和另一种新的外消旋黄烷醇混合物组分表现出了对人海拉卵巢肿瘤细胞系的细胞毒性作用(Sha et al.,2009)。

墨西哥蜂胶中的天然黄酮醇 1-phenylallyl moiety {(7″R)-8-[1-(4′-hydroxy-3′-methoxyphenyl) prop-2-en-1-yl] galangin}在营养缺乏培养条件下表现出很强的细胞毒性作用,同时会诱导PANC人胰腺癌细胞引发凋亡形态改变。同样,phenylpropanoid-substituted flavanol (2R,3S)-8-(4-phenylprop-2-en-1-one)-4′,7-dihydroxy 3,5-d imethoxyflavan 3 ol表现出对A549细胞和HT-1080细胞很强的细胞毒性作用,这种细胞毒性作用甚至比常用的临床抗肿瘤药物氟尿嘧啶还要有效(Li et al.,2010a)。

2.1.1 柯因

柯因(5,7-二羟黄酮)是一种具有抗炎、抗癌、抗过敏、抗焦虑和抗氧化等丰富生物学活性的黄酮类化合物,并且能阻碍细胞周期运行。然而,柯因阻碍癌症细胞生长和对细胞内相关信号通路的作用机制,我们目前还知之甚少。

Fu等(2007)研究发现,柯因可以抑制前列腺癌 DU145 细胞中缺氧诱因因子-1α(HIF-1α)和血管内皮生长因子的表达。柯因通过减少其稳定性来抑制由胰岛素引起的 HIF-1α 的表达,同时通过脯氨酰羟化增加了 HIF-1α 的泛素化和降解。此外,柯因干扰了 HIF-1α 和热休克蛋白 90 的反应。柯因还被发现可以通过 AKT 信号抑制 HIF-1α 的表达。而在小鼠 B16-F1 和人黑素瘤 A375 细胞系上,柯因可以通过合成和聚集细胞内血红素前体卟啉 IX 来减少黑素瘤细胞的增殖,并诱发各类黑素瘤细胞分化。

柯因在由二乙基亚硝胺引起的小鼠早期肝癌发展模型中的效果也得到了研究人员的证实。在癌症发生后,每周 3 次给小鼠喂食混有柯因的饲料,可以明显观察到小鼠肿瘤结节的减少和减小。同时,血清中谷草转氨酶(AST)、谷丙转氨酶(ALT)、碱性磷酸酶(ALP)、乳酸脱氢酶(LDH)和 γ-谷氨酰转移酶的活性显著降低。同时,COX-2 和 NF-κB p65 的表达也大幅度降低,并伴随着 p53、Bax 和 caspase3 蛋白质翻译量和转移量的增加。此外,研究人员还观察到 β 休止蛋白(在进行性肿瘤中扮演重要角色的一种蛋白)和抗凋亡 Marker Bcl-xL 也出现了明显下降(Patel,2015)。

柯因的构效关系揭示了其化学结构符合对于白血病细胞具有强效细胞毒作用的黄酮类化合物的要求。研究人员推测,修饰过的柯因或者联合治疗可能比未修饰柯因或者单一柯因的疗效更好。

2.1.2　槲皮素

槲皮素能够通过调整胰岛素样营养因子系统成分和引起细胞凋亡来降低非激素依赖性前列腺癌细胞的存活率。槲皮素还能抑制乳腺癌细胞、人肺癌和鼻咽癌细胞的增殖。Sugantha 等(2013)的前期实验表明,槲皮素可以导致细胞周期终止,并引起前列腺癌细胞的凋亡。槲皮素抑制了前列腺癌细胞系 PC-3 增殖和存活过程中的入侵、迁移行为及其信号分子。

Xing 等(2001)的实验初步证明了槲皮素可以显著下调前列腺特定基因 NKX3.1 的表达。NKX3.1 基因的表达与前列腺癌的侵袭表型有关。槲皮素可以抑制 ODC mRNA 的激素上调水平,而 ODC 是细胞增殖中合成多胺的十分重要的调控因素。

Maurya 和 Vinayak(2015)研究发现,槲皮素可以在 HepG2 细胞中下调 phosphop85α,此作用与通过减少络氨酸激酶活性导致 PI3K 失活的原理是一样的。槲皮素可以通过竞争抑制 P85α 的 ATP 连接位点抑制 PI3K 活性;槲皮素致使 PI3K 的失活也证明了它是降低 HepG2 细胞存活率的主要因素。

2.1.3　高良姜素

高良姜素(3,5,7-三羟黄酮)是在多种天然产物中被发现的一种黄酮类抗癌活性物质,在蜂胶中也占有十分重要的位置。高良姜素的抗氧化性能保护细胞不受自由基的伤害,并被认为对癌细胞有抑制作用。前期研究表明,高良姜素对人白血病细胞传代细胞的增殖有抑制作用,并能促进细胞凋亡。

高良姜素在 B16F10 小鼠黑素瘤细胞上的作用也被广泛研究,被认为可以破坏线粒体膜电位,促进细胞凋亡,降低肿瘤细胞存活性。此外,高良姜素还在一定时间内显著地降低了 phospho-p38 MAPK 的活性,并表现出了剂量依赖性(Zhu et al.,2014)。

2.2 萜烯类化合物

萜烯类化合物是蜂胶中另一大类主要活性成分。缅甸蜂胶中分离出来的 13 种环阿尔廷醇三萜类和 4 种异戊二烯类黄酮类化合物的抗癌活性已经得到证实(Li et al.,2009a)。一种环阿尔廷醇型三萜(3a,27-dihydroxycycloart-24E-en-26-oic acid)表现出有效的对抗 B16-BL6 黑素瘤细胞毒素活性,(2S)-5,7-dihydroxy-4′-methoxy-8,3′-diprenylflavanone 表现出很强的抑制人体肿瘤细胞系(肺腺癌 A549 细胞、子宫颈海拉细胞和纤维肉瘤)活性作用(Li et al.,2009b)。另一种缅甸蜂胶的甲醇提取物可以在营养剥夺条件下抑制人胰腺癌 PNAC-1 细胞的增殖。该提取物的活性追踪分离结果表明,它是一种坏阿尔廷醇三萜类化合物(22Z,24E)-3-oxocycloart-22,24-dien-26-oic acid,具有有效的随时间变化、剂量依赖的细胞毒性(Li et al.,2009b)。

Pratsinis 等(2010)研究发现,希腊蜂胶提取物和二萜类物质对抗 HT-29 人类直肠癌细胞有着很强的细胞毒活性,且对正常人类细胞没有影响。一种二萜化合物——泪杉醇,是其中分离发现的活性最强的物质,能有效阻碍癌细胞周期在 G2/M 期发生阻滞。

2.3 酚酸类化合物

2.3.1 咖啡酸苯乙酯

咖啡酸苯乙酯(CAPE)是杨树型蜂胶中的一种主要活性成分,已有大量实验研究了 CAPE 的生物学活性,包括抗氧化、抗炎、抗血管增生、抗癌细胞转移和抑癌活性等(Bankova et al.,1987)。Grunberger 等(1988)利用以色列蜂胶进行生物导向性研究,比较了其对大鼠/人的正常细胞与转移性的大鼠/人的黑素瘤及乳腺癌细胞系的细胞毒性作用,结果表明,CAPE 是蜂胶中主要的细胞生长抑制剂。相比于正常细胞,肿瘤细胞系对 CAPE 表现出了更高的敏感度。来自伊朗的蜂胶也表现出对肿瘤细胞但不是癌细胞的细胞毒性作用,此效果被认为与 CAPE 的存在有一定联系(Natarajan et al.,1996)。

在癌症、神经退行性疾病等慢性疾病中,氧化应激是细胞损伤的主要原因。CAPE 可作为一种有效的外源细胞保护剂和抗基因毒性药物来抵抗细胞氧化性损伤,并能作为一种新药先导来治疗由各种氧化应激诱导产生的疾病(Wang et al.,2008)。

通过细胞色素 p450(CYP)进行 CAPE 抗癌作用机制分析发现,CAPE 能影响化学致肝癌形成初期的二乙基亚硝酸代谢(Beltrán-Ramírez et al.,2008)。CAPE 改变了 CYP(如 CYP1A 1/2 和 CYP2B 1/20)在二乙基亚硝酸激活中的活性,这也从另外的角度解释了利用 CAPE 的保护作用来进行替代疗法的作用机制。

Jin 等(2008)也研究了 CAPE 在人体髓系白血病 U937 细胞中引发细胞凋亡的机制。DNA 片段化分析反映 CAPE 处理过的 U937 细胞中出现了典型的阶梯样寡聚核小体段。研究人员同时发现了细胞核浓缩现象,这也是典型的细胞凋亡的表现。随后,他们还观察到了细胞色素 C 的释放,Bcl-2 的减少和 Bax 表达量的上升以及 caspase-3、PARP 的激活与剪切。另外,Fas 蛋白(细胞死亡信号通路中的起始介导因子)、磷酸化的 Eif2a、CHOP(线粒体介导的凋亡通路中重要的信号因子)的表达量并没有明显的改变。因此,他们认为 CAPE 在 U937 细胞中引发的细胞凋亡主要是通过线粒体介导的细胞凋亡途径,并非死亡受体或者雌激素介导的细胞凋亡途径。

在肝癌中,CAPE 通过抑制基质金属蛋白酶 2 和 9 的表达来发挥抗转移作用,其机制可能是靶向干扰癌细胞中的核因子-κB。Lee 等(2008)利用 t-BHP 对 HegG2 细胞及大鼠肝脏的损伤模型研究了 CAPE 的保护作用，发现 CAPE 能明显降低 HepG2 细胞中由 t-BHP 引起的氧化损伤，并能剂量依赖地缓解 t-BHP 诱导产生的细胞毒性、减缓脂质过氧化作用和降低活性氧水平。CAPE 导致明显的 Bxpc-3 和 Panc-1 细胞活力的抑制，减缓脂质过氧化作用和降低活性氧水平。体内实验表明，在施用 t-BHP 之前进行 CAPE 预处理能显著且呈剂量依赖性地增加血清中肝脏标志酶的水平（通过检测血清中的肝酶标记物——丙氨酸氨基转移酶和门冬氨酸氨基转移酶），并减轻大鼠肝脏脂质过氧化水平。CAPE 对由 t-BHP 引起的肝中毒的保护作用，部分原因是清除活性氧的能力和保护 DNA 不受到氧化应激造成的伤害的功能。

CAPE 引起的人类胰脏癌细胞的细胞凋亡主要涉及诱导癌细胞凋亡蛋白酶和线粒体功能的失调，CAPE 会导致 BxPC-3 和 PNAC-1 细胞活力的显著下降。在 BxPC-3 细胞中，CAPE 会呈时间依赖性地增加亚二倍体的细胞比例，并能显著降低线粒体跨膜电位。同时，CAPE 还诱发了线粒体变化所引起的细胞凋亡，且在 DNA 电泳中未发现 DNA 碎片(Chen et al.，2008)。

CAPE 在 C6 神经胶质瘤细胞上具有诱发细胞周期停滞和抗增殖效果。用 CAPE 处理过的 C6 神经胶质瘤会导致其发生形态学变化，变换成为星形细胞型，同时神经胶质分化标识蛋白表达量增加，包括胶质细胞原纤维酸性蛋白(glial fibrillary acidic protein)和 s100β，并表现出了对 C6 胶质细胞侵袭的抑制作用(Lin et al.，2010)。

肿瘤坏死因子诱导凋亡相关配基(TRAIL/APO2L)，相关的肿瘤坏死因子是天然的抗肿瘤因子，可以特异性地引发癌细胞的凋亡，且对正常细胞不造成毒性作用。Szliszka 等(2011)利用蜂胶乙醇提取物及从蜂胶中分离的酚类成分，研究了它们与 TRAIL 联合使用对两种前列腺癌细胞（荷尔蒙敏感型 LNCaP 和荷尔蒙不敏型 DU-145）的影响。结果表明，与 TRAIL 协同使用下，芹菜素、山奈素、高良姜素和 CAPE 表现出最强细胞毒性作用；蜂胶提取物在前列腺癌细胞中 TRAIL 介导的细胞凋亡显著增强。研究结果推测 CAPE 在前列腺癌的化学预防中可能起着十分重要的作用。

2.3.2 阿替匹林 C

阿替匹林 C 作为巴西绿蜂胶特有的生物活性成分，最早是作为一种肿瘤抑制成分而受到广泛的关注的(Kimoto et al.，1996；Matsuno et al.，1997)。TRAIL 耐受前列腺癌细胞株在被 TRAIL 和阿替匹林 C 处理后，通过 MTT 和 LDH 测定，发现其具有细胞毒性。阿替匹林 C 能够增加 TRAIL-R2 的表达，降低 NF-κB 的活性。TRAIL 和阿替匹林 C 的联合作用使得 Caspase-8 和 Caspase-3 表达量显著增多，同时还引起线粒体膜电位混乱。这些研究结果表明，阿替匹林 C 能使得前列腺癌细胞对 TRAIL 介导的免疫预防敏感，证明了其在前列腺癌的生化预防方面有着一定的前景(Szliszka et al.，2012)。

Tani 等(2010)选择在 NF 肿瘤（小鼠移植瘤）中一种缺乏 NF2 快速生长的肿瘤细胞来评估阿替匹林 C 的治疗作用。结果发现，相较于对照组，经阿替匹林 C 处理过的小鼠，其体内 NF2 肿瘤的生长基本被抑制。

大约 50% 的人体结肠癌是由 RAS 变异导致 PAK1 的不正常活化造成的。Tomoki 等(2010)用阿替匹林 C 和绿蜂胶提取物处理患有结肠癌的小鼠，结果发现肿瘤面积显著减少（约减少 40%）。

3 蜂胶及其有效活性成分的抗肿瘤机制

3.1 诱导细胞凋亡

细胞凋亡在平衡细胞死亡与更新的过程中起着十分重要的作用,而肿瘤细胞的一部分成因也是由细胞的不断增殖和细胞凋亡的减少所造成的。Szliszka 等(2009)研究证实,蜂胶乙醇提取物可诱导海拉癌细胞发生凋亡,并表现出了剂量相关性。Syamsudin 等(2010)研究发现蜂胶乙醇提取物在 MCF-7 乳腺癌细胞系中具有更强的细胞毒性和促凋亡作用。

Samarghandian 等(2014)研究发现,蜂胶的主要活性成分柯因可以通过活化 Caspase-3 和 Caspase-9 诱发人肺上皮癌干细胞系(A549)的凋亡。另一项研究表明,在成神经瘤细胞(NGP 和 SK-N-AS)中,柯因可以通过诱发细胞凋亡来抑制肿瘤的生长,包括增加 DNA 修复酶和 Caspase-3,减少促生存蛋白 survivin 和 XIAP(X 染色体连锁的凋亡抑制蛋白)(Burke et al.,2012)。

蜂胶的另一成分咖啡酸苯乙酯(CAPE),同样也表现出了很强的抗肿瘤活性。在用 $10\mu M$ CAPE 处理人淋巴细胞性白血病细胞后,Avci 等(2011)观察到凋亡细胞数量增多;通过荧光显微镜还发现了大量的 JC-1 染料的聚集。这意味着在用 CAPE 处理后,CCRF-CEM 细胞息膜电位的消失致使促凋亡蛋白的释放。

阿替匹林 C 也可通过诱导人体癌细胞发生细胞凋亡从而发挥抗癌作用。Matsuno 等(1997)研究发现在人类和小鼠恶性肿瘤细胞中施用阿替匹林 C 后,肿瘤细胞在生长中产生的细胞毒性效应明显减轻,组织学分析还发现细胞凋亡、顿挫性核分裂及肿瘤大块性坏死;同时,$CD4^+/CD8^+$ T 细胞的比例、$CD4^+$ T 细胞的数量也有所上升。这说明阿替匹林 C 激活了免疫系统,从而直接发挥抗癌作用。

3.2 抑制肿瘤细胞增殖

蜂胶在特定的肿瘤细胞上可以表现出一定的抗增殖活性,主要通过阻滞细胞周期。目前已有研究机构希望通过蜂胶对癌细胞的生长抑制作用这一特点来作为癌症治疗的突破口。有研究表明,蜂胶可以通过调控细胞周期蛋白 D1、B1、细胞周期蛋白依赖激酶及 p21 的表达来抑制前列腺癌细胞的增殖。同时,古巴蜂胶在 MCF-7(雌激素受体阳性,ER^+)人类乳腺癌细胞系上表现出显著的抗增殖活性,而对 MDA-MB231 无显著副作用。此抗增殖效果与浓度、时间成依赖相关性,并且研究人员认为,诱导细胞凋亡作用也在其中起到了一定的作用。在剂量和时间相关的研究方式下,试验表明蜂胶可以使 MCF-7 细胞增长停滞在 G1 期(Popolo et al.,2009)。研究人员还从 Clusia rosea Jacq 植物树脂和古巴蜂胶中分离出低浓度的 nemorosone,一种多环二甲苯酮类萜,同样也表现出了对 MCF-7 的生长抑制作用,而不影响 MDA-MB-231 和 Lncap 的正常生长。因此,蜂胶有可能成为辅助雌激素受体拮抗对抗 ER^+ 乳腺癌的预防或者治疗的药物(Popolo et al.,2010)。

Li 等(2007)研究发现,巴西蜂胶的乙醇提取物可以有效抑制前列腺转移性癌细胞(DU145 和 PC-3 细胞)的生长。他们认为,巴西蜂胶乙醇提取物可以使前列腺癌细胞生长停滞在 G2 期,从而发挥抑制癌细胞增殖的作用。而蜂胶在膀胱癌患者身上提取的活体组织上的应用结果也证实了蜂胶的使用确实可以减少细胞分裂,这意味着蜂胶有可能用作抗有丝分裂和抗癌的药物。

3.3 抗血管增生

血管增生是肿瘤发展过程中十分重要的一部分，并被认为是治疗实体瘤的突破口。抑制血管增生不仅可以有效抑制肿瘤生长，还具有抗肿瘤迁移的作用。肿瘤主要通过释放血管生成因子（血管内皮生长因子 VEGF）诱发血管增生，借助新生血管来获取营养，达到生长、转移的目的（Vit et al.，2015）。

研究发现，100μg/L 柯因在体外可以抑制人乳腺癌细胞（MDA 细胞）和神经胶质瘤（U-343 和 U-118 细胞）的血管内皮生长因子 VEGF 的释放。在人前列腺癌细胞（DU145）体外实验条件下，柯因可以通过 AKT 信号通路抑制诱发 HIF-1α 表达和持续的 VEGF 释放的胰岛素，从而达到抗血管增生的效果。而在小鼠异种移植模型体内试验中，柯因也表现出了对 DU145 前列腺癌血管生成的抑制作用（Kasala et al.，2015）。

阿替匹林 C 不仅可以直接抑制肿瘤细胞生长，还可以通过抑制血管生成发挥抗癌作用。Ahn 等（2007）和 Ohta 等（2009）研究发现，阿替匹林 C 在抑制人类脐带内皮细胞（HUVECs）血管生成中有显著的量效关系（3.13～50μg/mL），抑制 HUVECs 的增殖过程也体现了这种量效关系；血管再生分析显示，阿替匹林 C 能显著减少体内新生血管的数量，这说明阿替匹林 C 具有很强的抗血管生长能力。

3.4 抑制信号转导通路的活化

3.4.1 NF-κB 信号通路

NF-κB 是一类重要的转录因子，可以诱导多种抗凋亡基因（如 Bcl-2、Bcl-xL、Mcl-1 和 c-FLIP）的表达。肿瘤细胞可以通过炎症刺激物的应答，包括炎性因子 TNF-α、白介素-1、生长因子、环境污染物、氧化应激等持续激活 NF-kB。

蜂胶成分柯因在人鼻咽癌细胞系 HCT-116 和人肝癌细胞系 HepG2 中可以显著增加 TNF-α 引起的细胞凋亡，从而抑制了 NF-κB 的活化，导致 c-FLIP-L 的下调（Li，2010b）。而 c-FLIP-L 是抗凋亡基因中十分重要的一环，具有阻碍 TNF caspases 活性的功能。

Shvarzbeyn 和 Huleihel（2011）研究表明，在 HTLV-1 感染和未感染的 T 细胞培养过程中，蜂胶乙醇提取物和 CAPE 都可以有效地抑制由 Tax 引起的 NF-κB 的转录活性。同时，CAPE 对其他启动子的 Tax 活性（如 HTLV-1 LTR 和 SRF）无作用，而蜂胶乙醇提取物则对这些启动子的 Tax 活性仍有强烈的抑制作用。这也说明了蜂胶中仍有其他活性组分在发挥作用，需要进一步的研究。

3.4.2 Wnt 信号通路

Wnt/β-catenin 信号通路是协助 NF-κB 发挥致癌作用的一类重要通路。有研究表明，NF-κB 和 Wnt/β-catenin 信号通路间的有效协作关系是通过糖原合成酶激酶 3β（GSK-3β）、Wnt/β-catenin 信号通路中的主要成分来进行调节的。GSK-3β 的激活可以使得细胞质中游离的 β-catenin 聚集，从而改变反式激活目标基因核物质，引起癌变。

Khan 等（2011）研究表明，在早期由 DEN 引起的小鼠肝癌模型中，柯因可以上调 GSK-3β 的表达，下调 casein kinase-2 的表达。因此，它能使 β-catenin 上的 Ser33/37 磷酸化，最后导致其降解并减少 Wnt 信号通路。

3.5 抗肿瘤转移

肿瘤的转移是肿瘤发展过程中标志性的一步，也是其高死亡率的原因之一。肿瘤发生转移的过程中包括了 ECM(用于调节基质金属蛋白酶 MMPs)的降解。Western blot(蛋白质印迹法)和 gelatin zymography analysis(基质金属蛋白酶明胶酶谱法)显示,CAPE 可以下调 MMP-2 的蛋白质表达和酶活性。CAPE 通过上调 TIMP-2 抑制 MMP-2 的表达和活性,同时通过减少点状黏附激酶(FAK)的磷酸化,下调 p38 胞外信号调节激酶分子信号通路(p38 MAPK)和 c-Jun 氨基末端激酶(JNKs)来减少细胞的迁移。这意味着 CAPE 有可能成为用于防止肿瘤迁移的化学成分。同样,蜂胶中的活性成分柯因也可以通过调节 MMP-10 抑制人三阴性乳腺癌细胞的转移(Kuo et al.,2015)。

有研究表明,蜂胶另一活性成分高良姜素也具有抗癌活性。高良姜素可以抑制多种人类肿瘤细胞增殖,并诱导其凋亡,包括乳腺癌、白血病、胰腺癌、胃癌、结肠癌和肝癌。Zhu 等(2014)的研究证明了高良姜素可以在 HNSCC 细胞中诱发细胞凋亡,同时伴随着 Bcl-2 和 Bcl-xL 蛋白表达量的降低,Bax 和 cleaved-caspase 3 表达量的增加。因此,推测蜂胶及其活性成分有可能通过此机制来抑制肿瘤的转移。而在肺转移小鼠模型中,高良姜素可以抑制 B16F10 黑素瘤细胞的增殖和转移。

3.6 对致癌因素的防治

有研究检测了巴西蜂胶甲醇提取物和乙醇提取物对雄性 Wistar Hannover(GALAS)小鼠体内氧化偶氮甲烷引起的异常隐窝灶的作用,通过分析后发现这两种提取物均在结肠癌初期有化学预防能力,可调控细胞的增殖(Yasui et al.,2008)。

而在日粮中加入蜂胶的主要活性成分柯因则可以在很大程度上减少小鼠癌前病变灶和 COX-2 的表达和核因子-κB 及 p65 的表达量,同时增加 p53,bax 和 Caspase3 mRNA 和蛋白质水平。同样地,研究人员也注意到了抑制蛋白水平和抗细胞凋亡蛋白 bcl-xl 表达量的降低。因此,推断柯因具有保肝药的效果,其化学预防活性与 p53 介导的细胞凋亡有关(Khan et al.,2011)。

蜂胶也由于在裸鼠实验模型上所表现出的抗光致癌效果而得到关注。悉尼蜂胶可以缓解由 UV 光线照射引起的皮肤炎症,抑制免疫反应和脂质过氧化作用。蜂胶表现出明显且呈剂量依赖性地对抗晒伤水肿,抑制接触性过敏反应和脂质过氧化过程。蜂胶对光致癌中,无论是白介素-10 的过度表达还是白介素-12 的减少,都有明显的抑制作用。此外,白介素-6 的上调减少和与血红素加氧酶的相关诱导性都预示着蜂胶在皮肤保护上有着很重要的作用(Cole et al.,2010)。

Tavares 等(2007a)研究了酒神菊树乙酸乙酯浸提物(Bd-EAE)对由阿霉素诱导的突变性的抑制作用,给实验组大鼠口服不同浓度的 Bd-EAE,再对其腹腔注射化疗剂阿霉素(DXR)。结果发现,相对于直接腹腔注射阿霉素的大鼠而言,口服 Bd-EAE 能显著减少微核多染性红细胞数量,抑制细胞突变的发生,而 Bd-EAE 本身不具致突变作用。HPLC 分析表明,Bd-EAE 样品中含大量阿替匹林 C,说明阿替匹林 C 可能是 Bd-EAE 具有抗突变性能的重要原因。他们在巴西绿蜂胶中也做了类似的实验,发现巴西绿蜂胶也具有抗染色体突变作用(Tavares et al.,2007b)。Azevedo 等(2011)在这些研究的基础上,对阿替匹林 C 抗化学诱变剂的作用进行了研究,采用微核实验和彗星实验方法,给小鼠服用阿替匹林 C,并验证其对施用 DXR 和甲基磺酸甲酯(MMS)小鼠的影响。结果表明,阿替匹林 C 在实验条件下对化学诱变剂诱导发生染色体 DNA 发生突变具有很好的保护作用,这对相关癌症的防治具有非常重要的意义。

小 结

虽然蜂胶在体外对不同肿瘤细胞有非常明显的细胞毒性作用,但蜂胶在动物或者人体上的应用仍需考虑其生物可利用性。此外,蜂胶在体内发挥抗癌作用除考虑蜂胶的化学预防或者治疗作用外,也需要与蜂胶的免疫调节活性相结合。

细胞免疫反应主要由激活的T淋巴细胞来完成,T淋巴细胞在治疗恶性肿瘤细胞中也扮演着十分重要的角色。同样,高频率出现的肿瘤浸润淋巴细胞与缓解肿瘤扩增密切相关,同时也有利于提高癌症病人总体存活率(如黑色素瘤和前列腺癌)。但是,免疫疗法至今没能解决任何一种肿瘤转移病例。此外,肿瘤还可能发展出多种机制来逃避体内和体外免疫疗法(Singer et al.,2011)。因此,围绕蜂胶及其组分的体内抗肿瘤和免疫调节活性研究还需要更加深入。

一些分离的组分同样也被认为可以解释蜂胶的抗肿瘤活性。但是,因为蜂胶的成分十分复杂,需要通过更多的体内外实验来进行相关检测,并阐述其活性成分相互间的协同作用。蜂胶的主要抗肿瘤机制包括促进细胞凋亡、诱导细胞周期阻滞和干扰细胞代谢途径,这也可以为研发新型诱使癌细胞死亡的靶性药物提供信息(Cotter,2009)。

参考文献

Ahn MR,Kunimasa K,Ohta T,Kumazawa S,Kamihira M,Kaji K,Uto Y,Hori H,Nagasawa H,Nakayama T (2007) Suppression of tumor-induced angiogenesis by Brazilian propolis:major component artepillin C inhibits in vitro tube formation and endothelial cell proliferation[J]. *Cancer Letters*,252(2):235-243.

Avcı CB,Gündüz C,Baran Y,Sahin F,Yilmaaz S,Dogan ZO,Saydam G (2011) Caffeic acid phenethyl ester triggers apoptosis through induction of loss of mitochondrial membrane potential in CCRF-CEM cells[J]. *Cancer Res Clin Oncol*,137(1):41-47.

Awale S,Li F,Onozuka H,Esumi H,Tezuka Y,Kadota S (2008) Constituents of Brazilian red propolis and their preferential cytotoxic activity against human pancreatic PANC-1 cancer cell line in nutrient-deprived condition[J]. *Bioorganic & Medicinal Chemistry*,16(1):181-189.

Azevedo Bentes Monteiro Neto M,Souza Lima IM,Furtado RA,Bastos JK,Silva Filho AA,Tavares DC (2011) Antigenotoxicity of artepillin C *in vivo* evaluated by the micronucleus and comet assays[J]. *Journal of Applied Toxicology*,31:714-719.

Bankova V (2005) Chemical diversity of propolis and the problem of standardization[J]. *Journal of Ethnopharmacology*,100(1/2):114-117.

Bankova V,Dyulgerov AL,Popov S and Marekov N (1987) A GC/MS study of the propolis phenolic constituents[J]. *Zeitschrift für Naturforschung. Section C,Biosciences*,42(1/2):147-151.

Beltrán-Ramírez O,Alemán-Lazarini L,Salcido-Neyoy M,Hernandez-Garcia S,Fattel-Fazenda S,Arce-Popoca E,Arellanes-Robledo J,Garcia-Roman R,Vazquez-Vazquez P,Sierra-Santoyo A Villa-Trevino S (2008) Evidence that the anticarcinogenic effect of caffeic acid phenethyl ester in the resistant hepatocyte model involves modifications of cytochrome P450[J]. *Toxicological Sciences*,104(1):100-106.

Burdock GA (1998) Review of the Biological PropertieReview of the Biological Properties and Toxicity of Bee Propolis (Propolis)[J]. *Food and Chemical Toxicology*,36 (1998):347-363.

Burke JF,Schlosser L,Chen H (2012) Chrysin induces growth suppression through apoptosis in neuroblastoma cells[J] *Journal of the American College of Surgeons*,215 (3):S70.

Chen MJ,Chang WH,Lin CC,Liu CY,Wang TE,Chu CH,Shih SC,Chen YJ (2008) Caffeic acid phenethyl ester induces apoptosis of human pancreatic cancer cells involving caspase and mitochondrial dysfunction[J] *Pancreatology*,8(6):566-576.

Cole N,Sou PW,Ngo A,Tsang KH,Severino JAJ,Arun SJ,Duke CC,Reeve VE (2010) Topical "Sydney" propolis protects against UV-radiation-induced inflammation,lipid peroxidation and immune suppression in mouse skin[J]. *Int Arch Allergy Immunol*, 152(2):87-97.

Cotter TG (2009) Apoptosis and cancer:the genesis of a research field[J]. *Nature Reviews Cancer*,9(7):501-507.

Cunha I,Sawaya A,Caetano FM (2004) Factors that influence the yield and composition of Brazilian propolis extracts[J]. *Journal of the Brazilian de Quimica*,15(6):964-970.

Demir S,Aliyazicioglu Y,Turan I,Misir S,Mentese A,Yaman SO,Akbulut K,Kilinc K, Deger O (2016) Antiproliferative and proapoptotic activity of turkish propolis on human lung cancer cell line[J]. *Nutrition and Cancer-an International Journal*,68 (1):165-172.

Dreesen O,Brivanlou AH (2007) Signaling pathways in cancer and embryonic stem cells [J]. *Stem Cell Review*,3(1):7-17.

Eroglu HE,Oezkul Y,Tatlisen A,Silici S (2008) Anticarcinogenic and antimitotic effects of Turkish propolis and mitomycin-C on tissue cultures of bladder cancer[J]. *Natural Production Research*,22(12):1060-1066.

Fu B,Xue J,Li Z,Shi X,Jiang BH,Fang J (2007) Chrysin inhibits expression of hypoxia-inducible factor-1 through reducing hypoxia-inducible factor-1 stability and inhibiting its protein synthesis[J]. *Molecular Cancer Therapeutics*,6(1):220-226.

Gardana C,Scaglianti M,Pietta P,Simonetti P (2007) Analysis of the polyphenolic fraction of propolis from different sources by liquid chromatography-tandem mass spectrometry [J]. *Journal of Pharmaceutical and Biomedical Analysis*,45(2007):390-399.

Grunberger D,Banerjee R,Eisinger K,Oltz EM,Efros L,Caldwell M,Estevez V,Nakanishi K (1988) Preferential cytotoxicity on tumor cells by caffeic acid phenethyl ester isolated from propolis[J]. *Experientia*,44(3):230-232.

Inoue K,Saito M,Kanai T,Tetsuya K,Shigematsu N,Uno T,Isobe K,Liu CH,Ito H (2008) Anti-tumor effects of water-soluble propolis on a mouse sarcoma cell line *in* vivo and *in vitro*[J]. *The American Journal of Chinese Medicine*, 36(3): 625-634.

Ishihara M, Naoi K, Hashita M, Itoh Y, Suzui M (2009) Growth inhibitory activity of ethanol extracts of Chinese and Brazilian propolis in four human colon carcinoma cell lines[J]. *Oncology Reports*, 22(2): 349-354.

Jasprica I, Mornar A, Debeljak Z, Smolcic-Bubalo A, Medic-Snric M, Mayer L, Romic Z, Bucan K, Balog T, Sobocanec S, Sverko V (2007) *In vivo* study of propolis supplementation effects on antioxidative status and red blood cells[J]. *Journal of Ethnopharmacology*, 110(3): 548-554.

Jin UH, Song KH, Motomura M, Suzuki I, Gu YH (2008) Caffeic acid phenethyl ester induces mitochondria-mediated apoptosis in human myeloid leukemia U937 cells[J]. *Molecular and cellular Biochemistry*, 310: 43-48.

Josipovic P, Orsolic N (2008) Cytotoxicity of polyphenolic/flavonoid compounds in a leukaemia cell culture[J]. *Arh Hig Rada Toksikol*, 59(4): 299-308.

Karikas GA (2010) Anticancer and chemopreventing natural products: some biochemical and therapeutic aspects[J]. *Journal of Buon*, 15(4): 627-638.

Kasala ER, Bodduluru LN, Madana RM, V AK, Gogoi R, Barua CC (2015) Chemopreventive and therapeutic potential of chrysin in cancer: mechanistic perspectives[J]. *Toxicol Letter*, 233(2): 214-225.

Khacha-ananda S, Tragoolpua K, Chantawannakul P, Tragoolpua Y (2016) Propolis extracts from the northern region of thailand suppress cancer cell growth through induction of apoptosis pathways[J]. *Investigational New Drugs*, 34(6): 707-722.

Khan MS, Devaraj H, Devaraj N (2011) Chrysin abrogates early hepatocarcinogenesis and induces apoptosis in N-nitrosodiethylamine-induced preneoplastic nodules in rats[J]. *Toxicology and Applied Pharmacology*, 251(2011): 85-94.

Khan MS, Halagowder D, Devaraj SN (2011) Methylated chrysin induces co-ordinated attenuation of the canonical Wnt and NF-κB signaling pathway and upregulates apoptotic gene expression in the early hepatocarcinogenesis rat model[J]. *Chemico-Biological Interactions*, 193(1): 12-21.

Kim DM, Lee GD, Aum SH, Kim HJ (2008) Preparation of propolis nanofood and application to human cancer (Retracted Article. See vol 32, pg 1135, 2009)[J]. *Biological and Pharmaceutical Bulletin*, 31(9): 1704-1710.

Kimoto T, Arai S, Aga M, Hanaya T, Kohguchi M, Nomura Y, Kurimoto M (1996). Cell cycle and apoptosis in cancer induced by the artepillin C extracted from Brazilian propolis[J]. *Gan To Kagaku Ryoho*, 23(13): 1855-1859.

Kouidhi B, Zmantar T, Bakhrouf A (2010) Anti-cariogenic and anti-biofilms activity of Tunisian propolis extract and its potential protective effect against cancer cells proliferation[J]. *Anaerobe*, 16(6): 566-571.

Kuo YY, Jim WT, Su LC, Chung CJ, Lin CY, Huo C, Tseng JC, Huang SH, Lai CJ, Chen BC, Wang BJ, Chan TM, Lin HP, Chang WSW, Chang CR, Chuu CP (2015) Caffeic acid phenethyl ester is a potential therapeutic agent for oral cancer[J]. *International Journal of Molecular Sciences*, 16(5):10748-10766.

Li H, Kapur A, Yang JX, Srivastava S, Mcleod DG, Paredes-Guzman JF, Daugsch A, Park YK, Rhim JS (2007) Antiproliferation of human prostate cancer cells by ethanolic extracts of Brazilian propolis and its botanical origin[J]. *International Journal of Oncology*, 31(3):601-606.

Li F, Awale S, Tezuka Y, Kadota S (2009a) Cytotoxic constituents of propolis from Myanmar and their structure-activity relationship[J]. *Biological and Pharmaceutical Bulletin*, 32(12):2075-2078.

Li F, Awale S, Zhang H, Tezuka Y, Esumi H, Kadota S (2009b) Chemical constituents of propolis from Myanmar and their preferential cytotoxicity against a human pancreatic cancer cell line[J]. *Journal of Natural products*, 72(7):1283-1287.

Li F, Awale S, Tezuka Y, Kadota S (2010a) Cytotoxicity of constituents from Mexican propolis against a panel of six different cancer cell lines[J]. *Natural Product Communications*, 5(10):1601-1606.

Li X, Huang Q, Ong CN, Yang XF, Shen HM (2010b) Chrysin sensitizes tumor necrosis factor-α-induced apoptosis in human tumor cells via suppression of nuclear factor-kappa B.[J]. *Cancer Letters*, 293(2010):109-116

Lin WL, Liang WH, Lee YJ, Chuang SK, Tseng TH (2010) Antitumor progression potential of caffeic acid phenethyl ester involving p75 NTR in C6 glioma cells[J]. *Chemico-Biological Interactions*, 188(3):607-615.

Mani F, Damasceno HCR, Novelli ELB, Martins EAM, Sforcin JM (2006) Propolis: Effect of different concentrations, extracts and intake period on seric biochemical variables [J]. *Journal of Ethnopharmacology*, 105(1-2):95-98.

Matsuno T, Jung SK, Matsumoto Y, Saito M, Morikawa J (1997) Preferential cytotoxicity to tumor cells of 3,5-diprenyl-4-hydroxycinnamic acid (artepillin C) isolated from propolis[J]. *Anticancer Research*, 17(5A):3565-3568.

Maurya AK, Vinayak M (2015) Anticarcinogenic action of quercetin by downregulation of phosphatidylinositol 3-kinase (PI3K) and protein kinase C (PKC) via induction of p53 in hepatocellular carcinoma (HepG2) cell line[J]. *Molecular Biology Reports*, 42(9):1419-1429.

Missima F, Pagliarone AC, Orsatti CL, Araujo JPJ, Sforcin JM (2010) The effect of propolis on Th1/Th2 cytokine expression and production by melanoma-bearing mice submitted to stress[J]. *Phytother Research*, 24(10):1501-1507.

Missima F, Pagliarone AC, Orsatti CL, Sforcin JM (2009) The effect of propolis on pro-inflammatory cytokines produced by melanoma-bearing mice submitted to chronic stress[J]. *Journal of ApiProduct and ApiMedical Science*, 1(2009):11-15.

Natarajan K, Singh S, Burke TR, Grunberger D, Aggarwal BB (1996) Caffeic acid phenethyl ester is a potent and specific inhibitor of activation of nuclear transcription factor NF-kappa B. [J]. *Proceedings of the National Academy of Sciences of the United States of America*, 93(17):9090-9095.

Ohta T, Ahn MR, Kunimasa K, Kumazawa S, Nakayama T, Kaji K, Uto Y, Hori H, Nagasawa H (2009) Correlation between antiangiogenic activity and antioxidant activity of various components from propolis [J]. *Molecular Nutrition & Food Research*, 53(5):643-651.

Orsíolic N, Sí ver L, Terzic S, Basíic I (2005) Peroral application of water-soluble derivative of propolis (WSDP) and its related polyphenolic compounds and their influence on immunological and antitumour activity [J]. *Veterinary Research Communications*, 29(2005):575-593.

Oršolii ć N, Basic I (2003) Immunomodulation by water-soluble derivative of propolis: a factor of antitumor reactivity [J]. *Journal of Ethnopharmacology*, 84(2-3):265-273.

Oršolic ć N, Knezevic AH, Šver L, Terzi ć S, Basic I (2004) Immunomodulatory and antimetastatic action of propolis and related polyphenolic compounds [J]. *Journal of Ethnopharmacology*, 94(2-3):307-315.

Oršolic N, Saranovic AB, Basic I (2006) Direct and indirect mechanism(s) of antitumour activity of propolis and its polyphenolic compounds [J]. *Planta Medica*, 72(1):20-27.

Ozkul Y, Eroglu HE, Ok E (2006) Genotoxic potential of Turkish propolis in peripheral blood lymphocytes [J]. *Die Pharmazie-An International Journal of Pharmaceutical Sciences*, 61(7):638-640.

Ozkul Y, Silici S, Eroglu E (2005) The anticarcinogenic effect of propolis in human lymphocytes culture [J]. *Phytomedicine*, 12(10):742-747.

Patel S (2015) Emerging adjuvant therapy for cancer: propolis and its constituents [J]. *Journal of Dietary Supplements*, Online:1-24.

Popolo A, Piccinelli AL, Morello S, Sorrentino R, Rubio OC, Rastrelli L, Aldo P (2011) Cytotoxic activity of nemorosone in human MCF-7 breast cancer cells [J]. *Canadian Journal of Physiology and Pharmacology*, 89(1):50-57.

Popolo A, Piccinelli LA, Morello S, Cuesta-Rubio O, Sorrentino R, Rastrelli L, Pinto A (2009) Antiproliferative activity of brown Cuban propolis extract on human breast cancer cells [J]. *Natural Product Communications*, 4(12):1711-1716.

Samarghandian S, Azimi Nezhad M, Mohammadi G (2014) Role of Caspases, Bax and Bcl-2 in Chrysin-Induced Apoptosis in the A549 Human Lung Adenocarcinoma Epithelial Cells [J]. *Anti-cancer Agents in Medicinal Chemistry*, 14(6):901-909.

Sforcin JM (2007) Propolis and the immune system: a review [J]. *Journal of Ethnopharmacology*, 113(1):1-14.

Sforcin JM, Bankova V (2011) Propolis: Is there a potential for the development of new drugs? [J]. *Journal of Ethnopharmacology*, 133(2):253-260.

Sforcin JM, Kaneno R, Funari S (2002) Absence of seasonal effect on the immunomodulatory action of Brazilian propolis on natural killer activity [J]. *Journal of Venomous Animals and Toxins*, 8(1):19-29.

Sforcin JM, Novelli E, Funari S (2002) Seasonal effect of Brazilian propolis on seric biochemical variables[J]. *Journal of Venomous Animals and Toxins*, 8(2):244-254.

Sha N, Guan SH, Lu ZQ, Guang-Tong C, Huang HL, Xie FB, Yue QX, Liu X, Guo DA (2009) Cytotoxic Constituents of Chinese Propolis[J]. *Journal of Natural Products*, 72(4):799-801.

Shvarzbeyn J, Huleihel M (2011) Effect of propolis and caffeic acid phenethyl ester (CAPE) on NF kappa B activation by HTLV-1 Tax activities[J]. *Retrovirology*, 90(3):108-115.

Singer K, Gottfried E, Kreutz M, Mackensen A (2011) Suppression of T-cell responses by tumor metabolites[J]. *Cancer Immunology Immunotherapy*, 60(3):425-431.

Sugantha Priya E, Selvakumar K, Bavithra S, Elumalai P, Arunkumar R, Raja Singh P, Brindha Mercy A, Arunakaran J (2013) Anti-cancer activity of quercetin in neuroblastoma: an *in vitro* approach[J]. *Neurological Sciences*, 35(2):163-170.

Syamsudin, Partomuan Simanjuntak. (2010) Apoptosis of human breast cancer cells induced by ethylacetate extracts of propolis. [J]. *American Journal of Biochemistry and Biotechnology*, 6(2):84-88.

Szliszka E, Czuba ZP, Domino M, Mazur B, Zydowicz G, Krol W (2009) Ethanolic extract of propolis (EEP) enhances the apoptosis-inducing potential of TRAIL in cancer cells [J]. *Molecules*. 14(2):738-754.

Szliszka E, Zydowicz G, Janoszka B, Dobosz C, Kowalczyk-Ziomek G, Krol W (2011) Ethanolic extract of Brazilian green propolis sensitizes prostate cancer cells to TRAIL-induced apoptosis[J]. *International Journal of Oncology*, 38(4):941-953.

Szliszka E, Zydowicz G, Mizgala E, Krol W (2012) Artepillin C (3,5-diprenyl-4-hydroxycinnamic acid) sensitizes LNCaP prostate cancer cells to TRAIL-induced apoptosis[J]. *International Journal of Oncology*, 41(3):818-828.

Tani H, Hasumi K, Tatefuji T, Hashimoto K, Koshino H, Takahashi S (2010) Inhibitory activity of Brazilian green propolis components and their derivatives on the release of cys-leukotrienes. [J]. *Bioorganic & Medicinal Chemistry*, 18(1):151-157.

Tavares DC, Lira WM, Santini CB, Takahashi CS, Bastos JK (2007b) Effects of propolis crude hydroalcoholic extract on chromosomal aberrations induced by doxorubicin in rats[J]. *Planta Medica*, 73(15):1531-1536.

Tavares DC, Resende FA, Alves JM, Munari CC, Senedese JM, Sousa JPB, Bastos JK (2007a) Inhibition of doxorubicin-induced mutagenicity by *Baccharis dracunculifolia* [J]. *Mutation Research-Genetic Toxicology and Environmental Mutagenesis*, 634(1-2):112-118.

Umthong S, Puthong S, Chanchao C (2009) Trigona laeviceps Propolis from Thailand: antimicrobial, antiproliferative and cytotoxic activities [J]. *American Journal of Chinese Medicine*, 37(5):855-865.

Valente MJ, Baltazar AF, Henrique R, Estevinho L, Carvalho M (2011) Biological activities of Portuguese propolis:Protection against free radical-induced erythrocyte damage and inhibition of human renal cancer cell growth in vitro[J]. *Food and Chemical Toxicology*,49(1):86-92.

Vit P,Huq F,Barth O,Campo M,Santos E. (2015) Use of Propolis in Cancer Research[J]. *British Journal of Medicine & Medical Research*,8(2):88-109.

Wang T,Chen L,Wu W,Long Y,Wang R (2008) Potential cytoprotection:antioxidant defence by caffeic acid phenethyl ester against free radical-induced damage of lipids,DNA,and proteins[J]. *Canadian Journal Physiology and Pharmacology*,86(5):279-287.

Xing N,Chen Y,Mitchell SH,Young CYF (2001) Quercetin inhibits the expression and function of the androgenreceptor in LNCaP prostate cancer cells[J]. *Carcinogenesis*,22(3):409-414.

Xuan H,Li Z,Yan H,Qing S,Wang K,He Q,Wang YJ,Hu FL. (2014) Antitumor activity of Chinese propolis in human breast cancer MCF-7 and MDA-MB-231 cells[J]. *Evidence-Based Complementary and Alternative Medicine*,2014(1):280120-11.

Xuan HZ,Li Z,Yan HY,Sang Q,Wang K,He QT,Wang YJ,Hu FL (2014) Antitumor activity of Chinese propolis in human breast cancer MCF-7 and MDA-MB-231 cells[J]. *Evidence-based Complementary and Alternative Medicine*,2014:280120.

Yasui Y,Miyamoto S,Kim M,Kohno H,Sugie S,Tanaka T (2008) Aqueous and ethanolic extract fractions from the Brazilian propolis suppress azoxymethane-induced aberrant crypt foci in rats[J]. *Oncology Reports*,20(3):493-499.

Zhu L,Luo Q,Bi J,Ding J,Ge S,Chen F (2014) Galangin inhibits growth of human head and neck squamous carcinoma cells *in vitro* and *in vivo*[J]. *Chemico-Biological Interactions*,224:149-156.

第五节　蜂胶的免疫调节作用

免疫是机体的一种保护性生理反应,其作用是识别"自己"和"非己",并排除抗原性"异物"包括病原微生物及其他物质,以及体内衰老细胞、突变细胞等,以维持机体内环境的平衡和稳定。蜂胶具有十分复杂的化学成分和广泛的生物学活性(胡福良,2005)。早在20世纪60年代,苏联喀山兽医学院就对蜂胶影响动物机体的免疫活性进行了观察。通过对小鼠、豚鼠、家兔等实验,证明单独应用蜂胶或配合抗原进入机体,均能促进机体发生免疫过程。随后,国内外许多学者不断对蜂胶的免疫调节活性及相关作用机制进行研究,并取得了许多新的进展。研究表明,蜂胶的免疫调节活性具有双向性,既可以通过活化巨噬细胞、刺激淋巴细胞的增殖和分化、促进抗体及免疫调节性细胞因子的产生达到免疫增强作用;同时,也可以通过抑制免疫刺激性细胞因子的释放起到抑制过度的免疫反应,从而维持机体正常的免疫水平。

1 蜂胶的免疫增强作用

蜂胶的免疫增强作用主要通过提高机体的非特异性和特异性免疫系统来实现,表现在促进免疫器官胸腺、脾脏及淋巴结的发育,增强体液免疫及细胞免疫反应,促进抗体的生成,增强巨噬细胞的吞噬能力和自然杀伤性细胞的活性,从而增强机体对外界病原菌、病毒的抗感染,识别和清除自身衰老组织细胞,杀伤和清除异常突变细胞,抑制恶性肿瘤生长等功能。

Dimov等(1991)研究了蜂胶水溶性物质的免疫调节作用,发现蜂胶可以抑制环磷酰胺对小鼠机体造成的损伤,提高其存活率;并进一步证明蜂胶通过活化巨噬细胞调节体内外C1q的产生,同时直接或间接作用于细胞因子从而调节补体-受体功能,最终起到增强机体非特异性免疫反应的作用。同时,该研究团队发现蜂胶水溶性物质对金黄色葡萄球菌和白色念珠菌感染也具有良好的免疫增强效果(Dimov et al.,1992)。

Sampietro等(2016)研究了15个来源于阿根廷北部的蜂胶的免疫调节活性,同时采用高良姜素、松属素作为标准活性物质作为参考,研究了它们对中性白细胞的趋化作用与吞噬能力的影响。有一半的阿根廷蜂胶在$40\mu g/mL$浓度时即表现出强于同浓度高良姜素、松属素的效果,对中性白细胞的趋化、吞噬作用提升明显。Saavedra等(2016)研究了智利蜂胶及智利蜂胶中的主要活性黄酮—松属素对巨噬细胞的影响,发现智利蜂胶及松属素均能有效抑制金属蛋白酶MMP-9的基因表达,并呈剂量依赖关系;但蜂胶的效果较松属素更强,提示蜂胶中其他多酚类化合物可能与松属素之间有协同作用。

Ivanovska等(1995)对蜂胶及其化学组分的体内免疫调节作用进行了一系列研究。结果发现,苯乙烯酸和咖啡酸对红细胞溶解有良好的抑制作用,并且对机体损伤造成的免疫功能低下以及本身原有的免疫功能缺乏都有很好的调节效果;肉桂酸则可以促进小鼠脾细胞在脂多糖(LPS)、植物凝集素及刀豆蛋白A(Con A)等分裂素的刺激下对胸腺嘧啶的吸收能力,同时还能提高血清中IL-1β的含量,从而促进小鼠脾细胞的增殖反应。研究还发现,体外条件下蜂胶尤其是蜂胶水提液更能够充分发挥其免疫增强功能,对革兰阴性菌感染的免疫调节主要是通过提高巨噬细胞活力,增强其吞噬功能,以保护宿主免受病原微生物的入侵。尽管体外条件下蜂胶水提液对革兰阴性菌无抑制作用,但它可诱导宿主产生非特异性免疫保护(Ivanovska et al.,1993,1995a,1995b)。Moriyasu等(1994)研究发现,蜂胶($0.2\sim 1.0mg/mL$)能够刺激活化小鼠腹腔巨噬细胞从而促进细胞因子IL-1及TNF-α的产生,同时还能抑制NO的产生。Tatefuji等(1996)研究表明,巴西蜂胶中6种咖啡酰奎宁酸类化合物均可通过促进小鼠体内巨噬细胞的迁移及扩散从而诱导巨噬细胞的活化,同时还能够刺激巨噬细胞产生淋巴细胞活化因子,并进一步调控淋巴细胞的转运。Banskota等(2000)用不同溶剂提取的巴西蜂胶处理体外培养的小鼠结肠癌26-L5细胞及人体HT-1080纤维肉瘤细胞,结果证明上述提取物均可以通过活化巨噬细胞产生TNF-α间接对恶性肿瘤细胞株产生细胞毒性。Orsi等(2005)系统研究了蜂胶醇提液对巨噬细胞抑菌作用的影响,体外细胞培养结果显示,蜂胶(5、10、$20\mu g/mL$)能促进巨噬细胞产生H_2O_2;而体内实验结果显示,250、$500\mu g/mL$蜂胶能增强巨噬细胞对IFN-γ的反应性能,促进H_2O_2及NO的释放,从而增强其抑菌活性。Orsatti等(2010)研究发现,蜂胶处理能增加腹腔巨噬细胞中TLR-2、TLR-4的表达以及IL-1β的产生;同时也能上调脾细胞中TLR-2、TLR-4的表达以及IL-1β、IL-6的产生。该结果表明,蜂胶主要是通过活化TLRs(TOLL样受体)的表达及促炎细胞因子的生成从而增强机体的非特异性免疫反应的。

Fan 等（2015）研究了蜂胶黄酮联合淫羊藿多糖脂质体（EPL）对肝脏库普弗细胞（Kupffer Cells，位于肝脏中的特殊巨噬细胞）的激活效果，结果发现 EPL 能显著诱导细胞趋化因子 RANTES 和 MCP-1 的分泌，同时能促进 NO 的分泌，并提高库普弗细胞的细胞吞噬能力。此外，EPL 处理还能提高库普弗细胞抗原呈递能力和激活淋巴细胞能力。该研究为 EPL 提高机体免疫力提供了新的证据。随后，Washio 等（2015）也发现蜂胶乙醇提取物能激活 C2C12 成肌细胞因子和趋化因子的产生。

人们曾经一度认为蜂胶的免疫增强活性只局限于巨噬细胞上。然而，随着研究的深入，人们发现蜂胶还能通过调控淋巴细胞及相关抗体、细胞因子的产生起到增强机体免疫反应的作用。Scheller 等（1988）就蜂胶对体液免疫的作用展开了研究，结果发现蜂胶醇提物可以促进小鼠体内的绵羊红细胞产生抗体；进一步研究表明，这一免疫刺激活性与巨噬细胞的活化直接相关，并通过促进细胞因子的产生而达到调控 B 细胞及 T 细胞的功能。此外，他们还观察到短期使用蜂胶具有更高水平的抗体含量。潘明（2007）研究了蜂胶提取物对荷瘤小鼠免疫系统的影响，将 S180 肉瘤移植到 BalB/C 纯种小鼠后再进行腹腔注射蜂胶提取物，结果显示，蜂胶提取物对 NK 细胞杀伤活性有明显的增强作用，且能显著提高淋巴细胞的增殖能力和 IL-2 的含量，证明蜂胶提取物能够显著提高荷瘤小鼠免疫系统的活性。另有研究发现，蜂胶还能够增强中老年大鼠的自然杀伤细胞活性和巨噬细胞吞噬功能，表明蜂胶可以提高中老年大鼠的免疫力（杨锋等，2007）。

李淑华等（2001）用单克隆抗体技术检测实验小鼠药物处理前后 T 细胞总数及亚群变化，发现蜂胶乙醇提取物能促进 Con A 诱导的淋巴细胞增殖，增加 T 细胞总数并调整 T 细胞亚群紊乱，表明 EEP 对免疫功能低下小鼠的细胞免疫反应具有免疫刺激和调节作用。于晓红等（2001）研究发现，蜂胶能够增强免疫功能低下小鼠的抗体生成细胞功能，提高溶血素含量，增强单核吞噬细胞功能。说明蜂胶有促进和调节机体体液免疫功能的作用和增加机体吞噬细胞的吞噬活性，并能刺激体内抗体的产生，抵抗多种疾病。Gao 等（2014）的研究结果表明，巴西绿蜂胶对免疫功能低下的老龄小鼠同样具有免疫增强作用，能显著提高其机体内腹腔巨噬细胞的吞噬活性、抗体生成及 IgG 的含量。胡箭卫和李旭涛（2003）研究发现，蜂胶本身不具有抗原特性，但能提高血清蛋白和丙种球蛋白的含量，并能提高白细胞和巨噬细胞的吞噬能力。

近年来，随着蜂胶在畜牧兽医上的广泛应用，通过对多种家畜、家禽等进行的一系列实验已经初步证明了蜂胶本身所具有的免疫增强作用。柴家前等（2002）研究发现，蜂胶及其提取物能够通过影响雏鸡免疫器官，如胸腺、法氏囊、脾脏的发育来提高机体的免疫能力，同时能显著提高 T 淋巴细胞的比率和活性；并能通过增强红细胞膜上的 C3b 受体活性来增强红细胞的免疫功能。Ahmed 等（2016）研究了中国蜂胶对 Ross 308 肉仔鸡法氏囊组织结构的影响，发现低剂量（100、250mg/kg）的蜂胶可以增加法氏囊褶以及法氏囊淋巴小结的大小，同时能够小幅度地增加滤泡间结缔组织数量；而高剂量的蜂胶（500、750mg/kg）则能够使法氏囊发生实质性变化，包括退化为淋巴滤泡，具体表现为囊腔形成、液化性坏死、滤泡间结缔组织的显著增加。该研究表明，高浓度的中国蜂胶能够诱导法氏囊的快速退化，从而进一步增强肉仔鸡的体液免疫。Daneshmand 等（2015）研究了蜂胶及益生素单独及混合使用对肉仔公鸡免疫反应的影响，结果显示，单独使用蜂胶及两者的混合使用均能显著增加肉仔鸡脾脏及法氏囊的相对重量,同时还能显著增加其抗新城疫病毒的抗体滴度。上述研究结果为蜂胶更好地用于畜禽日粮添加剂提供了合理的理论依据。

2 蜂胶的免疫抑制作用

正常情况下免疫系统只对侵入机体的外来物,如细菌、病毒、寄生虫以及移植物等产生反应,消灭或排斥这些异物。在某些因素影响下,机体组织或免疫系统本身会出现一些异常,导致过度的免疫反应,造成机体的损害。因此,为了抑制异常的免疫反应,维持机体健康,寻找合理有效、安全低毒的免疫抑制活性物质必不可少。目前,蜂胶的免疫抑制活性已经得到证实。在免疫抑制模型中,Ivanovska等(1995a)的体外实验结果显示,63～1000 $\mu g/mL$ 的蜂胶水提物能够抑制小鼠补体系统的经典及替代通路。Bratter等(1999)在研究蜂胶对人体促炎性细胞因子的作用过程中发现连续服用蜂胶胶囊(500mg)2周后,人外周血淋巴细胞中细胞因子TNF-α、IL-6 及 IL-8 的水平得以显著的上升,而血浆中促炎细胞因子 TNF-α、IL-1β、IL-6 及 IL-8 则无明显的改变。

李英华(2002)通过大鼠佐剂型关节炎、大鼠试验性胸膜炎、大鼠实验性肺损伤、大鼠急性关节炎、小鼠棉秋肉芽肿模型、小鼠腹腔毛细血管通透性试验以及小鼠S180实体瘤模型等一系列的试验,研究了蜂胶醇提液和水提液的抗炎机制以及免疫调节作用,发现蜂胶水提液和醇提液均能够降低炎症区域毛细血管的通透性,抑制炎性肿胀,对肺损伤有明显的保护作用,而且能够对抗水肿的产生,对慢性炎症和实体瘤模型有显著的抑制作用;并且能够降低胶叉莱胶致大鼠胸膜炎组织中前列腺素E2(PGE2)和总蛋白的含量,降低炎症组织中过氧化物代谢产物MDA的含量,减轻炎症反应发生程度;研究还发现,蜂胶能够对急性炎症渗出液中一氧化氮(NO)及溶菌酶含量的升高表现出抑制作用,能够对抗炎症模型内细胞因子IL-6的异常升高,对IL-2和IFN-γ影响不明显。说明蜂胶在发挥免疫调节作用时可能主要靠B淋巴系统发挥作用,而对T淋巴系统的影响不显著。

在对蜂胶抗炎作用研究过程中,Mirzoeva和Calder(1996)发现蜂胶提取液及其多酚类化合物不仅能够抑制NO的产生,减少PGE2的生成;同时也可以通过影响相关细胞因子的产生发挥其免疫抑制作用。Hu等(2005)研究发现,蜂胶醇提液及水提液均能显著抑制小鼠炎症部位IL-6水平的上升,而对IL-2及IFN-γ的水平无明显影响。Sa-Nunes等(2003)研究发现,体外条件下蜂胶可以抑制脾淋巴细胞的增殖反应,同时还能促进IFN-γ的生成。而此前You等(1998)研究指出,蜂胶中的黄酮类化合物对淋巴增生反应也具有相似的免疫抑制活性。Sforcin(2007)认为蜂胶抑制淋巴细胞增生的主要途径是通过影响一些调节性细胞因子如转化生长因子β(TGF-β)、IL-10的产生来实现的;而蜂胶的抗炎、抗新血管生成作用则主要是通过调控 $TGF-\beta_1$ 的生成来完成的。

近年来,众多研究结果表明,蜂胶还能显著抑制病理状态下免疫细胞产生IL-8及IL-17(de Moura et al.,2011;Skiba et al.,2011;Tanaka et al.,2012;Orsatti and Sforcin,2012)。此外,Orsatti等(2010)的研究结果表明,蜂胶对小鼠巨噬细胞的Toll样受体(Toll-like receptors,TLRs)2、4的表达也有相似的抑制作用;特别是在应激状态下,蜂胶通过下调小鼠TLR-2、TLR-4的mRNA表达水平,从而发挥免疫调节作用。TLRs在免疫系统,特别是天然免疫反应中发挥着重要作用,同时与细胞因子的产生、免疫细胞的激活也有着一定的联系(Steinhagen et al.,2011;Mills,2011)。Dantas等(2006)研究发现,蜂胶能够部分抑制感染动物体内淋巴细胞亚群 $CD4^+$、$CD8^+$ 中 $CD69^+$、$CD44^+$ 表达的上升,以及在 $CD8^+CD62L$ 中表达的下降,表明蜂胶对 T 细胞亚群效应器及其记忆功能的抑制作用。

此外,蜂胶对异体间组织器官移植过程中由于受体的 T 细胞活化而引起的免疫排斥反应也有调节作用。在混合白细胞实验模型中,Chan 等(2013)研究发现,巴西绿蜂胶主要通过刺激 $CD14^+$ 单核细胞,从而对人体休眠性外周单核血细胞进行免疫调节,T 细胞是受其影响的一种主要的外周单核血细胞;同时还发现,巴西绿蜂胶以剂量依赖性方式抑制了外周血淋巴细胞的增殖,而对 T 细胞没有表现出明显的刺激作用。这一结果表明,巴西绿蜂胶对休眠性和活化后的 T 细胞可能有着不同的作用。此外,巴西绿蜂胶对 T 细胞的免疫抑制作用也暗示了其潜在的抗炎活性。此前,Cheung 等(2011)在类似的混合白细胞反应模型中发现,巴西绿蜂胶在该反应过程中同时抑制了 T 细胞的增殖和活化。进一步研究发现,巴西绿蜂胶对 $CD4^+T$ 细胞的抑制作用部分是由于增殖过程中 T 细胞的选择性诱导凋亡作用,而不是调节性 T 细胞的诱导作用。然而,在同样的刺激物重复刺激后,巴西绿蜂胶的这种抑制作用是可逆的。巴西绿蜂胶还能抑制 T 细胞的活化和细胞因子的产生。

3 蜂胶及其化学成分的免疫佐剂活性

蜂胶本身虽不具有抗原特性,但它具有良好的免疫增强作用。将蜂胶配合抗原注入机体,能够有效激活机体的免疫系统,增强补体功能,增加免疫细胞数量,促进抗体的产生,从而提高机体的特异性和非特异性免疫力。

早在 20 世纪 80 年代末,沈志强等(1989)就提出将蜂胶用作免疫增强剂,并研制出禽霍乱蜂胶灭活疫苗,开创了蜂胶免疫佐剂研究的新领域。目前,蜂胶作为免疫佐剂已经成功地应用于禽霍乱灭活疫苗、猪副嗜血杆菌灭活疫苗、鸡新城疫疫苗以及鸡多杀性巴氏杆菌、大肠埃希菌和肺炎克雷伯菌的三价细菌疫苗等多个领域(邱昌庆和董恩娜,1996;张莉等,2004;赵恒章等,2009;金天明等,2009)。沈志强等研制的禽霍乱蜂胶菌苗产生抗体快速、高效,保持免疫期时间长,可达 6 个月之久,并且保护率近 100%。同时,这种疫苗可全面刺激机体免疫器官的发育和全面调动机体的免疫防护系统,尤其是提高了细胞免疫力(沈志强,1995)。赵恒章等(2009)以油乳剂、蜂胶和铝胶为佐剂,按一定比例配制成巴氏杆菌的灭活苗。经试验证明,该疫苗安全、有效,其中油乳剂灭活苗的保护期和蜂胶灭活苗的保护期相当,但蜂胶灭活苗的抗体水平上升速度比油乳剂灭活苗快且保护率高。Nassar 等(2012)将埃及蜂胶乙醇提物用于灭活的多杀巴斯德菌(*Pasteurella muitocida*)疫苗中,并比较了单独施用蜂胶、单独施用细菌疫苗及添加蜂胶提取物的细菌疫苗三者对兔的保护作用。在受到多杀巴斯德菌感染后,采用蜂胶作为佐剂的灭活细菌疫苗的试验组能有效地增强兔的特异性免疫与非特异性免疫反应,从而显著地减少了兔的死亡率并缓解了由细菌感染引起的临床组织病理学变化。Fischer 等(2007a)研究了蜂胶对小鼠的免疫辅佐活性,结果显示,小鼠接种 SuHV-1 后再配合使用氢氧化铝及蜂胶提取物能够显著提升其体内的抗体含量水平,而单独使用蜂胶提取物则不能诱导抗体的产生;两者结合使用可以增强小鼠体内的细胞免疫反应。这些研究结果表明,蜂胶作为免疫佐剂对机体的免疫反应具有极显著的增强效果。此外,多个研究证实蜂胶作为强效的免疫佐剂比传统的疫苗佐剂更具优势,其毒性要明显低于传统疫苗佐剂,且作用效果更迅速、高效、持久。

作为蜂胶中主要的活性成分——黄酮类化合物的免疫佐剂活性近来也得以广泛的研

究。张宝康等(2006)以蜂胶黄酮为免疫增强剂,探讨其对免疫雏鸡血清抗体效价和外周血T淋巴细胞增殖以及对培养的鸡脾脏淋巴细胞增殖的影响。结果显示,蜂胶黄酮能显著提高雏鸡的血清抗体效价,促进外周血淋巴细胞增殖,且有一定的量效和时效关系;在体外,也能促进鸡脾脏T淋巴细胞增殖,以低浓度的效果较好。Fischer 等(2007b)研究发现,巴西绿蜂胶中多种活性成分尤其是黄酮类化合物能够显著增加免疫肉牛血清中中和性抗体的含量,同时也促进了高效价抗体的产生。然而,由于黄酮类化合物极弱的水溶性以及易氧化的特性使其难以得到广泛的应用。

4 蜂胶中主要的免疫调节因子

蜂胶广泛的生物学活性与其自身复杂的化学成分密不可分。蜂胶中含有丰富的黄酮、酚酸及萜烯类化合物以及微量的维生素、矿物元素、酶类、氨基酸及脂肪酸等。大量研究证明,蜂胶中众多化学成分对机体的免疫系统均具有广泛的调节作用。沈志强等(1989)研究发现,蜂胶中的 VA 是一种中枢作用佐剂,可以通过免疫中枢刺激抗体应答的增强;而 VC 和 VE 则能够通过增加抗体的产生,从而增强机体体液免疫(沈志强和杨永福,1989)。冯俊鹏(1997)研究发现,蜂胶中的 Zn 也可刺激机体的免疫力。缺 Zn 对 T 淋巴细胞的免疫机能会有很大的损害,将导致吞噬细胞明显变化,失去其多型性和伪足而成为平滑的圆形细胞,从而使免疫机能下降。此外,蜂胶中重要的活性成分——黄酮类化合物松属素、高良姜素、山柰素以及酚酸类化合物——p-香豆酸、咖啡酸及其酯类、咖啡酰奎宁酸类化合物均能促进抗体的产生,使血清总蛋白和丙种球蛋白的含量增加,并能增强白细胞和巨噬细胞的吞噬能力,从而增强机体的非特异性和特异性免疫反应。同时,这些多酚类化合物还具有抑制促炎细胞因子、花生四烯酸及前列腺素产生的作用,从而起到免疫抑制的作用。Búfalo 和 Sforcin(2015)研究发现,不同浓度的咖啡酸对人单核细胞具有双向的免疫调节活性,并指出这一免疫调节活性主要是通过作用于 Toll 样受体(TLRs)来实现的。

在众多免疫学调节因子中,蜂胶中两种重要的特征性活性成分咖啡酸苯乙酯(CAPE)和阿替匹林 C(Artepillin C)因其强效的免疫调节功效而备受人们的关注。CAPE 几乎存在于所有温带地区的杨树型蜂胶中,是一种具有极强生物学活性的化合物。CAPE 属于咖啡酸衍生物,首次由 Hashimoto 等在 1988 年合成,随后多个科学家展开了对 CAPE 生物学活性的研究,其免疫调节作用也已得以证实(Hashimoto et al.,1988)。Park 等(2004)连续 14d 给予雌性 BALB/c 小鼠口服不同浓度的 CAPE,结果发现,小鼠体内相应的抗体含量及免疫调控细胞因子 IL-2、IL-4 的含量均有所升高,同时也促进了 T 细胞的增殖作用。而 CAPE 的抗炎作用主要是通过抑制刺激性免疫细胞的活性来完成的。Marquez 等(2004)用 CAPE 培养经 SEB(葡萄球菌肠毒素 B)和 PHA(植物血凝素)处理过的人单核外周血细胞,结果发现,CAPE 通过抑制活化的 T 细胞内 IL-2 基因转录及合成,从而抑制 T 细胞的增殖;同时,它也可以抑制 T 细胞活化的两个关键转录因子 NF-κB 和 NFAT(活化 T 细胞核因子)的 DNA 转录及两者间的结合。体外研究发现,CAPE 还可以抑制巨噬细胞(NR 8383)产生 NF-κB 和 TNF-α,显著促进 DNA 断裂;在结肠上皮细胞(SW620)的研究中也得到了类似结果,说明 CAPE 的免疫调节作用机制是通过抑制 NF-κB 而降低促炎症细胞因子的产生,以及减少巨噬细胞的凋亡。研究者还指出这一作用机制很可能与 CAPE 对 PG-PE 诱导的结肠炎的减弱作用相关。此外,Wang 等在研究 CAPE 对人树突状细胞作用的过程中发现,体外条件下 CAPE 可以抑制经脂多糖(LPS)诱导的人树突状细胞的成熟分化(Wang et al., 2009,2010)。

阿替匹林 C(3,5-二异戊烯基-4-羟基肉桂酸)是一种低分子量的酚酸类化合物。Aga 等(1994)在巴西绿蜂胶中首次分离得到阿替匹林 C 并测定了其抗菌活性;随后,Kimoto 等(1996)在研究白血病细胞模型时发现了阿替匹林 C 的抗癌活性,并证实这一活性是通过对体内外白血病细胞的诱导性凋亡作用来完成的。然而,奇怪的是,Shimizu 等(2004)在检测蜂胶的抗氧化活性过程中发现阿替匹林 C 可以保护人肝癌细胞株 HepG2 免受氧化损伤。

由于阿替匹林 C 只存在于巴西绿蜂胶中,目前对其免疫学活性的研究还远远少于 CPAE。在人与小鼠混合白血病实验模型中,Kimoto 等(1998)发现阿替匹林 C 主要通过增加体内 $CD4^+/CD8^+$ 的比率以及 T 淋巴细胞的总数来进一步增强 T 细胞的细胞毒性,从而间接杀伤癌细胞。Paulino 等(2008)在研究阿替匹林 C 在小鼠体内的抗炎效果、吸收作用和生物利用率的过程中发现,阿替匹林 C 可以抑制巨噬细胞 NF-κB 的活性以及炎性因子 PGE2 和 NO 的产生,从而遏制卡拉胶诱导的大鼠爪子水肿的进一步恶化。Tani 等(2010)用人工合成的阿替匹林 C 处理鼻炎患者的外周单核细胞,结果发现,阿替匹林 C 可以抑制中性粒细胞向炎症区域的游走。同时,多个大鼠实验模型研究结果表明,阿替匹林 C 对生物体没有明显的毒性作用,因此可以将其用作抗过敏药物。

5 影响蜂胶免疫调节作用的因素

蜂胶是由蜜蜂采集到的植物树脂与其自身分泌物混合而成的树脂状固形物,其化学成分及生物学活性受蜂种、植物来源、地理位置及采胶季节等因素的影响。其中,植物来源和地理位置是影响蜂胶化学成分的主要因素。Chan 等(2013)检测了巴西、新西兰、中国和塔斯曼尼亚 4 个不同地区蜂胶醇提物对人休眠性外周单核血细胞的免疫学活性影响,结果显示,除了巴西绿蜂胶外,其余 3 种蜂胶都没有表现出对该细胞明显的增殖作用。不过,有趣的是,巴西绿蜂胶适度的刺激作用是通过增强细胞内线粒体的活性而完成的,而新西兰、中国和塔斯曼尼亚蜂胶却表现出以剂量和时间依赖性方式诱导细胞死亡,这一抑制作用主要源于蜂胶的细胞毒性。这一结果表明,不同地区、不同来源的蜂胶具有不同的生物学活性,同时与前期 Orsi 等(2005)的研究结果一致。巴西绿蜂胶和其他 3 种蜂胶提取物的唯一不同就在于他们成分上的差异,巴西绿蜂胶是产自热带地区的蜂胶,其主要的化学成分是异戊烯基肉桂酸及其衍生物、咖啡酰奎宁酸及二萜酸类化合物;而新西兰、中国和塔斯曼尼亚蜂胶均来自温带地区。Bankova(2005)曾报道来自温带地区的不同蜂胶含有相似的化学成分——B 环无取代基的黄酮类化合物,多数属杨树型蜂胶。最近,Conti 等(2015)比较研究了巴西、墨西哥、古巴 3 个不同地区蜂胶样对人单核细胞促炎及抗炎细胞因子的调节作用,发现巴西蜂胶能够同时刺激 TNF-α 以及 IL-10 的产生;古巴蜂胶能够促进 TNF-α 的产生,然而抑制 IL-10 的产生;而墨西哥蜂胶与之相反。进一步检测发现这 3 种蜂胶发挥活性的主要化学成分分别为阿替匹林 C、异黄酮及松属素。

季节是影响胶源植物化学组成的物候学因素,同时也是影响蜂胶化学成分及生物学活性的因素。Simões-Ambrosio 等(2010)研究发现,蜂胶的免疫调节活性随着季节的变化而有显著的差异。此外,提取加工方法也是影响蜂胶化学成分及生物学活性的一个重要因素。薛晓丽(2007)比较了乙醇提取、索氏提取、超声波提取、超临界流体提取等 4 种提取工艺的粗提取物中总黄酮含量,结果表明,不同提取工艺对蜂胶中黄酮类物质的提取有较大的影响。超声波提取法所得的粗提物中黄酮类物质含量最高,具有更强的生物学活性。

小　结

免疫反应由各种细胞和它们分泌的可溶性分子所介导,并通过各种反馈而进行调节。各种免疫细胞与可溶性免疫介质以及各种细胞因子之间相互刺激和制约,形成复杂的免疫调节系统。蜂胶可以对非特异性免疫系统以及特异性免疫系统的多个环节、多个途径起作用。同时,蜂胶对免疫系统的调节作用具有双向性,既可以通过活化巨噬细胞、刺激单核淋巴细胞的增殖和分化、促进抗体的产生达到免疫增强作用,又可以通过抑制免疫刺激性细胞因子的释放起到免疫抑制作用。

蜂胶在使用上相对安全,无毒副作用,然而,蜂胶的有效成分及含量随着胶源植物、地理位置和提取方法的变化而变化,这将直接影响其免疫调节活性的强弱,使其在使用上无法达到完全的标准化；另一方面,蜂胶中两种重要的生物活性成分——CAPE 和阿替匹林 C(Artepillin C)是蜂胶发挥免疫调节作用的关键性成分,也是目前研究得最为深入的成分,但蜂胶的化学成分极为复杂,不同活性物质的作用机制不尽相同,而且它们之间是否存在协同或拮抗作用不得而知。因此,今后还需要全面系统地研究蜂胶有效成分对其生物学活性的影响,为探明蜂胶免疫调节的作用机制,建立蜂胶的化学标准化提供理论依据。

参考文献

Aga H,Shibuya T,Sugimoto T,KurimotoM,Nakajima S (1994) Isolation and identification of antimicrobial compounds in Brazilian propolis[J]. *Bioscience, Biotechnology, and Biochemistry*, 58: 945-946.

Ahmed OHB, Mahmoud UT, Mahmoud MA, El-Bab MRF (2016) Histomorphological changes associated with different doses of Chinese propolis in the bursa of fabricius of chickens[J]. *Journal of Advanced Veterinary Research*, 6(1): 1-6.

Bankova V (2005) Recent trends and important developments in propolis research[J]. *Evidence-based Complementary and Alternative Medicine*, 2(1): 29-32.

Banskota AH, Tezuka Y, AdnyanaIK, Midorikawaa K, Matsushigea K, Messageb D, HuertasbAAG, Kadota S (2000) Cytotoxic, hepatoprotective and free radical scavenging effects of propolis from Brazil, Peru, the Netherlands and China[J]. *Journal of Ethnopharmacology*, 72(1): 239-246.

Bratter C, Tregel M, Liebenthal C, Volk HD (1999) Prophylaktische wirkungen von propolis zur immunstimulation: eine klinische pilotstudie[J]. *Forschende Komplementarmedizin*, 6: 256-260.

Búfalo MC, Sforcin JM (2015) The modulatory effects of caffeic acid on human monocytes and its involvement in propolis action[J]. *Journal of Pharmacy and Pharmacology*, 67(5): 740-745.

Chan GCF, Cheung KW, Sze DMY (2013) Theimmunomodulatory and anticancer properties of propolis[J]. *Clinical Reviews in Allergy and Immunology*, 44(3):262-273.

Cheung KW, Sze DM, Chan WK, Deng RX, Tu W, Chan GC (2011) Brazilian green propolis and its constituent, Artepillin C inhibits allogeneic activated human CD4 T cells expansion and activation[J]. *Journal of Ethnopharmacology*, 138(2):463-471.

Conti BJ, Santiago KB, Búfalo MC, Herrera YF, Alday E, Velazquez C, Sforcin JM (2015) Modulatory effects of propolis samples from Latin America (Brazil, Cuba and Mexico) on cytokine production by human monocytes[J]. *Journal of Pharmacy and Pharmacology*, 67(10):1431-1438.

Daneshmand A, Sadeghi GH, Karimi A, Vaziry A, Ibrahim SA (2015) Evaluating complementary effects of ethanol extract of propolis with the probioticon growth performance, immune response and serum metabolites in male broiler chickens[J]. *Livestock Science*, 178(2015):195-201.

Dantas AP, Olivieri BP, Gomes FHM, de Castro SL (2006) Treatment of *Trypanosomacruzi*-infected mice with propolis promotes changes in the immune response[J]. *Journal of Ethnopharmacology*, 103:187-193.

de Moura SAL, Ferreira MAND, Andrade SP, Reis MLC, Noviello MDL, Cara DC (2011) Braziliangreenpropolis inhibits inflammatory angiogenesis in amurine sponge model[J]. *Evidence-Based Complementary and Alternative Medicine*, 2011:182703.

DimovVB, Ivanovska ND, Bankova V, Popov SS (1992) Immunomodulatory action of propolis: IV. Prophylactic activity against Gram-negative infections and adjuvant effect of the water-soluble derivative[J]. *Vaccine*, 10(12):817-823.

DimovVB, IvanovskaND, ManolovaN, Bankova V, Nikolov N, Popov SS (1991) Immunomodulatory action of propolis. Influence on anti-infection protection and macrophage function[J]. *Apidologie*, 22:155-162.

Fan YP, Ren MM, Hou WF, Guo C, Tong DW, Ma L, Zhang WM, He MM (2015) The activation of Epimedium polysaccharide-propolis flavone liposome on Kupffer cells[J]. *Carbohydrate Polymers*, 133:613-623.

Fischer G, Cleff MB, DummezharLA, Paulinoc N, Paulinoc AS, de Oliveira Vilelab C, Camposb FS, Storchb T, Vargasb GDA, de Oliveira Hübnerb S, Vidor T (2007b) Adjuvant effect of green propolis on humoral immune response of bovines immunized with bovine herpesvirus type 5[J]. *Veterinary Immunology and Immunopathology*, 116(1):79-84.

Fischer G, Conceicao FR, Leite FPL, Dummera LA, Vargasb GDA, de Oliveira Hübnerb S, Dellagostina OA, Paulinod N, Paulinod AS, Vidor T (2007a) Immunomodulation produced by a green propolis extract on humoral and cellular responses of mice immunized with SuHV-1[J]. *Vaccine*, 25:1250-1256.

Gao WN, Wu JQ, Wei JY, Pu LL, Guo CJ, Yang JJ, Yang M, Luo HJ (2014) Brazilian green propolis improves immune function in aged mice[J]. *Journal of Clinical Biochemistry and Nutrition*, 55(1):7-10.

Hashimoto T, Tori M, Asakawa Y (1988) Synthesis of two allergenic constituents of propolis and poplar bud excretion[J]. *FürNaturforschung C*, 43(5-6): 470-472.

Hu FL, Hepburn HR, Li YH, Chen ML, Radloffd SE, Dayae S (2005) Effects of ethanol andwater extracts of propolis (bee glue) on acuteinflammatory animal models[J]. *Journal of Ethnopharmacology*, 100(3): 276-283.

Ivanovska ND, Dimov VB, Bankova VS, Popov SS (1995a) Immunomodulatory action of propolis. VI. Influence of a water soluble derivative on complement activity *in vivo*[J]. *Journal of Ethnopharmacology*, 47(3): 145-147.

Ivanovska ND, Dimov VB, Pavlova S, BankovaV, Popov SS (1995b) Immunomodulatory action of propolis: V. Anticomplementary activity of a water-soluble derivative[J]. *Journal of Ethnopharmacology*, 47: 135-143.

Ivanovska ND, Stefanova Z, Valeva V, Neychev H (1993) Immunomodulatory action of propolis: VII. A comparative study on cinnamic and caffeic acid lysine derivatives[J]. *ComptesRendus de l'AcademieBulgare des Sciences*, 46(10): 115-117.

Kimoto T, Arai S, Aga M, Hanaya T, Kohguchi M, Nomura Y, Kurimoto M (1996) Cell cycle and apoptosis in cancer induced by the Artepillin C extracted from Brazilian propolis[J]. *GanTo Kagaku Ryoho. Cancer Chemotherapy*, 23(13): 1855-1859.

Kimoto T, Arai S, Kohguchi M, Aga M, Nomura Y, Micallef MJ, Mito K (1998) Apoptosis and suppression of tumor growth by Artepillin C extracted from Brazilian propolis[J]. *Cancer Detection and Prevention*, 22(6): 506-515.

Marquez N, Sancho R, Macho A, Calzado MA, Fiebich BL, Muñoz E (2004) Caffeic acid phenethyl ester inhibits T-cell activation by targeting both nuclear factor of activated T-cells and NF-κB transcription factors[J]. *The Journal of Pharmacology and Experimental Therapeutics*, 308(3): 993-1001.

Mills KHG (2011) TLR-dependent T cell activation inautoimmunity[J]. *Nature Reviews Immunology*, 11(12): 807-822.

Mirzoeva OK, Calder PC (1996) The effect of propolis and itscomponents on eicosanoid production during theinflammatory response[J]. *Prostaglandins Leukotrienes Essent Fatty Acids*, 55(6): 441-449.

MoriyasuJ, AraiS, MotodaR, Kurimoto M (1994) *In vitro* activation of mouse macrophage by propolis extract powder[J]. *Biotherapy*, 8: 364-365.

Nassar SA, Mohamed AH, Soufy H, Nasr SM, Mahran KM (2012) Immunostimulant effect of Egyptian propolis in rabbits[J]. *The Scientific World Journal*, 2012: 901516.

Orsatti CL, Missima F, Pagliarone AC, Bachiega TF, Búfalo MC, AraújoJr JP, Sforcin JM (2010) Propolisimmunomodulatory action *in vivo* on Toll-like receptors 2 and 4 expression and on pro-inflammatory cytokines production in mice[J]. *Phytotherapy Research*, 24(8): 1141-1146.

Orsatti CL, Missima F, Pagliarone AC, Bachiega TF, Búfalo MC, AraújoJr JP, Sforcin JM (2010) Propolisimmunomodulatory action *in vivo* on toll-like receptors 2and 4 expression and on pro-inflammatory cytokinesproduction in mice[J]. *Phytotherapy Research*,24(8):1141-1146.

Orsatti CL,Sforcin JM (2012) Propolisimmunomodulatoryactivity on TLR-2 and TLR-4 expression by chronicallystressed mice[J]. *Natural Product Research*,26(5):446-453.

Orsi RO,Sforcin JM,Funari SRC,Bankova V (2005) Effects of Brazilian and Bulgarian propolis on bactericidal activity of macrophages against *Salmonella Typhimurium*[J]. *International Immunopharmacology*,5(2):359-368.

Park JH,Lee JK,Kim HS,Chunga ST,Eoma JH,Kima KA,Chunga SJ,Paikb SY,Oh HY (2004) Immunomodulatory effect of caffeic acid phenethyl ester in Balb/c mice[J]. *International Immunopharmacology*,4(3):429-436.

Paulino N,Abreu SRL,Uto Y,Koyama D,Nagasawa H,Hori H,Dirsch VM,Vollmar AM, ScreminA,Bretz WA (2008) Anti-inflammatory effects of a bioavailable compound, Artepillin C,in Brazilian propolis[J]. *European Journal of Pharmacology*,587(1):296-301.

Saavedra N,Cuevas A,Cavalcante MF,Dorr FA,Saavedra K,Zambrano T,Abdalla DSP, Salazar LA (2016) Polyphenols from chilean propolis and pinocembrin reduce mmp-9 gene expression and activity in activated macrophages[J]. *Biomed Research International*, 2016:6505383.

Sampietro DA,Vattuone MMS,Vattuone MA (2016) Immunomodulatory activity of apis mellifera propolis from the north of argentina[J]. *Lwt-Food Science and Technology*, 70:9-15.

Sa-Nunes A,Faccioli LH,Sforcin JM (2003) Propolis:Lymphocyte proliferation and IFN-gamma production[J]. *Journal of Ethnopharmacology*,87(1):93-97.

Scheller S,Gazda G,Pietsz G,Gabrys J,Szumlas J,Eckert L,Shani J (1988) The ability of ethanol extract of propolis to stimulate plaque formation in immunized mouse spleen cells[J]. *Pharmacological Research Communications*,20:323-328.

Sforcin JM (2007) Propolis and the immune system:A review[J]. *Journal of Ethnopharmacology*, 113(1):1-14.

Shimizu K,Ashida H,Matsuura Y,Kanazawa K (2004) Antioxidative bioavailability of Artepillin C in Brazilian propolis[J]. *Archives of Biochemistry and Biophysics*,424(2):181-188.

Simões-Ambrosio LMC, Gregório LE, Sousa JPB, Figueiredo-Rinhela ASG, Azzolinia AECS,Bastosb JK,Lucisano-Valim YM (2010) The role of seasonality on the inhibitory effect of Brazilian green propolis on the oxidative metabolism of neutrophils[J]. *Fitoterapia*, 81(8):1102-1108.

Skiba M,Szliszka E,Kunicka M,Król W (2011) Effect of ethanolextract of propolis (EEP) on interleukin 8 release byhuman gastric adenocarcinoma cells (AGS) infected with*Helicobacter pylori*[J]. *Central European Journal of Immunology*,36(2):65-69.

Steinhagen F,Kinjo T,Bode C,Klinman DM (2011) TLR-based immuneadjuvants[J].

Vaccine,29(17):3341-3355.

Tanaka M,Okamoto Y,Fukui T,Masuzawa T (2012) Suppression of interleukin 17 production by Brazilian propolis in mice with collagen-induced arthritis[J]. *Inflammopharmacology*,20(1):19-26.

Tani H,Hasumi K,Tatefuji T,Hashimotoa K,Koshinob H,Takahashi S (2010) Inhibitory activity of Brazilian green propolis components and their derivatives on the release of cys-leukotrienes[J]. *Bioorganic and Medicinal Chemistry*,18(1):151-157.

Tatefuji T,Izumi N,Ohta T,Arai S,Ikeda M,Kurimoto M (1996) Isolation and indentification of compounds from Brazilian propolis which enhance macrophage spreading and mobility[J]. *Biological Pharmaceutical Bulletin*,19:966-970.

Wang LC,Chu KH,Liang YC,Lin YL,Chiang BL (2010) Caffeic acid phenethyl ester inhibits NF-κB and protein kinase B signalling pathways and induces caspase-3 expression in primary human $CD4^+$ T cells[J]. *Clinical and Experimental Immunology*,160(2):223-232.

Wang LC,Lin YL,Liang YC,Yang YH,Lee JH,Yu HH,Wu WM,Chiang BL (2009) The effect of caffeic acid phenethyl ester on the functions of human monocyte-derived dendritic cells[J]. *BMC Immunology*,10(1):39.

Washio K,Kobayashi M,Saito N,Amagasa M,Kitamura H (2015) Propolis ethanol extract stimulates cytokine and chemokine production through NF-κB activation in C2C12 myoblasts[J]. *Evidence-based Complementary and Alternative Medicine*.

You KM,Son KH,Chang HW,Kang SS,Kim HP (1998) Vitexicarpin,a flavonoid from the fruits of *Vitex rotundifolia*,inhibits mouse lymphocyte proliferation and growth of cell lines in vitro[J]. *Planta Medica*,64:546-550.

柴家前,王玲,庞昕,沈志强(2002)纳米蜂胶颗粒对鸡免疫功能的影响[J].畜牧兽医学报,33(4):412-416.

冯俊鹏(1997)蜂产品的营养机制与人体免疫系统的关系[J].中国养蜂,(5):24-26.

胡福良(2005)蜂胶药理作用研究[M].杭州:浙江大学出版社.

胡箭卫,李旭涛(2003)蜂胶应用于畜牧业的研究现状及发展前景[J].中兽医医药杂志,6:42-43.

金天明,谷禹,马文芝,周淑云,李富桂,郭亮,张建斌(2009)禽霍乱、大肠杆菌病和克雷伯菌病多价蜂胶三联灭活疫苗的研制[J].黑龙江畜牧兽医,8:99-100.

李淑华,于晓红,于英君,杨宝华,张德山(2001)蜂胶对免疫功能低下模型鼠细胞免疫功能的影响[J].中国药理学报,29(3):38-39.

李英华(2002)蜂胶的抗炎免疫作用及其机制的研究[D].杭州:浙江大学.

潘明(2007)蜂胶提取物对荷瘤小鼠肿瘤免疫系统的影响研究[J].时珍国医国药,18(2):415-416.

邱昌庆,董恩娜(1996)鸡新城疫—减蛋综合征蜂胶佐剂灭活二联苗的研究[J].中国兽医科技,26(12):3-5.

沈志强(1995)禽霍乱蜂胶灭活疫苗研究与应用总结[J].中国畜禽传染病,4:27-30.

沈志强,杨永福(1989)佐剂新秀——蜂胶[J].中国畜禽传染病,(5):55-57.

薛晓丽（2007）提取工艺对蜂胶总黄酮含量的影响[J].吉林农业科技学院学报,16(3):3-4.

杨锋,戴关海,潘慧云,任佩天,江克翊（2007）蜂胶增强蜂王浆生物活性作用的实验研究[J].中国中医药科技,14(1):42-43.

于晓红,李淑华,李树伟,于英君,张德山（2001）蜂胶对小鼠免疫功能影响的实验研究[J].中医药信息,18(4):53-54.

张宝康,王德云,孔祥峰,胡元亮（2006）蜂胶黄酮增强免疫作用的研究[J].扬州大学学报,27(3):20-22.

张莉,尹燕博,崔尚金,简子健,邓普辉,刘尚高（2004）甲壳素、油乳剂和蜂胶对鸡新城疫疫苗免疫调节作用的研究[J].中国预防兽医学报,26(2):136-138.

赵恒章,赵坤,余燕,李国旺,马金友,张慧辉（2009）副猪嗜血杆菌蜂胶苗的制备及免疫效果试验[J].生物技术,19(6):72-74.

第六节　蜂胶的降血糖和降血脂作用

　　蜂胶降血糖和降血脂作用一直被认为是蜂胶最主要的生物学活性之一。近年来,世界各国的学者采用不同的分析方法对蜂胶进行了广泛的研究,发现蜂胶具有良好的降血糖和降血脂作用。糖尿病不仅发病率高,而且并发症多,可引起心、脑、肾、眼、神经、血管、消化道等全身器官和系统的慢性疾病,死亡率仅次于肿瘤和心血管疾病,已成为目前威胁人类生命的第三大类疾病。此外,高脂血症是造成心血管疾病的主要原因之一。近年来,我国高脂血症的发病率一直呈上升状态,而高脂血症又是引起动脉粥样硬化的危险因素之一,对患者的生活质量和生命安全造成了极大的危害。然而,目前针对糖尿病和高脂血症的治疗方法并不能达到根治的目的,对其发病机制也没有完全阐明。近年来,人们在天然产物资源方面寻求具有降血糖、降血脂及防治并发症的活性成分取得了很大进展,而蜂胶正是研究的热点之一。下面对蜂胶及其几种重要活性成分的降血糖和降血脂作用进行综述,并对其可能的分子机制进行探讨,旨在为蜂胶活性成分的深度开发提供参考。

1　蜂胶的降血糖作用

1.1　不同地理来源蜂胶的降血糖作用

　　不同地理来源蜂胶因胶源植物不同,其黄酮、酚酸及萜烯类物质含量差异较大,但总体上均具有良好的降血糖作用。Zhu 等(2011a)研究了中国蜂胶和巴西绿蜂胶对链脲佐菌素(STZ)诱导的 1 型糖尿病大鼠的影响,结果发现中国蜂胶和巴西率蜂胶均能显著地抑制糖尿病大鼠体重的下降及血糖的升高。与对照组相比,中国蜂胶处理组的大鼠糖化血红蛋白含量下降 8.4%,总胆固醇含量下降 16.6%。同时,中国蜂胶和巴西绿蜂胶均使大鼠的血液、肝、肾的氧化应激反应得到了不同程度的提高,丙氨酸转氨酶和天冬氨酸转氨酶含量及微蛋白尿排泄率的下降证明蜂胶有益于恢复大鼠的肝肾功能。Li 等(2011)观察了中国杨属型蜂胶胶囊对实验型 2 型糖尿病(T2DM)大鼠体重、空腹血糖(FBG)、空腹血清胰岛素(FINS)、胰岛素作用指数(IAI)、血清三酰甘油(TG)、总胆固醇(TC)、高密度脂蛋白胆固醇(HDL-C)及低密度脂蛋白胆固醇(LDL-C)的影响。实验结果发现,蜂胶胶囊可显著降低 2

型糖尿病大鼠 FBG 和 TG 水平,提高 IAI 值,但对体重、TC、HDL-C 及 LDL-C 无显著影响,表明蜂胶胶囊具有改善 2 型糖尿病大鼠血糖、调节脂代谢及提高机体胰岛素敏感性的作用。赵丽婷等(2016)采用口服巴西蜂胶对 2 型糖尿病患者进行干预,检测其生化指标、氧化/抗氧化指标和炎症因子水平,进一步探讨蜂胶对糖尿病患者能量代谢和氧化应激的干预作用及其作用机制。结果表明,巴西蜂胶可显著增加血浆总多酚含量,提示蜂胶中的多酚类成分可以被机体有效吸收,增强机体抗氧化能力;同时,蜂胶可显著降低 2 型糖尿病患者的血清 TNF-α 和 LDH 活性,显著升高血清 IL-1β 和 IL-6,提示巴西蜂胶具有降低炎性反应的作用。Mahani 等(2013)对印度尼西亚爪哇和苏威拉西的蜂胶乙醇提取物(EEP)进行 1 型糖尿病大鼠降血糖研究,结果表明,EEP 的降血糖作用达到标准胰岛素的降血糖效果,且当注射 100mg/kg 剂量的苏威拉西蜂胶时效果最佳。Rifa'i 等(2014)研究了印尼蜂胶 EEP 对 S961 肽诱导的糖尿病小鼠 T 细胞的免疫功能、细胞表面分子以及体内 γ 干扰素的产生的影响,发现 EEP 能有效降低糖尿病小鼠的血糖浓度并抑制糖尿病小鼠效应 T 细胞表达 CD62L 分子的数量。同时,蜂胶促进了 $CD4^+CD25^+$ T 细胞的分裂以及效应细胞的增殖,抑制了糖尿病小鼠体内 γ 干扰素的产生。此外,Hadi(2014)研究发现,伊拉克蜂胶对四氧嘧啶诱导的糖尿病兔子有显著的降血糖效果,并改善血液指标。

1.2 蜂胶结合其他物质的降血糖作用

研究表明,酸奶及酸奶制品中的乳酸菌对癌症、感染、胃肠道功能紊乱、哮喘有着一定治疗和预防作用(Gauffin et al.,2002;Meydani et al.,2000),同时,酸奶在控制血糖的上升及糖尿病并发症方面具有良好作用(Bijvoet et al.,1996)。Bukhari 等(2012)就此研究了酸奶与蜂胶结合对 STZ 诱导的糖尿病雄性大鼠体内的血糖和血脂的作用,发现酸奶结合蜂胶饲喂可降低血清总胆固醇、三酰甘油 LDL、VLDL 的含量,并升高 HDL、LDL-C/HDL-C 和 HDL-C/TC% 水平,且蜂胶含量较高组效果更为显著。据此推断,酸奶和蜂胶减少了葡萄糖的吸收、抑制 α-葡萄糖苷酶的活性以及恢复胰岛 β 细胞的功能,这可能与动脉粥样硬化发病机制的基因表达调控有关,表明酸奶与蜂胶结合在降低患心血管疾病风险上有巨大的潜在医疗价值。

铬能促进胰岛素的分泌,且它的络合物能提高机体对胰岛素的敏感性(Sharma et al.,2011),如吡啶甲酸铬(Trent et al.,1995)、组氨酸铬(Dogukan et al.,2010)等铬的有机络合物对糖脂代谢均具有良好效果。Wu 等(2012)将三价铬的苹果酸络合物 $Cr_2(LMA)_3$ 与蜂胶结合饲喂 ICR 小鼠,2 周后测量小鼠的血糖水平、肝糖原水平,以及天冬氨酸转氨酶、谷丙转氨酶和碱性磷酸酶的活性。结果表明,$Cr_2(LMA)_3$ 与蜂胶结合比两者单用天冬氨酸转氨酶、谷丙转氨酶和碱性磷酸酶的活性具有更显著的效果。Zhang 等(2005)则观察了蜂胶与吡啶甲酸铬的复合剂对高血糖模型小鼠的空腹血糖和糖耐量的影响,试验结果显示,蜂胶与吡啶甲酸铬具有协同作用,能有效降低高血糖模型小鼠的空腹血糖含量,并提高机体对葡萄糖的糖耐量。

1.3 蜂胶对糖尿病并发症的作用

糖尿病肾病是糖尿病患者最重要的并发症之一,目前已成为终末期肾脏病的第二位原因。Oršolić 等(2012)用蜂胶水提物和醇提物给四氧嘧啶诱导的糖尿病瑞士白化小鼠按每天 50mg/kg 的剂量连续注射 7d,发现空泡化细胞数明显减少,空化程度降低,并提高了糖尿病小鼠的脂肪酸代谢自我修复能力，证明蜂胶能有效降低糖尿病对肝肾的损伤，这可能与蜂胶的抗氧化作用与解毒作用有关。Agawany 等（2012）用 STZ 诱导雄性维斯塔糖尿病大鼠，对小鼠肾脏进行石蜡包埋切片染色观察，发现蜂胶有良好的控制血糖作用，起到了降低肾小球的血糖滤过率而延迟了肾病的发生及改善发病后症状的作用。

　　肝在糖代谢中占有重要作用，有 30%～60% 的葡萄糖在胃肠道被吸收，在肝脏以糖原的形式存在。当血糖运输出现紊乱后，糖尿病患者的肝病发生率升高，且肝病是 2 型糖尿病死亡的原因之一。Mahmoud 等（2013b）就蜂胶对 STZ 诱导的糖尿病大鼠的保肝作用进行研究，发现糖尿病大鼠的肝切片表现出细胞质中脂质增多、免疫淋巴细胞浸润、枯否氏细胞增殖，而蜂胶处理组明显降低 STZ 带来的毒副作用，同时，线粒体嵴减少、基质浓缩、粗面内质网增生并分裂成更小的片段，表明蜂胶能有效降低 STZ 诱导的糖尿病带来的早期肝损伤。此外，Zhu 等（2011b）还比较了中国蜂胶与巴西绿蜂胶对 STZ 诱导的糖尿病大鼠肝肾损伤的作用效果，结果发现，中国蜂胶处理后的糖尿病大鼠比未处理的糖尿病大鼠的糖化血红蛋白含量低 7.4%，血清中的超氧化物歧化酶（SOD）显著增加；而巴西绿蜂胶处理后的糖尿病大鼠体内血清 SOD 显著增加，丙二醛和一氧化氮合成酶迅速降低。除此之外，血清中丙氨酸转氨酶（ALT）、天冬氨酸转氨酶（AST）和微量白蛋白含量降低，谷胱甘肽过氧化物酶含量增加抑制丙二醛的产生，这都证明了蜂胶能通过抑制脂质的过氧化及增强抗氧化酶的活性来防止肝肾损伤。

　　糖尿病足是糖尿病患者的慢性并发症，也是糖尿病患者致死致残的主要原因之一。统计表明，仅美国在 2007 年用于糖尿病病足的医疗花费就达 400 亿美元（Driver et al.，2010）。Al-Hariri 等（2011）用蜂胶处理 STZ 诱导的糖尿病大鼠，并对其血糖和骨盐含量进行分析研究，与对照组相比，蜂胶处理组的糖尿病大鼠的空腹血糖含量、血清过氧化物含量显著降低，血浆胰岛素、股骨灰中钙、磷、镁的含量升高。因此，蜂胶对血糖的平衡及骨盐的矿化具有显著作用。Al-Saeed 等（2015）除了检测糖尿病足的大鼠血液、尿液及骨样本以外，还检测了护骨素的表达基因，发现蜂胶处理组不仅使空腹血糖水平显著下降，提高了血清中骨钙素、钙和磷的含量，还提高了护骨素基因的表达。而 Temesio 等（2012）认为蜂胶能有效治疗糖尿病是因其抗菌、抗炎、抗氧化和免疫调节功能综合作用的结果。

1.4　几种活性成分的降血糖作用

1.4.1　类黄酮化合物

黄酮类化合物是蜂胶的主要功效成分。蜂胶中的黄酮类化合物主要有黄酮及黄酮醇、二氢黄酮及二氢黄酮醇、异黄酮及二氢异黄酮、查耳酮及二氢查耳酮、新黄酮类化合物。

（1）白杨素

白杨素具有抗菌、消炎、防止血脑血管疾病和抗癌等活性，在中国蜂胶中含量丰富。Premalatha 等（2012）研究了白杨素对 STZ 诱导的糖尿病大鼠肾病的影响，通过检验大鼠的血液和尿液样本，发现患有糖尿病肾病的大鼠体内血糖、尿素、血清肌酐酸总尿蛋白、尿液尿素、肌酸酐均比正常大鼠的含量高，肾小球的滤过率显著降低，注射白杨素后，各指标恢复正常，表明白杨素对 STZ 诱导的糖尿肾病大鼠具有保护作用。

（2）山奈酚和槲皮素

山奈酚和槲皮素同属于蜂胶中的黄酮醇类化合物（张翠平等，2009），由于具有较多酚羟基,因此抗氧化活性很高,对人体的保健作用尤为重要。Fang 等(2008)发现山奈酚和槲皮素能显著改善 2 型糖尿病成熟 3T3-L1 脂肪细胞对葡萄糖的吸收;进一步研究表明山奈酚和槲皮素在过氧化物酶受体 γ(PPARγ)的报告基因序列试验中具有辅助激动剂活性,然而这两者只有同时添加激动剂(罗格列酮),才能抑制 3T3-L1 脂肪细胞的分化,且在 PPARγ 受体过度表达的糖脂化处理的巨噬细胞中能显著降低 NO 的产生,效果比罗格列酮更好,这表明山奈酚和槲皮素从多个方面调节血糖,其中就包括作为 PPARγ 受体辅助激动剂。

(3)柚皮素

柚皮素($4'$,5,7-三羟基黄酮)是蜂胶中的一种二氢黄酮类化合物,具有抗菌、消炎、抗癌、解痉和利胆等作用,也有文献指出柚皮素可能是造成酶促反应的抑制剂(Ghosal et al.,1996)。Annadurai 等(2012)用柚皮素处理链脲佐菌素-烟酰胺诱导的糖尿病大鼠 21d 后,发现糖尿病大鼠的空腹血糖、糖化血红蛋白、丙二醛含量显著降低,血清胰岛素明显升高,胰腺的酶促及非酶促抗氧化活性均显著增强,且血清中 ALT、AST、ALT 及 LDH 的活性降低,进一步的组织病理学研究表明,柚皮素对糖尿病大鼠的胰腺组织具有一定保护作用。

1.4.2 酚酸类化合物

自然界中酚酸类化合物主要有以下两类:一是 C6-C1 型,基本骨架是苯甲酸,如没食子酸;二是 C6-C3 型,基本骨架是苯丙酸,如肉桂酸、咖啡酸、对香豆酸等。此外,还有从巴西绿蜂胶中分离出的特有的绿原酸类化合物。蜂胶中的酚酸类化合物有 100 多种(张翠平等,2013)。

(1)肉桂酸类化合物

咖啡酸(3,4-二羟基肉桂酸)具有抗菌、抗病毒、提高神经中枢兴奋性等药理学活性;肉桂酸作为一种食品添加剂已广泛应用于医药、美容、农药及有机合成等方面。Huang 等(2012)对咖啡酸和肉桂酸调节 2 型糖尿病小鼠的肝糖生成及糖异生机制进行研究,结果显示咖啡酸和肉桂酸能增加糖原合成酶的表达,减少 2 型糖尿病小鼠肝细胞中糖原合酶激酶的产生及 Ser641 的磷酸化,抑制 FL83B 肝细胞核中肿瘤细胞的表达,并认为咖啡酸和肉桂酸改善葡萄糖代谢主要是通过促进糖原的生成及抑制肿瘤坏死因子处理后的肝细胞的糖异生。

Yoon 等(2013)研究发现一定剂量下的 p-香豆酸(对羟基肉桂酸)能促进已分化的 L6 骨骼肌细胞中蛋白激酶的磷酸化,增加乙酰辅酶 A 羧化酶(ACC)的磷酸化、肉毒碱棕榈酰转移酶-1(CPT-1)mRNA 及过氧化物酶体增殖物激活受体 α 的表达,抑制十八烯酸三酰甘油的积累,增强已经分化的 L6 骨骼肌细胞中 2-脱氧葡萄糖的吸收。结果表明,p-香豆酸能通过活化 L6 骨骼肌细胞的蛋白激酶调节葡萄糖和脂类代谢,对于改善或治疗糖脂代谢紊乱具有潜在效果。

(2)绿原酸

绿原酸类化合物主要存在于忍冬科忍冬属、菊科蒿属等植物中,而蜂胶中发现的绿原酸类化合物均从巴西蜂胶中分离得到。Ong 等(2013)通过观察绿原酸对 Lepr$^{db/db}$ 小鼠的葡萄糖耐量、胰岛素敏感性、肝糖原异生、脂质代谢、骨骼肌葡萄糖摄入量的影响,发现绿原酸能通过活化的蛋白激酶提高糖脂代谢能力。Karthikesan 等(2010)将四氢姜黄素与绿原酸结合,研究其对 STZ-烟酰胺(NA)诱导的 2 型糖尿病大鼠氧化应激的保护作用,糖尿病大鼠口服四氢姜黄素 80mg/kg、绿原酸 5mg/kg 45d 后,空腹血糖、糖化血红蛋白、硫代巴比妥酸反应物等指标都得到改善,且二者结合使用的效果比单一使用效果更强。

(3)咖啡酸苯乙酯(Caffeic acid phenethyl ester,CAPE)

咖啡酸苯乙酯是咖啡酸的一种脂类衍生物,是杨树型蜂胶的一种代表性成分。Hassan等(2014)连续给动脉粥样硬化小鼠注射CAPE 30mg/kg/day 6周,测量小鼠血糖水平、血清胰岛素水平、肿瘤坏死因子、糖基化终产物、主动脉血红蛋白的表达量和胶原沉积水平;从KCl、去氧肾上腺素两个模型中发现CAPE减轻了血压的收缩和舒张,增强了血管的收缩性,抑制了肿瘤坏死因子,减少了主动脉血红蛋白的表达量和胶原沉积。

(4)阿替匹林C

阿替匹林C(Artepillin C,3,5-二异戊烯基-4-羟基肉桂酸)是巴西绿蜂胶中的主要酚酸类物质之一,也是其特有的标志性成分。Choi等(2011)研究发现阿替匹林C能诱导3T3-L1脂肪细胞的分化、过氧化物酶体增殖物激活受体和其靶基因(如P2)的表达以及脂联素和葡萄糖转运蛋白增加,在成熟的3T3-L1脂肪细胞中,阿替匹林C显著增强了对基质和胰岛素刺激下的血糖吸收,且在PI3K抑制剂的联合使用下,效果更强,有效地降低了2型糖尿病风险。

2 蜂胶的降血脂作用

2.1 蜂胶的降血脂作用

近年来,国内外研究发现蜂胶具有显著的调节血脂作用,其被誉为"血管清道夫"。Koya-Miyata等(2009)在给饮食诱导的肥胖小鼠连续喂食蜂胶提取物10d后,发现小鼠的体重,内脏脂肪组织重量,肝脏重量和血清三酰甘油、胆固醇、未脂化脂肪酸含量均下降,并抑制了脂肪酸合成相关酶的mRNA表达,表明蜂胶提取物能通过抑制或调节与脂肪酸合成相关的基因表达来缓和高脂饮食诱导的高脂血症。Kwon等(2001)研究发现,对高胆固醇喂食的大鼠喂以蜂胶水提物后,大鼠血清和肝组织中的低密度脂蛋白胆固醇、丙二醛、三酰甘油水平降低,高密度脂蛋白胆固醇水平升高,表明蜂胶水提物具有一定的降胆固醇作用。Fang等(2013)在Kwon等(2001)的研究基础上,测定了血清中白细胞介素6和白细胞介素17、内皮素、诱导性一氧化氮合酶、血管内皮生长因子的含量,认为蜂胶醇提物可能是通过调节胆固醇水平、炎症反应,抑制内皮素和血管内皮生长因子,保护血管内皮细胞以达到抑制动脉粥样硬化的形成的目的。此外,程清洲等(2013)观察了鄂西地区蜂胶提取物对高脂血症模型大鼠血脂的作用,发现蜂胶能调节血脂及抗脂质过氧化作用,预防和治疗高脂血症。

2.2 蜂胶结合其他物质的降血脂作用

西洋参具有增强中枢神经系统功能,保护心血管系统(扩张血管、降低血脂、改善血液流变性),提高免疫力等功效。童晔玲等(2013)观察了蜂胶西洋参软胶囊对高脂血症模型大鼠血脂及脂蛋白的影响,与高脂水对照组、高脂油对照组相比,低、中、高组均能明显降低高脂血症大鼠血清三酰甘油含量,中、高剂量组能明显降低血清总胆固醇含量,表明蜂胶西洋参软胶囊具有辅助降血脂功能。

银杏叶因含有丰富的黄酮和银杏内酯等活性成分,具有抗氧化、调节血脂、防止动脉硬化、改善血液流变性等功效而被国内外学者广泛研究,特别是以银杏为主要原料的调节血脂保健品和药品也逐渐上市。胎盘具有提高免疫力、补气血的药理活性,对此,李冀宏等

(2002)用银杏叶和胎盘配伍辅以蜂胶来防治动脉粥样硬化,发现复方高剂量组的胆固醇含量显著低于高脂对照组,其降血脂作用明显。

百里醌是黑种草种子挥发油的主要成分,Nader 等(2010)研究发现,灌胃蜂胶和百里醌能显著降低饲喂高胆固醇饲料组总胆固醇、低密度脂蛋白胆固醇、三酰甘油等含量,提高高密度脂蛋白胆固醇和谷胱甘肽含量,可通过抗氧化机制降低动脉粥样硬化的风险。

2.3 几种活性成分的降血脂作用

2.3.1 黄酮类化合物

(1)白杨素

Anandhi 等(2013)采用平菇提取物(主要成分为白杨素)和白杨素对 Triton WR-1339 诱导的高胆固醇大鼠的抗高胆固醇和抗氧化作用进行研究,发现大鼠口服平菇提取物和白杨素后均降低了血糖、血脂参数、肝标记酶(丙氨酸氨基转移酶、天门冬氨酸氨基转移酶、碱性磷酸酶和乳酸脱氢酶)水平,并提高了还原型谷胱甘肽酶、维生素 C 和维生素 E 水平,且口服白杨素的效果比口服平菇提取物的效果更显著,结果表明白杨素对 WR-1339 诱导的高胆固醇大鼠具有抗高胆固醇作用。Zarzecki 等(2014)也对白杨素对 Triton WR-1339 诱导的高胆固醇小鼠的抗高胆固醇进行研究,得到了类似的效果。

(2)松属素

Sang 等(2012)研究了松属素和辛伐他汀组合对 $apoE$ 基因缺陷小鼠的抗动脉粥样硬化作用,结果发现,该组合处理降低了血清总胆固醇、三酰甘油、低密度脂蛋白胆固醇水平,提高了一氧化氮水平和超氧化物歧化酶活性,抑制了内皮素和血管内皮生长因子的表达,且松属素和辛伐他汀组合的抗动脉粥样硬化效果比单一使用辛伐他汀的效果更佳。

2.3.2 酚酸类化合物

(1)肉桂酸类化合物

目前,关于肉桂酸类化合物有助于降血脂的报道较少。Yeh 等(2009)测试了咖啡酸、阿魏酸和香豆酸 3 种酚酸的降血脂和抗氧化能力,连续 6 周喂食含 3% 胆固醇的饲料并分别添加 0.2% 咖啡酸、阿魏酸和香豆酸 3 种酚酸后,发现均降低了血浆脂质和肝脏胆固醇水平,大鼠体内的 SOD 和 GSH 活性提高,抗氧化能力增强,促进了酸性和中性甾醇的排泄。

(2)绿原酸

王建辉等(2012)开展了杜仲绿原酸对高脂高胆固醇诱导的高血脂模型小鼠脂质代谢的影响研究,饲喂高脂饲料并灌胃杜仲绿原酸 4 周后,发现杜仲绿原酸显著降低了小鼠血清 TC、TG、LDL-C 水平,动脉硬化指数和冠心指数,肝脏 TC、TG 含量也显著降低,血清和肝脏中 MDA 生成下降,抗氧化酶活性增加,因而具有良好的调节脂质代谢作用。梁秀慈(2013)研究了绿原酸对长期高糖高脂饲料饲养 SD 大鼠脂代谢的影响,并通过检测生理生化及相关基因的表达,发现绿原酸能有效改善长期高糖高脂喂养下 SD 大鼠形成的脂代谢紊乱,减少内脏脂肪的形成,减缓脂肪肝的形成。

3 蜂胶调节血糖和血脂作用的可能机制

糖尿病是以高血糖为特征的代谢性疾病,高血糖则是由胰岛素分泌缺陷或其他生物作用受损,或者两者兼有引起的。胰岛素作用于细胞,可引起各种各样的生理效应,包括糖类和脂类代谢、蛋白合成以及细胞生长。此外,高血脂也是引起糖尿病的一个重要病因,肝脏脂肪积累是造成胰岛素抵抗的一个关键因素(Kim et al.,2000)。肥胖和糖尿病都会引起游离脂肪酸含量上升,降低胰岛素对肝葡萄糖生成的抑制作用(Hawkins et al.,2003;Shah et al.,2003)。Özcan 等(2004)研究发现肥胖会引起内质网应激,该应激会导致 JNK 超活化以及 IRS-1 丝氨酸磷酸化而失活,导致胰岛素信号抑制,表明高血糖和高血脂具有紧密的联系。

近年来,研究发现中草药提取物以及植物分离化合物对胰腺具有多重功效,如β细胞增殖、胰岛素合成及分泌,说明药用植物对于治疗糖尿病相关的胰岛素抵抗和缺乏具有潜在作用(Hawkins et al.,2003),而蜂胶正是研究热点之一。蜂胶的多酚类物质含量丰富,每类物质所具有的特定结构决定了它们的生物学活性与作用途径。例如,黄酮类物质的基本结构由两个苯环结构中间连接一个 3 碳单元形成的含氧杂环构,它发挥调节血糖作用主要是通过调节系列靶点分子以及信号通路(Hajiaghaalipour et al.,2015)。目前,蜂胶中已有报道具有显著降血糖、血脂作用的白杨素、柚皮素、山柰酚、木樨草素、绿原酸、咖啡酸、表儿茶素、没食子酸等则是通过不同通路进行调节。

目前,从胰岛素调节通路和非胰岛素调节通路的角度研究天然产物对血糖和血脂作用途径已取得了一定进展。下面主要从胰岛素调节和非胰岛素调节通路来阐述蜂胶调节血糖作用可能的分子机制。

3.1 通过 IRS-PI3K 通路调节血糖和血脂的作用

3.1.1 通过 IRS-PI3K 通路调节血糖的作用

胰岛素与胰岛素受体结合,活化的胰岛素受体进一步激活胰岛素受体底物(IRS),从而启动胰岛素信号转导通路。IRS-2 功能受损会严重影响外围胰岛素信号传导和胰岛 β 细胞功能,IRS-2 缺失小鼠因肝脏和骨骼肌的胰岛素耐受性和 β 细胞缺乏补偿胰岛素耐受性的能力而使葡萄糖内稳态严重失衡(Withers et al.,1998)。另一重要的胰岛素底物受体(IRS-1),在其自身酪氨酸磷酸化后,结合并激活含有 SH2 结构域的 PI3K 等蛋白,从而激活胰岛素通路。IRS-1 缺失小鼠表现出子宫内生长降低、葡萄糖耐受度受损以及胰岛素/胰岛素样生长因子依赖的葡萄糖吸收下降(Ellchi et al.,1994)。

(1)靶向 IRS 而调节血糖的作用

在正常细胞中,IRS-1 是主要的结合并激活 PI3K 的蛋白,而 IRS-2 需要在更高的胰岛素浓度下才能结合并激活 PI3K。胰岛素抵抗和 β 细胞功能紊乱是引发 2 型糖尿病的重要因素(Abdul-Ghani et al.,2006),而 IRS2 在增强 β 细胞功能和数量上具有重要作用。IRS2 的 pH 和 PTB 结构域可与磷酸化的胰岛素受体结合,从而使 IRS2 自身磷酸化,磷酸化的 IRS2 激活 PI3K,进而调控细胞生长、β 细胞蛋白合成等(Withers et al.,1998)。Park 等(2008)研究发现,地黄、人参等提取物可刺激 IRS2 的表达,改善 IRS2 的敏感性。有研究表明,激活的 IRS2 可改善葡萄糖敏感性和 β 细胞增殖(Hennige et al.,2003),从而发挥调节血糖的作用。

(2)靶向 PI3K 及其负性调节子(PTEN)调节血糖的作用

磷脂酰肌醇 3 激酶蛋白家族由 3 个成员组成,分别是Ⅰ型、Ⅱ型和Ⅲ型,其中对Ⅰ型的研究较多。Ⅰ型由酪氨酸激酶受体(ⅠA)和 G 蛋白偶联受体(ⅠB)组成,ⅠA 由 p85α/

p85β/p55 亚基组成,而ⅠB 由 p101/p84/p87PIKAP 亚基组成(Cantley,2002)。Akt 蛋白(丝氨酸/苏氨酸激酶)又叫蛋白激酶 B(PKB),是 PI3K 的一个主要的下游效应子。在 PIP_3 的作用下,PDK1 磷酸化 Akt 蛋白催化环的 Thr-308 位点,从而激活 Akt 蛋白(Mora et al.,2004),进一步磷酸化 Ser-473 位点,从而使 Akt 完全激活(Bayascas et al.,2005)。在肌肉和脂肪细胞中,Akt 促进葡萄糖转运子 GLUT4 的膜转移,增强细胞对葡萄糖的吸收(Thong et al.,2005;Virkamäki et al.,1999)。同时,Akt 可磷酸化糖原合酶激酶 3(GSK3)而使其失活,进而促进糖原合成(Cohen et al.,2001),表儿茶素和可可酚提取物提升了 GSK3 亚基的磷酸化水平,而降低了 p-GS 的水平,从而促进糖原合成,同时,表儿茶素和可可酚提取物可系列地激活 IRS-1/IRS-2、PI3K/Akt 信号通路,调节葡萄糖代谢。二者亦可提升葡萄糖转运子 2(GLUT2)的含量(Cordero-Herrera et al.,2013)。GLUTs 在调节血糖水平上发挥重要作用,胰岛素通过诱导 GLUT4 转运到质膜而刺激组织(肌肉组织和脂肪组织)的葡萄糖转运(Birnbaum,1989)。然而,并不是所有的植物提取物都是依赖于一致的信号通路发挥调节血糖和血脂的功能。Prasad 等(2010)研究表明,没食子酸通过激活 PI3K 蛋白方式促进 GLUT4 的膜转移,进而促进葡萄糖的吸收,然而,这并不依赖于 Akt 的激活。

PI3K 催化 3,4,5-三磷酸磷脂酰肌醇(PIP3)的产生,PTEN 是重要的脂类磷酸酶,可使 PIP3 脱磷酸而失活(Lee et al.,1999;Maehama et al.,1998)。在 3T3L1 细胞中过表达 PTEN 会抑制葡萄糖的吸收以及 GLUT4 的膜转移(Nakashima et al.,2000),Butler 等(Butler et al.,2002)系统饲喂 db/db 和 ob/ob 小鼠反义核苷酸,显著降低了小鼠 PTEN 的表达,也使小鼠血糖浓度趋于正常水平。卫晓怡等(2013)研究表明,矢车菊素-3-葡萄糖苷可以改善 KKAy 糖尿病小鼠胰岛素敏感性,抑制肝脏、肌肉和内脏脂肪组织 PTEN 表达而增强 Akt 磷酸化,从而改善小鼠胰岛素抵抗。

(3) 靶向 IPF1/PDX1 及其负性调节子(FOXO1)调节血糖的作用

同源域因子 IPF1/PDX1 最初是在早期鼠胰腺间叶原基细胞中表达(Leonard et al.,1993)。IPF1/PDX1 被认为调节多种胰岛内分泌基因的表达,如胰岛素、生长激素抑制素、葡糖激酶、胰岛淀粉样多肽和葡萄糖转运子 2(GLUT2)(Ohlsson et al.,1993;Serup et al.,1995;Waeber et al.,1996)。亦有研究指出,缺失 IPF1/PDX1 小鼠不能正常形成胰岛(Jonsson et al.,1994)。Ahlgren 等(1998)研究表明,IPF1/PDX1 通过正向调节胰岛素和胰岛淀粉样多肽的表达以及抑制胰高血糖素表达的方式维持 β 细胞特征。他们还发现 IPF1/PDX1 可呈剂量依赖性地调节 GLUT2 的表达;降低 IPF1/PDX1 的活性可导致 2 型糖尿病的发生。Zhang 等(2013)研究表明,山柰酚可通过增强 PDX1/cAMP/PKA/CREB 信号级联通路改善胰岛 β 细胞的存活及其功能,促进胰岛素的分泌与合成。1 型糖尿病和 2 型糖尿病都会引起胰岛 β 细胞的严重降低,极大地降低了机体调节血糖的能力。Soto 等(2014)研究显示,水飞蓟素能提升 PDX1 和胰岛素基因表达水平,并促进胰腺组织的 β 细胞增殖,促使血胰岛素水平上升和血糖水平下降。叉头状转录因子 1(FOXO1)是胰岛素通路重要的负性调节因子。在胰岛素应答组织中,FOXO1 是 FOXO 家族含量最丰富的亚基,它调节葡萄糖-6-磷酸酶(Nakae et al.,2001),抑制 PDX1 的表达(Kitamura et al.,2002),并逆转胰岛素抑制肝葡萄糖生成和促进 β 细胞增殖的作用(Nakae et al.,2002)。

3.1.2 通过 IRS-PI3K 通路调节血脂的作用

载脂蛋白 B(ApoB)在肝脏内质网中将磷脂质、胆固醇、胆固醇酯和三酸甘油酯装配成极低密度脂蛋白(VLDL)(Fisher and Ginsberg,2002),中性脂质通过分泌途径而募集,并经微粒体甘油三酸酯转运蛋白(MTP)介导后在内质网内腔中积累,进一步转移到 ApoB(Gordon and Jamil,2000;Kulinski et al.,2002;Wang et al.,1999)。柚苷配基可抑制 MTP 的活性,从而降低中性脂质在内质网中的积累以及 apoB 脂化和分泌,而低密度脂蛋白受体(LDLr)的表达能调节网格 apoB 的分泌(Borradaile et al.,2003a;Borradaile et al.,2002)。Borradaile 等(2003b)研究表明,柚苷配基可通过 PI3K 介导的通路促进固醇调控元件结合蛋白(SREBP-1)的表达,进而促进 LDLr 的表达。此过程不依赖于 IRS-1 的激活。

极低密度脂蛋白(VLDL)是由肝细胞合成并分泌的一类富含甘油三酸酯微粒,该过程受到胰岛素的调节(Kamagate et al.,2008a)。VLDL 的过多积累会引发高甘油三酯血症,而受胰岛素调节的 MTP 对 VLDL 生成的促进作用是该过程的限速步骤(Kamagate et al.,2008b)。Kamagate 等(2008b)研究表明,FOXO1 可结合并激活 MTP 启动子活性,增强 MTP 的表达,进而促进 VLDL 的生成以及提升血浆甘油三酸酯水平。因此,FOXO1 可作为蜂胶调节血糖和血脂的作用的潜在靶点。

3.2 通过 PPARs 转录因子调节血糖和血脂的作用

过氧化物酶体增殖物激活受体(PPARs)是属于配体激活型核受体超家族的转录因子,主要分为 PPAR-α、PPAR-β/δ 和 PPAR-γ 三个类型,其中 PPAR-α 和 PPAR-γ 是关键的脂质和葡萄糖代谢的调节因子,调控多种调节脂质和葡萄糖代谢的基因(Wahli et al.,1995)。已有大量的文献报道了多酚类物质促进 PPARs 的表达而调控葡萄糖和脂质代谢的作用(Huang et al.,2006;Jung et al.,2006;Kumar et al.,2009;Mahmoud et al.,2013a)。Shin 等(2013)运用体外实验和动物实验研究乌梅水果乙醇提取物的抗糖尿病作用,发现其可通过促进 PPARγ 的活性与表达抑制高脂肪诱导的体重增加、脂肪积累以及葡萄糖水平,高效液相色谱分析表明,乌梅水果提取物的主要成分是绿原酸、咖啡酸、芦丁、木樨草素、油苷等黄酮、酚酸类物质。Sharma 等(2008)研究发现,富含黄酮类化合物的乌墨蒲桃种子提取物通过上调 PPARα 和 PPARγ 的表达而有效改善糖原合成、葡萄糖内稳态酶活性,提升高密度脂蛋白含量,降低低密度脂蛋白和三酸甘油酯含量。更有研究发现巴西红蜂胶乙醇提取物增强 PPARγ 的转录活性,极大地诱导脂肪前体细胞分化为脂肪细胞(Iio et al.,2010)。PPARγ 的激活亦可恢复胰岛肌内质网 Ca^{2+} ATP 酶 2(SERCA2)的含量,并阻止 β 细胞功能紊乱(Kono et al.,2012)。

3.3 通过 AMPK 调节血糖和血脂的作用

$5'$-AMP 激活蛋白激酶(AMPK)信号通路在调节能量代谢方面发挥着重要作用。AMPK 是异源三聚体蛋白,由一个催化亚基 α 以及两个调节亚基 β 和 γ 构成(Schimmack et al.,2006)。AMPK 的激活依赖于催化亚基 α 上苏氨酸 172 的磷酸化,活化后的 AMPK 降低能量储存并激发能量的产生。白藜芦醇通过激活 AMPK 信号通路而促进葡萄糖的吸收、改善小鼠的能量代谢并延长其寿命(Baur et al.,2006;Breen et al.,2008)。近年来,AMPK 已成为治疗胰岛素抵抗和糖尿病的颇具吸引力的药物靶点。Jin 等(2013)研究表明,崖爬藤黄酮提取物能通过激活 AMPK 信号通路有效增强胰岛素抵抗细胞的葡萄糖吸收。大量研究表明,富含多酚类的天然提取物都能通过激活 AMPK 信号通路有效促进葡萄糖的吸收、缓解 1 型和 2 型糖尿病(Kang et al.,2010;Ong et al.,2012;Zygmunt et al.,2010)。同时,

AMPK 的激活也能有效调节脂肪代谢。Lee 等(2012)研究发现,光甘草定可激活 AMPK 信号通路,从而缓解肥胖症以及高脂血症状。杧果干能促进肝脏 AMPK 的表达并增强其活性,加速血浆游离脂肪酸的分解代谢而降低血浆游离脂肪酸的含量(Niu et al.,2012)。

3.4 抗氧化调节血糖和血脂的作用

有研究指出,表儿茶素通过调节细胞的氧化还原状态而调控编码葡萄糖异生作用的酶以及蛋白酪氨酸磷酸化的基因的表达,结果表明氧化还原状态对治疗糖尿病有一定的作用(Waltner-Law et al.,2002)。前期研究亦指出茶叶黄酮可抑制巨噬细胞介导的低密度脂蛋白的氧化,从而调节 LDL 的代谢(Ishikawa et al.,1997)。蜂胶作为天然药物,已有大量研究表明其具有强的抗氧化活性,可调节机体的氧化还原状态(张江临等,2013)。Lee 等(2013)研究结果表明,DMA 能通过 Nrf2 通路发挥抗氧化活性,进而缓解丙酮醛诱导的胰岛细胞损伤、提升胰岛素水平而改善糖尿病。因此,抗氧化活性是蜂胶调节血糖和血脂的关键因素之一。

参考文献

Abdul-Ghani MA, Tripathy D, DeFronzo RA (2006) Contributions of β-cell dysfunction and insulin resistance to the pathogenesis of impaired glucose tolerance and impaired fasting glucose[J]. *Diabetes Care*, 29(5):1130-1139.

Ahlgren U, Jonsson J, Jonsson L, Simu K, Edlund H (1998) β-Cell-specific inactivation of the mouseIpf1/Pdx1 gene results in loss of the β-cell phenotype and maturity onset diabetes[J]. *Genes & Development*, 12(12):1763-1768.

Al-Hariri M, Eldin TG, Abu-Hozaifa B, Elnour A (2011) Glycemic control and anti-osteopathic effect of propolis in diabetic rats[J]. *Diabetes, Metabolic Syndrome and Obesity: Targets and Therapy*, 4:377.

Al-Saeed HF, Mohamed NY (2015) The possible therapeutic effects of propolis on osteoporosis in diabetic male rats[J]. *Nature and Science*, 13(3):136-140.

Anandhi R, Annadurai T, Anitha TS, Muralidharan AR, Najmunnisha K, Nachiappan V, Geraldine P (2013) Antihypercholesterolemic and antioxidative effects of an extract of the oyster mushroom, Pleurotus ostreatus, and its major constituent, chrysin, in Triton WR-1339-induced hypercholesterolemic rats[J]. *Journal of Physiology and Biochemistry*, 69(2):313-323.

Annadurai T, Muralidharan AR, Joseph T, Hsu M, Thomas P, Geraldine P (2012) Antihyperglycemic and antioxidant effects of a flavanone, naringenin, in streptozotocin-nicotinamide-induced experimental diabetic rats[J]. *Journal of Physiology and Biochemistry*, 68(3):307-318.

Baur JA, Pearson KJ, Price NL, Jamieson HA, Lerin C, Kalra A, Sinclair DA (2006) Resveratrol improves health and survival of mice on a high-calorie diet[J]. *Nature*, 444(7117):337-342.

Bayascas JR, Alessi DR (2005) Regulation of Akt/PKB Ser473 phosphorylation[J].

Molecular Cell, 18(2): 143-145.

Bijvoet AG, Kroos MA, Pieper FR, de Boer HA, Reuser AJ, van der Ploeg AT, Verbeet MP (1996) Expression of cDNA-encoded human acid α-glucosidase in milk of transgenic mice[J]. *Biochimica et Biophysica Acta (BBA)-Gene Structure and Expression*, 1308(2): 93-96.

Birnbaum MJ (1989) Identification of a novel gene encoding an insulin-responsive glucose transporter protein[J]. *Cell*, 57(2): 305-315.

Borradaile NM, de Dreu LE, Barrett PHR, Huff MW (2002) Inhibition of hepatocyte apoB secretion by naringenin enhanced rapid intracellular degradation independent of reduced microsomal cholesteryl esters[J]. *Journal of Lipid Research*, 43(9): 1544-1554.

Borradaile NM, de Dreu LE, Barrett PHR, Behrsin CD, Huff MW (2003a) Hepatocyte apoB-containing lipoprotein secretion is decreased by the grapefruit flavonoid, naringenin, via inhibition of MTP-mediated microsomal triglyceride accumulation[J]. *Biochemistry*, 42(5): 1283-1291.

Borradaile NM, de Dreu LE, Huff MW (2003b) Inhibition of net HepG2 cell apolipoprotein B secretion by the citrus flavonoid naringenin involves activation of phosphatidylinositol 3-kinase, independent of insulin receptor substrate-1 phosphorylation[J]. *Diabetes*, 52(10): 2554-2561.

Breen DM, Sanli T, Giacca A, Tsiani E (2008) Stimulation of muscle cell glucose uptake by resveratrol through sirtuins and AMPK[J]. *Biochemical and Biophysical Research Communications*, 374(1): 117-122.

Bukhari HM, Abdelghany AH, Nada IS, Header EA (2012) Effect of yoghurt pillared with propolis on hyperglycemic rats[J]. *Egyptian Journal of Hospital Medicine*, 49: 691-704.

Butler M, McKay RA, Popoff IJ, Gaarde WA, Witchell D, Murray SF, Monia BP (2002) Specific inhibition of PTEN expression reverses hyperglycemia in diabetic mice[J]. *Diabetes* 51(4): 1028-1034.

Cantley LC (2002) The phosphoinositide 3-kinase pathway[J]. *Science*, 296(5573): 1655-1657.

Choi SS, Cha BY, Iida K, Lee YS, Yonezawa T, Teruya T, Woo JT (2011) Artepillin C, as a PPARγ ligand, enhances adipocyte differentiation and glucose uptake in 3T3-L1 cells[J]. *Biochemical Pharmacology*, 81(7): 925-933.

Cohen P, Frame S (2001) The renaissance of GSK3[J]. *Nature Reviews Molecular Cell Biology*, 2(10): 769-776.

Cordero-Herrera I, Martín MA, Bravo L, Goya L, Ramos S (2013) Cocoa flavonoids improve insulin signalling and modulate glucose production via Akt and AMPK in HepG2 cells[J]. *Molecular Nutrition & Food Research*, 57(6): 974-985.

Dogukan A, Tuzcu M, Juturu V, Cikim G, Ozercan İ, Komorowski J, Sahin K (2010) Effects of chromium histidinate on renal function, oxidative stress, and heat-shock proteins in

fat-fed and streptozotocin-treated rats[J]. *Journal of Renal Nutrition*, 20(2): 112-120.

Driver VR, Fabbi M, Lavery LA, Gibbons G (2010) The costs of diabetic foot: the economic case for the limb salvage team[J]. *Journal of Vascular Surgery*, 52(3): 17S-22S.

El Agawany A, Meguid EMA, Khalifa H, El Harri M (2012) Propolis Effect on Rodent Models of Streptozotocin-Induced Diabetic Nephropathy[J]. *Journal of American Science*, 8(12).

Ellchi A, Lipes MA, Patti ME, Jens CB, Haag B, Johnson RS, Kahn. CR (1994) Alternative pathway of insulin signalling in mice with targeted disruption of the IRS-1 gene[J]. *Nature*, 372: 186-190.

Fang XK, Gao J, Zhu DN (2008) Kaempferol and quercetin isolated from Euonymus alatus improve glucose uptake of 3T3-L1 cells without adipogenesis activity[J]. *Life Sciences*, 82(11): 615-622.

Fang Y, Sang H, Yuan N, Sun H, Yao S, Wang J, Qin S (2013) Ethanolic extract of propolis inhibits atherosclerosis in ApoE-knockout mice[J]. *Lipids Health Dis*, 12: 123-128.

Fisher EA, Ginsberg HN (2002) Complexity in the secretory pathway: the assembly and secretion of apolipoprotein B-containing lipoproteins[J]. *Journal of Biological Chemistry*, 277(20): 17377-17380.

Gauffin CP, Agüero G, Perdigón G (2002) Immunological effects of yogurt addition to a re-nutrition diet in a malnutrition experimental model[J]. *Journal of Dairy Research*, 69(2): 303-316.

Ghosal A, Satoh H, Thomas PE, Bush E, Moore D (1996) Inhibition and kinetics of cytochrome P4503A activity in microsomes from rat, human, and cdna-expressed human cytochrome P450[J]. *Drug Metab Dispos*, 24(9): 940-947.

Gordon DA, Jamil H (2000) Progress towards understanding the role of microsomal triglyceride transfer protein in apolipoprotein-B lipoprotein assembly[J]. *Biochimica et Biophysica Acta (BBA)-Molecular and Cell Biology of Lipids*, 1486(1): 72-83.

Hadi A-HA (2014) Study the effect of iraqi propolis extract on hematological parameters in alloxan-induced diabetic rabbits[J]. *Mirror of Research in Veterinary Sciences and Animals*, 3(2): 1-10

Hajiaghaalipour F, Khalilpourfarshbafi M, Arya A (2015) Modulation of Glucose Transporter Protein by Dietary Flavonoids in Type 2 Diabetes Mellitus[J]. *International Journal of Biological Sciences*, 11(5): 508.

Hassan NA, El-Bassossy HM, Mahmoud MF, Fahmy A (2014) Caffeic acid phenethyl ester, a 5-lipoxygenase enzyme inhibitor, alleviates diabetic atherosclerotic manifestations: effect on vascular reactivity and stiffness[J]. *Chemico-Biological Interactions*, 213: 28-36.

Hawkins M, Tonelli J, Kishore P, Stein D, Ragucci E, Gitig A, Reddy K (2003) Contribution of elevated free fatty acid levels to the lack of glucose effectiveness in type 2 diabetes[J].

Diabetes,52(11):2748-2758.

Hennige AM,Burks DJ,Ozcan U,Kulkarni RN,Ye J,Park S,White MF (2003) Upregulation of insulin receptor substrate-2 in pancreatic β cells prevents diabetes[J]. *Journal of Clinical Investigation* 112(10):1521.

Huang DW,Shen SC (2012) Caffeic acid and cinnamic acid ameliorate glucose metabolism via modulating glycogenesis and gluconeogenesis in insulin-resistant mouse hepatocytes. *Journal of Functional Foods*,4(1):358-366.

Huang TH-W,Peng G,Li GQ,Yamahara J,Roufogalis BD,Li Y (2006) Salacia oblonga root improves postprandial hyperlipidemia and hepatic steatosis in Zucker diabetic fatty rats:activation of PPAR-α[J]. *Toxicology and Applied Pharmacology*,210(3):225-235.

Iio A,Ohguchi K,Inoue H,Maruyama H,Araki Y,Nozawa Y,Ito M (2010) Ethanolic extracts of Brazilian red propolis promote adipocyte differentiation through PPARγ activation[J]. *Phytomedicine*,17(12):974-979.

Ishikawa T,Suzukawa M,Ito T,Yoshida H,Ayaori M,Nishiwaki M,Nakamura H (1997) Effect of tea flavonoid supplementation on the susceptibility of low-density lipoprotein to oxidative modification[J]. *The American journal of clinical nutrition* 66(2):261-266.

Jin MN,Shi GR,Tang SA,Qiao W,Duan HQ (2013) Flavonoids from Tetrastigma obtectum enhancing glucose consumption in insulin-resistance HepG2 cells via activating AMPK[J]. *Fitoterapia*,90:240-246.

Jonsson J,Carlsson L,Edlund T,Edlund H (1994) Insulin-promoter-factor 1 is required for pancreas development in mice[J]. *Nature*,371(6498):606-609.

Jung UJ,Lee MK,Park YB,Kang MA,Choi MS (2006) Effect of citrus flavonoids on lipid metabolism and glucose-regulating enzyme mRNA levels in type-2 diabetic mice[J]. *The International Journal of Biochemistry & Cell Biology*,38(7):1134-1145.

Kamagate A,Dong HH (2008a) FoxO1 integrates insulin signaling to VLDL production [J]. *Cell Cycle*,7(20):3162-3170.

Kamagate A,Qu S,Perdomo G,Su D,Kim DH,Slusher S,DongHH (2008b). FoxO1 mediates insulin-dependent regulation of hepatic VLDL production in mice[J]. *The Journal of Clinical Investigation*,118(6):2347.

Kang C,Jin YB,Lee H,Cha M,Sohn ET,Moon J,Kim E (2010) Brown alga Ecklonia cava attenuates type 1 diabetes by activating AMPK and Akt signaling pathways[J]. *Food and Chemical Toxicology*,48(2):509-516.

Karthikesan K,Pari L,Menon VP (2010) Protective effect of tetrahydrocurcumin and chlorogenic acid against streptozotocin-nicotinamide generated oxidative stress induced diabetes[J]. *Journal of Functional Foods*,2(2):134-142.

Kim JK,Gavrilova O,Chen Y,Reitman ML,Shulman GI (2000) Mechanism of insulin resistance in A-ZIP/F-1 fatless mice[J]. *Journal of Biological Chemistry*,275(12):

8456-8460.

Kitamura T, Nakae J, Kitamura Y, Kido Y, Biggs WH, Wright CV, Accili D (2002) The forkhead transcription factor Foxo1 links insulin signaling to Pdx1 regulation of pancreatic β cell growth[J]. *The Journal of Clinical Investigation*, 110(12):1839-1847.

Kono T, Ahn G, Moss DR, Gann L, Zarain-Herzberg A, Nishiki Y, Evans-Molina C (2012) PPAR-γ activation restores pancreatic islet SERCA2 levels and prevents β-cell dysfunction under conditions of hyperglycemic and cytokine stress[J]. *Molecular Endocrinology*, 26(2):257-271.

Koya-Miyata S, Arai N, Mizote A, Taniguchi Y, Ushio S, Iwaki K, Fukuda S (2009) Propolis prevents diet-induced hyperlipidemia and mitigates weight gain in diet-induced obesity in mice[J]. *Biological and Pharmaceutical Bulletin*, 32(12):2022-2028.

Kulinski A, Rustaeus S, Vance JE (2002) Microsomal triacylglycerol transfer protein is required for lumenal accretion of triacylglycerol not associated with ApoB, as well as for ApoB lipidation[J]. *Journal of Biological Chemistry*, 277(35):31516-31525.

Kumar R, Balaji S, Uma T, Sehgal P (2009) Fruit extracts of Momordica charantia potentiate glucose uptake and up-regulate Glut-4, PPARγ and PI3K[J]. *Journal of Ethnopharmacology*, 126(3):533-537.

Kwon MS, Han ZZ, Park SY, Choi YS, Lim H, Kim SD, Kim HC (2001) Effect of water-extracted propolis on the accumulation of cholesterol induced by high-cholesterol diet in the rat [C]. Proceedings of the 37th International Apicultural Congress. 1-8.

Lee B-H, Hsu W-H, Hsu Y-W, Pan T-M (2013) Dimerumic acid protects pancreas damage and elevates insulin production in methylglyoxal-treated pancreatic RINm5F cells[J]. *Journal of Functional Foods*, 5(2):642-650.

Lee J-O, Yang H, Georgescu M-M, Di Cristofano A, Maehama T, Shi Y, Pavletich, NP (1999) Crystal structure of the PTEN tumor suppressor: implications for its phosphoinositide phosphatase activity and membrane association[J]. *Cell*, 99(3):323-334.

Lee J-W, Choe SS, Jang H, Di Cristofano A, Maehama T, Shi Y, Pavletich NP (2012) AMPK activation with glabridin ameliorates adiposity and lipid dysregulation in obesity[J]. *Journal of Lipid Research*, 53(7):1277-1286.

Leonard J, Peers B, Johnson T, Ferreri K, Lee S, Montminy M (1993) Characterization of somatostatin transactivating factor-1, a novel homeobox factor that stimulates somatostatin expression in pancreatic islet cells[J]. *Molecular Endocrinology*, 7(10):1275-1283.

Li YJ, Chen ML, Xuan HZ, Hu F (2011) Effects of encapsulated propolis on blood glycemic control, lipid metabolism, and insulin resistance in type 2 diabetes mellitus rats[J]. *Evidence-based Complementary and Alternative Medicine*, 2012:1-8.

Maehama T, Dixon JE (1998) The tumor suppressor, PTEN/MMAC1, dephosphorylates the lipid second messenger, phosphatidylinositol 3, 4, 5-trisphosphate[J]. *Journal of Biological Chemistry*, 273(22):13375-13378.

Mahani M, Jannah I, Harahap E, Salman M, Habib N (2013) Antihyperglycemic Effect of

Propolis Extract from Two Different Provinces in Indonesia[J]. *International Journal on Advanced Science,Engineering and Information Technology*,3(4):1-4.

Mahmoud AM,Ahmed OM,Abdel-Moneim A,Ashour MB(2013a) Upregulation of PPARγ mediates the antidiabetic effects of citrus flavonoids in type 2 diabetic rats[J]. *International Journal of Bioassays*,2(5):756-761.

Mahmoud MF,Sakr SM(2013b) Hepatoprotective effect of bee propolis in rat model of streptozotocin-induced diabetic hepatotoxicity:light and electron microscopic study [J]. *Life Science Journal*,10(4).

Meydani SN,Ha W-K(2000) Immunologic effects of yogurt[J]. *The American Journal of Clinical Nutrition*,71(4):861-872.

Mora A,Komander D,van Aalten DM,Alessi DR(2004) PDK1,the master regulator of AGC kinase signal transduction[J]. *Seminars in Cell & Developmental Biology*,15 (2):161-170.

Nader MA,El-Agamy DS,Suddek GM(2010) Protective effects of propolis and thymoquinone on development of atherosclerosis in cholesterol-fed rabbits[J]. *Archives of Pharmacal Research*,33(4):637-643.

Nakae J,Biggs WH,Kitamura T,Cavenee WK,Wright CV,Arden KC,Accili D(2002) Regulation of insulin action and pancreatic β-cell function by mutated alleles of the gene encoding forkhead transcription factor Foxo1[J]. *Nature Genetics*,32(2): 245-253.

Nakae J,Kitamura T,Silver DL,Accili D(2001) The forkhead transcription factor Foxo1 (Fkhr) confers insulin sensitivity onto glucose-6-phosphatase expression[J]. *Journal of Clinical Investigation*,108(9):1359.

Nakashima N,Sharma PM,Imamura T,Bookstein R,Olefsky JM(2000) The tumor suppressor PTEN negatively regulates insulin signaling in 3T3-L1 adipocytes[J]. *Journal of Biological Chemistry*,275(17):12889-12895.

Niu Y,Li S,Na L,Feng R,Liu L,Li Y,Sun C(2012) Mangiferin decreases plasma free fatty acids through promoting its catabolism in liver by activation of AMPK[J]. *PLoS One*,7(1):e30782.

Ohlsson H,Karlsson K,Edlund T(1993) IPF1,a homeodomain-containing transactivator of the insulin gene[J]. *The EMBO Journal*,12(11):4251.

Ong KW,Hsu A,Tan BKH(2013) Anti-diabetic and anti-lipidemic effects of chlorogenic acid are mediated by ampk activation[J]. *Biochemical Pharmacology*,85(9):1341-1351.

Ong KW,Hsu A,Tan BKH(2012) Chlorogenic acid stimulates glucose transport in skeletal muscle via AMPK activation:a contributor to the beneficial effects of coffee on diabetes[J]. *PLoS ONE*,7(3):e32718.

Oršolić N,Sirovina D,Končić MZ,Lacković G,Gregórovic G(2012) Effect of Croatian propolis on diabetic nephropathy and liver toxicity in mice[J]. *BMC Complementary and Alternative Medicine*,12(1):117.

Özcan U, Cao Q, Yilmaz E, Lee AH, Iwakoshi NN, Özdelen E, Hotamisligil GS (2004) Endoplasmic reticulum stress links obesity, insulin action, and type 2 diabetes[J]. *Science*, 306(5695):457-461.

Park SM, Hong SM, Sung SR, Lee JE, Kwon DY (2008) Extracts of Rehmanniae radix, Ginseng radix and Scutellariae radix improve glucose-stimulated insulin secretion and β-cell proliferation through IRS2 induction[J]. *Genes & nutrition*, 2(4):347-351.

Prasad CV, Anjana T, Banerji A, Gopalakrishnapillai A (2010) Gallic acid induces GLUT4 translocation and glucose uptake activity in 3T3-L1 cells[J]. *FEBS Letters*, 584(3):531-536.

Premalatha M, Parameswari C (2012) Renoprotective effect of chrysin (5,7 dihydroxy flavone) in streptozotocin induced diabetic nephropathy in rats[J]. *International Journal of Pharmacy and Pharmaceutical Sciences*, 4(3):241-247.

Rifa'i M, Widodo N (2014) Significance of propolis administration for homeostasis of $CD4^+$ $CD25^+$ immunoregulatory T cells controlling hyperglycemia[J]. *Springer Plus*, 3(1):1-8.

Sang H, Yuan N, Yao S, Li F, Wang J, Fang Y, Qin S (2012) Inhibitory effect of the combination therapy of simvastatin and pinocembrin on atherosclerosis in apoE-deficient mice[J]. *Lipids Health Dis*, 11(166):10.1186.

Schimmack G, DeFronzo RA, Musi N (2006) AMP-activated protein kinase: role in metabolism and therapeutic implications[J]. *Diabetes, Obesity and Metabolism*, 8(6):591-602.

Serup P, Petersen HV, Pedersen EE, Edlund H, Leonard J, Petersen JS, Madsen OD (1995) The homeodomain protein IPF-1/STF-1 is expressed in a subset of islet cells and promotes rat insulin 1 gene expression dependent on an intact E1 helix-loop-helix factor binding site[J]. *Biochem J*, 310:997-1003.

Shah P, Vella A, Basu A, Basu R, Adkins A, Schwenk WF, Rizza RA (2003) Elevated Free Fatty Acids Impair Glucose Metabolism in Women Decreased Stimulation of Muscle Glucose Uptake and Suppression of Splanchnic Glucose Production During Combined Hyperinsulinemia and Hyperglycemia[J]. *Diabetes*, 52(1):38-42.

Sharma B, Balomajumder C, Roy P (2008) Hypoglycemic and hypolipidemic effects of flavonoid rich extract from Eugenia jambolana seeds on streptozotocin induced diabetic rats[J]. *Food and Chemical Toxicology*, 46(7):2376-2383.

Sharma S, Agrawal RP, Choudhary M, Jain S, Goyal S, Agarwal V (2011) Beneficial effect of chromium supplementation on glucose, HbA 1 C and lipid variables in individuals with newly onset type-2 diabetes[J]. *Journal of Trace Elements in Medicine and Biology*, 25(3):149-153.

Shin EJ, Hur HJ, Sung MJ, Park JH, Yang HJ, Kim MS, Hwang JT (2013) Ethanol extract of the Prunus mume fruits stimulates glucose uptake by regulating PPAR-γ in C2C12 myotubes and ameliorates glucose intolerance and fat accumulation in mice fed a high-fat diet[J]. *Food Chemistry*, 141(4):4115-4121.

Soto C, Raya L, Juárez J, Pérez J, González I (2014) Effect of Silymarin in Pdx-1 expression and the proliferation of pancreatic β-cells in a pancreatectomy model[J]. *Phytomedicine*, 21(3):233-239.

Temesio P, Ross N, Alvarez R (2012) Topical treatment with propolis dressings of poor healing foot ulcers in diabetic patients[J]. *Journal of Analytical Atomic Spectrometry*, 2012, 19:788-795.

Thong FS, Dugani CB, Klip A (2005) Turning signals on and off: GLUT4 traffic in the insulin-signaling highway[J]. *Physiology*, 20(4):271-284.

Trent L, Thieding-Cancel D (1995) Effects of chromium picolinate on body composition [J]. *The Journal of Sports Medicine and Physical Fitness*, 35(4):273-280.

Virkamäki A, Ueki K, Kahn CR (1999) Protein-protein interaction in insulin signaling and the molecular mechanisms of insulin resistance[J]. *Journal of Clinical Investigation*, 103(7):931.

Waeber G, Thompson N, Nicod P, Bonny C (1996) Transcriptional activation of the GLUT2 gene by the IPF-1/STF-1/IDX-1 homeobox factor[J]. *Molecular Endocrinology*, 10(11):1327-1334.

Wahli W, Braissant O, Desvergne B (1995) Peroxisome proliferator activated receptors: transcriptional regulators of adipogenesis, lipid metabolism and more[J]. *Chemistry & Biology*, 2(5):261-266.

Waltner-Law ME, Wang XL, Law BK, Hall RK, Nawano M, Granner DK (2002) Epigallocatechin gallate, a constituent of green tea, represses hepatic glucose production[J]. *Journal of Biological Chemistry*, 277(38):34933-34940.

Wang Y, Tran K, Yao Z (1999) The activity of microsomal triglyceride transfer protein is essential for accumulation of triglyceride within microsomes in McA-RH7777 cells a unified model for the assembly of very low density lipoproteins[J]. *Journal of Biological Chemistry*, 274(39):27793-27800.

Withers DJ, Gutierrez JS, Towery H, Burks DJ, Ren JM, Previs S, White MF (1998) Disruption of IRS-2 causes type 2 diabetes in mice[J]. *Nature*, 391(6670):900-904.

Wu X-Y, Li F, Zhao T, Mao GH, Li J, Qu HY, Yang LQ (2012) Enhanced anti-diabetic activity of a combination of chromium (III) malate complex and propolis and its acute oral toxicity evaluation[J]. *Biological Trace Element Research*, 148(1):91-101.

Yeh YH, Lee YT, Hsieh HS, Hwang DF (2009) Dietary caffeic acid, ferulic acid and coumaric acid supplements on cholesterol metabolism and antioxidant activity in rats [J]. *Journal of Food and Drug Analysis*, 7(2):123-132.

Yoon S-A, Kang S-I, Shin H-S, Kang S-W, Kim J-H, Ko H-C, Kim S-J (2013) p-Coumaric acid modulates glucose and lipid metabolism via AMP-activated protein kinase in L6 skeletal muscle cells[J]. *Biochemical and Biophysical Research Communications*, 432 (4):553-557.

Zarzecki MS, Araujo SM, Bortolotto VC, de Paula MT, Jesse CR, Prigol M (2014) Hypolipidemic action of chrysin on Triton WR-1339-induced hyperlipidemia in female C57BL/6 mice[J].

Toxicology Reports,1:200-208.

Zhang J,Wang HY,Zhang GX,Zhang Y (2005) Effect of propolis compounds on blood glucose[J]. *Henan Journal of Preventive Medicine*,5:005.

Zhang Y,Zhen W,Maechler P,Liu D (2013) Small molecule kaempferol modulates PDX-1 protein expression and subsequently promotes pancreatic β-cell survival and function via CREB[J]. *The Journal of Nutritional Biochemistry*,24(4):638-646.

Zhu W,Chen M,Shou Q,Li Y,Hu F (2011a) Biological activities of Chinese propolis and Brazilian propolis on streptozotocin-induced type 1 diabetes mellitus in rats[J]. *Evidence-Based Complementary and Alternative Medicine*,Volume 2011,Article ID 468529.

Zhu W,Li YH,Chen ML,Hu FL (2011b) Protective effects of Chinese and Brazilian propolis treatment against hepatorenal lesion in diabetic rats[J]. *Human & Experimental Toxicology*,30(9):1246-1255.

Zygmunt K,Faubert B,MacNeil J,Tsiani E (2010) Naringenin,a citrus flavonoid,increases muscle cell glucose uptake via AMPK[J]. *Biochemical and Biophysical Research Communications*,398(2):178-183.

程清洲,周威,瞿永华,朱珍斌,吴东方(2013)蜂胶乙醇提取物调节大鼠血脂的效果研究[J].辽宁中医杂志,40(4):810-811.

李冀宏,杨秀珍(2002)胎盘银杏叶蜂胶复方制剂对高脂大鼠血清总胆固醇,MDA含量及脾重的影响[J].齐齐哈尔医学院学报,23(12):1321-1322.

梁秀慈(2013)绿原酸对长期摄取高糖高脂饮食SD大鼠脂代谢的影响[D].长沙:湖南农业大学.

童晔玲,杨锋,戴关海,任泽明,竹剑平,王波波(2013)蜂胶西洋参软胶囊降血脂功能的实验研究[J].中华中医药学刊,31(3):569-570.

王建辉,刘永乐,李赤翎,俞健,李向红,王发祥,李艳(2012)杜仲绿原酸对高脂模型小鼠降血脂作用研究[J].食品工业科技,33(15):360-362.

卫晓怡,白晨,崔琳琳,陆红,蒲立柠(2013)矢车菊素-3-葡萄糖苷抑制KKAy糖尿病小鼠PTEN表达[J].食品科学,34(13):280-284.

张翠平,胡福良(2009)蜂胶中的黄酮类化合物[J].天然产物研究与开发,21(6):1084-1084.

张翠平,王凯,胡福良(2013)蜂胶中的酚酸类化合物[J].中国现代应用药学,30(1):102-105.

张江临,王凯,胡福良(2013)蜂胶的抗氧化活性及其分子机制研究进展[J].中国中药杂志,38(16):2645-2652.

赵丽婷(2016)蜂胶对2型糖尿病患者能量代谢和氧化应激的干预作用及其机制研究[D].北京:中国人民解放军军事医学科学院.

第七节 蜂胶的保肝作用

肝脏是人体中最大的腺体,也是最大的实质性脏器,犹如一个巨大的化工厂,通过各种复杂的生化反应,将流经肝脏的血液中各种物质进行代谢转化以满足生命活动所需。而解毒作用作为肝脏重要的生理功能之一,能将随血液进入肝脏的各种有毒物质通过肝细胞内的一些非特异性酶系的催化完成其代谢转化,使这些毒性物质转化为低毒,甚至无毒,或是可溶性物质,从而随尿液和胆汁排出体外(陈世贵,1990)。然而,作为各种毒性物质的"解毒器",肝脏自然也成了机体最容易受到损伤的脏器之一。各种毒物及其毒性代谢物通过损伤细胞膜,改变离子通透性,变性蛋白、脂,阻碍核酸合成,使细胞色素 P450 酶(CYP450)活性异常激活或抑制,并由此引起大量自由基产生等多种方式造成肝细胞的损伤(彭飞等,2012;沈杰,1990)。由毒物及其毒性代谢造成的大量活性自由基的产生及脂质过氧化是引起肝损伤最主要的原因之一,因为这些活性氧产物能造成膜脂、蛋白和 DNA 的损伤(朱润芝等,2010)。氧化应激性损伤通常也是各种不利因素造成肝损伤的最主要的途径。因此,研究者通常将蜂胶及其保肝成分的保肝作用与其抗氧化活性联系在一起,并以此开展了大量研究。

保肝活性作为蜂胶重要的生物学活性之一,一直受到人们的重视。通过对蜂胶及其活性成分保肝作用研究,发现蜂胶的保肝作用与其抗氧化活性紧密联系在一起,这主要是因为蜂胶中大量的黄酮类及酚类物质能通过直接清除自由基、增强机体抗氧化能力等方式改善肝的氧化应激性损伤。同时,蜂胶的保肝作用也可能与其他功效有一定关系,如抗炎活性或直接结合毒性物质(例如螯合毒性离子)等能力有关。下面就近年来国内外学者对蜂胶及其活性成分保肝作用的相关研究进行综述,以期为蜂胶产品的开发利用和保肝活性的深入研究提供参考。

1 蜂胶的保肝作用

1.1 蜂胶对 CCl_4 致肝损伤的保护作用

四氯化碳(CCl_4)是经典的实验性肝损伤模型常用到的毒剂之一,进入体内能引起实验动物肝的损伤和坏死。它能通过肝细胞的 CYP450 代谢为更有毒性的三氯甲基自由基($CCl_3\cdot$)和过氧化三氯甲基自由基($OOCCl_3\cdot$),进而引起脂质过氧化,导致抗氧化酶活性下降,细胞膜和细胞器膜的损伤,钙离子稳态被打破;并且,高浓度 CCl_4 本身具有的溶酶作用也会导致肝细胞的坏死(张欣和党双锁,2009)。因此,用 CCl_4 诱导的肝损伤模型是研究蜂胶保肝作用的常见模型之一。

El-Khatib 等(2002)给大鼠预先连续口服蜂胶水提物(APE)14d,之后通过腹腔注射 CCl_4 诱导肝损伤,CCl_4 注射 1d 后,分离出大鼠的肝细胞和准备肝匀浆物用来评估肝损伤。通过对分离的肝细胞研究发现,口服 APE 后乳酸脱氢酶(LDH)的流失和脂质过氧化减少,且胞内谷胱甘肽(GSH)水平得以维持,从而证明了 APE 的保肝作用;利用肝匀浆物也发现了类似结果。研究结果表明,蜂胶体内保肝作用可能与其抗氧化作用相关,通过其抗氧化作用维持胞内 GSH 含量,而 GSH 含量的稳定有利于减少细胞的脂质过氧化,从而保护细胞膜的完整性,减少 LDH 的流失。

Bhadauria 等(2007)研究表明,蜂胶能呈持续时间依赖性方式保护 CCl_4 诱导的急性肝损伤。应用 CCl_4 后,血清中转氨酶和碱性磷酸酶的活性急剧升高,而肝组织中的 GSH 含量大幅下降并伴随着显著增加的脂质过氧化,肝肾组织中糖原含量和碱性磷酸酶、三磷腺苷酶(ATP 酶)及琥珀酸脱氢酶活性显著下降,而总蛋白量和酸性磷酸酶活性有所增加。当加入蜂胶治疗后 6、12、24h,这些指标变化明显降低,且保护作用在蜂胶作用 24h 后表现最强。

Bhadauria 等(2008a)再次通过实验验证了蜂胶能对抗 CCl_4 诱导的肝损伤,使由 CCl_4 导致的肝的一些重要生化参数的变化恢复正常。例如,阻止血清中转氨酶、碱性磷酸酶、乳酸脱氢酶、γ-谷氨酰转肽酶、尿素和尿酸含量的升高,恢复肝微粒体药物代谢酶活性,显著抑制脂质过氧化和增加肝肾中谷胱甘肽含量,并使糖、脂、蛋白含量向正常水平靠近。他们分析认为这与蜂胶的抗氧化活性有密切关系。肝细胞内质网 CPY450 能通过脱卤作用将 CCl_4 转变为三氯甲基自由基($CCl_3\cdot$)和三氯甲基过氧化氢自由基($CCl_3O_2\cdot$)。当 GSH 的量不足以清除 CCl_4 毒性自由基时,这些自由基可以攻击或共价结合微粒体脂和蛋白,造成脂质过氧化,进而导致肝细胞的坏死。有研究表明,存在于蜂胶中的几种黄酮能增加 γ-谷氨酰胺半胱甘酸合成酶的表达,从而促进 GSH 的合成(Myhrstad et al.,2002)。因此,蜂胶可能通过直接中和活性氧,增强内源性抗氧化系统防御能力,增强 GSH 的稳定性和其合成速率的方式促进 GSH 的含量趋向正常水平,从而清除毒性自由基。CCl_4 作用后,糖、脂、蛋白含量的明显变化反映出肝受损后对碳水化合物的代谢障碍,而蜂胶能呈剂量依赖恢复糖原含量的原因可能与增加肝实质细胞再生有关。此外,蜂胶中丰富的矿物元素可能有助于调节一些金属离子依赖性酶的活性。另外,蜂胶中的咖啡酸苯乙酯(CAPE)可能通过供氢给 $CCl_3\cdot$ 和 $CCl_3O_2\cdot$,生成 $HCCL_3$ 和 $HOOCCl_3$,而其本身转变为邻醌 CAPE 或 CAPE 与自由基的结合物(如图 3.7)。

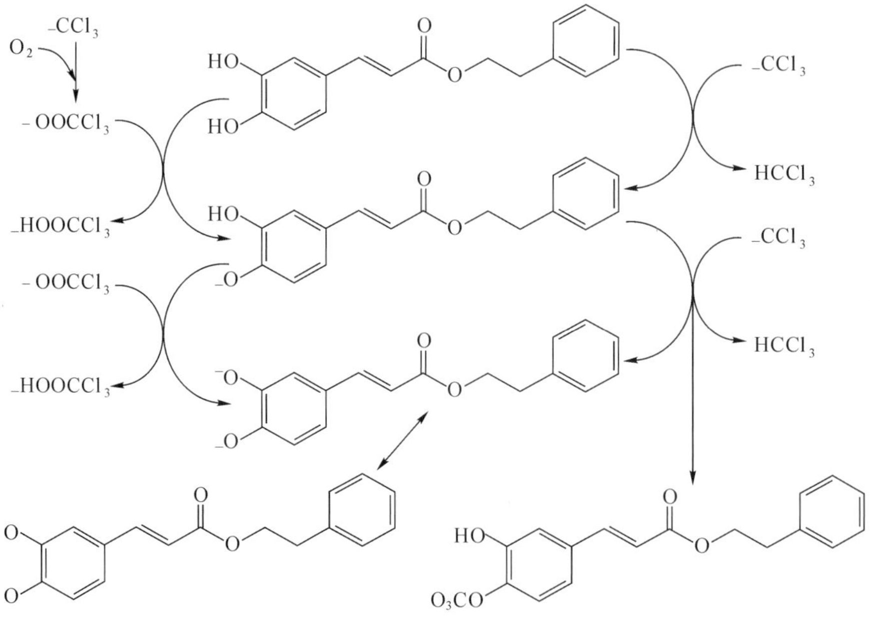

图 3.7 CAPE 与 CCl_4 代谢自由基可能发生的化学反应

Fig. 3.7 A possible mechanism of reaction of CAPE with $CCl_3\cdot$ radical and/or $CCl_3O_2\cdot$ radical

杜夏(2013)利用 L-02 细胞建立的 CCl_4 致肝损伤体外模型对蜂胶中保肝活性成分及机制进行了初步探索。结果表明,EEP 的保肝活性要远远强于 WEP,而且不同浓度乙醇(40%、70%、95%)提取出来的蜂胶中主要保肝活性成分不同。对 EEP 保肝活性机制的研究发现,蜂胶能通过其抗氧化活性和抑制 CYP2E1 减轻 CCl_4 诱导的氧化应激,抑制凋亡相关蛋白 Caspase 3 的活化,从而减少细胞凋亡。

1.2 蜂胶对药物性肝损伤的保护作用

药物性损伤是常见的肝损伤类型之一。正常情况下,药物进入肝脏能通过肝脏的解毒作用最终排出体外。正常情况下这些药物经肝解毒而产生的代谢转化物、亲电子基、自由基和氧基等有害活性物质能通过与谷胱甘肽、葡萄糖醛酸等结合而解毒,并不造成肝的损伤。而在有些情况下,药物的大量摄入必然加重肝脏的负担,一旦超出肝脏的代偿能力,毒物必然会蓄积在肝内,并通过多种方式损伤肝细胞(徐鑫和屈彩芹,2008)。

大量的研究表明,蜂胶能通过恢复肝中 GSH 水平、降低脂质过氧化等多种方式改善各种药物性肝损伤。Nirala 和 Bhadauria(2008)研究了蜂胶提取物对乙酰氨基酚(AAP,对乙酰氨基酚)诱导的肝的氧化应激和功能障碍的疗效。结果表明,AAP 高剂量给药后,大鼠血清中转氨酶、碱性磷酸酶、乳酸脱氢酶和血清胆红素增加,而血红蛋白和血糖水平下降。肝组织氧化应激水平也发生了明显变化,表现为 GSH 含量、MDA 含量、CYP 酶活性的变化。此外,组织中 ATP 酶、酸性和碱性磷酸酶活性,总蛋白、糖原、胆固醇的含量也发生明显改变,而蜂胶能改善这些生化指标的变化,表明蜂胶对药物性肝损伤具有保护作用。这种作用可能与蜂胶中酚类成分能改善机体氧化应激防御能力有关,如清除自由基、维持 GSH 的水平。GSH 同时也是清除醋氨酚(APAP)毒性代谢产物 N-乙酰-P-苯醌亚胺(NAPQI)的重要物质。Nirala 等(2008)再次通过 AAP 诱导的肝损伤模型研究发现蜂胶能改善 SOD、过氧化氢酶(CAT)的活性,进一步证实了蜂胶可以通过抗氧化及改善机体抗氧化防御能力改善 APAP 导致的肝损伤。Seo 等(2003)研究表明,蜂胶可能通过抑制 I 相药物代谢酶的活性而增强 II 相药物代谢酶活性的方式改善 APAP 诱导肝损伤。Siess 等(1996)研究发现蜂胶及其黄酮提取物能提高肝脏内一些药物代谢酶如脱乙基酶、脱甲基酶、胱甘肽 S 转移酶等的活性。

Badr 等(2011)将 Swiss 白化鼠腹腔注入 2.5×10^6 的艾氏腹水癌细胞,再用甲氨蝶呤治疗,用来研究埃及蜂胶对甲氨蝶呤诱导的肝肾功能障碍。结果表明,蜂胶能改善癌细胞植入及加入甲氨蝶呤治疗后的 SOD、CAT 及 GSH 的水平,并降低 MDA 的含量,同时改善以血清中天冬氨酸转氨酶(AST)、丙氨酸转氨酶(ALT)活性,总蛋白、白蛋白含量等为指标的病理学变化,表明了蜂胶对抗癌药诱导的肝功能障碍的保护作用。

Türkez 等(2013)探究了蜂胶在减轻四氯二苯并-p-二噁英(TCDD,除草剂中一种剧毒杂质)诱导的大鼠肝毒性的有效性。结果表明,高剂量的 TCDD 应用后,肝组织抗氧化酶活性降低并出现严重的组织病理学变化,同时肝细胞微核率增加。蜂胶加入后,能阻止肝抗氧化酶活性的抑制,降低肝细胞微核率,从而削弱 TCDD 的肝毒性。其体外实验的结果与之一致,即蜂胶能降低 TCDD 诱导的肝细胞的总氧化应激水平,提高其抗氧化能力,并减少 TCDD 诱导的肝细胞微核率和 DNA 损伤(Türkez et al.,2012)。

Abdelsameea 等(2013)还研究了蜂胶对阿托伐他汀(降血脂药,能降低胆固醇水平,治疗冠心病)诱导的肝毒性的抑制作用。在给实验鼠口服剂量为 20mg/kg 和 80mg/kg 的阿托伐他汀1h之前,提前将蜂胶以50mg/kg和100mg/kg的剂量经口给药予实验鼠,连续用药30d。试验结束后对血清中ALT、AST和肝匀浆物中SOD、CAT的水平进行评估,并进行组织病理学观察。结果表明,阿托伐他汀呈剂量依赖性地显著升高ALT、AST、SOD和CAT的水平,并造成肝细胞的变性。蜂胶能呈剂量依赖性方式显著降低这些检测指标的变化,并阻止阿托伐他汀诱导的肝组织结构的变化。他们同样认为蜂胶的这种保肝作用可能与其抗氧化作用有关。此外,还有人报道了蜂胶对异烟肼(抗结核药)、环孢霉素A、杀虫剂氰戊菊酯诱导的肝毒性也有一定的改善作用(Humayun et al.,2014;Seven et al.,2014;Al-Amoudi,2015)。

由此可见,蜂胶对各种药物所致的肝损伤都有一定的改善作用,这种保肝活性可能通过清除自由,恢复GSH水平及抗氧化酶活性,甚至调节一些药物代谢酶的方式得以实现。

1.3 蜂胶对金属离子致肝损伤的保护作用

随着工业化进程的不断推进,环境污染日益严重,加上人们环保意识的薄弱,近年来关于金属性中毒事件特别是镉、铅、汞等重金属中毒事件的相关报道屡见不鲜。这些重金属物质通过污染的水体、农作物、鱼类以及工业废气、化工制品等多种方式进入人体,通过消化道、呼吸道及皮肤吸收,再通过血液运输进入各种脏器和组织。当它们在体内的蓄积量超过机体清除能力时,便会通过变性生物大分子、损伤DNA、引发变态反应、造成氧化应激等方式损伤机体(朱玉真等,1997;袭著革等,2003;容丽萍等,2014)。这些金属离子往往具有多器官毒性。肝作为血流量较多且具有解毒作用的器官,是各种重金属离子最易蓄积的器官之一,因此容易受到损伤。此外,对身体有害的不只是那些非必需金属元素,一些机体必需的轻金属或重金属元素如镁、铝、铜、镍等超过机体耐受范围也会对机体不利。因此,不少学者研究了蜂胶对金属离子所致的肝损伤的保护作用。

Nirala等(2008)研究了由没食子酸和蜂胶及两者配伍对毒性金属离子铍诱导的肝损伤的保护作用,实验结束后评估反映肝肾病变的生化指标。血清中用来评估的指标包括天冬氨酸转氨酶(AST)、丙氨酸转氨酶(ALT)、乳酸脱氢酶(LDH)、γ-谷氨酰转肽酶(γ-GT)、总胆红素、肌酸酐、白蛋白和尿素;肝肾组织用来评估的指标包括糖原、总蛋白、脂质过氧化、谷胱甘肽、三酰甘油、总胆固醇的含量、酸性及碱磷酸酶、ATP酶、6-磷酸葡萄糖酶及琥珀酸脱氢酶的活性;再利用微粒体评估微粒体蛋白、微粒体脂质过氧化及微粒体药物代谢酶活性,这些指标主要用来反映肝肾细胞及其细胞器结构和功能的变化。结果发现,与正常鼠比较,接触铍的对照组大鼠各种理化指标都发生了显著的变化,能很好地反映由金属离子铍诱导的肝肾功能障碍,而蜂胶和没食子酸治疗后能将这些理化指标的变化幅度明显降低,两者配伍用药效果更为显著,使这些理化指标恢复到接近正常水平。组织病理学观察结果同样证实了两者对肝肾的保护作用。分析认为,毒性金属离子铍进入机体能通过引起肝肾组织内细胞及其细胞器的损伤及一些重要的酶活性的降低,并引起大量活性氧的产生,而造成肝肾器官组织结构和功能的失常,而没食子酸和蜂胶可能通过螯合金属离子铍、清除自由基、调节酶活性等多种方式发挥对肝肾的保护作用。

Türkez等(2010)将雄性Sprague Dawley大鼠分为4组:对照组、氯化铝组[氯化铝($AlCl_3$)加入量为34mg/kg]、蜂胶组(50mg/kg)和$AlCl_3$加蜂胶组,经口给药,时间为30d,试验结束后,麻醉小鼠并分离肝细胞进行微核肝细胞(MNHEPs)计数。此外,还对血清中酶的水平和肝的组织学变化进行了分析。结果表明,$AlCl_3$刺激后,MNHEPs、碱性磷酸酶、转氨酶(AST 和 ALT)和 LDH 的量上升,而且还出现了比较严重的病理学损伤如中央静脉阻塞、脂质堆积、淋巴细胞浸润等,而单独应用蜂胶则没表现出上述的副作用;结合蜂胶治疗能显著减轻由 $AlCl_3$ 诱导的毒性作用。

Bhadauria 等(2008b)的实验结果表明,蜂胶提取物治疗能抑制氯化汞($HgCl_2$,5mg/kg)诱导的肝的脂质过氧化和氧化型谷胱甘肽的形成,增加肝中还原型谷胱甘肽水平,显著降低肝的一些标志性酶在血清中的含量,并恢复一些抗氧化酶的活性,表明蜂胶能通过抗氧化防御能力对抗汞诱导的肝中毒。

综上所述,蜂胶对一些金属离子诱导的肝损伤也有一定的保护作用,这种作用可能与蜂胶具有清除自由基、调节抗氧化酶活性甚至螯合金属离子的能力有关。

1.4 蜂胶对酒精性肝损伤的保护作用

酒精性肝病种类较多,如酒精性脂肪肝和肝纤维化、酒精性肝炎、肝硬化等。大量长期的酗酒会加重肝脏代谢负担,代谢酒精的过程中,可致脂代谢紊乱,此外线粒体呼吸量功能亢进,大量耗氧,使某些部位的肝细胞缺氧坏死,并且脂质过氧化的增强,谷胱甘肽的减少以及自由基和毒性代谢产物乙醛的大量形成等共同构成肝损伤的病因(森本道雄,1997)。

Kolankaya 等(2002)研究了蜂胶对酒精诱导的血清中脂变化的调节和肝损伤的作用。EEP 以 200mg/kg 的剂量通过灌胃给药给雄性大鼠,连续服用 15d,再用等体积 20%的酒精诱导肝损伤 15d 后,评估血清中脂水平、肝酶活及其他生化参数。结果表明,酒精诱导的肝损伤会使高密度脂蛋白(HDL)水平下降和低密度脂蛋白(LDL)水平增加,而结合蜂胶的治疗会使这两种脂蛋白变化幅度显著减小。蜂胶的治疗还能降低胆固醇和三酰甘油的水平。与酒精刺激组比较,蜂胶的治疗能降低 ALP 和 AST 的酶活性,但 LDH 酶活性增加。

Chen 等(2008)通过酒精诱导的体内、体外肝损伤模型,研究酒精刺激下组织型转谷氨酰胺酶(tTG)表达的具体的分子机制以及 tTG 与肝纤维化疾病的关系,并探究蜂胶成分阻止体外 tTG 表达和体内抗肝纤维化的有效性。结果表明,ERK1/2 和 PI3K/Akt 信号通路参与了酒精诱导的 NF-κB 依赖性转录,这些信号通路可能参与到酒精诱导的 tTG 表达的激活。该研究还表明,tTG 可能是参与酒精诱导的肝纤维的重要因素之一,而蜂胶中的主要成分松属素能抑制酒精诱导的 tTG 的表达和酶活性,组织病理学观察也发现蜂胶能明显减轻硫代乙酰胺诱导的肝纤维化,进一步证实了蜂胶及其成分的保肝作用。

1.5 蜂胶对半乳糖胺致肝损伤的保护作用

利用 D-氨基半乳糖(D-GalN)诱导的肝损伤模型,是近年来研究肝病常用的实验性肝损伤模型之一。D-GalN 在肝内的代谢造成尿苷酸的大量消耗,并由此引发一系列病理变化,如核酸、蛋白、糖原合成受阻,引起磷脂代谢障碍、膜损伤加重,破坏钙离子稳态,损伤细胞器,加速氧自由基产生等,从而致肝损伤逐渐加重(张欣和党双锁,2009)。

Rodríguez 等(1997)利用 1000mg/kg 剂量的半乳糖胺诱导的肝炎模型,研究古巴红蜂胶的保肝作用。在用半乳糖胺诱导 30min 之前,给大鼠口服不同浓度(10、50、100mg/kg)的EEP。结果表明,EEP 能通过其抗炎和抗氧化作用对抗半乳糖胺诱导的肝损伤,抑制血清中 ALT 活性和 MDA 浓度的升高,阻止半乳糖胺诱导的肝细胞的细胞膜、细胞核及细胞器的损伤。

Banskota 等(2001)利用半乳糖苷/TNF-α 诱导的小鼠肝实质细胞死亡体外模型,研究蜂胶提取物的保肝作用，结果表明，蜂胶 MeOH 提取物的保肝作用强于蜂胶水提取物，发挥保肝作用的物质是包括黄酮在内的酚类物质。他们分析认为，这可能主要与蜂胶提取物的抗氧化活性有关，即蜂胶提取物通过清除 TNF-α 诱导的活性氧，阻碍 TNF-α 介导的信号转导，从而保护细胞免受损伤。

El-Mahalaway 等(2015)利用 D-GalN 及脂多糖(LPS)诱导的鼠肝炎模型，并结合组织学和免疫组化染色法对蜂胶保肝效果进行了研究，结果表明，蜂胶治疗能明显改善由 D-GalN 及 LPS 引起的肝细胞的坏死、变性，中央静脉内皮细胞增生、扩张充血，以及胶原纤维显著增加等一系列肝炎症性特征。

1.6 蜂胶对糖尿病老鼠肝的保护作用

肝脏作为糖脂代谢重要器官，在糖尿病状态下能产生脂肪肝、脂肪性肝炎、肝纤维化等肝脏病，这些慢性肝脏疾病的发生可能与氧化应激、内质网应激密切相关(吴悠等，2011)。因此，降血糖血脂同时缓解肝的氧化应激和内质网应激来减少细胞损伤，对改善糖尿病肝损伤有重要意义。大量的实验研究发现蜂胶对糖尿病鼠的肝有一定保护作用，这可能与蜂胶抗氧化、调节血糖血脂的活性有很大关系。

Zhu(2011)等研究了中国和巴西蜂胶对 STZ 诱导的糖尿病大鼠肝肾损伤的保护作用。结果表明，中国蜂胶治疗使糖尿病大鼠糖化血红蛋白降低 7.4%，并且显著增加血清中 SOD 的水平。巴西蜂胶治疗增加血清中的 SOD，并降低 MDA 和氮合成酶的水平，血清 AST 和 ALT 的减少和尿微量白蛋白的减少说明了蜂胶对肝肾功能的改善。他们还观察到中国蜂胶和巴西蜂胶能显著提高肝肾谷胱甘肽过氧化物酶(GSH-px)水平并抑制 MDA 的产生，表明蜂胶能通过抑制脂质过氧化和增加抗氧化酶活性的方式来保护肝肾。Mahmoud 和 Sakr (2013)通过组织学和超微结构观察的结果表明，蜂胶在糖尿病发生早期阶段，能够降低糖尿病的破坏性进程，对受损的肝细胞有明显的修复作用，表明了其对琏尿霉素诱导的高血糖大鼠肝的保护作用。此外，Babatunde 等(2015)研究表明，尼日利亚蜂胶乙醇提取物能通过降低血清中 MDA 含量和升高 SOD 含量的方式改善高血糖诱导的大鼠肝脏和胰腺氧化应激，而起到保肝护胰作用。

1.7 蜂胶对其他不利因素引起的肝损伤的保护作用

蜂胶除了对以上因素所致的肝损伤有一定的改善作用外，对其他不利因素诱导的肝损伤也有一定的保护作用。

Nakamura 等(2012)对比研究了巴西蜂胶和 VE 对水浸-束缚应激(WIRS)诱导的肝氧化损伤的保护作用。在对大鼠水浸-束缚 30min 前给禁食大鼠口服巴西绿蜂胶乙醇提取物(BPEE)10、50、100mg/kg，VE250mg/kg。大鼠接受 WIRS 6h 后使血清中丙氨酸转氨酶活性、天冬氨酸转氨酶活性、脂质过氧化水平、氮氧化合物含量(NOx)及髓过氧化酶活性升高，使肝的抗坏血酸含量及 SOD 活性下降。而提前给药 BPEE(50、100g/kg)或 VE 能降低 WIRS 引起的肝损伤，即降低脂质过氧化、NOx 含量和髓过氧化物酶(MPO)活性，并升高抗坏血酸含量和 SOD 活性。50mg/kg 剂量 BPEE 的保护作用强于 1000mg/kg 剂量，与 VE (250mg/kg)的保护作用接近。上述结果说明了 BPEE 能减轻 WIRS 诱导的肝损伤，而这种作用可能与其抗炎和抗氧化活性有关。

2 蜂胶中的保肝成分

2.1 咖啡酸苯乙酯(CAPE)

咖啡酸苯乙酯(CAPE)是公认的具有良好保肝效果的蜂胶成分,因此关于 CAPE 保肝作用的研究非常多。相关结果表明,CAPE 能通过抗炎、抗氧化等方式改善由各种因素导致的肝损伤。这些研究按肝损伤模型的诱导因素分为如下几种。

(1)CCl_4 诱导的肝损伤

Kus 等(2004)将 24 只大鼠分成 3 组:组 I 为对照组,组 II 大鼠每隔一天注射一次 CCl_4,时间为 1 个月,组 III 大鼠每隔一天注射一次 CCl_4 和 CAPE,实验结束后,收集血样,检测血清中天冬氨酸转氨酶、丙氨酸转氨酶、碱性磷酸酶及胆红素的含量和肝丙二醛含量。结果表明,CAPE 能显著降低经 CCl_4 诱导后的血清中这些指标的水平以及肝 MDA 的含量,同时组织病理学观察结果表明,CAPE 能减轻 CCl_4 诱导的肝脏病变。

Sirag 等(2011)每天给大鼠口服 2.5mg/kg 剂量的 CCl_4,持续 8 周,以诱导肝的损伤模型。CCl_4 诱导同时每天给予 10mg/kg 的 CAPE 治疗 8 周作为实验对照组,结束后对反映肝脏病变常见的生化指标进行检测。结果表明,CAPE 的治疗能使血清中 AST、ALT 和 ALP 的水平趋于正常,同时降低肝脂质过氧化并显著改善肝的 GST、GSH 及磷脂含量。此外,CAPE 的治疗还能改善 CCl_4 所致的基因毒性,抑制 CCl_4 诱导的 α-平滑肌肌动蛋白(α-SMA)表达,表明 CAPE 对 CCl_4 所致的肝纤维化有一定的保护作用。Colakoglu 等(2011)通过电子显微镜对肝组织进行观察,结果表明,CAPE 能降低由 CCl_4 导致的肝细胞细胞器膜和核膜的损伤,再次证实了 CAPE 对 CCl_4 致肝损伤的保护作用。

(2)药物性肝损伤

Albukhari 等(2009)研究了 CAPE 对乳腺癌治疗药物三苯氧胺(TAM,它莫昔芬)诱导的肝损伤的保护作用。TAM 诱导后,血清中丙氨酸转氨酶、天冬氨酸转氨酶、碱性磷酸酶含量升高,肝谷胱甘肽含量降低,氧化型谷胱甘肽含量增多,脂质过氧化水平提高。此外,谷胱甘肽还原酶(GR)、谷胱甘肽过氧化物酶(GPx)、超氧化物歧化酶(SOD)和过氧化氢酶(CAT)活性降低,并且肝组织中 TNF-α 含量升高。在 CAPE 的治疗下,血清中被检酶的活性有所下降,阻止谷胱甘肽含量的下降及氧化型谷胱甘肽的积累,降低脂质过氧化水平,并恢复 GR、GPx、SOD、CAT 的活性和抑制 TNF-α 的升高。总之,CAPE 能通过保护细胞膜的完整性,抑制脂质过氧化,增强抗氧化酶及抑制肝脏炎症的方式保护大鼠对抗 TAM 诱导的肝脏毒性。CAPE 还能通过降低活性氮和恢复谷胱甘肽水平的方式对抗癌药物顺铂诱导的肝毒性,起到保肝作用(Kart,2010)。

(3)糖尿病性肝损伤

Yilmaz 等(2004)利用链佐星诱导的大鼠糖尿病模型,研究 CAPE 对糖尿病大鼠肝脏脂质过氧化水平和一些抗氧化酶(SOD、CAT、GSH-Px)活性的影响。结果表明,与对照组比较,糖尿病大鼠 MDA 含量及抗氧化酶的活性显著上升;CAPE 能降低糖尿病大鼠脂质过氧化水平,使 MDA 恢复至正常水平,但 CAPE 治疗组的 SOD 和 CAT 活性低于糖尿病大鼠非治疗组,分析认为这可能是由于 CAPE 的抗氧化作用降低了糖尿病大鼠肝脏中自由基含量,从而阻止了抗氧化酶活性的升高。

(4)电磁波诱导的肝氧化应激

Koyu等(2005)以肝组织中CAT、SOD、GSH-Px、黄嘌呤氧化酶(XO)的酶活性及脂质过氧化水平作为检测指标,利用Spraque-Dawley大鼠研究CAPE对1800Hz微波诱导的肝氧化应激的影响。结果表明,CAPE能通过上调CAT、SOD等抗氧化酶的活性,缓解微波诱导的肝氧化应激。他们后续的研究发现,CAPE能通过降低活性氧和增强抗氧化酶活性的方式减轻由一定强度电磁场诱导下的肝的氧化应激(Koyu et al.,2009)。

(5)肝缺血/再灌注性损伤

Saavedra-Lopes等(2008)通过手术制造鼠的肝缺血再灌注模型,将白化大鼠分为4组:假手术组(只进行解剖,不进行缺血再灌注处理)、缺血组(让肝缺血60min,再灌注之前将其杀掉)、缺血/再灌注组(缺血60min,然后让其再灌注6h,之后将其杀掉)、缺血/再灌注加CAPE组(缺血60min,再灌注6h之前腹腔注射CAPE $10\mu mol/kg$,治疗30min)。将血清中ALT、AST、组织中谷胱甘肽作为检测指标,并评估肝组织的病理学损伤指数;通过肝组织切片结合免疫组化等方法观察中性粒细胞的浸润状况、肝细胞凋亡状况、四羟壬烯醛加合物的生成(评估脂质过氧化程度)及NF-κB激活状况。结果表明,再灌注后ALT、AST水平显著增加,而CAPE治疗组的这两种酶在血清中水平降低很多;缺血再灌注过程中组织谷胱甘肽含量逐渐降低,而在CAPE治疗组得到部分恢复;缺血/再灌注组组织病理学损伤指数、凋亡指数、中性粒细胞浸润程度、四羟壬烯醛的生成量及NF-κB激活程度都要高于缺血组,而CAPE治疗组这些变化显著降低。研究结果表明,这种保肝作用可能与CAPE抑制NF-κB激活有关。

(6)冷刺激诱导的肝损伤

Ates等(2006)研究了CAPE对冷刺激诱导的肝氧化应激的影响。将24只雌性大鼠分为4组:对照组、CAPE治疗组、冷刺激组、CAPE治疗+冷刺激组。将抗氧化酶(CAT、SOD、GSH-Px)活性、总谷胱甘肽含量以及MDA水平作为检测指标。结果表明,冷刺激会显著降低这3种抗氧化酶活性及总谷胱甘肽含量,MDA含量升高。与冷刺激组相比,CAPE治疗的冷刺激组中这3种抗氧化酶活性及总谷胱甘肽水平显著升高,MDA含量降低;组织病理学观察发现,CAPE能减轻肝脏病变。总之,CAPE能通过调节抗氧化酶活性对抗冷刺激诱导的肝的氧化应激,表现为抑制脂质过氧化,减轻肝损伤。

Pekmez等(2007)将21只大鼠平均分成3组:对照组、烟草烟雾刺激组和烟草烟雾刺激+CAPE治疗组。利用烟草烟雾刺激60d,一天4次,每次30min,CAPE通过腹腔注入大鼠体内,试验结束后,评估大鼠肝组织氧化应激水平和血清中总胆红素水平及转氨酶活性。结果表明,烟草烟雾长时间刺激会提高血清中总胆红素水平及转氨酶活性和一些常见抗氧化酶活性和MDA含量,而加入CAPE组大鼠的这些参数明显降低,接近正常水平。

(7)LPS诱导的肝损伤

Korish和Arafa(2011)研究了CAPE对LPS诱导的内毒素血症、肝和神经损伤以及与此相关的系统性炎症反应。将50只大鼠分为3组:对照组、LPS组、LPS+CAPE组。实验结束后对血浆中不同的细胞因子(TNF-α、IL-1α、IL-1β、IL-6、IL-4、IL-10)和sICAM-1进行评估,同时对肝细胞和神经细胞的组织病理学变化进行评估。结果表明,LPS组表现为较高的炎症因子水平,反映出系统性炎症反应。另外,还观察到肝细胞坏死、凋亡、广泛性出血、炎性细胞浸润,以及脑星形胶质细胞肿胀。而CAPE的使用能降低炎症细胞因子水平和增加抗炎因子水平,这与肝脑组织中炎症细胞浸润减少的组织学观察结果一致,表明 CAPE 能够通过调节促炎与抗炎因子的平衡,抑制黏附分子表达的方式,减轻 LPS 诱导的系统性炎症反应及肝细胞、神经细胞的损伤。

(8)胆汁淤积型肝损伤

Coban 等(2010)将 21 只瑞士白化大鼠平均分成 3 组:对照组(也称为假手术组,只进行简单的剖腹手术,不进行胆道结扎)、胆道结扎组、胆道结扎 + CAPE 治疗组(CAPE 10mmol/kg,腹腔注射,一天 1 次,共 14d)。以此来研究 CAPE 对胆汁郁积型肝损伤的作用。结果发现,与胆道结扎组相比,CAPE 治疗组能降低血清中 γ-谷氨酰转移酶(GGT)、AST、ALT 水平,同时还能显著降低组织中 MDA 和髓过氧化物酶(MPO)的水平,而相比之下,CAPE 治疗显著提高了组织中谷胱甘肽的水平。此外,CAPE 治疗还能显著降低白介素(IL-1a 和 IL-6)水平。表明 CAPE 能对胆汁郁积型肝损伤有保护作用,且这种作用与 CAPE 的抗氧化、抗炎有一定关系。Esrefoglu 和 Ara(2010)的实验也表明,CAPE 能保护肝细胞对抗胆道胆汁郁积诱导的肝损伤,并认为这与 CAPE 的抗氧化作用有密切联系。

CAPE 对老年大鼠的肝也有一定的保护作用,表现为降低老龄鼠肝组织 MDA 的水平并增加老龄鼠肝组织 CAT 活性(Esrefoglu et al.,2012)。

2.2 其他成分

除了 CAPE,蜂胶中还发现了其他具有保肝作用的活性单体,主要是一些黄酮类化合物。

Banskota 等(2001)从巴西蜂胶甲醇提取物中分离出大量具有保肝活性的成分,主要是酚类化合物,特别是其中的黄酮类化合物,如桦木酚(Betuletol)、堪非醇 3,4′-二-O-甲醚(Ermanin)、山柰素(Kaempferide)、3,5,7-三羟基-8,4′-二甲氧基黄酮。这些活性成分能有效抑制 D-半乳糖胺(D-GalN)/肿瘤坏死因子 a(TNF-a)诱导的肝细胞死亡。并且,通过构效关系比较发现,1,2-双氢肉桂酸基团会增加异戊二烯化酚类物质的保肝活性,而异戊二烯基基团数目的增加可能会抑制异戊二烯化酚类物质的活性。此外,蜂胶提取物中分离出的半日花烷型双萜类化合物也表现出显著的保肝活性。Chen 等(2008)研究表明,蜂胶中的主要成分松属素能对抗酒精诱导的肝纤维化。刘爽等(2009)研究表明,槲皮素对大鼠原代肝细胞酒精性氧化损伤有防护作用,表现为升高酒精刺激下的肝细胞的抗氧化酶活性及 GSH 水平,降低肝细胞的脂质过氧化,并证明了槲皮素的这种保护效应可能通过血红素加氧酶-1(HO-1)介导。Chen(2010)研究表明,槲皮素还能显著降低酒精诱导的肝血浆中一些炎症因子的水平。有报道表明,蜂胶中的白杨素对酒精(Sathiavelu et al.,2009)、CCl_4(Anand et al.,2011)诱导的肝氧化应激有一定的保护作用。

杜夏等(2013)对从蜂胶中分离得到的 21 种化合物进行研究,发现其中山柰酚、3′,4′,5-三羟基-3,7-二甲氧基黄酮、咖啡酸苯乙酯、异鼠李素、槲皮素的保肝活性明显。各活性成分均能通过抗氧化作用改善自由基对肝细胞的损伤,并且大多数化合物的保肝活性具有浓度依赖性。并且,通过对蜂胶保肝活性成分的构效分析发现,蜂胶清除自由基的活性主要取决于黄酮类化合物 B 环 C3 和 C4 位上的酚羟基(图 3.8)。这与 Pérez-Alvarez 等(2001)的研究结果有相似之处。Pérez-Alvarez 等通过对 3,4-二羟基肉桂酸(3,4-Dihydroxycinnamic Acid,一种咖啡酸衍生物)及其类似物的保肝活性比较研究发现其基本结构(图 3.9)上的 C3 和 C4 位上的酚羟基很重要,尤其是 C4 位上的羟基。而 Bhadauria 和 Nirala(2009)的研究也表明,CAPE 能改善 CCl_4 诱导的肝损伤可能与 CAPE 的苯环 C_3 和 C_4 位上的羟基通过供氢清除 CCl_4 代谢的毒性自由基($CCl_3·$,$CCl_3O_2·$)有一定关系。这暗示了具有这种类似结构的成分很有可能具有良好的抗氧化和保肝活性。

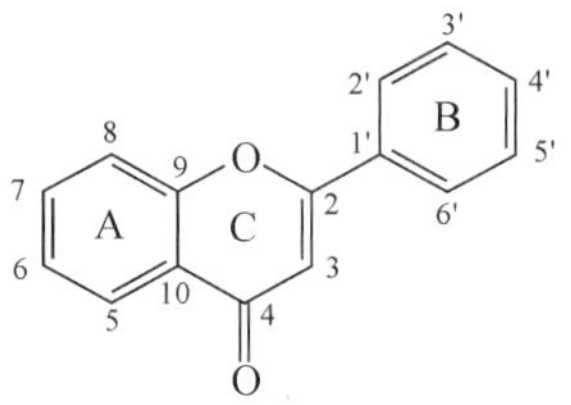

图 3.8　黄酮类化合物基本母核

Fig. 3.8　Flavonoid nucleus

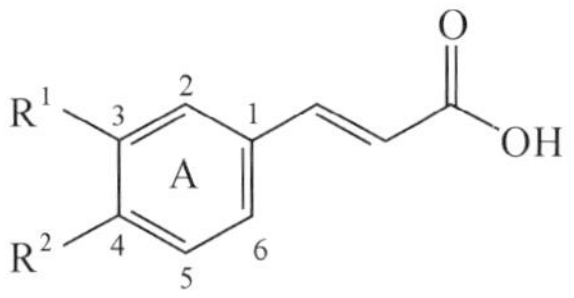

图 3.9　3,4-二羟基肉桂酸及其类似物基本结构

Fig. 3.9　Basic structure of 3,4-dihydroxycinnamicacid and its analogues

此外,有文献报道,台湾绿蜂胶中的 propolin G 能通过干扰 TGF-β 介导的 Smad2/3 信号通路,阻止肝的纤维化(Su,2014)。

小　结

已有大量实验证明蜂胶的保肝活性,然而对其机制的研究还不够深入。蜂胶化学成分复杂,药理活性广泛,因此其保肝作用的原因不能简单地归结于其抗氧化、抗炎活性,还可能涉及其他机制,如调节肝细胞的再生,与毒性物质之间的相互作用,促进肝内源性解毒,等等。同时,对蜂胶保肝活性单体成分的进一步分离鉴定,或许能分离出更为有效的保肝活性成分,有利于药物的开发及对单体成分保肝分子机制研究的深入。此外,蜂胶与其他药物配伍对肝损伤保护作用的研究相对较少。因此,可以对蜂胶与 1 种或几种保肝药物配伍保肝进行研究,以利于开发出更有效的保肝药物,使蜂胶得以最充分的利用。

参考文献

Abdelsameea AA, Mahgoub ML, Abdel Raouf SM (2013) Study of the possible hepatoprotective effect of propolis against the hepatotoxic effect of atorvastatin in albino rats[J]. *Journal of Medicine-Zagazig University*, 19(5):388-396.

Albukhari AA, Gashlan HM, El-Beshbishy HA, Nagy AA, Abdel-Naim AB (2009) Caffeic acid phenethyl ester protects against tamoxifen-induced hepatotoxicity in rats[J]. *Food and Chemical Toxicology*, 47(7):1689-1695.

Al-Amoudi(2015) Ameliorative role and antioxidant effect of propolis against hepatotoxicity offenvalerate in albino rats[J]. *Journal of Cytology & Histology*,6(1):doi:10.4172/2157-7099.1000303.

Anand KV,Anandhi R,Pakkiyaraj M,Geraldine P (2011) Protective effect of chrysin on carbon tetrachloride (CCl$_4$)-induced tissue injury in male Wistar rats[J]. *Toxicology and Industrial Health*,27(10):923-933.

Ates B,Dogru MI,Gul M,Erdogan A,Dogru AK,Yilmaz I,Yurekli M,Esrefoglu M (2006) Protective role of caffeic acid phenethyl ester in the liver of rats exposed to cold stress [J]. *Fundamental & Clinical Pharmacology*,20(3):283-289.

Babatunde IR,Abdulbasit A,Oladayo MI,Olasile OI,OlamideFR,Gbolahan BW (2015) Hepatoprotective and pancreatoprotective properties of the ethanolic extract of Nigerian propolis[J]. *Journal of Intercultural Ethnopharmacology*,4(2):102-108.

Badr MOT,Edrees NMM,Abdallah AAM,Hashem MA,El-Deen NAMN,Neamat-Allah ANF,Ismail HTH (2011) Propolis protects against methotrexate induced hepatorenal dysfunctions during treatment of ehrlich carcinoma[J]. *Journal of American Science*,7(12):313-319.

Banskota AH,Tezuka Y,Adnyana IK,Ishii E,Midorikawa K,Matsushige K,Kadota S (2001) Hepatoprotective and anti-helicobacter pylori activities of constituents from Brazilian propolis[J]. *Phytomedicine*,8(1):16-23.

Bhadauria M,Nirala SK,Shukla S (2007) Duration-dependent hepatoprotective effects of propolis extract against carbon tetrachloride-induced acute liver damage in rats[J]. *Advances in Natural Therapy*,24(5):1136-1145.

Bhadauria M,Nirala SK,Shukla S (2008a) Multiple treatment of propolis extract ameliorates carbon tetrachloride induced liver injury in rats[J]. *Food and Chemical Toxicology*,46(8):2703-2712.

Bhadauria M,Shukla S,Mathur R,Agrawal OP,Shrivastava S,Johri S,Joshi D,Singh V,Mittal D,Nirala SK (2008b) Hepatic endogenous defense potential of propolis after mercury intoxication[J]. *Integrative Zoology*,3(4):311-321.

Bhadauria M,Nirala SK (2009) Reversal of acetaminophen induced subchronichepatorenal injury by propolis extract in rats[J]. *Environmental Toxicology and Pharmacology*,27(1):17-25.

Chen CS,Wu CH,Lai YC,Lee WS,Chen HM,Chen RJ,Chen LC,Ho YS,Wang YJ (2008) NF-kappaB-activated tissue transglutaminase is involved in ethanol-induced hepatic injury and the possible role of propolis in preventing fibrogenesis[J]. *Toxicology*,246(2-3):148-157.

Chen X (2010) Protective effects of quercetin on liver injury induced by ethanol[J]. *Pharmacognosy Magazine*,22(6):135-141.

Coban S,Yildiz F,Terzi A,Al B,Ozgor D,Ara C,Polat A,Esrefoglu M (2010) The Effect of caffeic acid phenethyl ester (cape) against cholestatic liver injury in rats[J]. *Journal of Surgical Research*,159(2):674-679.

Colakoglu N, Kus I, Kukner A, Pekmez H, Ozan E, Sarsilmaz M (2011) Protective effects of CAPE on liver injury induced by CCL (4): An electron microscopy study[J]. *Ultrastructural Pathology*, 35(1):26-30.

El-Mahalaway AM, Selim AA, Mahboub FAR (2015) The potential protective effect of propolis on experimentally induced hepatitis in adult male albino rats; Histological and immunohistochemical study[J]. *Journal of Histology & Histopathology*, 2(1): doi: 10.7243/2055-091X-2-14.

El-Khatib AS, Agha AM, Mahran LG, Khayyal MT (2002) Prophylactic effect of aqueous propolis extract against acute experimental hepatotoxicity *in vivo* [J]. *Verlag der Zeitschrift für Naturforschung*, 57(3-4):379-385.

Esrefoglu M, Ara C. (2010) Beneficial effect of caffeic acid phenethyl ester (CAPE) on hepatocyte damage induced by bile duct ligation: An electron microscopic examination [J]. *Ultrastructural Pathology*, 34(5):273-278.

Esrefoglu M, Iraz M, Ates B, Gul M (2012) Melatonin and CAPE are able to prevent the liver from oxidative damage in rats: An ultrastructural and biochemical study[J]. *Ultrastructural Pathology*, 36(3):171-178.

Humayun F, Tahir M, Lone KP, Munir B, Ahmad A, Latif W (2014) Protective effect of ethanolic extract of propolis on isoniazid induced hepatotoxicity in male albino mice [J]. *Biomedica*, 30(2):85-91.

Kart A, Cigremis Y, Karaman M, Ozen H (2010) Caffeic acid phenethyl ester (CAPE) ameliorates cisplatin-induced hepatotoxicity in rabbit[J]. *Experimental and Toxicologic Pathology*, 62(1):45-52.

Kolankaya D, Selmanoglu G, Sorkun K, Salih B (2002) Protective effects of Turkish propolis on alcohol-induced serum lipid changes and liver injury in male rats[J]. *Food Chemistry*, 78(2):213-217.

Korish AA, Arafa MM. (2011) Propolis derivatives inhibit the systemic inflammatory response and protect hepatic and neuronal cells in acute septic shock[J]. *Brazilian Journal of Infectious Diseases*, 15(4):332-338.

Koyu A, Nazıroğlu M, Ozguner F, Yilmaz HR, Uz E, Cesur G (2005) Caffeic acid phenethyl ester modulates 1800MHz microwave-induced oxidative stress in rat liver [J]. *Electromagnetic Biology and Medicine*, 24(2):135-142.

Koyu A, Ozguner F, Yilmaz H, Uz E, Cesur G, Ozcelik N (2009) The protective effect of caffeic acid phenethyl ester (CAPE) on oxidative stress in rat liver exposed to the 900 MHz electromagnetic field[J]. *Toxicology and Industrial Health*, 25(6):429-434.

Kus I, Colakoglu N, Pekmez H, Seckin D, Ogeturk M, Sarsilmaz M (2004) Protective effects of caffeic acid phenethyl ester (CAPE) on carbon tetrachloride-induced hepatotoxicity in rats [J]. *Acta Histochemica*, 106(4):289-297.

Mahmoud MF, Sakr SM (2013) Hepatoprotective effect of bee propolis in rat model of streptozotocin-induced diabetic hepatotoxicity: light and electron microscopic study

[J]. *Life Science Journal*,10(4):2048-2054.

Myhrstad MCW,Carlsen H,Nordström O,Blomhoff R,Moskaug JØ (2002) Flavonoids increase the intracellular glutathione level by transactivation of the γ-glutamyl cysteine synthetase catalytical subunit promoter[J]. *Free Radical Biology and Medicine*,32(5):386-393.

Nakamura T,Ohta Y,Ohashi K,Ikeno K,Watanabe R,Tokunaga K,Harada N (2012) Protective effect of Brazilian propolis against hepatic oxidative damage in rats with water-immersion restraint stress[J]. *Phytotherapy Research*,26(10):1482-1489.

Nirala SK,Bhadauria M (2008) Propolis reverses acetaminophen induced acute hepatorenal alterations:A biochemical and histopathological approach[J]. *Archives of Pharmacal Research*,31(4):451-461.

Nirala SK,Li PQ,Bhadauria M,Guo GQ (2008) Combined effects of gallic acid and propolis on beryllium-induced hepatorenal toxicity[J]. *Integrative Zoology*,3(3):194-207.

Rodríguez S, Ancheta O, Ramos ME, Remíez D, Rojas E, González, González R (1997) Effects of Cuban red propolis on galactosamine-induced hepatitis in rats[J]. *Pharmacological Research*,35(1):1-4.

Pekmez H,Kus I,Colakoglu N,Ogeturk M,Ozyurt H,Turkoglu AO,Sarsilmaz M. (2007) The protective effects of caffeic acid phenethyl ester (CAPE) against liver damage induced by cigarette smoke inhalation in rats[J]. *Cell Biochemistry and Function*,25(4):395-400.

Pérez-Alvarez V,Bobadilla RA,Muriel P (2001) Structure-hepatoprotective activity relationship of 3,4-dihydroxycinnamic acid (caffeic acid) derivatives[J]. *Journal of Applied Toxicology*,21(6):527-531.

Saavedra-Lopes M,Ramalho FS,Ramalho LN,Andrade-Silva A,Martinelli AL,Jordão AA,Castro-e-Silva O,Zucoloto S (2008) The protective effect of CAPE on hepatic ischemia/reperfusion injury in rats[J]. *The Journal of Surgical Research*,150(2):271-277.

Sathiavelu J,Senapathy GJ,Devaraj R,Namasivayam N (2009) Hepatoprotective effect of chrysin on prooxidant-antioxidant status during ethanol-induced toxicity in female albino rats[J]. *Journal of Pharmacy and Pharmacology*,61:809-817.

Seo KW,Park M,Song YJ,Kim SJ,Yoon KR (2003) The protective effects of propolis on hepatic injury and its mechanism[J]. *Phytotherapy Research*,17(3):250-253.

Siess MH,Le Bon AM,CanivencLavier MC,AmiotMJ,Sabatier S,Aubert SY,Suschetet M (1996) Flavonoids of honey and propolis:Characterization and effects on hepatic drug-metabolizing enzymes and benzo[a]pyrene-DNA binding in rats[J]. *Journal of Agricultural and Food Chemistry*,44(8):2297-2301.

Sirag HM,Ibrahim HA,Amer AA,Salam TA,Samaka RM,El-Mowafy YA (2011) Ameliorative effect of caffeic acid phenethyl ester and candsartaincilexetil on CCl_4 induced hepatotoxicity in male rats[J]. *Annals of Biological Research*,2(4):503-515.

Seven İ,Baykalir BG,Seven PT,Dăgŏglu G (2014) The ameliorative effects of propolis against cyclosporine A induced hepatotoxicity and nephrotoxicity in rats[J]. *Source of* the Document Kafkas Universitesi Veteriner Fakultesi Dergisi,20(5):641-648.

Su KY,Hsieh CY,Chen YW,Chuang CT,Chen CT,Chen YL (2014) Taiwanese green propolis and propolin G protect the liver from the pathogenesis of fibrosis via eliminating TGF-β-induced smad 2/3 phosphorylation[J]. *Journal of Agricultural and Food Chemistry*,62(3):192-3201.

Türkez H,Yousef MI,Geyikoglu F (2010) Propolis prevents aluminium-induced genetic and hepatic damages in rat liver[J]. *Food and Chemical Toxicology*,48(10):2741-2746.

Türkez H,Yousef MI,Geyikoglu F (2012) Propolis protects against 2378-tetrachlorodibenzo-p-dioxin-induced toxicity in rat hepatocytes[J]. *Food and Chemical Toxicology*,50(6):2142-2148.

Türkez H,Geyikolu F,Yousef MI,Toar B,Vancelik S (2013) Propolis alleviates 2,3,7,8-Tetrachlorodibenzo-p-dioxin-induced histological changes,oxidative stress and DNA damage in rat liver[J]. *Toxicology and Industrial Health*,29(8):677-685.

Yilmaz HR,Uz E,Yucel N,Altuntas I,Ozcelik N (2004) Protective effect of caffeic acid phenethyl ester (CAPE) on lipid peroxidation and antioxidant enzymes in diabetic rat liver[J]. *Journal of Biochemical and Molecular Toxicology*,18(4):234-238.

Zhu W,Li YH,Chen ML,Hu FL (2011) Protective effects of Chinese and Brazilian propolis treatment against hepatorenal lesion in diabetic rats[J]. *Human & Experimental Toxicology*,30(9):1246-1255.

森本道雄(1997)酒精性肝损伤的最新研究进展[J].日本医学介绍,18(10):447-449.

陈世贵(1990)对肝脏解毒作用的理解[J].生物学通报,(11):15-16.

杜夏(2013)蜂胶保肝活性成分的分离及其保肝机制初探[D].北京:中国农业科学院.

刘爽,姚平,宋毅,许四元,李骏,刘烈刚(2009)槲皮素对酒精性肝损伤的防护作用及其与Ⅰ型血红素氧化酶关系的研究[J].营养学报,31(1):83-88.

沈杰(1990)几种常用肝毒剂致肝损伤机制的研究现况[J].生理科学进展,21(1):70-73.

彭飞,林丽,熊瑞芳(2012)肝损伤机制的研究[J].科技资讯,(33):236-237.

容丽萍,许园园,蒋小云(2014)重金属中毒与儿童肾损伤[J].中国当代儿科杂志,16(4):325-329.

徐鑫,屈彩芹(2008)药物性肝损伤机制[J].医学综述,14(5):747-749.

袭著革,晁福寰,孙咏梅,杨丹凤,张华山,李官贤,李媛(2003)金属离子介导活性氧引起DNA氧化损伤及机制研究[J].环境科学学报,23(5):662-667.

吴悠,丛晓东,张云(2011)糖尿病肝损伤相关机制及中药干预[J].中国民族民间医药,(21):16-17.

朱润芝,李京敬,谢超,郜尽,胡建军,袁运生,韩伟,俞雁(2010)过氧化作用与肝脏疾病[J]世界华人消化杂志,18(11):1134-1140.

张欣,党双锁(2009)实验性肝损伤机制研究进展[J].中国肝脏病杂志,1(2):60-62.

朱玉真,王立荣,孙应彪(1997)镍致大鼠肝损伤与脂质过氧化关系的研究[J].中国公共卫生学报,16(2):112-113.

第八节 蜂胶的促创(烧)伤修复作用

创伤广义上指由理化及生物因素所造成的机体损伤,包括刀器伤、挤压伤、冻伤等。烧烫伤属于创伤的一种,在人群中有着较高的发生率。人们从抗炎、抗菌、防腐等角度对创伤修复的机制进行了大量的研究(Farstvedt et al.,2004),并开发出许多临床上治疗局部创(烧)伤的药物,但多数药物的药理活性较单一,并且有可能对机体产生一定的副作用(陈炳等,2006;Olczyk et al.,2013)。而蜂胶作为一种能促进组织修复,具有抗炎、抗菌、抗氧化等多种生物学活性的天然产物,在促进创(烧)伤修复中具有明显的优势,取得了较理想的效果。下面对蜂胶在创(烧)伤修复作用上的研究进展及蜂胶促创伤修复作用的可能机制进行综述,对今后的研究方向进行展望,以期为蜂胶对创伤修复作用的进一步研究提供参考。

1 蜂胶对创(烧)伤组织的作用

经典的创伤修复分为炎症期、组织生长期和组织重塑期(Gurtner et al.,2008)。这三个时期在时空上并非严格分开,而是相互交叠。整个过程是在各种类型的修复细胞、细胞因子、细胞外基质分子及一些蛋白水解酶相互作用紧密协调下完成的。影响创伤修复的因素有很多,其中,伤口炎症反应程度、受细菌感染程度以及氧化损伤程度是3个非常重要的因素。因此,有人认为蜂胶抗炎、抗菌、抗氧化活性是其促进组织再生和创伤修复的重要原因(de Castro,2001;Ramos and Miranda,2007)。近年来,国内外研究者以大鼠和猪的烧伤和切割伤为主要模型开展了一些关于蜂胶创伤修复活性的研究,这些研究结果表明,蜂胶能通过调节细胞外基质(胶原)沉积、加速再上皮化、促进肉芽组织生长等多种方式促进创伤愈合。近年来蜂胶创伤修复作用的相关研究见表3.5。

表3.5 近年来蜂胶促创(烧)伤修复作用的相关研究

Table 3.5 Related research on propolis' effect on promoting (burn) wound healing in recent years

年份 Year	蜂胶来源 Propolis source	模型 Model	修复效果 Healing effect	给药方式 Administration method	参考文献 Reference
2005	土耳其	大鼠三度烧伤模型	抑制炎症反应,促进上皮再生和肉芽组织生长	外敷	Han et al.,2005
2008	澳大利亚	大鼠糖尿病全层皮肤损伤模型	抑制炎症反应,促进上皮再生,使中性粒细胞和巨噬细胞浸润正常化	外敷	McLennan et al.,2008
2008	土耳其	大鼠结肠切除术结合吻合术诱导的创伤模型	抑制炎症反应,促进成熟肉芽组织的形成,组织清创及胶原的合成	饲喂	Kilicoglu et al.,2008
2009	巴西	大鼠切割伤模型	抑制炎症反应,促进成纤维细胞增生,胶原纤维沉积和重塑	外敷	de Albuquerque-júnior et al.,2009

续表

年份 Year	蜂胶来源 Propolis source	模型 Model	修复效果 Healing effect	给药方式 Administration method	参考文献 Reference
2010	巴西	大鼠二度烧伤模型	抑制炎症反应,促进组织清创和胶原纤维产生	外敷	Pessolato et al., 2011
2010	伊朗	大鼠全层皮肤烧伤模型	抑制炎症反应,促进伤口收缩和修复	外敷	Khorasgani et al., 2010
2011	巴西	Swiss小鼠皮下植入海绵创伤模型	抑制炎症反应,减少胶原沉积,促进Ⅰ型胶原对Ⅲ型胶原的取代	灌胃	de Moura et al., 2011a
2012	波兰	家猪烧伤模型	烧伤早期刺激玻连蛋白、层连黏蛋白和硫酸乙酰肝素/肝素的增加,促进肉芽组织和上皮化及胶原沉积	外敷	Olczyk et al., 2012
2012	巴西	大鼠皮肤穿刺损伤模型	抗菌,促进再上皮化、成纤维细胞增殖及胶原生成	外敷	Berretta et al., 2012
2013	波兰	家猪烧伤模型	促进烧伤皮肤中软骨素/硫酸皮肤素和透明质酸积累,加速软骨素/硫酸皮肤素结构调整	外敷	Olczyk et al., 2013
2013	巴西	大鼠二度烧伤模型	抑制炎症反应,促进肉芽组织形成,加速上皮形成速率,并促进了Ⅰ型胶原更早取代Ⅲ型胶原	外敷	de Almeida et al., 2013
2015	沙特阿拉伯	糖尿病小鼠全层皮肤切除伤模型	显著加速伤口愈合,使糖尿病鼠IL-1、IL-6、TNF-α、MMP-9水平降低至接近正常,并能通过上调TGF-β/smad2,3信号通路促进胶原的表达,进而促进糖尿病鼠伤口的愈合。	外敷	Hozzein et al., 2015

1.1 蜂胶对再上皮化的作用

再上皮化是伤口边缘的角质形成细胞及残存的皮肤附属器官如毛囊、汗腺中的角质形成细胞通过增殖、迁移、分化以覆盖创面并最终恢复皮肤完整性和屏障功能的过程(Adam et al.,1999;Santoro and Gaudino,2005)。它是创伤修复过程中一个十分重要而复杂的生物学事件,其完成依赖于角质形成细胞的增殖、迁移和分化,而这些过程严格受到相应的生长因子、细胞外基质、蛋白水解酶及细胞表面整合素受体及其相互作用的调节。

近年来,大量有关蜂胶创伤修复作用的研究表明,蜂胶能够通过加速创面的再上皮化达到促进创伤修复的目的。Khorasgani等(2010)利用大鼠全层皮肤烧伤模型,通过组织病理学观察,并对伤口扩张、收缩以及再上皮化速率进行定性定量评估,对比研究了伊朗蜂胶和磺胺嘧啶银对鼠全层皮肤的烧伤愈合效果,结果表明蜂胶治疗组炎症反应最轻,伤口收缩和再上皮化速率最快。Han等(2005)采用大鼠三度烧伤模型,通过组织病理学观察对比了土耳其蜂胶和磺胺嘧啶银对烧伤大鼠的修复效果,结果表明蜂胶不仅具有良好的抗炎作用,还能促进再上皮化,从而加速烧伤皮肤的愈合,推测可能是蜂胶的抗炎作用促进了再上皮化过程。McLennan 等(2008)采用糖尿病全层皮肤创伤模型,研究蜂胶对糖尿病患者上皮闭合、伤口形态、细胞浸润以及血管密度的影响,结果表明蜂胶能明显改善糖尿病大鼠上皮闭合及再上皮化程度低的状况,说明蜂胶能改善糖尿病患者组织修复过程存在的某些缺陷。该研究还发现蜂胶具有有效抑制中性粒细胞浸润以及使糖尿病动物损伤组织巨噬细胞浸润正常化的作用,并以此推测炎症细胞浸润正常化和接下来的上皮闭合率之间可能存在某种关系。de Almeida 等(2013)的研究也观察到巴西蜂胶能明显提高再上皮化速率,并认为这可能是蜂胶中的咖啡酸类物质刺激了角质细胞的迁移,或者与巴西绿蜂胶能在一定条件下激活巨噬细胞免疫调节活性有关,因为激活的巨噬细胞能释放大量的成纤维细胞生长因子(FGF)以及上皮细胞增殖过程相关的一些细胞因子,进而促进再上皮化;也有可能是上皮化与纤维组织形成之间形成联系,从而促进烧伤修复。

总之,蜂胶有可能通过抗炎作用,减少炎症反应中一些细胞因子和炎性介质(自由基等)对再上皮化的抑制作用(Martin and Leibovich,2005);或促进一些促上皮化的生长因子的产生,从而起到间接的促上皮化作用;也有可能因为蜂胶中的某些活性成分对角质细胞有刺激作用(Sehn et al.,2009),直接促进了再上皮化。

1.2 蜂胶对肉芽组织形成的作用

肉芽组织是富含新生毛细血管的纤维结缔组织,由成纤维细胞、毛细血管及一定数量的炎性细胞等有形成分组成。具有抗感染,保护创面,填补创口及其他组织缺损,机化或包裹坏死、血栓、炎性渗出物及其他异物的作用。

大量研究表明,蜂胶通过刺激成纤维细胞和毛细血管内皮细胞增殖以及胶原基质沉积,从而促进肉芽组织生长。高畅等(2009)通过蜂胶对糖尿病大鼠创面新生血管和细胞增殖影响的研究,发现蜂胶能显著促进正常大鼠和糖尿病大鼠创面新生毛细血管生长、细胞增殖和肉芽组织的生长,加速创面愈合。Kilicoglu 等(2008)利用结肠切除术和断端吻合术诱导的大鼠创伤模型,通过透射电镜观察蜂胶经口给药后对大鼠结肠吻合的效果,超微结构的组织病理学分析显示蜂胶能加速创伤修复,刺激成熟肉芽组织的形成及胶原合成。de Almeida 等(2013)的实验也表明了蜂胶对肉芽组织形成和生长的促进作用。这些研究表明,蜂胶不仅能促进正常机体的创伤修复,同时能改善具有修复障碍的糖尿病患者的创伤修复。然而,也有研究发现蜂胶具有抗血管新生的作用(Song et al.,2002;de Moura et al.,2011a)。因此,蜂胶调节血管生成的具体机制还有待进一步研究。

1.3 蜂胶促进伤口收缩的作用

伤口收缩主要是由肉芽组织内肌成纤维细胞收缩引起的。肌成纤维细胞内含有平滑肌肌动蛋白(α-SMA),因而具有较强的收缩能力,此外,它还具有分泌大量胶原基质的能力(Gabbiani,2003)。创缘周围结缔组织内的成纤维细胞是它的主要来源,在趋化因子作用下迁移到创面,转变为肌成纤维细胞,这个转变过程受到细胞因子(如 TGF-β)和细胞外基质(纤连蛋白剪接变体 ED-A、胶原)的调节(Tateshita et al.,2001;Gabbiani,2003;Desmoulière et al.,2005)。

蜂胶广泛的生物学活性使其能有效促进伤口收缩。Khorasgani 等(2010)研究表明,蜂胶能降低损伤部位炎症反应,有效促进伤口收缩。Iyyam 等(2010)的研究也发现印度蜂胶具有明显地促进伤口收缩的作用,但其作用机制尚不是十分清楚。猜测可能是蜂胶上调炎症细胞 $TGF-\beta_1$ 的产生,由 $TGF-\beta_1$ 和它介导的基质生成进而促进了肌成纤维细胞的产生和伤口收缩。虽然有关蜂胶促进伤口收缩方面的报道较少,但有研究表明,一些富含黄酮类、多酚类及萜类的植物性成分具有良好的创伤修复活性,能有效促进伤口的收缩(Koca et al.,2009;Atiba et al.,2011;George et al.,2014)。因此,推测蜂胶中含有的黄酮类、酚类物质与伤口收缩之间可能存在关联。

1.4 蜂胶对细胞外基质生成的作用

皮肤组织缺损以后,为恢复其结构和功能的完整性,各种细胞外基质需要重新被合成和分泌(Schultz and Wysocki,2009),其中胶原蛋白是细胞外基质最主要的成分,尤其是Ⅰ型和Ⅲ型胶原成分。胶原纤维形成的一系列动态过程是皮肤修复的关键过程。皮肤损伤修复过程中,Ⅲ型胶原首先由成纤维细胞产生,用以引导肉芽形成过程中的成纤维细胞和内皮细胞的增殖与迁移,然后Ⅲ型胶原逐渐被Ⅰ型胶原取代,使真皮纤维结缔组织具有更强的张力和机械稳定性(Almeida et al.,2013)。细胞外基质对机体的重要性,不仅在于它能够作为组织和器官的结构成分,为细胞提供附着位点,更在于它能通过与细胞、细胞因子及其他成分相互作用,调节细胞的生长、分化和存活,因此,它们的再生对创伤修复有着极为重要的意义(Meredith et al.,1993;Schultz and Wysocki,2009)。

许多研究表明,蜂胶对细胞外基质的生成有调节作用。Albuquerque-Júnior 等(2009)通过实验证实了蜂胶对细胞外基质Ⅰ型胶原纤维和Ⅲ型纤维沉积的促进作用,进一步发现蜂胶能促进Ⅰ型胶原纤维对Ⅲ型胶原纤维的取代,据此,说明蜂胶能通过调节胶原沉积动态过程从而促进组织修复。Pessolato 等(2011)研究了蜂胶和羊膜对大鼠二度烧伤的再上皮化的影响,结果也表明了蜂胶对胶原生成的促进作用。由于实验时间只进行到 21d,组织修复尚未完全结束,胶原是否会继续生成直到修复末期并导致疤痕的产生还有待进一步研究。该实验还首次证明了蜂胶具有促进组织清创的效果,推测这可能是由于蜂胶刺激了巨噬细胞的吞噬活性。de Almeida 等(2013)采用大鼠二度烧伤模型,对含有两种蜂胶(巴西红蜂胶和巴西绿蜂胶)水醇提取物的胶原蛋白膜在促进真皮烧伤修复中的适用性进行了研究。结果表明,蜂胶控制组能促进胶原合成、沉积以及成熟的过程,这可能是由于蜂胶刺激成纤维细胞生长因子(FGF)的合成,因为成纤维细胞增生和其合成沉积胶原的过程有 FGF 的参与(Barrientos et al.,2008)。绿蜂胶组促进了Ⅰ型胶原对Ⅲ型胶原的取代,红蜂胶组促进胶原纤维束的逐渐交织,这进一步证实了 Albuquerque-Júnior 等(2009)的研究结果。Olczyk 等(2013)利用家猪作为模型,定性定量研究了蜂胶对烧伤组织细胞外基质Ⅰ型和Ⅲ型胶原积累的影响,结果表明,蜂胶能增加胶原及其成分表达,特别是在修复的最初阶段,进一步证实了蜂胶对胶原基质生成的调节作用。然而,de Moura 等(2011b)利用 Swiss 小鼠皮下植入海绵诱导创伤模型评估巴西绿蜂胶水提物的抗炎和创伤修复效果,结果表明蜂胶抑制修复早期胶原的合成,与其他报道(Kilicoglu et al.,2008;Olczyk et al.,2013)不太一致,这可能与组织特异性、蜂胶提取方式或来源有关。

蜂胶不仅能调节胶原基质的生成,而且对其他胞外基质的生成也有调节作用。Olczyk 等(2012)利用家猪建立烧伤模型,研究了蜂胶在烧伤修复过程中对玻连蛋白、层粘连蛋白、硫酸乙酰肝素/肝素表达的影响,结果表明,蜂胶治疗能在烧伤修复早期阶段显著刺激玻连蛋白、层粘连蛋白和硫酸乙酰肝素/肝素的表达,并使它们的含量最终趋向正常皮肤水平,这表明蜂胶能在不影响某些胞外基质在正常组织内的最终含量的情况下,加速组织的修复。该实验也同时观察到蜂胶促进胶原沉积的现象,与他人(Albuquerque-Júnior et al.,2009;Pessolato et al.,2011;de Almeida et al.,2013)的研究结果一致。Olczyk等(2013)还研究了蜂胶在烧伤修复过程中对骨素/硫酸皮肤素和透明质酸积累的影响,结果表明蜂胶能通过刺激肉芽组织生长和损伤闭合所需要的糖胺聚糖的积累,加速组织修复。而且,蜂胶能加速软骨素/硫酸皮肤素结构调整,使其结合生长因子在组织修复中发挥重要作用。这表明蜂胶不仅能促进细胞外基质的产生,还能通过加速其结构的调整,使修复朝有利的方向进行。

上述研究都表明了蜂胶对细胞外基质生成的调节作用,特别是对胶原生成的刺激作用。有研究表明,蜂胶及其活性成分能够上调人类外周血单核细胞(PBMC)和T淋巴细胞TGF-β_1的产生(Ansorge et al.,2003),而TGF-β_1是促进成纤维细胞增生及促进其分泌胶原等基质的十分重要的细胞因子(Gressner et al.,2002;Leask and Abraham,2004;Schiller et al.,2004;Faler et al.,2006)。据此,表明蜂胶可能通过影响一些参与胞外基质生成的细胞因子的合成,如TGF-β_1和FGF等(Barrientos et al.,2008;Ansorge et al.,2003),从而调节胞外基质的生成。

2 蜂胶促创(烧)伤修复作用可能的机制

如前所述,伤口炎症反应程度、受细菌感染程度以及氧化损伤程度是影响创伤修复3个非常重要的因素,因此有人认为蜂胶抗炎、抗氧化、抗菌活性,是其促进组织再生和创伤修复的重要原因(de Castro,2001;Ramos and Miranda,2007)。

2.1 蜂胶抗炎作用对创(烧)伤修复的影响

创伤发生后炎症反应立即发生,此时蜂胶的主要作用为抗感染,清除坏死组织,并释放一些趋化因子和细胞因子(TGF-β_1、bFGF、VEGF等),吸引一些修复细胞进入创面,从而调节创伤组织的修复(Martin and Leibovich,2005;Gurtner et al.,2008)。然而炎症反应对创伤修复的作用一直存在争议,关于炎症细胞在组织修复中的重要性开展了许多实验,这些实验结果大多表明无论是中性粒细胞还是巨噬细胞对创伤修复都不是绝对必要的,这些炎症细胞功能的缺失甚至会加速再上皮化和组织修复,减少瘢痕的产生(Simpson and Ross,1972;Egozi et al.,2003;Dovi et al.,2003;Martin et al.,2003)。但是,也有实验表明巨噬细胞功能缺失会使组织因为清创障碍影响创伤修复(Leibovich and Ross,1975)。Ashcroft等(1999)敲除小鼠的smad3基因,结果发现小鼠创面局部单核细胞浸润明显减少,而再上皮化速度加快,反映了炎症反应对再上皮化的抑制作用。总的说来,炎症反应能通过抗感染,组织清创,释放一些细胞因子,对创伤修复发挥一定的积极作用,但同时也会对组织修复的某些方面起抑制作用,例如再上皮化过程。所以,过度的炎症反应对创伤修复本身而言是弊大于利(Dovi et al.,2004)。

严重的烧伤和创伤后,大量变性坏死的组织、细菌的侵入,自由基的大量产生以及应激反应发生往往会引起过度的炎症反应(彭代智,2005)。大量研究蜂胶组织修复活性的体内实验都观察到蜂胶促进创伤修复的同时伴随着对创面局部炎症反应的抑制(见表3.5)。除了蜂胶,其他具有抗炎活性的植物成分对创伤修复的促进作用也有报道(Koca et al.,2009;George et al.,2014),其揭示了抗炎对创伤修复的积极作用,并且它们通常也含有黄酮及酚类物质。因此,蜂胶很有可能通过抗炎活性减少过度炎症反应对创伤修复的不利影响,从而

促进修复。

蜂胶的抗炎活性已经得到大量的研究证明和报道（Borrelli et al.，2002；胡福良等，2007；Paulino et al.，2008；Rebiai et al.，2011）。蜂胶这一生物学活性的发挥主要与蜂胶中含有大量的黄酮类抗炎物质如槲皮素、高良姜素等以及酚类抗炎物质如咖啡酸、阿魏酸、咖啡酸苯乙酯等有关（Guimarães et al.，2012）。这些抗炎活性成分能通过影响花生四烯酸代谢通路（Mirzoeva and Calder，1996）、L-精氨酸代谢通路抑制一些炎症介质如前列腺素 E2（PGE2）、一氧化氮（NO）等的产生（Tan-No et al.，2006），并能干扰 NF-κB 信号通路激活（Song et al.，2002），从而阻止某些炎症相关基因的转录，有效缓解炎症反应对细胞和组织造成的损伤。

2.2　抗氧化作用对创（烧）伤修复的影响

烧伤后，正常的氧化还原平衡被打破，局部组织缺血缺氧，加上大量的炎症因子及细菌的侵入，共同促进了烧伤组织的脂质过氧化反应和大量氧化物质的产生（Parihar et al.，2008）。这些活性氧类（ROS）和活性氮类（RNS）物质不只是帮助机体消灭细菌（Hampton et al.，1998），同时参与到血管通透性改变，细胞脂质过氧化，引起局部和系统性的炎症反应，降低巨噬细胞吞噬能力，损伤细胞 DNA 及引起花生四烯酸代谢反应在内的多个应激反应（Youn et al.，1992；Parihar et al.，2008），而这些应激反应进一步影响创伤修复。O'toole 等（1996）研究发现，过氧化氢在微摩尔浓度下就能抑制角质细胞迁移，而角质细胞的增生和迁移对再上皮化至关重要。由此可见，创伤后的氧化应激会因为对角质细胞的损伤而抑制再上皮化。总的说来，机体受剧烈损伤后大量产生的自由基会通过延长炎症反应时间，使蛋白酶解反应过度，同时抑制成纤维细胞和角质形成细胞的增生、迁移及细胞外基质的生成，并通过加速它们的衰老和凋亡等多种方式，阻碍创伤修复进程，并最终导致疤痕形成或修复障碍（Soneja et al.，2005）。因此，抗氧化治疗有利于减轻烧伤对患者的伤害。

蜂胶中丰富的黄酮、酚酸及萜烯类成分使其具有良好的抗氧化效果，所以，这很可能是蜂胶促进再上皮化和创伤修复的重要原因之一。这些抗氧化活性物质能通过直接清除自由基或提高细胞内一些抗氧化酶活性，甚至通过影响氧化应激相关转录调节因子（如 Nrf-2）的信号转导过程发挥抗氧化活性（张江临等，2012）。已有相关文献报道了蜂胶或其成分在烧伤组织中的抗氧化作用。Ocakci 等（2006）在蜂胶活性成分咖啡酸苯乙酯（CAPE）对氢氧化钠导致的小鼠食管烧伤的作用研究中表明 CAPE 组减弱了脂质过氧化，同时也促进抗氧化酶活性的恢复。王元元等（2012）在对蜂蜜、蜂胶对深Ⅱ度烫伤大鼠创面愈合影响的研究中，检测到蜂胶外用能显著提高大鼠血清中超氧化物歧化酶（SOD）含量。Olczyk 等（2013）研究了蜂胶对烧伤组织中自由基的作用，通过电子顺磁共振谱学检测蜂胶治疗组烧伤组织中自由基的浓度，结果表明蜂胶能使烧伤组织自由基浓度相对较低。这些研究表明，蜂胶及其活性成分可能通过促进创伤组织内一些抗氧化酶活性的增加或恢复，并促进其含量的增加以及对自由基的直接清除作用，保护细胞免受氧化损伤，从而发挥促进组织修复的作用。

2.3　抗菌作用对创（烧）伤修复的影响

细菌感染是烧伤患者常见的并发症，往往是造成创伤病理性修复的重要原因之一，同时也是对烧伤患者尤其是大面积严重烧伤患者造成生命威胁的主要因素之一。据报道，烧伤创面感染的病原微生物主要以金黄色葡萄球菌、铜绿假单胞菌、大肠埃希菌、鲍氏不动杆菌、肺炎克雷白杆菌等最为常见(Matsui et al.,2004;Ozyurt et al.,2007;罗念容等,2011;王宏等,2011)。近年来,由于抗生素的广泛应用,使得金黄色葡萄球菌、铜绿假单胞菌、鲍氏不动杆菌等致病菌对一些常见的抗菌药如头孢菌素类、大环内酯类、氨基糖苷类、氟喹诺酮类等表现出高耐药性(黎洪棉等,2005;于勇等,2006)。

蜂胶具有良好的抗菌活性,无一般抗生素的毒副作用,且不产生耐药性。蜂胶对革兰阳性菌的作用强于对革兰阴性菌的作用,尤其对金黄色葡萄球菌有较强的抗菌作用。此外,蜂胶对大肠杆菌、铜绿假单胞菌杆菌等阴性杆菌也有良好的抗菌效果(郭芳彬,2004)。由于地理位置、植物源甚至提取方式的差异,加之蜂胶成分极为复杂,各种实验结果可能并不完全一致(张芳英等,2011),但蜂胶的抗菌效果确实得到大量实验的证实(陈荷凤等,2001;申慧亭和靳月琴,2005;周建新等,2007)。Berretta 等(2012)通过肉汤大量稀释法用来确定蜂胶配方和提取物对烧伤中常见的微生物,如铜绿假单胞菌、克雷伯氏肺炎菌、大肠杆菌、金黄色葡萄球菌、表皮葡萄球菌的抗菌能力,并用大鼠穿刺损伤模型评估蜂胶的修复活性,结果表明,受试样品中蜂胶浓度达到3.6%时表现出最佳的抗菌和损伤修复活性。

蜂胶确切的抗菌机制尚不清楚,很可能是各种物质的协同作用。蜂胶中含有大量的黄酮类(松属素、高良姜素)、有机酸类及其脂类(龙胆酸、阿魏酸、咖啡酸、香豆素酸、咖啡酸酯等)及萜类物质,这些物质均具有良好的抗菌、杀菌作用(郭芳彬,2004;Wang et al.,2010)。蜂胶有效抑制微生物对创伤面的感染,抑制微生物对创伤面的损害,同时,降低了因感染而起的炎症,亦有利于维持创伤面的氧化还原平衡。总之,由于创伤修复是包括炎症反应、氧化应激、抗感染等多个事件在内的复杂过程,而这些事件之间存在密切的联系,能相互促进,放大彼此对修复的副作用。所以,蜂胶的各种生物学活性之间的相互协调、相互促进可能是其促进创伤修复最重要的原因。另外,蜂胶中含有的氨基酸、维生素、糖类物质(齐东梅,2008),可以为创伤组织提供局部营养,具有改善局部组织微环境的作用,这可能也是其促进组织修复的原因之一。需要指出的是,蜂胶成分复杂,不同提取方式、不同植物源的蜂胶活性成分差异很大,加之各研究采用的实验模型和给药方式和剂量也不尽相同,导致一些实验结果的不一致。未来的研究有必要在明晰蜂胶药效成分的基础上,综合比较研究不用提取方式、不同植物源蜂胶及其单体成分对创伤修复的效果,从而为蜂胶的临床应用提供参考。

小 结

从近年来对蜂胶促进创伤修复作用的研究报道来看,比起蜂胶抗炎、抗氧化方面的研究,蜂胶在创伤修复作用方面的研究相对较少,并且主要集中于蜂胶在组织病理学方面的作用和一些临床效果的观察上,而对于蜂胶促进创伤修复作用的内在分子机制却鲜有报道。这可能是因为创伤修复本身是一个十分复杂的过程,涉及的生物学机制和影响因素太多,加大了蜂胶在这一领域的研究难度。从近年来这些研究和报道发表的时间来看(表3.5),人们对于蜂胶的这一生物学活性的关注和研究在逐渐提升。随着人们对于创伤修复及其内在机制的研究越来越透彻,必然会带动蜂胶创伤修复作用研究的进一步深入。

参考文献

Adam J, Singer MD, Richard AF, Clark MD (1999) Cutaneous wound healing[J]. *The New England Journal of Medicine*, 341(10): 738-746.

Albuquerque-Júnior RLC, BarretoALS, Pires JA, Reis FP, Lima SO, Ribeiro MAG, Cardoso JC (2009) Effect of bovine type-Ⅰ collagen-based films containing red propolis on dermal wound healing in rodent model[J]. *International Journal of Morphology*, 27(4): 1105-1110.

Ansorge S, Reinhold D, Lendeckel U (2003) Propolis and some of its constituents down-regulate DNA synthesis and inflammatory cytokine production but induce TGF-β_1 production of human immune cells[J]. *Zeitschrift für Naturforschung*, 58C: 580-589.

Ashcroft GS, Yang X, Glick AB, Weinstein M, Letterio JL, Mizel DE, Anzano M, Greenwell-Wild T, Wahl SM, Deng C, Roberts AB (1999) Mice lacking smad3 show accelerated wound healing and an impaired local inflammatory response[J]. *Nature Cell Biology*, 1(5): 260-266.

Atiba A, Nishimura M, Kakinuma S, Hiraoka T, Goryo M, Shimada Y, Ueno H, Uzuka Y (2011) *Aloe vera* oral administration accelerates acute radiation-delayed wound healing by stimulating transforming growth factor-β and fibroblast growth factor production [J]. *The American Journal of Surgery*, 201(6): 809-818.

Barrientos S, Stojadinovic O, Golinko MS, Brem H, Tomic-Canic M (2008) Growth factors and cytokines in wound healing[J]. *Wound Repair Regen*, 16(5): 585-601.

Berretta AA, Nascimento AP, Bueno PC, de Oliveira LLV, Marchetti JM (2012) Propolis standardized extract (EPP-AF(R)), an innovative chemically and biologically reproducible pharmaceutical compound for treating wounds[J]. *International Journal of Medical Sciences*, 8(4): 512-521.

Borrelli F, Maffia P, Pinto L, Ianaro A, Russo A, Capasso F, Ialenti A (2002) Phytochemical compounds involved in the anti-inflammatory effect of propolis extract[J]. *Fitoterapia*, 73(Suppl1): S53-63.

de Almeida EB, Cordeiro Cardoso J, Karla de Lima A, de Oliveira NL, de Pontes-Filho NT, Oliveira Lima S, Leal Souza IC, de Albuquerque-Júnior RL (2013) The incorporation of Brazilian propolis into collagen-based dressing films improves dermal burn healing[J]. *Journal of Ethnopharmacology*, 147(2): 419-425.

de Castro SL (2001) Propolis: biological and pharmacological activity[J]. *Ann Rev Biomed Sci*, 3(1): 49-83.

de Moura SA, Ferreira MA, Andrade SP, Reis ML, de Lourdes NM, Cara DC (2011a) Brazilian green propolis inhibits inflammatory angiogenesis in a murine sponge model [J]. *Evidence-Based Complementary and Alternative Medicine*, 2011: 182703.

de Moura SA, Negri G, Salatino A, da Cunha LimaLD, DouradoLPA, Mendes JB, Andrade SP, Ferreira MA, Cara DC (2011b) Aqueous extract of brazilian green propolis: primary components, evaluation of inflammation and wound healing by using subcutaneous implanted sponges[J]. *Evidence-Based Complementary and Alternative Medicine*,2011:748283.

Desmoulière A, Chaponnier C, Gabbiani G (2005) Tissue repair, contraction, and the myofibroblast[J]. *Wound Repair and Regeneration*,13(1):7-12.

Dovi JV, He LK, DiPietro LA (2003) Accelerated wound closure in neutrophil-depleted mice[J]. *Journal of Leukocyte Biology*,73:448-455.

Dovi JV, Szpaderska AM, Dipietro LA (2004) Neutrophil function in the healing wound: adding insult to injury? [J]*Thromb Haemost*,92:275-280.

Egozi EI, Ferreira AM, Gamelli RL, Gamelli RL, Dipietro LA (2003) Mast cells modulate the inflammatory but not the proliferative response in healing wounds[J]. *Wound Repair Regen*,11(1):46-54.

Faler BJ, Macsata RA, Plummer D, Mishra L, Sidawy AN (2006) Transforming growth factor-β and wound healing[J]. *Perspectives in Vascular Surgery and Endovascular Therapy*,18(1):55-62.

Farstvedt E, Stashak TS, Othic A (2004) Update on topical wound medications[J]. *Clinical Techniques in Equine Practice*,3(2):164-172.

Gabbiani G (2003) The myofibroblast in wound healing and fibrocontractive diseases[J]. *The Journal of Pathology*,200(4):500-503.

George BP, Parimelazhagan T, Chandran R (2014) Anti-inflammatory and wound healing properties of *Rubus fairholmianus* Gard. Root—An *in vivo* study[J]. *Industrial Crops and Products*,54:216-225.

Gressner AM, Breitkopf K, Dooley S (2002) Roles of TGF-beta in hepatic fibrosis[J]. *Frontiers in Bioscience*,7:d793-807.

Guimarães NS, Mello JC, Paiva JS, Bueno PC, Berretta AA, Torquato RJ, Nantes IL, Rodrigues T (2012) Baccharis dracunculifolia, the main source of green propolis, exhibits potent antioxidant activity and prevents oxidative mitochondrialdamage[J]. *Food and Chemical Toxicology*,50(3-4):1091-1097.

Gurtner GC, Werner S, Barrandon Y, Michael T, Longaker MT (2008) Wound repair and regeneration[J]. *Nature*,453(7193):314-321.

Hampton MB, Kettle AJ, Winterbourn CC (1998) Inside the neutrophil phagosome: oxidants, myeloperoxidase and bacterial killing[J]. *American Society of Hematology*,92(9):3007-3017.

Han MC, Durmus AS. Karabulute E, Yaman I (2005) Effects of Turkish propolis and silversulfadiazine on burn wound healing in rats[J]. *Revue de Médecine Vétérinaire*,156(12):624-627.

Hozzein WN, Badr G, Al Ghamdi AA, Sayed A, Al-Waili NS, Garraud O (2015) Topical application of propolis enhances cutaneous wound healing by promoting TGF-Beta/Smad-mediated collagen production in a streptozotocin-induced type Ⅰ diabetic mouse model[J]. *Cellular Physiology and Biochemistry*,37(3):940-954.

Iyyam PS, Palsamy P, Subramanian S, Kandaswamy M(2010) Wound healing properties of

Indian propolis studied on excision wound-induced rats[J]. *Pharmaceutical Biology*, 48(11):1198-1206.

Khorasgani EM, Karimi AH, Nazem MR (2010) A comparison of healing effects of propolis and silver sulfadiazine on full thickness skin[J]. *Pakistan Veterinary Journal*, 30(2):72-74.

Kilicoglu SS, Kilicoglu B, Erdemli E (2008) Ultrastructural view of colon anastomosis under propolis effect by transmission electron microscopy[J]. *World Journal of Gastroenterology*, 14(30):4763-4770.

Koca U, Suntar IP, Keles H, Yesilada E, Akkol EK (2009) *In vivo* anti-inflammatory and wound healing activities of *Centaurea iberica* Trev. ex Spreng[J]. *Journal of Ethnopharmacology*, 126(3):551-556.

Leask A, Abraham DJ (2004) TGF-β signaling and the fibrotic response[J]. *Federation of American Societies for Experimental Biology*, 18(7):816-827.

Leibovich SJ, Ross R (1975) The role of the macrophage in wound repair[J]. *The American Journal of Pathology*, 78(1):71-100.

Martin P, D'Souza D, Martin J, Grose R, Cooper L, Maki R, McKercher SR (2003) Wound healing in the PU. 1 null mouse-tissue repair is not dependent on inflammatory cells[J]. *Current Biology*, 13(13):1122-1128.

Martin P, Leibovich SJ (2005) Inflammatory cells during wound repair: the good, the bad and theugly[J]. *Trends Cell Biol*, 5(11):599-607.

Matsui T, Ebuchi S, Fujise T, Abesundara KJ, Doi S, Yamada H, Matsumoto K (2004) Strong antihyperglycemic effects of water-soluble fraction of Brazilianpropolis and its bioactive constituent, 3, 4, 5-tri-o-caffeoylquinic acid[J]. *Biological and Pharmaceutical Bulletin*, 27(11):1797-1803.

McLennan SV, Bonner J, Milne S, Lo L, Charlton A, Kurup S, Jia J, Yue DK, Twigg SM (2008) The anti-inflammatory agent propolis improves wound healing in a rodent model of experimental diabetes[J]. *Wound Repair Regen*, 16:706-713.

Meredith JE, Fazeli B, Schwartz MA (1993) The Extracellular matrix as a cell survival factor[J]. *Molecular Biology of the Cell*, 4(9):953-961.

Mirzoeva OK, Calder PC (1996) The effect of propolis and its components on eicosanoid production during the inflammatory response[J]. *Prostaglandins Leukotrienes and Essential Fatty Acids*, 55(6):441-449.

Ocakci A, Kanter M, Cabuk M, Buyukbas S (2006) Role of caffeic acid phenethyl ester, an active component of propolis, against NAOH-induced esophageal burns in rats[J]. *International Journal of Pediatric Otorhinolaryngology*, 70(10):1731-1739.

Olczyk P, Komosinska-Vassev K, Winsz-Szczotka K, Kožma EM, Wisowski G, Stojko J, Klimek K, Olczyk K (2012) Propolis modulates vitronectin, laminin, and heparan sulfate/heparin expression during experimental burn healing[J]. *Journal of Zhejiang University-Science B (Biomedicine and Biotechnology)*, 13(11):932-941.

Olczyk P, Ramos P, Komosinska-Vassev K, Stojko J, Pilawa B (2013) Positive effect of propolis on free radicals in burn wounds[J]. *Evidence-Based Complementary and Alternative Medicine*, 2013:356737.

Olczyk P, Wisowski G, Komosinska-Vassev K, Stojko J, Klimek K, Olczyk M, Kozma EM (2013) Propolis modifies collagen types Ⅰ and Ⅲ accumulation in the matrix of burnt tissue[J]. *Evidence-Based Complementary and Alternative Medicine*, 2013:423809.

Olczyk P, Komosinska-Vassev K, Winsz-Szczotka K, Stojko J, Klimek K, Kozma EM (2013) Propolis induces chondroitin/dermatan sulphate and hyaluronic acid accumulation in the skin of burned wound[J]. *Evidence-Based Complementary and Alternative Medicine*, 2013:290675.

O'Toole EA, Goel M, Woodley DT (1996) Hydrogen peroxide inhibits human keratinocyte migration[J]. *Dermatologic Surgery*, 22(6):525-529.

Ozyurt B, Parlaktas BS, Ozyurt H, Aslan H, Ekici F, Atis O (2007) A preliminary study of the levels of testis oxidative stress parameters after MK-801-induced experimental psychosis model: Protective effects of CAPE[J]. *Toxicology*, 230(1):83-89.

Parihar A, Parihar MS, Milner S, Bhat S (2008) Oxidative stress and anti-oxidative mobilization in burn injury[J]. *Burns*, 34(1):6-17.

Paulino N, Abreu SR, Uto Y, Koyama D, Nagasawa H, Hori H, Dirsch VM, Vollmar AM, Scremin A, Bretz WA (2008) Anti-inflammatory effects of a bioavailable compound, Artepillin C, in Brazilian propolis[J]. *European Journal of Pharmacology*, 587(1-3):296-301.

Pessolato AG, Martins Ddos S, Ambrósio CE, Mançanares CA, de Carvalho AF (2011) Propolis and amnion reepithelialise second-degree burns in rats[J]. *Burns*, 37(7):1192-1201.

Ramos AFN, Miranda JL (2007) Propolis: A review of its anti-inflammatory and healing actions[J]. *Journal of Venomous Animals and Toxins Including Tropical Diseases*, 13(4):679-710.

Rebiai A, Lanez T, Belfar ML (2011) *In vitro* evaluation of antioxidant capacity of algerian propolis by spectrophotometrical and electrochemical assays[J]. *International Journal of Pharmacology*, 7(1):113-118.

Santoro MM, Gaudino G (2005) Cellular and molecular facets of keratinocyte reepithelization during wound healing[J]. *Experimental Cell Research*, 304(1):274-286.

Schiller M, Javelaud D, Mauviel A (2004) TGF-β-induced SMAD signaling and gene regulation: consequences for extracellular matrix remodeling and wound healing[J]. *Journal of Dermatological Science*, 35(2):83-92.

Schultz GS, Wysocki A (2009) Interactions between extracellular matrix and growth factors in wound healing[J]. *Wound Repair and Regeneration*, 17(2):153-162.

Sehn E, Hernandes L, Franco SL, Gonçalves CC, Baesso ML (2009) Dynamics of reepithelialisation and penetration rate of a bee propolis formulation during cutaneous wounds healing[J].

Analytica Chimica Acta, 635(1):115-120.

Simpson DM, Ross R (1972) The neutrophilic leukocyte in wound repair[J]. *Journal of Clinical Investigation*, 51(8):2009-2023.

Soneja A, Drews M, Malinski T (2005) Role of nitric oxide, nitroxidative and oxidative[J]. *Pharmacological Reports*, 57(suppl):108-119.

Song YS, Park EH, Hur GM, Ryu YS, Kim YM, Jin C (2002) Ethanol extract of propolis inhibits nitric oxide synthase gene[J]. *Journal of Ethnopharmacology*, 80(2-3):155-161.

Song YS, Park EH, Jung KJ, Jin C (2002) Inhibition of angiogenesis by propolis[J]. *Archives of Pharmacal Research*, 25(4):500-504.

Steffensen B, Hakkinen L, Larjava H (2001) Proteolytic events of wound-healing-coordinated interactions among matrix metalloproteinases (MMPs), integrins, and extracellular matrix molecules[J]. *Critical Reviews in Oral Biology Medicine*, 12(5):373-398.

Su KY, Hsieh CY, Chen YW, Chuang CT, Chen CT, Chen YL (2014) Taiwanese green propolis and propolin G protect the liver from the pathogenesis of fibrosis via eliminating TGF-beta-induced smad2/3 phosphorylation[J]. *Journal of Agricultural and Food Chemistry*, 62(14):3192-201.

Tan-No K, Nakajima T, Shoji T, Nakagawasai O, Niijima F, Ishikawa M, Endo Y, Sato T, Satoh S, Tadano T (2006) Anti-inflammatory effect of propolis through inhibition of nitric oxide[J]. *Biological and Pharmaceutical Bulletin*, 29(1):96-99.

Tateshita T, Ono I, Kaneko F (2001) Effects of collagen matrix containing transforming growth factor TGF-β_1 on wound contraction[J]. *Journal of Dermatological Science*, 27(2001):104-113.

Wang X, Stavchansky S, Kerwin SM, Bowman PD (2010) Structure-activity relationships in the cytoprotective effect of caffeic acid phenethyl ester (CAPE) and fluorinated derivatives: Effects on heme oxygenase-1 induction and antioxidant activities[J]. *European Journal of Pharmacology*, 635(1-3):16-22.

Youn YK, LaLonde X, Demling R (1992) The role of mediators in the response to thermal injury[J]. *World Journal of Surgery*, 16(1):30-36.

陈荷凤, 韩文辉, 李兵, 崔海辉 (2001) 蜂胶各种溶剂提取物的抑菌效果比较[J]. 食品研究与开发, 22(S1):18-19.

陈炯, 韩春茂, 林小玮, 唐志坚, 苏士杰 (2006) 纳米银敷料在修复Ⅱ度烧伤创面的应用研究[J]. 中华外科杂志, 44(1):50-52.

高畅, 贾军宏, 楚勤英, 唐志雄, 马虹颖, 李建华, 傅小玲, 王进伟 (2009) 蜂胶对糖尿病大鼠创面新生血管和细胞增殖的影响[J]. 中国医院用药评价与分析, 9(3):200-203.

郭芳彬 (2004) 蜂胶的抗菌作用[J]. 蜜蜂杂志, (3):10-12.

胡福良, 李英华, 朱威, 陈民利, 应华忠 (2007) 蜂胶提取液对大鼠急性胸膜炎的作用及其机制的研究[J]. 营养学报, 29(2):189-191.

黎洪棉, 梁自乾, 刘达恩, 蒙诚跃 (2005) 南宁地区某烧伤病房细菌生态学调查及耐药性分析

[J].中华烧伤杂志,21(2):107-110.
李晋辉,母得志(2007)整合素及其信号传导通路[J].医学分子生物学杂志,4(3):279-282.
罗念容,高华,周青峰(2011)烧伤患者病原菌感染及其耐药性调查[J].中国误诊学杂志,11(7):1761-1762.
彭代智(2005)烧伤后炎症反应的病因、分子机制及防治对策[J].中华烧伤杂志,21(6):405-409.
齐东梅(2008)蜂胶的药理作用和临床应用[J].首都医药,(4):44-45.
申慧亭,靳月琴(2005)蜂胶抗菌作用的实验室研究[J].长治医学院学报,19(4):253-254.
王宏,蔡金东,侯智慧(2011)烧伤感染的细菌学调查及其耐药性分析[J].中国冶金工业医学杂志,28(1):5-6.
王元元,黄云英,杜娟,沈丽,张德芹,王雪妮(2012)蜂蜜、蜂胶对深Ⅱ度烫伤大鼠创面愈合的影响[J].天津中医药大学学报,31(3):154-156.
于勇,盛志勇,柴家科,杨小强,常东,蒋伟(2006)抗菌药物使用与烧伤感染主要病原菌构成比变化的关系[J].解放军医学杂志,31(1):1-3.
张芳英,穆丽娟,杨继,杨树民,张旭东,张征章(2011)蜂胶提取物抗菌作用的研究进展[J].中国药房,22(11):1041-1043.
张江临,王凯,胡福良(2013)蜂胶的抗氧化活性及其分子机制研究进展[J].中国中药杂志,2013,38(16):2645-2652
周建新,姚明兰,岳文倩,潘海琼(2007)蜂胶的抗菌性及其影响因素的研究[J].食品与发酵工业,33(3):41-43.

第九节 蜂胶的其他生物学活性

1 抗胃溃疡

胃溃疡是消化系统常见病之一,在人群中有较高的发病率。胃溃疡的形成机制十分复杂,目前认为幽门螺杆菌(*Helicobacter pylori*)、药物、乙醇等各种损伤因素使胃黏膜受损,引起胃酸和胃蛋白酶对胃的自身消化,是导致消化性胃溃疡形成的主要原因(胡伏莲,2005)。

研究表明,蜂胶对各种因素所致的胃溃疡都有一定的治疗效果。金河奎等(1996)利用幽门结扎型、应激型、酒精型等5种实验性溃疡模型研究蜂胶乙醇提取液(EEP)的抗溃疡作用,结果表明,EEP能通过降低胃蛋白酶活性和胃酸分泌的方式对抗幽门结扎型胃溃疡,并促进醋酸型溃疡的愈合,也能减少其他几种因素致溃疡的面积,表明蜂胶提取物对各种因素所致的溃疡都有一定的改善作用。金春玉等(1996)研究表明,蜂胶石油醚提取物对醋酸型、应激型溃疡有明显的对抗作用,并能通过降低胃内游离酸度,增加PGE2和胃壁黏液含量对抗幽门结扎型溃疡;正丁醇萃取物对以上3种溃疡模型均无显著影响,但能降低胃蛋白酶活性,增加胃壁黏液含量,表明蜂胶抗溃疡作用的主要有效成分在石油醚萃取物中。de Barros等(2007)研究了巴西绿蜂胶对酒精型、吲哚美辛型等4种类型溃疡的保护作用,结果表明,

大鼠预先服用(50、25、0、500mg/kg)EEP便能明显降低酒精所致的胃溃疡的溃疡指数、总病变区域面积及病区与整个胃面积比。对于改善吲哚美辛型和应激型溃疡,绿蜂胶提取物的浓度需分别达到500mg/kg和250mg/kg;浓度达到500mg/kg时,蜂胶才能显著降低溃疡指数、总病变区域面积及病变区与整个胃面积比。而对于幽门结扎型溃疡,绿蜂胶浓度为250mg/kg和500mg/kg时,表现出抗分泌活性,使胃液体积、总酸度和pH值下降。进一步研究表明,巴西绿蜂胶的抗溃疡效果与其所含的咖啡酸、阿魏酸、p-香豆酸及肉桂酸成分有关(de Barros,2008)。幽门螺杆菌与胃溃疡发生特别是反复发作有着十分密切的联系,因此抑制幽门螺杆菌的生长对治疗胃溃疡有着十分重要的意义。有研究表明,EEP对幽门杆菌的生长有明显的抑制作用,并呈浓度依赖性抑制,而蜂胶水提物(WEP)的作用远不如EEP(陈炼,2009)。El-Ghazaly等(2011)研究了WEP对吲哚美辛诱导的受辐射大鼠和未受辐射大鼠胃溃疡的影响,结果表明,无论是否接受辐射,预先口服WEP都能有效对抗吲哚美辛诱导的胃溃疡,表现为胃酸分泌和消化活动减弱,黏蛋白分泌和PEG2水平增加,炎症因子(TNF-α和IL-1β)水平、脂质过氧化指标和MDA水平显著降低,表明蜂胶具有保护胃的作用。

2 抗运动性疲劳

运动性疲劳与氧化应激之间有着十分密切的关系。机体做剧烈运动和力竭运动时,体内会产生大量自由基。当这些自由基累积超过机体抗氧化防御系统的清除能力,便会造成脂质过氧化,影响细胞膜及线粒体结构和功能的完整性,使细胞无法发挥正常功能,导致疲劳(张陵,2006)。因此,补充外源性抗氧化剂或提升机体自身抗氧化防御能力有助于对抗运动性疲劳。蜂胶中富含黄酮和酚酸类化合物,具有良好的自由基清除活性,还能影响细胞内抗氧化相关信号通路的转导(Zhang et al.,2015)。

人们对蜂胶抗疲劳能力进行了研究。曹炜等(2004)的研究结果表明,蜂胶提取物可以显著延长小鼠游泳至力竭的时间,清除自由基,减轻力竭运动所致的氧化损伤,并促进力竭运动后小鼠物质和能量代谢的恢复,表明蜂胶提取物具有抗疲劳作用。潘燕等(2010)研究了蜂胶黄酮对疲劳小鼠心肌MDA含量、抗氧化酶活性以及ATP酶活性的影响,其结果同样表明蜂胶能延长小鼠的运动力竭时间,降低心肌MDA含量,并显著增强小鼠心肌抗氧化酶及ATPase活性,表明蜂胶能通过改善机体疲劳性氧化应激的方式抗疲劳。同时,这种作用还可能与蜂胶的强心作用有关,即蜂胶还可能通过增加心输出量,改善剧烈运动时心肌缺血缺氧状态的方式抵抗疲劳(王南州等,1999)。于洋等(2012)的研究也表明蜂胶能改善机体的运动源性氧化应激,表现为显著降低训练后小鼠的红细胞畸形率、心肌和血浆MDA含量,升高心肌和血浆SOD活性,心肌GSH-Px、Ca_2^+-ATP酶和红细胞Na^+-K^+-ATP酶活性。

3 神经保护作用

研究表明,蜂胶及其成分可以通过抗氧化、调节NO合成、抗凋亡及抗炎等多种方式起到神经保护作用。

Wei等(2004)研究表明,不论是在缺氧-缺血性脑损伤之前还是之后接受CAPE治疗,都能显著对抗新生大鼠由于缺氧-缺血所致的皮质、海马和丘脑的脑损伤。除了能阻止缺氧-

缺血诱导的凋亡调控因子 Caspase3 的激活，CAPE 还能抑制体内缺氧-缺血介导的诱导型一氧化氮合酶(iNOS)及 Caspase1 的表达，并能在体外强有力地抑制 NO 诱导的神经毒性。此外，CAPE 还能直接抑制分离的脑线粒体中细胞血素 C 的释放。因此，CAPE 可能通过直接阻碍缺氧-缺血诱导的神经元死亡通路或通过抗炎途径对抗神经元的死亡，从而发挥对新生鼠的神经保护作用。Shimazawa 等(2005)通过体外研究表明，蜂胶能显著降低由 H_2O_2 刺激 24h 或血清饥饿 48h 所致的神经元样细胞分化 PC12 细胞的细胞毒性，而其体内实验表明蜂胶能减轻缺血性神经元损伤，提示巴西绿蜂胶的这种神经保护作用可能与其清除自由基和抑制脂质过氧化的能力有关。Wei(2008)的研究结果表明，蜂胶中的 CAPE 能通过抑制 p38 磷酸化和 Caspase3 激活的方式对抗谷氨酸盐诱导的小脑颗粒神经元兴奋性神经毒性。覃燕飞等(2010)研究表明，蜂胶总黄酮能促进大鼠缺血再灌注后大脑皮质 bcl-2 表达增强，减弱凋亡相关基因 bax 与 Caspase-3 表达，从而能有效减少脑缺血/再灌注损伤引起的神经细胞凋亡，因此具有一定的神经保护作用。

神经细胞损伤通常与氧化应激及炎症密切相关。Barbosa 等(2016)利用巴西红蜂胶水醇提取物(HERP)研究了其神经保护效果，采用大鼠坐骨神经损伤模型，检测了大鼠神经损伤后的行为改变及中枢神经轴突数量。结果表明，与模型组相比，HERP 处理能有效加速大鼠神经细胞修复过程，增加有髓鞘轴突数量。da Silveira 等(2016)研究了一种富含萜烯类化合物(主要为三萜类化合物羽扇豆醇，二萜类化合物 β-香树素)蜂胶提取物的神经保护效果，在大鼠强迫游泳实验、迷宫实验、抑制性回避试验中，该蜂胶提取物能有效缓解大鼠焦虑症状，并降低焦虑大鼠血液中一氧化氮和丙二醛水平，但对总过氧化能力、超氧化物歧化酶和过氧化氢酶水平无显著影响。

Kurauchi 等(2012)研究表明，CAPE 能通过上调血红素氧合酶 1(HO1)和脑源性神经生长因子的机制保护黑质多巴胺能神经元。另外，还有体外研究表明，经中国蜂胶刺激后的牙髓细胞的条件培养液显示出神经保护和神经突延伸作用，具体表现为蜂胶呈剂量依懒性方式促进 DPCs 表达神经生长因子(nerve growth factor，NGF)，并且蜂胶刺激后的条件培养基能明显抑制过氧化氢诱导的细胞死亡，且这种保护效果要强于未用蜂胶刺激的条件培养基。此外，DPCs 的条件培养基能明显抑制衣霉素诱导的细胞死亡，且蜂胶刺激后的条件培养基能促进 PC12 细胞突起生长，效果强于未用蜂胶刺激的条件培养基的作用效果(Kudo et al.，2015)。

4 改善记忆力

学习和记忆是脑神经的高级生理活动，因此记忆相关的神经元之间信息传递不受干扰是保证良好记忆力的关键。任何导致脑神经元受损或阻碍脑神经元之间信息传递的因素都有可能导致记忆力的下降或丧失，如脑组织内的乙酰胆碱过少、脑组织的氧化应激性损伤、锌等必需元素的缺乏甚至脑组织的机械性损伤等。蜂胶对记忆力改善也是其神经保护作用的重要体现，可能与其抗氧化、抗凋亡活性等有关，甚至还有可能是蜂胶影响了大脑记忆区乙酰胆碱酯酶的活性。

王浩等(2008)通过 Y 型迷宫测试方法研究了 EEP 对化学药品所致的小鼠记忆障碍的效果，结果表明 EEP 能明显改善东莨菪碱、亚硝酸钠和三氯化铝所致小鼠学习记忆功能障碍，并猜测蜂胶对东莨菪碱所致记忆障碍的改善可能与蜂胶提高中枢胆碱能神经系统功能活动有关,而对亚硝酸钠和三氯化铝致学习记忆障碍的改善作用,可能与蜂胶抗氧化作用有关。陈娟等(2008)研究表明,WEP可能是通过抑制海马区域的乙酰胆碱酯酶的活性来减轻东莨菪碱引起的小鼠学习记忆损伤的。Miyazaki等(2015)通过体外实验研究表明巴西蜂胶乙醇提取物改善了高同型半胱氨酸引起的认知功能障碍。其体外实验结果表明巴西蜂胶乙醇提取物能呈剂量依赖性方式抑制由同型半胱氨酸引起的成神经细胞瘤SH-SY5Y细胞死亡和成胶质细胞瘤U-251MG细胞活性氧的产生。而水迷宫试验结果表明,巴西蜂胶乙醇提取物改善了高同型半胱氨酸引起的认知功能障碍,进一步研究表明这可能与蜂胶减轻淀粉样蛋白在大脑中的聚集有关。

5 提高畜禽免疫力和生产性能

将蜂胶作为饲料添加剂添加到畜禽日粮中改善畜禽体质和生产性能的研究在国内外有大量报道。早在1990年就有研究表明,将200mL 5%的蜂胶制剂添加到30kg的育肥猪饲料中可使育肥猪的生长速度提高22.83%,节省18.38%的饲料(彭和禄,1990)。后来彭和禄等又研究了蜂胶渣作为饲料添加剂对育肥猪生长的影响,结果表明,将蜂胶残渣以每头猪每日10g的量添加到育肥猪的基础日粮中,可以使育肥猪的平均日增重提高6.22%,每增重1kg可节约6.25%的饲料,这为蜂胶残渣的再利用也提供了一条新的思路(彭和禄,1993)。蜂胶对育肥猪生产性能的改善不只局限在提高畜禽的生长速度上,还能改善猪肉质量(彭和禄 1993)。此外,郭冬生等(2011)研究了蜂胶添加剂对母猪繁殖性能和仔猪腹泻率的影响。结果表明,在母猪基础日粮中添加蜂胶(0.07%)使仔猪断奶重平均提高0.62kg,成活率提高4.51%;在哺乳期和断奶前后7d期间仔猪腹泻率分别降低1.37%和2.96%。

Aguiar等(2014)研究了日粮中添加蜂胶提取物对奶牛乳产量、乳脂肪酸组成和乳抗氧化能力的影响,发现添加蜂胶会影响乳脂肪酸组成,显著影响共轭亚油酸比例。饲喂含高浓度蜂胶提取物日粮会显著增加单不饱和脂肪酸(MUFA)和多不饱和脂肪酸(PUFA)含量,降低饱和脂肪酸(SFA)比例。此外,饲喂蜂胶奶牛所产牛奶的抗氧化能力会显著增强。通过对奶牛消化功能和瘤胃代谢功能的研究,发现蜂胶能有效提升奶牛的消化功能,对奶牛瘤胃微生物的蛋白质代谢有着非常积极的作用(De Aguiar et al.,2014)。

国内外关于蜂胶对禽类生产性能的改善作用也有大量报道。Denli等(2005)曾报道在基础日粮中添加不同比例土耳其蜂胶均能有效提高鹌鹑的体重、饲料转化率,其中以每千克日粮中添加1g蜂胶的效果最好。家禽的生产性能受温度、湿度等外界条件的影响,持续的高温会造成家禽的热应激性损伤,表现为增重减缓、蛋品质下降、发病率及死亡率增加等多种症状,通常在高温的夏季比较常见。Seven等(2008)研究了蜂胶对高温刺激下蛋鸡生产性能的影响,结果表明,蜂胶能明显改善蛋鸡在热应激状态下的摄食量、饲料转换率、产蛋量、蛋重及蛋壳厚度,并能降低热应激状态下鸡的死亡率,由此说明蜂胶可以作为抗热应激制剂添加到畜禽饲料中。许合金等(2010)研究表明,日粮中添加75mg/kg和225mg/kg蜂胶能够显著降低蛋黄胆固醇含量;各个添加水平的蜂胶均能在蛋保存的各阶段不同程度地降低蛋黄MDA含量。日粮中添加225mg/kg蜂胶能够显著提高血清高密度脂蛋白水平,并显著降低血清总胆固醇水平;此外,日粮中添加150mg/kg和225mg/kg蜂胶能够显著提高血清T淋巴细胞转化率,添加150mg/kg蜂胶能够显著提高血液B淋巴细胞转化率。詹勋等(2010)研究表明,日粮中添加300g/t或500g/t蜂胶黄酮对黄羽肉种鸡的采食量、产蛋率、蛋重和料蛋比影响较小,但有降低种鸡死亡率,提高种蛋受精率和孵化率的效果。刘莹(2013)研究表明,一定剂量的槲皮素可通过调节蛋鸡营养物质代谢和相关内分泌激素水平来提高39~47周龄蛋鸡生产性能并改善蛋品质,推荐添加量为0.04%。Duarte等(2014)研究了饲料中添加蜂胶对肉鸡肠道形态和消化酶的影响,结果表明,饲喂蜂胶能显著提高7日龄肉鸡肌胃和大肠的相对重量,而蜂胶的最小有效添加浓度为278μg/kg;饲料中添加蜂胶也可显著提升21日龄肉鸡回肠隐窝深度和增强淀粉酶的活性。王留和张代(2016)研究了蜂胶提取物对蛋雏鸡免疫器官指数和腹腔巨噬细胞活性的影响,发现10日龄蛋雏鸡颈部皮下分别注射质量浓度为20mg/mL和40mg/mL的蜂胶乙醇浸提液0.2mL能显著提高蛋雏鸡胸腺、法氏囊、脾脏指数和腹腔巨噬细胞吞噬率,具有促进和调节机体免疫功能的作用,并能加强机体吞噬细胞的吞噬作用。陈佳亿等(2016)研究发现,饲粮中添加1.0%蜂胶残渣能改善黄羽肉鸡的屠宰性能和鸡肉品质。

奶牛乳腺炎是奶牛业最常见、危害最严重的疾病之一,它不仅影响产奶量,造成经济损失,而且影响牛奶的品质,危害人类的健康。Wang等(2016)采用奶牛乳腺上皮细胞(MAC-T细胞系)作为研究对象,利用多种乳腺炎病原物诱导建立细胞炎性损伤模型,研究了蜂胶对奶牛乳腺炎的防治作用。结果表明,蜂胶预处理能有效缓解各乳腺炎病原物诱导细胞活力的丧失,同时也伴随着促炎性细胞因子白介素6和肿瘤坏死因子基因表达量的降低。此外,与乳腺炎病原侵染过的细胞相比,蜂胶处理能显著上调细胞抗氧化基因HO-1、Txnrd-1和GCLM的表达。蜂胶及其多酚类活性成分(主要为咖啡酸苯乙酯和槲皮素)能有效抑制炎症相关转录因子NF-κB的激活,并提高细胞防御相关转录因子Nrf2-ARE的活力。Fiordalisi等(2016)也发现巴西蜂胶具有抗奶牛乳腺炎病原菌的效果,为蜂胶在防治奶牛乳腺炎上的应用提供了理论依据。

蜂胶对畜禽生产性能的改善可能是其抗氧化、抗病原微生物感染以及免疫调节等生物活性在畜禽体内综合作用的结果。随着抗生素的广泛使用,微生物耐药性问题和畜禽产品抗生素残留问题日益突出,因此寻找安全无害的天然替代物也变得日益重要。蜂胶作为一种动植物双源性天然产物,不仅安全,同时生物学活性广泛,具有显著的保健功能,作为饲料添加剂颇具优势。然而,由于蜂胶产量有限,所以在其实际生产应用中还要综合考虑成本等因素,如能以较少的用量换取丰厚的经济回报,将会拥有更广阔的应用前景。

此外,蜂胶还具有促进组织再生(Olczyk,2013;Sehn,2009)、抗辐射(Yalcin et al.,2016)、护肾(Baykara et al.,2015;Ulusoy et al.,2016)、镇痛(De Campos,1998)、对化疗副作用的缓解(Kumari et al.,2016),甚至治疗痔疮(徐传球,2009)、脱发(赵家明,2006)等诸多作用。

参考文献

Aguiar SC, Cottica SM, Boeing JS, Samensari RB, Santos GT, Visentainer JV, Zeoula LM (2014) Effect of feeding phenolic compounds from propolis extracts to dairy cows on milk production, milk fatty acid composition, and the antioxidant capacity of milk[J]. *Animal Feed Science and Technology*, 193:148-154.

Barbosa RA, Nunes T, da Paixao AO, Neto RB, Moura S, Albuquerque RLC, Candido EAF, Padilha FF, Quintans LJ, Gomes MZ (2016) Hydroalcoholic extract of red propolis

promotes functional recovery and axon repair after sciatic nerve injury in rats[J]. *Pharmaceutical Biology*, 54(6):993-1004.

da Silveira C, Fernandes LMP, Silva ML, Luz DA, Gomes ARQ, Monteiro MC, Machado CS, Torres YR, de Lira TO, Ferreira AG (2016) Neurobehavioral and antioxidant effects of ethanolic extract of yellow propolis[J]. *Oxidative Medicine and Cellular Longevity*, 2016:2906953.

De Aguiar SC, De Paula EM, Yoshimura EH, dos Santos WBR, Machado E, Valero MV, dos Santos GT, Zeoula LM (2014) Effects of phenolic compounds in propolis on digestive and ruminal parameters in dairy cows[J]. *Revista Brasileira De Zootecnia-Brazilian Journal of Animal Science*, 43(4):197-206.

de Barros MP, Lemos M, Maistro EL, Leite MF, Sousa JPB, Bastos JK, de Andraded SF (2008) Evaluation of antiulcer activity of the main phenolic acids found in Brazilian green propolis[J]. *Journal of Ethnopharmacology*, 120(3):372-377.

de Barros MP, Sousa JPB, Bastos JK, de Andrade SF (2007) Effect of Brazilian green propolis on experimental gastric ulcers in rats[J]. *Journal of Ethnopharmacology*, 110(3):567-571.

De Campos ROP, Paulino N, Da Silva CHM, SCREMIN A, CALIXTO JB (1998) Anti-hyperalgesiceffect of an ethanolicextract of propolis in mice and rats[J]. *Journal of Pharmacy and Pharmacology*, 50(10):1187-1193.

Denli M, Cankaya S, Silici S, Okan F, Uluocak AN (2005) Effect of dietary addition of Turkish propolis on the growth performance, carcass characteristics and serum variables of quail (*Coturnixcoturnix japonica*)[J]. *Asian-Australasian Journal of Animal Sciences*, 18(6):848-854.

Duarte CRA, Eyng C, Murakami AE, Santos TC (2014) Intestinal morphology and activity of digestive enzymes in broilers fed crude propolis[J]. *Canadian Journal of Animal Science*, 94(1):105-114.

El-Ghazaly MA, Rashed RRA, Khayyal MT (2011) Anti-ulcerogenic effect of aqueous propolis extract and the influence of radiation exposure[J]. *International Journal of Radiation Biology*, 87(10):1045-1051.

Ercis K, Aydoğan S, AtayogluAT. Silici S (2015) Effect of propolis on erythrocyte rheology in experimentalmercury intoxication in rats[J]. *Environmental Science and Pollution Research*, 22:12534-12543.

Fiordalisi SAL, Honorato LA, Loiko MR, Avancini CAM, Veleirinho MBR, Machado LCP, Kuhnen S (2016) The effects of brazilian propolis on etiological agents of mastitis and the viability of bovine mammary gland explants[J]. *Journal of Dairy Science*, 99(3):2308-2318.

Kudo D, Inden M, Sekine S, Tamaoki N, Iida K, Naito E, Watanabe K, Kamishina H. Shibata T. HozumiI (2015) Conditioned medium of dental pulp cells stimulated by Chinese propolis show neuroprotection and neurite extension *in vitro*[J]. *Neuroscience*

Letters, 589: 92-97.

Kumari S, Naik P, Vishma BL, Salian SR, Devkar RA, Khan S, Mutalik S, Kalthur G, Adiga SK (2016) Mitigating effect of indian propolis against mitomycin C induced bone marrow toxicity[J]. *Cytotechnology*, 68(5): 1789-1800.

Kurauchi Y, Hisatsune A, Isohama Y, Mishima S, Katsuki H (2012) Caffeic acid phenethyl ester protects nigral dopaminergic neurons via dual mechanisms involving haem oxygenase-1 and brain-derived neurotrophic factor[J]. *British Journal of Pharmacology*, 166(3): 1151-1168.

Miyazaki Y, Sugimoto Y, Fujita A, Kanouchi H (2015) Ethanol extract of Brazilian propolis ameliorates cognitive dysfunction and suppressed protein aggregations caused by hyperhomocysteinemia[J]. *Bioscience, Biotechnology, and Biochemistry*, 79(11): 1884-1889.

Olczyk P, Wisowski G, Komosinska-Vassev K, Stojko J, Klimek K, OlczykM, Kozma EM (2013) Propolis modifies collagen types Ⅰ and Ⅲ accumulation in the matrix of burnt tissue[J]. *Evidence-Based Complementary and Alternative Medicine*, 2013: 423809.

SehnE, Hernandes L, Franco SL, Goncalves CC, Baesso ML (2009) Dynamics of reepithelialisation and penetration rate of a bee propolis formulation during cutaneous wounds healing[J]. *Analytica Chimica Acta*, 635(1): 115-120.

Seven PT (2008) The effects of dietary Turkish propolis and Vitamin C on performance, digestibility, egg production and egg quality in laying hens under different environmental temperatures[J]. *Asian-Australasian Journal of Animal Sciences*, 21(8): 1164-1170.

Shimazawa M, Chikamatsu S, Morimoto N, Mishima S, Nagai H, Hara H (2005) Neuroprotection by Brazilian green propolis against *in vitro* and *in vivo* ischemic neuronal damage[J]. *Evidence-Based Complementary and Alternative Medicine*, 2(2): 201-207.

Ulusoy HB, Öztürk İ, Sönmez MF (2016) Protective effect of propolis on methotrexate-induced kidney injury in the rat[J]. *Renal Failure*, 38(5): 744-750.

Wang K, Jin XL, Shen XG, Sun LP, Wu LM, Wei JQ, Marcucci MC, Hu FL, Liu JX (2016) Effects of chinese propolis in protecting bovine mammary epithelial cells against mastitis pathogens-induced cell damage[J]. *Mediators of Inflammation*, 2016(3): Article ID 8028291.

Wei X, Zhao L, Ma ZZ, Holtzman DM, Yan C, Dodel RC, Hampel H, Oertel W, Farlow MR, Du YS (2004) Caffeic acid phenethyl ester prevents neonatal hypoxic-ischaemic brain injury[J]. *Brain*, 127(12): 2629-2635.

Wei X, Ma Z, Fontanilla CV, Zhao L, Xu ZC, Taggliabraci V, Johnstone BH, Dodel RC, Farlow MR, Du Y (2008) Caffeic acid phenethyl ester prevents cerebellar granule neurons (CGNs) against glutamate-induced neurotoxicity[J]. *Neuroscience*, 155(4): 1098-1105.

Yalcin CO, Aliyazicioglu Y, Demir S, Turan I, Bahat Z, Misir S, Deger O (2016) Evaluation of the radioprotective effect of turkish propolis on foreskin fibroblast cells[J]. *Journal*

　　of Cancer Research and Therapeutics,12(2):990-994.

Zhang JL,Cao XP,Ping S,Wang K,Shi JH,Zhang CP,Zheng HQ,Hu FL（2015）Comparisons of ethanol extracts of Chinese propolis (Poplar Type) and poplar gums based on the antioxidant activities and molecular mechanism[J]. *Evidence-Based Complementary and Alternative Medicine*,2015:307594.

陈佳亿,李四元,陈清华,赵蕾,欧阳谦,刘飞燕,高飒（2016）蜂胶残渣对28～56日龄黄羽肉鸡生长性能、屠宰性能及肌肉品质的影响[J].动物营养学报,28(5):1541-1548.

陈娟（2008）水溶性蜂胶减轻小鼠学习记忆损伤及侧脑室注射人的Opiorphin的镇痛作用和机制[D].兰州:兰州大学.

陈炼,廖旺娣,李国华,谢勇,吕农华,王崇文（2009）江西蜂胶对幽门螺杆菌的体外抑菌作用研究[J].江西医学院学报,49(12):27-32.

曹炜,尉亚辉,杨建雄,陈卫军（2004）蜂胶对增强小鼠力竭游泳耐力的影响[J].食品科学,25(11):273-276.

郭冬生,丁松林（2011）基础日粮添加蜂胶对母猪生产性能的影响[J].畜牧与饲料科学 32(12):3-4.

胡伏莲（2005）消化性溃疡发病机制的现代理念[J].中华消化杂志,25(3):189-190.

金春玉,朴世浩,张善玉,赵华（1996）蜂胶石油醚与正丁醇萃取物对大鼠实验性胃溃疡作用的比较观察[J].延边医学院学报,19(3):140-145.

金河奎,朴世浩,张善玉,张轶伦（1996）蜂胶对大鼠实验性胃溃疡的作用[J].延边医学院学报,19(1):14-17.

黎相广,张军民,詹曾王（2010）蜂胶黄酮对黄羽肉种鸡生产性能的影响[J].中国饲料,(13):12-15.

刘莹（2013）槲皮素对蛋鸡生产性能和蛋品质的影响[D].长春:东北农业大学.

彭和禄,李树荣,晋跃东,李世宗,王尚明（1990）蜂胶对肥育猪增重效果的研究[J]养蜂科技,(4):4-5.

彭和袜,李树荣,易嘉宾（1993）蜂胶残渣制剂对育肥猪增重效果的观察[J]养蜂科技,(2):6.

彭和禄,李树荣,易嘉宾,葛长荣（1993）蜂胶、蜂花粉对猪肉质影响的研究[J]养蜂科技,(4):9-10.

潘燕,彭彦铭（2010）蜂胶黄酮对疲劳小鼠心肌自由基代谢及ATP酶活性的影响[J].山东体育科技,32(4):30-32.

覃燕飞,毕丹东（2010）蜂胶总黄酮对大鼠缺血再灌注后大脑皮质神经细胞凋亡和caspase-3与bcl-2及bax表达的影响[J].中华临床医师杂志,4(7):1023-1027.

王浩,张庆乐,许利军,陈美华,高允生,朱玉云（2008）蜂胶醇提物对小鼠学习记忆的促进作用[J].中国蜂业,59(8):5-6,9.

王留,张代（2016）蜂胶提取物对蛋雏鸡免疫器官指数和腹腔巨噬细胞活性的影响[J].饲料与畜牧,(7):42-43.

王南舟,贾冬梅,黄宗锈,杨翔（1999）蜂胶复合制剂对心脏活动的影响[J].蜜蜂杂志,(12):3-4.

徐传球（2009）蜂胶治痔疮[J]. 蜜蜂杂志,(1):10.

许合金,张军民,王修启,赵青余（2010）蜂胶对蛋鸡生产性能、蛋品质和血液生化特性的影响[J]. 中国兽医学报,30(5):704-708.

于洋,韩晓燕,潘燕（2012）蜂胶对运动训练小鼠抗疲劳能力影响的研究[J]. 沈阳体育学院学报,31(2):86-89.

赵家明（2006）蜂胶与脱发[J]. 蜜蜂杂志,(4):29-30.

詹勋,曾佩玲,王修启,黎相广,张军民（2010）蜂胶黄酮对黄羽肉种鸡生产性能的影响[J]. 中国饲料,(13):12-15.

张陵,万宁（2006）氧自由基脂质过氧化反应所致运动性疲劳产生机制研究进展[J]. 中国实验诊断学,10(9):1104-1108.

赵青余,许张王（2010）蜂胶对蛋鸡生产性能、蛋品质和血液生化特性的影响[J]. 中国兽医学报,30(5):704-708.

朱玉云,王张,许陈高（2008）蜂胶醇提物对小鼠学习记忆的促进作用[J]. 中国蜂业,59(8):5-9.

第四章 蜂胶的质量控制与标准化

第一节 蜂胶化学成分的多样性

蜂胶的化学成分及药理活性的现代研究始于20世纪50年代,当时人们认为不同植物来源采集的蜂胶其化学成分可能不同,但是大部分蜂胶是由50%树脂(由类黄酮和相关的酚酸以及多酚类组成)、30%蜂蜡、10%挥发油、5%花粉和5%其他有机化合物组成(Marcucci,1995),这一观点至今仍被大多数人沿用。但是,最近10余年来,植物化学家和药理学家联合开展了不同地理来源蜂胶化学成分的大量研究,认为蜂胶化学成分十分复杂,是多成分、多靶点的复杂体系。"蜂胶"这个词并不特指哪类化学成分,蜂胶表现出的药理活性是多种成分作用于不同靶点的整合结果。这是蜂胶的优势,但同时也给蜂胶的质量控制与标准化造成极大的困难。

1 蜂胶化学成分的影响因素

蜂胶植物来源的多样性直接导致蜂胶化学组成的多样性和复杂性,除此之外,蜂胶的成分随着季节、光照、海拔、采集蜂种类以及蜂胶开采时植物源的可利用性不同而变化。蜂胶化学成分的多样性是其商品化和标准化的严重制约因素(Bankova et al.,2000)。

1.1 植物来源对蜂胶化学成分的影响

蜂胶的化学成分首先取决于其胶源植物。杨属型蜂胶的特征性成分是B环无取代基的类黄酮以及苯丙酸及其酯类,如松属素、短叶松素、高良姜素、柯因和咖啡酸苯乙酯(CAPE)(Bankova et al.,1992);酒神菊属型蜂胶主要成分为异戊烯苯丙类,如阿替匹林C和咖啡酰奎宁酸(dos santos pereira et al.,2003);委内瑞拉(Tomas-Barberan et al.,1993)、古巴(Hernandez et al.,2005)、巴西亚马逊(de Castro et al.,2011)和巴西东北部(Trusheva et al.,2006)地区的克鲁西属型蜂胶的特征性成分是异戊烯基苯甲酮;台湾岛(Huang et al.,2007)和冲绳岛(Kumazawa et al.,2008)地区的血桐属型蜂胶的特征性成分是香叶基黄酮类。从埃及(El-Bassuony et al.,2009)和肯尼亚(Petrova et al.,2010)的蜂胶中也鉴定出类似的成分,这是因为从非洲西部到南太平洋岛,有大约280种血桐属植物(Matsubayamshi et al.,2006);地中海东部地区,如希腊、克里特岛和土耳其地区的蜂胶中主要含有二萜(如异柏油酸)或蒽醌类(如大黄酚)成分(Kalogeropoulos et al.,2009)。此外,俄罗斯北部地区桦树型蜂胶的主要成分也是黄酮类,但与杨属型蜂胶的黄酮类组成不同(Popravko and Sokolov,1980;Bankova et al.,2000)。在巴西,除了来自酒神菊属的绿蜂胶外,研究较多的还有红蜂胶,其主要活性成分为异黄酮类化合物(Daugsch et al.,2008;Alencar et al.,

2007)。de Castro Ishida 等(2011)在巴西的马瑙斯(Manaus)地区发现了一种新的蜂胶类型,其主要成分为:polyprenylated benzophenones、7-epi-nemorosone、7-epi-clusianone、xanthochymol 和 gambogenone。

　　植物来源的复杂性决定了蜂胶化学成分的多样性。一个地区的蜂胶中可能含有多种植物源的成分,如希腊及附近地区的蜂胶中除含有杨属植物典型成分外,还有醌类化合物和其他不明植物来源的物质(Kalogeropoulos et al.,2009);巴西红蜂胶中除了含有一种黄檀植物(*Dalbergia ecastophyllum*)的典型成分外,还含有藤黄科的典型成分异戊二烯基苯甲酮等成分(Trusheva et al.,2007)。同时,不同地区的蜂胶植物来源又可能是相同的,如古巴红蜂胶和巴西红蜂胶中均含有异甘草素、甘草黄素、柚皮素、异黄酮、异黄烷和紫檀素,这些成分均来自黄檀植物的分泌物。但是,只在巴西红蜂胶中检测到异戊烯多异苯甲酮类(guttiferone E/xanthochymol)和成分奥氏叶黄素 A (oblongifolin A)。*Dalbergia ecastophyllum* 分泌物和红蜂胶呈现红色的色素是两个 C30 异构体,即新的 retusapurpurin B 和 retusapurpurin A。这说明热带地区不同的红蜂胶有相似的化学组成。*Dalbergia ecastophyllum* 是红蜂胶的主要植物来源,但是在巴西红蜂胶中含有其特有成分多异戊烯基二苯甲酮类(polyisoprenylated benzophenones),说明巴西红蜂胶还有其他的植物来源(Piccinelli et al.,2011)。

　　蜂胶中同种类型的成分可能来源于不同的植物,如希腊蜂胶中的二萜类可能来自柏科植物(Popova et al.,2010),巴西南部蜂胶中的二萜类可能来自松科植物(Bankova et al.,1996),而巴西中部和东南部蜂胶中的二萜类很可能来自酒神菊属植物(Missima et al.,2007);从巴西红蜂胶中发现的异黄酮和新类黄酮聚合物来源于黄檀植物 *Dalbergia ecastophyllum* 分泌的树脂(Daugsch et al.,2008);而尼泊尔蜂胶中的这两种成分可能来源于另一种黄檀类植物 *Dalbergia sisoo*(Shrestha et al.,2007)。

1.2　蜂种对蜂胶化学成分的影响

　　Silici 和 Kutluca(2005)比较了土耳其同一个蜂场饲养的高加索蜂(*Apis mellifera caucasica*)、卡尼鄂拉蜂(*A. m. carnica*)和安纳托利亚蜂(*A. m. anatoliaca*)3 种西方蜜蜂亚种采集的蜂胶的化学成分组成,在 3 种蜂胶样本中共检测到 48 种成分,其中共有成分 16 种。此外,卡尼鄂拉蜂和安纳托利亚蜂采集的蜂胶还有 6 种共有成分,高加索蜂和安纳托利亚蜂有 3 种共有成分,高加索蜂和卡尼鄂拉蜂有 1 种共有成分,造成 3 种蜂胶成分差异的主要原因可能是不同蜂种对植物树脂具有不同的采集偏好性。

1.3　季节对蜂胶化学成分的影响

　　季节对蜂胶化学组成也有较大影响。季节对中国蜂胶醇提物(EEP)的化学组成有显著影响,中国 EEP 中总酚、总黄酮、黄酮-黄酮醇和黄烷酮的含量均在 7—8 月为最高,而在 9 月份最低(Guo et al.,2011)。不同季节采集的巴西绿蜂胶中酚类化合物和萜类化合物的比例也会发生变化(Bankova et al.,1998;Teixeira et al.,2010)。通常巴西绿蜂胶以酚类化合物为主要成分,但在极端情况下三萜类化合物是巴西绿蜂胶的主要成分(Teixeira et al.,2006)。Valencia 等(2012)研究了季节对蜂胶化学组成的影响。他们从墨西哥索诺兰沙漠收集了不同季节蜜蜂采集的蜂胶样本,分析了这些样本间化学成分及生物学活性的差异,发现不同季节采集得到的样本基本化学成分类似,但春季采集的索诺兰蜂胶样本抗恶性肿瘤

细胞增生的效果最强,同时,这些蜂胶样本清除自由基能力都不是很强。这也说明蜂胶采集季节可能影响了蜂胶抗肿瘤细胞增生的效果。Isla 等(2012)对不同季节采集的阿根廷蜂胶也进行了类似的研究,结果发现,采自春夏季的蜂胶样本的总黄酮、总酚酸含量要高于其他季节采集的蜂胶样本。另外,夏季和秋季收集的蜂胶样本抗革兰阳性菌活力较强。

季节对蜂胶化学成分的影响可能是由分泌树脂的主要植物的物候变化引起的,也可能是由可提供的主要树脂减少的情况下,蜜蜂开始采集替代胶源的树脂造成的。

1.4 采集方式对蜂胶化学成分的影响

Stan 等(2011)对使用蜂胶采集器收集的 12 个蜂胶样本与蜂箱内直接采集的 12 个蜂胶样本进行了对比研究,发现采集方法对蜂胶中蜡含量和多酚类化合物的含量影响很大。采胶器收集的蜂胶蜂蜡含量低,多酚含量更高。

不同取胶方式会对蜂胶品质产生影响。王勇等(2014)采用塑料取胶器、尼龙纱网、覆布以及从框梁缝隙刮取等 4 种取胶方式采集蜂胶,分析不同取胶方式所得蜂胶样品的乙醇提取物含量、8 种黄酮含量和总黄酮含量。结果表明,取胶方式对蜂胶品质的影响很大:尼龙纱网方式收集所得蜂胶的品质最好,3 个品质指标均极显著地优于其他取胶方式;塑料取胶器和从框梁缝隙处刮取的蜂胶品质次之,两种取胶方式间无显著差异,但都显著地优于覆布取胶;采用覆布方式收集所得蜂胶的品质最差。

1.5 蜂胶与植物源相关性的研究

蜂胶中新成分的不断发现带动了人们对胶源植物化学成分的研究。在植物化学研究中,植物不同部位成分的分离鉴定通常要用大量的植物材料,由于植物树脂和顶芽的收集难度较大,因而相关研究很少。而蜜蜂能察觉到植物表面的化学信息(Simone-Finstrom and Spivak,2010),自然就可以收集到大量的树脂材料,从中分离出单体物质,便于发现新的化学成分。同时,对蜂胶生物学活性的研究促进了对胶源植物化学成分的研究。因此,有关杨属植物、酒神菊属植物以及血桐属植物的化学成分、药理活性及主要活性成分的研究报道日益增多。

酚苷是杨柳科植物的特征性成分(Thieme,1967)。1830 年,Braconnot 从欧洲山杨(*Populus tremula*)的树皮和叶中分离得到水杨苷,之后又从杨树的皮、叶、芽等部位分离出柳匍匐苷、特里杨苷、柳皮苷、白杨苷、云杉苷(对羟基苯乙酮-D-葡糖苷)、杨属灵、2′-苯甲酰水杨苷、毛果杨苷、大齿杨苷、去羟基大齿杨苷等几十种酚苷(王欣等,1999,2000;Kwon et al.,2009;Si et al.,2011a,b)。在杨属植物中还发现了许多黄酮苷类化合物。Pearl 和 Darling(1971)从 *P. deltoides bartr* 叶中分离出白杨素-7-O-葡萄糖苷。Crwaford(1994)从 *P. angustifolia* 中鉴定出槲皮素-3-O-葡萄糖苷、槲皮素-3,7-O-二葡萄糖苷、山奈酚-3-O-葡萄糖苷、杨梅素-3-O-葡萄糖苷、木樨草素-7-O-葡萄糖苷、木樨草素-7-O-葡萄二糖苷、芹菜素-7-O-二葡萄糖苷。林茂和李守珍(1993)从毛白杨叶中分离出芹菜素-7-O-(6″-O-p-羟基肉桂酰)-葡萄糖苷。王想想等(2007)从加拿大杨树芽中分离出高良姜素-7-O-葡萄糖苷、白杨素-7-O-葡萄糖苷。Neacsu 等(2007)从欧洲白杨中分离鉴定出二氢山奈酚 7-β-葡萄糖苷、柚皮素 7-O-β-葡萄糖苷。此外,还从杨属中发现了槲皮素-3-半乳糖苷、槲皮素-3-芸香糖苷、槲皮素-3-葡萄糖苷酸、山奈酚-3-鼠李糖二葡萄糖苷、野樱黄苷、新野樱苷等(王欣等,1999)。

有关蜂胶化学成分的研究很多,迄今为止尚未从蜂胶中发现过任何酚苷,而很多研究曾报道过杨树型蜂胶中含有酚苷的降解产物,如水杨酸及乙酰水杨酸等衍生物(Marcucci,1995)。蜂胶中已发现黄酮类化合物 139 种,发现的黄酮类化合物也多为苷元。杨属植物中发现的酚苷均以 β-葡萄糖苷形式连接,β-葡萄糖苷酶可催化水解结合于末端非还原性的 β-D-糖苷键,同时释放出配基与葡萄糖体。Zhang 等(2011a)在新鲜蜂胶的酶提取液中检测到 β-葡萄糖苷酶活力,并证实该酶能够水解芳香基 β-D-葡萄糖苷,与 Pontoh 和 Low(2002)从西方蜜蜂体内分离纯化的 β-葡萄糖苷酶性质相符。进一步研究发现,蜂胶中 β-葡萄糖苷酶提取液只能水解类黄酮单葡萄糖苷,对类黄酮多糖苷没有水解作用(Zhang et al.,2011b),这与蜂胶中已鉴定出的成分中除鼠李糖苷(Popova et al.,2010)、槲皮素-3-氧芸香糖苷(Bonvehi et al.,1994)、异鼠李素-3-氧芸香糖苷(Popova et al.,2010)和黄酮碳苷(Righi et al.,2011)等 4 种黄酮苷外,其余均为黄酮苷元的事实相符(张翠平和胡福良,2009)。这说明蜜蜂在采集和加工蜂胶过程中加入的腺体分泌物改变了原有树脂中的某些成分,为蜂胶及植物源化学成分、药理活性以及质量控制的深入研究提供了理论依据。

2 蜂胶中的生物学活性成分

围绕蜂胶生物学活性的相关研究一直都是国内外蜂胶研究者关注的热点,因为蜂胶的生物学活性是蜂胶及蜂胶产品开发以及在医药、保健食品、化妆品等领域应用的基础。蜂胶植物来源广泛,胶源植物不同导致蜂胶化学组成差异很大。10 余年来,几乎所有关于蜂胶生物学活性的研究都包括了所用蜂胶的化学特征,但遗憾的是,在任何蜂胶类型中都尚未建立起一致可接受的用于评价其生物学活性的化学质量标准。

2.1 不同类型蜂胶生物学活性比较

不同来源的蜂胶化学成分的差异使得人们期望不同类型蜂胶的生物学活性不同。然而,事实是所有研究过的各种类型的蜂胶都表现出抗细菌、抗真菌、抗病毒、抗氧化、抗炎、护肝和抗肿瘤等相似的生物学活性(Banskota et al.,2001)。Kujumgiev 等(1999)比较了保加利亚、阿尔巴尼亚、蒙古、埃及、巴西和加那利群岛蜂胶的抗细菌、抗真菌和抗病毒作用及其化学组成。结果表明,尽管不同地理来源蜂胶化学组成差异很大,但所有的蜂胶都表现出明显的抗细菌和抗真菌作用,而且大部分具有抗病毒作用。因此,他们认为各种化合物的协同作用是蜂胶生物学活性的根本。

不同来源的蜂胶表现出的活性强弱有异。Banksota 等(2000)比较了巴西、秘鲁、新西兰和中国蜂胶的细胞毒性、肝保护作用和自由基清除活性,结果发现,新西兰和中国蜂胶具有最强的细胞毒性,而几乎所有蜂胶均具有肝保护作用,所有蜂胶对 DPPH 自由基清除作用类似,只有秘鲁蜂胶样本的活性较弱。Hegazi 等(2000)研究发现,德国蜂胶抗白色葡萄球菌和大肠杆菌活性很强,而澳大利亚蜂胶对白色念珠菌活性更高。Salomao 等(2004)研究发现,巴西和保加利亚蜂胶对锥虫(*Trypanozoma cruzi*)和某些病原真菌具有活性,且保加利亚蜂胶对细菌的活性强于巴西蜂胶。Banskota 等(2000)研究发现,乌拉圭蜂胶的乙醇提取物是乳腺癌的一种抑制剂,而巴西蜂胶的 WEP 和 EEP 对乳腺癌的作用不明显,荷兰蜂胶和中国蜂胶抑制鼠结肠 26-L5 癌活性最强,而巴西蜂胶则没有作用。巴西、秘鲁、荷兰和中国蜂胶的甲醇提取液都比相应的乙醇提取液对鼠结肠 26-L5 癌和人类 HT-1080 纤维肉瘤的细胞毒性强。

中国蜂胶 EEP 对肥大细胞脱粒抑制力最强,得到抗过敏成分为白杨素、山奈酚及其衍

生物。白杨素具有抑制从抗原激活的 RBL-2H3 细胞的 IL-4 和 MCP-1 的生成作用 (Nakamura et al.,2010)。巴西绿蜂胶醇提物对患有过敏性鼻炎的病人外围白细胞 Cry j1 引起的半胱氨酰白三烯和组胺释放具有抑制作用;Artepillin C、baccharin 和山奈素是最重要的抗过敏活性成分,其中 Artepillin C 苯乙酯的抑制活性最强(Tani et al.,2010)。因此,中国蜂胶和巴西蜂胶的抗敏成分和抗敏机制均有所不同。

2.2 不同类型蜂胶生物学活性成分比较

不同来源的蜂胶都具有广泛的生物学活性,但来源不同的蜂胶所起作用的活性成分不同,因此,研究者一直致力于寻找特定蜂胶中对某种活性起关键作用的活性成分,以实现对蜂胶的质量控制。

抗氧化活性是蜂胶最主要的活性之一,蜂胶中含有丰富有效的抗氧化活性成分。黄酮类化合物和酚酸类衍生物是温带地区杨属型蜂胶中最丰富、最有效的抗氧化剂。Kumazawa 等(2004)比较了产自阿根廷、澳大利亚、巴西、保加利亚、智利、中国、匈牙利、新西兰、南非、泰国、乌克兰、乌拉圭、美国和乌兹别克斯坦的蜂胶的抗氧化活性和化学组成,结果发现,具有强抗氧化活性的蜂胶含有较高浓度的山奈酚和苯乙醇;阿根廷、澳大利亚、中国以及匈牙利的蜂胶中抗氧化活性比较强,其中主要的抗氧化组分为咖啡酸、槲皮素、山奈酚、CAPE 和咖啡酸肉桂酯;而南非和乌兹别克斯坦蜂胶的抗氧化活性很弱,泰国蜂胶几乎没有抗氧化活性。采用同样的方法,Ahn 等(2004)对日本和韩国不同地区的蜂胶抗氧化活性和化学组成进行了分析,结果发现蜂胶强抗氧化活性与高浓度的咖啡酸和 CAPE 相关。Yamauchi 等(1992)研究了中国、日本、巴西、美国蜂胶对甲基亚油酸自动氧化的抑制作用,发现咖啡酸酯苯甲基是中国蜂胶中抗氧化的活性成分之一。中国和韩国蜂胶中总酚类和类黄酮含量明显高于巴西蜂胶,抗氧化活性也高于巴西蜂胶(Kumazawa et al.,2004;Choi et al.,2006)。Ahn 等(2007)研究发现,中国蜂胶的抗氧化活性成分为咖啡酸、阿魏酸及 CAPE。日本秋田蜂胶中含有大量的抗氧化活性物质,如咖啡酸和 CAPE,其抗氧化活性很强(Hamasaka et al.,2004)。Chen 等(2004)比较了台湾不同地区蜂胶的自由基清除活性、细胞毒性和对人类黑色素瘤细胞凋亡诱导作用,结果表明,高浓度的 Propolins 具有凋亡诱导人类黑色素瘤细胞和抗自由基作用。

类黄酮及酚酸酯是温带地区蜂胶的主要抗菌活性成分(Burdock,1998)。Hegazi 等(2000)研究了欧洲蜂胶的化学成分及其抗菌活性,发现德国蜂胶以 CAPE 苯甲基阿魏酸和高良姜素为主,对金黄色葡萄球菌和大肠杆菌表现出高抗菌活性;奥地利蜂胶主要含有松属素和香豆酸,抗白色念珠菌效果好;咖啡酸苯甲基酯和松属素是法国蜂胶的主要成分,对各种细菌都有抑制作用,但效果不如德国和奥地利蜂胶好。保加利亚蜂胶比巴西蜂胶抗菌活性强(Salomao et al.,2004)。Salomao 等(2008)研究发现,巴西蜂胶抑菌活性与 p-香豆酸及其衍生物,特别是异戊烯化衍生物和咖啡酰奎宁酸类存在有关。

蜂胶对鼻咽癌、子宫颈癌、乳腺癌、肾癌、大肠癌等人癌细胞株均具有强烈的毒杀作用。已有多种抑制肿瘤细胞生长的有效成分被鉴定出来,主要是咖啡酸及肉桂酸类的衍生物和黄酮类化合物等(韩利文等,2007)。台湾蜂胶能引起人类黑色素瘤细胞凋亡,这与其富含 Propolins A、B、C、D、E、F 有关(Chen et al.,2004)。巴西绿蜂胶树脂的主要成分是肉桂酸,主要成分有异戊烯化的色原烷和苯丙素衍生物而不是类黄酮,最典型的成分是 Artepillin C,具有抗肿瘤和诱导凋亡活性(Pisco et al.,2006)。Sun 等(2012)对中国不同地理来源的 9

个蜂胶样本进行分析,从中分离出具有抗人 MDA-MB-231 乳房癌细胞的活性成分白杨素。Chen 等(2004)从台湾蜂胶中分离出 2 个新的细胞毒性异戊二烯基黄酮类化合物,这两种化合物对 3 个癌细胞系均表现出细胞毒性,也有潜在的自由基(DPPH)清除活性。Banskota 等(2000)从新西兰蜂胶中分离出具有抗癌细胞系增殖的活性成分:CAPE 和几个类似物也被 Nagaoka 等(2003)在新西兰蜂胶中发现,具有 NO 抑制活性。

2.3 蜂胶中化学成分与其生物学活性

近年来,随着蜂胶研究的深入,许多新类型的化学成分及药理活性得以研究。Ahn 等(2007)研究了蜂胶中的金合欢素、芹菜素、阿替匹林 C、CAPE、白杨素、p-香豆酸、高良姜素、山柰酚、松属素和槲皮素的抗血管生成和抗氧化活性,结果发现,CAPE 和槲皮素具有很强的抑制血管生成和内皮细胞增殖作用,同时还具有很强的抗氧化活性。阿替匹林 C、高良姜素、山柰酚的抗血管生成作用也很强,抗氧化活性略弱;相反,金合欢素、芹菜素、松属素具有相当强的抗血管生成活性,但抗氧化活性很低。巴西酒神菊属蜂胶中肉桂酸衍生物酒神菊素和 drupanin 能显著抑制 HIF-1α 蛋白和 HIF-1 的下游靶基因,如己糖激酶 2、葡萄糖转运蛋白 1 和血管内皮生长因子 A,对雏鸡的表达尿囊膜具有显著的抗血管生成作用。另一方面,黄酮类化合物 beturetol 和异樱花素能诱导 HIF-1 依赖的荧光素酶的活性和缺氧条件下 HIF-1 靶基因的表达,调节 HIF-1 的化合物为酒神菊素、beturetol、山柰素、异樱花素和 drupanin(Hattori et al.,2011)。

尼泊尔蜂胶中的新类黄酮具有抑制 NO 生成作用(Awale et al.,2005);台湾蜂胶中的 Propolins G 具有细胞凋亡蛋白酶依赖型凋亡作用(Huang et al.,2007);巴西东北部蜂胶中的异黄酮类具有细胞毒性(Awale et al.,2008);缅甸蜂胶中的环菠萝烷型三萜具有细胞毒性(Li et al.,2009a,b);埃及蜂胶中的异山梨醇 C 具有抗菌活性(El-Bassuony,2009,2010);墨西哥蜂胶中的 8-苯烯丙基高良姜素衍生物具有细胞毒性(Li et al.,2010);肯尼亚蜂胶中的香叶基二苯乙烯具有抑菌活性(Petrova et al.,2010)。Usia 等(2002)从中国蜂胶中分离出许多具有抗增殖活性的化合物,大部分是杨属型蜂胶中的已知成分,但也有 2 种新的黄酮类化合物:2-甲基丁酰基短叶松素和 6-肉桂基白杨素;古巴红蜂胶的主要成分 prenylated benzophenone nemorosone 对多个肿瘤细胞系均具有细胞毒性,也有自由基清除作用(Cuesta-Rubio et al.,2002)。

3 蜂胶中新成分的研究

蜂胶中的化学成分是其发挥生理药理作用的物质基础,对其进行深入的研究是蜂胶质量控制的关键,是揭示蜂胶作用机制的核心。随着蜂胶基础研究的深入开展,寻找"有效成分""主要成分"和"特征成分"并以这三种成分作为控制指标,才能切实起到质量控制的目的。

蜂胶研究仍是天然产物研究中的新兴领域。随着对蜂胶的类型、组成、植物来源和生物学活性的深入系统研究,将会分离鉴定出更多新的化合物,确定更多的蜂胶类型及胶源植物。

3.1 蜂胶中的典型特征性新成分

从 20 世纪 90 年代开始,蜂胶化学成分和药理活性的研究受到关注,新化合物不断被鉴定出来,其中很多化合物具有蜂胶典型特征性的化学结构。

杨树型蜂胶鉴定出的新化合物均为 B 环无取代基的类黄酮,例如,短叶松素 3-O 乙酸酯;墨西哥蜂胶中的白杨素和高良姜素苯烯丙基衍生物(Li et al.,2010);中国蜂胶中的 6-内桂基白杨素(Usia et al.,2002)等。

巴西绿蜂胶中鉴定出的新化合物有类似阿替匹林 C 的结构,如阿替匹林 C 的异戊烯基烯键水合物(Nafady et al.,2003)、阿替匹林 C 酯及阿替匹林 C 的杂环酚类似物(Tazawa et al.,1999)、阿替匹林 C 碳骨架多色烷(Banskota et al.,1998)等。

克鲁西属型蜂胶的特征性成分为异戊二烯基苯甲酮,如委内瑞拉蜂胶中的聚异丙烯基化二苯酮 nemorosone 和 18-ethoxy-17-hydroxy-17,18-dihydroxyscrobiculatone(Trusheva et al.,2004),古巴蜂胶中的 propolone D(Hernandez et al.,2005)。这些化合物都含有 C6C1 莽草酸和 C6 乙酸基团及几个甲羟戊酸衍生的异戊二烯单位。

血桐属型蜂胶是主要含有香叶基残基的类黄酮(Kumazawa et al.,2008)。研究者已经从台湾岛、冲绳岛、埃及和肯尼亚蜂胶中分离鉴定出许多此类成分(Petrova et al.,2010;Kumazawa et al.,2004;El-Bassuony et al.,2009)。

地中海东部地区,如希腊、克里特岛和土耳其地区的蜂胶,主要含有二萜类(如异柏油酸)或蒽醌类(如大黄酚)成分(Silici et al.,2005;Kalogeropoulos et al.,2009)。

3.2　蜂胶中新的特征性成分

产自巴西东北部的巴西红蜂胶中也被鉴定出多种新类黄酮和异黄酮类化合物(Awale et al.,2008)。在巴西亚马逊发现一种新类型的蜂胶(De castro Ishida et al.,2011),主要含有异戊二烯基苯甲酮。在巴西最南端还发现一种新的蜂胶类型,含有 1 个新的 Melliferone 衍生物、几个其他的三萜和苯丙素(Ito et al.,2001)。Athikomkulchai 等(2013)研究了泰国蜂胶的主要成分,并从泰国蜂胶甲醇提取物中发现了 2 种新苯丙烯黄烷酮类化合物:(7″S)-8-[1-(4′-hydroxy-3′-methoxyphenyl) prop-2-en-l-yl]-(2S)-pinocembrin 和 (E)-cinnarnyl-(E)-cinnamylidenate。Shimomura 等(2013)研究了韩国济州岛蜂胶的化学成分,并从中分离鉴定出 8 种新的查耳酮类化合物,同时也鉴定得到 19 种其他已知化合物,研究结果提示济州岛蜂胶具有独特化学组成。Popova 等(2013)基于阿曼蜂胶的化学成分鉴定出几种标志性成分,并发现阿曼蜂胶的主要胶源植物为尼姆树(*Azadiracta indicadica*)和金合欢属(*Acacia*)植物(主要是 *A. nilotica*),这为阿曼蜂胶的质量控制提供了基本依据。

日本学者研究发现,在冲绳岛和冈山采集的蜂胶与欧洲和巴西等地采集的蜂胶不同,其中含有异戊二烯基类黄酮,发现 *M. tanarius*(Kumazawa et al.,2008)和 *Rhus javanica var. chinensis*(Murase,2008)分别是这两个地区的植物胶源。Abu-Mellal 等(2012)利用从澳大利亚坎加鲁岛上收集获得的蜂胶乙酸乙酯提取物,从中分离鉴定出一种异戊烯基肉桂酸衍生物和 6 种异戊烯基羟基芪类化合物;同时从蜂胶提取物中鉴定出 8 种已知物质,并据此推断,坎加鲁岛蜂胶因具有异戊烯基羟基芪类化合物而可以被认为是一种独特类型的蜂胶。

3.3　蜂胶中非特征性新成分

蜂胶化学成分复杂,不同来源蜂胶化学组成差异很大。除了特征性成分,许多新类型的化合物也不断被鉴定出来。Lotti 等(2010)从墨西哥红蜂胶中鉴定出 3 种新成分:1-(30,40-dihydroxy-20-ethoxyphenyl)-3-(phenyl) propane、(Z)-1-(20-methoxy-40,50-dihydroxyphenyl)-2-

(3-phenyl) propene 和 3-羟基-5,6-二甲氧基黄烷。此外,1,3-二芳基丙烷和 1,3-二芳基丙烯碳骨架也在蜂胶中首次被发现。Li 等(2010)从墨西哥蜂胶中鉴定出 3 种新的类黄酮:(2R,3R)-3,5-二羟基-7-甲氧基黄烷酮 3-(2-甲基)丁酸盐、(7″R)-8-[1-(4′-羟基-3′-甲氧基苯基)丙烯基-2-反式-1-烷基]白杨素、(7″R)-8-[1-(4′-羟基-3′-甲氧基苯基) 丙烯基-2-反式-1-烷基]高良姜素。Trusheva 等(2010)从伊朗蜂胶中分离出 5 种新成分:prenylated coumarin suberosin 和 4 个萜酯 tschimgin (bornyl p-hydroxybenzoate)、tschimganin(bornyl vanillate)、ferutinin(ferutinol p-hydroxybenzoate)和 tefernin(ferutinol vanillate)。Popova 等(2010)采用标准品对照结合质谱解析,从希腊蜂胶中分离出 30 多种二萜类化合物;从肯尼亚蜂胶中分离出 2 种新的芳香基环烷类木脂素:四氢化爵床脂素 B 和 6-甲氧基山荷叶素,以及 4 种多酚 phyllamyricin C、macarangin、schweinfurthin A 和 schweinfurthin B。

Righi 等(2011)在巴西红蜂胶中发现一种碳苷,即 narigenin-8-C-hexoside,这是在蜂胶中发现的第一个碳苷。De Castro Ishida 等(2011)从巴西玛瑙斯市蜂胶中鉴定出 4 种新的多异戊二苯甲酮类化合物:7-epi-nemorosone、7-epi-clusianone、xanthochymol 和 gambogenone。Trusheva 等(2011)从印度尼西亚蜂胶中分离鉴定出 4 种新的烷基间苯二酚类化合物:5-pentadecyl-resorcinol、5-(8,Z,11,Z-heptadecadienyl)-resorcinol、5-(11,Z-heptadecenyl)-resorcinol 和 5-heptadecyl-resorcinol。

Inui 等(2012)从位于南太平洋所罗门群岛收集的蜂胶样本中鉴定出 4 种新化合物,并通过质谱联合核磁共振法确定了它们的结构,分别为 3 种异戊烯-黄酮类化合物 Solophenols B、C、D 和一种对称二苯代乙烯类化合物 Solomonin。Shi 等(2012)从中国武汉蜂胶甲醇提取物中分离鉴定出 5 种甘油酯类化合物:2-acetyl-1-coumaroyl-3-cinnamoylglycerol、(＋)-2-acetyl-1-feruloyl3-cinnamoylglycerol、(-)-2-acetyl-1-feruloyl-3-cinnamoylglycerol、2-acetyl-1,3-dicinnamoylglycerol 和 (-)-2-acetyl-1-(E)-feruloyl-3-(3″(X),16″)-dihydroxy-palmitoylglycerol。

3.4 蜂胶中新成分的鉴定方法

传统蜂胶成分研究主要采用提取分离方法,从复杂混合物中分离纯化出单体。采用这种方法,要确定一个化合物必须分离和纯化足够的(至少 5~10mg)纯物质(纯度大于 90%),用光谱和质谱等分析技术进行鉴定。这种方法需要经过复杂的分离提取过程,费力费时,而且由于在确定其结构之前对目标化合物的性质了解甚少,往往分离出来的成分并不是感兴趣的化合物。液相色谱质谱联用(LC-MS)是将提取物先经过 LC 分离,流份直接导入 MS 进行分析,根据采集的质谱图(一级或多级),可解析流份的部分结构,仅需几十分钟即可获得待测样品的大量化学信息,如化合物的可能结构以及相对含量(依采用的质谱类型而定)。ESI 和 APCI 源是目前蜂胶高效液相色谱质谱联用(HPLC-MS)主要采用的电离源,适用于蜂胶中大部分化合物的分析。ESI 和 APCI 均为软电离源,得到的质谱中,准分子离子常常是基峰,产生的碎片很少或者没有,能提供未知化合物的分子量信息,但缺少有关分子结构的信息。而蜂胶中主要活性成分类黄酮具有类似的母核结构,仅根据分子量很可能存在误判结构信息。根据文献记载,蜂胶中含有柚配基,但是 Gardana 等(2007)分析发现,柚配基与短叶松素具有相同的分子量、紫外吸收和类似的保留时间,采用液相色谱-电喷雾电离串联质谱(LC-ESI-MS)提供的分子量信息不能将两者分开。为了获得更加准确的结构信息,需采用多极质谱分析(MS^n, $n \geq 2$),用第一级质量分析器根据质荷比选择母离子,经过碰撞

诱导解离技术或其他活化方式使其裂解,然后进行第二级质谱分析,得到碎片离子的质谱图。他们采用 HPLC-MS" 分析发现,柚配基得到[m/z]为 151,而短叶松素产生典型的[m/z]为 253,因而认为蜂胶中没有柚配基存在,而是短叶松素。多极质谱分析增加了选择性,对于不明化合物的结构鉴定更加准确。

蜂胶化学成分种类繁多,含有许多在传统的分离过程中可能被忽略的微量或痕量成分,而 MS 具有高度的灵敏性,根据化合物的质荷比、提取离子色谱以及相关文献,很容易发现新化合物的存在。因此,HPLC-MS 在蜂胶新成分的分离鉴定中发挥着重要的作用。Falcão 等(2010)采用 HPLC-ESI(-)MS 从葡萄牙蜂胶中鉴定出 37 种酚类化合物,除了温带杨属型蜂胶常见的酚酸和类黄酮外,还发现几种新成分,通过质谱解析,其中 4 种是常见的类黄酮山奈酚以及橙皮素和白杨素的甲基化、酯化或羟基化衍生物,6 种是松属素或短叶松素的异戊烯基丙烯酸衍生物。Chang 等(2008)采用高效液相色谱串联大气压化学电离源质谱(HPLC-APCI-MS)在巴西绿蜂胶提取物中检测到肉桂酸及衍生物、类黄酮、苯甲酸和一些苯甲酸盐,以及常规研究中检测不到的不具有紫外吸收的非羟基化的芳香物、脂肪酸和酯等。Cuesta-Rubio 等(2007)采用 HPLC-ESI/MS 确定古巴棕色蜂胶的特征性成分为 nemorosone、scrobiculatones A 和 scrobiculatones B。采用 HPLC ESI/MS" 在古巴红蜂胶中鉴定出 isoliquiritigenin、liquiritigenin、formononetin、biochanin A、vestitol、neovestitol、7-O-metilvestitol、medicarpin、homopterocarpin、vesticarpan、3,8-dihydroxy-9-methoxypterocarpan 和 3-hydroxy-8,9-dimethoxypterocarpan,确定红蜂胶的特征性成分为 medicarpin 和 formononetin。Midorikawa 等(2001)采用 HPLC-APCI(-)-MS 建立蜂胶的总离子流色谱图,基于蜂胶中已鉴定或已分离出的类黄酮、二萜和酚类化合物等 41 种标准品和质谱鉴定,分析巴西、秘鲁、中国和荷兰蜂胶水提物和甲醇提取物中的成分组成。在巴西蜂胶水提物中均检测到二咖啡酰奎宁酸类,而在巴西蜂胶甲醇提取物中检测到二萜类、类黄酮和异戊烯化的酚类化合物。秘鲁蜂胶水提物和甲醇提取物均与巴西蜂胶总离子流图不同,未检测到对照化合物。在中国蜂胶水提物中检测到没食子酸、咖啡酸、香草醛、p-香豆酸和阿魏酸;甲醇提取物中有许多峰出现,但仅鉴定出肉桂酸。在荷兰蜂胶水提物和甲醇提取物中均检测到香兰醛、绿原酸、肉桂酸、p-香豆酸、咖啡酸、阿魏酸和白杨素,还检测到一些未知峰;在甲醇提取物中还检测到槲皮素、山奈酚、山奈素和桦木酚;而在水提物中还检测到松柏醛。

虽然 HPLC-MS 技术在蜂胶研究中具有很强的应用潜力,但由于缺少商品化质谱图库,质谱图的解析对专业人员的技术要求比较高,研究人员往往不能充分利用实验提供的信息解决具体研究中的问题,HPLC-MS 进行定量分析仍以与标准品对照为主,而很多标准品很难得到,因此,阐明蜂胶中各种成分的质谱裂解规律,积累 HPLC-MS 谱图库才可能更大限度地发挥 HPLC-MS 的作用。同时,应充分发挥多极质谱优势,从蜂胶中探究更多的未知成分。

第二节 蜂胶化学成分的分析方法

蜂胶的化学成分复杂,生理药理活性的发挥是多种成分协同作用的结果,而目前我国的蜂胶国标仍以总黄酮含量的高低来确定蜂胶的质量,这在一定程度上忽视了酚酸及挥发性成分的作用。现代各种检测技术尤其是色谱技术的不断发展,检测手段和能力的不断增强,为蜂胶中有效成分的分析研究提供了良好的条件。

1 蜂胶中类黄酮的分析方法

类黄酮是蜂胶中最丰富、活性最强的物质,其含量随蜂胶来源的不同而有很大变化,这些差异可能会影响蜂胶的生物学活性及临床作用。因此,蜂胶中类黄酮物质分析方法的研究,对蜂胶及其制品的质量控制有着十分重要的作用。

蜂胶中类黄酮分析通常采用以下的预处理:先冷冻脱水,将蜂胶原料粉碎成粉末,取一定量的粉末溶解于溶剂中(通常为1∶10,w/v),室温下放置24h(Ahn et al.,2004;Fu et al.,2005),或者在70℃下振荡提取30min(Park et al.,1998)。过滤,重复几次以使多酚提取完全。减压蒸干溶剂再溶解(Popova et al.,2004),或者离心获得上清液(Park et al.,1998;Bosio et al.,2000;Fu et al.,2005),然后采用以下几种方法进一步分析。

1.1 紫外分光光度比色法

紫外分光光度比色法具有重复性好、准确、简便、易掌握和不需要复杂仪器设备等优点,是最常规的分析方法。该方法可以确定蜂胶中总酚或总类黄酮(Nagy and Grancai,1996;Bonvehi et al.,1994;Woisky and Salatino,1998)、总二氢黄酮或二氢黄酮醇、总黄酮或黄酮醇(Nagy and Grancai,1996)的含量。

国家标准《蜂胶中总黄酮含量的测定方法 分光光度比色法》(GB/T 20574—2006)方法:以芦丁为对照品,用Al^{3+}作显色剂,于415nm测定吸光度来确定蜂胶中总黄酮含量。但是,由于蜂胶成分复杂,富含多种酚类物质,对吸光值会产生干扰,从而影响实验结果的准确性。而且不同类型的类黄酮与Al^{3+}作用形成的复合物的吸收波长与415nm有很大偏差。大部分天然黄酮和黄酮醇的吸收波长在390~440nm,而二氢黄酮与二氢黄酮醇的吸收波长在310~320nm。显然,在415nm处测定蜂胶中总黄酮含量时,没有检测到二氢黄酮和二氢黄酮醇,而其又是杨树型蜂胶的主要成分,因此,得到的总黄酮含量往往偏低。

Popova等(2004)提出以相同或相似化学结构的分组定量测定蜂胶中类黄酮含量的分光光度法,分别以高良姜素、松属素、松属素∶高良姜素(2∶1)作为对照品,测定蜂胶中总黄酮与总黄酮醇的含量、总二氢黄酮与二氢黄酮醇的含量、总多酚的含量,并与高效液相色谱法测定的结果进行对比,发现二者基本一致。因此,蜂胶中的活性成分分组定量看起来对评价蜂胶的生物学活性更合适。

蜂胶中类黄酮的定量有两种比色方法(Chang et al.,2002):一是用氯化铝法确定黄酮和黄酮醇含量,基于Al^{3+}与羰基和羟基间形成的复合物,然后在425nm处测定吸光度;二是用2,4-二硝基苯肼(DNP)法测定二氢黄酮和二氢黄酮醇的含量,在486nm处测定吸光度(Chang et al.,2002;Popova et al.,2004)。用以上两种方法测得的总黄酮含量与蜂胶中真实的总黄酮含量较为接近(Chang et al.,2002)。

1.2 薄层色谱(TLC)法

TLC分离和定性分析类黄酮,具有分离速度快、效率高、操作简便的优点,因此广泛应用于分离、鉴定类黄酮。在TLC方法中,固定相和恰当的流动相的选择取决于所研究的类黄酮的结构。硅胶是经典的最常用的固定相(Park et al.,2002;Jasprica et al.,2004),样本用不同的流动相洗脱:甲醇/水(55∶45,v/v)(Park et al.,2002),石油醚/乙酸乙酯(70∶30)

(Popova et al.,2005),氯仿/乙酸乙酯(60∶40)(Santos et al.,2002),甲苯/氯仿/丙酮(40∶25∶35)(Moreno et al.,2000),n-己烷/乙酸乙酯/乙酸(31∶14∶5)(Moreno et al.,2000;Medic-Saric et al.,2004),或者(60∶40∶3)(Moreno et al.,2000),氯仿/甲醇/甲酸(44.1∶3∶2.35)(Jasprica et al.,2004)。

目测法可以在短波长或长波长下进行,有时会喷上不同的试剂以帮助显色。检测波长通常为366nm(Park et al.,2002;Popova et al.,2005)。

1.3 气相色谱(GC)法

GC法可定量也可定性测定蜂胶中的类黄酮化合物,已被广泛采用(Hegazi et al.,2002;Negri et al.,2002;Popova et al.,2005),但是,通常需要衍生才能进行GC分析。近年来发展起来的高温高溶解性GC(HT-HRGC)(Pereira et al.,1998;2000;Park et al.,2002)分离复杂混合物,是确定常规GC柱不能洗脱的高分子量化合物的有效方法。

GC-MS是使用最广泛的方法,因为MS可以获得分子量和结构信息。但是,蜂胶中含有大量的即使衍生也不足以挥发进行GC-MS或HT-GC-MS直接分析的化合物(Sawaya et al.,2004)。表4.1给出了使用GC法时的温度范围、色谱柱特征、分析条件及鉴定的化合物。

表 4.1 蜂胶中类黄酮鉴定的GC条件

Table 4.1　GC conditions for identification of flavonoids in propolis

温度/℃	色谱柱	时间/min	检测器	衍生法	鉴定的化合物	文献
85～310	DB1柱 (30m×0.32mm i.d.)	85	MS	Pyridine+BSTFA	球松素查耳酮、六甲氧基黄酮、球松素、松属素、短叶松素、短叶松素-3-乙酸盐、白杨素、高良姜素、柚皮素	Hegazi et al.,2002
100～310	HP5-MS毛细管柱 (23m×0.25mm i.d.)	42	MS	Pyridine+BSTFA	松属素、短叶松素、短叶松素O-乙酸盐、白杨素、高良姜素	Bankova et al.,2002
100～310	HP5-MS毛细管柱 (23m×0.25mm i.d.)	42	MS	Pyridine+BSTFA	松属素、短叶松素、短叶松素3-乙酸盐、白杨素、高良姜素	Popova et al.,2005
50～285	CBP5柱 (30m×0.25mm i.d.)	55	MS	甲基化	短叶松素、山柰酚、芹菜素、异樱花素、松属素、短叶松素3-乙酸盐、白杨素、高良姜素、山柰素、柚木柯因	Park et al.,2002

1.4 高效液相色谱(HPLC)法

HPLC是目前最流行最可靠的类黄酮化合物的分析方法。NY 5136—2002、NY/T 629—2002和GB/T 19427—2003等均采用HPLC法测定蜂胶中高良姜素、槲皮素、芦丁、杨梅酮、莰菲醇、芹菜素、松属素、柯因等8种黄酮含量,以其总量来表示蜂胶中总黄酮的含量,通常这8种黄酮的峰面积之和是总峰面积之和的65%～75%。

HPLC法已广泛应用于蜂胶中的类黄酮化合物的分析。表4.2归纳了蜂胶分析中曾使用的 HPLC 色谱柱、流动相、洗脱类型、检测和提取方式等信息。近年来,指纹图谱技术已成为国际公认的控制中药或天然药物质量的最有效手段,也引起了蜂胶研究者的广泛关注。目前,HPLC 指纹图谱已用于蜂胶的真伪鉴定(周平等,2005),以及不同地理来源蜂胶的相关性研究(朱恩圆等,2005;Zhou et al.,2008)。

表 4.2 蜂胶中类黄酮鉴定的 HPLC 条件

Table 4.2 HPLC conditions for identification of flavonoids in propolis

色谱柱	流动相	洗脱	检测器	提取方式	鉴定的化合物	文献
Intersil 5 ODS-2 (25cm×0.46cm i.d.)	A. 水:乙酸(95:5,v/v); B. 甲醇	梯度	UV λ=290nm	70%乙醇室温下提取24h	黄酮和黄酮醇,黄烷酮和二氢黄酮醇	Popova et al.,2004
YMC Pack ODS-A (RP)	乙酸:甲醇:水(5:75:60,v/v/v)	等度	DAD λ=254nm	80%乙醇 70℃提取30min,离心	槲皮素、山柰酚、芹菜素、异鼠李素、鼠李素、松属素、樱花素、异樱花素、白杨素、金合欢素、高良姜素、山柰素、柚皮柯因	Park et al.,1998
YMC Pack ODS-A RP-18 (25cm×0.46cm,5μm)	A:水; B:甲醇	梯度	DAD λ=268nm	80%乙醇 70℃提取30min,离心	知叶松素、山柰酚、芹菜素、异樱花素、松属素、短叶松素3-乙酸盐、白杨素、高良姜素、山柰素、柚木柯因	Park et al.,2002
Superspher 100 RP-18 (12.5cm×0.4cm,4μm)	A. 甲醇:乙酸(50:50); B. 甲醇:乙酸(40:60); C. 乙腈	梯度	UV λ=254nm	95%乙醇室温提取7天,离子再溶解	松属素、高良姜素	Bosio et al.,2000
Capcell Pak ACR 120 C18 column (25cm×0.2cm i.d.,5μm)	A:0.1%甲酸:水; B:0.08%甲酸:乙腈	梯度	PAD λ=195～650nm MS(ESI)	室温下乙醇提取24h,离心	槲皮素、短叶松素5-甲基醚、芹菜素、山柰酚、短叶松素、白杨素、松属素、高良姜素、短叶松素3-乙酸盐、柚木柯因	Kumazawa et al.,2004b
Capcell Pak ACR 120 C18 column (25cm×0.2cm i.d.,5μm)	A:0.1%甲酸:水; B:0.1%甲酸:乙腈	梯度	PAD λ=195～650nm MS(ESI)	室温下乙醇提取24h,离心	短叶松素 5-甲基醚、芹菜素、山柰酚、短叶松素、白杨素、松属素、高良姜素、短叶松素3-乙酸盐、柚木柯因	Ahn et al.,2004
LiChrospher 100 RP-18 (11.9cm×0.4cm,5μm)	(1)A 甲酸; B:甲醇; (2)A':H₃PO₄ (pH 2.0); B:MeCN	梯度	PAD λ=268nm	乙醇醚剂	短叶松素、松属素、短叶松素 3-乙酸盐、白杨素、高良姜素、生松素 7-甲基醚、白杨素 7 甲基醚、高良姜素 7 甲基醚	Markham et al.,1996
YMC PACK ODS column (25cm×2cm)	0.1%三氟乙酸:乙腈:水(60:40)	等度	UV,MS,2D' NMR	乙醇室温下提取12h,浓缩	isonymphaeol-B、nymphaeol-A、nymphaeol-B,nymphaeol-C	Kumazawa et al.,2004b
Agilent ZORBAX Eclipse XDB C18 (4.6mm×150mm,3.5μm)	A:乙腈:甲醇(9:1); B:20mmoL⁻¹ 醋酸铵 (pH 3.5)	梯度	UV λ=360nm	甲醇室温提取30min		朱恩圆等,2005

续表

色谱柱	流动相	洗脱	检测器	提取方式	鉴定的化合物	文献
Hypersil RP-C18 (4.6mm×150mm,3μm)	甲醇：0.4% H_3PO_4 (65∶35)	等度	UV λ=360nm	95%乙醇超声		周平等，2005
Symmetry C18column (4.6mm×250mm i.d.5μm)	甲醇：0.4% H_3PO_4 (60∶40)	等度	UV,PDA λ=280nm	75%乙醇超声 4h		Zhou et al., 2008

从表 4.2 可以看出，HPLC 分析蜂胶全部是反相柱，通常为 C18 柱，检测系统全部为紫外检测器，波长范围 350～360nm。流动相通常是二元的，采用酸性的水系统，如乙酸、磷酸或甲酸(流动相 A)，这是由于黄酮类化合物带有酚羟基，在水中会部分解离，而未解离的羟基与固定相作用较强，加入酸可以抑制解离，减少拖尾。另一个是弱极性的流动相，如甲醇或乙腈，通常也是酸化的(流动相 B)。很少有有关于三元或四元流动相的报道。

1.5 毛细管电泳(CE)法

CE 是一种离子域和电离子在电场驱动下，在毛细管中按淌度和分配系数不同而进行高效快速分离的新技术。具有高效快速、分辨率高、成本低，最重要的是使用寿命长、易洗脱等优点。CE 按其电泳迁移技术的差异分为不同的种类，其中，在黄酮类化合物的分析中，最常用的为毛细管区带电泳(CZE)和毛细管胶束电泳(MEKC)。

毛细管区带电泳基于物质电荷和分子量的不同引起的电泳淌度不同而相互分离。胶束电泳是在缓冲溶液中加入表面活性剂(十二烷基硫酸钠，SDS)，不同物质根据其电迁移率及分配系数的不同而被分离。影响的变量有缓冲液的 pH、缓冲液浓度、分离电压和注射时间等，蜂胶中类黄酮分离的这些参数都被优化了，见表 4.3。

表 4.3　CE 鉴定蜂胶中类黄酮的参数
Table 4.3　CE parameters for identification of flavonoids in propolis

仪器参数						实验变量		鉴定的物质	文献
长度/cm	内径/μm	波长/nm	电压/kV	温度/℃	进样时间/s	缓冲液			
56	75	214	18	25	4	硼酸盐+0.5% 甲醇 (100 Mm, pH 9.5)		芦丁、白杨素、杨梅素、山柰酚、橘皮素、大豆黄素、染料木素、芹菜素、槲皮素、木樨草素、高良姜素	Fu et al., 2005
50	75	262	23	25	12	H_3BO_3-$Na_2B_4O_7$ (40～60mmol/L, pH 9.2)		芦丁、芹菜素、木樨草素、槲皮素	Cao et al., 2004

续表

仪器参数					实验变量	鉴定的物质	文献
长度/cm	内径/um	波长/nm	电压/kV	温度/℃ 进样时间/s	缓冲液		
50	50	254	15	25	四硼酸钠(30mmol/L,pH 9)	松属素、金合欢素、白杨素、儿茶素、柚皮素、高良姜素、木樨草素、山奈酚、芹菜素、杨梅素、槲皮素	Volpi,2004
56	50	200	30	25　2	硼酸盐/SDS+10%乙腈(v/v)(25/50mmol/L,pH 9.3)	松属素、白杨素、高良姜素	Hihorst et al.,1998

1.6 液质联用法(HPLC-MS)

类黄酮和多酚是蜂胶中最重要的生理活性物质,但不同来源的蜂胶中这些组分间含量差异很大,对于含量较低的,用HPLC法很难鉴定出来,而通过提高样品浓度、增加进样量来富集,又会导致蜂胶提取液浓度过高,引起仪器过载。同时,蜂胶成分复杂,不同活性成分相互干扰严重,很难分离。HPLC-MS利用MS选择性强的特点,根据待测成分选择离子质谱图峰面积进行定量,可以达到不同成分同时定量的目的,且灵敏度高。MS多采用ESI源和APCI源。ESI可以直接电离,将分子转移到质谱,从而扩大了MS的应用范围,可以分析许多新的热不稳定、高极性、高分子量的分子。APCI源质谱也可以得到蜂胶的特征指纹。采用不同的HPLC-MS法,蜂胶中大部分黄酮类化合物已被精确鉴定(表4.4)。

表4.4 HPLC-MS在蜂胶活性成分定量中的应用

Table 4.4 Application of HPLC-MS in determination of active constituents in propolis

分析方法	色谱条件	被测化合物	参考文献
HPLC-DAD-ESI/TQ	柱:ODS(150mm×2mm,3μm);流动相:(A) 0.05%甲酸和(B)乙腈,梯度洗脱;流速:0.2mL/min;	白杨素、柚木柯因、芹菜素、金合欢素、槲皮素、高良姜素、山奈酚、松属素、松属素、樱花素、异戊二烯	Medana et al.,2008
HPLC-DAD-MS	柱:C18(250mm×4.6mm,5μm);流动相:(A) 0.1%甲酸和(B)乙腈,梯度洗脱;流速:1.2mL/min;	白杨素、高良姜素、松属素、槲皮素、山奈酚、异鼠李素、樱花素、异樱花素、芹菜素、金合欢素、松属素-5,7-二甲醚、白杨素-7-甲醚	Gardana et al.,2007
HPLC-APCI-IT-MS	柱:C18(220×4.6mm,5μm);流动相:(A) 30mmol/L NaH$_2$PO$_4$(pH 3)和(B)乙腈,梯度洗脱;流速:1.5mL/min;	槲皮素、山奈酚、柚配基、白杨素、高良姜素和松属素	Pietta et al.,2002
HPLC-ESI/MS	柱:C18(150mm×4.6mm,4μm);流动相:(A) 0.25%乙酸和(B)甲醇,梯度洗脱;流速:0.5mL/min;	金合欢素、芹菜素、黄芩素、儿茶酚、白杨素、高良姜素、染料木质素、山奈酚、木樨草素、杨梅素、柚皮素、松属素、枸橘苷、槲皮素、香草醛	Volpi and Bergonzini,2006

2 蜂胶中有机酸类化合物的分析方法

目前,蜂胶的质量鉴别主要依赖于对其黄酮类成分的含量测定,但有机酸类也是蜂胶的主要活性成分,其含量的变化也会影响蜂胶质量。依据有机酸分子碳架来源不同,蜂胶中有机酸可分成 3 大类:脂肪族羧酸,如甲酸、乙酸;糖衍生的有机酸,如葡萄糖醛酸、半乳糖醛酸等;酚酸类化合物,如苯甲酸、肉桂酸、p-香豆酸等。研究表明,蜂胶中的有机酸,如苯甲酸、咖啡酸、阿魏酸、香草酸、肉桂酸、绿原酸等,具有很强的抗病原微生物和保护动物肝脏的作用,与蜂胶中黄酮类物质起协同作用。

蜂胶中有机酸分析方法较多,薄层色谱、气相色谱、毛细管电泳等方法都已用于蜂胶中有机酸的组成及定量分析。高效液相色谱法具有简便、快速且选择性好、准确度好的优点,现已广泛用于蜂胶中有机酸的定性定量研究中。

蜂胶中有机酸的常见分析方法见表 4.5。

表 4.5 蜂胶中有机酸的常见分析方法

Table 4.5 Common analytical methods of organic acids in propolis

检测方法	有机酸组成	提取方法	地理来源	参考文献
胶束动电毛细管色谱法	肉桂酸、烟酸	乙醇	中国	Lu 等,2004
HPLC	咖啡酸、阿魏酸、丁香酸、鞣花酸、p-羟基苯甲酸、p-香豆酸、没食子酸、抗坏血酸	乙醇	中国	Ahn et al.,2007
HPLC、HPLC-MS	咖啡酸、p-香豆酸、阿魏酸	乙醇	土耳其	Gülçin et al.,2010
LC-DAD-MS、LC-MS/MS	咖啡酸、p-香豆酸、阿魏酸、异阿魏酸、3,4-二甲基-咖啡酸、肉桂酸、香豆酸甲基酯、Cinnamilidenacetic acid、咖啡酸苄酯、咖啡酸异戊烯基酯、CAPE、p-香豆酸甲基丁烯酯、p-香豆酸苄酯、咖啡酸肉桂酯、p-香豆酸肉桂酯、p-甲氧基-肉桂酸肉桂酯、p-香豆酸酯	乙醇	意大利	Gardana et al.,2007
LC-DAD-MS、LC-MS/MS	咖啡酸异戊烯基酯	乙醇	阿根廷	Gardana et al.,2007
HPLC-UV/DAD	咖啡酸、p-香豆酸、阿魏酸、肉桂酸	乙醇	意大利	Pellati et al.,2011
CE	肉桂酸、咖啡酸	乙醇	意大利	Volpi et al.,2004
HPLC	咖啡酸、香豆酸、阿魏酸、肉桂酸、drupanin、阿替匹林 C、baccharin、2,2-二甲基-6-羧基乙烯基-2H-1-苯并吡喃酸(DCBEN)	乙醇	巴西	de Sousa et al.,2007
GC-MS	苯甲酸、p-香豆酸、3,4-二甲氧基肉桂酸、阿魏酸、异阿魏酸、咖啡酸	BSTFA 衍生		Prytzyk et al.,2003

续表

检测方法	有机酸组成	提取方法	地理来源	参考文献
CE	阿魏酸、咖啡酸	甲醇	中国	Cao et al., 2004
TLC	肉桂酸、o-香豆酸、m-香豆酸、p-香豆酸、咖啡酸、阿魏酸	乙醇	克罗地亚	Medic-Šaric et al., 2004
超高效液相色谱-二级管阵列	阿魏酸、咖啡酸苯乙酯	甲醇	中国	李熠等, 2007
反相高效薄层色谱（RPHPTLC）	香豆酸、阿魏酸、肉桂酸、1,1-二甲基烯丙基咖啡酸	乙醇	阿根廷	Isla et al., 2005
HPLC-DAD	没食子酸、p-香豆酸、肉桂酸、咖啡酸、阿魏酸、3,4-二甲氧基肉桂酸	乙醇	中国	徐元君等, 2010
HPLC-UV	咖啡酸、异阿魏酸和3,4-二甲氧基桂皮酸	蜂胶制剂	中国	迟家平等, 1997
氨基酸自动分析仪	天冬氨酸、苏氨酸、丝氨酸、谷氨酸、甘氨酸、丙氨酸、胱氨酸、缬氨酸、蛋氨酸、异亮氨酸、亮氨酸、酪氨酸、苯丙氨酸、赖氨酸、组氨酸、精氨酸、脯氨酸、色氨酸	柱后茚三酮法	中国、巴西	吴健全等, 2012
返滴定法	游离酸	弱碱溶解再酸化萃取	中国	周立东等, 2005
HPLC-UV	咖啡酸、p-香豆酸、反式阿魏酸、肉桂酸	95%乙醇	中国	索志荣等, 2008
HPLC	苯甲酸、咖啡酸、阿魏酸、香草酸、肉桂酸、绿原酸	提纯蜂胶液	中国	孟霞等, 2008
HPLC-ESI-MS/MS	咖啡酸、p-香豆酸、阿魏酸、3-羟基-4-甲氧基肉桂酸、苯甲酸、3,4-二甲氧基肉桂酸、肉桂酸、咖啡酸苯乙酯、苯甲酰肉桂酯	乙醇	中国	吴正双, 2011
HPLC-DAD-ESI-MS^n	咖啡酸、p-香豆酸、阿魏酸、肉桂酸	乙醇	中国	符军放, 2006
HPLC-DAD-ESI-MS^n	咖啡酸、p-香豆酸、肉桂酸	水	中国	符军放, 2006
HPLC-DAD-ESI-MS^n	p-香豆酸、阿魏酸、肉桂酸	CO_2	中国	符军放, 2006
HPLC-DAD	没食子酸、3,4-二羟基-苯甲酸、咖啡酸、阿魏酸、芥子酸、p-香豆酸、4-对羟基苯甲酸乙酯、o-肉桂酸	溶剂分步提取	中国、巴西、乌拉圭	Bonvehi and Coll, 1994
HPLC	阿魏酸	乙醇	中国	于敏等, 2005

续表

检测方法	有机酸组成	提取方法	地理来源	参考文献
HPLC-MS/MS	咖啡酸苯乙酯	甲醇	中国	李丽等,2014
HPLC-DAD	咖啡酸、p-香豆酸、肉桂酸、阿魏酸	乙醇	中国	于世锋等,2007
HPLC-UV	咖啡酸	乙醇	中国	周萍等,2013
HPLC-UV	香草酸、咖啡酸、p-香豆酸、阿魏酸、3-羟基-4-甲氧基肉桂酸、苯甲酸、3,4-二甲氧基肉桂酸、肉桂酸、咖啡酸苯乙酯、苯甲酰肉桂酯	75%乙醇	中国	赵亮亮等,2012
HPLC-UV/DAD	咖啡酸、阿魏酸、苯甲酸、肉桂酸、咖啡酸苯乙酯、亚桂皮乙酸、p-香豆酸、苯甲酸、异阿魏酸、咖啡酸、3,4-二甲氧基肉桂酸、肉桂酸肉桂酯、香草酸、4-甲氧基肉桂酸、咖啡酸苄酯、p-香豆酸苄酯、咖啡酸肉桂酯、水杨酸	75%乙醇	中国	罗照明等,2013
HPLC-UV	阿魏酸	蜂胶液/蜂胶胶囊	中国	赵静,2003
CE	对香豆酸、阿魏酸、咖啡酸	80%甲醇	中国	彭友元,2011
HPLC-UV/FL	莽草酸、奎宁酸、绿原酸、迷迭香酸、对羟基苯甲酸	水	斯洛伐克	Hroboňová et al.,2009
HPLC-UV	肉桂酸、1,1-二甲烯丙基咖啡酸	醇酊剂	新西兰	Markham et al.,1996
HPLC-UV	奎尼、莽草酸	甲醇	斯洛伐克	Hroboňová et al.,2007

3 蜂胶中挥发性成分的分析方法

蜂胶具有特异的辛辣芳香味,占蜂胶10%左右的芳香挥发性成分被认为对蜂胶的鉴别、品质控制及生物学活性的发挥具有不可替代的作用。蜂胶中的挥发性成分是一类数量巨大,结构种类繁多,生物学活性多样,具有很高药用价值的天然成分,包括醇、醛、酮、醚、酯、羧酸等含氧基团的化合物,其主要活性成分是萜烯类及其含氧衍生物。它们与蜂胶中的其他生物学活性物质协同,共同体现蜂胶的生物学及药理学活性(Gomez-Caravaca et al.,2006;Sahinler et al.,2005)。

蜂胶挥发性成分研究始于1974年,一开始只从蜂胶中鉴定出了苯甲酸、苯甲醇、香草醛和丁香油酚(Janas and bumba,1974)。大量研究表明,蜂胶中挥发性成分比酚酸和类黄酮化合物可变性更大。

3.1 蜂胶中挥发性成分的提取

在蜂胶挥发性成分研究中,样品的制备方法对检测结果有很大影响。常用于提取挥发性成分的方法主要有:水蒸气蒸馏法、有机溶剂提取法、超临界流体萃取法、同时蒸馏萃取法、固相微萃取法及顶空进样等。不同方法提取出的蜂胶挥发性成分组成差异较大。

水蒸气蒸馏法是提取蜂胶中挥发性成分最常用的方法,尽管该法工艺复杂,提取率低,但提取物中杂质少,萜烯类化合物相对含量较高;用低沸点的有机溶剂如乙醚、石油醚等连续回流或冷浸提取,提取液蒸馏或减压蒸馏除去溶剂,操作相对简单,得率高,但是该法所得到的产物杂质较多,需进一步提纯;二氧化碳超临界流体萃取法是一种新的提取分离技术,其萃取剂二氧化碳具有无毒、无味、不腐蚀、价格便宜、易于回收等优点,且临界温度接近于室温,临界压力处于中等压力,特别适用于高沸点、挥发度低的热敏性物质的提取,近年来在蜂胶挥发性成分研究中较常用,其取得产物主要为烷烃及芳香烃衍生物,萜烯类化合物含量也较高。同时蒸馏法将蒸馏和萃取融合到一起,有利于挥发物的提取,损失少,方法也简单;顶空进样和固相微萃取无须有机溶剂,操作简单、快速,所需样品量少,集萃取、浓缩、采样于一体,减少了痕量挥发性成分的损失,而且便于与气相色谱-质谱联用,因此,这两种方法在蜂胶挥发性成分分析中应用越来越广泛。

3.2 挥发性成分的检识

气相色谱法具有分离效率和灵敏度高,样品用量少,分析速度快等优点,现已广泛用于挥发油的定性和定量分析中。由于蜂胶挥发性成分复杂多样,很难找到标准品作为参照,蜂胶挥发性成分分析主要采用气相色谱质谱联用技术。

以下是不同地理来源的蜂胶的挥发性成分的分析结果(表 4.6)。

表 4.6 蜂胶挥发性成分提取分离方法

Table 4.6 Methods for extraction and separation of volatile components in propolis

温度/℃	色谱柱	提取方法	鉴定的主要化合物	地理来源	文献
35~280	RTX-5MS(30m×0.25mm×0.25μm)	静态顶空	α-蒎烯、β-蒎烯	巴西	Kaškoniene et al.,2014
100~240	HP5MS(30m×0.32mm×0.25μm)	水蒸气蒸馏	α-蒎烯、β-蒎烯、大根香叶烯D、δ-荜澄茄烯	巴西	Ioshida et al.,2010
55~250	HP5MS column(30m×0.25mm×0.25μm)	水蒸气蒸馏	橙花叔醇、反式石竹烯、香木兰烯、δ-杜松烯、匙叶桉油烯醇、蓝桉醇	巴西	Marostica Junior et al.,2008
60~240	ZB-5 MS column(30m×0.25mm×0.25μm)	水蒸气蒸馏	β-石竹烯、苯乙酮、芳樟醇、γ-榄烯、γ-杜松烯、γ-衣兰油烯	巴西	Oliveira et al.,2010
50~250	SE-54/peg-20M(25m×0.25mm×0.2μm)	水蒸气蒸馏	α-蒎烯、β-蒎烯、柠檬油精	巴西	Simionatto et al.,2012
60~240	DB-5(50m×0.25mm×0.25μm)	溶剂提取	α-蒎烯、石竹烯氧化物、β-蒎烯、α-可巴烯	巴西	Torres et al.,2008
35~280	DB-5(30m×0.25mm×0.25μm)	水蒸气蒸馏	β-石竹烯、(E)-橙花叔醇、γ-芹子烯	巴西	De Albuquerque et al.,2008

续表

温度/℃	色谱柱	提取方法	鉴定的主要化合物	地理来源	文献
60～280	SPB-1（25m×0.25mm×0.25μm）	水蒸气蒸馏	spatulenol、(2Z,6E)-法尼醇、异戊二烯基-苯乙酮、苯甲酸苄酯	巴西	Bankova et al.，1998a
60～240	DB-5（50m×0.25mm×0.25μm）	溶剂提取	(E)-石竹烯、α-可巴烯、α-蒎烯、石竹烯氧化物、δ-荜澄茄烯	巴西	Torres et al.，2008
60～240	DB-5（50m×0.25mm×0.25μm）	溶剂提取	(E)-石竹烯、α-古芸烯、δ-荜澄茄烯、α-可巴烯	巴西	Torres et al.，2008
60～240	DB-5（50m×0.25mm×0.25μm）	溶剂提取	(E)-石竹烯、α-古芸烯、β-芹子烯	巴西	Torres et al.，2008
60～240	DB-5（50m×0.25mm×0.25μm）	溶剂提取	1,8-桉树脑、表小茴香醇、4-萜烯醇、小茴香醇	巴西	Torres et al.，2008
80～185	HP5MS（30m×0.25mm×0.25μm）	乙醇提取	氢化肉桂酸、3-苯丙酸乙酯、反式-橙花叔醇	巴西	黄帅等，2013
80～185	HP-5MS（30m×0.25mm×0.25μm）	石油醚超声波辅助	长叶烯、β-桉叶醇、α-桉叶醇、β-丁子香烯、愈创木醇	巴西	李雅晶等，2011
50～250	HP5MS（30m×0.25mm×0.25μm）	水蒸气蒸馏	α-蒎烯、棕榈酸、反式马鞭草烯醇	墨西哥	Pino et al.，2006
80～300	HP5MS（30m×0.25mm×0.25μm）	水蒸气蒸馏	橙花叔醇、spatulenol、喇叭茶醇	加那利群岛	Bankova et al.，1998b
50～300	HP5MS（30m×0.25mm×0.25μm）	水蒸气蒸馏	橙花叔醇、spatulenol、β-cayophillene、喇叭茶醇、ledene、T-muurolol	加那利群岛	Bankova et al.，1998
55～250	HP5MS（30m×0.25mm×0.25μm）	水蒸气蒸馏	5,6,7,8-tetramethylbicyclo[4,1,0]hept-4-en-3-one、菖蒲二烯、表雪松醇	埃塞俄比亚	Haile et al.，2012
55～250	HP5MS（30m×0.25mm×0.25μm）	水蒸气蒸馏	去氢白菖烯、4-松油醇双环倍半水芹烯	埃塞俄比亚	Haile et al.，2012
35～280	RTX-5MS（30m×0.25mm×0.25μm）	静态顶空	α-蒎烯、β-蒎烯、柠檬油精	乌拉圭	Kaškoniene et al.，2014
35～280	RTX-5MS（30m×0.25mm×0.25μm）	静态顶空	桉油精、α-蒎烯、苯甲醛、β-蒎烯	爱沙尼亚	Kaškoniene et al.，2014
70～200	HP-101（25m×0.2mm×0.2μm）	水蒸气蒸馏	苯甲酸、苯甲醇、柠檬油精、β-蒎烯、γ-萜品烯	克罗地亚	Bořcić et al.，1996
60～280	HP5MS（30m×0.25mm×0.25μm）	水蒸气蒸馏	刺伯烯、α-蒎烯、泪柏醚、β-松油醇、α-桉油精、n-decanal、guaiol、δ-荜澄茄烯、α-muurolene、n-decanal、cedrol、n-nonanal、manool	希腊	Melliou et al.，2007

续表

温度/℃	色谱柱	提取方法	鉴定的主要化合物	地理来源	文献
40~280	HP-5（30m×0.25mm×1.0μm）	顶空动态和水蒸气蒸馏	苯甲酸、benzyl benzoate、benzyl salicylate、benzyl cinnamate	意大利南部	Pellati et al., 2013
30~200	FSC（30m×0.25mm×0.5μm）	顶空	6-methylhept-5-en-2-one、benzyl alcohol、isopentyl acetate、benzaldehyde	英国威尔士	Greenaway et al., 1989
60~220	FSC（60m×0.25mm×0.25μm）	顶空固相微萃取	phenyl ethyl alcohol、苯乙醇、decanal、ethyl benzoate、nonanal、cedrol	土耳其	Hames-Kocabas et al., 2013
60~220	FSC（60m×0.25mm×0.25μm）	顶空固相微萃取	雪松醇、α-没药醇、δ-荜澄茄烯、α-桉叶醇	土耳其	Hames-Kocabas et al., 2013
50~200	Supelcowax-10（30m×0.32mm×0.25μm）	水蒸气蒸馏	tricosane、hexacosane、palmitic acid、linalool、methyleugenol、palmitoleic acid（(Z)-hexadec-9-enoic acid）、geraniol、(Z)-ethylcinnamate、heneicosane	印度	Naik et al., 2013
35~280	RTX-5MS（30m×0.25mm×0.25μm）	静态顶空	3-甲基-3-丁烯-1-醇、3-甲基-2-丁烯-1-醇、4-penten-1-yl acetate、α-longipinene	中国	Kaškoniene et al., 2014
35~230	DB-Wax（30m×0.32mm×0.25μm）	动态顶空	乙酸、2-乙酸苯乙酯、萘	中国	Yang et al., 2010
40~290	HP5MS（60m×0.25mm×0.32μm）	固相微萃取	甲酸、乙酸、苯甲酸、苯甲醇、乙酸苯乙酯	中国	徐响等，2008
80~280	HP-5（30m×0.25mm×0.25μm）	微波辅助萃取	17-三十五碳烯、二十七烷、1-十九烯、二十九烷、正二十一烷、1-二十二烯、羊毛甾醇醋酸酯	中国	赵强等，2007
40~250	DB-1（30m×0.25mm×0.25μm）	动态顶空/电子鼻	乙酸、苯乙醇、3-甲基-3-丁烯-1-醇、3-甲基-2-丁烯-1-醇、2-甲基-2-丁烯-1-醇	中国	Cheng et al., 2013
40~250	DB-5（30m×250m×0.25μm）	固相微萃取	1-(1,5-二甲基-4-己烯)-4-甲基苯、八氢二甲基-2-(1-亚异丙基)萘、雪松烯、α-古芸烯、β-荜草烯	中国	程焕等，2012
40~250	DB-5（30m×250m×0.25μm）	动态顶空	乙酸、苯甲醇、3-甲基-3-丁烯-1-醇、	中国	程焕等，2012
40~290	HP5MS（60m×0.25mm×0.32μm）	固相微萃取	甲酸、乙酸、苯甲酸、苯甲醇	中国	徐响等，2008

续表

温度/℃	色谱柱	提取方法	鉴定的主要化合物	地理来源	文献
100~230	DB-1（30m×0.25mm×0.25μm）	乙醚索氏提取	桉叶醇,松香酸,愈创木醇	中国	王小平等,2008
80~240	AC-5（30m×0.25mm×0.25μm）	水蒸气蒸馏	苯甲醇,苯甲酸,雪松醇,β-桉叶醇	中国	曾晞等,2004
40~250	DB-1（30m×0.25mm×0.25μm）	动态顶空/电子鼻	乙酸,苯乙醇,甲苯,苯甲醛,2-甲基-2-丁烯-1-醇	中国	Cheng et al.,2013
35~250	HP5（30m×0.25mm×0.25μm）	同时蒸馏法	α-没药醇、2-甲基-3-丁烯-2-醇、3-甲基-2-丁烯-1-醇、奠	中国	付宇新等,2009
35~250	HP5（30m×0.25mm×0.25μm）	动态顶空	十七烷、菲、芳姜黄酮、1-(1,5-二甲基-4-己烯基)-4-甲基-苯、十八烷、1-甲氧基-4-(1-丙烯基)-苯、十六烷等	中国	付宇新等,2009
35~250	HP5（30m×0.25mm×0.25μm）	水蒸气蒸馏	3-甲基-2-丁烯-1-醇、苯乙醇、1,2,3,4,4a,5,6,7-八氢-α,α,4a,8-四甲基-2-萘甲醇、2-甲氧基-4-乙烯基苯酚、α-没药醇	中国	付宇新等,2009
40~250	DB-1（30m×0.25mm×0.25μm）	动态顶空/电子鼻	乙酸、1-(1,5-二甲基-4-己烯基)-4-甲基苯、1,2,3,4,4a,5,6,8a-octahydro-4a,8-dimethyl-2-(1-methylethenyl)-萘、苯乙醇	中国	Cheng et al.,2013
40~250	DB-1（30m×0.25mm×0.25μm）	动态顶空/电子鼻	乙酸、雪松烯、3-甲基-3-丁烯-1-醇、1-(1,5-dimethyl-4-hexenyl)-4-methyl-benzene、3-甲基-2-丁烯-1-醇	中国	Cheng et al.,2013
60~280	HP5MS（30m×0.25mm×0.25μm）	超临界萃取	2,3-二氧化-5,7-二羟基-2-苯基-4-氢-苯并吡喃-4-酮、烷烃、绿原酸	中国	韩玉谦等,2003
100~230	DB-1（30m×0.25mm×0.25μm）	固相微萃取	杜松烯,去氢白菖烯,愈创木烯,α-雪松烯,香木兰烯,α-芹菜烯,桉叶油醇,杜松烯醇,脱氢杜松油萜烯,愈创木醇	中国	贺丽苹等,2008
100~230	DB-1（30m×0.25mm×0.25μm）	乙酸乙酯	桉叶油醇、十四烯	中国	贺丽苹等,2008
60~250	DB-1（30m×0.25mm×0.25μm）	超临界萃取	亚油酸、烷烃	中国	李雅洁等,2006
100~230	DB-1（30m×0.25mm×0.25μm）	乙酸乙酯浸提	有机酸、烷烃、酚类物质	中国	李雅萍等,2007

续表

温度/℃	色谱柱	提取方法	鉴定的主要化合物	地理来源	文献
100~230	DB-1（30m×0.25mm×0.25μm）	固相微萃取	α-桉叶油醇、荜澄茄油烯醇、愈创木醇	中国	李雅萍等，2007
40~280	HP5MS（30m×0.25mm×0.25μm）	乙醚浸提	苯甲醇、苯乙醇、β-桉叶醇、肉桂醇、α-桉叶醇	中国	刘安洲等，2009
40~280	HP5MS（30m×0.25mm×0.25μm）	固相微萃取	苯甲醇、苯甲酸、苯乙醇、肉桂醇	中国	刘安洲等，2009
80~240	HP-12m×0.2mm	超临界萃取	烷烃及芳香烃衍生物	中国	刘映等，2000
40~250	DB-Wax（30m×0.25mm×0.25μm）	顶空固相萃取	苯甲醇、苯甲醛、苯乙醇、2-甲基-2-丁烯-1-醇、3-甲基-3-丁烯-1-醇、苯乙烯、醋酸	中国	延莎等，2012a
80~230	AC-5（30m×0.25mm×0.25μm）	水蒸气蒸馏乙醚萃取	雪松醇、邻苯二甲酸二丁酯、二二碳酸、苯甲醇、9,12-十八碳二烯酸甲酯、苯甲酸、4-羟基-2-甲基苯乙酮、	中国	曾晞等，2004
35~230	DB-Wax（30m×0.25mm×0.25μm）	固相微萃取	姜黄烯、雪松烯、苯乙醇、乙酸苯乙酯和柏木烯	中国	田文礼等，2012
40~270	HP-MS（30m×0.25mm×0.25μm）	索氏石油醚提取	α-桉叶醇、油酸、芹子烯、愈创醇、十六酸	中国	杨书珍等，2009
40~270	HP-MS（30m×0.25mm×0.25μm）	索氏正己烷提取	苯甲酸、愈创醇、芹子烯、3,4-二氢-2-萘甲酸、γ-桉叶酸、β-桉叶醇	中国	杨书珍等，2009
40~270	HP-MS（30m×0.25mm×0.25μm）	索氏二氯甲烷提取	3,4-二甲氧基肉桂酸、亚油酸、α-桉叶醇、苯乙酸、顺式-9-二十三烯、愈创醇、肉桂酸肉桂酯	中国	杨书珍等，2009
40~260	HP-MS（30m×0.25mm×0.25μm）	HS-SPME	3-甲基-2-丁烯醛、苯甲醛、α-柏木烯、β-马榄烯、β-桉叶醇、α-桉叶醇	中国	延莎等，2012b
40~260	HP-MS（30m×0.25mm×0.25μm）	HS-SPME	苯甲醇、壬醛、苯甲酸、α-紫穗槐烯、β-桉叶醇、α-桉叶醇	中国	延莎等，2012b
40~260	HP-MS（30m×0.25mm×0.25μm）	HS-SPME	苯乙烯、苯甲醛、桉叶油醇、苯甲醇、壬醛	中国	延莎等，2012b
40~260	HP-MS（30m×0.25mm×0.25μm）	HS-SPME	苯甲醛、苯甲醇	中国	延莎等，2012b

续表

温度/℃	色谱柱	提取方法	鉴定的主要化合物	地理来源	文献
80~280	HP-5MS(30m×0.25mm×0.25μm)	水蒸气蒸馏	17-三十五碳烯、二十七烷、1-十九烯、二十九烷、正二十一烷、1-二十二烯、Lanosta-8,24-dien-3β-ol,acetate、愈创木醇	中国	薄文飞,2009
80~185	HP-5MS (30m×0.25mm×0.25μm)	石油醚超声波辅助	肉桂酸、β-桉叶醇、α-桉叶醇、γ-桉叶醇	中国	李雅晶等,2011
35~230	DB-Wax (30m×0.25mm×0.25μm)	固相微萃取	苯甲醛、萘、2-戊基呋喃、2-甲基萘和可巴烯	马来西亚	田文礼等,2012

(1)亚洲蜂胶

中国研究者已经将很多方法用于蜂胶挥发性成分的分析。Yang 等(2010)对中国 23 个地区采集的蜂胶样本进行顶空进样分析,发现主要挥发性成分是芳香性活性成分:乙酸、2-乙酸苯乙酯和萘。徐响等(2008)采用固相微萃取联合 GC-MS 分析了北京和河北的蜂胶挥发性成分,发现乙酸、2-乙酸苯乙酯和苯乙醇是其主要活性成分。这些成分与中国栽培的杨树分泌的胶质中的挥发性成分相似(Cheng et al.,2013)。内蒙古蜂胶挥发油的主要成分是α-没药醇、2-甲基-3-丁烯-2-醇、3-甲基-2-丁烯-1-醇(付宇新,2009)。Kaškoniene 等(2014)也在中国蜂胶中发了现 3-甲基-3-丁烯-1-醇、3-甲基-2-丁烯-1-醇、4-戊烯-1-乙酸酯和α-长叶蒎烯。这些半松油精醇及其酯是典型的杨树代谢物,也在威尔士的杨树型蜂胶顶空挥发性成分中被发现(Greenaway,1989)。

经顶空固相微萃取与 GC-MS 分析,土耳其安大略省东北部蜂胶的挥发性成分主要是氧化烃、氧化倍半萜、芳香醇及其酯(Hames-Kocabas et al.,2013)。Naik 等(2013)采用水蒸气蒸馏法研究了印度蜂胶的挥发性成分,主要含有长链烷类(二十三烷、正二十六烷、二十七烷、二十一烷)、萜类(芳樟醇、甲基丁子香酚、香叶醇)和酚类(肉桂酸甲酯)。

(2)欧洲蜂胶

欧洲蜂胶的挥发性成分与中国蜂胶类似,主要是倍半萜,然后是芳香族化合物,如乙酸苄酯、苯甲酸苄酯和苯甲醇等。

希腊蜂胶挥发性成分中发现了较高比例的 α-蒎烯,在希腊柏杉中发现的主要精油成分是 α-蒎烯(Melliou et al.,2007;Milos et al.,2005),因此,认为希腊蜂胶的植物来源可能是柏杉。意大利南部的蜂胶也发现了较高比例的 α-蒎烯,针叶树种被认为是其植物来源(Pellati et al.,2013)。在爱沙尼亚的蜂胶样品的挥发物成分中也检测到高含量的单萜 α-蒎烯、β-蒎烯和桉油精(Kaškoniene et al.,2014)。

(3)南美蜂胶

目前研究最多的是巴西蜂胶,尤其是绿蜂胶或 Alecrim 蜂胶(Queiroga et al.,2008;Ferracini et al.,1995)。巴西绿蜂胶特征性挥发性成分是倍半萜,已经鉴定出的主要成分有:橙花叔醇、β-石竹烯、spatulenol 和 δ-荜澄茄烯。圣保罗、里约热内卢和皮奥伊州蜂胶中主要挥发性成分是石竹烯、spatulenol 和 δ-荜澄茄烯。而来自米纳斯吉拉斯州的绿蜂胶主要含有橙花叔醇、β-石竹烯和 γ-芹子烯。对巴西绿蜂胶乙醇提取物中的挥发性成分分析发现,橙花叔醇和苯丙烯酸是其主要的挥发性成分(黄帅等,2013)。Nunes 等(2012)采用顶空 GC/MS 证实了巴西绿蜂胶挥发油组成随季节变化并不明显(Bankova et al.,1998)。

巴西蜂胶植物来源广泛,皮奥伊州蜂胶中的主要成分是单萜类:α-and β-蒎烯、1,8-桉树脑和 4-萜品醇(Torres et al.,2008)。里约热内卢州蜂胶的主要成分是 β-蒎烯(Ioshida et al.,2010)。α-蒎烯和 β-蒎烯在很多巴西蜂胶中被发现,而且在乌拉圭蜂胶挥发性成分中的含量也较高。此外,乌拉圭蜂胶的特征之一是还含有高浓度的柠檬油精。

热带地区的生物多样性决定了蜂胶挥发性成分组成的多样性。阿根廷蜂胶挥发性成分特征是单萜类化合物含量很高,主要是邻伞花烃和柠檬油精,同样的挥发油特征性在矮灌木 *Larrea nitida* 中被发现,这种植物分泌物被证明是该地区蜂胶的植物源。在墨西哥尤卡坦半岛的蜂胶中,α-蒎烯、棕榈酸和反式马鞭草烯醇是最丰富的挥发性成分(Pino et al.,2006)。

(4)非洲蜂胶

埃塞俄比亚蜂胶挥发性成分中最主要的是氧化的单萜、倍半萜和氧化的脂肪烃,而加那利群岛蜂胶挥发性成分中,倍半萜(橙花叔醇、spatulenol、喇叭茶醇)和长链烃是其主要成分。

对蜂胶挥发性成分的提取及成分鉴定结果均表明,根据产地及提取方法的不同,蜂胶挥发性成分的化学组成及生物活性组分的含量有显著差异。

蜂胶植物来源广泛,挥发性成分组成具有特定植物源的化学成分特征,但蜂胶中挥发性成分研究还主要局限于杨树型蜂胶和巴西绿蜂胶,其他类型蜂胶中挥发性成分的研究还处于空白阶段。系统开展蜂胶挥发性成分研究,有望建立蜂胶挥发性成分标准化图谱,为实现蜂胶的化学成分标准化奠定基础。

第三节 蜂胶的质量控制与标准化

1 蜂胶的化学成分标准化

蜂胶植物来源广泛,胶源植物不同导致蜂胶化学组成差异很大。要实现蜂胶质量控制的标准化的确是相当困难的,主要是由于蜂胶中的成分复杂,各个成分之间存在着协同作用,以及药物主要活性成分与所含主要化合物不一定统一,即使是指纹图谱也存在很多问题。一般指纹图谱的制作都是经过多个产地蜂胶的收集比较,再给出一份较完整的特征图谱,但采收时期、季节、提取方法都会造成特征图谱的大不同。十多年来,几乎所有关于蜂胶生物学活性的研究都包括了所用蜂胶的化学特征(Bankova,2005),蜂胶化学成分标准化研究也成为研究热点,然而,至今尚未形成统一的蜂胶化学成分标准化评价系统。

1.1 按植物来源的标准化

蜜蜂采集植物不同部分的材料来生产蜂胶,植物源的化学成分决定了蜂胶的化学组成,每种植物来源有其典型的化学成分以及与之相适应的药理活性。因此,蜂胶植物来源的明确是进行蜂胶质量控制的源头和关键。典型的蜂胶类型及其主要的生物学活性成分已基本明确(见表 4.7)。当然,还有许多其他胶源植物和相应化学类型的蜂胶有待探索。例如,除绿蜂胶外,巴西还有 12 种蜂胶类型(Park et al.,2002;De Castro Ishida et al.,2011)。从表 4.7 可见,一种蜂胶具有的生物学活性的结论无法自动转换成另一种类型蜂胶的生物学活性;不同植物来源的蜂胶难以找到有效的共有活性成分指标。因此,对不同地理来源的蜂胶建立统一的化学成分标准化几乎是不可能的。

植物源的化学成分决定了蜂胶的化学组成,参照药用植物的方法:①如果活性部分是已知的可接受的,可以采用合适的方法定性;②如果活性部位未知或仍有争议,总提取物被认为是"活性部位",但必须有标志性化合物用于质量控制。目前,蜂胶主要以总黄酮含量为指标进行质量控制,缺少标志性化合物做指标,而不同植物源蜂胶具有各自的特征性成分,因此,按植物来源分别进行化学成分标准化是蜂胶质量控制的可行方向。

表 4.7 主要类型蜂胶的化学成分与其生物学活性
Table 4.7 The chemical composition and its biological activity of main types of propolis

蜂胶类型	地理来源	植物来源	主要成分	参考文献
杨属型	温带地区	杨属	黄烷酮、黄酮、咖啡酸苯乙酯、阿魏酸、咖啡酸、其他酚酸及其酯	Nagy et al.,1986;Greenaway et al.,1989;Bankova et al.,2000
酒神菊属型	巴西	酒神菊属	异戊二烯化 p-香豆酸、半日花烷型二萜类、克罗烷二萜、黄酮类、木脂素类、咖啡酰奎宁酸类	Marcucci et al.,2001;Park et al.,2002;Banskota et al.,2001;Kumazawa et al.,2003
克鲁西属型	古巴、委内瑞拉	克鲁西属	异戊二烯化苯甲酮类、聚异丙烯基化的二苯酮	Cuesta Rubio et al.,2002;Trusheva et al.,2004
黄檀属型	古巴、巴西、墨西哥	黄檀属	异黄酮类、紫檀素	Daugsch et al.,2007,2008;Alencar et al.,2007;López et al.,2014
血桐属型	日本冲绳、中国台湾、肯尼亚、埃及	血桐属	异戊二烯化的黄烷酮	Chen et al.,2004;Kumazawa et al.,2004,2008;Huang et al.,2007;El-Bassuony,2009;Petrova et al.,2010
地中海型	希腊、克里特岛、土耳其	地中海柏木	二萜酸	Silicis and Kutluca,2005;Kalogeropoulos,2009;Popova et al.,2010,2012
桦树型	俄罗斯	桦树属	黄酮、黄酮醇类(与杨树型不同)	Popravko and Sololov,1980
加那利群岛型	加那利群岛	未鉴定	木脂素类	Christov et al.,1999;Kujumgiev et al.,1999

(1)杨属型蜂胶

杨属型蜂胶是研究最早、最系统的蜂胶类型。长期以来,总黄酮含量被认为是杨属型蜂胶质量优劣的评价指标。然而,杨属型蜂胶以黄酮含量高为特征,涉及多种类型的黄酮类组分,因此,总黄酮检测方法也备受争议。Popova 等(2004)提出具有相同或相近的活性化合物分组分析更有利于杨属型蜂胶的标准化。他们认为,杨属型蜂胶的化学特征能用 3 个参数——总黄酮和黄酮醇、总二氢黄酮和二氢黄酮醇、总酚含量——实现特征化；同时建立和确定了快速、低成本的分光光度法来定量测定杨属型蜂胶中这3种主要活性成分，并将该方法与混合14种化合物代表杨属型蜂胶成分的HPLC方法进行比较研究，认为两种方法检测结果一致。用这一方法，他们分析了欧洲和中东不同地区的114个蜂胶样本，也测定了其抗菌活性，大部分样本表现出典型的"杨属型"蜂胶特征：黄酮和黄烷酮含量为8%±4%，二氢黄酮和二氢黄酮醇为6%±2%，总酚类为28%±9%，MIC为(211±132)μg/mL(Popova et al.，2007)。

现已明确温带地区杨属型蜂胶的化学特征是不含有B-环取代基的类黄酮，如松属素、短叶松素、高良姜素和白杨素，以及苯丙酸类及其酯类，如咖啡酸、阿魏酸及CAPE等(Salatino et al.，2011)。随着现代分析技术的发展，多个黄酮或酚酸类化合物同时测定成为蜂胶质量控制的主流。Popova等(2003)基于现有对杨树芽分泌物化学组成的了解(Nagy et al.，1986；Greenaway et al.，1990；Bankova et al.，2000)，选择7个酚类化合物为指标，建立了快速的TLC方法来区分杨属型蜂胶与其他类型蜂胶。但是，测定活性主成分的含量非常困难，因为杨属型蜂胶中超过25种单个酚类具有不同类型的生物学活性(Marcucci，1995；Banskota et al.，2001)。Gardana等(2007)采用HPLC-MS/MS同时分析了欧洲、中国、巴西和阿根廷蜂胶中多种类黄酮和多酚含量，结果表明，欧洲、中国和阿根廷蜂胶以多酚和类黄酮为主要成分，其中含量最多的为白杨素(2%～4%)、松属素(2%～4%)、短叶松素-乙酸盐和高良姜素(1%～2%)；并提出根据总黄酮来衡量蜂胶的质量，低于11%、11%～14%、14%～17%和>17%的蜂胶可分别归类为质量差、可以接受、好和高质量的蜂胶。

(2)酒神菊属型蜂胶

酒神菊属型蜂胶，俗称巴西绿蜂胶，是目前国际上最流行、研究最系统的蜂胶类型。酒神菊属型蜂胶采自巴西东南部和中西部地区，其主要植物来源是酒神菊属植物(*Baccharis dracunculifolia* DC)，主要成分为异戊烯苯丙类和 *p*-香豆酸的异戊烯基衍生物，而黄酮类化合物含量较低。由于缺少酒神菊属型蜂胶的质量评价方法，在生产和贸易中，酒神菊属型蜂胶的质量控制也大多参照杨属型蜂胶的标准，即以总黄酮含量作为质控指标。

目前，比较公认的是，在世界各种蜂胶类型中，仅酒神菊属型蜂胶含有阿替匹林C，因此，大多数学者认为，阿替匹林C是酒神菊属型蜂胶质量控制的有效指标。但王维(2010)采用ESI-MS/MS法对中国蜂胶进行分析，发现峰5的紫外色谱图中有两个吸收峰，最大吸收波长为222nm和307nm，表明峰5可能为含有桂皮酰基结构单元的化合物，并列出总离子流色谱图中峰5的EIC中的一级质谱和二级质谱。m/z 300.4为[M-H]离子峰，283.9为[M-H-OH]碎片离子峰，他们认为这些数据与Sawaya等(2004)报道的阿替匹林C的数据相一致。因此，认为中国吉林蜂胶中峰5为3,5-二异戊烯基-4-羟基桂皮酸(阿替匹林C)。但事实上，Sawaya等(2004)报道的巴西蜂胶中(FIG 1 Brazil-G，Brazil-B2)阿替匹林C的m/z分别为299、284、255、244、200、145，其结果并不一致。此外，吴健全等(2013)通过对中国不同产地蜂胶与巴西绿蜂胶的对比研究认为，中国山东、北京、江苏和新疆蜂胶中均检测到阿替匹林C的存在，含量分别为0.51%、0.34%、0.47%和0.47%，而巴西蜂胶中阿替匹林C含量为2.3%。上述研究认为中国蜂胶中含有阿替匹林C。

以阿替匹林C作为酒神菊属型蜂胶质量控制的指标关键是阿替匹林C是否仅存在于酒

神菊属型蜂胶中。除了上述研究认为中国蜂胶中也含有阿替匹林C外,未见其他有关非酒神菊属蜂胶中发现阿替匹林C的研究报道。为此,浙江大学科研团队对中国蜂胶中是否存在阿替匹林C进行了详细的论证。对中国蜂胶、酒神菊属型蜂胶和阿替匹林C标准品进行对比研究发现,阿替匹林C的ESI(+)二级质谱的m/z分别为301.7、283.7、245.6、227.5、177.4和69.2;ESI(+)二级质谱的m/z分别为255.7、200.5和145.3;酒神菊属型蜂胶中检测到的阿替匹林C的二级质谱图与之一致;而中国蜂胶在阿替匹林C出峰位置附近有个干扰峰存在,其ESI/MS(-)无响应,而ESI/MS(+)m/z 301.7与阿替匹林标准品一致,但是二级质谱的m/z与阿替匹林C的不同,说明该峰并不是阿替匹林C,但该峰极易被误认为是阿替匹林C。从以上研究结果基本证实,中国蜂胶中并不存在阿替匹林C。但采用HPLC或HPLC-MS方法,中国蜂胶中会出现阿替匹林C检测的干扰成分。因此,建立酒神菊属蜂胶中阿替匹林C的检测方法非常关键。

基于植物源对蜂胶进行标准化的方法是可行的,目前以总黄酮含量及多种黄酮或酚酸类化合物共同检测有望建立杨属型蜂胶的化学成分标准化方法;阿替匹林C也是目前公认的酒神菊属型蜂胶定性指标,建立多指标检测方法也有可能有助于建立该类型蜂胶化学成分标准化方法。但是,对于其他类型蜂胶,实现化学成分标准化还有很多工作要做。深入开展不同类型蜂胶化学成分及药理活性的对比研究,将特定的蜂胶类型与特定的生物学活性相联系,将更有效地发挥蜂胶的药理活性。

1.2 化学指纹图谱

蜂胶及制剂药理活性发挥的最大特征在于其成分的整体性和复杂性,而不同来源蜂胶的化学成分组成和含量差异显著,凭借对蜂胶中1个或2个化学成分进行定量,难以全面地反映蜂胶及其制剂的内在质量。指纹图谱是运用现代分析技术对化学信息以图形(图像)的方式进行表征并加以描述,经光谱或色谱测定而得到的组分群体的特征图谱。指纹图谱可起到辨别样品真伪、监测样品质量特征等重要作用,现已广泛地应用于中药材的质量控制中。目前,中药指纹图谱技术已涉及众多方法,包括薄层扫描(TLCS)、高效液相色谱法(HPLC)、气相色谱法(GC)和高效毛细管电泳法(HPCE)等色谱法以及紫外光谱法(UV)、红外光谱法(IR)、质谱法(MS)、磁共振法(NMR)和X-射线衍射法等光谱法,其中色谱方法为主流方法。如能将多维信息化学特征指纹图谱所显示的大量参数配合多个指标成分的量化信息,应用于不同来源蜂胶的化学成分标准化研究中,这将成为评价蜂胶质量优劣的有效指标。

(1)蜂胶的植物来源分析

蜂胶与植物来源成分对比研究最早采用气相色谱和质谱联用技术(Greenaway et al.,1987),但蜂胶中主要活性成分类黄酮和多酚类化合物没有足够的挥发性而不能直接用GC-MS分析,而HPLC-MS能更加全面地表征蜂胶与植物源间主要成分的异同。

Kumazawa等(2003)研究发现,巴西绿蜂胶与酒神菊属植物 *Baccharis dracunculifolia* 的HPLC图谱几乎一样,采用HPLC-ESI(-)-MS进行验证,发现二者乙醇提取物的化学组成一致,p-香豆酸的异戊烯化衍生物是其主要成分,从而确定巴西绿蜂胶的植物来源为 *B. dracunculifolia*。曹炜等(2007)采用高效液相色谱-二极管阵列检测器-多极质谱联用法,分析了陕西、河南、湖北、河北、辽宁、浙江、山东、甘肃和贵州等地的30个蜂胶样品以及2个不同产地杨树芽乙醇提取物的化学成分,结果发现蜂胶与杨树芽化学成分极其相似,松属素、白杨素和高良姜素是存在于蜂胶和杨树芽中的主要黄酮类化合物,证实中国蜂胶的主要植物来源为杨树芽。Chang 等(2008)首次将 HPLC-APCI-MS 应用于蜂胶指纹图谱研究中,作为 GC-MS 的有益补充,更加全面真实地反映出蜂胶的植物来源和地理来源。Maróstica Junior 等(2008)将 HPLC、HPTLC 和 GC-MS 等色谱技术相结合,对巴西绿蜂胶与其植物来源 *B. dracunculifolia* 的乙醇提取物和挥发油进行对比研究,确定了蜂胶的植物来源,同时,有利于找出巴西绿蜂胶的指标成分。Popova 等(2012)采用 GC-MS 确定了地中海柏木(*Cupressus sempervirens*)是地中海型蜂胶的主要植物来源,具有典型的二萜指纹图谱。

(2)蜂胶的分类

蜂胶植物来源多样,化学成分复杂,世界不同蜂胶很难形成统一的标准,目前关于蜂胶质量标准化的研究主要集中在采用不同的手段将蜂胶进行分类。应用 TLC、GC-MS、HPLC、HPLC-MS、HPCE 及 HPCE-MS 进行指纹图谱研究,建立指纹图谱库,采用化学计量学方法,研究不同来源蜂胶的相似性是蜂胶分类最有效的方法,是实现蜂胶化学成分标准化的有效途径。

Yang 等(2010)采用气相色谱-嗅辨仪-质谱联用方法分析了中国 23 个不同地区蜂胶的挥发性成分,发现中国蜂胶的主要挥发性成分为乙酸、2-苯乙基乙酸和萘。除了 4 个样本外,其余的样本以这 3 个成分作为主成分,并分成 3 组,蜂胶中挥发性成分与地理来源紧密相关。马海乐等(2011)采用 GC 建立蜂胶超临界 CO_2 萃取物(SEP)的指纹图谱,用于鉴别市售超临界 CO_2 萃取的蜂胶类产品。结果发现来自 9 省份的 16 批蜂胶样品超临界 CO_2 萃取物的 GC 图谱具有 9 个共有峰,其图谱信息构成蜂胶超临界 CO_2 萃取物的指纹图谱,与所得对照图谱的相似度均大于 0.8。然而,由于蜜蜂在蜂胶采集过程中出现跨区域或者跨树种采集的情况,此方法的聚类分析特征不明显。王蓓等(2011)建立了由 8 个共有峰及其相关信息组成的中国蜂胶超临界 CO_2 萃取物 GC-MS 指纹图谱。通过对 GC-MS 数据进行聚类分析,结果得到 2 大类别:第一类别地区所产蜂胶的超临界 CO_2 萃取物中有 61 种相似性组分;第二类别地区有 75 种相似性成分。结合我国地理区划特征,黑龙江和河北 2 省、云南和海南 2 省、山东、河南、湖北、湖南和四川 5 省的蜂胶超临界 CO_2 萃取物特征成分组成相近。Pellati 等(2013)首次将顶空固相微萃取-气相色谱质谱分析法(HS-SPME-GC)用于蜂胶挥发性物质的分析,从意大利不同区域采集获得的蜂胶中鉴定得到 99 种成分,这些成分主要包括苯甲酸及其酯类化合物。此项研究为蜂胶中的挥发性成分提供了指纹图谱,为全面理解蜂胶的化学成分提供了参考。

Cheng 等(2013)利用 GC-MS 联合电子鼻对蜂胶的地理来源进行了分析研究,首先利用动态顶空取样对蜂胶的主要成分进行分析,并利用电子鼻对蜂胶中 28 种气味相关化合物进行了分类,检测频率分析结果显示,此种方法能有效区分从中国 4 种不同地理来源收集得到的 12 个代表性样本,这也为中国蜂胶的地理溯源提供了参考依据。Sârbu 和 Moţ(2011)采用 TLC 对不同来源的罗马尼亚蜂胶进行分层模糊聚类和影像分析,结果表明,该方法可对蜂胶的植物来源和地理来源进行准确分析。

尽管色谱指纹图谱分析是一种十分可行的质量控制模式,但受重复性和成本的限制,难以普及。Bertrams 等(2013)采用 TLC 和 TLC-MS 分析德国蜂胶和杨树芽提取物中酚类化合物的组成,将德国蜂胶分成 3 种类型,确定黑杨杂交种和白杨是蜂胶的主要植物来源。

Morlock 等(2014)首次将 HPTLC 与 DART-MS 联合应用快速分析蜂胶,并进行多变量数据分析(PCA)、层序聚类分析(HCA)和线性判别分析(LDA)。该方法基于苯酚化合物图谱将德国 91 个蜂胶与其他地区蜂胶分开。Ristivojevic 等(2014)将 HPTLC 与影响分析和模式识别联合,用于 52 个塞尔维亚和 1 个克罗地亚蜂胶的指纹分析和分类。

朱恩圆等(2005)采用 RP-HPLC 梯度洗脱的方法对蜂胶总黄酮进行研究,对不同产地的蜂胶指纹图谱进行比较,选出 10 个稳定的且具有代表性的峰作为共有峰,建立了蜂胶的指纹图谱。Hernandez 等(2005)分析了 19 个古巴黄色蜂胶样本,根据 NMR、HPLC-PDA 定性和定量分析将这些样本分成 2 组:一组以富含三萜醇以及少量的多甲基黄酮为主;另一组以乙酰基三萜为主。Volpi 和 Bergonzini(2006)采用 HPLC-ESI/MS 对不同来源蜂胶的化学成分和特征图谱进行对比研究,发现阿根廷、意大利和西班牙的蜂胶乙醇提取液具有类似的总离子流图;相反,阿塞拜疆、中国、埃塞俄比亚和肯尼亚蜂胶具有特殊的总离子流图。Gardana(2007)采用 HPLC/DAD-MS/MS 从阿根廷、巴西和意大利蜂胶中鉴定出 60 种化合物,并发现 3 个地区蜂胶所含成分明显不同,阿根廷蜂胶与意大利蜂胶具有某些共有成分,但与巴西蜂胶没有共有成分。Cuesta-Rubio 等(2007)采用 HPLC-PDA、HPLC-MS 和 NMR 对古巴 11 个地区的 65 个蜂胶样品进行分析,发现古巴蜂胶可分为 3 类,富含聚异戊烯化二苯甲酮的棕色蜂胶,主要成分为异黄酮的红蜂胶,以及以脂肪族类化合物为主要成分的黄蜂胶。

Zhou 等(2008)对中国 10 个省 17 个地区的 120 种蜂胶样品建立了 HPLC 指纹图谱。根据色谱图的相似性以及芦丁、杨梅酮、槲皮素、山奈酚、芹菜素、松属素、白杨素和高良姜素的含量建立不同来源蜂胶的相关性,从而区分国内不同来源的蜂胶。沙娜等(2009)应用反相高效液相色谱法使蜂胶中各成分得到较好的分离,根据检测结果确定了 15 个共有指纹峰,并指认了其中的 14 个成分,分别为咖啡酸、对香豆酸、异阿魏酸、3,4-二甲氧基肉桂酸、5-甲氧基短叶松素、短叶松素、乔松素、咖啡酸卞酯、5-甲氧基高良姜素、白杨素、咖啡酸苯乙基酯、高良姜素、3-甲氧基高良姜素和咖啡酸桂皮酰酯,以建立蜂胶药材指纹图谱,对蜂胶进行质量控制。韩红祥等(2011)应用 HPLC-MS 联用技术研究不同基源蜂胶指纹图谱,发现 10 个产地蜂胶有 16 个共有峰。雷雨等(2012)采用共有峰率和变异峰率双指标序列分析、聚类分析法研究了蜂胶提取物的 HPLC 指纹图谱,并比较二者的差异,发现双指标序列分析法和聚类分析法都能反映蜂胶 HPLC 图谱的特征。前者能利用共有峰率和变异峰率指明任意两张样品图谱的差异,从而反映两个样品组分的差别,但样本量多、计算工作量大;后者能较快聚类大量的 HPLC 指纹图谱特征,并提供直观的分类图,但对图谱特征的鉴别能力不如前者。

Righi 等(2013)比较了来自巴西 6 个地区的 8 个代表性蜂胶样本乙醇提取物、氯仿提取物之间的化学成分差异。采用 HPLC/DAD/ESI/MS 和 GC-MS 检测方法,并根据化学成分将所取样本分为 2 类,一类主要是来自巴西皮奥伊州、戈亚斯州的巴西黑蜂胶,此类蜂胶主要特点是富含黄烷酮类、黄烷酮糖苷类化合物;另一类主要是来自巴伊亚州、米纳斯吉纳斯州、圣保罗州及巴拉那州,也是人们最为熟知的巴西绿蜂胶,此类蜂胶富含异戊二烯苯丙烷类化合物和咖啡奎宁酸类化合物;并首次从巴西蜂胶中鉴定出异戊二烯黄酮类化合物和夏佛塔苷类化合物。

Falcao 等(2013)对所收集的 40 个葡萄牙蜂胶样本采用二极管阵列检测色谱偶联电喷

雾多级串联质谱(LC-DAD-ESI-MS^n)技术进行化学成分的检测,从中发现了76种多酚类化合物,并可将所有蜂胶样本分为2大类:一类为普通温带蜂胶,其特征为富含杨树型蜂胶共有的酚类化合物,如黄酮、甲基黄酮、苯丙酸及其酯等;而另一类葡萄牙蜂胶比较特殊,含有槲皮素糖苷及山柰酚糖苷类化合物,其中很多成分在蜂胶中并未有过报道,并进一步鉴定出了山柰酚-二甲基乙醚,其可作为葡萄牙蜂胶的地理标志物。Kasote等(2014)利用超高液相色谱-电喷雾-质谱(UPLC-ESI-MS)分析了39个南非蜂胶样本,并将这39个样本与3个巴西蜂胶样本的化学成分进行对比,采用化学计量学方法对南非蜂胶的化学-地理格局进行了进一步分析。结果表明,南非蜂胶总体上可以分为两大类,且这两大类蜂胶的主要化学成分同巴西蜂胶截然不同,大部分南非蜂胶样本同温带区域蜂胶化学成分具有一定相似性,进而选定了15种酚酸类、黄酮醇类化合物作为南非蜂胶的代表性成分。

Sawaya等(2004)采用ESI-MS和ESI-MS/MS直接分析了欧洲、北美、非洲和巴西蜂胶的乙醇提取物。结果表明,ESI-MS提供了特征性指纹质谱,可将蜂胶按地理来源分组;化学计量多变量分析表明蜂胶ESI-MS指纹图谱方法是可靠的;在线ESI-MS/MS提供的特征性[M-H]-离子标志物增加了指纹图谱的灵敏度;通过与标准品对比,鉴定出8个标志性成分 p-香豆酸、3-甲氧基-4--羟基肉桂醛、2,2-二甲基-6-羧甲基-2H-1-苯并吡喃、3-苯基-4-羟基肉桂酸、白杨素、松属素、3,5-二苯基-4-羟基肉桂酸和二咖啡酰奎宁酸;ESI-MS,指纹图谱方法能区分不同的差异成分,筛选同样的来源,揭示不同地理来源蜂胶中特征性、极性和酸性化学成分。随后,他们又采用轻便超声喷雾电离源质谱(EASI-MS)对五大洲采集的样本进行指纹图谱鉴定,该方法简便、省时,EASI-MS数据的主成分分析分组与样本植物源一致,可以特征化蜂胶的地理来源(Sawaya et al.,2007,2010)。

López等(2014)采用电喷雾离子化质谱指纹图谱法对巴西不同地区采集的红蜂胶样本进行分析。根据化学组成和指标化合物对这些样本进行分类,巴西红蜂胶可以分为3类。Mot等(2010)用反射光谱区分39个罗马尼亚蜂胶样本的植物来源,发现在220～850nm的紫外可见光区,样本间存在差异。采用聚类分析、主成分分析和线性判别式分析,得到零阶系数的数字化数据,零阶系数归一化和一阶系数导数光谱证实该技术的可行性。根据植物源地理位置,样本分为两大类,主要来源于森林地区和草原地区。在第一组,根据植物主要类型,分成2个亚组,落叶的和树脂的,而后一组根据牧场的种类分成3个亚组。Cantarelli等(2011)建立了阿根廷不同来源蜂胶8种化学元素(Br、Co、Cr、Fe、Rb、Sb、Sm和Zn)含量的测定方法,通过该方法可以对不同来源的蜂胶进行分类。

Cai等(2012)采用近红外光谱法分析了中国不同地区蜂胶中黄酮类化合物的含量,并基于这个方法研究了蜂胶样本的地理来源情况,选取了180个中国不同地区蜂胶样本,并针对蜂胶中的黄酮类化合物建立了偏最小二乘模型,并利用主成分分析-马氏距离法对不同地理来源蜂胶样本进行归类,精确度达到了100%,同时也利用该方法对蜂胶样本进行了质量评价。Yang等(2016)提出一种以近红外光谱技术快速鉴别蜂胶品种的方法,以3种不同植物源的蜂胶(杨树型蜂胶、桦树型蜂胶和橡树型蜂胶)为研究对象,利用傅里叶变换近红外光谱仪对蜂胶的无水乙醇溶液进行光谱扫描,采用主成分分析结合马氏距离判别法和典型判别分析,分别建立了蜂胶品种的判别模型并对其性能进行检验。主成分分析结合马氏距离判别法与典型判别分析对蜂胶样品的分类效果均较好。近红外光谱技术结合化学计量学方法应用于蜂胶植物源的快速、准确鉴别具有一定的可行性和实用性。

Zhang 等(2014)采用多种现代分析化学联用技术,包括液相色谱(LC)-紫外检测(UV)-蒸发光散射检测(ELSD)联用、LC-高分辨率质谱(HRMS)联用、气质联用(GC-MS)、LC-二极管阵列检测器(DAD)-HRMS/MS 联用技术,对 22 个撒哈拉沙漠以南的非洲蜂胶样本的化学成分进行了全面分析。通过分析 LC-UV 和 ELSD 热图数据,并进行主成分分析后发现,这些非洲蜂胶样本并没有非常明显的地理界定。三萜类化合物被认为是大部分非洲蜂胶的主要成分,而其他一些蜂胶则主要归类为温带杨树型和地中海型蜂胶。此外,他们还发现尼日利亚蜂胶样本因其具有异戊二烯异黄酮类化合物而显著区别于其他蜂胶样本,该蜂胶样本的化学成分也同巴西红蜂胶十分类似。在尼日利亚蜂胶样本中还首次发现了芪类化合物,这也是首次在蜂胶中发现该类化合物。Pierini 等(2016)开发了一种基于数码图像技术的蜂胶地理溯源方法,对 6 个采集于阿根廷不同地区的毛胶进行数码图像分析,结合色彩分析模型(灰度 RGB 和 HSI)及多种变量分析模型,发现采用连续投影算法线性判别分析得到的强度直方图能 100%辨别得到蜂胶的地理来源数据,为蜂胶质量控制提供了新的方法。

1.3 蜂胶的多指标检测

蜂胶化学组成复杂,Bonvehi 等(1994)试图建立杨属型蜂胶抗菌活性和不同活性成分比例的相关性,发现没有单一化合物超过 Pearson-Lee 值。Kujumgiev 等(1999)认为蜂胶所表现的任何生物学活性都不可能用单一成分来表达。目前,蜂胶的质量控制主要是快速、准确地对蜂胶中多种化学成分同时定性定量分析。

Morlock 等(2014)首次将 HPTLC 和 DART-MS 联合使用用于对蜂胶化学成分差异方面的研究中。相关的统计学分析方法的应用显著降低了对蜂胶进行化学成分分析的复杂程度,节约分析时间,降低实验费用,而联合正交处理法对样本进行超快样本鉴定也获得了高水平的置信概率。HPTLC 结合后续选择性的色谱衍生化可为蜂胶中标志性成分提供有关极性、功能基团和光谱性能方面的信息,而 DART-MS 则可用于蜂胶主成分(酚酸类标志物)基本化学结构式的确定。他们成功地将该方法应用于对 91 个来自德国和其他地区蜂胶中酚类化合物成分的鉴定。

Kumazawa 等(2004b)对不同地理来源蜂胶中的咖啡酸、p-香豆酸、阿魏酸、3,4-二甲氧基肉桂酸、短叶松素-5 甲基醚、芹菜素、槲皮素、山柰酚、短叶松素、肉桂叉乙酸、白杨素、松属素、高良姜素、短叶松素-3-乙酸酯、咖啡酸苯乙酯、肉桂酰咖啡酯、柚木柯因、阿替匹林 C 等进行 HPLC-ESI/MS 定量分析,发现蜂胶抗氧化活性与其较高的多酚含量相关,确定了山柰酚、咖啡酸、阿魏酸和咖啡酸苯乙酯具有较高的抗氧化活性。李熠等(2007)采用超高效液相色谱法同时测定蜂胶中的黄酮类物质如芦丁、杨梅酮、桑色素、槲皮素、山柰酚、芹菜素、松属素、白杨素、高良姜素、金合欢素,酸类物质如阿魏酸以及脂类物质咖啡酸苯乙酯等 12 种成分的含量,提供了一种更为全面的评价蜂胶质量的新方法。Medana 等(2008)建立了 HPLC-ESI/MS 同时定性定量分析蜂胶中白杨素、柚木柯因、芹菜素、金合欢素、槲皮素、高良姜素、山柰酚、松属素、樱花素、异戊二烯、苯甲基和苯乙基咖啡酸的方法。

蜂胶药理活性的发挥取决于其化学成分的组成及含量。Volpi 和 Bergonzini(2006)采用 HPLC-ESI/MS,以金合欢素、芹菜素、黄芩素、儿茶酚、白杨素、高良姜素、染料木质素、山柰酚、木樨草素、杨梅素、柚皮素、松属素、枸橘苷、槲皮素和香草醛为标准品定性定量测定,结果表明,阿根廷、意大利和西班牙蜂胶中含有大量的松属素(分别占检测总黄酮的 49%、

48%和39%);中国、阿塞拜疆和埃塞俄比亚蜂胶中也含有大量的松属素(分别占检测总黄酮的63%,46%和62%)。然而,中国蜂胶中没有鉴定出染料木黄酮、山柰酚、芹菜素和白杨素;阿塞拜疆蜂胶中未鉴定出染料木黄酮、山柰酚、金合欢素和白杨素;埃塞俄比亚蜂胶中未鉴定出山柰酚和金合欢素;肯尼亚蜂胶中未鉴定出类黄酮;阿根廷、意大利和西班牙蜂胶比阿塞拜疆、中国、埃塞俄比亚和肯尼亚蜂胶含有更多的多酚。Pellati等(2011)采用HPLC-ESI-MS/MS对意大利市场的蜂胶醇提取中多种活性成分同时定性定量,包括咖啡酸、p-香豆酸、阿魏酸、异阿魏酸、3,4-二甲基咖啡酸、槲皮素、短叶松素-5甲基醚、槲皮素-3甲基醚、肉桂酸、白杨素-5甲醚、芹菜素、山柰酚、短叶松素、异鼠李素、木樨草素甲醚、槲皮素二甲醚、高良姜素-5甲醚、短叶松素-5甲醚-3-O-醋酸盐、肉桂叉乙酸、槲皮素-7甲醚、槲皮素二甲醚、咖啡酸苯乙酯、咖啡酸苯甲酯、白杨素、松属素、高良姜素、咖啡酸异戊二烯基酯、短叶松素-3-O醋酸盐、甲氧基白杨素、p-香豆素异戊二烯基酯、p-香豆素苯甲酯、咖啡酸肉桂酸酯、短叶松素-3-O丙酸酯、p-香豆素肉桂酯、短叶松素-3-O丁酸酯、短叶松素-3-O戊酸酯、短叶松素-3-O己酸酯、p-甲氧基肉桂酸肉桂酯。尽管市场上的样本色谱图类似,但是含量差别较大,总酚含量范围为0.17~16.67mg/mL,类黄酮为2.48~41.10mg/mL。

Shi等(2012)分析了中国蜂胶甲醇提取物的植物化学组成,并研究了其体外抗炎及自由基清除能力。他们采用超高效液相色谱法(UHPLC)鉴定了15个中国不同地区蜂胶样本中的11种化合物,采用超高效液相色谱联合飞行时间串联质谱法(UPLC/Q-TOF-MS)鉴定出中国蜂胶中的38种化合物。王冰等(2014)以河南南阳蜂胶为原料,通过RP-HPLC-DAD-ESI-MS和MS法对蜂胶乙酸乙酯提取物进行分离鉴定,首先以10种黄酮对照品为分析对象建立了RP-HPLC-DAD的分析方法,在此基础上,采用RP-HPIC-DAD-ESI-MS/MS对蜂胶乙酸乙酯提取物进行分析,发现河南蜂胶中可能含有槲皮素、芹菜素、山柰酚、异鼠李素、白杨素、松属素、高良姜素、短叶松素、短叶松素-3-乙酸酯、短叶松素-3-丙酸酯和短叶松素-3-异丁酸酯等11种黄酮类化合物以及对甲氧基肉桂酸肉桂酯和咖啡酸苯乙酯两种酚酸及其酯类化合物。罗照明等(2013)采用高效液相色谱方法对9种中国河南蜂胶样品中的多酚类物质进行分析,从75%乙醇提取物中鉴定出29种多酚类成分,对其中25种成分进行定量分析。结果表明,河南蜂胶中含有大量的酚酸、黄酮及酯类物质,酚酸及其酯类中含量较高的主要有咖啡酸苯乙酯、亚桂皮乙酸、p-香豆酸、苯甲酸、异阿魏酸、咖啡酸、3,4-二甲氧基肉桂酸,其中咖啡酸苯乙酯平均含量最高;黄酮中含量较高的主要有短叶松素-3-乙酸酯、白杨素、松属素、短叶松素及高良姜素,其中短叶松素-3-乙酸酯的平均含量最高。Castro等(2014)通过高效液相色谱偶联电喷雾质谱法(HPLC-ESI-MS/MS)鉴定了来自智利6个地区蜂胶样本中的30种酚酸类化合物,发现智利库拉卡维地区蜂胶CAPE含量最为丰富,而布因地区蜂胶中咖啡酸苄酯和槲皮素含量最为丰富;还发现短叶松素是6个智利蜂胶样本中均可以检测出的酚酸类化合物,可作为地理标志性成分用于智利地区蜂胶产品的标准化。

Bertelli等(2012)利用磁共振氢谱法(^{1}H-NMR)分析了蜂胶提取物中发挥生物活性功能的化合物,选择蜂胶中典型的12种酚类化合物作为参考化合物,分析了他们在蜂胶中的含量,并用HPLC-MS对实验结果进行确认。结果表明,采用磁共振氢谱法联合光谱分析法能很好地实时检测蜂胶中的酚类化合物。

高效毛细管电泳法与HPLC互补可以测定蜂胶中的芳香类化合物如酯类化合物、糖、糖醇类化合物、氨基酸、脂肪酸及黄酮类的含量。Volpi(2004)采用毛细管电泳技术(CZE)

同时检测了松属素、金合欢素、白杨素、芦丁、儿茶素、柚皮素、高良姜素、木樨草素、山柰酚、芹菜素、杨梅素、槲皮素、肉桂酸、咖啡酸和白藜芦醇。

多酚类化合物一直被认为是蜂胶中的主要生物学活性成分,但蜂胶中的其他植物来源成分如单宁类物质等,一直未受到研究人员的重视。Mayworm等(2014)通过对巴西7个地区的蜂胶样本检测发现,所有蜂胶样本中的原花青素反应均为阳性,提示单宁类化合物在巴西蜂胶中普遍存在。定量检测结果发现,蜂胶中单宁含量在0.6%~4.1%,单宁含量与蜂胶总酚酸含量呈显著正相关。红蜂胶、绿蜂胶中的单宁含量较高,而褐蜂胶中单宁含量较低,提示单宁含量可作为一种新的参数用于蜂胶的质量控制中。

尽管HPLC-MS已广泛用于不同来源蜂胶中多种成分的同时定性定量,但不同来源蜂胶中主要成分差异较大,因此,根据选定的几种成分仍不能反映蜂胶的来源及整体的质量或药效。建立多种成分定量结合指纹图谱是有效的蜂胶分类方法,是实现蜂胶化学成分标准化的有效途径。

1.4　蜂胶成分标准化指标成分

为使质量控制既能充分体现蜂胶特点,又能符合蜂胶功效的要求,蜂胶的质量控制指标应包括具有整体生物学功能的共同成分和具有特定生物活性的功效成分。以蜂胶中的标志物为依据进行分类将使蜂胶的分类不再受地域所限,对反映蜂胶的质量优劣更具代表性。标志物是指在某种类型蜂胶中含量比较多,而且比较稳定的已知物质,例如,杨属型蜂胶中的CAPE,巴西绿蜂胶中的阿替匹林C。

(1)咖啡酸苯乙酯(CAPE)

CAPE,化学名为3-(3′,4′-二羟基苯基)-2-丙烯酸苯乙醇酯,分子量为284.31,其分子结构式见图4.1。

图4.1　CAPE的分子结构式

CAPE是蜂胶中报道的第一个主要活性成分(Bankova et al.,1992),也是蜂胶中药理活性研究得最多的活性成分,已发现具有多种生物学活性,包括抑制核因子κB、抑制环氧合酶2活性和表达、抑制TREK-2钾通道激活、预防回肠Th2免疫反应、抑制细胞增殖、诱导细胞周期阻滞和凋亡、减少淤积的肠道损伤、预防腺瘤性息肉和结肠癌等(Aviello et al.,2010)。自从1987年,Grunberger等报道了从蜂胶中提取CAPE以来,CAPE便是蜂胶中研究最热门的成分。1990年,Greenaway和Whatley采用GS-MS从白杨树幼芽中分离出CAPE。1993年,Sud'ina等使用酯化过程实现了CAPE的全合成。2000年,Bankova等报道,CAPE是欧洲、亚洲和北美洲蜂胶的特征成分之一。Kumazawa(2004b)研究发现,产自中国的蜂胶中咖啡酸苯乙酯含量比较高,可达到15~29mg/g,其次是产自乌拉圭、匈牙利、乌兹别克斯坦的蜂胶中CAPE含量也相对较高,而产自巴西的蜂胶中基本不含CAPE。但是,CAPE是否是杨属型蜂胶的特征性成分还有待深入研究。

(2)阿替匹林 C(Artepillin C)

阿替匹林 C,化学名为 3-[4-羟基-3,5-二(3-甲基-2-丁烯基)苯基]-2-(E)-丙烯酸或 3,5-二异戊烯基-4-羟基肉桂酸,分子量为 300.40,是一种 2,4,6-三取代苯酚类物质。它也是巴西蜂胶中发现的一种主要的酚酸,其分子结构式如图 4.2 所示。

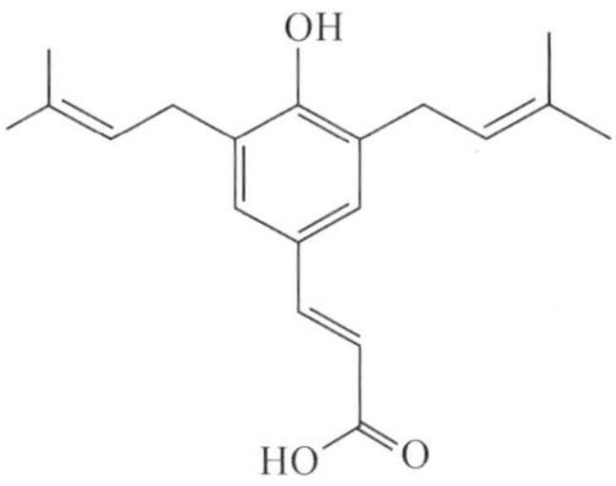

图 4.2　Artepillin C 的分子结构式

阿替匹林 C 是巴西绿蜂胶的代表性成分,首次由 Bohlmann 和 Jakupovic 于 1979 年从菊花(*Flourensia heterolepis*)中分离出来,但其他种属如 *Baccharis*(Bohlmann et al.,1981)和 *Relhania*(Tsichritzis et al.,1990),特别是产自巴西东南部的巴西绿蜂胶的主要胶源植物酒神菊树 *Baccharis dracunculifolia* DC.(Compositae)中都含有大量的阿替匹林 C 成分(de Sousa et al.,2009;Park et al.,2004)。随着对巴西绿蜂胶研究的逐步深入,针对阿替匹林 C 在蜂胶质量控制、生物学活性、分离纯化以及人工合成等方面的研究也得到了较为广泛和深入的开展。2002 年,阿替匹林 C 的全合成完成(Uto et al.,2002)。

由于阿替匹林 C 是巴西绿蜂胶的代表性成分,阿替匹林 C 的有无是确定巴西绿蜂胶真实性的主要依据,而且其含量高低是评价巴西绿蜂胶质量优劣的主要指标。Kumazawa 等(2004)对阿根廷、澳大利亚、巴西、保加利亚、智利、中国、匈牙利、新西兰、南非、泰国、乌克兰、乌拉圭、美国和乌兹别克斯坦等地的蜂胶进行检测,结果仅在巴西蜂胶中检测到阿替匹林 C,而且含量较高,在乙醇提取物中的含量达 43.9mg/g。但是,阿替匹林 C 在巴西蜂胶中的含量受地理来源和植物来源影响较大。Gardana 等(2007)建立了同时测定 57 种类黄酮或酚酸的 HPLC-MS/MS 法,并对欧洲、中国和阿根廷的蜂胶进行检测,发现类黄酮含量的高低是评价欧洲、中国和阿根廷蜂胶质量好坏的重要指标。但是,巴西绿蜂胶中类黄酮含量相对较低,而酚酸含量较高,特别是阿替匹林 C 含量是目前评价巴西绿蜂胶质量优劣的主要指标。韩利文等(2008)建立了反相高效液相色谱法测定蜂胶中阿替匹林 C 含量的方法,为巴西绿蜂胶的质量评价提供了更加简单易行的方法。Matsuda 和 de Almeida-Muradian(2008)测定了巴西不同地区收集的 33 个蜂胶样本中检测到其阿替匹林 C 的含量,结果发现样本间的含量差异较大,含量范围从 0% 至 11% 不等。从巴西东南部采集的样本中阿替匹林 C 的含量最为丰富(通常在 5%~11%),而从东北部采集的样本中却基本检测不到阿替匹林 C,这是因为巴西绿蜂胶的主要胶源植物酒神菊在巴西东北部基本没有分布。此外,巴西绿蜂胶中阿替匹林 C 含量受季节影响明显,春夏季采集的绿蜂胶中阿替匹林 C 含量明显高于秋冬季采集的蜂胶(Simoes-Ambrosio et al.,2010)。

除了 CAPE 和阿替匹林 C,蜂胶中其他成分还未被系统研究过。因此,蜂胶中更多具有典型特征的质控指标成分还有待深入研究与开发。

1.5 蜂胶活性指纹图谱

蜂胶中多种化学成分的综合检测以及蜂胶指纹图谱技术,已被国内外研究者广泛用于蜂胶化学成分分析、真伪鉴别、质量评价中,已成为蜂胶化学成分及质量标准研究的方向。但是,蜂胶指纹图谱所面临的局限越来越明显,它反映的仅是蜂胶中的化学信息,与药效活性信息无关。这些化学信息所对应的物质类群并不完全等同于蜂胶的药效物质基础,因此,不能实现通过控制蜂胶中各化学成分的"量"进而控制蜂胶的"质"。尽管目前蜂胶药理活性的研究都离不开蜂胶中主要化学成分的鉴定与相应活性的研究,但化学成分的检测与药理活性的研究是分离的,迄今为止,蜂胶发挥药理活性的物质基础及基于活性对蜂胶进行质量评价还未见报道。

"活性指纹图谱"是指与药理活性及化学成分相关联的指纹图谱,建立在有效部位且有效成分明确的基础上,以有代表性的有效成分来反映蜂胶特征性的指纹图谱。蜂胶标准化可以把特定化学成分的蜂胶与特定的生物学活性联系起来,以更有效地使用蜂胶。因此,采用多指标成分定量结合指纹图谱的质量控制模式结合蜂胶的生物学活性评价,建立蜂胶生物活性指纹图谱是未来蜂胶质量控制发展的方向。

2 蜂胶与杨树胶鉴别研究

蜂胶生产受蜂种、采胶方式、季节、气候条件以及生产过程等多种因素的影响,产量相对有限。蜂胶掺杂使假现象愈演愈烈,真蜂胶的价值难以体现,蜂农、蜂胶加工企业及消费者的权益难以保障。

温带地区公认的蜂胶植物来源是杨属及其杂交属的芽苞分泌物。蜂胶原料的紧俏和利益的驱动,促使不法厂家用杨树芽熬制出杨树胶掺入到蜂胶中或直接以杨树胶冒充蜂胶,并加入类黄酮类物质,人为地提高总黄酮含量以达到以次充好的目的。

杨属植物是杨属型蜂胶的植物来源,近年来,蜂胶与杨树胶二者的对比研究成为蜂胶质量控制研究的热点。研究者尝试通过液相色谱法、气相色谱质谱法、红外光谱法、热分析法等现代分析分离技术以及蜂胶与杨树胶的化学成分对比研究等多种方法来鉴别蜂胶与杨树胶。

2.1 液相色谱法(HPLC)

潘建国等(2002)采用 HPLC 在蜂胶中检测到 10-羟基-2-癸烯酸(10-HDA),含量大约为 0.22%。10-HDA 是由蜜蜂上颚腺分泌的一种蜂王浆中所特有的高效生物活性物质,蜜蜂有可能在采胶的过程中加入了该物质。然而他们仅测定了 3 个样本,是否所有蜂胶中都含有 10-HDA、含量多少等,都有待进一步研究。盛文胜等(2008)用 HPLC 对蜂胶与杨树胶中肉桂酸含量进行研究,发现真假蜂胶中肉桂酸的含量存在差异。但是,由于不同地区采集的蜂胶即使同一成分含量差异也较大,因此还要进行大量取样才能确定其可行性。

液相色谱指纹图谱近年在蜂胶与杨树胶鉴别方面发挥了很大作用。周萍等(2005)测定了蜂胶、杨树胶和掺假蜂胶的 HPLC 指纹图谱,发现蜂胶含有 2 个杨树胶不含有的特征峰,杨树胶中含有 1 个蜂胶中不含有的特征峰。他们对不同产地来源的 62 个蜂胶样品建立 HPLC 指纹图谱,将蜂胶图谱分为 6 个区域,提出了以第四区的 1 号峰、2 号峰、5 号峰为蜂胶鉴别的主要依据,结合 6 个区域的指纹图谱进行真伪鉴别(周萍等,2009)。李樱红等

(2014)建立了区分蜂胶与杨树胶的 HPLC 指纹图谱,以对照模板为参照,通过《中药指纹图谱相似度计算软件》计算,对蜂胶、杨树胶、蜂胶中掺杨树胶、蜂胶中掺外源性黄酮类化合物、既非蜂胶又非杨树胶等样品的相似度范围进行了限定。该方法可以鉴别出杨树胶掺假量 40% 以上的样品。

张翠平等(2011)采用 HPLC 建立区分蜂胶与杨树胶的指纹图谱,8 个不同地区蜂胶的 HPLC 指纹图谱十分相似,与杨树胶表现出完全不同的指纹图谱特征。其中,蜂胶中 2 个共有峰是杨树胶中没有的特征指纹峰,而 3 个不同来源的杨树胶中有 3 个共有峰是蜂胶中没有的特征指纹峰,且这 5 个特征峰明显,相互间没有干扰,易鉴别。

王小平等(2009)研究发现,蜂胶与杨树胶主要化学成分基本相同,只是相对含量有所不同,并建立了蜂胶指纹图谱分析方法,确定有 13 个共有峰,其中 4 个成分为已知成分,而 10 批蜂胶中 8 号和 10 号共有峰的峰面积甚微,而在杨胶中这两个峰均较高,认为这两个峰的明显程度用来进行蜂胶与杨树胶的鉴别。但该方法所需样本量小,需收集不同产地大量的样品进行系统规范研究。

刘莉敏等(2014)对 9 个国内外蜂胶样品 75% 热乙醇提取液用超高效液相色谱四级杆飞行时间质谱联用仪(UPLC-Q-TOF-MS)进行了一级阳离子化$[M+H]^+$总离子流色谱(TIC)和总质谱(色谱和质谱指纹)的采集分析。从 TIC 指纹中发现,2 个广东成品蜂胶色谱指纹与其余 7 个极为不同,判断为异常蜂胶;9 个蜂胶样品均未检出芦丁和槲皮素。总质谱指纹主成分分析结果表明,2 个异常蜂胶的判别与色谱指纹结果一致。由于色谱指纹图谱以"整体性"和"模糊性"为特点,在没有具体指标成分参考条件下,在不同检测条件下图谱很难完全重复,这给方法的推广应用带来困难。

Zhang 等(2014)采用 RP-HPLC 开发出了一种同时分析中国蜂胶 12 种黄酮类和 8 种酚酸类化合物的方法,并成功将其应用于中国蜂胶的质量控制。分析结果显示,香草酸、芦丁、杨梅酮和木樨草素在所有分析的蜂胶和树胶样本中均未被检出;而咖啡酸、阿魏酸和 p-香豆酸仅在蜂胶中被检出,而在杨树胶样本中并未发现,提示这几种化合物可作为区分蜂胶与杨树胶的标志性成分。此外,杨树胶的黄酮类化合物图谱同蜂胶相近,短叶松素、松属素、短叶松素-3-O-乙酸酯、白杨素和高良姜素是其中的主要成分。因此,此方法可将咖啡酸、阿魏酸和 p-香豆酸用于区分蜂胶和杨树胶的定性指标,通过测定短叶松素、松属素、短叶松素-3-O-乙酸酯、白杨素和高良姜素含量的多少,在一定程度上可判断中国蜂胶质量的优劣。

2.2 气相色谱与质谱联用(GC-MS)

蜂胶中含有大量的挥发性成分,关于蜂胶与杨树胶中挥发性成分的对比研究也较多。余兰平等(2006)采用 GC-MS 研究发现,蜂胶中芳香酸、醇及酯类物质的含量明显高于杨树胶,杨树胶中总黄酮和萜烯类化合物的含量略高于蜂胶。蜂胶和杨树胶中苯甲醇、3,4-二甲氧基肉桂酸、3-苯基-2,3-二甲基环丙烯、2,6-二羟基-4-甲氧基查耳酮和柯因的含量有比较大的差异,反油酸、1-(3-甲氧苯基)乙酮、4-乙酰基苯甲酸苯基甲基酯只在蜂胶中存在,并且含量在 1% 以上,4-羟基-3-甲基苯乙酮和佛手油烯只在杨树胶中存在。

程焕等(2012)分别采用固相微萃取法(SPME)、动态顶空(DHS)对蜂胶和杨树树胶挥发性成分进行提取,经 GC-MS 法结合计算机检索对其挥发性成分进行分析和鉴定。结果发现,蜂胶中的酯类物质(乙酸-3-甲基-3-丁烯-1-醇酯、3-甲基-2-丁烯-1-醇甲酸酯、乙酸-3-甲基-3-丁烯-1-醇酯、3-甲基-2-丁烯酸-2-苯乙酯、壬酸乙酯、月桂酸乙酯、棕榈酸乙酯)、醇类物质

(3-戊烯-2-醇、3-甲基-2-丁烯-1-醇、α-桉叶醇)、萜烯类物质(β-荜澄茄烯、雪松烯)、烯烃类物质[八氢二甲基-2-(1-亚异丙基)萘]在杨树胶中均未检测到。而杨树胶中醇类物质(＋/－)-α-红没药醇、烯烃类物质α-姜黄烯 2,6,6,9-四甲基-三环[5.4.0.0(2,8)]-9-十一烯也未在蜂胶中检测到。

王向平等(2012)以气相色谱-质谱联用仪分别对蜂胶和杨树胶的化学组分进行了分离鉴定,蜂胶中共鉴定出 75 种物质,杨树胶中共鉴定出 71 种化合物。蜂胶和杨树胶中的 2,6-二羟基-4-甲氧基查耳酮、高良姜素和柚木柯因的含量有明显差异。3,4-二甲氧基肉桂酸、棕榈酸、1-二十六烷醇、1-十九烯只在蜂胶中存在并且含量均高于 2%,愈创木醇、1-甲基-4-(6-甲基-5-庚烯-2-基)苯、2-苯基苯并咪唑在杨树胶中存在而在蜂胶中不存在。因此,可通过检测愈创木醇、1-甲基-4-(6-甲基-5-庚烯-2-基)苯、2-苯基苯并咪唑是否存在于蜂胶中来鉴别蜂胶中是否掺入了杨树胶,而可通过检测 2,6-二羟基-4-甲氧基查耳酮、柚木柯因和高良姜素的含量来确定蜂胶的真伪和掺假程度。

电子鼻是根据仿生学原理,由传感器阵列和自动化模式识别系统所组成,是一种新颖的分析、识别和检测复杂气味和大多数挥发性成分的仪器,已经用于食品的等级划分与新鲜度检测。延莎等(2012a)用电子鼻对来自 17 个省份的 71 个蜂胶样品及其杨树胶进行测定,通过主成分分析,可以较好地区分蜂胶和杨树胶。董捷等(2008)利用电子鼻对中国 14 个省市的共 52 个掺假蜂胶样品进行分析,通过电子鼻对蜂胶芳香特征的响应实验,可以得出电子鼻对蜂胶的芳香成分有明显的响应,并且每一个传感器对蜂胶的响应各不相同。应用电子鼻的几类传感器,针对几类特定的物质(而不是几种物质)进行分析,并对所得的结果应用主成分分析方法(PCA)进行处理,能够鉴别掺假蜂胶。但是在该实验中,也存在几个地区的蜂胶样品难以区分的情况,且该实验中并未取不同品种蜂胶来做对比。因此,对于利用电子鼻鉴别真假蜂胶技术还有待进一步深入研究和完善。

延莎等(2012b)采用气质与嗅闻仪结合的方法,分析了 3 种蜂胶样品和杨树胶的气味活性成分,共鉴别出 48 种气味活性成分,包括酯类、醛类、醇类和酸类化合物。结果表明,形成蜂胶的气味活性成分中具有更多的花香、果香物质,如梨醇酯、糠醛、苯甲醛等化合物,使蜂胶具有更为清香、柔和的总体气味特征,而杨树胶呈现出的整体气味较蜂胶更为刺鼻、尖酸,可以考虑从气味特征上对蜂胶与杨树胶进行识别。

由于不同来源的蜂胶中挥发性成分差异较大,至今还未形成有效的基于蜂胶中挥发性成分的蜂胶与杨树胶的鉴别方法。

2.3 薄层色谱法(TLC)

Tang 等(2014)采用 TLC 来区分蜂胶和杨树胶。结果表明,采用相似度、分层群聚、k-means 聚类、神经式网络、支持向量机等分析手段,能够完全将蜂胶与杨树胶区分开。

2.4 红外光谱法

Wu 等(2008)使用傅里叶变换红外光谱和二维红外光谱联用对不同地区蜂胶及杨树芽分泌物进行分析,发现它们之间具有相似的红外谱图,二者的差异在于长链烃基化合物的不同,包括长链烷烃、长链烷基酯和长链烷基醇。蜂胶提取物比杨树芽提取物中长链烷基类化合物更多,这些化合物中碳原子呈 Z 形排列,而杨树芽提取物中是不规则的。由于红外光谱图具有整体性、特征性和模糊性的特点,可以作为复杂混合体系鉴别与质量监控的快速、有效的方法。红外光谱法是应用于蜂胶真伪鉴别较为可行的一种技术,但其实验结果受样品处理的影响容易产生假象。此外,龚上佶(2011)研究发现,树胶与蜂胶的光谱图差异显著。所有蜂胶在基团频率区存在两个特征峰区域,分别是 2848.53~2849.08 和 2916.76~2917.74(为烷烃的特征峰),而树胶在这两个区域不存在特征峰;杨树胶特征峰集中在 1150~1300(为醇、酚及醚类的特征峰)和 1550~1650(为亚硝基、酰胺及烯烃的特征峰)两个区域,而蜂胶在 1800~4650 间存在很多特征峰。这两方面的差异可用于区别蜂胶与树胶。Xu 等(2013)利用傅里叶变换近红外光谱法(FT-NIR)对中国蜂胶中掺入杨树胶进行检测。首先针对树胶掺假建立了偏最小二乘数学模型,并采用偏最小二乘回归方程对树胶含量进行定量分析,其结果也证实,采用 FT-NIR 光谱结合化学计量学方法可以对中国蜂胶中树胶掺假进行快速检测。

2.5 热分析法

付阳等(2006)采用热分析法对蜂胶和杨树胶的 DSC-TG 曲线的特征区间进行比对,结果发现,TG 曲线分为两个阶段:100~300 度为第一阶段,是小分子的分解,蜂胶的失重率为 47.72%,杨树胶的失重率为 32.69%;340~700 度为第二阶段,为大分子的降解,蜂胶的失重率为 47.29%,杨树胶的失重率为 54.86%。而 DSC 曲线中,蜂胶在 333.0 度有一个吸收峰,在 503.0 度有一个放热峰。杨树胶在 314 度有一个吸收峰,在 478.0 度有一个放热峰。通过比较发现,在第一失重阶段,蜂胶的小分子有机物含量比杨树胶高 15%,而第二阶段,蜂胶的失重较为缓慢,为黄酮类混合物的降解,杨树胶具有典型的高分子聚合物的特征。从两者的残碳率来看,蜂胶几乎为零,杨树胶为 7.48%。但该方法并未形成有效的检测方法得以应用。

2.6 以水杨苷为指标

杨属植物的特征性成分是酚苷,以水杨苷、柳皮苷、2′-苯甲酸水杨苷和特里杨苷含量较为丰富(Thieme,1967;Pearl and Darling,1971)。有关蜂胶化学成分的研究很多,迄今为止尚未有关于蜂胶中存在任何酚苷的报道,但很多研究曾报道过蜂胶中含有水杨酸及乙酰水杨酸等衍生物(Marcucci,1995;Krol et al.,1996)。杨属植物中发现的酚苷均以 β-葡萄糖苷形式连接,β-葡萄糖苷酶可催化水解结合于末端非还原性的 β-D-糖苷键,同时释放出配基与葡萄糖体。Zhang 等(2011a)在进行蜂胶中酶活性研究中首次在新鲜蜂胶的酶提取液中检测到 β-葡萄糖苷酶活力,并证实该酶能够水解芳香基 β-D-葡萄糖苷,与 Pontoh 等(2002)从西方蜜蜂体内分离纯化得到的 β-葡萄糖苷酶性质相符。因此,推测蜜蜂在采集蜂胶以及蜂巢内传递蜂胶的过程中已将酚苷水解。

杨属植物有 100 多种,主要分布在温带和亚热带地区。水杨苷是酚苷的基本结构单元,是更高级的水杨酸盐类衍生物的主要降解产物之一,在所有研究过的杨属植物的皮、叶、花、芽中均有分布(Clausen et al.,1989)。Zhang 等(2011b)用高效液相色谱法优化了杨树芽中水杨苷的测定方法,在蜂场周围分布较多的北京杨和加拿大白杨的叶子和芽中均检测到水杨苷存在;在随机购买的 11 个杨树胶样本中也均检测到水杨苷的存在;而在蜜蜂采集回蜂箱内 5d、10d、15d、20d 及 30d 的蜂胶样本中均未检测到水杨苷的存在。以上事实支持蜜蜂在采集加工蜂胶过程中已将水杨苷水解,而在杨树胶制作过程中水杨苷稳定存在的结论,证明以水杨苷为参照鉴别蜂胶与杨树胶的可行性。除了水杨苷特征峰,还有一个峰 A 在所有杨树胶样本中存在,而在蜂胶中不存在,该峰经鉴定为邻苯二酚(Huang et al.,2014),它在所研究的杨属植物的芽、叶、花、皮中均有分布,其物理性质极其稳定,可在甲醇、酸性、碱性以及中性介质中均能稳定存在。在存在多酚氧化酶的条件下,可被氧化为 1,2-苯醌,但其所需的氧化酶应具有较强的特异性,即可以氧化具有邻位二羟基的多酚类化合物。根据相关研究表明,多酚氧化酶存在于植物组织中,天然状态下无活性。蜜蜂在采集和制造蜂胶过程中,将植物组织破坏并激活多酚氧化酶的活性,使邻苯二酚氧化成邻苯二醌,使其在蜂胶中消失(黄帅,2015)。从收集到的国内不同来源的蜂胶样本中均未检测到水杨苷或邻苯二酚,证实了以水杨苷或邻苯二酚为代表的液相色谱指纹图谱方法鉴别蜂胶与杨树胶是可行的。目前,以水杨苷为指标的蜂胶中杨树胶的鉴别方法已成为行业标准和国家标准,用以区分蜂胶与杨树胶。

2.7 以叶绿素为指标

周立东等(2007)利用光谱法对比蜂胶与杨树叶、杨树芽中叶绿素含量的差别,发现测定的 11 个蜂胶样品中均不含叶绿素,杨树叶提取物中叶绿素含量较高,杨树芽 50% 乙醇提取物中含有叶绿素,但含量较低,杨树芽 95% 乙醇提取物中未能检出叶绿素,杨树胶中也未检测到叶绿素。该方法可以鉴别直接掺入杨树叶或杨树芽粉末的假蜂胶,以及各种含叶绿素杂质的掺假蜂胶,但不能排除其中掺有杨树芽高极性溶剂提取物的可能。

杨茂森等(2012)根据相似相溶原理,利用极性较低有机溶剂对蜂胶及杨树(芽)胶中叶绿素类物质进行溶解、分离萃取,通过分光光度法进行定量对比检测,发现蜂胶中叶绿素平均含量为 6.73mg/100g,提纯杨树(芽)胶中叶绿素平均含量为 111.2mg/100g,两种不同胶体叶绿素含量差异十分明显;在蜂胶中按不同比例掺入杨树(芽)胶进行回收检测试验,发现杨树(芽)胶掺入量与叶绿素的增加量呈线性关系。

3 蜂胶的安全性研究

蜂胶已在食品、药品、化妆品行业得到广泛应用。如果蜂胶的安全不能得到保障,将会危害人类健康。随着食品安全问题日益受到人们的重视,蜂胶本身是否无毒、农药残留、重金属是否超标、蜂胶过敏、人为添加西药等问题已成为蜂胶安全性研究的热点。

3.1 蜂胶的毒性试验

急性毒性是指动物一次或 24h 内多次接受一定剂量的受试物,在短期内出现的毒性反应。迄今为止,关于蜂胶的急性毒性研究已多不胜数,然而,由于提取、加工、给药方式的不同和蜂胶来源各异,蜂胶的急性毒性致死剂量(LD)值差异很大。Gritsenko 等(1976)研究发现,蜂胶对小鼠的半数致死量(median lethal dose,LD_{50})为 2050mg/kg,100% 致死剂量(completely lethal dose,LD_{100})为 2750mg/kg。Ghisalberti(1979)报道指出,蜂胶对小鼠的 LD_{50} 为 700mg/kg,猫可以耐受 100mg/kg 蜂胶乙醇提取物(EEP)的皮下给药。Kleinrok 等(1977)观察到腹腔注射 EEP(100~2000mg/kg)能够以剂量依赖效应降低小鼠的自主运动,并指出大鼠的阈剂量为 1~400mg/kg,小鼠的阈剂量为 10~100mg/kg。Dobrowolski(1991)对 10 只小鼠(雌雄各半)口服 700mg/kg 的蜂胶,经过 48h 观察,无小鼠死亡现象。Arvouet 等(1993)报道蜂胶提取物对小鼠口服的 LD_{50} 超过了 7340mg/kg;而日本食品分析中心通过对巴西和中国产 EEP 进行急性毒性实验,得出经口服 LD_{50} 是 3600mg/kg(陈东海和闫德斌,1999)。Burdock(1998)研究发现,200~5000mg/(kg·d)大剂量口服蜂胶,没有引起实验动物死亡,但考虑到缺乏相应的长期毒性研究,推荐人体安全使用剂量应为1.4mg/(kg·d)或大约70mg/d。Okamoto等(2015)报道了巴西绿蜂胶类似的雌激素样活性,发现蜂胶可以结合人类雌激素受体(estrogen receptors,ERs),同时能诱导雌激素响应基因的表达;进一步的研究发现,蜂胶能有效提升切除卵巢大鼠的子宫净重及腔上皮厚度,并能诱导乳腺导管细胞增殖。由于植物雌激素对机体的内分泌系统具有双向调节作用,因此,对于蜂胶的安全性使用也需要更加全面的评估。

有关蜂胶的长期系统性毒性研究目前还少有报道。Kakehashi等(2016)采用实验周期长达104周的大鼠实验研究了巴西绿蜂胶对大鼠的致瘤性影响。近2年的实验结果表明,日粮中添加0.5%和2.5%蜂胶不会造成大鼠肿瘤病变,也没有对任何器官造成组织病理学变化,同时蜂胶组大鼠的存活率高于对照组,推测可能与蜂胶的抗炎、抗突变活性有关。Cuesta等(2005)研究发现,连续6个月用高剂量(10g/kg)的蜂胶处理金头海鲷(*Sparus aurata*)不会引起任何损伤。Mani等(2006)对不同浓度、不同提取溶剂及不同时期(90d、150d)处理下蜂胶对大鼠血清生化指标的影响展开了系统性研究。结果显示,上述不同处理组大鼠机体均未见任何异常及组织病变,也证实了长期使用蜂胶不会对机体造成心肌损伤。Jasprica等(2007)研究发现,连续1个月口服蜂胶粉对人体内氧化-还原状态及红细胞参数无异常影响。Kashkooli等(2011)随后也报道了长期喂食高剂量的EEP(9g/kg/d)对虹鳟鱼(*Oncorhynchus mykiss*)的生长及血清指标无任何实质性的影响。在鱼食中连续8周分别添加0、0.5、1.5、4.5和9g/kg的蜂胶,所有剂量组中,生长参数和血清总蛋白、白蛋白、球蛋白、低密度脂蛋白胆固醇、高密度脂蛋白胆固醇、三酰甘油、谷氨酸、谷氨酸丙酮酸转氨酶谷草转氨酶、碱性磷酸酶、乳酸脱氢酶等指标均未发生显著变化。因此,蜂胶是一种无毒物质,可长期用于虹鳟鱼的饲料添加。

因此,鉴于蜂胶毒性研究结果及LD_{50}的不一致性,有必要进一步完善蜂胶的毒物学效应评价体系。

为了确保食用安全,研究人员对蜂胶软胶囊进行了毒理学安全性评价,采用急性毒性试验、遗传毒性试验(Ames试验、小鼠骨髓微核试验、小鼠精子畸形试验)和30d喂养试验进行评价。蜂胶软胶囊急性毒性分级属无毒级、无遗传毒性,最大无损害作用剂量大于1.5g/kg,相当于人体推荐摄入量的100倍,说明这些蜂胶软胶囊属安全性保健食品(冯丁山等,2010;傅颖等,2010;谢玮等,2011)。

童晔玲等(2012)对蜂胶洋参软胶囊的食用安全性进行了毒理学评价,结果表明,雌雄大、小鼠经口急性毒性最大耐受量均大于20.0g/kg,Ames试验、小鼠骨髓细胞微核试验、小鼠精子畸形试验3项遗传毒性试验结果均为阴性。30d喂养试验大鼠的生长发育、血液学、生化、脏体比及组织病理学未见异常变化,属安全性保健食品。

尽管多年来对蜂胶毒性的研究表明,蜂胶对人体没有任何毒副作用。但蜂胶大剂量使用时会对机体产生损伤。Pereira等(2008)报道,Swiss小鼠外周血中EEP的含量高于1000mg/kg会对其产生遗传毒性作用并表现出剂量依赖效应,这一发现与之前Tavares等(2006)的研究结果相一致。da Silva等(2015)的巴西红蜂胶水醇提取液的急性毒性研究结果显示,口服剂量达到300mg/kg时无致死效应,然而小鼠表现出明显的中毒迹象,表明其$LD_{50} \geq 300$mg/kg;而亚急性毒性研究结果显示,200mg/kg的口服剂量对雄性小鼠具有最显著的毒性作用。考虑到该类型蜂胶中丰富的异黄酮类化合物的存在，所以推测可能是其雌激素样作用所造成的机体毒性。因此，一次性过量摄入蜂胶会引起血细胞的诱变。当然，所有制成的蜂胶产品正常服用剂量不会达到这个限度。

此外，蜂胶引起急性毒性的个例也有发现。Li 等（2005）曾报道一例由巴西蜂胶诱导的急性肾功能衰竭患者，该患者连续服用蜂胶2周后出现呼吸急促、尿量过少、腿部水肿等现象，而停用后肾功能得以改善，继续服用蜂胶后病症再度恶化；二次停用后肾功能又继续恢复到正常水平。这也是首次关于蜂胶对人体急性毒性的报道，然而由于蜂胶化学成分的复杂性和多变性，目前还不清楚具体的诱因。

3.2 蜂胶中的重金属

蜂胶本身无毒，但在蜂胶的生产过程中有很多可能被重金属污染的环节，如胶源中的植物成分来源于污染严重的地区、使用铁纱收集蜂胶等。如此一来，消费者长期食用或使用这些蜂胶产品会造成重金属在体内富集，危害健康。因此，如何更准确、快速、便捷地检测蜂胶产品中重金属含量就显得尤为重要。

王益民等（2008）采用石墨炉原子吸收光谱法测定蜂胶中的铅含量。Pierini 等（2003）利用电化学分析法研究了阿根廷蜂胶（毛胶）中铅的含量，他们选择铋膜玻璃碳电极代替传统的水银膜电极，减轻了电化学分析的毒性。研究结果发现很多阿根廷蜂胶中铅含量很高，同电感耦合等离子体原子发射光谱法（ICP-AES）获得的结果一致，这也为蜂胶重金属检测提供了一种新的方法。

王强等（2011）采用原子荧光法测定蜂胶中砷的含量，采用湿法消化和干法灰化两种不同方法对蜂胶进行前处理，结果发现湿法消解处理后的砷回收率显著大于干法灰化。

俎志平（2013）用微波消解-石墨炉原子吸收法测定蜂胶胶囊壳中铬的含量，采用微波消解法对蜂胶胶囊壳进行消解，在波长357.9nm下，通过石墨炉原子吸收进行测定，结果发现5个批次的20份蜂胶胶囊中铬含量均不超标。

Gong 等（2012）采用电感耦合等离子体原子发射光谱法测定微波消解后31个中国蜂胶样本、1个美国蜂胶样本中的15种重金属（钙、铝、镁、钾、铁、钠、锌、锰、锶、铬、镍、砷、镉和铅）含量。结果发现，蜂胶中这些重金属的平均含量一般为：钙 1449.1mg/kg；铝 971.2mg/kg；镁 552.5mg/kg；钾 1137.4mg/kg；铁 1188.4mg/kg；钠 497.5mg/kg；锌 160.8mg/kg；锰 36.93mg/kg；铅 19.92mg/kg；锶 8.70mg/kg；镉 0.60mg/kg。而砷、铜、铬、镍在很多样本中都未发现。

此外，蜂胶中的放射性微粒也引起了人们的普遍关注，这些有害微粒主要富集在土壤中，从而污染植物、昆虫及其他相关产品，并进一步危害人类健康。目前，在蜂胶样中已检测出的天然放射性粒子有钾（^{40}K）、铍（^{7}Be）及铯（^{137}Cs），不过定量检测发现，这些微粒的含量一般低于人体的耐受范围（Orsi et al.，2006），然而长期的富集作用则会对人体产生组织病变及基因变异现象。

张修景（2014）建立了微波消解-分子荧光猝灭法测定蜂胶中痕量硒的方法。蜂胶样品经微波消解后，用盐酸将消化液中的六价硒全部还原成四价硒。再与碘化物和罗丹明B发生显色反应；当蜂胶中的硒元素与罗丹明B形成离子缔合物时，罗丹明B溶液的荧光猝灭，硒浓度在 $0\sim0.100\mu g/mL$ 线性范围内荧光猝灭值与四价硒的浓度成正比。其线性回归方程是 $y=3536.1x+4.76$，线性相关系数为0.9995，加标回收率在96.8%~103.0%，相对标准偏差($n=4$)$<4.2\%$，检测限为 $0.0010\mu g/mL$。

高丽红(2014)研究了蜂胶中总砷的分析方法，采用微波消解法对蜂胶样品进行前处理，采用氢化物发生-原子荧光法对蜂胶中的总砷含量进行测定。测定结果：总砷的线性范围为 $0\sim90\mu g/L$，相关系数为 0.9995，方法回收率为 $96.60\%\sim101.75\%$，RSD 为 1.08%。

蜂胶中去除重金属的方法也层出不穷。高振中等(2009)研究了改性天然沸石与超临界萃取法分离蜂胶中的铅，发现两种方法除铅效果明显，改性天然沸石除铅效果更好，除铅后蜂胶中的钙、铁、镁等含量也随之降低。张洛红等(2011)将改性制备的纤维素黄原酸盐用于蜂胶中铅的吸附脱除，筛选出以废弃灯芯草为原料制备的纤维素黄原酸盐为最佳吸附材料，不仅能获得较高的铅去除率，而且对蜂胶中黄酮类物质的影响不大，这为降低蜂胶中重金属铅含量提供了一种新的技术方法。随后，他们继续研究自制的灯芯草纤维素黄原酸盐用于吸附去除蜂胶中重金属铅的可行性和适应性，考察了灯芯草纤维素黄原酸盐对蜂胶中铅的去除效果，探讨了静态吸附反应中铅离子浓度、吸附时间、初始溶液 pH 和吸附温度 4 个因素对铅去除率的影响，确定最佳吸附反应条件；并利用 HPLC 对脱铅处理前后蜂胶中黄酮类物质的成分和含量进行比较。结果表明，灯芯草黄原酸盐对铅具有良好的选择性，而对蜂胶中黄酮类物质的影响不大。这既为控制蜂胶中重金属铅含量提供了新的处理技术，同时还为废弃中草药灯芯草的资源化及深度利用创造了新的思路(张洛红等，2012)。李波等(2013)采用自制的灯芯草纤维素黄原酸盐用于吸附蜂胶中的重金属镉，采用原子吸收分光光度法测定了蜂胶中镉含量，并考察了灯芯草纤维素黄原酸盐对蜂胶中镉的去除效果，探讨静态吸附反应中镉离子初始浓度、吸附时间、溶液 pH 值和吸附温度这 4 种因素对镉去除率的影响，并确定最佳吸附反应条件。结果表明，当向 10mL 含镉浓度为 $200\mu g/L$ 的蜂胶溶液中加入 0.1g 灯芯草纤维素黄原酸盐，并调节蜂胶液的 pH 值为 $4.0\sim4.5$，恒温($30℃$)振荡吸附 5min 时，该吸附剂对蜂胶中镉离子的去除率为 52.5%。这一结果也说明灯芯草纤维素黄原酸盐用于吸附去除蜂胶中重金属镉是可行的，为控制蜂胶中重金属镉含量提供了新的处理技术。

3.3 蜂胶中的农兽药残留

由于蜂胶生产与现代农业生产密切相关，而农业生产中使用的各类杀虫剂、农药也对蜂胶产品质量造成严重影响，其中最为突出的问题就是蜂胶产品中农兽药残留问题。根据蜂产品生产的相关规定，蜜蜂在养殖过程中不得使用违禁兽药，但蜂农仍可能使用氯霉素、磺胺类、硝基呋喃类等抗生素来预防和治疗蜜蜂疾病，导致抗生素在蜂胶产品中的残留。

3.3.1 蜂胶中氯霉素残留的检测

氯霉素是一种高效广谱的抗生素，对各种好氧和厌氧微生物都有活性。由于该药物成本低，见效快，曾被用于家畜和家禽疾病的预防和治疗。若将氯霉素用于治疗蜜蜂疾病，则会导致其在蜂产品中残留。由于蜂胶化学成分复杂，应用气相色谱、气相色谱质谱、ELISA 等方法分析蜂胶中的氯霉素，存在检出限难以达到要求、背景干扰或样品净化不完全以及衍生化难等问题。近年来，欧盟对进口动物源产品中的氯霉素的检出限量不断降低，而蜂胶中氯霉素残留的检测又是动物源产品检测中的难点。因此，在蜂胶中检测 ppb 级甚至更低浓度水平的氯霉素残留对仪器灵敏度、重现性与选择性的要求非常高，并且需要良好的样品前处理手段来净化浓缩目标组分。

Mochizuki 等(2008)建立了检测蜂胶乙醇提取物中的氯霉素的 LC-ESI/MS" 方法。前处理为在蜂胶提取物中加水,然后用氯化钠盐析去除蜂胶提取物中的蜂蜡。以醋酸铵和乙腈为流动相,该方法的检测限和定量限分别为 0.05 和 0.15ng/g,回收率为 111.2%。Bononi 等(2008)建立了蜂胶制成品中氯霉素检测的 LC/MS 方法,采用含水乙醇或乙二醇进行提取,然后用乙酸乙酯稀释,以甲砜霉素为内标定量,该方法的最低检测限为 0.05ng/g。用该方法对意大利市场上流通的 8 个知名品牌的蜂胶产品中的氯霉素进行检测,发现在其中 2 个蜂胶产品中检出氯霉素,浓度分别为 0.70ng/g 和 0.85ng/g。周萍等(2010)利用液相色谱串联质谱对蜂胶中的氯霉素残留进行测定,采用 10% 高氯酸提取,用 1mol NaOH 调节溶液 pH=10.5,并用乙酸乙酯进行反萃取,不再经过固相萃取,大大节约了检测成本,样品前处理方法简单快捷,加快了检测速度,该方法的检测限为 0.1g/kg。张晓燕等(2012)建立了用 HPLC-MS/MS 测定蜂胶中氯霉素残留的方法。样品用水提取后,以醋酸铅溶液作为沉淀剂除去样品中的大部分黄酮类成分,用液-液萃取的方式提取样品中的氯霉素残留,最后用 HPLC-MS/MS 对样品进行定性、定量分析。该方法能除去蜂胶中的大部分黄酮类成分,减少了干扰,可以用于蜂胶中氯霉素残留的测定。尽管欧盟等国家相关法规明确禁止在动物产品中检出氯霉素,但是蜂胶及制剂中确实存在氯霉素残留的风险,因此,这些方法的建立将为规范蜂胶产品、保证其质量安全提供依据。

3.3.2 蜂胶中氟胺氰菊酯残留的检测

氟胺氰菊酯是蜂群中最广泛应用的杀螨剂。王祥云等(2009)建立了蜂胶中氟胺氰菊酯和氟氯苯氰菊酯残留的测定方法。以石油醚为溶剂提取样品,采用凝胶渗透色谱法净化样品,采用气相色谱法测定氟胺氰菊酯和氟氯苯氰菊酯残留量。结果表明,匀浆提取的提取效果远远好于震荡提取的提取效果。石油醚的提取效果比石油醚-丙酮混合液的提取效果好。与丙酮-石油醚(9∶1)和石油醚-乙酸乙酯(95∶5)相比,石油醚-乙酸乙酯(98∶2)的净化效果最理想。氟胺氰菊酯和氟氯苯氰菊酯的回收率均在 69.0%~96.4%,相对标准偏差均在 1.27%~7.69%,表明该方法的准确度和精密度均能达到残留分析的要求。氟胺氰菊酯和氟氯苯氰菊酯的检测限均为 0.05mg/kg。

3.3.3 蜂胶中四环素类抗生素残留的检测

在养蜂生产过程中,四环素类抗生素通常被用来治疗或预防蜜蜂美洲幼虫腐臭病或欧洲幼虫腐臭病,易在蜂产品中残留。Zhou 等(2009)首次采用 HPLC 法测定蜂胶中的 4 种四环素类抗生素的残留。蜂胶超声波提取后,采用 Oasis HLB 固相萃取和弱阳离子交换柱去除水溶性和脂溶性类黄酮、芳香酸、萜类化合物、蜡和花粉残渣。4 种四环素的定量线为 100~150ng/g,回收率为 61.9%~88.5%,RSD 为 4.80%~13.2%。30 个分析样品中检测到痕量四环素,稍高于欧洲委员会设定的最高残留量。

3.3.4 蜂胶中磺胺类药物残留的检测

磺胺类药物在畜牧生产中应用十分广泛。张睿等(2014)建立了同时检测蜂胶中 16 种磺胺及 2 种林可胺类药物残留的方法,以 1moL/L 盐酸为提取液,经阳离子固相萃取小柱净化富集后,进行高效液相色谱-串联质谱分析。采用 Aglient Polaris C18 色谱柱,以甲醇和 0.1% 甲酸为流动相梯度洗脱,质谱模式为电喷雾正离子检测。此方法前处理较简单,采用磺胺类药物、林可胺类药物内标进行定量后,磺胺和林可胺类药物的线性范围均为 1.0~

50μg/L，相关系数均在 0.99 以上，方法定量限为 10.0μg/kg，在 10、20 和 40μg/kg 3 个水平做添加回收，回收率范围为 69.5%～114.6%，相对标准偏差小于 10%。

3.3.5 蜂胶中硝基呋喃类药物残留的检测

硝基呋喃类药物是一种广谱抗生素，对大多数革兰阳性菌和革兰阴性菌、真菌和原虫等病原体均有杀灭作用。杨雯筌等（2013）建立了蜂胶中硝基呋喃类代谢物液相色谱-串联质谱检测方法。将蜂胶样品经固相萃取、衍生、乙酸乙酯提取后进行质谱分析，在 1.0、2.0、5.0μg/kg 3 个添加水平下，硝基呋喃类代谢物的平均回收率为 92.6%～99.3%，日内相对标准偏差小于 10%，日间相对标准偏差小于 15%。在 0.5～20ng/L 范围内呈良好的线性（$r>0.99$），检测限为 0.25μg/kg，定量限为 1.0μg/kg，为蜂胶中硝基呋喃类代谢物的检测提供了分析确证方法。

3.3.6 蜂胶中有机磷农药残留的检测

有机磷农药是用于防治植物病虫害的含有机磷农药的有机化合物。Pérez-Parada 等（2011）建立了基于基质固相扩散技术，通过气相色谱与火焰光度检测器以及质谱联用技术痕量分析蜂胶酊剂（乙醇提取物）中有机磷农药（蝇毒磷、毒死蜱和乙硫磷）的方法，对乌拉圭不同蜂场 1800 多个蜂胶样本的检测结果显示，绝大多数蜂胶样中存在杀螨剂蝇毒磷、毒死蜱的残留，而很少检测到乙硫磷的存在；这 3 种物质的检出率及含量水平与之前 Mullin 等（2010）在蜂蜡中的检测结果相一致。Gloria 等（2011）基于基质固相扩散技术，通过气相色谱与质谱、选择离子监测模式，对蜂胶中 5 种有机磷农药（敌敌畏、二嗪磷、马拉硫磷、甲基对硫磷和蝇毒磷）进行了分析。Medina-Dzul 等（2014）采用基质分散固相萃取法优化了蜂胶毛胶中有机磷农药的提取方法。由于蜂胶本身化学成分非常复杂，他们首先选取不同比例的二元溶剂混合物（主要是乙腈和另一种非极性溶剂）对蜂胶中的有机磷类农药进行选择性洗脱，采用气质联用仪对提取得到的有机磷类农药和主要干扰化合物进行鉴定。结果表明，蜂胶中有机磷农药提取方法的最优实验条件为：1mL 蜂胶提取液配合 8mL 乙腈，二氯甲烷混合液（混合液中有机溶剂比为 1∶3）。

3.3.7 蜂胶中多环芳香碳水化合物残留的检测

Dobrinas 等（2008）采用 GC-MS 对罗马尼亚地区 15 个蜂胶样本中多环芳香碳水化合物进行分析，该类化合物的最高含量达到 618ng/g，低于人体推荐的最大耐受量（成人，8.4μg/d；儿童，7.4μg/d）（Falco et al.，2003）。Santos 等（2008）用基质固相分散处理，采用 GC-MS 确定蜂胶中的噻嗪酮、四氯二苯砜、乙烯菌核利和联苯菊酯残留。Moret 等（2010）采用高效液相色谱结合荧光检测法对意大利蜂胶中二氢苊、莹蒽、萘、苯并蒽、苯并[b]荧蒽、苯并芘、苯并[k]荧蒽、稠二萘、苊、蒽、苯并[ghi]苝、芴、菲、二苯并蒽、茚并(1,2,3-cd)芘、芘等 13 种多环芳烃进行检测，大约一半蜂胶中检测到苯并芘浓度大于 2μg/kg，即超出食品强化剂的规定限。

3.4 蜂胶中人为添加药物的检测

近年来，蜂胶产业发展迅猛，其中有不少蜂胶保健品具有辅助调节血糖的功能。个别不法厂商为了提高蜂胶的降血糖效果，在蜂胶保健品中违法添加西药。因此，对蜂胶中人为添加药物进行检测很有必要。

朱明达等（2010）采用超高效液相色谱-电喷雾串联四极杆质谱仪（UPLC-ESI-MS/MS）

同时测定蜂胶保健品中14种活性成分:阿魏酸、香豆酸、咖啡酸、王浆酸、芦丁、槲皮素、杨梅酮、桑色素、山柰酚、芹菜素、松鼠素、白杨素、高良姜素、金合欢素;9种违禁降糖西药:瑞格列奈、那格列奈、甲苯磺丁脲、妥拉磺脲、格列吡嗪、格列齐特、格列本脲、格列苯脲以及格列喹酮。蜂胶保健品样品用甲醇稀释,超声波提取,样品溶液经高速离心后过滤。该方法简便、有效、灵敏,为评价蜂胶保健品质量提供了新的检测方法。

滕军和况磊(2014)建立了同时测定蜂胶中非法添加吡格列酮和罗格列酮等2种化学合成降糖药物的方法。采用薄层系统对蜂胶进行双向展开、分离、显色及刮板,提取后在紫外区扫描测定其吸收曲线;用薄层色谱法和分光光度法同时进行定性分析。结果在选定条件下,蜂胶中的黄酮类物质被有效分离,2种合成降糖药物均被检出,检出限为 $2\mu g/mg$,整个分析时间不超过 2h。该方法操作简单、快速准确,可用于检测添加了上述合成降糖药物的蜂胶及蜂胶质量的快速筛查。

3.5 蜂胶引起的过敏

尽管蜂胶具有多种生理药理活性,但也有可能引起一些过敏症状。据 Corazza 等(2014)的一项调查结果显示,在所涉及的14种天然植物源性产品中,蜂胶是最常见的过敏原,包括引起养蜂人产生疹性变应性接触性皮炎和诱发易过敏人群产生过敏症状。随着蜂胶应用人群的增长,其致敏率有明显上升的趋势。蜂胶会引发有过敏体质的人出现系统性接触性皮炎,表现出严重的瘙痒症、红斑丘疹,脸、颈部、手臂、腹部和大腿浮肿等症状。

Rudzki 和 Grzywa(1987)研究发现,蜂胶过敏症主要是直接接触引起的,包括表皮接触、皮下注射、静脉注射、肌肉注射及口腔接触,机体组织一般会出现皮炎、红疹、水肿等现象,手是最频发部位。蜂胶的接触性过敏反应表现出剂量依赖效应。Ledón 等(2002)研究发现,蜂胶能够以剂量依赖性方式诱导豚鼠产生轻微至中等程度的接触性过敏反应,使其机体出现红疹现象。Budimir 等(2012)采用问卷调查的方式调查了波兰蜂农对蜂胶的过敏情况。558 份有效问卷的分析结果显示,蜂农对蜂胶过敏反应发生率为 21.97%,接触性过敏症发生率为 3.05%。在 404 位使用蜂胶作为治疗性药物的蜂农中,仅有 5 位蜂农出现过敏症状。这项调查结果也说明蜂农在收集蜂胶时可能发生过敏反应,但使用蜂胶造成的过敏症并不普遍。

Walgrave 等(2005)调查发现,部分养蜂者口服蜂胶后表现出过敏性唇炎、口腔炎、口周湿疹、嘴唇水肿、口腔痛及呼吸困难等症状。随着蜂胶口服制剂形式的多样化,相应的过敏反应也日趋增多。Orsi 等(2005)报道高剂量的蜂胶提取物($300\mu g/mL$)可以直接活化机体肥大细胞,促进炎性介质的释放,从而引发人体的过敏反应。Cho 等(2011)报道了韩国一位36 岁女性在口服蜂胶液一段时间后,身体多处部位出现严重的红斑状丘疹、皮肤斑块及水肿现象。Ramien 和 Pratt(2012)还报道了首例口服蜂胶粉诱发人体产生固定性药疹。此外,口服蜂胶还可能会在肠道吸收过程中引起过敏反应。

研究表明,对蜂胶过敏通常是次生性的,是由于对某些物质尤其是对杨树新芽分泌物和秘鲁香脂(Myroxylonpereirae)有原发敏感性的人,体内发生交叉反应而引起的(Rudzki 和 Grzywa,1987)。Rajpara 等(2009)研究发现,蜂胶与蜂蜡(7.2%)、混合香精(10.9%)、松香(27.2%)、秘鲁香脂(40%)均可发生体内交叉过敏反应。De Groot 等(1994)认为,对蜂胶过敏主要是对树的芽孢成分过敏。现在多数学者认为,蜂胶中的过敏原主要是咖啡酸及其酯类衍生物,即"LB-1"(1,1-咖啡酸二甲烯丙酯)和 3-甲基-2-丁烯咖啡酸酯(54%)、3-甲基-3-丁烯咖啡酸酯(28%)、2-甲基-2-丁烯咖啡酸酯(4%)、CAPE(8%)、咖啡酸(1%)、苄基咖啡酸盐(1%)(Hausen et al.,1987;Acciai et al.,1990;deGroot et al.,1994;Gardana and Simonetti,2011)。此外,异戊烯基咖啡酸盐、苯基异阿魏酸盐、肉桂酸苄酯、水杨酸苄酯以及黄酮类化合物柚木柯因也被认为是蜂胶中的第二类过敏原,然而其致敏作用相对较弱(Hausen,2005;Aliboni et al.,2010;Marcucci,1995)。上述致敏性化合物主要存在于温带地区的杨树型蜂胶中,而其他类型蜂胶中几乎不含有这些物质,也很少有相关过敏现象的报道。Hausen 等(1992)研究发现,萘醌类、丁子香酚、甲基异丁香酚及榄香素可能是古巴红蜂胶中的过敏原。Karsten(2001)则认为蜂胶中还含有其他未知的、不溶于水的非黄酮类致敏性物质。

Gardana 等(2011)建立了超高效液相色谱法确定蜂胶原料及制品中过敏原咖啡酸及其酯类的含量。欧洲蜂胶($n=8$)和亚洲蜂胶($n=3$)中含量较大的过敏原依次为:苄基咖啡酸＞3-甲基-2-丁烯基-咖啡酸＞CAPE＞3-甲基-3-丁烯基咖啡因＞咖啡酸＞2-甲基-2-丁烯基咖啡酸。采用该方法分析了蜂胶含水乙醇提取物($n=6$)和片剂($n=6$),发现这几种致敏原含量也较高。而在巴西红蜂胶($n=1$)和绿蜂胶($n=1$)中均未检测到这些过敏原。Aliboni 等(2011)采用气质联用定量分析了意大利中部地区蜂胶中两个引起过敏的酯类:水杨酸苄酯和肉桂酸苄酯。

Gardana 等(2012)利用一种特异性的瑞士乳酸杆菌(*Lactobacillus helveticus*),使蜂胶发生生物转化,去除蜂胶中的过敏原类物质——咖啡酸酯。此方法也基本不影响蜂胶的总黄酮含量,且处理过后的蜂胶抗菌活性也没有明显的下降。这也是首次关于利用细菌生物转化方法去除蜂胶中过敏原的报道。

4 蜂胶的质量标准

蜂胶的质量标准是指根据蜂胶的来源、物理化学特性、卫生指标制定的蜂胶品质的具体条例,是蜂胶产品和质量控制的依据。由于蜂胶的植物来源多样,化学成分复杂,不同来源的蜂胶难以形成统一标准,给蜂胶的质量控制带来很大困难。

4.1 蜂胶质量标准现状

4.1.1 国内蜂胶标准

我国在 1992 年就建立了适用于生产、收购、运输和销售的蜂胶行业标准,规定了蜂胶的质量要求、检验方法、标志、包装、运输和贮存。《中华人民共和国药典》(一部)自 2005 年出版以来,收录了蜂胶的判别方法,包括理化性质鉴别和液相色谱法。2009 年,《蜂胶》国家标准(GB/T 24283—2009)正式颁布实施,2016 年又进行了修订。

现行的这些蜂胶标准(表 4.8)内容上仍限于感观鉴别、理化检测和氧化时间测定。有效成分的含量测定方面,主要是指总黄酮含量高低,或者几个黄酮类化合物表征蜂胶质量的质控方法,难以体现蜂胶的整体质量和疗效,无法表征蜂胶的物质基础和化学成分的整体性和复杂性,从而难以真正有效地控制蜂胶质量。

表 4.8 蜂胶相关的现行国家及行业标准

Table 4.8 Current national and industry standards of propolis

标准名称	标准编号	颁发部门	类别	检测项目
蜂胶	SB/T 10096—1992	商业部	行业标准	状态、颜色、气味、结构、硬度、95%乙醇提取物、杂质和蜂蜡含量、碘值、氧化时间、酚类化合物、黄酮类化定性反应合物
蜂胶	NY/T629—2002	农业部	行业标准	状态、气味、色泽、滋味、75%乙醇提取物、蜂蜡和75%乙醇不溶物、氧化时间、总黄酮含量、铅含量
蜂胶	《中华人民共和国药典》2005,2010版（一部）	国家药典委员会	国家药品标准	来源、性状、鉴别、检查项（干燥失重、总灰分、酸不溶性成分、氧化时间）、浸出物以及含量测定（白杨素和高良姜素）
蜂胶	GB/T 24283—2009	国家质检总局	国家标准	色泽、状态、气味、滋味、结构、乙醇提取物含量、总黄酮、氧化时间
蜂胶中芦丁、杨梅酮、槲皮素、莰菲醇、芹菜素、松属素、苛因、高良姜素含量的测定方法 液相色谱-串联质谱检测法和液相色谱-紫外检测法	GB/T 19427—2003	中华全国供销合作总社	国家标准	黄酮类化合物
蜂胶中阿魏酸含量的测定方法 液相色谱-紫外检测法	GB/T 23196—2008	国家质检总局	国家标准	阿魏酸
蜂胶中总黄酮含量的测定方法 分光光度比色法	GB/T 20574—2006	国家质检总局	国家标准	黄酮类化合物
地理标志产品 饶河（东北黑蜂）蜂蜜、蜂王浆、蜂胶、蜂花粉	GB/T 19330—2008	国家质检总局	国家标准	状态、气味、色泽、滋味、总黄酮含量、75%乙醇提取物含量、蜂蜡和75%乙醇不溶物、氧化时间、卫生指标
蜂胶中铅的测定 微波消解-石墨炉原子吸收分光光度法	GB/T 23870—2009	国家质检总局	国家标准	铅
蜂胶中杨树胶的检测方法 反相高效液相色谱法	GH/T 1081—2012	中华全国供销合作总社	行业标准	杨树胶及掺杂物
蜂胶真实性鉴别方法 高效液相色谱指纹图谱法	GH/T 1087—2013	中华全国供销合作总社	行业标准	杨树胶

(1) 蜂胶质量评价

现行国内蜂胶标准中对质量评价主要是通过总黄酮含量来评价，但各标准间总黄酮含量测定的方法不一，所测数据与实际含量可能存在较大的偏差。

《蜂胶中总黄酮含量的测定方法 分光光度比色法》(GB/T 20574—2006)的方法原理,是黄酮类化合物在氯化铝-醋酸钾溶液体系下反应,反应溶液在415nm波长处有最大吸收峰,以芦丁为对照品,用Al^{3+}为显色剂,该法主要检测的是蜂胶中黄酮醇类化合物,测定的总黄酮含量偏低。

国家标准《蜂胶》(GB/T 24283—2009)的方法原理,是用聚酰胺树脂吸附、洗脱后,以芦丁为对照品,在360nm波长处测定吸光值,检测在此波长下有吸收的黄酮类化合物。该方法引自《保健食品检验与技术评价规范》(2003)中规定的保健食品中总黄酮的测定方法,是蜂胶申报保健食品时官方认定的黄酮类检测方法。但该方法也存在以下问题:(1)虽然苯是非极性物质,但它同样能够洗脱某些弱极性物质,而蜂胶中的白杨素、高良姜素、松属素是弱极性的黄酮,通过对苯洗脱液的HPLC检测,发现这3种物质有吸收峰出现,表明这些弱极性的黄酮被苯洗脱了,而使测定结果出现偏差;(2)聚酰胺树脂有粉末和颗粒状两种,2005版《中国药典》中要求的吸附剂的直径为0.07~0.15nm。颗粒状的聚酰胺树脂在装柱时存在的空隙比较大,并且装柱时松紧不一,在洗脱时会造成洗脱不完全,进而导致测定结果有偏差(杨琴 等,2008)。

在《保健食品中总黄酮的测定方法》(征求意见稿)中,用Al^{3+}在碱性条件下显色,在510nm波长处测定吸光度,该方法主要检测黄酮类化合物,所得蜂胶中总黄酮含量值也较高。

高效液相色谱法是目前普遍采用的测定物质有效成分含量的检测方法。GB/T 19427—2003规定了蜂胶中8种黄酮的高效液相色谱法和高效液相-质谱联用的检测方法,其主要检测的是芦丁、杨梅酮、槲皮素、茨菲醇、芹菜素、松属素、苛因、高良姜素的含量,以这8种黄酮的总含量评价蜂胶的总黄酮含量。但蜂胶中含有130种多种黄酮类化合物,因此,所测黄酮含量相对值偏低。

由于缺乏蜂胶药效物质基础研究,仅以总黄酮含量为质量评价指标,不能全面反映蜂胶的内在品质,造成蜂胶制成品药效不稳定,质量的不齐和药效的不稳定成为阻碍蜂胶深入研究与开发的瓶颈。

(2)蜂胶真实性评价

由于温带地区蜂胶的植物来源主要以杨属和其杂交属的芽苞分泌物为主,所以在这些地区采集的蜂胶的化学成分与杨树芽的成分极其相似。市场上出现了在蜂胶中掺入杨树胶(杨树芽提取物)或直接以杨树胶冒充蜂胶的现象。杨树胶在颜色、气味、形状上都与蜂胶十分相似,以杨树胶冒充蜂胶始终困扰着蜂胶行业的健康发展。2010年11月21日,中央电视台曝光假蜂胶事件后,2010年11月24日,国家食品药品监督管理局下发通知,对蜂胶产品进行大规模检查,并向社会公布有关信息,推动了整个蜂产业的规范、良性发展,其生产经营秩序也有了明显好转,但打假并未形成长效机制,杨树胶仍大量充斥蜂胶市场。而且近几年,大量的澳大利亚、新西兰等进口蜂胶产品充斥中国市场,据中国蜂产品协会调查发现,在两国境内和中国市场销售的所谓澳大利亚和新西兰的蜂胶产品原料大都是来自中国大陆的杨树胶。GH/T 1081—2012《蜂胶中杨树胶检测方法》和GH/T 1087—2013《蜂胶真实性鉴别方法》相继出台,基本遏止了杨树胶冒充蜂胶的现象。然而,好景不长,2014年市场上又出现了蜂胶中掺入松树枝芽提取物、人为水解以去除水杨苷的杨树胶等现象,蜂胶真实性评价仍任重道远。

(3)蜂胶中铅残留检测

蜂胶本身无毒,但在蜂胶的生产过程中有很多可能被重金属污染的环节,会使部分蜂胶铅含量过量。GB/T 23870—2009 规定了蜂胶中铅的测定方法。目前养蜂生产中,以尼龙纱代替金属纱网或使用无污染集胶器,从源头控制,显得尤其重要。蜂胶中的铅以无机状态存在,通过乙醇等有机溶剂提取以及针对性的除铅工艺,可以除去铅等重金属,达到食品卫生的要求。

4.1.2 国外蜂胶标准

(1)日本

日本是蜂产品的生产小国,消费大国,将蜂产品作为维护公众身体健康的产业来发展。据 2013 年日本公布的保健食品排行榜,蜂胶位列第 13 位,年销售额达 120 亿日元。

日本的蜂胶质量控制与管理以日本蜂胶协议会为主导,其出台了蜂胶质量控制标准,对蜂胶的有害菌、农兽药残留、重金属含量等进行了规定(表 4.7)。

表 4.7 日本蜂胶质量控制标准

项目		乙醇提取液	颗粒、片剂	软胶囊
规格成分		>8%(W/V)	8%以上	8%以上
槲皮素		检出	检出	检出
紫外吸光度		275~315nm	275~315nm	275~315nm
乙醇浓度		>50%	—	—
细菌总数		—	5×10^4 个/g 以下	3×10^3 个/g 以下
大肠杆菌		—	阴性	阴性
过酸化物值		—	—	5meq/kg 以下
酸值		—	—	15 以下
残留农药	异狄氏剂或狄氏剂	不得检出		
	六氯苯(BHC)	$<0.2\times10^{-6}$		
	DDT	$<0.2\times10^{-6}$		
	对硫磷(马拉硫磷和杀螟松)	均$<0.2\times10^{-6}$		
多氯联苯		不得检出		
砷		$<2\times10^{-6}$		
铅		$<20\times10^{-6}$		
n-正己烷		—		
四环素族		不得检出		
二甘醇		—		

(2)巴西

巴西蜂胶因其外观与成分等与世界其他地区的蜂胶有较大差别,且作用明确,药理研究充分而备受世人关注。巴西蜂胶的质量标准也主要是感官和理化指标要求(表 4.8 和表 4.9)。

表 4.8　巴西蜂胶质量控制的感官要求

项目	特征
颜色	有黄色、棕色、绿色及其他颜色,颜色依源植物而变化
稠度(室温)	随源植物不同带有不同程度韧性的固体
香气	具有源植物的树脂和软树胶的香气特征
味	有源植物的浓郁的树脂芳香
状态	呈现不均匀的颗粒

表 4.9　巴西蜂胶质量控制的理化指标要求

项目	蜂胶分级		
	低	中	高
总黄酮含量	<1.0%(m/m)	≥1.0%～2.0%(m/m)	>2.0%(m/m)
干燥失重	<8%(m/m)		
灰分	<5%(m/m)		
蜡	<25%(m/m)		
酚醛树脂	≥5%(m/m)		
黄酮	≥0.5%(m/m)		
氧化活性	≥22 秒		
乙醇中的溶解度	≥35%(m/m)		
可见紫外吸收光谱	200nm～400nm 呈现主要黄酮类化合物的特征峰		
醋酸铅	阳性		
氢氧化钠	阳性		
芽孢杆菌孢子	不得检出		

4.2　蜂胶质量标准展望

近年来,无论是蜂胶的基础研究还是应用研究,都有了快速的发展,发明专利的申请与授权数量逐年递增;每年发表的有关蜂胶的研究论文数量众多,研究领域广泛,研究成果丰硕,但缺乏完整的质量标准体系,这也是国内外蜂胶研究开发中所面临的一个十分迫切的问题。

蜂胶中多种化学成分的综合检测以及蜂胶指纹图谱的建立和完善,已被国内外研究者广泛用于蜂胶化学成分、真伪鉴别、质量评价中,但是,蜂胶指纹图谱所面临的局限越来越明显,它反映的仅是蜂胶中的化学信息,与药效活性信息无关。这些化学信息所对应的物质类群并不完全等同于蜂胶的药效物质基础,不能实现通过控制蜂胶中各化学成分的"量"进而控制蜂胶的"质"。尽管蜂胶药理活性的研究都离不开蜂胶中主要化学成分的鉴定与相应活性的研究,但化学成分的检测与药理活性的研究是分离的,迄今为止,蜂胶发挥药理活性的物质基础及基于活性对蜂胶进行质量评价还未见报道。"活性指纹图谱"是指与药理活性及化学成分相关联的指纹图谱,建立在有效部位且有效成分明确的基础上,以有代表性的有效成分来反映蜂胶特征性的指纹图谱。

蜂胶及其制剂的质量标准应该是建立在大量坚实的基础研究之上,包括蜂胶植物来源的明确、提取生产工艺的统一以及制剂工艺过程的研究等。化学成分组成是蜂胶发挥药效作用的物质基础,它的深入研究是蜂胶质量控制的关键,是揭示蜂胶作用机制的核心。应加大蜂胶的基础研究,寻找"有效成分""主要成分"和"特征成分",以这3种成分作为控制指标,并且这3种成分应达到一定的量值和比例,才能切实起到质量控制的目的。另外,在色谱指纹图谱的研究中,也要对这3种成分进行归属。在现行质量控制方法的基础上,采用多指标成分定量结合活性指纹图谱的质量控制模式是蜂胶质量控制标准研究的方向。

参考文献

Abu-Mellal A, Koolaji N, Duke RK, Tran VH, Duke CC (2012) Prenylated cinnamate and stilbenes from Kangaroo Island propolis and their antioxidant activity[J]. *Phytochemistry*, 77:251-259.

Acciai MC, Ginanneschi M, Bracci S, Sertoli A (1990) Studies of the sensitizing properties of propolis[J]. *Contact Dermatitis*, 23(4):274-275.

Ahn MR, Kumazawa S, Hamasaka T, Bang KS, Nakayama T (2004) Antioxidant activity and constituents of propolis collected in various areas of Korea[J]. *Journal of Agricultural and Food Chemistry*, 52(24):7286-7292.

Ahn MR, Kumazawa S, Usui Y, Nakamura J, Matsuka M, Zhu F, Nakayama T (2007) Antioxidant activity and constituents of propolis collected invarious areas of China[J]. *Food Chemistry*, 101:1383-1392.

Ahn MR, Kumazawa S, Hamasaka T, Bang KS, Nakayama T (2004) Antioxidant activity and constituents of propolis collected in various areas of Korea[J]. *Journal of Agricural and Food Chemistry*, 52(24):7286-7292.

Ahn MR, Kumazawa S, Usui Y, Nakamura J, Matsuka M, Zhu F, Nakayama T (2007) Antioxidant activity and constituents of propolis collected in various areas of China[J]. *Food Chemistry*, 101:1383-1392.

Alencar SM, Oldoni TLC, Castro ML, Cabral ISR, Costa-Neto CM, Cury JA, Rosalen PL, Ikegaki M (2007) Chemical composition and biological activity of a new type of Brazilian propolis:Red propolis[J]. *Journal of Ethnopharmacology*, 113:278-283.

Aliboni A, D'Andrea A, Massanisso P (2010) Propolis specimens from different locations of central Italy:chemical profiling and gas chromatography-mass spectrometry (GC-MS) quantitative analysis of the allergenic esters benzyl cinnamate and benzyl salicylate[J]. *Journal of Agricultural and Food Chemistry*, 59(1):282-288.

Arvouet GA, Lejeune B, Bastide P, Pourrat A, Legret P (1993) Propolis extract Part 6:Subacute toxicity and cutaneous primary irritation index[J]. *Journal de Pharmacie de Belgique*, 48:165-170.

Athikomkulchai S; Awale S, Ruangrungsi N, Ruchirawat S, Kadota S (2013) Chemical constituents of Thai propolis[J]. *Fitoterapia*, 88:96-100.

Aviello G, Scalisi C, Fileccia R, Capasso R, Romano B, Izzo AA, Borrelli F (2010) Inhibitory effect of caffeic acid phenethyl ester, a plant-derived polyphenolic compound, on rat intestinal contractility[J]. *European Journal of Pharmacology*, 640(1-3):163-167.

Awale S, Li F, Onozuka H, Esumi H, Tezuka Y, Kadota S (2008) Constituents of Brazilian red propolis and their preferential cytotoxic activity against human pancreatic PANC-1 cancer cell line in nutrient-deprived condition[J]. *Bioorganic & Medicinal Chemistry*, 16(1):181-189.

Awale S, Shrestha SP, Tezuka Y, Ueda JY, Matsushige K, Kadota S (2005) Neoflavonoids and related constituents from Nepalese propolis and their nitric oxide production inhibitory activity[J]. Journal of Natural Products, 68(6):858-64.

Bankova V, Boudourova-Krasteva G, Popov S, Sforcin J, Funari SC (1998) Seasonal variations in essential oil from Brazilian propolis[J]. *Journal of Essential Oil Research*, 1998, 10:693-696.

Bankova V, De Castro SL, Marcucci MC (2000) Propolis: recent advances in chemistry and plant origin[J]. *Apidologie*, 31(1):3-15.

Bankova V, Dyulgeroy A, Popov S, Evstatieva L, Kuleva L, Pureb O, Zamjansan Z (1992) Propolis produced in Bulgaria and Mongolia: phenolic compounds and plant origin[J]. *Apidologie*, 23(1):79-85.

Bankova V, Goudourova-Krasteva G, Popov S, Sforcinb JM, Cunha Funarib SR (1998) Seasonal variations in essential oil from Brazilian propolis[J]. *Journal of Essential Oil Research*, 10:693-696.

Bankova V, Marcucci MC, Simova S, Nikolova N, Kujumgiev A, Popov S (1996) Antibacterial diterpenic acids from Brazilian propolis[J]. *Zeitschrift für Naturforschung C*, 51(5-6):277-280.

Bankova V, Popova M, Bogdanov S, Sabatini AG (2002) Chemical composition of European propolis: expected and unexpected results[J]. *Zeitschrift für Naturforschung C*, 57(5-6):530-533.

Bankova V (2005) Chemical diversity of propolis and the problem of standardization[J]. *Journal of Ethnopharmacology*, 100(1):114-117.

Banskota AH, Tezuka Y, Adnyana IK, Ishii E, Midorikawa K, Matsushige K, Kadota S (2001) Hepatoprotective and anti-helicobacter pylori activities of constituents from Brazilian propolis[J]. *Phytomedicine*, 8(1):16-23.

Banskota AH, Tezuka Y, Adnyana IK, Midorikawa K, Matsushige K, Message D, Huertas AAG, Kadota S (2000) Cytotoxic, hepatoprotective and free radical scavenging effects of propolis from Brazil, Peru, the Netherlands and China[J]. *Journal of Ethnopharmacology*, 72(1-2):239-246.

Banskota AH, Tezuka Y, Prasain JK, Matsushige K, Saiki I, Kadota S (1998) Chemical constituents of Brazilian propolis and their cytotoxic activities[J]. *Journal Natural*

Product,61(7):896-900.

Basista K M, Filipek B (2012) Allergy to propolis in Polish beekeepers[J]. *Postepy Dermatologii i Alergologii*,29 (6):440-445.

Bertelli D,Papotti G,Bortolotti L,Marcazzan GL,Plessi M (2012) Smultaneous identification of health-relevant compounds in propolis extracts Phytochemical [J]. *Analysis*, 23 (3): 260-266.

Bertrams J,Kunz N,Muller M,Kammerer D,Stintzing FC (2013) Phenolic compounds as marker compounds for botanical origin determination of German propolis samples based on TLC and TLC-MS[J]. *Journal of Applied Botany and Food Quality*,86 (1):143-153.

Bohlmann F,akupovic J (1979) New sesquiterpene acids, sesquiterpene diol, flavanones and other aromatic-compounds from flourensia-heterolepis. 198. naturally occurring terpene derivatives[J]. *Phytochemistry*,18 (7),1189-1194.

Bohlmann F, Zdero C, Grenz M, Dhar AK, Robinson H, King RM (1981) Naturally-occurring terpene derivatives 307. 5 Diterpenes and other constituents from 9 *Baccharis* species[J]. *Phytochemistry*,1981,20 (2):281-286.

Bononi M,Tateo F (2008) Liquid chromatography/tandem mass spectrometry efficiency of propolis extract against mitochondrial stress induced by antineoplasic agents (doxorubicin and vinblastin) in rats. analysis of chloramphenicol in propolis extracts available on the Italian market[J]. *Journal of Food Composition and Analysis*,21:84-89.

Bonvehi JS,Coll FV,Jorda RE (1994) The composition, active components and bacteriostatic activity of propolis in dietetics[J]. *Journal of the American Oil Chemists Society*,71(5): 529-532.

Borčić I,Radonic A,Grzunov K (1996) Comparison of the volatile constituents of propolis gathered in different regions of Croatia[J]. *Flavour and Fragrance Journal*, 11: 311-313.

Bosio K,Avanzini C,D'Avolio A,Ozino O,Savola D (2000) *In vitro* activity of propolis against Streptococcus pyogenes[J]. *Letters in Applied Microbiology*,31 (2):174-177.

Budimir V,Brailo V,Alajbeg I,Vuićevíc Boras V,Budimir J (2012) Allergic contact cheilitis and perioral dermatitis caused by propolis case report[J]. *Acta Dermatovenerologica Croatica*,20 (3):187-190.

Burdock GA (1998) Review of the biological properties and toxicity of bee propolis[J]. *Food and Chemical Toxicology*,36 (4):347-363.

Cai R,Wang S,Meng Y,Meng QG,Zhao WJ (2012) Rapid quantification of flavonoids in propolis and previous study for classification of propolis from different origins by using near infrared spectroscopy[J]. *Analytical Methods*,4(8):2388-2395.

Cantarelli MÁ, Camiña JM, Pettenati EM, Marchevsky EJ, Pellerano RG (2011) Trace mineral content of Argentinean raw propolis by neutron activation analysis (NAA): Assessment of geographical provenance by chemometrics[J]. *Food Science Technology*, 44

(1):256-260.

Cao YH, Wang Y, Yuan Q (2004) Analysis of flavonoids and phenolic acid in propolis by capillary electrophoresis[J]. *Chromatographia*, 59(1-2):135-140.

Castro C, Mura F, Valenzuela G, Figueroa C, Salinas R, Zuniga MC, Torres JL, Fuguet E, Delporte C (2014) Identification of phenolic compounds by HPLC-ESI-MS/MS and antioxidant activity from Chilean propolis[J]. *Food Research International*, 64:873-879.

Chang CC, Yang MH, Wen HM, Chem JC (2002) Estimation of total flavonoid content in propolis by two complementary colorimetric methods[J]. *Journal of Food and Drug Analysis*, 10(3):178-182.

Chang R, Pilo-Veloso D, Morais SAL, Nascimento EA (2008) Analysis of a Brazilian green propolis from Baccharis dracunculifolia by HPLC-APCI-MS and GC-MS[J]. *Brazilian J Pharmacog*, 18(4):549-556.

Chen CN, Weng MS, Wu CL, Lin JK (2004) Comparison of radical scavenging activity, cytotoxic effects and apoptosis induction in human melanoma cells by taiwanese propolis from different sources[J]. *Evid Based Complement Alternat Med*, 1(2) 175-185.

Cheng H, Qin ZH, GUo XF, GUO XF, Hu XS, Wu JH (2013) Geographical origin identification of propolis using GC-MS and electronic nose combined with principal component analysis[J]. *Food Research International*, 51(2):813-822.

Cho E, Lee JD, Cho SH (2011) Systemic contact dermatitis from propolis ingestion[J]. *Annals of Dermatology*, 23(1):85-88.

Choi YM, Noh DO, Cho SY, Suh HJ, Kim KM, Kim JM (2006) Antioxidant and antimicrobial activities of propolis from several regions of Korea[J]. *LWT-Food Science and Technology*, 39(7):756-761.

Christov R, Trusheva B, Popova M, Bankova V, Bertrand M (2005) Chemical composition of propolis from Canada, its antiradical activity and plant origin[J]. *Natural Product Research*, 19(7):673-678.

Clausen TP, Rrichardt PB, Bryant JP, Werner RA, Post K, Frisby K (1989) Chemical model for short-term induction in quaking aspen (*Populus tremuloides*) foliage against herbivores[J]. *Journal of Chemical Ecology*, 15(9):2335-2346.

Corazza M, Borghi A, Gallo R, Schena D, Pigatto P, Lauriola MM, Guarneri F, Stingeni L, Vincenzi C, Foti C, Virgili A (2014) Topical botanically derived products: use, skin reactions, and usefulness of patch tests: A multicentre Italian study[J]. *Contact Dermatitis*, 70(2):90-97.

Crawford DJ (1994) A morphological and chemical study of populus acuminate Rydberg [J]. *Brittonia*, 26(1):74-89.

Cuesta A, Rodrıguez A, Esteban MA, Meseguer J (2005) *In vivo* effects of propolis: A honeybee product on gilthead seabream innate immune responses[J]. *Fish and Shelfish Immunology*, 18(1):71-80.

Cuesta-Rubio O, Frontana-Uribe BA, Ramírez-Apan T, Cárdenas C (2002) Polyisoprenylated benzophenones in Cuban propolis: Biological activity of nemorosone[J]. *Zeitschrift für Naturforschung C*, 57(3-4): 372-378.

Cuesta-Rubio O, Piccinelli AL, Fernandez MC, Hernández IM, Rosado A, Rastrelli L (2007) Chemical characterization of Cuban propolis by HPLC-PDA, HPLC-MS, and NMR: the brown, red, and yellow Cuban varieties of propolis[J]. *Journal of Agricutural and Food Chemistry*, 55(18): 7502-7509.

da Silva RO, Andrade VM, Rêgo ESB, Dória GAA, dos Santos Lima B, da Silva FA, de Souza Araújo AA, de Albuquerque Júnior RLC, Cardoso JC, Gomes MZ (2015) Acute and sub-acute oral toxicity of Brazilian red propolis in rats[J]. *Journal of Ethnopharmacology*, 170: 66-71.

Daugsch A, Moraes CS, Fort P, Park YK (2008) Brazilian red propolis chemical composition and botanical origin[J]. *Evid Based Complement Alternat Med*, 5(4): 434-441.

De Albuquerque IL, Alves LA, Lemos TLG, Dorneles CA, de Morais MO (2008) Constituents of the essential oil of Brazilian green propolis from Brazil[J]. *Journal of Essential Oil Research*, 20(5): 414-415.

De Castro Ishida VF, Negri G, Salatino A, Bandeira MFCL (2011) A new type of Brazilian propolis: Prenylated benzophenones in propolis from Amazon and effects against cariogenic bacteria[J]. *Food Chemistry*, 125(3): 966-972.

De Sousa JP, da Silva Filho AA, Bueno PC, Gregorio LE, Furtado NA, Jorge RF, Bastos JK (2009) A validated reverse-phase HPLC analytical method for the quantification of phenolic compounds in *Baccharis dracunculifolia*[J]. *Phytochem Analysis*, 20(1): 24-32.

De Sousa JP, Bueno PCP, Gregorio LE, da Silva Filho AA, Furtado NAJC, de Sousa ML, Bastos JK (2007) A reliable quantitativemethod for the analysis of phenolic compounds in Brazilian propolis by reverse phase high performance liquid chromatography[J]. *Journal of Separation Science*, 30(16): 2656-2665.

DeGroot AC, Weyland JW, Nater JP (1994) Unwanted effects of cosmetics and drugs used in dermatology [M]. Elsevier.

Dobrinas S, Birghila S, Coatu V (2008) Assessment of polycyclic aromatic hydrocarbons in honey and propolis produced from various flowering trees and plants in Romania[J]. *Journal of Food Composition and Analysis*, 21(1): 71-77.

Dobrowolski JW (1991) Antibacterial, antifungal, antiamoebic, antiinflammatory and antipyretic studies on propolis bee products[J]. *Journal of Ethnopharmacology*, 35(1): 77-82.

Dos Santos TFS, Aquino A, Dorea HS, Navickiene S (2008) MSPD procedure for determining buprofezin, tetradifon, vinclozolin, and bifenthrin residues in propolis by gas chromatography-mass spectrometry[J]. *Analytical and Bioanalytical Chemistry*, 390(5): 1425-1430.

El-Bassuony AA,AbouZid S (2010) A new prenylated flavanoid with antibacterial activity from propolis collected in Egypt[J]. *Natural Product Communication*,5(1):43-5.

El-Bassuony AA (2009) New prenilated compound from egyptian propolis with antimicrobial activity[J]. *Rev Latinoam Quim*,37:85-90.

Falcao S,Tomas A,Vale N,Gomes P,Freire C,Vilas-Boas M (2013) Phenolic quantification and botanical origin of Portuguese propolis[J]. *Industrial Crops and Products*,49:805.

Falcao SI,Vale N,Gomes P,Domingues MR,Freire C,Cardoso SM,Vilas-Boas M (2013) Phenolic profiling of Portuguese propolis by LC-MS spectrometry:uncommon propolis rich in flavonoid glycosides[J]. *Phytochem Anal*,24(4):309-318.

Falco G,Domingo JL,Llobet JM,Teixidó A,Casas C,Müller L (2003) Polycyclic aromatic hydrocarbons in foods:human exposure through the diet in Catalonia, Spain[J]. *Journal of Food Protection*,66:2325-2331.

Ferracini VL,Paraiba LC,Filhob HFL,da SilvaAG,Nascimento LR,Marsaioli AG (1995) Essential oils of seven Brazilian *Baccharis species*[J]. *Journal of Essential Oil Research*,7(4):355-367.

Fu SH,Yang MH,Wen HM,Chem JC (2005) Analysis of flavonoids in propolis by capillary electrophoresis[J]. *Journal of Food and Drug Analysis*,13(1):43-50.

Gardana C,Barbieri A,Simonetti P,Guglielmetti S (2012) Biotransformation strategy to reduce allergens in propolis[J]. *Applied and Environmental Microbiology*,78(13):4654-4658.

Gardana C,Scaglianti M,Pietta P,Simonetti P (2007) Analysis of the polyphenolic fraction of propolis from different sources by liquid chromatography-tandem mass spectrometry [J]. *Journal of Pharmaceutical and Biomedical Analysis*,45:390-399.

Gardana C,Simonetti P (2011) Evaluation of allergens in propolis by ultra-performance liquid chromatography/tandem mass spectrometry[J]. *Rapid Commun Mass Spectrom*,25:1675-1682.

Ghisalberti EL (1979) Propolis:A review[J]. *Bee World*,60:59-84.

Gloria M,Acosta-Tejada G,Medina-Peralta M,Yolanda B,Moguel-Ordóñez,Muñoz-Rodríguez D (2011) Matrix solid-phase dispersion extraction of organophosphorus pesticides from propolis extracts and recovery evaluation by GC/MS[J]. *Analytical and Bioanalytical Chemistry*,400(3):885-891.

Gomez-Caravaca AM,Gomez-Romero M,Arraez-Roman D,Segura-Carretero A,Fernandez-Gutierrez A (2006) Advances in the analysis of phenolic compounds in products derived from bees[J]. *Journal of Pharmaceutical and Biomedical Analysis*,41(4):1220-1234.

Gong S,Luo L,Gong W,Gao YY,Xie MY (2012) Multivariate analyses of element concentrations revealed the groupings of propolis from different regions in China[J]. *Food Chemistry*,134(1):583-588.

Greenaway W,Scaysbrok T,whatley FR (1987) The analysis of bud exudate of Populus euramericana, and of propolis, by gas chromatography-mass spectrometry[J]. *Proceedings Royal Society*, 232(1268): 249-272.

Greenaway W, Scaysbrook T, Whatley FR (1987) Headspace volatiles from propolis[J]. *Flavour and Fragrance Journal*, 4(4): 173-175.

Greenway W, Scaysbrook T, Whatley FR (1990) The composition and plant origins of propolis: A report of work at Oxford[J]. *Bee Word*, 71(3): 107-118.

Gritsenko VI, Tikhonov OI, Priakhin OR (1976) Study of the polysaccharide preparation propolis[J]. *Farmatsevtychnyi Zhurnal*, (3): 92-93.

Gulcin I, Bursal E, Sehitog lu MH, Bilsel M, Goren (2010) Polyphenol contents and antioxidant activity of lyophilized aqueous extract of propolis from Erzurum[J]. *Turkey Food and Chemical Toxicology*, 48: 2227-2238.

Guo XL, Luo LP, Xu YJ, Chen B, Fu YX (2011) Chemical components and biological activity of chinese propolis from different seasons[J]. *Food Science*, 32(17): 141-146.

Haile K, Kebede T, Dekebo A (2012) A comparative study of volatile components of propolis (bee glue) collected from Haramaya University and Assela beekeeping centers, Ethiopia[J]. *Bulletin of the Chemical Society of Ethiopia*, 26: 353-360.

Hamasaka T, Kumazawa S, Fujimoto T, Nakayama T (2004) Antioxidant Activity and Constituents of Propolis Collected in Various Areas of Japan[J]. *Food Science and Technology*, 10 (1): 86-92.

Hames-Kocabas EE, Demirci B, Uzel A, Demirci F (2013) Volatile composition of anatolianpropolis by headspace-solid-phase microextraction (HS-SPME), antimicrobial activity against food contaminants and antioxidant activity[J]. *Journal of Medicinal Plants Research*, 7: 2140-2149.

Hattori H, Okuda K, Murase T, Shigetsura Y, Narise K, Semenza GL, Nagasawa H (2011) Isolation, identification, and biological evaluation of HIF-1-modulating compounds from Brazilian green propolis[J]. *Bioorganic & Medicinal Chemistry*, 19 (18): 5392-5401.

Hausen BM, Evers P, Stuwe HT, König WA, Wollenweber E (1992) Propolis allergy (IV) Studies with sensitizers from propolis and constituents common to propolis, poplar buds and balsam of Peru[J]. *Contact Dermatitis*, 26(1): 34-44.

Hausen BM, Wollenweber E, Senff H, Post B (1987) Propolis allergy ⅰ: Origin properties usage and literature review[J]. *Contact Dermatitis*, 17(3): 163-170.

Hausen BM (2005) Evaluation of the main contact allergens in propolis (1995 to 2005)[J]. *Dermatitis*, 16(3): 127-129.

El Hady FKA, Hegazi AG (2002) Egyptian propolis: 2. Chemical composition, antiviral and antimicrobial activities of East Nile Delta propolis[J]. *Zeitschrift für Naturforschung C*, 57(3-4): 386-394.

Hegazi AG, Hady FKAE, Abd Allah FAM (2000). Chemical composition and antimicrobial activity of European propolis[J]. *Zeitschrift für Naturforschung C*, 55(1-2): 70-75

Hernandez IM, Fernandez MC, Cuesta-Rubio O, Piccinelli AL, Rastrelli L (2005) Polyprenylated benzophenone derivatives from Cuban propolis[J]. *Natural Product*, 68(6):931-934.

Hilhorst MJ, Somsen GW, de Jong (1998) Potential of capillary electrophoresis for the profiling of propolis[J]. *Journal of High Resolution Chromatography*, 21(11): 608-612.

Hrobonova K, Lehotay J, Cizmarik J (2009) Determination of Organic Acids in Propolis by HPLC Using Two Columns with an On-Line SPE System[J]. *Journal of Liquid Chromatography & Related Technologies*, 32(1):25-135.

Hrobonova K, Lehotay J (2007) Determination of Quinic and shikimic acidsin products derived from bees and their preparates by HPLC[J]. *Journal of Liquid Chromatography & Related Technologies*, 30(17):2635-2644.

Huang S, Zhang CP, Li GQ, Sun YY, Wang K, Hu FL (2014) Identification of catechol as a new marker for detecting propolis adulteration[J]. *Molecules*, 19(7):10208-10217.

Huang WJ, Huang CH, Wu CL, Lin JK, Chen YW, Lin CL, Chuang SE, Huang CY, Chen CN (2007) Propolin G, a prenylflavanone, isolated from Taiwanese propolis, induces caspase-dependent apoptosis in brain cancer cells[J]. *Journal of Agricultural and Food Chemistry*, 55(18):7366-7376.

Inui S, Hosoya T, Shimamura Y, Masuda S (2012) Solophenols B-D and Solomonin: New prenylated polyphenols isolated from propolis collected from the Solomon Islands and their antibacterial activity[J]. *Journal of Agricultural and Food Chemistry*, 60(47): 11765-11770.

Ioshida MDM, Young MCM, Lago JHG. Chemical composition and antifungal activity of essential oil from Brazilian propolis[J]. *Journal of Essential Oil Bearing Plants*, 2010,13(5):633-637.

Ioshida MDM, Young MCM, Lago JHG (2010) Chemical composition and antifungal activity of essential oil from Brazilian propolis[J] *Journal of Essential Oil Bearing Plants*, 13(5):633 - 637.

Isla MI, Dantur Y, Salas A, Danert C, Zampini C, Arias M, Ordonez R, Maldonado L, Bedascarrasbure E, Nieva Moreno MI (2012) Effect of seasonality on chemical composition and antibacterial and anticandida activities of Argentine propolis: Design of a topical formulation[J]. *Natural Product Communications*, 7(10):1315-1318.

Isla MI, Paredes-Guzman JF, Nieva-Moreno MI, KOO H, Park YK (2005) Some chemical composition and biological activity of Northern Argentine propolis[J]. *Journal of Agricultural and Food Chemistry*, 53(4):1166-1172.

Ito J, Chang FR, Wang HK, Park YK, Ikegaki M, Kilgore N, Lee KH (2001) Anti-AIDS agents. 48. (1) Anti-HIV activity of moronic acid derivatives and the new melliferone-related triterpenoid isolated from Brazilian propolis[J]. *Journal Natural Product*, 64 (10):1278-1281.

Janas K, Bumba V (1997) Contribution to composition of beeswax propolis[J]. *Pharmazie*, 29: 544-545.

Jasprica I, Mornar A, Debeljak Z, Smolcic-Bubalo A, Medic-Saric M, Mayer L, Romic Z, Bucan K, Balog T, Sobocanec S, Sverko V (2007) In vivo study of propolis supplementation effects on antioxidative status and red blood cells[J]. *Journal of Ethnopharmacology*, 110(3): 548-554.

Jasprica I, Smolcic-Bubalo A, Momar A, Medic-Saric M (2004) Investigation of the flavonoids in Croatian propolis by thin-layer chromatography[J]. *Journal of Planar Chromatography*, 17(2): 95-101.

Kakehashi A, Ishii N, Fujioka M, Doi K, Gi M, Wanibuchi H (2016) Ethanol-extracted brazilian propolis exerts protective effects on tumorigenesis in wistar hannover rats [J]. *PLoS ONE*, 11(7): e0158654.

Kalogeropoulos N, Konteles SJ, Troullidou E, Mourtzinos I, Karathanos VT (2009) Chemical composition, antioxidant activity and antimicrobial properties of propolis extracts from Greece and Cyprus[J]. *Food Chemistry*, 116(2): 452-461.

Karsten M (2001) Propolis-current and future medical uses[J]. *American Bee Journal*, (7): 507-510.

Kashkooli OB, Dorcheh EE, Mahboobi-Soofiani N, Samie A (2011) Long-term effects of propolis on serum biochemical parameters of rainbow trout (*Oncorhynchus mykiss*) [J]. *Ecotoxicol Environm Safety*, 74(3): 315-318.

Kaškoniene V, Kaškonas P, Maruška A, Kubiliene L (2014) Chemometric analysis of volatiles of propolis from different regions using static GC-MS[J]. *Central European Journal of Chemistry*, 12(6): 736-746.

Kasote D, Suleman T, Chen W, Sandasi M, Viljoen A, Van Vuuren SF (2014) Chemical profiling and chemometric analysis of South African propolis[J]. *Biochemical Systematics and Ecology*, 55: 156-163.

Kleinrok Z, Borzecki Z, Scheller S, Matuga W (1997) Biological properties and clinical application of propolis. X. Preliminary pharmacological evaluation of ethanol extract of propolis (EEP)[J]. *Arzneimittel-Forschung*, 28(2): 291-292.

Krol W, Scheller S, Czuba Z, Matsuno T, Zydowicz G, Shani J, Mos M (1996) Inhibition of neutrophils' chemiluminescence by ethanol extract of propolis (EEP) and its phenolic components[J]. *Journal of Ethnopharmacology*, 55(1): 19-25.

Kujumgiev A, Tsvetkova I, Serkedjieva Y, Bankova V, Christov R, Popov S (1999) Antibacterial, antifungal and antiviral activity of propolis of different geographic origin [J]. *Journal of Ethnopharmacology*, 64(3): 235-240.

Kumazawa S, Goto H, Hamasaka T, Fukumoto S, Fujimoto T, Nakayama T (2004) A new prenylated flavonoid from propolis collected in Okinawa, Japan[J]. *Bioscience Biotechnology and Biochemistry*, 68(1): 260-262.

Kumazawa S, Hamasaka T, Nakayama T (2004) Antioxidant activity of propolis of various geographic origins[J]. *Food Chemistry*, 84(3): 329-339.

Kumazawa S, Nakamura J, Murase M, Miyagawa M, Ahn MR, Fukumoto S (2008). Plant origin of Okinawan propolis: honeybee behavior observation and phytochemical analysis[J]. *Naturwissenschaften*, 95(8): 781-786.

Kumazawa S, Yoneda M, Shibata I, Kanaeda J, Hamasaka T, Nakayama T (2003) Direct evidence for the plant origin of Brazilian propolis by the observation of honeybee behavior and phytochemical analysis[J]. *Chemical and Pharmaceutical Bulletin*, 51(6): 740-742.

Kumazawa S, Hamasaka T, Nakayama T (2004b) Antioxidant activity of propolis of various geographic origins[J]. *Food Chemistry*, 84(3): 329-339.

Kwon DJ, Bae SY (2009) Phenolic glucosides from bark of *Populus alba glandulosa* (*Salicaceae*)[J]. *Biochemical Systematics and Ecology*, 37(2): 130-132.

Li F, Awale S, Tezuka Y, Kadota S (2009b) Cytotoxic constituents of propolis from Myanmar and their structure-activity relationship[J]. *Biological & Pharmaceutical Bulletin*, 32(12): 2075-2078.

Li F, Awale S, Zhang H, Tezuka Y, Esumi H, Kadota S (2009a). Chemical constituents of propolis from Myanmar and their preferential cytotoxicity against a human pancreatic cancer cell line[J]. *Journal of Natural Products*, 72(7): 1283-1287.

Li F, Awale S, Tezuka Y, Esumi H, Kadota S (2010) Study on the constituents of Mexican propolis and their cytotoxic activity against PANC-1 human pancreatic cancer cells[J]. *Journal of Natural Products*, 73(4): 623-627.

Li YJ, Lin JL, Yang CW, Yu CC (2005). Acute renal failure induced by a Brazilian variety of propolis[J]. *American Journal of Kidney Diseases*, 46(6): 125-129.

López BG, Schmidt EM, Eberlin MN, Sawaya AC (2014) Phytochemical markers of different types of red propolis[J]. *Food Chemistry*, 146: 174-180.

Lotti C, Fernandez MC, Piccinelli AL, Cuesta-Rubio O, Hernandez IM, Rastrelli L (2010) Chemical constituents of red Mexican propolis[J]. *Journal of Agriculture and Food Chemistry*, 58(4): 2209-2213.

Lu Y, Wu C, Yuan Z (2004) Determination of hesperetin, cinnamic acid and nicotinic acid in propolis with micellar electrokinetic capillary chromatography[J]. *Fitoterapia*, 75(3): 267-276.

Mani F, Damasceno HCR, Novelli ELB, Martins EAM, Sforcin JM (2006) Propolis: effect of different concentrations, extracts and intake period on seric biochemical variables[J]. *Journal of Ethnopharmacology*, 105(1): 95-98.

Marcucci MC, Ferreres F, Garcia-Viguera C, Bankova B, Castro S, Dantas AP (2001) Phenolic compounds from Brazilian propolis with pharmacological activities[J]. *Journal of Ethnopharmacology*, 74(2): 105-12.

Marcucci MC (1995) Propolis: chemical composition, biological properties and therapeutic activity[J]. *Apidologie*, 26(2): 83-99.

Markham KR,Mitchell KA,Wilkins AL,Daldy JA,Lu Y (1996) HPLC and GC-MS identification of the major organic constituents in New Zeland propolis[J]. *Phytochemistry*,42(1):205-211.

Marostica Junior RM,Daugsh A,Moraes CS,Queiroga CS,Pastore GM,Park YK (2008) Comparison of volatile and polyphenolic compounds in Brazilian green propolis and its botanical origin *Baccharis dracunculifolia*[J]. *Cienc Tecnol Aliment*,28(1):178-181.

Martos I,Cossentini M,Ferreres F,Tomas-Barberan FA (1997) Flavanoid composition of Tunisian honeys and propolis[J]. *Journal of Agricural and Food Chemistry*,45(8):2824-2829.

Matsubayamshi H,Lagan P,Sukor JRA (2006) Utilization of Macaranga trees by the Asian elephants (*Elephas maximus*) in Borneo[J]. *Mammal Study*,31(2):115-118.

Matsuda AH,de Almeida-Muradian LB (2008) Validated method for the quantification of artepillin-C in Brazilian propolis[J]. *Phytochem Anal*,19(2):179-183.

Matsuno T,Jung SK,Matsumoto Y,Saito M,Morikawa J (1997) Preferential cytotoxicity to tumor cells of 3,5-diprenyl-4-hydroxycinnamic acid (artepillin C) isolated from propolis[J]. *Anticancer Research*,17(5A):3565-3568.

Mayworm MAS,Lima CA,Tomba ACB,Fernandes-Silva CC,Salatino MLF,Salatino A (2014) Does propolis contain tannins [J]. *Evidence-based Complementary and Alternative Medicine*,Article ID 613647.

Medana C,Carbone F,Aigotti R,Appendino G,Baiocchi C (2008) Selective analysis of phenolic compounds in propolis by HPLC-MS/MS[J]. *Phytochemical Analysis*,19(1):32-39.

Medic-Saric M,Jasprica L,Mormar A,Smolcic-Bubalo A,Golja P (2004) Quantitative analysis of flavonoids and phenolic acids in propolis by two-dimensional thin layer chromatography[J]. *Journal of Planar Chromatography-modern Tlc*,17(100):459-463.

Medina-Dzul K,Munoz-Rodriguez D,Moguel-Ordonez Y,Carrera-Figueiras C (2014) Application of mixed solvents for elution of organophosphate pesticides extracted from raw propolis by matrix solid-phase dispersion and analysis by GC-MS[J]. *Chemical Papers*,68(11):1474-1481.

Melliou E,Stratis E,Chinou I (2007) Volatile constituents of propolis from various regions of Greece-antimicrobial activity[J]. *Food Chemistry*,103(2):375-380.

Midorikawa K,Banskota AR,Tezuka Y,Nagaoka T,Matsushige K,Message D,Huertas AAG,Kadota S (2011) Liquid chromatography-mass spectrometry analysis of propolis [J]. *Phytochemistry Analysis*,12(6):366-373.

Milos M,Radonic A,Mastelic J (2005) Seasonal variations of essential oil composition of cupressussempervirens L[J]. *Journal of Essential Oil Research*,17:160-165.

Missima F,Silva-Filho AA,Nunes GA,Bueno PC,de Sousa JP,Bastos JK,Sforcin JM (2007) Effect of *Baccharis dracunculifolia* D.C (*Asteraceae*) extracts and its isolated compounds on macrophage activation[J]. *Pharmacy and Pharmacology*, 59(3): 463-468.

Mochizuki N, Aoki E, Suga K, Ishii R, Horie M (2008) Analysis of chloramphenicol in propolis extract by LC/MS/MS[J]. *Shokuhin Eiseigaku Zasshi*, 9(6): 399-402.

Moreno MIN, Isla MI, Sampietro AR, Vattuone MA (2000) Comparison of the free radical-scavenging activity of propolis from several regions of Argentina[J]. *Journal of Ethnopharmacology*, 71(1): 109-114.

Moret S, Purcaro G, Conte L (2010) Polycyclic aromatic hydrocarbons (PAHs) levels in propolis and propolis-based dietary supplements from the Italian market[J]. *Food Chemistry*, 122(1): 333-338.

Morlock G, Ristivojevic P, Chernetsova ES (2014) Combined multivariate data analysis of high-performance thin-layer chromatography fingerprints and direct analysis in real time mass spectra for profiling of natural products like propolis[J]. *Journal of Chromatography A*, 1328: 104-112.

Morlock GE, Ristivojevic P, Chernetsova ES (2014) Combined multivariate data analysis of high-performance thin-layer chromatography fingerprints and direct analysis in real time mass spectra for profiling of natural products like propolis[J]. *Journal of Chromatography A*, 1328: 104-12.

Mot AC, Soponar F, Sarbu C (2010) Multivariate analysis of reflectance spectra from propolis: Geographical variation in Romanian samples[J]. *Talanta*, 81(3): 1010-1015.

Mullin CA, Frazier M, Frazier JL, Ashcraft S, Simonds R, vanEngelsdorp D, Pettis JS (2010) High levels of miticides and agrochemicals in North American apiaries: implications for honey bee health[J]. *PLoS One*, 5(3): e9754.

Murase M, Kato M, Sun A, Ono T, Nakamura J, Sato T, Kumazawa S (2008) *Rhus javanica* var. *chinensis* as a new plant origin of propolis from Okayama, Japan[J]. *Bioscience, Biotechnology, and Biochemistry*, 72(10): 2782-2784.

Nafady AM, El-Shanawany MA, Mohamed MH, Hassanean HA, Nohara T, Yoshimitsu H, Ono M, Sugimoto H, Doi S, Sasaki K, Kuroda H (2003) Cyclodextrin-enclosed substances of Brazilian propolis[J]. *Chem Pharm Bull (Tokyo)*, 51(8): 984-985.

Nagaoka T, Banskota AH, Tezuka Y, Midorikawa K, Matsushige K, Kadota S (2003) Caffeic acid phenethyl ester (CAPE) analogues: Potent nitric oxide inhibitors from the Netherlands propolis[J]. *Biological and Pharmaceutical Bulletin*, 26(4): 487-491.

Nagy E, Papay V, Litkei G, Dinya Z (1986) Investigation of the chemical constituents, particularly the flavonoid components, of propolis and *Populi* gemma by the GC/MS method[J]. *Stud Org Chem (Amsterdam)*, 23: 223-232.

Nagy M, Grancai D (1996) Colorimetric determination of flavanones in propolis[J]. *Pharmazie*, 51(2): 100-101.

Naik DG, Vaidya HS, Namjoshi TP (2013) Essential oil of Indian propolis: chemical composition and repellency against the honeybee apis florea[J]. *Chemistry & Biodiversity*, 10(4):

649-657.

Nakamura R, Nakamura R, Watanae K, Oka K, Ohta S, Mishima S (2010) Effects of propolis from different areas on mast cell degranulation and identification of the effective components in propolis[J]. *International Immunopharmacology*, 10(9):1107-1112.

Neacsu M, Eklund PC, Sjoholm RE, Pietarinen SP, Ahotupa MO, Holmbom BR, Willfor SM (2007) Antioxidant flavonoids from knotwood of jack pine and European aspen[J]. *Holz Roh Werkst*, 65(1):1-6.

Nunes CA, Guerreiro M (2012) Characterization of Brazilian green propolis throughout the seasons by headspace GC/MS and ESI-MS[J]. *Journal of the Science of Food and Agriculture*, 92(2):433-438.

Okamoto Y, Tobe T, Ueda K, Takada T, Kojima N (2015) Oral administration of Brazilian propolis exerts estrogenic effect in ovariectomized rats[J]. *Journal of Toxicological Sciences*, 40(2):235-242.

Oliveira AP, França HS, Custeer RM, Teixeira LA, Rocha LM (2010) Chemical composition and antibacterial activity of Brazilian propolis essential oil[J]. *Journal of Venomous Animals and Toxins including Tropical Diseases*, 16(1):121-130.

Orsi RO, Funari SRC, Barbattini R, Giovani C, Frilli F, Sforcin JM, Bankova V (2006) Radionuclides in honeybee propolis (*Apis mellifera* L.)[J]. *Bulletin of Environmental Contamination and Toxicology*, 76(4):637-640.

Orsi RO, Sforcin JM, Funari SRC, Gomes JC (2005) Effect of propolis extract onguinea pig lungmast cell[J]. *Journal of Venomous Animals and Toxins including Tropical Diseases*, 11(1):76-83.

Park YK, Alencar SM, Aguiar C (2002) Botanical origin and chemical composition of Brazilian propolis[J]. *Journal of Agricural and Food Chemistry*, 50(9):2502-2506.

Park YK, Ikegaki M (1998) Preparation of water and ethanolic extracts of propolis and evaluation of the preparations[J]. *Bioscience Biotechnology and Biochemistry*, 62(1):2230-2232.

Pearl I A, Darling S F (1971) Hot water phenolic extractives of the bark and leaves of diploid Populus tremuloides[J]. *Phytochemistry*, 10(2):483-484.

Pearl I, Darling SF (1971) Studies of the hot water extractives of the bark and leaves of populus deltoids bartr[J]. *CanadianJournal of Chemistry*, 49(1):49-55.

Pellati F, Orlandini G, Pinetti D, Benvenuti S (2011) HPLC-DAD and HPLC-ESI-MS/MS methods for metabolite profiling of propolis extracts[J]. *Journal of Pharmaceutical and Biomedical Analysis*, 55(5):934-948.

Pellati F, PioPrencipe F, Benvenuti S (2013) Headspace solid-phase microextraction-gas chromatography-mass spectrometry characterization of propolis volatile compounds [J]. *Journal of harmaceutical and Biomedical Analysis*, 84:103-111.

Pereira AD, de Andrade SF, de Oliveira Swerts MS, Maistro EL (2008) First *in vivo* evaluation of the mutagenic effect of Brazilian green propolis by comet assay and micronucleus test[J]. *Food and Chemical Toxicology*, 46(7):2580-2584.

Pereira ADS, Pinto AC, Cardoso JN, Neto FRD, Ramos HFS, Dellamora-Ortix GM, Santos EP (1998) Application of high temperature high resolution gas chromatography to crude extracts of propolis[J]. *Journal of High Resolution Chromatography*, 21(7):396-400.

Pereira AS, Pereira AFM, Trugo LC, Neto AFR (2003) Distribution of quinic acid derivatives and other phenolic compounds in Brazilian propolis[J]. *Zeitschrift für Naturforschung C*, 58(7-8):590-593.

Perez-Parada A, Colazzo M, Besil N, Geis-Asteggiante L, Rey F, Heinzen H (2011) Determination of coumaphos, chlorpyrifos and ethion residues in propolis tinctures by matrix solid-phase dispersion and gas chromatography coupled to flame photometric and mass spectrometric detection[J]. *Journal of Chromatogram A*, 1218(34):5852-5857.

Petrova A, Popova M, Kuzmanova C, Tsvetkova I, Naydenski H, Muli E, Bankova V (2010) New biologically active compounds from Kenyan propolis[J]. *Fitoterapia*, 81(6):509-514.

Piccinelli AL, Lotti C, Campone L, Cuesta-Rubio O, Campo Fernandez M, Rastrelli L (2011) Cuban and Brazilian red propolis: botanical origin and comparative analysis by high-performance liquid chromatography-p Photodiode array detection/electrospray ionization tandem mass spectrometry[J]. *Journal of Agricural and Food Chemistry*, 59(12):6484-6491.

Pierini GD, Granero AM, Nezio MSD, Centurion ME, Zon MA, Fernandez H (2013) Development of an electroanalytical method for the determination of lead in Argentina raw propolis based on bismuth electrodes[J]. *Microchemical Journal*, (106):102-106.

Pierini GD, Fernandes DDS, Diniz P, de Araujo MCU, Di Nezio MS, Centurion ME (2016) A digital image-based traceability tool of the geographical origins of argentine propolis [J]. *Microchemical Journal*, 128:62-67.

Pietta PG, Gardana C, Pietta AM (2002) Analytical methods for quality control of propolis [J]. *Fitoterapia*, 73:S7-S20.

Pino JA, Marbot R, Delgado A, Zumarraga C, Sauri E (2006) Volatile constituents of propolis from honey bees and stingless bees from Yucatan[J]. *Journal of Essential Oil Research*, 18(1):53-56.

Pisco L, Kordian M, Peseke K, Feist H, Michalik D, Estrada E, Carvalho J, Hamilton G, Rando D, Quincoces J (2006) Synthesis of compounds with antiproliferative activity as analogues of prenylated natural products existing in Brazilian propolis[J]. *European Journal of Medicinal Chemistry*, 41(3):401-407.

Pontoh J, Low N H (2002) Purification and characterization of β-glucosidase from honey bees (*Apis mellifera*)[J]. *Insect Biochemistry and Molecular Biology*, 32(6):

679-690.

Popova M, Bankova V, Butovska D, Petkov V, Nikolova-Damyanova B, Sabatini AG, Marcazzan GL, Bogdanov S (2004) Validated methods for the quantification of biologically active constituents of poplsr-type propolis[J]. *Phytochemical Analysis*, 15(4): 235-240.

Popova M, Bankova V, Bogdanov S, Tsvetkova I, Naydenski C, Marcuzzan L, Sabatini AG (2007) Chemical characteristics of poplar type propolis of different geographic origin[J]. *Apidologie*, 38: 306-311.

Popova M, Dimitrova R, Al-Lawati HT, Tsvetkova I, Najdenski H, Bankova V (2013) Omani propolis: chemical profiling antibacterial activity and new propolis plant sources[J]. *Chemistry Central Journal*, 7((1): 158.

Popova M, Silici S, Kaftanoglu O, Bankova V (2005) Antibacterial activity of Turkish propolis and its qualitative and quantitative chemical composition[J]. *Phytomedicine*, 12(3): 221-228.

Popova M, Trusheva B, Cutajar S, Antonova D, Mifsud D, Farrugia C, Bankova V (2012) Identification of the plant origin of the botanical biomarkers of Mediterranean type propolis[J]. *Nat Prod Commun*, 7(5): 569-570.

Popova MP, Graikou K, Chinou I, Bankova VS (2010) GC-MS profiling of diterpene compounds in mediterranean propolis from Greece[J]. *Journal of Agricural and Food Chemistry*, 58(5): 3167-3176.

Popravko SA, Sololov IV (1980) Plant sources of propolis[J]. *Pchelovodstvo*, (2): 28-29.

Prytzyk E, Dantas AP, Salamo K, Pereira AS, Bankova VS, DeCastro SL, Neto FRA (2003) Flavonoids and trypanocidal activity of Bulgarian propolis[J]. *Journal of Ethnopharmacology*, 88(2): 189-193.

Queiroga CL, Bastos JK, de Sousa JPB, de Magalhães PM (2008) Comparison of the chemical composition of the essential oil and the water soluble oil of *Baccharis dracunculifolia* DC (Asteraceae)[J]. *Journal of Essential Oil Research*, 20(2): 111-114.

Rajpara S, Wilkinson MS, King CM, Gawkrodger DJ, English JS, Statham BN, Green C, Sansom JE, Chowdhury MMU, Horn HL, Ormerod AD (2009) The importance of propolis in patch testing-amulticentre survey[J]. *Contact Dermatitis*, 61(5): 287-290.

Ramien ML, Pratt MD (2012) Fixed drug eruption to ingested propolis[J]. *Dermatitis*, 23(4): 173-175.

Righi AA, Alves TR, Negri G, Marques LM, Breyer H, Salatino A (2011) Brazilian red propolis: unreported substances, antioxidant and antimicrobial activities[J]. *Journal of Scienes and Food Agriculture*, 91(13): 2363-2370.

Righi AA, NegriG, SalatinoA (2013) Comparative chemistry of propolis from eight Brazilian localities[J]. *Evidence-Based Complementary and Alternative Medicine*, Article ID 267878.

Rudzki E,Grzywa Z (1987) Primary and secondary allergy to propolis[J]. *Przegl Dermatol*,74(1):11-14.

Sahinler N,Kaftanoglu O (2005) Natural product propolis:Chemical composition[J]. *Natural Product Research*,19(2):183-188.

Salatino A,Fernandes-Silva CC,Righi AA,Salatino MLF (2011) Propolis research and the chemistry of plant products[J]. *Nature Product Reports*,28(5):925-936.

Salomao AP,Dantas1 CM,Borba LC,Campos DG,Machado FR,Aquino Neto,de Castro SL (2004) Chemical composition and microbicidal activity of extracts from Brazilian and Bulgarian propolis[J]. *Letters in Applied Microbiology*,38(2):87-92.

Salomao K,Pereira PRS,Campos LC,Borba CM,Cabello pH,Marcucci MC,de Castro SL (2008) Brazilian propolis:correlation between chemical composition and antimicrobial activity[J]. *Evidence-Based Complementary and Alternative Medicine*,5(3):317-324.

Santos FA,Bastos EMA,Uzeda M,Carvalho MAR,Farias LM,Moreira ESA,Braga FC (2002) Antibacterial activity of Brazilian propolis and fractions against oral anaerobic bacteria[J]. *Journal of Ethnopharmacol*,80(1):1-7.

Sarbu CS, Mot AC (2011) Ecosystem discrimination and fingerprinting of Romanian propolis by hierarchical fuzzy clustering and image analysis of TLC patterns[J]. *Talanta*,85(2):1112-1117.

Sawaya ACHF,Abdelnur PV,Eberlin MN,Kumazawa S,Ahn MR,Bang KS,Nagaraja N,Bankova VS,Afrouzan H (2010) Fingerprinting of propolis by easy ambient sonic-spray ionization mass spectrometry[J]. *Talanta*,81(1-2):100-108.

Sawaya ACHF,da Silva Cunha IB,Marcucci MC,Aidar DS,Silva ECA,Carvalho CAL,Eberlin MN (2007) Electrospray ionization mass spectrometry fingerprinting of propolis of native Brazilian stingless bees[J]. *Apidologie*,38(1):93-103.

Sawaya ACHF,Tomazela DM,Cunha IBS,Bankova VS,Marcucci MC,Custodio AR,Eberlin MN (2004) Electrospray ionization mass spectrometry fingerprinting of propolis[J]. *Analyst*,129(8):739-744.

Shi H,Yang H,Zhang X,Sheng Y,Huang H,Yu L (2012) Isolation and characterization of five glycerol esters from wuhan propolis and their potential anti-inflammatory properties[J]. *Journal of Agricultural and Food Chemistry*,60(40):10041-10047.

Shi H,Yang H,Zhang X,Yu LL (2012) Identification and quantification of phytochemical composition and anti-inflammatory and radical scavenging properties of methanolic extracts of Chinese propolis[J]. *Journal of Agricultural and Food Chemistry*,60(50):12403-12410.

Shimomura K,Sugiyama Y,Nakamura J,Ahn MR,Kumazawa S (2013) Component analysis of propolis collected on Jeju Island,Korea[J]. *Phytochemistry*,93:222-229.

Shrestha SP,Narukawa Y,Takeda T (2007) Chemical constituents of Nepalese propolis:isolation of new dalbergiones and related compounds[J]. *Natural Medicines*,61(1):73-76.

Si CL, Lu YY, Zhang Y, Xu J, Qin PP, Sun RC, Ni YH (2010) Antioxidative low molecular weight extractives from triploid populus tomentosa xylem[J]. *Bioresources*, 6(1): 232-242.

Si CL, Xu J, Kim JK, Bae YS, Liu PT, Liu Z (2011a). Antioxidant properties and structural analysis of phenolic glucosides from bark of *Populus ussuriensis* Kom[J]. *Wood Science and Technology*, 45:5-13.

Sillici S, Kutluca S (2005) Chemical composition and antibacterial activity of propolis collected by three different races of honeybees in the same region[J]. *Journal of Ethnopharmacol*, 99(1):69-73.

Simionatto E, Facco JT, Morel AF, Giacomelli SR, Linares CEB (2012) Chiral analysis of monoterpenes in volatile oils from propolis[J]. *Journal of the Chilean Chemical Society*, 57(3):1240-1243.

Simoes-Ambrosio LM, Gregorio LE, Sousa JP, Figueiredo-Rinhel AS, Azzolini AE, Bastos JK, Lucisano-Valim YM (2010) The role of seasonality on the inhibitory effect of Brazilian green propolis on the oxidative metabolism of neutrophils[J]. *Fitoterapia*, 81(8):1102-1108.

Simone-Finstrom M, Spivak M (2010) Propolis and bee health: the natural history and significance of resin use by honey bees[J]. *Apidologie*, 41(3):295-311.

Stan L, Marghitas LA, Dezmirean D (2011) Influence of collection methods on propolis quality[J]. *Bulletin UASVM Animal Science and Biotechnologies*, 68(1-2):278-282.

Sud'ina GF, Mirzoeva OK, Pushkareva MA, Korshunova GA, Sumbatyan NV, Varfolomeev SD (1993) Caffeic acid phenethyl ester as a lipoxygenase inhibitor with antioxidant properties[J]. *FEBS Letters*, 329(1):21-24.

Sun LP, Chen AL, Hung HC, Chien YH, Huang JS, Huang CY, Chen YW, Chen CN (2012) Chrysin: A histone deacetylase 8 inhibitor with anticancer activity and a suitable candidate for the standardization of chinese propolis[J]. *Journal of Agricural and Food Chemistry*, 60(47):11748-11758.

Tang TX, Guo WY, Xu Y, Zhang SM, Xu XJ, Wang DM, Zhao ZM, Zhu LP, Yang DP (2014) Thin-layer chromatographic identification of Chinese propolis using chemometric fingerprinting[J]. *Phytochem Analysis*, 25(3):266-272.

Tani H, Hasumi K, Tatefuji T, Hashimoto K, Koshino H, Takahashi S (2010) Inhibitory activity of Brazilian green propolis components and their derivatives on the release of cys-leukotrienes[J]. *Bioorg Med Chem*, 18(1):151-157.

Tavares DC, Barcelos GRM, Silva LF, Tonin CCC, Bastos JK (2006) Propolis induced genotoxicity and antigenotoxicity in Chinese hamster ovary cells[J]. *Toxicology in Vitro*, 20(7):1154-1158.

Tazawa S, Warashina T, Noro T (1999) Studies on the constituents of brazilian propolis Ⅱ[J]. *Chemical & Pharmaceutical Bullettin*, 47(10):1388-1392.

Teixeira EW, Message D, Negri G, Salatino A, Stringheta PC (2010) Seasonal variation, chemical composition and antioxidant activity of Brazilian propolis samples[J]. *Evidence-Based Complementary and Alternative Medicine*, 7(3):307-315.

Teixeira EW, Message D, Negri G, Salatino A (2006) Bauer-7-en-3b-yl acetate: A major constituent of unusual samples of Brazilian propolis[J]. *Química Nova*, 29(2): 245-246.

Thieme H (1976) Phenolic glycosides of the genus *Populus*[J]. *Planta Medica*, 15:35-40.

Tomas-Barberan FA, Garcia-Viguera G, Vit-Olivier P, Ferreres F, Tomas Premte F (1993) Phytochemical evidence for the botanical origin of tropical propolis from Venezuela *Populus*[J]. *Phytochemistry*, 34(1):191-196.

Torres RNS, Lopes JAD, MoitaNeto JM, Cito AM, Das GL (2008) The volatile constituents of propolis from Piauí[J]. *Química Nova*, 31(3):479-485.

Trusheva B, Popova M, Bankova V, Simova S, Marcucci MC, Miorin PL, da Rocha Pasin F, Tsvetkova I (2006) Bioactive constituents of Brazilian red propolis[J]. *Evidence-Based Complementary and Alternative Medicine*, 3(2):249-254.

Trusheva B, Popova M, Koendhori EB, Tsvetkova I, Naydenski C, Bankova V (2011) Indonesian propolis: chemical composition, biological activity and botanical origin[J]. *Nature Product Research*, 25(6):606-613.

Trusheva B, Popova M, Naydenski H, Tsvetkova I, Rodriguez JG, Bankova V (2004) New polyisoprenylated benzophenones from Venezuelan propolis[J]. *Fitoterapia*, 75(7-8): 683-689.

Trusheva B, Todorov I, Ninova M, Najdenski H, Daneshmand A, Bankova V (2010) Antibacterial mono- and sesquiterpene esters of benzoic acids from Iranian propolis [J]. *Chemistry Central Journal*, 4(1):8.

Tsichritzis F, Jakupovic J (1990) Diterpenes and other constituents from *Relhania* Species [J]. *Phytochemistry*, 29(10):3173-3187.

Usia T, Banskota AH, Tezuka Y, Midorikawa K, Matsushige K, Kadota S (2002) Constituents of Chinese propolis and their antiproliferative activities[J]. *Journal of Natural products*, 65(5):673-676.

Uto Y, Hirata A, Fujita T, Takubo S, Nagasawa H, Hori H (2002) First total synthesis of artepillin C established by o,o'-diprenylation of p-halophenols in water[J]. *Journal of Organic Chemistry*, 67(7):2355-2357.

Valencia D, Alday E, Robles-Zepeda R, Garibay-Escobar A, Galvez-Ruiz JC, Salas-Reyes M, Jimenez-Estrada M, Velazquez-Contreras EV, Hernandez J, Velazquez C (2012) Seasonal effect on chemical composition and biological activities of Sonoran propolis [J]. *Food Chemistry*, 131(2):645-651.

Volpi N, Bergonzini G (2006) Analysis of flavonoids from propolis by on-line HPLC-electrospray mass spectrometry[J]. *Journal of Pharmaceutical and Biomedical Analysis*, 42(3):354-361.

Volpi N (2004) Separation of flavonoids and phenolic acids from propolis by capillary zone electrophoresis[J]. *Electrophoresis*, 25(12):1872-1878.

Walgrave SE, Warshaw EM, Glesne LA (2005) Allergic contact dermatitis from propolis [J]. *Dermatitis*, 16(4):209-215.

Woisky RG, Salatino A (1998) Analysis of propolis: some parameters and procedures for chemical quality control[J]. *Journal of Apicultural Research*, 37(2):99-105.

Wu YW, Sun SQ, Zhao J, Zhou Q (2008) Rapid discrimination of extracts of Chinese propolis and poplar buds by FT-IR and 2D IR correlation spectroscopy[J]. *Journal of Molecular Structure*, 883:48-54.

Xu L, Yan SM, Cai CB, Yu XP (2013) Untargeted detection and quantitative analysis of poplar balata (PB) in Chinese propolis by FT-NIR spectroscopy and chemometrics[J]. *Food Chemistry*, 141(4):4132-4137.

Yamauchi, Kato, Oida, Kanaeda, Ueno (1992) Benzyl Caffeate, an antioxidative compound isolated from propolis [J]. *Bioscience, Biotechnology, and Biochemistry*, 56(8):1321-1322

Yang C, Luo L, Zhang H, Yang X, Lv Y, Song H (2010) Common aroma-active components of propolis from 23 regions of China[J]. *Journal of the Science of Food and Agriculture*, 90(7):1268-1282.

Yang J, Chen LZ, Xue XF, Wu LM, Li Y, Zhao J, Wu ZB, Zhang YN (2016) A feasibility study on the discrimination of the propolis varieties based on near infrared spectroscopy[J]. *Spectroscopy and Spectral Analysis*, 36(6):1717-1720.

Zhang CP, Zheng HQ, Hu FL (2011b) Extraction, partial characterization and storage stability of β-glucosidase from propolis[J]. *Journal of Food Science*, 76(1):75-79.

Zhang C, Zheng HQ, Liu G, Hu FL (2011b) Development and validation of HPLC method for determination of salicin in poplar buds: Application for screening of counterfeit propolis[J]. *Food Chemistry*, 127(1):345-350.

Zhang CP, Huang S, Wei WT, Ping S, Sheng XG, Li YJ, Hu FL (2014) Development of high-performance liquid chromatographic for quality and authenticity control of Chinese propolis[J]. *Journal of Food Science*, 79(7):C1315-C1322.

Zhang CP, Zheng HQ, Hu FL (2011a) Extraction, partial characterization and storage stability of β-glucosidase from propolis[J]. *Journal of Food Science*, 76(1):75-79.

Zhang CP, Zheng HQ, Liu G, Hu FL (2011a) Development and validation of HPLC method for determination of salicin in poplar buds: Application for screening of counterfeit propolis[J]. *Food Chemistry*, 127(1):345-350.

Zhang T, Omar R, Siheri W, Mutairi SA, Clements C, Fearnley J, Edrada-Ebel RA, Watson D (2014) Chromatographic analysis with different detectors in the chemical characterisation and dereplication of African propolis[J]. *Talanta*, 120:181-190.

Zhou JH, Li Y, Zhao J, Xue XF, Wu LM, Chen F (2008) Geographical traceability of propolis by high-performance liquid-chromatography fingerprints[J]. *Food Chemisty*, 108(2):749-759.

Zhou JH, Xue XF, Li Y, Zhang JZ, Chen F, Wu LM, Chen LZ, Zhao J (2009) Multiresidue determination of tetracycline antibiotics in propolis by using HPLC-UV detection with ultrasonic-assisted extraction and tow-step solid phase extraction[J]. *Food Chemistry*, 115(3):1074-1080.

薄文飞(2009)蜂胶萜类成分及生物活性研究[D].济南:山东轻工业学院.

曹炜,符军放,索志荣,陈卫军,郑建斌(2007)蜂胶与杨树芽提取物成分的比较研究[J].食品与发酵工业,33(7):162-166.

曾晞,卢玉振,牟兰,张长庚(2004)GC-MS法分析比较贵州不同产地蜂胶挥发油化学成分[J].生命科学仪器,(2):28-29.

陈东海,闫德斌(1999)蜂胶的安全性—急性毒性实验[J].养蜂科技,(3):29-31.

程焕,秦子涵,胡小松,吴继红(2012)固相微萃取/动态顶空-气相色谱-质谱联用法对蜂胶与杨树胶挥发性成分的分析[J].食品安全质量检测学报,3(1):1-9.

迟家平,薛秉文,徐大庆,李琪(1997)RP-HPLC法测定蜂胶益胃胶囊中咖啡酸异阿魏酸和3′,4′-二甲氧基桂皮酸的含量[J].人民军医药学专刊,13(4):243-245.

董捷(2008)电子鼻对不同地域的蜂胶气味测定的初步研究[J].食品科学,29(10):468-470

冯丁山,郑定仙,黄业宇,王湛,张晓昕(2010)蜂胶胶囊的毒理学安全性评价[J].中国热带医学,10(4):454-455.

符军放(2006)中国蜂胶中酚类化合物的色谱分析方法研究[D].西安:西北大学.

付阳,任志勇,朱琰,余守志(2006)热分析快速鉴别蜂胶的新方法研究[C].中国化学会第十三届全国化学热力学和热分析学术会议论文摘要集.106.

付宇新,徐元君,陈滨,丽艳,罗丽萍(2009)气相色谱/质谱法分析内蒙古蜂胶挥发性成分[J].分析化学,37(5):745-748.

傅颖,梅松,刘冬英,来伟旗,陈建国,王茵(2010)蜂胶软胶囊毒理学安全性研究[J].中国卫生检验杂志,(7):1697-1700.

高丽红(2014)氢化物发生-原子荧光法测定蜂胶中的总砷[J].河南预防医学杂志,(3):013.

高振中,李宁,杨志岩,常彧,宋诗莹,韩英素(2009)改性天然沸石与超临界萃取法分离蜂胶中的铅[J].食品研究与开发,30(1):1-4.

龚上佶(2011)中国不同地区蜂胶元素含量测定及红外光谱分析[D].南昌:南昌大学.

韩红祥,王维,陈雪松,董雪莲,董金香,樊美玲,邱智东(2011)应用HPLC-MS联用技术研究不同基源蜂胶指纹图谱[J].吉林中医药,31(12):87-89.

韩利文,刘可春,王雪,侯海荣,王思锋(2007)蜂胶中抗肿瘤成分的研究[J].中国蜂业,58(5):11-13.

韩利文,刘可春,王思锋,王雪,侯海荣(2008)HPLC法测定蜂胶中阿替匹林C的含量[J].中国药事,22(4):312-314.

韩玉谦,隋晓,冯晓梅,管华诗(2003)超临界CO_2萃取蜂胶有效成分的研究[J].精细化工,20(7):421-424.

贺丽苹,李雅萍,陈玉芬,潘建国,郑尧隆,林励,卢占列(2008)SPME-GC/MS联用技术分析蜂胶的挥发性成分[J].中国蜂业,59(4):36-37.

黄帅,卢媛媛,张翠平,胡福良(2013)巴西绿蜂胶乙醇提取前后挥发性成分的分析比较[J].食品与生物技术学报,(7):680-685.

黄帅(2015)基于邻苯二酚以及多指标指纹图谱为指标的中国蜂胶真伪鉴别方法的研究[D].杭州:浙江大学.

雷雨,庄红林,毕玉芬,戴云(2010)蜂胶 HPLC 指纹图谱的 2 种分析方法[J].云南民族大学学报(自然科学版),(1):13-17.

李波,张洛红,李莹(2013)灯芯草纤维素黄原酸盐的制备及吸附蜂胶中镉的效果分析[J].黑龙江畜牧兽医,(15):128-130.

李丽,吴俐勤,章虎,莫卫民,王方莉,陈志民,钱鸣蓉(2014)高效液相色谱-串联质谱法测定蜂胶中咖啡酸苯乙酯[J].理化检验-化学分册,50(1):54-57.

李雅洁,凌建亚,邓勇(2006)蒙山蜂胶超临界二氧化碳萃取挥发性组分的气相色谱-质谱联用分析[J].时珍国医国药,17(10):1975-1976.

李雅晶 胡福良 陆旋 詹忠根(2011)蜂胶中挥发性成分的微波辅助提取工艺研究及中国蜂胶、巴西蜂胶挥发性成分比较[J].中国食品学报,11(5):93-99.

李雅萍,贺丽苹,陈玉芬,卢占列,郑尧隆,潘建国(2007)SPME-GC/MS 联用技术分析蜂胶中挥发性成分的研究[J].现代食品科技,23(7):78-80.

李熠,赵静,薛晓锋,周金慧(2007)超高效液相色谱法同时测定蜂胶中的 12 种活性成分[J].色谱,25(6):857-860.

李樱红,周萍,罗金文,周明昊,陶巧凤(2014)蜂胶与杨树胶 HPLC 指纹图谱的建立及应用[J].药物分析杂志,34(2):349-354.

林茂,李守珍(1993)毛白杨化学成分的研究[J].药学学报,28(6):437-444.

刘安洲,杜安华,王晓(2009)泰山蜂胶挥发性成分检测[J].食品与发酵工艺,35(5):163-166.

刘莉敏,郭军,米智慧,王竹,向雪松(2014)蜂胶乙醇提取物 UPLC-Q-TOF-MS 指纹特征初探[J].食品科学,35(18):96-99.

刘映,许静芬,史庆龙,葛发欢(2000)蜂胶挥发性成分的超临界 CO_2 萃取及 GC-MS 分析[J].中药材,23(9):547-549.

罗照明,董捷,赵亮亮,张红城(2013)河南蜂胶中多酚类物质成分分析[J].食品科学,34(10):139-143.

马海乐,李倩,赵杰文,骆琳,何荣海,王振斌(2011)蜂胶超临界 CO_2 萃取物 GC 指纹图谱及聚类分析[J].江苏大学学报(自然科学版),32(1):1-5.

孟霞,彭敬东,刘绍璞(2008)高效液相色谱法测定蜂胶液中的 6 种有机酸[J].西南大学学报,30(9):66-70.

潘建国,王开发,段怡,李立群(2002)蜂胶中 10-羟基-癸烯酸的含量测定[J].中药材,25(7):505-506.

彭友元(2011)毛细管电泳电化学检测法测定蜂胶中的黄酮和酚酸[J].分析试验室,30(3):54-57.

沙娜,黄慧莲,张金强,李萍,果德安(2009)蜂胶的 HPLC 指纹图谱研究[J].中国药科大学学报,40(2):144-146.

盛文胜,杜青桃,李树岚,喻建辉,汪玲（2008）蜂胶与杨树胶中八种黄酮与肉桂酸含量的测定[J].中国蜂业,59(3):7-8.

索志荣,曹炜,秦海燕（2008）蜂胶浸膏中7种多酚化合物的高效液相色谱法分析[J].食品科学,29(8):496-498.

滕军,况磊（2014）蜂胶中两种合成降糖药物及蜂胶质量的快速鉴定[J].实验与检验医学,32(4):372-373.

田文礼,赵亚周,方小明,高凌宇,彭文君（2012）蜂胶醇提物的挥发性成分分析[J].现代食品科技,28(4):456-461.

童晔玲,任泽明,戴关海,王波波,杨锋（2012）蜂胶洋参软胶囊毒理学安全性研究[J].中国中医药科技,(2):143-145.

王蓓,马海乐,赵杰文,骆琳（2011）国产蜂胶超临界CO_2萃取物GC-MS指纹图谱与聚类分析[J].中国食品学报,11(6):183-192.

王冰,陈洋,李宝丽,朱宇轩,唐翠娥,黄艳春,张缔,刘睿（2014）蜂胶乙酸乙酯提取物的反相高效液相色谱-质谱分离鉴定[J].食品科学,35(2):186-190.

王强,常海军,吴洪斌,任彦荣,李贵节（2011）不同前处理方法对检测蜂胶总砷含量的影响-采用氢化物原子荧光技术(Ha-AFS)[J].食品研究与开发,32(8):77-80.

王维（2010）不同产地蜂胶的质量评价研究[D].长春:长春中医药大学.

王祥云,章虎,罗志强,李文丹,吴俐勤（2009）气相色谱法检测蜂胶中的氟胺氰菊酯和氟氯苯氰菊酯残留[J].安徽农业科学,37(22):10361-10362.

王想想,周立东,南垚,孟庆伟（2007）加拿大杨树芽化学成分的研究[J].中医药现代化,9(6):64-67.

王向平,品斌,刘伟忠,严凤仙,王荣（2012）气相色谱-质谱法分析蜂胶与杨树胶化学组分及鉴别研究[J].湖北农业科学,51(2):382-384.

王小平,林励,潘建国,林芳花,卢占列（2008）甘肃蜂胶中化学成分的气相色谱-质谱联用分析[J].时珍国医国药,19(3):559-560.

王小平,林励,白吉庆（2009）HPLC指纹图谱法鉴别蜂胶和树胶[J].陕西农业科学,55(3):133-134.

王欣,王强,徐国钧,徐珞珊（1999）杨属植物化学成分及药理活性研究进展[J].天然产物研究与开发,11(1):65-74.

王欣,王强,汪红（2000）山杨的化学成分研究[J].中草药,31(12):891-892.

王益民,杨仕军,毛小江（2008）石墨炉原子吸收光谱法测定蜂胶中的铅[J].理化检测-化学分册,44(5):419-420.

王勇,张中印,陈芳,张金振,薛晓锋,魏月,吴黎明（2014）不同取胶方式对蜂胶品质影响的比较研究[J].食品科学,(10):52-56.

吴健全,高蔚娜,赵丽婷,唐振闯,罗海吉,郭长江（2012）不同产地蜂胶中氨基酸含量的比较[J].氨基酸和生物资源,34(4):17-19.

吴健全,高蔚娜,韦京豫,薄玲玲,焦昌娅,郭长江（2013）不同产地蜂胶成分含量的比较[J].中国食物与营养,19(7):62-65.

吴正双（2011）蜂胶提取物中酚类化合物分析及其抗氧化和抗肿瘤活性研究[D].广州:华

南理工大学.

谢玮,郭婕,卢连华,薛薇(2011)某种蜂胶软胶囊的毒理学安全性评价[J].预防医学论坛,17(3):237-240.

徐响,董捷,李洁(2008)固相微萃取与GC-MS法分析蜂胶中挥发性成分[J].食品工业科技,29(5):57-60.

徐元君,罗丽萍,丽艳,陈滨,付宇新,高荫榆(2010)HPLC测定两种中国蜂胶醇提物化学组成[J].林产化学与工业,30(2):61-66.

延莎,张红城,董捷(2012a)电子鼻对蜂胶质量的判别[J].食品科学,33(20):201-205.

延莎,张红城,董捷(2012b)蜂胶及杨树胶关键气味活性成分研究[J].食品科学,33(4):157-161.

杨茂森,刘泊,王鑫,张华,蒲晓亚(2012)应用检测叶绿素含量鉴别蜂胶和杨树(芽)胶的研究[J].中国蜂业,63(2):81-85.

杨琴,朱美玲,郦宏岩,马海燕,毛海波,赵淑云(2008)蜂胶提取物及蜂胶保健食品的质量标准探讨及质量安全控制[J].蜜蜂杂志,(1):10-14.

杨书珍,彭丽桃,江海,艾倩倩,王萍,潘思轶,范刚(2009)不同溶剂提取蜂胶挥发油的成分分析[J].天然产物研究与开发,21(B10):337-340.

杨雯筌,许蔚,殷耀,张晓燕,陈惠兰,辛志宏,沈崇钰,张睿(2013)LC-MS/MS检测蜂胶中硝基呋喃类代谢物的残留[J].分析试验室,32(3):47-50.

于敏,弥宏,焦连庆(2005)蜂胶总黄酮中阿魏酸与总酚酸的含量测定[J].中药材,28(12):1117-1118.

于世锋,符军放,陈卫军,王毕妮,曹炜(2007)高效液相色谱法测定蜂胶中4种酚酸的含量[J].中国蜂业,58(9):8-10.

余兰平,盛文胜(2006)气相色谱-质谱法分析蜂胶和杨树胶的化学成分[J].蜜蜂杂志,(6):3-5.

张翠平,胡福良(2009)蜂胶中的黄酮类化合物[J].天然产物研究与开发,21(6):1084-1090.

张翠平,胡福良(2011)蜂胶与杨树胶HPLC指纹图谱鉴别[J].中国食品学报,11(1):222-225.

张洛红,李莹,仝攀瑞(2012)灯芯草纤维素黄原酸盐用于脱除蜂胶中铅的研究[J].食品工业科技,(1):226-229.

张洛红,李莹,仝攀瑞(2011)改性废弃灯芯草吸附去除蜂胶中的铅[J].西安工程大学学报,25(4):503-508.

张睿,张晓燕,陈磊,刘艳,殷耀,丁涛,朱文君,吴斌,陈惠兰(2014)液相色谱-串联质谱法同时测定蜂胶中的磺胺及林可胺类药物残留[J].分析试验室,(12):1420-1424.

张晓燕,张睿,许蔚,黄娟,刘艳,吴斌,陈磊,丁涛,沈崇钰,陈惠兰(2012)高效液相色谱-串联质谱法测定蜂胶中的氯霉素[J].色谱,(3):314-317.

张修景(2014)微波消解-分子荧光猝灭法测定蜂胶中痕量硒[J].食品与发酵工业,(3):208-210.

赵静(2003)利用HPLC测定蜂胶中阿魏酸含量[J].现代科学仪器,(1):65-66.

赵亮亮,王光新,陈平,董捷,张红城(2012)高效液相色谱法分析北方部分地区蜂胶醇提取物成分[J].食品科学,33(18):143-148.

赵强,张彬,周武,郭志芳(2007)微波辅助萃取-GC/MS联用分析蜂胶挥发油[J].精细化工,24(12):1192-1203.

周立东,南垚,孙兰,黄红钢,郭伽(2005)返滴定法测定蜂胶中游离酸含量[J].中国养蜂,56(5):10-11.

周立东,郑莲香,南垚,王想想,李熠,赵静(2007)蜂胶与杨树叶和杨树芽中叶绿素含量的比较研究[J].中国蜂业,58(6):5-7.

周平,章征天,胡福良,余秀珍(2005)蜂胶HPLC指纹图谱真伪初探[J].蜜蜂杂志,(8):5-6.

周萍,章征天,胡福良,余秀珍(2005)蜂胶HPLC指纹图谱真伪鉴别初探[J].蜜蜂杂志,(8):5-6.

周萍,陈建清,胡福良,胡元强,邵巧云(2009)不同产地蜂胶HPLC指纹图谱测定及真伪判定[J].中国蜂业,(10):5-8.

周萍,胡福良,徐权华,王立华(2010)高效液相色谱串联质谱测定蜂胶中氯霉素药物残留量[J].中国蜂业,61(9):42-44.

周萍,邵巧云,徐权华,黄帅,胡福良(2013)高效液相色谱法测定蜂胶中咖啡酸及8种黄酮类化合物的含量[J].蜜蜂杂志,(4):4-6.

朱恩圆,窦玉玲,魏东芝,王峥涛,卢艳花(2005)蜂胶HPLC指纹图谱及质量控制[J].中国中药杂志,30(18):1423-1425.

朱明达,马微,陈冬东,彭涛,李淑娟,赵广华,唐英章(2010)液相色谱-电喷雾串联质谱法同时测定蜂胶保健品中的活性成分和降糖西药[J].分析化学,38(2):169-174.

俎志平(2013)微波消解-石墨炉原子吸收法测定蜂胶中重金属铬[J].河南预防医学杂志,24(2):109-110.

第五章　蜂胶有效成分的提取与产品加工

蜂胶的化学成分十分丰富和复杂,除富含黄酮类、酚酸类和萜类化合物三大类活性成分外,还含有维生素、氨基酸、木脂素类及脂肪酸类等其他生物学活性成分,种类达数百种之多(Bankova,2005;Silva et al.,2008;de Groot,2013)。蜂胶原料中含有大量的活性因子,但同样也存在一些无明显药用价值的成分甚至有毒成分;加上蜂胶具有低温硬脆、高温变黏的质地特异性,不溶于水,部分溶于乙醇等有机溶剂,因此,在蜂胶使用前,必须经过加工提取,将其有效因子提取出来,并使其转化为易被人体吸收的形式(Burdock,1998)。同时,由于蜂胶中不同生物学活性成分的理化特性各异,因而在蜂胶加工过程中需考虑不同加工方式对不同生物学活性成分的影响。

随着对蜂胶生物学活性的深入研究及分析手段的提高,蜂胶中的各类组分及在蜂胶活性发挥中的作用会得到进一步明晰。当前,蜂胶经提取加工后开发为蜂胶产品主要以黄酮类化合物作为活性物质参照指标,对蜂胶中的酚酸类化合物、萜类化合物及其他生物学活性组分是否保留及保留量顾及甚少,而蜂胶生物学活性的发挥是蜂胶中各种活性物质协同发挥作用的结果。因此,对蜂胶中不同类型化合物进行提取、分离以及药理活性的深入研究,将为蜂胶的进一步研究开发及蜂胶产品的深加工奠定基础。

第一节　蜂胶有效成分的提取工艺

蜂胶成分极为复杂,影响因素诸多,产地来源、胶源植物、采收季节、蜂种、气候、提取方式等不同,都会使蜂胶的成分在质量和数量上产生显著差异(Toreti et al.,2013;Bankova et al.,2014;Huang et al.,2014;Czyzewska et al.,2015)。准确可靠的提取方法有助于蜂胶中活性物质的充分利用以及推进蜂胶标准化的发展。

由于蜂胶具有成分复杂、低温硬脆、高温黏稠、不溶于水、气味独特、滋味苦涩等特性,人们在利用蜂胶时需将蜂胶进行加工,去除蜂胶中的蜂蜡等无效及妨碍人体吸收的成分。目前用于蜂胶产品加工制备的原料绝大多数为用不同溶剂提取后的蜂胶提取物。由于溶剂及提取方法不同,这些提取物根据挥发性来分,主要分为2类:一类为蜂胶中的非挥发性物质,另一类为蜂胶中的挥发性物质。蜂胶中挥发性及非挥发性成分的全面有效提取对蜂胶生物学活性的发挥起着至关重要的作用。

1　蜂胶中非挥发性有效成分的提取工艺

蜂胶中的非挥发性提取物质是蜂胶产品加工的主要原料,这类提取物中包括了蜂胶中的黄酮类化合物、酚酸类化合物及氨基酸等活性物质。在溶剂选择方面,目前,国内外用于药品和保健食品的蜂胶几乎全部采用乙醇或者水作提取溶剂。有些化妆品采用乙二醇作溶剂,以提高水乳剂的溶解度。蜂胶的丙酮提取物也被用在洗发剂和洗涤剂中。其他一些溶剂,如乙醚、醋酸、苯、2%的氢氧化钠以及氨水等,也被用于蜂胶的提取。

1.1 蜂胶中非挥发性有效成分的不同溶剂提取法

按照提取溶剂的不同,蜂胶中非挥发性有效成分提取方法有以下几种。

(1)乙醇提取法

乙醇是目前国内外蜂胶有效成分提取中最常使用的溶剂。使用乙醇溶解的方法对蜂胶进行提取,操作简单,成本较低,一次性投入少,易为一些中小企业所采用。乙醇提取主要利用黄酮作为监控指标来评价整个蜂胶提取工艺的优劣(刘冀等,2011)。乙醇提取法的原理主要是利用黄酮类化合物与其他物质极性不同。选用乙醇进行萃取可以提取黄酮类化合物,还可以起到分离苷和配基或极性和非极性配基的目的(迟家平等,2002;罗登林等,2006)。

冷浸法是乙醇提取中操作最为简单的一种方法,目前蜂胶加工生产厂商大多采用此法。操作过程为(工艺流程见图 5.1):首先将蜂胶切成小片或碾磨成细粉,以增加与乙醇接触的表面积,提高溶解程度。若蜂胶太黏难以粉碎,可将其放入冰柜内冷冻数小时后再进行粉碎。然后将乙醇和蜂胶倒入容器内,稍作搅拌,使乙醇浸没蜂胶,以后每天搅拌 1～2 次,将混合物避光放置 1～2 周,或更久后,将液体过滤,过滤后的渣滓再次浸泡,提取。为减少溶剂用量,降低成本,提取时可先采用高浓度乙醇,然后根据用途稀释提取液。高浓度的提取液可减少蒸干的过程。

滤渣→→→反复提取
↑   ↓
粗蜂胶→冷冻→粉碎→乙醇提取→过滤→滤液→真空浓缩→蜂胶膏→保藏

图 5.1 蜂胶的提纯工艺流程

Fig. 5.1 Technological flow sheet of propolis extraction

乙醇浸提过程中,蜂胶乙醇提取物的活性成分受浸渍时间长短、温度高低、乙醇溶液浓度等因素的影响(吕武清等,2014)。通常认为蜂胶在乙醇中浸渍时间越长,则溶解成分越多,一般在 2 周内随时间的延长而升高,工业上大规模生产一般要浸泡 7d 以上甚至数月(侯加强,1993;胡福良,2000)。有报道说,蜂胶在浸泡 30d 后其总黄酮含量还会缓慢增加,但浸渍时间太长其提取物的含量并不显著增加;乙醇浓度不同,提取物的成分有差别,所提物质的溶解性亦有差异(郑艳萍,2008)。60%的乙醇提取的蜂胶液中所含异樱花素、槲皮素、山奈酚的量最高;70%乙醇提取的乔松素、樱花素含量最多;80%的乙醇提取的山奈甲黄素、刺槐素、异鼠李素含量较多;60%和 80%的乙醇提取液对微生物的抑制作用最强;70%和 80%的乙醇提取液的抗氧化作用最强;80%的乙醇提取液对透明质酸酶的抑制作用强。常规提取采用的乙醇浓度为 95%～100%,也有利用较低浓度乙醇,如 70%乙醇进行提取的工艺(沈海涛,2002)。程伟贤等(2006)研究发现,用 70%的乙醇浸泡蜂胶 36h 的提取率较高,而王启发等(2004)通过试验确定最佳工艺为乙醇质量分数为 80%,溶剂与蜂胶的质量配比(液固比)为 8∶1,搅拌下提取温度为 40℃,提取时间为 6h,在此条件下提取液中黄酮类化合物的质量分数可达 21.5%。乙醇冷浸法进行蜂胶提取,该方法较简单,并且所需温度不高,蜂胶中某些易失活的活性物质不易受到损失,但是提取率受时间限制较大。

考虑蜂胶溶解性差、易结块的物理特点,人们在蜂胶乙醇冷浸法加工提取的过程中,对一些工艺过程进行了改进,增加了回流提取或搅拌等过程。回流浸提法就是用回流装置对其进行加热,其操作与冷浸法相类似,这种方法所需时间较短,但由于操作温度较高容易使某些黄酮类物质失去活性。搅拌法则在乙醇浸渍的基础上再加以连续的搅拌,以增加溶解效率,缩短浸提时间。吕武清等(2014)就蜂胶超声、回流、浸渍、浸渍加搅拌4种提取工艺对蜂胶浸膏得率及浸膏中总黄酮、白杨素、松属素和高良姜素质量的影响进行了比较研究。结果发现,回流和超声提取的得膏率、黄酮含量均低于浸渍(加搅拌)提取;经观察,回流和超声提取是由于加热的原因,蜂胶在提取时溶液形成胶状,影响了浸出,超声提取发热程度不及回流提取,所以得膏率、黄酮含量高于回流提取;而浸渍(加搅拌)提取无发热过程,蜂胶中的成分为溶解过程,因此,提取效果最佳。在乙醇浓度方面,95%乙醇提取得膏率,总黄酮、白杨素、松属素和高良姜素质量分数显著高于75%和90%乙醇。研究得到最佳工艺为:蜂胶冷藏后粉碎,加5倍量95%乙醇浸渍(加搅拌,50r/min)提取4h,过滤,回收乙醇,浓缩、干燥。得蜂胶浸膏率超过50%,其中总黄酮的量为57.65%,白杨素、松属素、高良姜素等的含量也较高。

相对于其他溶剂来说,乙醇的毒性比较小,萃取液的组分范围与回收率都较好,比较经济可行,为蜂胶提取的最主要的方法(林贤统等,2008)。

(2)水提取法

由于蜂胶的物理特点,天然蜂胶在水中的溶解度非常小,若以黄酮类化合物作为提取率衡量指标,目前水提法制得的蜂胶提取物的得率仅在3%~10%。研究发现,采用合适的蜂胶水提取工艺,其提取率虽然远比醇提取法低,但蜂胶水提物仍具有多种生物学活性,如抗肿瘤、增强免疫等,在调节血脂、血糖、抗炎免疫等效果上与醇提物比较基本无显著差异(Hu et al.,2005)。利用水作溶剂提取蜂胶可以改善传统蜂胶乙醇提取物所附带的酒精刺激性气味及部分对乙醇过敏人群不能耐受蜂胶醇提物产品等弊端,更易于在制药、化妆品以及食品上得到应用。因而,开发一种切实可行的工艺,提高水作为溶剂的蜂胶提取效率,受到了研究人员的关注。

目前用水进行蜂胶提取主要有2类方法:一类是直接法,即直接用水提取;一类是间接法,即以水作为主要提取溶剂,但在蜂胶处理过程中利用酶等对蜂胶进行预处理后再用水提取。

直接法进行水提与乙醇提取法基本相同,只是溶剂变为水,提取率同样也受浸提时间和温度的影响。用水浸渍蜂胶数天或水煮的方法可获得蜂胶的水提取物,其加工、过滤等方法均与乙醇提取法相同。水提取物中黄酮等活性成分的含量要大大低于乙醇提取物中的,且提取效率较低,但提取物具有一定的抗细菌和抗真菌的作用。通过改变水作为溶剂的料液比、提取温度、提取时间等可以增加蜂胶水提物的得率。王锐等(2004)将新鲜蜂胶粉碎后置于蒸馏水(1:20,w/w)于室温条件下搅拌48h;过滤;滤渣加水反复提取、过滤;合并滤液,在15000r/min条件下离心60min;合并上清液;经冻干处理后得水溶性蜂胶成分。经测定,所提取的水溶性蜂胶含3种黄酮成分。Takeshi等(2003)将蜂胶在一定量蒸馏水中混匀,在20℃振荡1d,提取液在28000r/min条件下离心30min,收集上清。残渣在同样条件下再次提取,合并上清,用蒸馏水透析,透析液抵押冻干,蜂胶充分水提,得率达13%。陈滨(2010)

将10g蜂胶在300mL蒸馏水中混匀,60℃水浴提取7h,提取液在28000r/min条件下离心30min,收集上清。残渣在同样条件下再次提取,合并上清,于60℃真空减压旋转蒸发仪中浓缩,得到固态蜂胶水提物,蜂胶水提平均得率为(6.0±1.6)%。郝胤博等(2012)通过比较中国不同地区蜂胶水提物的抑菌活性,得到优化的蜂胶水提物制备工艺:取4.0g蜂胶,在120mL蒸馏水中混匀,在60℃下水浴提取7h,提取液离心30min,收集上清液,残渣在同样条件下再提取1次,合并提取液,于60℃真空旋转蒸发仪中浓缩,得到固态蜂胶水提物。使用时蒸馏水复溶,配制成所需浓度待用。该方法可以提高蜂胶的利用率。

用水作溶剂直接对蜂胶进行提取还可将蜂胶中的蜂蜡有效去除。蜂蜡作为蜂胶中的惰性物质,有碍蜂胶中的活性物质在人体内的吸收,因此,蜂胶产品的纯度和蜂蜡的去除率有直接关系。魏强华等(2007)采用蜂胶水提除蜡工艺去除粗蜂胶中的蜂蜡,并通过均匀设计法优化蜂胶水提除蜡工艺。发现在蜂胶水提除蜡工艺中,水提温度是显著因素,而水提时间、液固比不是显著因素。综合考虑经济技术要求,蜂胶水提除蜡工艺最佳条件为:水提温度70℃、水提时间30min、液固比为6。马海芳等(2014)也研究了蜂胶水提除蜡的工艺,发现选择75℃水浴加热再冷却可以达到有效分离蜂蜡的目的。

间接水提法进行蜂胶的提取,常用酶等对蜂胶进行预处理,以达到提高蜂胶在水中有效溶出的目的。酶解法制备水溶性蜂胶的原理是选用合适的酶对蜂胶进行处理,从而破坏蜂胶的结构,产生局部的坍塌、溶解、疏松,加快有效成分溶出,提高提取效率,缩短提取时间。为提高蜂胶在水中的溶解性,制备水溶性蜂胶,何新益等(2009)研究了酶解法制备水溶性蜂胶的制备工艺。以水溶液中黄酮提取率为指标,研究了果胶酶、蛋白酶和脂酶对蜂胶酶解效果的影响,选定脂酶作为作用酶,通过正交试验优化了酶用量、作用时间、酶反应温度和pH值。结果表明,酶解法制备水溶性蜂胶的最佳提取工艺为酶的用量为11%、作用时间2.25h、酶反应温度40℃和pH值为5.5;并采用1,1-二苯基苦基苯肼(DPPH)自由基清除率测定方法评价酶解法制备的水溶性蜂胶、蜂胶醇提取液和维生素C(阳性对照)的抗氧化能力。结果表明,两种蜂胶提取液对DPPH自由基的清除率随着浓度的增大而增强,水溶性蜂胶的清除率强于维生素C,接近于醇提取物,水溶性蜂胶、蜂胶醇提取物、维生素C的半数抑制浓度分别为0.069、0.061和0.126mg/mL。酶解法由于反应特异性强、条件温和、提取时间短、提取率高、绿色节能等特点,具有较大的应用潜力。

(3)乳化法

蜂胶中的活性成分众多,极性各不相同,黄酮等主要活性成分不溶于水,在油中溶解度也较差,可溶解于乙醇等有机溶剂中。考虑到尽可能多地将蜂胶中的活性成分加以利用,人们研究了蜂胶的乳化工艺,通常采用有机溶剂和乳化剂来制备蜂胶溶液,再将蜂胶溶解于水中。蜂胶乳化可以降低有机溶剂的用量,尽可能多地保留蜂胶中的活性成分,提高蜂胶的水溶效率,丰富蜂胶产品的形式。因此,安全性高、性质稳定的蜂胶乳化工艺值得关注。

蜂胶的乳化工艺,主要是根据乳剂形成的理论,选择适宜的乳化剂,考察其浓度、搅拌速度、乳化温度等参数,以纯化水作为无机相,最终获得蜂胶在水中的稳定乳化物。乳化剂的应用可以考虑多种乳化剂相复合。乳化剂复合使用更有利于降低界面张力,界面张力越低,界面吸附作用增强,分子定向排列更加紧密,界面膜增强,防止液滴的聚集倾向,有利于乳浊液的稳定。复合乳剂往往要比单一乳剂具有更好的表面活性。

何新益等(2008)研究发现,0.1mg/mL明胶和甘油体系对蜂胶黄酮类化合物增溶效果

最佳。最佳的乳化条件为:蜂胶于100倍质量体积的0.1mg/mL明胶和甘油溶液中(1:4,v/v)进行乳化,乳化温度为70℃,所得蜂胶水溶液中黄酮类化合物的含量为0.6977mg/mL。陈崇羔等(1999)采用Tween-20、Tween-60、Tween-80和span-80等4种表面活性剂,按一定浓度梯度配置水溶液,以作为蜂胶的增溶剂。结果表明,20%Tween-20水溶液与80%Tween-60水溶液以及60%Tween-20水溶液与40%Tween-60水溶液的组配液,对蜂胶水溶的增容效果最好;并确定当上述增溶剂用量不小于蜂胶用量的7.2倍和8.4倍时,蜂胶可任意加水稀释均保持澄清。王嘉军等(2008)选择具有天然、广谱抗微生物活性的蜂胶醇溶液为油相,硝酸银水溶液为水相,表面活性剂吐温-80为乳化剂,按照纳米乳剂形成的理论,以适当的比例混合后,室温下温和搅拌制备成水包油型(O/W)蜂胶-银离子复合纳米乳剂。结果发现,制成的复合纳米乳剂含蜂胶浓度为20g/L,总黄酮为5.61g/L,银离子浓度为50μg/L。含体积分数6.7%复合纳米乳剂的溶液与大肠杆菌接触作用10min后,菌体产生融合,胞内容物开始溶解;作用15min时完全融合,胞内容物完全溶解;作用30min时融合菌体出现裂解。用上述浓度的复合纳米乳液与金黄色葡萄球菌接触作用10min后,菌体内含物产生溶解,核质破裂;作用15min时,胞壁出现部分裂解;作用30min时,胞内容物完全溶解,胞壁损坏。表明蜂胶-银离子复合纳米乳剂通过菌体融合的方式杀灭革兰阴性菌,而对革兰阳性菌则主要使其胞内容物溶解,以裂解胞壁的方式致使细菌死亡。

杨锋等(2006)将蜂胶加入95%乙醇水中,充分搅拌使其溶解后,边搅拌边加入50~100℃的热水,温浴,得蜂胶-乙醇-水混合物,混合物中乙醇含量为20%~60%,冷却至室温,静置,过滤;滤液回收乙醇至无醇味,过滤,得滤液;将滤渣和沉渣重复上述醇提水提过程,获得滤液合并,80℃浓缩或用水调至相对密度1.010~1.040;冷却,过滤,即为蜂胶水溶液。该方法确定了制备蜂胶水溶液的最佳工艺条件,采用该工艺条件提取的蜂胶水溶液不含任何有机溶剂,黄酮含量高达8mg/mL,比单纯水提取率(1mg/mL)提高7倍以上,而且生产工艺简单,成本低廉。胡福良等将粉碎后的蜂胶按料液比15:100(W/V)加入纯净水中,控制温度在80℃,加入少量天然绿色表面活性剂进行增溶,加热12h后超声波处理30min,过滤,得到蜂胶水提液。通过动物实验发现,在相同剂量的情况下,蜂胶水提液的抗炎和抗肿瘤效果与醇提液相比差异不显著,甚至优于醇提液(胡福良等,2003;Hu et al.,2005;胡福良等,2005b)。由于水提蜂胶易被机体吸收,副作用小且容易调味,极具开发潜力。研究发现,蜂胶水提物比蜂胶醇提取物具有更好的抗氧化作用,可以更好地抑制一些酶的活性,被认为是天然的抗氧化剂。

(4)乙二醇提取法

乙二醇提取法与乙醇提取法相似,只是溶剂由乙醇换成乙二醇,而且料液比不能太小(蜂胶的浓度不能超过10%),若在真空条件下提取效果会更好。与乙醇相比,乙二醇的缺点是蒸发时需要较高的温度,这对蜂胶提取物中的挥发性物质会造成破坏。乙二醇的价格通常比食用酒精低,且乙二醇易溶解于洗发水中,因此乙二醇很受化妆品生产者的欢迎。虽然乙二醇在外用时没有毒性,但成人内服的剂量不得超过1.5g/d(胡福良,2000)。

(5)乙酸乙酯提取法

将蜂胶用乙酸乙酯溶解形成浸膏状,静置数小时后,再用5%碳酸氢钠溶液提取,得到蜂胶乙酸乙酯提取物和蜂胶酸性物质,再将蜂胶酸性物质用盐酸酸化后,先用乙醚提取,再用乙酸乙酯提取,最后用正丁醇提取可得到正丁醇提取物。

(6)二氯甲烷加热提取法

取一定量的经冷藏、研碎后的蜂胶,加入一定量的二氯甲烷,充分搅拌,进行加热、回流、搁置数小时后,进行过滤,让溶剂挥发后,用乙醇热回流提取、过滤、去蜡,即可得到蜂胶抽提物。

(7)乙醇石油醚双相溶剂萃取法

将粉碎后的天然蜂胶粉置于萃取器中,加入90%乙醇水4倍,60%~90%石油醚1倍(m/v),在40℃恒温下搅拌萃取,如此重复3次,合并萃取液分层,冷却后过滤,减压浓缩,将脂溶性物质合并即得较纯的蜂胶。石油醚、乙醇和蜂胶的容积重量比在1∶4∶1范围内,40℃温度下,提取率较高,成分较全面。但因设备和工艺较单溶剂提取法复杂,大规模生产还需一定的过渡阶段。

(8)氢氧化钠溶剂提取法

将粉碎后的天然蜂胶加入1%~2%的氢氧化钠溶液(10倍),充分搅拌,使之溶解后立即过滤,再加入5%盐酸水溶液酸化蜂胶析出,经水洗后注模成型。这种方法工艺尚比较难以掌握,在用碱液溶解及酸化过程中都会出现时间过长、温度过高等情况,使析出的蜂胶呈炉渣样物质。分析证明,虽然氢氧化钠溶液对蜂胶有效成分有良好的溶解性,但酸化后并不能使所有成分还原,而出现了部分异物化,部分皂化呈酸性不可逆反应(林贤统等,2008)。叶静凌(2008)通过增加料液比、调节pH值,并结合超声等手段,利用氢氧化钠提取蜂胶取得了良好的效果。

(9)油提取法

取10g蜂胶加入200g橄榄油,或杏仁油,或100mL上等亚麻籽油,或100g黄油(其他可食用油参照上述标准)中。将它们放入锅中用文火加热约10min,油温不超过50℃,并不停地搅拌,过滤后密闭置暗处存放,最好能冷藏。

(10)硼高分子电解质法

硼高分子电解质法是将阴离子型高分子化合物-硼高分子电解物的稀释精制水溶液与含有蜂胶物的纤维素充分混溶,高分子电解物的稀释精制水溶液可不断地分解蜂胶,然后再加入乙醇和以海藻为原料的纤维素,从而可提取出含有有效成分的亲水性凝胶体,经提炼后的结晶具有很强的除菌力,而且用这种方法得到的提取物具有很好的活性(张伟,2000)。

1.2 蜂胶非挥发性成分提取新技术

由于蜂胶的特性,人们对蜂胶中非挥发性活性成分的提取在溶剂选择方面进行了以上探索,蜂胶有效成分的提取率得到了很大的提高,但问题依然存在。各类提取溶剂都有其优缺点,如单纯以水作为溶剂,安全性高,但提取效率低;单纯以乙醇作为溶剂,提取率高,但提取物具有刺激性气味及部分酒精过敏的人群不能耐受;用有机溶剂(如聚乙二醇、甘油等)和乳化剂(吐温-80、磷脂)来制备蜂胶溶液,能提高蜂胶成分在水中的溶解度,但所用的物质大部分为非常用物质,国家对这些物质的添加也有严格的规定和限制,同时这些物质的添加降低了蜂胶的生物学活性,增加了产品成本,也增加了蜂胶产生毒副作用的可能性;以氢氧化钠、碳酸钠制备水溶性蜂胶,往往存在蜂胶浸出率较低、造成资源浪费,提取过程中往往需要高温、容易破坏蜂胶中某些成分的活性和提取液的pH值呈碱性、影响蜂胶的稳定性和人体的吸收等问题。因此,在蜂胶的提取加工过程中,除了在按照原料用途对提取溶剂进行严格选择的基础上,在传统浸提法外,一些新的提取加工手段尤其是高新技术,被不断应用到蜂

胶的加工中。以下是蜂胶非挥发性成分提取中采用的一些技术手段。

(1)超声波辅助萃取技术

超声波是频率高于20kHz,并且不引起听觉的弹性波。利用超声振动能量可改变物质组织结构、状态、功能或者加速这些改变的过程。现普遍认为超声波的空化效应、热效应和机械作用是超声技术协助提取的三大依据。超声波在提取中可以有效破碎细胞壁或者包埋结构的外层,促进物质扩散,释放出内容物,提高提取率。超声波提取技术能避免高温高压对有效成分的破坏,但它对容器壁的厚薄及容器放置位置要求较高,否则会影响原材料浸出效果。目前用于实际生产的是一种连续逆流超声提取设备。其工艺采用浸渍法或重浸渍法动态连续试提取,蜂胶运动方向与溶液的流动方向相反,药材和溶剂在连续不断地进入提取器,同时药渣和提取液连续不断地排出提取器。此法提取率高,要求提取液浓度高,适用于大规模生产线的提取段工序。除具有连续逆流提取设备的优点外,利用超声波空化作用,可加速有效成分浸出,大幅度缩短提取时间;由于连续逆流超声提取温度低,可最大限度地保护天然物的有效成分。近年来,超声技术广泛应用于提取植物中的生物碱、苷类等生物活性物质,也是提取蜂胶黄酮类化合物的一种重要方法。

超声波辅助法提取蜂胶黄酮比单纯的溶剂回流提取条件优越,大大缩短了提取时间;液料比也有所降低,节省了溶剂用量;而且超声波提取不需要加热,最大限度地保护了蜂胶中的活性成分,提取物中总黄酮含量也有较大提高。这可能是由于蜂胶颗粒在超声波的空化和机械作用下被破碎成碎片,增大了蜂胶颗粒与溶剂的接触表面积,同时也增强了溶剂向蜂胶细胞的渗透作用,故可明显加速溶质中有效成分的提取过程,而超声波的热效应可使蜂胶颗粒内部温度很快升高,加速有效成分的溶解,因此超声波辅助提取比单纯的溶剂提取效果要好。

王春玲等(2003)采用超声波技术对蜂胶黄酮的提取工艺进行了研究,经响应面优化浸泡时间、乙醇浓度、超声处理时间和液料比4个影响因素,得到最佳提取条件为:乙醇浓度89%,浸泡时间11h,超声处理时间13min,液料比12∶1。李维莉等(2003)比较了水浴和超声波辅助提取2种方法提取蜂胶中黄酮类物质的效果,结果表明,超声法以20倍原料重的75%乙醇水溶液超声20min浸提为最佳,且优于水浴法,黄酮提取率高达45.6%。付英娟(2007)以蜂胶为提取原料,采用超声波提取技术对蜂胶中黄酮类化合物进行提取,采用了响应曲面法建立了超声波提取蜂胶黄酮的二次多项式数学模型,考察了乙醇浓度、提取时间、提取功率和料液比对超声波蜂胶中总黄酮含量的影响,优化出提取蜂胶中总黄酮的工艺参数为乙醇浓度79.51%,提取时间19.31min,提取功率538.28W,液料比39.48∶1,且预测可获得最大的黄酮提取量为5.859%。刘元法等(2004)采用超声波技术在提取功率750W,料液比1∶15,提取时间15min的条件下,可以获得高于醇提蜂胶的提取率和总黄酮含量,是一种高效的蜂胶提取技术。王春玲和高玲美(2012)研究发现,采用超声波提取法,能够提高蜂胶中黄酮类化合物的提取率,缩短提取时间。超声提取率只需浸泡1h即可,而常规的浸泡提取方法则需要浸泡40h以上,最大得率只有87.5%,比超声提取率低了将近2个百分点。杨林莎和王东(2005)以蜂胶总黄酮含量为指标,通过乙醇冷浸法、乙醇温浸提取法、乙醇超声提取法对蜂胶提取方法进行比较,结果表明,利用乙醇超声提取法所得提取率为16.23%,高于其他提取方法。潘秋月等(2011)通过单因素实验研究了蜂胶黄酮的超声波提取工艺,并对蜂胶中黄酮类化合物的清除活性氧自由基的能力进行测定。结果表明,蜂胶黄酮的最佳提取工艺为75%乙醇作为溶剂,料液比为1:10,提取时间为20min,超声波提取2次,温度为60℃,此时黄酮提取含量最高,达到5.046mg/g。曾林晖等(2016)在选择超声波频率的基础上,分别研究了提取时间、超声波功率、超声波占空比和液面高度对蜂胶总黄酮提取率的影响。结果表明,20kHz聚能超声波对蜂胶黄酮的提取效果最好。其他因素对蜂胶总黄酮得率的影响大小顺序为超声波占空比>液面高度>提取时间>超声波功率,若采用最佳的超声波提取工艺参数,蜂胶黄酮得率达到(39.06±0.48)mg/g。此方法提取出的蜂胶黄酮有较强的DPPH自由基清除能力和FRAP抗氧化能力。

(2)微波辅助萃取技术

微波技术是利用微波能量对物料加热达到要求的一种新技术,吸收微波的各个分子能按微波的频率高速往返运动,相互碰撞,彼此摩擦而产生热量,已应用于加热、干燥、杀菌灭酶、医疗等领域。利用微波技术可使溶剂分子发生高速震荡,增大了溶剂进入溶质内部的能力,使黄酮类化合物从基质上脱离,从而加速了向溶剂扩散的速度,缩短了提取时间,是一种提取蜂胶黄酮省时、高效的方法。

吴玉敏等(2007a,2007b)利用微波辅助萃取法提取蜂胶中黄酮类化合物,考察了溶剂浓度、微波功率、提取时间、料液比对蜂胶黄酮提取率的影响,优化的工艺条件为:乙醇浓度80%,微波提取功率80W,提取时间1.5min,料液比1:4。与溶剂浸提法相比,微波辅助法具有提取效率高、时间短、节省溶剂用量等优点。唐坤等(2008)以蜂胶总黄酮含量为检测指标,采用微波-超声波技术,通过正交试验确定最佳提取工艺条件。结果显示,蜂胶的最佳提取工艺为:加入所提蜂胶15倍量的80%乙醇,在微波功率600W下处理5min,置于70℃温水浴中超声提取2次,每次提取20min。微波-超声法的提取效率比单纯的超声法显著,微波磁控管能产生超高频快速震荡,使物质分子能迅速吸收电磁波的能量而受热,因此蜂胶能在短时间内受热融化成黏稠细颗粒,这样非常有利于乙醇的浸入提取;而且超声波的频率一般高于20kHz,功率在50~500W,通过强烈的空化效应、机械振动和热效应来增大物质分子运动频率和速度,加速蜂胶的溶出。余晶晶和童群义(2013)采用超声微波协同法萃取蜂胶中的黄酮类化合物,在单因素和正交实验的基础上得出了最佳提取工艺:乙醇浓度65%,料液比1:15,时间150s,微波功率175W。在该提取条件下,蜂胶提取液中总黄酮含量为30.64%,蜂胶提取率为70.38%。与其他提取方法相比,超声微波协同萃取在提取效果上具有明显的优势。郭华等(2001)采用微波技术,对蜂胶黄酮的提取工艺进行了改进性研究,优选出微波提取的最佳工艺:乙醇浓度80%,微波处理时间15min,恒温时间2d。提取时间较常规方法缩短一半,提取率有很大提高。

(3)高压加工技术

高压加工技术是指在较低温度(通常低于100℃)的条件下,对生物材料施加100~800MPa的流体冷等静压力,从而使生物材料发生物理或化学的变化而得到新产品的一种技术。高压提取在室温条件下进行,不会使有效成分因受热而发生变性,而且提取时间非常短,高压设备的操作简单,节约能源并环保。

席军(2005)将蜂胶的高压提取方法与室温浸泡提取方法进行比较,发现高压提取的组分和室温浸泡提取的组分完全一样;而且在高达600MPa压力作用下,黄酮类化合物组分的提取率也没有减少,说明压力对蜂胶黄酮类化合物组分没有破坏作用,其最优提取工艺参数为:提取压力450~500MPa、乙醇浓度65%~75%、料液比1:28~1:35。Zhang等(2005)

比较了高压提取(500MPa,1min)、乙醇常温浸提(乙醇浓度75%,料液比1∶35,时间7d)和热回流提取(乙醇浓度95%,料液比1∶4,85℃,4h)3种方法,结果表明高压提取法的提取率(5.10%±0.14%)显著高于其他2种方法(4.70%±0.21%和4.56%±0.11%)。王娜等(2009)以蜂胶黄酮得率为指标,采用响应面分析法优化超高压提取工艺条件,考察乙醇浓度、料液比、保压时间和压力对蜂胶黄酮得率的影响。结果表明,蜂胶黄酮最佳提取工艺条件为:乙醇浓度85.9%,液料比46.75∶1,最佳压力357.5MPa,保压时间2min,此工艺条件下提取蜂胶黄酮得率为12.38%。

(4)超临界CO_2流体萃取技术

超临界流体技术在近几十年来发展迅速,它是一种非热提取技术,解决了传统提取方法存在的一系列问题,特别是在分离纯化方面有着特殊的优势,在食品、医药、化工、材料科学、环境科学、分析技术等领域已经得到广泛的应用。二氧化碳因其临界温度和临界压力低(31.06℃,7.39MPa),对中、低分子量和非极性的天然产物有较强的亲和力,而且具有无色、无味、无毒、不易燃、不易爆、低膨胀性、低黏度、低表面张力、易于分离、价廉、易制得高纯气体等特点,是应用最为广泛的超临界流体。超临界CO_2是亲脂性的,它只能有选择地提取脂类物质。在医药和食品工业中,超临界流体技术主要用于中草药有效成分包括挥发性油、生物碱、苷类、香豆素类、萜类的提取。

蜂胶中的黄酮类化合物等有效成分在超临界CO_2流体中的溶解度极低,需要加入乙醇作为夹带剂;而且该方法设备投资较高,蜂胶的提取率也较低。该法与乙醇浸泡法和水浸泡法相比,提取物的极性是较低的,蜂胶的香气成分以及二萜等成分几乎100%可以提取出来,而黄酮类化合物却仅有20%~30%被提取出来,而极性较高的糖类、有机酸几乎没有被提取出来(卢媛媛等,2013)。研究表明,超临界提取的蜂胶液具有非常强的抗肿瘤和抗过敏作用,动物实验也表明超临界提取的蜂胶制品比乙醇提取物的以上作用要强5~10倍(Takara et al.,2007)。

谷玉洪等(2006)采用超临界CO_2并用乙醇作夹带剂萃取蜂胶中的黄酮类有效成分,萃取物中仅含有少量树脂、蜂蜡等亲脂性成分,萃取物处理方法简单方便,萃取的最佳条件为:粉碎粒度20目,萃取压力35MPa,萃取温度40℃,蜂胶与夹带剂用量比例为1∶2。韩玉谦等(2003b)采用四因素三水平的正交设计研究了超临界CO_2萃取蜂胶有效成分的萃取工艺,并用气质联用法分析了蜂胶提取物的组成,结果表明超临界CO_2萃取蜂胶有效成分最佳工艺条件为:萃取温度55℃,萃取压力30MPa,CO_2循环量0.6kg/g,携带剂用量5%。曾志将等(2006a,2006b)研究得出的超临界CO_2萃取蜂胶的最佳条件为:萃取温度46℃、萃取压力25MPa、分离温度为50℃、分离压力为7MPa、萃取时间为3h;并发现随着乙醇含量的提高,蜂胶CO_2超临界萃取率随之增大,但萃取蜂胶中黄酮含量在5%乙醇时最高,为3.09%;他们还发现,超临界萃取是一种去除蜂胶原料中铅的有效方法。巴西蜂胶中含有3,5-异戊二烯-4-羟基肉桂酸(阿替匹林C),是重要的生理活性成分之一。Lee等(2007)采用超临界CO_2提取纯化了巴西蜂胶中的阿替匹林C,提取温度和夹带剂乙酸乙酯添加量是影响阿替匹林C提取率和产物纯度的主要因素,添加6%的乙酸乙酯将提取率由不添加时的3.7%增加到13.9%,超临界CO_2萃取的阿替匹林C纯度均在40%,远高于乙酸乙酯索氏提取法的16.9%。Chen等(2009a)在萃取压力20.7MPa,萃取温度323K时,以质量分数为6%的乙酸乙酯作为夹带剂,对巴西蜂胶进行提取,亦发现提取所得阿替匹林C对HL-60、 colo205等肿瘤细胞有抑制作用。De Zordi等(2014)采用中心组合设计法,利用超临界CO_2流体技术提取意大利蜂胶中的多酚类物质。结果发现,超临界提取产物与以乙醇为溶剂的超声提取产物在化学成分上存在差异,为得到富含黄酮类化合物的蜂胶提取物,超临界提取可作为原料预处理步骤,以便进一步利用乙醇作为溶剂加以提取。Catchpole等(2004)用超临界CO_2萃取技术对蜂胶酊进行了提取,在蜂胶酊质量分数为10%,萃取温度333K(约60℃),萃取压力275~300bar(27.5~30MPa)条件下可以得到2部分萃取物,一种富含黄酮类化合物,其中黄酮类化合物的含量为20%~35%,另一种主要为蜂胶挥发油。Wang等(2004)报道,先将蜂胶与乙醇以1:10的比例混合放置24h,然后进行超临界CO_2萃取,萃取参数为温度60℃,压力分别为20、15、10、5MPa,得到4部分产物(R、F1、F2、F3),进行总黄酮测定及抗氧化分析,结果表明,R、F1、F2和F3的总黄酮含量分别为137、90、83和76mg/g,并且R的抗氧化作用最强。Biscaia和Ferreira(2009)的研究结果表明,利用5%的乙醇作为夹带剂对蜂胶进行超临界提取,得率可达24.8%。Machado等(2016)比较了利用超临界CO_2萃取法和乙醇提取巴西不同类型蜂胶(红蜂胶、绿蜂胶、棕蜂胶),结果表明,传统乙醇提取得到的蜂胶提取物的活性更强,而利用超临界CO_2萃取法获得的蜂胶提取物中阿替匹林C和对香豆素含量最高。

以上研究表明,萃取温度、压力、助溶剂的含量对黄酮类化合物的提取率有重要影响,其中萃取压力的影响最大。萃取压力稍有变化,超临界CO_2流体的性质随之改变,导致提取物中黄酮化合物的含量也发生改变。超临界CO_2流体是非极性,而大多数黄酮类化合物为极性,因此较难溶于纯的超临界CO_2流体,需借助一些极性助溶剂。超临界提取蜂胶中最常用的助溶剂是乙醇溶液,其含量通常为5%~10%。

此外,不同来源的蜂胶原料,由于含纯胶量和所含黄酮量不同,也会影响蜂胶超临界CO_2流体提取率和黄酮化合物等蜂胶非挥发性活性成分的种类及提取率。赵淑云等(2006)比较了不同产地蜂胶超临界CO_2萃取物中的8种黄酮含量,发现各产地蜂胶的超临界CO_2萃取物中一个显著的共性是松属素的含量较其他的组分高出较多,而松属素具有抗菌、消炎、局部镇痛的效果,说明蜂胶的超临界CO_2萃取物有较强的抗菌、消炎活性。云南样品中槲皮素及杨梅酮含量较高,而总黄酮及其他6种黄酮含量均明显低于黑龙江、河北和四川的样品。

刘伟等(2007)采用分光光度法与HPLC法测定了蜂胶超临界CO_2萃取物中总黄酮的含量,并与蜂胶乙醇提取物作对比。结果表明,HPLC法测定的超临界蜂胶CO_2萃取物中8种黄酮和总黄酮的含量均低于蜂胶乙醇提取物,分光光度法与HPLC法测定结果基本一致。吉挺等(2008)对醇提蜂胶、超临界蜂胶及醇提超临界蜂胶残渣进行了抗氧化作用比较,结果表明,醇提蜂胶的抗氧化作用最强,醇提超临界蜂胶残渣次之,而超临界蜂胶的抗氧化作用最弱。由此表明,超临界CO_2萃取技术虽然可以从蜂胶中提取出黄酮类化合物,但是与醇提蜂胶相比,超临界萃取物中黄酮含量较低,而大多黄酮类化合物都以残渣的形式流失,而黄酮类化合物是蜂胶起抗氧化作用的主要成分之一,导致超临界蜂胶在抗氧化方面的效果不明显。因此,超临界技术在蜂胶非挥发性有效成分提取上的应用及超临界萃取产物的生物学活性有待进一步研究证明。

(5)微胶囊技术

微胶囊技术是一项具有很高实用价值并被广泛应用的新技术,它采用壁材(变性淀粉、麦芽糊精、环糊精、乳糖、纤维素、大豆蛋白、乳清蛋白、酪蛋白酸钠、石蜡、蜂蜡、单甘酯、卵磷脂等)与芯材(功能因子)混合,通过喷雾、挤压等工艺在外层形成一层连续而薄薄的包裹层,制备成微胶囊成品,其直径一般在5~200μm。功能因子的微胶囊化避免了外界温度、湿度、氧气、紫外线等因素的影响,显著提高了功能因子的稳定性、有效保护了活性成分的活性,而在壁材被破坏时,芯材又能被释放出来,充分发挥其生理活性。

游海等(2002)利用喷雾干燥法,采用大米液体麦芽糊精为微胶囊壁材对蜂胶活性成分提取物进行了包埋,得到了包埋率高(95%以上)的固体的蜂胶微囊粉末,大大提高了产品的保质期。张英宣(2012)研究发现,以阿拉伯树胶和糊精以1∶1比例混合作为壁材,固形物含量为20%,芯材与壁材比例为1∶4,进料量20mL/min,进风压力为0.2MPa,微胶囊化蜂胶中总黄酮的效率最高。Bruschi等(2003)采用明胶包埋蜂胶提取物,考察喷雾干燥条件和配方对产品品质的影响,在有无甘露醇时微胶囊的粒径分别为2.50和2.70μm,包埋率分别为39%和41%,包埋后产品可掩盖蜂胶特殊的气味,并减少了乙醇的用量,并对金黄色葡萄球菌(*Staphylococcus aureus*)保持了很好的抑制作用。俞益芹等(2010)以类黄酮的微胶囊化效率为考察指标,得到蜂胶微胶囊化的最佳工艺为以阿拉伯胶与β-环糊精作为壁材,其比例为1∶1,芯材与壁材的比例为1∶3,固形物浓度为30%,单甘酯为0.2%;最佳喷雾干燥工艺条件为进料流量40mL/min,进风温度180℃,出风温度60℃,在此工艺条件下微胶囊化蜂胶提取物的效率可达到93.51%。

吉挺等利用冷冻干燥技术对醇提超临界萃取蜂胶剩余物进行微胶囊粉末的制备,利用麦芽糊精作为壁材进行包埋,包埋后制备出的微胶囊产品粉质细腻、质地疏松,包埋率可达到69.87%(吉挺等,2010;李文艳等,2010)。他们还利用大豆分离蛋白、麦芽糊精为壁材,超临界蜂胶与超临界紫苏籽油为芯材进行蜂胶和紫苏籽油混合微胶囊的制备,得到最佳工艺为大豆分离蛋白∶麦芽糊精=1.5∶1,壁材∶芯材=2.5∶1,固形物浓度为15%,稳定剂添加量5g/L;混合微胶囊粉末制备的工艺参数为乳化温度60℃,均质时间6min,均质转速4000r/min;用冷冻干燥法制备的微胶囊包埋率达到67.63%(李文艳和吉挺,2010a,2010b)。

胡福良等(2005a)将2%~10%的蜂胶乙醇溶液加入到0.5%~5%的环糊精及其衍生物溶液中,加入聚乙二醇400溶液,超声振动15~45min后用微孔滤膜进行抽滤,滤液用0.1μm的微孔滤膜过滤,滤渣真空冷冻干燥后得到一种亚微米蜂胶微胶囊。进一步改良工艺,通过制备蜂胶提取液、制备水溶性玉米蛋白、用果胶与明胶制备胶质溶液,以改性玉米蛋白、明胶和果胶为壁材,以醇溶性的蜂胶为芯材,再通过聚合反应和冷冻干燥制备获得蜂胶微胶囊。将毛蜂胶粉碎过筛,加入V/V为95%的乙醇水,搅拌浸提24h,过滤,旋转蒸发除去液体,制成蜂胶乙醇提取物;将玉米蛋白用水溶解,制备W/V为2%~8%的玉米蛋白水溶液,用0.1mol/L NaOH调pH为10.0,然后用盐酸将溶液调节到pH为8.0;将果胶和明胶按W/W为1∶1的比例制备成100mL浓度为2%~6%的胶质溶液W/V,用1mol/L盐酸调pH为4.0。将蜂胶乙醇提取液加入到制备好的玉米蛋白水溶液中,充分混匀,然后慢慢加入胶质溶液,此过程在磁力搅拌器上进行,保持温度为35~45℃;聚合反应结束后,将混合物放在冰箱内4℃下冷却过夜,冷却的混合物置于-18℃冷冻,然后冷冻干燥,得粉末状蜂胶微胶囊(胡福良和张翠平,2011)。蜂胶微胶囊可广泛应用于食品、饮料、保健食品及营养强化食品等领域。

(6)超细微粒及纳米技术

化学工程技术领域的发展进步使得粒径在纳米范围的微胶囊被制备出来。纳米粒子的尺寸效应、表面效应和宏观量子隧道效应可提高其溶解性和在体内的生理作用。李雅晶等将蜂胶乙醇溶液用β-环糊精包埋,得到的纳米蜂胶的载药量为(29.65±0.01)%,包埋率为(98.82±0.04)%,平均粒径为220.1nm。动物实验结果表明,纳米蜂胶具有明显降低血清胆固醇的作用(李雅晶等,2007;李雅晶和胡福良,2007)。

超临界流体制粒是一门新兴的制备微细颗粒的技术,其制备的超细微粒特别是纳米粒子与传统工艺(粉碎、研磨、干喷等)相比,具有工艺流程简单、设备少、周期短、有机溶剂残留少、颗粒细小和粒径分布窄等特点(孟庆伟和回闯,2007)。超临界流体抗溶剂(supercritical antisolvent,SAS)技术在蜂胶纳米微粒的制备中已有应用。Wu 等(2009)利用超临界流体抗溶剂技术制备蜂胶微粒,并以阿替匹林 C 作为其活性成分衡量指标,研究发现,阿替匹林 C 的纯度随 CO_2 流速的下降而增高,当蜂胶浓度超过 27mg/mL 时,提取所得蜂胶微粒主体粒径增大。Chen 等(2009b)利用超临界流体抗溶剂技术,在压力 20MPa,温度 328 K 时,以乙酸乙酯作为夹带剂制备蜂胶亚微粒体,制备所得微粒体 DHCA 的含量达 306mg/g;研究结果表明,CO_2 的流速及体积对蜂胶亚微粒体的粒径均有影响。

2 蜂胶挥发性有效成分的提取工艺

蜂胶所含挥发性成分主要为在蜂胶中约占10%的芳香挥发油。传统的蜂胶提取工艺中,由于溶剂及提取方法的限制,蜂胶中的挥发性成分往往得不到充分利用。乙醇浸提所废弃的蜂胶渣依然存在浓郁的蜂胶特有香气(吉挺等,2008)。研究发现,蜂胶原胶、蜂胶乙醇提取物及蜂胶乙醇提取物残渣中挥发性成分含量相似,组成成分亦相近,可见,蜂胶挥发性成分在蜂胶传统提取工艺中被忽略并大量废弃了(黄帅等,2013;刘嘉等,2013)。蜂胶中挥发性成分如萜烯类化合物等形成了蜂胶特殊香气,被发现具有多种生物学活性,是蜂胶活性发挥不可或缺的一部分。研究蜂胶挥发性成分有助于进一步分析蜂胶功效成分的药用价值,并为蜂胶的标准化及开发利用提供依据。因此,在蜂胶加工提取中必须重视对蜂胶挥发性成分的提取与利用。

2.1 蜂胶中挥发性成分的生物活性

对植物组织的挥发性成分研究发现,很多挥发油具有驱虫、镇痛、抗菌、消炎、增强免疫等作用。挥发油的应用也很广泛,不仅应用在香料工业中,还在食品工业及化学工业上有重要的应用价值,近年来随着"回归大自然"热潮的掀起,利用精油的芳香疗法成为热点。对蜂胶中挥发性成分的生物活性研究主要集中在抑菌活性、抗氧化活性及其应用上。

Melliou 等(2007b)使用改良的水蒸气蒸馏法研究了希腊5个地区的蜂胶挥发性成分,体外抑菌实验显示,蜂胶挥发性物质对革兰阳性菌(金黄色葡萄球菌、表皮葡萄球菌)和革兰阴性菌(绿脓假单胞菌、大肠杆菌、肺炎克雷伯氏菌、阴沟肠杆菌)共6种细菌及白色念珠菌、热带假丝酵母、光滑念珠菌等3种真菌有抑制作用。Bankova 等(1999)研究了巴西三地蜂胶挥发油的抗菌活性,结果表明,蜂胶挥发油对葡萄球菌具有抑制作用,对大肠杆菌生长无影响,其中一种挥发油对大肠杆菌有活性可能是含有独特成分的缘故。Kujumgiev 等(1999)还研究发现,蜂胶提取物(包括挥发性成分)有抗滤过性病原体活性作用。

Atungulu 等(2007)利用同时蒸馏-萃取法得到蜂胶挥发性物质,研究其在大米储藏过程中对大米脂质过氧化及水解作用的影响。结果发现,用蜂胶挥发油处理的米糠油的过氧化物值要低于用迷迭香油溶液及迷迭香水溶液处理的样品;蜂胶挥发油能够抑制α-亚麻酸和亚油酸的聚集,并能对糙米表面的需氧菌、霉菌及酵母的生长等起到抑制作用;蜂胶挥发油还能抑制大米储藏过程中二氧化碳释出,从而延长大米的储藏期。他们还发现,蜂胶用不同溶剂提取后的挥发性成分均有清除自由基的作用,但抗氧化效果强弱不一,其中,经蜂胶无水乙醇提取物挥发性成分处理的大米抗氧化损伤能力最弱,而经正己烷提取的蜂胶挥发性成分处理的大米抗氧化损伤能力最强(Gregory et al.,2008)。赵强等(2008)采用超临界CO_2萃取和冷冻分离所得蜂胶挥发油能有效抑制羟自由基和超氧阴离子自由基诱导的化学发光,其IC_{50}分别约为1.3和0.75mg/mL。Li等(2012)研究发现,蜂胶挥发性成分具有抗焦虑作用,对束缚应激小鼠HPA轴基础水平有影响,对动物的焦虑行为具有显著的改善效果。

蜂胶中挥发性成分在抑菌、抗氧化及食物防腐保藏方面已表现出令人兴奋的效果。从蜂胶中有效提取挥发性成分的技术手段、蜂胶挥发性成分中具有生物活性物质的鉴定及分离以及蜂胶挥发性成分在蜂胶质量控制、蜂胶生物学活性发挥中的进一步利用,应引起关注。

2.2 蜂胶挥发性成分提取方法

由于挥发油具有常温下易挥发,能随水蒸气蒸馏且不溶于水的特点,目前常用于提取获得挥发油的方法主要有:水蒸气蒸馏法、有机溶剂浸提法、超临界流体萃取法、同时蒸馏-萃取法、固相微萃取法、酶辅助提取法、超声提取法、微波萃取法及半仿生提取法等。蜂胶不同于常规用于挥发油提取的植物组织,具有低温硬脆、高温黏稠的质地特异性,目前对蜂胶挥发性成分的提取主要有以下几种方法。

(1)水蒸气蒸馏提取

作为挥发油提取的经典方法,水蒸气蒸馏法具有所需设备简单、操作容易且成本低廉等特点,在挥发性成分的提取中被广泛采用。根据蜂胶产地不同及水蒸气蒸馏时间等工艺参数不同,蜂胶挥发性成分的得率也不尽相同。曾晞等(2004)对贵州兴仁、德江两地蜂胶进行水蒸气蒸馏提取,所得蜂胶挥发油得率分别为1.15%和1.32%。郭伽和周立东(2000)将北京蜂胶进行水蒸馏法提取后,得到质量比为1.1%的挥发性成分。付宇新等(2009)将内蒙古蜂胶用水蒸气蒸馏法提取5h后,所得蜂胶挥发性物质得率为0.26%。王小平等(2007)将蜂胶于水中浸泡过夜再进行水蒸气蒸馏,也获得了蜂胶挥发性物质。徐响等(2010)采用水蒸气蒸馏法对蜂胶挥发油进行提取,从125g蜂胶粉末中提取得到1.427g蜂胶挥发油。Kusumoto等(2001)将巴西蜂胶用550g水蒸气蒸馏5h,馏出物经乙醚提取,无水硫酸钠干燥得到了蜂胶挥发油,得率为0.34%。Melliou等(2007a)采用水蒸气蒸馏法对希腊5个地区的蜂胶挥发性成分进行了提取,提取时间为3h,5个地区蜂胶的挥发性成分得率分别为0.05%、0.10%、0.04%、0.08%和0.03%。阿布力克木·吾甫尔等(2011)对新疆蜂胶中的挥发油利用水蒸气蒸馏法进行了提取,并发现其对直肠癌HCT-116细胞具有明显的增殖抑制作用。

(2)有机溶剂提取

溶剂提取技术可将挥发性物质转变成溶液状态,同时得到富集。杨文超等(2012)用石油醚、正己烷、乙醚、四氯化碳4种溶剂提取蜂胶挥发油,并对其进行了成分分析。王小平等

(2009a)采用溶剂提取法对河南、山东、江西、内蒙古、甘肃等 5 产地蜂胶的挥发性成分进行了提取。提取装置为索式提取器,提取溶剂为乙醚,经过脱水、除溶剂处理后得到河南、山东、江西、内蒙古等 4 产地蜂胶挥发性物质得率分别为 3.70%、3.00%、3.73% 和 3.74%。他们还尝试了用超声辅助有机溶剂提取的方法提取蜂胶中的挥发性物质,将蜂胶浸泡于乙醚中进行超声处理,滤液除溶剂后得到蜂胶挥发性成分;同时采用有机溶剂冷浸法,将蜂胶加入乙醚中,室温浸泡 1d,滤液除溶剂后即得到蜂胶挥发性成分(王小平等,2009b)。卫永第等(1996)将蜂胶用乙醚冷浸 24h 后于 50℃ 水浴中加热回流 3h,蜂胶挥发油得率为 2.1%。

(3)超临界 CO_2 萃取

超临界 CO_2 萃取可直接适用于分子量较小、脂溶性、热敏性、易挥发性成分的萃取,用这种技术提取的挥发油具有防止氧化、热解及提高品质的优点。蜂胶中的弱极性及非极性成分,如脂类、萜类、有机酸类、醇类等物质可利用超临界流体 CO_2 进行萃取。但超临界流体 CO_2 萃取对蜂胶中黄酮类化合物的提取率较低,萃取后得到的蜂胶中的黄酮主要为高良姜素、柯因等一些弱极性物质,一般蜂胶中黄酮的萃取率为萃取物中的比例为 6%~7%。De Zordi 等(2014)认为超临界 CO_2 流体技术可以作为蜂胶中亲脂性成分的有效提取方法或作为醇提法提取蜂胶中黄酮类物质的前处理方法。超临界 CO_2 流体萃取技术对蜂胶中挥发性成分的提取有着独特的优势。目前,已经有一些学者利用超临界 CO_2 流体萃取技术对蜂胶挥发性成分的提取进行了一些探索。

韩玉谦等(2003a)用正交实验对超临界 CO_2 萃取蜂胶挥发性成分的条件进行了优化,得到最佳工艺条件为:萃取温度 55℃、萃取压力 30MPa、CO_2 循环量 0.6kg/g、携带剂量 5%。徐响等(2008b)应用超临界 CO_2 萃取工艺提取蜂胶挥发性物质,在萃取压力 33.4MPa,温度 45.4℃,二氧化碳流速 16.8L/h,萃取蜂胶 120min 时萃取物的得率达 19.62%。他们还进一步利用超声强化超临界 CO_2 萃取方式对蜂胶中的挥发性成分进行了提取,参数为超声波频率 20kHz,超声功率 100W,萃取温度 55℃,萃取压力 30MPa,并与同条件下常规超临界 CO_2 萃取所得产物的成分进行了比较(徐响等,2009)。刘映等(2000)在萃取压力为 30MPa,萃取温度 45℃ 下对蜂胶挥发性成分进行萃取,采用三级分离,分离柱压力 11.5MPa、温度 60℃,解析Ⅰ压力 10.5MPa、温度 60℃,解析Ⅱ压力 7.5MPa、温度 46℃,萃取时间 5h,蜂胶挥发性成分得率为 8%。赵强等(2007)在 CO_2 流量为 50kg/h,萃取压力为 30MPa,萃取温度为 55℃,萃取时间为 4h 的工艺条件下,蜂胶挥发油得率为 1.5%;用 GC-MS 分析其成分,共分析鉴定出 51 种成分,其中 17-三十五碳烯、1-二十二烯、二十七碳烷、α-桉叶油醇和羊毛甾醇醋酸酯等物质相对含量较高,萜烯类及其衍生物种类较多且总相对含量较高。李雅洁等(2006)在压力 20Mpa、萃取池温度 50℃、解析温度 40℃、静态萃取 10min、动态萃取 75min、CO_2 流量 1.0L/min 的工艺条件下得到了黄色透明油状的蒙山蜂胶挥发油,并分析鉴定出 44 种成分,含量较高的组分为亚油酸、正二十六碳烷、二十九碳烷、棕榈酸、二十四碳烷、亚油酸甲酯等。Catchpole 等(2004)用超临界 CO_2 萃取技术对蜂胶酊进行了提取,得到两部分萃取物,一种富含黄酮类化合物,另一种主要为蜂胶挥发油。徐响等(2010)将 30MPa、50℃、15L/h 条件下得到的超临界 CO_2 萃取物进行 GC-MS 分析,共鉴定出 44 种物质,萃取物中含量较高的为二十六烷、二十九烷等烷烃类物质,以及苯甲酸、苯乙醇、肉桂醇、肉桂醛、愈创木醇类物质等。他们进一步比较了传统水蒸气蒸馏法与超临界 CO_2 萃取法得到的挥发油在组成和抑菌活性上的差异,采用超临界 CO_2 萃取得到 1.968g 蜂胶挥发油,而采用水蒸气

蒸馏法得到 1.427g 挥发油；GC-MS 检测结果表明，水蒸气蒸馏法取得 34 种物质，其中二甲基丁酸、苯甲醛、柠檬烯、桉叶油醇、苯甲醇、苯乙醇、乙酸苯甲酯、苯甲酸、2-甲氧基-4-乙烯苯酚、愈创醇和甘菊环族化合物等的相对含量较高，而超临界 CO_2 萃取得到 37 种物质，相对含量较高的主要成分为苯甲醇、甘菊环族及萘甲醇类化合物、愈创醇、癸醛、部分烃类、异香橙烯环氧化物和肉桂酸肉桂酯等；抑菌实验表明，两种方法提取的挥发油具有抑菌作用，而超临界法萃取的挥发油抑菌效果强于水蒸气蒸馏法。

从以上结果可以看出，超临界 CO_2 萃取与水蒸气蒸馏法制备的蜂胶挥发油的化学组成存在差异，超临界法在提取率及挥发油成分上略显优势。并且水蒸气蒸馏法提取过程温度较高，容易对挥发油成分造成破坏，而超临界萃取法萃取温度较低，能萃取出较多的非极性化合物。因此，这也可能是超临界萃取法比水蒸气蒸馏法抑菌效果较好的原因之一。

不同实验条件，包括超临界萃取的设备、工艺条件、蜂胶原料等因素，使得不同研究结果所获得的工艺参数差异巨大。萃取压力和温度对挥发油的提取具有显著影响，而不同蜂胶原料中挥发性成分种类和含量存在差异，导致相应的提取条件发生变化。同时对这些实验获得的成分进行分析，可以发现超临界 CO_2 流体萃取技术对蜂胶中烷烃类、酯类的萃取比较明显，可能原因是超临界 CO_2 流体本身适用于非极性成分萃取，并且也可能与其杨树型植物来源和混入的蜂蜡有关。值得注意的是，虽然不同工艺条件下获得的挥发油成分种类与相对含量差异巨大，但大多挥发油中含有较多的萜烯类化合物，如桉醇、愈创木醇等，可见萜烯类化合物是挥发油中最主要的活性物质。

(4) 固相微萃取技术

固相微萃取（solid phasemicroextraction，SPME）是在固相萃取基础上发展起来的方法，其操作时间短、样品用量少、无须萃取溶剂、重现性好、方便与气相色谱、液相色谱联用，适于分析蜂胶中挥发性物质。贺丽苹等（2008）采用固相微萃取法对山东、北京、河北蜂胶的挥发性成分进行了富集及成分鉴定。徐响等（2008a）采用 SPME 联合 GC-MS 分析技术分析了蜂胶中挥发性成分，结果表明，中国不同地区蜂胶中挥发性成分存在差异，但均含有萜烯类、醇、醛、酮、酯等，其中甲酸、乙酸、苯甲酸、苯甲醇、乙酸苯乙酯是蜂胶主要的挥发性成分。

(5) 其他提取方法

目前报道的对蜂胶中挥发性成分的提取方法还包括同时蒸馏萃取法及微波辅助萃取法等。付宇新等（2009）采用同时蒸馏萃取法提取蜂胶，得到淡黄色具芳香气味的蜂胶挥发油。他们还采用动态顶空进样的方法将蜂胶的挥发性成分进行吸附及解析，用于成分鉴定。Griffths 等（2007）也利用同时蒸馏-萃取法，以乙醚为溶剂、提取时间为 2h，得到了蜂胶的挥发性物质，得率为 0.09%。赵强等（2007）采用微波辅助萃取工艺提取蜂胶的挥发性成分，在以石油醚为溶剂，萃取温度 30℃，按蜂胶 10g、石油醚 110mL 的比例进行萃取，萃取时间 32min，蜂胶挥发油提取率为 13.83%，提取效率较常规溶剂萃取法增加 1.51 倍。李雅晶等（2011a）利用微波辅助萃取工艺对中国蜂胶、巴西蜂胶中的挥发性成分以石油醚为溶剂进行了提取，得到巴西蜂胶挥发性成分得率为 11.21%，中国蜂胶挥发性成分得率为 9.73%。利用超声辅助萃取法得到中国山东蜂胶挥发油得率达 8.83%（李雅晶等，2011b）。

第二节 蜂胶产品的加工与应用

蜂胶原料(毛胶)不能直接利用,因此对于蜂胶产品加工来说,蜂胶原料的预处理及其有效成分的获取是其产品加工必须突破的重要技术瓶颈。蜂胶中有效成分提取技术的长足发展使蜂胶产品的开发及加工获得了重大机遇。适宜的产品形式能进一步让蜂胶为大众所利用。目前,蜂胶产品已广泛应用于医药、食品、营养保健、日化、饲料等领域,涉及的剂型包括酊剂、胶囊、片剂、乳膏等。

1 蜂胶产品的应用

随着蜂胶研究的不断深入,蜂胶的成分进一步明晰,生物学活性得以验证。蜂胶产品市场的拓展及消费者认可度的提升使得大量蜂胶新产品涌现出来。鉴于蜂胶中所含的活性成分所发挥的特殊生物学效应,蜂胶除广泛应用于医疗保健领域外,研究人员还依据蜂胶的各种特性将其应用范围进一步拓展至食品保鲜、畜禽水产养殖、口腔保健、免疫佐剂及化妆品等各领域。

1.1 蜂胶产品在食品保鲜中的应用

由于具有高效、广谱的抑菌作用,蜂胶可作为保鲜剂应用于各种食品中。

在农副产品保鲜方面,Ozdemir 等(2010)将红宝石葡萄柚采摘后立即在不同浓度蜂胶乙醇提取液(1%、5%、10%)中沾一下,然后在8℃、相对湿度90%条件下储存6个月。结果发现,5%的蜂胶可以有效预防真菌性腐烂;用5%蜂胶乙醇提取液处理的红宝石葡萄柚可以在8℃下成功储存5个月。柑橘青霉病和绿霉病是由 *Penicillium digitatum* 和 *Penicillium italicum* 引起的,主要危害贮藏期的果实。Yang 等(2010)研究发现,中国蜂胶乙酸乙酯提取物能强烈抑制病菌菌丝生长和诱导菌丝突出的异常形态改变,呈浓度依赖性,能对病原菌孢子萌发产生不利影响;能减少 *P. digitatum* 和 *P. italicum* 引起的伤口感染和自然感染的水果腐烂,而对柑橘类水果的整体质量没有不利影响。他们还发现,蜂胶提取物处理柑橘48h 后人工接种意大利青霉,该柑橘所表现出的抗病性最强。蜂胶提取物处理增加了柑橘果皮中酚类物质的含量,提高了柑橘果皮中苯丙氨酸解氨酶(PAL)、过氧化物酶(POD)、多酚氧化酶(PPO)、几丁质酶(CHT)与诱导抗病性密切相关的酶的活性(杨书珍等,2010)。因此,蜂胶提取物可作为一种天然抗真菌剂来控制柑橘青霉病和绿霉病,降低果实贮藏期间的发病率,对柑橘表现出良好的保鲜效果。

陈小利等(2011a,2011b)研究了蜂胶对冷藏红富士苹果的保鲜效果,发现蜂胶处理能显著降低果实的失重率,延缓果皮色泽 a^*、C^* 值及花色素苷的下降,提高果实蜡质层、果皮和果肉中的类黄酮和总酚含量,抑制果实红色的消退;在贮藏前期(0~150d),蜂胶处理对降低果实的腐烂指数和丙二醛含量,提高过氧化物酶活性,抑制多酚氧化酶活性,延缓果实褐变速度的效果较明显,但贮藏后期(150~180d),蜂胶作用的效果与对照相比差异不显著。因此,蜂胶对短期冷藏红富士苹果的保鲜效果较好。俞益芹和张焕新(2011)以质量浓度0.04g/L 单甘酯、0.1g/L 山梨醇、0.12g/L $CaCl_2$、0.1g/L 蜂胶复配后的保鲜剂对枇杷进行涂膜处理,常温贮藏,可以有效降低枇杷失水率,抑制呼吸强度,较大限度地减少维生素 C 和可滴定酸和可溶性总糖等营养成分的损失,12d后,枇杷仍保持良好的感官品质。刁春英等(2013a)以雪花梨果实为试材,研究了室温条件下不同浓度蜂胶乙醇提取物涂膜处理对雪花梨果实保鲜效果的影响。发现不同浓度的蜂胶提取物(2%、4%、6%)均可有效降低果实的呼吸速率,维持产品的硬度,降低产品的失重率,较好地维持产品的可滴定酸含量和维生素C含量,其中6%浓度蜂胶效果最好,对雪花梨果实可溶性固形物的含量影响不大。同时也发现蜂胶提取物对草莓室温保鲜效果良好,可以显著降低果实的呼吸强度、失重率和腐烂率,维持果实原有品质,延缓果实成熟衰老(刁春英等,2013b)。吴雪丽等(2013)将蜂胶提取液用于扇贝保鲜,发现蜂胶提取液在扇贝冷藏保鲜过程中能有效地抑制微生物的生长和蛋白质的降解,减缓腐败变质,延长货架期。张楠和任战军(2014)研究了陕北蜂胶对香蕉的保鲜作用。利用蜂胶乙醇提取液制备不同浓度的涂膜剂分别作用于香蕉,测定香蕉中维生素C含量、失重率、呼吸强度、总糖含量、可滴定酸含量和腐烂指数的变化及香蕉感官属性变化,来反映蜂胶涂膜剂对香蕉保鲜的最佳浓度和最长保鲜期。结果表明,各浓度蜂胶乙醇提取液对香蕉均有不同程度的保鲜作用,且浓度为1%的蜂胶乙醇提取液保鲜效果最佳。此外,还有研究发现蜂胶具有延缓油脂氧化的作用(王大红和陈明胜,2014)。

低温肉制品普遍存在货架期短,易发生腐败变质的问题。徐世明等(2011)对胀袋西式火腿中的优势腐败菌进行分离,并通过16S-rRNA基因序列分析对分离菌株进行鉴定,综合分析确定这些菌株分别为枯草芽孢杆菌、短小芽孢杆菌与生孢梭状芽孢杆菌。蜂胶对分离菌株及菌株混合液均有明显抑菌效果。因此,可将蜂胶作为保鲜剂进行低温肉制品综合保鲜,以有效延长低温肉制品保质期,在一定程度上解决低温肉制品的腐败变质问题。孟良玉等(2011)研究了不同浓度蜂胶提取液对鲜猪肉的保鲜效果,发现在蜂胶提取液浓度为0.015mg/mL时采用涂抹法,对猪肉的保鲜效果最好,在37℃条件下可保鲜贮存24h,在0~4℃条件下可保鲜贮存12d以上。

抗菌包装膜是活性包装材料的一种,它主要是以天然大分子物质为载体,添加一定的抗菌剂而制成的薄膜,与普通包装相比,抗菌包装对确保食品安全以及延长食品货架期的作用更为突出。利用具有抑菌活性的天然产物作为抑菌剂制备抗菌包装薄膜是近年来抗菌包装研究的一大亮点。Mascheroni等(2010)研究发现,将蜂胶添加到聚乳酸膜中制成活性包装,直接接触液体介质时,蜂胶中的某些活性成分(酚酸)可以释放到包装的食物中,而类黄酮等物质仍留在聚合体中起作用。Pastor等(2010)还用羟丙基甲基纤维素和不同浓度的蜂胶乙醇提取物制成了抗真菌的可食薄膜。

蜂胶的抗菌作用及优良的成膜性使蜂胶在食品保鲜方面有很大的应用空间。

1.2 蜂胶产品在畜禽养殖中的应用

蜂胶具有免疫增强活性,作为绿色饲料添加剂饲喂畜禽已有不少文献资料报道。

在产蛋鸡的基础日粮中加入3g/kg的蜂胶,可显著增加血清IgG和IgM的浓度,显著降低外周血T淋巴细胞百分率;显著增加红细胞计数(红细胞)而不影响血红蛋白、红细胞压积值,总白细胞(白细胞)和白细胞计数差异。研究结果表明,在基础日粮中加入3g/kg的蜂胶可能对产蛋鸡体液免疫功能产生积极影响(Cetin et al.,2010)。经蜂胶处理的产蛋母鸡具有较高浓度的天然抗体。蜂胶处理组(50mg/kg)中异嗜白细胞以及异嗜淋巴细胞比显著下降。绵羊红细胞免疫后,IgG水平明显增加,可用于增加对疫苗的抗原特异性抗体反应(Freitas et al.,2011)。

李香子等(2011)研究发现,蜂胶饲料添加对猪屠宰率、平均背膘厚及眼肌面积没有显著影响,但可明显提高瘦肉率;降低猪肉中的胆固醇含量,还可显著提高肉中的人体必需氨基酸及亚油酸含量。因此,蜂胶可用于生产功能性畜产品,满足人们对低胆固醇畜产品的需求。樊兆斌等(2013)研究了蜂胶对断奶仔猪抗病力、血液生化指标及肠道结构的影响,结果证实饲喂蜂胶能够显著提高仔猪的抗病力,使排粪异常率和呼吸异常率分别下降7.03%和3.38%,成活率提高10%;能够显著提高血清中总蛋白、白蛋白和球蛋白的质量分数,降低血清中三酰甘油的水平,并能明显改善仔猪肠道结构。

Seven等(2012)研究了日粮中添加蜂胶对肉鸡在由铅诱导氧化应激饲养条件下的生产性能、营养素利用率和胴体品质改变的影响。结果表明,日粮中过多添加铅会对肉鸡生长造成不利影响,而日粮中蜂胶的添加改善了肉鸡生长状况、养分利用率和胴体品质。Aygun等(2012)研究了蜂胶对日本鹌鹑(*Coturnix coturnix japonica*)蛋蛋重损失率、孵化率及蛋壳微生物活性的影响,同时也研究了蜂胶对日本鹌鹑雏鸟的生长影响。发现相较于对照组(70%乙醇、苯扎氯铵),蜂胶喷雾能显著降低蛋重损失率,并能有效抑制微生物的活力。同时,蜂胶喷雾在蛋孵化率、胚胎死亡率、雏鸡生长指标上与对照组没有显著差异。这说明相较于通常鹌鹑蛋孵化中使用的化学消毒剂,蜂胶也可以达到令人满意的效果。Ozkok等(2013)的研究证实饲料中添加蜂胶能有效提升产蛋母鸡的生长性能及鸡蛋品质,且无任何不利影响。

Hashem等(2013)研究发现,蜂胶能有效缓解炎热季节对公兔精液质量、氧化应激、血液生化指标所造成的不利影响。Morsy等(2013)也证实蜂胶用于饲料添加剂能对发情期、发情后期母羊的生长及健康状态有着良好的促进作用。巴西马林加州立大学的科研人员研究了日粮中添加蜂胶提取物对奶牛乳产量、乳脂肪酸组成和乳抗氧化能力的影响,发现日粮中添加蜂胶对奶牛干物质摄入、乳产量、饲料转化率、乳固形物和体细胞指数并没有显著影响,而添加蜂胶会影响乳脂肪酸组成,显著影响共轭亚油酸比例。饲喂含高浓度蜂胶提取物日粮会显著增加单不饱和脂肪酸(MUFA)和多不饱和脂肪酸(PUFA)含量,降低饱和脂肪酸(SFA)比例(Aguiar et al.,2014a)。此外,饲喂蜂胶奶牛所产牛奶的抗氧化能力会显著增强。通过对奶牛消化功能和瘤胃代谢功能的研究,结果发现蜂胶能有效提升奶牛的消化功能,对奶牛瘤胃微生物的蛋白质代谢有着非常积极的作用(Aguiar et al.,2014b)。

巴西另一科研团队研究了饲料中添加蜂胶对肉鸡肠道形态和消化酶的影响。实验共使用了1020只雄性肉鸡,并根据蜂胶的添加量(0~500μg/g)将这些肉鸡分为6个处理,每组5个重复,每个实验单位34羽。蜂胶添加从1日龄开始一直持续到21日龄,21日龄后开始使用相同的完全饲料。结果发现,所有处理组中的肉鸡生长性能并没有显著差异,但42d后这些肉鸡的屠宰率与蜂胶添加量呈显著线性上升关系。同时发现,饲喂蜂胶能显著提高7日龄肉鸡肌胃和大肠的相对重量,而蜂胶的最小有效添加浓度为278μg/g;饲料中添加蜂胶也可显著提升21日龄肉鸡回肠隐窝深度和增强淀粉酶的活性(Duarte et al.,2014)。此外,蜂胶在肉鸡生产中的应用也得到了沙特阿拉伯阿卜杜勒阿齐兹国王大学Attia等和巴西马林加州立大学Eyng等的关注(Attia et al.,2014;Eyng et al.,2014)。

1.3 蜂胶产品在水产养殖中的应用

膳食补充蜂胶乙醇提取物能显著加速虹鳟鱼(*Oncorhynchus mykiss*)的生长速度,增加饲料和蛋白质实用率,增加血浆超氧化物歧化酶、溶菌酶、谷胱甘肽过氧化物酶和过氧化氢

酶的活性等,降低血浆中丙二醛浓度和三酰甘油浓度,降低血浆中谷草转氨酶和谷丙转氨酶活性,但增加了肝谷草转氨酶和丙氨酸氨基转移酶的活性。这说明蜂胶乙醇提取物是虹鳟鱼潜在的生长促进剂、保肝剂和免疫刺激剂(Deng et al.,2011)。Bae 等(2012)研究了饲料中添加蜂胶对幼鳗(*Anguilla japonica*)生长性能、免疫反应、疾病抵抗力及身体成分的影响,并评价了蜂胶作为幼鳗饲料添加剂的生物利用率。结果表明,对于幼鳗来说,蜂胶不同添加剂量所产生的效果不完全相同,作为生长促进剂而言蜂胶的最佳添加量为0.25%～0.5%,而用作免疫增强及防病抗病目的的最适添加量应为0.5%～1%。Yonar 等(2012)研究发现,蜂胶对农药毒死蜱造成鲤鱼(*Cyprinus carpio carpio*)的毒性有一定的保护作用,并能缓解鲤鱼遭受到的氧化应激状况。Kelestemur 和 Seven(2012)研究发现,饲料中添加蜂胶能有助于维持幼年虹鳟鱼在缺氧应激条件下血电解质的平衡。Segvic-Bubic 等(2013)研究了蜂胶对低水温应激条件下的鲈鱼(*Dicentrarchus labrax*)生长的影响,发现饲料中添加2.5g/kg 的蜂胶能有效提高鲈鱼特定生长率和饲料转化率。低水温应激会导致血清中三酰甘油、葡萄糖和皮质醇的含量上升,但饲喂蜂胶可以显著降低血糖及皮质醇的水平,这也说明饲料中添加蜂胶能有效缓解低水温对鲈鱼生长造成的不利影响。Orun 等(2013)的研究也证实蜂胶对草甘膦引起的鲤鱼(*Cyprinus carpio*)急慢性毒性症状有着很好的保护效果。

1.4 蜂胶产品在口腔保健中的应用

蜂胶具有良好的抗菌活性,对多种口腔疾病均有良好的预防、治疗作用,口腔保健也是蜂胶应用的重要领域。Saavedra 等(2011)研究发现,蜂胶对与龋病发生直接相关的病原菌 *Lactobacillus fermentum* 具有抑制作用。Pereira 等(2011)研究表明,含有5%(W/V)巴西绿蜂胶的无酒精漱口水可有效降低牙菌斑和牙龈炎,对口腔软硬组织没有副作用。Dodwad 和 Kukreja(2011)研究表明,蜂胶可替代化学漱口水保护牙龈健康。10%和30%的蜂胶凝胶没有降低体外牙本质水分传导,但可部分去除牙本质小管(de Carvalho et al.,2011)。玻璃离子水门汀是由可析出离子的氟铝硅酸盐玻璃粉和聚丙烯酸水溶液组成的牙体粘结、修复材料。加入巴西绿蜂胶乙醇提取物或冻干粉可明显增加玻璃离子水门汀的吸水性。因此,蜂胶可用于口腔保健,不会引起明显的副作用,而且牙本质敏感性患者也可选用。

病原微生物在口腔中的大量繁殖是引发口腔疾病的重要因素,而蜂胶具有良好的抗菌活性,因此,近年来也被广泛用于口腔保健中。Venero 等(2012)比较了5%蜂胶酊和传统治疗方法对牙槽炎的治疗效果,结果发现相较于传统治疗方法,使用蜂胶酊可以更快、更有效地治疗牙槽炎。Coutinho(2012)将蜂胶用于牙龈冲洗,用于牙周病的辅助治疗方法,可以显著减少口腔厌氧细菌的数量。而 Madhavan 等(2012)的研究也证实蜂胶对治疗牙本质过敏也有良好的疗效。Tanasiewica 等(2012)研究发现,蜂胶牙膏(添加3%巴西蜂胶提取物)能有效减少牙菌斑,从而对牙龈炎、牙周炎具有很好的预防保健效果。Skaba 等(2013)对含有巴西蜂胶醇提物的蜂胶药膏的口腔保健效果进行了短期临床研究。通过对32名成年病人边缘牙周组织的调查发现,含3%蜂胶药膏能有效去除牙菌斑及缓解边缘牙周组织的病变症状。崔艺兰和刘泓(2013)比较了不同脱敏剂与蜂胶对牙本质小管的封闭性,发现实验中选用的脱敏剂与蜂胶均具有封闭牙本质小管的效能。魏广治等研究发现,蜂胶与蒙脱石粉联合应用对口腔溃疡有着很好的治疗效果(魏广治和沈兰花,2013;魏广治和薛峰,2013)。台湾众泰卫生科学与技术研究所 Hwu 和 Lin(2014)采用荟萃分析法(Meta-Analysis)来评价蜂胶对口腔健康的效果。选取了7个电子数据库,检索了1969—2012年发表的相关文献,采用纳入和排除标准对数据进行收集,采用澳大利亚循证护理中心(Joanna Briggs Institute)开发的荟萃统计分析和回顾工具来评估这些文章的质量与效用。结果发现,从1997 到 2011 年间有 194 位参与者具有可萃取的数据。荟萃分析结果表明,虽然蜂胶对减少牙菌斑有一定效果,但并不具统计学意义,对于口腔感染和口腔炎症方面的结果也并没有显著差异性。这说明虽然蜂胶在促进口腔健康上有许多非常有前景的发现,但考虑到发表文章中可利用数据并不是很多,且差异性很大,未来的研究需要进一步从完善实验设计与数据统计方面出发,为蜂胶在口腔健康方面的应用提供更多有效而可信的数据。

1.5 蜂胶在免疫佐剂中的应用

佐剂是一种非特异性免疫增强剂,当与抗原一起注射或预先注入机体时,可增强机体对抗原的免疫应答或改变免疫应答类型。很多研究已经证实了蜂胶可以作为一种免疫佐剂发挥独特的免疫调节作用(El Sayed and Tarek,2012)。鸡新城疫(ND)在世界范围内广泛流行,对养鸡业的危害非常严重。近年来,又出现了非典型 ND 的流行,目前防治 ND 最有效的方法是疫苗接种。而在生产中由于种种原因疫苗接种后常出现免疫失败。淫羊藿多糖与蜂胶黄酮组合能增效体外淋巴增生,增强鸡 ND 疫苗的免疫效应(Fonseca et al.,2011)。蜂胶作为佐剂一般是以蜂胶乙醇提取物、蜂胶水提液的形式加入的。柴家前等(2010)以纳米蜂胶颗粒为佐剂的新城疫灭活疫苗能促进鸡免疫器官的生长、发育和成熟,促进免疫细胞的分化和增殖,使血液循环中 T 淋巴细胞增多,使腹腔巨噬细胞活性增强。蜂胶疫苗产生抗体早,可以获得早期保护。Fan 等(2012)采用淫羊藿多糖-蜂胶黄酮制备脂质体,将其用于新城疫疫苗佐剂。结果发现,淫羊藿多糖-蜂胶黄酮脂质体能显著提高抗体滴度,促进淋巴细胞增生,提高血清中干扰素 γ 和白介素 6 的含量。同时,脂质体组具有最佳保护效果和最低的新城疫发病率和致死率,说明脂质体能显著增强淫羊藿多糖-蜂胶黄酮的免疫佐剂效果。王宇航等(2012)利用蜂胶作为佐剂制成禽霍乱蜂胶佐剂灭活疫苗,将其与同批菌液制成的禽霍乱氢氧化铝胶灭活疫苗进行免疫效果比较试验。结果显示,用蜂胶佐剂制成的禽霍乱灭活疫苗免疫效果优于禽霍乱氢氧化铝胶灭活疫苗。王晓丽等(2012)利用实验室自制的兔病毒性出血症、多杀性巴氏杆菌病二联蜂胶灭活疫苗,对健康易感兔进行最小免疫剂量测定,结果表明该疫苗的最小免疫剂量为 0.5mL/只。

1.6 蜂胶在化妆品中的应用

氧化是肌肤衰老的最大威胁,饮食不健康、日晒、压力、环境污染等都能让肌肤自由基泛滥,从而产生面色黯淡、缺水等氧化现象。预防皮肤老化是化妆品的主要目的之一。巴西绿蜂胶和褐色蜂胶提取物的局部预处理,可阻止紫外线引起的体内谷胱甘肽(GSH)耗竭,说明蜂胶对抗皮肤氧化应激有应用前景(Fonseca et al.,2011)。蜂胶提取物和维生素 E 醋酸酯配方的霜剂在 60℃下放置 28d,逐渐变暗和 pH 值略微降低,但呈现出稳定的流变性,鉴于稳定性和受试者偏好,二者联合应用制成的霜剂在化妆品中的应用更有前景(Gonçalves et al.,2011)。

2 蜂胶产品加工工艺

鉴于蜂胶预处理手段以及有效成分提取技术的不断改进,应用范围的不断扩展,蜂胶的精深加工以及产品的多元化成为蜂胶研究开发的重要方向。对蜂胶产品进行深加工,开发各种产品形式,能够达到扩大蜂产品应用范围,增强应用效果,提高使用价值,提升市场竞争力的目的。为促进蜂胶行业持续健康发展,蜂胶产品的加工工艺研究与创新必不可少。

2.1 内服型蜂胶产品加工工艺

内服型蜂胶产品主要用于药品及保健食品领域,以蜂胶提取物为原料或以蜂胶与其他生物学活性成分相配伍,进行产品制备,以达到相应的预防或治疗目的。目前内服型蜂胶产品形式主要有蜂胶口服液、蜂胶片剂、蜂胶胶囊等。

(1) 蜂胶口服液加工工艺

蜂胶口服液是目前国内外蜂胶产品的主要形式之一,分为酒精型和无醇型。酒精型蜂胶口服液以食用酒精作溶剂,将蜂胶配制成不同浓度的溶液后进行灌装或与其他成分配伍后进行灌装。醇溶型蜂胶口服液对蜂胶原料预处理过程简单,黄酮类化合物含量较高,便于自制(震声,2001;邓希尧,2002;赵家明,2005)。

醇溶型蜂胶口服液基本工艺流程为:

纯蜂胶→冷冻→粉碎→按比例加入食用酒精→溶解→过滤→分装→贴标→成品

玄红专和顾美儿(2006)选择80%乙醇提取粗蜂胶,经过滤、浓缩制成纯蜂胶,而后进行蜂胶口服液的制备,通过按比例加入食用酒精,经搅拌、溶解等过程制得浓度为25%、30%和50%的蜂胶口服液。王振山(2010)用饮用纯净水和食用酒精以不同比例配合萃取蜂胶,固液分离后,合并滤液,调节有效成分含量至标准要求,所得产品蜂胶总黄酮含量为26mg/mL。

由于含乙醇的产品在老人、儿童以及酒精过敏者和肝病患者等特殊人群的使用中有诸多限制,因而研究者设计开发了无醇型蜂胶口服液产品。

无醇型蜂胶口服液基本工艺流程为:

纯蜂胶→水溶工艺预处理或可分散蜂胶微粒制备→配料(食用甘油和蜂蜜)→混合→分装→贴标→检验→成品

孙丽萍等(2007)发明了一种制备固体水溶性蜂胶提取物的方法。取蜂胶原料(毛胶)粉碎,加含水乙醇溶液提取2~4次,提取液合并滤过,减压回收乙醇得浓缩液,静置得到沉淀和上清液,沉淀水洗后干燥即得到固体浅色蜂胶提取物,该提取物可方便用于蜂胶口服液的制备。张红城等(2014)将蜂胶提取物、溶剂油、表面活性剂和注射用水按重量比为(1~10):(5~12):(10~40)混合,高纯水余量,乳剂中乳滴的平均粒径为15~90nm,得到蜂胶纳米乳口服液。葛飞等(2014)公开了一种高雄山虫草菌丝体蜂胶口服液的配方及其制备方法,该高雄山虫草菌丝体蜂胶口服液的配方按体积比包括:菌丝体水溶液过滤液:蜂胶纳滤溶液=1:2,产品中其余成分按照常规口服液的配方原则加入。该高雄山虫草菌丝体蜂胶口服液的制备方法包括以下步骤:利用固态发酵方式发酵生产高雄山虫草菌丝体;高雄山虫草菌丝体水溶液的制备;通过超声波提取,超临界CO_2萃取、浓缩技术,将高雄山虫草固态发酵菌丝体与优质蜂胶进行调配;加入辅料,进行调配,然后灌装、检验、包装,最后制备完成。毛丽珍和徐世芳(1998)将蜂胶和银耳等原料按照一定的配方,经特殊工艺制备成乳白色胶状液体。该产品富含黄酮类化合物和蛋白质。通过紫外分光光度法能够简便快捷的对口服液中的总黄酮含量进行测定,该方法的回收率高,重现性好,方法简便,数据可靠,对产品的质量控制是切实可行的。

(2) 蜂胶片加工工艺

蜂胶片剂的加工参照药品片剂的生产工艺。片剂的优点在于剂量准确,体积小,携带便利,服用方便,适于大量生产,便于运输和贮存;利用包衣技术可遮盖不良气味,避免对胃肠道的刺激,物料加工成片剂后,光线、空气、水分和灰尘等因素对其影响很小。

蜂胶片剂的生产要添加部分赋形剂。赋形剂在片剂中有五大功能:黏性,使片剂成型;吸湿性,使片剂崩解;润滑性,使颗粒具有流动性;吸收性,可使液体药物成为固态;稀释性,使小剂量的药物制成片剂。

常用的吸收剂与稀释剂有淀粉、糊精、蔗糖粉、乳糖、氢氧化铝、硫酸钙、碳酸钙和碳酸氢钙等。常用的黏合剂和崩解剂有明胶、羟甲纤维素、羟丙基纤维素、微晶纤维素、麦芽糖醇、聚乙烯醇等。常用的润滑剂有硬脂酸镁、滑石粉、三硅酸镁、二甲硅油、聚乙二醇等。

蜂胶片剂的基本工艺流程为:

拟定处方→准备物料→粉碎、过筛→称量、混合→制软材、制颗粒(干法和湿法)→干燥→制粒、总混→压片、包衣

江苏连云港蜂疗医院研制的主治高脂血症的蜂胶片,该产品经脱蜡后添加其他辅料压制成片,每片含黄酮不少于3mg(相对于原蜂胶0.1g)。配方包括:蜂胶粉(或浸膏)80kg,淀粉70kg,硫酸钙空白颗粒170kg,氢氧化铝粉9kg,硬脂酸镁2kg。将原料按比例混匀,加适量润湿剂,手捏可成团为好,切勿过量,再制粒,干燥,压片,或挂糖衣制得。陈坤等(2008)以蜂胶酒精提取物为原料;将蜂胶提取物与微粉硅胶混合,然后冷冻至$-20\sim0^\circ C$,粉碎温度保持在$-20\sim0^\circ C$,然后再采用等量递增法与其他物料混合,然后加入体积分数为80%~90%的乙醇溶液,进行湿法制粒,过14~16目筛,于38~42℃温度下干燥,至水分重量含量达到4%~6%时,冷却至室温,加入薄荷脑,混合后压片,制成蜂胶咀嚼片。石聚彬和石聚领(2007)制备红枣蜂胶含片,将蜂胶与碾磨后的红枣以(2~4):1的比例混合,造粒,之后将其干燥压片,最后经包装得到红枣蜂胶含片。含片既具有良好的口感,又具有红枣和蜂胶的药用保健功能,便于储存,携带方便。周斌等将蜂胶冷冻、粉碎除杂,再以常压常温乙醇重复浸泡3次后,冷冻离心除去固体杂质,取乙醇上清液,最后,在1~5℃的环境温度下分离出分子量在4000D以下的蜂胶乙醇提取物,经乙醇蒸馏回收得到干粉,将蜂胶粉与蜂王浆冻干粉、异麦芽糖醇、硬脂酸镁和薄荷脑等混合后制得防治酒精性脂肪肝蜂胶口含片。为丰富产品形式,改进工艺将蜂胶冷冻、粉碎除杂,再以常压常温乙醇浸泡后的超声波提取,冷冻离心除去固体杂质,取醇溶液,最后,在1~5℃的环境温度下分离出分子量在5000D以下的蜂胶乙醇提取物,经乙醇蒸馏回收得到采用乙醇从蜂胶中提取的分子量小于5000D的、固形物含量大于90%的生物活性干燥物,将其与蜂王浆冻干粉、异麦芽糖醇、硬脂酸镁和包衣粉等混合制得蜂胶解酒肠溶片(周斌和叶满红,2012a,2012b)。徐水荣和卢媛媛(2012)发明了一种葛根素、松花粉和蜂胶复合片剂,包括如下质量分数的组分:破壁松花粉,40%~50%;蜂胶粉,10%~20%;葛根素,20%~30%;蔗糖,2%~5%;糊精,2%~5%;羧甲基淀粉钠,1%~3%;微晶纤维素,1%~3%。其制备方法包括主要原材料混合、黏合剂制、湿法制粒、片剂制作等步骤。该发明以葛根素、松花粉和蜂胶为主要原料,无毒性和无副作用,兼具保肝护肝保健功效;通过制粒后的片剂,具有服用携带方便、质量稳定等特点。

(3) 蜂胶胶囊加工工艺

蜂胶胶囊分为硬胶囊和软胶囊两种。

蜂胶硬胶囊是指将蜂胶制成粉体,并装入硬胶囊壳中的一种蜂胶产品,具有服用方便、无酒精刺激味等优点,该产品在国内外都比较流行。

蜂胶硬胶囊基本工艺流程为:

纯蜂胶→用乙醇溶解制成50%蜂胶液→加入吸收剂→低温真空干燥→粉碎→填充胶囊→分装→成品→消毒、检验→贴标→入库

姜德勇(2001)发明了一种蜂胶抗病毒抗氧化活性提取物(提纯脱铅蜂胶粉)及其制造方法,并将其用于蜂胶胶囊的制备。将原料蜂胶在-4~0℃的温度下预冻后,粉碎至40目以下,室温在8~15r/min搅拌速度下,用大于等于95%质量分数的乙醇提取8~12h,蜂胶粉与等于大于95%质量分数的乙醇提取溶剂的质量比为1:(3.5~4.5)后用微孔精密过滤工艺过滤,精密过滤后,用对含氯有机物具有选择性吸附的树脂进行脱除,再用活性陶土吸附剂进行吸附脱除致敏物质,接着用对重金属具有选择性吸附的离子交换树脂进行选择性吸附,脱除重金属,后常温提取以去除热乙醇溶出的可溶性的树脂类非活性成分,最后用孔径为100nm的微孔精密过滤工艺进行精制过滤,减压干燥粉碎后得到蜂胶粉。利用蜂胶粉制成硒灵蜂胶粉胶囊、蜂胶粉胶囊、蜂胶灵芝杰胶囊。周萍(2011)将蜂胶、环状糊精和聚甘油脂肪酸酯,按一定的比例及工艺步骤加工,进行蜂胶胶囊的制备,制得的蜂胶粉胶囊的组分科学,经过环状糊精进行包埋,添加的聚甘油脂肪酸酯可增加蜂胶的水溶性,有利于提高人体对产品的吸收。该制备工艺科学、简单,制备过程中不会破坏蜂胶粉营养成分。

蜂胶软胶囊是将纯蜂胶与载体混合,然后填充到软胶囊中的一种产品。该剂型是目前市场上占有率最高的剂型,技术的关键是软胶囊的制备和载体的选择。

蜂胶软胶囊工艺流程为:

纯蜂胶→熔化→用甘油和乳化剂乳化→填充胶囊→成品分装→贴标→成品→入库

马建中(2010)发明了一种高含量蜂胶软胶囊的制备方法。该制备方法为:以质量份配比计,取高纯度蜂胶浸膏20~50份,水浴加热至50℃并保温;另取食用植物油20~50份,加热至50℃时,在搅拌下,缓缓加入到高纯度蜂胶浸膏中,与此同时,用循环水冷却,并不断搅拌,待温度降至40℃时,停止搅拌,得到均质性好的流体状蜂胶软胶囊内容物,经压丸、成型、洗丸、干燥,制得高含量的蜂胶软胶囊。所述高纯度蜂胶浸膏制备方法为:将粗蜂胶在温度为4℃环境下冷藏2h以上,经粉碎机粉碎成粗粉;以质量比计,向粗粉中加入粗粉10倍量的95%乙醇回流提取,过滤,将提取液减压回收乙醇,得浸膏,将此浸膏在4℃环境温度冷藏2h以上,再粉碎成浸膏粗粉;以质量比计,向浸膏粗粉中加入浸膏粗粉10倍量的95%乙醇并使其在常温下浸渍溶解,不时搅拌,使浸膏粗粉充分溶解,静置,过滤,滤液减压回收乙醇,再浓缩至相对密度为1.30,此密度为50℃测得;经用紫外分光光度计测得蜂胶总黄酮含量达28%~40%,得高纯度蜂胶浸膏。郑浩亮(2010)用丙二醇溶解蔗糖酯,边加热边搅拌,溶解至澄清透明的溶液后再加入单辛酸甘油酯,溶解至澄清透明的溶液后加入事先粉碎的蜂胶,不断搅拌至蜂胶全部溶解,得棕红透明的蜂胶液,将此液冷却,搅拌状态下,加入大豆磷脂,得到棕红透明的亲水性蜂胶软胶囊内容物。用以制备蜂胶软胶囊。戴关海等(2011)将蜂胶、西洋参提取物、三七提取物和大豆油组成按质量比为5:(1~5):(1~5):(10~20)比例配伍,制备蜂胶软胶囊,能有效降低血脂并提高人体免疫功能。田文礼等(2007)提纯蜂胶粉碎、过筛,然后将蜂胶粉、食用油及助悬剂在温度为5~45℃通过超微粉碎机进行湿法超微粉碎,在粉碎的同时达到分散乳化的目的,再把料液进行抽真空处理,最后压丸、定型、洗丸、

干燥得到油溶蜂胶软胶囊。由于在低温下进行超微粉碎,避免了蜂胶中挥发油组分的挥发,提高了蜂胶在人体的生物利用率。

陈璇等(2013)测定了蜂胶鱼油软胶囊对小鼠免疫功能的影响,蜂胶鱼油软胶囊主要由蜂胶和鱼油等成分组成。其中蜂胶占30%,鱼油占64.8%,内容物为褐色油膏状。结果发现其具有增强非特异和体液免疫功能的作用。蔡烈涛和吴俊(2011)采用压制法制备蜂胶软胶囊,并采用正交试验方法,以崩解时限为指标,以定型时间、干燥温度、相对湿度、干燥时间为因素进行正交试验,最终确定最佳工艺为:定型时间12h,干燥温度25℃,相对湿度30%,干燥时间24h。刘华(2012)以天然蜂胶为原料,辅以聚乙二醇400、甘油、纯化水发挥协同成型作用,软胶囊皮制备中明胶、甘油、纯化水的配方最佳比例为1∶0.45∶1,胶囊内容物所用的蜂胶用量约0.5g/d,制备了增强免疫力、辅助降血糖的双功能蜂胶软胶囊。王凤忠(2011)将蜂胶和王浆制备成具备一定保健功能的蜂胶王浆软胶囊(以聚乙二醇为分散基质的软胶囊),并在制备过程中对蜂胶原料中的重金属铅的含量进行了质量控制。

(4)蜂胶滴丸加工工艺

滴丸剂是将固体或液体药物与基质加热熔融成溶液、混悬液或者乳液后,滴入不相溶的冷凝液中,滴丸剂是固体分散体的一种形式。蜂胶滴丸剂既可供内服、外用和局部使用,亦可制成缓控释制剂,是一种开始引人注目并有良好发展前景的剂型。

王平和孙建伟(2012)以蜂胶为主要原料,制备蜂胶滴丸,蜂胶滴丸制法:按每100g蜂胶滴丸计,加入泊洛沙姆70g,蜂胶提取物20g,葡萄籽油10g,在60~95℃水浴中融化物料,并搅拌均匀;预热滴丸设备,滴头温度50~70℃,冷凝柱内冷凝液二甲硅油的温度梯度40~25℃→5~0℃;待滴丸设备预热结束,将已经熔融的物料加入滴罐中,通过滴头,以30~60滴/min的速度滴入冷凝液中;由滴丸机的出口将成型的滴丸取出,脱去表面的冷凝液,干燥后制得。

(5)蜂胶颗粒加工工艺

颗粒剂是将药物与适宜的辅料配合而制成的颗粒状制剂,可以直接吞服,也可以冲入水中饮入,应用和携带比较方便,溶出和吸收速度较快。蜂胶常与其他药用成分配伍,制成颗粒剂应用。

杨汝伟等(2014)公开了一种辅助调节血糖的复方蜂胶颗粒制剂配方。其配方包括以下组分:蜂胶粉0.6质量份,雄蜂幼虫粉0.5质量份,维生素B 0.61质量份,葡萄糖酸锌0.1质量份,2%/kg的富硒酵母0.25质量份,2%/kg的富铬酵母1质量份,中药提取物8质量份;所述的中药提取物是由人参、霜桑叶、生地、黄芪、肉桂经粉碎、加水、加热、提取、浓缩、制粒而成。这一辅助调节血糖的复方蜂胶颗粒制剂配方,可调节血糖、血脂,起到抗氧化,避免对胰岛细胞造成伤害,促进胰岛β细胞再生,消除机体胰岛素抵抗的作用,从而使机体糖代谢趋于正常,从而逐渐取代依靠服用降糖西药或者胰岛素来控制血糖,使身体状况得到良好的改善。戴关海等(2008)观察了桑胶颗粒(主要由蜂胶、地骨皮、桑叶、山茱萸等组成)对自发性2型糖尿病GK大鼠降血糖作用,结果发现桑胶颗粒具有较好的降低糖尿病模型大鼠血糖作用,其作用机制可能是与桑胶颗粒能促进胰岛细胞增加胰岛素分泌有关。

2.2 外用型蜂胶产品加工工艺

外用型蜂胶产品主要用于药品,剂型包括蜂胶软膏、蜂胶酊、蜂胶喷雾剂等。

(1) 蜂胶软膏加工工艺

软膏剂是将药物与适宜基质均匀混合而制成的具有一定稠度的半固体外用制剂。由于蜂胶本身的物理特点及生物学活性，蜂胶软膏被加以制备并进行了应用。蜂胶膏有止痒、镇痛、促进创面愈合及良好的消炎作用，适用于久治不愈的创伤、皮肤瘙痒和慢性湿疹等症。每支软管装乳膏 30g 或 50g，味芳香，不溶于水。具有释药快、穿透力强、易清洗、不污染衣物等特点。

程瑛等(2013)制备了一种蜂胶花粉复方按摩膏，制备方法如下：①量取硬脂酸、鲸蜡醇、单硬脂酸甘油酯、精制水等，混合搅拌，保持水温，得混合液 a；②量取硼砂，用蒸馏水溶解后保持水温，加入到混合液 a 中，形成混合液 b；③量取丙二醇、丙三醇，加热，加入到混合液 b 中，搅拌，形成混合液 c；④取精制蜂胶，打碎至小块，加入酒精，得蜂胶乙醇溶液；⑤取蜂花粉，加入酒精和蒸馏水混合，搅拌，浸泡，合并两次上清液，置于水浴锅中加热浓缩，控制水温；⑥将蛇床子等中药粉碎，加水煎煮，合并两次滤出液；⑦取蜂胶花粉及中草药混合液加热，关闭水浴锅，缓慢加入蜂花粉液、蜂胶液和中药液，搅拌，温度下降时，加入适量对羟基苯甲酸乙酯等物，搅拌，冷却，制得复方按摩膏成品。陈英华(2011)将寄生、炭灰、穿山龙、蜂胶、天南星、南蛇藤根和洋金花复合配伍制成蜂胶综合膏药。成分配比为：寄生 20 份、炭灰 10 份、穿山龙 10 份、蜂胶 20 份、天南星 15 份、南蛇藤根 8 份、洋金花 0.04 份。蜂胶综合膏药能有效地治疗及治愈许多骨科类疾病，外用可化解骨刺、瘀血坏死、疑难的抽筋、腰腿疼、腰椎间盘突出、风湿筋骨疼、骨髓、肩周炎等。王维人(1993)制备了一种蜂胶冻疮软膏，由蜂胶、鲜橘皮、生姜、紫草、红花、当归、日本灵芝草等活血化瘀中药、防腐剂硼砂及赋形剂液状石蜡配制而成。该冻疮膏对冻疮具有治疗及预防复发的效果。

(2) 复方蜂胶酊加工工艺

酊剂是将生药浸在酒精里或把化学药物溶解在酒精里而成的药剂，由于蜂胶在酒精中较易溶解，因此，蜂胶酊剂是一种常见的蜂胶产品类型，将蜂胶与其他药用成分相配伍制得的酊剂可获得更广泛的用途。

蒋云云等(2011)将浓度为 0.02g/mL 的蜂胶液 5%～30%、茶树精油 0.1%～0.5%、薄荷精油 0.1%～0.5%及乙醇复配制备蜂胶酊。制得的外用蜂胶酊的抗菌消炎效果显著，具有良好的可溶性，提高了产品的纯度，避免本产品含重金属和其他杂质，使其颜色鲜亮纯净透明。另外，此蜂胶酊可有效脱除蜂胶中的致敏成分，减少使用时过敏现象的发生。杨汝伟(2001)制备了一种治疗口腔溃疡的外用制剂。该制剂主要由蜂胶、达克罗宁粉、蜂王浆、薄荷脑组成，其组分含量为 25～35 份蜂胶，0.2～0.6 份薄荷脑，0.5～1.5 份达克罗宁粉和 2～6 份蜂王浆，上述组分在混合溶剂中浸渍后制成酊剂。该酊剂中有效蜂胶物质含量为 5%～15%。该酊剂无毒，使用安全，擦于患处，复发性口腔溃疡 3～7d 即可痊愈，治愈率为 100%。对小儿鹅口疮、牙周炎、龋齿引发的牙痛以及放射治疗引起的放射损伤性舌炎及口腔溃疡也有良好的治疗作用。阮德荣(2005)将鲜蜂胶 50g，鲜细辛根 50g，65 度纯粮白酒 500mL 放入茶色瓶中密封，每天摇动一次，一周后可得蜂胶细辛酊。

(3) 蜂胶喷雾剂加工工艺

喷雾剂是用压缩空气或惰性气体作动力，用非金属喷雾器将药液喷出的剂型。

解放军第 302 医院制备了一种新型的抗感染性皮肤疾病的复方蜂胶喷雾剂。复方蜂胶喷雾剂的主要原料为精制蜂胶和甘草酸单胺盐，产品中黄酮含量为 3.26mg/mL。经解放军第 302 医院门诊部皮肤性病科对 15 例患有不同皮肤病(皮肤湿疹、干性湿疹、顽固性皮肤瘙痒症、过敏性皮肤瘙痒、股癣、轻二度皮肤烫伤等)进行随机初步疗效观察,总有效率为 95.6%,治愈率为 68.6%,其中对湿疹、过敏性皮肤瘙痒、股癣尤为有效(蔡光明,2006)。

2.3 化妆品型蜂胶产品加工工艺

根据蜂胶的生物学特性,蜂胶特别适合应用于香皂、香波等化妆品中。因为蜂胶具有抗菌、消炎、止痛、止痒、防腐、除臭、排除毒素、促进再生、增强 SOD 活性等多种作用,在化妆品中加入适量的蜂胶有利于增强它们保护皮肤、清除体表污垢、预防外在感染以及其他医疗保健功效。已经走向市场的蜂胶化妆品包括含蜂胶的洗发水、药皂、冷膏、发膏、护肤露、面霜、面膜、牙膏、沐浴露、漱口液、脱发剂等,显示出良好的发展前景(胡福良等,2000;胡福良等,2004)。

(1)蜂胶唇膏加工工艺

刘昊涅(2012)制备了一种蜂胶唇部去角质软膏,其成分包括乳木果精油、蜂胶蜜、维生素 B、甘草提取物、木糖醇、去离子水、丙三醇、卡波姆、十六烷基三甲基氯化铵、VPEG-40 氢化蓖麻油、葡糖糖苷;各组分的质量配比为乳木果精油 10~20 份,蜂胶蜜 10~20 份,维生素 B 11.5 份,甘草提取物 3~5 份,木糖醇 10~15 份,去离子水 30~50 份,丙三醇 4~6 份,卡波姆 3~5 份,十六烷基三甲基氯化铵 2~3 份,VPEG-40 氢化蓖麻油 2~5 份,葡糖糖苷 8~10 份。该产品能够保温和有效去除双唇因干燥、缺水造成的角质堆积,有效修复干燥,舒展唇纹,持久润泽,令双唇柔软滋润,饱满光滑,唇色鲜艳娇嫩。

(2)蜂胶护肤油加工工艺

蜂胶配制的护肤油是具有医疗作用的化妆品,有抗菌、消炎、止痒、止痛的功效,还可促进组织再生,对多种皮肤病有疗效。

参考配方(含量,%)

水貂油:2~4;蜂胶乙醇液:1~3;甘油三硬脂酸酯:2~4;二乙烯干醇硬脂酸酯:2~4;对-羟基苯甲酸甲酯:0.1~0.3;对-羟基苯甲酸丙酯:0.05~0.15;羊毛脂:2~4;甘油:1~3;甘菊油:1~2;麦胚油:1~3;香料:适量;三乙醇胺:0.7~0.9;水加至 100。

需要特别注意的是,蜂胶乙醇液含量以 1%~3%为宜,超过 3%,对皮肤有轻微的刺激;低于 1%,达不到杀菌效果。

加工方法

将脱臭的水貂油、羊毛脂、甘油三硬脂酸酯、二乙烯甘醇硬脂酸酯、对-羟基苯甲酸丙酯放入带有加热管和搅拌器的搪瓷反应器中,加热至 80~85℃,搅拌至完全熔化。按处方量,将水加入另一搪瓷反应器中,加热至 80~85℃,加入甘油、对-羟基苯甲酸甲酯、三乙醇胺,搅拌 30min;然后加入前面已熔化的油溶性材料,在 65~70℃温度下乳化 10~15min;之后降温至 50~60℃,边搅拌边加入麦胚油、甘菊油、蜂胶乙醇液;继续乳化 30min,降温至 30~40℃;加入香料,搅拌,冷却至 20~25℃。将制品移入包装车间,静置 24h,分装。

(3)蜂胶护肤霜加工工艺

参考配方

A:羊毛脂 0.5%,白蜂蜡 10.0%,18 号白油 27.0%;

B:硼砂 0.5%、去离子水 50.5%;

C:10%蜂胶油溶液 2.0%,香精、防腐剂适量。

制作方法

将油相 A 部分加热至 85℃,水相 B 部分加热至 90℃,灭菌 20min,冷却至 85℃。水相加入油相中进行乳化搅拌,冷却至 50℃时加蜂胶油溶液,40℃时加香精,35℃时停止搅拌,经分析合格后包装。

陈颖(2010)在温度 60℃,时间 2h,超声次数 1 次(20min),料液比 1∶1,乙醇浓度 95%的工艺条件下对蜂胶进行了提取。经过实验确定出蜂胶润肤霜的配方为:橄榄油 11.5g,硬脂酸 3.0g,司盘-80 4.5g,吐温-80 1.0g,蒸馏水 20mL,海藻酸钠 0.5g,薰衣草精油 600μL,蜂胶乳化液 1mL。并确定了润肤霜的制作工艺为:油相在 60℃下乳化 60min、水相在 90℃下乳化 30min,之后双相以 2090r/min 搅拌乳化 15min。最终研制出膏体细腻、均匀一致,有光泽感,香气舒适怡人,涂覆感好的蜂胶润肤霜。许应强(2014)将蜂胶、荷花粉、花生衣、水蛭、松仁和老蜂蜜配伍,制备蜂胶美容霜。制备方法:取水蛭浸泡于老蜂蜜中直至水蛭溶化得到含蛭蜂蜜,备用;取蜂胶于 0～5℃冷冻 20～30min 后,破碎为粗粒状,与荷花粉、花生衣及松仁一起粉碎成过 200～300 目筛的混合粉,备用;将混合粉与含蛭蜂蜜混合均匀,必要时加入常规量的防腐剂混合均匀,包装即得。制得的蜂胶美容霜有抗菌、美容养颜、促进细胞再生、改善脸部微循环及保护皮肤的作用。

(4)蜂胶防晒水加工工艺

参考配方

A:单对氨基苯甲酸甘油酯 2.0%、乙醇 65.0%、甘油 10.0%、丙二醇蓖麻醇酸酯 10.0%、5%蜂胶酊液 1.0%;

B:去离子水 12.0%;

C:香精和色素适量。

制作方法

充分溶解 A 成分,在搅拌下慢慢加入 B,最后加适量的香精和色素。

本防晒水敷在皮肤上,不会有油腻的感觉,且能形成保护膜,具有特殊的遮光防晒作用。

(5)蜂胶祛臭液(喷雾型)加工工艺

参考配方

A:10%蜂胶乙醇溶液 2.0%、水合氧化铝(丙二醇复体)10.0%、豆蔻酸异丙酯 2.0%、磷酸三油醇酯 3.0%、乙醇 87.0%、香精;

B:喷射液配比:A 料 35%,F11∶F12(22∶43)。

制作方法

A 料混合后过滤,取 A 料按 B 配比加入喷雾器中,装好喷嘴,然后压进氟利昂喷剂即成。

此喷剂具有除臭、抗菌和抑汗的作用。

(6)蜂胶调理香波加工工艺

参考配方

A:月桂醇硫酸酯三乙醇胺盐 40.0%、两性取代的咪唑啉 8.0%、羊毛醇聚氧乙烯醚 1.0%、月桂酰二乙醇胺 4.0%、羟丙基纤维素 1.0%、去离子水 45.0%;

B:5%蜂胶酊溶液 1.0%;

C:香精和色素适量。

制作方法

A 组加热至 60℃,缓慢搅拌,待降温至 50℃ 时加入 B,最后加适量香精和色素。

此香波除具有止痒、去头屑作用外,同时还能改善发质和梳理性,使头发具有光泽和柔软感。

(7) 蜂胶须后水加工工艺

参考配方

乙醇 50.0%、去离子水 43.9%、硼酸 2.0%、山梨醇 3.0%、薄荷脑 0.1%、10% 蜂胶乙醇溶液 1.0%、香料和色素适量。

制作方法

将山梨醇溶解于去离子水中;将色素以外的其他成分溶解于乙醇;把水相加到醇体系中混合,再加入色素后过滤。

此须后水能抑制剃须造成的刺激,使脸部产生清凉感,同时有较强的抑菌和收敛作用。

(8) 蜂胶香皂加工工艺

参考配方

无水钠皂 86%、蜂胶浸膏 1%、护肤剂 1%、增效剂 2%~2.5%、辅助添加剂 3.5%~4%、香精、色素、防腐剂适量。

制作方法

在无水钠皂中加入蜂胶浸膏以及各种添加剂,按香皂生产工艺加工。

王彬(2004)制备了一种保健蜂胶蜜皂,它是由下述组分按质量份数比组成:皂基:蜂胶:蜂蜜:地肤子:白鲜皮:全蝎:香精=(92~95):(0.5~2):(1~5):(0.15~0.6):(0.1~0.7):(0.05~0.2):(0.01~0.04)。该产品中含有的重要成分能有效抑制细菌的生长,对肌肤具有美容和保健功效;蜂胶具有广谱抗菌、消炎的作用,将各组分进行组合后,使得各组分功效产生协同作用,从而能够有效清洁肌肤,对肌肤具有保护作用;成本低,使用安全,即使有过敏性肤质的人也能使用,它采用中药成分,具有抗过敏的效果;加工制造工艺简单,流程短,各组分来源易得,具有广泛的市场前景。杨珍(2014)制作了一种蜂胶汉方美白皂。成分组成包括:橄榄油 15~25 份、棕榈油 15~25 份、椰子油 5~15 份、蜂胶 5~15 份、中药复方 5~15 份、纯净水 5~15 份、茶树精油 4~6 份、维生素 E 10~20 份、迷迭香精油 2~3 份、甜杏仁油 2~3 份、有机玫瑰纯露 2~3 份。其中,所述中药复方是由阿胶 19~21 份、龙眼肉 14~16 份、甘草 19~21 份、薏仁 19~21 份、黄芩 14~16 份、枸杞子 14~16 份、当归 19~21 份、木通 11~13 份、桃花 14~16 份、桑叶 14~16 份、荷叶 11~13 份、菊花 14~16 份、柳叶 19~21 份、槐树皮 9~11 份、香椿树 4~6 份组成。该产品能够促进脸部血液循环和新陈代谢,恢复肌肤紧实弹性,保湿滋养,美白淡斑,增强肌肤抵抗力。沈新荣(2011)将椰子油 10~15 份、牛油 20~30 份、乙醇 16~22 份、32% 的 NaOH 16~23 份、保湿剂 10~15 份、蜂胶浸膏 0.5~2 份、香精 0.5~2 份和水 4~8 份配方制成一种蜂胶香皂。蜂胶香皂具有抑菌抗菌、消炎、止痛、止痒、除臭等功效,对皮肤瘙痒、轻微皮肤炎症有一定的改善效果。

2.4 洁口型蜂胶产品加工工艺

研究发现,蜂胶对多种口腔细菌有抑制作用,因而蜂胶产品在口腔卫生和保健中的应用广泛,产品多样,主要产品有蜂胶牙膏、蜂胶洁口剂、蜂胶气雾剂、蜂胶喷剂和蜂胶漱口液等。

(1)蜂胶牙膏加工工艺

蜂胶有消炎、杀菌和镇痛等作用,对口腔疾病有较好的疗效。将蜂胶添加到牙膏里,可以防止牙病,有利牙齿保健,因而国内外部分生产企业开发了蜂胶牙膏。

参考配方(质量分数,%)

蜂胶5、无水磷酸氢钠20、碳酸钙20、十二醇硫酸钠2.5、羟甲纤维素钠5.0、香精1~2、甜味剂1.0、甘油40、去离子水余量。

工艺流程

首先在螺旋桨搅拌下,将羟甲纤维素钠、蜂胶分散于甘油中,搅拌10min,以达到足够的分散度,少数块在以后的加热中熔化;然后加入碳酸钙和甜味剂并搅拌5min,再加入水搅拌30min,在搅拌下将分散体于水浴中加热至60℃保持5min;然后加入去离子水搅拌,将热胶水移至一般拌和机中,加入磷酸氢钠搅拌5min,直至膏体细致光滑;再加入香料拌和1分钟,从配方中取出一部分水溶解十二醇硫酸钠,可以在水浴中加热帮助溶解,将十二醇硫酸钠溶液缓慢搅拌入膏体中,最后将膏体坯移入真空脱气机中,于86.65~101.3kPa环境中脱气20min,脱气时缓慢搅拌,防止过多气泡形成,然后将膏体灌入软管。

陈万金等(2013)公开了一种可食性天然蜂胶抑菌牙膏及其制备方法,原料的质量配比为:蜂胶提取物0.2%~0.8%、摩擦剂15%~20%、保湿剂50%~70%、黏合剂0.5%~1.0%、发泡剂2.0%~3.0%、甜味剂0.1%~0.3%、调节剂0.5%~3.5%、防腐剂0.1%~0.3%、香精1.0%~1.2%、去离子水余量。这种牙膏利用蜂胶天然的抗菌防腐性,无毒无刺激性,代替三氯生、甲醛、苯甲酸钠、尼泊金酯等应用于牙膏中,减少合成抗菌剂的使用,具有良好的抑菌效果,且可避免安全隐患。宋繁华(2013)将0.6%~1.0%蜂胶、0.4%~0.8%D-泛醇、30%~40%二氧化硅、8%~16%甘油、12%~16%山梨醇、1.6%~2.4%月桂酰肌氨酸钠、0.1%~0.3%羟乙基纤维素、0.4%~0.8%羟甲纤维素混合制得牙膏。制得的蜂胶保健牙膏对于牙齿具有较好的保护作用,而且还能够滋养牙龈,预防牙龈炎症的发生,在使用的过程中还会使口腔感觉舒适。洪滔(2005)在牙膏基质中配入鱼腥草浸膏、蜂胶粉,经加工制成鱼腥草蜂胶牙膏,它具有杀菌、消炎、固齿的作用,同时又具有抗氧化、抗自由基、改善循环、延缓衰老的作用。李俊等(2011b)制备了一种蜂胶牙膏,质量配比为:蜂胶粉0.03%~3.2%、天然碳酸钙25%~46%、二氧化硅5%~15%、甘油3%~20%、聚乙二醇-400 1%~10%、焦磷酸钠0.2%~1.0%、黄原胶0.2%~1.5%、羧甲基纤维素0.1%~1.5%、山梨醇3%~25%、糖精钠0.1%~1%、十二烷基硫酸钠1%~5%、羟苯乙酯钠0.05%~0.5%、尼泊金丙酯钠0.02%~2%、香精0.3%~3%、三氯生0.01%~0.3%、柠檬酸锌0.05%~2%和水余量。此蜂胶牙膏对于牙齿具有较好的保护作用,而且还能够滋养牙龈,预防牙龈炎症的发生。他们另将三七和蜂胶配伍,制备牙膏。原料的质量分数为:三七提取物0.%~1.5%、蜂胶粉0.03%~3.2%、天然碳酸钙25%~46%、二氧化硅5%~15%、甘油3%~20%、聚乙二醇-400 1%~10%、焦磷酸钠0.2%~1.0%、黄原胶0.2%~1.5%、羧甲基纤维素0.1%~1.5%、山梨醇3%~25%、糖精钠0.1%~1%、十二烷基硫酸钠1%~5%、羟苯乙酯钠0.05%~0.5%、尼泊金丙酯钠0.02%~2%、香精0.3%~3%、三氯生0.01%~0.3%、柠檬酸锌0.05%~2%和水余量。含有三七提取物的天然护理牙膏,对牙龈炎、牙龈出血、口腔溃疡等有显著疗效。

(2) 蜂胶漱口液加工工艺

曹月秀(2007)以配方(质量分数)山梨醇 8%～15%、聚氧乙烯山梨醇单月桂酸酯 0.5%～2%、甘油 10%～20%、柠檬酸 0.1%～0.3%、蜂胶乙醇溶液(10%质量浓度)1%～3%、香精 0.2%～0.4%、去离子水 70%～80%，制备蜂胶漱口液。制得的漱口液在蜂胶和乙醇的协同作用下，可杀灭口腔中的病菌，并能消除口臭和口腔的各种炎症；产品中加入柠檬酸、乙醇、香精等材料，含漱时，口感清爽，无任何异味。

(3) 蜂胶气雾剂加工工艺

蜂胶气雾剂是指将蜂胶与适宜的溶剂装于具有密闭容器中制成的澄明液体，使用时将内容物呈雾粒喷出的一种外用的消炎喷剂。适用于多种皮肤科疾病、口腔科各种黏膜和齿龈疾患、口腔和咽部真菌损害以及促进拔牙后创面愈合等。使用方法：视炎症程度每日喷雾 1～3 次。瓶装的蜂胶气雾剂是深黄色的透明液体，有浓厚的蜂胶香味，为无菌的稳定剂型。蜂胶气雾剂的配方如下。

参考配方

蜂胶 3.6g，乙醇 48g，甘油 8.4g，氟氯烷 40g，香料适量。

制作方法

按配方将蜂胶在特定的条件下与甘油、氟氯烷等混合，搅拌均匀。用以上原料可制成蜂胶气雾剂 100g，在温度为 30℃、湿度为 70% 的条件下装于密闭的喷雾器中。

蜂胶喷剂在生产和贮存期间应符合下列有关规定：蜂胶喷剂应在清洁、避菌环境下配制；可按需要加入适量增稠剂等附加剂，皮肤和黏膜用蜂胶喷剂应无不良刺激性；蜂胶喷剂的容器不应与内容物发生理化反应；气雾剂阀门调节系统中的弹簧、阀杆、定量杯和橡胶圈等组成部件均不应与药液发生理化作用，其尺寸精度和溶胀性必须符合要求，吸入气雾剂所用定量阀门每次喷射应能释放出均匀的液体；气雾剂需用适宜方法进行漏气和爆破检查，确保安全使用；气雾剂应放置于阴暗处保存，并避免暴晒、受热、敲打和撞击。

(4) 蜂胶口腔溃疡膜

参考配方

蜂胶 5g，维生素 A 4 万国际单位，羧甲基纤维素 20g，达克洛宁 0.5g，冰片 0.5g，95% 乙醇适量，甜味剂适量。

制作方法

将蜂胶、冰片溶于乙醇中，然后加入羧甲基纤维素，将维生素 A、达克洛宁液搅匀，使成糊剂，倒在涂有液状石蜡油的玻板上，做成 4cm×25cm 厚度均匀一致的薄膜，干燥后切成适宜的大小和形状备用。

2.5 其他类型蜂胶产品加工工艺

随着蜂胶研究的进一步深入、蜂胶市场的拓展，蜂胶产品的类型越来越多，除上述几类外，目前研究开发的蜂胶产品还有蜂胶口香糖、蜂胶酒、蜂胶蜜、蜂胶饮料，等等。

(1) 蜂胶口香糖加工工艺

蜂胶口香糖味香可口，无苦涩味，可保护牙齿、清洁口腔、清除口臭以及防治口腔疾病和牙周病。健康人也可食用。

参考配方

蜂胶浸膏 20g，糖粉 500g，活性小麦谷朊(面筋纯度 70%)100g，蒸馏水 400～800mL，香精适量。

制作方法

将上述原料混合,搅匀,压成薄片,切成一定大小的长方块,每块重 2~5g,烘干并保持水分在不挥发掉的条件下加热处理 1h 即成。冷却后用纸包装。

付中民(2009)在工艺参数为乙醇浓度 80%、提取温度 60℃、提取时间 3h、超声波处理时间 20min 的条件下,对蜂胶进行提取。其次,为了解决蜂胶本身口感差、易黏附等应用瓶颈,对提取后的蜂胶以 β-环状糊精为壁材进行包埋,包埋后的产物口感有了很大的改善,且不会黏附。最后将包埋后的蜂胶以不同比例添加到口香糖配料中制成口香糖,并在口感及抑菌能力等方面对该口香糖进行测试分析,确定出蜂胶口香糖配方中蜂胶包合物的添加量以 3.00% 为宜。该口香糖抑制变形链球菌的能力好,且其口感有较大程度的改善。

(2)蜂胶蜜加工工艺

蜂胶蜜是指蜂胶经特殊处理后混入蜂蜜的一种新型产品,在国内外均有生产,在很大程度上改变了蜂胶的感观质量,而且服用方便,深受老年人和妇女的喜爱。

工艺流程

蜂胶蜜的加工工艺流程如下:

纯蜂胶→加入熔化→乳化
↓
蜂蜜→粗过滤→精过滤→浓缩→混合→搅拌→罐装→封口→巴氏杀菌→分段冷却→贴标签→成品检验→入库

操作要点

①蜂胶蜜所用原料:主要以深色蜜为主,如枣花蜜、荞麦蜜和荆条蜜等,在加工蜂胶蜜前应该对蜂蜜进行除杂、破晶和浓缩处理。

②蜂胶的处理:加工蜂胶蜜的最关键技术是使蜂胶能够很好地溶入蜂蜜中而且长时间不分层。因此蜂胶的表面活性剂的选择就显得特别重要,可以利用环糊精和聚乙二醇将蜂胶进行处理。

③搅拌:为了使蜂胶乳状液与蜂蜜很好地融为一体,必须将蜂蜜加热到 60℃,边搅拌边加入蜂胶。

④罐装:采用棕色玻璃瓶包装,玻璃瓶在使用前应充分清洗和消毒。

⑤巴氏消毒:将封好口的蜂胶蜜置入常压杀菌锅中,在 100℃ 下杀菌 15min。

⑥分段冷却:玻璃瓶装蜂胶蜜在刚杀完菌后应及时分段冷却,冷却温度分别为 70℃、50℃ 和 25℃。

⑦成品检验入库。

梁霭(2000)公开了一种对皮肤及过敏性、感冒、癌症、外伤等疾病具有显著疗效的蜂胶蜜配方,它是由纯度 95% 以上、重金属含量不大于 5μg/g 的蜂胶膏 0.7%~0.8%,酒精度数 60°~62° 的高粱酒 3%~10%,纯正蜂蜜 89.2%~96.3% 组成。其制备方法为,先在蜂胶膏上加入一定量的食用酒精,浸 3~7d,使蜂胶膏完全溶解,再加入高粱酒后,加热炖 30min,使酒精挥发去除后,加入纯正蜂蜜,搅拌均匀,制得蜂胶蜜。

(3)蜂胶酒加工工艺

由于蜂胶中基本不含糖类,而且具有很强的杀菌效果,因而蜂胶多在酿制后期加入,目前所开发的蜂胶酒多是由一定比例的蜂胶和基酒配合而成的(季文静等,2007)。

李士光的"一种蜂胶酒及其配制生产工艺"将蜂胶研磨后与基酒混合,经陈化处理后精制成蜂胶酒。张亚州等的"蜂胶酒"以蜂胶与高浓度纯粮白酒或食用酒精配合而成。Kim Won Jin 的"含蜂胶提取物的白兰地酒的制备方法"以葡萄为原料发酵蒸馏,并加入醇提蜂胶,制成蜂胶酒。而他的"蜂胶果酒的制备"则是将橘子、柠檬、杧果、苹果等多种水果粉碎后溶于醇提蜂胶中发酵,经离心后与水提蜂胶混合,再通过离心、过滤、陈酿获得蜂胶果酒。于帮春制备了一种蜂胶酒。方法为:选取不透明、其颜色呈黄褐、棕褐、灰褐、暗绿、黑色固体蜂胶,放入冷冻室在零下 30℃进行冷冻 8h,使其变硬、变脆,将其碾碎,达到 5mm 以下的颗粒,放入到有 70%～75%配置好食用酒精溶液的罐体中,将罐体置于避光、18～25℃条件处,每日搅动 5～7 次,使蜂胶中的物质溶解,1 周后停止搅动,使之自然沉降,2 周后从沉降液出口放出沉降液,进入到沉降罐罐体中,将沉降罐罐体置入 0～30℃的环境中自然沉降 10h 后,从蜂胶液成品出口放出蜂胶酒。许应强(2013)进行了蜂胶酒的制备,其中含有蜂胶、蜂蜜以及白酒,白酒的度数在 30°以上。蜂胶与酒的质量比为 0.35∶1～0.35∶15,蜂蜜与酒的质量比为 0.3∶1～0.3∶15。齐永增和程国珍(2007)制备了一种蜂胶配制酒。配制酒的配方及比例为:0.1%～1%天然蜂胶,0.01%～0.05%天然调味料提取的乙醇流浸膏,3%～5%蒸馏酒,食用乙醇和水余量调节配制酒的酒精度为 25%～50%。蜂胶配制酒由于采用葛根与山楂、丁香等调味料的复方组合,并且采用制药技术中的流浸膏提取方式,有效改善了蜂胶配制酒的辛辣口感和风味,使其绵软醇和,提高了酒的档次。尤其是利用了葛根等调味料中的药用成分,还可以突出葛根黄酮、葛根素等多种成分对心脑血管的保健作用,可缓解饮酒症状。

(4)蜂胶饮料加工工艺

陈立阁(2004)研究了以蜂胶、蜂蜜为原料开发营养饮料的生产工艺。通过正交实验、均匀实验和感观评定的方法确定,该营养饮料配方为:蜂胶 0.35%,蜂蜜 32%,柠檬酸 0.9%。最适合的杀菌工艺为:温度为 65℃,时间为 15min。制得的蜂胶营养饮料产品呈鲜亮的柠檬黄色,澄清,不但具有良好的感官性状,而且其理化指标、微生物指标均符合国家软饮料质量标准,是一种老少皆宜的营养型保健饮料。

参考文献

Aguiar SC, Cottica SM, Boeing JS, Samensari RB, Santos GT, Visentainer JV, Zeoula LM (2014a) Effect of feeding phenolic compounds from propolis extracts to dairy cows on milk production, milk fatty acid composition, and the antioxidant capacity of milk[J]. *Animal Feed Science and Technology*, 193:148-154.

Aguiar SCD, Paula EMD, Yoshimura EH, Santos WBRD, Machado E, Valero MV, Zeoula LM (2014) Effects of phenolic compounds in propolis on digestive and ruminal parameters in dairy cows[J]. *Revista Brasileira De Zootecnia-Brazilian Journal of Animal Science*, 43:197-206.

Attia YA, Al-Hamid AA, Ibrahim MS, Al-Harthi MA, Bovera F, Elnaggar AS (2014) Productive performance, biochemical and hematological traits of broiler chickens supplemented with propolis, bee pollen, and mannan oligosaccharides continuously or intermittently[J]. *Livestock Science*,164:87-95.

Atungulu G,Miura M,Atungulu E,Atungulu,E. ,Satou Y,Suzuki K (2007) Activity of gaseous phase steam distilled propolis extracts on peroxidation and hydrolysis of rice lipids[J]. *Journal of Food Engineering*,80(3):850-858.

Aygun A,Sert D,Copur G (2012) Effects of propolis on eggshell microbial activity, hatchability,and chick performance in Japanese quail (*Coturnix coturnix japonica*) eggs[J]. *Poultry Science*,91(4):1018-1025.

Bae JY,Park GH,Lee JY,Okorie OE,Bai SC (2012) Effects of dietary propolis supplementation on growth performance, immune responses, disease resistance and body composition of juvenile eel,anguilla japonica[J]. *Aquaculture International Journal of Pharmaceutics*,20:513-523.

Bankova V (2005) Chemical diversity of propolis and the problem of standardization[J]. *Journal of Ethnopharmacology*,100:114-117.

Bankova V,Christov R,Popov S,Marcucci MC,Tsvetkova I,Kujumgiev A (1999) Antibacterial activity of essential oils from brazilian propolis[J]. *Fitoterapia*,70:190-193.

Bankova V,Popova M,Trusheva B (2014) Propolis volatile compounds:Chemical diversity and biological activity:A review[J]. *Chemistry Central Journal*,8:28.

Biscaia D,Ferreira SRS (2009) Propolis extracts obtained by low pressure methods and supercritical fluid extraction[J]. *The Journal of Supercritical Fluids*,51:17-23.

Bruschi ML,Cardoso MLC,Lucchesi MB,Gremião MPD (2003) Gelatin microparticles containing propolis obtained by spray-drying technique:Preparation and characterization[J]. *International Journal of Pharmaceutics*,264:45-55.

Burdock GA (1998) Review of the biological properties and toxicity of bee propolis (propolis)[J]. *Food Chemistry and Toxicology*,36:347-363.

Catchpole OJ,Grey JB,Mitchell KA (2004) Supercritical antisolvent fractionation of propolis tincture[J]. *Journal of Supercritical Fluids*,29:97-106.

Catchpole OJ,Grey JB,Mitchell KA,Lan JS (2004) Supercritical antisolvent fractionation of propolis tincture[J]. *The Journal of Supercritical Fluids*,29:97-106.

Cetin E,Silici S,Cetin N,Guclu BK (2010) Effect s of diets containing different concent rations of propolis on hematological and immunological variables in laying hens[J]. *Poultry Science*,89:1703.

Chen CR,Lee YN,Lee MR,Chang CMJ (2009a) Supercritical fluids extraction of cinnamic acid derivatives from brazilian propolis and the effect on growth inhibition of colon cancer cells[J]. *Journal of the Taiwan Institute of Chemical Engineers*,40:130-135.

Chen CR,Shen CT,Wu JJ,Yang HL,Hsu SL,Chang CMJ (2009b) Precipitation of sub-micron particles of 3,5-diprenyl-4-hydroxycinnamic acid in brazilian propolis from supercritical carbon dioxide anti-solvent solutions[J]. *The Journal of Supercritical Fluids*,50:176-182.

CoutinhoA (2012) Honeybee propolis extract in periodontal treatment: A clinical and microbiological study of propolis in periodontal treatment[J]. *Indian Journal of Dental Research*, 23:294.

Czyzewska U, Kononczuk J, Teul J, Dragowski P, Pawlak-Morka R, Surazynski A, Miltyk W (2015) Verification of chemical composition of commercially available propolis extracts by gas chromatography-mass spectrometry analysis[J]. *Journal of Medicinal Food*, 18:584-591.

De Groot AC (2013) Propolis: A review of properties, applications, chemical composition, contact allergy, and other adverse effects[J]. *Dermatitis*, 24:263-282.

De Zordi N, Cortesi A, Kikic I, Moneghini M, Solinas D, Innocenti G (2014) The supercritical carbon dioxide extraction of polyphenols from propolis: A central composite design approach [J]. *The Journal of Supercritical Fluids*, 95:491-498.

Deng J, An Q, Bi B, Wang Q, Kong L, Tao L, Zhang X (2011) Effect of ethanolic extract of propolis on growth performance and plasma biochemical parameters of rainbow trout (*Oncorhynchus mykiss*)[J]. *Fish Physiology Biochemistry*, 37(4):959-967.

Dodwad V, Kukreja BJ (2011) Propolis mouthwash: A new beginning[J]. *Indian Soc Periodontol*, 15:121-125.

DuarteCRA, Eyng C, Murakami AE, Santos TC (2014) Intestinal morphology and activity of digestive enzymes in broilers fed crude propolis[J]. *Canadian Journal of Animal Science*, 94(1):105-114.

El Sayed H, Ahmad TA. (2012) The use of propolis as vaccine's adjuvant[J]. *Vaccine*, 31:31-39.

El Sayed H, El Ashry, Tarek A, Ahmad (2012) The use of propolis as vaccine's adjuvant [J]. *Vaccine*, 31:31-39.

Eyng C, Murakami AE, Duarte CRA, Santos TC (2014) Effect of dietary supplementation with an ethanolic extract of propolis on broiler intestinal morphology and digestive enzyme activity[J]. *Journal of Animal Physiology and Animal Nutrition*, 98(2):393-401.

Fan Y, Hu Y, Wang D, Guo Z, Zhao X, Guo L, Nguyen TL (2010) Epimedium polysaccharide and propolis flavone can synergist ically stimulate lymphocyte proliferat ion in vitro and enhance the immune responses to ND vaccine in chickens[J]. *International Journal of Biological Macromolecules*, 47(2):87-92.

Fan Y, Wang D, Liu J, Hu Y, Zhao X, Han G, Chang S (2012) Adjuvanticity of epimedium polysaccharide-propolis flavone on inactivated vaccines against AI and ND virus[J]. *International Journal of Biological Macromolecules*, 51(5):1028-1032.

Fonseca YM, Marquele-Oliveira F, Vicentini FT, Furtado NAJ, Sousa JPB, Lucisano-Valim YM, Fonseca MJV (2011) Evaluation of the potential of brazilian propolis against UV-induced oxidative stress[J]. *Evidence-Based Complementary and Alternative Medicine*, 2011:863917.

Freitas JA,Vanat N,Pinheiro JW,Balarin MR,Sforcin JM,Venancio EJ (2011) The effects of propolis on antibody production by laying hens[J]. *Poultry Science*,90:1227-1233.

Gonçalves GMS,Srebernich SM,Souza JADM (2011) Stability and sensory assessment of emulsions containing propolis extract and/or tocopheryl acetate[J]. *Brazilian Journal of Pharmaceutical Sciences*,47(3):585-592.

Gregory AG,Toshitaka U,Fumihiko T (2008) Effect of vapors from fractionated samples of propolis on microbialand oxidation damage of rice during storage[J]. *Journal of Food Engineering*,88:341-352.

Griffiths A,Makoto M,Elizabeth A,Yoshinori S,Koichi S (2007) Activity of gaseous phase steam distilled propolis extracts on peroxidation and hydrolysis of rice lipids[J]. *Journal of Food Engineering*,80(3):850-858.

Hashem NM,El-Hady AA,Hassan O (2013) Effect of vitamine or propolis supplementation on semen quality,oxidative status and hemato-biochemical changes of rabbit bucks during hot season[J]. *Livestock Science*,157(2):520-526.

Hu F,Hepburn HR,Li Y,Chen M,Radloff SE,Daya S (2005) Effects of ethanol and water extracts of propolis (bee glue) on acute inflammatory animal models[J]. *Journal of Ethnopharmacology*,100:276-283.

Huang S,Zhang CP,Wang K,Li GQ,Hu FL (2014) Recent advances in the chemical composition of propolis[J]. *Molecules*,19:19610-19632.

Hwu YL,Lin FY (2014) Effectiveness of propolis on oral health: A meta-analysis[J]. *Journal of Nursing Research*,22:221-230.

Kelestemur GT,Seven I (2012) Effect of propolis on blood electrolyte levels of juvenile rainbow trout (oncorhynchus mykiss w. 1792) under hypoxic stress[J]. *Indian Journal of Animal Research*,46(3):231-235.

Kujumgiev A,Tsvetkova I,Serkedjieva Y,Bankova V,Christov R,Popov S (1999) Antibacterial,antifungal and antiviral activity of propolis of different geographic origin [J]. *Journal of Ethnopharmacology*,64:235-240.

Kusumoto T,Miyamoto TR,Doi S,Sugimoto H,Yamada H (2001) Isolation and structures of two new compounds from the essential oil of Brazilian propolis[J]. *Chemical & Pharmaceutical Bulletin*,49:1207-1209.

Lee YN,Chen CR,Yang HL,Lin CC,Chang CMJ (2007) Isolation and purification of 3,5-diprenyl-4-hydroxycinnamic acid (Artepillin c) in Brazilian propolis by supercritical fluid extractions[J]. *Separation and Purification Technology*,54:130-138.

Li YJ,Xuan HZ,Shou QY,Zhan ZG,Lu X,Hu FL (2012) Therapeutic effects of propolis essential oil on anxiety of restraint-stressed mice[J]. *Human & Experimental Toxicology*,31:157-165.

Machado BAS,Silva RPD,Barreto GD,Costa SS,da Silva DF,Brandao HN,da Rocha JLC,Dellagostin OA,Henriques JAP,Umsza-Guez MA (2016) Chemical composition and biological activity of extracts obtained by supercritical extraction and ethanolic extraction of brown, green and red propolis derived from different geographic regions in Brazil[J]. *PLoS One*, 11(1): e0145954.

Madhavan S, Nayak M, Shenoy A, Shetty R, Prasad K (2012) Dentinal hypersensitivity: A comparative clinical evaluation of cpp-acpf, sodium fluoride, propolis, and placebo[J]. *Journal of Conservative Dentistry*, 15: 315-318.

Mascheroni E, Guillard V, Nalin F, Mora L, Piergiovanni L (2010) Diffusivity of propolis compounds in polylactic acid polymer for the development of anti-microbial packaging films[J]. *Journal of Food Engineering*, 98(3): 294-301.

Melliou E, Stratis E, Chinou I (2007) Volatile constituents of propolis from various regions of greece-antimicrobial activity[J]. *Food Chemistry*, 103: 375-380.

MorsyAS, Abdalla AL, Soltan YA, Sallam SM, El-Azrak KEDM, Louvandini H, Alencar SM (2013). Effect of Brazilian red propolis administration on hematological, biochemical variables and parasitic response of Santa Inês ewes during and after flushing period[J]. *Tropical Animal Health and Production*, 45(7), 1609-1618.

Nagai T, Inoue R, Inoue H, Suzuki N (2003) Preparation and antioxidant properties of water extract of propolis[J]. *Food Chemistry*, 80(1): 29-33.

Orun I, Dogru MI, Erdogan K, Dogru A, Ongun A, Yuksel, E, Talas ZS (2013) Effects of acute and chronic exposure to glyphosate on common carp (*Cyprinus carpio* L.) hematological parameters: The beneficial effect of propolis[J]. *Fresenius Environmental Bulletin*, 22: 2504-2509.

Ozdemir AE, Candir EE, Kaplankiran M, Soylu EM, Şahinler N, Gül A (2010). The effects of ethanol-dissolved propolis on the storage of grapefruit cv. Star Ruby[J]. *Turkish Journal of Agriculture and Forestry*, 34(2), 155-162.

Ozkok D, Iscan K, Silici S (2013) Effects of dietary propolis supplementation on performance and egg quality in laying hens[J]. *Journal of Animal and Veterinary Advances*, 12(2): 269-275.

Pastor C, Sanchez-Gonzalez L, Chafer M, Chiralt A, Gonzalez-Martinez C (2010) Physical and antifungal properties of hydroxypropylmet hylcellulose based films containing propolis as affected by moisture content[J]. *Carbohydrate Polymers*, 82: 1174-1183.

Pereira EMR, da Silva JLDC, SilvaFF, De Luca MP, Lorentz TCM, Santos VR (2011) Clinical evidence of the efficacy of a mouthwash containing propolis for the control of plaque and gingivitis: A phase II study[J]. *Evidence-based Complementary and Alternative Medicine*, 2011: 750249.

Saavedra N, Barrientos L, Herrera CL, Alvear M, Montenegro G, Salazar L(2011) Effect of chilean propolis on cariogenic bacteria lactobacillus fermentum[J]. *Cienciae Investigación Agraria*, 38(1): 117-125.

Sales-Peres SHDC, Carvalho FND, Marsicano JA, Mattos MC, Pereira JC, Forim MR, Silva MFGF (2011) Effect of propolis gel on the *in vitro* reduction of dentin permeability [J]. *Journal of Applied Oral Science*, 19(4): 318-323.

Segvic-Bubic T, Boban J, Grubisic L, Trumbic Z, Radman M, Percic M, Coz-Rakovac R (2013) Effects of propolis enriched diet on growth performance and plasma biochemical parameters of juvenile European sea bass (*Dicentrarchus labrax* L.) under acute low-temper ature stress[J]. *Aquaculture Nutrition*, 19(6): 877-885.

Seven I, Aksu T, Seven PT (2012) The effects of propolis and vitamin C supplemented feed on performance, nutrient utilization and carcass characteristicsin broilers exposed to lead[J]. *Livestock Science*, 148(1): 10-15.

Silva BB, Rosalen PL, Cury JA, Ikegaki M, Souza VC, Esteves A, Alencar SM (2008) Chemical composition and botanical origin of red propolis, a new type of Brazilian propolis[J]. *Evidence-Based Complementary and Alternative Medicine*, 5(3): 313-316.

Skaba D, Morawiec T, Tanasiewicz M, Mertas A, Bobela E, Szliszka E, Krol W (2013) Influence of the toothpaste with brazilian ethanol extract propolis on the oral cavity health[J]. *Evidence-Based Complementary and Alternative Medicine*, 2011: 750249.

Takara K, Fujita M, Matsubara M, Minegaki T, Kitada N, Ohnishi N, Yokoyama T (2007) Effects of propolis extract on sensitivity to chemotherapeutic agents in hela and resistant sublines[J]. *Phytother Res*, 21(9): 841-846.

Tanasiewicz M, Skucha-Nowak M, Dawiec M, Krol W, Skaba D, Twardawa H (2012) Influence of hygienic preparations with a 3% content of ethanol extract of brazilian propolis on the state of the oral cavity[J]. *Advances in Clinical and Experimental Medicine*, 21(1): 81-92.

Toreti VC, Sato HH, Pastore GM, Park YK (2013) Recent progress of propolis for its biological and chemical compositions and its botanical origin[J]. *Evidence-Based Complementary and Alternative Medicine*, 2013: 697390.

Troca VBPB, Fernandes KBP, Terrile AE, Marcucci MC, Andrade FBD, Wang L (2011) Effect of green propolis addition to physical mechanical properties of glass ionomer cements[J]. *Journal of Applied Oral Science*, 19(2): 100-105.

Venero AVB, García LMD, González LA (2012) Treatment of dental alveolitis with 5% propolis tincture[J]. *Revista Cubana de Farmacia*, 46: 97-104.

Wang BJ, Lien YH, Yu ZR (2004) Supercritical fluid extractive fractionation-study of the antioxidant activities of propolis[J]. *Food Chemistry*, 86(2): 237-243.

Wu JJ, Shen CT, Jong TT, Young CC, Yang HL, Hsu SL, Shieh CJ (2009) Supercritical carbon dioxide anti-solvent process for purification of micronized propolis particulates and associated anti-cancer activity[J]. *Separation and Purification Technology*, 70(2): 190-198.

Yang S, Peng L, Cheng Y, Chen F, Pan S (2010) Control of citrus green and blue molds by Chinese propolis[J]. *Food Science and Biotechnology*, 19(5): 1303-1308.

Yonar ME, Yonar SM, Ural MS, Silici S, Dușukcan M (2012) Protective role of propolis in chlorpyrifos-induced changes in the haematological parameters and the oxidative/

antioxidative status of *Cyprinus carpio carpio*[J]. *Food and Chemical Toxicology*, 50(8):2703-2708.

Zhang SQ, Xi J, Wang CZ (2005) High hydrostatic pressure extraction of flavonoids from propolis[J]. *Journal of Chemical Technology and Biotechnology*, 80:50-54.

阿布力克木·吾甫尔,依米提·热合曼,吐尔逊娜依·阿布都热依木,阿尔孜古丽·吐尔逊,木塔力甫·艾买提(2011)新疆蜂胶挥发油对结直肠癌 hct-116 细胞增生、周期及凋亡的影响[J].世界华人消化杂志,19(14):1469-1475.

蔡光明(2006)复方蜂胶喷雾剂中总黄酮的含量测定[J].解放军药学学报,22(1):60-62.

蔡烈涛,吴俊(2011)正交试验优选蜂胶软胶囊的最佳压丸工艺[J].中国中医药信息杂志,18(8):49-50.

曹月秀(2007)蜂胶漱口液[P].CN200710032149.8.

曾林晖,邓泽元,余修亮,李红艳(2016)蜂胶黄酮的超声波提取工艺优化及其抗氧化活性研究[J].食品工业科技,37(12):295-300.

曾晞,卢玉振,牟兰,张长庚(2004)GC-MS 法分析比较贵州不同产地蜂胶挥发油化学成分[J].生命科学仪器,(2):28-29.

曾志将,樊兆斌,谢国秀,颜伟玉(2006a)蜂胶 CO_2 超临界萃取研究[J].江西农业大学学报,28(5):769-771.

曾志将,杨明,杨新跃,周银平,刘志勇(2006b)CO_2 超临界和乙醇提取蜂胶对大鼠降血脂效果[J].江西农业大学学报,28(3):433-435.

柴家前,庞昕,刘金华(2010)鸡新城疫纳米蜂胶灭活疫苗免疫作用机制[J].中国兽医学报,30(9):1151-1155.

陈滨(2010)中国不同地区蜂胶水提物化学组成及生物活性[D].南昌:南昌大学.

陈崇羔,邱忠平,沈利平.1999.蜂胶水溶工艺的研究(一)——蜂胶增溶剂的筛选.中国养蜂,50(5):3-5.

陈坤,林德祥,邵飞(2008)蜂胶咀嚼片及生产工艺[P].CN200810203254.8.

陈立阁(2004)蜂胶营养饮料的研制[J].吉林工程技术师范学院学报(工程技术版),20(3):12-15.

陈万金,陈良熊,王毕璟,吴世杰(2013)一种可食性天然蜂胶抑菌牙膏及其制备方法[P].Vol. CN2013103199173.

陈小利,任小林,吕燕荣,徐义杰,魏敏(2011a)1-MCP 和蜂胶对冷藏苹果品质的影响[J].西北农林科技大学学报(自然科学版),39(5):126-132.

陈小利,任小林,蒲飞,吕燕荣,徐义杰(2011b)蜂胶涂膜对红富士苹果贮藏品质和生理活性的影响[J].食品与发酵工业,37(7):230-234.

陈璇,任泽明,杨锋,戴关海,童晔玲,王波波(2013)蜂胶鱼油软胶囊对小鼠免疫功能的影响研究.中国现代医生,51(14):7-9.

陈英华(2011)蜂胶综合膏药[P].CN201110256199.0.

陈颖(2010)蜂胶抗氧化润肤霜的研制[D].福州:福建农林大学.

程伟贤,陈鸿雁,张义平,谭理想,高树鹏,古昆(2006)蜂胶中黄酮成分及提取方法研究[J].云南大学学报(自然科学版),28(S1):25-27.

程瑛,毛玉花,刘彩云,逯彦国(2013)蜂胶花粉复方按摩膏的制备方法[P].CN201310396549.2.

迟家平,张锋,姜莉(2002)国外蜂胶研究回顾[C].中国药学会学术年会论文集,67-70.

崔艺兰,刘泓(2013)不同脱敏剂与蜂胶对牙本质小管封闭性的实验研究[J].中国美容医学, 22(1):62-64.

戴关海,童晔玲,陆拯,杨锋(2008)桑胶颗粒对 gk 糖尿病模型大鼠降血糖作用的实验研究 [C].第九届亚洲养蜂大会论文摘要集,229.

戴关海,童晔玲,杨锋,竹剑平,邱汝民(2011)一种蜂胶中药组合物[P].CN201110075192.9.

邓希尧(2002)蜂胶口服液治肾囊肿有效[J].中国养蜂,53(4):28.

刁春英,闫洪波,刘月英(2013a)蜂胶提取物对雪花梨的保鲜效果[J].北方园艺,(18): 136-138.

刁春英,高秀瑞,张玲(2013b)蜂胶提取物对草莓室温保鲜效果的研究[J].食品科技,38(1): 248-252,256.

樊兆斌,李立山,王春强(2013)蜂胶对断奶仔猪抗病力、血液生化指标及肠道结构的影响 [J].饲料研究,(1):54-56.

付英娟(2007)蜂胶中有效成分的提取及应用研究[D].杨陵:西北农林科技大学.

付宇新,徐元君,陈滨,丽艳,罗丽萍(2009)气相色谱/质谱法分析内蒙古蜂胶挥发性成分 [J].分析化学,37(5):745-748.

付中民(2009)蜂胶口香糖的研制[D].福州:福建农林大学.

葛飞,桂琳,陶玉贵,张慧敏,龚倩,石贝杰(2014)一种高雄山虫草菌丝体蜂胶口服液的配方 及其制备方法[P].CN2014100206358.

谷玉洪,罗濛,徐飞,赵余庆(2006)超临界 CO_2 提取蜂胶中总黄酮的工艺研究[J].中草药,37 (3):380-382.

郭华,叶暾昊,李次力(2001)微波对蜂胶黄酮提取率影响的研究[J].江苏食品与发酵,(2): 10-12.

郭伽,周立东(2000)北京蜂胶挥发油的化学成分研究[J].中国蜂业,51(1):9.

国家食品药品监督管理总局(2015)数据查询.Available:http://app1.sfda.gov.cn/datasearch/face3/dir.html.

韩玉谦,隋晓,冯晓梅,管华诗(2003)超临界 CO_2 萃取蜂胶有效成分的研究[J].精细化工,20 (7):422-424.

郝胤博,吴学志,罗丽萍,郭夏丽,张茜,戴喜末,陈滨(2012)中国不同地区蜂胶水提物的抑菌 活性[J].食品工业科技,33(10):101-104.

何新益,王磊,符绳慧(2008)乳化法制备水溶性蜂胶的研究[J].食品科技,33(9):86-88.

何新益,符绳慧,王磊(2009)水溶性蜂胶的酶解制备工艺优化[J].农业工程学报,25(2): 280-284.

贺丽苹,李雅萍,陈玉芬(2008)SPME-GC/MS 联用技术分析蜂胶的挥发性成分[J].中国蜂 业,59(4):36-37.

洪滔(2005)鱼腥草蜂胶牙膏[P].CN200510052078.9.

侯加强(1993)如何提高蜂胶酊的质量[J].养蜂科技,(6):61.

胡福良(2000)蜂胶的提取方法[J].养蜂科技,(5):28-29,40.

胡福良,李英华,朱威(2000)国外系列蜂胶化工产品及其专利[J].蜜蜂杂志,(12):13-14.

胡福良,李英华,陈民利,应华忠,朱威(2003)蜂胶醇提液和水提液对急性炎症动物模型的作用[J].浙江大学学报(农业与生命科学版),29(4):444-448.

胡福良,朱威,李英华(2004)蜂胶香皂的加工技术[J].养蜂科技,(2):39-40.

胡福良,李英华,朱威(2005a)亚微米蜂胶微胶囊的制备方法[P].CN200510049746.2.

胡福良,李英华,朱威,陈民利,应华忠(2005b)不同方法提取的蜂胶液中总黄酮含量的测定及抗肿瘤与抗炎作用研究[J].中国食品学报,5(3):11-15.

胡福良,张翠平(2011)一种蜂胶微胶囊的制备方法及应用[P].CN201110339839.4.

黄帅,卢媛媛,张翠平,胡福良(2013)巴西绿蜂胶乙醇提取前后挥发性成分的分析比较[J].食品与生物技术学报,32(7):680-685.

吉挺,李文艳,吴宏安,陈国宏(2008)不同提取方法下蜂胶抗氧化作用的比较[J].中国蜂业,59(10):33-34.

吉挺,李文艳,岑宁,吴宏安(2010)醇提超临界蜂胶剩余物的性质及其微胶囊粉末的制备[J].食品科学,31(20):256-259.

季文静,郑火青,胡福良(2007)蜂产品酒的开发现状与理念[C].2007年全国蜂产品市场信息交流会暨中国(桂林)蜂业博览会论文集,91-97.

姜德勇(2001)蜂胶抗病毒抗氧化活性提取物及其分子包合物制剂[P].CN01141763.3.

蒋云云,邵兴军,毛日文(2011)一种外用蜂胶酊剂及其制备方法[P].CN201110329230.9.

李俊,罗珊珊,苏莎,杨立民(2011a)一种含有三七和蜂胶的牙膏及其制备方法[P].CN201110241479.4.

李俊,罗珊珊,苏莎,杨立民(2011b)一种蜂胶牙膏及其制备方法[P].CN201110241483.0.

李维莉,马银海,彭永芳(2003)蜂胶黄酮提取方法研究[J].食品科学,24(5):100-101.

李文艳,吉挺(2010a)蜂胶和紫苏籽油混合微胶囊的制备工艺[J].食品工业科技,31(9):242-245.

李文艳,吉挺(2010b)冷冻干燥法制备蜂胶紫苏微胶囊粉末[J].江苏农业科学,(3):380-382.

李文艳,吉挺,岑宁(2010)超临界蜂胶残渣微胶囊化配方的研究[J].食品科技,35(6):136-139.

李香子,张敏,李成云,耿春银,严昌国(2011)蜂胶对猪胴体品质及肉品质的影响[J].畜牧科学,(4):68-70.

李雅洁,凌建亚,邓勇(2006)蒙山蜂胶超临界二氧化碳萃取挥发性组分的气相色谱-质谱联用分析[J].时珍国医国药,17(10):1975-1976.

李雅晶,冯磊,胡福良,张挺,陈民利(2007)纳米蜂胶对实验性高脂血症大鼠脂质代谢的影响[J].中国药学杂志,42(12):903-906.

李雅晶,胡福良(2007)纳米蜂胶的体外药剂学性质研究[J].中国蜂业,58(7):5-6.

李雅晶,胡福良,陆旋,詹忠根(2011a)蜂胶中挥发性成分的微波辅助提取工艺研究及中国蜂胶、巴西蜂胶挥发性成分比较[J].中国食品学报,11(5):93-99.

李雅晶,黄美珍,胡福良(2011b)均匀设计法优化超声波提取蜂胶精油工艺条件的研究[J].农产品加工(学刊),(2):60-62.

梁霭(2000)蜂胶蜜及其制备方法[P].CN00114163.5.

林贤统,朱威,胡福良(2008)不同溶剂提取蜂胶的得率及其提取物的抗氧化性[J].蜜蜂杂志,28(7):3-5.

刘昊湿(2012)一种蜂胶唇部去角质软膏[P].CN201210526583.2.

刘华(2012)双功能富莱欣牌蜂胶软胶囊的研制[J].食品科技,37(1):89-93.

刘嘉,姜楠,李雅晶,张翠平,胡福良(2013)巴西绿蜂胶原胶、乙醇提取液及蜂胶渣中挥发性成分的分析与比较[J].中国蜂业,64(1-3中):50-54.

刘伟,李岂凡,张彬,周武,侯宗福(2007)用分光光度法测定蜂胶超临界CO_2萃取物中总黄酮含量[J].南昌大学学报(理科版),31(3):276-278.

刘䶮,罗宇倩,郭辉,钱俊青(2011)蜂胶黄酮提取纯化的工艺研究[J].中国蜂业,62(1):6-10.

刘映,许静芬,史庆龙(2000)蜂胶挥发性成分的超临界CO_2萃取及GC-MS分析[J].中药材,23(9):547-549.

刘元法,王兴国,金青哲(2004)超声波技术提取蜂胶黄酮类功能性物质的研究[J].食品科学,25(6):35-39.

卢媛媛,魏文挺,胡福良(2013)超临界CO_2流体萃取技术在蜂胶提取中的应用[J].食品工业科技,34:364-368.

罗登林,丘泰球,卢群(2006)超声波技术及应用(Ⅲ)——超声波在分离技术方面的应用[J].日用化学工业,36(1):46-49.

吕武清,文萍,姚剑平,范婷婷,胡棠洪,俞建辉,李淑岚(2014)蜂胶提取物制备工艺研究[J].中草药,45(6):791-794.

马海芳,王昌利,史亚军(2014)蜂胶除蜡工艺研究[J].中南药学,12:448-451.

马建中(2010)一种高含量蜂胶软胶囊的制备方法[P].CN201010230004.0.

毛丽珍,徐世芳(1998)蜂胶口服液中黄酮类化合物的测定[J].中草药,29(4):231-232.

孟良玉,蔡文倩,卢佳琨,兰桃芳,丁文姝,励建荣(2011)蜂胶提取液对猪肉保鲜效果的研究[J].食品工业科技,32(12):155-157,161.

孟庆伟,回闯(2007)超临界流体药物微粒化技术的研究进展[J].化学试剂,29(4):212-216.

潘秋月,李英华,周晓红,杨凤杰(2011)蜂胶黄酮的超声波提取工艺研究[J].蜜蜂杂志,(12):10-12.

齐永增,程国珍(2007)一种蜂胶配制酒及制备方法[P].CN200710139302.7.

阮德荣(2005)蜂胶细辛酊的配制[J].中国养蜂,(4):23.

沈海涛(2002)蜂胶提取工艺和抗氧化活性研究[D].杭州:浙江大学.

沈新荣(2011)一种蜂胶香皂及其制作方法[P].CN201110275306.4.

石聚彬,石聚领(2007)红枣蜂胶含片及其生产方法[P].CN200710180484.2.

宋繁华(2013)蜂胶保健牙膏[P].CN 2013103746936.

孙丽萍,张智武,田文礼(2007)具有生物活性的水溶性蜂胶提取物和浅色蜂胶提取物[P].CN200710111189.1.

唐坤,李标,蔡应繁,夏晨燕,王伯初(2008)微波-超声波提取蜂胶的工艺研究[J].食品研究与开发,29(10):74-77.

田文礼,彭文君,韩胜明,高凌宇,张杨(2007)油溶蜂胶软胶囊及其制备方法[P].CN200710302091.4.

王彬(2004)一种保健蜂胶蜜皂[P].CN200410100463.1.

王春玲,张玉军,刘建平,陈杰镕(2003)超声波对蜂胶中有效成分的提取率影响的研究[J].郑州工程学院学报,24(4):28-33.

王春玲,高玲美(2012)超声波与溶剂法提取蜂胶中黄酮类化合物的效果[J].食品研究与开发,33(6):77-80.

王大红,陈明胜(2014)蜂胶在食用油脂中的抗氧化作用[J].武汉职业技术学院学报,13(3):91-93.

王凤忠(2011)蜂胶王浆软胶囊制备-功能评价及全程质量控制体系研究[D].北京:中国农业科学院.

王嘉军,张伟,王全立(2008)蜂胶-银离子复合纳米乳剂杀菌机制的研究[J].中国消毒学杂志,25(1):1-3.

王娜,叶萌祺,王春会,石颖,张建新(2009)超高压提取蜂胶黄酮的工艺优化[J].西北农业学报,18(3):300-303.

王平,孙建伟(2012)蜂胶滴丸及制备方法[P].CN201210454923.5.

王启发,程青芳,许冰(2004)从蜂胶中提取黄酮类化合物的研究[J].淮海工学院学报(自然科学版),13(2):60-62.

王锐,陇源,徐培涵(2004)一种蜂胶有效成分提取方法及用途[P].CN200410025836.

王维人(1993)蜂胶冻疮软膏[P].CN93112399.2.

王小平,陈玉芬,李雅萍(2007)蜂胶化学成分的提取方法研究[J].现代食品科技,23(6):73-77.

王小平,林励,潘建国,刘晓涵,卢占列(2009a)不同产地蜂胶挥发油成分的GC-MC比较分析[J].药物分析杂志,29(1):86-90.

王小平,林励,肖凤霞(2009b)GC-MC分析不同产地蜂胶的乙醚提取物[J].华西药学杂志,24(4):383-385.

王晓丽,程志伟,王永明(2012)兔病毒性出血症、多杀性巴氏杆菌病二联蜂胶灭活疫苗最小免疫剂量测定的试验[J].上海畜牧兽医通讯,(4):31-31.

王宇航,王艳娟,陈昨含(2012)禽霍乱蜂胶与氢氧化铝胶灭活疫苗免疫效果比较试验[J].吉林农业科技学院学报,21(2):4-6,118.

王振山(2010)一种水溶解蜂胶萃取液的制备方法[P].CN101081233B.

卫永第,安占元,丁长江,阎吉昌,张宏(1996)色质联用法分析蜂胶挥发油成分[J].分析测试学报,15(4):67-69.

魏广治,沈兰花(2013)蜂胶与蒙脱石粉联合应用治疗小儿口腔溃疡的临床研究[J].中国校医,27(8):628.

魏强华,邓桂兰,李桂锋(2007)均匀设计法优化蜂胶水提除蜡工艺[J].食品工业科技,28(7):117-118.

吴雪丽,申亮,刘红英(2013)蜂胶提取液对扇贝保鲜效果的研究[J].食品科技,38(7):166-169.

吴玉敏,何锦风,蒲彪(2007a)蜂胶总黄酮含量快速测定方法的研究[J].农产品加工(学刊),(2):71-72.

吴玉敏,何锦风,蒲彪,魏雅萍(2007b)微波辅助萃取蜂胶中黄酮类化合物的研究[J].现代食品科技,23(5):50-52.

席军(2005)高压加工技术在蜂胶黄酮类化合物提取中的应用研究[D].长春:吉林大学.

徐世明,赵瑞连,徐冬雪,宋维娟(2011)蜂胶对肉制品中主要腐败菌的抑菌效果研究[J].肉类研究,25(2):5-8.

徐水荣,卢媛媛(2012)一种葛根素、松花粉和蜂胶复合片剂及其制备方法[P].CN201210493317.4.

徐响,董捷,李洁(2008a)固相微萃取与GC-MC法分析蜂胶中挥发性成分[J].食品工业科技,29(5):57-60.

徐响,孙丽萍,董捷(2008b)超声强化对蜂胶超临界CO_2萃取物组成的影响[J].农产品加工学刊,(11):4-7.

徐响,孙丽萍,董捷(2009)响应面法优化蜂胶超临界二氧化碳萃取工艺的研究[J].食品科学,30(8):86-89.

徐响,董捷,丁小宇,杨佳林,孙丽萍(2010)不同方法萃取蜂胶挥发油组成及抑菌作用的研究[J].食品科学,31(3):60-63.

许应强(2013)一种蜂胶酒及其制备方法[P].CN201310497695.4.

许应强(2014)蜂胶美容霜[P].CN201410475663.9.

玄红专,顾美儿(2006)蜂胶口服液的制备及其黄酮类化合物的测定[J].食品研究与开发,27(3):55-56.

杨锋,戴关海,江克翊(2006)蜂胶水溶液制备工艺的研究[J].蜜蜂杂志,(5):3-4.

杨林莎,王东(2005)蜂胶提取工艺的研究[J].中国现代应用药学,22(2):135-137.

杨汝伟(2001)用于治疗口腔溃疡的蜂胶制剂及其配制方法[P].CN01107701.8.

杨汝伟,孙凤桂,李有志(2014)一种辅助调节血糖的复方蜂胶颗粒制剂配方[P].CN201410820993.7.

杨书珍,彭丽桃,潘思铁(2010)蜂胶提取物处理对柑橘诱导抗病性的影响[J].食品科学,31(8):275-279.

杨文超,蒋金龙,吴珍红,缪晓青(2012)蜂胶挥发油成分比较[J].中国蜂业,(Z1):58-63.

杨珍(2014)蜂胶汉方美白皂[P].CN201410452819.1.

叶静凌(2008)超声波结合碱性离子水提取蜂胶总黄酮工艺的研究[D].福州:福建农林大学.

游海,郑为完,高荫榆(2002)蜂胶微胶囊粉末的研究[J].食品科学,23(1):61-64.

余晶晶,童群义(2013)超声微波协同萃取蜂胶中黄酮类物质的研究[J].食品工业科技,34(4):314-317.

俞益芹,张焕新,唐劲松(2010)喷雾干燥法制备蜂胶微胶囊的研究[J].食品工业科技,31(1):301-304.

俞益芹,张焕新(2011)蜂胶涂膜对枇杷的保鲜效果[J].安徽农业科学,39(32):20018-20020.

张红城,董捷,胡浩(2014)一种蜂胶纳米乳口服液及其制备方法[P].CN 2014100201405.

张楠,任战军(2014)陕北蜂胶对香蕉的保鲜作用研究[J].现代农业科技,(3):290-293.

张伟(2000)新型蜂胶提取物用途广[J].养蜂科技,(2):23-24.

张英宣(2012)蜂胶微胶囊化工艺的研究[J].农产品加工(学刊),(10):79-80.

赵家明(2005)自制蜂胶口服液的方法[J].中国养蜂,56(10):21.

赵强,张彬,周武,郭志芳(2007)微波辅助萃取-GC/MS联用分析蜂胶挥发油[J].精细化工,24(12):1192-1196.

赵强,张彬,李岂凡(2008)蜂胶挥发油抗氧化性能及其成分研究[J].天然产物研究与开发,20(1):82-86.

赵淑云,马海燕,朱美玲,毛海波,房柱,邵兴军(2006)不同产地蜂胶超临界CO_2萃取物8种黄酮比较研究[J].中国蜂业,57(11):8-10.

震声(2001)自制蜂胶口服液的方法及应用注意事项[J].中国养蜂,52(2):29.

郑浩亮(2010)一种亲水性蜂胶软胶囊的内容物配方及制备方法[P].CN201010126306.3.

郑艳萍(2008)蜂胶乳化工艺及抑菌性研究[D].福州:福建农林大学.

周斌,叶满红(2012a)用于防治酒精性脂肪肝的蜂胶乙醇提取物、制备方法及在生产口含片中的应用[P].CN201210500237.7.

周斌,叶满红(2012b)用于解酒的蜂胶乙醇提取物、制备方法及在生产肠溶片中的应用[P].CN201210500238.1.

周萍(2011)一种亲水性蜂胶粉胶囊组合物及其生产方法[P].CN201110456323.8.

第二篇

蜂胶生物学活性研究

第六章 蜂胶抗氧化作用及其机制研究

抗氧化活性是蜂胶重要的生物学活性之一,一直是国内外蜂胶研究的热点。然而,对不同植物来源蜂胶发挥抗氧化作用的分子机制的相关研究较少。本研究首先对杨树型蜂胶乙醇提取物(EECP)和杨树胶乙醇提取物(EEPG)的化学成分、抗氧化活性及其分子机制进行比较,验证蜂胶的化学成分及活性与植物来源的相关性。然后,比较研究杨树型蜂胶乙醇提取物(EECP)、桉树型蜂胶乙醇提取物(EEEP)和酒神菊属型蜂胶乙醇提取物(EEBGP)的化学成分、抗氧化活性及其分子机制,为进一步开发蜂胶产品、寻找更多的具有良好药理活性的潜在蜂胶类型提供参考。

第一节 三种植物来源蜂胶和杨树胶乙醇提取物成分及其抗氧化活性

大量研究已经表明,蜂胶的生物学活性与其化学成分密切相关。早期研究表明,蜂胶的化学成分与植物来源相关,而不同地理来源的植物可能会因为植物种类、各种地理环境等因素造成植物的化学成分具有较大差异。蜂胶的抗氧化活性是蜂胶的主要活性之一,并能对蜂胶的其他生物学活性产生一定影响。因此,我们采集三个不同地理来源的三种植物来源的蜂胶以及中国的杨树胶为样本,对中国杨树胶乙醇提取物(EEPG)、中国杨树型蜂胶乙醇提取物(EECP)、巴西酒神菊属型蜂胶(巴西绿蜂胶)乙醇提取物(EEBGP)和澳大利亚桉树型蜂胶乙醇提取物(EEEP)的化学成分进行比较研究,并运用化学分析方法对它们的抗氧化活性进行初步的评估,为进一步比较研究不同植物来源蜂胶的抗氧化活性及其分子机制奠定物质基础。

1 实验材料

1.1 材料与试剂

中国杨树型蜂胶采自中国山东境内蜂场,杨树胶购自市场,桉树型蜂胶采自澳大利亚(南澳)蜂场,酒神菊属型蜂胶(巴西绿蜂胶)由蜂乃宝本铺(南京)保健品有限公司提供。

DPPH、ABTS、Trolox、松鼠素、没食子酸、槲皮黄酮、咖啡酸、p-香豆素、阿魏酸、白藜芦醇、芹黄素、山奈酚、柯因、高良姜素、CAPE 等标准品购自 Sigma 公司;无水乙醇、甲醇等均为国产分析纯试剂;以及液氮、三氯化铝、福林酚试剂、2,4-二甲基苯肼、PMS、NADH、NBT、亚铁氰化钾、三氯乙酸等。

1.2 主要仪器与设备

旋转蒸发器 RE-2000A(上海亚荣生化仪器厂);循环水式多用真空泵 SHB-ⅢA(上海豫

康教仪器设备有限公司);Spectra Max M5 多功能连续光谱酶标仪[美谷分子仪器(上海)有限公司];HX-200 型高速中药粉碎机(浙江省永康市溪岸五金药具厂);冰箱(荣事达电冰箱有限公司);高压灭菌锅[施都凯仪器设备(上海)有限公司];电热鼓风干燥箱(上海博讯实验有限公司);电子天平(北京赛多利斯仪器系统有限公司);微量移液器(美国 Eppendorf 公司);高速冷冻离心机(上海实维实验仪器技术有限公司)。

2 实验方法

2.1 蜂胶乙醇提取物的制备

蜂胶原胶一直贮存在-20℃冰箱。称取一定量(大于50g)的蜂胶原胶,然后用粉碎机粉碎,将粉碎后的蜂胶原胶粉末装于封口袋中。准确称取 50g(M)蜂胶原胶粉末溶于 500mL(分三次加入乙醇:第一次 200mL,第二次 150mL,第三次 150mL)95%的乙醇溶液中,40℃超声 3h,然后用滤纸(用前称重,M1)过滤。将残渣收集(滤纸和残渣一起放入烧杯中),再加入 150mL 95%乙醇,40℃超声 3h,然后滤纸过滤;以同样的方法再超声过滤一次,将残渣收集(包括滤纸)烘干后称重(M2),将三次过滤所得的乙醇提取液收集起来,4℃放置过夜,然后减压过滤除蜂蜡,将滤液 50℃减压旋转蒸发至少许乙醇时停止蒸发,倒入培养皿中,放入烘箱中 50℃烘干,然后放入-20℃冷冻,取下蜂胶提取物,用 100%的乙醇配成 20mg/mL 的贮备液,细胞实验中,乙醇终浓度不超过 0.1%。计算得率:原胶提取率=[(M+M1)-M2]/M×100%。

2.2 植物化学分析方法

2.2.1 蜂胶成分含量的测定

总黄酮含量(TFC)测定(Yang et al.,2011):加 2%(w/v)三氯化铝(150μL)在 96 孔板中,再加入乙醇样品溶液(150μL,0.3mg/mL),混匀,在室温孵育 15min,435nm 下测吸光值,设置一个重复。总黄酮含量用不同浓度的芦丁作标准曲线,样品总黄酮含量表示为芦丁(μg)/样品干重(mg)。

总酚酸含量(TPC)测定(Yang et al.,2011):将 10μL 乙醇样品溶液(10μL 乙醇,空白对照)加入到 450μL 的蒸馏水中,随后加入 10μL 福林酚试剂并振荡 3min,然后加入 30μL 2%(w/v)碳酸钠溶液,振荡混匀并放置 3h,取 200μL 加入 96 孔板在 760nm 处测吸光值,设立空白对照与复孔。以没食子酸建立标准曲线,总酚酸含量表示为没食子酸(μg)/样品干重(mg)。

黄烷酮与黄烷酮醇含量(FDC)测定(Miguel et al.,2010):将 40μL 的蜂胶溶液(空白对照、标准品)加入 80μL 的 DNP 溶液中[50mg DNP 溶于 100μL 96%的硫酸(v/v)],用甲醇稀释到 5mL),然后将混合溶液在 50℃加热 50min。室温冷却,用甲醇配置的 10%(w/v)的 KOH 将混合物稀释到 400μL,混合均匀,取 20μL 用甲醇稀释到 1mL。在 486nm 处测吸光值,以松属素建立标准曲线,表示为松属素(mg)/蜂胶(g)。

高效液相色谱法:

(1)对 EECP 和 EEPG 的 11 种有效成分进行高效液相色谱分析,安捷伦科技 HPLC 系统由真空除气器 G1322A、四进制泵 G1311A、自动进样器 G1329A、可编程的可变波长检测器(VWD)G1314B 和恒温柱隔间 G1316A 组成。分离条件为:Agilent Eclipse XDB-C18 柱

(4.6mm×150mm,5μm)为30℃,流动相流速为1.0mL/min,其包含(C)氰化甲烷,(D) 0.4%的醋酸和梯度洗脱:0～40min,5%～25%(B);40～45min,25%～35%(B);55～ 60min,35%～40%(B);80～90min,40%～5%(B);90～100min,5%(B)。上样量5μL,上样浓度1mg/mL,结果表示为平均值±SD($n=3$)。

(2)对EECP、EEBGP和EEEP的22种有效成分进行高效液相色谱分析(Zhang et al.,2014),系统为赛分科技(美国)的HP-C18柱(150mm×4.6mm,5μm)。流动相为1.0%(v/v)水溶醋酸(A),甲醇(B);1.0mL/min的流速进行梯度洗脱(33℃):15%～40%(B) 0～30min,40%～55%(B) 30～65min,55%～62%(B) 65～70min,100%(B) 70～85min。上样量5μL,上样浓度10mg/mL,结果表示为平均值±SD($n=3$)。

2.2.2 蜂胶自由基清除能力的测定

ABTS自由基清除活性(Yang et al.,2011):将ABTS粉末溶于超纯水,配制成7mmol/L的ABTS溶液。将过硫酸钾粉末溶于水,配置成140mmol/L的过硫酸钾水溶液。将7.5mL 7mmol/L ABTS溶液和132μL 140mmol/L过硫酸钾水溶液避光反应16h后生成,该溶液为前一天配置,使用前用乙醇稀释到吸光值在732nm处为0.7左右(稀释25倍左右)。在96孔板中加入100μL ABTS,再加入50μL样品,黑暗放置10min,然后在734nm波长下测吸光值(Ai),平行设一个复孔。以乙醇代替样品为对照(A0,乙醇+ABTS),以蒸馏水代替ABTS为阴性对照(Ai0,蒸馏水+样品),清除率=[1-(Ai-Ai0)/A0]×100%。

DPPH自由基清除活性(Yang et al.,2011):精密称取DPPH标准品约15mg,用乙醇定容至15mL,即得DPPH标准工作液(1mg/mL),低温(4℃)、避光保存备用。分别精密称取待测样品和维生素C阳性对照品约25.0mg,用乙醇定容至25mL,即为1mg/mL的样品储备溶液和对照品储备液。样品测试前用无水乙醇稀释成所需浓度。取稀释液120μL并加入120μL DPPH工作液,同时以不加DPPH(120μL无水乙醇代替DPPH)的供试品溶液各浓度作为对照以消除供试品本身颜色对测试结果的干扰,并设DPPH阴性对照(以120μL无水乙醇代替供试品),每组平行设2个复孔。每孔100μL加入到96孔板各孔中,室温避光反应30min后,酶标仪波长517nm处测定吸光度,使其读数在0.2～0.8。据此将每个样品储备溶液稀释成6～8个浓度梯度的样品溶液。DPPH清除率(%)=(A0-AS)/A0×100%,式中:AS,加入样品提取液的DPPH溶液的吸光度;A0,加入无水乙醇的DPPH溶液的吸光度。以样品溶液的浓度为横坐标、DPPH自由基清除率为纵坐标作图,求线性方程,并计算清除50% DPPH所需样品及维生素C对照品的浓度,即半抑制浓度IC50,比较样品清除DPPH自由基活性。

还原力(RP)测定(Guo et al.,2011):在1.5mL管中先加入312.5μL磷酸缓冲液、312.5μL 1%铁氰化钾溶液,再取125μL样品(阴性对照:将样品改为125μL的乙醇,其他不变,本底对照为只加蒸馏水)加入1.5mL管中。混合,50℃孵育20min,加入312.5μL 10%的三氯乙酸溶液,混合,离心2000r/min 10min。取上清液(1mL),加入312.5μL蒸馏水和62.5μL 0.1%的氯化铁溶液。然后,在700nm处测吸光值。阴性对照:将样品改为125μL的乙醇,其他不变。

溶液配制:

(1)2%的铁氰化钾溶液50mL,使用时需稀释1倍,1.00g铁氰化钾加50mL水。

(2)20% 三氯乙酸溶液:10.00g三氯乙酸固体加50mL蒸馏水。

(3)0.1%的三氯化铁溶液:0.083g 六水合三氯化铁(三氯高铁)加 50mL 蒸馏水。

(4)0.2μmoL/L 的二水合磷酸二氢钠:3.12g 二水合磷酸二氢钠加 100mL 水。

(5)0.2μmoL/L 十二水合磷酸氢二钠:7.162g 十二水合磷酸氢二钠加 100mL 水。

(6)取 93.75mL 0.2μmoL/L 的二水合磷酸二氢钠溶液和 56.25mL 的十二水合磷酸氢二钠溶液混合均匀,然后调 pH 至 6.6。

超氧阴离子自由基清除活性(SRSA)测定(Song et al.,2011):依次加入 20μL 样品,300μL 468μmoL/L 的 NADH,300μL 150μmoL/L 的 NBT,最后加入 300μL 60μmoL/L 的 PMS 溶液触发产生超氧离子,反应 10min,在 560nm 处测定吸光值,结果以自由基清除率表示。

氧自由基吸收能力(ORAC)测定(Shi et al.,2012):将荧光素稀释成 8.163×10^{-8} mol/L 的工作液,加入 96 孔板(每孔 225μL),然后加入 30μL(样品、标准品、空白)振荡混匀,用酶标仪 37℃孵育 20min,然后加入 0.36mol/L AAPH 25μL,振荡,然后迅速测定(485nm 和 535nm,每分钟测一次,测 2h)。

溶液配制:

(1)0.36mol/L AAPH(分子质量:271.19)溶液:称取 2.44071g AAPH 溶于 25mL 75mmol/L pH 7.4 的磷酸缓冲液中。

(2)75mmol/L pH7.4 磷酸缓冲液:磷酸盐缓冲液的配制为取 13%的磷酸氢二钾水溶液 79.9mL,与 10.2%的磷酸二氢钾水溶液 20.1mL 混合,加到 890mL 的去离子水中。用磷酸氢二钾或磷酸二氢钾水溶液调节缓冲液 pH 值,使其达到 7.4,定容到 1L,最终缓冲液浓度为 75mmol/L。

13%的二水合磷酸氢二钾水溶液:称取磷酸氢二钾 17.03g 溶于 100mL 蒸馏水。

10.2%的磷酸二氢钾水溶液:称取磷酸二氢钾 10.2g 溶于 100mL 蒸馏水。

(3)8.163×10^{-8} mol/L 荧光素溶液:用 75mmol/L pH 7.4 的磷酸盐缓冲液配制成 1.6009×10^{-4} mol/L 的贮备液,待用。

3 结果与讨论

3.1 三种植物来源蜂胶的提取率

中国杨树型蜂胶、巴西绿蜂胶和澳大利亚桉树型蜂胶的提取率相近,分别为 57.42%、58.06%和 56.36%。根据蜂胶国家标准(GB/T 24283—2009)对蜂胶原胶的等级划分(蜂胶原胶优等品为蜂胶乙醇提取物的提取率不小于 60%)可知,试验用中国杨树型蜂胶原胶、巴西绿蜂胶原胶(酒神菊属型蜂胶)和澳大利亚桉树型蜂胶原胶都与优等品接近。

3.2 EECP 和 EEPG 的主要成分比较

多酚类物质有良好的抗氧化活性,蜂胶中含有丰富的多酚类物质。对 EECP 和 EEPG 的总黄酮、总酚酸以及黄烷酮和黄烷酮醇含量进行测定,结果如表 6.1 所示,EECP 的多酚类物质含量明显高于 EEPG。而且,EECP 的总酚酸、黄烷酮和黄烷酮醇含量显著高于 EEPG,而两者的黄酮含量没有差异。

表 6.1 EECP 和 EEPG 的总酚酸(TPC)、总黄酮(TFC)以及黄烷酮和黄烷酮醇(FDC)的含量

Table 6.1 Total phenolic contents (TPC), total flavonoid contents (TFC) and flavanone and dihydroflavonol contents (FDC) in EECP and EEPG

样品	TPC(mg GAE/g)	TFC(mg RE/g)	FDC(mg NE/g)
EECP	192.80±10.85**	297.24±10.32 NS	229.64±7.05**
EEPG	121.81±8.83	297.09±10.66	164.14±4.95

注:数据表示为平均数±标准差($n=3$);NS 表示差异不显著,* 表示差异显著($P<0.05$),** 表示差异极显著($P<0.01$);GAE,没食子酸等量;RE,芦丁等量;NE,柚苷配基等量。

3.3 EECP、EEEP 和 EEBGP 的主要成分比较

运用化学分析方法对三种植物来源蜂胶乙醇提取物的总酚酸、总黄酮以及黄烷酮和黄烷酮醇含量进行测定,结果见表 6.2。结果表明,EECP 的多酚类物质含量最高;EEEP 的总酚酸含量最高,但是总黄酮含量最低,而 EEBGP 总黄酮含量最高。有研究表明,酚酸类物质是巴西绿蜂胶的主要活性物质(Bankova et al.,1995),而黄酮被认为是欧洲蜂胶(一般为杨树型蜂胶)的主要活性物质(Hegazi et al.,2000)。从该结果可知,三种植物来源蜂胶的成分具有一定差异,这为进一步探讨三种植物来源蜂胶的抗氧化活性及其分子机制提供了物质基础。

表 6.2 EECP、EEEP、EEBGP 的总酚酸(TPC)、总黄酮(TFC)、黄烷酮和黄烷酮醇(FDC)的含量

Table 6.2 Total phenolic contents (TPC), total flavonoid contents (TFC) and flavanone and dihydroflavonol contents (FDC) in EECP,EEEP and EEBGP

样品	TPC(mg GAE/g)	TFC(mg RE/g)	FDC(mg NE/g)
EECP	192.80±10.85A	297.24±10.32a	229.64±7.05a
EEEP	217.70±4.99Ba	55.76±4.17b	118.21±8.58b
EEBGP	135.07±1.11Bb	229.68±7.84c	82.36±5.82c

注:数据表示为平均数±标准差($n=3$);相同的字母表示差异不显著($P>0.05$),不同的大写字母表示差异显著($P<0.05$),不同的小写字母表示差异极显著($P<0.01$);GAE,没食子酸等量;RE,芦丁等量;NE,柚苷配基等量。

3.4 高效液相色谱法(HPLC)测定 EECP、EEPG、EEBGP 和 EEEP 的化合物含量

蜂胶的抗氧化活性与其有效的化合物含量密切相关。有研究指出,杨树型蜂胶和杨树胶的化合物成分差异不明显(Bankova et al.,1992;Isidorov and Vinogorova,2003),同时,亦有研究发现水杨苷能在杨树胶中检出,而在杨树型蜂胶中未检出(Zhang et al.,2011)。因此,我们用 HPLC 方法测定 EECP 和 EEPG 中 11 种主要物质的含量,这 11 种物质在杨树胶和杨树型蜂胶的相关研究文献中都有报道。11 种化合物的 HPLC 色谱图如图 6.1 所示,其含量如表 6.3 所示。结果表明,EECP 的 11 种化合物的总含量为 EEPG 的两倍多。其中,芹黄素、柯因、松属素、高良姜素和咖啡酸苯乙酯在 EECP 和 EEPG 中都有检出,而白藜芦醇、槲皮黄酮和山奈酚未检出。更有趣的是,咖啡酸、p-香豆酸和阿魏酸只在 EECP 中检出,但其含量较低。由结果可知,EECP 和 EEPG 的成分既有一定的差异,也有一定的相似性,其差异性主要体现在物质的含量上,而相似性主要体现在成分基本一致。

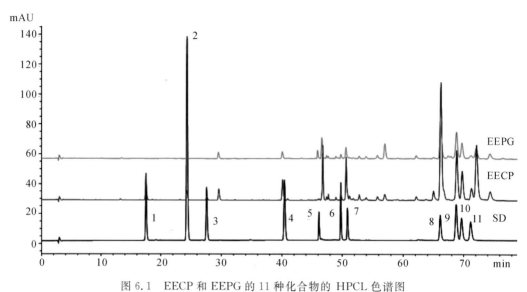

图 6.1 EECP 和 EEPG 的 11 种化合物的 HPCL 色谱图

Fig. 6.1 HPLC chromatograms of 11 compounds in EECP and EEPG

1.咖啡酸;2.p-香豆酸;3.阿魏酸;4.白藜芦醇;5.槲皮黄酮;6.芹黄素;7.山奈酚;8.柯因;9.松属素;10.高良姜素;11.咖啡酸苯乙酯;SD:标准品。

表 6.3 EECP 和 EEPG 中 11 种化合物的含量

Table 6.3 Contents of 11 compounds in EECP and EEPG

化合物	保留时间（min）	EECP（g/100g 提取物）	EEPG（g/100g 提取物）
咖啡酸	17.52	0.35±0.00	—
p-香豆酸	24.32	0.08±0.00	—
阿魏酸	27.71	0.07±0.00	—
白藜芦醇	40.53	—	—
槲皮黄酮	46.24	—	—
芹黄素	49.89	0.12±0.00	0.07±0.00
山奈酚	50.98	—	—
柯因	66.44	2.33±0.06	0.96±0.02
松属素	69.10	1.22±0.04	0.69±0.01
高良姜素	70.04	1.11±0.05	0.65±0.01
咖啡酸苯乙酯	71.75	0.58±0.04	0.22±0.00
总含量		5.85±0.20	2.59±0.05

注:结果表示为平均值±标准差($n=3$);"—"表示未检出。

以上成分分析结果表明,蜂胶的成分与植物来源有一定的关系。那么,不同植物来源蜂胶的单体成分是否存在差异呢?我们进一步用 HPLC 方法比较 EECP、EEBGP 和 EEEP 的 22 种单体成分的含量,其色谱图如图 6.2 所示,单体成分含量如表 6.4 所示。

从标准品色谱图(图 6.2A-a、B-a)可知,该方法能有效对 22 种单体成分进行区分。香草酸、芦丁、绿原酸、山奈素和阿替匹林 C 在 EECP 中未检出,而其他单体成分在 EECP 中含量明显高于 EEBGP 和 EEEP，EEEP 中检测出 10 种物质，而咖啡酸、p-香豆酸、阿魏酸、绿原酸、山奈素和阿替匹林 C 都未检出（表 6.4）。EEBGP 中也检出 10 种物质，然而，EEBGP 中有检出且含量较高的咖啡酸、p-香豆酸、阿魏酸和绿原酸、山奈素和阿替匹林 C，尤其是 p-香豆酸、山奈素和阿替匹林 C，但 3,4-二甲氧基肉桂酸、肉桂酸、木樨草素、芹黄素、短叶松素-3-乙酸酯、柯因、高良姜素在 EEBGP 中未检出（表 6.4）。更有趣的是，EEEP 中检出的物质在 EECP 中都存在，且含量低于 EECP，但 EECP 中含有的某些物质在 EEEP 中未检出（表 6.4）。此外，槲皮黄酮、柯因和高良姜素在 EEBGP 的峰属于干扰峰（图 6.2A-b），因为我们在另外的方法中对 EEBGP 的这三种物质进行鉴定，但并未检出（结果未给出）。

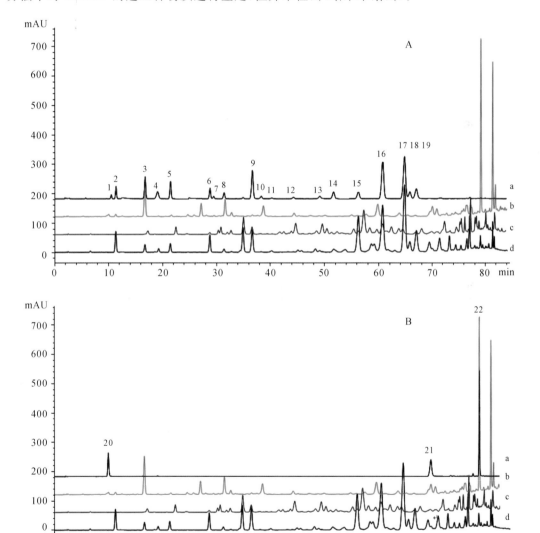

图 6.2　EEBGP、EEEP 和 EECP 中 22 种化合物的 HPLC 色谱图

Fig. 6.2　HPLC chromatograms of 22 compounds in EEBGP, EEEP and EECP

1. 香草酸；2. 咖啡酸；3. p-香豆酸；4. 阿魏酸；5. 异阿魏酸；6. 3,4-二甲氧基肉桂酸；7. 芦丁；8. 肉桂酸；9. 短叶松素；10. 柚皮素；11. 槲皮黄酮；12. 木樨草素；13. 山奈酚；14. 芹黄素；15. 松鼠素；16. 短叶松素-3-乙酸酯；17. 柯因；18. 咖啡酸苯乙酯；19. 高良姜素；20. 绿原酸；21. 山奈素；22. 阿替匹林 C；a. 标准品；b. EEBGP；c. EEEP；d. EECP。

以上结果表明,EEEP 的成分与 EECP 相似,但其含量低于 EECP,然而,EEBGP 与 EECP 和 EEEP 的成分差异较大。EECP、EEBGP 和 EEEP 之间的成分差异性与相似性为进一步研究三种植物来源蜂胶的抗氧化能力及其分子机制的相似性与差异性奠定了物质基础。

表 6.4 EECP、EEBGP 和 EEEP 中 22 种化合物的含量
Table 6.4 Contents of 22 compounds in EECP, EEBGP and EEEP

化合物	EEBGP		EEEP		EECP	
	保留时间/min	含量/(mg/g)	保留时间/min	含量/(mg/g)	保留时间/min	含量/(mg/g)
香草酸	—	—	—	—	—	—
咖啡酸	11.417	1.47±0.43	—	—	11.417	13.35±3.71
p-香豆酸	16.847	18.01±1.05	—	—	16.851	2.70±0.75
阿魏酸	19.347	0.40±0.08	—	—	19.342	3.14±0.57
异阿魏酸	—	—	—	—	21.489	3.29±1.02
3,4-二甲氧基肉桂酸	—	—	28.84	1.99±0.76	28.835	12.59±2.56
芦丁	—	—	—	—	—	—
肉桂酸	—	—	31.466	0.15±0.05	31.464	0.87±0.01
短叶松素	36.711	1.75±0.60	36.711	9.62±0.83	36.707	28.23±5.13
柚皮素	38.348	2.49±0.16	38.345	1.79±0.37	38.345	0.77±0.02
槲皮黄酮	—	—	—	—	40.299	5.90±0.24
木樨草素	—	—	44.329	7.29±0.18	44.325	1.16±0.53
山奈酚	47.988	0.99±0.24	—	—	47.984	1.85±0.28
芹黄素	—	—	51.301	3.96±1.03	51.302	8.03±1.67
松鼠素	55.818	1.13±0.84	55.819	4.80±0.82	55.817	44.55±4.26
短叶松素-3-乙酸酯	—	—	60.243	22.94±1.24	60.24	97.49±6.89
柯因	—	—	64.281	7.22±0.94	64.286	66.89±6.04
咖啡酸苯乙酯	—	—	—	—	65.275	24.43±8.06
高良姜素	—	—	66.454	3.08±1.67	66.453	22.34±5.04
绿原酸	10.069	8.98±0.86	—	—	—	—
山奈素	69.604	11.18±8.31	—	—	—	—
阿替匹林 C	79.17	107.70±10.86	—	—	—	—

注:结果表示为平均值±标准差($n=3$);"—"表示未检出。

3.5 EECP、EEPG、EEBGP 和 EEEP 的自由基清除活性及还原力比较

植物化学方法被广泛应用于评价天然产物的抗氧化能力,尤其是 ABTS 法和 DPPH 法。然而,也有学者指出单一地运用植物化学方法不能很好地评价多功能食物的抗氧化活性(Edwin et al.,2000)。因此,本研究选用多个指标比较不同植物来源蜂胶以及杨树胶的抗氧化活性,结果见表 6.5 和表 6.6。表 6.5 的结果表明,EECP 的 ABTS、SRSA 和 RP 活性极显著高于 EEPG,EECP 的 ORAC 值显著高于 EEPG,而两者具有相当的 DPPH 自由基清除活性。

第六章 蜂胶抗氧化作用及其机制研究

表 6.5 EECP 和 EEPG 的自由基清除能力和还原力

Table 6.5 Free radical scavenging activities and reducing power of EECP and EEPG

样品	DPPH(IC_{50})	ABTS(IC_{50})	RP/ mmol TE/g	SRSA/ mmol TE/g	ORAC/ mmol TE/g
EECP	32.35±2.84 NS	40.5±2.38**	2.08±0.08**	1.52±0.03**	9.25±0.85*
EEPG	31.83±2.68	55.4±1.49	1.55±0.05	0.67±0.02	6.43±0.46

注：数据表示为平均数±标准差($n=3$)；NS 表示差异不显著，* 表示差异显著（$P<0.05$），** 表示差异极显著（$P<0.01$）；DPPH 和 ABTS 结果表示为 IC50（$\mu g/mL$）；FRAP、SRSA 和 ORAC 结果表示为毫摩尔 Trolox 等量(TE)每克样品。

进一步比较三种植物来源蜂胶的自由基清除活性和还原力(表 6.6)，发现 EEEP 具有最强的 DPPH 自由基清除活性和 ABTS 自由基清除活性，而 EEBGP 和 EECP 差异不显著。同时，EEEP 具有最强的还原力，EEBGP 最弱。然而，EEEP 对超氧阴离子的清除活性显著低于 EECP 和 EEBGP。另一方面，对于总氧吸收能力的强弱顺序依次为：EECP=EEEP＞EEBGP。

有学者指出，植物化学方法（DPPH、FRAP 和 ABTS）是基于相似的氧化还原体系（SPLET），他们认为选择其中的一种作为衡量抗氧化活性的指标也是被广泛接受的(Dejian et al.，2005)。而且，ORAC 分析（HAT）方法被食品和保健品行业广泛认为是检测抗氧化能力的有效手段。但是，氧化应激在机体内是一个相当复杂的过程，植物化学方面不能模拟体内复杂的环境，因此，这些结果只能在一定程度上比较 EECP、EEEP 和 EEBGP 的抗氧化活性。

表 6.6 EECP、EEEP 和 EEBGP 的自由基清除活性和还原力

Table 6.6 Free radical scavenging activities and reducing power of EECP, EEEP and EEBGP

样品	DPPH (IC50)	ABTS (IC50)	RP (mmoL TE/g)	SRSA (mmoL TE/g)	ORAC (mmoL TE/g)
EECP	32.35±2.84a	40.5±2.38a	2.08±0.08a	1.52±0.03a	9.25±0.85a
EEEP	19.55±1.28b	20.0±0.31b	2.70±0.08b	0.96±0.04b	7.73±0.76a
EEBGP	43.85±0.54a	38.2±0.33a	1.53±0.05c	1.65±0.07a	5.60±0.53c

注：数据表示为平均数±标准差($n=3$)；相同的字母表示差异不显著（$P>0.05$），不同的大写字母表示差异显著（$P<0.05$），不同的小写字母表示差异极显著（$P<0.01$）；DPPH 和 ABTS 结果表示为 IC50（$\mu g/mL$）；FRAP、SRSA 和 ORAC 结果表示为毫摩尔 Trolox 等量(TE)每克样品。

小 结

本研究采用植物化学方法和高效液相色谱法（HPLC）首先对中国杨树型蜂胶乙醇提取物（EECP）和杨树胶乙醇提取物进行比较，并对中国杨树型蜂胶乙醇提取物（EECP）、巴西绿蜂胶乙醇提取物（EEBGP）和澳大利亚桉树型蜂胶乙醇提取物（EEEP）作成分分析，然后以植物化学方法测定 EECP、EEPG、EEBGP 和 EEEP 的自由基清除活性和还原力。结果表明：蜂胶的成分与植物来源有一定关系，但在含量上具有差异。EEEP 和 EECP 的总体成分

具有一定的相似性,但EEEP的有效成分种类少于EECP,而EEBGP与EECP和EEEP的成分具有较大的差异。进一步的抗氧化活性分析结果表明,EEEP通过电子转移途径(ET)的抗氧化活性最强;而通过氢原子转移途径(HAT)的抗氧化活性与EECP相似,但强于EEBGP;EEBGP的抗氧化活性最弱。

第二节 不同植物来源蜂胶和杨树胶在RAW 264.7中的抗氧化活性及其分子机制的比较

关于蜂胶抗氧化活性的体内、体外实验的报道已有很多,然而,目前还鲜有关于蜂胶在生物体内是如何发挥抗氧化作用的报道。了解蜂胶发挥抗氧化活性的具体机制,不仅能丰富蜂胶在抗氧化活性方面的理论,更有利于研究蜂胶药理活性作用靶点,为研究蜂胶对疾病的治疗作用提供理论依据。

核转录因子(Nrf2)是一个重要的氧化还原敏感因子,它能被多酚类物质激活而加强机体的抗氧化防御系统。因此,本研究以RAW264.7小鼠巨噬细胞为细胞模型,首先研究EECP、EEPG、EEBGP和EEEP的ROS自由基清除活性。然后,比较研究EECP、EEPG、EEBGP和EEEP对抗氧化相关基因表达的作用效果,并进一步探讨它们对MAPKs和Nrf2的调节作用。

1 实验材料

1.1 材料与试剂

RAW264.7小鼠巨噬细胞,浙江大学生命科学研究院夏总平实验室馈赠;中国杨树型蜂胶采自中国山东境内蜂场,杨树胶购自市场,澳大利亚桉树型蜂胶采自澳大利亚(南澳)蜂场,酒神菊属型蜂胶(巴西绿蜂胶)由蜂乃宝本铺(南京)保健品有限公司提供。

兔抗小鼠抗体:HO-1、GCLM、TrxR1、GCLC、β-tubulin、p38、Erk1 Phospho/Erk2 Phospho、JNK1 phospho/JNK2 phospho/JNK3 phospho、JNK1、AKt1、AKt1 phospho(美国Epitomics公司);Phospho-p38(Thr180/Tyr182)抗体(美国Cell Signaling公司);Nrf2抗体(美国Abcam公司);SB203580,SP600125,LY294002和PD98059抑制剂(美国Selleckchem公司);山羊二抗(美国Sigma公司);HRP二抗(美国Epitomics公司);Alexa Fluor 488标记抗兔二抗(杭州联科生物技术有限公司)。

DMEM高汤培养基、PBS(1×)(杭州科易生物技术有限公司);一次性滤器(美国Pall公司);培养板(6孔、24孔、48孔、96孔)[科晶(宁波)生物技术有限公司];2′7′-dichlorodihydrofluorescein(DCHF)(美国Sigma公司);蛋白分子量marker(美国Thermo scientific公司);PVDF膜(德国Millipore公司);RNApure超纯总RNA快速提取试剂盒(艾德莱生物技术有限公司);SYBR Premix Ex Taq™(日本Takara公司);PBST缓冲液、Tris-甘氨酸-SDS电泳缓冲液(上海百赛生物科技有限公司);TEMED、Tris缓冲液(1mol/L,pH8.0)、β-巯基乙醇(生工生物科技有限公司);过硫酸铵(北京索莱宝科技有限公司);Tris-盐酸缓冲液(1.5 M,pH8.8)(上海双螺旋生物科技有限公司);RIPA裂解液(强)(碧云天生物技术公司);40% Acr-Bis(39∶1)(碧云天生物技术公司);Nonodet P-40(生工生物科技有限公司);BCA 蛋白浓度测定试剂盒(WB0123)(上海威奥生物科技有限公司);细胞计数(CCK-8)试剂盒(同仁化学研究所)。

1.2 主要溶液与试剂配方

DCHF 贮存液:将 DCHF 溶于 DMSO,配成 100mmol/L 的贮存液。

DCHF 使用液:DCHF 贮存液 $2\mu L$,溶于 1mL 细胞培养液。

10%过硫酸铵(w/v):现配现用。

封闭液 100mL:脱脂奶粉 5g,加 PBST(1×)定容至 100mL。

PBST(×1)1000mL:PBST(5×)200mL,加三蒸水 800mL。

Tris-甘氨酸-SDS 电泳缓冲液(1×):Tris-甘氨酸-SDS 电泳缓冲液(5×)200mL,加三蒸水 800mL。

1× AP 显色缓冲液:1mmol/L Tris-Cl pH9.5 20mL,500mmol/L 氯化钠 20mL,50mmol/L 氯化镁 20mL,加三蒸水至 200mL。

免疫印迹转膜缓冲液(10×):144.13g 甘氨酸,250mL 1mmol/L Tris-盐酸缓冲液,pH8.0,加三蒸水至 1000mL。

免疫印迹转膜缓冲液(1×):100mL 10×转膜缓冲液,200mL 甲醇,700mL 三蒸水。

NBT/NCBI 显色液:1g NCIP+100mg NBT,溶于 10mL 67% DMSO(v/v),$-20℃$ 保存,显色时,取 $200\mu L$ 显色液加入 10mL 1×AP 显色缓冲液中。

20% SDS 贮存液:20g SDS 溶于 100mL 三蒸水。

5×SDS 上样缓冲液:1mL 0.5mmol/L pH6.8 Tris-盐酸缓冲液,2mL 20% SDS,6mL 甘油,0.5mL 5% 溴酚蓝溶液,加三蒸水至 10mL,分装成 10 管,贮存于$-20℃$,使用时每毫升贮存液加入 $30\mu L$ β-巯基乙醇。

溴酚蓝贮存液:0.5g 溴酚蓝溶于 10mL 50%的无水乙醇。

1.3 主要仪器与设备

HX-200 型高速中药粉碎机(浙江省永康市溪岸五金药具厂);CO_2 培养箱(日本 SANYO 公司);洁净工作台(苏州安康空气技术有限公司);倒置相差显微镜(日本 Nikon 公司);激光扫描共聚焦显微镜(德国 Leica 公司);流式细胞仪(美国 BD 公司);GL-802A 型微型台式真空泵(海门市其林贝尔仪器制造有限公司);电热恒温水槽(上海精宏实验设备有限公司);冰箱(荣事达电冰箱有限公司);CO_2 钢瓶(杭州今工特种气体有限公司);液氮罐(亚洗牌橡塑机器有限公司);高压灭菌锅[施都凯仪器设备(上海)有限公司];电热鼓风干燥箱(上海博讯实验有限公司);电子天平(北京赛多利斯仪器系统有限公司);DYY-10C 型电泳仪电源(北京六一仪器厂);DYY-6D 型电泳仪电源(北京六一仪器厂);微量移液器(美国 Eppendorf 公司);$-80℃$ 超低温冰箱(中科美菱);凝胶成像分析系统(培清科技);高速冷冻离心机(上海实维实验仪器技术有限公司);Spectra Max M5 多功能连续光谱酶标仪[美谷分子仪器(上海)有限公司]Perfection V300 Photo 扫描仪(日本爱普生公司)。

2 实验方法

2.1 细胞培养和蜂胶毒性分析

将小鼠巨噬细胞 RAW264.7 复苏于含有 10%胎牛血清的 DMEM 高糖培养基的细菌培

养皿中,在37℃ 5% CO_2 饱和湿度的培养箱中培养32h,然后以1∶3传代,之后每24h传代一次。为保证实验中细胞处于良好状态,所用细胞传代次数均为7代。

检测细胞存活率:待细胞传至第2代后,将细胞以 $1×10^5 \sim 5×10^5$ 种于96孔板(96孔板四周孔内以PBS代替,防止边缘效应),每孔100μL,培养24h后,加入1μL不同浓度的蜂胶溶液(空白对照组为1μL无水乙醇),继续培养24h后,每孔加入5μL CCK-8试剂,继续培养2h,然后在450nm测定吸光度。其中,本底为不加CCK-8试剂的正常细胞组。

存活率=[OD(蜂胶组)−OD(本底)]/[OD(空白对照组)−OD(本底)]×100%,空白对照组存活率为100%。

2.2 ROS含量测定

将RAW264.7细胞以 $1×10^6$ 左右接种于12孔板,培养24h,加入1μL不同浓度的蜂胶溶液培养0.5h,然后以300μmol/L H_2O_2 处理细胞13h。除去培养基,以暖的PBS清洗2遍,再以含有终浓度为200μmol/L DCHF-DA培养30min。除去培养基,用冷的PBS清洗2遍,用胰酶消化并收集于1.5mL离心管中,2500r/min离心5min,加入500μL PBS清洗一次,离心,再加入700μL PBS重悬,在流式细胞仪上测定ROS含量。

ROS清除率=1−[ROS(蜂胶+H_2O_2处理组)−ROS(空白组)]/[ROS(H_2O_2处理组)−ROS(空白组)]。

2.3 实时荧光定量聚合酶链式反应(qRT-PCR)

将培养基除去,根据RNApure超纯总RNA快速提取试剂盒的说明,提取总RNA,并用Nano Drop分光光度计测定提取的总RNA浓度,定量1μg RNA进行cDNA的合成(primeScript^{TM} RT试剂盒)。最后,以合成的cDNA为模板,用SYBR Premix Ex Taq^{TM} 试剂进行qRT-PCR。反应条件为:95℃ 30s,95℃ 5s和60℃ 30s,然后溶解曲线分析为95℃ 15s,50℃ 15s和95℃ 15s。每孔反应体系为7μL:水2.85μL,前后引物各0.25μL,SYBR Premix Ex Taq^{TM} 试剂3.5μl,模板0.15μL。PCR产物以核酸电泳进行分析,引物为生工生物科技有限公司合成。数据采用 $2^{-\Delta\Delta Ct}$ 法进行计算,引物序列如表6.7。

表6.7 qRT-PCR引物序列
Table 6.7　Primer sequences of qRT-PCR

Primer	Sense primers	Antisense primers
HO-1	5′-ACATTGAGCTGTTTGAGGAG-3′	5′-TACATGGCATAAATTCCCACTG-3′
TrxR1	5′-AGGATTTCTGGCTGGTATCG-3′	5′-CTCGCTGTTTGTGGATTGAG-3′
GCLM	5′-CTGACATTGAAGCCCAGGAT-3′	5′-GTTCCAGACAACAGCAGGTC-3′
GCLC	5′-GATGATGCCAACGAGTCTGA-3′	5′-GACAGCGGAATGAGGAAGTC-3′
GAPDH	5′-GAGAAACCTGCCAAGTATGATGAC-3′	5′-TAGCCGTATTCATTGTCATACCAG-3′

2.4 蛋白质印迹(Western blot)

2.4.1 细胞总蛋白提取

(1)弃去培养基中细胞培养液,用PBS洗两遍,甩干。

(2)加入蛋白裂解液(RIPA或NP40),充分裂解10min(冰上)。

(3)用刮刀刮去蛋白,装入相应的编号 eppendorf 管中。

(4)裂解物在 4℃条件下 16000r/min 离心 10min,取上清液(取 10μl,根据 BCA 试剂盒说明书测定蛋白浓度,),以 1∶4(上清液∶5×SDS 上样缓冲液)体积加入 5×SDS 上样缓冲液,充分混匀。

(5)95℃条件下煮 10min,然后贮存于-80℃。

2.4.2 Western blot 检测 HO-1、GCLM、TrxR1 以及 MAPKs 蛋白表达

(1)制 SDS-PAGE 胶,10%分离胶,5%浓缩胶。

(2)根据蛋白定量结果,每孔上样为 30~50μg。

(3)恒压 90V 跑浓缩胶,当溴酚蓝前沿刚进入分离胶后,恒压 120V 跑胶。

(4)湿法转膜 3h。

(5)用封闭液室温下封闭 1h。

(6)用 PBST 溶液漂洗 2~3 次,每次 5~10min。

(7)用封闭液稀释抗体,HO-1,1∶2000;GCLM,1∶1000;TrxR1,1∶1000;β-tubulin,1∶1000;二抗,1∶10000;p38,1∶1000;p-p38,1∶1000;Akt1,1∶1000;p-Akt1,1∶1000;p-Erk,1∶1000;JNK,1∶2000;p-JNK,1∶2000。以稀释后的抗体室温下孵育含有目标蛋白的 PVDF 膜 1h。

(8)回收抗体,用 PBST 缓冲液漂洗含有目标的蛋白的 PVDF 3 次,每次 5~10min。

(9)再以稀释后的二抗孵育含有目标蛋白的 PVDF 膜 1h。

(10)用 PBST 缓冲液漂洗含有目标蛋白的 PVDF 膜 3 次,每次 5~10min。

(11)NBT/NCPI 显色(2~3min),水漂洗终止显色,室温晾干。

(12)用 Perfection V300 Photo 扫描仪扫描。

2.5 免疫细胞化学法检测 Nrf2 蛋白核移位

(1)弃 12 孔板中培养基,用暖的 PBS(1×)清洗 2 遍。

(2)加入固定液(甲醇∶丙酮,1∶1),在室温条件下固定 30min。

(3)PBS(1×)清洗 3 遍,每次 5min。

(4)以含有 0.5% TRITON 的 PBST 溶液孵育 30min。

(5)去除 PBST 溶液,加入 10%的山羊血清(1×PBS 稀释)在室温下孵育 30min。

(6)去除孵育液,加入 1∶100 的 Nrf2 抗体(1×PBS 稀释)在 4℃孵育过夜,然后取出在 37℃孵育 30min。

(7)用 PBS 洗 2 遍,每次 5min,以 Alexa fluor 488 标记的二抗(1∶500,1×PBS 稀释)在 37℃条件下孵育 1h。

(8)1×PBS 清洗 3 遍,每次 5min,然后与 DAPI 孵育 5min,再以 1×PBS 清洗 3 遍。

(9)在激光共聚焦显微镜下观察 Nrf2 的位置并拍照。

2.6 统计学分析

所有结果表示为平均值±SD($n=3$),采用 t 检验或单因素方差分析中的 Student-Newman-Keules 方法对结果进行统计学分析。*($P<0.05$)为差异显著,代表处理组与空白组的数理统计;♯($P<0.05$)为差异显著,代表处理组与处理组的数理统计。

3 结果与讨论

3.1 EECP、EEPG、EEBGP 和 EEEP 对 RAW264.7 细胞的毒性

为了进一步研究蜂胶的抗氧化活性及其分子机制,用 RAW264.7 细胞建立氧化应激模型。EECP、EEPG、EEEP 和 EEBGP 对 RAW264.7 细胞的毒性作用以 CCK-8 方法测定,结果如图 6.3 所示。结果表明,在一定浓度范围内,所有样品对 RAW264.7 细胞没有毒性,但是,不同蜂胶的安全浓度不一样。其中,EEEP 的毒性效果最大(3.75μg/mL),EEBGP 的毒性效果最低(100μg/mL),EECP(5μg/mL)的毒性也要大于 EEPG(15μg/mL)的毒性。以下实验都在安全浓度范围内进行。

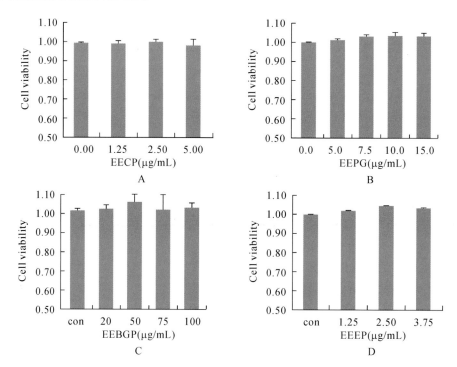

图 6.3 EECP、EEPG、EEBGP 和 EEEP 对 RAW264.7 细胞的细胞活性效果
Fig. 6.3 Effects of EECP, EEPG, EEBGP and EEEP on cell viabilities of RAW264.7 cells
在 96 孔板中培养 RAW264.7 细胞 24h 后,以所示浓度的 EECP、EEPG、EEBGP 和 EEEP 处理 24h,用 CCK-8 试剂盒测定细胞活力(%)。结果表示为平均值±标准差($n=3$)。

3.2 EECP、EEPG、EEBGP 和 EEEP 对 RAW264.7 细胞中 H_2O_2 诱导的 ROS 的清除活性

3.2.1 EECP 和 EEPG 对 RAW264.7 细胞中 H_2O_2 诱导产生的 ROS 的清除活性

蜂胶能通过有效抑制活性氧(ROS)的产生而保护机体免受氧化损伤(Tetsuro et al., 2012)。运用流式细胞术检测 EECP 和 EEPG 对 RAW264.7 细胞中 ROS 含量进行测定,结果如图 6.4 所示。从图 6.4 可知,H_2O_2 可有效刺激 RAW264.7 细胞产生大量的 ROS,使细胞处于氧化应激状态,而 EECP 和 EEPG 都能有效降低细胞中 ROS 含量,并且使细胞中的 ROS 含量低于正常状态时的含量。同时,EECP 和 EEPG 也可以在一定程度上抑制正常

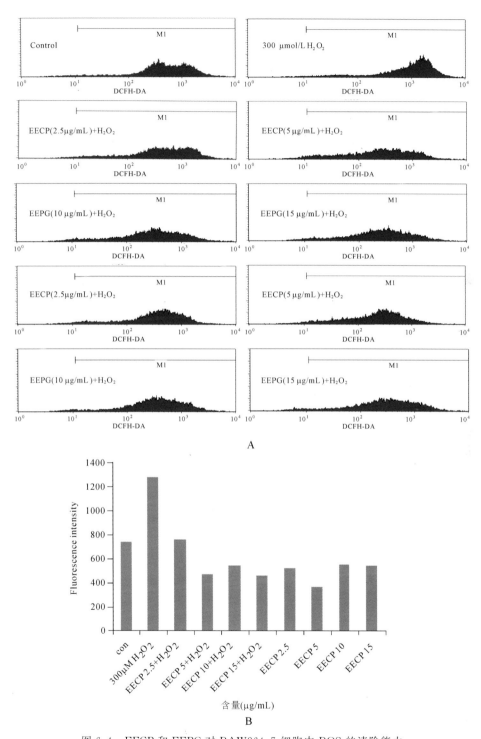

图 6.4　EECP 和 EEPG 对 RAW264.7 细胞中 ROS 的清除能力

Fig. 6.4　The ROS elimination effects of EECP and EEEP in RAW264.7 cells

A:流式细胞术测定 ROS 含量的直观结果图。B:每条柱状图代表细胞中 ROS 含量的具体数值,此图为辅助的直观反应 ROS 含量变化,因此无标注标准差。

RAW264.7 细胞的 ROS 产生。EECP 和 EEPG 在最大有效浓度下,对 H_2O_2 诱导的 ROS 清除效果差异不明显,但是在正常细胞里,EECP 的清除活性要强于 EEPG。

3.2.2 EECP、EEEP 和 EEBGP 对 RAW264.7 细胞中 H_2O_2 诱导的 ROS 的清除活性

运用流式细胞术进一步比较三种植物来源蜂胶对 ROS 的清除活性,结果表示为对 ROS 的抑制率(如图 6.5B),从图 6.5 可知,在蜂胶最大耐受浓度下,EECP 的清除活性最强,抑制率达到了 60%;其次为 EEBGP;EEEP 最弱。尽管 EEEP 的自由基清除活性和还原力较强,但在 RAW264.7 细胞中对 ROS 的清除活性最弱,这也在一定程度上反映了植物化学方法不能很准确地反映天然化合物的抗氧化能力,只能在一定程度上发挥指示作用。EECP(5μg/mL)的 ROS 清除活性显著高于 EEEP(3.75μg/mL),而两者的浓度很接近。很多研究指出,ROS 具有重要的调节功能,ROS 被过多的清除会激活细胞信号通路进而引发慢性疾病(Finley et al.,2011)。本研究发现,EEBGP 和 EECP 都能有效地减少 H_2O_2 诱导正常细胞产生的 ROS,而 EEEP 相对较弱,但三者都能降低正常细胞的 ROS 含量。最重要的是,机体持续产生的 ROS 需要被抗氧化系统调节,从而使 ROS 发挥正常功能,减少氧化损伤(Halliwell,2011),而 EECP、EEEP 和 EEBGP 能改善调节过程。

3.3 EECP、EEPG、EEBGP 和 EEEP 对抗氧化相关基因(HO-1、TrxR1、GCLM、GCLC)表达的影响

3.3.1 EECP 和 EEPG 对抗氧化相关基因(HO-1、TrxR1、GCLM、GCLC)在 mRNA 水平表达影响的比较

天然产物不仅以直接清除自由基的方式发挥抗氧化活性,还能通过激活抗氧化相关基因的表达进而发挥抗氧化作用(Finley et al.,2011)。正如我们所预测的,EECP 和 EEPG 都能明显地激活抗氧化基因的表达,如图 6.6 所示。EECP 和 EEPG 能浓度依赖性和时间依赖性地激活 HO-1、GCLM 和 TrxR1 的表达,但是对 GCLC 的作用不是很明显。从图可知,在最大安全浓度下,对 EECP 和 EEPG 的作用效果进行比较具有一定的合理性,统计分析结果(图 6.6A,C)表明 EECP 对 HO-1 和 GCLM 的作用效果要强于 EEPG,但是 EEPG 能更持久地激活 HO-1 和 GCLM 的表达。同时,低浓度的 EECP(1.25μg/mL、2.5μg/mL)的激活效果要显著低于高浓度 EECP(5μg/mL)的激活效果,EEPG 则随着浓度的递增显示出较为平缓的增长趋势。

3.3.2 EECP、EEEP 和 EEBGP 对抗氧化相关基因(HO-1、TrxR1、GCLM 和 GCLC)在 mRNA 水平表达影响的比较

三种不同植物来源蜂胶(EECP、EEEP、EEBGP)处理 RAW264.7 细胞不同的时间,不同的浓度处理 RAW264.7 细胞相同时间,结果如图 6.7 所示。从图中可知,与 EECP 和 EEBGP 相比,EEEP 对 HO-1、TrxR1、GCLM 和 GCLC 的作用效果很弱,而 EECP 和 EEBGP 能在很大程度上激活 HO-1、TrxR1 和 GCLM 基因的表达,但对 GCLC 的作用效果也不明显,激活的方式都显示出时间依赖性和浓度依赖性。血红素加氧酶(HO)可将氧化的亚铁血红素降解为胆绿、游离铁和一氧化碳而减少氧化损伤(Kikuchi et al.,2005)。与 EECP 相比,EEEP 和 EEBGP 对抗氧化基因的激活最强时间要滞后,而且 EEBGP 表现出更持久的激活效果。HO-1 在机体内发挥着重要作用,HO-1 能够保护内皮细胞免受氧化损伤并且降低前炎症因子的产生(Jeong et al.,2006),HO-1 的缺失会导致严重且持久的内皮损伤(Kawashima et al.,2002)。

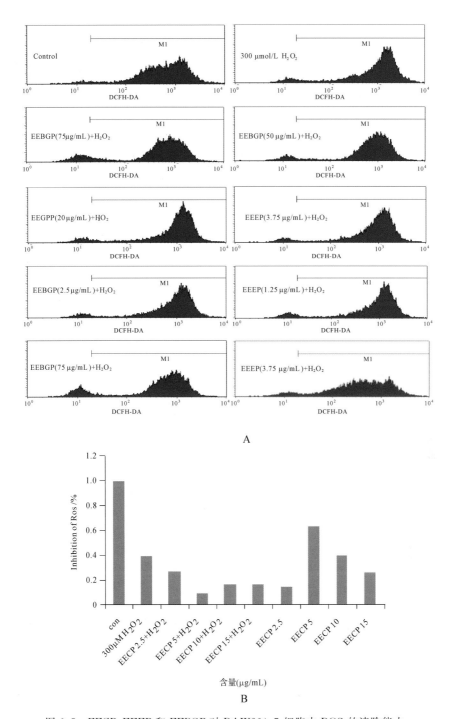

图 6.5 EECP、EEEP 和 EEBGP 对 RAW264.7 细胞中 ROS 的清除能力
Fig. 6.5 ROS elimination effects of EECP, EEEP and EEBGP in RAW264.7 cells

A：流式细胞术测定 ROS 含量的直观结果图。B：每条柱状图代表蜂胶对细胞中 ROS 产生量的抑制率（％），此图为辅助的直观反应 ROS 含量变化，故此无标注标准差。

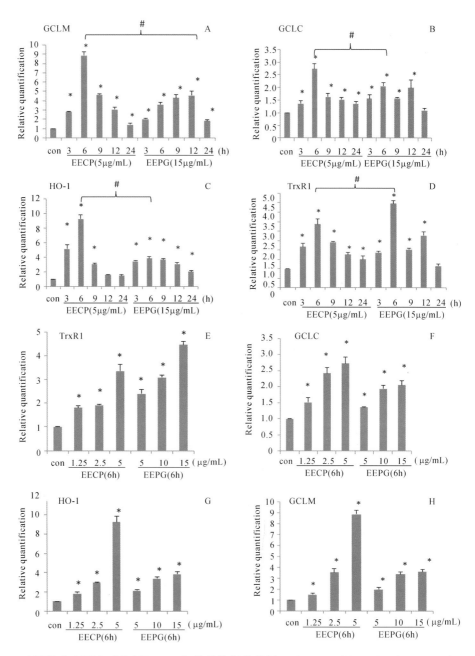

图 6.6 EECP 和 EEPG 对 RAW264.7 细胞的抗氧化基因（HO-1、GCLM、TrxR1 和 GCLC）在 mRNA 水平表达的作用

Fig. 6.6 Determine the expressions of antioxidant genes (HO-1, GCLM, TrxR1 and GCLC) at mRNA level regulated by EECP and EEPG

A~D：测定 EECP 和 EEPG 对抗氧化基因表达的时间依赖关系，以 EECP（5μg/mL）、EEPG（15μg/mL）处理细胞相应的时间。E~H：用相应浓度的 EECP 和 EEPG 处理 RAW264.7 细胞 6h。结果以内参基因 GAPDH 为标准，并表示为平均值±标准差（$n=3$）。

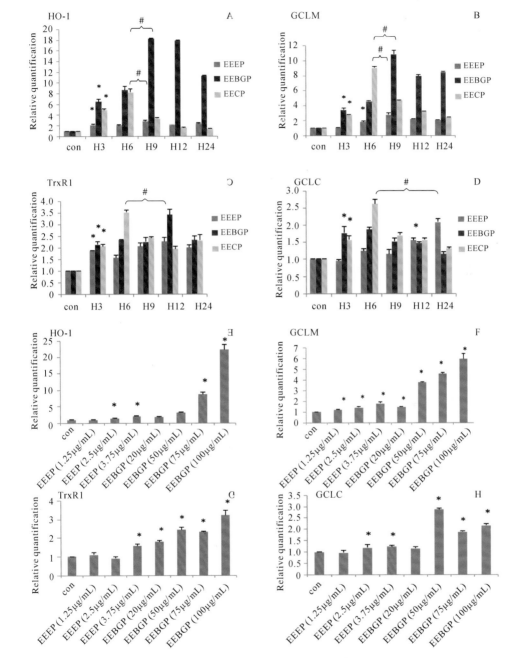

图 6.7 EECP、EEEP 和 EEBGP 对 RAW264.7 细胞的抗氧化基因（HO-1、GCLM、TrxR1 和 GCLC）在 mRNA 水平表达的作用

Fig. 6.7 Determine the expressions of antioxidant genes (HO-1, GCLM, TrxR1 and GCLC) at mRNA level regulated by EECP, EEEP and EEBGP

A～D：EECP (5μg/mL)、EEEP (3.75μg/mL) 和 EEBGP(75μg/mL) 处理 RAW264.7 细胞相应的时间。E～H：用相应浓度的 EEEP 和 EEBGP 处理 RAW264.7 细胞 6h。结果以内参基因 GAPDH 为标准，并表示为平均值±标准差 ($n=3$)。

另一方面,硫氧还蛋白还原酶(TrxR)、硫氧还蛋白(Trx)和 NADP(H)组成硫氧还蛋白系统,该系统能有效调节多个氧化应激相关的紊乱(Huang et al.,2014)。与 HO-1 和 GCLM 的表达量相比,三种植物来源蜂胶对 TrxR1 的激活效果较弱,但也有明显的激活效果。TrxR 能催化多种抗氧化化合物的产生,如维生素 C、含硒物质、硫辛酸和辅酶 Q(Jonas and Arner,2001)。另有文献表明,氨溴索(ambroxol)能加强细胞清除 ROS 的能力,而这种能力会因 Trx 的缺少而明显被削弱(Huang et al.,2014)。更有趣的是,有文献指出,虽然 TrxR1 的 mRNA 表达上调,但是其活性却降低了,这可能是由于 TrxR1 蛋白的合成缺乏硒代半胱氨酸(Fusako et al.,2011)。因此,我们推测蜂胶与硒的复合制剂具有更强的药理学活性,这有待进一步的研究证实。

谷胱甘肽系统是另一个重要的细胞内氧化还原状态调节系统,谷氨酸盐-半胱氨酸连接酶(GCL)是谷胱甘肽合成的限速酶,它由催化作用的亚基(GCLC)和修饰作用的亚基(GCLM)组成(Levonen et al.,2004)。图 6.7 结果表明,在 EECP、EEEP 和 EEBGP 的作用下,GCLM 的表达量明显增加,而其中 EEEP 的作用效果最弱,EEBGP 的作用效果最强。与 HO-1 和 TrxR1 的效果一致,EECP 的作用最强时间点在 6h 左右,而 EEEP 和 EEBGP 在 9h 左右。然而,GCLC 的表达量只是轻微的上调,几乎没有什么变化。因此,EECP、EEEP 和 EEBGP 主要通过激活 GCLM 而激活 GSH。临床研究揭示了冠状动脉疾病与血清中低 GSH 含量相关(Morrison et al.,1999)。同时,巨噬细胞的细胞内 GSH 含量决定了细胞对氧化脂肪和其他前氧化剂的应答反应(Bea et al.,2003)。

3.4 EECP、EEPG、EEEP 和 EEBGP 对抗氧化相关基因(HO-1、TrxR1、GCLM)在蛋白水平表达的作用

不同浓度的蜂胶和树胶处理 RAW264.7 细胞不同的时间,然后收集细胞蛋白,用 western blotting 检测 HO-1、TrxR1 和 GCLM 在蛋白水平的表达情况。结果如图 6.8 所示。

由图 6.8A~B 可知,EECP 和 EEPG 都能有效增强 GCLM、HO-1 和 TrxR1 的表达,且呈现出时间依赖性和浓度依赖性,其中,对 HO-1 的激活效果最明显。而且图 6.7A 显示出 EEPG 的激活效果要滞后于 EECP。

进一步对 EECP、EEBGP 和 EEEP 的作用效果进行比较,发现 EEBGP 促进 GCLM、HO-1 和 TrxR1 在蛋白水平表达的效果最显著,其次是 EECP,而 EEEP 能轻微地促进 GCLM 和 TrxR1 的表达,对 HO-1 效果不明显(数据未显示)(图 6.8)。另外,EEBGP 的激活效果明显滞后于 EECP。EEEP 对 GCLM 的激活效果也滞后于 EECP,但是对 TrxR1 的效果则与 EECP 相似。从图 6.8A、C、E 可知,EEBGP 对 GCLM 的激活效果要显著强于 EECP 和 EEEP。EECP、EEBGP 和 EEEP 对 HO-1、GCLM 和 TrxR1 在蛋白水平的激活效果与 mRNA 水平一致。

3.5 EECP、EEPG、EEEP 和 EEBGP 对 MAPKs 在抗氧化基因表达上的调节作用

3.5.1 EECP、EEPG、EEEP 和 EEBGP 对 MAPKs 蛋白表达的影响

促分裂原激活蛋白激酶(MAPKs)参与多个信号通路的调节。本研究运用 western blot 检测蜂胶乙醇提取物对 MAPKs(p38/p-p38、Erk/p-Erk、JNK1/p-JNK1、AKt/p-Akt)蛋白表达的调节效果。结果如图 6.9 所示。

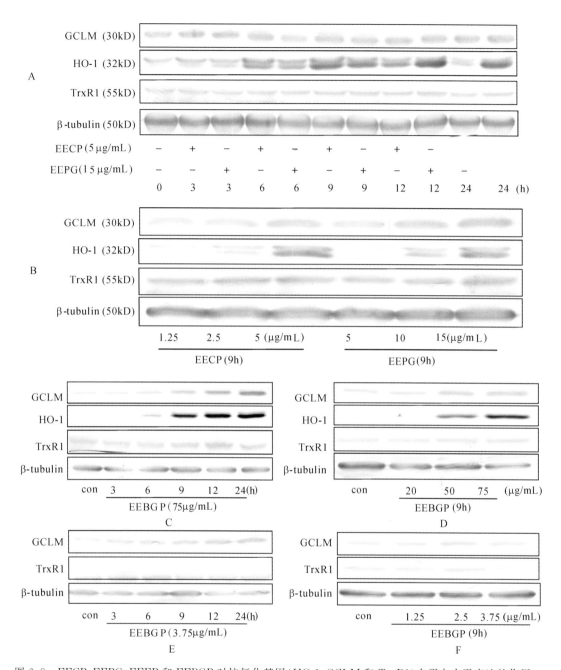

图 6.8　EECP、EEPG、EEEP 和 EEBGP 对抗氧化基因（HO-1、GCLM 和 TrxR1）在蛋白水平表达的作用

Fig. 6.8　Determine the expressions of antioxidant genes (HO-1, GCLM and TrxR1) at protein level regulated by EECP, EEPG, EEEP and EEBGP

由图 6.9 可知，EECP、EEPG、EEBGP 和 EEEP 都能加强 p-p38 和 p-Erk 的表达，而且 EECP、EEPG、EEBGP 和 EEEP 对 p-Erk 的刺激作用强于 p-p38。另外，EEEP 对 p-Erk 的刺激作用弱于 EECP 和 EEBGP。但它们对 JNK/p-JNK 和 Akt/p-Akt 的表达未表现出促进作用。

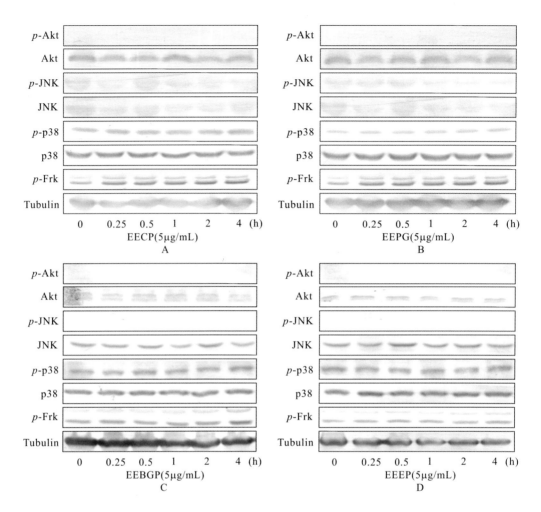

图 6.9 EECP、EEPG、EEEP 和 EEBGP 对 MAPKs 蛋白表达调节作用

Fig. 6.9 Determine the expressions of MAPKs protein regulated by EECP, EEPG, EEEP and EEBGP

3.5.2 EECP、EEPG、EEEP 和 EEBGP 对 MAPKs 在 HO-1、TrxR1 和 GCLM 基因表达上的调节作用

采用 MAPKs 的特定激酶抑制剂(LY294002、Akt/p-Akt 抑制剂;SP600125、JNK/p-JNK 抑制剂;SB203580、p38/p-p38 抑制剂;PD98059、Erk/p-Erk 抑制剂)抑制激酶的表达,进而检测 HO-1、TrxR1 和 GCLM 的蛋白表达,从而验证 p38/p-p38 和 Erk/p-Erk 激酶是否参与蜂胶调节 HO-1、TrxR1 和 GCLM 基因表达的过程。结果如图 6.10 所示。

从图 6.10 A～C 可知,抑制剂能有效抑制 EECP、EEPG、EEBGP 和 EEEP 对 p-p38 和 p-Erk 的激活。为了观察抑制剂对 RAW264.7 细胞的作用效果(避免抑制剂造成的假阳性),我们在未加蜂胶样品处理 RAW264.7 细胞的情况下,以抑制剂处理 RAW264.7 细胞,然后测定 HO-1、GCLM 和 TrxR1 的蛋白表达效果。结果如图 6.10D 所示。发现 SP600125、LY294002、PD98059 均对 HO-1、GCLM 和 TrxR1 的表达无影响,而 p38/p-p38 的抑制剂 SB203580 处理 RAW264.7 细胞后,HO-1、GCLM 和 TrxR1(TrxR1 结果未展示)的蛋白表达量明显升高。有研究指出,SB203580 是 p38α 和 p38β 亚基的特定抑制剂,而且能有效提高 HepG2 细胞中 ARE 报告基因的活性(Keum et al., 2006)。

为了验证 p38/p-p38 和 p-Erk 的作用,我们进一步检测抑制剂处理细胞后,EECP、EEPG、EEEP 和 EEBGP 对 HO-1、GCLM 和 TrxR1 基因在蛋白水平表达的作用。结果如图 6.10E～H 所示。由图可知,在 Erk/p-Erk 的抑制剂 PD98059 作用下,EECP、EEPG、EEEP 和 EEBGP 对 HO-1、GCLM 和 TrxR1 蛋白表达的促进效果明显减弱,由此可知,EECP、EEPG、EEEP 和 EEBGP 通过上调 Erk/p-Erk 激酶激活 HO-1、GCLM 和 TrxR1 的表达(TrxR1 的效果不是很明显,条带亮度低,故此未展示)。

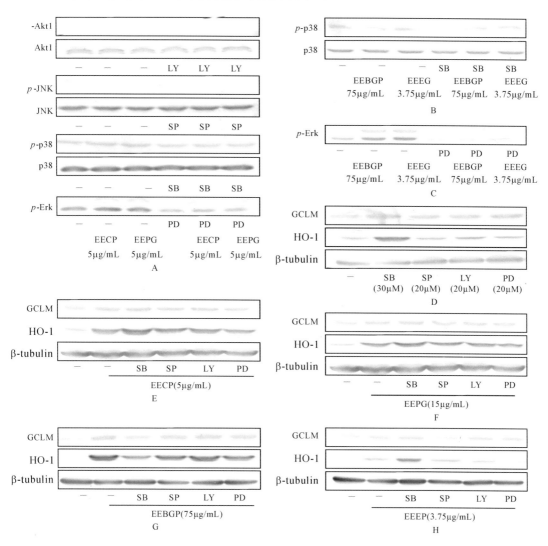

图 6.10 EECP、EEPG、EEEP 和 EEBGP 对 MAPKs 蛋白在 HO-1、TrxR1 和 GCLM 蛋白表达上的调节作用

Fig. 6.10 Determine the regulation effects of MAPKs on the expression of HO-1, TrxR1 and GCLM stimulated by EECP, EEPG, EEEP and EEBGP

D～H:抑制剂(30μM SB,20μM SP,20μM LY,20μM PD)预处理 RAW264.7 细胞 0.5h,然后以 EECP、EEPG、EEBGP 和 EEEP 处理细胞 5h,换新培养基继续培养 4h,收集细胞总蛋白。

然而,p38/p-p38 抑制剂(SB203580)抑制 p-p38 的表达后,不仅不能抑制 EECP、EEPG 和 EEEP 对 HO-1、GCLM 和 TrxR1 的激活效果,其表达量反而增加(图 6.10E、F、H)。这可能是由于 SB203580 会激活 HO-1、GCLM 和 TrxR1 蛋白表达,而 EECP、EEPG 和 EEEP 并不能通过上调 p-p38 蛋白表达而促进 HO-1、GCLM 和 TrxR1 蛋白表达。有趣的是,SB203580 能显著抑制 EEBGP 对 HO-1、GCLM 和 TrxR1 蛋白的激活效果(图 6.10G)。在人乳腺上皮细胞中,表没食子酸(epigallocatechin gallate)能通过激活 PI3K 和 ERK 激酶的途径诱导 Nrf2 调控的抗氧化酶的表达(Na et al.,2008)。然而,绿原酸能抑制醋氨酚(acetaminophen)诱导的 ERK1/2、JNK、p38 激酶磷酸化,进而增强抗氧化防御系统,保护肝脏(Ji et al.,2013)。相反,三白草酮(sauchinone)通过 p38 激酶和 Nrf2/ARE 通路上调 HO-1 的表达,进而抑制叔丁基氢过氧化物(t-butyl hydroperoxide)诱导的活性氧的产生(Jeong et al.,2010)。同时,血根碱(sanguinarine)通过激活 p38 激酶/Nrf2 通路增强抗氧化防御系统(Vrba et al.,2012)。而且,过表达 p38α、β 和 δ 亚基抑制 HO-1 的表达,而过表达 p38γ 亚基激活 HO-1 的表达(Kietzmann et al.,2003)。由此可见,p38/p-p38 的调节作用极其复杂,其具体发挥抑制作用还是激活作用可能与细胞或机体所处的应激环境密切相关。

3.6 EECP、EEPG、EEEP 和 EEBGP 对 Nrf2 核转移的影响

核因子 E2 相关蛋白 2(Nrf2)是氧化还原敏感转录因子,Nrf2 激活后从 Keap1-Nrf2 复合物上释放出来并进入细胞核内并与 DNA 上的抗氧化反应原件结合,进一步诱导抗氧化和细胞防御相关基因的表达(Kensler et al.,2007)。为了检测 EECP、EEPG、EEEP 和 EEBGP 对 Nrf2 的核转移效果,以 EECP、EEPG、EEEP 和 EEBGP 处理 RAW264.7 细胞 4h,然后用免疫荧光激光共聚焦显微镜观察 Nrf2 的位置,并拍照。结果如图 6.11 所示。在正常情况下,荧光主要聚集于细胞质,呈环状(图 6.11),当 EECP、EEPG、EEEP 和 EEBGP 处理后,荧光主要聚集于细胞核,呈点状(图 6.11)。Srivastava 等在研究 Nrf2 的核移位效果时,得到了相似的免疫荧光结果(Srivastava et al.,2013)。有文献指出,三七总皂苷(panax notoginseng saponins)通过 Nrf2 通路激活 HO-1 和 GSTP1(glutathione S-transferase pi 1) 的表达,进而改善 H_2O_2 诱导的细胞凋亡(Zhou et al.,2014)。同时,安卡黄素通过 Nrf2 通路调节氧化应激状态,改善肺损伤和呼吸道炎症(Hsu et al.,2012)。而且,大量研究已证明 p38 和 Erk 激酶后磷酸化,进一步激活 Nrf2,进而促进抗氧化基因的表达(Jiang et al.,2014;Zipper and Mulcahy,2000;Zipper and Mulcahy,2003)。

小 结

本研究采用细胞实验,采用流式细胞术、qRT-PCR、western blotting 以及免疫细胞荧光方法分别对 EECP、EEPG、EEBGP 和 EEEP 的 ROS 清除能力、基因(HO-1、GCLM、GCLC、TrxR1 和 MAPKs)表达以及转录因子 Nrf2 的核转移效果进行了分析,发现 EECP、EEPG 和 EEEP 主要通过 Erk/p-Erk-Nrf2-抗氧化基因(HO-1、GCLM、GCLC 和 TrxR1)发挥抗氧化作用,而 EEBGP 主要通过 p38/p-p38 和 Erk/p-Erk-Nrf2-抗氧化基因(HO-1、GCLM、GCLC 和 TrxR1)发挥抗氧化作用。同时,EECP、EEPG、EEBGP 和 EEEP 还可以直接清除 ROS。

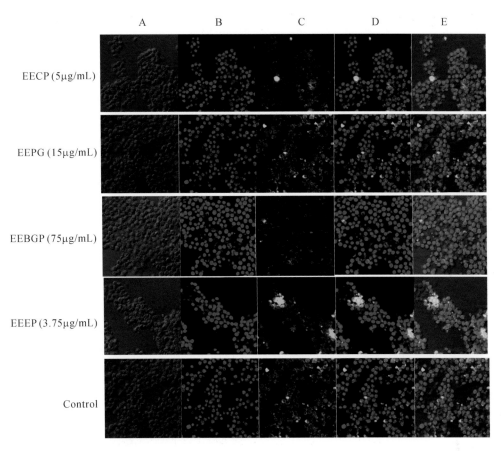

图 6.11 EECP、EEPG、EEBGP 和 EEEP 对 RAW264.7 细胞中 Nrf2 的核转移效果

Fig. 6.11 Nucleus translocation effects of Nrf2 stimulated by EECP, EEPG, EEBGP and EEEP in RAW264.7 cells

A:RAW264.7 细胞形态;B:DAPI 染细胞核图;C:Alexa Fluor 488 标记的抗兔抗体染色 RAW264.7 细胞的 Nrf2 图片;D:DAPI 染色图和 Alexa Fluor 488 染色图的融合图;E:DAPI 图、Alexa Fluor 488 图和细胞形态图的融合图。

参考文献

Abu-Mellal A, Koolaji N, Duke RK, Tran VH, Duke CC (2012) Prenylated cinnamate and stilbenes from Kangaroo Island propolis and their antioxidant activity[J]. *Phytochemistry*, 77:251-259.

Amoros M, Lurton E, Boustie J, Girre L, Sauvager F, Cormier M (1994) Comparison of the anti-herpes simplex virus activities of propolis and 3-methyl-but-2-enyl caffeate[J]. *Journal of Natural Products*, 57(5):644-647.

Attia AA, ElMazoudy RH, El-Shenawy NS (2012) Antioxidant role of propolis extract against oxidative damage of testicular tissue induced by insecticide chlorpyrifos in rats [J]. *Pesticide Biochemistry and Physiology*, 103(2):87-93.

Azevedo Bentes Monteiro Neto M, de Souza Lima IM, Furtado RA, Bastos JK, da Silva Filho AA, Tavares DC (2011) Antigenotoxicity of Artepillin C *in vivo* evaluated by the micronucleus and comet assays[J]. *Journal of Applied Toxicology*, 31(8):714-719.

Bankova V, Dyulgerov A, Popov S, Evstatieva L, Kuleva L, Pureb O, Zamjansan Z (1992) Propolis produced in Bulgaria and Mongolia: Phenolic compounds and plant origin[J]. *Apidologie*, 23(1):79-85.

Bankova V, Christov R, Kujumgiev A, Marcuccic MC, Popov S (1995) Chemical composition and antibacterial activity of Brazilian propolis[J]. *Zeitschrift für Naturforschung*, 50c(3-4):167-172.

Banskota A, Tezuka Y, Adnyana I, Ishii E, Midorikawa K, Matsushige K, Kadota S (2001) Hepatoprotective and anti-Helicobacter pylori activities of constituents from Brazilian propolis[J]. *Phytomedicine*, 8(1):16-23.

Banskota AH, Tezuka Y, Adnyana IK, Midorikawa K, Matsushige K, Message D, Huertas AA, Kadota S (2000) Cytotoxic, hepatoprotective and free radical scavenging effects of propolis from Brazil, Peru, the Netherlands and China [J]. *Journal of Ethnopharmacology*, 72(1):239-246.

Bea F, Hudson FN, Chait A, Kavanagh TJ, Rosenfeld ME (2003) Induction of glutathione synthesis in macrophages by oxidized low-density lipoproteins is mediated by consensus antioxidant response elements[J]. *Circulation Research*, 92(4):386-393.

Bufalo MC, Candeias JM, Sforcin JM (2009) *In vitro* cytotoxic effect of Brazilian green propolis on human laryngeal epidermoid carcinoma (HEp-2) cells[J]. *Evidence-Based Complementary and Alternative Medicine*, 6(4):483-487.

Capucho C, Sette R, de Souza Predes F, de Castro Monteiro J, Pigoso AA, Barbieri R, Dolder MAH, Severi-Aguiar GD (2012) Green Brazilian propolis effects on sperm count and epididymis morphology and oxidative stress[J]. *Food and Chemical Toxicology*, 50(11):3956-3962.

Cigut T, Polak T, Gašperlin L, Raspor P, Jamnik P (2011) Antioxidative activity of propolis extract in yeast cells[J]. *Journal of Agricultural and Food Chemistry*, 59(21):11449-11455.

Dejian H, Boxin O, Prior RL (2005) The chemistry behind antioxidant capacity assays[J]. *Journal of Agricultural and Food Chemistry*, 53:1841-1856.

Edwin NF, Meyer AS (2000) The problems of using one-dimensional methods to evaluate multifunctional food and biological antioxidants[J]. *Journal of the Science of Food and Agriculture*, 80:1925-1941.

Eggler AL, Gay KA, Mesecar AD (2008) Molecular mechanisms of natural products in chemoprevention: Induction of cytoprotective enzymes by Nrf2[J]. *Molecular Nutrition & Food Research*, 52(S1):S84-S94.

Enis Yonar M, Mişe Yonar S, Silici S (2011) Protective effect of propolis against oxidative stress and immunosuppression induced by oxytetracycline in rainbow trout (*Oncorhynchus mykiss*, W.)[J]. *Fish & Shellfish Immunology*, 31(2):318-325.

Enis Yonar M, Yonar SM, Ural MŞ, Silici S, Düşükcan M (2012) Protective role of propolis in chlorpyrifos-induced changes in the haematological parameters and the oxidative/antioxidative status of *Cyprinus carpio carpio*[J]. *Food and Chemical Toxicology*, 50 (8):2703-2708.

Erdogan S, Ates B, Durmaz G, Yilmaz I, Seckin T (2011) Pressurized liquid extraction of phenolic compounds from Anatolia propolis and their radical scavenging capacities[J]. *Food and Chemical Toxicology*, 49(7):1592-1597.

Finley JW, Kong AN, Hintze KJ, Jeffery EH, Ji LL, Lei XG (2011) Antioxidants in foods: state of the science important to the food industry[J]. *Journal of Agricultural and Food Chemistry*, 59(13):6837-6846.

Fonseca YM, Marquele-Oliveira F, Vicentini FT, Furtado NAJ, Sousa JPB, Lucisano-Valim YM, Fonseca MJV (2011) Evaluation of the potential of Brazilian propolis against UV-induced oxidative stress[J]. *Evidence-Based Complementary and Alternative Medicine*, 2011:863917.

Fusako U, Akio Y, Fujimura M (2011) Post-transcriptional defects of antioxidant selenoenzymes cause oxidative stress under methylmercury exposure[J]. *Journal of Biological Chemistry*, 286:6641-6649.

Gazzani G, Daglia M, Papetti A (2012) Food components with anticaries activity[J]. *Current Opinion in Biotechnology*, 23(2):153-159.

Gregoris E, Fabris S, Bertelle M, Grassato L, Stevanato R (2011) Propolis as potential cosmeceutical sunscreen agent for its combined photoprotective and antioxidant properties[J]. *International Journal of Pharmaceutics*, 405(1):97-101.

Guo X, Chen B, Luo L, Zhang X, Dai X, Gong S (2011) Chemical compositions and antioxidant activities of water extracts of Chinese propolis[J]. *Journal of Agricultural and Food Chemistry*, 59(23):12610-12616.

Halliwell B (2011) Free radicals and antioxidants-quo vadis? [J] *Trends in Pharmacological Sciences*, 32(3):125-130.

Hayacibara MF, Koo H, Rosalen PL, Duarte S, Franco EM, Bowen WH, Ikegaki M, Cury JA (2005) *In vitro* and *in vivo* effects of isolated fractions of Brazilian propolis on caries development[J]. *Journal of Ethnopharmacology*, 101(1):110-115.

Hegazi AG, Abd El Hady FK, Abd Allah FA (2000) Chemical composition and antimicrobial activity of European propolis[J]. *Zeitschrift Für Naturforschung*, 55c(1-2):70-75.

Hernandez J, Goycoolea FM, Quintero J, Acosta A, Castañeda M, Dominguez Z, Robles R, Vazquez-Moreno L, Velazquez EF, Astiazaran H (2007) Sonoran propolis: Chemical composition and antiproliferative activity on cancer cell lines[J]. *Planta Medica*, 73 (14):1469-1474.

Hsu WH, Lee BH, Huang YC, Hsu YW, Pan TM (2012) Ankaflavin, a novel Nrf-2 activator for attenuating allergic airway inflammation[J]. *Free Radical Biology & Medicine*, 53(9):1643-1651.

Huang J, Xu J, Tian L, Zhong L (2014) A thioredoxin reductase and/or thioredoxin system-based mechanism for antioxidant effects of ambroxol[J]. *Biochimie*, 97: 92-103.

Isidorov VA, Vinogorova VT (2003) GC-MS analysis of compounds extracted from buds of *Populus balsamifera* and *Populus nigra*[J]. *Zeitschrift Für Naturforschung C*, 58(5/6): 355-360.

Jeong GS, Oh GS, Pae HO, Jeong SO, Kim YC, Shin MK, Seo BY, Han SY, Lee HS, Jeong JG, Koh JS, Chung HT (2006) Comparative effects of curcuminoids on endothelial heme oxygenase-1 expression: Ortho-methoxy groups are essential to enhance heme oxygenase activity and protection[J]. *Experimental and Molecular Medicine*, 38(4): 393-400.

Jeong GS, Lee DS, Li B, Byun E, Kwon DY, Park H, Kim YC (2010) Protective effect of sauchinone by upregulating heme oxygenase-1 via the P38 MAPK and Nrf2/ARE pathways in HepG2 cells[J]. *Planta Medica*, 76(1): 41-47.

Ji L, Jiang P, Lu B, Sheng Y, Wang X, Wang Z (2013) Chlorogenic acid, a dietary polyphenol, protects acetaminophen-induced liver injury and its mechanism[J]. The *Journal of Nutritional Bochemistry*, 24(11): 1911-1919.

Jiang G, Hu Y, Liu L, Cai J, Peng C, Li Q (2014) Gastrodin protects against MPP^+-induced oxidative stress by up regulates heme oxygenase-1 expression through p38 MAPK/Nrf2 pathway in human dopaminergic cells[J]. *Neurochemistry International*, 75: 79-88.

Jonas N, Arner ESJ (2001) Reactive oxygen species, antioxidants, and the mamalian thioredoxin system[J]. *Free Radical Biology & Medicine*, 31(11): 1287-1312.

Jorge R, Furtado N, Sousa J, da Silva Filho A, Gregório Junior L, Martins C, Soares A, Bastos J, Cunha W, Silva M (2008) Brazilian propolis: Seasonal variation of the prenylated *p*-coumaric acids and antimicrobial activity[J]. *Pharmaceutical Biology*, 46(12): 889-893.

Kamiya T, Izumi M, Hara H, Adachi T (2012) Propolis suppresses $CdCl_2$-induced cytotoxicity of COS7 cells through the prevention of intracellular reactive oxygen species accumulation[J]. *Biological and Pharmaceutical Bulletin*, 35(7): 1126-1131.

Kawashima A, Oda Y, Yachie A, Koizumi S, Nakanishi I (2002) Heme oxygenase-1 deficiency: The first autopsy case[J]. *Human Pathology*, 33(1): 125-130.

Kawashima T, Manda S, Uto Y, Ohkubo K, Hori H, Matsumoto K-i, Fukuhara K, Ikota N, Fukuzumi S, Ozawa T (2012) Kinetics and mechanism for the scavenging reaction of the 2, 2-diphenyl-1-picrylhydrazyl radical by synthetic artepillin c analogues[J]. *Bulletin of the Chemical Society of Japan*, 85(8): 877-883.

Kensler TW, Wakabayashi N, Biswal S (2007) Cell survival responses to environmental stresses via the Keap1-Nrf2-ARE pathway[J]. *Annual Review of Pharmacology and Toxicology*, 47: 89-116.

Keum YS,Yu S,Chang PP,Yuan X,Kim JH,Xu C,Han J,Agarwal A,Kong AN (2006) Mechanism of action of sulforaphane: Inhibition of p38 mitogen-activated protein kinase isoforms contributing to the induction of antioxidant response element-mediated heme oxygenase-1 in human hepatoma HepG2 cells[J]. *Cancer Research*,66(17):8804-8813.

Kietzmann T,Samoylenko A,Immenschuh S (2003) Transcriptional regulation of heme oxygenase-1 gene expression by MAP kinases of the JNK and p38 pathways in primary cultures of rat hepatocytes[J]. *The Journal of Biological Chemistry*,278(20):17927-17936.

Kikuchi G,Yoshida T,Noguchi M (2005) Heme oxygenase and heme degradation[J]. *Biochemical and Biophysical Research Communications*,338(1):558-567.

Laskar RA,Sk I,Roy N,Begum NA (2010) Antioxidant activity of Indian propolis and its chemical constituents[J]. *Food Chemistry*,122(1):233-237.

Levites Y,Weinreb O,Maor G,Youdim MB,Mandel S (2001) Green tea polyphenol (−)-epigallocatechin-3-gallate prevents N-methyl-4-phenyl-1, 2, 3, 6-tetrahydropyridine-induced dopaminergic neurodegeneration[J]. *Journal of Neurochemistry*,78(5):1073-1082.

Levonen AL,Landar A,Ramachandran A,Ceaser EK,Dickinson DA,Zanoni G,Morrow JD,Darley-Usmar VM (2004) Cellular mechanisms of redox cell signalling: role of cysteine modification in controlling antioxidant defences in response to electrophilic lipid oxidation products[J]. *Biochemical Journal*,378(2):373-382.

Li F,Awale S,Zhang H,Tezuka Y,Esumi H,Kadota S (2009) Chemical constituents of propolis from Myanmar and their preferential cytotoxicity against a human pancreatic cancer cell line[J]. *Journal of Natural Products*,72(7):1283-1287.

Li Y,Chen M,Xuan H, Hu F (2011) Effects of encapsulated propolis on blood glycemic control,lipid metabolism,and insulin resistance in type 2 diabetes mellitus rats[J]. *Evidence-Based Complementary and Alternative Medicine*,2012:981896.

Lima B,Tapia A,Luna L,Fabani MP,Schmeda-Hirschmann G,Podio NS,Wunderlin DA,Feresin GE (2009) Main flavonoids,DPPH activity,and metal content allow determination of the geographical origin of propolis from the province of San Juan (Argentina)[J]. *Journal of Agricultural and Food Chemistry*,57(7):2691-2698.

López-Alarcón C,Denicola A (2013) Evaluating the antioxidant capacity of natural products:A review on chemical and cellular-based assays[J]. *Analytica Chimica Acta*,763(1):1-10.

Majiene D,Trumbeckaite S,Pavilonis A,Savickas A,Martirosyan D (2007) Antifungal and antibacterial activity of propolis[J]. *Current Nutrition & Food Science*,3(4):304-308.

Mavri A,Abramovič H,Polak T,Bertoncelj J,Jamnik P,Smole Možina S,Jeršek B (2012) Chemical properties and antioxidant and antimicrobial activities of Slovenian propolis[J]. *Chemistry & Biodiversity*,9(8):1545-1558.

Miguel MG, Nunes S, Dandlen SA, Cavaco AM, Antunes MD (2010) Phenols and antioxidant activity of hydro-alcoholic extracts of propolis from Algarve, South of Portugal[J]. *Food and Chemical Toxicology*, 48(12): 3418-3423.

Morrison JA, Jacobsen DW, Sprecher DL, Robinson K, Khoury P, Daniels SR (1999) Serum glutathione in adolescent males predicts parental coronary heart disease[J]. *Circulation*, 100(22): 2244-2247.

Moţ AC, Silaghi-Dumitrescu R, Sârbu C (2011) Rapid and effective evaluation of the antioxidant capacity of propolis extracts using DPPH bleaching kinetic profiles, FT-IR and UV-vis spectroscopic data[J]. *Journal of Food Composition and Analysis*, 24(4): 516-522.

Moura SALd, Negri G, Salatino A, Lima LDdC, Dourado LPA, Mendes JB, Andrade SP, Ferreira MAND, Cara DC (2011) Aqueous extract of Brazilian green propolis: Primary components, evaluation of inflammation and wound healing by using subcutaneous implanted sponges[J]. *Evidence-Based Complementary and Alternative Medicine*, 2011: 748283.

Na HK, Kim EH, Jung JH, Lee HH, Hyun JW, Surh YJ (2008) (−)-Epigallocatechin gallate induces Nrf2-mediated antioxidant enzyme expression via activation of PI3K and ERK in human mammary epithelial cells[J]. *Archives of Biochemistry and Biophysics*, 476(2): 171-177.

Nakajima Y, Shimazawa M, Mishima S, Hara H (2007) Water extract of propolis and its main constituents, caffeoylquinic acid derivatives, exert neuroprotective effects *via* antioxidant actions[J]. *Life Sciences*, 80(4): 370-377.

Nordberg J, Arnér ES (2001) Reactive oxygen species, antioxidants, and the mammalian thioredoxin system[J]. *Free Radical Biology and Medicine*, 31(11): 1287-1312.

Oldoni TLC, Cabral IS, d'Arce MA, Rosalen PL, Ikegaki M, Nascimento AM, Alencar SM (2011) Isolation and analysis of bioactive isoflavonoids and chalcone from a new type of Brazilian propolis[J]. *Separation and Purification Technology*, 77(2): 208-213.

Özeren M, Sucu N, Tamer L, Aytacoglu B, Bayrı Ö, Döndaş A, Ayaz L, Dikmengil M (2005) Caffeic acid phenethyl ester (CAPE) supplemented St. Thomas' hospital cardioplegic solution improves the antioxidant defense system of rat myocardium during ischemia-reperfusion injury[J]. *Pharmacological Research*, 52(3): 258-263.

Ozyurt B, Parlaktas BS, Ozyurt H, Aslan H, Ekici F, Atis Ö (2007a) A preliminary study of the levels of testis oxidative stress parameters after MK-801-induced experimental psychosis model: Protective effects of CAPE[J]. *Toxicology*, 230(1): 83-89.

Ozyurt H, Ozyurt B, Koca K, Ozgocmen S (2007b) Caffeic acid phenethyl ester (CAPE) protects rat skeletal muscle against ischemia-reperfusion-induced oxidative stress[J]. *Vascular Pharmacology*, 47(2): 108-112.

Park YK, Ikegaki M (1998) Preparation of water and ethanolic extracts of propolis and evaluation of the preparations[J]. *Bioscience, Biotechnology and Biochemistry*, 62

(11):2230-2232.

Petrova A,Popova M,Kuzmanova C,Tsvetkova I,Naydenski H,Muli E,Bankova V (2010) New biologically active compounds from Kenyan propolis[J]. *Fitoterapia*,81(6):509-514.

Potkonjak NI,Veselinovic DS,Novakovic MM,Gorjanovic SŽ,Pezo LL,Sužnjevic DŽ (2012) Antioxidant activity of propolis extracts from Serbia:A polarographic approach [J]. *Food and Chemical Toxicology*,50(10):3614-3618.

Qian YP,Shang YJ,Teng QF,Chang J,Fan GJ,Wei X,Li RR,Li HP,Yao XJ,Dai F (2011) Hydroxychalcones as potent antioxidants:Structure-activity relationship analysis and mechanism considerations[J]. *Food Chemistry*,126(1):241-248.

Righi AA,Alves TR,Negri G,Marques LM,Breyer H,Salatino A (2011) Brazilian red propolis:unreported substances,antioxidant and antimicrobial activities[J]. *Journal of the Science of Food and Agriculture*,91(13):2363-2370.

Sforcin J (2007) Propolis and the immune system:A review[J]. *Journal of Ethnopharmacology*,113(1):1-14.

Shi H,Yang H,Zhang X,Yu LL (2012) Identification and quantification of phytochemical composition and anti-inflammatory and radical scavenging properties of methanolic extracts of Chinese propolis[J]. *Journal of Agricultural and Food Chemistry*,60(50):12403-12410.

Shimizu K,Ashida H,Matsuura Y,Kanazawa K (2004) Antioxidative bioavailability of artepillin C in Brazilian propolis[J]. *Archives of Biochemistry and Biophysics*,424(2):181-188.

Silva V,Genta G,Möller MaN,Masner M,Thomson L,Romero N,Radi R,Fernandes DC,Laurindo FR,Heinzen H (2011) Antioxidant activity of Uruguayan propolis. *In vitro* and cellular assays[J]. *Journal of Agricultural and Food Chemistry*,59(12):6430-6437.

Simões-Ambrosio L,Gregório L,Sousa J,Figueiredo-Rinhel A,Azzolini A,Bastos J,Lucisano-Valim Y (2010) The role of seasonality on the inhibitory effect of Brazilian green propolis on the oxidative metabolism of neutrophils[J]. *Fitoterapia*,81(8):1102-1108.

Song GL,Du Q (2011) Antioxidant activity comparation of polysaccharides from nine traditional edible fungi in China[J]. *Biomedical Engineering and Informatics*,4:1196-1200.

Srivastava S,Alfieri A,Siow RC,Mann GE,Fraser PA (2013) Temporal and spatial distribution of Nrf2 in rat brain following stroke:Quantification of nuclear to cytoplasmic Nrf2 content using a novel immunohistochemical technique[J]. *The Journal of Physiology*,591(14):3525-3538.

Sulaiman GM,Sammarrae KWA,Ad'hiah AH,Zucchetti M,Frapolli R,Bello E,Erba E,D'Incalci M,Bagnati R (2011) Chemical characterization of Iraqi propolis samples and assessing their antioxidant potentials[J]. *Food and Chemical Toxicology*, 49(9): 2415-2421.

Tetsuro K, Misato I, Hirokazu H, Adachi T (2012) Propolis suppresses CdCl2-induced cytotoxicity of COS7 cells through the prevention of intracellular reactive oxygen species accumulation[J]. *Biological and Pharmaceutical Bulletin*, 35(7): 1126-1131.

Thirugnanasampandan R, Raveendran SB, Jayakumar R (2012) Analysis of chemical composition and bioactive property evaluation of Indian propolis[J]. *Asian Pacific Journal of Tropical Biomedicine*, 2(8): 651-654.

Valencia D, Alday E, Robles-Zepeda R, Garibay-Escobar A, Galvez-Ruiz JC, Salas-Reyes M, Jiménez-Estrada M, Velazquez-Contreras E, Hernandez J, Velazquez C (2012) Seasonal effect on chemical composition and biological activities of Sonoran propolis[J]. *Food Chemistry*, 131(2): 645-651.

Valente MJ, Baltazar AF, Henrique R, Estevinho L, Carvalho M (2011) Biological activities of Portuguese propolis: Protection against free radical-induced erythrocyte damage and inhibition of human renal cancer cell growth *in vitro*[J]. *Food and Chemical Toxicology*, 49(1): 86-92.

Varì R, D'Archivio M, Filesi C, Carotenuto S, Scazzocchio B, Santangelo C, Giovannini C, Masella R (2011) Protocatechuic acid induces antioxidant/detoxifying enzyme expression through JNK-mediated Nrf2 activation in murine macrophages[J]. *The Journal of Nutritional Biochemistry*, 22(5): 409-417.

Velazquez C, Navarro M, Acosta A, Angulo A, Dominguez Z, Robles R, Robles-Zepeda R, Lugo E, Goycoolea F, Velazquez E (2007) Antibacterial and free-radical scavenging activities of Sonoran propolis[J]. *Journal of Applied Microbiology*, 103(5): 1747-1756.

Vrba J, Orolinova E, Ulrichova J (2012) Induction of heme oxygenase-1 by Macleaya cordata extract and its constituent sanguinarine in RAW264.7 cells[J]. *Fitoterapia*, 83(2): 329-335.

Wang K, Ping S, Huang S, Hu L, Xuan H, Zhang C, Hu F (2013) Molecular mechanisms underlying the *in vitro* anti-inflammatory effects of a flavonoid-rich ethanol extract from Chinese propolis (poplar type)[J]. *Evidence-Based Complementary and Alternative Medicine*, 2013: 127672.

Wang X, Stavchansky S, Kerwin SM, Bowman PD (2010) Structure-activity relationships in the cytoprotective effect of caffeic acid phenethyl ester (CAPE) and fluorinated derivatives: Effects on heme oxygenase-1 induction and antioxidant activities[J]. *European Journal of Pharmacology*, 635(1): 16-22.

Xuan H, Zhao J, Miao J, Li Y, Chu Y, Hu F (2011) Effect of Brazilian propolis on human umbilical vein endothelial cell apoptosis[J]. *Food and Chemical Toxicology*, 49(1): 78-85.

Yang H, Dong Y, Du H, Shi H, Peng Y, Li X (2011) Antioxidant compounds from propolis collected in Anhui, China[J]. *Molecules*, 16(4): 3444-3455.

Yeh CT, Ching LC, Yen GC (2009) Inducing gene expression of cardiac antioxidant enzymes by dietary phenolic acids in rats[J]. *The Journal of Nutritional Biochemistry*, 20(3):163-171.

Zhang CP, Zheng HQ, Liu G, Hu FL (2011) Development and validation of HPLC method for determination of salicin in poplar buds: Application for screening of counterfeit propolis[J]. *Food Chemistry*, 127(1):345-350.

Zhang CP, Huang S, Wei WT, Ping S, Shen XG, Li YJ, Hu FL (2014) Development of high-performance liquid chromatographic for quality and authenticity control of Chinese propolis. *Journal of Food Science*, 79(7):C1315-1322.

Zhou LZH, Johnson AP, Rando TA (2001) NFκB and AP-1 mediate transcriptional responses to oxidative stress in skeletal muscle cells[J]. *Free Radical Biology and Medicine*, 31(11):1405-1416.

Zhou N, Tang Y, Keep RF, Ma X, Xiang J (2014) Antioxidative effects of *Panax* notoginseng saponins in brain cells[J]. *Phytomedicine*, 21(10):1189-1195.

Zhu W, Li YH, Chen ML, Hu FL (2011) Protective effects of Chinese and Brazilian propolis treatment against hepatorenal lesion in diabetic rats[J]. *Human & Experimental Toxicology*, 30(9):1246-1255.

Zipper LM, Mulcahy RT (2000) Inhibition of ERK and p38 MAP kinases inhibits binding of Nrf2 and induction of GCS genes[J]. *Biochemical and Biophysical Research Communications*, 278(2):484-492.

Zipper LM, Mulcahy RT (2003) Erk activation is required for Nrf2 nuclear localization during pyrrolidine dithiocarbamate induction of glutamate cysteine ligase modulatory gene expression in HepG2 cells[J]. *Toxicological Sciences*, 73(1):124-134.

胡福良,李英华,陈民利,应华忠,朱威(2003)蜂胶醇提液和水提液对急性炎症动物模型的作用[J].浙江大学学报(农业与生命科学版),29(4):444-448.

胡福良,李英华,朱威,陈民利,应华忠(2005)蜂胶对大鼠佐剂性关节炎的作用及其机制的研究[J].中国药学杂志,40(15):1146-1148.

胡福良,李英华,朱威,陈民利,应华忠(2007)蜂胶提取液对大鼠急性胸膜炎的作用及其机制的研究[J].营养学报,29(2):189-191.

第七章 蜂胶对炎症性疾病及炎症微环境的影响及其作用机制研究

炎症通过影响机体微环境中多种细胞与因子的相互作用，参与调控机体多种生理、病理过程。尽管炎症本身并不被认为是癌症等复杂疾病的起始致病因素，但当前已有越来越多的研究发现炎症在很多复杂疾病（如癌症、心脑血管疾病、代谢性疾病等）的发生发展进程中起重要作用，利用体内动物模型或体外细胞实验，探讨炎症疾病发生发展过程，开发新型高效低毒的抗炎药物已成为当前研究人员关注的热点内容。

蜂胶是一种历史十分悠久的传统天然药物，其良好的抗炎效果，也是蜂胶最早受到关注的药理学活性之一，但对其抗炎机制等仍缺乏系统性研究。此外，由于蜂胶的化学组成与其植物来源、地理来源密切相关，这两大因素是否会对蜂胶生物学活性的发挥产生直接影响，目前仍缺乏直接证据。

基于前期研究基础，本研究利用细菌脂多糖诱导建立小鼠急性肺损伤模型、硫酸葡聚糖诱导建立大鼠溃疡性结肠炎模型，研究了蜂胶对这两种代表性炎症性疾病的影响，利用不同体外细胞实验模型研究了蜂胶对炎症微环境的影响，并探讨蜂胶可能存在的抗炎机制，为蜂胶在治疗炎症性疾病中的临床应用提供理论依据。

第一节 中国蜂胶及其植物来源杨树芽提取物对细菌脂多糖诱导小鼠急性肺损伤的影响研究

由于蜜蜂采集的植物芽苞和树木渗出的胶状物来自于不同地区的不同植物，导致其所得的蜂胶在化学组成和生物学活性上存在很大的差异（Toreti et al.，2013）。在欧洲和亚洲，包括中国和其他温带地区，蜂胶的主要植物来源是杨树（*Populus* spp.）（Wu et al.，2008；Sforcin et al.，2011）。这些区域的蜂胶具有相似的化学特点，即富含黄酮类、酚酸及其酯类化合物（Toreti et al.，2013）。杨树芽提取物和杨树型蜂胶也具有相似的化学组成，但其在具体数量上可能存在着差异。杨树型蜂胶和杨树芽主要的活性成分是柯因、高良姜素、松属素、槲皮素、山柰酚和一些酚酸（CAPE，*p*-香豆酸等）（Rubiolo et al.，2013）。

研究发现，源自酒神菊属植物（*Baccharis*）的巴西绿蜂胶在化学组成上与其亲本植株非常相似，它们均富含戊烯对香豆酸，主要为阿替匹林C和香草素。此外，体内和体外的比较研究发现，巴西绿蜂胶和酒神菊树的提取物均展现出显著的抗氧化（Guimaraes et al.，2012）、抗菌（da Silva Filho et al.，2008）、抗溃疡（Lemos et al.，2007）、抗利什曼病（da Silva Filho et al.，2009）、抗突变（Munari et al.，2008）、免疫调节（Missima et al.，2007）和抗炎（dos Santos et al.，2010；Cestari et al.，2011）效果。然而迄今为止，对于中国杨树型蜂胶提取物及其植物来源杨树芽的生物活性的比较研究非常有限。因此,我们从中国杨树型蜂胶和杨树芽(*Populus×Canadensis*)乙醇提取物的生物活性比较出发,研究了二者体外自由基清除活力,并通过体内试验研究了这两种提取物在细菌脂多糖诱导下对小鼠急性肺损伤的治疗效果,为蜂胶对炎症性疾病的治疗提供生物学证据。

1 实验材料与方法

1.1 试剂

DPPH、ABTS、维生素 E、细菌脂多糖(*Escherichia coli* 0111:B4)、没食子酸、槲皮素及高效液相色谱分析中所用的标准品,均购自美国 Sigma 公司。福林酚试剂、格里斯(Griess)试剂、亚硝酸钠和地塞米松购自上海生工生物技术有限公司。其他化学试剂均为分析纯,购自上海生工生物技术有限公司。

1.2 样品的收集和提取

中国蜂胶样本于 2010 年夏天取自山东省的意大利蜂(*Apis mellifera ligustica*)蜂群。其主要的植物源为杨树(*Populus* spp.),属于杨柳科。称取 100g 蜂胶,采用研磨机研磨成粉状,用 1L 95%(v/v)乙醇提取蜂胶样本,40℃超声 3h。超声后的上清液用 4 号滤纸过滤取出残留杂质。将残留杂质重新采用 95%(v/v)乙醇浸润超声 3h 收集过滤清液,重复如上操作,共提取 3 次。之后,集中收集全部的滤液采用 50℃减压旋转蒸发仪蒸发回收。最后,采用烘箱烘烤干燥直至提取物恒重,得到蜂胶乙醇提取物,将提取物保存于 -20℃冰箱直至使用。加拿大杨树芽提取物购自市场,使用前溶于 1L 95%(v/v)乙醇进行提取,提取步骤同中国蜂胶。动物实验中,1g 中国蜂胶乙醇提取物(CP)和杨树芽乙醇提取物(PB)采用 0.5%的黄芪胶溶解,得到 25mg/mL 的贮备液后备用。CP 和 PB 样本中总黄酮和总酚酸的含量分别采用福林酚法和氯化铝比色法进行检测(Ahn et al.,2007)。总黄酮和总酚酸分别采用没食子酸当量和槲皮素当量来表示。

1.3 高效液相色谱分析

蜂胶中水杨苷的检测参照我们团队前期报道的检测方法,用于鉴别、区分本研究中所使用的中国蜂胶乙醇提取物和杨树芽乙醇提取物(Zhang et al.,2011)。黄酮类和酚酸类化合物定性定量检测采用安捷伦高效液相色谱分析,系统为 Agilent Eclipse XDB-C18 色谱柱(4.6mm×150mm,5μm),柱温为 30℃,流动相为乙腈(A)和 4%乙酸(B),流动速度 1.0mL/min。梯度洗脱程序如下:0~40min,5%~25%(A);40~45min,25%~35%(A);55~60min,35%~40%(A);80~90min,40%~100%(A)。检测波长为 280nm。所有的样本均采用 0.22μm 滤膜过滤,进样量为 5μL。

1.4 中国蜂胶乙醇提取物和杨树芽乙醇提取物的体外抗氧化活性检测

1.4.1 DPPH 自由基清除活性检测

样本的 DPPH 自由基清除活性检测参考文献(Ahn et al.,2007),并进行一些改进。反应前,将样本进行梯度稀释,浓度梯度在 10~150μg/mL。在 96 孔板中将 100μL 样本和 100μL 50μg/mL 的 DPPH 溶液混合,避光孵育 30min,检测 517nm 波长下的吸光度。所有的测定均有 3 个重复。样本的自由基清除活力表示为 IC50 值,即清除 50% DPPH 自由基所需的浓度。

1.4.2 ABTS自由基清除活性检测

ABTS自由基清除活性检测采用改良的ABTS阳离子自由基消色反应(Shi et al., 2012)。中国蜂胶乙醇提取物和杨树芽乙醇提取物溶液处理同DPPH检测。ABTS反应液采用7mmol/L的ABTS水溶液15mL与140mmol/L的过硫酸钾反应制得,该溶液在使用前室温避光放置16h。检测开始前,ABTS工作液采用甲醇稀释,使其在室温平衡后在734nm波长下波长下吸光度为0.70±0.02。96孔板中反应体系为:50μL样本和100μL ABTS工作液。反应液混合均匀后室温避光稳定10min后检测734nm吸光度。所有的测定均有3个重复。样本的自由基清除活力表示为IC50值,即清除50%ABTS自由基所需的浓度。

1.5 动物实验

1.5.1 实验动物

雄性ICR小鼠(6~8周龄,18~22g)购自浙江中医药大学动物实验研究中心,8只/笼饲养于光照(12h昼夜循环)和温度(20~23℃)稳定的房间。供应空气为过滤无病原菌空气,相对湿度50%。小鼠自由采食饮水,饮食供应为商业化无污染饲料。实验动物质量认证和试验设施执照分别为SYXK(浙)2008-0115和SCXK(沪)2008-0016。对实验中使用的动物均按实验动物使用的"3R"原则给以人道主义关怀。

1.5.2 动物分组和急性肺损伤诱导

经过1周的适应期,小鼠随机分为7个组($n=8$):①空白对照组,该组小鼠采用溶剂对照处理;②模型组,该组小鼠尾静脉注射1mg/kg细菌脂多糖;③阳性对照组,该组小鼠在LPS刺激前1h灌胃2mg/kg地塞米松;④CP低剂量组,该组小鼠口服25mg/kg中国蜂胶乙醇提取物后尾静脉注射LPS;⑤CP高剂量组,该组小鼠口服100mg/kg中国蜂胶乙醇提取物后尾静脉注射LPS;⑥PB低剂量组,该组小鼠口服25mg/kg杨树芽乙醇提取物后尾静脉注射LPS;⑦PB高剂量组,该组小鼠口服100mg/kg杨树芽乙醇提取物后尾静脉注射LPS。中国蜂胶、杨树芽处理组小鼠连续3d灌胃CP或PB,而其他组(空白组、模型组、阳性药物对照组)的小鼠则灌胃溶剂对照(PBS)。在第3天口服完最后一次CP或PB后1h,对所有的动物(除了空白组)经尾静脉注射细菌脂多糖来小鼠急性肺损伤,注射脂多糖3h后处死小鼠取样(Choi et al., 2013)。

1.5.3 小鼠肺部组织病理学检查

小鼠肺部样本采用10%福尔马林固定,并采用石蜡包埋,切片采用苏木精-伊红染色法对肺部组织进行染色。切片在尼康光学显微镜(Nikon eclipse 80i)下观察并拍照(Zhu et al., 2011)。

1.5.4 小鼠血清中炎症因子的检测

获得小鼠全血后,采集装入离心管中,静置过夜,血液凝固后离心(3000r/min,10min),得到上清液即为血清,分装冻存于-80℃冰箱中备用。采用BD公司(美国)微量样本多指标流式蛋白定量技术Cytometric Bead Array (CBA)检测小鼠血清中炎症因子(IL-6、IL-10、MCP-1、IFN-γ、TNF和IL-12p70)。操作步骤根据试剂盒说明书进行,血清进行25倍稀释。检测结果采用BD公司(美国)FCAP Array软件进行分析。流式检测设备为BD FACSCalibur流式细胞仪。

1.6 数据分析

文中数据采用独立实验中特定重复的平均值±标准偏差表示。数据的统计比较采用非配对学生 t 检验或者 Student-Newman-Keules 方法进行单向方差分析,P 值小于 0.05 即定义为差异显著。所有的数据统计均采用 SPSS 17.0 统计。

2 结果与分析

2.1 中国蜂胶乙醇提取物和杨树芽乙醇提取物的化学成分分析

通过高效液相色谱,我们对中国蜂胶乙醇提取物或杨树芽乙醇提取物的主要化学成分进行了分析。中国蜂胶乙醇提取物和杨树芽乙醇提取物的色谱图见图 7.1,其定量结果见表 7.1。比色法测定得到的中国蜂胶乙醇提取物和杨树芽乙醇提取物的总黄酮含量分别为 124.92±9.74mg/g 和 126.23±8.46mg/g(槲皮素当量);总酚酸含量为 233.98±

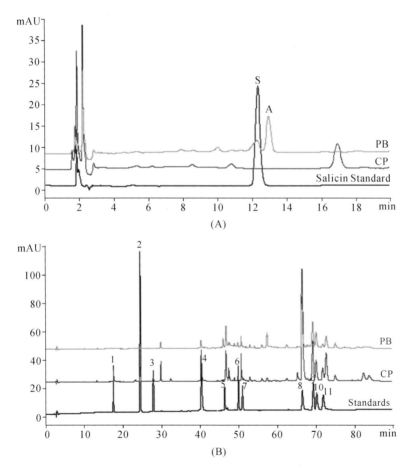

图 7.1 中国蜂胶乙醇提取物(CP)和杨树芽乙醇提取物(PB)高效液相色谱图谱
Fig 7.1 HPLC chromatograms for CP and PB.

A:CP 和 PB 中的水杨苷;B:CP(红线)和 PB(绿线)中的其他酚酸和类黄酮代表峰。S,水杨苷;A,邻苯二酚;(1)咖啡酸;(2)p-香豆酸;(3)阿魏酸;(4)白藜芦醇;(5)槲皮素;(6)芹菜素;(7)山奈酚;(8)白杨素;(9)乔松素;(10)高良姜素;(11)咖啡酸苯乙酯.

0.84mg/g 和 145.54±5.89mg/g（没食子酸当量）。我们的前期研究表明，水杨苷仅存在于杨树芽提取物中，而蜂胶无法检测到。图 7.1A 的结果清楚地区分了 CP 和 PB。此外，我们分析了杨树型蜂胶中最具代表性的 13 种黄酮及酚酸类化合物，其色谱峰如图 7.1B 所示，其相对含量见表 7.1。定性定量结果表明，中国蜂胶乙醇提取物的酚酸类化合物含量、种类较杨树芽乙醇提取物更为丰富。

表 7.1 中国蜂胶乙醇提取物（CP）和杨树芽乙醇提取物（PB）中主要的黄酮类和酚酸类化合物含量

Table 7.1 Major phenolic acids and flavonoids presented in CP and PB[a]

Peak No.	Compounds	Retention time (min)	Contents of extract(g/100g)	
			CP	PB
1	Caffeic acid	17.54	0.53±0.01	ND
2	p-coumaric acid	24.36	0.44±0.00	ND
3	Ferulic acid	27.75	0.18±0.00	ND
4	Resveratrol	40.53	1.45±0.01	0.38±0.00
5	Quercetin	46.24	0.14±0.01	1.10±0.02
6	Apigenin	49.89	0.42±0.09	0.54±0.01
7	Kaempferol	50.78	3.10±0.10	1.85±0.23
8	Chrysin	66.56	10.11±0.12	4.17±0.15
9	Pinocembrin	69.16	3.65±0.16	2.88±0.12
10	Galangin	70.04	3.21±0.24	2.83±0.21
11	CAPE	71.76	2.33±0.19	1.23±0.12

a：表中的值为平均值±标准差（$n=3$）。ND：未检测到。

2.2 中国蜂胶乙醇提取物和杨树芽乙醇提取物的体外抗氧化活性

如表 7.2 所示，中国蜂胶乙醇提取物和杨树芽乙醇提取物均表现出良好的 DPPH/ABTS 自由基清除能力。中国蜂胶乙醇提取物对 DPPH 自由基清除 IC50 值为 15.49±0.59μg/mL，低于杨树芽乙醇提取物（IC50＝28.69±1.52μg/mL），与阳性对照维生素 E 接近；中国蜂胶乙醇提取物对 ABTS 自由基清除 IC50 值为 36.66±1.82μg/mL，其自由基清除能力同样强于杨树芽乙醇提取物（IC50＝55.63±0.78μg/mL）和阳性对照维生素 E。

表 7.2 中国蜂胶乙醇提取物和杨树芽乙醇提取物中总黄酮、总酚酸含量 DPPH/ABTS 自由基清除活性

Table 7.2 Total phenolic content (TPC), Total flavonoid content (TFC) and DPPH/ABTS free-radical scavenging activities of CP and PB

Sample	TPC (mg GAE/g)	TFC (mg QE/g)	IC50(μg/mL)	
			DPPH-scavenging activity	ABTS-scavenging activity
CP	233.98±0.84	124.92±9.74	15.49±0.59	36.66±1.82
PB	145.54±5.89	126.23±8.46	28.69±1.52	55.63±0.78
α-Tocopherol	—	—	17.80±1.11	56.43±5.75

注：表中的值为平均值±标准差（$n=3$）。

2.3 中国蜂胶乙醇提取物和杨树芽乙醇提取物对细菌脂多糖诱导小鼠急性肺损伤模型的影响

与对照组相比,小鼠注射细菌脂多糖能显著诱导血清中 6 种炎症因子水平。小鼠灌胃高剂量中国蜂胶乙醇提取物和杨树芽乙醇提取物(100mg/kg 体重)能显著降低 IL-6、IL10、MCP-1、IFN-γ、TNF-α 和 IL-12p70 的含量($P<0.05$)。且二者除了对 IL-10 和 IFN-γ 调控水平有差异外,对 IL-6 等其他几个炎性因子的影响均没有明显差异($P>0.05$)。中国蜂胶乙醇提取物高剂量组显著抑制了 IL-10 的释放,但杨树芽乙醇提取物高剂量组小鼠 IL-10 水平显著上升(表 7.3)。这表明中国蜂胶乙醇提取物和杨树芽乙醇提取物均能有效抑制小鼠促炎症因子的释放,但对特定的炎症因子的调控却不完全相同。

图 7.2 表明了中国蜂胶乙醇提取物和杨树芽乙醇提取物对细菌脂多糖诱导小鼠急性肺损伤模型肺部病理组织化学的影响。苏木精-伊红染色结果表明,LPS 注射小鼠(1mg/kg,3h)肺部发生明显的肺部损伤症状,肺泡壁增厚,肺部水肿出血,并出现炎性细胞浸润(图 7.2B)。如图 7.2B 和 E 所示,100mg/kg 体重单位的中国蜂胶乙醇提取物和杨树芽乙醇提取物处理会显著地抑制 LPS 引起的组织变化,其效果与阳性药物地塞米松(2mg/kg)相类似(图 7.2C)。

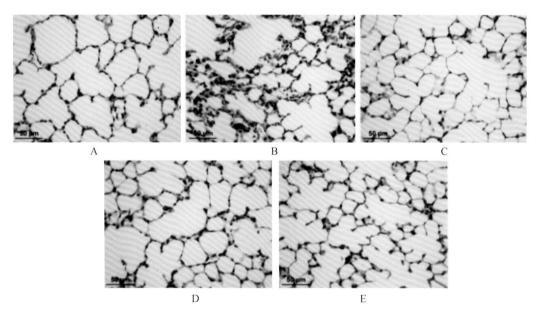

图 7.2　中国蜂胶及其植物来源杨树胶对细菌脂多糖诱导小鼠急性肺损伤模型肺部病理组织化学的影响

Fig 7.2　Effects of CP and PB on LPS-induced lung histopathologic changes

从注射细菌脂多糖 3h 后的每个实验组收集到的肺组织进行组织学评价。A:标准组;B:LPS 诱导肺急性损伤模型组;C:脂多糖刺激小鼠并以 2mg/kg 地塞米松处理组;D,E:小鼠注射脂多糖并分别以 100mg/kg 蜂胶和杨树芽处理组。用苏木精-伊红对肺部组织进行染色(放大倍数:200 倍)。

3　讨论

细菌脂多糖的发病机制很大程度上可以归因于炎症反应系统的调控紊乱,特别是组织器官的炎性损伤和各种炎症因子的过度蓄积(Choi et al.,2013)。我们的前期研究结果表明,蜂胶乙醇提取物在各种体内炎症模型中均具有较好的抗炎效果(Hu et al.,2005),同时还

表 7.3 中国蜂胶乙醇提取物和杨树芽乙醇提取物对脂多糖诱导小鼠急性炎症模型血清炎症因子的作用

Table 7.3 Effect of CP and PB on serum levels of inflammatory cytokines in LPS-challenged mice

Group	Dose (mg/kg)	IL-6 (ng/mL)	IL-10 (pg/mL)	MCP-1 (ng/mL)	IFN-γ (ng/mL)	TNF-α (pg/mL)	IL-12p70 (pg/mL)
Standard	—	ND	ND	ND	ND	ND	ND
Model	—	96.00±10.66^a	237.15±0.59^a	98.27±10.69^a	5.22±0.32^a	1897.57±93.31^a	100.60±8.08^a
CP	25	103.27±2.54^a	245.18±5.22^a	87.13±3.12^{a,b}	7.58±0.16^b	2706.41±36.23^b	97.29±3.39^{a,b}
CP	100	70.15.00±8.58^b	107.99±3.40^b	82.41±5.04^{b,c}	10.16±1.05^c	1303.18±113.43^c	67.56±8.00^b
PB	25	88.28±2.18^b	186.16±13.51^c	83.23±0.01^{a,b,c}	3.98±0.09^{a,d}	1981.88±93.87^a	114.19±25.06^a
PB	100	68.77±3.00^b	302.12±4.91^d	73.03±2.45^{b,c}	3.36±0.05^d	1263.79±74.56^{c,d}	61.91±1.69^b
Dexamethasone	2	52.05±7.37^b	180.22±11.53^c	67.93±1.18^c	2.69±0.64^d	850.82±58.44^d	6.61±0.95^c

注:图中的值为平均值±标准差($n=8$)。采用 Student-Newman-Keuls 方法对数据结果进行单因素方差分析,字母标志不同意味着差异显著($P<0.05$)。ND.未检出。

发现中国蜂胶可以显著地抑制炎症器官中 IL-6 的增加，而不能影响 IL-2 和 IFN-γ 的水平（Hu et al.，2005）。Debbache-Benaida 等（2013）研究发现白杨提取物可以在角叉菜胶引起的小鼠足肿胀模型中起到抗炎的效果。而 Machado 等（2012）则应用 LPS 诱导的小鼠模型发现巴西绿蜂胶提取物具有抗炎的效果，可以降低促炎因子（IL-6、TNF）并增加抗炎因子（IL-10、TGF-β）的产生。

为了研究中国蜂胶乙醇提取物和杨树芽乙醇提取物对急性炎症的保护作用，我们利用细菌脂多糖诱导建立了小鼠急性肺损伤模型，在了解了二者植物化学成分差异与体外抗氧化活性区别后，发现中国蜂胶乙醇提取物和杨树芽乙醇提取物均可通过抑制一些特定炎症因子的产生而在体内发挥抗炎效果，这也可能与中国蜂胶乙醇提取物和杨树芽乙醇提取物中富含的黄酮和酚酸类化合物有关。更有趣的是，口服高剂量（100mg/kg）的杨树芽提取物可以有效地抑制 IL-6、MCP-1、TNF-α 和 IL-12p70 产生，效果与中国蜂胶蜂胶提取物类似。IL-10 在机体应对炎症阶段就有重要的抗炎活性（Berbaum et al.，2008）。我们的研究发现，中国蜂胶乙醇提取物和杨树芽乙醇提取物可以有效地抑制单核巨噬细胞分泌的一些重要的单核因子，包括 IL-6、MCP-1 和 TNF-α 等。这提示中国蜂胶乙醇提取物和杨树芽乙醇提取物中的多酚类化合物也参与调节 T 细胞的淋巴因子的分泌，但二者对这一因子的调控作用模式非常不同。低剂量的中国蜂胶乙醇提取物上调了循环中的 IL-10 的水平，高剂量中国蜂胶乙醇提取物却抑制了 IL-10 的分泌；而杨树芽乙醇提取物却产生了完全相反的效果。由此，我们推测二者在抗炎的有效性上与其免疫调节活性密切相关（Machado et al.，2012）。

总之，本研究证实了中国蜂胶和杨树芽乙醇提取物具有相似的化学组成，但其各种化合物的相对含量有很大不同。中国蜂胶乙醇提取物和杨树芽乙醇提取物对细菌脂多糖诱导的急性肺损伤有着良好的预防效果。同时，本研究也证实杨树芽提取物具有良好的抗炎效果和自由基清除能力，提示杨树芽乙醇提取物也是一种具有开发利用潜力的天然产物。需要注意的是，尽管我们发现中国蜂胶和杨树芽乙醇提取物均可以发挥体内抗炎效果，但也不能排除二者是通过与不同的分子靶标相互作用而发挥抗炎效果的，这需通过后续的实验进一步加以研究证实。

小 结

本研究结果表明，中国杨树型蜂胶及杨树芽提取物的主要植物化学成分存在相似性，但中国杨树型蜂胶提取物的酚酸类化合物含量、种类较杨树芽提取物更为丰富，体外自由基清除能力也更强。利用细菌脂多糖诱导建立小鼠急性肺损伤模型，发现高剂量中国蜂胶提取物和杨树芽提取物灌胃小鼠（100mg/kg）均可抑制血清中炎症因子（白介素-6、单核细胞趋化蛋白-1、肿瘤坏死因子-α 和白介素-12p70）的产生并能显著缓解肺部炎性损伤症状。

第二节　中国蜂胶对硫酸葡聚糖诱导大鼠溃疡性结肠炎的影响研究

炎症性肠病（Inflammatory bowel disease，IBD）是一类以肠道炎症为主要特征的疾病，主要包括克罗恩病（Crohn's disease，CD）和溃疡性结肠炎（Ulcerative colitis，UC）。炎症性肠病的发病原因目前尚未完全明确，临床上炎症性肠病常表现为反复发作而治愈难度很大。

目前普遍认为其发病机制主要与宿主因素和外界因素之间复杂的相互作用密切相关。炎症性肠病发生的主要因素包括肠道微环境紊乱、免疫失调、外界环境改变、宿主的遗传因素等。据统计,全球炎症性肠病患者约 500 万人,其中美国约有 140 万人,欧洲约有 300 万人(Burisch et al.,2015)。

溃疡性结肠炎在临床较为常见,且近年来在我国的发病率呈上升趋势。该病病程长,病变范围位于结直肠的黏膜层,主要的病理表现为结直肠充血、水肿、溃疡、糜烂。当前,治疗溃疡性结肠炎的主要策略是抑制肠黏膜炎症,使用的药物主要是抗炎药,包括糖皮质激素、水杨酸类、抗生素类、免疫抑制剂如环磷酰胺等。尽管这些药物治疗方法在很多患者中取得了积极的治疗效果,但仍有相当数量的患者临床症状得不到完全改善,因此针对溃疡性结肠炎的治疗仍需研究开发新的药物治疗方法。

目前,建立溃疡性结肠炎动物模型的方法有化学药诱导法、构建基因缺失动物法、细胞移植法和免疫诱导法等四类。葡聚糖硫酸钠(Dextran sodium sulfate,DSS)是一种常见的硫酸多糖,也是构建溃疡性结肠炎最常用的一种化学药,它可破坏肠黏膜屏障并导致肠道菌群的失调,该模型因其良好的重现性被科研人员广泛采用。

前一节的研究已经发现中国蜂胶具有良好的预防由细菌脂多糖诱导的小鼠急性肺损伤症状,提示蜂胶对于预防细菌感染引起的急性炎症有着良好的干预作用。此外,有研究利用巴西蜂胶水醇提取物发现其对 2,4,6-trinitrobenzenesulfonic acid(TNBS)诱导的大鼠肠炎有着良好的预防作用(Gonçalves et al.,2013)。而 TNBS 诱导的炎症主要与 Th1 介导的免疫炎症有关,TNBS 灌肠会导致肠上皮细胞的蛋白质结构变化,形成自身抗原,引起免疫反应最终导致肠炎,这与 DSS 诱导的溃疡性结肠炎有一定的区别。本研究利用 DSS 诱导的大鼠溃疡性结肠炎模型,研究中国蜂胶乙醇提取物对 DSS 诱导大鼠肠道炎症的影响。

1 实验材料与方法

1.1 实验动物

雄性 SPF 级 SD 大鼠(45 只,8 周龄,体重 190±20g)购自澳大利亚阿德莱德大学动物实验中心。动物饲养于澳大利亚联邦科工组织食品营养研究所动物房,培养环境为 12h 光照+12h 黑暗交替,温度 23℃,相对湿度 50%～70%,自由采食和饮水。该实验严格遵守澳大利亚联邦科工组织食品营养研究所动物实验中心实验动物福利规章制度,并经澳大利亚联邦科工组织食品营养研究所实验动物研究伦理学委员会批准执行。

1.2 试剂

硫酸葡聚糖(分子量 36～50kDa)购于澳大利亚 MP Biomedical 公司;日粮各组分购自阿德莱德大学动物实验中心;气相色谱检测用氢氧化钠、磷酸、磺基水杨酸、二乙醚、挥发性脂肪酸标准品均购自美国 Sigma 公司。

1.3 日粮及动物分组

大鼠在经过 1 周的适应性饲养后,随机分为 5 组:空白对照组($n=5$);模型组($n=10$);蜂胶低剂量组($n=10$,日粮中添加 0.1% 蜂胶);蜂胶中剂量组($n=10$,日粮中添加 0.2% 蜂胶);蜂胶高剂量组($n=10$,日粮中添加 0.3% 蜂胶)。日粮配方参考美国 AIN-93M 标准配方(表 7.4)。

表 7.4　动物日粮配方
Table 7.4　Composition of control diet

Composition	Diet(g/kg)
Casein	250
Cornstarch	350
Sucrose	100
Fat Blend (Canola and Palm oils)	200
Wheat bran	50
L-Cystine	3
Choline bitartrate	2.5
Vitamins (AIN 93)	10
Minerals	35
Tert-butyl hydroquinol	0.014

1.4　溃疡性结肠炎的诱导

实验大鼠经过 1 周实验日粮饲喂后,开始溃疡性结肠炎的诱导。除空白对照组饮用蒸馏水外,其余 40 只大鼠均饮用含 3% DSS 的水,持续 7d。7d 后替换为正常蒸馏水,经过 7d 恢复期后处死大鼠,取盲肠内容物、近端结肠组织进行肠道短链脂肪酸的检测和肠道组织病理学检测。

1.5　大鼠疾病活动指数(Disease activity index,DAI)评估

DSS 饮水开始后直到实验结束,进行大鼠疾病活动指数(DAI)评估。每日观察 2 次,按表 7.5 标准根据体重下降、大便性状、隐血情况和动物总体状况评分相加,得出每只大鼠的疾病活动指数。

表 7.5　疾病活动指数
Table 7.5　Disease activity index (DAI) observations

Weight loss(%)	Stool consistency	Rectal bleeding	Overall condition[1]	Score[2]
No weight loss	Normal	No observable blood	Normal	0
0.1%～5%	Loose stool	Small amount of blood in some stool	Some signs of poor condition	1
5%～10%	Mild diarrhoea	Blood in stool regularly seen	Moderately poor condition	2
>10%	Diarrhoea	Blood in all stool	Very poor condition	3

[1] 整体状况包括活力、蜷缩、姿势和行为。
[2] 得分解释:0 代表正常,1 代表轻微程度,2 代表中等程度,3 代表严重。

1.6 盲肠内容物短链脂肪酸(SCFA)的测定

将预处理的盲肠内容物样品(500mg)加入 2mL 庚酸(Heptanoic acid)内参溶液,2000g/min 离心 10min(4℃),后吸取上清液 150μL,转移至预冷过的玻璃试管内。加入 50μL 10%的磺基水杨酸、3mL 乙醚溶液,震荡混匀。200g 离心 2min(4℃)后使用巴斯德管转移乙醚层(上清液)至新的玻璃管中,加入 50μL 氢氧化钠(0.2mol/L)溶液,震荡混匀再离心(200g,2min)。离心后弃去上清乙醚层,加入 2mL 乙醚溶液后,再离心。小心吸取上层乙醚后,将样本用氮吹仪吹干,以去除剩余乙醚后加入 30μL 1mol/L 磷酸酸化后迅速转移至预冷过的气相色谱管中并密封上机检测。采用 Agilent 6890N 气相色谱仪测定样品中乙酸、丙酸、异丁酸、丁酸、异戊酸、戊酸、己酸的含量。

1.7 大鼠结肠病理学检查

大鼠宰杀后,取 1.5cm 远端结肠组织样本采用 10%福尔马林固定,并采用石蜡包埋,切片采用苏木精-伊红染色法对结肠组织进行染色。切片在尼康光学显微镜(Nikon eclipse 80i)下观察并拍照。

1.8 数据分析

文中数据采用独立实验中特定重复的平均值±标准偏差表示。数据的统计比较采用非配对学生 t 检验或者 Student-Newman-Keules 方法进行单向方差分析。P 值小于 0.05 即定义为差异显著。所有的数据统计均采用 SPSS 17.0 统计。

2 结果与分析

2.1 蜂胶对溃疡性结肠炎大鼠日增重及器官指数的影响

蜂胶对溃疡性结肠炎大鼠体重指标的影响结果见图 7.3。由图可见,随着大鼠日龄的增加,大鼠体重呈现逐渐上升的趋势,但与对照组相比,可以发现在第 7 天 DSS 饮水开始后,大鼠体重上升趋势明显减缓,日粮中加入 0.1%蜂胶体重下降最为明显,日粮添加 0.2%蜂胶、0.3%蜂胶处理组与 DSS 组相比,没有显著差异。

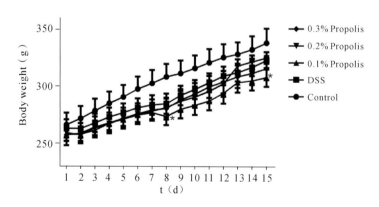

图 7.3 蜂胶对溃疡性结肠炎大鼠体重指标的影响

Fig 7.3 Effects of propolis on the body weight changes of colitis changes

蜂胶对溃疡性结肠炎大鼠器官指数的影响结果见表 7.6。结果表明,除 0.3%蜂胶处理组外,各剂量蜂胶处理并没有对大鼠各器官指数(肝脏指数、肾脏指数、体脂蓄积比)产生显著影响。与对照组相比,DSS 处理能显著上调脾脏指数,但 0.2%蜂胶处理组的脾脏指数较DSS 组显著上升,说明蜂胶具有一定的免疫调节作用。

表 7.6 蜂胶对溃疡性结肠炎大鼠器官指数的影响
Table 7.6 Effects of propolis on organ indices changes of colitis rats

	Treatment group				
	DSS	0.1% Propolis	0.2% Propolis	0.3% Propolis	Control
Total fat pad	2.08±0.38^a	1.84±0.29^a	2.3±0.68^{a,b}	2.04±0.29^a	2.72±0.56^b
Liver	4.05±0.35^a	3.85±0.22^{a,b}	4.31±0.58^a	4.05±0.28^a	3.55±0.25^b
Spleen	0.27±0.04^a	0.27±0.04^a	0.44±0.42^b	0.25±0.03^a	0.23±0.01^a
Kidney	0.75±0.05	0.74±0.05	0.77±0.12	0.74±0.04	0.71±0.02

2.2 蜂胶对溃疡性结肠炎大鼠疾病活动指数的影响

蜂胶对 DSS 诱导的大鼠疾病活动指数的影响结果见图 7.4。由图可见,与正常对照组相比,模型组大鼠从第 4 天开始出现明显的溃疡性结肠炎症状,包括精神状态下降、腹泻、便血等,DSS 开始处理后第 8 天 DAI 指数达到高峰。与 DSS 模型组相比,0.2%和 0.3%蜂胶处理即对大鼠疾病活动指数有一定影响。造模第 7 天起,0.3%蜂胶处理组即能显著降低大鼠疾病活动指数。

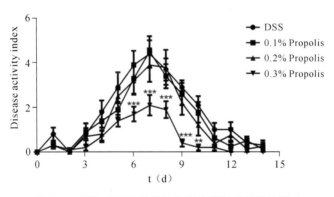

图 7.4 蜂胶对 DSS 诱导的大鼠疾病活动指数的影响

Fig 7.4 Effects of Propolis on DAI induced by DSS in rats. Disease activity index (DAI) scores in rats

2.3 蜂胶对溃疡性结肠炎大鼠结肠形态及组织学损伤的影响

各组大鼠结肠组织切片如图 7.5 所示,光镜结果表明,与对照组完整结肠黏膜结构相比,DSS 可对肠黏膜造成严重影响,黏膜和黏膜下层多处可见溃疡及炎性渗出物,并伴随着水肿和大量中性粒细胞浸润。与 DSS 组相比,0.3%蜂胶干预可使大鼠结肠上述病理症状有着明显程度的减轻。结肠黏膜腺体排列整齐,肉眼少见水肿,基本未见溃疡和炎性渗出物。

 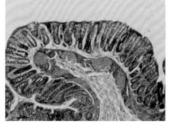 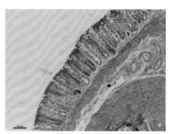

DSS　　　　　　　　　0.3% Propolis　　　　　　　　Control

图 7.5　各组大鼠结肠组织切片图

Fig 7.5　Histological analysis of DSS treated rats

(A)空白对照喂养的大鼠的结肠部位显示正常结构(放大 400 倍);(B)葡聚糖硫酸钠处理小鼠的结肠组织显示地下室炎症和中度破坏;(C)0.3%蜂胶处理的小鼠的结肠组织出现炎症和轻微损伤,损伤程度明显低于 DSS 组(放大 400 倍)。

2.4　蜂胶对溃疡性结肠炎大鼠盲肠总短链脂肪酸 SCFAs 产生的影响

蜂胶对溃疡性结肠炎大鼠盲肠总短链脂肪酸产生的影响结果见表 7.7。与对照组相比,DSS 大鼠盲肠内容物中乙酸、丙酸和总短链脂肪酸的含量均有所上升,但丁酸含量有所降低($P<0.05$)。在蜂胶组中,0.1%蜂胶处理显著增加了盲肠内容物总短链脂肪酸的含量,但 0.2%和 0.3%蜂胶处理均明显降低了盲肠内容物总短链脂肪酸的含量,表明大鼠肠道代谢产生的短链脂肪酸受摄入蜂胶剂量的影响。结合大鼠疾病活动指数结果,我们推测溃疡性结肠炎严重程度与肠道盲肠内短链脂肪酸有一定负相关联系。

表 7.7　蜂胶对溃疡性结肠炎大鼠盲肠总短链脂肪酸产生的影响

Table 7-7　Effects of propolis on SCFAs production in DSS induced colitis rats

Component (μmol/L)	Treatment group				
	DSS	0.1% Propolis	0.2% Propolis	0.3% Propolis	control
Acetate	81.9±18.1[ab]	89.5±15[a]	66.8±11.6[c]	74.0±9.4[bc]	60.8±9.9[c]
Propionate	20.9±3.7[a]	19.5±2.2[ab]	17.2±2.2[b]	19.0±2.6[ab]	17.4±1.7[b]
Butyrate	13.4±2.9	12.6±3.7	12.1±2.7	12.7±1.9	15.1±2.8
Valerate	2.8±0.5	2.8±0.6	2.8±0.4	3.0±0.3	3.2±0.2
Caproate	0.3±0.2	0.3±0.3	0.4±0.3	0.2±0.1	0.2±0.2
Total VFA	119.2±23.3[a]	124.7±11.8[a]	99.3±14.3[b]	108.8±12.9[b]	96.7±12.5[b]

3　讨论

在正常情况下,机体肠道中出现的轻微炎症是对细菌、食品中感染原的正常生理反应。但当炎症介质持续存在并不断刺激肠道情况下,机体会诱发导致炎症性肠病,主要包括克罗恩病和溃疡性结肠炎(Melmed et al.,2013)。二者尽管临床表现类似,但在发病机制和预后方面存在一些明显的差异。克罗恩病最常见的病变部位是从回肠末端到大肠,而溃疡性结肠炎往往是影响结肠和直肠的黏膜层。从临床病因角度来看,遗传因素、肠道感染、免疫因素是溃疡性结肠炎的发病发展的主要诱因。溃疡性结肠炎患者肠道上皮细胞会出现明显的慢性炎症样改变。临床治疗仍然以传统抗炎药水杨酸进行对应症处理为主,同时也有利用糖皮质激素(但泼尼松)以及免疫抑制剂(环磷酰胺)联合使用对溃疡性结肠炎进行治疗,但这些方法都可能会造成机体免疫系统的失调,甚至导致肠道慢性炎症的加剧(Ben-Horin et al.,2013)。

我们在前一个实验中已经发现蜂胶能显著缓解由细菌脂多糖诱导的小鼠急性肺损伤症状,并显著抑制血清中相关促炎症因子的释放,病理切片观察结果也证实了蜂胶对炎性细胞浸润有着良好的改善效果。由于前人针对IBD的研究发现细菌脂多糖也是慢性肠炎发展的重要诱导因素之一,我们因此推测蜂胶对肠炎的预防治疗上可能也有一定效果(Morgan et al.,2012)。在本实验中,SD大鼠自由饮用含DSS溶液3d后,模型组大鼠开始出现溃疡性结肠炎的典型特征,第7d撤去DSS饮水后,结肠炎症状达到高峰,出现明显的腹泻、便血、食欲不振等症状。病理切片结果也表明,DSS大鼠结肠组织出现肠黏膜损伤和炎症细胞浸润,说明动物模型造模成功。日粮中添加0.3%蜂胶对大鼠的保护作用明显,表现为溃疡性结肠炎相关症状包括腹泻、隐血、便血的症状得到明显缓解、DAI分值显著低于模型组和结肠组织病理损伤明显减轻。

短链脂肪酸,又称挥发性脂肪酸,是一组由5个及5个以下的碳原子组成的饱和脂肪酸。和长链脂肪酸不同,短链脂肪酸主要是由厌氧细菌或酵母菌代谢过程中产生的,可以被机体吸收利用,是重要的体内代谢能源(Smith et al.,2013)。乙酸主要参与心、脑、肌肉、肾脏等器官代谢;结肠吸收丙酸后机体可直接用于肝脏代谢,并能调节肝脏中胆固醇的合成;丁酸是肠上皮细胞重要的能量来源,并发挥一定免疫调节作用,对维持肠道内环境稳定和预防结肠癌等方面起着重要作用(Chang et al.,2014)。我们的研究也发现,0.3%蜂胶处理能显著提高肠道挥发性脂肪酸的含量,提示蜂胶可能通过改变肠道菌群代谢来影响肠炎过程。

Aslan等(2007)利用4%乙酸诱导小鼠肠炎模型首次发现蜂胶抵抗肠炎效果,其效果可与美沙拉嗪(Mesalamine)相媲美(Aslan et al.,2007)。Okamoto等(2013)研究发现巴西蜂胶乙醇提取物能有效缓解TNBS诱导的小鼠肠炎症状,发现巴西蜂胶能抑制肠炎过程中Th1型淋巴细胞的分化,并抑制相关炎症因子的分泌,从而达到缓解肠炎的效果;Gonçalves等(2013)也利用TNBS诱导的大鼠肠炎模型证实巴西蜂胶水醇提取物能显著降低肠组织炎症过程产生的MPO酶的含量,同时对肠黏膜修复过程有着显著的促进效果并提升杯状细胞的数目。需要注意的是,本研究中所使用的DSS诱导肠炎模型与上述两种化学药物诱导的肠炎模型有一定区别。前期有研究发现TNBS诱导的肠炎反应主要是通过增强Th1—Th17(以IL-12和IL-17的上升为标志)并会朝慢性炎症方向转变,而DSS诱导的肠炎通常伴随着Th1—Th17介导的炎症反应向Th2介导的炎症反应转变(以IL-4和IL-10上升、TNF-α和IL-6的下降为标志)。此外,免疫细胞(T细胞和B细胞)的缺失并不会对DSS诱导肠炎症状产生影响,说明先天性免疫对于DSS诱导结肠炎具有更为重要的意义。同时,研究者也认为,DSS对肠上皮细胞具有显著的细胞毒性作用,会导致肠道黏膜屏障功能的丧失,进而诱导食物、细菌相关过敏原、毒素进入血液,引起全身性炎症反应。另一方面,由于中国蜂胶与巴西蜂胶的化学成分存在相当大的差异,二者生物学活性的发挥并不能完全等同。本研究首次为中国蜂胶抵御慢性溃疡性肠炎提供了实验证据。

小　结

本研究通过建立3%硫酸葡聚糖饮水成功诱导建立大鼠溃疡性结肠炎模型,发现日粮中添加0.3%蜂胶提取物对大鼠溃疡性结肠炎的保护作用最为明显,能显著降低大鼠疾病活动指数、有效缓解结肠炎性病理改变,并增加了盲肠内容物总短链脂肪酸的含量。

第三节　中国蜂胶对肠道上皮细胞Caco-2屏障功能的影响及其机制研究

肠道黏膜的屏障功能主要由肠道腔侧的肠道上皮细胞(Intestinal epithelial cells,IECs)及肠道上皮细胞之间的紧密连接(Tight junctions,TJ)所构成(Peterson et al.,2014)。这种屏障的存在使必需营养物质通过跨细胞转运的方式通过肠黏膜,并且阻止了肠腔内的大分子及微生物穿过肠黏膜扩散。紧密连接的功能紊乱和破坏时常伴随着肠道黏膜屏障的破坏,肠黏膜屏障的破坏会导致有害异物或毒素渗透到内腔,进而激活肠道的免疫系统,引发炎症性肠道疾病(Inflammatory bowel disease,IBD)(Neurath et al.,2012)。尽管现在针对IBD存在很多的药物疗法,其主要成分是抗炎药物,然而有相当一部分的病人在采用这些疗法后仍然不能达到完全缓解的效果。研究表明,增加肠道屏障的完整性对于应对IBD具有潜在的治疗价值(Peterson et al.,2014；Kosińska et al.,2013)。

紧密连接的存在对于维持肠道物理屏障完整性具有重要意义(Chasiotis et al.,2012)。紧密连接主要由大量的跨膜蛋白(例如occludin、claudins和junctional adhesion molecule)所构成,它可以与细胞质内的诸如zonulaoccludens(ZO-1、ZO-2、ZO-3)等适应性蛋白发生相互作用。Occludin和ZO-1之间的协同互作在正常TJ结构和肠道屏障功能的维持中起到了关键性的作用(Ulluwishewa et al.,2011)。当前一些研究结果证实紧密连接结构的渗透性和紧密连接蛋白受到一些细胞内信号通路的调节,例如AMP-activated protein kinase (AMPK)信号通路和mitogen-activated protein kinase(MAPK)级联反应。这些信号通路的激活则取决于各种各样生理反应的需要和病理学刺激的调节(González-Mariscal et al.,2008)。一些应用人肠道上皮细胞系Caco-2的体外研究强调指出,这些信号通路在调节TJ的渗透性及TJ蛋白表达等方面发挥了重要作用(Wang et al.,2013；Beutheu et al.,2013)。为了评价TJ结构的完整性,科学家采用了电子显微镜,包括透射电镜(Transmission electron microscopy,TEM)和扫描电镜(Scanning electron microscopy,SEM)来观察细胞间的空隙以及相邻IEC之间的潮位结果(Capaldo et al.,2014；Finotti et al.,2015)。此外,物质通过旁细胞途径穿过TJ结构的渗透性可以通过检测跨细胞电阻值(Transepithelial electrical resistance,TER)或者示踪物流量来检测,示踪物可采用钌红、FITC-右旋糖酐、3H-甘露醇等物质(Danielsen et al.,2013；Hering et al.,2014)。

越来越多的研究者对这种通过增加肠道屏障功能来抵御IBD的新型疗法产生兴趣。当前研究已经发现,诸如绿茶、咖啡、浆果、葡萄及一些水果和蔬菜中含有丰富的多酚类化合物,这类富含多酚类化合物食物的大量摄入,对完整的肠道屏障维持具有积极作用,且可以增加诸如occludin、claudin-1和ZO-1等TJ相关蛋白的表达丰度(Kosińska et al.,2013)。蜂胶提取物也含有丰富的植物多酚,前一节的动物实验研究已经证实蜂胶可以强有力地抵御包括结肠炎在内的炎症性疾病。广泛的研究证实了蜂胶的抗菌活性,并有研究提出蜂胶在调节肠道微生物菌群中发挥了一定的作用(Roquetto et al.,2015),但是蜂胶对于肠道黏膜屏障的有益作用却尚不明确。为此,我们研究了中国蜂胶对肠道上皮细胞Caco-2屏障功能的影响及其作用机制。

1 实验材料与方法

1.1 试剂

荧光素黄、alkaline phosphatase (AP)-conjugated secondary antibody (anti-rabbit IgG)、叔丁基过氧化氢(tert-butyl hydroperoxide,tBHP),2,4,6-三硝基苯磺酸(2,4,6-trinitrobenzenesulfonic acid,TNBS)均购自 Sigma-Aldrich(St. Louis,MO,USA)。兔抗anti-ZO-1 和 anti-occludin 购自 Abcam(Cambridge,MA,USA)。特异性信号通路的抑制剂 orsomorphin(AMPK 信号通路抑制剂)、PD98059(ERK1/2 信号通路抑制剂)、SB203580(p38 信号通路抑制剂)和 LY294002(Akt 信号通路抑制剂)购自 Selleck Chemicals(Houston,TX,USA)。其他化学试剂均为分析纯,购自上海生工。

1.2 蜂胶样品收集、提取和植物化学成分分析

同本章第一节。

1.3 细胞培养及细胞活力检测

人肠道上皮细胞系 Caco-2 源自中国科学院上海生命科学研究院细胞资源中心,将其培养在 37℃ 和 5% CO_2 的恒温细胞培养箱中,培养基配方为 90%(V/V)高糖 Dulbecco's modified Eagle's medium(Pierce Hyclone,Fremont,CA,USA)和 10%(V/V)热灭火胎牛血清(Gibco,Carlsbad,CA,USA),并添加 100U/mL 青霉素和 100μg/mL 链霉素。为了检测肠道细胞的屏障功能,细胞接种于 Transwell 小室(12mm diameter,0.4μm pore size;Costar,Cambridge,MA,USA)中聚碳酸酯膜上,接种培养 14d,使细胞增殖和单细胞层形成。培养过程中每天换液,每 2 天检测跨膜电阻值。细胞活力采用商品化试剂细胞活力 CCK-8 检测试剂盒(Cell counting kit-8,Dojindo,Kumamoto,Japan)。

1.4 肠道屏障功能测定

肠道的屏障功能主要通过检测 TER 值和测定荧光黄的非定向流速来评估(Suzuki et al.,2009)。指定的药物加入到小室内并开始计时,加入的时间点定义为 0。

跨膜电阻(Transepithelial electrical resistance,TER)用电阻仪 Millicell-ERS(Bedford,MA,USA)测定,在使用前,两个电极先用 75% 酒精泡 5min 消毒,再用 D-Hanks 液冲洗电极,然后将电阻仪的两极插入小室内外侧检测,读取稳定后的电阻值,每孔从不同方向插入电极,测定 3 次求平均值。首先检测未接种细胞的孔阻值,而后检测待测定的细胞孔,待测定细胞孔的阻值需减去接种细胞孔阻值。

跨细胞渗透性则通过检测荧光黄从顶端侧向基底侧转移的量反应,在加入药物孵育一段时间后,HBSS 清洗细胞,而后将上室液体更换为 100μmol/L 的荧光素黄,下室则更换为 HBSS。孵育特定时间后,取 200μL 基底侧的培养基,采用 96 孔板检测荧光素钠的量,并补充 200μL 新鲜预热的 HBSS 至下室,最后计算每个时间点的累计转运量。

1.5 基因的信使RNA表达丰度检测

主要步骤包括提取细胞内的总RNA,反转录为cDNA,接着采用荧光定量PCR的方法进行检测。

细胞总RNA的提取采用RNA提取试剂盒提取,具体操作步骤如下:

(1)将Caco-2细胞接种于细胞培养板中,经过生长和相应处理后,于待检测点吸尽细胞培养基并每孔加入1mL RNA lysis buffer,采用细胞刮刀刮取细胞并用移液枪加入1.5mL离心管,剧烈震荡15s,并保存于−80℃冰箱;

(2)室温解冻冻存RNA样本,提取前先标记好提取和检测过程中所需的离心管和纯化柱,待完全溶解后,加入200μL氯仿溶液,剧烈震荡15s后室温静置3min;

(3)12000r/min,4℃离心15min,溶液分为3层,小心准确定量吸取上层清液(约400μL)移入标记好的离心管中,并记录好体积,并将离心机调至25℃;

(4)向上述清液中加入一半体积的无水乙醇,颠倒混匀10次,用移液枪将液体移入纯化柱上柱中,12000r/min,25℃离心1min,弃滤出下层液;

(5)向纯化柱上层加入500μL去蛋白液,12000r/min,25℃离心1min,弃滤出下层液;

(6)向纯化柱上层加入500μL漂洗液(初次使用需加入42mL无水乙醇),12000r/min,25℃离心1min,弃滤出下层液;

(7)重复第(6)步操作;

(8)将纯化柱移入新的收集管中,13000r/min,25℃离心2min,取出纯化柱放入标记好的1.5mL RNA收集管中;

(9)向纯化柱中心的纯化膜上小心加入30μL 70℃预热的DEPC水,并室温静置3min;

(10)12000r/min,25℃离心1min,弃纯化柱,枪头轻柔吹打混匀,分装保存于−80℃冰箱。

而后进行反转录反应,将样本RNA在冰上解冻,并采用Nanodrop检测RNA浓度,采用1%琼脂糖凝胶电泳100V 15min,若RNA呈现清晰的三条带,则将其稀释为200ng/μL按照反转录试剂盒对RNA进行反转录,在冰上配置反转录反应液,体系见表7.8。

表7.8 反转录体系

Figure 7.8 Reverse transcription system

试剂	体积(μL)	终浓度
RNA	2.5	50ng/μL
5×PrimeScript® Buffer	2	1×
PrimeScript® RT Enzyme Mix	0.5	—
Oligo dT Primer(50μmol/L)	0.5	25 pmol/L
Random 6 mers(100μmol/L)	0.5	50 pmol/L
RNase-free H_2O	4	—

荧光定量 PCR 反应采用 SYBR premix EX Taq 在 96 孔板上进行,其每孔反应体系 20μL,具体用量如表 7.9。

表 7.9 荧光定量 PCR 反应体系
Table 7.9 Real-time PCR system

试剂	用量(μL)	终浓度
SYBR® Premix Ex Taq ™ II(2×)	10	1×
PCR Forward Primer(10μmol/L)	0.4	0.2μmol/L
PCR Reverse Primer(10μmol/L)	0.4	0.2μmol/L
ROX Reference Dye or Dye II(50×)	0.4	1×
RT 反应液(1∶10 稀释 cDNA 溶液)	2	
dH$_2$O	6.8	
Total	20	

反应条件时两步法 PCR 反应,包括:95℃预变形 30s;40 个 PCR 循环,95℃变性 5s,60℃退火延伸 34s,于 60℃时收集荧光信号绘制扩增曲线;添加溶解曲线程序。每个样本设置 4 个复孔,并设置 RNase-free H$_2$O 空白对照孔排除试剂污染。基因表达变化倍数采用 $2^{-\triangle\triangle Ct}$ 法计算,其计算公式为 $2^{(对照组目的基因平均Ct-对照组看家基因平均Ct)-(待检样品目的基因Ct-待检样品看家基因平均Ct)}$。

试验中所采用的引物序列为 Occludin:5'-GAGGTTTAGATTAGATTTCCGAC-3'(F)、5'-CACAACAAACTCCTTAGAACAAT-3'c(R);ZO-1:5'-AGATGAACGGGCTACGC-3'(F)、5'-GGAGACTGCCATTGCTTG-3'(R);GAPDH:5'-AGGGATGATGTTCTGGAGAG-3'(F)、5'-TCAAGATCATCAGCAATGCC-3'(R)。

1.6 蛋白质印迹分析(Western Blotting)

(1)分离胶的配制:采用 10% 分离胶,每块胶版需要 5mL 分离胶,静置 1h 凝胶,配方见表 7.10;

表 7.10 SDS-PAGE 的 10% 分离胶配方
Table 7.10 The formula of 10% separating gel for SDS-PAGE

10% 分离胶各成分	体积(mL)	所占体积比(%)
超纯水	2.425	48.5
1.5mol/L Tris (pH 8.8)	1.25	25
20% SDS	0.025	0.5
10% 过硫酸铵 e	0.05	1
TEMED	0.002	—
40% 丙烯酰胺	1.25	25

(2)浓缩胶的配置:采用 5% 浓缩胶,每块胶版需要 3mL 浓缩胶,静置 30min 凝胶,配方见表 7.11;

表 7.11 SDS-PAGE 的 5％浓缩胶配方

Table 7.11　The formula of 5％ stackinggel for SDS-PAGE

5％浓缩胶中各成分	体积(mL)	所占体积比(％)
超纯水	1.83	61
0.5mol/L Tris (pH 6.8)	0.75	25
20％ SDS	0.015	0.5
10％过硫酸铵	0.03	1
TEMED	0.003	—
40％丙烯酰胺	0.375	12.5

（3）电泳：加入电泳缓冲液，恒压 90V 跑胶约 30min，当蓝色进入分离胶时，将电压调至 120V 跑胶约 1h，待蓝色接近下边缘时停止电泳；

（4）转膜：甲醇活化 PVDF 膜至少 1min，然后从下到上依次是海绵、滤纸、PVDF 膜、凝胶、滤纸、海绵的顺序，然后放入转膜仪中，加入转膜缓冲液，转膜条件为 300mA 恒流 3h；

（5）封闭：转完膜后将 PVDF 膜取出放入 5％的脱脂牛奶中室温封闭 2h；

（6）一抗杂交：将封闭后的膜直接放入对应的配置好的一抗中，4℃水平震荡孵育过夜；

（7）漂洗：取出第（6）步孵育后的 PVDF 膜，用 1× 的 TBST 震荡漂洗，每次 10min，漂洗 3 次；

（8）二抗杂交：将 PVDF 膜放入对应的配置好的二抗中，室温震荡孵育 2h；

（9）漂洗：取出第（8）步孵育后的 PVDF 膜，用 1× 的 TBST 震荡漂洗，每次 10min，漂洗 3 次；

（10）显色：将 PVDF 膜放入 10mL BCIP/NBT 染色工作液中，室温避光孵育 5～30min 或更长时间（可长达 24h），直至显色至预期深浅。去除 BCIP/NBT 染色工作液，用蒸馏水洗涤 1～2 次即可终止显色反应；

（11）数据保存：将晾干的 PVDF 膜进行拍照或扫描后保存。

1.7　免疫荧光分析

（1）在细胞对数生长期将细胞以 $1×10^5$/mL 的浓度接种到激光共聚焦小皿中，待生长和相应处理后用于检测目的蛋白；

（2）用冰上预冷的 PBS 清洗细胞 3 次，每次 5min，总计 15min；

（3）甲醇和丙酮等量混合制备固定液，而后缓慢滴加到共聚焦小皿中，RT 静置 15min。用冰上预冷的 PBS 清洗细胞 3 次，每次 5min，总计 15min；

（4）用 0.5％ PBS-Triton 常温通透 10min 后，用冰上预冷的 PBS 清洗细胞 3 次，每次 5min，总计 15min；

（5）10％山羊血清常温封闭 30min 后，去除山羊血清；

（6）加入稀释的一抗，置于湿盒中 4℃孵育过夜；

（7）用冰上预冷的 PBS 清洗细胞 3 次，每次 5min，总计 15min，而后加入 1∶500 稀释比例的 FITC-conjugated goat anti-rabbit IgG 二抗，37℃避光孵育 1h；

（8）用冰上预冷的 PBS 清洗细胞 3 次，每次 5min，总计 15min；加入 DAPI 染色液，染色 5min；

(9)用冰上预冷的 PBS 清洗细胞 3 次,每次 5min,总计 15min;最后用 PBS 覆盖待检细胞,采用激光共聚焦显微镜观察并拍照记录。

1.8 透射电镜分析

将 Caco-2 细胞接种于 Transwell 小室中,待细胞处理完后用 PBS 清洗 2 遍后,加入 2.5%的戊二醛+1%四氧化锇覆盖小室膜,4℃固定过夜,PBS 清洗 3 遍后,分别用梯度浓度丙酮脱水,每个梯度中静置 10min,脱水完毕,用纯酒精脱水 3 次,每次 30min,脱水后刮下细胞,根据说明书(SPI-EM, Division of Structure Probe, Westchester, NY, USA)用 Epon 812 树脂包埋细胞,连续切出细胞 70nm 超薄切片,再用乙酸双氧铀和碱性柠檬酸铅染色 15min,置冷冻干燥机内真空干燥后用透射电镜观察(H-7650, Hitachi, Tokyo, Japan)。

1.9 数据分析

数据采用独立实验中特定重复的平均值±标准偏差表示。所有的数据统计均采用 SPSS 16.0(SPSS Inc., Chicago, IL, USA)统计。数据的统计比较采用 Duncan's multiple range tests 方法进行单向方差分析。P 值小于 0.05 即定义为差异显著。

2 结果与分析

2.1 中国蜂胶对分化的 Caco-2 细胞单层跨膜电阻及对荧光黄渗透率的影响

为了评价 PPE 对肠道上皮细胞的通透性的影响,我们检测了 Caco-2 单细胞层的 TER 和 LY 流量。结果如图 7.6 所示,在加入中等浓度(25μg/mL)或者高浓度(50μg/mL)的 PPE 处理 2h 后,肠道细胞的紧密连接时间和浓度依赖性有所加强。50μg/mL PPE 处理细胞 36h 后,细胞的 TER 值达到最大值,较试验初始值上调了 32.2%。而中等浓度 25μg/mL 蜂胶对上皮细胞的 TJ 通透性则温和很多,处理 24h 后 TER 较试验初始值仅上调了 20.1%(图 7.6A)。LY 流量所代表的旁细胞途径通透性也与 TER 值表现出相匹配的趋势,加入中等浓度(25μg/mL)或者高浓度(50μg/mL)蜂胶处理导致 LY 渗透量的显著下降(图 7.6B)。而且加入中等浓度(25μg/mL)或者高浓度(50μg/mL)的蜂胶处理对于 Caco-2 细胞是安全的,因为这两种浓度的处理不会引起细胞活力的显著下降。由于发现 50μg/mL 的蜂胶处理表现出最佳的效果,故后续研究中采用 50μg/mL 蜂胶处理 Caco-2 细胞。

2.2 中国蜂胶对紧密连接蛋白基因表达、紧密连接蛋白分布及对紧密连接超微结构的影响

应用荧光定量 PCR 对主要的紧密连接基因、ZO-1 和 occludin 的基因表达丰度进行了定量检测。结果如图 7.7 所示,50μg/mL 蜂胶处理引起了 ZO-1 和 occludin 的表达显著上升。50μg/mL 蜂胶处理 3h 后细胞内的 ZO-1 表达开始发生变化,至处理 24h 达到最大值,最大值较对照组的基因表达丰度上调了 48.9%。蜂胶对 occludin 的上调作用表现出一定时间上的滞后,但是上调作用强于对 ZO-1 的作用,检测结束时蜂胶对 occludin 的上调幅度达 4.7 倍。

采用激光共聚焦显微镜观测了蜂胶对紧密连接蛋白定位的作用。结果如图 7.8 和图 7.9 所示,无论是对照组还是蜂胶处理组,绿色荧光所代表的 occludin 和 ZO-1 蛋白都明显适量且正确地定位于细胞间接触部位。然而,采用公认的氧化物 tBHP 处理细胞则发现绿色荧光大幅减弱,并在细胞膜表面形成了不连续的环状。

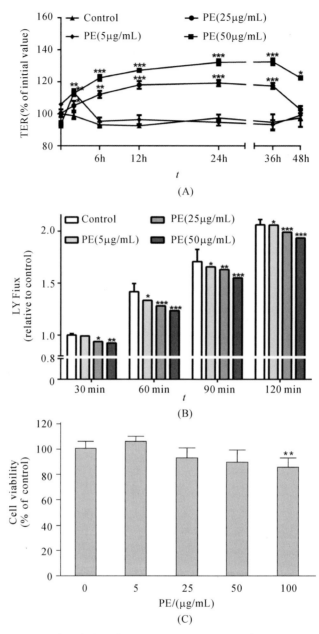

图 7.6 中国蜂胶对分化的 Caco-2 细胞肠道屏障功能的影响

Fig 7.6 Effects of propolis extracts (PE) on intestinal TJ permeability in Caco-2 cell monolayers.

(A)蜂胶提取物处理 Caco-2 细胞单层的跨膜电阻。细胞接种于 Transwell 小室中并培养 14d，形成 Caco-2 单细胞层。在指定时间点用 PE 处理 Caco-2 细胞单层后测量跨膜电阻。与初始每个单层细胞的 TER 值相比百分比的变化来表示跨膜电阻的变化。数据用平均值±标准差($n=3$)来表示。* $P<0.05$、** $P<0.01$ 和 *** $P<0.001$ 表示与每个时间点的控制值相比的显著性差异。(B)PPE 处理的 Caco-2 细胞单层的单向荧光素黄流量。LY 流量在 PE 处理 Caco-2 单层后 120min 测定。值以 LY 渗透百分比的平均值±标准偏差 ($n=3$)表示。* $P<0.05$、** $P<0.01$ 和 *** $P<0.001$ 表示与对照组相比的显著性差异。(C)细胞活力是在 PE 处理 48h 后用 CCK-8 试剂盒测定分析。结果为相对于对照细胞%表示，数据采用 3 个独立实验中特定重复 3 次的平均值±标准偏差表示。

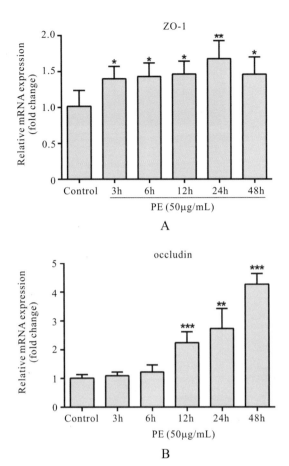

图 7.7　中国蜂胶对紧密连接蛋白基因表达的影响

Fig 7.7　Effect of PE treatment on the mRNA expression of ZO-1 and occludin

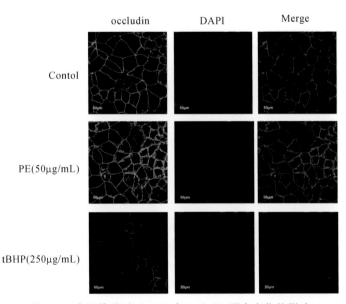

图 7.8　中国蜂胶对 Caco-2 中 occludin 蛋白定位的影响

Fig 7.8　Confocal microscopy images of occludin proteins in confluent Caco-2 cells

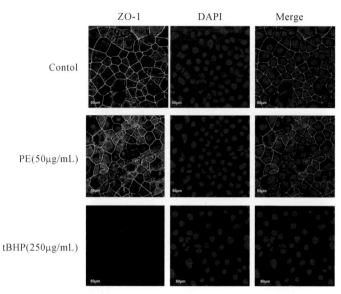

图 7.9 中国蜂胶对 Caco-2 中 ZO-1 蛋白定位的影响

Fig 7.9 Confocal microscopy images of ZO-1 proteins in confluent Caco-2 cells

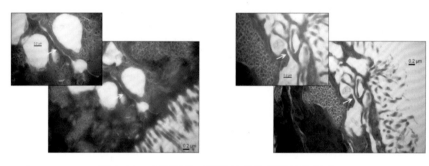

图 7.10 中国蜂胶对紧密连接的超微结构影响

Fig 7.10 Effect of PE treatment on the morphological ultrastructure of tight junction in Caco-2 cell monolayers

2.3 中国蜂胶对 Caco-2 细胞 AMPK 和 MAPK 信号通路的影响

为了探明个别信号转导通路在 PE 调控的通透性中所起的作用,我们检测了蜂胶对 AMPKα、Akt、ERK1/2 和 p38 磷酸化的作用。结果如图 7.11A 所示,蜂胶处理的 Caco-2 细胞中 ERK1/2、AMPKα、Akt 和 p38 的磷酸化水平显著上升,且分别于 15min、45min、120min 和 120min 达到峰值。这一结果促使我们采用特异性的抑制剂来抑制这些信号转导,探索其对 TJ 通透性的影响。因此,我们采用 Dorsomorphin(特异性 AMPK 信号通路抑制剂)、LY294002(特异性 Akt 抑制剂)、PD98059(特异性 ERK1/2 抑制剂)和 SB203580(特异性 p38 抑制剂)分别处理 Caco-2 单细胞层而后加入 $50\mu g/mL$ PPE 处理。如图 7.11B 所示的 TER 值可以看出,不论是 AMPK 信号通路抑制剂 Dorsomorphin 还是特异性 ERK1/2 抑制剂 PD98059 均抵消了蜂胶引起 TER 阻值的上升,而特异性 Akt 抑制剂 LY294002 则不会影响蜂胶对阻值的上调作用。有趣的是,应用 SB203580 特异阻断 p38 的激活进一步促进了蜂胶对 TER 值的上调幅度,这一现象提示 p38 在蜂胶调控的 TJ 通透性中所起的作用与其他信号通路有所不同。

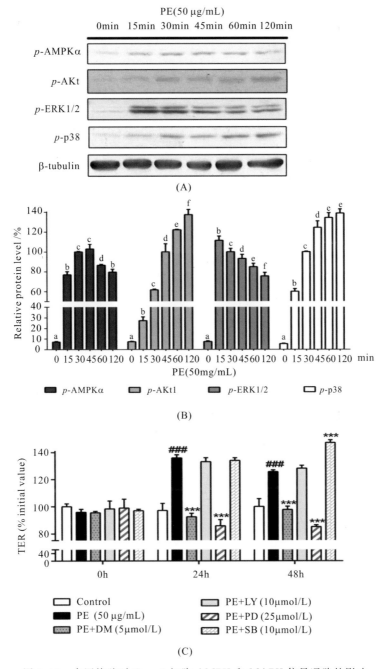

图 7.11 中国蜂胶对 Caco-2 细胞 AMPK 和 MAPK 信号通路的影响

Fig 7.11 Effect of PE treatment on AMPK and ERK signaling in Caco-2 cell monolayers

(A)Caco-2 细胞融合于 6 孔板后，在特定期间用 PE(50μg/mL)处理。收集完整的细胞溶解产物，并进一步进行免疫印迹分析。特定的抗体被用来检测 phospho-AMPKα、phospho-Akt、phospho-ERK1/2 和 p38 的表达。β-微管蛋白被作为对照。具有代表性的印迹图表示三个独立的实验。(B)相应的谱带的强度用光密度法测定并利用 β-微管蛋白进行归一化。三个独立的实验的平均值如图。(C)Caco-2 单层细胞接种于 Transwell 小室中培养 14d。PE 处理细胞前被选择性抑制剂进行预处理 1h。与每个单层细胞初始 TER 值相比来表示 TER 的百分比变化。值以平均值±标准偏差($n=3$)表示。* $P<0.05$，** $P<0.01$ 和 *** $P<0.001$ 表示与对照组相比的显著性差异程度。

3 讨论

众所周知,完整的肠道屏障功能对于健康肠道的维持和抵御病原物入侵具有重要意义。人肠道上皮细胞 Caco-2 细胞广泛用于体外评价肠道屏障功能的研究中。我们的研究发现,蜂胶用于体外模拟的肠道单细胞层 Caco-2 中可以显著提升肠道紧密性,主要表现在显著增加了 TER 值,并降低 LY 流量。当前仅有少量物质被发现具有促进肠道屏障功能的作用,这类物质包括诸如 TGF-β_1、IL-7 和 EGF 等的细胞蛋白(González-Mariscal et al.,2008)、糖皮质激素(Woo et al.,1999)和一些膳食营养素(氨基酸、多酚类化合物、短链脂肪酸及多糖类化合物)等(Ulluwishewa et al.,2011)。而很多物质则会破坏肠道的屏障功能,例如氧化剂、细菌毒素和炎性因子(Martin-Martin et al.,2010)。在这些肠道保护性物质中,相当一部分富含多酚类化合物的食物被证实具有增强肠道屏障的功能,例如绿茶、可可粉和水果等(Liu et al.,2013;Parkar et al.,2008;Kosińska et al.,2012),还有一些诸如槲皮素、山奈酚、杨梅酮等多酚类化合物单体也发现具有这种促进肠道紧密连接的作用(Suzuki et al.,2009,2011)。而这些多酚类化合物,例如槲皮素、山奈酚也可以在蜂胶样本中检测到。而且紧密联系系统是一个由胞质蛋白和跨膜蛋白整合而成的完整的系统,紧密连接蛋白的表达丰度和分布模式是肠道屏障功能的主要决定因素。因此,蜂胶对于肠道屏障功能的提升作用可能一方面由于它富含多种植物多酚,另一方面能够提高紧密连接蛋白 ZO-1 和 occludin 的信使 RNA 表达分度,并使其合理有序地分布到细胞的接触部位。

一系列的蛋白参与了紧密连接的形成和分布,而这些蛋白受到很多信号,例如激酶、磷酸酶和 G 蛋白等的调节(Kosińska et al.,2013)。越来越多的研究结果证实不同信号通路的相互作用参与调节紧密连接的蛋白的组装和紧密连接的形成。AMPK 可以调节一些重要的膜转运蛋白(Hallows,2005),它参与了细胞内的 ATP 代谢稳态,是一种代谢敏感性激酶。近期的一些研究表明,AMPK 可以促进肠道紧密连接的形成。细胞内的信号转导通路 MAPK 家族是一种典型的细胞内信号转导途径(Gehart et al.,2010),胞外信号调节蛋白激酶(ERK)、N 段 c-Jun 蛋白激酶(JNK)及 p38 亚族是 MAPK 家族最典型的三个成员(Gehart et al.,2010)。我们应用人的 Caco-2 细胞研究发现,PPE 处理可以引起 ERK1/2 和 p38 的磷酸化,而对 JNK 没有作用。这一结果与我们前期将 PPE 用于小鼠的 RAW264.7 巨噬细胞实验结果相一致,而与在神经元样细胞 PC12 和纤维原细胞上的相关研究的结果则不同(Murase et al.,2013;Kano et al.,2008)。MAPK 通路对于紧密连接的调节作用是错综复杂的,已有研究者发现了一些相互矛盾的结果。Kinugasa 等(2000)发现 ERK1/2 的激活引起了肠道上皮细胞中的 TER 值上升,并上调了 TJ 蛋白 claudin-1 和 claudin-2 的信使 RNA 表达丰度。也有研究者在肾脏上皮细胞 LLC-PK1 中发现了相似的结果,然而肿瘤坏死因子 TNF-α 作为一种 IBD 中重要的促炎因子,在增加肠道通透性的同时会快速地激活 ERK1/2 并引起肌球蛋白轻链激酶依赖的肠道紧密连接的开放(Al-Sadi et al.,2013)。Wu 等(2013)在一种氯吡格雷诱导的肠道损伤模型中发现紧密连接蛋白的下调可能与 p38 的激活有关。本研究中我们发现,利用 dorsomorphin 和 PD98059 抑制 AMPK 和 ERK1/2 显著降低了 PPE 引起的 TER 上升程度。有趣的是,抑制 p38 则表现出相反的效果,TER 值不降反而更高地上升。这些结果提示,p38 信号通路的激活在 Caco-2 细胞通透性调节中起到了负面作用。因此,本研究提出了直接的证据证实了蜂胶激活了肠道上皮细胞中的 ERK1/2 和

AMPK 信号通路。

然而值得一提的是,蜂胶调节的屏障功能的改变与细胞毒性作用并无关联。一些研究已经提出 TER 值的下降可能是源于细胞数量的下降,因为细胞数量的减少意味着细胞的密度和单细胞层中细胞间的空隙增加(Martin-Martin et al.,2010)。我们应用免疫荧光染色检测 occludin 和 ZO-1 蛋白空间表达特性时显示相似的紧密连接模式,细胞密度在对照组和蜂胶处理组中并无明显差异。不仅如此,我们应用透射电镜首次发现蜂胶处理的细胞具有正常的组织结构,而不是如 tBHP 处理组那样出现肠道损伤。因此,我们推测蜂胶对 TER 的上升和肠道屏障功能的增强并不是通过改变细胞的数量或者细胞的大小来实现的。

综上所述,蜂胶可以增强 Caco-2 单细胞层的屏障功能,而且蜂胶对肠道屏障功能的调节作用至少部分是受 AMPK 和 ERK1/2 激活所调节,而 p38 的激活则抑制了蜂胶对肠道紧密连接的上调作用。这些结果对于诸如蜂胶之类的富含多酚食物对肠道健康的有益作用提供了新的分子机制,对于人类 IBD 的预防和治疗具有重要意义。肠道微环境也是机体微环境的重要组成部分,本实验也从细胞水平为蜂胶治疗炎症性肠炎提供了新的证据。

小 结

本研究结果表明,中国蜂胶能显著提高肠上皮细胞屏障功能。具体表现为能提高 Caco-2 肠上皮细胞单层跨膜电阻阻值,并能降低荧光黄渗透率。荧光定量 PCR 结果表明,蜂胶处理能提高紧密连接蛋白 ZO-1 和 occludin 的基因表达丰度;蛋白质印迹结果表明,蜂胶能使肠道上皮 Caco-2 细胞 AMPK-α、AKT、ERK1/2 和 p38 蛋白发生磷酸化。采用特异性的蛋白抑制剂处理发现中国蜂胶对肠道屏障功能的调节作用受 AMPK 和 ERK1/2 信号通路激活所调节,而 p38 的激活则抑制了蜂胶对肠道屏障功能的促进作用。

第四节 中国蜂胶和巴西蜂胶对巨噬细胞免疫调节作用及其机制研究

蜂胶的化学组分随着其植物来源的地理位置差异而发生变化,这一特点反映了一个地域植物的多样性(Bankova,2005)。中国和巴西是世界蜂胶原料的主要出产国,然而中国蜂胶和巴西蜂胶的植物来源却大相径庭。大部分中国蜂胶的胶源植物为杨树(*Populus* spp.),而巴西绿蜂胶主要的胶源植物为一种生长于巴西东南部的植物酒神菊树(*Baccharis dracunculifolia* DC,Asteraceae)(Park et al.,2004)。通过对蜂胶的植物化学研究发现蜂胶中的多酚类化合物主要出现在醇溶性组分(Ethanol-soluble fraction)中(Cuevas et al.,2014)。中国蜂胶中主要的多酚类化合物包括黄酮和类黄酮酯类,如柯因、高良姜素、咖啡酸苯乙酯等。而巴西绿蜂通常主要由诸如咖啡酸和 prenylated *p*-coumaric acids(阿司匹林 C 和巴卡林)等的酚酸类所组成(Sforcin et al.,2011)。尽管这两种蜂胶是世界上最常见的蜂胶类型,但有关它们的比较研究却非常少。

越来越多的研究证实炎症和非传染性疾病存在关联,各种各样的癌症、类风湿性关节炎、动脉粥样硬化、炎性肠病和糖尿病等方面的研究均提到这一规律(Baker et al.,2011)。核转录因子 NF-κB(Nuclear factor kappa B)在免疫和炎症反应中起到了至关重要的作用,

它可以调控大量基因的转录激活,这些基因不仅编码了诸如白介素 IL-1β、IL-6、IL-10 和肿瘤坏死因子(TNF-α)等细胞因子,也编码了像单核细胞趋化蛋白-1(MCP-1)和 IL-8 之类的趋化因子,还编码了一些诱导酶类,如环氧酶、诱导性一氧化氮合酶等(Chawla et al.,2011;Jakus et al.,2013)。大量的炎症调节因子可以诱导 NF-κB 的激活。

众所周知,巨噬细胞在免疫系统和炎症过程中扮演着重要作用。LPS(脂多糖)刺激的巨噬细胞可产生多种炎症调节因子,包括 NO、IL-1β 和 IL-6。这些调节因子的过量产生将破坏免疫内环境的稳态,引起一些炎症性疾病,包括脓血性休克、动脉硬化和癌症等。

基于我们前期的研究发现,结合当前公认的蜂胶抗炎效果,本研究采用巨噬细胞体外实验比较、确认中国蜂胶和巴西蜂胶的抗炎效果,并且更深层次地探索蜂胶对炎症微环境中具有核心作用的巨噬细胞炎症反应的调控机制。

1 实验材料与方法

1.1 试剂

脂多糖(LPS)(*Escherichia coli* 0111:B4)、ROS 探针 2′,7′-dichlorofluorescin diacetate(DCFH-DA)、碱性磷酸酶标记兔二抗,均购自生工生物工程(上海)股份有限公司;色谱分析用到的化合物购自 Sigma 公司(St. Louis,MO,USA);特异性抗 IkappaB-alpha(IκBα)、phosphor-IkappaB-alpha(p-IκBα)、NF-κB-p65、β-tubulin、β-actin 一抗购自 Epitomics 公司(Burlingame,CA,USA);特异性识别辣根过氧化物酶的结合 HA 标签(Horseradish peroxidase conjugated HA tag)和组蛋白 H3(histone H3)的一抗及 HRP 标记的抗兔二抗(HRP-conjugated secondary antibody,anti-rabbit IgG)购自华安生物技术有限公司(杭州);重组人源 TNF-α 和 IL-1β 购自 Peprotech 公司(Rocky Hill,USA);胎牛血清(FBS)购自 GIBCO 公司(Gibco Laboratories,NY,USA);细胞活性检测试剂盒购自同仁化学研究所(Dojindo,Kumamoto,Japan);转染试剂聚乙烯亚胺(polyethylenimine,PEI)购自 Polysciences 公司(美国);Dual-Luciferase® Reporter Assay System 和 ARE-Luc 质粒及 Renila 质粒购自 Promega 公司(Invitrogen Life Technologies,Carlsbad,USA);泛素特异性抗体购自 Santa Cruz Biotechnology(CA,USA)。

1.2 蜂胶样品收集、提取和植物化学成分分析

中国蜂胶样本收集、提取方法参照本章第一节。

巴西蜂胶样本来自蜂乃宝本铺(南京)保健食品有限公司,采集于巴西米纳斯吉拉斯州,胶源植物为酒神菊树(*Baccharis dracunculifolia* DC.)。巴西蜂胶样本提取方法同中国蜂胶,采用高效液相色谱法分析其主要多酚类化合物。

1.3 细胞培养及细胞活力检测

鼠白血病巨噬细胞系 RAW 264.7、人胚胎肾细胞 HEK293T、稳定转染表达白介素 1(IL-1)受体的 HEK293T 细胞(293T-C6)及赫拉细胞(Hela)由浙江大学生命科学研究院夏总平教授馈赠。基本培养基配方为 90% DMEM 高糖培养基,10% 胎牛血清,100U/mL 青霉素,100μg/mL 链霉素。培养环境为 5% CO_2,37℃ 和 100% 湿度细胞培养箱。

细胞活力检测应用 CCK-8 检测试剂盒,将细胞接种于 96 孔板中,经过相应处理后进行 CCK-8 检测。CCK-8 的具体操作方法为:种板时将 96 孔板外围 36 个孔加入 100μL PBS,其

他孔每孔加入 $100\mu L$ 浓度为 $1\times10^5/mL$ 的细胞悬液,并经过生长和相应处理后,于待检测时间点,采用 PBS 轻柔清洗细胞两遍,而后加入 90% DMEM 高糖培养基和 10% CCK-8 检测液。将细胞放回细胞培养箱孵育 2h 后,采用酶标仪检测每孔在 450mm 波长下的吸光度,然后进行分析计算。

1.4 基因的信使 RNA 表达丰度检测

参见本章第一节。小鼠引物列表如下(表 7.12):

表 7.12 小鼠引物序列

Table 7.12　Primer sequences for mice

Primer	Primer sequence	Product size(bp)	GenBank accession No.
IL-1β	F:5′-CCAACAAGTGATATTCTCCATGAG-3′ R:5′-ACTCTGCAGACTCAAACTCCA-3′	239	NM_008361.3
IL-6	F:5′-CTCTGCAAGAGACTTCCATCC-3′ R:5′-GAATTGCCATTGCACAACTC-3′	210	NM_031168.1
IL-10	F:5′-CTATGCTGCCTGCTCTTACTG-3′ R:5′-CAACCCAAGTAACCCTTAAAGTC-3′	221	NM_010548.2
HO-1	F:5′-ACATTGAGCTGTTTGAGGAG-3′ R:5′-TACATGGCATAAATTCCCACTG-3′	241	NM_010442.2
MCP-1	F:5′-AAGAAGCTGTAGTTTTTGTCACCA-3′ R:5′-TGAAGACCTTAGGGCAGATGC-3′	155	NM_011333.3
TNF-α	F:5′-CCACGCTCTTCTGTCTACTG-3′ R:5′-ACTTGGTGGTTTGCTACGAC-3′	169	NM_013693.2
iNOS	F:5′-TTTCCAGAAGCAGAATGTGACC-3′ R:5′-AACACCACTTTCACCAAGACTC-3′	294	NM_010927.3
COX-2	F:5′-GAAATATCAGGTCATTGGTGGAG-3′ R:5′-GTTTGGAATAGTTGCTCATCAC-3′	237	NM_011198.3
GAPDH	F:5′-GAGAAACCTGCCAAGTATGATGAC-3′ R:5′-TAGCCGTATTCATTGTCATACCAG-3′	212	NM_008084.2

1.5 瞬时转染及报告基因检测

具体步骤如下:

(1) HEK 293T 或者 293T C6 细胞以 1×10^5/孔的密度接种到 12 孔细胞培养板中,每孔细胞培养基的体积为 1mL,转染前培养基更换为 90%DMEM+10%FBS;

(2) 试验前配置转染复合物,每孔转染物的体系为 30ng 萤火虫荧光素酶报告质粒(pGL4.2-3×NF-κB-Luc)+5ng 海肾荧光素酶报告质粒(pRL-TK)+470ng 空载质粒 pcDNA3.1,转染试剂为聚乙烯亚胺(polyethylenimine,PEI);

(3) 转染后培养 24h 后加入相应的试验处理,处理 12h 后检测荧光素酶活性;

(4) 去除培养板中的培养基,加入 PBS 清洗细胞两遍,吸尽 PBS 液;

(5) 加入 100μL 1×PLB 细胞裂解,摇床上轻摇 15min 后,收集细胞裂解液;

(6) 4℃,12000r/min 离心 10min,取上清液;

(7) 向酶标板中预先加好 50μL LAR Ⅱ试剂,而后小心加入 20μL 细胞裂解液,每个样本 3 个复孔;

(8) 吹打混匀后,用酶标仪检测化学发光强度,即为萤火虫荧光素酶的活性;

(9) 继续向检测孔中加入 100μL Stop & Glo® Reagent,用酶标仪检测化学发光强度,即为海肾荧光素酶活性;

(10) 检测结果采用(8)、(9)中数值的比值计算。

1.6 融合 TRAF6-gyrase B 蛋白稳定细胞系的构建

插入了 TRAF6-gyrase B 的嵌合重组质粒主要由 T6RZC 构成,其中 TRAF6C 段的 1~358 残基被替换成微生物转旋酶 B 的 N 段片段,而后克隆到了哺乳动物表达质粒 pEF-IRES-P 中。然后将该质粒转染 Hela 细胞,通过 puromycin(1μg/mL)筛选稳定嵌合重组到 Hela 细胞克隆。这样构建的细胞系中 TRAF6 的寡聚化和后续 NF-κB 的激活是通过库马霉素 A(2μmol/L)诱导的。

1.7 总蛋白提取及核蛋白的分离提取

1.8 蛋白质印迹

参照本章第一节。

1.9 体外泛素化检测

参照 Deng 等(2000)报道的泛素化检测体系,检测蜂胶抑制 NF-κB 激活的生化机制。采用 Gibco bac-to-bac 杆状病毒表达系统在昆虫 Sf9 细胞中表达 TRAF6-gyrase B 融合蛋白(T6RZC)和带 His6 标签的 E1。His6-Ubc13,His 6-Uev2 和泛素则从大肠杆菌中纯化。该试验体系为 E1(85nmol/L)、Ubc13(3.9μmol/L)、Uev2(3.9μmol/L)、T6RZC(20nmol/L)、泛素(10μmol/L),加入不同浓度的 PPE 于 30℃孵育 30min,反应缓冲液配方为 50mmol/L Tris-HCl(pH 7.5)、2mmol/L ATP、5mmol/L $MgCl_2$ 和 0.5mmol/L DTT。在反应开始前,以上反应混合物先在冰上孵育 15min。反应采用变性缓冲液终止,而后采用泛素特异性抗体 P4D1(Santa Cruz Biotechnology,CA,USA)对其中的泛素进行蛋白质印迹分析。

1.10 一氧化氮及活性氧检测

NO 在细胞上清中的浓度检测采用 Griess 比色反应方法检测硝酸盐浓度。其中标准曲线应用亚硝酸盐构建。试验前首先配置如下反应液:

对氨基苯磺酰胺(磺胺类,Sulfonamides)溶液(Griess A 试剂):取 2.37mL 85%磷酸,加蒸馏水 47.63mL,配成 50mL 4%的磷酸溶液,称取 0.5g 对氨基苯磺酰胺,溶于 50mL 4%的磷酸溶液中,超声溶解。现用现配。

萘乙二胺盐酸盐溶液(Griess B 试剂):取 0.1g N-(1-萘基)乙二胺二盐酸盐溶于 100mL 纯净水。

标准曲线的配制:先配制 50mmol/L 的 $NaNO_2$ 贮备液,若需配制 10mL 贮备液,则称取 34.5mg $NaNO_2$,加 10mL DMEM 溶解。再将得到的贮备液用 DMEM 稀释成不同的梯度浓度,如 0、5、10、15、20、30、40、50μmol/L,来绘制标准曲线。

检测时取适量样本至离心管中,加入等量 Griess A 试剂,颠倒混匀,可见溶液褪色。然后吸取 100μL 混合液加入酶标板的孔中。最后加入 50μL Griess B 试剂,轻轻晃动酶标板,并在 3～5min 内测其在 540nm 波长下的吸光度值。在检测完成后采用标准品浓度和吸光度值构建标准曲线,计算样本中的 NO 浓度。

1.11 细胞内活性氧检测

细胞内 ROS 的检测采用过氧化物敏感的荧光探针标记结合流式细胞仪进行检测,具体操作方法为:将细胞以 1×10^5/mL 接种于 6 孔细胞培养板中,经过生长和相应处理后,于待检测时间点,采用 PBS 轻柔清洗细胞两遍,而后避光加入含 10μmol/L carboxy-H2DCF-DA 探针的 DMEM 高糖溶液。然后将细胞放回细胞培养箱避光孵育 30min 后,消化清洗后重悬于 PBS 中制备单细胞悬液,采用流式细胞仪检测 2×10^4 个细胞的平均荧光强度,荧光的激发波长为 488nm,发射波长为 530nm。

1.12 炎性细胞因子检测

收集细胞培养基,其中炎性细胞因子的检测方法参见本章第一节。

1.13 数据分析

数据采用独立实验中特定重复的平均值±标准偏差表示。所有的数据统计均采用 SPSS 16.0(SPSS Inc., Chicago, IL, USA)统计。数据的统计比较采用 Duncan's multiple range tests 方法进行单向方差分析。P 值小于 0.05 即定义为差异显著。

2 结果与分析

2.1 蜂胶预处理对脂多糖诱导小鼠巨噬细胞关键炎性介质基因表达的影响

通过 CCK-8 检测细胞活力发现高达 10μg/mL CP 或者 50μg/mL 的 BP 对于 RAW 264.7 没有毒性。与 LPS 刺激组相比,低至 5μg/mL 的 CP 仍然可以显著下调促炎细胞因子 IL-1β 和 IL-6 的表达($P<0.05$)(图 7.12A 和 B)。而 50μg/mL BP 也表现出相同的效果。CP 和 BP 对 MCP-1 的信使 RNA 表达的调节作用也遵循相似的规律,即抑制 LPS 诱导的 MCP-1 的表达上调(图 7.12E)。而且,CP 和 BP 预处理对 IL-10 的调节作用存在不同,试验中用到的不同浓度的 CP 均可以抑制 IL10 的表达,而 BP 浓度只有在浓度高达 30μg/mL 和 50μg/mL 才能抑制 IL-10 的表达,甚至当 BP 浓度较低为 5μg/mL 和 10μg/mL 时促进了 IL-10 的表达。TNF-α 的 mRNA 水平只能被 50μg/mL 的 BP 所抑制,而 CP 则促进了 TNF-α 的 mRNA 表达。此外,CP 和 BP 均抑制了 iNOS 的表达上调,10μg/mL 的 CP 处理组中 iNOS 仅为 LPS 组的 16%,而 50μg/mL BP 处理组则是 LPS 组的 23%。COX-2 也是炎症的重要调节因子,两种蜂胶可以时间和剂量依赖性地显著抑制 COX-2 的表达(图 7.12H)。

HO-1 及其副产物在炎症消减中发挥着重要作用,我们的结果表明,PPE 可以显著地上调 LPS 激活的巨噬细胞中 HO-1 的表达,且表现出明显的剂量依赖性,CP 对 HO-1 的激活作用强于 BP。

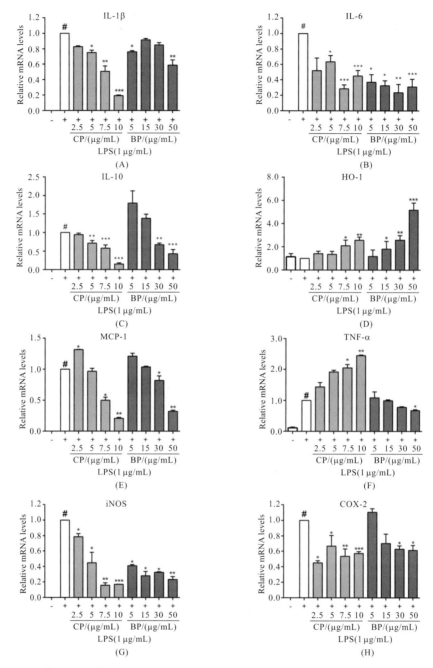

图 7.12 中国、巴西蜂胶预处理对脂多糖诱导小鼠巨噬细胞关键炎性因子基因表达的影响

Fig 7.12 Effects of Chinese and Brazilian propolis pretreatments on the mRNA expressions of key inflammatory-mediators and cytokine genes in LPS-stimulated RAW 264.7 cells

注:小鼠白血病来源巨噬细胞 RAW 264.7 细胞经 1μg/mL 的脂多糖刺激后,诱导小鼠巨噬细胞炎性因子表达,并通过 PPE 预处理来研究蜂胶对炎性因子表达的影响。通过 qRT-PCR 和归一化 GAPDH 定量 IL-1β (A)、IL-6 (B)、IL-10 (C)、HO-1 (D)、MCP-1 (E)、TNF-α (F)、iNOS (G) 和 COX-2 (H) 的 mRNA 水平。LPS 刺激组的基因表达水平设为1。数据显示了三个实验组平均值±标准差。相比较于 LPS 处理组,* 表示 $P<0.05$,** 表示 $P<0.01$ 和 *** 表示 $P<0.001$;相比较于对照组,# 表示 $P<0.01$。

2.2 蜂胶预处理对脂多糖诱导小鼠巨噬细胞炎性介质释放的影响

蜂胶对 NO 及炎症因子产生的作用表现出时空依赖性的规律,但是 CP 和 BP 的作用却不尽相同。如图 7.13 所示,LPS 处理 3h 后细胞培养基的 NO 水平显著地不断上升直至实

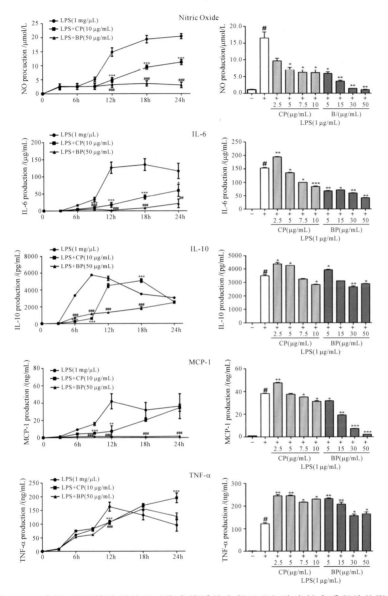

图 7.13 中国、巴西蜂胶预处理对脂多糖诱导小鼠巨噬细胞炎性介质释放的影响

Fig 7.13 Effects of Chinese and Brazilian propolis pretreatments on inflammatory mediator release by RAW 264.7 cells stimulated with LPS

注:用 10μg/mL 中国蜂胶提取液(CP)或 50μg/mL 巴西蜂胶提取液(BP)处理小鼠巨噬细胞 1h,培养一段时间后,用 1.0μg/mL 的脂多糖诱导小鼠巨噬细胞发生严重反应。相比较于脂多糖诱导细胞,CP 组细胞 * 表示 $P<0.05$,** 表示 $P<0.01$ 和 *** 表示 $P<0.001$;# 表示 $P<0.05$,## 表示 $P<0.01$ 和 ### 表示 $P<0.001$。用不同浓度的 CP 或者 BP 预处理 RAW 264.7 细胞,并加入 1.0μg/mL 的脂多糖诱导小鼠吞噬细胞,时间持续 24h。通过格里斯反应分析测定 NO,流式细胞仪测定其他炎性细胞因子。数据显示了 3 个实验组的平均值±标准差。相比较于脂多糖处理组,* 表示 $P<0.05$,** 表示 $P<0.01$,*** 表示 $P<0.001$;相比较于对照组 # 表示 $P<0.01$。

验结束，而 BP 和 CP 显著地抑制了这一趋势，且 BP 的效果强于 CP。在 LPS 处理 24h 后 NO 的产量是 $16.5\pm1.8\mu mol/L$，$50\mu g/mL$ BP 处理组、$10\mu g/mL$ CP 处理组则分别为 $10.5\pm1.8\mu mol/L$ 和 $6.2\pm0.8\mu mol/L$，二者均低于 LPS 组，且 BP 组更低；LPS 刺激可以显著地提高 IL-6 和 MCP-1 的产量，而蜂胶对这两种细胞因子的调节作用随蜂胶的预处理时间及浓度的不同而发生变化；$50\mu g/mL$ 的 BP 预处理对 IL-6 和 MCP-1 的过量产生表现出最强的抑制效果；LPS 刺激 RAW 264.7 时 IL-10 分泌不断上升并于处理后 9h 后达到最大值，而后开始缓慢下降，蜂胶预处理对于 IL-10 的分泌的影响却是大相径庭，CP 预处理推迟了这一变化进程，致使 IL-10 的峰值延迟到 LPS 处理 18h 才出现，BP 预处理随着时间延长 IL-10 的产量逐渐缓慢上升，但是全程均低于 LPS 处理组。

结果还表明，在低剂量的 CP（2.5 和 $5\mu g/mL$）和 BP（$5\mu g/mL$）显著地促进了 LPS 处理 24h 后的 IL-10 的分泌，而高剂量处理则抑制了 IL-10 的分泌。同时，也得到一些出乎意料的结果，即 CP 促进了 TNF-α 的分泌而 BP 则没有显著的作用。不仅如此，在我们的检测中没有检测到 IFN 和 IL-12p70，这与前人的研究结果一致（Berbaum et al.，2008），提示这些细胞因子不是巨噬细胞中持续表达的，且也不能被 LPS 所激发产生。

2.3　蜂胶预处理对脂多糖诱导巨噬细胞氧化应激的影响

结果如图 7.14 所示，LPS 处理 RAW 264.7 细胞显著提高了 DCF 阳性的细胞比率，通过检测荧光强度可见本实验中所用的各个浓度的蜂胶处理均可以抑制 ROS 的产生，而 ROS 阳性细胞数则在各个浓度与 LPS 间不存在显著差异。

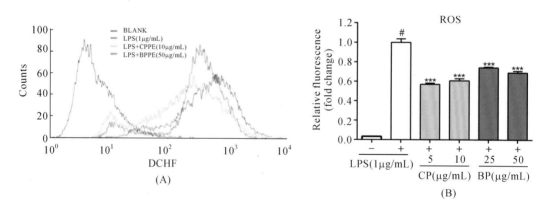

图 7.14　中国、巴西蜂胶预处理对脂多糖诱导巨噬细胞氧化应激的影响

Fig 7.14　Inhibition of ROS generation by PPE in LPS-stimulated RAW 264.7 cells

注：小鼠巨噬细胞 RAW 264.7 经不同浓度的 PE 孵育 1h，用 $1\mu g/mL$ 的脂多糖诱导 24h。避光加入 $10\mu mol/L$ 的 DCFH-DA 于细胞中，避光孵育 30min 后采用流式细胞仪检测细胞平均 DCF 荧光强度。（A）三个实验组中未经处理细胞的流式细胞直方图，细胞经 $10\mu g/mL$ 的中国蜂胶提取液（CP）或者 $50\mu g/mL$ 的巴西蜂胶提取液（BP）预处理 1h 后，加入 $1\mu g/mL$ 的脂多糖（LPS）溶液，暴露 24h。（B）采用流式细胞仪检测 DCF 荧光强度。数据显示了 3 个实验组的平均值±标准差。相比较于脂多糖处理组，* 表示 $P<0.05$，** 表示 $P<0.01$ 和 *** 表示 $P<0.001$；相比较于对照组 # 表示 $P<0.01$。

2.4　蜂胶对核因子-κB 激活的影响

通过 3×NF-κB-Luc 质粒转染构建的报告基因系统，检测了蜂胶预处理对 TNF-α 和 IL-1β 处理来诱导 NF-κB 激活的作用。细胞转染 24h 后采用不同浓度的蜂胶处理，而后加入

10ng/mL TNF-α 或 IL-1β,结果如图 7.15A 所示,蜂胶剂量依赖性地抑制了 NF-κB 报告基因表达活性。

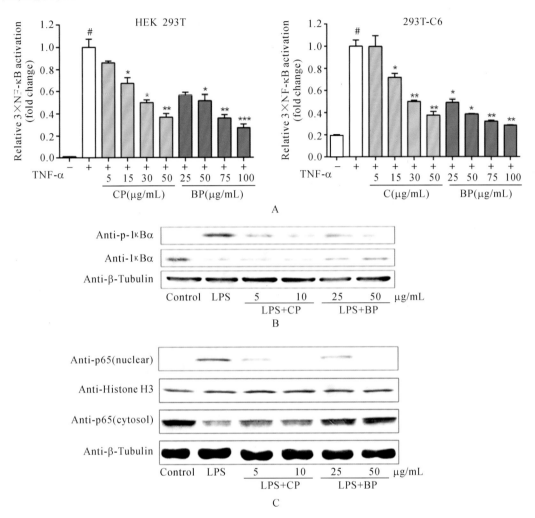

图 7.15　中国、巴西蜂胶预处理对核因子-κB 激活的影响

Fig 7.15　Effects of CP and BP on NF-κB activation

A:通过 TNF-α 和 IL-1β 分别诱导 HEK 293T 细胞和 293T C6 细胞后,CP 和 BP 蜂胶对 NF-κB 激活作用的影响。293T 或 293T-C6 细胞被萤火虫荧光素酶报告质粒(pGL4.2-3×NF-κB-Luc)和海肾荧光素酶报告质粒(pRL-TK)转染,培养 24h。使用不同浓度的 CP(0~50μg/mL,灰条)或 BP(0~100μg/mL,白条)处理细胞,培养 1h。通过 TNF-α(10ng/mL)刺激 HEK 293T 细胞或 IL-1β(10ng/mL)刺激 293-C6 细胞 12h 后,利用评估报告检测 NF-κB 的激活作用。采用 tk-Renilla 荧光素酶活性归一萤火虫荧光素酶和海肾荧光素酶。数据显示了 3 个实验组的平均值±标准差。相比较于 TNF-α 处理细胞或 IL-1β 处理细胞,* $P<0.05$,** $P<0.01$ 和 *** $P<0.001$,具有显著的统计学差异。相比较于控制组,# $P<0.01$。B:PE 对脂多糖诱导小鼠吞噬细胞产生磷酸化和 IκB-α 降解作用的影响。使用给定浓度的 PE 处理或不处理 RAW264.7 细胞 1h,1μg/mL 的 LPS 诱导 30min。通过磷酸化或非磷酸化 IκB-α 产生特异性抗体,并利用特定抗体检测细胞蛋白表达。β-微管蛋白作为对照。C:PE 对脂多糖诱导的 NF-κB 在小鼠吞噬细胞 RAW 264.7 中核转移的作用。按上述方法处理 RAW 264.7 细胞,NF-κB-p65 亚基抗体免疫印迹胞质提取物和核提取物,采用蛋白质印迹分析。组蛋白 H3 作为核提取物的对照,β-微管蛋白作为胞质蛋白的对照。3 个实验组的免疫印迹显示相似的实验结果。

在巨噬细胞炎症过程中，LPS 诱导的 NF-κB 激活过程主要包括 IκB-α 的磷酸化，NF-κB-p65 亚基的降解及核转位。我们应用特异性的抗体对 IκB-α、NF-κB-p65 的修饰及分布进行了蛋白质印迹检测，结果如图 7.15B 所示，BP 抑制了 LPS 诱导细胞内 IκB-α 的磷酸化和降解，CP 则抑制了 IκB-α 的磷酸化，但是对 LPS 诱导其发生降解却没有作用。更重要的是，BP 和 CP 预处理均引起核内 NF-κB-p65 大幅下降，这提示它们具有抑制 LPS 诱导的 NF-κB-p65 在 RAW 264.7 中核转位的作用。

2.5 蜂胶对 TRAF6 泛素连接酶活力及多聚泛素链形成的影响

为了进一步阐明蜂胶是如何调控 NF-κB 的激活，我们检测了蜂胶处理是否影响细胞 TRAF6 的自泛素化过程。基于 Wang 等(2001)的前期研究，我们建立了一个稳定的 Hela 细胞系(HeLa-T6RZC)。该细胞系中 TRAF6 的 C 端结构域被细菌促旋酶 B 所替换，这样使得该细胞内 TRAF6 的寡聚化不再依赖上游的 IRAK 信号而可以直接通过库马霉素处理细胞即可诱导激活 NF-κB。实验结果确定所构建的 HeLa-T6RZC 对 2 μmol 库马霉素处理非常敏感，库马霉素处理 5min 即可引起 IκBα 的磷酸化并伴随着 IκBα 蛋白快速的降解，而且这一过程中伴随着 TRAF6 多聚泛素化的积累。结果还发现，蜂胶处理可以推迟细胞内 TRAF6 多聚泛素化的形成。

基于这一结果，我们进一步检测了蜂胶是否影响直接泛素化酶的活性，应用了一个无细胞系统，严格模拟 TRAF6 的泛素连接酶的活性环境，这一系统中包括从 Sf9 细胞中纯化的 TRAF6 蛋白、E1、E2(Ubc13/Uev2)、泛素、ATP 及缓冲液。结果如图 7.16 所示，100~400 μg/mL CP 处理剂量依赖性地显著抑制了 T6RZC 催化多聚泛素链形成的能力，同时，在无细胞体系中所用的 CP 浓度高于细胞培养体系中所使用的量，200 μg/mL 或 400 μg/mL BP 处理也可以产生相似的抑制效果。这一结果提示 PPE 抑制了 TRAF6 依赖的泛素链的形成，破坏了 TRAF6 的泛素连接酶活性，进而抑制了 NF-κB 的激活。

3 讨论

大量研究证实，多酚类化合物具有强大的抗氧化功能，因此可被用作预防和治疗急性和慢性炎症的天然替代品(Muschietti et al.,2001;Joseph et al.,2016)，而蜂胶富含这类多酚类化合物。大量在中国蜂胶中发现的主要的多酚类化合物在不同的试验中通过体外和体内研究模型均证实具有抗炎性能，例如柯因(Hougee et al.,2005)、高良姜素(Jung et al.,2014)、山奈酚(Hämäläinen et al.,2007)和咖啡酸苯乙酯(dos Santos et al.,2013)等。一些关于巴西绿蜂胶的研究也证实了巴西绿蜂胶的抗炎效果主要源自于其中富含的多酚类化合物，特别是咖啡酸(Bufalo et al.,2013)和阿替匹林 C(Paulino et al.,2008;Szliszka et al.,2013)。

我们实验室前期研究已经发现 CP 显著地下调了弗氏完全佐剂(Freund's complete adjuvant,FCA)诱导的关节炎大鼠中的 IL-6(Hu et al.,2005)。其他体内试验也表明，BP 显著地抑制了三硝基苯磺酸(2,4,6-trinitrobenzene sulfonic acid,TNBS)诱导的结肠炎中的 IFN-γ 和 IL-12 的产生，而 TNBS 诱导的结肠炎模型是广为认可的炎性肠病研究模型(Okamoto et al.,2013)。Sforcin 等研究发现，BP 不影响小鼠脾细胞中包括 IL-4 和 IL-10 在内的抗炎因子的表达和产生，但是抑制诸如 IL-6、IL-1 和 IFN-γ 等促炎因子的产生(Orsatti et al.,2010)。源自特定区域的不同蜂胶提取物其化学组成也不同，因而在特定的

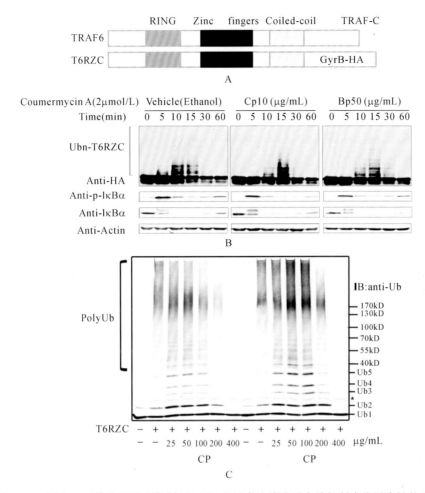

图 7.16　中国、巴西蜂胶预处理抑制 TRAF6 的泛素连接酶活力并抑制多聚泛素链的形成

Fig 7.16　PPE inhibit the E3 activity of TRAF6 by controlling its auto-ubiquitination

A：TRAF6 蛋白（上部）的嵌合结构，其 TRAF6 的 C 端结构被人工寡聚促旋酶 B 所替换（T6RZC，下部）。B：稳定表达 TRAF6（T6RZC）的 HeLa 细胞体系中加入 2μmol 的库马霉素作为刺激，用含有/不含有 10μg/mL 的 CP 或 50μg/mL 的 BP 预处理 1h。收集细胞提取物，使用特定抗体做免疫印迹。β-微管蛋白作为加载控制，3 个独立实验的代表印迹具有相似的实验结果。C：50mmol/L Tris-HCl（pH 7.5），2mmol/L ATP，5mmol/L $MgCl_2$ 和 0.5mmol/L DTT 配成缓冲液，不同浓度的 PE 加入到缓冲体系中在 30℃ 条件下孵育纯化的 T6RZC（20nmol/L），E1（85nmol/L），E2（Ubc13，3.9μmol/L/Uev2，3.9μmol/L）和泛素（10μmol/L）30min。加入 SDS 缓冲液停止反应，利用 SDS-PAGE 实现分离，通过泛素蛋白印迹进行分析。代表性重复实验具有相似的实验结果。＊非特异性条带。

免疫细胞中表现出不同的免疫调节活性（Chan et al.，2013）。我们的研究结果发现，CP 和 BP 对于 IL-10 的调节作用确实存在不同，这很可能是由它们的多酚组成成分不同所致（Lin et al.，2012；Bueno-Silva et al.，2013）。因而我们猜想 BP 可能降低了炎症细胞特别是活化的巨噬细胞中炎性因子的释放来抵御炎症，而 CP 则可以刺激抗炎因子产量的增加来代偿性地抵御炎症。

　　巨噬细胞对于宿主防御病原入侵具有重要作用。在败血症发生发展过程中，巨噬细胞通过产生大量的促炎调节因子和黏附蛋白来激活免疫系统（Martinez et al.，2007），能够抑制炎症调节因子产生的物质可作为潜在免疫调节和抗炎特效治疗药物（Haque et al.，2011；

Jakus et al.,2013)。本研究中,我们发现蜂胶可强有力地调节活化的巨噬细胞中的各种炎症调节因子。蜂胶剂量和时间依赖性地抑制了 RAW264.7 中 LPS 诱导的 NO 的产生,并且抑制了 IL-6 和 MCP-1 的基因表达丰度和蛋白分泌量。我们前期研究发现,富含黄酮的中国蜂胶可以抑制 IL-1β 的基因表达丰度和蛋白分泌量(Wang et al.,2013)。近期关于巴西绿蜂胶的一些研究也证实巴西绿蜂胶的体外抗炎活性归因于其具有丰富的多酚类物,特别是咖啡酸(Bufalo et al.,2013)和阿替匹林 C(Paulino et al.,2008;Szliszka et al.,2013)。

LPS 诱导的炎症反应可以改变 ROS 的产生,进而激活 MAPK 和 PKC,进一步激活 NF-κB 等转录因子。我们团队早期研究发现,适宜浓度的中国蜂胶和巴西蜂胶预处理可以抑制 HUVEC 细胞中的 ROS 过量产生(Xuan et al.,2011)。本研究中,我们发现蜂胶不仅在 LPS 刺激中表现出 ROS 清除活性,而且促进了 HO-1 的信使 RNA 表达。HO-1 是一种重要的抗氧化酶,我们的结果提示蜂胶可能参与调节一些重要的自由基敏感转录因子,例如 Nrf-2。

蜂胶在 NF-κB 的激活中起着重要的作用,并可能是通过调节 NF-κB 进而引起下游促炎基因表达下调(Bufalo et al.,2013;Wang et al.,2013),然而仍然缺乏蜂胶对这一激活过程的具体调节机制的相关研究。我们知道,蜂胶中含有多种多酚类物质,CP 中包括咖啡酸苯乙酯(Natarajan et al.,1996)、白藜芦醇(Jakus et al.,2013)、芹菜素(Nicholas et al.,2007)、山奈酚和槲皮素等(Hämäläinen et al.,2007),BP 中包含阿替匹林 C(Paulino et al.,2008;Szliszka et al.,2013)、咖啡酸(Bufalo et al.,2013)及其衍生物、肉桂酸及其衍生物等(Sawaya et al.,2011)。我们的研究进一步发现了这两种蜂胶对 NF-κB 的激活具有较强的抑制作用。在 LPS 诱导的脓毒血症整小鼠模型中,我们发现小鼠肺部存在明显的 NF-κB-p65 表达,通过口服 PPE 可以显著地下调其中的 NF-κB,这一结果坚定了我们的假设,即蜂胶可能是通过下调 NF-κB 来发挥抗炎活性的。试验中采用的两种蜂胶均剂量依赖性地抑制了 TNF-α 或 IL-1β 诱导的 293T 细胞中的 NF-κB 活性。通过进一步检测发现,在 LPS 诱导的 RAW264.7 中蜂胶剂量依赖性地抑制 NF-κB 活性主要是通过抑制 IκBα 的磷酸化并抑制 p65 的核转位。我们的结果显示只有巴西蜂胶可以抑制 IκBα 的降解。

LPS 可以激活两条下游炎症信号通路,这种激活均是依赖于 LPS 结合 TLR4 受体进而通过 MyD88 传递信号而实现的。MyD88 则招募白介素 1 受体相关激酶 1 和 6(interleukin-1 receptor associated kinase,IRAK-1,IRAK-6)和 TRAF6,进而诱发典型的 NF-κB 激活反应(Xia et al.,2012)。TRAF6 是连接 TLR4 受体和 NF-κB 转录因子的重要蛋白,大量基础研究证实 TRAF6 的激活主要依赖于位点特异性的非降解的 Lys63 连接的自泛素化反应(Lamothe et al.,2007;Yin et al.,2009)。TRAF6 可以通过其 C 端域对上游信号做出应答反应。而其 N 端 RING 域包括一个不断扩大的泛素连接酶家族,并作为 E3s 起作用。这一结构对于 TRAF6 调控的 IKK-NF-κB 激活是必需的。本研究中用到的 HeLa-T6RZC、TRAF6 依赖的 IKK-NF-κB 激活是通过库马霉素所诱导的,而不依赖于上游的 IRAK 信号。PE 处理的 HeLa-T6RZC 细胞表现出之后的泛素化形成过程,这一结果促使我们进一步探究蜂胶对 TRAF 泛素连接酶的作用。体外激酶测试清楚地显示试验中用到的蜂胶破坏了泛素链的形成,确定了蜂胶对 NF-κB 激活的调节作用可能与其对 TRAF6 的泛素化过程调节直接相关。

但是值得注意的是,我们在体外激酶检测中蜂胶的有效浓度远高于细胞试验中的有效浓度,猜测蜂胶中能够调节 TRAF6 的有效成分在蜂胶中含量较低。我们检测了蜂胶中的多酚类化合物发现并不是所有的组分都有调节 NF-κB 激活的能力,也并不是所有的组分可以抑制 TNF-α 诱导的 NF-κB 激活。一些蜂胶中发现的多酚类化合物,如柯因、高良姜素和山柰酚甚至增加了 NF-κB 报告基因系统中荧光素酶的活性。而且不像大量药物化合物特异性的作用于某一个受体或者信号通路,多酚类化合物具有多靶点的作用特点(Koeberle et al.,2014)。多酚类物质根据其组分的不同可能发挥特异或非特异的作用机制。前期结果证实一些多酚诸如白藜芦醇(Jakus et al.,2013)、黄芩苷(Hou et al.,2012)、夫拉平度(Haque et al.,2011)等可以通过直接调节 TRAF6 缓解 LPS 诱导的炎症反应。有趣的是,我们发现 CAPE 和阿替匹林 C 这两种潜在活性成分可以强力抑制 NF-κB 报告基因系统中 NF-κB 的激活,这与前人的研究结果相一致(Natarajan et al.,1996;Paulino et al.,2008;Szliszka et al.,2013)。我们也发现即使 CAPE 和阿替匹林 C 用到最高浓度也不能影响泛素化链的形成过程,这一结果提示这两种组分对 NF-κB 的调节是不依赖于 TRAF6 的。因此,我们认为 CP 和 BP 可能是参与调节了 TRAF6 上游的信号转导蛋白,例如 MyD88 或者 IRAK 等。显然,后续研究应着眼于蜂胶的主要抗炎组分及其作用机制研究。

综上所述,我们应用体外试验证实富含多酚的蜂胶提取物尽管地理来源和成分组成不尽相同,但均表现出明显的抗炎效果。这种抗炎效果主要表现为抑制 LPS 诱导的巨噬细胞中促炎症因子的表达和分泌,抑制 NO 和 ROS 的生成。而产生这种效果主要是通过抑制 TRAF6 的泛素化链形成,进而干扰了 NF-κB 激活过程。这一结果丰富了蜂胶提取物对炎症微环境中炎症关键细胞的调节机制研究,为蜂胶抗炎机制研究提供了参考。

小 结

本研究利用巨噬细胞体外炎症实验模型,采用不同地理来源蜂胶探索了其对炎症微环境中具有核心作用的巨噬细胞炎症反应和对炎症相关信号通路的调控作用及其机制。结果发现,中国蜂胶和巴西蜂胶尽管化学成分存在显著差异,但二者均能显著抑制细菌脂多糖诱导小鼠巨噬细胞炎症因子的表达与释放,同时也能缓解细胞内活性氧(ROS)的累积。在调节细胞炎症信号通路方面,报告基因结果表明,蜂胶能显著促使炎症过程中核心转录因子NF-κB 的激活,但两种蜂胶调节机制并不完全相同。巴西蜂胶能同时抑制脂多糖诱导巨噬细胞内的 IκB-α 蛋白的磷酸化和降解,中国蜂胶能抑制 IκB-α 的蛋白的磷酸化但对其炎症过程中发生降解并没有显著抑制作用。进一步研究表明,蜂胶是通过抑制 TRAF6 依赖的泛素链的形成进而影响 TRAF6 的泛素连接酶活性从而抑制了 NF-κB 的激活的。

参考文献

Al-Sadi R,Guo SH,Ye DM,Ma Thomas Y(2013)TNF-α modulation of intestinal epithelial tight junction barrier is regulated by ERK1/2 activation of Elk-1. *The American Journal of Pathology*,183(6):1871-1884.

Aslan A,Temiz M,Atik E,Polat G,Sahinler N,Besirov E,Aban N,Parsak CK(2007)Effectiveness of mesalamine,propolis in experimental colitis. *Advances in Therapy*,24(5):1085-1097.

Baker Rebecca G, Hayden Matthew S, Ghosh S (2011) NF-κB, inflammation, and metabolic disease. *Cell Metabolism*, 13(1): 11-22.

Bankova V (2005) Chemical diversity of propolis and the problem of standardization[J]. *Journal of Ethnopharmacology*, 100(1): 114-117.

Ben-Horin S, Waterman M, Kopylov U, Yavzori M, Picard O, Fudim E, Awadie H, Weiss B, Chowers Y (2013) Addition of an immunomodulator to infliximab therapy eliminates antidrug antibodies in serum and restores clinical response of patients with inflammatory bowel disease. *Clinical Gastroenterology and Hepatology*, 11(4): 444-447.

Berbaum K, Shanmugam K, Stuchbury G, Wiede F, Korner H, Munch G (2008) Induction of novel cytokines and chemokines by advanced glycation endproducts determined with a cytometric bead array. *Cytokine*, 41(3): 198-203.

Beutheu S, Ghouzali I, Galas L, Déchelotte P, Coëffier M (2013) Glutamine and arginine improve permeability and tight junction protein expression in methotrexate-treated Caco-2 cells. *Clinical Nutrition*, 32(5): 863-869.

Bueno-Silva B, Alencar SM, Koo H, Ikegaki M, Silva G VJ, Napimoga MH, Rosalen PL (2013) Anti-inflammatory and antimicrobial evaluation of neovestitol and vestitol isolated from Brazilian red propolis. *Journal of Agricultural and Food Chemistry*, 61(19): 4546-4550.

Bufalo MC, Ferreira I, Costa G, Francisco V, Liberal J, Cruz MT, Lopes MC, Batista MT, Sforcin JM (2013) Propolis and its constituent caffeic acid suppress LPS-stimulated pro-inflammatory response by blocking NF-kappaB and MAPK activation in macrophages. *Journal of Ethnopharmacology*, 149(1): 84-92.

Burisch, Munkholm P (2015) The epidemiology of inflammatory bowel disease. *Scandinavian Journal of Gastroenterology*, 50(8): 942-951.

Capaldo CT, Farkas AE, Hilgarth RS, Krug SM, Wolf MF, Benedik JK, Fromm M, Koval M, Parkos C, Nusrat A (2014) Proinflammatory cytokine-induced tight junction remodeling through dynamic self-assembly of claudins. *Molecular Biology of The Cell*, 25(18): 2710-2719.

Cestari SH, Bastos JK, Di Stasi LC (2011) Intestinal anti-inflammatory activity of Baccharis dracunculifolia in the trinitrobenzenesulphonic acid model of rat colitis. *Evidence-Based Complementary and Alternative Medicine*.

Chan GCF, Cheung KW, Sze DMY (2013) The immunomodulatory and anticancer properties of propolis. *Clinical Reviews in Allergy & Immunology*, 44(3): 262-273.

Chang PV, Hao LM, Offermanns S, Medzhitov R (2014) The microbial metabolite butyrate regulates intestinal macrophage function via histone deacetylase inhibition. *Proceedings of the National Academy of Sciences*, 111(6): 2247-2252.

Chasiotis H, Kolosov D, Bui P, Kelly SP (2012) Tight junctions, tight junction proteins and paracellular permeability across the gill epithelium of fishes: a review. *Respiratory Physiology & Neurobiology*, 184(3): 269-281.

Chawla A, Nguyen KD, Goh YPS (2011) Macrophage-mediated inflammation in metabolic disease. *Nature Reviews Immunology*, 11(11): 738-749.

Choi SJ, Shimomura K, Kumazawa S, Ahn MR (2013) Antioxidant properties and phenolic composition of propolis from diverse geographic regions in Korea. *Food Science and Technology Research*, 19(2): 211-222.

Cuevas A, Saavedra N, Cavalcante MF, Salazar L A, Abdalla DSP (2014) Identification of microRNAs involved in the modulation of pro-angiogenic factors in atherosclerosis by a polyphenol-rich extract from propolis. *Archives of Biochemistry and Biophysics*, 557: 28-35.

da Silva Filho AA, de Sousa JP, Soares S, Furtado NA, Andrade e Silva ML, Cunha WR, Gregorio LE, Nanayakkara NP, Bastos JK (2008) Antimicrobial activity of the extract and isolated compounds from Baccharis dracunculifolia D. C. (Asteraceae). *Zeitschrift Für Naturforschung*, C 63(1-2): 40-46.

da Silva Filho AA, Resende DO, Fukui MJ, Santos FF, Pauletti PM, Cunha WR, Silva ML, Gregorio LE, Bastos JK, Nanayakkara NP (2009) *In vitro* antileishmanial, antiplasmodial and cytotoxic activities of phenolics and triterpenoids from Baccharis dracunculifolia D. C. (Asteraceae). *Fitoterapia*, 80(8): 478-482.

Danielsen Erik Michael, Hansen Gert H, Rasmussen K, Niels-Christiansen LL (2013) Permeabilization of enterocytes induced by absorption of dietary fat. *Molecular Membrane Biology*, 30(3): 261-272.

Deng L, Wang C, Spencer E, Yang LY, Braun A, You JX, Slaughter C, Pickart C and Chen ZJ (2000) Activation of the IκB kinase complex by TRAF6 requires a dimeric ubiquitin-conjugating enzyme complex and a unique polyubiquitin chain. *Cell*, 103(2): 351-361.

dos Santos DA, Fukui Mde J, Dhammika Nanayakkara NP, Khan SI, Sousa JP, Bastos JK, Andrade SF, da Silva Filho AA, Quintao NL (2010) Anti-inflammatory and antinociceptive effects of Baccharis dracunculifolia DC (Asteraceae) in different experimental models. *Journal of Ethnopharmacology*, 127(2): 543-550.

dos Santos JS, Monte-Alto-Costa A (2013) Caffeic acid phenethyl ester improves burn healing in rats through anti-inflammatory and antioxidant effects. *Journal of Burn Care & Research*, 34(6): 682-688.

Gehart H, Kumpf S, Ittner A, Ricci R (2010) MAPK signalling in cellular metabolism: Stress or wellness? *EMBO Reports*, 11(11): 834-840.

Gonçalves CCM, Hernandes L, Bersani-Amado CA, Franco SL, Silva JF de Souza, Natali MRM (2013) Use of propolis hydroalcoholic extract to treat colitis experimentally induced in rats by 2,4,6-Trinitrobenzenesulfonic acid. *Evidence-based Complementary and Alternative Medicine*, 2013.

González-Mariscal L, Tapia R, Chamorro D (2008) Crosstalk of tight junction components with signaling pathways. *Biochimica et Biophysica Acta (BBA)-Biomembranes*, 1778(3): 729-756.

Guimaraes NS, Mello JC, Paiva JS, Bueno PC, Berretta AA, Torquato RJ, Nantes IL, Rodrigues T (2012) Baccharis dracunculifolia, the main source of green propolis, exhibits potent antioxidant activity and prevents oxidative mitochondrial damage. *Food and Chemical Toxicology*, 50(3-4):1091-1097.

Hallows KR (2005) Emerging role of AMP-activated protein kinase in coupling membrane transport to cellular metabolism. *Current Opinion In Nephrology and Hypertension*, 14(5):464-471.

Hämäläinen M, Nieminen R, Vuorela P, Heinonen M, Moilanen E (2007) Anti-inflammatory effects of flavonoids: Genistein, kaempferol, quercetin, and daidzein inhibit STAT-1 and NF-kappa B activations, whereas flavone, isorhamnetin, naringenin, and pelargonidin inhibit only NF-kappa B activation along with their inhibitory effect on iNOS expression and NO production in activated macrophages. *Mediators of Inflammation*, 45673.

Haque A, Koide N, Noman ASM, Odkhuu E, Badamtseren B, Naiki Y, Komatsu T, Yoshida T, Yokochi T (2011) Flavopiridol inhibits lipopolysaccharide-induced TNF-α production through inactivation of nuclear factor-κB and mitogen-activated protein kinases in the MyD88-dependent pathway. *Microbiology and Immunology*, 55(3):160-167.

Hering NA, Richter JF, Fromm A, Wieser A, Hartmann S, Günzel D, Bücker R, Fromm M, Schulzke JD, Troeger H (2014). TcpC protein from E. coli Nissle improves epithelial barrier function involving PKCζ and ERK1/2 signaling in HT-29/B6 cells. *Mucosal Immunology*, 7(2):369-378.

Hou JC, Wang J, Zhang P, Li D, Zhang CX, Zhao HP, Fu JH, Wang B, Liu JX (2012) Baicalin attenuates proinflammatory cytokine production in oxygen-glucose deprived challenged rat microglial cells by inhibiting TLR4 signaling pathway. *International Immunopharmacology*, 14(4):749-757.

Hougee S, Sanders A, Faber J, Graus YMF, van den Berg Wim B, Garssen J, Smit HF, Hoijer MA (2005) Decreased pro-inflammatory cytokine production by LPS-stimulated PBMC upon in vitro incubation with the flavonoids apigenin, luteolin or chrysin, due to selective elimination of monocytes/macrophages. *Biochemical Pharmacology*, 69(2):241-248.

Hu F, Hepburn HR, Li Y, Chen M, Radloff SE, Daya S (2005) Effects of ethanol and water extracts of propolis (bee glue) on acute inflammatory animal models. *Journal of Ethnopharmacology*, 100(3):276-283.

Jakus PB, Kalman N, Antus C, Radnai B, Tucsek Z, Gallyas Jr F, Sumegi B, Veres B (2013) TRAF6 is functional in inhibition of TLR4-mediated NF-κB activation by resveratrol. *Journal of Nutritional Biochemistry*, 24(5):819-823.

Joseph SV, Edirisinghe I, Burton-Freeman BM (2016) Fruit polyphenols: a review of anti-inflammatory effects in humans. *Critical Reviews in Food Science and Nutrition*, 56(3):419-444.

Jung YC, Kim ME, Yoon JH, Park PR, Youn Hwa-Young, Lee H W, Lee JS (2014) Anti-inflammatory effects of galangin on lipopolysaccharide-activated macrophages via ERK and

NF-κB pathway regulation. *Immunopharmacology and Immunotoxicology*, 36（6）：426-432.

Kano Y, Horie N, Doi S, Aramaki F, Maeda H, Hiragami F, Kawamura K, Motoda H, Koike Y, Akiyama J (2008) Artepillin C derived from propolis induces neurite outgrowth in PC12m3 cells via ERK and p38 MAPK pathways. *Neurochemical Research*, 33(9)：1795-1803.

Kinugasa T, Sakaguchi T, Gu X, Reinecker H (2000) Claudins regulate the intestinal barrier in response to immune mediators. *Gastroenterology*, 118(6)：1001-1011.

Koeberle A, Werz O (2014) Multi-target approach for natural products in inflammation. *Drug Discovery Today*, 19(12)：1871-1882.

Kosińska A, Andlauer W (2012) Cocoa polyphenols are absorbed in Caco-2 cell model of intestinal epithelium. *Food Chemistry*, 135(3)：999-1005.

Kosińska A, Andlauer W (2013) Modulation of tight junction integrity by food components. *Food Research International*, 54(1)：951-960.

Lamothe B, Besse A, Campos AD, Webster WK, Wu H, Darnay BG (2007) Site-specific Lys-63-linked tumor necrosis factor receptor-associated factor 6 auto-ubiquitination is a critical determinant of IκB kinase activation. *Journal of Biological Chemistry*, 282(6)：4102-4112.

Lemos M, de Barros MP, Sousa JP, da Silva Filho AA, Bastos JK, de Andrade SF (2007) Baccharis dracunculifolia, the main botanical source of Brazilian green propolis, displays antiulcer activity. *Journal of Pharmacy and Pharmacology*, 59(4)：603-608.

Lin HP, Jiang SS, Chuu CP (2012) Caffeic acid phenethyl ester causes p21 induction, Akt signaling reduction, and growth inhibition in PC-3 human prostate cancer cells. *PLoS one*, 7(2)：e31286.

Liu XB, Wang ZH, Wang P, Yu B, Liu YH, Xue YX (2013) Green tea polyphenols alleviate early BBB damage during experimental focal cerebral ischemia through regulating tight junctions and PKCalpha signaling. *BMC Complementary and Alternative Medicine*, 13(1)：187.

Machado JL, Assunçao AKM, da Silva MCP, Reis AS, Costa GC, Arruda DS, Rocha BA, Vaz MM, Oliveira LL, Paes AMA, Guerra RNM (2012) Brazilian green propolis：anti-inflammatory property by an immunomodulatory activity. *Evidence-based Complementary and Alternative Medicine*, 157652.

Martin-Martin N, Ryan G, McMorrow T, Ryan MP (2010) Sirolimus and cyclosporine a alter barrier function in renal proximal tubular cells through stimulation of ERK1/2 signaling and claudin-1 expression. *American Journal of Physiology-Renal Physiology*, 298(3)：F672-F682.

Martinez FO, Sica A, Mantovani A, Locati M (2007) Macrophage activation and polarization. *Frontiers in Bioscience*, 13：453-461.

Melmed GY, Siegel CA, Spiegel BM, Allen JI, Cima Robert, Colombel JF, Dassopoulos T,

Denson LA, Dudley-Brown S, Garb A (2013) Quality indicators for inflammatory bowel disease: development of process and outcome measures. *Inflammatory Bowel Diseases*, 19(3): 662-668.

Missima F, da Silva Filho AA, Nunes GA, Bueno PC, de Sousa JP, Bastos JK, Sforcin JM (2007) Effect of Baccharis dracunculifolia D. C. (Asteraceae) extracts and its isolated compounds on macrophage activation. *Journal of Pharmacy and Pharmacology*, 59(3): 463-468.

Morgan Xochitl C, Tickle Timothy L, Sokol Harry, Gevers Dirk, Devaney Kathryn L, Ward Doyle V, Reyes Joshua A, Shah Samir A, LeLeiko Neal, Snapper Scott B (2012) Dysfunction of the intestinal microbiome in inflammatory bowel disease and treatment. *Genome Biology*, 13(9): R79.

Munari CC, Resende FA, Alves JM, de Sousa JPB, Bastos JK and Tavares DC (2008) Mutagenicity and antimutagenicity of baccharis dracunculifolia extract in chromosomal aberration assays in Chinese hamster ovary cells. *Planta Medica*, 74(11): 1363-1367.

Murase H, Shimazawa M, Kakino M, Ichihara K, Tsuruma K, Hara H (2013) The effects of Brazilian green propolis against excessive light-induced cell damage in retina and fibroblast cells. *Evidence-based Complementary and Alternative Medicine*, 2013.

Muschietti L, Gorzalczany S, Ferraro G, Acevedo C, Martino V (2001) Phenolic compounds with anti-inflammatory activity fromEupatorium buniifolium. *Planta Medica*, 67(743): 743-744.

Natarajan K, Singh S, Burke TR, Grunberger D and Aggarwal BB (1996) Caffeic acid phenethyl ester is a potent and specific inhibitor of activation of nuclear transcription factor NF-kappa B. *Proceedings of the National Academy of Sciences*, 93(17): 9090.

Neurath MF, Travis Simon PL (2012) Mucosal healing in inflammatory bowel diseases: a systematic review. *Gut*, 61(11): 1619-1635.

Nicholas C, Batra S, Vargo MA, Voss OH, Gavrilin MA, Wewers MD, Guttridge DC, Grotewold E, Doseff AI (2007) Apigenin blocks lipopolysaccharide-induced lethality *in vivo* and proinflammatory cytokines expression by inactivating NF-κB through the suppression of p65 phosphorylation. *The Journal of Immunology*, 179(10): 7121-7127.

Okamoto Y, Hara T, Ebato T, Fukui T and Masuzawa T (2013) Brazilian propolis ameliorates trinitrobenzene sulfonic acid-induced colitis in mice by inhibiting Th1 differentiation. *International Immunopharmacology*, 16(2): 178-183.

Orsatti CL, Missima F, Pagliarone AC, Bachiega TF, Búfalo MC, Araújo Jr JP, Sforcin JM (2010) Propolis immunomodulatory action in vivo on toll-like receptors 2 and 4 expression and on pro-inflammatory cytokines production in mice. *Phytotherapy Research*, 24(8): 1141-1146.

Park YK, Paredes-Guzman JF, Aguiar CL, Alencar SM, Fujiwara FY (2004) Chemical constituents in Baccharis dracunculifolia as the main botanical origin of southeastern Brazilian propolis. *Journal of Agricultural and Food Chemistry*, 52(5): 1100-1103.

Parkar Shanthi G,Stevenson David E,Skinner Margot A (2008) The potential influence of fruit polyphenols on colonic microflora and human gut health. *International Journal of Food Microbiology*,124(3):295-298.

Paulino N,Abreu SRL,Uto Y,Koyama D,Nagasawa H,Hori H,Dirsch VM,Vollmar AM,Scremin A, Bretz WA (2008) Anti-inflammatory effects of a bioavailable compound, Artepillin C,in Brazilian propolis. *European Journal of Pharmacology*,587(1-3):296-301.

Peterson LW,Artis D (2014) Intestinal epithelial cells:regulators of barrier function and immune homeostasis. *Nature Reviews Immunology*,14(3):141-153.

Roquetto AR,Monteiro NES,Moura CS,Toreti VC,de Pace F,dos SA,Park YK,Amaya-Farfan J (2015) Green propolis modulates gut microbiota, reduces endotoxemia and expression of TLR4 pathway in mice fed a high-fat diet. *Food Research International*,76:796-803.

Rubiolo P,Casetta C,Cagliero C,Brevard H,Sgorbini B,Bicchi C (2013) *Populus nigra* L. bud absolute:A case study for a strategy of analysis of natural complex substances. *Analytical and Bioanalytical Chemistry*,405(4):1223-1235.

Sawaya ACHF,Cunha IBS,Marcucci Maria Cristina (2011) Analytical methods applied to diverse types of Brazilian propolis. *Chemistry Central Journal*,5(1):27.

Sforcin JM,Bankova V (2011). Propolis:is there a potential for the development of new drugs? *Journal of Ethnopharmacology*,133(2):253-260.

Smith Patrick M,Howitt Michael R,Panikov N,Michaud M,Gallini CA,Bohlooly YM,Glickman JN,Garrett WS (2013) The microbial metabolites,short-chain fatty acids,regulate colonic Treg cell homeostasis. *Science*,341(6145):569-573.

Suzuki T,Hara H (2009) Quercetin enhances intestinal barrier function through the assembly of zonnula occludens-2,occludin,and claudin-1 and the expression of claudin-4 in Caco-2 cells. *The Journal of Nutrition*,139(5):965-974.

Suzuki T, Tanabe S, Hara H (2011) Kaempferol enhances intestinal barrier function through the cytoskeletal association and expression of tight junction proteins in Caco-2 cells. *The Journal of Nutrition*,141(1):87-94.

Szliszka E,Mertas A,Czuba ZP,Krol W (2013) Inhibition of inflammatory response by Artepillin C in activated RAW264.7 macrophages. *Evidence-based Complementary and Alternative Medicine*.

Toreti VC,Sato HH,Pastore GM,Park YK (2013) Recent progress of propolis for its biological and chemical compositions and its botanical origin. *Evidence-Based Complementary and Alternative Medicine*,2013:697390.

Ulluwishewa D, Anderson RC, McNabb WC, Moughan PJ, Wells JM, Roy Nicole C (2011) Regulation of tight junction permeability by intestinal bacteria and dietary components. *The Journal of Nutrition*,141(5):769-776.

Wang X,Valenzano MC,Mercado JM,Zurbach E P,Mullin JM (2013) Zinc supplementation modifies tight junctions and alters barrier function of CACO-2 human intestinal epithelial

layers. *Digestive Diseases and Sciences*,58(1):77-87.

Woo PL,Ching D,Guan Y,Firestone GL (1999) Requirement for Ras and phosphatidylinositol 3-kinase signaling uncouples the glucocorticoid-induced junctional organization and transepithelial electrical resistance in mammary tumor cells. *Journal of Biological Chemistry*,274(46):32818-32828.

Wu YW,Sun SQ,Zhao J,Li Y,Zhou Q (2008) Rapid discrimination of extracts of Chinese propolis and poplar buds by FT-IR and 2D IR correlation spectroscopy. *Journal of Molecular Structure*,883-884(1-3):48-54.

Wu HL,Gao X,Jiang ZD,Duan ZT,Wang SK,He BS,Zhang ZY,Xie HG (2013) Attenuated expression of the tight junction proteins is involved in clopidogrel-induced gastric injury through p38 MAPK activation. *Toxicol*,304:41-48.

Xia MZ,Liang YL,Wang H,Chen X,Huang YY,Zhang ZH,Chen YH,Zhang C,Zhao M, Xu DX (2012) Melatonin modulates TLR4-mediated inflammatory genes through MyD88-and TRIF-dependent signaling pathways in lipopolysaccharide-stimulated RAW264.7 cells. *Journal of Pineal Research*,53(4):325-334.

Xuan HZ,Zhao J,Miao JY,Li YJ,Chu YF,Hu FL (2011) Effect of Brazilian propolis on human umbilical vein endothelial cell apoptosis. *Food and Chemical Toxicology*,49(1):78-85.

Yin Q,Lin SC,Lamothe B,Lu M,Lo YC,Hura G,Zheng LX,Rich Rebecca L,Campos Alejandro D and Myszka David G (2009) E2 interaction and dimerization in the crystal structure of TRAF6. *Nature Structural &Molecular Biology*,16(6):658-666.

Zhu W,Li YH,Chen ML,Hu FL (2011) Protective effects of Chinese and Brazilian propolis treatment against hepatorenal lesion in diabetic rats. *Human & Experimental Toxicology*,30(9):1246-1255.

第八章 蜂胶对糖尿病的作用及其机制研究

国内外大量的研究表明,蜂胶具有显著的调节血脂血糖活性,对糖尿病具有预防和治疗的作用。我们先前的研究也证明蜂胶能有效降低糖尿病模型大鼠血糖水平,调节糖尿病大鼠体内脂质代谢和蛋白质代谢(玄红专,2003)。

在本研究中,我们利用链佐星构建1型糖尿病(T1DM)大鼠模型,链佐星+高脂饲料构建2型糖尿病(T2DM)大鼠模型,通过研究大鼠尿液、血液和肝肾生化指标及肝肾组织病理变化,比较中国蜂胶和巴西蜂胶改善糖尿病大鼠糖代谢、脂质代谢、蛋白质代谢、氧化应激和肝肾功能的效果。同时测定T2DM大鼠肾组织内促炎细胞因子IL-2、IL-6、TNF-α及MCP-1 mRNA表达,测定PKC mRNA表达以及TGF-$β_1$表达量,测定大鼠肾脏血流动力学指标,探讨蜂胶改善糖尿病肾病可能的作用机制。

第一节 蜂胶改善T1DM大鼠效果的研究

1型糖尿病(T1DM)又称为胰岛素依赖型糖尿病,主要发生在儿童及青少年中,但是也可以发生在任何年龄阶段。发病原因多由于胰岛β细胞发生细胞介导的自身免疫性损伤,患者多具有特征性的自身免疫抗体存在,如胰岛细胞自身抗体(ICA)、胰岛素自身抗体(IAA)、谷氨酸脱羧酶自身抗体(GAD)及酪氨酸磷酸酶自身抗体IA-2等(Graves and Eisenbarth,1999)。

T1DM患者一般终生需要服用胰岛素,如果发生中断则可能出现酮症并危及生命。胰岛素药物效果肯定,机制明确,但是长期治疗费用昂贵,同时容易产生低血糖,少数服用者还会产生过敏、水肿及胰岛素抗药性等现象(陈灏珠,2001)。临床上采用口服药结合胰岛素的方式治疗T1DM患者。

本实验以中国杨树型蜂胶和巴西酒神菊属型绿蜂胶为研究对象,系统地比较不同来源蜂胶对T1DM大鼠控制血糖、调节脂质代谢和保护肝肾功能的效果,同时通过研究大鼠血液及肝肾组织氧化应激水平,探讨蜂胶改善糖尿病的作用机制。

1 材料与方法

1.1 蜂胶溶液的制备

蜂胶原料:中国杨树型蜂胶产自华北地区,巴西绿蜂胶(酒神菊属型)从巴西进口,由杭州蜂之语蜂业股份有限公司提供。

蜂胶醇提物制备:将蜂胶和95%乙醇按1∶5混合后,浸提5h,再超声提取2h,过滤,去滤渣,滤液在45℃真空干燥箱内真空抽提5h,去除液体,得到干燥纯蜂胶。

蜂胶溶液制备:提纯后的蜂胶和PEG400及PEG2000以1∶1∶1比例混合,加入蒸馏水,获得10mg/mL和5mg/mL的蜂胶水溶液,4℃避光保存,实验期间每周制备一次。

1.2 动物实验

1.2.1 实验动物

品种,SD大鼠;级别,清洁级;性别,雄性;来源,中国科学院上海实验动物中心;生产许可证,SCXK(沪)2003-0003。

1.2.2 实验条件

动物实验在浙江中医药大学动物实验研究中心进行。对实验中使用的动物均按实验动物使用的"3R"原则给以人道主义关怀。

SPF级屏障系统大鼠实验饲养室,温度(23±1)℃,湿度50%～70%,光照150～200Lx,12h明暗交替(6:00～18:00 PM),噪音<50dB,使用许可证为SYXK(浙)2003-0003。

饮水:自来水过滤灭菌,置于高压灭菌的饮水瓶中自由饮用。

饲料:Co^{60}辐照灭菌大鼠全价营养颗粒饲料。

喂养方式:自由饮食,大鼠饲养笼中给予充足的水和饲料,每笼饲养5只大鼠。造模后,每只大鼠称重、标记编号,每日加饲料3次(分别于9:00、15:00、21:00添加),每日加水5次(分别于6:30、9:30、14:30、18:30、21:30添加)。

1.2.3 T1DM模型大鼠实验

(1)动物造模

取体重为230～280g的雄性SD大鼠150只,禁食不禁水15h,用0.1mmol/L柠檬酸缓冲液(pH=4.2),将链佐星(STZ,Sigma公司产品)配成2%的溶液,按50mg/kg的剂量尾静脉注射制造糖尿病模型。另外选取8只正常大鼠,作为正常对照组。7d后取尾血测血糖和尿糖,筛选血糖在15～27mmol/L糖尿病大鼠48只用于实验。

(2)分组和给药

将筛选的糖尿病大鼠按血糖值、体重随机分组,每组8只。各组分别给予相应的药物,每天2次,连续给药8周。具体给药剂量如下:

A1组:中国杨树型蜂胶,10mg/100g大鼠体重;

A2组:中国杨树型蜂胶,5mg/100g大鼠体重;

B1组:巴西绿蜂胶,10mg/100g大鼠体重;

B2组:巴西绿蜂胶,5mg/100g大鼠体重;

阳性药物组(Positive):拜糖平,1mg/100g大鼠体重;

模型组(Model)和正常组(Normal):给予含有PEG的蒸馏水,剂量为1mL/100g大鼠体重。

(3)实验进程

①造模后,每周测定一次大鼠体重,尾部取血测定血糖值;

②造模后,每2周进行一次代谢试验,测定采食量、饮水量、排尿量,测定尿液中尿肌酐、尿蛋白、尿糖含量;

③每2周空腹眼眶后静脉取血,分离血清,测定大鼠甘油三酯(TG)、胆固醇(TC)、高密度脂蛋白胆固醇(HDL-C)、低密度脂蛋白胆固醇(LDL-C)、尿素氮(BUN)、Scr、总蛋白

(TP)、谷丙转氨酶(ALT)和谷草转氨酶(AST);

④实验第8周最后一次取血,用半自动生化仪测定血清一氧化氮(NO)、一氧化氮合酶(NOS)、超氧化物歧化酶(SOD)、CAT、GSH-px 和丙二醛(MDA),同时用多功能全定量特种蛋白金标检测仪测定 HbA_{1c};

⑤实验第8周,最后一次代谢实验收集尿液后,用全自动生化仪测定尿液中白蛋白含量;

⑥8周实验结束,处死动物,取大鼠肾脏和肝脏测定 SOD、CAT、GSH-px 和 MDA 含量;

⑦取大鼠肾脏和肝脏进行 HE 染色作病理学常规检查。

1.3 生化指标测定

1.3.1 主要仪器

BT-815A 半自动生化分析仪(上海三科仪器有限公司);NYCOCARD READER Ⅱ多功能全定量蛋白金标检测仪器(挪威 AXIS SHIELD);OneTouch^R Ultra 血糖仪[强生(上海)医疗器材有限公司];日立 7020 全自动生化分析仪(日立公司)。

1.3.2 主要试剂

糖化血红蛋白(HBA1C)测定试剂盒(日本和光纯药工业株式会社);胆固醇(TC)测定试剂盒(上海申能德赛技术诊断有限公司);三酰甘油(TG)测定试剂盒(上海申能德赛技术诊断有限公司);高密度脂蛋白(HDL)测定试剂盒(上海申能德赛技术诊断有限公司);低密度脂蛋白(LDL)测定试剂盒(上海复星长征医学科学有限公司);(SOD)测定试剂盒(南京建成生物工程研究所);丙二醛(MDA)测定试剂盒(南京建成生物工程研究所);尿素氮(BUN)测定试剂盒(上海申能德赛技术诊断有限公司);肌酐(CREA)测定试剂盒(上海申能德赛技术诊断有限公司);总蛋白(TP)测定试剂盒(上海申能德赛技术诊断有限公司);白蛋白液体(ALB)试剂盒(上海申能德赛技术诊断有限公司);谷丙转氨酶(ALT)测定试剂盒(上海申能德赛技术诊断有限公司);谷草转氨酶(AST)测定试剂盒(上海申能德赛技术诊断有限公司);拜糖平(拜耳医药保健有限公司);链脲佐菌素(STZ)(ALEXIS CORPORATION 公司);一氧化氮(NO)测定试剂盒(南京建成生物工程研究所);一氧化氮合酶(NOS)测定试剂盒(南京建成生物工程研究所)。

1.4 组织病理学观察

1.4.1 材料和方法

(1)主要材料

多聚甲醛,规格:500g/瓶,含量≥95%。由天津市化学试剂研究所生产。配制方法:称取多聚甲醛 40g,置于烧瓶中,加入 500～800mL 0.1mol/L 磷酸盐缓冲液(PBS),加热至 60℃,持续搅拌,使粉末全部溶解,滴加 1N NaOH,使溶液清亮,最后定容至 1000mL,按用量需要分批配制。

磷酸盐缓冲液(ZLI-9062),2000mL/袋,溶液终浓度为 0.01mol/L,pH 7.2～7.4,室温保存,由北京中杉金桥生物技术有限公司生产。

柠檬酸盐缓冲液(ZLI-9065),2000mL/袋,溶液终浓度为 0.01mol/L,pH 6.0,室温保存,由北京中杉金桥生物技术有限公司生产。

氨水,含量 25%～28%,规格:500mL/瓶。由杭州长征化工厂生产。0.5%稀氨水配制(用于 HE 染色返蓝):0.5mL 氨水,加入到 100mL 蒸馏水中。

无水乙醇:含量≥99.7%,规格为500mL/瓶,由杭州化学试剂有限公司生产。95%乙醇:分析纯,规格为500mL/瓶,由安徽安特生物化学有限公司生产。80%、70%、60%乙醇配制:分别取95%乙醇84.2mL、73.6mL、63.1mL加入蒸馏水至100mL。0.5%盐酸酒精溶液配制(用于染色分化):取70%乙醇100mL,加入0.5mL盐酸。

(2)主要仪器

STP120脱水机、AP280-2包埋机、HM335E切片机(MICROM公司)

Nikon eclipse 80i 显微镜(Nikon公司)

CCD相机 DS－Fi1(500万像素)(Nikon公司)

1.4.2 HE染色

1)取材组织块,经固定后,常规石蜡包埋,4μm切片。

2)切片常规用二甲苯脱蜡,经各级乙醇至水洗:二甲苯(Ⅰ)5min→二甲苯(Ⅱ)5min→100%乙醇2min→95%的乙醇1min→80%乙醇1min→75%乙醇1min→蒸馏水洗2min。

3)苏木素染色5min,自来水冲洗。

4)盐酸乙醇分化30 s(提插数下)。

5)自来水浸泡15min或温水(约50℃)浸泡5min。

6)置伊红液2min。

7)常规脱水、透明、封片:95%乙醇(Ⅰ)min→95%乙醇(Ⅱ)1min→100%乙醇(Ⅰ)1min→100%乙醇(Ⅱ)1min→二甲苯(I)1min→二甲苯(Ⅱ)1min→中性树胶封固。

1.5 数据处理

实验数据用 SPSS 16.0 软件统计,采用单因素方差分析法和事后多重比较 LSD 法进行分析,结果以 $\bar{x} \pm S.D.$ 表示,$P<0.05$ 表示差异显著。

2 结果与分析

2.1 蜂胶改善T1DM大鼠血糖含量和体重的效果

实验结果见表8.1至表8.3和图8.1至图8.3。

由表8.1和图8.1可见,给药前各组糖尿病大鼠体重无显著差异,但是低于正常组大鼠($P<0.01$),这表明链佐星注射降低了大鼠体重。给药期间中国蜂胶、巴西蜂胶和阳性药均能不同程度抑制大鼠体重下降,高剂量组效果优于低剂量组,呈现量效关系。A1组大鼠(饲喂高剂量中国蜂胶)体重从第1周到第8周明显高于模型组大鼠($P<0.05$ 或者 $P<0.01$),B1组大鼠(饲喂高剂量巴西蜂胶)体重在第5到8周明显高于模型组大鼠($p<0.05$),B2组大鼠(饲喂低剂量巴西蜂胶)体重在第2、第6和第7周明显高于模型组大鼠($P<0.05$)。

表8.1 蜂胶改善糖尿病大鼠体重的效果

Table 8.1 Effects of propolis on body weight of diabetic rats

Week	Body weight/g						
	Model	A1	A2	B1	B2	Positive	Normal
0	269.4±13.5	275.0±18.4	269.4±15.2	270.0±16.1	268.1±16.5	271.0±14.1	329.0±27.0**
1	260.8±13.9	283.8±26.8*	271.0±17.3	269.4±18.3	271.9±24.3	280.5±21.5	337.8±28.2**

续表

Week	Body weight/g						
	Model	A1	A2	B1	B2	Positive	Normal
2	249.0±20.6	285.9±26.7**	262.5±18.4	271.5±18.0	272.9±25.4*	274.6±27.1*	354.8±31.5**
3	258.4±25.1	294.1±25.6*	271.4±19.6	278.1±24.5	282.4±24.7	282.3±24.8	361.3±47.5**
4	256.0±29.3	290.8±30.3*	269.5±18.2	277.1±21.3	280.6±27.4	289.9±24.6*	386.3±34.6**
5	262.9±31.6	316.4±41.4**	275.9±22.5	294.0±24.3*	286.6±29.3	298.6±28.2*	407.1±35.4**
6	260.5±33.5	309.1±24.4*	279.3±23.4	309.4±39.2**	296.3±28.8*	301.4±30.0*	427.1±39.2**
7	263.6±32.1	316.9±27.5*	288.0±20.3	308.5±31.2**	299.4±27.2*	312.6±30.5**	445.6±41.6**
8	260.6±25.1	305.8±28.6**	283.6±17.4	293.8±29.7*	285.6±31.5	311.4±33.0**	458.9±44.5**

结果以平均值±标准差表示，$n=8$。与模型对照组比：* $P<0.05$，** $P<0.01$。

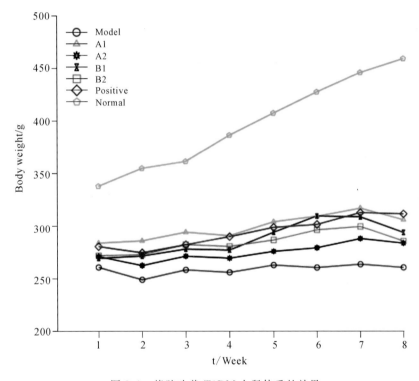

图 8.1　蜂胶改善 T1DM 大鼠体重的效果

Fig. 8.1　Effects of propolis on body weight of diabetic rats

由表 8.2 和图 8.2 可见，给药前各组糖尿病大鼠空腹血糖（FBG）含量无显著差异，但是显著高于正常组大鼠（$P<0.01$），这表明大鼠造模成功。给药期间中国蜂胶、巴西蜂胶和阳性药均能不同程度地抑制大鼠血糖升高。A1 组大鼠（饲喂高剂量中国蜂胶）血糖从第 1 周到第 5 周、第 8 周明显低于模型组大鼠（$P<0.05$ 或者 $P<0.01$），A2 组大鼠（饲喂低剂量中国蜂胶）血糖在第 5 周明显低于模型组大鼠（$P<0.05$），B1 组大鼠（饲喂高剂量巴西蜂胶）血糖在第 3 周到第 5 周、第 7 周到第 8 周明显低于模型组大鼠（$P<0.05$ 或 $P<0.01$），B2 组大

鼠(饲喂低剂量巴西蜂胶)血糖在第 5 周明显低于模型组大鼠($P<0.01$)。

表 8.2 蜂胶改善糖尿病大鼠空腹血糖的效果
Table 8.2 Effects of propolis on FBG in diabetic rats

Week	Fast blood glucose/(mmol/L)						
	Model	A1	A2	B1	B2	Positive	Normal
0	20.81±3.59	20.56±3.46	20.69±3.38	20.80±3.19	20.47±3.23	20.15±3.22	4.88±0.47**
1	21.01±4.92	11.16±6.82**	18.34±5.73	15.96±5.97	17.15±6.33	12.73±6.94**	4.90±0.36**
2	26.63±9.41	17.49±8.91*	21.62±7.11	19.31±10.11	23.88±5.40	22.20±8.23	5.11±0.42**
3	26.44±4.59	20.78±6.65*	23.66±5.56	20.79±5.22*	24.76±2.69	20.75±5.51*	4.88±0.56**
4	31.15±3.27	23.40±8.67*	31.35±2.01	25.00±9.13*	30.90±3.15	26.05±8.44	5.75±0.44**
5	26.61±4.52	19.04±5.13**	20.11±3.99*	18.60±4.96*	17.35±4.17**	16.25±6.33**	4.76±0.49**
6	22.42±9.57	23.99±6.58	26.49±4.12	20.00±8.49	24.96±5.43	22.19±8.11	5.24±0.62**
7	26.64±5.41	26.28±3.12	26.25±3.17	21.30±4.58*	22.40±5.31	24.15±4.23	4.97±0.52**
8	29.25±5.60	19.99±8.30**	26.59±5.56	18.75±9.11**	28.62±4.48	20.49±7.41**	5.00±0.76**

结果以平均值±标准差表示,$n=8$。与模型对照组比:* $P<0.05$,** $P<0.01$。

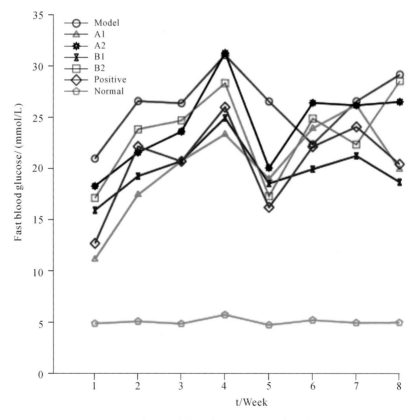

图 8.2 蜂胶改善糖尿病大鼠空腹血糖的效果
Fig. 8.2 Effects of propolis on fast blood glucose in diabetic rats.

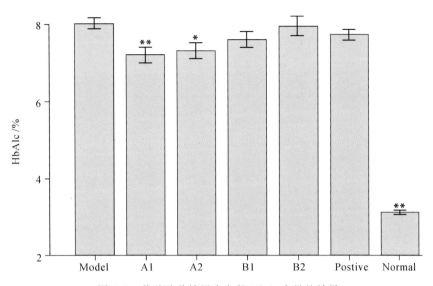

图 8.3　蜂胶改善糖尿病大鼠 HbA_{1c} 含量的效果

Fig. 8.3　Effects of propolis on HbA_{1c} in diabetic rats

由图 8.3 可见,实验第 8 周,模型组大鼠空腹 HbA_{1c} 含量[(8.00±0.35)%]显著高于正常组大鼠 HbA_{1c} 含量[(3.09±0.11)%]($P<0.01$)。与模型组大鼠相比,中国蜂胶(包括高剂量组和低剂量组)能显著降低糖尿病大鼠 HbA_{1c} 含量($P<0.01$,$P<0.05$),其中 A1 组大鼠 HbA_{1c} 含量比模型组约低 0.67%,A2 组大鼠 HbA_{1c} 含量比模型组约低 0.59%。阳性药和巴西蜂胶对于大鼠 HbA_{1c} 含量影响不显著($P>0.05$)。

由表 8.3 可见,模型组大鼠 24h 尿糖含量明显高于正常组大鼠($P<0.01$)。与模型组大鼠相比,阳性药能在第 2、6 及第 8 周显著减轻尿糖排放($P<0.05$,或 $P<0.01$),中国蜂胶和巴西蜂胶对于糖尿病大鼠尿糖含量影响不显著,低剂量巴西蜂胶在第 4 周显著增加尿糖排放($P<0.05$)。

表 8.3　蜂胶改善糖尿病大鼠尿糖的效果

Table 8.3　Effects of propolis on urine glucose in diabetic rats

Group	Urine glucose/(mmol/24h)			
	Week 2	Week 4	Week 6	Week 8
Model	6.48±0.91	6.64±1.06	8.83±1.46	9.61±1.75
A1	7.49±1.42	7.50±1.38	8.88±1.26	9.60±1.08
A2	7.36±0.89	7.51±1.32	8.87±1.80	9.48±1.70
B1	6.97±1.49	7.66±1.33	9.05±1.58	8.58±1.52
B2	7.23±1.01	8.16±0.81*	8.08±1.08	8.72±1.57
Positive	3.69±1.91**	6.86±1.33	6.81±1.75**	7.64±3.14*
Normal	0.043±0.036**	0.006±0.005**	0.015±0.009**	0.037±0.075**

注:结果以平均值±标准差表示,$n=8$。与模型对照组比:* $P<0.05$,** $P<0.01$。

2.2 蜂胶改善 T1DM 大鼠脂质代谢紊乱的效果

实验结果见表 8.4 至表 8.7。

表 8.4 蜂胶改善糖尿病大鼠 TC 含量的效果
Table 8.4 Effects of propolis on TC level in diabetic rats

Group	Total cholesterol/(mmol/L)			
	Week 2	Week 4	Week 6	Week 8
Model	1.85±0.48	2.45±0.71	1.76±0.42	1.86±0.33
A1	1.54±0.19	1.73±0.25*	1.76±0.24	1.57±0.22*
A2	1.62±0.28	2.38±0.86	1.72±0.33	1.88±0.40
B1	1.66±0.32	1.96±0.41	1.58±0.23	1.66±0.28
B2	1.68±0.20	2.19±0.93	1.63±0.24	1.84±0.25
Positive	1.80±0.20	1.92±0.46	1.73±0.23	1.75±0.27
Normal	1.49±0.23*	1.62±0.23*	1.58±0.22	1.71±0.21

注:结果以平均值±标准差表示,$n=8$。与模型对照组比:* $P<0.05$,** $P<0.01$。

由表 8.4 可见,模型组大鼠 TC 含量高于正常组大鼠,其中第 2 周和第 4 周差异明显($P<0.05$),第 6 周和第 8 周差异不显著($P>0.05$),这表明链佐星诱导引发 TC 代谢紊乱。与模型组大鼠相比,高剂量中国蜂胶和高剂量巴西蜂胶能降低糖尿病大鼠 TC 含量。A1(饲喂高剂量中国蜂胶)TC 含量在第 4 周和第 8 周明显低于模型组($P<0.05$),这表明中国蜂胶能改善大鼠脂质代谢。

由表 8.5 可见,模型组大鼠 LDL-C 含量与正常组大鼠相比无显著差异,链佐星诱导没有引发 LDL-C 代谢紊乱。与模型组大鼠相比,中国蜂胶(高剂量组)在第 4 周显著降低大鼠 LDL-C 含量($P<0.05$),巴西蜂胶(低剂量组)和阳性药在第 6 周显著降低大鼠 LDL-C 含量($P<0.05$)。

表 8.5 蜂胶改善糖尿病大鼠 LDL-C 含量的效果
Table 8.5 Effects of propolis on LDL-C level in diabetic rats

Group	LDL-C (mmol/L)			
	Week 2	Week 4	Week 6	Week 8
Model	0.39±0.09	0.41±0.17	0.36±0.08	0.36±0.02
A1	0.35±0.10	0.25±0.08*	0.32±0.11	0.30±0.12
A2	0.36±0.09	0.39±0.20	0.29±0.06	0.30±0.08
B1	0.38±0.10	0.32±0.07	0.29±0.06	0.32±0.05
B2	0.36±0.10	0.42±0.38	0.26±0.06*	0.32±0.09
Positive	0.39±0.09	0.30±0.08	0.28±0.07*	0.30±0.09
Normal	0.43±0.11	0.36±0.01	0.32±0.07	0.34±0.05

注:结果以平均值±标准差表示,$n=8$。与模型对照组比:* $P<0.05$,** $P<0.01$。

表 8.6 蜂胶改善糖尿病大鼠 HDL-C 含量的效果

Table 8.6　Effects of propolis on HDL-C level in diabetic rats

Group	HDL-C/(mmol/L)			
	Week 2	Week 4	Week 6	Week 8
Model	0.99±0.23	1.02±0.22	0.78±0.10	0.96±0.21
A1	0.90±0.13	0.88±0.10	0.88±0.11	0.83±0.12
A2	0.97±0.17	0.99±0.26	0.87±0.16	0.90±0.14
B1	0.96±0.09	0.87±0.15	0.83±0.17	0.84±0.17
B2	1.05±0.16	0.95±0.11	0.85±0.16	0.91±0.13
Positive	0.92±0.14	0.93±0.22	0.91±0.17	0.92±0.17
Normal	0.92±0.14	0.93±0.13	0.87±0.15	0.97±0.16

注：结果以平均值标准差表示，$n=8$。与模型对照组比：* $P<0.05$，** $P<0.01$。

由表 8.6 可见，模型组大鼠 HDL-C 含量与正常组大鼠相比无显著差异，链佐星诱导没有引发 HDL-C 代谢紊乱。各给药组大鼠 HDL-C 含量与模型组含量相比无显著差异（$P>0.05$）。

由表 8.7 可见，模型组大鼠 TG 含量与正常组大鼠相比无显著差异，链佐星诱导引发 TG 代谢紊乱。与模型组大鼠相比，高剂量中国蜂胶和高剂量巴西蜂胶能降低糖尿病大鼠 TG 含量，但是效果不显著。A1（饲喂高剂量中国蜂胶）TG 含量在第 2 周明显低于模型组（$P<0.05$）。

表 8.7 蜂胶改善糖尿病大鼠 TG 含量的效果

Table 8.7　Effects of propolis on TG level in diabetic rats

Group	Triglyeride (mmol/L)			
	Week 2	Week 4	Week 6	Week 8
Model	2.48±1.05	3.64±1.78	1.87±0.72	1.90±0.47
A1	1.49±0.54*	2.29±0.96	1.85±0.75	1.75±0.65
A2	1.99±0.59	3.60±1.32	2.06±0.75	2.66±1.47
B1	1.91±0.71	2.99±1.74	1.52±0.45	1.62±0.44
B2	1.92±0.62	3.42±2.09	1.80±0.57	2.24±0.57
Positive	2.13±0.79	2.80±1.67	1.97±0.65	1.91±0.86
Normal	0.81±0.13**	0.86±0.17**	1.01±0.28*	1.11±0.27*

注：结果以平均值±标准差表示，$n=8$。与模型对照组比：* $P<0.05$，** $P<0.01$。

2.3　蜂胶改善 T1DM 大鼠血液氧化应激的效果

实验结果见表 8.8。正常组大鼠 NO、NOS 和 MDA 含量明显低于模型组大鼠（$P<0.01$），SOD 和 CAT 含量明显高于模型组大鼠（$P<0.01$），GSH-px 含量差异不显著。阳性药和蜂胶对于 NO 含量改变不显著；A1 组大鼠 MDA 含量明显低于模型组大鼠（$P<0.01$）；

A2 组大鼠 SOD 含量和 MDA 含量显著高于模型组大鼠($P<0.05$);B1 组大鼠 SOD 含量显著高于模型组大鼠($P<0.01$ 或 $P<0.05$),NOS 含量明显低于模型组大鼠($P<0.01$);B2 组大鼠 SOD 含量显著高于模型组大鼠($P<0.01$),MDA 和 NOS 含量明显低于模型组大鼠($P<0.01$)。

表 8.8　蜂胶改善糖尿病大鼠血清氧化应激、NO 和 NOS 的效果
Table 8.8　Effects of propolis on serum oxidative stress, NO and NOS level in diabetic rats

Group	NO/(μmol/L)	NOS/(U/mL)	SOD/(U/mL)	CAT/(U/mL)	GSH-px/(μmol/L)	MDA/(nmol/L)
Model	19.40±9.04	46.93±4.83	39.42±14.30	9.65±0.83	687.88±48.29	5.15±0.55
A1	14.70±2.96	46.98±3.21	44.46±11.66	9.97±1.04	682.35±48.89	3.61±0.81**
A2	16.84±9.11	43.00±6.60	49.96±9.60*	10.40±2.59	695.69±73.77	6.70±1.00
B1	16.25±2.69	36.47±6.38**	54.53±3.41**	11.04±1.07	663.38±80.86	4.80±2.11
B2	16.48±5.30	38.06±3.07**	54.42±3.23**	10.35±2.41	677.80±54.79	3.59±0.66**
Positive	19.13±5.60	36.97±3.35**	57.16±4.07**	9.60±1.10	674.76±44.08	2.50±0.59**
Normal	10.41±1.59**	35.30±6.04**	53.23±5.51**	12.69±2.11**	692.22±35.85	3.23±0.41**

注:结果以平均值±标准差表示,$n=8$。与模型对照组比:* $P<0.05$,** $P<0.01$。

2.4　蜂胶改善 T1DM 大鼠肝脏功能和氧化应激的效果

实验结果见表 8.9 至表 8.11。

由表 8.9 可见,模型组大鼠 ALT 含量明显高于正常组大鼠($P<0.01$),链佐星诱导引发肝功能损伤。与模型组大鼠相比,中国蜂胶、巴西蜂胶和拜糖均能降低糖尿病大鼠 ALT 含量,但是阳性药和低剂量巴西蜂胶效果不显著($P>0.05$)。A1 组大鼠 ALT 含量在第 2 周、第 6 周和第 8 周明显低于模型组($P<0.05$ 或者 $P<0.01$),A2 组 ALT 含量在第 6 周和第 8 周明显低于模型组($P<0.01$,$P<0.05$),B1 ALT 含量第 6 周和第 8 周明显低于模型组($P<0.01$)。

表 8.9　蜂胶改善糖尿病大鼠血清 ALT 的效果
Table 8.9　Effects of propolis on serum ALT level in diabetic rats

Group	ALT/(IU/L)			
	Week 2	Week 4	Week 6	Week 8
Model	151.33±19.78	111.17±18.48	125.00±38.10	128.67±51.06
A1	111.67±11.36**	94.67±18.14	99.33±20.06*	93.00±18.75*
A2	132.83±14.13	104.33±18.09	89.50±9.05**	95.83±16.08*
B1	138.83±29.03	98.17±35.68	86.00±9.94**	85.83±21.78**
B2	159.00±25.24	108.17±22.34	102.67±11.52	105.33±28.56
Positive	140.00±11.93	110.50±17.05	107.17±23.24	123.83±34.61
Normal	76.12±14.39**	51.67±5.35**	42.50±5.32**	45.83±5.91**

注:结果以平均值±标准差表示,$n=8$。与模型对照组比:* $P<0.05$,** $P<0.01$。

由表 8.10 可见,模型组大鼠 AST 含量在第 2 周和第 4 周正常组大鼠无显著差异,在第 6 周和第 8 周明显高于正常组大鼠($P<0.01$)。与模型组大鼠相比,中国蜂胶、巴西蜂胶和拜糖平能降低糖尿病大鼠 AST 含量,但是阳性药效果不显著($P>0.05$)。与模型组大鼠相比,中国蜂胶(含高剂量和低剂量)在第 6 周和第 8 周显著降低糖尿病大鼠 AST 含量($P<0.05$ 或者 $P<0.01$),巴西蜂胶在第 6 周和第 8 周显著降低糖尿病大鼠 AST 含量($P<0.01$)。

表 8.10 蜂胶改善糖尿病大鼠血清 AST 的效果

Table 8.10 Effects of propolis on serum AST level in diabetic rats

Group	AST/(IU/L)			
	Week 2	Week 4	Week 6	Week 8
Model	160.83±29.29	133.50±30.42	209.83±77.10	250.17±65.67
A1	135.83±8.28	131.50±28.85	134.00±9.51**	195.50±23.74*
A2	140.33±16.56	114.83±20.74	133.17±19.79**	182.83±52.42*
B1	156.67±14.51	143.00±52.41	125.50±21.17**	172.50±13.28**
B2	163.00±41.29	127.83±32.15	145.67±41.92**	179.33±35.89**
Positive	170.50±23.19	129.83±21.27	187.00±46.78	226.67±64.85
Normal	130.17±12.86	108.50±30.16	124.83±22.21**	127.67±7.99**

注:结果以平均值±标准差表示,$n=8$。与模型对照组比:* $P<0.05$,** $P<0.01$。

表 8.11 蜂胶改善糖尿病大鼠肝脏氧化应激的效果

Table 8.11 Effects of propolis on hepatic oxidative stress in diabetic rats

Group	SOD/(U/mL)	CAT/(U/mL)	GSH-px/(μmol/L)	MDA/(nmol/L)
Model	69.33±6.90	83.28±6.28	265.18±27.86	2.67±0.66
A1	79.68±10.66	91.73±9.83	345.93±15.61**	2.04±0.34
A2	78.54±17.86	94.88±18.56	443.02±60.07**	2.25±0.61
B1	91.07±16.49**	97.43±18.11*	318.29±12.75*	1.86±0.83*
B2	74.30±7.17	81.92±10.35	345.76±46.98**	1.96±0.20*
Positive	76.23±8.42	86.35±6.93	306.3±48.67	1.83±0.61*
Normal	84.77±9.51*	96.32±9.17	310.20±24.65*	1.80±0.38*

注:结果以平均值±标准差表示,$n=8$。与模型对照组比:* $P<0.05$,** $P<0.01$。

由表 8.11 可见,正常组大鼠抗氧化酶 SOD 和 GSH-px 含量明显高于模型组大鼠($P<0.05$),CAT 含量与模型组差异不显著,而 MDA 含量明显低于模型组大鼠($P<0.05$)。拜糖平对于糖尿病大鼠肝脏氧化应激无改善效果,中国蜂胶能有效增加肝脏 GSH-px 含量($P<0.01$);B1 组大鼠肝脏各抗氧化酶含量显著高于模型组大鼠($P<0.01$ 或者 $P<0.05$),而 MDA 含量明显低于模型组大鼠($P<0.05$);B2 组大鼠肝脏 GSH-px 含量显著高于模型

组大鼠($P<0.01$),而 MDA 含量明显低于模型组大鼠($P<0.05$)。

2.5　蜂胶改善 T1DM 大鼠肾脏功能和氧化应激的效果

实验结果见表 8.12 至表 8.16 和图 8.4。

由表 8.12 可见,模型组大鼠 BUN 含量明显高于正常组大鼠($P<0.01$),链佐星诱导提升大鼠 BUN 含量。中国蜂胶、巴西蜂胶和阳性药能降低糖尿病大鼠 BUN 含量,其中低剂量蜂胶组效果不显著($P>0.05$),A1 组大鼠 BUN 含量在第 2 周明显低于模型组($P<0.05$),B1 组 BUN 含量在第 6 周、第 8 周明显低于模型组($P<0.05$,$P<0.01$)。

表 8.12　蜂胶改善糖尿病大鼠血清 BUN 的效果
Table 8.12　Effects of propolis on BUN level in diabetic rats

Group	BUN/(mmol/L)			
	Week 2	Week 4	Week 6	Week 8
Model	15.05±2.87	14.10±1.87	15.69±2.84	14.19±2.41
A1	11.29±3.05*	13.56±3.17	14.39±2.80	12.66±3.59
A2	12.09±2.38	13.25±1.95	12.16±2.28	12.74±3.35
B1	12.53±2.6108	12.55±2.46	11.40±1.92*	9.23±1.89**
B2	14.43±3.35	13.16±2.56	14.00±1.92	12.56±3.31
Positive	10.94±4.34*	12.23±2.78	11.09±2.24*	10.50±3.02**
Normal	7.44±2.15**	6.95±1.13**	6.26±0.74**	4.14±0.70**

注:结果以平均值±标准差表示,$n=8$。与模型对照组比:* $P<0.05$,** $P<0.01$。

由表 8.13 可见,模型组大鼠 SCr 含量与正常组大鼠相比无显著差异。各给药组大鼠 SCr 含量与模型组相比无显著差异($P>0.05$)。

表 8.13　蜂胶改善糖尿病大鼠血清 SCr 的效果
Table 8.13　Effects of propolis on serum SCr in diabetic rats

Group	SCr/(μmol/L)			
	Week 2	Week 4	Week 6	Week 8
Model	44.6±1.2	69.0±2.2	56.5±2.2	57.8±1.2
A1	49.1±4.9	67.3±0.7	53.5±0.8	56.3±1.5
B2	50.9±5.9	66.9±3.8	55.8±1.0	58.4±1.3
A1	47.9±1.3	68.3±1.2	53.8±0.8	56.3±1.2
B2	42.9±2.4	68.0±1.1	52.9±1.0	61.4±1.0
Positive	44.1±2.2	68.6±1.0	53.0±0.5	59.3±1.4
Normal	45.9±1.1	67.6±1.1	56.5±2.2	57.3±1.9

注:结果以平均值±标准差表示,$n=8$。与模型对照组比:* $P<0.05$,** $P<0.01$。

由表 8.14 可见,模型组大鼠 CCR 与正常组大鼠相比明显偏高($P<0.01$)。与模型组相比,蜂胶对糖尿病大鼠 CCR 无显著影响,阳性药能在第 2 周和第 4 周显著降低 CCR 值。

表 8.14　蜂胶改善糖尿病大鼠 CCR 的效果

Table 8.14　Effects of propolis on CCR in diabetic rats

Group	CCR/(mL/min)			
	Week 2	Week 4	Week 6	Week 8
Model	2.02±0.59	2.67±0.94	2.20±1.03	2.02±0.53
A1	2.43±0.37	2.89±0.64	2.48±0.64	1.97±0.25
A2	2.36±0.40	2.62±1.11	2.43±0.76	1.90±0.24
B1	1.99±0.25	2.41±0.44	2.46±0.41	1.97±0.29
B2	2.22±0.59	2.59±0.31	2.34±0.48	1.62±0.61
Positive	1.42±0.94*	2.50±0.49	2.35±0.47	1.08±0.54**
Normal	0.72±0.26**	0.90±0.51**	0.96±0.25**	0.81±0.43**

注:结果以平均值±标准差表示,$n=8$。与模型对照组比:* $P<0.05$,** $P<0.01$。

由表 8.15 可见,模型组大鼠 24h 尿蛋白含量明显高于正常组大鼠($P<0.01$ 或 $P<0.05$)。与模型组大鼠相比,中国蜂胶对于大鼠尿蛋白含量影响不显著,B1 组大鼠尿蛋白含量在第 8 周显著降低($P<0.05$),B2 组大鼠尿蛋白含量在第 6 和第 8 周显著降低。

表 8.15　蜂胶改善糖尿病大鼠尿蛋白含量的效果

Table 8.15　Effects of propolis on urine protein in diabetic rats

Group	Urine protein/(mg/24h)			
	Week 2	Week 4	Week 6	Week 8
Model	8.19±5.36	17.19±7.93	19.60±8.59	67.98±29.75
A1	9.24±2.14	21.25±7.15	22.61±8.59	55.07±21.87
A2	8.67±2.65	21.36±7.14	20.30±8.95	62.15±6.65
B1	7.58±1.99	18.50±8.09	17.51±11.77	50.33±15.46*
B2	6.41±1.92	21.49±8.09	10.17±5.43*	44.88±12.71**
Positive	4.40±2.28**	20.24±4.84	9.35±5.51**	31.49±9.97**
Normal	5.11±1.77*	10.07±4.51*	10.26±4.48**	18.36±5.50**

注:结果以平均值±标准差表示,$n=8$。与模型对照组比:* $P<0.05$,** $P<0.01$。

由图 8.4 可见,模型组大鼠 24h 尿液白蛋白含量明显高于正常组大鼠($P<0.01$)。与模型组大鼠相比,巴西蜂胶和中国蜂胶能有效降低大鼠尿液中白蛋白含量($P<0.05$)。

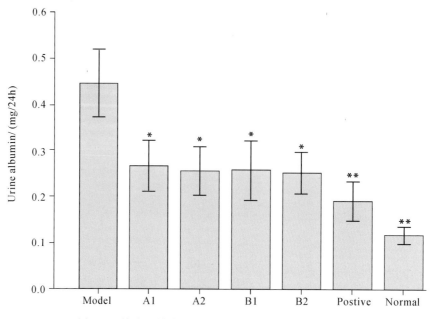

图 8.4 蜂胶改善糖尿病大鼠尿液白蛋白排泄量的效果

Fig. 8.4 Effects of propolis on urine albumin in diabetic rats

由表 8.16 可见,正常组大鼠抗氧化酶 GSH-px 和 MDA 含量明显低于模型组大鼠($P<$ 0.01),SOD 和 CAT 含量与模型组相比略高,但差异不显著。拜糖平和巴西蜂胶(包括高剂量组和低剂量组)能显著提升 CAT 含量($P<0.01$ 或者 $P<0.05$),降低 GSH-px 和 MDA 含量($P<0.01$);A1 组大鼠肾脏中 CAT 含量显著高于模型组大鼠($P<0.01$),而 MDA 含量明显低于模型组大鼠($P<0.05$);A2 组大鼠肾脏 GSH-px 含量显著高于模型组大鼠($P<$ 0.01),而 MDA 含量明显低于模型组大鼠($P<0.01$)。

表 8.16 蜂胶改善糖尿病大鼠肾脏氧化应激的效果

Table 8.16 Effects of propolis on renal oxidative stress in diabetic rats

Group	SOD/ (U/mL)	CAT/ (U/mL)	GSH-px/ (μmol/L)	MDA/ (nmol/L)
Model	80.34±2.42	30.69±2.48	729.80±70.47	2.84±0.68
A1	78.26±5.00	37.40±2.07**	654.27±86.96	2.09±0.24**
A2	80.47±7.36	32.09±7.27	550.70±76.39**	1.77±0.20**
B1	83.25±5.26	37.21±1.53**	639.82±122.22*	1.83±0.21**
B2	82.78±5.09	39.48±2.32**	557.12±37.43**	2.07±0.22**
Positive	88.57±7.95*	41.65±6.03**	613.95±72.60**	2.08±0.17**
Normal	84.70±3.23	33.94±2.76	552.37±39.40**	2.02±0.22**

注:结果以平均值±标准差表示,$n=8$。与模型对照组比:* $P<0.05$,** $P<0.01$。

2.6 肝脏和肾脏病理切片观察

结果见图 8.5 和图 8.6。

由图 8.5 可见,链佐星诱导对糖尿病大鼠肝脏造成伤害,中国蜂胶、巴西蜂胶和阳性药均能改善肝脏病变。

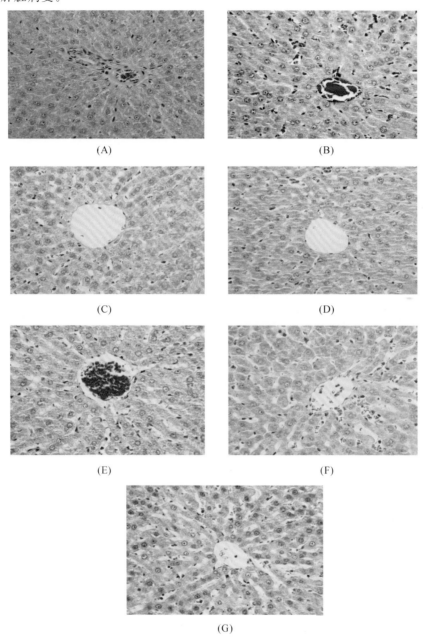

图 8.5　蜂胶对 T1DM 大鼠肝脏病理变化的影响(HE×400)

Fig 8.5　The histological analysis of liver in experimental rats(HE×400)

A:正常组大鼠肝组织各小叶结构清晰,肝细胞形态正常,胞质丰富,呈现条索状排列,小叶中央静脉结构完好,周围肝细胞索呈放射状排列;B:模型组大鼠肝组织结构清晰,出现大量肝细胞浊肿,肝窦扩张充血,双核细胞增多;C:A1组大鼠肝组织结构清晰,部分肝细胞浊肿;D:A2组大鼠肝组织结构清晰,局部肝细胞轻度浊肿;E:B1组大鼠肝组织结构清晰,局部肝细胞轻度浊肿;F:B2组大鼠肝组织结构清晰,局部肝细胞轻度浊肿;G:阳性药组大鼠肝组织结构清晰,局部肝细胞轻度浊肿。

由图 8.6 可见,链佐星诱导对糖尿病大鼠肾脏造成较为严重的损伤,中国蜂胶、巴西蜂胶和阳性药均能有效改善肾脏病变,且高剂量蜂胶效果优于低剂量蜂胶。

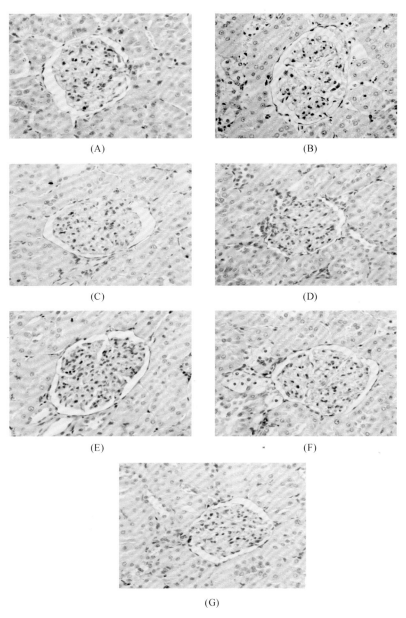

图 8.6　蜂胶对 T1DM 大鼠肾脏病理变化的影响(HE×400)

Fig 8.6　The histological analysis of kidney in experimental rats (HE×400)

A:正常组大鼠肾被膜、皮质、髓质结构清晰,皮质部肾小体和肾小管结构完好,未见肾小球增大;B:模型组大鼠肾脏,肾小球体积增大,系膜细胞增生,部分肾小管上皮细胞空泡化,出现肾小管上皮管型;C:A1 组大鼠肾肾小球体积轻度增大,系膜细胞增生;D:A2 组大鼠肾脏肾小球体积轻度增大,系膜细胞增生;E:B1 组大鼠肾脏肾小球体积轻度增大,系膜细胞增生;F:1 组大鼠肾脏肾小球体积轻度增大,系膜细胞增生,部分肾小管上皮细胞空泡化;G:阳性药组大鼠肾脏肾小球体积轻度增大,系膜细胞增生。

3 讨论

3.1 蜂胶对 T1DM 大鼠血糖的影响

在糖尿病治疗过程中,控制血糖水平是治疗的首要目标,持续性高血糖是引起糖尿病并发症的主要因素,强化血糖控制有利于减缓 T1DM 患者糖尿病肾病、糖尿病眼病及糖尿病性神经病发作及发展进程(The Diabetes Control and Complications Trial Research Group,1993)。HbA_{1c}是一种用于反应血糖控制的有效检测指标,是血红蛋白在高血糖环境下经非酶途径糖化产生。大型研究调查表明,当 T1DM 患者血液中 HbA_{1c} 含量下降时,其病情呈现明显的好转趋势(Larsen et al.,1990)。El-Sayed 等(2009)证明巴西蜂胶能显著降低糖尿病大鼠血糖含量。我们团队以前的实验发现中国蜂胶醇提物和水提物能降低糖尿病大鼠空腹血糖含量(Hu et al.,2005)。本实验进一步验证了中国蜂胶和巴西蜂胶控制血糖的效果,而中国蜂胶还能显著降低糖尿病大鼠血液中 HbA_{1c} 含量,且呈现量效关系。

在本实验中,我们发现高剂量中国蜂胶的降血糖效果相当于或者优于拜糖平。阳性药拜糖平是一种 α-葡萄糖苷酶抑制剂,能竞争性抑制位于小肠的各种 α-葡萄糖苷酶,使淀粉类分解为葡萄糖的速度减慢,从而减缓肠道内葡萄糖的吸收,降低餐后高血糖,主要用于 T2DM 治疗,但是临床上也用于辅助治疗 T1DM(Neuser et al.,2005)。拜糖平配合胰岛素治疗,能减少 T1DM 患者胰岛素用量,并可稳定血糖(Juntti-Berggren et al.,2000)。Matsui 等(2004)研究发现巴西蜂胶水溶性提取物能降低 SD 大鼠餐后血糖的升高,这种活性可能和其抑制麦芽糖酶活性有关。高海琳和刘富海(2000)将 50 例 T2DM 患者和 3 例 T1DM 患者随机分为蜂胶治疗组和对照组,进行了为期 2 年的临床疗效观察。发现服蜂胶后空腹及餐后 2h 血糖水平明显下降,原应用胰岛素治疗的病人在服用蜂胶后均不同程度地减少了胰岛素用量。这表明中国蜂胶和巴西蜂胶可能是通过发挥 α-葡萄糖苷酶抑制剂的效果实现或者部分实现降血糖效果。

在本实验中,我们采用链佐星诱导 T1DM 大鼠模型。链佐星通过损伤细胞 DNA 产生毒性,与葡萄糖结构相似,能由葡萄糖转运蛋白 GLUT2 转运而不能被其他葡萄糖载体识别,而在胰腺-细胞中有较高含量的 GLUT2,因而造成胰腺损伤(Schnedl et al.,1994)。Coskun 等(2005)研究了槲皮素(蜂胶中一种常见的黄酮类成分)保护糖尿病大鼠胰腺 β 细胞的效果。预先用槲皮素饲喂 3d 后,再用链佐星诱导大鼠糖尿病,然后再用槲皮素饲喂糖尿病大鼠 4 周。实验结果表明,槲皮素能降低患病大鼠血糖含量,增加胰岛素含量,降低红细胞和胰腺中 MDA 含量,降低血液中 NO 含量,恢复胰腺中抗氧化酶含量。此外,胰腺免疫组化检测表明,槲皮素能保护大部分朗格罕氏胰岛 β-细胞,减少细胞水肿、脱粒、降解及坏死。这表明蜂胶有可能从保护胰腺组织、恢复胰岛素分泌角度降低糖尿病大鼠血糖水平。

3.2 蜂胶对 T1DM 大鼠血液氧化应激的改善效果

ROS 和 RNS 在人体中有重要的生理功能,能帮助机体消灭病原微生物,参与信号传导等,但是当生物体内 ROS/RNS 过量则出现氧化应激,攻击体内细胞,破坏细胞膜、蛋白质、脂质及 DNA 等,对人体造成严重伤害(Valko et al.,2007)。高血糖能通过多种途径刺激糖尿病患者过度产生 ROS 和 NO,这些途径包括线粒体途径、葡萄糖自氧化、脂氧化酶、NAD(P)H 氧化酶、细胞色素 C P450(CYP 450)、黄嘌呤氧化酶(XO)和 NOS 等(Valko et al.,2007)。在这些途径中,高血糖刺激线粒体过度产生 ROS 被认为是糖尿病患者体内氧化应激的主要因素,当线粒体内 MnSOD 超量表达时则减少 ROS 产生,同时抑制上述几条途径的激活(Brownlee,2005)。

生物体内氧化应激有多种平衡体系,其中抗氧化酶是其中重要的一环,当生物体内氧化应激的时候,抗氧化酶 SOD、CAT 和 GSH-px 会出现不同程度的改变。动物实验和临床应用均表明,糖尿病模型动物或者患者服用抗氧化剂能降低患者血液中氧化应激水平,降低 MDA 含量,提升抗氧化酶含量(Rahimi et al.,2005)。蜂胶具有显著的抗氧化活性,能清除自由基,提升抗氧化酶活性,抑制生物体内脂质过氧化(Banskota et al.,2001;Khalil,2006)。健康人群服用蜂胶一段时间后,能改善服用者 MDA 和 SOD 含量(Jasprica et al.,2007)。我们先前的实验结果以及 El-Sayed 的研究均表明蜂胶能改善 T1DM 大鼠氧化应激水平(Hu et al.,2005;El-Sayed et al.,2009)。本实验中中国蜂胶和巴西蜂胶均能不同程度改善糖尿病大鼠血液氧化应激,巴西蜂胶的效果优于中国蜂胶,而且巴西蜂胶能有效抑制 NOS 活性,但是没有表现出剂量效应。这些实验结果表明蜂胶能通过抑制糖尿病大鼠氧化应激而延缓糖尿病进程。

蜂胶改善糖尿病大鼠氧化应激的效果,可能和蜂胶抑制高血糖激发活性氧途径有关。Song 等(2002)研究发现蜂胶醇提物能抑制巨噬细胞 RAW 264.7 中 iNOS 和 NO 表达,并且证明蜂胶抑制 iNOS mRNA 表达的效果可能是通过抑制 NF-κB 实现。Seo 等(2003)研究发现在动物实验和体外细胞培养中蜂胶能抑制醋氨酚诱导下小鼠肝脏细胞中细胞色素 CP450 含量增加。CAPE 能阻止烧伤大鼠血液中和脊髓炎大鼠大脑中 SOD 含量下降,抑制 XO 和 MDA 含量增加(Hosnuter et al.,2004;Ilhan et al.,2004)。Alyane 等(2008)用抗癌药阿霉素引发雌性 WISTAR 大鼠心脏过氧化损伤和线粒体功能失调,观察蜂胶改善心脏损伤的效果。实验结果表明,抗癌药引发组织线粒体呼吸控制率(RCV)及氧化磷酸化效率(P/O)下降,线粒体肿大,超氧化物阴离子和 MDA 含量上升。预先给大鼠饲喂蜂胶,能显著抑制线粒体 MDA 和超氧阴离子产生,恢复 RCV 和 P/O,降低线粒体肿胀程度。这些实验结果表明,蜂胶有可能通过影响 iNOS、CYP 450、XO 和线粒体途径抑制糖尿病大鼠发生氧化应激。

3.3 蜂胶对 T1DM 大鼠脂质代谢的影响

T1DM 患者的脂质代谢紊乱一般表现为 LDL-C 和 HDL-C 含量正常而 TG 含量显著升高,当情况恶化也会出现 LDL-C 升高而 HDL-C 下降。这种紊乱主要由高血糖所引发,控制高血糖能帮助 T1DM 患者恢复脂质代谢水平(O'Brien et al.,1998)。从本研究结果(表 8.4 至表 8.7)可以发现,正常组大鼠和模型组大鼠之间 LDL-C 和 HDL-C 含量差异不显著,但是模型组大鼠三酰甘油含量明显偏高,链佐星诱导的糖尿病大鼠呈现典型的 T1DM 脂质代谢紊乱症状,这从另一侧面表明 T1DM 模型组建成功。

脂质代谢紊乱容易造成人体动脉硬化并导致各类慢性心血管疾病,降低血液中胆固醇含量能帮助减少冠心病发生率(Jain et al.,2007)。有研究指出,载脂蛋白 E 基因缺乏小鼠存在高血脂及动脉硬化现象,蜂胶中的活性成分 CAPE 能不通过降低胆固醇含量而延缓小鼠动脉硬化过程,这种活性可能与 CAPE 抑制小鼠动脉中 NF-κB 活性及改善小鼠氧化应激有关(Hishikawa et al.,2005)。LDL-C 氧化产生一系列氧化副产品,这些成分和巨噬细胞进入动脉壁,积聚产生含有大量胆固酯的泡沫细胞,进而导致动脉硬化(Matsuura et al.,2008)。蜂胶是一种强抗氧化剂,具有抑制 LDL-C 氧化的活性(Abd El-Hady et al.,2007)。这些实验结果结合在一起,表明蜂胶有可能通过降低血液 TC 含量、抑制 LDL-C 氧化和抑制 NF-κB 活性来减少糖尿病患者动脉硬化发生。

3.4 蜂胶保护 T1DM 大鼠肝脏的效果

与正常人群肝脏相比,T1DM 患者肝脏内源性葡萄糖含量显著增加,其中只有 25%~45%合成肝糖原,这可能与胰岛素缺乏导致血糖升高、肝糖降解和糖异生作用增加有关(Roden and Bernroider,2003)。持续高血糖刺激肝脏组织氧化应激,肝脏内过多的 ROS 能诱导肝脏细胞死亡、炎症应答和肝脏纤维化等各类慢性肝脏疾病(Novo and Parola,2008)。Sugimoto 等(2005)研究发现高血糖能通过刺激 NADPH 氧化酶产生过量的 ROS,刺激肝脏星形细胞增生及胶原形成,进而诱导肝脏纤维化。目前已经有实验指出抗氧化剂有助于肝脏纤维化治疗,但是在不同动物模型中并不完全一致(Novo and Parola,2008)。

蜂胶对于各类因素诱导的肝脏功能损伤均表现出良好的效果。Zhao 等(2009)研究发现蜂胶能降低小鼠汞中毒产生的氧化应激并改善小鼠肝脏功能。Liu 等(2004)研究发现蜂胶能改善益康唑(一种抗微生物药)诱导大鼠肝功能受损,减少肝脏氧化应激水平。Ates 等(2006)研究发现 CAPE 能抑制因冷冻应激所致肝脏脂质过氧化及抗氧化酶活性的下降,阻止肝组织恶化。在本实验中我们发现,高血糖已经影响大鼠肝脏功能,中国蜂胶和巴西蜂胶能显著降低血清 ALT 和 AST 含量,而阳性药效果不显著,这表明两种蜂胶均能恢复糖尿病大鼠肝脏功能。对于肝脏氧化应激水平检测发现,中国蜂胶显著提升肝脏 GSH-px,但是对于脂质过氧化影响不显著,而巴西蜂胶能全面恢复肝脏各类抗氧化酶含量,抑制 MDA 产生。肝脏组织切片 HE 染色结果表明,和正常组大鼠相比,模型组大鼠肝脏病变程度并不严重,而各给药组大鼠肝组织形态基本维持正常。这些实验结果表明蜂胶有可能通过降低肝脏氧化应激水平来恢复肝脏功能。

3.5 蜂胶保护 T1DM 大鼠肾脏的效果

糖尿病肾病是一种微血管疾病,早期表现出肾脏肥大和肾小球滤过率增高,随后 GBM 增厚和 ECM 增多,病变发展过程可以发现尿液中白蛋白含量逐渐增多,病变继续发展出现肾小球硬化症状,最后形成肾衰竭(Girach et al.,2006)。T1DM 患者中 15%~25%患有糖尿病肾病,T2DM 患者中 30%~40%患有糖尿病肾病(Østergaard et al.,2005),约有 10%~20%的糖尿病患者死于肾衰竭(WHO,2008)。在本实验中我们测定了血肌酐、尿素氮、肌酐清除率、尿总蛋白及尿液白蛋白水平,以反映蜂胶保护糖尿病大鼠肾脏的效果。实验结果表明,BUN、尿液总蛋白、尿液白蛋白含量呈现显著的上升趋势,而肾脏病理组织 HE 染色结果显示模型组大鼠肾脏肾小管上皮细胞空泡化,出现肾小管上皮管型,这表示糖尿病大鼠开始出现肾病症状。中国蜂胶和巴西蜂胶均能降低尿液中白蛋白含量,同时巴西蜂胶能抑制 BUN 和尿液总蛋白升高,这表明两种蜂胶均能改善糖尿病大鼠肾脏病变,而巴西蜂胶的效果更为全面。观察蜂胶组大鼠肾脏组织切片发现,蜂胶能减少肾小管上皮细胞空泡化现象,这表明蜂胶能改善肾组织病变。

糖尿病病因复杂,血糖的严重程度和持续时间是引起糖尿病并发症的关键因素,但是控制高血糖并不能杜绝糖尿病的发生,因为多种因素共同作用促进其发作,这些因素主要包括:多元醇通道、AGEs、PKC 激活、氧化应激、炎症、肾脏血流动力学、细胞因子以及某些酶类物质等(Wolf,2004)。在本实验中,中国蜂胶和巴西蜂胶被证明能有效降低 T1DM 大鼠血糖,这可能是蜂胶抑制糖尿病发展的一种作用途径。氧化应激是促进糖尿病发展的另一

重要病因,体内过多的 ROS 能促进 AGEs 的生成,激发多元醇通道、蛋白激酶 C 途径以及己糖胺通道(Brownlee,2001);氧化应激能刺激肾小球系膜细胞分泌 TGF-β_1、纤溶酶原激活物抑制剂-1 和细胞外基质蛋白,促进系膜增生,同时激活各类信号途径如 MAPK 和转录因子 NF-κB 和 AP-1,促进细胞因子和生长因子表达(Ha and Lee,2001)。通过肾脏组织氧化应激水平检测发现,两种蜂胶有效提升肾脏 CAT,抑制肾脏脂质过氧化并降低 GSH-px 含量。这表明蜂胶有可能通过改善糖尿病大鼠肾脏氧化应激水平而保护肾脏组织。

小 结

本研究系统地比较了中国杨树型蜂胶和巴西绿蜂胶对 T1DM 大鼠控制血糖、调节脂质代谢和保护肝肾功能的效果及其作用机制,结果表明:(1) T1DM 大鼠体重比正常大鼠明显下降,中国蜂胶和巴西蜂胶能减缓 T1DM 大鼠体重下降,呈现剂量效应。在相同剂量下比较,中国蜂胶效果略优于巴西蜂胶;(2) T1DM 大鼠空腹血糖含量和 HbA_{1c} 值比正常大鼠明显增加,中国蜂胶和巴西蜂胶能降低 T1DM 大鼠的空腹血糖含量,呈现剂量效应,而且在相同剂量下比较,中国蜂胶效果略优于巴西蜂胶。中国蜂胶能降低 HbA_{1c} 水平,呈现剂量效应;(3) T1DM 大鼠出现脂质代谢紊乱现象,高剂量中国蜂胶在实验第 4 周能显著降低血清 LDL-C 含量,并在第 8 周能明显降低 TC 水平。低剂量巴西蜂胶在实验第 6 周明显降低血清 LDL-C 含量;(4) T1DM 大鼠出现血清氧化应激现象,中国蜂胶和巴西蜂胶能在不同程度抑制 T1DM 大鼠血清氧化损伤,提升抗氧化酶含量,抑制脂质过氧化,巴西蜂胶改善氧化应激的效果要比中国蜂胶更为全面。巴西蜂胶还能有效降低血清 NOS 含量;(5) T1DM 大鼠肝脏和肾脏功能受损,中国蜂胶和巴西蜂胶对 T1DM 大鼠肝脏和肾脏功能具有保护作用,能抑制肝肾组织结构破坏,降低组织氧化应激水平,巴西蜂胶改善氧化应激的效果要比中国蜂胶更为全面。

第二节 蜂胶改善 T2DM 大鼠效果的研究

2 型糖尿病(T2DM)又称非胰岛素依赖型糖尿病,占糖尿病患者发病类型总数的 90%,主要发生在成年人中,其临床特征为起病比较缓和、隐蔽,无明显症状,难以估计发病时间,常以糖尿病的大血管或微血管病变为首发症状而就诊。T2DM 患者在基因缺陷基础上以及外界环境因素(如肥胖、缺乏锻炼、酒精和烟)影响下,出现胰岛素抵抗及 β 细胞功能受损所导致胰岛素分泌障碍,导致血糖升高,高血糖进一步破坏 β 细胞功能和增加胰岛素抵抗,而 T2DM 患者所表现出脂质代谢紊乱也会破坏 β 细胞功能而造成病情加剧(Leahy,2005)。

根据 T2DM 发病机制,市场上出现一些治疗药物,如磺酰脲类药物、双胍类药物、α-葡萄糖苷酶抑制剂、噻唑烷二酮和非磺酰类促胰岛素分泌剂等,其中磺酰脲类药物可与胰岛 β 细胞表面受体结合,促进胰岛素释放,并强化胰岛素和受体集合;α-葡萄糖苷酶能抑制小肠上皮细胞表面的 α-糖苷酶,延缓碳水化合物的吸收,降低餐后血糖。这些药物具有较好的治疗效果,但是这类药物大部分针对单途径治疗而不是多靶位治疗,部分药物存在较大的副作用(陈灏珠,2001)。筛选安全性高、疗效好的天然药物,降低医疗成本,替代或者部分替代目前糖尿病的治疗药物,是一个值得研究的方向。

蜂胶和桑树叶混合后给 T2DM 患者服用,能有效降低病人空腹血糖含量和糖化血红蛋白含量(Murata et al.,2004)。同时蜂胶能改善大鼠胰岛素抵抗(Zamami et al.,2007),促进 T2DM 的Ⅰ、Ⅱ级糖尿病患者伤口愈合(楚勤英等,2008)。这些结果说明蜂胶能帮助改善 T2DM,但是缺乏系统的研究。蜂胶胶源植物的差异是造成蜂胶化学成分差异最主要的因素,这同时影响蜂胶的生物学活性(Bankova,2005)。本实验以中国杨树型蜂胶和巴西酒神菊属型绿蜂胶为研究对象,系统地比较不同来源蜂胶对 T2DM 大鼠改善糖代谢、胰岛素、脂质代谢和保护肝肾功能的效果,通过研究大鼠血液及肝肾组织氧化应激水平,探讨蜂胶抗糖尿病的作用机制。

1 材料与方法

1.1 蜂胶溶液的制备

同本章第一节 1.1。

1.2 动物实验

1.2.1 实验动物

同本章第一节 1.2.1。

1.2.2 实验条件

同本章第一节 1.2.2。

1.2.3 T2DM 模型大鼠实验

(1)动物造模

取 SPF 级体重为 200±20g 的雄性 SD 大鼠 100 只,饲喂高脂饲料 1 个月后,禁食不禁水 16h,用 0.1mmol/L 柠檬酸缓冲液(pH=4.2),将链脲佐菌素(STZ)配成质量分数为 2% 的溶液,按 50mg/kg STZ 的剂量尾静脉注射,1 周后测定血糖值和体重,选取血糖值＞20mmol/L、体重 450±50g 的大鼠 32 只用于试验。

高脂饲料配方:10%猪油,10%白糖,10%蛋黄粉,0.25%胆固醇,69.75%基础饲料,由浙江省医学科学院配置。

(2)分组和给药

将糖尿病大鼠随机分组,每组 8 只,各组分别给予相应的药物,每日 2 次,连续给药 8 周。另取 8 只雄性 SD 大鼠作正常对照组。具体给药剂量如下:

A 组:中国杨树型蜂胶,5mg/100g 大鼠体重;

B 组:巴西绿蜂胶,5mg/100g 大鼠体重;

阳性药物组(Positive):卡司平,1mg/100g 大鼠体重;

正常组(Normal)和模型组(Model):给予含有 PEG 的蒸馏水,剂量为 1mL/100g。

(3)实验进程

①实验期间,糖尿病大鼠饲喂高脂饲料,正常大鼠饲喂普通饲料;

②实验开始后,每周测定一次大鼠体重和空腹血糖值;

③造模后,每 2 周进行一次代谢试验,测定采食量、饮水量、排尿量,测定尿液中尿肌酐、尿总蛋白、尿糖含量;

④每 2 周空腹尾静脉取血,分离血清,测定大鼠 TG、TC、HDL-C、LDL-C、BUN、Scr、TP、Alb、ALT、AST;

⑤每4周测定大鼠糖化血红蛋白含量;

⑥实验第8周,动物禁食12h后用戊巴比妥麻醉,腹主动脉取血,测定HbA_{lc},并测定血液中NO、NOS、SOD、MDA、CAT、GSH-px和胰岛素含量,胰岛素采用酶联免疫法进行测定;

⑦处死大鼠后取肝脏组织,做HE染色观察组织病理变化;

⑧处死大鼠后取肾组织,进行HE染色和PAS观察组织病理变化。

1.3 生化指标测定

1.3.1 主要仪器

BT-815A半自动生化分析仪(上海三科仪器有限公司);NYCOCARD READER Ⅱ多功能全定量特种蛋白金标检测仪器(挪威 AXIS SHIELD);OneTouch^R Ultra血糖仪[强生(上海)医疗器材有限公司];日立7020全自动生化分析仪(日立公司);多功能酶标仪 VarioScan Flash(美国热电 Thermo)。

1.3.2 主要试剂

糖化血红蛋白(HbA_{lc})测定试剂盒(日本和光纯药工业株式会社);胆固醇(TC)测定试剂盒(上海申能德赛技术诊断有限公司);三酰甘油(TG)测定试剂盒(上海申能德赛技术诊断有限公司);高密度脂蛋白(HDL)测定试剂盒(上海申能德赛技术诊断有限公司);低密度脂蛋白(LDL)测定试剂盒(上海复星长征医学科学有限公司);超氧化物歧化酶(SOD)测定试剂盒(南京建成生物工程研究所);丙二醛(MDA)测定试剂盒(南京建成生物工程研究所);尿素氮(BUN)测定试剂盒(上海申能德赛技术诊断有限公司);肌酐(CREA)测定试剂盒(上海申能德赛技术诊断有限公司);总蛋白(TP)测定试剂盒(上海申能德赛技术诊断有限公司);白蛋白(ALB)液体试剂盒(上海申能德赛技术诊断有限公司);谷丙转氨酶(ALT)测定试剂盒(上海申能德赛技术诊断有限公司);谷草转氨酶(AST)测定试剂盒(上海申能德赛技术诊断有限公司);链脲佐菌素(STZ)(ALEXIS CORPORATION公司);一氧化氮(NO)测定试剂盒(南京建成生物工程研究所);一氧化氮合酶(NOS)测定试剂盒(南京建成生物工程研究所);卡司平(杭州中美华东制药有限公司);大鼠胰岛素(iNS)酶联免疫分析试剂盒(美国R&D公司进口分装)。

1.4 组织病理学观察

1.4.1 材料和方法

(1)主要材料

同本章第一节1.4.1。

(2)主要仪器

同本章第一节1.4.1。

1.4.2 实验方法

(1)HE染色

同本章第一节1.4.2。

(2)PAS染色

Schiff试剂配制:将碱性品红0.5g加入到100mL沸水中,持续煮沸5min,并随时搅拌,待冷却到50℃时过滤到棕色瓶中,加1N HCl 10mL,冷却至25℃时加入1g $NaHSO_3$,此时需很好地振荡,避光过夜。次日取出(呈淡黄色),加0.25g活性炭剧烈振荡1min。过滤后即得Schiff试剂。避光低温保存。

①1%过碘酸水溶液染色 10min,使含有乙二醇的化合物氧化成醛基;
②蒸馏水换洗数次,洗去过碘酸;
③Schiff 试剂反应 30min,使醛基和 Schiff 试剂结合,形成紫红色产物;
④0.5%亚硫酸氢钠(NaHSO$_3$)处理 3 次,每次 2min;
⑤流水冲洗 10min,再用蒸馏水冲洗一次;
⑥Harris 苏木精复染 1min,使胞核着色;
⑦流水充分冲洗;
⑧1%盐酸酒精分化;
⑨经 95%酒精Ⅰ、Ⅱ,100%酒精Ⅰ、Ⅱ脱水,每次 1~2min;
⑩二甲苯透明,DPX 封片。

1.5 数据处理

实验数据用 SPSS 16.0 软件统计,采用单因素方差分析法和事后多重比较 LSD 法进行分析,结果以平均值±标准差表示,$P<0.05$ 表示差异显著。

2 结果与分析

2.1 蜂胶改善 T2DM 大鼠体重、糖代谢、胰岛素和蛋白质代谢的效果

实验结果见图 8.7 至图 8.10 和表 8.17 至表 8.19。

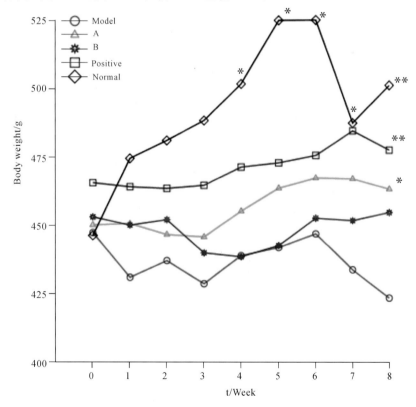

图 8.7 中国蜂胶和巴西蜂胶改善 T2DM 大鼠体重的效果

Fig. 8.7 Effects of propolis on body weight of diabetic rats

由图 8.7 可见,给药前及正常组大鼠和糖尿病大鼠体重无显著差异。给药阶段正常组大鼠体重高于模型组大鼠体重,但是实验前 3 周各组差异不显著($P>0.5$),从第 4 周到第 8 周显著高于模型组大鼠体重($P<0.05$ 或者 $P<0.01$)。蜂胶和阳性药有抑制大鼠体重下降的趋势,但是效果不显著。A 组大鼠(饲喂中国蜂胶)体重在第 8 周明显高于模型组大鼠体重($P<0.05$)。

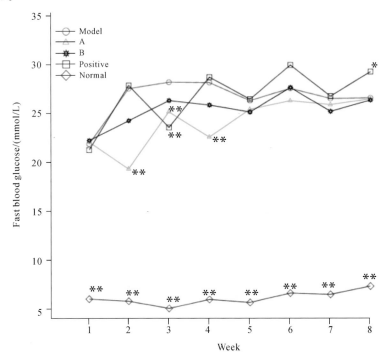

图 8.8 蜂胶改善糖尿病大鼠空腹血糖的效果

Fig. 8.8 Effects of propolis on fast blood glucose in diabetic rats

由图 8.8 可见,整个给药期间各组糖尿病大鼠空腹血糖含量显著高于正常组大鼠($P<0.01$),这表明大鼠造模成功。A 组大鼠(饲喂中国蜂胶)血糖含量从第 1 周到第 8 周低于模型组大鼠血糖含量,其中第 2 周到第 4 周效果显著($P<0.01$);B 组大鼠(饲喂巴西蜂胶)血糖含量与模型组大鼠差异不显著;阳性药组大鼠(饲喂卡司平)血糖含量在第 3 周明显低于模型组大鼠($P<0.01$),在第 8 周明显高于模型组大鼠($P<0.05$)。

由图 8.9 可见,模型组大鼠第 4 周和第 8 周 HbA_{1c} 含量分别为 $6.29\pm0.26\%$ 和 $8.33\pm0.38\%$,显著高于正常组大鼠($P<0.01$)。与模型组大鼠相比,中国蜂胶和巴西蜂胶在实验第 4 周能显著降低糖尿病大鼠 HbA_{1c} 含量($P<0.01$),其中 A 组大鼠 HbA_{1c} 含量比模型组约低 0.54%,B 组大鼠 HbA_{1c} 含量比模型组约低 0.64%;实验第 8 周中国蜂胶显著降低大鼠 HbA_{1c} 含量($P<0.01$),A 组大鼠 HbA_{1c} 含量比模型组约低 0.44%。

由图 8.10 可见,模型组大鼠胰岛素含量与正常组大鼠无显著差异。阳性药和蜂胶对于糖尿病大鼠胰岛素含量无明显影响。

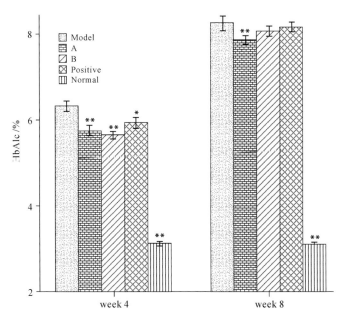

图 8.9 蜂胶改善糖尿病大鼠 HbA_{1c} 含量的效果

Fig. 8.9 Effects of propolis on HbA_{1c} in diabetic rats

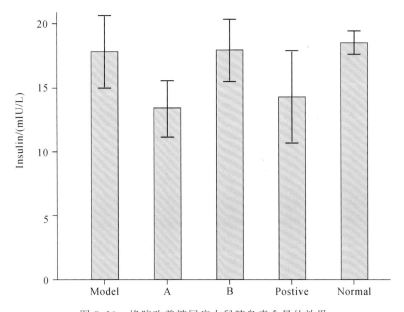

图 8.10 蜂胶改善糖尿病大鼠胰岛素含量的效果

Fig. 8.10 Effects of propolis on insulin in diabetic rats

由表 8.17 可见,模型组大鼠尿糖含量与正常组大鼠相比明显偏高($P<0.01$)。与模型组大鼠相比,整个实验期间蜂胶和阳性药能有效降低糖尿病大鼠尿糖含量($P<0.01$ 或者$P<0.05$)。

表 8.17　蜂胶改善糖尿病大鼠尿糖的效果

Table 8.17　Effects of propolis on urine glucose in diabetic rats

Week	Urine glucose/(mmol/24h)				
	Model	A	B	Positive	Normal
2	10.2339±1.2112	8.1072±2.6176*	6.8834±1.0297*	6.2162±1.4712**	0.0099±0.0070**
4	11.0316±0.5639	9.1814±1.3893**	8.7928±1.6719*	8.4599±1.2325*	0.0106±0.0128**
6	9.5185±1.5014	7.4889±1.7312*	7.0547±1.0112**	7.1350±1.9591*	0.0053±0.0036**
8	11.3625±1.8087	7.5558±2.4356**	9.3439±1.3340*	7.6773±1.4641**	0.0039±.0019**

注：结果以平均值±标准差表示，$n=8$。与模型对照组比：$^* P<0.05$，$^{**} P<0.01$。

由表 8.18 可见，正常组大鼠血清总蛋白含量高于模型组大鼠，实验第 4 周到实验第 6 周差异显著（$P<0.01$ 或者 $P<0.05$）。与模型组大鼠相比，阳性药效果不显著，蜂胶能抑制血清总蛋白流失，巴西蜂胶在第 4 周和第 6 周显著增加总蛋白含量（$P<0.05$，$P<0.01$），中国蜂胶在第 6 周显著增加总蛋白含量。

表 8.18　蜂胶改善糖尿病大鼠血清总蛋白的效果

Table 8.18　Effects of propolis on serum TP in diabetic rats

Week	Serum Alb/(g/L)				
	Model	A	B	Positive	Normal
2	66.91±2.16	68.43±4.18	67.14±2.80	67.81±2.56	68.60±2.60
4	66.58±1.99	69.70±4.16	71.07±3.56*	68.51±3.89	73.63±3.56**
6	65.44±2.37	70.19±1.97**	70.06±4.20**	67.44±1.86	69.69±3.72*
8	67.96±2.18	69.31±2.59	68.74±1.63	67.18±2.52	68.63±1.92

注：结果以平均值±标准差表示，$n=8$。与模型对照组比：$^* P<0.05$，$^{**} P<0.01$。

由表 8.19 可见，实验第 4 周到实验第 8 周模型组大鼠血清白蛋白含量明显低于正常组大鼠含量（$P<0.01$）。与模型组大鼠相比，阳性药和巴西蜂胶对血清白蛋白含量无显著影响，中国蜂胶在第 4 周和第 6 周显著增加白蛋白含量（$P<0.05$）。

表 8.19　蜂胶改善糖尿病大鼠血清 Alb 的效果

Table 8.19　Effects of propolis on serum Alb in diabetic rats

Week	Serum Alb/(g/L)				
	Model	A	B	Positive	Normal
2	31.39±1.37	29.16±7.36	31.08±1.60	31.16±1.24	34.27±1.06
4	31.06±1.39	34.50±2.16*	32.55±2.31	31.56±0.96	37.90±6.74**
6	31.55±1.68	34.53±2.78*	32.54±2.89	31.83±1.94	35.87±2.75**
8	31.36±1.10	32.60±2.22	31.13±2.06	31.03±0.92	33.83±0.64**

注：结果以平均值±标准差表示，$n=8$。与模型对照组比：$^* P<0.05$，$^{**} P<0.01$。

2.2 蜂胶改善 T2DM 大鼠脂质代谢紊乱的效果

实验结果见表 8.20 至表 8.22 和表 8.23。

由表 8.20 可见,模型组大鼠 TC 含量与正常组大鼠相比明显偏高($P<0.05$ 或者 $P<0.01$)。蜂胶和阳性药对于糖尿病大鼠 TC 含量影响不显著($P>0.05$)。

表 8.20 蜂胶改善糖尿病大鼠 TC 含量的效果

Table 8.20　Effects of propolis on TC level in diabetic rats

Week	Total cholesterol/(mmol/L)				
	Model	A	B	Positive	Normal
2	12.07±8.70	9.87±8.88	7.52±3.81	10.12±9.97	1.78±0.20*
4	15.31±6.22	12.31±7.72	11.89±4.85	13.66±13.60	2.01±0.16**
6	13.94±8.67	13.20±10.59	14.82±11.11	14.18±13.29	1.90±0.27*
8	13.48±8.43	18.34±13.49	20.32±12.98	13.66±12.42	1.65±0.16*

注:结果以平均值±标准差表示,$n=8$。与模型对照组比:* $P<0.05$,** $P<0.01$。

由表 8.21 可见,实验期间正常组大鼠血清 LDL-C 含量明显低于模型组大鼠($P<0.01$),这表明 T2DM 大鼠存在脂质代谢紊乱。与模型组大鼠相比,蜂胶和阳性药能不同程度降低大鼠 LDL-C 含量,A 组大鼠 LDL-C 含量在第 4 周和第 6 周显著低于模型组大鼠($P<0.05$,$P<0.01$),B 组大鼠 LDL-C 含量在第 4 周显著降低模型组大鼠($P<0.05$)。

表 8.21 中国蜂胶改善糖尿病大鼠 LDL-C 含量的效果

Table 8.21　Effects of propolis on LDL-C level in diabetic rats

Week	LDL-C/(mmol/L)				
	Model	A	B	Positive	Normal
2	6.63±5.57	4.43±2.24	4.42±2.19	3.93±1.72	0.42±0.09**
4	12.73±1.82	8.84±4.64*	8.67±2.80*	6.55±2.11**	0.45±0.08**
6	8.23±2.80	6.90±2.44**	9.75±2.42	6.92±3.04**	0.35±0.05**
8	8.54±3.34	12.18±6.45	12.46±5.33	7.27±2.79	0.25±0.05**

注:结果以平均值±标准差表示,$n=8$。与模型对照组比:* $P<0.05$,** $P<0.01$。

由表 8.22 可见,模型组大鼠 HDL-C 含量与正常组大鼠相比无显著差异。各给药组大鼠 HDL-C 含量与模型组含量相比无显著差异($P>0.05$)

表 8.22 中国蜂胶改善糖尿病大鼠 HDL-C 含量的效果

Table 8.22　Effects of propolis on HDL-C level in diabetic rats

Week	HDL-C/(mmol/L)				
	Model	A	B	Positive	Normal
2	1.03±0.13	1.15±0.20	1.18±0.24	1.16±0.17	1.09±0.14
4	1.00±0.21	1.35±0.25	1.03±0.04	1.09±0.19	1.14±0.22

续表

Week	HDL-C/(mmol/L)				
	Model	A	B	Positive	Normal
6	0.95±0.16	1.01±0.13	0.91±0.16	1.14±0.20	1.08±0.17
8	0.99±0.11	1.04±0.15	0.99±0.08	0.95±0.09	0.97±0.06

注：结果以平均值±标准差表示，$n=8$。与模型对照组比：* $P<0.05$，** $P<0.01$。

由表 8.23 可见，第 2 周到第 4 周模型组大鼠 TG 含量与正常组大鼠相比明显偏高（$P<0.01$），第 6 周到第 8 周模型组大鼠 TG 含量与正常组大鼠相比偏高，但是差异不显著。蜂胶和阳性药在第 2 周能有效降低糖尿病大鼠 TG 含量（$P<0.01$）。

表 8.23　中国蜂胶改善糖尿病大鼠 TG 含量的效果

Table 8.23　Effects of propolis on TG in diabetic rats

Week	Triglyeride/(mmol/L)				
	Model	A	B	Positive	Normal
2	4.78±2.10	3.89±1.95	4.11±3.23	4.17±2.58	1.23±0.40**
4	5.70±0.31	3.95±0.89**	3.61±0.53**	3.15±1.05**	1.50±0.39**
6	5.38±3.16	5.98±3.13	7.15±4.84	6.43±5.06	2.20±0.66*
8	3.97±2.62	6.88±5.00	6.72±3.97	5.46±3.45	2.68±0.84*

注：结果以平均值±标准差表示，$n=8$。与模型对照组比：* $P<0.05$，** $P<0.01$。

2.3　蜂胶改善 T2DM 大鼠血液氧化应激的效果

实验结果见表 8.24。

由表 8.24 可见，正常组大鼠血清 MDA 含量明显低于模型组大鼠（$P<0.05$），GSH-px 含量明显高于模型组大鼠（$P<0.05$），SOD 和 CAT 含量略高于模型组大鼠，NO 和 NOS 略低于模型组大鼠，但差异不显著。与模型组大鼠相比，阳性药和蜂胶能改善糖尿病大鼠氧化应激水平，对于血清 NO 和 NOS 含量改变不显著，其中中国蜂胶能明显提升 CAT 和 GSH-px 含量（$P<0.01$ 或者 $P<0.05$），降低 MDA 含量（$P<0.05$）；巴西蜂胶能有效降低 MDA 含量（$P<0.05$）。

表 8.24　中国蜂胶改善糖尿病大鼠血清氧化应激、NO 和 NOS 的效果

Table 8.24　Effects of propolis on serum oxidative stress, NO and NOS level in diabetic rats

Group	Model	A	B	Positive	Normal
NO/(μmol/L)	37.41±2.08	44.41±10.09	39.11±10.58	38.79±5.05	33.11±4.00
NOS/(U/mL)	29.94±6.20	36.87±13.11	41.66±20.46	39.32±16.54	25.52±1.40
SOD/(U/mL)	39.08±5.75	39.49±12.97	33.02±12.55	34.51±14.22	44.35±1.10
CAT/(U/mL)	10.38±4.73	15.18±6.81	18.97±9.78*	18.65±9.88*	13.92±5.36

续表

Group	Model	A	B	Positive	Normal
GSH-PX/(μmol/L)	183.62±98.38	519.41±262.97**	289.54±175.42	192.55±152.90	321.20±26.73*
MDA/(nmol/L)	13.42±1.38	11.09±1.78*	10.73±1.60*	11.93±3.21	10.70±1.34*

注:结果以平均值±标准差表示,n=8。与模型对照组比: * $P<0.05$, ** $P<0.01$。

2.4 蜂胶改善 T2DM 大鼠肝脏功能的效果

实验结果见表 8.25 和表 8.26。

由表 8.25 可见,实验期间正常组大鼠 ALT 含量明显低于模型组大鼠($P<0.01$)。与模型组大鼠相比,阳性药对于糖尿病大鼠 ALT 含量无显著影响,巴西蜂胶第 8 周显著降低大鼠血清 ALT 含量($P<0.05$),中国蜂胶在第 2 周和第 8 周显著降低大鼠血清 ALT 含量($P<0.01$)。

表 8.25 中国蜂胶改善糖尿病大鼠血清 ALT 的效果

Table 8.25 Effects of propolis on serum ALT level in diabetic rats

Week	ALT/(IU/L)				
	Model	A	B	Positive	Normal
2	109.67±10.03	70.50±37.24**	89.83±10.21	88.43±15.80	58.88±11.95**
4	109.12±9.40	108.50±11.39	101.14±10.43	99.60±15.85	57.88±17.67**
6	94.00±9.54	95.33±8.39	88.17±15.36	91.33±15.29	45.62±13.53**
8	121.38±37.46	81.75±20.26**	96.29±13.71*	100.60±13.78	45.00±16.36**

注:结果以平均值±标准差表示,n=8。与模型对照组比: * $P<0.05$, ** $P<0.01$。

由表 8.26 可见,实验第 2 周和第 8 周正常组大鼠 AST 含量与模型组大鼠相比无显著差异,第 4 周到第 6 周正常组大鼠 AST 含量明显高于模型组大鼠($P<0.01$)。与模型组大鼠相比,巴西蜂胶和阳性药对于糖尿病大鼠 AST 含量无显著影响,中国蜂胶在第 8 周显著降低大鼠血清 AST 含量($P<0.01$)。

表 8.26 中国蜂胶改善糖尿病大鼠血清 AST 的效果

Table 8.26 Effects of propolis on serum AST level in diabetic rats

Week	AST/(IU/L)				
	Model	A	B	Positive	Normal
2	94.75±26.16	103.50±22.93	100.00±19.10	85.62±7.35	100.29±9.52
4	80.50±9.32	105.29±23.23	91.75±22.39	99.00±12.81	121.43±47.64**
6	69.75±9.98	88.20±40.87	91.57±18.36	92.43±15.06	111.57±31.42**
8	102.75±30.50	102.33±29.21	72.50±13.50*	88.00±32.97	94.00±9.59

注:结果以平均值±标准差表示,n=8。与模型对照组比: * $P<0.05$, ** $P<0.01$。

2.5 蜂胶改善 T2DM 大鼠肾脏功能和氧化应激的效果

实验结果见表 8.27 至表 8.30 和图 8.11。

由表 8.27 可见，与模型组大鼠相比，从实验第 2 周到第 6 周，正常组大鼠和给药组大鼠 BUN 含量与模型组大鼠相比无显著差异。实验第 8 周正常组大鼠 BUN 含量显著低于模型组大鼠（$P<0.01$），中国蜂胶、巴西蜂胶和阳性药能有效降低大鼠 BUN 含量（$P<0.01$）。

表 8.27 蜂胶改善糖尿病大鼠血清 BUN 的效果
Table 8.27 Effects of propolis on BUN level in diabetic rats

Week	BUN/(mmol/L)				
	Model	A	B	Positive	Normal
2	6.83±0.99	6.56±1.51	6.41±2.96	7.30±2.35	6.64±0.79
4	8.40±1.53	8.07±2.38	8.16±2.65	10.54±1.85	7.19±1.51
6	6.21±1.35	4.84±1.53	6.53±1.24	7.50±0.31	6.73±1.85
8	12.53±1.64	7.69±3.63**	7.03±1.79**	8.14±1.42**	5.70±0.95**

注：结果以平均值±标准差表示，$n=8$。与模型对照组比：* $P<0.05$，** $P<0.01$。

由表 8.28 可见，模型组大鼠血肌酐含量与正常组大鼠相比无明显差异，蜂胶和阳性药对于糖尿病大鼠血液 Scr 含量无显著影响。

表 8.28 蜂胶改善糖尿病大鼠血清 SCr 的效果
Table 8.28 Effects of propolis on serum SCr level in diabetic rats

Week	SCR/(umol/L)				
	Model	A	B	Positive	Normal
2	67.62±5.95	60.80±15.06	66.88±15.49	69.40±7.70	67.88±4.61
4	72.50±6.85	72.8±8.37	70.50±4.63	72.75±6.50	70.75±4.71
6	67.88±10.82	70.60±15.08	75.12±9.67	74.75±5.42	76.17±2.93
8	74.12±11.04	72.75±7.87	71.62±4.03	72.00±3.42	68.88±3.44

注：结果以平均值±标准差表示，$n=8$。与模型对照组比：* $P<0.05$，** $P<0.01$。

由表 8.29 可见，模型组大鼠 CCR 与正常组大鼠相比明显偏高（$P<0.01$ 或者 $P<0.05$）。蜂胶对于糖尿病大鼠 CCR 无显著影响，阳性药能在第 4 和第 6 周显著降低 CCR 值（$P<0.05$，$P<0.01$）。

表 8.29 蜂胶改善糖尿病大鼠 CCR 的效果
Table 8.29 Effects of propolis on CCR in diabetic rats

Week	CCR/(mL/min)				
	Model	A	B	Positive	Normal
2	2.81±0.73	4.46±1.97	3.63±1.99	2.13±1.38	1.07±0.76*
4	2.28±0.58	2.16±1.31	2.10±0.60	1.44±0.26*	0.77±0.17**

续表

Week	CCR/(mL/min)				
	Model	A	B	Positive	Normal
6	2.33±0.44	2.23±0.69	1.91±0.52	1.47±0.43**	0.92±0.60**
8	1.51±0.33	1.86±1.22	2.13±0.79	1.22±0.40	0.75±0.26*

注：结果以平均值±标准差表示，$n=8$。与模型对照组比：* $P<0.05$，** $P<0.01$。

由图 8.11 可见，模型组大鼠血尿液白蛋白含量与正常组大鼠相比显著偏高（$P<0.01$ 或者 $P<0.05$）。与模型组大鼠相比，蜂胶和阳性药能降低尿液白蛋白排泄率，但是效果不显著。第 8 周 A 组大鼠尿液白蛋白含量明显低于模型组大鼠（$P<0.05$）。

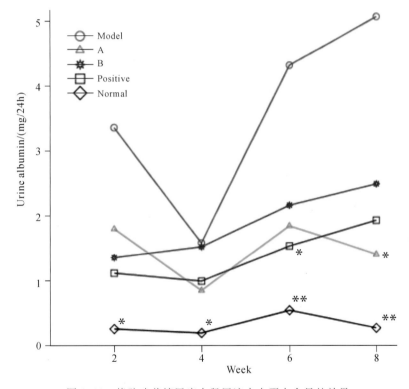

图 8.11　蜂胶改善糖尿病大鼠尿液中白蛋白含量的效果

Fig. 8.11　Effects of propolis on urine albumin in diabetic rats

由表 8.30 可见，正常组大鼠肾脏 NO、NOS 和 MDA 含量明显低于模型组大鼠（$P<0.01$ 或者 $P<0.05$），SOD 和 CAT 含量显著高于模型组（$P<0.01$），GSH-px 略高于模型组大鼠，但差异不显著，这表明大鼠肾脏存在氧化应激现象。与模型组大鼠相比，蜂胶和阳性药均能改善糖尿病大鼠氧化应激水平，减少 NO 和 NOS 含量，其中中国蜂胶能显著提升大鼠肾脏 CAT 含量，巴西蜂胶能有效增加 GSH-px 含量（$P<0.05$），两种蜂胶同时能降低 NO、NOS 和 MDA 含量（$P<0.01$ 或者 $P<0.05$）。

表 8.30 蜂胶改善糖尿病大鼠肾脏氧化应激、NO 和 NOS 的效果

Table 8.30 Effects of propolis on renal oxidative stress, NO and NOS in diabetic rats

Group	Model	A	B	Positive	Normal
SOD/(U/mL)	115.87±19.34	106.99±14.33	134.56±26.05	108.64±19.20	195.88±79.27**
CAT/(U/mL)	111.89±13.2	136.12±21.36*	111.78±24.01	117.74±13.13	182.17±46.34**
GSH-px (μmol/L)	413.94±141.01	618.79±250.41	670.60±259.88*	843.81±251.71**	588.90±147.44
MDA/(nmol/L)	2.94±1.06	2.28±0.23**	1.98±0.31**	1.99±0.38**	2.24±0.46*
NO/(μmol/L)	187.54±74.18	81.04±20.82**	74.94±36.17**	70.36±59.49**	47.83±7.18**
NOS/(U/mL)	30.39±13.79	19.74±7.28*	18.23±6.42**	20.71±6.17*	17.91±4.61**

注：结果以平均值±标准差表示，$n=8$。与模型对照组比：* $P<0.05$，** $P<0.01$。

2.6 肝脏和肾脏病理切片观察

结果见图 8.12 至图 8.14。

由图 8.12 可见，链佐星加高脂饲料诱导对糖尿病大鼠肝脏造成严重的伤害，中国蜂胶和巴西蜂胶能明显改善肝脏病变，效果优于阳性药。

由图 8.13 和图 8.14 可见，T2DM 大鼠肾脏呈现出较为严重的病变，中国蜂胶、巴西蜂胶和阳性药均能有效改善肾脏病变。

3 讨论

3.1 蜂胶改善 T2DM 大鼠糖代谢、胰岛素和蛋白质代谢的效果

研究人员常用小鼠、大鼠、仓鼠、兔和狗等动物，通过长期选择性培养获得糖尿病自发性动物/基因缺陷型动物，或者通过物理损伤/药物手段进行诱导，组建糖尿病动物模型（Rees and Alcolado，2005）。在这些动物模型中，利用 STZ 诱导 T1DM 以及 STZ+高脂饲料诱导的 T2DM 动物模型因其组建方便、费用低及模型可靠受到广泛的关注（Rees and Alcolado，2005；Zhang et al.，2008）。本实验中我们通过饲喂高脂饲料，获得肥胖大鼠，再单次尾静脉注射 STZ 以诱发 SD 大鼠形成 T2DM。STZ 能通过损伤胰腺细胞 DNA 造成胰腺损伤而使胰岛素分泌减少（Schnedl et al.，1994）。利用链佐星制造 T2DM 大鼠模型，需要避免胰岛素合成和分泌能力的完全耗竭，成为 T1DM 模型。高脂饲料则能通过增加血液脂肪酸而干扰胰岛素信号途径增加胰岛素抵抗，并诱导胰腺 β 细胞凋亡而减少胰岛素分泌（McGarry，2002）。通过测定糖尿病大鼠空腹胰岛素发现，模型组大鼠空腹胰岛素与正常组相比无显著差异，各给药组对于糖尿病大鼠无显著影响，而模型组和各给药组大鼠血糖含量则远高于正常大鼠，这表明糖尿病大鼠可能存在胰岛素抵抗伴胰岛素分泌不足，利用本模型测定蜂胶改善 T2DM 大鼠的效果具有可靠的生物学价值。

第八章 蜂胶对糖尿病的作用及其机制研究

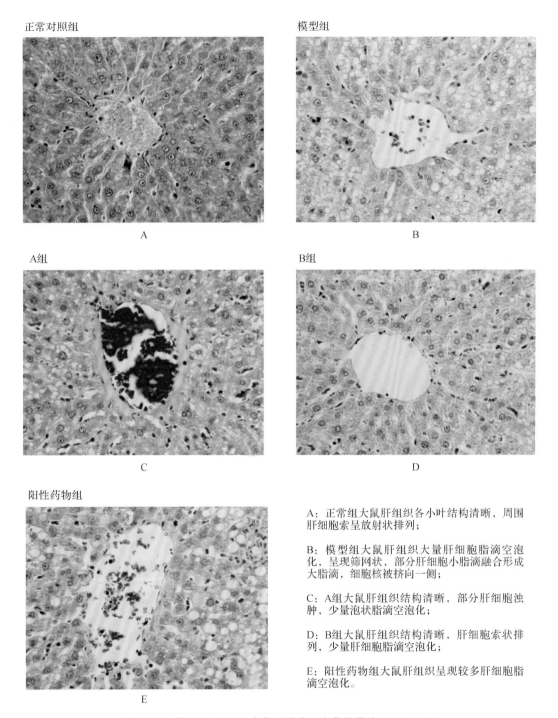

A：正常组大鼠肝组织各小叶结构清晰，周围肝细胞索呈放射状排列；

B：模型组大鼠肝组织大量肝细胞脂滴空泡化，呈现筛网状，部分肝细胞小脂滴融合形成大脂滴，细胞核被挤向一侧；

C：A组大鼠肝组织结构清晰，部分肝细胞浊肿，少量泡状脂滴空泡化；

D：B组大鼠肝组织结构清晰，肝细胞索状排列，少量肝细胞脂滴空泡化；

E：阳性药物组大鼠肝组织呈现较多肝细胞脂滴空泡化。

图 8.12　蜂胶对 T2DM 大鼠肝脏病理变化的影响（HE×400）

Fig. 8.12　The histological analysis of liver in experimental rats （HE×400）

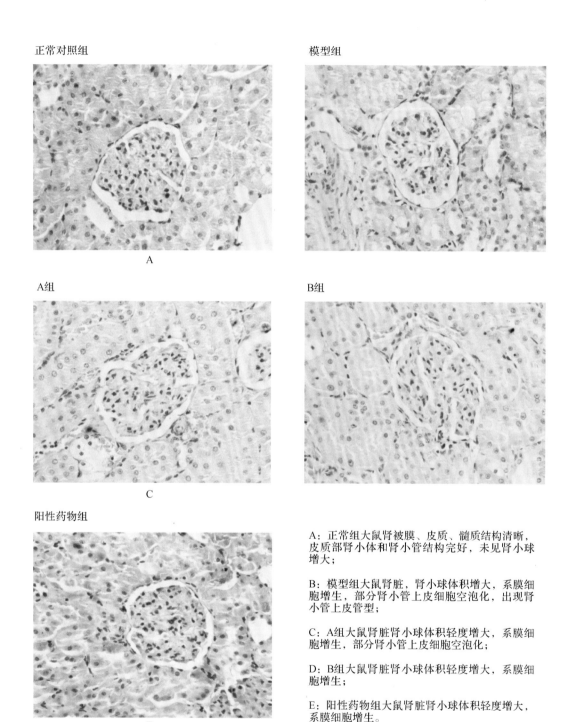

图 8.13 蜂胶对 T2DM 大鼠肾脏病理变化的影响（HE×400）

Fig. 8.13 The histological analysis of kidney in experimental rats (HE×400)

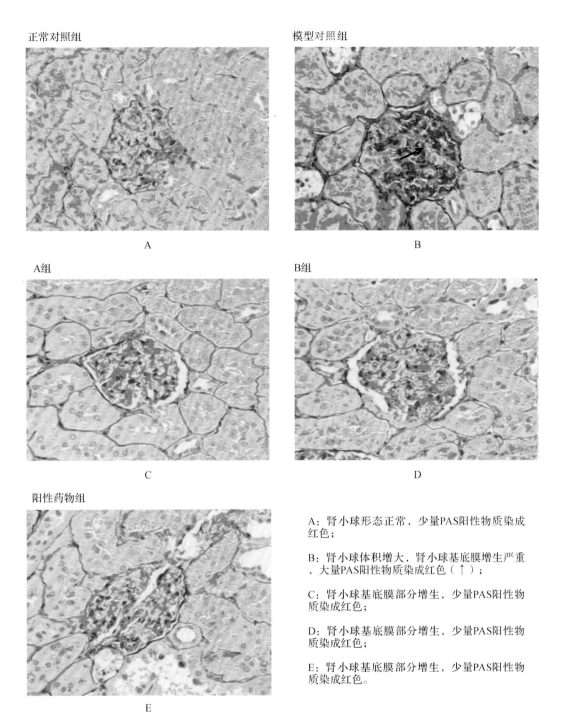

图 8.14 蜂胶对于 T2DM 大鼠肾脏病理变化的影响（PAS×400）

Fig. 8.14 The histological analysis of kidney in experimental rats (PAS×400)

控制血糖水平是治疗 T2DM 的首要目标,HbA_{1c} 是衡量 T2DM 患者血糖控制的理想指标。在进行血糖强化治疗后,当 HbA_{1c} 含量控制在 7% 左右,患者死亡率及微血管并发症发生率显著下降,但是对于大血管并发症无明显效果(The UK Prospective Diabetes Study Group,1998a)。而当 HbA_{1c} 下降到 6.5% 时,与常规治疗相比(HbA_{1c} 值为 7.3%),T2DM 患者大血管并发症和微血管并发症发生率下降 10%,而 DN 发作率下降 21%(The Advance Collaborative Group et al.,2008)。在本实验中,我们发现巴西蜂胶能降低糖尿病大鼠血糖含量,但是效果不明显;中国蜂胶在实验前 4 周能显著降低血糖含量,而在实验后 4 周则不能显著降低血糖含量;阳性药卡司平在第 2 周能显著降低血糖含量,但是在第 8 周促进血糖升高。从尿糖检测情况发现,蜂胶和阳性药在整个实验周期能有效降低尿糖含量。从 HbA_{1c} 分析发现,阳性药能降低糖尿病大鼠 HbA_{1c} 含量,但是效果不明显;A 组大鼠(饲喂中国蜂胶)和 B 组大鼠(饲喂巴西蜂胶)HbA_{1c} 含量在第 4 周比模型组大鼠低 8.6% 和 10.2%,效果显著;在实验第 8 周 A 组大鼠(饲喂中国蜂胶)HbA_{1c} 含量比模型组大鼠低 5.3%,差异明显。这些实验结果表明,阳性药和蜂胶能帮助控制大鼠血糖,但是随病情发展效果下降。

对于蜂胶改善 T2DM 血糖的作用机制目前尚不明确。Matsui 等(2004)认为巴西蜂胶水溶性提取物可能通过抑制葡萄糖苷酶活性降低 SD 大鼠餐后血糖的升高。Coskun 等(2005)研究发现蜂胶中的槲皮素能恢复糖尿病大鼠胰腺 β 细胞功能,调节糖尿病大鼠血糖含量。Zamami 等(2007)研究发现蜂胶能改善果糖诱导造成的大鼠胰岛素抵抗。这些研究从不同侧面分析了蜂胶控制血糖的作用机制,但其确切机制还需要进一步展开研究。

T2DM 中,胰岛素抗性使得蛋白质合成减少,分解增加,人体处于负氮平衡,导致病人消瘦、乏力、组织修复能力和抵抗能力下降,易于遭受各种感染(陈灏珠,2001)。在本实验中,模型组大鼠血清总蛋白和白蛋白含量显著低于正常大鼠,而中国蜂胶和巴西蜂胶能显著抑制大鼠总蛋白减少,这说明中国蜂胶能改善 T2DM 大鼠蛋白质代谢紊乱。

3.2 蜂胶改善 T2DM 大鼠脂质代谢紊乱的效果

T2DM 患者脂质代谢紊乱表现为 TG 升高和 HDL-C 含量减少,LDL-C 含量不变但是转化成小而密的 LDL,更加容易导致动脉粥样硬化发生(Ginsberg,2006)。脂质代谢紊乱被认为是促进 T2DM 患者病情发展的一种独立因素,主要因为脂质代谢紊乱能通过破坏胰腺 β 细胞功能减少胰岛素分泌,同时增加胰岛素抵抗,促进血糖升高(McGarry,2002)。

给 T2DM 患者服用富含多酚的绿茶提取物和石榴提取物,能降低患者血液中 LDL 含量,增加 HDL 含量(Fenercioglu et al.,2009)。蜂胶中含有丰富的酚类物质(Huang et al.,2014),能改善 T1DM 大鼠模型中脂质代谢紊乱(Hu et al.,2005;El-Sayed et al.,2009),也能改善金属铝中毒而出现血脂升高(Newairy et al.,2009)。在本实验中我们发现,与正常组大鼠相比较,模型组大鼠 HDL-C 含量变化不显著,但是 TC、TG 和 LDL-C 含量则显著增加。蜂胶和阳性药对于 TC 和 HDL-C 含量影响不明显,但是在实验前期能有效帮助降低 LDL-C 和 TG 含量,而在实验后期效果下降。这可能是与在实验过程中糖尿病大鼠一直被饲喂高脂饲料,导致持续性高血脂有关。

T2DM 患者因具有高血糖、高血脂、炎症反应及胰岛素抵抗等因素,比非糖尿病病人更容易产生心血管疾病,单纯的降低血糖或者使用他汀类药物降低血脂都不能满足治疗需求,需要更为全面的治疗方案(Erdmann,2005)。在本实验中,我们发现蜂胶能在 T2DM 前期控制血糖和降低血脂。蜂胶具有显著的抗炎活性(Wang et al.,2014),其活性成分 CAPE 能通过抑制 NF-κB 活性而延缓小鼠动脉硬化过程(Hishikawa et al.,2005)。LDL-C 氧化是动脉硬化的前兆(Matsuura et al.,2008),蜂胶能抑制 LDL-C 氧化(Abd El-Hady et al.,2007)。这些实验结果表明,蜂胶有可能在 T2DM 发展前期,通过抗氧化、抗炎、降血糖和降血脂而延缓 T2DM 患者心血管疾病发生。

3.3 蜂胶改善 T2DM 大鼠氧化应激的效果

高血糖刺激线粒体产生过多的 ROS 被认为是糖尿病患者血管并发症共同的上游调节因子(Brownlee,2005)。在微血管并发症中,高血糖刺激细胞中线粒体产生过多 ROS,ROS 损伤细胞 DNA 而激活多聚(ADP-核糖)聚合酶(PARP),PARP 通过级联反应降低磷酸甘油醛脱氢酶(GAPDH)活性,GAPDH 活性下降激活 PKC 途径、AGEs 途径、多元醇通路和已醣胺途径,并进一步激活 NF-κB(Brownlee,2001)。在大血管并发症中,排除高血糖和脂质代谢紊乱因素后,胰岛素抵抗使脂肪细胞进一步释放游离脂肪酸(FFA),FFA 进入大血管内皮细胞后被线粒体释放的 ROS 氧化,氧化后的 FFA 刺激线粒体过度产生 ROS,ROS 则进一步激活下游途径(同高血糖作用途径)(Brownlee,2005)。在本实验中,我们发现中国蜂胶和巴西蜂胶对于糖尿病大鼠血清 NO 和 NOS 含量无显著影响,但是能明显降低 MDA 含量;中国蜂胶能显著提升 CAT 含量,巴西蜂胶能显著提升 GSH-px 含量,这些结果表明蜂胶能抑制 T2DM 大鼠氧化应激。

氧化应激是糖尿病并发症发作的上游因子,从理论上来说改善糖尿病患者氧化应激水平能帮助控制糖尿病并发症发生,抗氧化剂能改善糖尿病患者氧化应激水平,但是对于能否改善糖尿病并发症存在争论。小范围临床实验表明,抗氧化剂能帮助改善糖尿病患者并发症状;大范围临床研究表明,糖尿病患者服用抗氧化剂一段时间后其并发症没有得到明显改善(Johansen et al.,2005)。简单的清除自由基不能直接帮助糖尿病并发症,可能还存在其他影响因素,需要进一步展开研究。

3.4 蜂胶保护 T2DM 大鼠肝脏和肾脏的效果

高血糖、肥胖及胰岛素抵抗等多种因素能引发 T2DM 患者肝脏损伤,形成非酒精性脂肪肝,并进一步发展为肝硬化及肝细胞癌(Garcia-Compean et al.,2009)。T2DM 患者胰岛素抵抗促进脂肪组织释放游离脂肪酸(FFA),FFA 在肝脏内积累生成更多 TG,胰岛素抵抗同时降低肝脏极低密度脂蛋白分泌效率,这导致肝细胞脂肪堆积而变性;肝脏内 FFA 增加促使线粒体氧化应激,过多的 ROS 诱导肝细胞炎症和坏死,肝细胞炎症和坏死激活肝脏星状细胞产生更多胶原、结缔组织生长因子并积累胞外基质,进而向肝纤维化发展(Garcia-Compean et al.,2009)。在本实验中,我们发现模型组大鼠血清 ALT 含量显著高于正常大鼠而 AST 含量在实验中期明显偏低,中国蜂胶和巴西蜂胶能有效降低糖尿病血清 ALT 含量,但是对于 AST 含量影响不显著,这表明两种蜂胶均能恢复糖尿病大鼠肝脏功能。肝脏病理观察发现,模型组大鼠肝细胞浊肿,且大泡状脂滴空泡化,已有脂肪肝产生,但尚未产生肝纤维化,蜂胶和阳性药能减少糖尿病大鼠肝脏脂肪化程度,蜂胶效果优于阳性药。

T1DM 肾病与 T2DM 肾病在基本病理生理学机制上表现一致(Parving,2001)。与 T1DM 相比,T2DM 增加了高血压、肥胖、脂质代谢紊乱以及动脉硬化所导致的缺血性肾病等因素,更容易导致 DN 的发生(Wolf,2004)。本实验中我们发现 T2DM 大鼠 BUN、CCR 及尿白蛋白排泄率明显高于正常大鼠,观察其肾脏病理发现模型组大鼠肾脏、肾小球体积增大,系膜细胞增生,肾小球基底膜增生严重,部分肾小管上皮细胞空泡化,出现肾小管上皮管型,大量 PAS 阳性物质染成红色,这说明 T2DM 模型鼠已经出现 DN 症状。与模型组比较,蜂胶和阳性药能显著降低 BUN 和尿液白蛋白含量,阳性药卡司平还能明显降低肾脏 CCR 值,观察其肾脏病理发现,蜂胶和阳性药可减轻肾脏病变,对 T2DM 引起的肾损伤具有保护作用。观察肾脏组织氧化应激水平,我们发现蜂胶能有效抑制肾脏内脂质过氧化,不同程度恢复抗氧化酶活性,这表明蜂胶可能能通过抑制肾脏氧化应激而延缓肾病病变。NO 参与 DN 发展过程,在 DN 发展早期肾脏内 NO 含量增加,促进肾小球高过滤及蛋白尿产生;随 DN 病情发展,肾脏内 NO 和 NOS 受到多种途径抑制而含量下降(Prabhakar,2004)。在本实验中,我们发现蜂胶和阳性药能显著减少肾脏 NO 和 NOS 含量,这表明蜂胶可能通过减少 NO 产生而降低尿液白蛋白含量及肾小球滤过率。

小 结

本研究系统地比较了中国杨树型蜂胶和巴西绿蜂胶对 T2DM 大鼠改善糖代谢、胰岛素、脂质代谢和保护肝肾功能的效果及其作用机制,结果表明:(1)中国蜂胶和巴西蜂胶能减缓 T2DM 大鼠体重下降,但是效果不显著,中国蜂胶在实验第 8 周明显抑制实验大鼠体重下降;(2)中国蜂胶能在实验前期降低 T2DM 大鼠的空腹血糖含量,巴西蜂胶效果不显著。中国蜂胶在实验第 4 周和第 8 周能明显降低 HbA_{1c} 水平,巴西蜂胶在实验第 4 周能明显降低 HbA_{1c} 水平;(3)中国蜂胶和巴西蜂胶能改善 T2DM 大鼠蛋白质代谢紊乱,中国蜂胶能显著抑制糖尿病大鼠血清总蛋白和白蛋白下降,巴西蜂胶能显著抑制血清总蛋白下降;(4)中国蜂胶和巴西蜂胶对 T2DM 大鼠胰岛素含量无显著影响;(5)中国蜂胶能在实验前期降低 T2DM 大鼠血清 TG,而 LDL 含量增加,在实验后期效果不显著;(6)中国蜂胶和巴西蜂胶能明显降低 T2DM 大鼠血清 MDA 含量,中国蜂胶能显著提升 GSH-px 含量,巴西蜂胶能显著提升 CAT 含量;(7)中国蜂胶和巴西蜂胶能保护 T2DM 大鼠肝脏和肾脏功能,抑制肝肾组织结构破坏,改善肾脏组织氧化应激水平,减少肾组织中 NO 和 NOS 含量。

第三节　蜂胶对 T2DM 大鼠肾病的作用机制研究

糖尿病发病机制十分复杂,多种因素共同作用促进糖尿病并发症发作。这些因素主要包括:多元醇通路的激活、糖基化终端产物 AGEs 生成、PKC 激活、氧化应激、炎症、血流动力学改变、细胞因子以及某些酶类物质等(Wolf,2004;Kanwar et al.,2008)。此外,糖尿病大血管并发症和微血管并发症之间存在相互促进作用,当患者存在大血管并发症时,会比无并发症的糖尿病患者更容易产生糖尿病(Charles and Selam,2005)。

在我们前面的实验中,蜂胶已经表现出改善 T1DM 和 T2DM 模型大鼠肾脏的功能,这种保护作用可能与蜂胶降低血糖、改善肾脏组织氧化应激和降低 NO 含量有关。但是对于改善氧化应激能否治愈糖尿病各类血管并发症存在争议,大型前瞻性研究指出单纯补充抗氧化剂并不能帮助患者改善并发症(Johansen et al.,2005)。这表明蜂胶对 T2DM 大鼠肾脏的保护作用可能存在多种作用途径。

研究证实,各种黄酮类物质如槲皮素、芦丁等能抑制醛糖还原酶(AR)活性,减少山梨醇产生(闫泉香,2004;刘英华等,2006),并且这种抑制活性和黄酮结构有着密切的关联(Suzen and Buyukbingol,2003)。蜂胶中含有丰富的黄酮类成分(Bankova et al.,2000),传统中药制剂蜂贝化瘀软胶囊(含蜂胶)能降低糖尿病大鼠血液 AR 活性(金光香等,2005),蜂胶对糖尿病保护效果可能与蜂胶抑制 AR 活性的效果有关。

高血压是促进糖尿病发展的另一重要因素,改善 T2DM 患者血压能显著减少尿液微量白蛋白排放。血压正常的糖尿病患者的 GFR 比高血压患者低将近 50%(The UK Prospective Diabetes Study,1998b)。巴西蜂胶中黄酮类提取物能改善自发性高血压大鼠主动脉压(Maruyama et al.,2009)。这表明蜂胶改善高血压也可能是抑制糖尿病发展的作用途径。

多种炎症介质介入糖尿病发展过程,包括趋化因子、黏附分子和促炎细胞因子如 IL-1、IL-6、IL-8 和 TNF-α 等,这些炎症物质的表达和糖尿病的发展呈现显著的正相关性(Mora and Navarro,2006;Ruster and Wolf,2008)。蜂胶具有显著的抗炎活性,能抑制炎症细胞向炎症部位转移(Aslan et al.,2007),减少炎症部位 MCP-1、ICAM-1 和多种促炎细胞因子如 IL-1β、IL-6、IL-8 和 TNF-α 等的表达(Khayyal et al.,2003;Shin et al.,2009)。蜂胶的抗炎活性可能是蜂胶保护糖尿病的作用机制之一。

总之,蜂胶可能通过多种途径改善糖尿病,但是其确切的作用机制尚不明确。在本实验中,我们通过分析 T2DM 大鼠肾脏炎症反应、血流动力学、PKC 及 TGF-β_1 的表达,对蜂胶保护肾脏组织的作用机制进行深入探讨。

1 材料与方法

1.1 实验进程

1)在 T2DM 大鼠实验第 8 周,用激光多普勒血流仪测定大鼠肾脏血流动力学;
2)实验第 8 周测定 T2DM 大鼠血清 C 反应蛋白(CRP);
3)实验结束后,取大鼠肾组织,测定肾组织中相关指标:
- 采用 ELISA 法测定肾脏炎症因子 TNF-α、IL-2 和 IL-6 含量;
- 用 RT-PCR 方法分析 PKC 和 MCP-1mRNA 表达情况;
- 免疫组化测定肾脏 TGF-β_1 表达情况。

1.2 生化指标测定

1.2.1 主要仪器

日立 7020 全自动生化分析仪(日立公司);多功能酶标仪 VarioScan Flash(美国 Thermo 公司);激光多普勒血流仪(英国 Moor 公司)。

1.2.2 主要试剂

C 反应蛋白(CRP)试剂盒(上海复星长征医学科学有限公司);大鼠肿瘤坏死因子-α(TNF-α)ELISA 试剂盒(美国 R&D 公司进口分装);大鼠白介素-2(IL-2)ELISA 试剂盒(美国 R&D 公司进口分装);大鼠白介素-6(IL-6)ELISA 试剂盒(美国 R&D 公司进口分装)。

1.3 组织病理学观察

1.3.1 主要试剂

一抗:兔抗 TGF-β_1,规格:0.1mL/支,Santa Cruz 进口分装,由中杉金桥生物技术有限公司提供。

二抗:兔两步法试剂盒(PV-6001),规格:18mL/支,中杉金桥生物技术有限公司生产。

DAB 显色试剂盒(ZLI9032,20×),规格:3mL/支×3 支,中杉金桥生物技术有限公司生产。

1.3.2 主要仪器

STP120 脱水机、AP280-2 包埋机、HM335E 切片机(MICROM 公司);Nikon eclipse 80i 显微镜(Nikon 公司);CCD 相机 DS-Fi1(500 万象素)(Nikon 公司);图像分析软件:Carl Zeiss Imaging Systems(Carl Zeiss 公司)。

1.3.3 免疫组化染色

采用两步法操作方法:

1)切片置于 60℃烤箱中烘烤 2h;

2)切片脱蜡至水;

3)蒸馏水洗 2min;

4)高压热修复:高压锅内放适量自来水,将修复液倒入烧杯内(修复液体积视切片多少而定),将烧杯置于高压锅内煮沸(高压锅不需盖紧),将切片放入装有修复液的烧杯内,盖紧高压锅,加热至设定压力冒气后持续 2min,自来水冷却;

5)3% H_2O_2 溶液阻断过氧化物酶,10min;

6)PBS 洗 3 次,每次 3min;

7)滴加一抗,37℃孵育 60min,稀释比例均为 1∶100;

8)PBS 洗×3 次,每次 3min;

9)滴加二抗,37℃孵育 45min;

10)PBS 洗 3 次,每次 3min;

11)DAB 显色 2min;

12)复染、透明后封片。

1.3.4 图像分析

每张切片所要分析的部位在显微镜下拍摄 3~5 张照片,进行图像分析。光源亮度:恒定电压 5.5V(显微镜卤素灯电源电压 0~12V 可调);放大倍数:目镜 400×,物镜 40×;自动白平衡:关;手动曝光,曝光时间:1/90 s;图像分辨率:1280×960=1228800;阳性面积:照片内被染成棕黄色的物质的面积总和,反映阳性表达的多少;平均光密度:阳性表达区域的光密度平均值,反映表达部位的阳性强度;

总光密度又称积分光密度,总光密度=∑(阳性表达部位面积×该部位的平均)×光密度,反映阳性表达的总量,阳性指数=总光密度/相应肾小球截面积。

1.4 mRNA 测定

采用 RT-PCR 法测定单核细胞趋化蛋白-1(MCP-1)和蛋白激酶 C(PKC)mRNA。

1.4.1 材料和试剂

(1)实验器具

①移液枪:1mL、200μL、20μL、2.5μL;

②枪头:1mL、200μL、20μL;

③EP管:5mL、1.5mL、0.2mL;

④试剂瓶:100mL、250mL各2个备用,一个装无水乙醇,另一个装DEPC水;

⑤量筒:50mL、250mL;

⑥烧杯:250mL、500mL、1000mL;

⑦架子:10mL、1.5mL、20μL各2个;

⑧铝制饭盒:2个;

⑨大瓷缸:2个;

⑩锡箔纸:一卷;

⑪牛皮纸:一卷;

⑫剪刀10把,镊子10把。

(2)试剂

TRIZOL(GIBCO)、反转录试剂盒(Promega A3500)、DEPC(Geneview)、DNA染料(Biotium)、琼脂糖(BIOWEST)、dNTP、Taq酶、Marker DL2000,均购自TaKaRa公司。

1.4.2 引物合成

(1)MCP-1

正义:5′-CACCTGCTGCTACTCATTCACT-3′

反义:5′-GTTCTCTGTCATACTGGTCACTTC-3′

GAPDH

正义:5′-TCCCTCAAGATTGTCAGCAA-3′

反义:5′-AGATCCACAAACGGATACATT-3′

(2)PKC

正义:5′-ACCCTCAGTGGAATGAGTCCTTCACGT-3′

反义:5′-TTAGATGGCTGCTTCCTGTCTTCTGAA-3′

GAPDH

正义:5′-TCCCTCAAGATTGTCAGCAA-3′

反义:5′-AGATCCACAACGGATACATT-3′

1.4.3 实验方法

(1)实验器具的处理与准备

①塑料制品:(包括枪头、EP管、匀浆管等)先将DEPC水从容量瓶中倒入瓷缸中,将塑料制品逐个浸泡其中,其中小枪头需要吸管打入DEPC水,过夜,然后高压,再烤干备用,实验前将枪头等放入吸头台,再高压一次(EP管)。

②玻璃制品:泡酸过夜,冲洗干净,蒙锡纸烤干备用(DEPC水泡)(洗净后先泡1‰DEPC过夜,再烤干)。

③剪刀、镊子:先洗净后,再高压(不需要泡DEPC)。

(2)试剂配制

①DEPC水:吸出1mL DEPC放在1000mL双蒸水中配成1‰ DEPC水,放在1000mL容量瓶中静置4h备用。

②75%乙醇:用无水乙醇+DEPC水配,然后放于-20℃冰箱保存(其中DEPC水需先高压)。

③50×TAE电泳缓冲液(贮存液):242g Tris,57.1mL冰乙酸,100mL 0.5mol/L EDTA(pH8.0)。

④1×TAE电泳缓冲液:将50×TAE电泳贮存液稀释50倍。

(3)Trizol法抽提总RNA

肾皮质组织总RNA提取按Trizol试剂盒说明书操作。取100mg肾皮质组织,加1mL TRIzol试剂,研磨匀浆直至组织完全裂解,无团块,混匀器上颠倒混匀2~3min,室温静置5min。加0.2mL氯仿,颠倒混匀10次,室温静置5min,4℃、12000r/min离心15min,吸取上层水相(约400μL)于另一1.5mL EP管中,加等体积异丙醇(400μL),温柔颠倒混匀,室温静置10min,4℃、12000r/min离心20min,弃上清,加1mL预冷的75%乙醇(用DEPC水配),4℃、12000r/min离心10min,弃上清,沉淀室温干燥5~10min(不能完全干燥),加入40μL DEPC水溶解,取出10μL RNA用多功能酶标仪检测RNA含量。

(4)RT-PCR两步法

第一步:反转录反应。

取10μg RNA,加DEPC水至10μL,混匀,离心,70℃孵育10min,立即冰水浴,稍离心,按以下顺序加:

①$MgCl_2$(25mmol/L) 4μL

②Reverse Transcription 10×Buffer 2μL

③dNTP 2μL

④Recombinant Rnasin Ribonuclease Inhibitor 0.5μL

⑤AMV reverse transcriptase 15U

⑥Oligo(DT) 0.5μg

⑦加DEPC水至20μL。混匀,离心,42℃ 15min,95℃ 5min(破坏MLV),4℃保存。

第二步:PCR反应。

以反转录合成的cDNA为模板用MCP-1、PKC或GADPH引物经PCR方法进行扩增。总体积20μL,其中10× Buffer 2μL,$MgCl_2$ 25mmol/L 1.4μL,dNTP 2.5mmol/L 1.5μL,上游引物15 pmol/μL 0.15μL,下游引物15 pmol/μL 0.15μL,Taq酶5U/μL 0.15μL,cDNA模板1μL。最后加DEPC水至20μL。

MCP-1(GAPDH)扩增条件为:94℃预变性4min;94℃变性60 s,56℃退火60 s,72℃延伸60 s,35个循环;72℃延伸10min,4℃保存。

PKC扩增条件为:94℃预变性4min;94℃变性30 s,65℃退火45s,72℃延伸60 s,39个循环;72℃延伸10min,4℃保存。

GAPDH扩增条件为:94℃预变性4min;94℃变性30 s,60℃退火30 s,72℃延伸30 s,30个循环;72℃延伸10min,4℃保存。

(5) PCR 产物电泳

① 1.0%琼脂糖凝胶的配制：0.4g 琼脂糖+40mL 电泳缓冲液，微波炉加热熔化，加入 Gelred 0.4μL，充分混匀，将温热的凝胶倒入已置好梳子的胶膜中，在室温下放置 30~45min 后进行电泳。

② 水平电泳：将 10μL PCR 产物和 1μL 溴酸兰混匀，用移液枪吸到琼脂糖凝胶加样孔中，进行水平电泳鉴定，设置稳压电源 80V，电泳 50min，UVP 凝胶成像系统中拍照，以 MCP-1/GAPDH、PKC/GAPDH 条带光密度比值表示 MCP-1 和 PKC mRNA 的相对表达量。

1.5 数据处理

实验数据用 SPSS 16.0 软件统计，采用单因素方差分析法和事后多重比较 LSD 法进行分析，结果以平均值±标准差表示，$P<0.05$ 表示差异显著。

2 结果与分析

2.1 蜂胶改善 T2DM 大鼠肾脏炎症因子及血清 CRP

实验结果见图 8.15 至图 8.18。

由图 8.15 可见，模型组大鼠血清 CRP 含量与正常组大鼠相比显著偏高（$P<0.05$），这表明 T2DM 大鼠存在炎症反应。与模型组大鼠相比，阳性药和中国蜂胶能降低糖尿病大鼠血清 CRP 含量但是，效果不显著，巴西蜂胶能有效降低 A 组大鼠 CRP 含量（$P<0.05$）。

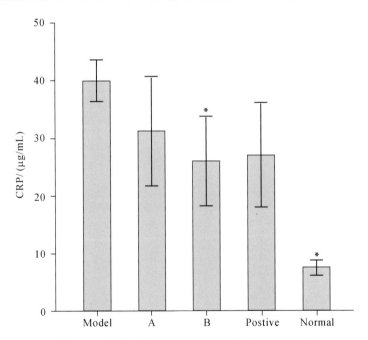

图 8.15　蜂胶改善 T2DM 大鼠血清 CRP 含量效果

Fig. 8.15　Effects of propolis on serum CRP in diabetic rats

由图 8.16 可见，模型组大鼠 IL-2 含量与正常组大鼠相比显著偏高（$P<0.05$）。与模型组大鼠相比，阳性药和巴西蜂胶对于肾脏 TNF 含量效果不显著，中国蜂胶能有效降低 A 组大鼠肾脏 IL-2 含量（$P<0.05$）。

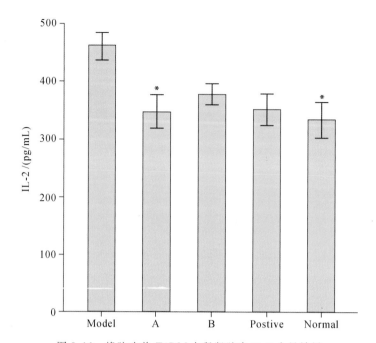

图 8.16 蜂胶改善 T2DM 大鼠肾脏中 IL-2 含量效果

Fig. 8.16 Effects of propolis on renal IL-2 level in diabetic rats

注:结果以平均值±标准差表示,$n=8$。与模型对照组比:* $P<0.05$,** $P<0.01$。

由图 8.17 可见,模型组大鼠 IL-6 含量与正常组大鼠相比显著偏高($P<0.05$),这表明大鼠肾脏存在炎症反应。与模型组大鼠相比,巴西蜂胶和阳性药对于肾脏 IL-6 含量效果不显著,中国蜂胶能有效降低 A 组大鼠肾脏 IL-6 含量($P<0.05$)。

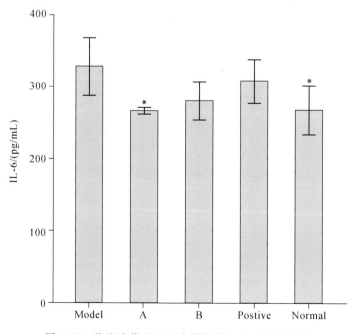

图 8.17 蜂胶改善 T2DM 大鼠肾脏中 IL-6 含量效果

Fig 8.17 Effects of propolis on renal IL-6 level in diabetic rats

由图 8.18 可见,模型组大鼠 TNF-α 含量与正常组大鼠相比显著偏高($P<0.01$)。与模型组大鼠相比,阳性药对于肾脏 TNF-α 含量效果不显著,巴西蜂胶和中国蜂胶能有效降低 A 组大鼠肾脏 TNF-α 含量($P<0.01$)。

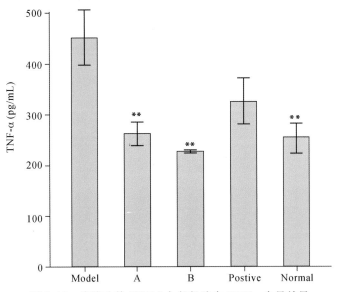

图 8.18　蜂胶改善 T2DM 大鼠肾脏中 TNF-α 含量效果

Fig. 8.18　Effects of propolis on renal TNF-α level in diabetic rats

2.2　蜂胶改善 T2DM 大鼠肾脏 MCP-1 和 PKC mRNA 表达

采用 RT-PCR 方法测定肾脏 MCP-1 和 PKC mRNA 表达,以 MCP-1/GAPDH、PKC/GAPDH 电泳条带光密度比值表示 MCP-1 和 PKC mRNA 的相对表达量。MCP-1 能趋化白细胞向炎症部位定向移动,促进炎症发展。实验结果见图 8.19。

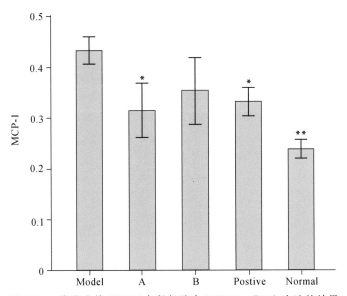

图 8.19　蜂胶改善 T2DM 大鼠肾脏中 MCP-1 mRNA 表达的效果

Fig. 8.19　Effects of propolis on the expression of renal MCP-1 mRNA in diabetic rats

由图 8.19 可见，模型组大鼠肾脏 MCP-1mRNA 与正常组大鼠相比显著偏高（$P<0.01$）。与模型组大鼠相比，巴西蜂胶对于糖尿病大鼠肾脏 MCP-1 表达影响不显著，阳性药和中国蜂胶能有效降低大鼠肾脏 MCP-1 表达量（$P<0.05$）。

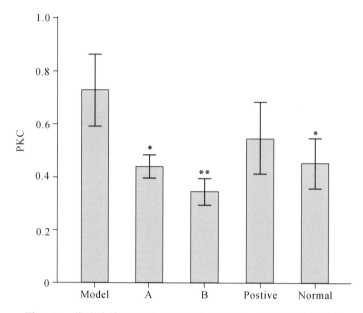

图 8.20　蜂胶改善 T2DM 大鼠肾脏中 PKC mRNA 表达的效果

Fig. 8.20　Effects of propolis on the expression of renal PKC mRNA in diabetic rats

高血糖激活肾脏 PKC，PKC 能影响肾脏血流动力学，同时通过刺激多种途径促进 DN 发展。实验结果见图 8.20。

由图 8.20 可见，模型组大鼠肾脏 PKC mRNA 与正常组大鼠相比显著偏高（$P<0.05$）。与模型组大鼠相比，阳性药对于糖尿病大鼠肾脏 PKC mRNA 表达影响不显著，巴西蜂胶和中国蜂胶能有效降低大鼠肾脏 MCP-1 表达（$P<0.01$，$P<0.05$）。

2.3　蜂胶改善 T2DM 大鼠肾脏血流动力学

持续高血糖能引发机体肾脏血流动力学异常，血流动力学异变促进糖尿病发展。实验第 8 周，各组大鼠用 3% 戊巴比妥钠（45mg/kg）麻醉后，用激光多普勒血流仪检测肾脏的血流动力学变化，通过计算肾血浆流量（RPF）、肾小球滤过率（GFR）、肾血管阻力（RVR）及滤过分数（FF）分析蜂胶保护糖尿病大鼠肾脏功能的效果。结果见表 8.31。

由表 8.31 可见，正常组大鼠 RPF 值显著高于模型组大鼠，而 PRV 值和 FF 值明显低于模型组大鼠（$P<0.05$，$P<0.01$）。与模型组大鼠相比，巴西蜂胶对于肾脏功能指标影响不显著，中国蜂胶能有效提升 RPF 值，降低 PVR 和 FF 值（$P<0.05$）。

表 8.31 蜂胶改善糖尿病大鼠肾脏血流动力学的效果
Table 8.31 Effects of propolis on renal hemodynamics in diabetic rats

Group	Model	A	B	Positive	Normal
肾血流量(RPF)/(mL/min/100g)	0.138±0.110	0.449±0.344*	0.181±0.214	0.440±0.481*	0.406±0.354*
肾小球滤过率(GFR)	0.336±0.234	0.318±0.270	0.299±0.389	0.353±0.204	0.274±0.193
肾血管阻力(RVR)	231.5±99.8	131.0±106.2*	245.1±118.8	107.9±63.0*	115.7±72.9*
滤过分数(FF)	2.72±2.26	1.36±0.84*	1.76±1.56	1.96±1.80	1.06±0.48**

注:结果以平均值±标准差表示,$n=8$。与模型对照组比:* $P<0.05$,** $P<0.01$。

2.4 蜂胶改善 T2DM 大鼠肾脏 TGF-β_1 表达

TGF-β_1 是肾脏细胞外基质合成和纤维化的主要因素,其过度表达是 DN 重要的病理生理特征。我们采用免疫组化染色法研究糖尿病大鼠肾脏 TGF-β_1 表达,利用图像分析系统对 TGF-β_1 表达的阳性面积、平均光密度、阳性指数进行分析。结果见表 8.32 和图 8.21。图像分析结果显示,正常对照组 TGF-β_1 表达较低,模型对照组与正常组比较阳性面积、总光密度和阳性指数均有显著差异($P<0.05$ 或 $P<0.01$)。与模型对照组比,B 组和阳性对照组大鼠肾脏 TGF-β_1 表达总光密度、阳性指数和阳性面积显著降低($P<0.01$,$P<0.05$)。

表 8.32 蜂胶改善糖尿病大鼠肾脏 TGF-β_1 表达的效果
Table 8.32 Effects of propolis on the expression of renal TGF-β_1 in diabetic rats

Group	阳性面积($\times 10^2 \mu m^2$)	平均光密度	总光密度	阳性指数($\times 10^{-3}$)
Model	14.545±10.600	0.175±0.006	392.930±227.580	2.081±1.035
Normal	3.988±2.322*	0.172±0.004	78.89±48.62**	0.614±0.382**
A	6.641±3.439	0.178±0.007	213.33±174.64	1.394±1.326
B	5.564±2.817*	0.176±0.006	124.92±83.42**	0.737±0.486**
Positive	1.808±1.536*	0.172±0.006	34.22±29.99**	0.196±0.160**

注:结果以平均值±标准差表示,$n=8$。与模型对照组比:* $P<0.05$,** $P<0.01$。

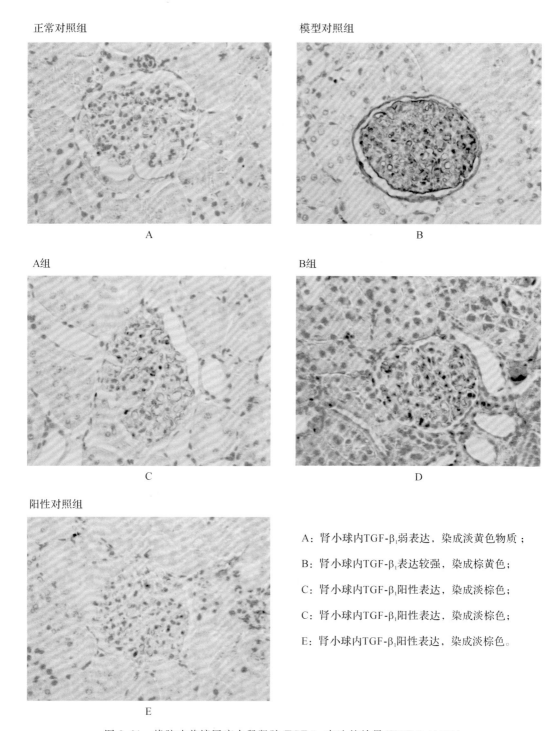

图 8.21 蜂胶改善糖尿病大鼠肾脏 TGF-β_1 表达的效果(TGF-β_1×400)

Fig. 8.21 Effects of propolis on the expression of renal TGF-β_1 in diabetic

3 讨论

3.1 蜂胶改善 T2DM 大鼠肾脏炎症的效果

糖尿病患者及各类糖尿病动物模型肾脏组织切片观察均表明,肾脏组织存在炎症细胞浸润现象,而且这种浸润是以单核细胞为主(Noronha et al.,2002)。糖尿病患者肾脏炎症发病机制还不是十分明确,但是多种因素促进炎症发展。高血糖能刺激肾小球系膜细胞产生 ROS 和 PKC,级联激活 NF-κB,并上调 MCP-1 蛋白和 mRNA 表达(Ha et al.,2002)。Mezzano 等(2003)研究发现血管紧张素Ⅱ(AngⅡ)/肾素激活促进 T2DM 患者肾小管细胞内 NF-κB 表达,并进一步上调 MCP-1 和 RANTES(淋巴细胞趋化因子)表达。肾素受体能单独刺激糖尿病大鼠肾脏内 TNF-和 IL-β,促进肾脏炎症发展(Matavelli et al.,2010)。

MCP-1 对单核细胞具有趋化活性,激活单核细胞和巨噬细胞向炎症部位转移。Mezzano 等(2004)分析了 11 位 T2DM 患者肾脏组织病理切片,结果发现 NF-κB 明显激活,主要表达在肾小管上皮细胞内,同时 MCP-1(单核细胞/单核巨噬细胞趋化蛋白)和 RANTES(淋巴细胞趋化因子)mRNA 表达也显著增加,主要表达在肾小管细胞,且 NF-κB 激活与趋化因子表达呈正相关性;蛋白尿是促进炎症因子表达的重要因素,其表达量与 NF-κB、趋化因子表达呈现正相关性。Morii 等(2003)研究发现 T2DM 患者尿白蛋白含量越高则尿液中 MCP-1 和 N-乙酸氨基葡萄糖苷酶(NAG,肾小管损伤产物)含量也越高,提示大量蛋白尿能刺激肾小管表达释放 MCP-1 而加快 DN 的进程。这些结果表明,蛋白尿激活肾小管上皮细胞分泌 NF-κB,而 NF-κB 激活则上调趋化因子表达,促进炎症细胞浸润。

IL-6 和 TNF-α 在糖尿病炎症发展过程具有重要的生理意义。肾组织切片观察表明,IL-6 含量与肾小球基底膜增厚呈现密切的正相关性(Dalla Vestra et al.,2005)。当肾髓质中编码 IL-6 的基因表达量增加的时候,尿液中 IL-6 的排泄量也随之增加(Navarro et al.,2006)。Choudhary 和 Ahlawat(2008)测定 T2DM 患者治疗前后尿液微量白蛋白和血清 CRP 及 IL-6 含量变化发现,蛋白尿和血清 CRP 及 IL-6 存在显著的正相关性,血清 CRP 和 IL-6 含量是可用以衡量肾组织损伤程度的指标。肾脏内 TNF-α 含量随糖尿病病情发展而显著增加,两者呈现密切的正相关性(Sugimoto et al.,1999)。TNF-α 能破坏肾小球滤过屏障而促进尿液白蛋白排泄,降低肾脏血流量和滤过率,引发炎症细胞浸润,调节炎症分子分泌,诱导肾脏细胞凋亡,引发肾脏内钠滞留和肾肥大现象(Navarro and Mora-Fernandez,2006)。

血清 CRP 是人体炎症标志物。在本实验中,我们发现 T2DM 大鼠血清 CRP 含量比正常大鼠明显升高,说明糖尿病大鼠存在炎症现象。蜂胶和阳性药均能降低血清 CRP 含量,其中巴西蜂胶效果显著。观察肾脏促炎因子和 MCP-1 mRNA 表达发现,模型组大鼠肾脏 IL-2、IL-6、TNF-α 及 MCP-1 mRNA 含量比正常组大鼠明显增加,这表明糖尿病大鼠存在炎症反应。蜂胶和阳性药能抑制肾脏炎症反应,其中中国蜂胶能有效抑制糖尿病大鼠肾脏中 IL-2、IL-6 和 TNF-α 含量增加,巴西蜂胶能有效抑制糖尿病大鼠肾脏中 TNF-α 含量增加及 MCP-1 mRNA 表达。

NF-κB 是一个多功能核转录因子,在炎症过程中起到重要的调节作用,激活后可以促进各类细胞因子、黏附因子、趋化因子、趋氧化酶-2(Cox-2)及一氧化氮合酶(NOS)等表达及基因转录(Tak and Firestein,2001)。蜂胶中的活性成分 CAPE 能通过抑制 NF-κB 活性而降

低下调促炎细胞因子表达和炎症细胞浸润(Paulino et al.,2006;Jung et al.,2008)。本实验中,蜂胶改善 T2DM 大鼠肾脏炎症反应的效果,可能和 CAPE 抑制 NF-κB 活性有关。此外,蛋白尿能刺激肾小管上皮分泌 NF-κB,蜂胶能降低尿液中白蛋白含量,这可能有助于抑制肾脏炎症发生。NO 具有促进炎症发展的生理效应(Cirino et al.,2006),能在糖尿病早期促进尿微量白蛋白排泄(Prabhakar,2004),蜂胶降低肾组织中 NO 和 NOS 含量的效果,可能有助于抑制肾脏炎症发生。

3.2 蜂胶改善 T2DM 大鼠肾脏血流动力学的效果

肾脏血流动力学异变是促进糖尿病发生发展的重要原因。高血糖/糖尿病环境下,高灌注引发肾脏高压力,而高压力进一步诱发高滤过(Ruggenenti et al.,2001;Hostetter,2003)。肾脏血流动力学异变能促进系膜基质扩张和 GBM 增厚,破坏正常的滤过屏障,增加肾脏生长因子合成和释放,刺激糖尿病发展(Hostetter,2003;Wolf et al.,2003)。多种血管活化激素诱发肾脏血流动力学异变,包括血管收缩因子血管紧张素 II-肾素系统(RAS)、内皮素、尾加压素 II 和抗利尿激素等,血管舒张因子如血管缓激肽、利钠因子、前列腺素和 NO 等(Forbes et al.,2007;Maric,2008)。

对于糖尿病的研究最初比较强调肾脏血流动力学异常对于糖尿病发展的影响,利用 RAS 抑制剂降低糖尿病模型动物血压而减缓糖尿病发展(Anderson and Vora,1995)。高血糖/糖尿病环境刺激人体和各组织器官分泌 RAS,RAS 能造成肾脏组织血压升高而诱发肾脏血流动力学异常;但是 RAS 不仅能增加肾小球内压力,而且直接独立刺激系膜细胞基质蛋白合成,抑制基质降解酶活性,同时刺激 TGF-$β_1$ 分泌(Leehey et al.,2000)。市场上的抑制剂,如血管紧张素转化酶抑制剂(ACEi)、血管紧张素受体阻滞剂(ARBs)和直接肾素抑制剂(DRIs),已经被证明能有效改善肾脏血流动力学异常以及肾组织结构恶化,延缓糖尿病发展(Estacio,2009)。

本实验中发现,与模型组大鼠比较,正常组大鼠肾脏血流量明显偏高而肾脏血管阻力值和过滤分数明显降低,肾小球滤过率偏低,但是差异不显著;中国蜂胶能有效改善 T2DM 大鼠肾脏血流动力学异常现象,升高肾脏血流量,并降低肾脏血管阻力值和过滤分数,巴西蜂胶也呈现恢复各项指标的趋势,但是效果不明显。这表明蜂胶能帮助恢复肾脏血流动力学异常现象,延缓糖尿病发展。

对于蜂胶改善肾脏血流动力学异常作用的途径尚不明确,巴西蜂胶及其黄酮类提取物能改善自发性高血压大鼠主动脉压(Mishima et al.,2005;Maruyama et al.,2009),但是能否改善糖尿病大鼠肾脏动脉压效果不明确。NO 具有血管舒张活性,在糖尿病发展早期肾脏内 NO 含量增加,促进肾小球高过滤及蛋白尿产生(Prabhakar,2004),蜂胶能降低肾组织内 NO 和 NOS 含量,这可能有助于帮助恢复肾脏血流动力学。此外,有报道指出抑制 T1DM 大鼠血液中 COX-2 含量能显著恢复其 GFR 值。蜂胶中活性成分白杨素和 CAPE 分别能抑制 COX-2 在巨噬细胞和胃上皮细胞中的表达(Abdel-Latif et al.,2005;Woo et al.,2005),这可能是蜂胶的另外一条作用途径。

3.3 蜂胶改善 T2DM 大鼠肾脏 PKC mRNA 表达的效果

在糖尿病环境下,高血糖由葡萄糖转运蛋白-1(GLUT1,非胰岛素依赖性)转运进入肾细胞;在肾细胞内高血糖通过刺激细胞氧化应激、糖酵解获得 DAG 及 AGEs 生成,激活各类

PKC 同工酶(Meier et al.,2009)。PKC 主要通过以下 4 条途径促进糖尿病发展(Meier et al.,2009):① 激活 NADPH,产生更多 ROS;② 促进 TGF-β_1 表达,促进细胞外基质扩张,增加肾小球基底膜厚度,最终导致组织纤维化;③ 通过作用于血管紧张素、PGE_2/PGI_2、NO 和内皮激素-1(ET-1),引发肾脏血流动力学异常,改变肾血流量和 GFR;④ 促进血管内皮生长因子(VEGF)产生,增加血管通透性和蛋白尿流出量。PKC 作为糖尿病并发症的治疗靶位引起广泛的兴趣,PKC β 特异性抑制剂 LY333531 能阻止 T2DM 模型 *db/db* 小鼠渐进性肾小球基质增厚(Koya et al.,2000);利用链佐星注射转基因(mRen 2)27 大鼠后得到糖尿病模型,LY333531 能抑制该模型大鼠肾脏肾小球硬化和肾管间质纤维化,降低肾脏 TGF-β_1 表达以及大鼠蛋白尿量,但是对于高血糖和高血压无改善效果(Kelly et al.,2003);临床应用中均发现,阻止 T2DM 患者服用 LY333531 后能明显减少尿液白蛋白量,恢复血清肌酐含量(Tuttle et al.,2005),降低尿液中 TGF-β_1 含量(Gilbert et al.,2007)。

本实验中我们发现,与模型组相比,正常组 PKC 表达明显偏低,而中国蜂胶和巴西蜂胶则能有效抑制 T2DM 大鼠肾脏 PKC mRNA 表达。这表明中国蜂胶和巴西蜂胶能通过抑制肾脏 PKC 表达而减缓糖尿病发展。有研究指出高血糖能刺激肾脏细胞产生大量的 ROS 而激活 PKC,补充抗氧化剂能阻断 PKC 激活(Ha et al.,2008)。在本实验中,我们发现蜂胶能改善糖尿病大鼠肾脏组织氧化应激,这可能是蜂胶抑制 PKC mRNA 表达的作用途径之一。

3.4 蜂胶改善 T2DM 大鼠肾脏 TGF-β_1 的效果

TGF-β 可以刺激肾脏细胞外基质表达增强,并通过抑制细胞外基质降解蛋白酶合成、激活蛋白酶抑制剂如纤溶酶原活化抑制剂-1(PAI-1)而减少细胞外基质降解,并通过上调整联蛋白而增强细胞基质作用,其过度表达被认为是糖尿病肾脏肥大及细胞外基质扩增最为关键的因素(Ziyadeh,2004)。

在糖尿病/高血糖环境下,多条途径能激活 TGF-β_1 表达。高血糖能通过激活 PKC 途径而促进糖尿病模型大鼠肾脏 TGF-β_1 表达,而 PKC 抑制剂能阻止这个过程(Kelly et al.,2003);高血糖刺激产生的 ROS 能促进人系膜细胞中 TGF-β_1 和及 ECM 扩增(Iglesias-De La Cruz et al.,2001),并能通过激活人系膜细胞和糖尿病大鼠肾小球中 TGF-β_1 增加 PAI-1 表达,而抗氧化剂牛磺酸能缓减这个过程(Lee et al.,2005);高血糖能刺激肾组织表达 12/15 脂氧化酶,并级联激活 12(S)-HETE,HETE 则能刺激系膜细胞中 TGF-β_1 表达(Kim et al.,2005);高血糖刺激下的已糖胺积累能通过激活 PKC 途径和 p38 MAPK 途径,增加人系膜细胞中 TGF-β_1 表达(Burt et al.,2003);高血糖能激活 ERK 和 p38 MAPK 而增加系膜细胞和肾小球中 TGF-β_1 表达(Kang et al.,2001;Toyoda et al.,2004),这类激活可能是通过 PKC 或者 ROS 级联激活实现(Wilmer et al.,2001;Toyoda et al.,2004)。

本实验中我们发现,从图像分析系统分析,模型组大鼠肾脏 TGF-β_1 表达明显高于正常组大鼠,蜂胶和阳性药能降低大鼠肾脏 TGF-β_1 表达总吸光度、平均吸光度和阳性指数,其中巴西蜂胶和阳性药效果显著,而中国蜂胶效果不显著,巴西蜂胶效果优于中国蜂胶而与阳性对照组相当。这些结果表明,蜂胶能改善 T2DM 大鼠肾脏 TGF-β_1 表达,巴西蜂胶效果优于中国蜂胶而与阳性药相当。

对于蜂胶改善肾脏 TGF-β_1 表达的作用机制尚不明确,从 TGF-β_1 激活途径分析,蜂胶降血糖、改善肾脏组织氧化应激而减少 PKC 和 ROS 的效果,可能是蜂胶发挥作用的途径。

蜂胶及其各类活性能通过抑制单核细胞和 T 淋巴细胞中 ERK/MAPK 信号途径的调节免疫细胞分化及细胞因子分泌（Ansorge et al.，2003）。而 ERK/MAPK 途径激活促进肾脏细胞中 TGF-β_1 表达，这可能是蜂胶的另外一条作用途径。糖尿病大鼠肾脏巨噬细胞浸润被认为是 TGF-β 表达增加的另外一个原因（Okada et al.，2003）。与普通糖尿病模型小鼠相比，ICAM-1 基因敲除糖尿病小鼠肾脏 TGF-β_1、胶原蛋白 Ⅳ 表达显著下降，同时伴有 ECM 和蛋白尿降低（Okada et al.，2003）。蜂胶通过抗炎活性减少炎症细胞浸润，这可能是蜂胶降低 T2DM 大鼠肾脏 TGF-β_1 表达的另外一种作用途径。

小 结

本研究通过分析 T2DM 大鼠肾脏炎症反应、血流动力学、PKC 及 TGF-β_1 的表达，对蜂胶保护肾脏组织的作用机制进行了探讨，结果表明：(1)中国蜂胶和巴西蜂胶均能降低 T2DM 大鼠血清 CRP 含量，巴西蜂胶效果显著，这表明蜂胶能改善 T2DM 大鼠炎症现象；(2)中国蜂胶和巴西蜂胶均能抑制 T2DM 大鼠肾脏炎症，中国蜂胶能有效降低糖尿病大鼠肾脏中 IL-2、IL-6 和 TNF-α 含量，巴西蜂胶能有效抑制糖尿病大鼠肾脏中 TNF-α 含量增加及 MCP-1 mRNA 表达；(3)中国蜂胶能明显改善 T2DM 大鼠肾脏血流动力学异常现象，恢复肾脏血流量，并降低肾脏血管阻力值和过滤分数，巴西蜂胶效果不明显；(4)中国蜂胶和巴西蜂胶能明显抑制 T2DM 大鼠肾脏 PKC mRNA 表达；(5)免疫组化染色和图像分析结果显示，模型组大鼠 TGF-β_1 表达明显高于正常组。中国蜂胶和巴西蜂胶均能改善 T2DM 大鼠肾脏病理变化，巴西蜂胶效果优于或相当于阳性对照。

参考文献

Abd El-Hady F K，Hegazi AG，Wollenweber E（2007）Effect of Egyptian propolis on the susceptibility of LDL to oxidative modification and its antiviral activity with special emphasis on chemical composition[J]. *Zeitschrift für Naturforschung C*，62(9-10)：645-655.

Abdel-Latif MM，Windle HJ，Homasany BS，Sabra K，Kelleher D（2005）Caffeic acid phenethyl ester modulates *Helicobacter* pylori-induced nuclear factor-kappa B and activator protein-1 expression in gastric epithelial cells[J]. *British Journal of Pharmacology*，146(8)：1139-1147.

Alyane M，Kebsa LB，Boussenane H，Rouibah H，Lahouel M（2008）Cardioprotective effects and mechanism of action of polyphenols extracted from propolis against doxorubicin toxicity [J]. *Pakistan Journal of Pharmaceutical Sciences*，21(3)：201-209.

Anderson S，Vora JP（1995）Current concepts of renal hemodynamics in diabetes[J]. *Journal of Diabetes and its Complications*，9(4)：304-307.

Ansorge S，Reinhold D，Lendeckel U（2003）Propolis and some of its constituents down-regulate DNA synthesis and inflammatory cytokine production but induce TGF-beta1 production of human immune cells[J]. *Zeitschrift für Naturforschung C*，58(7-8)：580-589.

Aslan A, Temiz M, Atik E, Polat G, Sahinler N, Besirov E, Aban N, Parsak CK (2007) Effectiveness of mesalamine and propolis in experimental colitis[J]. *Advances in Therapy*, 24(5):1085-1097.

Ates B, Dogru MI, Gul M, Erdogan A, Dogru AK, Yilmaz I, Yurekli M, Esrefoglu M (2006) Protective role of caffeic acid phenethyl ester in the liver of rats exposed to cold stress [J]. *Fundamental Clinical Pharmacology*, 20(3):283-289.

Bankova V (2005) Chemical diversity of propolis and the problem of standardization[J]. *Journal of Ethnopharmacology*, 100(1-2):114-117.

Bankova V, De Castro SL, Marcucci MC (2000) Propolis: recent advances in research on chemistry and plant origin[J]. *Apidologie*, 31:3-15.

Banskota AH, Tezuka Y, Kadota S (2001) Recent progress in pharmacological research of propolis[J]. *Phytotherapy Research*, 15(7):561-571.

Brownlee M (2001) Biochemistry and molecular cell biology of diabetic complications[J]. *Nature*, 414(6865):813-820.

Brownlee M (2005) The pathobiology of diabetic complications: a unifying mechanism[J]. *Diabetes*, 54(6):1615-1625.

Burt DJ, Gruden G, Thomas SM, Tutt P, Dell'Anna C, Viberti GC, Gnudi L (2003) P38 mitogen-activated protein kinase mediates hexosamine-induced TGF beta1 mRNA expression in human mesangial cells[J]. *Diabetologia*, 46(4):531-537.

Charles MA, Selam JL (2005) Cyclic relationships between diabetic nephropathy and cardiovascular risk factors[J]. *Metabolic Syndrome and Related Disorders*, 3(3):203-212.

Choudhary N, Ahlawat RS (2008) Interleukin-6 and C-reactive protein in pathogenesis of diabetic nephropathy: new evidence linking inflammation, glycemic control, and microalbuminuria [J]. *Iranian Journal of Kidney Diseases*, 2(2):72-79.

Cirino G, Distrutti E, Wallace JL (2006) Nitric oxide and inflammation[J]. *Inflammation Allergy-Drug Targets*, 5(2):115-119.

Coskun O, Kanter M, Korkmaz A, Oter S (2005) Quercetin, a flavonoid antioxidant, prevents and protects streptozotocin-induced oxidative stress and beta-cell damage in rat pancreas[J]. *Pharmacological Research*, 51(2):117-123.

Dalla Vestra M, Mussap M, Gallina P, Bruseghin M, Cernigoi AM, Saller A, Plebani M, Fioretto P (2005) Acute-phase markers of inflammation and glomerular structure in patients with type 2 diabetes[J]. *Journal of the American Society of Nephrology*, 16(Suppl 1):78-82.

El-Sayed el-SM, Abo-Salem OM, Aly HA, Mansour AM (2009) Potential antidiabetic and hypolipidemic effects of propolis extract in streptozotocin-induced diabetic rats[J]. *Pakistan Journal of Pharmaceutical Sciences*, 22(2):168-174.

Erdmann E (2005) Diabetes and cardiovascular risk markers[J]. *Current Medical Research and Opinion*, 21(Suppl 1):21-28.

Estacio RO (2009) Renin-angiotensin-aldosterone system blockade in diabetes: role of direct renin inhibitors[J]. *Postgraduate Medical Journal*, 121(3):33-44.

Fenercioglu AK, Saler T, Genc E, Sabuncu H, Altuntas Y (2009) The effects of polyphenol containing antioxidants on oxidative stress and lipid peroxidation in type 2 diabetes mellitus without complications[J]. *Journal of Endocrinological Investigation*, 33(2): 118-124.

Forbes JM, Fukami K, Cooper ME (2007) Diabetic nephropathy: where hemodynamics meets metabolism[J]. *Experimental and Clinical Endocrinology Diabetes*, 115(2):69-84.

Hu FL, Hepburn HR, Xuan HZ, Chen ML, Daya S, Radloff SE (2005) Effects of propolis on blood glucose, blood lipid and free radicals in rats with diabetes mellitus[J]. *Pharmacological Research*, 51(2):147-152.

Garcia-Compean D, Jaquez-Quintana JO, Gonzalez-Gonzalez JA, Maldonado-Garza H (2009) Liver cirrhosis and diabetes: risk factors, pathophysiology, clinical implications and management[J]. *World Journal of Gastroenterology*, 15(3):280-288.

Gilbert RE, Kim SA, Tuttle KR, George L, Bakris GL, Toto RD, McGill JB, Haney DJ, Kelly DJ, Anderson PW (2007) Effect of ruboxistaurin on urinary transforming growth factor-beta in patients with diabetic nephropathy and type 2 diabetes[J]. *Diabetes Care*, 30(4):995-996.

Ginsberg HN (2006) REVIEW: Efficacy and mechanisms of action of statins in the treatment of diabetic dyslipidemia[J]. *The Journal of Clinical Endocrinology Metabolism*, 91(2): 383-392.

Girach A, Manner D, Porta M (2006) Diabetic microvascular complications: can patients at risk be identified? A review[J]. *International Journal of Clinical Practice*, 60(11): 1471-1483.

Graves PM, Eisenbarth GS (1999) Pathogenesis, prediction and trials for the prevention of insulin-dependent (type 1) diabetes mellitus[J]. *Advanced Drug Delivery Reviews*, 35 (2-3):143-156.

Ha H, Hwang IA, Park JH, Lee HB (2008) Role of reactive oxygen species in the pathogenesis of diabetic nephropathy[J]. *Diabetes Research and Clinical Practice*, 82 (Suppl 1):42-45.

Ha H, Lee HB (2001) Oxidative stress in diabetic nephropathy: basic and clinical information[J]. *Current Diabetes Reports*, 1(3):282-287.

Ha H, Yu MR, Choi YJ, Kitamura M, Lee HB (2002) Role of high glucose-induced nuclear factor-kappa B activation in monocyte chemoattractant protein-1 expression by mesangial cells[J]. *Journal of the American Society of Nephrology*, 13(4):894-902.

Hishikawa K, Nakaki T, Fujita T (2005) Oral flavonoid supplementation attenuates atherosclerosis development in apolipoprotein E-deficient mice[J]. *Arteriosclerosis, Thrombosis, and Vascular Biology*, 25(2):442-446.

Hosnuter M, Gurel A, Babuccu O, Armutcu F, Kargi E, Işikdemir A (2004) The effect of CAPE on lipid peroxidation and nitric oxide levels in the plasma of rats following thermal injury[J]. *Burns*, 30(2):121-125.

Hostetter TH (2003) Hyperfiltration and glomerulosclerosis[J]. *Seminars in Nephrology*, 23(2):194-199.

Huang S, Zhang CP, Wang K, Li GQ, Hu FL (2014) Recent advances in the chemical composition of propolis[J]. *Molecules*, 19:19610-19632.

Iglesias-De La Cruz MC, Ruiz-Torres P, Alcami J, Díez-Marqués L, Ortega-Velázquez R, Chen S, Rodríguez-Puyol M, Ziyadeh FN, Rodríguez Puyol D (2001) Hydrogen peroxide increases extracellular matrix mRNA through TGF-beta in human mesangial cells[J]. *Kidney International*, 59(1):87-95.

Ilhan A, Akyol O, Gurel A, Armutcu F, Iraz M, Oztas E (2004) Protective effects of caffeic acid phenethyl ester against experimental allergic encephalomyelitis-induced oxidative stress in rats[J]. *Free Radical Biology Medicine*, 37(3):386-394.

Jain KS, Kathiravan MK, Somani RS, Shishoo CJ (2007) The biology and chemistry of hyperlipidemia[J]. *Bioorganic Medicinal Chemistry*, 15(14):4674-4699.

Jasprica I, Mornar A, Debeljak Z, Smólcic-Bubalo A, Medic-Saric M, Mayer L, Romic Z, Búcan K, Balog T, Sobocanec S, Sverko V (2007) *In vivo* study of propolis supplementation effects on antioxidative status and red blood cells[J]. *Journal of Ethnopharmacology*, 110(3):548-554.

Johansen JS, Harris AK, Rychly DJ, Ergul A (2005) Oxidative stress and the use of antioxidants in diabetes: linking basic science to clinical practice[J]. *Cardiovascular Diabetology*, 4(1):5.

Jung WK, Lee DY, Choi YH, Yea SS, Choi I, Park SG, Seo SK, Lee SW, Lee CM, Kim SK, Jeon YJ, Choi IW (2008) Caffeic acid phenethyl ester attenuates allergic airway inflammation and hyperresponsiveness in murine model of ovalbumin-induced asthma[J]. *Life Science*, 82(13-14):797-805.

Juntti-Berggren L, Pigon J, Hellstrom P, Holst JJ, Efendic S (2000) Influence of acarbose on post-prandial insulin requirements in patients with type 1 diabetes[J]. *Diabetes, Nutrition Metabolism*, 13(1):7-12.

Kang S, Adler S, Lapage J, Natarajan R (2001) p38 MAPK and MAPK kinase 3/6 mRNA and activities are increased in early diabetic glomeruli[J]. *Kidney International*, 60(2):543-552.

Kelly DJ, Zhang Y, Hepper C, Gow RM, Jaworski K, Kemp BE, Wilkinson-Berka JL, Gilbert RE (2003) Protein kinase C beta inhibition attenuates the progression of experimental diabetic nephropathy in the presence of continued hypertension[J]. *Diabetes*, 52(2):512-518.

Kanwar YS, Wada J, Sun L, Xie P, Wallner EI, Chen S, Chugh S, Danesh FR (2008) Diabetic nephropathy: mechanisms of renal disease progression[J]. *Experimental*

Biology and Medicine (*Maywood*), 2008, 233(1):4-11.

Khalil ML (2006) Biological activity of bee propolis in health and disease [J]. *Asian Pacific Journal of Cancer Prevention*, 7(1):22-31.

Khayyal MT, el-Ghazaly MA, el-Khatib AS, Hatem AM, de Vries PJ, el-Shafei S, Khattab MM (2003) A clinical pharmacological study of the potential beneficial effects of a propolis food product as an adjuvant in asthmatic patients [J]. *Fundamental Clinical Pharmacology*, 17(1):93-102.

Kim Y, Xu Z, Reddy M, Li SL, Lanting L, Sharma K, Adler SG, Natarajan R (2005) Novel interactions between TGF-{beta} 1 actions and the 12/15-lipoxygenase pathway in mesangial cells [J]. *Journal of the American Society of Nephrology*, 16(2):352-362.

Koya D, Haneda M, Nakagawa H, Isshiki K, Sato H, Maeda S, Sugimoto T, Yasuda H, Kashiwagi A, Ways DK, King GL, Kikkawa R (2000) Amelioration of accelerated diabetic mesangial expansion by treatment with a PKC beta inhibitor in diabetic db/db mice, a rodent model for type 2 diabetes [J]. *Federation of American Societies for Experimental Biology* (*FASEB J*), 14(3):439-447.

Larsen ML, Horder M, Mogensen EF (1990) Effect of long-term monitoring of glycosylated hemoglobin levels in insulin-dependent diabetes mellitus [J]. *New England Journal of Medicine*, 323(15):1021-1025.

Leahy JL (2005) Pathogenesis of type 2 diabetes mellitus [J]. *Archives of Medical Research*, 36(3):197-209.

Lee EA, Seo JY, Jiang Z, Yu MR, Kwon MK, Ha H, Lee HB (2005) Reactive oxygen species mediate high glucose-induced plasminogen activator inhibitor-1 up-regulation in mesangial cells and in diabetic kidney [J]. *Kidney International*, 67(5):1762-1771.

Leehey D. J., Singh A. K., Alavi N., Singh R (2000) Role of angiotensin II in diabetic nephropathy [J]. *Kidney International*, *Suppl*, 77:S93-98.

Liu CF, Lin CH, Lin CC, Lin YH, Chen CF, Lin CK, Lin SC (2004) Antioxidative natural product protect against econazole-induced liver injuries [J]. *Toxicology*, 196(1-2):87-93.

Maric C (2008) Vasoactive hormones and the diabetic kidney [J]. *Scientific World Journal*, 8:470-485.

Maruyama H, Sumitou Y, Sakamoto T, Araki Y, Hara H (2009) Antihypertensive effects of flavonoids isolated from brazilian green propolis in spontaneously hypertensive rats [J]. *Biological and Pharmaceutical Bulletin*, 32(7):1244-1250.

Matavelli LC, Huang J, Siragy HM (2010) (Pro)renin receptor contributes to diabetic nephropathy through enhancing renal inflammation [J]. *Clinical and Experimental Pharmacology and Physiology*, 37(3):277-282.

Matsui T, Ebuchi S, Fujise T, Abesundara KJ, Doi S, Yamada H, Matsumoto K (2004) Strong antihyperglycemic effects of water-soluble fraction of Brazilian propolis and its bioactive constituent, 3, 4, 5-tri-O-caffeoylquinic acid [J]. *Biological and Pharmaceutical*

Bulletin, 27(11): 1797-1803.

Matsuura E, Hughes GR, Khamashta MA (2008) Oxidation of LDL and its clinical implication [J]. *Autoimmunity Reviews*, 7(7): 558-566.

McGarry JD (2002) Banting lecture 2001: dysregulation of fatty acid metabolism in the etiology of type 2 diabetes[J]. *Diabetes*, 51(1): 7-18.

Meier M, Menne J, Haller H (2009) Targeting the protein kinase C family in the diabetic kidney: lessons from analysis of mutant mice[J]. *Diabetologia*, 52(5): 765-775.

Mezzano S, Aros C, Droguett A, Burgos ME, Ardiles L, Flores C, Schneider H, Ruiz-Ortega M, Egido J (2004) NF-kappaB activation and overexpression of regulated genes in human diabetic nephropathy[J]. *Nephrology Dialysis Transplantation*, 19(10): 2505-2512.

Mezzano S, Droguett A, Burgos ME, Ardiles LG, Flores CA, Aros CA, Caorsi I, Vio CP, Ruiz-Ortega M, Egido J (2003) Renin-angiotensin system activation and interstitial inflammation in human diabetic nephropathy[J]. *Kidney International Suppl*, 86: S64-70.

Mishima S, Yoshida C, Akino S, Sakamoto T (2005) Antihypertensive effects of Brazilian propolis: identification of caffeoylquinic acids as constituents involved in the hypotension in spontaneously hypertensive rats[J]. *Biological and Pharmaceutical Bulletin*, 28(10): 1909-1914.

Mora C, Navarro JF (2006) Inflammation and diabetic nephropathy[J]. *Current Diabetes Reports*, 6(6): 463-468.

Morii T, Fujita H, Narita T, Koshimura J, Shimotomai T, Fujishima H, Yoshioka N, Imai H, Kakei M, Ito S (2003) Increased urinary excretion of monocyte chemoattractant protein-1 in proteinuric renal diseases[J]. *Renal Failure*, 25(3): 439-444.

Murata K, Yatsunami K, Fukuda E, Onodera S, Mizukami O, Hoshino G, Kamei T (2004) Antihyperglycemic effects of propolis mixed with mulberry leaf extract on patients with type 2 diabetes[J]. *Alternative Therapies In Health And Medicine*, 10(3): 78-79.

Navarro JF, Milena FJ, Mora C, León C, García J (2006) Renal pro-inflammatory cytokine gene expression in diabetic nephropathy: effect of angiotensin-converting enzyme inhibition and pentoxifylline administration[J]. *American Journal of Nephrology*, 26(6): 562-570.

Navarro JF, Mora-Fernandez C (2006) The role of TNF-alpha in diabetic nephropathy: pathogenic and therapeutic implications[J]. *Cytokine Growth Factor Reviews*, 17(6): 441-450.

Neuser D, Benson A, Bruckner A, Goldberg RB, Hoogwerf BJ, Petzinna D (2005) Safety and tolerability of acarbose in the treatment of type 1 and type 2 diabetes mellitus[J]. *Clinical Drug Investigation*, 25(9): 579-587.

Newairy AS, Salama AF, Hussien HM, Yousef MI (2009) Propolis alleviates aluminium-induced lipid peroxidation and biochemical parameters in male rats[J]. *Food and Chemical Toxicology*, 47(6):1093-1098.

Noronha I, Fujihara C, Zatz R (2002) The inflammatory component in progressive renal disease—are interventions possible? [J]. *Nephrology, Dialysis, Transplantation*, 17(3):363-368.

Novo E, Parola M (2008) Redox mechanisms in hepatic chronic wound healing and fibrogenesis[J]. *Fibrogenesis & Tissue Repair*, 1(1):5.

O'Brien T, Nguyen TT, Zimmerman BR (1998) Hyperlipidemia and diabetes mellitus[J]. *Mayo Clinic Proceedings*, 73(10):969-976.

Okada S, Shikata K, Matsuda M, Ogawa D, Usui H, Kido Y, Nagase R, Wada J, Shikata Y, Makino H (2003) Intercellular adhesion molecule-1-deficient mice are resistant against renal injury after induction of diabetes[J]. *Diabetes*, 52(10):2586-2593.

Østergaard J, Hansen TK, Thiel S, Flyvbjerg A (2005) Complement activation and diabetic vascular complications[J]. *Clinica Chimica Acta*, 361(1-2):10-19.

Parving HH (2001) Diabetic nephropathy: prevention and treatment[J]. *Kidney International*, 60(5):2041-2055.

Paulino N, Teixeira C, Martins R, Scremin A, Dirsch VM, Vollmar AM, Abreu SR, de Castro SL, Marcucci MC (2006) Evaluation of the analgesic and anti-inflammatory effects of a Brazilian green propolis[J]. *Planta Medica*, 72(10):899-906.

Prabhakar SS (2004) Role of nitric oxide in diabetic nephropathy[J]. *Seminars in Nephrology*, 24(4):333-344.

Rahimi R, Nikfar S, Larijani B, Abdollahi M (2005) A review on the role of antioxidants in the management of diabetes and its complications[J]. *Biomedicine Pharmacotherapy*, 59(7):365-373.

Rees DA, Alcolado JC (2005) Animal models of diabetes mellitus[J]. *Diabetic Medicine*, 22(4):359-370.

Roden M, Bernroider E (2003) Hepatic glucose metabolism in humans—its role in health and disease[J]. *Best Practice Research. Clinical Endocrinology Metabolism*, 17(3):365-383.

Ruggenenti P, Schieppati A, Remuzzi G (2001) Progression, remission, regression of chronic renal diseases[J]. *The Lancet*, 357(9268):1601-1608.

Ruster C, Wolf G (2008) The role of chemokines and chemokine receptors in diabetic nephropathy[J]. *Frontiers in Bioscience*, 13:944-955.

Schnedl WJ, Ferber S, Johnson JH, Newgard CB (1994) STZ transport and cytotoxicity. Specific enhancement in GLUT2-expressing cells[J]. *Diabetes*, 43(11):1326-1333.

Seo KW, Park M, Song YJ, Kim SJ, Yoon KR (2003) The protective effects of propolis on hepatic injury and its mechanism[J]. *Phytotherapy Research*, 17(3):250-253.

Shin EK, Kwon HS, Kim YH (2009) Chrysin, a natural flavone, improves murine inflammatory bowel diseases[J]. *Biochemical and Biophysical Research Communications*, 381(4):

502-507.

Song YS, Park EH, Hur GM, Ryu YS, Kim YM, Jin C (2002) Ethanol extract of propolis inhibits nitric oxide synthase gene expression and enzyme activity[J]. *Journal of Ethnopharmacology*, 80(2-3): 155-161.

Sugimoto H, Shikata K, Wada J, Horiuchi S, Makino H (1999) Advanced glycation end products-cytokine-nitric oxide sequence pathway in the development of diabetic nephropathy: aminoguanidine ameliorates the overexpression of tumour necrosis factor-alpha and inducible nitric oxide synthase in diabetic rat glomeruli[J]. *Diabetologia*, 42(7): 878-886.

Sugimoto R, Enjoji M, Kohjima M, Tsuruta S, Fukushima M, Iwao M, Sonta T, Kotoh K, Inoguchi T, Nakamuta M (2005) High glucose stimulates hepatic stellate cells to proliferate and to produce collagen through free radical production and activation of mitogen-activated protein kinase[J]. *Liver International*, 25(5): 1018-1026.

Suzen S, Buyukbingol E (2003) Recent studies of aldose reductase enzyme inhibition for diabetic complications[J]. *Current Medicinal Chemistry*, 10(15): 1329-1352.

Tak PP, Firestein GS (2001) NF-kappaB: a key role in inflammatory diseases[J]. *Journal of Clinical Investigation*, 107(1): 7-11.

The Advance Collaborative Group (2008) Intensive blood glucose control and vascular outcomes in patients with type 2 diabetes[J]. *New England Journal of Medicine*, 358(24): 2560-2572.

The Diabetes Control and Complications Trial Research Group (1993) The effect of intensive treatment of diabetes on the development and progression of long-term complications in insulin-dependent diabetes mellitus[J]. *New England Journal of Medicine*, 329(14): 977-986.

The UK Prospective Diabetes Study (UKPDS) Group (1998a) Intensive blood-glucose control with sulphonylureas or insulin compared with conventional treatment and risk of complications in patients with type 2 diabetes (UKPDS 33)[J]. *The Lancet*, 352(9131): 837-853.

The UK Prospective Diabetes Study (UKPDS) Group (1998b) Tight blood pressure control and risk of macrovascular and microvascular complications in type 2 diabetes: UKPDS 38[J]. *British Medical Journal (BMJ)*, 317(7160): 703-713.

Toyoda M, Suzuki D, Honma M, Uehara G, Sakai T, Umezono T, Sakai H (2004) High expression of PKC-MAPK pathway mRNAs correlates with glomerular lesions in human diabetic nephropathy[J]. *Kidney International*, 66(3): 1107-1114.

Tuttle KR, Bakris GL, Toto RD, McGill JB, Hu K, Anderson PW (2005) The effect of ruboxistaurin on nephropathy in type 2 diabetes[J]. *Diabetes Care*, 28(11): 2686-2690.

Wang K, Zhang JL, Ping S, Ma QX, Chen X, Xuan HZ, Shi JH, Zhang CP, Hu FL (2014) Anti-inflammatory effects of ethanol extracts of Chinese propolis and buds from poplar (*Populus×canadensis*)[J]. *Journal of Ethnopharmacology*, 155: 300-311.

Valko M, Leibfritz D, Moncol J, Cronin MT, Mazur M, Telser J (2007) Free radicals and antioxidants in normal physiological functions and human disease[J]. *International Journal of Biochemistry and Cell Biology*, 39(1): 44-84.

WHO (2008) Fact sheet N°312, Diabetes. from http://www.who.int/mediacentre/factsheets/fs312/en/index.html.

Wilmer WA, Dixon CL, Hebert C (2001) Chronic exposure of human mesangial cells to high glucose environments activates the p38 MAPK pathway[J]. *Kidney International*, 60(3): 858-871.

Wolf G. New insights into the pathophysiology of diabetic nephropathy: from haemodynamics to molecular pathology[J]. *Eur J Clin Invest*, 2004, 34(12): 785-796.

Wolf G, Butzmann U, Wenzel UO (2003) The renin-angiotensin system and progression of renal disease: from hemodynamics to cell biology[J]. *Nephron Physiology*, 93(1): 3-13.

Woo KJ, Jeong YJ, Inoue H, Park JW, Kwon TK (2005) Chrysin suppresses lipopolysaccharide-induced cyclooxygenase-2 expression through the inhibition of nuclear factor for IL-6 (NF-IL6) DNA-binding activity[J]. *FEBS Lett*, 579(3): 705-711.

Zamami Y, Takatori S, Koyama T, Goda M, Iwatani Y, Doi S, Kawasaki H (2007) Effect of propolis on insulin resistance in fructose-drinking rats[J]. *Yakugaku Zasshi*, 127(12): 2065-2073.

Zhang M, Lv XY, Li J, Xu ZG, Chen L (2008) The characterization of high-fat diet and multiple low-dose streptozotocin induced type 2 diabetes rat model[J]. *Experimental Diabetes Research*, 2008(704045): 1-9.

Zhao JQ, Wen YF, Bhadauria M, Nirala SK, Sharma A, Shrivastava S, Shukla S, Agrawal OP, Mathur R (2009) Protective effects of propolis on inorganic mercury induced oxidative stress in mice[J]. *Indian Journal of Experimental Biology*, 47(4): 264-269.

Ziyadeh FN (2004) Mediators of diabetic renal disease: the case for tgf-Beta as the major mediator[J]. *Journal of the American Society of Nephrology*, 15 (Suppl 1): S55-57.

陈灏珠, 林果为, 王吉耀 (2013) 实用内科学[M]. 第14版. 北京: 人民卫生出版社.

楚勤英, 唐志雄, 王进伟, 傅小玲, 蒋虹, 贾军宏 (2008) 蜂胶治疗2型糖尿病ⅠⅡ级糖尿病足的临床观察[J]. 河北医学, 14(3): 263-265.

高海琳, 刘富海 (2000) 蜂胶在糖尿病综合治疗中的作用[J]. 北京医学, 22(2): 115-116.

金光香, 宫海民, 段文卓, 王红艳 (2005) 蜂贝化瘀胶囊对糖尿病大鼠糖脂代谢的调整及微血管并发症的防治[J]. 中国临床康复, 9(027): 96-98.

刘英华, 薛长勇, 欧阳红, 张荣欣, 张永 (2006) 茶多酚对大鼠晶状体醛糖还原酶的抑制作用[J]. 实用预防医学, 13(5): 1162-1164.

玄红专 (2003) 蜂胶对糖尿病的作用及其机制的研究[D]. 杭州: 浙江大学.

闫泉香 (2004) 黄酮类醛糖还原酶抑制剂的活性研究[J]. 中药药理与临床, 20(2): 9-11.

第九章 蜂胶对血管内皮细胞的作用及其机制研究

国内外大量的研究及我们团队前期动物实验结果表明,蜂胶具有降血糖(玄红专,2013;朱威,2011)、降血脂(詹耀锋,2014)、抗炎症(李英华,2002)等生物学活性,但是作用机制还不完全清楚。血管内皮细胞对维持血管正常功能非常重要,内皮细胞损伤或功能异常引起多种炎症和退行性疾病,如糖尿病、动脉粥样硬化、高血压、心肌缺血和肿瘤等。因此,调控血管内皮细胞凋亡,维持其正常的功能引起研究者的重视。目前关于蜂胶对血管内皮细胞的作用及其研究机制还较少。因此,本研究通过在体外建立不同的血管内皮细胞损伤模型,即去除血清和生长因子诱导血管内皮细胞凋亡、脂多糖(LPS)诱导血管内皮细胞损伤以及氧化性低密度脂蛋白(ox-LDL)诱导血管内皮细胞损伤,研究蜂胶影响血管内皮细胞的作用及其可能的机制,同时为研究蜂胶辅助治疗糖尿病、心血管疾病以及炎症提供理论依据。

第一节 去除血清和生长因子条件下蜂胶对血管内皮细胞的影响

1 去除血清和生长因子条件下巴西蜂胶对血管内皮细胞的影响

巴西蜂胶的种类繁多,产业化开发的主要有巴西绿蜂胶和巴西红蜂胶。巴西绿蜂胶(Brazilian green propolis),又称酒神菊属型蜂胶(Baccharis-type propolis),是蜜蜂从巴西东南部地区酒神菊属植物(*Baccharis*)上采集的树脂经加工而成的一种蜂胶类型,富含香豆酸衍生物,其中阿替比林C(Artipillin C)是其特有的活性成分(Daugsch et al.,2008)。巴西红蜂胶是蜜蜂从巴西东北部沿海地区紫檀属植物[*Dalbergia ecastophyllum*(L)Taub.(Leguminosae)]采集的红色树脂,经加工而成的一种红色树胶,富含异黄酮类物质。目前,关于巴西红蜂胶的研究还较少(Awale et al.,2008;Li et al.,2008)。由于地域差异,巴西蜂胶的化学成分与其他地区的蜂胶化学成分相差很大,因而生物学活性也存在一定的差异,如巴西绿蜂胶具有特殊的抗龋齿(Salomao et al.,2009)、杀锥虫(Ayres et al.,2007;Machado et al.,2007)、抗溃疡(Chen et al.,2009;de Barros et al.,2007;Lemos et al.,2007)、抗氧化(Fonseca et al.,2011)、免疫调节(Missima and Sforcin,2008;Moura et al.,2009)、神经保护(Messerli et al.,2009;Nakajima et al.,2009)、抗肿瘤(Bufalo et al.,2009;Ribeiro et al.,2006)以及抗微生物(Salomao et al.,2008)等活性。

Kunimasa等(2009)研究发现,巴西绿胶通过抑制细胞外存活信号ERK1/2,活化Caspase3诱导血管内皮细胞凋亡。血管内皮细胞在血管正常发育中起重要作用,血管内皮细胞损伤或功能异常引起多种炎症和退行性疾病的发生,如糖尿病、动脉粥样硬化、高血压、败血症及肿瘤等(Heitzer et al.,2001;Higashi et al.,2009)。调控血管内皮细胞凋亡,维持其正常生理活动是防止血管疾病的重要方面。因此,影响血管内皮细胞凋亡的信号通路受到研究者的重视。在体外,Miao 等(1997a)研究发现膜整连蛋白 β4、PC-PLC、p53 和 ROS 在调控血管内皮凋亡方面发挥着重要作用。

我们团队前期的动物实验结果表明,巴西绿胶能够降低血糖和血脂,预防 1 型糖尿病和 2 型糖尿病的发生(Zhu et al.,2011),但巴西绿胶保护糖尿病的分子机制目前研究还不完全清楚。糖尿病的发生与内皮细胞损伤有关,而目前关于巴西绿蜂胶对血管内皮细胞的影响研究较少。因此,在本实验中,通过去除血清和生长因子诱导 HUVECs 凋亡,研究不同浓度的巴西蜂胶对 HUVECs 的影响,并进一步分析巴西绿蜂胶对膜整连蛋白 β4、PC-PLC、p53 和 ROS 以及线粒体膜电位的影响,为阐明巴西绿胶影响血管内皮细胞的分子机制提供新的实验依据。

1.1 材料和方法

1.1.1 主要试剂

M199 培养基和胎牛血清(美国 Hyclone 公司);碱性生长因子(中国 EssexBio 公司);L-α-卵磷脂、磺酰罗丹明 B(SRB)、DCHF 探针(美国 Sigma 公司);脂多糖(LPS)(Escherichia coli 055:B5)、JC-1 探针(美国 Invitrogen 公司);p53、β-actin(美国 Santa Cruz 公司);辣根过氧化物酶(HRP)标记的二抗(南京中杉金桥公司)。

1.1.2 主要设备与仪器

HPLC-MS 分析系统(德国 Agilent 公司);超净工作台(苏州净化设备厂);低温冰箱、CO_2 培养箱(日本 SANYO 公司);倒置相差显微镜(日本 Nikon 公司);酶标仪(美国 Perkin Elmer 公司);Mini-Protein Ⅱ 垂直板电泳仪(美国 BIO-RAD 公司);电泳转移槽(北京六一仪器厂);共聚焦显微镜(德国 Leica 公司)。

1.1.3 实验方法

(1)蜂胶样品的制备

巴西绿蜂胶,产自巴西 Minas Gerais 州,主要的植物来源是(*Baccharis dracunculifolia* DC.)。巴西绿蜂胶经冷冻、粉碎、无水乙醇浸提、减压过滤,滤液在 40℃旋转蒸发仪中浓缩至恒重,4℃保存。使用时溶解在无水乙醇中,分别配成不同浓度的巴西蜂胶乙醇溶液(EEBP)。制备的巴西绿蜂胶颜色为棕色。

(2)HPLC-MS 分析 EEBP 的化学组分

为了分析所使用的巴西蜂胶的化学组分,本实验采用 HPLC-MS 系统进行了检测。HPLC 系统:Agilent 1200(德国),C18 柱(2.1mm×150mm i.d.,5μm)。巴西蜂胶浓度,2mg/mL;进样体积,10μL。流动相包括:0.1% 甲酸(A)和 0.1% 乙腈(B),梯度洗脱 20%~80% B(0~60min),流速 0.4mL/min。UV 吸收光谱:195nm。MS 分析系统是 Agilent 6510(德国)。

(3)人脐静脉血管内皮细胞(HUVECs)培养

HUVECs 由山东大学发育与细胞生物学实验室馈赠。HUVECs 培养在添加 20% 胎牛血清和 70ng/mL 碱性生长因子的 M199 培养基中。孵育条件为:37℃,5% CO_2 饱和湿度培养箱中。每隔一天换液。倒置相差显微镜下观察细胞形态。

(4) 实验分组

待 HUVECs 长满后,将细胞分为3组:正常组、溶剂对照组和实验组。正常组的细胞在正常培养条件下培养;对照组细胞在去除血清和 FGF-2 的培养液中加入酒精溶剂培养,酒精浓度<0.1%(v/v),培养液中乙醇浓度<0.1%不会影响细胞的正常生长;实验组细胞在去除血清和 FGF-2 的同时用不同浓度(12.5、25 和 50 $\mu g/mL$)的巴西蜂胶处理。

1.1.4 检测指标

(1) 细胞存活率的测定

将长势良好的血管内皮细胞种植到96孔板,37℃、5% CO_2 孵育,至细胞长满。按照实验分组,分别处理细胞12h和24h后,弃去细胞培养液,用10%三氯乙酸4℃固定1h,然后用 SRB 染色10min,晾干,100mmol/L Tris 碱溶解,在540nm处测吸光值。

$$细胞存活率(\%)=(处理组 OD 值/对照组 OD 值)\times 100\%$$

(2) 吖啶橙染色观察细胞核凝集和片段化

将处于对数生长期的细胞种植到24孔板上,待细胞长满,移弃各孔中原有培养液。向实验组加入不同浓度的含蜂胶的培养原液;处理结束后,弃各孔中原有培养液,1×PBS 轻轻冲洗细胞,加入吖啶橙染液在常温下作用1min,置于倒置荧光显微镜下观察并拍照。

(3) Hoechst33258 染色检测细胞凋亡

将处于对数生长期的细胞种植到24孔板上,待细胞长满,移弃各孔中原有培养液。向实验组加入不同浓度的含蜂胶的培养原液;处理结束后,弃各孔中原有培养液,加入 Hoechst33258 染色液,放入二氧化碳培养箱中孵育15min;弃24孔板中染色液,加入1×PBS 缓冲液200μL,在荧光显微镜下观察、拍照,紫外光激发。正常的细胞核看上去是均匀的蓝色,凋亡细胞的细胞核表现为非常明亮的蓝色,以至于发白色光,形状变化包括收缩、破碎等。

(4) PC-PLC 活性的检测

消化法收集不同处理的细胞,以 L-α-卵磷脂作为 PC-PLC 的底物,在660nm处测定吸光值。

(5) western blotting 检测 p53 蛋白的表达

用蛋白裂解液收集不同处理组细胞,测定蛋白浓度。每组细胞分别经12% SDS-PAGE 电泳、转印、杂交、ECL 显色、得到 p53、β-actin 蛋白杂交带。蛋白相对量通过 Quantity One 软件分析。

(6) 免疫细胞化学法检测膜整连蛋白β4的表达

待处理结束后,弃24孔板中细胞培养液,4%多聚甲醛固定15min;加入正常血清封闭液液,室温封闭20min;弃封闭液,加入一抗(1∶100),湿盒中4℃过夜;弃一抗,用0.1mol/L PBS 缓冲液洗3次,加入二抗(1∶200),37℃恒温箱中反应1h,待恢复到室温后,弃二抗,用0.1mol/L PBS 缓冲液洗3次,在激光扫描共聚焦显微镜下观察、分析结果。

(7) ROS 的检测

将处于对数生长期的细胞种植到24孔板上,待细胞长满后,分别经蜂胶处理不同时间后,弃24孔板中培养液,用1×PBS 冲洗细胞,将 DCHF 应用液代替细胞培养液,37℃孵育30min;弃 DCHF 染液,用1×PBS 冲洗细胞两次,再加入 M199 原液,在激光扫描共聚焦显微镜下观察、分析结果。

(8) JC-1 染色检测细胞线粒体膜电位

将处于对数生长期的细胞种植到 24 孔板上,待细胞长满后,分别经蜂胶处理不同时间后,弃 24 孔板中培养液,用 1×PBS 冲洗细胞,加入 JC-1 的 M199 原液,37℃ 孵育 15min;弃 JC-1 染液,用 M199 原液轻轻冲洗细胞两次,每次 5min。在激光扫描共聚焦显微镜下观察、分析结果。

1.1.5 统计学分析

采用 SPSS v11.5 软件包统计分析。计量资料以均数±标准差($\bar{x}\pm S$)表示,采用 t 检验进行统计学分析。$P<0.05$ 为差异显著。

1.2 实验结果

1.2.1 HPLC-MS 分析巴西蜂胶的化学组分

通过 HPLC-MS 分析,我们检测了巴西蜂胶的主要组分(图 9.1),分别为咖啡酸、p-香豆酸和阿替比林 C(Artepillin C)。

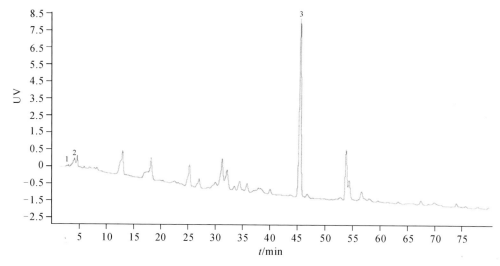

图 9.1 巴西蜂胶乙醇提取物 MS 图谱

Fig. 9.1 MS chromatogram of ethanol-extracted Brazilian propolis (EEBP)

1:咖啡酸;2:p-香豆酸;3:阿替比林 C。
1:caffeic acid;2:p-coumaric acid;3:artepillin C.

1.2.2 巴西蜂胶对 HUVECs 形态和存活率的影响

去除血清和 FGF-2 培养 12h 和 24h 后,细胞会逐渐变圆,细胞膜出泡形成凋亡小体,最终细胞脱离培养皿底部并凋亡。经不同浓度的巴西蜂胶处理 6h 后,SRB 分析表明,12.5μg/mL 巴西蜂胶显著地提高细胞存活率($P<0.01$);25 和 50μg/mL 巴西蜂胶与对照组相比差异不显著。在 12h、50μg/mL 的巴西蜂胶处理组细胞存活率与对照组相比显著下降($P<0.01$)。在 24h、25 和 50μg/mL 的巴西蜂胶处理组细胞存活率与对照组相比显著下降($P<0.01$),表明 25 和 50μg/mL 的巴西蜂胶促进 HUVECs 细胞的凋亡(图 9.2)。

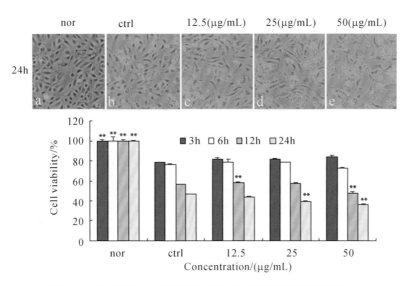

图 9.2 巴西蜂胶对 HUVECs 细胞形态和细胞存活率的影响

Fig. 9.2 Effect of EEBP on cell morphology and viability of human umbilical vein endothelial cells (HUVECs)

nor:正常组细胞;ctrl:对照组细胞;12.5μg/mL:巴西蜂胶 12.5μg/mL 组;25μg/mL:巴西蜂胶 25μg/mL 组;50μg/mL:巴西蜂胶 50μg/mL 组。下同。

1.2.3 巴西蜂胶对细胞核凝集和片段化的影响

为了进一步确定高浓度的巴西蜂胶的促凋亡功效,通过吖啶橙染色和 hoechst 33258 染色法观察了巴西蜂胶对细胞核凝集和片段化的影响。结果表明,在去除血清和 FGF-2 的条件下,高浓度的巴西蜂胶促进染色质浓缩和细胞核片段化(图 9.3)。

1.2.4 巴西蜂胶对膜整连蛋白 β4 表达的影响

如图 9.4 所示,经不同浓度的巴西蜂胶处理 3h 和 6h 后,膜整连蛋白 β4 的表达显著下调($P<0.01$)。在 12h,不同浓度的巴西蜂胶处理对膜整连蛋白的表达没有显著性影响。然而,在 24h,25 和 50μg/mL 组巴西蜂胶显著上调膜整连蛋白 β4 的表达($P<0.01$)。这表明高浓度的巴西蜂胶诱导的 HUVECs 凋亡中,膜整连蛋白 β4 是一个重要的信号分子。

1.2.5 巴西蜂胶对 PC-PLC 的影响

如图 9.5 所示,经不同浓度的巴西蜂胶处理 HUVECs 24h 后,25 和 50μg/mL 组巴西蜂胶显著降低了 PC-PLC 的表达($P<0.01$,$P<0.05$)。

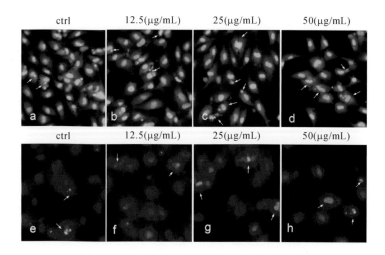

图 9.3 巴西蜂胶对 HUVECs 细胞核的影响

Fig. 9.3 Effect of EEBP on nuclear fragmentation of HUVECs

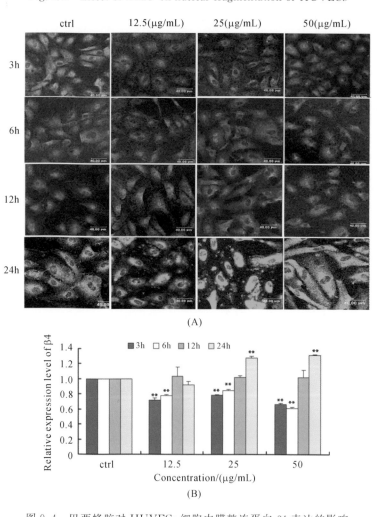

图 9.4 巴西蜂胶对 HUVECs 细胞内膜整连蛋白 β4 表达的影响

Fig. 9.4 Effect of EEBP on integrin β4 expression level in HUVECs

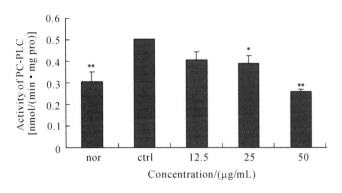

图 9.5 巴西蜂胶对 HUVECs 细胞内 PC-PLC 活性的影响

Fig. 9.5 Effect of EEBP on the activity of PC-PLC in HUVECs

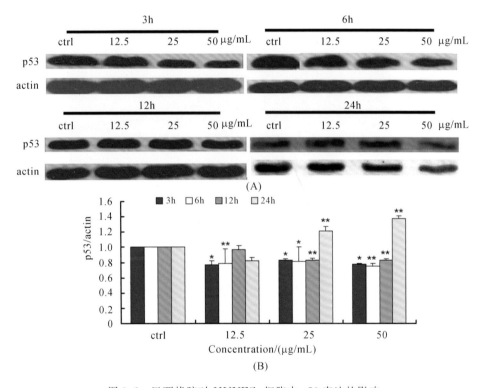

图 9.6 巴西蜂胶对 HUVECs 细胞内 p53 表达的影响

Fig. 9.6 Effect of EEBP on the expression of p53 in HUVECs

1.2.6 巴西蜂胶对 p53 蛋白表达的影响

如图 9.6 所示,经不同浓度的巴西蜂胶处理 HUVECs 3h 和 6h 后,p53 蛋白的表达与对照组相比显著下调($P<0.01$,$P<0.05$),但在 24h,25 和 50 μg/mL 组巴西蜂胶显著上调 p53 蛋白的表达($P<0.01$)。

1.2.7 巴西蜂胶对 ROS 的影响

为了进一步探讨 ROS 是否参与到高浓度巴西蜂胶诱导的 HUVECs 凋亡中,我们检测了不同浓度巴西蜂胶处理的 HUVECs 中 ROS 的水平。如图 9.7 所示,在 3h、50 μg/mL 组巴西蜂胶显著降低 ROS 的水平（$P<0.01$），在 6h 和 12h，12.5、25 和 50μg/mL 组巴西蜂胶显著降低细胞中 ROS 的水平（$P<0.01$，$P<0.05$），然而，在 24h，50μg/mL 组巴西蜂胶显著提高细胞中 ROS 的水平（$P<0.01$）。

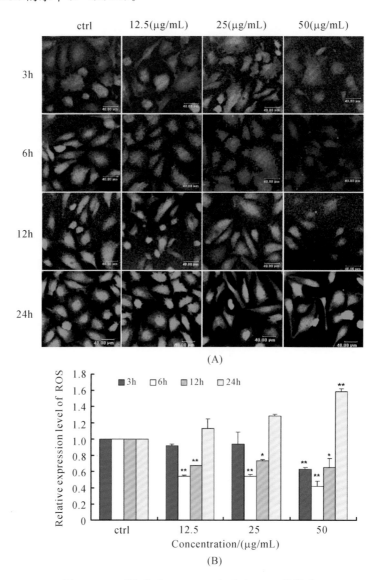

图 9.7　巴西蜂胶对 HUVECs 细胞内 ROS 的影响

Fig. 9.7　Effect of EEBP on reactive oxygen species (ROS) level in HUVECs

1.2.8　巴西蜂胶对线粒体膜电位的影响

如图 9.8 所示，从 3h 一直到 24h，50μg/mL 巴西蜂胶处理的细胞中线粒体膜电位水平显著下降（$P<0.01$，$P<0.05$）；从 12h 到 24h，25μg/mL 巴西蜂胶处理的细胞中线粒体膜电位水平显著下降（$P<0.01$，$P<0.05$）。

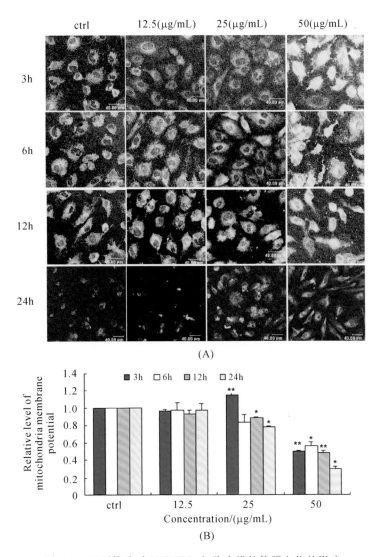

图 9.8 巴西蜂胶对 HUVECs 细胞内线粒体膜电位的影响
Fig. 9.8 Effect of EEBP on changes of mitochondrial membrane potential level in HUVECs

1.3 讨 论

本实验主要研究了在去除血清和 FGF-2 的条件下,巴西蜂胶对 HUVECs 细胞的影响,并且分析了巴西蜂胶对膜整连蛋白 β4、p53、ROS 和线粒体膜电位水平的影响。本实验研究结果首次发现,在去除血清和 FGF-2 的条件下,高浓度的巴西蜂胶通过上调膜整连蛋白 β4、p53 和 ROS 以及降低 PC-PLC 活性和线粒体膜电位水平诱导 HUVECs 的凋亡,而且巴西蜂胶对 HUVECs 的影响具有时间和剂量依赖性。研究表明,主要存在两种起始凋亡的信号,一种是细胞死亡受体介导的凋亡信号通路,另一种是线粒体介导的凋亡信号通路。我们的研究初步显示,在去除血清和 FGF-2 条件下,高浓度巴西蜂胶诱导 HUVECs 凋亡可能是通过线粒体介导的途径实现的。

膜整连蛋白 β4 是一种重要的膜蛋白,主要存在于半桥粒中,是介导细胞外基质黏附的

细胞表面受体家族,在基膜和细胞骨架的中间纤维中提供坚固的支撑。许多研究表明,膜整连蛋白 β4 参与多种信号转导,如细胞黏附、分化、迁移、存活以及凋亡(Laferriere et al.,2004)。最近研究表明,在大多数肿瘤细胞中膜整连蛋白 β4 的 mRNA 表达升高(Feng et al.,2004)。Miao 等(1997a)研究表明,膜整连蛋白 β4 在 HUVECs 凋亡中发挥重要作用。在本实验我们首次发现,巴西蜂胶影响膜整连蛋白 β4 的表达。在 3h 和 6h,不同浓度的巴西蜂胶显著地降低膜整连蛋白 β4 的表达,在 24h,25 和 50 $\mu g/mL$ 的巴西蜂胶显著上调了膜整连蛋白 β4 的表达。由此可以看出,不同浓度的巴西蜂胶对膜整连蛋白 β4 的影响不同,低浓度的巴西蜂胶能够降低膜整连蛋白 β4 的表达,而高浓度的巴西蜂胶则能够上调膜整连蛋白 β4 的表达。膜整连蛋白 β4 可能参与到高浓度的巴西蜂胶诱导的 HUVECs 的凋亡。

PC-PLC 是磷脂酶 C 家族中的重要一员,能特异性水解磷脂酰胆碱产生磷酸胆碱和二酰基甘油(DAG)。磷酸胆碱和 DAG 作为第二信使,参与哺乳动物多种细胞反应,如细胞生长、分化、老化以及凋亡等(Miao et al.,1997b)。苗俊英等发现抑制 PC-PLC 活性能够抑制血管内皮细胞凋亡(Miao et al.,1997b),但赵静等报道黄樟素氧化物上调 Fas、膜整连蛋白 β4、p53 并降低 PC-PLC 活性和 ROS 的水平,诱导血管内皮细胞凋亡(Zhao et al.,2005),这些研究结果表明,PC-PLC 活性的升高和降低在不同的凋亡信号中发挥的功效可能不同。我们的研究发现,高浓度的巴西蜂胶降低 PC-PLC 的活性,但诱导内皮细胞凋亡,这一结果与赵静等的研究结果一致。

p53 肿瘤抑制蛋白是膜整连蛋白和 PC-PLC 调控的下游信号分子,参与 HUVECs 凋亡的信号通路,p53 蛋白的表达及活性增强能诱导细胞凋亡。线粒体途径是 p53 抗肿瘤应答的主要途径之一。p53 在多种细胞类型包括肿瘤细胞、成纤维细胞以及血管内皮细胞的凋亡中发挥重要作用(Speidel,2010;Speidel et al.,2006)。我们的结果表明,在 3h 和 6h,巴西蜂胶处理的 HUVECs 中,p53 蛋白伴随着膜整连蛋白 β4 表达下降而下降;而在 24h,高浓度的巴西蜂胶(25,50 $\mu g/mL$)处理的血管内皮细胞中 p53 蛋白显著上调,而同时膜整连蛋白 β4 的表达也显著上调。这表明,膜整连蛋白 β4 在调节 HUVECs 凋亡中与 p53 蛋白相关,尽管它们之间的关系还不是很清楚,需要进一步研究。

大量的研究表明,ROS 在凋亡信号通路发挥重要作用。以前的研究表明,细胞内 ROS 水平过度升高诱导细胞凋亡,引发多种疾病的发生。本实验研究表明,在 6h 和 12h,不同浓度的巴西蜂胶处理的 HUVECs 中,ROS 的水平显著下降,但是在 24h,50 $\mu g/mL$ 的巴西蜂胶显著地上调 ROS 的水平,说明不同浓度的巴西蜂胶对 ROS 的影响具有时间依赖性。有报道指出,膜整连蛋白 β4 在 HUVECs 中参与调控 ROS 的水平。我们的研究结果表明,膜整连蛋白 β4 的表达水平在 3h 显著下调,而 ROS 的水平在 6h 显著下调。我们推测,在巴西蜂胶处理的 HUVECs 中,膜整连蛋白 β4 可能调控 ROS 的水平。线粒体是细胞内 ROS 的重要来源,而且在调控细胞死亡过程中发挥重要作用。反过来,ROS 水平的升高能降低线粒体膜电位,导致细胞色素 C 和凋亡诱导因子的释放,诱导细胞凋亡。在本研究中,高浓度的巴西蜂胶显著地降低线粒体膜电位,破坏线粒体的稳定性,结果显示高浓度的巴西蜂胶诱导 HUVECs 凋亡主要是通过提高 ROS 水平并降低线粒体膜电位途径实现的。

另外,值得注意的是,尽管长期以来人们将巴西蜂胶作为一种传统药物,但 Banskota 等指出,由于巴西蜂胶含有苯并呋喃衍生物,因而具有细胞毒性(Banskota et al.,2000)。对巴西绿蜂胶进行的体外毒性试验也发现,低剂量的巴西绿蜂胶具有抗突变的功效,但是高剂量的巴西绿蜂胶具有致突变的功效。Pereira等报道,过度服用巴西绿蜂胶引起小鼠的血细胞产生了突变功效(Pereira et al.,2008)。此外,Munari等建议巴西绿蜂胶主要植物来源(*Baccharis dracunculifolia*)提取物最有效的使用剂量是 12.5μg/mL(Munari et al.,2010)。通过本次实验,我们的研究结果也表明,在去除血清和 FGF-2 的条件下,高剂量的巴西蜂胶能够诱导 HUVECs 凋亡,而低剂量(12.5μg/mL)的巴西蜂胶不会诱导 HUVECs 凋亡,相反具有一定的保护功效。因此,我们认为,巴西蜂胶的浓度对血管内皮细胞的影响非常大,而且巴西蜂胶对细胞的毒性具有时间依赖性。内皮细胞凋亡在许多疾病中发挥着重要的作用,如动脉粥样硬化以及肿瘤等。内皮细胞凋亡不利于动脉粥样硬化疾病的治疗,然而却促进肿瘤细胞的凋亡,因而,我们推断高剂量的巴西蜂胶(25 和 50μg/mL)可以诱导 HUVECs 细胞凋亡,这有利于抑制肿瘤细胞的生长。而在去除血清和 FGF-2 的条件下,低剂量的巴西蜂胶如 12.5μg/mL 不会诱导 HUVECs 凋亡,反而在一定时间内促进 HUVECs 生长,下调膜整连蛋白 β4、p53 和 ROS 水平,稳定线粒体膜电位。因此,低剂量的巴西蜂胶可以作为辅助治疗心血管疾病如动脉粥样硬化药物的替代剂。这就提示我们,今后在巴西蜂胶的应用中应特别重视其浓度问题。

此外,阿替比林 C 是巴西蜂胶中一种最主要的酚酸类物质,具有多种生物学活性,如抗细菌、抗病毒、抗氧化、诱导凋亡、抗炎以及抗肿瘤等(Hoshida et al.,1997;Paulino et al.,2008;Pontin et al.,2008)。但目前关于阿替比林 C 对血管内皮细胞的研究较少。阿替比林 C 是否影响血管内皮细胞的功能,巴西蜂胶影响膜整连蛋白 β4 和 PC-PLC 的信号通路是否是由其主要组分阿替比林 C 在发挥主要功效呢?因此,阿替比林 C 在血管内皮细胞中的作用,以及对膜整连蛋白 β4 及其下游信号分子 PC-PLC、p53、ROS 和线粒体膜电位的影响需要进一步研究。

总之,本实验研究了在去除血清和 FGF-2 条件下,巴西蜂胶对 HUVECs 的影响,以及膜整连蛋白 β4、p53、ROS 和线粒体膜电位的变化。本实验首次发现,膜整连蛋白 β4 可能通过调控 PC-PLC、p53 和 ROS 来参与巴西蜂胶对 HUVECs 凋亡信号通路的调控。

2 去除血清和生长因子条件下中国蜂胶对血管内皮细胞的影响

由于植物来源不同,巴西蜂胶与中国蜂胶化学成分相差很大。中国蜂胶的主要成分是黄酮类化合物(Usia et al.,2002),而且中国蜂胶与巴西蜂胶一样具有广泛的生物学活性,如抗细菌、抗真菌、抗病毒(Nolkemper et al.,2010)、抗氧化(Fuliang et al.,2005)、抗炎症(Hu et al.,2005)、抗肿瘤(Ishihara et al.,2009)等。还有研究表明,中国蜂胶对神经细胞显示了新的生物学活性,如中国蜂胶及其主要组分柯因能够抑制衣霉素诱导的神经细胞的死亡(Izuta et al.,2008)。中国蜂胶的乙醇提取物还能促进脊髓损伤后运动功能的恢复,提高脑源性神经营养因子的产生(Kasai et al.,2011),这为中国蜂胶的研究和应用提供了新的方向。

在上一实验(去除血清和生长因子条件下巴西蜂胶对血管内皮细胞的影响)中,我们发现高浓度的巴西蜂胶通过上调膜整连蛋白 β4、p53 以及 ROS 的水平,降低线粒体膜电位诱导 HUVECs 凋亡,说明高浓度的巴西蜂胶对 HUVECs 具有细胞毒性。而目前关于中国蜂胶对 HUVECs 的影响的研究较少,对膜整连蛋白 β4 信号通路的影响还未有报道。因此,本研究的重点是进一步研究在去除血清和生长因子条件下中国蜂胶对 HUVECs 的影响,并比较中国蜂胶与巴西蜂胶对 HUVECs 影响的差异,以及中国蜂胶对膜整连蛋白 β4、PC-PLC、p53、ROS 以及线粒体膜电位的影响,以阐明中国蜂胶对血管内皮细胞影响的作用机制,为中国蜂胶的应用提供实验依据。

2.1 材料和方法

2.1.1 主要试剂

同本章 1.1.1。

2.1.2 主要设备与仪器

同本章 1.1.2。

2.1.3 实验方法

(1)蜂胶样品的制备

中国蜂胶取自中国山东省,主要植物来源为白杨树(*Populus* spp.);制备方法同巴西蜂胶的制备,制备的中国蜂胶(EECP)于 4℃下贮存。

(2)HPLC-MS 分析 EECP 的化学组分

分析方法同巴西蜂胶。

(3)人脐静脉血管内皮细胞(HUVECs)培养

方法同前。

(4)实验分组

待 HUVECs 长满后,将细胞分为 3 组:正常组、溶剂对照组和实验组。正常组的细胞是在正常培养条件下培养;对照组细胞是在去除血清和 FGF-2 的培养液中加入酒精溶剂培养,酒精浓度<0.1% (v/v),培养液中乙醇浓度<0.1%不会影响细胞的正常生长;实验组细胞是在去除血清和 FGF-2 的同时用不同浓度(6.25、12.5 和 25μg/mL)的中国蜂胶处理。

2.1.4 检测指标

(1)细胞存活率的测定

将长势良好的血管内皮细胞种植到 96 孔板,37℃、5% CO_2 孵育,至细胞长满。按照实验分组,分别处理细胞 12h 和 24h 后,弃细胞培养液,用 10%三氯乙酸于 4℃下固定 1h,然后用 SRB 染色 10min,晾干,100mmol/L Tris 碱溶解,在 540nm 波长处测吸光值。

细胞存活率(%)=(处理组 OD 值/对照组 OD 值)×100%

(2)PC-PLC 活性的检测

消化法收集不同处理的细胞,方法同前。

(3)western blotting 检测 p53 蛋白的表达

用蛋白裂解液收集不同处理组细胞,方法同前。

(4)免疫细胞化学法检测膜整连蛋白 β4 的表达。

方法同前。

(5)ROS 的检测

方法同前。

(6)JC-1 染色检测细胞线粒体膜电位

方法同前。

2.1.5 统计学分析

采用 SPSS v11.5 软件包统计分析。计量资料以均数±标准差($\bar{x}\pm S$)表示,采用 t 检验进行统计学分析。$P<0.05$ 为差异显著。

2.2 结果

2.2.1 中国蜂胶的化学组分

首先,我们通过 HPLC-MS 分析了中国蜂胶中的化学成分,主要包括:5-甲基酯短叶松素、短叶松素、阿魏酸、生松素、3-乙酸短叶松素、柯因、咖啡酸苯乙酯、高良姜、柚木柯因(图9.9)。

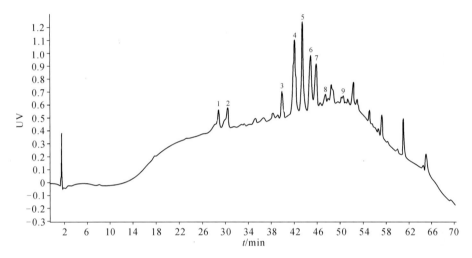

图 9.9 中国蜂胶乙醇提取物 MS 图谱

1:5-甲基酯短叶松素;2:短叶松素;3:阿魏酸;4:生松素;5:3-乙酸短叶松素;6:柯因;7:咖啡酸苯乙酯;8:高良姜;9:柚木柯因。

Fig. 9.9 MS chromatogram of ethanol-extracted Chinese propolis (EECP)

1:5-methylether;2:Pinobanksin;3:Ferulic acid;4:Pinocembrin;5:Pinobanksin 3-acetate;6:Chrysin;7:Caffeic acid phenethyl ester;8:Galangin;9:Tectochrysin

2.2.2 中国蜂胶对 HUVECs 存活率的影响

去除血清和 FGF-2 后,HUVECs 逐渐变圆,并脱离皿底,发生凋亡。在去除血清和 FGF-2 的条件下,中国蜂胶(6.25、12.5 和 25μg/mL)分别处理 HUVECs 3、6、12 和 24h,通过 SRB 法检测了中国蜂胶对 HUVECs 存活率的影响。结果表明,中国蜂胶(6.25 和 12.5μg/mL)在 12h 显著地提高了 HUVECs 的存活率($P<0.05$,$P<0.01$);然而,25μg/mL 的中国蜂胶对 HUVECs 存活率没有显著影响(图9.10)。

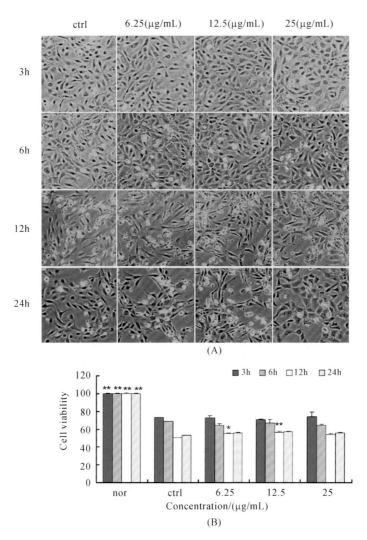

图 9.10 中国蜂胶对 HUVEC 细胞形态和细胞存活率的影响

Fig. 9.10 Effect of EECP on cell morphology and viability of HUVECs

注：nor：正常组细胞；ctrl：对照组细胞；6.25μg/mL：中国蜂胶 6.25μg/mL 组；12.5μg/mL：中国蜂胶 12.5μg/mL 组；25μg/mL：中国蜂胶 25μg/mL 组。下同。

2.2.3 中国蜂胶对膜整连蛋白 β4 的影响

如图 9.11 所示，6.25μg/mL 的中国蜂胶处理 HUVECs 6h 后，膜整连蛋白 β4 的表达与对照组相比显著下调（$P<0.01$）；12.5 和 25μg/mL 的中国蜂胶分别处理 HUVECs 6h 和 12h 后，膜整连蛋白 β4 表达显著下调（$P<0.05$）；而在 24h，中国蜂胶对膜整连蛋白 β4 的表达影响不显著。

2.2.4 中国蜂胶对 PC-PLC 活性的影响

为了探讨中国蜂胶对 HUVECs 凋亡细胞中 PC-PLC 活性的影响，我们检测了 PC-PLC 的活性，结果如图 9.12 所示。中国蜂胶（6.25、12.5 和 25μg/mL）处理 HUVECs 6h 和 12h 后，PC-PLC 的活性显著下降（$P<0.05$，$P<0.01$）。

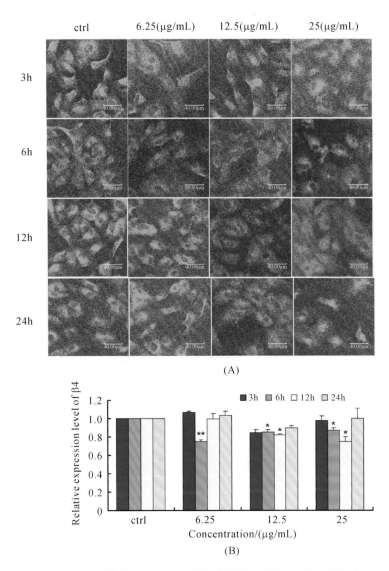

图 9.11 中国蜂胶对 HUVEC 细胞内膜整连蛋白 β4 表达的影响
Fig. 9.11 Effect of EECP on integrin β4 expression level in HUVECs

2.2.5 中国蜂胶对 p53 蛋白表达的影响

如图 9.13 所示,6.25μg/mL 的中国蜂胶处理 HUVECs 6h 后,p53 蛋白的表达与对照组相比显著下降($P<0.01$);12.5 和 25μg/mL 的中国蜂胶处理 HUVECs 6、12、24h 后 p53 蛋白的表达显著下调($P<0.05,P<0.01$)。结果表明,中国蜂胶可以抑制去除血清和 FGF-2 所引起的 p53 蛋白表达上调。

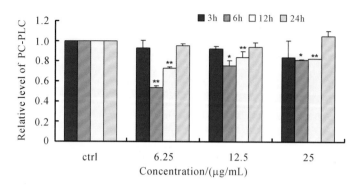

图 9.12 中国蜂胶对 HUVECs 细胞内 PC-PLC 活性的影响
Fig. 9.12 Effect of EECP on the activity of PC-PLC in HUVECs

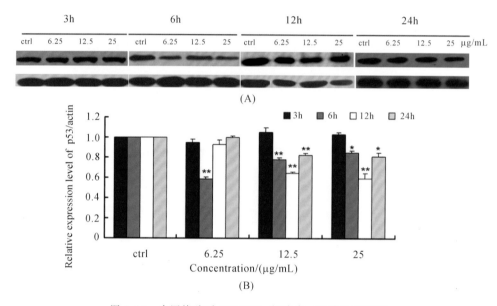

图 9.13 中国蜂胶对 HUVECs 细胞内 p53 表达的影响
Fig. 9.13 Effect of EECP on the expression of p53 in HUVECs

2.2.6 中国蜂胶对 ROS 的影响

在去除血清和 FGF-2 的条件下,测定了中国蜂胶对 ROS 的影响,结果如图 9.14 所示,在 6、12 和 24h,12.5μg/mL 的中国蜂胶与对照组相比显著降低 ROS 的水平($P<0.01$);在 3、6、12 和 24h,25μg/mL 的中国蜂胶与对照组相比极其显著地降低 ROS 的水平($P<0.01$);然而,6.25μg/mL 的中国蜂胶对 ROS 的水平没有显著影响。

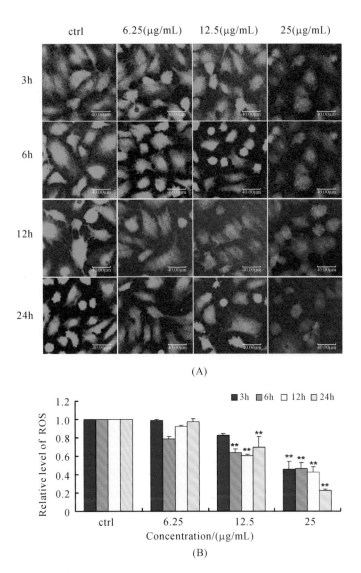

图 9.14 中国蜂胶对 HUVECs 细胞内 ROS 的影响

Fig. 9.14 Effect of EECP on ROS level in HUVECs

2.2.7 中国蜂胶对线粒体膜电位的影响

为了探讨在去除血清和 FGF-2 的条件下,中国蜂胶是否影响线粒体膜电位,我们用 JC-1 检测了中国蜂胶对线粒体膜电位的影响。结果如图 9.15 所示。6.25 和 12.5μg/mL 的中国蜂胶对线粒体膜电位没有显著影响;然而,25μg/mL 的中国蜂胶从 3h 开始一直到 24h 都显著降低线粒体膜电位($P<0.05, P<0.01$)。结果表明,在去除血清和生长因子的条件下,高浓度的中国蜂胶破坏线粒体膜电位。

2.3 讨论

上一实验研究了在去除血清和 FGF-2 的条件下巴西蜂胶对血管内皮细胞的影响,发现高剂量的巴西蜂胶诱导血管内皮细胞凋亡。内皮细胞凋亡或损伤以及功能异常引发多种疾病,如高血压、高血脂、糖尿病以及心肌缺血等,因此,高剂量的巴西蜂胶损伤内皮细胞,具有

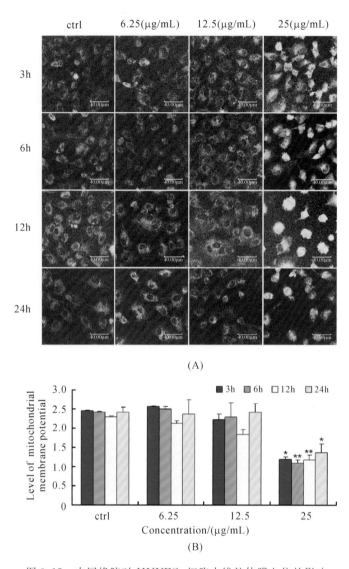

图 9.15 中国蜂胶对 HUVECs 细胞内线粒体膜电位的影响

Fig. 9.15 Effect of EECP on changes of mitochondrial membrane potential level in HUVECs

细胞毒性,不利于心血管等疾病的治疗。在本实验中,我们降低了中国蜂胶的浓度,评价了在去除血清和 FGF-2 的条件下中国蜂胶对 HUVECs 的影响,并进一步分析了细胞中膜整连蛋白 β4、PC-PLC、p53、ROS 和线粒体膜电位的变化。实验结果表明,中国蜂胶(6.25 和 12.5μg/mL)能够通过下调膜整连蛋白 β4、p53 蛋白表达,抑制 PC-PLC 的活性,并部分降低 ROS 的水平来抑制 HUVECs 的凋亡。高浓度的中国蜂胶(25μg/mL)尽管对 HUVECs 的存活率没有显著影响,而且能够下调膜整连蛋白 β4、p53 蛋白表达,抑制 PC-PLC 的活性,但会过度降低 ROS 的水平,损伤线粒体膜电位,因此高浓度的中国蜂胶具有潜在的细胞毒性。

膜整连蛋白 β4 是细胞信号转导过程中的一个关键的膜蛋白,并且参与血管内皮细胞的凋亡过程。上一实验研究发现,高浓度的巴西蜂胶上调膜整连蛋白 β4 的表达诱导内皮细胞凋亡。本实验结果表明,在去除血清和 FGF-2 的条件下,中国蜂胶(6.25、12.5 和 25μg/mL)处理 HUVECs 6h 和 12h 后均下调了膜整连蛋白 β4 的表达，因而，可以看出膜整连蛋白 β4 可能参与中国蜂胶抑制 HUVECs 凋亡的信号通路。

PC-PLC 参与血管内皮细胞的凋亡信号通路，但 PC-PLC 活性的高低在凋亡中的作用报道并不一致。苗俊英等指出，抑制 PC-PLC 的活性抑制了内皮细胞的凋亡，而赵静等研究也指出，黄樟素氧化物抑制 PC-PLC 的活性，却促进了内皮细胞凋亡。本实验对 PC-PLC 活性影响的研究发现，中国蜂胶（6.25、12.5 和 25μg/mL）处理 HUVECs 6h 和 12h 后均显著降低了 PC-PLC 的活性，表明中国蜂胶可能是通过降低 PC-PLC 活性来参与抑制 HUVECs 凋亡信号通路，这一结果与巴西蜂胶的研究结果不同，高浓度的巴西蜂胶抑制 PC-PLC 的活性并促进内皮细胞凋亡，而中国蜂胶抑制 PC-PLC 的活性的同时抑制了内皮细胞凋亡。因此，我们推断巴西蜂胶和中国蜂胶对 PC-PLC 活性的影响与细胞存活率之间的关系可能不大。

p53 是 PC-PLC 调控血管内皮细胞凋亡的下游信号分子，苗俊英等研究表明，在去除血清和 FGF-2 的条件下，p53 蛋白在 HUVECs 凋亡中发挥重要作用，而且抑制 PC-PLC 活性后能够抑制 p53 蛋白的表达；此外，另一项研究也指出，PC-PLC 通过调控 p53 来调控凋亡（Zhao et al.，2005）。本实验研究结果表明，中国蜂胶（6.25、12.5 和 25μg/mL）处理 HUVECs 6h、12h 和 24h 后，p53 蛋白的表达伴随着 PC-PLC 活性显著下降而下调。

大量研究证实 ROS 参与凋亡信号通路。中国蜂胶除了来自云南地区的蜂胶外，均含有大量的酚酸类物质，因而具有较强的抗氧化活性。上一实验的研究结果显示，高浓度的巴西蜂胶处理的细胞中 ROS 水平升高，破坏线粒体膜电位，诱导内皮细胞凋亡。在本研究中，12.5μg/mL 的中国蜂胶在 6h、12h 和 24h 显著地降低 ROS 的水平。而 25μg/mL 的中国蜂胶从 3h 一直到 24h 均极度降低 ROS 水平，同时，线粒体膜电位也受到破坏，以致线粒体膜电位显著下降，这表明 ROS 的过度下调也会破坏线粒体膜电位，也可能是由于线粒体膜电位的损伤导致 ROS 过度下调。由此可以看出，高浓度的巴西蜂胶和中国蜂胶均破坏线粒体膜电位，但是高浓度的巴西蜂胶的破坏作用是由 ROS 水平过高导致，而高浓度的中国蜂胶的破坏作用则是由 ROS 极度降低所致。综上考虑，应当指出细胞内 ROS 的平衡对于细胞的存活非常重要，过低的 ROS 和过高的 ROS 都具有潜在的促凋亡的功效。

与相同浓度的巴西蜂胶相比，低浓度的巴西蜂胶与低浓度的中国蜂胶（12.5μg/mL）对 HUVECs 的影响作用相似，均表现出对 HUVECs 的保护功效，而高浓度的巴西蜂胶（25μg/mL）在与同浓度的中国蜂胶在相同处理条件下，降低细胞存活率，上调膜整连蛋白 β4 和 p53 蛋白表达，抑制 PC-PLC 的活性，降低 ROS 水平，破坏线粒体膜电位。因此，高浓度的中国蜂胶和巴西蜂胶在抑制 PC-PLC 活性，降低 ROS 水平以及破坏线粒体膜电位水平上结果一致，但是需要指出，高浓度的中国蜂胶对 ROS 的降低是过度的降低，而同浓度的巴西蜂胶对 ROS 的降低是适度的。因而，我们认为不管是中国蜂胶还是巴西蜂胶在浓度为 12.5μg/mL 时对 HUVECs 起到保护功效，高于该浓度蜂胶将对 HUVECs 产生细胞毒性，促进细胞凋亡。

此外，中国蜂胶的化学组分比巴西蜂胶的化学组分复杂，其中槲皮素、CAPE、堪菲醇、高良姜、金合欢素、柯因、芹菜素、生松素等均具有广泛的生物学活性，特别是目前对 CAPE 的研究较多。那么，中国蜂胶对膜整连蛋白 β4 和 PC-PLC 活性的影响是中国蜂胶中的哪一种或者哪几种主要组分在发挥作用，目前还不清楚，因此有必要深入研究，这也是为了更好地应用中国蜂胶所必需的。

总之，本实验研究结果表明，在去除血清和生长因子的条件下，中国蜂胶对 HUVECs 抗凋亡功效可能部分是由于下调膜整连蛋白 β4 的表达，抑制 PC-PLC 活性、ROS 的水平以及 p53 蛋白的表达而发挥作用。由于 12.5μg/mL 的中国蜂胶还能显著地降低 ROS 的水平，因而 12.5μg/mL 的中国蜂胶抗凋亡功效要好于 6.25μg/mL 的中国蜂胶。然而，25μg/mL 的中国蜂胶由于极度地降低 ROS，损伤线粒体膜电位，是一种潜在的 HUVECs 凋亡诱导剂，因此，在中国蜂胶使用过程中也应注意浓度的选择。

第二节 蜂胶对脂多糖诱导的血管内皮细胞的影响

前面我们研究了在去除血清和生长因子条件下，蜂胶对血管内皮细胞的影响，并指出 12.5μg/mL 的巴西蜂胶和中国蜂胶对 HUVECs 起到保护作用。然而，去除血清和生长因子容易导致血管渗透性增加，诱发炎症反应。由于蜂胶具有较好的抗炎活性，但蜂胶抗炎功效的作用机制还不完全清楚，而且，蜂胶在炎症条件下对血管内皮细胞的影响较少，所以有必要深入研究。

脂多糖(LPS)是一种重要的促炎剂，靶向内皮细胞。TLRs 在各种免疫细胞中大量的表达，能够识别保守的病原相关分子，如 LPS、细菌或分枝杆菌多肽、病毒 RNA 和 DNA，因而在天然免疫和获得性免疫中发挥重要作用。由于 TLR2 和 TLR4 对微生物发挥重要作用，因而得到广泛的研究。TLR2 主要识别革兰阳性菌以及真菌，而 TLR4 主要识别革兰阴性菌 LPS，因而 TLR4 是 LPS 信号通路中的关键分子，在炎症信号通路中 LPS 与 TLR4 结合，进而活化下游信号级联反应，促进转录因子 NF-κB 和 AP-1 的转录和活化，促进和启动炎性相关因子的转录(Pagliarone et al.,2009)。

在炎症过程中，LPS 活化 PC-PLC，而 PC-PLC 又促进多种炎症的发生。研究指出，药理性阻断 PC-PLC 活性能抑制动脉粥样硬化的进展及促进斑块的稳定，而目前普遍认为动脉粥样硬化疾病是一种慢性炎症(Zhang et al.,2010)。而且，最近的报道表明，100ng/mL LPS 能够使细胞内 PC-PLC 活性升高，促进血管内皮细胞中 IL-8 和单核细胞趋化因子(MCP-1)的产生，阻断 PC-PLC 的活性，显著地抑制 LPS 诱导的 IL-8 和 MCP-1 的产生(Zhang et al.,2011)。

近年来，蜂胶的抗炎功效得到一定的研究。我们团队前期动物实验研究结果也显示，蜂胶具有显著的抗炎功效，然而蜂胶抑制炎症的机制目前尚不完全清楚。TLR4 和 PC-PLC 与炎症关系密切，而蜂胶在炎症介质诱导下对 TLR4 以及 PC-PLC 的影响还未有报道。因此，有必要研究 LPS 诱导的血管内皮细胞损伤条件下，蜂胶对 TLR4 表达和 PC-PLC 活性的影响，及其对下游信号分子如 NF-κB p65、p53、ROS 水平以及线粒体膜电位的影响，旨在探讨在炎症条件下，蜂胶对血管内皮细胞的影响及其机制。

1 材料和方法

1.1 主要试剂

M199 培养基和胎牛血清(美国 Hyclone 公司)；碱性生长因子(中国 EssexBio 公司)；L-α-卵磷脂、磺酰罗丹明 B(SRB)、DCHF 探针(美国 Sigma 公司)；脂多糖(LPS)(Escherichia coli 055：B5)、JC-1 探针(美国 Invitrogen 公司)；p53、β-actin(美国 Santa Cruz 公司)；辣根

过氧化物酶(HRP)标记的二抗(南京中杉金桥公司);NO 试剂盒(南京建成生物生物工程研究所)。

1.2 主要设备与仪器

同本章第一节 1.2。

1.3 实验方法

(1)蜂胶样品的制备

中国蜂胶取自中国山东省,主要植物来源为白杨树(*Populus* spp.)。粗蜂胶经冷冻、粉碎、无水乙醇室温浸提、减压过滤、旋转蒸发、干燥至恒重,于 4℃ 贮存。

(2)人脐静脉血管内皮细胞(HUVECs)培养

HUVECs 由山东大学发育与细胞生物学实验室提供。HUVECs 培养在添加 20% 胎牛血清和 70ng/mL 碱性生长因子的 M199 培养基中进行。孵育条件为:37℃、5% CO_2 饱和湿度培养箱中。每隔一天换液。倒置相差显微镜下观察细胞形态。

(3)实验分组

实验分为 3 组:空白对照组(0.5%血清+M199 培养基)(control)、LPS 组(0.5%血清+M199 培养基+100ng/mL LPS)(LPS)、蜂胶组(0.5%血清+M199 培养基+100ng/mL LPS+12.5μg/mL 蜂胶)(12.5μg/mL)。

1.4 检测指标

(1)细胞存活率的测定

将长势良好的血管内皮细胞种植到 96 孔板中,37℃、5% CO_2 孵育,至细胞长满。按照实验分组,分别处理细胞 12h 和 24h 后,弃去细胞培养液,用 10%三氯乙酸于 4℃下固定 1h,然后用 SRB 染色 10min,晾干,100mmol/L Tris 碱溶解,在 540nm 波长处测吸光值。

细胞存活率(%)=(处理组 OD 值/对照组 OD 值)×100%

(2)细胞培养液中 NO 的测定

收集不同处理的细胞培养液,按照试剂盒的步骤测定 NO 的活性。

(3)细胞中 PC-PLC 活性的测定

消化法收集不同处理的细胞,以 L-α-卵磷脂作为 PC-PLC 的底物,在 660nm 波长处测定吸光值。

(4)Western blotting 检测 TLR4、NF-κB p65、p53 的表达

用蛋白裂解液收集不同处理组细胞,测定蛋白质浓度。每组细胞分别经 12% SDS-PAGE 电泳、转印、杂交、ECL 显色,得到 TLR4、NF-κB p65、p53、β-actin 蛋白杂交带。蛋白相对分子质量通过 Quantity One 软件分析。

(5)细胞中 ROS 的测定

细胞经不同处理 12h、24h 后,弃 24 孔板中培养液,用 M199 原液轻轻冲洗细胞 2 次,然后弃 M199 原液,加入含有 2μL/mL DCHF 荧光探针的 M199 原液,37℃ 孵育 30min,孵育结束后弃 DCHF 染液,用 M199 原液轻轻冲洗细胞 2 次,置于激光扫描共聚焦显微镜下观察,分析结果。

(6)细胞中线粒体膜电位的测定

细胞经不同处理 24h 后,弃 24 孔板中培养液,用 M199 原液轻轻冲洗细胞 2 次,加入含

有 2μL/mL JC-1 探针的 M199 原液,37℃ 孵育 15min,孵育结束后弃 JC-1 染液,用 M199 原液轻轻冲洗细胞 2 次,置于激光扫描共聚焦显微镜下观察,分析结果。

1.5 统计学分析

采用 SPSS v11.5 软件包统计分析。计量资料统计结果以均数±标准差($\bar{x}\pm S$)表示,采用 t 检验进行统计学分析。$P<0.05$ 为差异显著。

2 结果

2.1 蜂胶对 LPS 诱导的血管内皮细胞存活率的影响

如图 9.16 所示,在 0.5% 血清、100ng LPS 诱导条件下,12.5μg/mL 蜂胶分别处理血管内皮细胞 12h 和 24h 后,通过 SRB 法检测了细胞存活率。结果表明,在 12h 和 24h 蜂胶组细胞存活率与 LPS 组相比差异不显著。

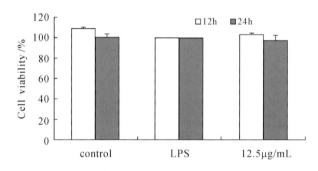

图 9.16 蜂胶对 LPS 诱导的血管内皮细胞存活率的影响

Fig. 9.16 Effect of propolis on cell viability of VEC induced by LPS

注:control:对照组;LPS:LPS 处理组;12.5μg/mL:中国蜂胶 12.5μg/mL 组,下同。

2.2 蜂胶对 LPS 诱导的血管内皮细胞液中 NO 的影响

如图 9.17 所示,在 LPS 处理 12h 后,蜂胶组与 LPS 组相比 NO 含量差异不显著,而在 LPS 处理 24h 后,对照组和蜂胶处理组与 LPS 组相比,NO 含量显著下降($P<0.01$)。

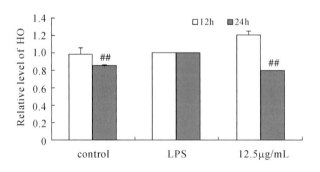

图 9.17 蜂胶对 LPS 诱导的血管内皮细胞液中 NO 的影响

Fig. 9.17 Effect of propolis on nitric oxide of VEC induced by LPS

2.3 蜂胶对 LPS 诱导的血管内皮细胞中 PC-PLC 活性的影响

如图 9.18 所示,LPS 处理 12h 后,对照组和蜂胶处理组与 LPS 组相比,PC-PLC 活性显著下降($P<0.05$, $P<0.01$),而在 LPS 处理 24h 后,蜂胶处理组与 LPS 组相比差异不显著。

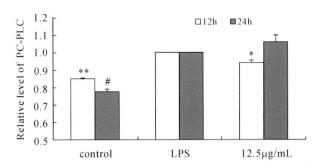

图 9.18　蜂胶对 LPS 诱导的血管内皮细胞中 PC-PLC 活性的影响

Fig. 9.18　Effect of propolis on PC-PLC activity in VEC induced by LPS

2.4　蜂胶对 LPS 诱导的血管内皮细胞中 TLR4、NF-κB p65、p53 表达的影响

如图 9.19 所示,LPS 处理 12h 和 24h,蜂胶处理组与 LPS 组相比降低了 TLR4 的表达($P<0.05$, $P<0.01$);在 LPS 处理 12h 后,蜂胶处理组 NF-κB p65 的表达与 LPS 组相比略有下降($P<0.05$);而 LPS 处理 12h 和 24h 后,蜂胶处理组与 LPS 组相比 p53 表达显著下降($P<0.05$, $P<0.01$)。

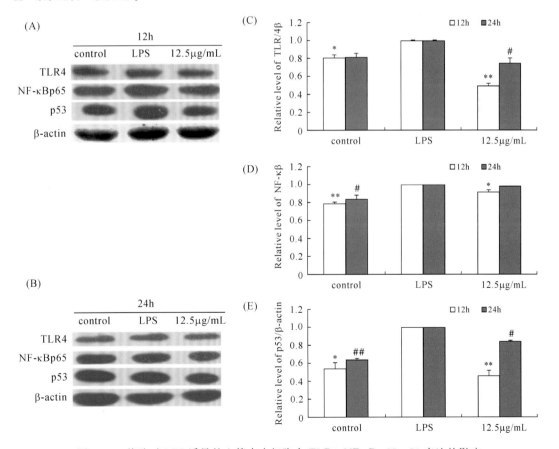

图 9.19　蜂胶对 LPS 诱导的血管内皮细胞中 TLR4、NF-κB p65、p53 表达的影响

Fig. 9.19　Effect of propolis on expression of TLR4, NF-κB p65 and p53 in VEC induced by LPS

2.5 蜂胶对 LPS 诱导的血管内皮细胞中 ROS 的影响

如图 9.20 所示，LPS 处理 12h 后，蜂胶组 ROS 水平与 LPS 组相比差异不显著；在 LPS 处理 24h 后，蜂胶组与 LPS 组相比 ROS 水平显著下降（$P<0.01$）。

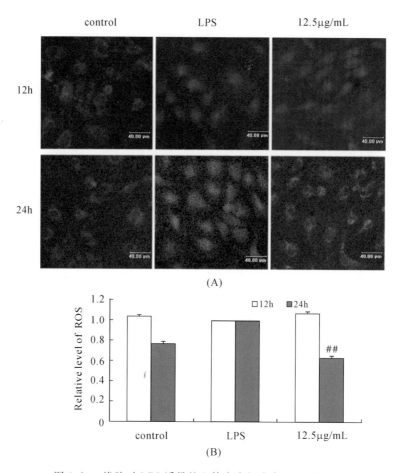

图 9.20　蜂胶对 LPS 诱导的血管内皮细胞中 ROS 的影响

Fig. 9.20　Effect of propolis on ROS level in VEC induced by LPS

2.6 中国蜂胶对 LPS 诱导的血管内皮细胞中线粒体膜电位的影响

如图 9.21 所示，LPS 处理 24h 后，蜂胶处理组的线粒体膜电位水平与 LPS 组相比差异不显著。

3 讨 论

蜂胶具有较好的抗炎功效，然而，其抑制炎症的作用机制还不完全清楚。本实验结果表明，中国蜂胶能够降低 TLR4 的表达和 PC-PLC 活性，进而抑制其下游信号分子 NF-κB p65、p53 的表达和 ROS 的水平，以及抑制 NO 的释放，发挥抗炎功效。

TLR4 是介导先天免疫和炎症反应的跨膜受体，是 LPS 信号转导途径中的关键分子，其主要功能是作为细菌细胞壁脂多糖的转导受体。因此，LPS 的识别及跨膜信号转导位于整个信号转导通路的最上游，对其所介导的下游信号分子起决定性的作用。TLR4 信号通路

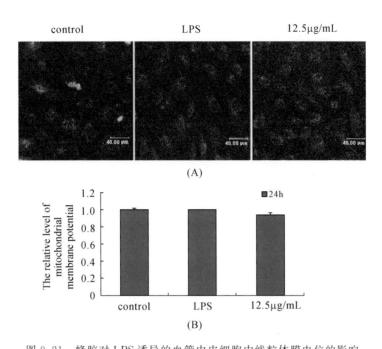

图 9.21 蜂胶对 LPS 诱导的血管内皮细胞中线粒体膜电位的影响
Fig. 9.21 Effect of propolis on mitochondrial membrane potential in VEC induced by LPS

存在 MyD88 依赖途径和 MyD88 非依赖途径。MyD88 依赖途径主要介导 NF-κB 的活化和细胞因子产生,而 MyD88 的非依赖途径主要负责 LPS 诱导的干扰素诱导蛋白-10、干扰素调节基因-1 表达。本研究结果显示,中国蜂胶抑制 LPS 诱导的 HUVECs 中 TLR4 蛋白表达的上调,并抑制下游分子信号分子 NF-κB p65 的表达。因此,中国蜂胶可能通过抑制 TLR4 的表达发挥抗炎功效,而且中国蜂胶阻断细胞内信号 TLR4 的表达可能是 MyD88 依赖的途径。

NF-κB 是 TLR4 下游信号通路中的重要分子。NF-κB 调控免疫、炎症以及增殖等细胞基因的转录。在静息状态,NF-κB 以无活性的形式与抑制性蛋白 IκB 结合,在 LPS 等刺激的作用下,IκB 发生磷酸化并降解,NF-κB 活化,并从细胞质进入细胞核,促进和启动炎性相关因子的转录。本实验结果显示,中国蜂胶能够降低 NF-κB p65 的表达,发挥抗炎功效。而 NF-κB p65 表达的下调除了与蜂胶抑制其上游信号分子 TLR4 的表达有关,还与蜂胶的抗氧化活性有关,因为细胞内 ROS 活性升高,导致 NF-κB 转录活性增强,而本实验研究结果表明,蜂胶能够降低 LPS 诱导的血管内皮细胞中 ROS 水平,进而也抑制了 NF-κB p65 的活化。

PC-PLC 与多种炎症反应密切相关,而且有研究指出,PC-PLC 今后可作为检测炎症的一个指标(Zhang et al.,2010)。LPS 可以活化 PC-PLC,引发一系列信号转导,加剧炎症反应。如在人肺泡巨噬细胞中,LPS 可以活化 PC-PLC,进而活化 NF-κB,释放 TNF-α 和 IL-6 参与炎症反应(Carter et al.,1998);在人单核细胞中,LPS 诱导脂筏中鞘磷脂水解产生神经酰胺、PKC-δ 磷酸化和 TLR4 复合体的装配,激活下游 MAPK 信号通路,释放细胞因子。在此过程中细胞表面的 CD14 与 IL-4 结合,激活 PC-PLC,产生神经酰胺,开启 JAK1-STAT6 信号转导通路(Cuschieri et al.,2006;Zamorano et al.,2003)。在小鼠的巨噬细胞

和小胶质细胞中,LPS 激活 PC-PLC,活化下游 PKC,提高 COX2 mRNA 的表达参与炎症反应(Akundi et al.,2005;Lee et al.,2003)。本实验的研究结果显示,LPS 诱导的血管内皮细胞中 PC-PLC 活性升高,经不同浓度的蜂胶处理后,PC-PLC 的活性降低,这与我们前面的研究一致。因此,本实验结果进一步说明,抑制 PC-PLC 活性参与蜂胶的抗炎功效。

NO 由 NOS 合成,在生物体内作为一种反应性极强的自由基,在体内具有多种生理作用。LPS 长期作用于内皮细胞,iNOS 增加使 NO 自由基含量升高。高浓度的 NO 可以竞争性结合超氧阴离子,生成 ONOO$^-$,从而对细胞造成损伤(Higashi et al.,2009)。本实验结果表明,蜂胶处理 HUVECs 24h 后,可以降低 LPS 诱导的血管内皮细胞中 NO 含量的升高,降低超氧阴离子对内皮细胞的损伤,进而保护内皮细胞。

此外,LPS 长期作用于内皮细胞,NO 含量升高也会导致细胞内 ROS 的升高。作为细胞内重要的信号分子 ROS,在 LPS 诱导的血管内皮细胞损伤中,蜂胶显著地降低了 ROS 的活性,进而保护内皮细胞。

总之,本实验首次研究了中国蜂胶对 PC-PLC 活性和 TLR4 的表达的影响,表明在 LPS 诱导的炎症反应中,中国蜂胶通过降低 PC-PLC 的活性和 TLR4 的表达,进而抑制了 NF-κB p65 和 p53 的表达和 ROS 的活性,以及降低 NO 的产生发挥抗炎功效。此外,在 LPS 诱导的条件下,中国蜂胶处理不影响 LPS 诱导的血管内皮细胞的存活率,表明中国蜂胶不是通过抑制内皮细胞凋亡发挥抗炎机制,其对内皮细胞的保护机制还需要进一步研究。

第三节 蜂胶对 ox-LDL 诱导的血管内皮细胞的影响

血管内皮损伤或功能异常时,血液中的单核细胞通过内皮间隙在内膜下转化为巨噬细胞。巨噬细胞通过 A 型清道受体吞噬大量氧化修饰的低密度脂蛋白(ox-LDL),导致细胞内脂质堆积,形成泡沫细胞,泡沫细胞形成是动脉粥样硬化形成的早期事件。动脉粥样硬化是一种炎症性疾病(Eldika et al.,2004;Montecucco and Mach,2009;Ross,1999),与内皮细胞损伤有直接关系(Szmitko et al.,2003)。炎症反应使大量淋巴细胞和巨噬细胞聚集在斑块内,这些细胞分泌金属蛋白酶,基质金属蛋白酶溶解斑块的纤维成分,从而使斑块的纤维帽变薄,易于破裂最终导致血栓的形成(Choy et al.,2001)。在此过程中,细胞间黏附分子-1(ICAM-1)、血管细胞黏附分子-1(VCAM-1)、单核细胞趋化因子-1(MCP-1)、血凝素样氧化低密度脂蛋白受体-1(LOX-1)、IL-1、IL-6、IL-8、ROS 以及 NF-κB 水平都会升高。Zhang 等(2010)报道,药理性阻断 PC-PLC 活性,能够抑制动脉粥样硬化的进展,并且稳定已存在的斑块的稳定性,而且抑制 PC-PLC 活性可以降低 ox-LDL 诱导的血管内皮细胞中 VCAM-1、ICAM-1 和 MCP-1 的表达。我们团队组前期的动物实验表明,蜂胶醇提液和水提液均能有效抑制高脂血症 SD 大鼠血清中的三酰甘油(TG)、总胆固醇(TC)、低密度脂蛋白(LDL-C)、谷丙转氨酶(GPT)、谷草转氨酶(GOT)的升高,并且能有效抑制肝脏组织中的 TC、TG、MDA 的升高,改善高脂血症大鼠的脂质代谢(Hu,2004b)。通过前面两部分的研究,我们发现巴西蜂胶和中国蜂胶均能够降低 PC-PLC 的活性,而且最近有报道指出,膜联蛋白 A7(ANXA7)是 PC-PLC 的一种内源性调控剂(Li et al.,2013)。由于血脂异常伴随着血管内皮细胞的损伤,在此过程中 ox-LDL 发挥重要作用,那么巴西蜂胶和中国蜂胶是否影响 ox-LDL 能够诱导血管内皮细胞损伤,在 ox-LDL 诱导的血管内皮细胞中,两种类型的蜂胶是否影响 PC-PLC 的活性及其内源调控剂,目前还未有报道,因而有必要在体外研究蜂胶对 ox-LDL 诱导血管内皮细胞损伤的影响,探讨蜂胶对血脂调控及动脉粥样硬化的作用机制。

Ox-LDL 损伤血管内皮细胞,破坏内皮细胞的完整性,根据报道,$45\mu g/mL$ 的 ox-LDL 能够诱导 HUVECEs 凋亡(Liu et al.,2009),因此本实验在选用 $45\mu g/mL$ 的 ox-LDL 诱导 HUVECs 的基础上,进一步研究中国蜂胶和巴西蜂胶对 ox-LDL 诱导的血管内皮损伤的影响。

1 材料和方法

1.1 主要试剂

中国蜂胶取自中国山东省,主要植物来源为白杨属(*Populus* spp.);巴西蜂胶(巴西绿蜂胶),取自巴西 Minas Gerais 州,主要的植物来源是酒神菊树(*Baccharis dracunculifolia* DC.)。DMEM 培养基、胎牛血清(美国 Gibco 公司);胰酶、吖啶橙、MTT、L-α-卵磷脂、DCHF 探针(sigma 公司,美国);ox-LDL(1.5mg/mL)(北京 Union-Biology 公司)、JC-1 探针、驴抗兔 Alexa Fluor-488(Invitrogen 公司,美国);ANXA7、β-actin(Santa Cruz 公司,美国);辣根过氧化物酶(HRP)标记的二抗(南京中杉金桥公司)。

1.2 主要设备与仪器

Heal Force 生物安全柜、Heal Force CO_2 培养箱(力康生物医疗科技控股有限公司);TE2000S 倒置荧光显微镜(日本 Nikon 公司);MK3 酶标仪(芬兰雷勃公司);AR2140 电子分析天平(美国奥豪斯公司);Milli-Q Synthesis 超纯水(美国密理博公司);激光扫描共聚焦显微镜(日本 Olympus FV1200)。

1.3 实验方法

(1)蜂胶样品的制备

巴西蜂胶和中国蜂胶的制备方法同前,于4℃贮存。

(2)人脐静脉血管内皮细胞(HUVECs)培养

HUVECs 由泰山医学院动脉粥样硬化研究所馈赠。HUVECs 细胞培养在添加10%胎牛血清的 DMEM 培养基中进行。孵育条件为:5% CO_2、37℃、饱和湿度 CO_2 培养箱中。倒置相差显微镜下观察细胞形态。

(3)实验分组

实验分为四组:空白对照组(3.5%血清+DMEM 培养基)(control)、ox-LDL 组($45\mu g/mL$ ox-LDL+DMEM 培养基)、EECP 组($45\mu g/mL$ ox-LDL+DMEM 培养基+$12.5\mu g/mL$ EECP)和 EEBP 组($45\mu g/mL$ ox-LDL+DMEM 培养基+$12.5\mu g/mL$ EEBP)。

1.4 检测指标

(1)细胞存活率的测定

将长势良好的血管内皮细胞种植到96孔板中,37℃、5% CO_2 孵育,至细胞长至80%。按照实验分组分别处理细胞12h 和24h 后,分别加入 MTT 继续培养4h,570nm 波长处测吸光值。ox-LDL 组细胞存活率设为 100%。

细胞存活率(%)=(处理组 OD 值/ox-LDL 组 OD 值)×100%

(2)吖啶橙染色检测细胞核

将长势良好的 HUVECs 细胞种植到共聚焦专用小皿中,37℃、5% CO_2 孵育,根据不同的处理 24h 后,加入吖啶橙(AO)染液作用 5min,置于激光共聚焦显微镜下观察细胞核。

(3)细胞中 PC-PLC 活性的测定

采用消化法收集不同处理的细胞,以 L-α-卵磷脂作为 PC-PLC 的底物,在 660nm 波长处测定吸光值。

(4)Western blotting 检测 ANXA7 的表达

用蛋白裂解液收集不同处理组细胞,测定蛋白浓度。每组细胞分别经 12% SDS-PAGE 电泳、转印、杂交、ECL 显色,得到 ANXA7、β-actin 蛋白杂交带。蛋白相对量通过 Quantity One 软件分析。

(5)免疫荧光法检测 ANXA7 和 NF-κB p65 水平

细胞经不同处理 24h 后,4% 多聚甲醛固定 15min,驴血清室温封闭 20min,分别加入 ANXA7 和 NF-κB p65 一抗(1∶100),4℃过夜,0.1mol/L PBS 缓冲液洗 3 遍,加入 FITC 标记的二抗,37℃孵育 1h,0.1mol/L PBS 缓冲液洗 3 遍,激光聚焦显微镜下观察。

(6)细胞中 ROS 的测定

细胞经不同处理 24h 后,弃培养液,用 DMEM 原液轻轻冲洗细胞 2 次,然后弃 DMEM 原液,加入含有 2μl/mL DCHF 荧光探针的 DMEM 原液,37℃孵育 30min,孵育结束后弃 DCHF 染液,用 DMEM 原液轻轻冲洗细胞 3 次,置于激光扫描共聚焦显微镜下观察并分析结果。

(7)细胞中线粒体膜电位的测定

细胞经不同处理 24h 后,弃培养液,用 DMEM 原液轻轻冲洗细胞 2 次,然后弃 DMEM 原液,加入含有 2μl/mL JC-1 荧光探针的 DMEM 原液,37℃孵育 15min,孵育结束后弃 DCHF 染液,用 DMEM 原液轻轻冲洗细胞 3 次,置于激光扫描共聚焦显微镜下观察并分析结果。

1.5 统计学分析

采用 SPSS v11.5 软件包统计分析。计量资料统计结果以均数±标准差($\bar{x}\pm S$)表示,采用 t 检验进行统计学分析。$P<0.05$ 为差异显著。

2 结果

2.1 EECP 和 EEBP 对 ox-LDL 诱导的 HUVEC 细胞存活率的影响

Ox-LDL 是内皮细胞损伤的主要诱因。MTT 检测结果表明,ox-LDL 显著抑制细胞存活率,经 12.5μg/mL 的 EECP 和 EEBP 分别处理 12 和 24h 后与 ox-LDL 组相比细胞存活率显著上调(**,$P<0.01$;图 9.22A)。

2.2 EECP 和 EEBP 对 ox-LDL 诱导的 HUVEC 细胞凋亡的影响

我们进一步检测了 ox-LDL 诱导的内皮细胞凋亡,经 AO 染色后,ox-LDL 组细胞呈现出明显的凋亡现象(图 9.22B),而 EECP 和 EEBP 显著地降低细胞的凋亡率(**$P<0.01$;图 9.22C)。

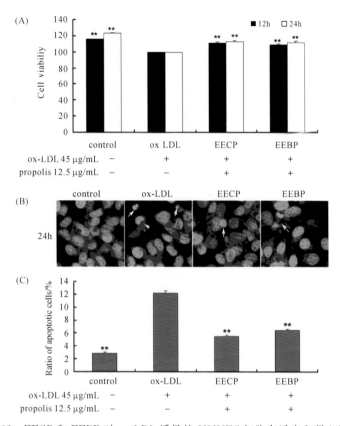

图 9.22 EECP 和 EEBP 对 ox-LDL 诱导的 HUVEC 细胞存活率和凋亡的影响

Fig.9.22 Effect of EECP and EEBP on cell viability and apoptosis in ox-LDL-stimulated HUVECs

注：control:对照组；ox-LDL:氧化型低密度脂蛋白组；EECP:中国蜂胶组；EEBP:巴西蜂胶组；* $P<0.05$，** $P<0.01$，下同。

2.3 EECP 和 EEBP 对 ox-LDL 诱导的 HUVEC 细胞中 PC-PLC 活性的影响

Ox-LDL 处理血管内皮细胞 24h 后细胞中 PC-PLC 活性升高，经不同浓度的 EECP 和 EEBP 处理均显著地降低了细胞中 PC-PLC 的活性（* $P<0.05$，** $P<0.01$；图 9.23）。

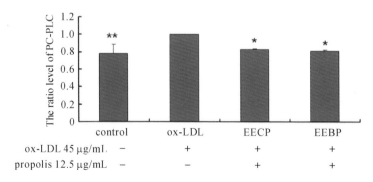

图 9.23 EECP 和 EEBP 对 ox-LDL 诱导的 HUVEC 细胞中 PC-PLC 活性的影响

Fig.9.23 Effect of EECP and EEBP on PC-PLC activity in ox-LDL-stimulated HUVECs

2.4　EECP 和 EEBP 对 ox-LDL 诱导的 HUVEC 细胞中 ANXA7 水平的影响

ANXA7 是 PC-PLC 内源性的调控因子。为了进一步研究 PC-PLC 和 ANXA7 的关系,我们检测了 EECP 和 EEBP 对 ox-LDL 诱导的 HUVEC 细胞中 ANXA7 表达和分布的影响。Western blotting 结果表明,在 12h,EEBP 显著地升高了 ANXA7 的表达;在 24h,两种蜂胶均显著地提高 ANXA7 的表达(** $P<0.01$;图 9.24)。而且,免疫荧光分析表明,EECP 和 EEBP 中 ANXA7 的荧光密度比 ox-LDL 组显著升高,而且出现点状聚集现象(图 9.24A)。

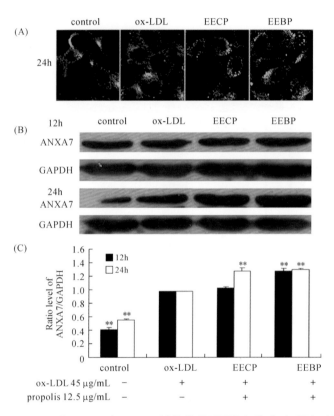

图 9.24　EECP 和 EEBP 对 ox-LDL 诱导的 HUVEC 细胞中 ANXA7 的影响

Fig. 9.24　Effect of EECP and EEBP on ANXA7 level in ox-LDL-stimulated HUVECs

2.5　EECP 和 EEBP 对 ox-LDL 诱导的 HUVEC 细胞中 NF-κB p65 水平的影响

Ox-LDL 诱导 HUVEC 细胞中 NF-κB 的活性。通过荧光检测分析,EECP 和 EEBP 均显著地降低 NF-κB p65 水平,而且两种蜂胶军显著地抑制 NF-κB p65 从细胞质进入细胞核(** $P<0.01$;图 9.25)。

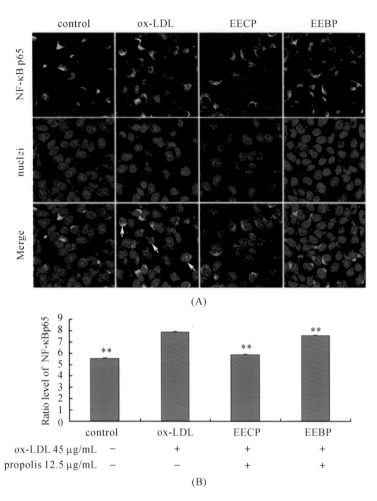

图 9.25　EECP 和 EEBP 对 ox-LDL 诱导的 HUVEC 细胞中 NF-κB p65 水平的影响
Fig. 9.25　Effect of EECP and EEBP on NF-κB p65 level in ox-LDL-stimulated HUVECs

2.6　EECP 和 EEBP 对 ox-LDL 诱导的 HUVEC 细胞中 ROS 的影响

如图 9.26 所示,ox-LDL 处理血管内皮细胞 24h 后,ox-LDL 组细胞中 ROS 水平升高,经 EECP 和 EEBP 处理后显著地降低了细胞中 ROS 的水平($^*P<0.05$)。

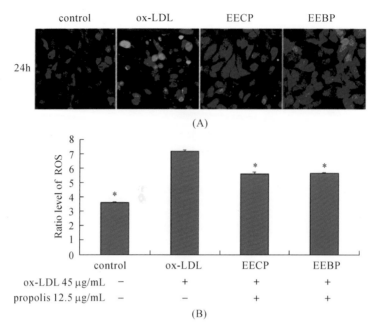

图9.26 EECP和EEBP对ox-LDL诱导的HUVEC细胞中ROS的影响
Fig. 9.26 Effect of EECP and EEBP on ROS level in ox-LDL-stimulated HUVECs

2.7 EECP和EEBP对ox-LDL诱导的HUVEC细胞中线粒体膜电位的影响

Ox-LDL降低线粒体膜电位。12.5 μg/mL EECP和EEBP处理细胞与ox-LDL组相比,显著升高线粒体膜电位($**P<0.01$;图9.27)。

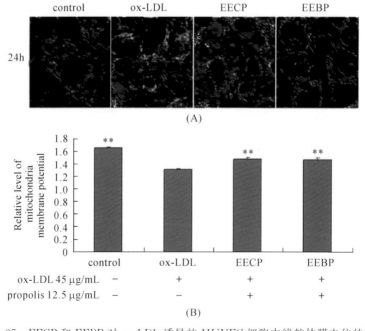

图9.27 EECP和EEBP对ox-LDL诱导的HUVEC细胞中线粒体膜电位的影响
Fig. 9.27 Effect of EECP and EEBP on mitochondria membrane potential in ox-LDL-stimulated HUVECs

小 结

目前的研究结果表明,在 ox-LDL 诱导的血管内皮细胞损伤中,12.5μg/mL 的中国蜂胶和巴西蜂胶均显著地提高细胞存活率,降低细胞凋亡率,升高 ANXA7 的表达,降低 PC-PLC 的活性。而且两种蜂胶均能通过降低 ROS 和升高线粒体膜电位保护 ox-LDL 诱导的内皮损伤。此外,两种蜂胶均能够抑制 NF-κB p65 的活性,抑制其从细胞质进入细胞核。以上很可能是蜂胶保护血管内皮细胞预防动脉粥样硬化的机制之一。我们的研究结果也表明,对 ox-LDL 诱导的血管内皮细胞损伤,中国蜂胶和巴西蜂胶表现出相似的生物学活性。两种类型的蜂胶均可以作为预防动脉粥样硬化的选择替代品。然而,蜂胶调控动脉粥样硬化的作用机制还需要进一步深入研究。

参考文献

Akundi RS, Candelario-Jalil E, Hess S, Hull M, Lieb K, Gebicke-Haerter PJ, Fiebich BL (2005) Signal transduction pathways regulating cyclooxygenase-2 in lipopolysaccharide-activated primary rat microglia[J]. *Glia*, 51(3):199-208.

Awale S, Li F, Onozuka H, Esumi H, Tezuka, Y, Kadota S (2008) Constituents of Brazilian red propolis and their preferential cytotoxic activity against human pancreatic PANC-1 cancer cell line in nutrient-deprived condition[J]. *Bioorg Med Chem*, 16(1):181-189.

Ayres DC, Marcucci MC, Giorgio S (2007) Effects of Brazilian propolis on Leishmania amazonensis[J]. *Mem Inst Oswaldo Cruz*, 102(2):215-220.

Banskota AH, Tezuka Y, Midorikawa K, Matsushige K, Kadota S (2000) Two novel cytotoxic benzofuran derivatives from Brazilian propolis[J]. *J Nat Prod*, 63(9):1277-1279.

Bufalo MC, Candeias JM, Sforcin JM (2009) *In vitro* cytotoxic effect of Brazilian green propolis on human laryngeal epidermoid carcinoma (HEp-2) cells[J]. *Evid Based Complement Alternat Med*, 6(4):483-487.

Carter AB, Monick MM, Hunninghake GW (1998) Lipopolysaccharide-induced NF-kappaB activation and cytokine release in human alveolar macrophages is PKC-independent and TK-and PC-PLC-dependent[J]. *Am J Respir Cell Mol Biol*, 18(3):384-391.

Chen YJ, Huang AC, Chang HH, Liao HF, Jiang CM, Lai LY, Chan JT, Chen YY, Chiang J (2009) Caffeic acid phenethyl ester, an antioxidant from propolis, protects peripheral blood mononuclear cells of competitive cyclists against hyperthermal stress[J]. *J Food Sci*, 74(6):H162-167.

Choy JC, Granville DJ, Hunt DW, McManus BM (2001) Endothelial cell apoptosis: biochemical characteristics and potential implications for atherosclerosis[J]. *J Mol Cell Cardiol*, 33(9):1673-1690.

Cuschieri J, Billgren J, Maier RV (2006) Phosphatidylcholine-specific phospholipase C (PC-PLC) is required for LPS-mediated macrophage activation through CD14[J]. *J Leukoc*

Biol,80(2):407-414.

Daugsch A,Moraes CS,Fort P,Park YK (2008) Brazilian red propolis—chemical composition and botanical origin[J]. *Evid Based Complement Alternat Med*,5(4):435-441.

de Barros MP,Sousa JP,Bastos JK,de Andrade SF (2007) Effect of Brazilian green propolis on experimental gastric ulcers in rats[J]. *J Ethnopharmacol*,110(3):567-571.

Eldika N,Yerra L,Chi DS,Krishnaswamy G (2004) Atherosclerosis as an inflammatory disease:implications for therapy[J]. *Front Biosci*,9:2764-2777.

Feng C,Ye C,Liu X,Ma H,Li M (2004) Beta4 integrin is involved in statin-induced endothelial cell death[J]. *Biochem Biophys Res Commun*,323(3):858-864.

Fonseca YM,Marquele-Oliveira F,Vicentini FT,Furtado NA,Sousa JP,Lucisano-Valim YM,Fonseca MJ (2011) Evaluation of the potential of Brazilian propolis against UV-induced oxidative stress[J]. *Evid Based Complement Alternat Med*,2011.

Fuliang HU,Hepburn HR,Xuan H,Chen M,Daya S,Radloff SE (2005) Effects of propolis on blood glucose, blood lipid and free radicals in rats with diabetes mellitus[J]. *Pharmacol Res*,51(2):147-152.

Heitzer T,Schlinzig T,Krohn K,Meinertz T,Munzel T (2001) Endothelial dysfunction, oxidative stress,and risk of cardiovascular events in patients with coronary artery disease[J]. *Circulation*,104(22):2673-2678.

Higashi Y,Noma K,Yoshizumi M,Kihara Y (2009) Endothelial function and oxidative stress in cardiovascular diseases[J]. *Circ J*,73(3):411-418.

Hoshida Y,Tsukuma H,Yasunaga Y,Xu N,Fujita MQ,Satoh T,Ichikawa Y,Kurihara K,Imanishi M,Matsuno T,Aozasa K (1997) Cancer risk after renal transplantation in Japan[J]. *Int J Cancer*,71(4):517-520.

Hu F,Hepburn HR,Li Y,Chen M,Radloff SE,Daya S (2005) Effects of ethanol and water extracts of propolis (bee glue) on acute inflammatory animal models[J]. *J Ethnopharmacol*,100(3):276-283.

Ishihara M,Naoi K,Hashita M,Itoh Y,Suzui M (2009) Growth inhibitory activity of ethanol extracts of Chinese and Brazilian propolis in four human colon carcinoma cell lines[J]. *Oncol Rep*,22(2):349-354.

Izuta H,Shimazawa M,Tazawa S,Araki Y,Mishima S,Hara H (2008) Protective effects of Chinese propolis and its component,chrysin,against neuronal cell death via inhibition of mitochondrial apoptosis pathway in SH-SY5Y cells[J]. *J Agric Food Chem*,56(19):8944-8953.

Kasai M,Fukumitsu H,Soumiya H,Furukawa S (2011) Ethanol extract of chinese propolis facilitates functional recovery of locomotor activity after spinal cord injury[J]. *Evid Based Complement Alternat Med*,2011.

Kunimasa K,Ahn MR,Kobayashi T,Eguchi R,Kumazawa S,Fujimori Y,Nakano T,Nakayama T,Kaji K,Ohta T (2009) Brazilian propolis suppresses angiogenesis by inducing apoptosis in tube-forming endothelial cells through inactivation of survival signal ERK1/2[J]. *Evid Based Complement Alternat Med*, 870753.

Laferriere J, Houle F, Huot J (2004) Adhesion of HT-29 colon carcinoma cells to endothelial cells requires sequential events involving E-selectin and integrin beta4[J]. *Clin Exp Metastasis*, 21(3):257-264.

Lee SC, Han JS, Seo JK, Cha YN (2003) Modulation of cyclooxygenase-2 expression by phosphatidylcholine specific phospholipase C and D in macrophages stimulated with lipopolysaccharide[J]. *Mol Cells*, 15(3):320-326.

Lemos M, de Barros MP, Sousa JP, da Silva Filho AA, Bastos JK, de Andrade SF (2007) Baccharis dracunculifolia, the main botanical source of Brazilian green propolis, displays antiulcer activity[J]. *J Pharm Pharmacol*, 59(4):603-608.

Li F, Awale S, Tezuka Y, Kadota S (2008) Cytotoxic constituents from Brazilian red propolis and their structure-activity relationship[J]. *Bioorg Med Chem*, 16(10):5434-5440.

Li H, Huang S, Wang S, Zhao J, Su L, Zhao B, Miao J (2013) Targeting annexin A7 by a small molecule suppressed the activity of phosphatidylcholine-specific phospholipase C in vascular endothelial cells and inhibited atherosclerosis in apolipoprotein E-/-mice [J]. *Cell Death Dis*, 4:e806.

Liu X, Zhao J, Xu J, Zhao B, Zhang Y, Zhang S, Miao J (2009) Protective effects of a benzoxazine derivative against oxidized LDL-induced apoptosis and the increases of integrin beta4, ROS, NF-kappaB and P53 in human umbilical vein endothelial cells[J]. *Bioorg Med Chem Lett*, 19(10):2896-2900.

Machado GM, Leon LL, De Castro SL (2007) Activity of Brazilian and Bulgarian propolis against different species of *Leishmania*[J]. *Mem Inst Oswaldo Cruz*, 102(1):73-77.

Messerli SM, Ahn MR, Kunimasa K, Yanagihara M, Tatefuji T, Hashimoto K, Mautner V, Uto Y, Hori H, Kumazawa S, Kaji K, Ohta T, Maruta H (2009) Artepillin C (ARC) in Brazilian green propolis selectively blocks oncogenic PAK1 signaling and suppresses the growth of NF tumors in mice[J]. *Phytother Res*, 23(3):423-427.

Miao JY, Araki S, Kaji K, Hayashi H (1997a) Integrin beta4 is involved in apoptotic signal transduction in endothelial cells[J]. *Biochem Biophys Res Commun*, 233(1):182-186.

Miao JY, Kaji K, Hayashi H, Araki S (1997b) Suppression of apoptosis by inhibition of phosphatidylcholine-specific phospholipase C in vascular endothelial cells[J]. *Endothelium*, 5(4):231-239.

Missima F, Sforcin JM (2008) Green Brazilian propolis action on macrophages and lymphoid organs of chronically stressed mice[J]. *Evid Based Complement Alternat Med*, 5(1):71-75.

Montecucco F, Mach F (2009) Atherosclerosis is an inflammatory disease[J]. *Semin Immunopathol*, 31(1):1-3.

Moura SA, Ferreira MA, Andrade SP, Reis ML, Noviello MD, Cara DC (2009) Brazilian green propolis inhibits inflammatory angiogenesis in a murine sponge model[J]. *Evid Based Complement Alternat Med*.

Munari CC, Alves JM, Bastos JK, Tavares DC (2010) Evaluation of the genotoxic and antigenotoxic potential of Baccharis dracunculifolia extract on V79 cells by the comet assay[J]. *J Appl Toxicol*, 30(1):22-28.

Nakajima Y, Shimazawa M, Mishima S, Hara H (2009) Neuroprotective effects of Brazilian green propolis and its main constituents against oxygen-glucose deprivation stress, with a gene-expression analysis[J]. *Phytother Res*, 23(10):1431-1438.

Nolkemper S, Reichling J, Sensch KH, Schnitzler P (2010) Mechanism of herpes simplex virus type 2 suppression by propolis extracts[J]. *Phytomedicine*, 17(2):132-138.

Pagliarone AC, Orsatti CL, Bufalo MC, Missima F, Bachiega TF, Junior JP, Sforcin JM (2009) Propolis effects on pro-inflammatory cytokine production and Toll-like receptor 2 and 4 expression in stressed mice[J]. *Int Immunopharmacol*, 9(11):1352-1356.

Paulino N, Abreu SR, Uto Y, Koyama D, Nagasawa H, Hori H, Dirsch VM, Vollmar AM, Scremin A, Bretz WA (2008) Anti-inflammatory effects of a bioavailable compound, Artepillin C, in Brazilian propolis[J]. *Eur J Pharmacol*, 587(1-3):296-301.

Pereira AD, de Andrade SF, de Oliveira Swerts MS, Maistro EL (2008) First *in vivo* evaluation of the mutagenic effect of Brazilian green propolis by comet assay and micronucleus test[J]. *Food Chem Toxicol*, 46(7):2580-2584.

Pontin K, Da Silva Filho AA, Santos FF, Silva ML, Cunha WR, Nanayakkara NP, Bastos JK, de Albuquerque S (2008) *In vitro* and *in vivo* antileishmanial activities of a Brazilian green propolis extract[J]. *Parasitol Res*, 103(3):487-492.

Ribeiro LR, Mantovani MS, Ribeiro DA, Salvadori DM (2006) Brazilian natural dietary components (annatto, propolis and mushrooms) protecting against mutation and cancer[J]. *Hum Exp Toxicol*, 25(5):267-272.

Ross R (1999) Atherosclerosis is an inflammatory disease[J]. *Am Heart J*, 138(5 Pt 2):S419-420.

Salomao K, de Souza EM, Henriques-Pons A, Barbosa HS, de Castro SL (2009) Brazilian green propolis: effects in vitro and in vivo on *Trypanosoma cruzi* [J]. *Evid Based Complement Alternat Med*.

Salomao K, Pereira PR, Campos LC, Borba CM, Cabello pH, Marcucci MC, de Castro SL (2008) Brazilian propolis: correlation between chemical composition and antimicrobial activity[J]. *Evid Based Complement Alternat Med*, 5(3):317-324.

Sha N, Guan SH, Lu ZQ, Chen GT, Huang HL, Xie FB, Yue QX, Liu X, Guo DA (2009) Cytotoxic constituents of chinese propolis[J]. *J Nat Prod*, 72(4):799-801.

Speidel D (2010) Transcription-independent p53 apoptosis: an alternative route to death [J]. *Trends Cell Biol*, 20(1):14-24.

Speidel D, Helmbold H, Deppert W (2006) Dissection of transcriptional and non-transcriptional p53 activities in the response to genotoxic stress[J]. *Oncogene*, 25(6):940-953.

Szmitko PE,Wang CH,Weisel RD,Jeffries GA,Anderson TJ,Verma S (2003) Biomarkers of vascular disease linking inflammation to endothelial activation: part II[J]. *Circulation*,108(17):2041-2048.

Usia T,Banskota AH,Tezuka Y,Midorikawa K,Matsushige K,Kadota S (2002) Constituents of Chinese propolis and their antiproliferative activities[J]. *J Nat Prod*,65(5):673-676.

Zamorano J,Rivas MD,Garcia-Trinidad A,Qu CK,Keegan AD (2003) Phosphatidylcholine-specific phospholipase C activity is necessary for the activation of STAT 6[J]. *J Immunol*,171(8):4203-4209.

Zhang L,Li HY,Li H,Zhao J,Su L,Zhang Y,Zhang SL,Miao JY (2011) Lipopolysaccharide activated phosphatidylcholine-specific phospholipase C and induced IL-8 and MCP-1 production in vascular endothelial cells[J]. *J Cell Physiol*,226(6):1694-1701.

Zhao J,Miao J,Zhao B,Zhang S (2005) Upregulating of Fas,integrin beta4 and P53 and depressing of PC-PLC activity and ROS level in VEC apoptosis by safrole oxide[J]. *FEBS Lett*,579(25):5809-5813.

Zhang L,Zhao J,Su L,Huang B,Wang L,Su H,Zhang Y,Zhang S,Miao J (2011) D609 inhibits progression of preexisting atheroma and promotes lesion stability in apolipoprotein e-/-mice:a role of phosphatidylcholine-specific phospholipase in atherosclerosis[J]. *Arterioscler Thromb Vasc Biol*,30(3):411-418.

Zhu W,Li YH,Chen ML,Hu FL (2010) Protective effects of Chinese and Brazilian propolis treatment against hepatorenal lesion in diabetic rats[J]. *Hum Exp Toxicol*,30(9):1246-1255.

胡福良,詹耀锋,陈民利,应华忠,朱威.(2004).蜂胶对高脂血症大鼠血液和肝脏脂质的影响[J].浙江大学学报(农业与生命科学版),30(5):510-514.

詹耀锋(2004)蜂胶对高脂血症SD大鼠的作用及其机制的研究[D].杭州:浙江大学.

玄红专(2003)蜂胶对糖尿病的作用及其机制的研究[D].杭州:浙江大学.

李英华(2002)蜂胶的抗炎免疫作用及其机制的研究[D].杭州:浙江大学.

朱威(2010)中国蜂胶和巴西蜂胶改善糖尿病大鼠的效果及对糖尿病大鼠肾脏的作用机制[D].杭州:浙江大学.

第十章 蜂胶促创伤修复机制研究

　　创伤修复是一个极为复杂的过程,是在各种细胞及其分泌的细胞因子和各种酶类的协同作用下完成的。修复的速度及好坏与致伤因素、伤口的面积和深浅、受感染程度以及机体本身的健康程度等有重要关系,因为这些因素将会进一步影响到伤口炎症反应及氧化应激程度,进而影响到整个修复过程。相对于蜂胶其他生物学活性上的研究,蜂胶在创伤修复上的研究要落后很多,而且迄今为止未见相关的体外实验对蜂胶这一生物学活性进行报道。因此,蜂胶作为众多被报道具有促进创伤修复作用的天然药物之一,尽管其促进创伤修复的效果已被相关的体内研究证实(Han et al.,2005;Pessolato et al.,2011),但具体的作用机制还不明晰,只是简单地将其与蜂胶的其他活性如抗炎、抗氧化及抗菌等基础活性联系起来。然而,蜂胶如何通过其基础活性的发挥加速创伤的修复,以及是否还有可能通过其他方式促进修复,如直接刺激某些修复细胞的增殖、迁移甚至相关功能基因的表达,目前还不清楚。这也是目前植物性成分促创伤修复研究的普遍特点,即作用效果明显,但作用机制含糊。因此,本研究通过建立体外实验模型,初步考察了蜂胶乙醇提取物对小鼠成纤维细胞增殖、迁移以及胶原等基因表达的影响;此外,考虑到氧化应激是影响创面修复的一个极为重要因素,本研究还将深入考察蜂胶的抗氧化作用及其对创伤修复的意义,以期为蜂胶促创伤修复的进一步研究及蜂胶产品的开发提供参考。考虑到蜂胶的化学成分主要受植物源影响,而成分的差异有可能会导致作用方式及效果的不同,本研究采用两种不同来源的蜂胶提取物,分别为中国蜂胶乙醇提取物(EECP)及巴西绿蜂胶乙醇提取物(EEBP)。

第一节 两种植物来源蜂胶提取物的成分分析

　　物质基础决定其生物学活性的发挥。蜂胶生物学活性广泛,正是由于其拥有极为复杂的化学成分。目前研究者已从蜂胶中分离鉴定出二十余类、四百多种天然成分,主要包括类黄酮化合物、酚酸类化合物、醛酮类化合物、萜类化合物、维生素、氨基酸、木脂素以及脂肪酸等(黄帅等,2013)。然而,受地理源、植物源、采胶季节及蜂种等多种因素的影响,不同来源的蜂胶的物质组成会存在一定的差异,而这成分的差异可能会导致不同来源蜂胶药理活性的差异。以杨树型植物为主要植物源的中国蜂胶和以酒神菊属植物为主要植物源的巴西绿蜂胶是目前市场上较为常见的两种蜂胶,并且关于巴西蜂胶体内促创伤修复效果的报道最多。因此,本研究采用中国蜂胶乙醇提取物(EECP)和巴西蜂胶乙醇提取物(EEBP)两种不同来源蜂胶乙醇提取物进行比较研究它们的化学组成及作用差异。

1 实验材料

1.1 材料和试剂

蜂胶来源：中国蜂胶样本为采自中国山东的杨树型蜂胶，主要植物来源为杨树（*Populus*）；巴西蜂胶样本惠赠于蜂乃宝本铺（南京）保健食品有限公司，主要植物来源为酒神菊树属（*Baccharis*）。

主要试剂：香草酸、咖啡酸、*p*-香豆酸、阿魏酸、异阿魏酸、3,4-二甲氧基肉桂酸、芦丁、肉桂酸、短叶松素、柚皮素、槲皮素、木樨草素、山柰酚、芹菜素、松属素、短叶松素-3-乙酸酯、白杨素、CAPE、高良姜素、绿原酸、山柰素、阿替匹林 C（均为标准品纯度≥98%），购自美国 Sigma 公司；无水乙醇和95%乙醇、乙酸等均为国产分析纯试剂；甲醇为色谱纯；色谱级超纯水；福林酚试剂、没食子酸、三氯化铝等。

1.2 主要仪器和设备

旋转蒸发仪 RE-2000A（上海亚荣生化仪器厂）；电冰箱（荣事达电冰箱有限公司）；干燥箱 GZX-9246 MBE（上海博讯实验有限公司）；分析天平 AL104（瑞士梅特勒-托利多仪器有限公司）；超声波清洗器 SK20GT（上海科导超声仪器有限公司）；循环水式真空泵 SHB-III（巩义市予华仪器有限责任公司）；高速中药粉碎机 HX-200（浙江省永康市溪岸五金药具厂）；高压灭菌锅［施都凯仪器设备（上海）有限公司］；多功能连续光谱酶标仪 Spectre Max M5［美谷分子仪器（上海）有限公司］；微量移液器（德国 Eppendorf 公司）；超纯水机（杭州永洁达净化科技有限公司）；高效液相色谱仪 Agilent 1200—内含真空脱气机 G1322A，四进制泵 G1311A，自动进样器 G1329A，可变波长检测紫外检测器（VWD）G1314B 以及恒温箱 G1316A 组成（安捷伦科技有限公司）。

2 实验方法

2.1 蜂胶乙醇提取物的制备

将存储于−20℃冰箱的蜂胶原胶取出适量，用高速中药粉碎机将其粉碎，然后称取适量的蜂胶粉末，按料液比 1∶10 的比例将其溶于 95% 的乙醇溶液中，超声提取 3h 后抽滤。将获得的残渣用同样的方法再重复提取 2 次，然后弃去残渣并合并 3 次提取得到的滤液。将获得的蜂胶滤液置于 4℃ 下制冷，待蜂胶中析出大量絮状物，便可通过抽滤除蜡，反复多次制冷除蜡干净后，滤液经 50℃ 旋转蒸发，最后将获得的蜂胶流浸膏倒在培养皿里的锡箔纸上，放入干燥箱中 50℃ 烘干，每隔一段时间进行称量，达到恒重后取出，放入封口袋中置于−20℃ 冰箱中冻存。待需要时再取出少量溶于无水乙醇中制成 20mg/mL 储备液。

2.2 蜂胶化学成分分析及总黄酮总酚含量的测定

2.2.1 高效液相色谱（HPLC）法成分分析

对 EECP 和 EEBP 的 24 种单体物质的含量进行 HPLC 分析，系统为赛分科技（美国）Sepax HP-C$_{18}$ 柱（150mm×4.6mm，5μm）。柱温为 33℃，检测波长为 280mm，流动相为 1%（*v/v*）的水溶冰醋酸-甲醇。以 1.0mL/min 的流速用甲醇进行梯度洗脱：0～30min 15%～40%，30～65min 40%～55%，65～70min 55%～62%，70～85min 62%～100%。进样量为

5μL，上样浓度为 5mg/mL。

2.2.2 总黄酮和总酚酸含量的测定

总黄酮含量测定：参照 Yang 等（2011）的方法并加以改进，利用芦丁作为标准品。让乙醇样品溶液（EECP 和 EEBP 分别为 0.25mg/mL 和 0.4mg/mL）和 2%（w/v）的三氯化铝（$AlCl_3$）在 1.5mL 离心管中等比例混匀，室温下反应 15min 后，取出 300μL 加入 96 孔酶标板的 1 个酶标孔中，每个样品 2 个复孔。以去离子水等体积替换 2% $AlCl_3$ 作为样本本底对照，然后于 435nm 波长处测定吸光值，最后根据芦丁标准曲线将样品中总黄酮含量表示为芦丁/样品干重（μg/mg）。

总酚酸含量测定：参照 Yang 等（2011）的方法并稍加改进，利用没食子酸制作标准曲线。100μL 0.4mg/mL 的乙醇样品溶液与 100μL 福林酚试剂混合并振荡 3min，然后加入 300μL 2%（w/v）碳酸钠溶液，振荡混匀并放置 3h，取 200μL 加入 96 孔板在 760nm 波长处测吸光值，设立空白对照与复孔。总酚酸含量用没食子酸（μg）/样品干重（mg）表示。

2.3 数据统计分析

结果为 3 次独立重复实验的结果，表示为平均值±标准偏差（$n=3$），利用 SPASS 16.0 软件的 t 检验对数据结果进行差异显著性分析。

3 结果与讨论

3.1 高效液相色谱（HPLC）法对 EECP 和 EEBP 成分分析结果

基于之前的研究，我们采用 HPLC 法对蜂胶中常见的 24 种化学成分在 EECP 和 EEBP 中的含量进行了分析比较（Kumazawa et al.，2003；Zhang et al.，2014）。分析结果如图 10.1 和表 10.1 所示。结果表明，EECP 和 EEBP 的化学成分种类及含量存在明显差异。EECP 含有非常丰富的黄酮类化合物，其中以松属素、短叶松素-3-乙酸酯、白杨素的比例为最高。EEBP 以酚酸类化合物为主，其中以阿替匹林 C 的含量最高。两者共有的成分为咖啡酸、p-香豆酸、山奈酚。

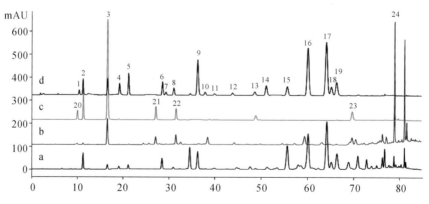

图 10.1 EECP、EEBP 以及混合标准品的 HPLC 图谱

Fig 10.1 HPLC chromatograms of EECP, EEBP and the standard solution

注：a. EECP 图谱；b. EEBP 图谱；c. 混合标准品 1；d. 混合标准品 2。检测波长 280nm。1. 香草酸；2. 咖啡酸；3. p-香豆酸；4. 阿魏酸；5. 异阿魏酸；6. 3,4-二甲氧基肉桂酸；7. 芦丁；8. 肉桂酸；9. 短叶松素；10. 柚皮素；11. 槲皮素；12. 木樨草素；13. 山奈酚；14. 芹菜素；15. 松属素；16. 短叶松素-3-乙酸酯；17. 柯因；18. 咖啡酸苯乙酯；19. 高良姜素；20. 绿原酸；21. 3,5-二咖啡酰奎宁酸；22. 4,5-二咖啡酰奎宁酸；23. 山奈酚；24. 阿替匹林 C。

表 10.1　EECP 和 EEBP 中 24 种单体物质的含量
Table 10.1　Contents of 24 monomer ingredients in EECP and EEBP

成分名称	出峰时间/min	EECP/(mg/g)	EEBP/(mg/g)
香草酸	10.41	—	—
咖啡酸	11.18	7.42±2.34	4.91±2.19
p-香豆酸	16.53	1.71±0.49	30.48±1.27
阿魏酸	19.02	1.49±0.21	—
异阿魏酸	21.17	2.61±0.09	—
3,4-二甲氧基肉桂酸	28.50	7.93±1.19	—
芦丁	29.19	—	—
肉桂酸	31.05	0.52±0.09	—
短叶松素	36.22	14.74±1.89	—
柚皮素	38.24	0.59±0.12	—
槲皮素	40.17	3.14±0.38	—
木樨草素	44.64	2.59±0.68	—
山奈酚	47.82	2.14±0.32	1.50±0.41
芹菜素	51.19	4.10±0.98	—
松属素	55.61	26.73±6.71	—
短叶松素-3-乙酸酯	60.08	53.53±5.29	—
白杨素	64.18	37.81±3.13	—
CAPE	65.22	12.30±5.71	—
高良姜素	66.36	12.03±2.64	—
绿原酸	10.00	—	10.72±0.25
3,5-二咖啡酰奎宁酸	27.11	—	49.21±1.34
4,5-二咖啡酰奎宁酸	31.55	—	36.90±4.05
山奈素	69.27	—	32.73±0.09
阿替匹林 C	79.21	—	104.71±10.39

注：结果表示为平均值±标准差（$n=3$）；"—"表示低于检测限。

3.2　EECP 和 EEBP 主要化学成分比较

酚酸和黄酮是蜂胶中两大主要成分，是蜂胶很多药理活性发挥的物质基础，特别是蜂胶的抗氧化活性（Yang et al.，2011）。而蜂胶抗氧化活性的发挥与蜂胶其他生物学活性的发挥有着密切联系，例如蜂胶的保肝活性（Banskota et al.，2001；El-Khatib et al.，2002）。氧化应激通常作为直接或间接诱因参与机体多种病理过程的发生，对于创伤修复来说亦是如此。此外，有报道表明，某些包括黄酮在内的酚类成分具有促进角质细胞及成纤维细胞增殖、调节胶原基质生成的作用(Stipcevic et al., 2006; Singh et al., 2014)。

因此,我们对 EECP 和 EEBP 的总酚及总黄酮含量进行了测定,结果如表 10.2 所示。结果表明,无论是总酚还是总黄酮含量,EECP 都显著高于 EEBP,与之前的 HPLC 成分分析结果具有一致性。两者在总酚和总黄酮含量的差异也为后续验证两者在作用效果上的差异提供了物质依据。

表 10.2 EECP 和 EEBP 的总酚(TPC)和总黄酮(TFC)含量

Table 10.2 Total phenolic contents (TPC) and total flavonoid contents (TFC) of EECP and EEBP

样品	TPC/(mg GAE/g)	TFC/(mg RE/g)
EECP	182.44±4.18**	258.43±4.75**
EEBP	112.55±4.83	174.02±3.21

注:数据表示为平均值±标准差($n=3$);** $P<0.01$,表示差异极显著;GAE 表示没食子酸等价值;RE 表示芦丁等价值。

小 结

通过利用高效液相色谱(HPLC)法及植物化学方法对 EECP 和 EEBP 的化学成分种类和含量进行了分析比较,分析结果显示,EECP 和 EEBP 在其化学成分种类和含量上有明显的差异,EECP 中酚类物质种类更为丰富,尤其是其中的黄酮类成分,而 EEBP 中主要含有一些酚酸类成分;无论是在总酚还是在总黄酮含量上,EECP 都显著高于 EEBP。

第二节 蜂胶对 L929 细胞增殖、迁移及胶原等基因表达的影响

现有的体内实验研究结果表明,蜂胶具有促进再上皮化、促进肉芽组织形成及调节细胞外基质沉积等多种效果,而这些生物学事件的发生离不开修复细胞的增殖、迁移以及特定功能基因的表达,成纤维细胞是其中最为关键的细胞类型之一,与肉芽组织的形成、伤口的收缩及胞外基质的生成密不可分。因此,本研究以小鼠皮下成纤维细胞 L929 为研究对象,综合考查了 EECP 和 EEBP 对其增殖、迁移以及胶原等基因表达的影响。

1 实验材料

1.1 材料和试剂

L929 细胞来源:小鼠皮下成纤维细胞 L929 购自上海中科院细胞典藏中心。

蜂胶来源:同本章第一节 1.1。

主要试剂和耗材:DMEM 高糖培养基(100 U/mL 青霉素,100 U/mL 链霉素)、PBS 磷酸缓冲液(1×)、0.25%胰酶(含 EDTA)、澳洲胎牛血清(美国 Gibco 公司);RNA 快速提取试剂盒(艾德莱生物科技有限公司);PCR 引物[生工生物工程(上海)有限公司];反转录试剂盒 PrimeScript™ RT-PCR Kit、荧光染料 SYBR Premix EX Taq™(日本 Takara 公司);细胞增殖与活性检测试剂盒 CCK8 试剂盒(日本同仁化学研究所)

一次性移液枪头、细胞培养板、细胞培养皿、细胞冻存管(美国 Coring 公司);一次性针头式滤器(美国 Pall 公司,0.22μm 孔径);一次性灭菌枪头(美国 Kirgen 公司);进口灭菌离心管(美国 Crystalgen 公司);PCR 单管、PCR 八联管(美国 Axygen 公司)。

1.2 主要仪器与设备

CO_2 细胞培养箱(美国 Thermo 公司);高压灭菌锅[施都凯仪器设备(上海)有限公司];多功能连续光谱酶标仪 Spectrc Max M5[美谷分子仪器(上海)有限公司];超净工作台(苏州安康空气技术有限公司);倒置相差显微镜 ECLIPSE TS100(日本 Nikon 公司);电热恒温水槽(上海精宏实验设备有限公司);电冰箱(荣事达电冰箱有限公司);超低温冰箱(中科美菱低温科技股份有限公司);CO_2 钢瓶(杭州今工特种气体有限公司);液氮罐(亚西牌橡塑机器有限公司);微量移液器(德国 Eppendorf 公司);Drummond XP2 电动移液器(美国 Drummond 公司);梯度 PCR 仪 Tgradient(德国 Biometra 公司);微型台式真空泵 GL-802A(江苏海门其林贝尔仪器制造有限公司);5417 R 小型台式高速冷冻离心机(德国 Eppendorf 公司);NanoDrop 2000 超微量分光光度计(美国 Thermo 公司);ABI StepOnePlus™ 实时荧光定量 PCR 系统(美国应用生物系统公司);涡旋仪 Vortex-Genie 2(美国 Scientific Industries 公司)。

2 实验方法

2.1 细胞培养

将小鼠皮下成纤维细胞 L929 复苏于含双抗和 10% 胎牛血清的高糖培养基 DMEM 中,然后置于含 5% CO_2、37℃ 的培养箱中培养。细胞近乎长满后传代,一般 2~3d 进行一次传代,传到第 3 代后开始用于实验。

2.2 CCK8 法检测蜂胶对 L929 细胞增殖和活力的影响

Cell Counting Kit-8,简称 CCK-8,其中含有 WST-8[中文名:2-(2-甲氧基-4-硝基苯基)-3-(4-硝基苯基)-5-(2,4-二磺酸苯)-2H-四唑单钠盐],它在电子载体 1-甲氧基-5-甲基吩嗪鎓硫酸二甲酯(1-Methoxy PMS)的作用下被细胞中的脱氢酶还原为黄色甲䐶产物(Formazan dye),生成的甲䐶物的量与活细胞数目成正比。因此,可被用作细胞增殖和毒性分析。

将 L929 细胞用含 10% 胎牛血清的 DMEM 调成 $5×10^4$ 个/mL 的密度接种于 96 孔细胞培养板中,每个孔 100μL 细胞(最外一圈的孔只加培养基,不作为测定孔用),种板 24h 后,细胞贴壁稳定,每孔加入等体积不同浓度的蜂胶处理,继续培养 24h 后检测。同时设有空白对照(含细胞,但不加任何处理)、本底对照(只是同样体积的培养液,不含细胞)以及溶剂对照(蜂胶溶剂无水乙醇),每组设 3 个复孔。检测试剂为 CCK8 试剂,每孔 5μL,反应 2h 后于 450nm 波长处测定吸光值。

细胞存活率 = [OD(蜂胶组) − OD(本底)] / [OD(空白对照组) − OD(本底)] ×100%

2.3 划痕实验观察蜂胶对 L929 细胞迁移的影响

将 L929 细胞用含 10% 胎牛血清的 DMEM 调成 $5×10^4$ 个/mL 的密度接种于 6 孔细胞培养板(之前已用记号笔在 6 孔板背后画上细线,用作定位),每孔 2mL。待细胞近乎长满后,弃去旧培养基,用 37℃ 预热的 PBS 洗 2 遍后,每孔加入 2mL 不含胎牛血清的 DMEM,饥饿培养 24h,然后利用 10μL 灭菌枪头枪尖和直尺在孔内划出划痕,每个孔左右方向及前后方向各 3 条划痕,共 9 个交点。弃去旧培养基及细胞碎片,用预热的 PBS 洗 2 遍后,重新换上 2mL 含有不同蜂胶浓度的无血清 DMEM 继续培养,然后分别于加药后 6、12、24h 对细胞迁徙状况进行拍照。由于无血清培养条件下,细胞对蜂胶毒性的耐受范围更低,经预实验后,确定采用安全浓度为 1、2μg/mL 的 EECP 和 5、10μg/mL 的 EEBP 处理细胞并观察。

2.4 蜂胶对 L929 胶原等基因 mRNA 表达的影响

2.4.1 样本采集

将 L929 细胞用含 10% 胎牛血清的 DMEM 调成 5×10^4 个/mL 的密度接种于 12 孔细胞培养板,每孔 1mL,培养 24h 后,细胞状态稳定,于不同时间点加入蜂胶(终浓度为 EECP 10μg/mL,EEBP 60μg/mL)样品处理,每个时间点 2 个复孔,使细胞接受药物作用 3、6、9、12、24h 后收获细胞。吸取培养液,PBS 洗 2 遍,然后每个孔加入 500μL TRIzol 裂解液,轻轻吹打几次后移至 1.5mL 无菌无 RNA 酶的离心管中,存于 -80℃ 冰箱,用于下一步的 RNA 提取及反转录实验。

2.4.2 总 RNA 提取和反转录

利用 RNA 快速提取试剂盒对样品的总 RNA 进行提取。大致步骤为:样本解冻后,向已裂解的细胞中加入氯仿(200μL/mL TRIzol),高速涡旋 15s 后,室温下静置 2~3min 出现分层,然后以 12000r/min 的转速于 4℃ 离心 15min。转移上清于 1.5mL 无菌无 RNA 酶离心管中,转移过程中勿触及中间层,并记录体积 V_1。向离心管中加入 $1/2\ V_1$ 无水乙醇,混匀后移入吸附柱中,然后高速离心,再经过去蛋白及漂洗过程后,最后再用无 RNA 酶水将 RNA 洗脱出来。对 RNA 浓度进行测定,然后各样品统一定量到 1μg 后利用反转录试剂进行反转录,然后将获得的 cDNA 存于 -20℃ 备用。

2.4.3 实时荧光定量 PCR 反应

引物设计:在 Genbank 数据库中查找到关于小鼠 I 型和 III 型胶原基因 *COL1A2*、*COL3A1* 及其生长因子 *TGF-β₁* 和基质金属蛋白酶 *MMP-2* 的基因信息,然后利用引物设计软件 Primer Premier 6.0 设计出相应的引物(表 10.3),管家基因 *GAPDH* 的引物参照他人的设计(Wang et al., 2010),由生工生物工程(上海)有限公司负责合成。

表 10.3 实时荧光定量 PCR 引物序列

Table 10.3 Sequences of primers for RT-PCR

Genes	Sense Sequences	Antisense Sequences
COL1A2	5′-CAGGTGCCAGAGGACTTGTTG-3′	5′-GTTGAACCACGATTGCCAGGAG-3′
COL3A1	5′-ACACCTGCTCCTGTGCTTCCT-3′	5′-ACCCATTCCTCCCACTCCAGAC-3′
MMP-2	5′-CCCCGATGCTGATACTGA-3′	5′-CTGTCCGCCAAATAAACC-3′
TGF-β₁	5′-ATTCCTGGCGTTACCTTGG-3′	5′-AGCCCTGTATTCCGTCTCCT-3′
GAPDH	5′-GAGAAACCTGCCAAGTATGATGAC-3′	5′-TAGCCGTATTCATTGTCATACCAG-3′

引物溶解:首先将含引物粉末的小管高速离心 3~5min,转速为 10000r/min,然后轻轻拧开盖子,然后根据引物的量向其中加入适量的 1×TAE,配成终浓度为 100μmol/L 的引物储备液,然后涡旋混匀,存于 -20℃,使用前冰上融化,并用 1×TAE 稀释 10 倍。

反应体系:每个反应设 3 个复孔,每孔 7μL 体系,其中 cDNA 0.15μL、上下游引物各 0.25μL、水 2.85μL、SYBR Green Master Mix 3.5μL。

PCR 扩增条件:95℃预变性 30s,扩增 40 个循环,每个循环 95℃变性 5s,60℃退火 30s。

溶解曲线:65~95℃,每上升 0.3℃采集一次荧光信号。

2.5 数据统计分析

结果为 3 次独立重复实验的结果,表示为平均值±标准差($n=3$),利用 SPASS 16.0 软件的 t 检验对数据结果进行差异显著性分析。$P<0.05$ 为差异显著,用 * 表示;$P<0.01$ 为差异极显著,用 ** 表示。

3 结果与讨论

3.1 EECP 和 EEBP 对 L929 细胞增殖和活力的影响

细胞的增殖是创伤修复过程中最重要的生物学事件之一。角质形成细胞通过增殖帮助促进创面的再上皮化;血管内皮细胞通过增殖促进肉芽组织中新血管的形成;而成纤维细胞的增殖对真皮结构的重建则起着非常重要的作用,细胞外基质的生成、伤口的收缩以及其他生物学事件的发生发展都与成纤维细胞的存在密不可分。很多具有创伤修复作用的药物成分,都被证明具有良好的促修复细胞增殖的作用。桑科榕属植物 *Ficus asperifolia* Miq. 和锦葵科亚洲棉 *Gossypium arboreum* L. 是存在于非洲西部的两种具有治疗创伤效果的民间药材,有体外实验研究表明,它们的水提物具有刺激人真皮成纤维细胞增殖的作用(Annan and Houghton,2008)。此外,其他的一些促进伤口愈合的植物性成分也被报道具有刺激修复细胞增殖的作用(Zippel et al.,2009;Gescher and Deters,2011)。

因此,本研究考察察了 EECP 和 EEBP 两种蜂胶提取物对小鼠皮下成纤维细胞 L929 的细胞增殖的影响,结果如图 10.2 所示。由图 10.2 可见,两种蜂胶提取物对 L929 都没有生长刺激作用,并且当提取物的浓度超过一定范围会产生细胞毒性,表现为抑制细胞的生长。由此可见,蜂胶乙醇提取物并不能直接刺激成纤维细胞的增殖以实现其促创伤修复效果,它的作用很可能通过其他途径实现。例如,某些修复成分被报道虽然不能通过直接刺激成纤维细胞增殖促进修复,但有可能通过刺激其迁移、其他修复细胞增殖,抗菌,抗氧化等途径帮

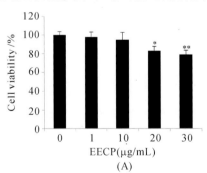
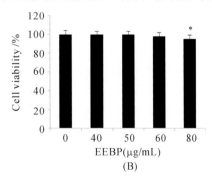

图 10.2 EECP 和 EEBP 对 L929 细胞活力的影响

Fig 10.2 Effects of EECP and EEBP on viability of L929 cells

注:以图中所示浓度 EECP 和 EEBP 处理 L929 细胞 24h 后,利用 CCK-8 试剂检测细胞活力(%),数据结果表示为平均值±标准差($n=3$)。* 表示 $P<0.05$,即与未处理组比较差异显著;** 表示 $P<0.01$,即与未处理组比较差异极显著。

助伤口修复(Steenkamp et al.,2004;Gescher and Deters,2011;Balekar et al.,2012)。另外,从图 10.2 中可以看出,EECP 在其浓度为 20μg/mL 时,已表现出细胞毒性,存活率在 83% 左右;EEBP 在浓度为 80μg/mL 时才表现出明显的细胞生长抑制。由此可见,EECP 比 EEBP 具有更强的细胞毒性,这可能与 EECP 含有更多的黄酮性成分有关。这提示蜂胶在临床用药时需要慎重考虑其合适的用药浓度。接下来的实验将建立在两种蜂胶的安全浓度范围内:EECP≤10μg/mL,EEBP≤60μg/mL。

3.2 EECP 和 EEBP 对 L929 细胞迁移的影响

细胞的迁移是整个创伤修复过程中另一个重要的生物学事件,尤其对创面的再上皮化过程尤为关键。伤口边缘及残存的皮肤附件上的角质形成细胞通过迁移、增殖及分化逐渐覆盖创口,重建表皮完整性,抵抗感染。而成纤维细胞、血管内皮细胞也是首先在各种因素共同作用下迁移进入创面,才能通过进一步的增殖、分化及分泌生长因子等方式促进肉芽组织的形成,填补组织缺损。因此,对修复细胞迁移能力的考察也是研究一些药物促创伤修复效果及方式重要的参考指标之一。划痕实验是体外研究药物促伤口修复效果常用到的一种实验手段,用于考察细胞运动,是对体内伤口愈合的简单模拟,便于操作。其原理是:当细胞长满单层时,在培养板底部人为刮出一条不含细胞的空白区域,称为"划痕",通过划痕弥合的快慢,考察细胞迁移的能力。一些促创伤修复药物被证实具有良好的促修复细胞迁移的能力(Fronza et al.,2009;Balekar et al.,2012)。

本实验利用划痕实验考察蜂胶对 L929 迁移的影响。为避免细胞增殖带来的干扰,采用无血清条件下观察药物的刺激作用,此时细胞对蜂胶耐受的毒性范围也变小,因此,EECP 浓度采用 1 或 2μg/mL 的安全作用浓度,EEBP 采用 5 或 10μg/mL 的安全作用浓度。分别于药物作用后 6、12、24h 观察划痕的弥合情况,并拍照,结果见图 10.3。由图可见,两种蜂胶醇提物对细胞的迁移都没有明显的促进作用。

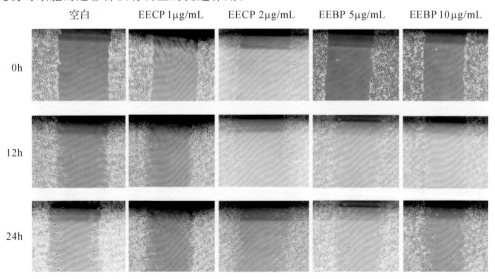

图 10.3 EECP 和 EEBP 对 L929 细胞迁移的影响

Fig 10.3 Effects of EECP and EEBP on migration of L929 cells

注:利用划痕实验观察 EECP 和 EEBP 对 L929 细胞迁移的影响。在 L929 细胞单层上制造划痕,然后加入如图所示浓度的 EECP 和 EEBP 刺激,并与刺激后 0、12、24h 用倒置相差显微镜拍照保存(放大倍数为 10)。

3.3 EECP 和 EEBP 对 L929 胶原等基因 mRNA 表达的影响

胶原蛋白是动物体内含量最丰富的一类蛋白,也是构成细胞外基质最主要的成分,主要存在于结缔组织中。皮肤中最主要的胶原类型是Ⅰ型和Ⅲ型胶原,它们特有三股螺旋结构使其具有较强的抗张能力,从而赋予皮肤韧性。皮肤组织缺损以后,为恢复皮肤结构和功能的完整性,成纤维细胞会在趋化因子的作用下迁移进入创面,然后开始增殖、分化并合成和分泌大量以胶原为主的细胞外基质(ECM)(Schultz et al.,2009),从而填补组织的缺损。创伤修复过程的完成还涉及各种细胞因子以及各种酶的共同参与及相互作用。TGF-β 是这个修复过程中最为关键的生长因子之一,它可以由巨噬细胞、表皮细胞以及成纤维细胞等多种细胞通过各种方式分泌,在调节修复细胞增殖、分化以及胶原等基质生成中起着非常关键的作用(Barrientos et al.,2008)。除了生长因子,基质金属蛋白酶家族也是调控整个修复过程非常关键的因素,它们可以通过调节炎症反应、降解细胞外基质促进细胞的迁移以及胞外基质的重塑(Xue et al.,2006)。因此,调节这些参与创伤修复调控的关键因子的表达和代谢有可能是一些促修复药物促进伤口恢复的重要原因之一。例如,有研究表明促创伤修复草药积雪草含有的三萜类化合物积雪草皂苷和羟基积雪草苷,能促进人皮肤成纤维细胞 $TGF-\beta_1$ 表达,并激活相应的 $TGF-\beta_1/smad3$ 信号通路,从而促进成纤维细胞表达Ⅰ型胶原、Ⅲ型胶原 mRNA 表达(Wu et al.,2012)。而之前的研究表明,蜂胶具有促进肉芽组织形成、调节胞外基质生成的作用,这暗示蜂胶有可能通过调节某些生长因子甚至基质金属蛋白酶的表达,或直接刺激成纤维细胞中胶原等胞外基质基因的表达,从而促进创伤修复。因此,本研究考察了两种蜂胶提取物对 L929 成纤维细胞Ⅰ型胶原(COL1A2)、Ⅲ型胶原(COL3A1)、生长因子($TGF-\beta_1$)以及基质金属蛋白酶(MMP-2)表达的影响,结果见图 10.4。

从图 10.4(A)、(B)、(C)、(D)中可以看出,两种蜂胶提取物对 COL1A2,COL3A1 的 mRNA 表达都没有明显的刺激作用;从图 10.4(C)、(F)中可以看出,EECP 和 EEBP 对 $TGF-\beta_1$ 表现出一定的抑制作用,而 $TGF-\beta_1$ 则是调节胶原表达非常关键的生长因子,不过两种蜂胶对 $TGF-\beta_1$ 的表达抑制并没有进一步影响到 L929 胶原 mRNA 的表达[如图 10.4(A)、(B)、(C)、(D)],原因尚不清楚。但是这些结果与蜂胶某些体内修复实验报道存在差异,例如 Albuquerque-júnior 等(2009)和 Pessolato 等(2011)的体内实验结果表明,蜂胶能够促进创伤修复早期创面胶原等基质的沉积。猜测造成这种差异的可能的原因有以下几种:(1)蜂胶通过抗炎、抗氧化等基础活性的发挥使创面更快地从炎症期过渡到组织生长期,这使得在修复的某个阶段,同时对控制组和蜂胶治疗组取样观察,会出现蜂胶治疗组表现出比控制组更轻的炎症反应及更多的胶原沉积。(2)成纤维细胞并不是 $TGF-\beta_1$ 的唯一来源,这表明蜂胶仍有可能通过刺激其他细胞 $TGF-\beta_1$ 的表达和分泌,进一步影响到创面胶原基因的表达。例如,Ansorge 等(2003)的实验表明蜂胶能够刺激人类外周血单核细胞 $TGF-\beta_1$ 的表达,而单核巨噬细胞一直被认为是创面修复过程 $TGF-\beta_1$ 等多个生长因子的重要来源(Adam and Richard,1999)。此外,$TGF-\beta_1$ 也不是调节成纤维细胞胶原基因表达的唯一一个生长因子,如碱性成纤维细胞因子(bFGF)也能调节成纤维细胞胶原的表达(Barrientos et al.,2008)。因此,蜂胶是否对其他重要生长因子的表达产生影响有待于进一步研究。总的来说,通过当前的结果可以推测,蜂胶促进体内修复过程中胶原基质的沉积很可能并不是由于直接刺激成纤维细胞 ECM 表达,而是通过其他方式间接促进 ECM 的表达和沉积。

最后,通过图 10.4(E)、(F)可以看出,EECP 和 EEBP 都能刺激 L929 细胞基质金属蛋白

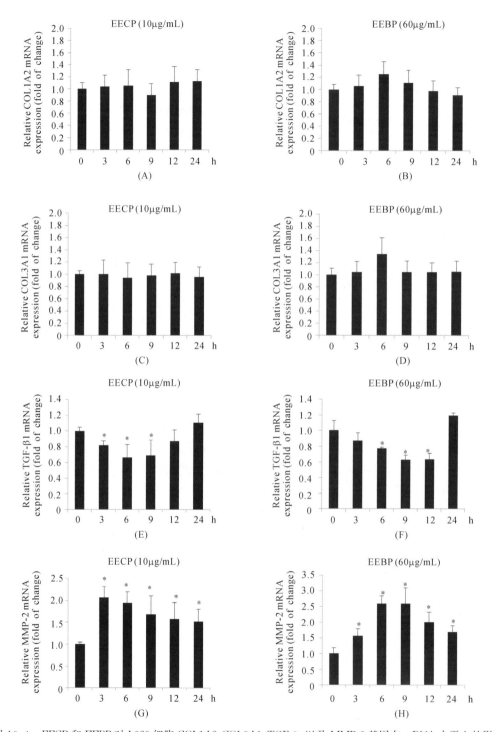

图 10.4 EECP 和 EEBP 对 L929 细胞 COL1A2、COL3A1、TGF-β_1 以及 MMP-2 基因在 mRNA 水平上的影响

Fig 10.4 Effects of EECP and EEBP on mRNA expression of COL1A2, COL3A1, TGF-β_1 and MMP-2 expression in L929 Cells

注：为探究 EECP 和 EEBP 对 L929 细胞修复相关基因表达的影响，用 10μg/mL 的 EECP 处理 L929 细胞相应的时间。利用 qRT-PCR 技术对 EECP 作用不同时间后几个修复相关基因(COL1A2、COL3A1、TGF-β_1 及 MMP-2)的 mRNA 相对表达量进行检测。结果表示为平均值±标准差($n=3$)，* 表示 $P<0.05$，即蜂胶组与未处理组比较差异显著。

酶 MMP-2 的表达。在蜂胶作用 3h 后 MMP-2 的表达上调就能到达显著水平,当蜂胶作用超过 12h 后,MMP-2 的上调开始减弱,但直到蜂胶作用 24h 后,MMP 的表达水平仍然高出正常水平。MMP 是一个大家族,主要负责 ECM 的降解与重塑,对创伤修复过程组织清创、细胞迁移、血管化、再上皮化、肉芽组织的重塑等多个生物学过程起非常重要的调节作用(Xue et al.,2006)。MMP-2 是其中非常重要的一种,参与修复的各个过程,在早期可以帮助降解变性的胞外基质,降低炎症反应及促进再上皮化;中期可以帮助降解基底膜以促进血管内皮细胞迁移及新血管形成(Xue et al.,2006)。它还能促进基质中 TGF-β_1 的释放(McCawley and Matrisian 2001),而 TGF-β_1 是调节 ECM 沉积及重塑的关键生长因子。此外,创伤发生后,MMP-2 将持续稳定地上调,几乎贯穿于整个修复过程,表明 MMP-2 还参与到修复后期漫长的基质降解及重塑过程(Elizabeth Fin et al.,1992;Ågren,1994)。因此,蜂胶对创面再上皮化、肉芽组织生长及修复前期 ECM 沉积的促进作用可能与它对 MMP-2 的表达刺激作用有一定关系。蜂胶对 L929 细胞 TGF-β_1 的表达抑制[如图 10.4(C)、(D)]和其对 MMP-2 的表达上调[如图 10.4(E)、(F)]暗示了蜂胶治疗创伤修复可能有助于抑制瘢痕的形成,起到抗纤维化的作用。

鉴于创伤修复的复杂性,蜂胶是否能够通过影响其他细胞因子、生长因子及蛋白酶在创面的表达及作用发挥,从而促进伤口更好更快地愈合,仍需要进一步研究。

小 结

通过考察 EECP 和 EEBP 对 L929 细胞增殖、迁移以及相关功能基因表达的影响,结果表明两者都没有明显的增殖及促进迁移的效果;相反,当蜂胶浓度超过一定范围时,反而表现出抑制细胞增殖的效果,提示临床用药时需要特别注意蜂胶的安全作用浓度范围。EECP 和 EEBP 都不能影响 L929 细胞 I 和 III 胶原基因在体外的表达,而且两种蜂胶虽然成分存在差异,但都能轻微抑制 L929 细胞 TGF-β_1 的表达并刺激其 MMP-2 的表达,这表明蜂胶促进创面修复早期胶原基质沉积的效果有可能是因为蜂胶通过其他途径减少了阻碍创面胶原生成的不利因素,而不是通过直接刺激成纤维细胞胶原表达而发挥作用。此外,蜂胶对促胶原生成因子 TGF-β_1 的表达抑制和基质代谢酶 MMP-2 表达刺激暗示蜂胶可能具有防止创面胶原过度沉积起到抗瘢痕形成的作用。

第三节　蜂胶对 H_2O_2 致 L929 细胞损伤的保护作用及可能机制

ROS(活性氧类)是活细胞代谢过程中产生的各种活性氧的总称,包括活性氧自由基,如超氧阴离子($\cdot O_2^-$)、羟基自由基($\cdot OH^-$)等,以及非自由基类的过氧化物,如过氧化氢(H_2O_2)。一方面,这些小分子物质在可以充当第二信使调节胞内正常的生化反应;另一方面,当其过量产生时,会因其强氧化性造成细胞膜的脂质过氧化甚至 DNA 的损伤,即所谓的氧化应激。正常情况下,机体可以通过自身的抗氧化防御系统将体内的活性氧含量维持在一个较低的水平内,使机体正常的氧化还原系统维持一个良好的动态平衡。但在某些病理状态下这种平衡会被打破,导致 ROS 的过度产生和蓄积,从而加重病情的发展。机体受到严重创伤后,如烧烫伤,局部组织缺血缺氧,加上大量的炎症因子及细菌的侵入,共同促进了烧伤组织的脂质过氧化反应和大量氧化物质的产生(Parihar et al.,2008),造成创面的氧化应激,进而阻碍修复的进程。另外,氧化应激也被认为是造成其他修复障碍性疾病的重要原因,如慢性下肢静脉溃疡(James et al.,2003)。因此,抗氧化治疗有助于加速创伤的修复。而蜂胶的抗氧化活性已被大量的研究证实,包括其自由基清除能力、还原力以及细胞抗氧化能力(张江临等,2013;Zhang et al.,2015)。本实验利用 H_2O_2 诱导的 L929 氧化应激模型,深入探索蜂胶及其抗氧化作用对创伤修复的意义。

1 实验材料

1.1 材料与试剂

L929 细胞来源:同本章第二节 1.1。

蜂胶来源:同本章第一节 1.1。

主要试剂和耗材:DMEM 高糖培养基(100 U/mL 青霉素,100 U/mL 链霉素)、磷盐酸缓冲液(1×)、0.25%胰酶(含 EDTA)、澳洲胎牛血清(美国 Gibco 公司);RNA 快速提取试剂盒(艾德莱生物科技有限公司);反转录试剂盒 PrimeScript™ RT-PCR Kit,荧光染料 SYBR Premix EX TaqTM(日本 Takara 公司);细胞增殖与活性检测试剂盒 CCK-8 试剂盒(日本同仁化学研究所);蛋白 Marker(美国 Thermo 公司);磷酸酶抑制剂、蛋白酶抑制剂(美国 Roche 公司);信号通路特异性抑制剂(SB203580、PD98059、SP600125、LY294002,美国 Selleck 公司);40%丙烯酰胺、NP-40 裂解液(碧云天生物技术研究所);BCA 蛋白浓度测定试剂盒(上海威奥生物科技有限公司);5×Tris-甘氨酸-SDS 电泳缓冲液(上海百赛生物技术有限公司);10×TBST,pH 8.8,1.5mol/L Tris-Cl 缓冲液,pH 6.8 1mol/L Tris-Cl 缓冲液(上海双螺旋生物科技有限公司);PCR 引物、30%过氧化氢 H_2O_2、TEMED、β-巯基乙醇、甘氨酸、脱脂奶粉、甲醇、Tris 缓冲液(1mol/L,pH 8.0)、十二烷基硫酸钠 SDS、过硫酸铵等试剂[生工生物工程(上海)有限公司]。

一次性移液枪头、细胞培养板、细胞培养皿、细胞冻存管、细胞刮刀(美国 Coring 公司);一次性针头式滤器(美国 Pall 公司,0.22μm 孔径);一次性灭菌枪头(美国 Kirgen 公司);进口灭菌离心管(美国 Crystalgen 公司);PCR 单管、PCR 八联管(美国 Axygen 公司);PVDF 膜(美国 Millipore 公司);其他均为普通国产耗材,如一次性手套等。

1.2 主要仪器与设备

CO_2 细胞培养箱(美国 Thermo 公司);高压灭菌锅(施都凯仪器设备(上海)有限公司);多功能连续光谱酶标仪 Spectre Max M5(美谷分子仪器(上海)有限公司);超净工作台(苏州安康空气技术有限公司);倒置相差显微镜 ECLIPSE TS100(日本 Nikon 公司);电热恒温水槽(上海精宏实验设备有限公司);电冰箱(荣事达电冰箱有限公司);超低温冰箱(中科美菱低温科技股份有限公司);CO_2 钢瓶(杭州今工特种气体有限公司);液氮罐(亚西牌橡塑机器有限公司);微量移液器(德国 Eppendorf 公司);Drummond XP2 电动移液器(美国 Drummond 公司);梯度 PCR 仪 Tgradient(德国 Biometra 公司);微型台式真空泵 GL-802A(江苏海门其林贝尔仪器制造有限公司);5417 R 小型台式高速冷冻离心机(德国 Eppendorf 公司);NanoDrop 2000 超微量分光光度计(美国 Thermo 公司);ABI StepOnePlus™ 实时荧

光定量 PCR 系统(美国应用生物系统公司);流式细胞仪(美国 BD 公司);涡旋仪 Vortex-Genie 2(美国 Scientific Industries 公司);双垂直电泳槽 DYCZ-24DN、DYCZ-40D 型转印电泳仪、DYY-6D 型电泳仪电源(北京市六一仪器厂)。

1.3 主要试剂和配方

10%过硫酸铵:过硫酸铵粉末与水按质量体积比为1:10的比例混合,现配现用。

5%脱脂牛奶:一定量的脱脂奶粉溶于1×TBST(质量体积比为5%)。

5×蛋白上样缓冲液:将0.5mol/L pH6.8 的 Tris-盐酸缓冲液、20%SDS、5%溴酚蓝溶液、甘油、水按照2:4:1:12:1的体积比混合制成,以1mL/管的量分装于1.5mL离心管中,存于-20℃,待用时每管加入 30μL 的 β-巯基乙醇。

NBT/BCIP 底物显色液:准确称量200mg 的 NBT 粉末和100mg 的 BCIP 粉末,共溶于67%的 DMSO 中,涡旋混匀,避光存于-20℃。适用于按1:50的体积比与 AP(1×)混合。

AP 显色缓冲液(1×):100mmol/L Tris-Cl,50mmol/L NaCl,5mmol/L $MgCl_2$;pH9.5。

Western 转膜缓冲液(1×):25mmol/L Tris-Cl,192mmol/L 甘氨酸,20%甲醇;pH8.0。

Tris-甘氨酸-SDS 电泳缓冲液(1×):由5×Tris-甘氨酸-SDS 电泳缓冲加水稀释而成。

TBST(1×):由10×TBST 加水稀释而成。

2 实验方法

2.1 自由基及总还原力测定

DPPH 自由基清除能力测定:参照 Yang 等(2011)的方法,并稍加改进。首先称取适量的 DPPH,将其定溶于无水乙醇中配成1mg/mL的储备液,4℃避光保存,使用时用无水乙醇进一步稀释成0.1mg/mL的工作液浓度。另配制1mg/mL的蜂胶乙醇提取物储备溶液,使用时再用无水乙醇稀释一系列的浓度梯度。取上述稀释液100μL加入100μL 0.1mg/mL DPPH 工作液在96孔酶标板的小孔中避光反应,同时本底对照,即以不加 DPPH(以100μL 无水乙醇代替 DPPH)的样品溶液各浓度作为对照以消除样品本身颜色对测试结果的干扰,并设 DPPH 阴性对照(以100μL 无水乙醇代替样品),每组平行设2个复孔。室温避光反应30min 后,在517nm 波长处测定吸光度。

清除率(%)=[1-(A1-A2)/A0]×100%

A0 是指 DPPH 本身的吸光度,即100μL 无水乙醇+100μLDPPH

A1 是指样品和 DPPH 反应后的吸光度,即100μL 样品液+100μLDPPH

A2 是指样品本底吸光度,即100μL 样品液+100μL 无水乙醇

结果是以样品溶液浓度为横坐标、DPPH 自由基清除率为纵坐标的散点图,求线性方程,并计算清除50% DPPH 所需样品的浓度,即半抑制浓度 IC_{50},比较样品清除 DPPH 自由基活性。

ABTS 自由基清除能力测定:参照 Yang 等(2011)的方法,并稍加改进。首先称取适量的 ABTS 粉末溶于超纯水中,配制成7mmol/L 的 ABTS 溶液。再称取适量的过硫酸钾粉末,将过硫酸钾粉末溶于水配置成140mmol/L 的过硫酸钾水溶液。然后将7.5mL 7mmol/L 的 ABTS 溶液和132μL 140mmol 过硫酸钾水溶液避光反应16h,该溶液为前一天配置,使用前用无水乙

醇稀释到吸光值在732nm波长处为0.7左右。使用前将样品储备液(1mg/mL)稀释成一系列浓度梯度。取上述稀释液50μL和ABTS工作液100μL在96孔酶标板的小孔中暗光反应。同时设本底对照,即以不加ABTS(以100μL无水乙醇代替ABTS)的样品溶液各浓度作为对照以消除样品本身颜色对测试结果的干扰,并设ABTS阴性对照(以50μL无水乙醇代替样品),每组平行设2个复孔。室温避光反应10min后,在734nm波长处测定吸光度。样品对ABTS自由基清除率计算按下面的公式:

$$清除率(\%) = [1-(A1-A2)/A0] \times 100\%$$

A0指ABTS本身的吸光度,即50μL无水乙醇+100μL ABTS

A1指样品和ABTS反应后的吸光度,即50μL样品液+100μL ABTS

A2指样品本底吸光度,即50μL样品液+100μL无水乙醇

结果是以样品溶液浓度为横坐标、ABTS自由基清除率为纵坐标的散点图,求线性方程,并计算清除50% ABTS所需样品的浓度,即半抑制浓度IC_{50},比较样品清除ABTS自由基活性。

总还原力测定:参照Moreira等(2008)的方法,并稍加改进,该方法也称为普鲁士蓝法(曾军和石国荣,2008)。

溶液配制:

(1)磷酸缓冲液的配制:称取3.1202g二水合磷酸二氢钠(相对分子质量156.01),定溶于100mL容量瓶中,配制成0.2mol/L磷酸二氢钠溶液。称取3.5814g十二水合磷酸氢二钠(相对分子质量358.14),定溶于50mL容量瓶中,配制成0.2mol/mL磷酸氢二钠溶液。取62.5mL磷酸二氢钠溶液与37.5mL磷酸氢二钠溶液混合成0.2mol/L的磷酸缓冲液,并用pH计将缓冲液的pH调至6.6。

(2)铁氰化钾溶液的配制:精确称取0.2413g六氰合铁(Ⅲ)酸钾固体,用蒸馏水溶解,配制成1%的铁氰化钾溶液,避光保存。

(3)三氯乙酸溶液的配制:精确称取2.1121g三氯乙酸固体,用蒸馏水溶解,配制成10%的三氯乙酸溶液,避光保存。

(4)三氯化铁溶液的配制:精确称取0.0322g三氯化铁固体,用蒸馏水溶解,配制成0.1%的三氯化铁溶液,避光保存。

具体操作步骤:

先取312.5μL pH6.6的磷酸缓冲液和312.5μL的1%的铁氰化钾溶液一起加入1.5mL离心管中,再取125μL样品溶液混入其中,涡旋混匀后于50℃水浴锅中反应20min后迅速冷却,加入312.5μL 10%的三氯乙酸,混合均匀,然后以2000r/min的速度离心10min,取上清液900μL,加入蒸馏水312.5μL和0.1%的三氯化铁62.5μL涡旋混匀。取96孔的酶标板,每个孔加180μL混合液,每组2个复孔。在700nm处测定吸光值,吸光值越大则样品的还原能力越强。其中以Trolox作为标准品,阴性对照是将样品改为125μL的乙醇,其他不变。

2.2 不同浓度H_2O_2对L929细胞存活率的影响

将L929细胞用含10%胎牛血清的DMEM调成5×10^4个/mL的密度接种于96孔细胞培养板中,每个孔100μL细胞(最外一圈的孔只加培养基,不作为测定孔用),种板24h后,细胞贴壁稳定,每孔加入不同浓度的H_2O_2刺激,终浓度为0.3~0.9mmol/L,继续培养24h

后检测。同时设有空白对照(含细胞,但不加任何处理)、本底对照(只是同样体积的培养液,不含细胞)以及溶剂对照(无菌去离子水),每组设 3 个复孔。检测试剂为 CCK-8 试剂,每孔 5μL,反应 2h 后于 450nm 波长处测定吸光值。

细胞存活率=[OD(刺激组)-OD(本底)]/[OD(空白对照组)-OD(本底)]

2.3 蜂胶提取物(EECP 和 EEBP)存在条件下 H_2O_2 对 L929 细胞存活率的影响

将 L929 以 $5×10^4$ 个/mL 的密度接种于 96 孔板 24h 后,先加入不同浓度蜂胶提取物(EECP 为 5、7.5、10μg/mL,EEBP 为 40、50、60μg/mL)预孵育 3h,再给予细胞以 800μmol/L 终浓度的 H_2O_2 刺激 24h 后,利用 CCK-8 法试剂盒检测细胞存活率。实验分成三大组:空白对照组、H_2O_2 单独刺激组、蜂胶与 H_2O_2 共孵育组。每组设 3 个复孔,实验重复不少于 3 次。每次检测完后,在倒置显微镜下观察细胞形态变化,并拍照保存(放大倍数为 10×)。

2.4 蜂胶提取物(EECP)作用下 H_2O_2 对胶原基因表达的影响

将 L929 细胞用含 10%胎牛血清的 DMEM 调成 $5×10^4$ 个/mL 的密度接种于 12 孔细胞培养板中,培养 24h 后,提前加入不同浓度 EECP 预孵育 3h,再加入 H_2O_2 单独或与蜂胶共孵育,每个处理设有 2 个复孔,实验分成三大组:空白对照组、H_2O_2 单独刺激组、蜂胶与 H_2O_2 共孵育组。由于过高浓度的过氧化氢,可能会造成细胞 DNA 大量断裂和损伤,不利于实验重复,因此 H_2O_2 作用浓度选择为 600μmol/L。另外,预实验结果表明 H_2O_2 刺激 L929 细胞 12h,其胶原基因 mRNA 表达明显降低,因此本实验中加入 H_2O_2 刺激时间 12h 后,收集样本。先 PBS 洗 2 遍,然后每个孔加入 500μL TRIzol 裂解液,轻轻吹打几次后移至 1.5mL 无菌无 RNA 酶的离心管中,存于 -80℃,用于下一步的 RNA 提取、反转录以及 qRT-PCR 实验。

2.5 胞内 ROS 含量的测定

将 L929 细胞用含 10%胎牛血清的 DMEM 调成 $5×10^4$ 个/mL 的密度接种于 12 孔细胞培养板中,培养 24h 后,提前加入不同浓度蜂胶提取物预孵育 3h,再加入 600μmol/L H_2O_2 单独或与蜂胶共孵育,每个处理设有 2 个复孔,实验分成四大组:空白对照组、H_2O_2 单独刺激组、蜂胶单独作用组、蜂胶与 H_2O_2 共孵育组。H_2O_2 刺激时间 12h 后,弃去旧培养基,用 PBS 洗 2 遍,换新的培养基后,加入终浓度为 100μmol/L 的 DCHF-DA 继续培养 30min。然后再次弃去培养基,并用 PBS 洗 2 遍,加入胰酶短暂消化后收集于 1.5mL 离心管中,1000r/min 离心 1min,弃去上清,再用 PBS 漂洗及离心 2 遍后,将细胞重悬于 400μL PBS 中,转移至流式细胞管中,在流式细胞仪上测定胞内 ROS 含量,注意整个过程尽量避光操作。测定结果选取了一次具有代表性结果,并以相对荧光强度表示。

2.6 蜂胶对 L929 抗氧化相关基因 mRNA 及蛋白表达的影响

将 L929 细胞用含 10%胎牛血清的 DMEM 调成 $5×10^4$ 个/mL 的密度接种于细胞培养板中,其中 12 孔板每孔 1mL 细胞,用于 mRNA 样本的收集,6 孔板每孔 2mL 细胞,用于蛋白样本的收集,培养 24h 后,细胞状态稳定,于不同时间点加入蜂胶(终浓度为 EECP 10μg/mL)样品处理,使细胞接受药物作用 3、6、9、12、24h 后收获细胞,用于下一步的 mRNA 或蛋白提取。

2.7 实时荧光定量 PCR 反应

小鼠抗氧化相关基因(*HO-1*、*GCLM*、*GCLC*)及管家基因(*GAPDH*)的引物参照他人的设计(Zhang et al.,2015;Wang et al.,2011),由生工生物科技有限公司负责合成(表 10.4)。

表 10.4 实时荧光定量 PCR 引物

Table 10.4 Sequences of primers for qRT-PCR

Genes	Sense primers	Antisense primers
HO-1	5′-ACATTGAGCTGTTTGAGGAG-3′	5′-TACATGGCATAAATTCCCACTG-3′
GCLM	5′-CTGACATTGAAGCCCAGGAT-3′	5′-GTTCCAGACAACAGCAGGTC-3′
GCLC	5′-GATGATGCCAACGAGTCTGA-3′	5′-GACAGCGGAATGAGGAAGTC-3′
GAPDH	5′-GAGAAACCTGCCAAGTATGATGAC -′	5′-TAGCCGTATTCATTGTCATACCAG -3′

反应体系与之前的一致,即每个反应 7μL 体系,每个反应设 3 个复孔。

PCR 扩增条件也与之前的一致,即 95℃预变性 30s,扩增 40 个循环,每个循环 95℃变性 5s,60℃退火 30s。

溶解曲线:65～95℃,每上升 0.3℃采集一次荧光信号。

2.8 蛋白免疫印记(western blotting)

细胞蛋白质提取:

(1)细胞样本处理完毕,弃去培养基,迅速用冰 PBS 洗 2 遍,吸尽培养孔里的 PBS;

(2)洗完后,立即将置于冰盒上,然后迅速向培养板中加入 NP40(内含蛋白酶和磷酸酶抑制剂),6 孔板,NP40 80μL/孔,微微晃匀;

(3)用干净的细胞刮刀,刮取细胞,最后尽可能将细胞全部收集至 1.5mL 的离心管中;

(4)高速涡旋 5s 后,放在冰盒中静置 10min;

(5)4℃条件下高速离心 10min,转速为 16000r/min;

(6)取上清,转入新的 1.5mL 离心管中,并记录体积 V_1;

(7)从 V_1 中取出 10μL,转入新的离心管中,用于后来的蛋白浓度测定;

(8)将剩下的蛋白上清液和 5×蛋白上样缓冲液混合(体积比为 4∶1);

(9)再次涡旋混匀,然后在 95℃条件下煮沸 10min;

(10)将处理好的样品和预留的 10μL 蛋白上清一同存入-80℃超低温冰箱中,以作后用。另外,为避免反复冻融给样品带来影响,可根据实际需要进行分装存储。

蛋白浓度测定:

利用 BCA 蛋白定量试剂盒对蛋白浓度进行测定,测定基本原理:碱性条件下,蛋白将 Cu^{2+} 还原为 Cu^+,Cu^+ 可以被 BCA(Bicinchoninic acid)试剂螯合形成紫色的络合物,其最高吸收峰的吸收波长为 562nm。吸光度强度与蛋白浓度成正比。利用 5mg/mL 的 BSA 蛋白标准品建立标准曲线,经换算得到待测蛋白的浓度。操作步骤按说明书进行,最后将蛋白样品统一定量后上样。

SDS-PAGE 电泳:

(1)10%分离胶的配制,分离胶各成分的比例如表 10.5,以 5mL 为例:

表 10.5　SDS-PAGE 的 10%分离胶配方

Table 10.5　Formula of 10% separating gel for SDS-PAGE

10% 分离胶各成分	5mL 分离胶中各成分体积(mL)	各成分所占体积比/%
H$_2$O	2.425	48.5
1.5mol/L Tris (pH 8.8)	1.25	25
20% SDS	0.025	0.5
10% ammonium persulfate	0.05	1
TEMED	0.002	—
40% acrylamide mix	1.25	25

注："—"表示不计入其中。

(2) 5%浓缩胶的配制,浓缩胶中各成分的比例如表 10.6,以 3mL 为例:

表 10.6　SDS-PAGE 的 10%浓缩胶配方

Table 10.6　Formula of 5% stacking gel for SDS-PAGE

5% 浓缩胶中各成分	3ML 浓缩胶中各成分体积(mL)	各成分所占体积比/%
H$_2$O	1.83	61
0.5mol/L Tris (pH 6.8)	0.75	25
20% SDS	0.015	0.5
10% ammonium persulfate	0.03	1
TEMED	0.003	—
40% acrylamide mix	0.375	12.5

注："—"表示不计入其中。

(3) 电泳:在 1×SDS-Tris-甘氨酸缓冲液中进行电泳,跑浓缩胶时,保持恒压 90V,当溴酚蓝进入分离胶时,可将电压调至 120V,直到溴酚蓝跑至底部,停止电泳,开始湿法转膜。

(4) 转膜:将 PVDF 膜放入甲醇中浸泡不少于 1min,然后按海绵—滤纸—凝胶—PVDF 膜—滤纸—海绵顺序制作"转印三明治",然后移入电泳槽中,加足预冷的转膜缓冲液,在 300mA 的恒流条件下湿转 3h 后停止。

(5) 封闭:将转膜后的 PVDF 膜放入 5%的脱脂牛奶中于 4℃于封闭过夜或常温下封闭 2h。

(6) 一抗杂交:首先用 5%脱脂牛奶稀释后一抗孵育封闭后的 PVDF 膜,常温下杂交 2h 即可。

(7) 漂洗:用 1×的 TBST 漂洗孵育了一抗后的 PVDF 膜,漂洗 3 次,时间分别为 15、5、5min。

(8) 二抗杂交:将上述漂洗后的 PVDF 膜再用 5%脱脂牛奶稀释后的二抗孵育,室温下杂交 2h 即可。

(9) 漂洗:用 1×的 TBST 漂洗孵育了二抗后的 PVDF 膜,漂洗 3 次,时间分别为 15、5、5min。

(10) 显色:用配置的 NBT/NCIP 显色工作液(现配现用)在常温下对 PVDF 膜显色 2~3min 后终止显色,然后漂洗、晾干。

(11) 数据保存:将晾干的 PVDF 膜进行拍照或扫描后保存。

2.9 结果统计分析

结果为 3 次独立重复实验的结果,表示为平均值±标准差($n=3$),利用 SPASS 16.0 软件的 T 检验对数据结果进行差异显著性分析。$P<0.05$ 为差异显著,用 * 或 # 表示;$P<0.01$ 为差异极显著,用 ** 表示。

3 结果与讨论

3.1 EECP 和 EEBP 自由基清除力及总还原力

自由基清除能力和还原能力是评价蜂胶及其他植物化学成分抗氧能力最常用的指标。为更好地反映出受试样品的抗氧化能力,本实验综合采用 DPPH 和 ABTS 自由基清除法以及用于还原力评估的普鲁士蓝法对 EECP 和 EEBP 的抗氧化能力进行评估和比较。结果如表 10.7 所示。

表 10.7 EECP 和 EEBP 的自由基清除活性及还原力

Table 10.7 Radical-scavenging activities and reducing power of EECP and EEBP

Sample	DPPH(IC_{50})	ABTS(IC_{50})	RP(mmol Trolox/g)
EECP/($\mu g/mL$)	47.71±1.34**	110.28±0.63*	1.73±0.09**
EEBP/($\mu g/mL$)	82.11±0.49	120.63±1.09	1.10±0.04

注:DPPH 和 ABTS 自由基清除能力结果用 IC_{50}($\mu g/mL$)表示,即自由基清除率为 50% 时的蜂胶浓度。RP 表示还原力,结果换算为毫摩尔 Trolox 等量每克蜂胶样品。数据结果表示为平均值±标准差($n=3$),* 表示 $P<0.05$,即差异显著;** 表示 $P<0.01$,即差异极显著。

从表 10.7 可以看出,不同抗氧化方法反映出抗氧化能力之间存在一定差异,这可能是不同抗氧化方法测定原理之间的差异造成的。从 DPPH 自由基清除能力的 IC_{50} 值测定结果可看出,EECP 的 DPPH 自由基清除能力远远强于 EEBP;而就 ABTS 自由基清除效果而言,EECP 稍强于 EEBP,两者在 ABTS 自由基清除能力上的差异不如它们在 DPPH 自由基清除能力上的差异明显。根据总还原力的测定结果可看出,EECP 的总还原力也要明显强于 EEBP。总体看来,EECP 的抗氧化能力明显强于 EEBP,并且这种差异与之前研究发现两者在总黄酮及总酚含量上的差异(本章第一节表 10.2)十分一致,这也进一步证实了包括黄酮在内酚类成分是决定蜂胶抗氧化能力的主要成分,是蜂胶抗氧化作用的物质基础。

3.2 不同浓度 H_2O_2 作用下 L929 细胞的存活率

H_2O_2 属于 ROS 的一种,参与机体内多种生理及病理过程。由于其具有膜渗透性、扩散性以及相对稳定性,常被用来诱导体外氧化应激模型,借以研究 ROS 介导的细胞氧化损伤及相关信号通路反应。创伤修复早期,由于组织缺氧,加上剧烈炎症反应甚至伤口感染等因素,包括 H_2O_2 在内的 ROS 会大量产生,它们一方面帮助创面抵抗感染,而另一面会加重创面氧化应激程度,进而影响整个修复的进程。因此,本实验利用 H_2O_2 诱导 L929 细胞损伤模型,研究 EECP 对成纤维细胞氧化应激性损伤的影响及可能机制,以期为蜂胶促创伤修复作用提供进一步的理论依据。

为考察不同浓度 H_2O_2 对 L929 细胞活力的影响,我们将 0.3～0.9mmol/L 的 H_2O_2 分别作用于 L929 细胞,24h 后观察细胞状态并测定其活力,结果如图 10.5 所示。由图可见,H_2O_2 对细胞活力的损伤具有浓度依赖性,当 H_2O_2 浓度小于 0.6mmol/L 时,细胞能保持 60% 以上的存活率,并且显微镜下观察发现,细胞数量虽明显少于正常组,但并未出现大量的细胞死亡(图片未展示),说明当刺激浓度低于 0.6mmol/L 时,H_2O_2 影响细胞活力的主要方式是抑制其增殖;当 H_2O_2 浓度为 0.7mmol/L 时,细胞活力为 18.7% 左右,镜下观察发现细胞出现大量收缩、变圆及脱落等凋亡及坏死样特征(图片未展示)。而当 H_2O_2 浓度达到 0.8mmol/L 时,仅剩下 8% 左右的细胞存活。

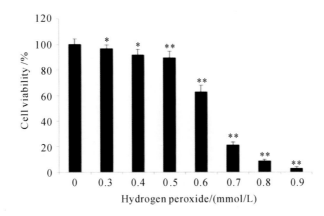

图 10.5　H_2O_2 对 L929 细胞活力的影响

Fig 10.5　Effects of H_2O_2 on viability of L929 cells

注:以图中所示的不同浓度 H_2O_2 处理 L929 细胞 24h 后,利用 CCK-8 试剂检测细胞活力(%),数据结果表示为平均值±标准差($n=3$)。* 表示 $P<0.05$,即 H_2O_2 处理组与未处理组比较差异显著;** 表示 $P<0.01$,即 H_2O_2 处理组与未处理组比较差异极显著。

3.3　EECP 和 EEBP 对 H_2O_2 作用下的 L929 死亡率的影响

通过之前的研究结果可知,EECP 和 EEBP 都具有较强的还原力,并能有效清除自由基,且 EECP 表现出更优越的抗氧化能力,因此推测 EECP 和 EEBP 都能帮助 L929 细胞对抗 H_2O_2 诱导的氧化应激性损伤,并且 EECP 的保护作用要远强于 EEBP。为证实我们的推测,本实验观察和测定了不同浓度 EECP(5、7.5、10μg/mL)和 EEBP(40、50、60μg/mL)对高浓度 H_2O_2 作用下 L929 细胞存活率的影响。首先用不同浓度 EECP 和 EEBP 对 L929 细胞进行预处理 3h,再加入 0.8mmol/L 的 H_2O_2 共孵育或单独孵育 24h,然后测定细胞的活力,观察细胞状态并拍照保存。

从图 10.6(b) 和 (d) 中可以看出,单独用 0.8mmol/L H_2O_2 刺激 L929 细胞 24h 后,细胞存活率降到 8% 左右,而提前孵育蜂胶,能明显降低细胞的死亡率,并且蜂胶浓度越大,这种保护效果越好。当 EECP 的作用浓度为 10μg/mL 时,能将细胞存活率能提升到 90.03%,而 EEBP 在最大浓度 60μg/mL 时,能将细胞存活率的提升到 66.22%,两者的保护作用都具有明显剂量依赖性效应。从图 10.6(a) 和 (c) 中细胞形态学结果中也可看出,当单独用 0.8mmol/L 的 H_2O_2 刺激 L929 细胞 24h 后,镜下观察发现大量细胞出现收缩、变圆及脱落

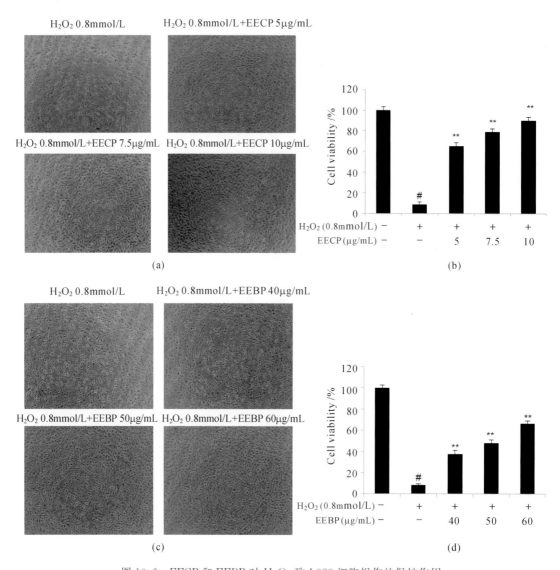

图 10.6 EECP 和 EEBP 对 H_2O_2 致 L929 细胞损伤的保护作用

Fig 10.6 Protective effects of EECP and EEBP on H_2O_2-treated L929 cells

注：用如图所示浓度的 EECP 和 EEBP 对 L929 细胞预处理 3h，再加入终浓度为 0.8mmol/L 的 H_2O_2 孵育 24h 后，利用 CCK-8 试剂盒检测 L929 细胞活力（%）。(a)和(c)分别为 EECP 和 EEBP 阻止 H_2O_2 致 L929 细胞死亡的具有代表性的图片；(b)和(d)分别表示 EECP 和 EEBP 提升 H_2O_2 作用下 L929 细胞活力检测结果，数据结果表示为平均值±标准差（$n=3$）。#表示 $P<0.05$，即 H_2O_2 处理组与未处理组比较差异显著；** 表示 $P<0.01$，即蜂胶预处理组与 H_2O_2 处理组比较差异极显著。

等凋亡及坏死样特征，而蜂胶对 H_2O_2 诱导的细胞损伤具有明显的剂量依赖性保护效果，提升细胞存活率并使其维持正常形态。将 EECP 和 EEBP 的保护效果比较发现，EECP 要明显强于 EEBP。这和之前研究的抗氧化测定结果十分一致，表明蜂胶抗氧化能力越强，其保护效果越好。鉴于 EECP 对 H_2O_2 致 L929 细胞损伤的保护效果更佳，以下将以 EECP 为例，进一步探索蜂胶对成纤维细胞 L929 氧化应激性损伤的保护作用及可能机制。

3.4 EECP 对 H_2O_2 作用下 L929 胶原基因 mRNA 表达的影响

胶原蛋白作为皮下结缔组织重要组成成分,为皮肤提供抗张能力,使皮肤保持弹性;因此胶原的合成、沉积及重塑对于创伤的修复具有十分重要的意义。然而在某些病理状态下会出现胶原的过度生成或生成不足,并最终导致纤维化疾病或修复障碍。有研究表明,ROS 及其诱导的氧化应激能够上调或下调某些成纤维细胞的胶原生成,进而参与疾病的发生发展(Tanaka et al.,1993;Nieto et al.,1999;Siwik et al.,2001)。

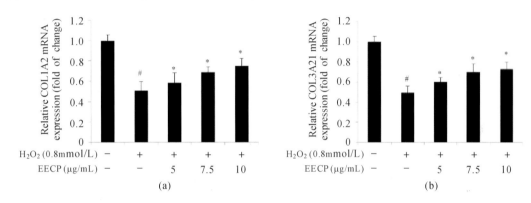

图 10.7　EECP 对 H_2O_2 作用下 L929 细胞胶原 mRNA 表达的影响

Fig10.7　Effects of EECP on COL1A2 and COL3A1 mRNA expression in H_2O_2-stimulated L929 cells

注:以图所示的蜂胶浓度对 L929 细胞预处理 3h 后,加入终浓度为 0.6mmol/L 的 H_2O_2 共孵育 12h。(a)和(b)分别表示Ⅰ型胶原 α2 链基因(COL1A2)和Ⅲ型胶原 α1 链基因(COL3A1)的相对 mRNA 表达量,检测方法为实时荧光定量 PCR(qRT-PCR)。结果表示为平均值±标准差($n=3$)。#表示 $P<0.05$,即 H_2O_2 处理组与未处理组比较差异显著;** 表示 $P<0.01$,即蜂胶预处理组与 H_2O_2 处理组比较差异极显著。

本实验进一步探究了高浓度 H_2O_2 作用下 L929 胶原表达变化以及 EECP 对 H_2O_2 诱导的 L929 细胞胶原 mRNA 表达变化的影响。考虑到过高浓度的 H_2O_2 会造成细胞的大量死亡及核酸片段化,不利于实验重复,本实验选用 0.6mmol/L 剂量的 H_2O_2 进行刺激。结果如图 10.7 所示,0.6mmol/L 的 H_2O_2 对 L929 刺激 12h 后,Ⅰ型胶原的 $α_2$ 链基因(COL1A2)的 mRNA 相对表达量只有正常组的 0.51 倍,而 EECP 存在时,能明显提升 H_2O_2 作用下 L929 细胞Ⅰ型胶原的表达量,并且具有浓度依赖性,当 EECP 作用浓度为 10μg/mL 时,COL1A2 的 mRNA 相对表达量可以提升到正常组的 0.75 倍;而对于Ⅲ型胶原的 $α_1$ 链基因(COL3A1)的 mRNA 表达也表现出相似的规律,当只有 0.6mmol/L 的 H_2O_2 存在时,COL3A1 的 mRNA 表达只有正常组的 0.49 倍,而提前预孵育 EECP 能明显提升 COL3A1 的表达量,并同样表现为 EECP 浓度越大,提升效果越明显;当 EECP 浓度为最大浓度 10μg/mL 时,COL3A1 的 mRNA 表达量可提升到正常组的 0.73 倍。以上结果表明,蜂胶能帮助 L929 抵御 H_2O_2 对其胶原基因表达的抑制,这进一步证实了蜂胶可能通过减弱氧化应激等不利因素对修复过程的影响,从而帮助创面修复。

3.5 EECP 对 H_2O_2 诱导的胞内 ROS 水平的影响

H_2O_2 引起细胞损伤的一个重要机制是它可以通过芬顿反应(Fenton reaction)产生氧化性更强的 ROS(如羟基自由基·OH),进而引发细胞的氧化应激损伤,如细胞膜的脂质过氧化、线粒体及 DNA 损伤(Farber,1994;Imlay,2003)。此外,ROS 不只是作为一种毒性物质存在,它同样可以发挥信号分子的作用,通过调节细胞特定的信号通路反应影响细胞的生命活动状态(Veal et al.,2007;D'Autreaux and Toledano,2007)。之前的实验结果表明,高浓度 H_2O_2 能抑制 L929 细胞胶原基因的表达,以及造成细胞的生长抑制或死亡,而在 EECP 存在条件下,能明显阻止这些变化的发生,猜测这可能是由于 EECP 帮助清除了胞内过量产生的 ROS,从而减低了 H_2O_2 诱导的氧化应激性损伤。为证实这一猜测,我们首先将 L929 细胞用不同浓度的 EECP 预孵育 3h,再加入 0.6mmol/L 的 H_2O_2 刺激 12h 后,利用 DCFH-DA 探针检测了胞内 ROS 水平。DCFH-DA 是一种氧化敏感性荧光探针,本身没有荧光且能自由穿过细胞膜,进入细胞后可被胞内的酯酶水解生成不能通透细胞膜的 DCFH,进而被胞内的 ROS 氧化成有荧光的 DCF,再通过流式细胞术检测胞内 DCF 的荧光就可以知道细胞内活性氧的水平。

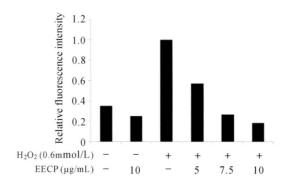

图 10.8　EECP 对 H_2O_2 作用下 L929 细胞内 ROS 含量变化的影响

Fig 10.8　Effects of EECP on intracellular ROS level change induced by H_2O_2

注:以图所示的蜂胶浓度对 L929 细胞预处理 3h,加或不加入终浓度为 0.6mmol/L 的 H_2O_2 孵育 12h。利用 DCFH-DA 探针结合流式细胞术对 L929 细胞内 ROS 含量进行检测。结果从 3 次独立重复实验中选取的最具代表性的结果,并用相对荧光强度来表示。

结果如图 10.8 所示,0.6mmol/L 的 H_2O_2 刺激 12h 后,胞内 ROS 水平几乎增加到控制组的 3 倍,而提前预孵育 EECP 可抑制胞内 ROS 水平的上升,蜂胶浓度越高作用越明显,甚至可将刺激后细胞内的 ROS 降到低于控制组水平。另外,从图中可以看出,EECP 单独作用时也能将胞内 ROS 水平降到略低于正常水平,说明细胞正常呼吸代谢过程中也会产生一定的 ROS,而蜂胶的存在减少了这些 ROS 在细胞内的蓄积。EECP 中丰富的黄酮及酚酸类物质可能是它能够有效清除胞内过量 ROS 的主要原因,即 EECP 通过这些物质抗氧化能力帮助清除 ROS,这一点从之前的抗氧化能力评估结果中也可以得到证实(见表 10.7)。由此可见,蜂胶及其成分对细胞氧化应激性损伤具有的重要意义。

3.6　EECP 对抗氧化基因表达的影响

3.6.1　EECP 对抗氧化基因在 mRNA 水平上表达的影响

皮肤受伤后早期炎症期会有大量炎症细胞迁入创面,通过释放活性氧帮助创面抵御感染;然而不断增加的 ROS 也会给创面带来负面影响,如抑制修复细胞的增生迁移甚至造成组织损伤。因此,为防止这些 ROS 在创面的过度蓄积,一些具有抗氧化作用的酶及小分子会在创伤发生后迅速增加(auf dem Keller et al.,2006)。HO-1 和 GSH 是其中比较重要的两种参与调节创面氧化应激的影响因素。血红素加氧酶(HO)是亚铁血红素分解代谢为

CO、Fe^{2+}、胆红素途径中的限速酶,具有明显的细胞保护作用,HO-1 是其中一种(Schafer and Werner,2008)。GSH 则是一种重要的内源性抗氧化剂,谷胱甘肽半胱氨酸连接酶(GCL)是其合成的限速酶,由催化亚基(GCLC)和修饰亚基(GCLM)组成(Seelig et al.,1984)。我们团队之前的研究表明,蜂胶乙醇提取物具有刺激小鼠巨噬细胞 Raw264.7 的 HO-1、GCLC 及 GCLM 等抗氧化相关基因的表达(Zhang et al.,2015),表明蜂胶还具有加强细胞抗氧化防御的作用。因此,我们猜测 EECP 对 L929 细胞的保护作用还可能与其加强 L929 抗氧化防御能力有关。本实验考察了 EECP(10μg/mL)对 L929 细胞作用不同时间后,*HO-1*、*GCLM* 和 *GCLC* 基因表达的变化。结果见图 10.9。

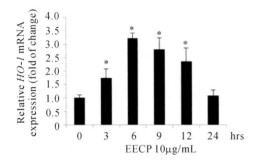
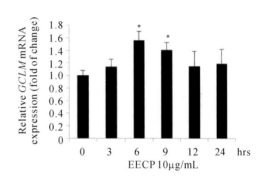
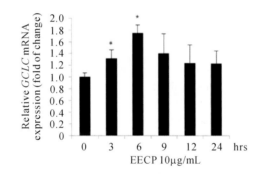

图 10.9　EECP 对 L929 细胞抗氧化相关基因(*HO-1*、*GCLC*、*GCLM*)mRNA 表达的影响

Fig 10.9　Effects of EECP on mRNA expression of antioxidant-related genes (*HO-1*、*GCLC*、*GCLM*) in L929 cells

注:为探究 EECP 刺激时间与抗氧化相关基因 mRNA 表达效应的关系,用 10μg/mL 的 EECP 处理 L929 细胞相应的时间。利用 qRT-PCR 技术对 EECP 作用不同时间后抗氧化相关基因(*HO-1*、*GCLC*、*GCLM*)的 mRNA 相对表达量进行检测。结果表示为平均值±标准差($n=3$)。* 表示 $P<0.05$,即蜂胶组与未处理组比较差异显著。

从图 10.9 可见,EECP 能明显刺激 L929 细胞抗氧化相关基因在 mRNA 水平上的表达,当 EECP 刺激 L929 细胞 6h 后,几种抗氧化相关基因的 mRNA 表达量达到最大,然后逐渐恢复正常。这与我们团队之前的实验结果也十分一致(Zhang et al.,2015)。由此可以看出,EECP 能够有效抑制 H_2O_2 作用下 L929 胞内 ROS 的水平的提升,不仅是因为 EECP 自身的自由基清除能力及还原力,还与其加强细胞的抗氧化防御能力有关。

3.6.2　EECP 对抗氧化基因在蛋白水平上表达的影响

为进一步证实 EECP 对 L929 细胞抗氧化基因的表达刺激作用,我们接着考察了 EECP(10mg/mL)对 L929 细胞作用不同时间后 *HO-1* 和 *GCLM* 基因在蛋白水平上表达变化。结果见图 10.10。由图可见，$HO\text{-}1$ 在基础状态下的表达量非常低，EECP 作用后能逐渐增强 $HO\text{-}1$ 在蛋白水平上的表达，在作用后的 6~9h 内表达量达到最大，然后逐渐恢复至正常水平。而对于 $GCLM$ 而言，其在基础状态已有一定量的表达，这可能有助于维持胞内正常的氧化还原平衡；EECP 作用后，$GCLM$ 的蛋白表达量进一步增强，在刺激 6~24h 内始终维持一个较高的表达量。需要指出的是，HO-1 及 GSH 在创面的表达对于修复具有十分重要的意义，例如有文献报道抑制谷胱甘肽的合成会降低创口破裂强度（Adamson et al.，1996）；Grochot-Przeczek 等（2009）研究表明，野生型小鼠伤后 2~3d，$HO\text{-}1$ 的表达量达到最大，抑制 $HO\text{-}1$ 活性会阻碍伤口的愈合；同样地，当小鼠敲出 $HO\text{-}1$ 基因后会出现修复障碍如再上皮化及新血管形成受阻；此外，他们还发现糖尿病小鼠伤口修复障碍很可能与伤后 $HO\text{-}1$ 的延迟上调有关。由此可见，蜂胶促创伤修复作用很可能与其促进修复细胞抗氧化基因的表达有很大关系。

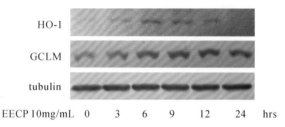

图 10.10　EECP 对 L929 细胞抗氧化相关基因（$HO\text{-}1$、$GCLM$）蛋白表达的影响

Fig 10.10　Effects of EECP on proteins expression of antioxidant-related genes（$HO\text{-}1$，$GCLM$）in L929 cells

注：为探究 EECP 刺激时间与抗氧化相关基因蛋白表达效应的关系，用 $10\mu g/mL$ 的 EECP 处理 L929 细胞相应的时间。利用 Western-blotting 技术对 EECP 作用不同时间后抗氧化相关基因（$HO\text{-}1$、$GCLC$、$GCLM$）的蛋白表达量进行检测。图片为 3 次独立重复实验中挑出的具有代表性的结果。

3.7　EECP 对 H_2O_2 作用下 MAPK 及 PI3K/AKT 信号通路变化的影响及意义

ROS 致细胞死亡的途径不仅包括直接损伤维持细胞生命活动的大分子物质，它还可以通过调节胞内特定的激酶反应，激活细胞的死亡通路。MAPK 及 PI3K/Akt 激酶信号通路被报道在调节 ROS 诱导的凋亡或坏死通路中发挥着重要的作用（Sakon et al.，2003；Matsuzawa and Ichijo，2005；McCubrey et al.，2006；Liu et al.，2007）。Lee 等（2003）研究表明，H_2O_2 诱导的 L929 细胞的氧化应激性凋亡与它对 Erk 通路的持续激活有关；Wu 等（2014）的研究表明 L929 细胞的坏死与 PI3K-Akt-mTOR 通路的激活有关；而 Cheng 等（2008）的研究则表明 Erk 和 JNK 信号通路参与调节 TNF-α 诱导的 L929 细胞凋亡及自噬性死亡。因此，我们猜测 EECP 能降低高浓度 H_2O_2 作用下 L929 细胞的死亡率，很可能是它调节了 H_2O_2 作用下 MAPK 及 PI3K/Akt 信号通路的变化；为验证此推论，我们考察了 H_2O_2 单独或与 EECP 共孵育时对 L929 细胞 MAPK 及 PI3K/Akt 信号通路中目标蛋白磷酸化水平的影响，然后又进一步利用特定的激酶抑制剂，考察了各信号通路在调节 L929 细胞氧化应激性死亡的过程中所发挥的作用。结果如图 10.11 所示。

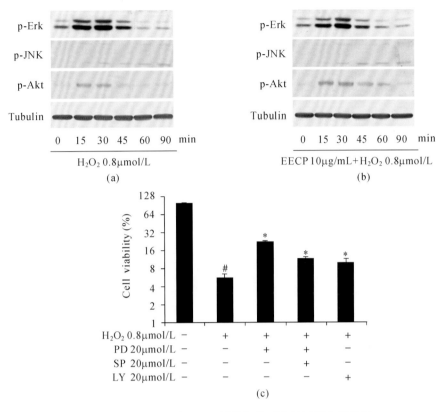

图 10.11 EECP 阻止 H_2O_2 诱导细胞死亡的可能分子机制

Fig 10.11 Possible molecular mechanisms underlying the preventive effects of EECP on L929 cells death induced by H_2O_2

注：为探明 EECP 阻止 H_2O_2 诱导细胞死亡的可能分子机制，首先检测了 EECP 对 H_2O_2 作用下 MAPK（ERK、JNK）蛋白及 AKT 蛋白磷酸化水平变化的影响。(a)是指单独用 0.8mmol/L H_2O_2 处理 L929 细胞相应时间。(b)是指提前用 EECP（10μg/mL）预处理 3h 再加入 0.8mmol/L H_2O_2 处理相应时间。Western-blotting 方法被用来检测相应蛋白的磷酸化水平。其次，通过特异性的通路抑制剂来探究 MAPK（ERK、JNK）以及 PI3K/AKT 信号通路在 H_2O_2 诱导 L929 细胞死亡过程中所发挥的作用。(c)是指提前用抑制剂（PD98059、SP600125、LY294002）处理细胞 30min，然后加或不加 0.8mmol/L H_2O_2 再培养 24h，再利用 CCK-8 试剂检测细胞活力（%）。数据结果表示为平均值±标准差（$n=3$）。# $P<0.05$，表示与未处理组比较差异显著；* $P<0.05$，表示与 H_2O_2 处理组比较差异显著。

从图 10.11（a）可以看出，0.8mmol/L H_2O_2 刺激 L929 细胞 15min 时就能明显增强 MAPK 通路中 Erk 及 JNK 的磷酸化水平和 PI3K/Akt 通路中 Akt 蛋白的磷酸化水平，表明了 H_2O_2 对 L929 的 ERK、JNK 及 PI3K/Akt 信号通路具有较强的激活作用，这和 Lee 等（2003）的研究结果具有一致性。而图 10.11（b）表明，EECP 能明显抑制 H_2O_2 对 Erk 磷酸化水平的增强，这表明 EECP 能够抑制 H_2O_2 对 ERK 信号通路的激活。然而有趣的是，EECP 并不能有效抑制 H_2O_2 对 JNK 及 PI3K/Akt 信号通路的激活，相反 EECP 能进一步加强 H_2O_2 对 PI3K/Akt 信号通路的激活，具体原因尚不清楚，有待于进一步研究。

我们又利用特定的激酶抑制剂考察了这几条信号通路在调节 L929 细胞氧化应激性死亡的过程中所发挥的作用。从图 10.11（c）的细胞活力检测结果中可以看出，ERK、JNK 及 PI3K/Akt 信号通路阻断剂都能明显降低 H_2O_2 作用下 L929 细胞的死亡率，尤其以 ERK 通

路阻断剂的作用效果最明显,表明这3条通路的激活在调节H_2O_2诱导的L929细胞死亡过程中发挥着重要作用。ERK通路阻断剂降低高浓度H_2O_2作用下细胞死亡率可能与它抑制L929氧化应激性凋亡有关(Lee et al.,2003);JNK抑制剂降低高浓度H_2O_2作用下细胞死亡率可能与它抑制L929细胞自噬性死亡有关(Cheng et al.,2008),而PI3K抑制剂降低高浓度H_2O_2作用下细胞死亡率很可能与它抑制L929细胞的坏死有关。之前的结果表明,EECP能显著降低H_2O_2作用下的L929细胞死亡率,猜测这很有可能与EECP能有效抑制H_2O_2作用下ERK通路过度激活有关,或者说抑制ERK通路的过度激活至少可以部分解释EECP对L929细胞氧化应激性死亡的保护作用。另外,结合他人的研究结果来看,EECP很可能通过抑制ERK通路的过度激活从而阻止L929的氧化应激性凋亡(Lee et al.,2003;Cheng et al.,2008)。至于EECP增强H_2O_2对PI3K/Akt信号通路激活的具体原因及其对EECP降低L929氧化应激性死亡过程中所具有的意义尚不清楚,猜测EECP有效抑制L929细胞氧化应激性死亡可能是它综合调节H_2O_2诱导的MAPK及PI3K/Akt通路激活的结果。

小　结

本研究通过考察EECP和EEBP抗L929细胞氧化应激性损伤效果发现,EECP比EEBP具有更好的保护效果,这可能与EECP含有更高含量的酚类有关,使得EECP具有更强抗氧化能力,从而有效地抑制胞内ROS含量过度增加。从H_2O_2对L929细胞活力及其胶原基因表达的影响可以看出氧化应激对创伤修复的不利影响,而蜂胶乙醇提取物良好的保护效果则为蜂胶促创伤修复效果提供了强有力的证据。另外,从蜂胶对L929细胞抗氧化相关基因的表达刺激作用可以看出,蜂胶的抗氧化能力不止依靠自身的自由基清除能力及还原力,它还能发挥信号分子作用,通过加强细胞的抗氧化防御帮助修复细胞抵御氧化应激性损伤。总的说来,我们的研究表明蜂胶可以通过自身抗氧化作用及加强细胞抗氧化防御的方式,抑制胞内ROS的过度蓄积,从而减弱由氧化应激引起的细胞损伤。由此可见,蜂胶的抗氧化作用对创伤修复所具有的意义。

参考文献

Adamson B,Schwarz D,Klugston P,Gilmont R,Perry L,Fisher J,Lindblad W,Rees R (1996) Delayed repair:the role of glutathione in a rat incisional wound model[J]. *Journal of Surgical Research*,62(2):159-164.

Ågren MS (1994) Gelatinase activity during wound healing[J]. *British Journal of Dermatology*,131:634-640.

Albuquerque-júnior RLC,Barreto ALS,Pires JA,Reis FP,Lima SO,Ribeiro MAG,Cardoso JC (2009) Effect of bovine type-I collagen-based films containing red propolis on dermal wound healing in rodent model[J]. *International Journal of Morphology*,27(4):1105-1110.

Annan K,Houghton PJ (2008) Antibacterial,antioxidant and fibroblast growth stimulation of aqueous extracts of *Ficus asperifolia* Miq. and *Gossypium arboreum* L.,wound-healing plants of Ghana[J]. *Journal of Ethnopharmacology*,119(1):141-144.

Ansorge S, Reinhold D, Lendeckel U (2003) Propolis and some of its constituents down-regulate DNA synthesis and inflammatory cytokine production but induce TGF-β_1 production of human immune cells[J]. *Zeitschrift für Naturforschung*, 58C: 580-589.

auf dem Keller U, Kümin A, Braun S, Werner S (2006) Reactive oxygen species and their detoxification in healing skin wounds[J]. *Journal of Investigative Dermatology Symposium Proceedings*, 11(1): 106-11.

Banskota AH, Tezuka Y, Adnyana IK, Ishii E, Midorikawa K, Matsushige K, Kadota S (2001) Hepatoprotective and anti-Helicobacter pylori activities of constituents from Brazilian propolis[J]. *Phytomedicine*, 8(1): 16-23.

Balekar N, Katkam NG, Nakpheng T, Jehtae K, Srichana T (2012) Evaluation of the wound healing potential of *Wedelia trilobata* (L.) leaves[J]. *Journal of Ethnopharmacology*, 141(3): 817-824.

Barrientos S, Stojadinovic O, Golinko MS, Brem H, Tomic-Canic M (2008) Growth factors and cytokines in wound healing[J]. *Wound Repair and Regeneration*, 16(5): 585-601.

Berretta AA, Nascimento AP, Bueno PC, de Oliveira LLV, Marchetti JM (2012) Propolis standardized extract (EPP-AF(R)), an innovative chemically and biologically reproducible pharmaceutical compound for treating wounds[J]. *International Journal of Medical Sciences*, 8(4): 512-521.

Blakytny R, Jude E (2006) The molecular biology of chronic wounds and delayed healing in diabetes[J]. *Diabetic Medicine*, 23(6): 594-608.

Borrelli F, Maffia P, Pinto L, Ianaro A, Russo A, Capasso F, Ialenti A (2002) Phytochemical compounds involved in the antiinflammatory effect of propolis extract[J]. *Fitoterapia*, 73 (Suppl 1): S53-63.

Cheng Y, Qiu F, Tashiro S, Onodera S, Ikejima T (2008) ERK and JNK mediate TNF alpha-induced p53 activation in apoptotic and autophagic L929 cell death[J]. *Biochemical and Biophysical Research Communications*, 376(3): 483-488.

D'Autreaux B, Toledano MB (2007) ROS as signalling molecules: mechanisms that generate specificity in ROS homeostasis[J]. *Nature Reviews Molecular Cell Biology*, 8(10): 813-824.

de Almeida EB, Cordeiro Cardoso J, Karla de Lima A, de Oliveira NL, de Pontes-Filho NT, Oliveira Lima S, Leal Souza IC, de Albuquerque-Júnior RL (2013) The incorporation of Brazilian propolis into collagen-based dressing films improves dermal burn healing[J]. *Journal of Ethnopharmacology*, 147(2): 419-425.

de Castro SL (2001) Propolis: biological and pharmacological activity. Therapeutic uses of this bee-product[J]. *Annual Review of Biomedical Science*, 3(1): 49-83.

de Moura SA, Negri G, Salatino A, da Cunha Lima LD, Dourado LPA, Mendes JB, Andrade SP, Ferreira MA, Cara DC (2011a) Aqueous extract of Brazilian green propolis: primary components, evaluation of inflammation and wound healing by using subcutaneous implanted sponges[J]. *Evidence-Based Complementary and Alternative Medicine*, 2011: 748283.

de Moura SA, Ferreira MA, Andrade SP, Reis ML, de Lourdes NM, Cara DC (2011b) Brazilian green propolis inhibits inflammatory angiogenesis in a murine sponge model [J]. *Evidence-Based Complementary and Alternative Medicine*, 2011:182703.

Desmoulière A, Chaponnier C, Gabbiani G (2005) Tissue repair, contraction, and the myofibroblast[J]. *Wound Repair and Regen*, 13(1):7-12.

Dovi JV, He LK, DiPietro LA (2003) Accelerated wound closure in neutrophil-depleted mice[J]. *Journal of Leukocyte Biology*, 73:448-455.

Dovi JV, Szpaderska AM, DiPietro LA (2004) Neutrophil function in the healing wound: adding insult to injury? [J]. *Thrombosis and Haemostasis*, 92:275-280.

Egozi EI, Ferreira AM, Gamelli RL, Gamelli RL, Dipietro LA (2003) Mast cells modulate the inflammatory but not the proliferative response in healing wounds[J]. *Wound Repair and Regeneration*, 11(1):46-54.

Elizabeth Fini MGM, Matsubara M (1992) Collagenolytic/gelatinolytic enzymes in corneal wound healing[J]. *Acta Ophthalmologica*, 70(S202):26-33.

El-Khatib AS, Agha AM, Mahran LG, Khayyal MT(2002) Prophylactic effect of aqueous propolis extract against acute experimental hepatotoxicity *in vivo* [J]. *Verlag der Zeitschrift für Naturforschung*, 57(3-4):379-385.

Epstein FH, Singer AJ, Clark RAF(1999) Cutaneous wound healing[J]. *New Engl J Med*, 341(10):738-746.

Faler BJ, Macsata RA, Plummer D, Mishra L, Sidawy AN (2006) Transforming growth factor-β and wound healing[J]. *Perspect Vascr Surg Endovasc Ther*, 18(1):55-62.

Farber JL. (1994) Mechanisms of cell injury by activated oxygen species[J]. *Environ Health Persp* 102(S10):17-24.

Farstvedt E, Stashak TS, Othic A(2004) Update on topical wound medications[J]. *Clin Tech Equine Pract*, 3(2):164-172.

Fronza M, Heinzmann B, Hamburger M, Laufer S, Merfort I (2009) Determination of the wound healing effect of *Calendula* extracts using the scratch assay with 3T3 fibroblasts[J]. *J Ethnopharmacol*, 126(3):463-467.

Fuliang HU, Hepburn HR, Xuan H, Chen M, Daya S, Radloff SE(2005) Effects of propolis on blood glucose, blood lipid and free radicals in rats with diabetes mellitus[J]. *Pharmacol Res*, 51(2):147-152.

Gabbiani G. (2003) The myofibroblast in wound healing and fibrocontractive diseases[J]. *J Pathol*, 200(4):500-503.

George BP, Parimelazhagan T, Chandran R (2014) Anti-inflammatory and wound healing properties of Rubus fairholmianus Gard. Root—An *in vivo* study[J]. *Ind Crop Prod*, 54:216-225.

Gescher K, Deters AM. (2011) *Typha latifolia* L. fruit polysaccharides induce the differentiation and stimulate the proliferation of human keratinocytes *in vitro*[J]. *J Ethnopharmacol*, 137(1):352-358.

Gressner AM WR, Breitkopf K, Dooley S(2002)Roles of TGF-beta in hepatic fibrosis[J]. *Frontiers in Bioscience Front in Biosci*, 7: d793-807.

Grochot-Przeczek A, Lach R, Mis J, Skrzypek K, Gozdecka M, Sroczynska P, Dubiel M, Rutkowski A, Kozakowska M, Zagorska A, Walczynski J, Was H, Kotlinowski J, Drukala J, Kurowski K, Kieda C, Herault Y, Dulak J, Jozkowicz A (2009) Heme oxygenase-1 accelerates cutaneous wound healing in mice[J]. *PLoS One*, 4(6): e5803.

Guimarães NSS, Guimarães NSS, Mello JC, Paiva JS, Bueno PCP, Berretta AA, Torquato RJ, Nantes IL, Rodrigues Tiago(2012)*Baccharis dracunculifolia*, the main source of green propolis, exhibits potent antioxidant activity and prevents oxidative mitochondrial damage[J]. *Food Chem Toxicol*, 50(3-4): 1091-1097.

Gurtner GC, Werner S, Barrandon Y, Michael T, Longaker MT(2008)Wound repair and regeneration[J]. *Nature*, 453(7193): 314-321.

Hampton MB, Kettle AJ, Winterbourn CC (1998) Inside the neutrophil phagosome: oxidants, myeloperoxidase and bacterial killing[J]. *Am Soc Hematol*, 92(9): 3007-3017.

Han MC, Durmus AS. Karabulute E, Yaman I (2005) Effects of Turkish propolis and silversulfadiazine on burn wound healing in rats[J]. *Revue Méd Vét*, 156(12): 624-627.

Imlay JA(2003)Pathways of oxidative damage[J]. *Annu Rev Microbiol*, 57: 395-418.

Iyyam Pillai S, Palsamy P, Subramanian S, Kandaswamy M(2010)Wound healing properties of Indian propolis studied on excision wound-induced rats[J]. *Pharm Biol*, 48(11): 1198-1206.

James TJ, Hughes MA, Cherry GW, Taylor RP(2003) Evidence of oxidative stress in chronic venous leg ulcers[J]. *Wound Rep Reg*, 11: 172-176.

Khorasgani EM, Karimi AH, Nazem MR(2010)A comparison of healing effects of propolis and silver sulfadiazine on full thickness skin[J]. *Pak Vet J*, 30(2): 72-74.

Kilicoglu SS, Kilicoglu B, Erdemli E(2008)Ultrastructural view of colon anastomosis under propolis effect by transmission electron microscopy[J]. *World J Gastroentero*, 14(30): 4763-4770.

Koca U, Suntar IP, Keles H, Yesilada E, Akkol EK (2009) *In vivo* anti-inflammatory and wound healing activities of *Centaurea iberica* Trev. ex Spreng[J]. *J Ethnopharmacol*, 126(3): 551-556.

Kumazawa S, Yoneda M, Shibata I, Kanaeda J, Hamasaka T, Nakayama T (2003) Direct evidence for the plant origin of Brazilian propolis by the observation of honeybee behavior and phytochemical analysis. *Chem Pharm Bull*, 51(6): 740-742.

Leask A, Abraham DJ (2004) TGF-β signaling and the fibrotic response[J]. *J Federa Am Soc Exp Biol*, 18(7): 816-827.

Lee YJ, Cho HN, Soh JW, Jhon GJ, Cho CK, Chung HY, Sangwoo B, Lee SJ, Lee YS (2003). Oxidative stress-induced apoptosis is mediated by ERK1/2 phosphorylation [J]. *Exp Cell Res*, 291(1): 251-266.

Leibovich SJ, Ross R. 1975. The role of the macrophage in wound repair[J]. *Am J Pathol*,

78(1):71-100.

Liu CL, Xie LX, Li M, Durairajan SSK, Goto S, Huang JD(2007) Salvianolic acid B inhibits hydrogen peroxide-induced endothelial cell apoptosis through regulating PI3K/Akt signaling[J]. *PLoS One*, 2(12):e1321.

Martin P(1997) Wound healing—aiming for perfect skin regeneration[J]. *Science*, 276(5309):75-81.

Martin P, D'Souza D, Martin J, Grose R, Cooper L, Maki R, McKercher SR(2003) Wound healing in the PU.1 null mouse-tissue repair is not dependent on inflammatory cells[J]. *Curr Biol*, 13(13):1122-1128.

Martin P, Leibovich SJ(2005) Inflammatory cells during wound repair: the good, the bad and the ugly[J]. *Trends Cell Biol*, 5(11):599-607.

Matsui T, Ebuchi S, Fujise T, Abesundara KJ, Doi S, Yamada H, Matsumoto K(2004) Strong antihyperglycemic effects of water-soluble fraction of Brazilian propolis and its bioactive constituent, 3,4,5-tri-o-caffeoylquinic acid[J]. *Biol Pharm Bull*, 27(11):1797-1803.

Matsuzawa A, Ichijo H(2005) Stress-responsive protein kinases in redox-regulated apoptosis signaling[J]. *Antioxid Redox Sign*, 7:472-481.

McCawley LJ, Matrisian LM(2001) Matrix metalloproteinases: they're not just for matrix anymore! [J]. *Curr opin cell biol*, 13:534-540.

McCubrey JA, LaHair MM, Franklin RA(2006) Reactive oxygen species-induced activation of the MAP kinase signaling pathways[J]. *Antioxid Redox Sign*, 8:1775-1789.

McLennan SV, Bonner J, Milne S, Lo L, Charlton A, Kurup S, Jia J, Yue DK, Twigg SM(2008) The anti-inflammatory agent propolis improves wound healing in a rodent model of experimental diabetes[J]. *Wound Repair Regen*, 16:706-713.

Meredith JE, Fazeli B, Schwartz MA(1993) The Extracellular matrix as a cell survival factor[J]. *Mol Biol Cell*, 4(9):953-961.

Mirzoeva OK, Calder PC(1996) The effect of propolis and its components on eicosanoid production during the inflammatory response[J]. *Prostag Leukotr Ess*, 55(6):441-449.

Moreira L, Dias LG, Pereira JA, Estevinho L(2008) Antioxidant properties, total phenols and pollen analysis of propolis samples from Portugal[J]. *Food Chem Toxicol*, 46(11):3482-3485.

Nieto N, Friedman SL, Greenwel P, Cederbaum AI(1999) CYP2E1-mediated oxidative stress induces collagen type I expression in rat hepatic stellate cells. *Hepatology*, 30(4):987-996.

Ocakci A, Kanter M, Cabuk M, Buyukbas S(2006) Role of caffeic acid phenethyl ester, an active component of propolis, against NAOH-induced esophageal burns in rats[J]. *Int J Pediatr Otorhi*, 70(10):1731-1739.

Olczyk P, Komosinska-Vassev K, Winsz-Szczotka K, Koźma EM, Wisowski G, Stojko J, Klimek K, Olczyk K(2012) Propolis modulates vitronectin, laminin, and heparan sulfate/heparin expression during experimental burn healing[J]. *J Zhejiang Univ-Sci B (Biomed & Biotechnol)*, 13(11):932-941.

Olczyk P, Wisowski G, Komosinska-Vassev K, Stojko J, Klimek K, Olczyk M, Kozma EM (2013a) Propolis modifies collagen types Ⅰ and Ⅲ accumulation in the matrix of burnt tissue[J]. *Evid-Based Compl Alt*, 2013:423809.

Olczyk P, Komosinska-Vassev K, Winsz-Szczotka K, Stojko J, Klimek K, Kozma EM (2013b) Propolis induces chondroitin/dermatan sulphate and hyaluronic acid accumulation in the skin of burned wound[J]. *Evid-Based Compl Alt*, 2013:290675.

Olczyk P, Ramos P, Komosinska-Vassev K, Stojko J, Pilawa B (2013c) Positive effect of propolis on free radicals in burn wounds[J]. *Evid-Based Compl Alt*, 2013:356737.

O'Toole EA, Goel M, Woodley DT (1996) Hydrogen peroxide inhibits human keratinocyte migration[J]. *Dermatol Surg*, 22(6):525-529.

Ozyurt B, Parlaktas BS, Ozyurt H, Aslan H, Ekici F, Atis O (2007) A preliminary study of the levels of testis oxidative stress parameters after MK-801-induced experimental psychosis model: protective effects of CAPE[J]. *Toxicology*, 230(1):83-89.

Parihar A, Parihar MS, Milner S, Bhat S (2008) Oxidative stress and anti-oxidative mobilization in burn injury[J]. *Burns*, 34(1):6-17.

Paulino N, Abreu SR, Uto Y, Koyama D, Nagasawa H, Hori H, Dirsch VM, Vollmar AM, Scremin A, Bretz WA (2008) Anti-inflammatory effects of a bioavailable compound, Artepillin C, in Brazilian propolis[J]. *Eur J Pharmacol*, 587(1-3):296-301.

Pessolato AG, Martins Ddos S, Ambrósio CE, Mançanares CA, de Carvalho AF (2011) Propolis and amnion reepithelialise second-degree burns in rats[J]. *Burns*, 37(7):1192-1201.

Ramos AFN, Miranda JL (2007) Propolis: A review of itsanti-inflammatory and healing actions[J]. *J Venom Anim Toxins*, 13(4):679-710.

Rebiai A, Lanez T, BelfarML (2011) *In vitro* evaluation of antioxidant capacity of algerian propolis by spectrophotometrical and electrochemical assays[J]. *Int J Pharmacol*, 7(1):113-118.

Sakon S, Xue X, Takekawa M, T Sasazuki. Okazaki T, Kojima Y, Piao JH, Yagita H, Okumura K, Doi T, Nakano H (2003) NF-κB inhibits TNF-induced accumulation of ROS that mediate prolonged MAPK activation and necrotic cell death[J]. *EMBO J*, 22(15):3898-3909.

Santoro MM, Gaudino G (2005) Cellular and molecular facets of keratinocyte reepithelization during wound healing[J]. *Exp Cell Res*, 304(1):274-286.

Schafer M, Werner S (2008) Oxidative stress in normal and impaired wound repair[J]. *Pharmacol res*, 58(2):165-171.

Schultz GS, Wysocki A (2009) Interactions between extracellular matrix and growth factors in wound healing[J]. *Wound Repair and Regen*, 17(2):153-162.

Schiller M, Javelaud D, Mauviel A (2004) TGF-β-induced SMAD signaling and gene regulation:

consequences for extracellular matrix remodeling and wound healing[J]. *J Dermatol Sci*,35(2):83-92.

Seelig GF,Simondsen RP,Meister A(1984)Reversible dissociation of gamma-glutamylcysteine synthetase into two subunits[J]. *J Biol Chem*,259(15):9345-9347.

Sehn E,Hernandes L,Franco SL,Gonçalves CC,Baesso ML(2009)Dynamics of reepithelialisation and penetration rate of a bee propolis formulation during cutaneous wounds healing[J]. *Analytica Chimica Acta*,635(1):115-120.

Shirai M,Yamanishi R,Moon JH,Murota K,Terao J(2002)Effect of quercetin and its conjugated metabolite on the hydrogen peroxide-induced intracellular production of reactive oxygen species in mouse fibroblasts[J]. *Biosci Biotechnol Biochem*,66(5):1015-1021.

Simone-Finstrom M,Spivak M(2010)Propolis and bee health:the natural history and significance of resin use by honey bees[J]. *Apidologie*,41(3):295-311.

Simpson DM,Ross R(1972)The neutrophilic leukocyte in wound repair[J]. *J Clin Invest*,51(8):2009-2023.

Singh D,Singh D,Choi SM,Zo SM,Painuli RM,Kwon SW,Han SS(2014)Effect of extracts of terminalia chebula on proliferation of keratinocytes and fibroblasts cells:An alternative approach for wound healing[J]. *Evid-based Compl Alt*,2014:701656.

Siwik DA,Pagano PJ,Colucci WS(2001)Oxidative stress regulates collagen synthesis and matrix metalloproteinase activity in cardiac fibroblast[J]. *Am J Physiol Cell Physiol*,280:C53-C60.

Soneja A,Drews M,Malinski T(2005)Role of nitric oxide,nitroxidative and oxidative[J]. *Pharmacol Rep*,57(Suppl):108-119.

Song YS,Park EH,Jung KJ,Jin C(2002a)Inhibition of angiogenesis by propolis[J]. *Arch Pharm Res*,25(4):500-504.

Song YS,Park EH,Hur GM,Ryu YS,Kim YM,Jin C(2002b)Ethanol extract of propolis inhibits nitric oxide synthase gene[J]. *J Ethnopharmacol*,80(2-3):155-161.

Steenkamp V,Mathivha E,Gouws MC,van Rensburg CEJ(2004)Studies on antibacterial, antioxidant and fibroblast growth stimulation of wound healing remedies from South Africa[J]. *J Ethnopharmacol*,95(2-3):353-357.

Steffensen B,Hakkinen L,Larjava H(2001)Proteolytic events of wound-healing-coordinated interactions among matrix metalloproteinases (MMPs),integrins,and extracellular matrix molecules[J]. *Crit Rev Oral Biol M*,12(5):373-398.

Stipcevic T,Piljac J,Berghe DV(2006)Effect of different flavonoids on collagen synthesis in human fibroblasts[J]. *Plant Food Hum Nutr*,61(1):29-34.

Su KY,Hsieh CY,Chen YW,Chuang CT,Chen CT,Chen YL(2014)Taiwanese green propolis and propolin G protect the liver from the pathogenesis of fibrosis via eliminating TGF-beta-induced Smad2/3 phosphorylation[J]. *J Agr Food Chem*,62(14):3192-3201.

Tanaka H,Okada T,Konishi H,Tsnji T(1993)The effect of reactive oxygen species on the biosynthesis of collagen[J]. *Arch Dermatol Res*, 285:352-355.

Tan-No K, Nakajima T, Shoji T, Nakagawasai O, Niijima F, Ishikawa M, Endo Y, Sato T, Satoh S, Tadano T(2006) Anti-inflammatory effect of propolis through inhibition of nitric oxide[J]. *Biol Pharm Bull*, 29(1):96-99.

Tateshita T, Ono I, Kaneko F(2001) Effects of collagen matrix containing transforming growth factor TGF-β_1 on wound contraction[J]. *J Dermatol Sci*, 27(2001):104-113.

Veal EA, Day AM, Morgan BA(2007) Hydrogen peroxide sensing and signaling[J]. *Mol Cell*, 26(1):1-14.

Villegas LF, Marçalo A, Martin J, Fernández ID, Maldonado H, Vaisberg AJ, Hammond GB (2001) (+)-*epi*-α-Bisbolol is the wound-healing principle of *Peperomia galioides*: investigation of the *in vivo* wound-healing activity of related terpenoids[J]. *J Nat Prod*, 64:1357-1359.

Wang K, Ping S, Huang S, Hu L, Xuan H, Zhang C, Hu F(2013) Molecular mechanisms underlying the *in vitro* anti-inflammatory effects of a flavonoid-rich ethanol extract from Chinese propolis (poplar type)[J]. *Evid-based Compl Alt*, 2013:127672.

Wang X, Stavchansky S, Kerwin SM, Bowman PD(2010) Structure-activity relationships in the cytoprotective effect of caffeic acid phenethyl ester (CAPE) and fluorinated derivatives: effects on heme oxygenase-1 induction and antioxidant activities[J]. *Eur J Pharmacol*, 635(1-3):16-22.

Wu F, Bian D, Xia Y, Gong Z, Tan Q, Chen J, Dai Y(2012) Identification of major active ingredients responsible for burn wound healing of *Centella asiatica* herbs[J]. *Evid-based Compl Alt*, 2012:848093.

Wu YT, Tan HL, Huang Q, Ong CN, Shen HM(2014) Activation of the PI3K-Akt-mTOR signaling pathway promotes necrotic cell death via suppression of autophagy[J]. *Autophagy*, 5(6):824-834.

Xue M, Le NTV, Jackson CJ, Jackson CJ(2006) Targeting matrix metalloproteases to improve cutaneous wound healing[J]. *Expert Opin Ther Targets*, 10(1):143-155.

Yang H, Dong Y, Du H, Shi H, Peng Y, Li X(2011) Antioxidant compounds from propolis collected in Anhui, China[J]. *Molecules*, 16(4):3444-3455.

Youn YK, LaLonde X, Demling R(1992) The role of mediators in the response to thermal injury[J]. *World J Surg*, 16(1):30-36.

Zhang C, Huang S, Wei W, Ping S, Shen X, Li Y, Hu F(2014) Development of high-performance liquid chromatographic for quality and authenticity control of Chinese propolis. *J Food Sci*, 79(7):C1315-1322.

Zhang J, Cao X, Ping S, Wang K, Shi J, Zhang C, Zheng H, Hu F(2015) Comparisons of ethanol extracts of Chinese propolis (poplar type) and poplar gums based on the antioxidant activities and molecular mechanism[J]. *Evid-based Compl Alt*, 2015:307594.

Zippel J, Deters A, Hensel A(2009) Arabinogalactans from Mimosa tenuiflora (Willd.) Poiret bark as active principles for wound-healing properties: specific enhancement of dermal fibroblast activity and minor influence on HaCaT keratinocytes[J]. *J Ethnopharmacol*, 124(3):391-396.

陈荷凤,韩文辉,李兵,崔海辉(2001)蜂胶各种溶剂提取物的抑菌效果比较[J].食品研究与开发,22(S1):18-19.

陈炯,韩春茂,林小玮,唐志坚,苏士杰(2006)纳米银敷料在修复Ⅱ度烧伤创面的应用研究[J].中华外科杂志,44(1):50-52.

高畅,贾军宏,楚勤英,唐志雄,马虹颖,李建华,傅小玲,王进伟(2009).蜂胶对糖尿病大鼠创面新生血管和细胞增殖的影响[J].中国医院用药评价与分析,9(3):200-203.

郭芳彬(2004)蜂胶的抗菌作用[J].蜜蜂杂志,(3):10-12.

黄帅,张翠平,胡福良(2013)2008—2012年蜂胶化学成分研究进展[J].天然产物研究与开发,25:1146-1153,1165.

胡福良,李英华,朱威,陈民利,应华忠(2007)蜂胶提取液对大鼠急性胸膜炎的作用及其机制的研究[J].营养学报,29(2):189-191.

胡晓,林亲录(2007)蜂胶的保健功能及应用进展[J].农产品加工学报,(5):31-38.

黎洪棉,梁自乾,刘达恩,蒙诚跃(2005)南宁地区某烧伤病房细菌生态学调查及耐药性分析[J].中华烧伤杂志,21(2):107-110.

李晋辉,母得志(2007)整合素及其信号传导通路[J].医学分子生物学杂志,4(3):279-282.

李雅晶,胡福良,冯磊(2005)蜂胶在食品保鲜中的应用及机制[J].天然产物研究与开发,17:108-12.

罗念容,高华,周青峰(2011)烧伤患者病原菌感染及其耐药性调查[J].中国误诊学杂志,11(7):1761-1762.

彭代智(2005)烧伤后炎症反应的病因、分子机制及防治对策[J].中华烧伤杂志,21(6):405-409.

申慧亭,靳月琴(2005)蜂胶抗菌作用的实验室研究[J].长治医学院学报,19(4):253-254.

齐东梅(2008)蜂胶的药理作用和临床应用[J].首都医药,(4):44-45.

王宏,蔡金东,侯智慧(2011)烧伤感染的细菌学调查及其耐药性分析[J].中国冶金工业医学杂志,28(1):5-6.

王元元,黄云英,杜娟,沈丽,张德芹,王雪妮(2012)蜂蜜、蜂胶对深Ⅱ度烫伤大鼠创面愈合的影响[J].天津中医药大学学报,31(3):154-156.

于勇,盛志勇,柴家科,杨小强,常东,蒋伟(2006)抗菌药物使用与烧伤感染主要病原菌构成比变化的关系[J].解放军医学杂志,31(1):1-3.

曾军,石国荣(2008)天然产物抗氧化活性的测定方法和原理[J].安徽农学通报,14(22):35-36.

张芳英,穆丽娟,杨继,杨树民,张旭东,张征章(2011)蜂胶提取物抗菌作用的研究进展[J].中国药房,22(11):1041-1043.

张江临,王凯,胡福良(2013)蜂胶的抗氧化活性及其分子机制研究进展[J].中国中药杂志,38(16):2645-2652.

周建新,姚明兰,岳文倩,潘海琼(2007)蜂胶的抗菌性及其影响因素的研究[J].食品与发酵工业,33(3):41-43.

第三篇

蜂胶质量控制研究

第十一章　蜂胶中 β-葡萄糖苷酶的研究

　　蜂胶化学成分复杂,目前已从蜂胶中分离鉴定出的化学成分有黄酮类、萜烯类、醌类、酯类、醇类、醛类、酚类、有机酸类,还有大量的氨基酸类、酶类、维生素类、多糖及多种微量元素等。黄酮类化合物是植物中最丰富的一类多酚类化合物,天然存在的黄酮类化合物多以 β-D-葡萄糖苷形式存在。例如,黄酮醇类多为 3-O-糖苷,也有的为 7-O-糖苷和 4′-O-糖苷(Fossen et al.,1998),而黄酮和二氢黄酮及异黄酮类多为 7-O-糖苷(Day,1998)。然而,蜂胶中已分离鉴定出的黄酮类化合物大都以游离黄酮苷元形式存在。因此,很多学者推测蜜蜂在采集蜂胶的过程中加入了自身分泌物,使黄酮苷水解成黄酮苷元(Bonvehi et al.,1994;Banskota et al.,2001;Pietta et al.,2002;Lu et al.,2004)。

　　β-葡萄糖苷酶,又称 β-D-葡萄糖苷水解酶,别名龙胆二糖酶、纤维二糖酶和苦杏仁苷酶,属于水解酶类。它可催化水解结合于末端非还原性的 β-D-糖苷键,同时释放出配基与葡萄糖体(Webb,1992)。β-葡萄糖苷酶广泛存在于自然界中,可来源于植物和微生物,也可来源于动物(Esen,1993)。Gilliam 等(1988)在西方蜜蜂的中肠和后肠中发现 β-葡萄糖苷酶的活性。Pontoh 等(2002)从西方蜜蜂的消化道、蜜囊、咽下腺中均分离纯化出 β-葡萄糖苷酶,而且分离出的是相同的酶。因此,他们推断 β-葡萄糖苷酶产生于蜜蜂的咽下腺,在蜜蜂取食时分泌到口器里,传递到蜜囊中,然后再转移到巢房和消化道中。在蜂蜜中已发现 β-葡萄糖苷酶的活性(Low et al.,1986),蜂蜜中该酶的活性与 β-O-寡糖的连接方式有关(Low et al.,1988)。

　　那么,蜂胶中也存在 β-葡萄糖苷酶吗? 其性质如何? 为了探究这些问题,我们对蜂胶中 β-葡萄糖苷酶进行检测,对其性质及其在蜂胶成分转化中的作用以及储存条件对酶活力的影响进行研究,为 β-葡萄糖苷酶在蜂胶质量控制中的应用提供理论依据。

第一节　蜂胶中 β-葡萄糖苷酶的检测

　　β-葡萄糖苷酶活性的测定方法主要有三种:一是 Barush 和 Swiain 法,以水杨苷作底物,酶解产物用 4-氨基安替比林作显色剂,使释放出来的水杨醇显色,再用分光光度法比色测定;二是荧光法,利用伞形酮(7-羟基香豆素)与 4-甲基伞形酮具强烈荧光的特点,将它们生为无荧光的底物,以此测定;三是以对-硝基 β-D-葡萄糖苷(p-NPG)为底物进行酶解,底物水解后释放出来的对-硝基苯酚可用分光光度法比色测定。由于第三种方法简单、灵敏度高、反应活性大,所以大多数实验采用 p-NPG 作为底物的分光光度法测定酶活性。

　　本研究采用对-硝基 β-D-葡萄糖苷为底物的方法对蜂胶中的 β-葡萄糖苷酶活性进行测定,并对影响该酶活性的条件进行优化,包括 PVPP 的加入量、反应底物的浓度、缓冲液 pH、反应温度和时间等。采用最优化的条件,探讨了不同储存时间和温度条件下 β-D 葡萄糖苷酶活力的变化；对蜂胶中 β-D 葡萄糖苷酶的底物特异性进行研究。

1 材料与方法

1.1 材料和仪器

9 个新鲜蜂胶样本从浙江（杭州、长兴、平湖、绍兴、江山）、安徽（蒙山）、河北（玉田）、天津（蓟县）、山西（阳泉）的蜂箱中采集，采集时间为 2008 年 11 月至 2009 年 8 月。所有的蜂胶样本收集后均立即于 -18℃ 冷冻储存。

此外，分别从市场上购得来自山东、山西、河北、内蒙古、四川、安徽、江西等地的 7 个蜂胶样本。

p-硝基苯酚-β-D-葡萄糖苷（p-NPG）（北京凯森莱医药科技有限公司），对硝基苯酚（上海国药），聚乙烯吡咯烷酮（PVPP）（杭州南杭化工有限公司），UV2550 紫外可见分光光度计（岛津），低温高速离心机（Himac CR22G），分析天平（Denver TB-215 D）。

1.2 粗酶液的制备及 PVPP 的加入

冷冻过的蜂胶捣碎，过 30 目筛。取蜂胶粉末 3.0g 加入 25mL 不同 pH 值的柠檬酸-磷酸氢二钠缓冲液，然后加入不同量的 PVPP（0、0.3、0.6、0.9、1.2、1.5g）及少许石英砂。在冰浴上研磨成糊状。10000r/min 4℃ 离心 15min，上清液转移到另一只离心管中，14000r/min 4℃ 二次离心 30min，取上清液定容至 25mL，即为蜂胶 β-葡萄糖苷酶粗酶提取液。

1.3 β-葡萄糖苷酶活力的测定

取 0.5mL 粗酶液，加入 0.5mL 不同底物浓度的 p-NPG。反应混合液在不同温度的水浴锅中温育不同时间，加入 2.5mL 1mol/L Na_2CO_3 终止反应。冷却至室温后转移置比色皿中于分光光度计在 405nm 波长处测定吸光度。空白对照为反应混合液立即在 100℃ 水浴中加热 5min 使酶灭活，然后加入 1mol/L Na_2CO_3 终止反应。

1.4 酶反应条件的优化

1.4.1 pH 值和温度对酶活力的影响

首先优化缓冲液 pH 值对酶提取及测定的影响，柠檬酸-磷酸氢二钠缓冲液 pH 值分别为 4.0、5.0、6.0、7.0，反应温度为 37℃，p-NPG 用与提取时相同 pH 值的缓冲液溶解，浓度为 10mmol/L，反应时间为 1.5h。

在温度对酶活力影响测定中，柠檬酸-磷酸氢二钠缓冲液 pH6.0，酶活反应温度分别设为 27℃、37℃、47℃、57℃、67℃、77℃，p-NPG 用 pH6.0 的柠檬酸-磷酸氢二钠缓冲液溶解，浓度为 10mmol/L，反应时间为 1.5h。

1.4.2 底物浓度和反应时间对酶活力的影响

p-NPG 用 pH 6.0 的柠檬酸-磷酸氢二钠缓冲液溶解，浓度范围为 5~40mmol/L，间隔为 5mmol/L，反应条件为 37℃ 反应 1.5h。

为了测定准确，使吸光度在准确范围内（0.100~0.900），在缓冲液 pH 6.0、57℃、30mmol/L 条件下，对不同反应时间的吸光度进行检测。在反应不同时间（0.5~4.5h），从反应混合物中取出 1mL，加入 2.5mL 1mol/L Na_2CO_3 终止反应。

1.5 不同蜂胶样本中 β-葡萄糖苷酶活性的测定

采用上述最优化条件，对 9 个从蜂箱中新采集的蜂胶样本和 7 个随机买的蜂胶样本的

β-葡萄糖苷酶活力进行测定。酶活力单位为每分钟催化形成一个微摩尔对硝基苯酚所需要的酶量为一个酶活力单位。蜂胶中酶活力为每克蜂胶具有的酶活力单位（unit/g）。

2 结果与讨论

2.1 PVPP加入量对β-葡萄糖苷酶活力的影响

提取蜂胶粗酶液时，加入不同量的PVPP对酶活力影响很大。不加PVPP时，基本检测不到β-葡萄糖苷酶活力。随着PVPP量的增加，酶活性显著提高，当加到蜂胶量的20%时，酶活性达到最高，继续加入PVPP，酶活力不但没有增加，反而略有降低（图11.1）。这主要是因为蜂胶富含多酚类化合物，在提取过程中阻碍酶活力的发挥。加入PVPP吸附多酚，使酶活力得到发挥，但是过量的PVPP也会吸附酶，从而使酶活力降低。

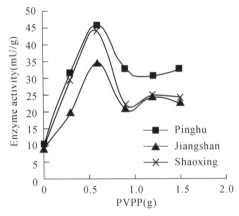

图 11.1　PVPP加入量对蜂胶中β-葡萄糖苷酶活性的影响

Fig. 11.1　Effect of PVPP on β-glucosidase activity

2.2 缓冲液pH值及温度对β-葡萄糖苷酶活力的影响

缓冲液的pH值对蜂胶中β-葡萄糖苷酶的提取影响很大，随pH值增加，酶活性显著提高，pH值为6.0时酶活力最高。pH值4和7时，吸光度降到0.1以下（图11.2）。不同来源蜂胶样本提取的β-葡萄糖苷酶表现出相同的变化规律。

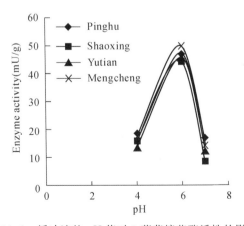

图 11.2　缓冲液的pH值对β-葡萄糖苷酶活性的影响

Fig. 11.2　Effect of pH on β-glucosidase activity

温度对蜂胶中β-葡萄糖苷酶活性的影响显著。在27～57℃时,随温度升高,酶活力显著升高,温度再升高,酶活力明显降低。表明测定该酶的适宜温度为37～67℃,最适温度为57℃(图11.3)。不同来源蜂胶样本提取的β-葡萄糖苷酶表现出相同的变化规律。

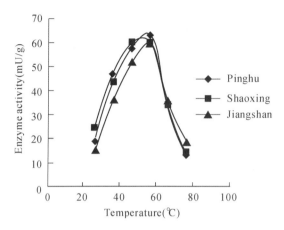

图11.3　温育温度对β-葡萄糖苷酶活性的影响

Fig. 11.3　Effect of incubation temperature on β-glucosidase activity

2.3　底物浓度对β-葡萄糖苷酶活力的影响

底物浓度在0～25mmol/L内,随底物浓度的增加,所测定的β-葡萄糖苷酶活力明显增加,底物浓度进一步增加,当底物浓度大于30mmol/L时,所测定的β-葡萄糖苷酶活力基本不增加(图11.4)。因此,测定蜂胶中β-葡萄糖苷酶活性时底物浓度应大于25mmol/L。

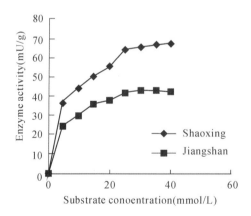

图11.4　底物浓度对β-葡萄糖苷酶活性的影响

Fig. 11.4　Effect of substrate concentration on β-glucosidase activity

2.4　温育时间对β-葡萄糖苷酶活力的影响

在57℃,底物浓度为30mmol/L的条件下,在4.5h内的酶促反应呈现线性关系,在0.5～3h内,吸光度在0.1～0.9之间,实验准确度最高,反应时间超过3.5h时,吸光度大于0.1,超过分光光度计准确范围。温育时间对β-葡萄糖苷酶活力影响见图11.5。因此,后续实验中我们选择温育时间为1.5h。

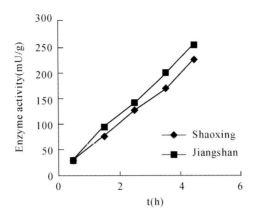

图 11.5　温育时间对酶活力的影响

Fig. 11.5　Effect of incubation time to the activity of β-Glucosidase

综上所述,测定蜂胶中 β-葡萄糖苷酶活性的最佳条件为:PVPP 加入量为蜂胶量的 20%,柠檬酸磷酸氢二钠缓冲液 pH 值为 6.0,反应温度为 57℃,底物浓度大于 25mmol/L,反应时间在 3h 内。

2.5　不同来源蜂胶中的 β-葡萄糖苷酶活性

不同来源蜂胶样本中 β-葡萄糖苷酶活性变化很大(表 11.1)。新鲜蜂胶与随机购买的蜂胶样本酶活力存在显著差异。新鲜蜂胶中 β-葡萄糖苷酶的活力均较高,吸光度均在分光光度计的准确范围内(0.26～0.58 abs),说明蜂箱中的蜂胶中普遍含有 β-葡萄糖苷酶,而且该酶活力较高;从市场随机购得的蜂胶除了四川蜂胶酶活力较高(0.12 abs)外,其余蜂胶样本中检测不到 β-葡萄糖苷酶活力,吸光度远远低于分光光度计的准确检测范围(0.01～0.05 abs)。由此可见,β-葡萄糖苷酶活力可能随储存时间延长而下降。

蜂胶是蜜蜂逐渐采集回来的天然的树脂产品,人为采收蜂胶前蜂胶至少已在蜂箱内累积超过一个月,而新鲜蜂胶样本中均检测到酶活力,说明 β-葡萄糖苷酶可能是蜜蜂在蜂胶加工过程中不断加入的。中国蜂胶的采集季节大约为 5 月到 11 月,我们随机买的蜂胶样本均为当年采集的蜂胶,根据所得结果说明,蜂胶在经过半年的储存后酶活力已基本全部丧失。

表 11.1　蜂胶中的 β-葡萄糖苷酶活力

Table 11.1　β-Glucosidase activity in propolis

新鲜蜂胶		储存蜂胶	
来源	酶活力(mU/g)	来源	酶活力(mU/g)
蓟县	116.9±5.5	四川	24.6±5.1
江山	95.8±3.9	河北	10.1±1.2
长兴	94.2±4.8	合肥	7.1±1.0
平湖	93.1±5.2	山东	6.1±1.6
蒙城	90.1±4.6	山西	5.9±1.3
绍兴	86.9±4.5	内蒙古	2.9±1.8

续表

新鲜蜂胶		储存蜂胶	
来源	酶活力（mU/g）	来源	酶活力（mU/g）
阳泉	78.7±3.9	江西	2.9±1.5
杭州	71.6±4.1		
玉田	54.0±4.4		

注：表中酶活力为一式三份的平均值±标准差（$\bar{x}\pm SD$）。

第二节　蜂胶中 β-葡萄糖苷酶的性质及其在成分转化中的作用

β-葡萄糖苷酶按其底物特异性可以分为 3 类：第一类是能水解烃基-β-葡萄糖苷或芳香基-β-葡萄糖苷的酶，此类 β-葡萄糖苷酶能水解的底物有纤维二糖、对硝基苯-β-D-葡萄糖苷等；第二类是只能水解烃基-β-葡萄糖苷的酶，这类 β-葡萄糖苷酶能水解纤维二糖等；第三类是只能水解芳香基-β-葡萄糖苷的酶，这类酶能水解对硝基苯-β-D-葡萄糖苷等类似物（Terra and Ferreira，1994）。Pontoh 和 Low（2002）从西方蜜蜂中分离出的 β-D 葡萄糖苷酶属于第三类。

本研究对蜂胶中 β-葡萄糖苷酶的底物特异性、对类黄酮单糖苷及二糖苷的水解作用进行研究，并探讨苷元结构和糖基的连接部位对酶水解作用的影响。

1　材料与方法

1.1　材料与试剂

蜂胶样本分别采自河北、安徽和浙江，从蜂箱内刮取，－18℃冷冻保存。

试剂：

甲醇，色谱纯：Merck 公司

二甲基亚砜（DMSO）：Sigma 公司

水为色谱级：杭州永洁达超纯水机自制

磷酸、柠檬酸、磷酸氢二钠均为分析纯

聚乙烯吡咯烷酮（PVPP）：杭州南杭化工有限公司

槲皮素 3-O-葡萄糖苷：Sigma 公司

芦丁：Sigma 公司

柚配基 7-O-葡萄糖鼠李糖苷：Sigma 公司

槲皮素：Sigma 公司

柚配基：Sigma 公司

槲皮素 7-O-葡萄糖苷：深圳美荷尔蒙生物科技有限公司

芹菜素 7-O-葡萄糖苷：深圳美荷尔蒙生物科技有限公司

木樨草素 7-O-葡萄糖苷：深圳美荷尔蒙生物科技有限公司

第十一章 蜂胶中 β-葡萄糖苷酶的研究

p-硝基苯酚-β-D-葡萄糖醛酸:Sigma 公司
纤维二糖:Sigma 公司
苦杏仁苷:Sigma 公司
葡萄糖氧化酶:Sigma 公司
过氧化物酶:Sigma 公司
p-硝基苯酚-β-D-半乳糖苷:北京凯森莱医药科技有限公司
p-硝基苯酚-β-D-纤维二糖苷:北京凯森莱医药科技有限公司
龙胆二糖:北京凯森莱医药科技有限公司
对硝基苯酚:上海国药

1.2 主要仪器

Agilent 1200,真空脱气器 G 1322A,四元泵 G1311A,紫外检测器(VWD)G1314B,恒温箱 G1316A。
高温冷冻离心机:Himac CR22G
分析天平:Denver TB-215 D
超纯水机:杭州永洁达净化科技有限公司
UV2550 紫外可见分光光度计:岛津
低温高速离心机:Himac CR22G
分析天平:Denver TB-215 D

1.3 色谱条件

色谱柱:Agilent Eclipse XDB-C18(4.6mm×250mm,5μm)
流动相:甲醇-0.5% H_3PO_4,梯度洗脱程序见表 11.2
流速:0.8mL/min
进样量:10μL
柱温:25℃
检测波长:柚配基 7-O-葡萄糖鼠李糖苷及苷元柚配基为 280nm 波长;槲皮素 3-O-葡萄糖苷、槲皮素 7-O-葡萄糖苷、芹菜素 7-O-葡萄糖苷、木樨草素 7-O-葡萄糖苷、芦丁及苷元柚配基为 350nm 波长。

表 11.2 类黄酮苷分析的梯度洗脱程序
Table 11.2 Gradient elution program for flavonoid glucosides analysis

	0min	18min	30min	31min	40min	41min	46min
B%(甲醇)	45	55	55	100	100	45	45

1.4 类黄酮苷底物的制备

几种类黄酮苷底物分别用 DMSO 溶解(最终反应液中 DMSO 浓度不高于 0.3%(v/v)),用柠檬酸-磷酸氢二钠缓冲液(pH 6.0)制成浓度为 200μmol/L 的母液。然后用柠檬酸-磷酸氢二钠缓冲液(pH 6.0)分别稀释到所需浓度。

1.5 对照品的制备

不同浓度的黄酮苷和相应苷元标准品分别用 50mL 甲醇溶解,然后用柠檬酸-磷酸氢二钠缓冲液(pH 6.0)定容至100mL。绘制每种类黄酮苷的标准曲线。

1.6 酶提取

冷冻过的蜂胶捣碎,过30目筛,混匀。取蜂胶粉末3.0g,加入液氮研碎,加入少量石英砂,0.6g PVPP,然后加入25mL柠檬酸-磷酸氢二钠缓冲液(pH 6.0)研磨成糊状。10000r/min 4℃离心15min,上清液转移到另一只离心管中,14000r/min 4℃离心30min,取上清液定容至25mL。

1.7 酶的底物特异性

蜂胶中β-葡萄糖苷酶对p-NP-β-D-单糖苷、p-NP-β-D-二糖苷和二糖的底物特异性。底物用pH6.0的柠檬酸-磷酸氢二钠缓冲液溶解,浓度为5mmol/L,反应温度为57℃,反应时间1.5h。测定从p-NP-β-D-半乳糖苷、p-NP-β-D-葡萄糖醛酸、p-NP-β-D-纤维二糖水解下来的p-NP。由于苦杏仁苷、纤维二糖和龙胆二糖中没有p-NP结构,故采用葡萄糖氧化酶法测定(Otero et al.,2003)。

1.8 酶对黄酮苷水解作用研究

取0.25mL粗酶液,分别加入0.25mL 100μmol/L不同的类黄酮苷底物。反应混合液在37℃的水浴锅中温育不同时间,加入0.5mL甲醇终止反应。空白对照为反应混合液立即在100℃水浴中加热5min使酶灭活,然后加入0.5mL甲醇终止反应。反应结束后,混合液离心,过0.22μm滤膜,然后用HPLC分析。

1.9 Km值和Vmax的计算

不同浓度的芹菜素7-O-葡萄糖苷、木樨草素-7-O-葡萄糖苷分别与不同来源的等量(0.5mL)蜂胶β-葡萄糖苷酶提取液混合,在37℃的水浴锅中温育60min,然后加入0.5mL甲醇终止反应。反应结束后,混合液离心,过0.22μm滤膜,然后用HPLC分析。

2 结果

2.1 蜂胶中β-葡萄糖苷酶的底物特异性

3个不同产地蜂胶中提取的β-葡萄糖苷酶具有相同的底物特异性(表11.3)。蜂胶中的β-葡萄糖苷酶只能水解芳基β-葡萄糖苷和β-半乳糖苷,对二糖苷(p-硝基苯酚-β-D-纤维二糖苷和苦杏仁苷)和有β-连接方式的二糖(纤维二糖和龙胆二糖)没有活力。

表11.3 蜂胶中β-葡萄糖苷酶的底物特异性

Table 11.3 Activity of β-glucosidase towards various glycosides

底物	酶活力(mU/g)		
	平湖	长兴	江山
p-硝基苯酚-β-D-半乳糖苷	28.0±4.2	31.0±3.7	35.4±3.6
p-硝基苯酚-β-D-葡萄糖醛酸	0	0	0
p-硝基苯酚-β-D-纤维二糖苷	0	0	0
苦杏仁苷	0	0	0
纤维二糖	0	0	0
龙胆二糖	0	0	0

注:表中数据为一式两份样本重复三次测定的平均值±标准差($\bar{X}$±SD)。

2.2 蜂胶中 β-葡萄糖苷酶对不同类黄酮苷底物的水解初速度和水解率

蜂胶中 β-葡萄糖苷酶水解木樨草素 7-O-葡萄糖苷、芹菜素-7-O-葡萄糖苷、槲皮素-7-O-葡萄糖苷、槲皮素-3-O-葡萄糖苷不同时间的色谱图分别见图 11.6、图 11.7、图 11.8 和图 11.9。如图所示，蜂胶中 β-葡萄糖苷酶能够将木樨草素 7-O-葡萄糖苷、芹菜素-7-O-葡萄糖苷、槲皮素-7-O-葡萄糖苷和槲皮素-3-O-葡萄糖苷水解去葡萄糖，分别生成其苷元木樨草素、芹菜素和槲皮素。蜂胶中 β-葡萄糖苷酶水解木樨草素 7-O-葡萄糖苷、芹菜素-7-O-葡萄糖苷、槲皮素-7-O-葡萄糖苷、槲皮素-3 O-葡萄糖苷在 180min 内均呈线性反应；而二糖苷芦丁、柚配基 7-O-葡萄糖鼠李糖苷在 180min 内不被该酶水解。

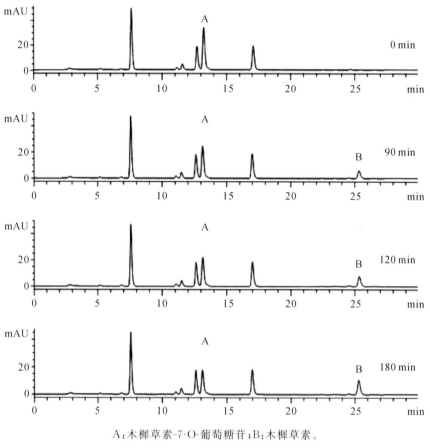

A：木樨草素-7-O-葡萄糖苷；B：木樨草素。
A：Luteolin 7-O-glucoside；B：Luteolin.

图 11.6 蜂胶中酶对木樨草素-7-O-葡萄糖苷不同水解时间的 HPLC 图

Fig. 11.6 HPLC chromatogram of Luteolin-7-O-glucoside after incubation with β-glucosidase from propolis

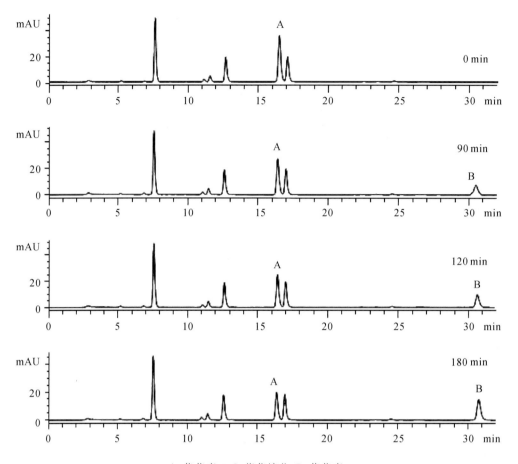

A:芹菜素-7-O-葡萄糖苷;B:芹菜素。
A:Apigenin 7-O-glucoside;B:Apigenin。
图 11.7 蜂胶中酶对芹菜素-7-O-葡萄糖苷不同水解时间的 HPLC 图
Fig. 11.7 HPLC chromatogram of Apigenin-7-O-glucoside after incubation with β-glucosidase from propolis

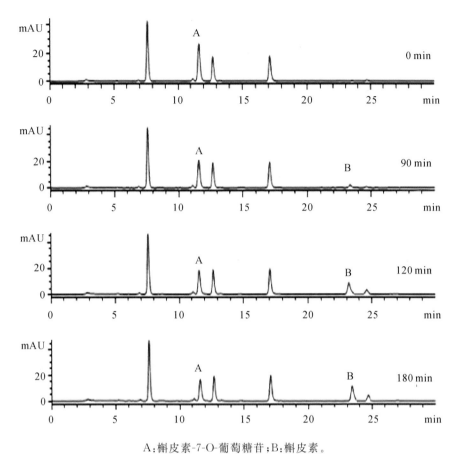

A：槲皮素-7-O-葡萄糖苷；B：槲皮素。
A：Quercetin 7-O-glucoside；B：Quercetin。

图 11.8 蜂胶中酶对槲皮素-7-O-葡萄糖苷不同水解时间的 HPLC 图

Fig. 11.8 HPLC chromatogram of Quercetin-7-O-glucoside after incubation with β-glucosidase from propolis

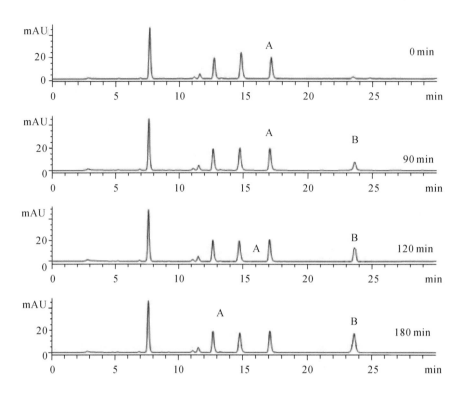

图 11.9 蜂胶中酶对槲皮素-3-O-葡萄糖苷不同水解时间的 HPLC 图
A:槲皮素-3-O-葡萄糖苷;B:槲皮素

Fig. 11.9 HPLC chromatogram of Quercetin-3-O-glucoside after incubation with β-glucosidase from propolis
A:Quercetin 3-O-glucoside;B:Quercetin

不同类黄酮葡萄糖苷的水解率和水解程度依类黄酮苷元的结构和糖取代基的位置而不同。由于糖取代基位置不同,槲皮素 3-O-葡萄糖苷的水解速度远大于槲皮素 7-O-葡萄糖苷。不同类黄酮苷元、木樨草素 7-O-葡萄糖苷水解速度大于芹菜素-7-O-葡萄糖苷,大于槲皮素 7-O-葡萄糖苷(表 11.4)。

表 11.4 蜂胶中酶提取物对不同黄酮苷的水解初速度和水解程度

Table 11.4 Initial rates and extent of hydrolysis of flavonoid glycosides with β-glucosidase from propolis

底物	初速度 (μmol/min g)	不同时间水解率(%)		
		90min	120min	180min
槲皮素-O-葡萄糖苷	3.80	28.64	35.77	49.55
槲皮素-O-葡萄糖苷	1.60	20.59	27.63	41.96
芦丁	0	0	0	0
柚配基 7-O-葡萄糖鼠李糖苷	0	0	0	0
芹菜素-O-葡萄糖苷	2.10	22.19	28.65	43.20
木樨草素 7-O-葡萄糖苷	3.24	25.59	32.72	46.47

2.3 蜂胶中 β-葡萄糖苷酶对不同类黄酮苷的 K_m 和 V_{max}

蜂胶中 β-葡萄糖苷酶对芹菜素 7-O-葡萄糖苷、木樨草素 7-O-葡萄糖苷的 K_m 分别为 $(12.93\pm1.73)\mu mol$ 和 $(19.49\pm2.76)\mu mol$，不同来源蜂胶中的 β-葡萄糖苷酶催化这两种类黄酮苷的 K_m 相似，说明不同地区蜂胶中 β-葡萄糖苷酶相同或作用相似。但是，不同来源蜂胶提取出的 β-葡萄糖苷酶对这两种类黄酮苷的 V_{max} 相差较多（表 11.5）。

表 11.5 蜂胶中 β-葡萄糖苷酶对芹菜素 7-O-葡萄糖苷、木樨草素-7-O-葡萄糖苷的 K_m 和 V_{max}

Table 11.5 K_m and V_{max} of apigenin 7-glucoside and luteolin-7-O-glucoside by β-glucosidase from propolis

底物	K_m ($n=3$)	$V_{max}/(\mu mol/min\cdot g)$ ($n=3$)
芹菜素-O-葡萄糖苷	12.93±1.73	1.91—3.80
木樨草素-7-O-葡萄糖苷	19.49±2.76	3.54—5.66

3 讨 论

蜂胶中的 β-葡萄糖苷酶只能水解芳基 β-葡萄糖苷和 β-半乳糖苷，对二糖苷（p-硝基苯酚-β-D-纤维二糖苷和苦杏仁苷）和有 β-连接方式的二糖（纤维二糖和龙胆二糖）没有活力。这与 Pontoh 和 Low（2002）在蜜蜂中肠和蜜囊中纯化出的 β-葡萄糖苷酶的特性一致，均对纤维二糖没有活力。但是，这并不能确定蜂胶中的 β-葡萄糖苷酶来源于蜜蜂，因为 β-葡萄糖苷酶来源广泛，各种有机体（植物、动物、真菌和细菌）均可分泌该酶（Esen，1993）。而且在蜂胶的主要植物源杨属植物中也曾发现 β-葡萄糖苷酶（Juntheikki and Julkunen-Tiitto，2000）。

尽管从蜂胶中提取 β-葡萄糖苷酶的过程中可能也提取出其他的糖苷酶，但是蜂胶 β-葡萄糖苷酶粗酶提取液仅对 p-硝基苯酚-β-D-葡萄糖苷和 p-硝基苯酚-β-D-半乳糖苷有水解作用，而对 p-硝基苯酚-β-D-纤维二糖苷、苦杏仁苷、纤维二糖和龙胆二糖等芳基 β-二糖苷以及 β-二糖均没有活力。这些结果与假说蜂胶中的类黄酮苷在蜂胶采集和加工过程中被 β-葡萄糖苷酶水解一致，同时也为蜂胶中存在异鼠李素-3-O-芸香糖苷（Popova et al.，2009）和芦丁（Bonvehi and Coll，2000）提供了一种解释。

自然界中的类黄酮多数以糖苷形式存在，最常见的糖为葡萄糖和鼠李糖，其次为二糖，如芸香糖等（Rice-Evans et al.，1997）。因此，根据实验结果，期望有更多的二糖苷化的类黄酮能从蜂胶中分离鉴定出来，为蜂胶化学成分研究提供了理论基础。

尽管本文所用酶液为 β-葡萄糖苷酶粗酶提取液，其中可能含有其他水解酶，但是蜂胶中 β-葡萄糖苷酶粗酶液只能够水解不同类型的类黄酮单葡萄糖苷，但对黄酮二糖苷没有水解作用。这些结果既肯定了蜜蜂在采集蜂胶的过程中将类黄酮单糖苷水解，又从理论上解释了蜂胶中检测到的芦丁和异鼠李素葡萄糖鼠李糖苷。

第三节 储存条件对蜂胶中葡萄糖苷酶活力的影响

类黄酮类化合物相对比较稳定,但在空气中与氧接触会被氧化分解。王永刚等(2005)发现田基黄中二氢槲皮素-7-O-鼠李糖苷在空气中易被氧化,生成槲皮素-7-O-鼠李糖苷。陈伟光和盛静(2006)研究不同存储时间的杭白菊总黄酮含量的变化趋势,发现杭白菊中总黄酮含量随存储时间的延长逐渐降低。孙君明等(2004)研究发现大豆种子随着储藏时间的增加异黄酮含量逐渐降低,储藏半年的大豆异黄酮含量降低较小,平均仅为12.9%,而储藏1年后平均降低57.1%。但室温、4℃和-10℃储存对异黄酮含量的影响不显著。

蜂胶从蜂箱中采集后基本在室温下自然储存,Bonvehi和Coll(2000)也曾报道新鲜蜂胶和储存蜂胶的主要活性成分(多酚和类黄酮)及生物学活性存在明显差异。除了主要活性成分类黄酮外,蜂胶中还含有约10%的芳香挥发油(Burdock,1998)。蜂胶中挥发油含量虽然不高,但成分相当复杂,发挥着许多药理活性,如抑菌(Trusheva et al.,2010;Popova et al.,2009)、抗真菌(Ota et al.,2001)等。许多研究发现,香紫苏、香薄荷、丁香、豆蔻、芫荽和墨角兰中的挥发性成分在储存过程中容易损失,影响其药理活性(Misharina et al.,2003)。

现行的蜂胶国家质量标准仅规定蜂胶中总黄酮含量,未对蜂胶的新鲜程度加以规定,这对蜂胶化学成分的稳定性及产品稳定性不利。本研究以蜂胶β-葡萄糖苷酶活力为指标,探讨其是否随储存温度和时间呈现规律性变化。

1 材料与方法

以从蓟县、阳泉、玉田和蒙城取的新鲜蜂胶为样本,测定储存温度和时间对酶活力的影响。每个蜂胶样本研碎,混匀,然后分别分成三等份,分别储存于-20℃、4℃和室温下密闭保存。分别于储存的第3、6、10、15、20、30、45、60、75和90d从每组样本中取两份按照优化的条件测定酶活力,重复测定取平均值。

酶活力测定方法同本章第一节。

2 实验结果

随着储存温度和储存时间的增加,β-葡萄糖苷酶活力下降。当蜂胶样本储存于-20℃和4℃ 90d时,没有明显的酶活力降低($P>0.37$,LSD)(表11.6)。但是,蜂胶储存在4℃时,酶活力下降呈现出一阶导数(表11.7)。当蜂胶储存于室温时,仅储存10d,酶活力就明显下降($p=0.03$,LSD),在储存期间,β-葡萄糖苷酶活力下降呈现指数降低(图11.6)。在前45d,呈现线性的半对数降低,表明呈现一阶导数反应。储存60d后,几乎检测不到β-葡萄糖苷酶活力(吸光度<0.1 abs)。

表 11.6 蜂胶在－20℃和 4℃储存时的 β-葡萄糖苷酶活力随时间的变化

Table 11.6 Variation of β-glucosidase activity in cold storage conditions.

时间/d	β-葡萄糖苷酶活力(mU/g)							
	蓟县		阳泉		玉田		蒙城	
	4℃	－20℃	4℃	－20℃	4℃	－20℃	4℃	－20℃
6	115.4±5.4	116.0±4.6	79.4±3.7	79.9±4.2	54.6±4.9	54.2±3.6	89.8±5.8	90.3±5.2
10	113.6±5.9	114.6±5.5	78.1±4.6	78.9±5.1	53.6±4.6	55.0±3.9	87.5±5.3	90.7±4.7
15	112.2±4.8	118.8±4.8	77.9±5.8	78.5±5.4	52.4±3.4	54.4±4.9	87.0±4.6	89.3±4.6
20	111.6±5.4	115.8±5.2	74.7±3.7	78.6±4.2	50.1±5.1	55.0±4.5	85.7±5.3	88.7±3.9
30	109.0±4.6	118.6±3.9	75.5±4.8	78.1±5.4	46.6±5.0	53.0±5.3	80.3±3.7	89.7±4.8
45	108.8±3.9	115.2±4.6	71.3±5.7	77.8±3.9	44.0±3.7	53.6±5.0	78.3±5.8	87.6±5.2
60	105.8±4.3	116.4±5.1	68.5±5.6	74.1±4.7	43.0±5.6	53.8±3.9	75.4±5.2	88.5±5.6
75	102.8±5.0	115.0±4.7	64.9±5.7	73.5±3.8	42.2±3.5	54.2±4.3	69.6±4.0	87.5±4.4
90	102.6±3.5	115.6±4.3	63.9±4.5	73.2±3.7	41.0±4.6	53.8±4.0	67.8±4.6	87.9±4.9

注:表中数据为一式两份样本重复三次测定的平均值±标准差($\overline{X}\pm SD$)。

表 11.7 蜂胶储存于 4℃时 β-葡萄糖苷酶呈现一阶导数反应

Table 11.7 First order reactions for β-glucosidase activity degradation in different propolis samples at 4℃

	蓟县	阳泉	玉田	蒙城
k	－0.1568	－0.1844	－0.1652	－0.2615
R^2	0.9558	0.9746	0.9124	0.9841

注:表中数据为一式两份样本重复三次测定的平均值±标准差($\overline{X}\pm SD$)。

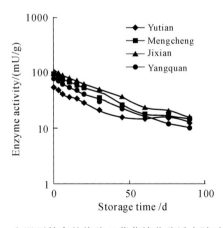

图 11.10 室温下储存的蜂胶 β-葡萄糖苷酶活力随时间的变化

Fig. 11.10 Variation of β-glucosidase activity according to the storage duration at room temperature

小 结

蜂胶中的 β-葡萄糖苷酶活力随储存温度的升高、储存时间的延长而降低,因此,可以作为蜂胶新鲜度的一个判断指标。但是,不同的蜂胶样本中 β-葡萄糖苷酶活力差异较大。在我们的实验条件下,如果从蜂胶中检测不到酶活力,说明这个蜂胶在室温下至少已储存 2 个月。作为新鲜度判断的指标,蜂胶中 β-葡萄糖苷酶活力范围的界定还有待进一步扩大样本确定。蜂胶中的活性成分,如多酚、类黄酮和挥发性成分等,在室温储存过程中可能降解,因此,β-葡萄糖苷酶活力的降低与蜂胶储存过程中化学成分变化的相关性还有待进一步研究。

参考文献

Banskota AH, Tezuka Y, Kadota SH (2001) Recent progress in pharmacological research of propolis[J]. *Phytother Research*, 15:561-571.

Bonvehi JS, Coil FV, Jorda RE (1994) The composition, active components and bacteriostatic activity of propolis in dietetics[J]. *Journal of the American Oil Chemists' Society*, 71(5):529-532.

Bonvehi SJ, Coll VF (2000) Study on propolis quality from China and Uruguay[J]. *Zeitschrift für Naturforschung*, 55c:778-784.

Burdock G (1998) Review of the biological properties and toxicity of bee propolis (propolis)[J]. *Food and Chemical Toxicology*, 36(4):347-363.

Day AJ, Dupont MS, Ridley S, Rhodes M, Rhodes MJ, Morgan MR, Williamson G (1998) Deglycosylation of flavonoid and isoflavonoid glycosides by human small intestine and liver beta-gulcosidase activity[J]. *FEBS Letters*, 436:71-75.

Esen A (1993) β-glucosidase:Overview. Biochemistry and Molecular Biology. ACS Symposium Series 533 [M]. Washington, DC:American Chemical Society, 1-14.

Fossen T, Pedersen AT, Andersen YM (1998) Flavonoids from red onion (*Allium cepa*)[J]. *Phytochemistry*, 47(2):281-285.

Gilliam M, Lorenz BJ, Richardson GV (1988) Digestive enzymes and microorganisms in honey bees, *Apis mellifera*: influence of streptomycin, age, season and pollen[J]. *Microbios*, 55:95-114.

Juntheikki M, Julkunen-Tiitto R (2000) Inhibition of β-glucosidase and esterase by tannins from Betula, Salix and Pinus species[J]. *Journal of Chemical Ecology*, 26(5):1151-1165.

Low NH, Nelson DL, Sporns P (1988) Carbohydrate analysis of Western Canadian honeys and their nectar sources to determine the origin of honey oligosaccharides[J]. *Journal of Apicultural Research*, 27:245-251.

Low NH, Vong KV, Sporns P (1986) A new enzyme, β-glucosidase, in honey[J]. *Journal of Apicultural Research*, 25:178-181.

Lu Y, Wu C, Yuan Z (2004) Determination of hesperetin, cinnamic acid and nicotinic acid in propolis with micellar electrokinetic capillary chromatography[J]. *Fitoterapia*, 75:267-

276.

Misharina T, Polshkov A, Ruchkina E, Medvedeva I (2003) Changes in the composition of the essential oil of marjoram during storage[J]. *Applied Biochemistry and Microbiology*, 39(3): 311-316.

Ota C, Unterkircher C, Fantinato V, Shimizu MT (2001) Antifungal activity of propolis on different species of Candida[J]. *Mycoses*, 44(9-10): 375-378.

Pietta PG, Gardana C, Pietta AM (2002) Analytical methods for quality control of propolis [J]. *Fitoterapia*, 73(1): S7-S20.

Pontoh J, Low NH (2002) Purification and characterization of β-glucosidase from honey bees (*Apis mellifera*)[J]. *Insect Biochemistry and Molecular Biology*, 32: 679-690.

Popova MP, Chinou IB, Marekov IN, Bankova VS (2009) Terpenes with antimicrobial activity from Cretan propolis[J]. *Phytochemistry*, 70: 1262-1271.

Rice-Evans C, Miller N, Paganga G (1997) Antioxidant properties of phenolic compounds [J]. *Trends in Plant Science*, 2(4): 152-159.

Trusheva B, Todorov I, Ninova M, Najdenski H, Daneshmand A, Bankova V (2010) Antibacterial mono-and sesquiterpene esters of benzoic acids from Iranian propolis[J]. *Chemistry Central Journal*, 4: 8.

Webb EC (1992) *Enzyme Nomenclature*. Recommendation of the Nomenclature Committee of the International Union of Biochemistry and Molecular Biology [M]. San Diego, CA: Academic Press.

陈伟光, 盛静(2006)不同存储时间和存储方法对杭白菊总黄酮含量的影响[J]. 时珍国医国药, 8: 1483-1484.

孙君明, 韩粉霞, 丁安林(2004)储藏温度与时间对大豆子粒中异黄酮含量的影响[J]. 大豆科学, 4: 245-248.

王永刚, 吴钉红, 杨立伟, 李沛波, 苏薇薇(2005)田基黄中一种不稳定黄酮的研究[J]. 中药材, 6: 468-469.

第十二章 蜂胶与杨树胶鉴别研究

温带地区蜂胶的植物来源主要以杨属和其杂交属的芽苞分泌物为主(Bankova et al.,2000)。在这些地区采集的蜂胶的化学成分与杨树芽的成分极其相似(Greenaway et al.,1987;曹炜等,2007)。由于蜂胶的生产受到蜂种、采胶方式以及胶源情况的影响,产量相对较低,处于供不应求状况。近年来,市场上出现了在蜂胶中掺入杨树胶(杨树芽提取物)的现象。杨树胶在颜色、气味、形状上都与蜂胶十分相似,而我国现有的蜂胶质量标准无法区分蜂胶与杨树胶,质量监管部门及加工、贸易企业无法在蜂胶生产和流通过程中进行有效的监管和控制。因此,如何高效准确地鉴定蜂胶的质量和真伪显得十分重要和迫切。

第一节 HPLC 指纹图谱鉴别蜂胶与杨树胶

在蜂胶的真假鉴别技术方面,许多研究者多年来进行了不懈的探索,试图通过液相色谱法(周萍等,2005,2009;盛文胜等,2008;王小平等,2009)、液相色谱串联质谱法(曹炜等,2007)、气相色谱质谱法(余兰平和盛文胜,2006)等来找出蜂胶的特征指纹图谱,以进一步控制蜂胶质量,防止掺假现象的发生,但未有成熟的鉴别杨树胶与蜂胶的方法。

由于以往鉴别蜂胶真假的研究主要针对蜂胶的醇溶性成分,而醇提法提取出的成分大多数为蜂胶与杨树胶的共有成分,而且干扰峰很多,因此,难以将蜂胶与杨树胶区分开。本研究主要对蜂胶及杨树胶中水溶性成分进行 HPLC 对比研究,以建立蜂胶真伪品的指纹图谱,为蜂胶质量控制及鉴别提供依据。

1 材料与方法

1.1 材料

不同来源的 3 个杨树胶样本均从市场购得,其中 Y_1 为未提纯过的杨树芽水熬物,Y_2 和 Y_3 均为提纯过的杨树芽水熬物。

不同地理来源的 8 个毛蜂胶样本由辽宁、内蒙古、山西、河北、山东、安徽、浙江和福建的蜂场提供。采集回来的蜂胶和杨树胶样本分别粉碎,混匀保存。

1.2 仪器与试剂

Agilent 1200,真空脱气器 G1322A,四元泵 G1311A,紫外检测器(VWD) G1314B,恒温箱 G1316A。

分析天平:Denver TB-215 D

超纯水机:杭州永洁达净化科技有限公司

甲醇:色谱纯,Merk 公司

磷酸:分析纯

水,色谱级:杭州永洁达高纯水机自制

1.3 试验方法

1.3.1 色谱条件

色谱柱:Agilent Eclipse XDB-C18 (4.6mm×250mm,5μm)

柱温:30℃

流动相:甲醇:0.5% H_3PO_4 =14:86

流速:0.6mL/min

进样量:10μL(定量环)

检测波长:268nm

1.3.2 样品制备

蜂胶供试品溶液的制备:分别取不同来源的蜂胶 10g,分别用 80mL 95%乙醇超声溶解,过滤,用真空干燥法除去乙醇,加入少量水,用氯仿萃取,除去水溶部分的残留氯仿,用流动相定容到 25mL,混匀,用 0.22μm 滤膜过滤。

杨树胶供试品溶液的制备:分别取不同来源的杨树胶 2.5g,分别用 20mL 95%乙醇超声溶解,过滤,用真空干燥法除去乙醇,加入少量水,用氯仿萃取,除去水溶部分的残留氯仿,用流动相定容到 25mL,混匀,用 0.22μm 滤膜过滤。

掺杨树胶的蜂胶供试品溶液的制备:分别取掺入 10%、20%、30%杨树胶的蜂胶各 10g,分别用 80mL 95%乙醇超声溶解,过滤,用真空干燥法除去乙醇,加入少量水,用氯仿萃取,除去水溶部分的残留氯仿,用流动相定容到 25mL,混匀,用 0.22μm 滤膜过滤。

1.3.3 方法学考察

(1)精密度试验

取同一份掺入 30%杨树胶的蜂胶样品溶液,连续进样 5 次,检测指纹图谱,考察特征色谱峰相对保留时间和相对峰面积的一致性,结果表明,各特征色谱峰保留时间的 RSD 均在 0.2%以内,相对峰面积的 RSD 均在 1.1%以内。

(2)稳定性试验

取同一份掺入 30%杨树胶的蜂胶样品溶液,分别于 0、3、6、9、12、24h 进样,测得各特征色谱峰的相对保留时间的 RSD 均在 0.5%以内,相对峰面积 RSD 均在 1.2%以内,表明样品溶液在 24h 内基本稳定。

(3)重复性试验

取同一批掺入 30%杨树胶的蜂胶样品溶液 5 份,按供试品溶液的制备方法,分别制成供试品溶液,结果表明,测得各特征色谱峰的相对保留时间的 RSD 均在 0.15%以内,相对峰面积 RSD 均在 3%以内。

2 结 果

2.1 蜂胶与杨树胶指纹图谱对比模型

按 1.3.1 的色谱条件对蜂胶和杨树胶进行 HPLC 分析,分别建立了蜂胶与杨树胶

HPLC 指纹特征图谱(图 12.1 和图 12.2)。

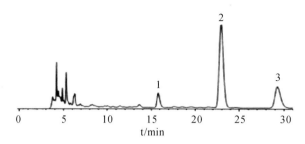

图 12.1　蜂胶 HPLC 指纹特征峰

Fig. 12.1　HPLC fingerprint of propolis

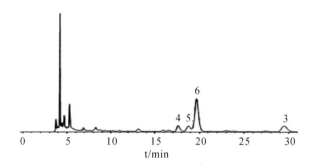

图 12.2　杨树胶 HPLC 指纹特征峰

Fig. 12.2　HPLC fingerprint of poplar tree gum

经比较发现,蜂胶与杨树胶的 HPLC 指纹图谱存在显著差异,特征性明显,特征峰附近没有干扰峰存在。在 HPLC 指纹图谱中,13～31min 内蜂胶和杨树胶有一个明显的共有成分(3 号峰),存在 5 个特征差异峰。其中,蜂胶中 1 号和 2 号两个明显的特征峰是杨树胶中不具有的,而且 2 号峰峰面积较大,特征性强;而杨树胶中 4、5 和 6 号 3 个特征峰是蜂胶中没有的,而且 6 号峰峰面积较大,特征性强。6 个特征峰的保留时间值见表 12.1。

表 12.1　蜂胶与杨树胶 HPLC 指纹图谱特征峰的保留时间

Table 12.1　Retention time of HPLC fingerprint from propolis and poplar tree gum　　min

	1	2	3	4	5	6
蜂胶	15.90	23.05	29.06			
杨树胶			29.09	17.63	18.77	19.69

2.2　不同产地蜂胶指纹图谱比较

按 2.3.1 色谱条件对不同来源的蜂胶进行测定,经对比发现,尽管蜂胶产地不同,但是该方法测得的各地蜂胶指纹图谱极其相似,其中 1、2、3 号峰为各地蜂胶的共有峰,但各共有峰的峰面积不同(图 12.3)。在所有蜂胶样本中,2 号峰峰面积均较大,1 号峰峰面积较小,但不同产地蜂胶 HPLC 图谱中均未出现 4、5、6 号特征峰。

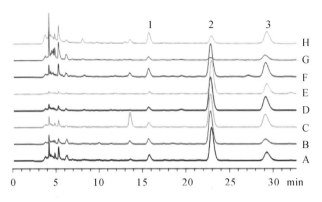

图 12.3　不同来源蜂胶 HPLC 指纹图谱
A:内蒙古;B:辽宁;C:山西;D:河北;E:山东;F:安徽;G:浙江;H:福建
Fig. 12.3　HPLC fingerprint of propolis from different geographic origins.
A:Inner Mongolia;B:Liaoning;C:Shanxi;D:Hebei;E:Shandong;F:Anhui;G:Zhejiang;H:Fujian

2.3　不同来源杨树胶指纹图谱比较

按供试品溶液的制备方法,分别取不同来源的杨树胶样品制成供试品溶液,取 10μL 进样,按前述色谱条件测定。不同来源及不同加工方法的杨树胶指纹图谱极其相似,其中 3、4、5、6 号峰为各杨树胶的共有特征峰,但彼此间峰面积不同。未提纯过的杨树芽水熬物(Y_1)指纹图谱中 6 号峰的峰面积明显高于提纯过的杨树芽水熬物(图 12.4)。

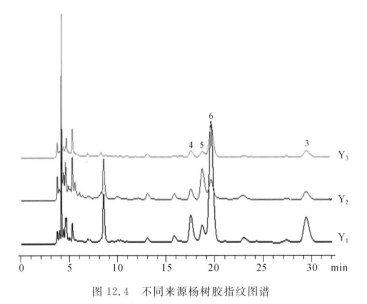

图 12.4　不同来源杨树胶指纹图谱
Fig. 12.4　HPLC fingerprint of poplar tree gum from different geographic origins

2.4　掺入不同比例杨树胶的蜂胶色谱图

按供试品溶液的制备方法,分别取掺入 10%、20% 和 30% 杨树胶的蜂胶样品制成供试品溶液,取 10μL 进样。按前述色谱条件测定,结果见图 12.5。掺入不同比例杨树胶的蜂胶溶液中,掺假 10% 杨树胶的可以检出 1、2、3、5、6 号特征峰,掺假 20% 和 30% 的均可检出 1、

2、3、4、5、6号特征峰。随掺入杨树胶比例的增加,4、5、6号峰峰面积明显增加,2号特征峰的峰面积减少,而1、3号峰峰面积变化不大。

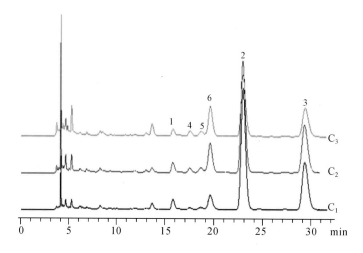

图 12.5　掺杨树胶的蜂胶指纹图谱

C_1:掺加10%杨树胶的蜂胶;C_2:掺加20%杨树胶的蜂胶;C_3:掺加30%杨树胶的蜂胶

Fig. 12.5　HPLC fingerprint of proplis mixed with poplar tree gum

C_1:Propolis mixed with 10% poplar tree gum;C_2:Propolis mixed with 20% poplar tree gum;C_3:Propolis mixed with 30% poplar tree gum

3　讨　论

本实验以毛蜂胶10g和杨树胶2.5g为基准样,用95%乙醇进行提取,该比例下蜂胶提取物远大于杨树胶提取物,而杨树胶中3个明显的特征峰在蜂胶中不存在,说明用这3个特征峰鉴定真假蜂胶具有可行性。但是,由于蜂胶成分复杂,不同产地蜂胶成分间差异很大,杨树胶的提取制作工艺也不尽相同,因此,本研究所提供的蜂胶与杨树胶间特异组分仅能定性区分蜂胶与杨树胶,以及蜂胶中是否掺有杨树胶,对于掺假多少很难进行定量。

在 Y_1、Y_2 两个杨树胶样本中也有较小的1号和2号峰存在,由于这些样本是从市场随机购买的,因此,可能是由这些杨树胶在加工过程中掺有部分蜂胶造成的。但这两个峰不影响鉴定样本中是否有杨树胶存在,只需在实验所述色谱条件下,看是否有4、5、6号峰或者仅有6号峰出现即可认为样本中有无杨树胶存在。

本研究定性地给出了蜂胶与杨树胶差异的特征图谱,这些特征峰具体是什么物质有待进一步分离鉴定。

小　结

通过不同产地蜂胶样本与不同来源杨树胶样本进行对比指纹图谱研究发现,在本实验建立的条件下,蜂胶与杨树胶HPLC指纹图谱存在显著差异,特征峰明显,且没有干扰峰存在。蜂胶中2号峰和杨树胶中6号峰峰面积较大,特征性最明显。如果所测样本图谱中同时具有2号峰和6号峰,且峰面积较大则认为是掺有杨树胶的蜂胶;如果图谱中有6号峰而未检出2号峰则认为是纯的杨树胶样本。本实验确定的液相色谱条件基线平稳,峰形对称,分离度较好,保留时间较短,对掺杨树胶的蜂胶样检出灵敏度高。因此,本研究建立的HPLC色谱条件可用于区分蜂胶与杨树胶,以及蜂胶中是否掺入杨树胶。

第二节 以水杨苷和邻苯二酚为参照的蜂胶与杨树胶鉴别方法

1 以水杨苷为参照

蜂胶中所含化学成分有三个来源:蜜蜂采集回来的植物分泌物、蜜蜂代谢的分泌物以及在蜂胶采集加工过程中加入的物质(Ghisalberti,1979;Marcucci,1995)。蜜蜂加入的腺体或酶分泌物可能改变了原始树脂中的某些成分(Bonvehí et al.,1994;Burdock,1998),可能引起蜂胶与植物源分泌的树脂存在成分差异。

杨属植物的特征性成分是酚苷,以水杨苷、柳皮苷、2′-苯甲酸水杨苷和特里杨苷含量较为丰富(Thieme,1967;Pearl and Darling,1971),典型酚苷结构见图12.6。有关蜂胶化学成分的研究很多(Greenaway et al.,1987;Marcucci,1995;Bankova et al.,2000;Usia et al.,2002;Bankova,2005),迄今为止尚未有蜂胶中存在任何酚苷的报道,很多研究曾报道过蜂胶中含有水杨酸及乙酰水杨酸等衍生物(Marcucci,1995;Krol et al.,1996)。杨属植物中发现的酚苷为β-葡萄糖苷形式,我们在蜂胶中也检测到β-葡萄糖苷酶活力,因此,推测蜜蜂在采集蜂胶以及蜂巢内传递蜂胶的过程中已将酚苷水解。

杨属植物有100多种,主要分布在温带和亚热带地区(王欣等,1999)。水杨苷是酚苷的基本结构单元,是更高级的水杨酸盐类衍生物的主要降解产物之一,在所有研究过的杨属植物的皮、叶、花、芽中均有分布(Clausen et al.,1989),而且水杨苷是市场上唯一商品化的酚苷,价格便宜,易购得。因此,我们选取水杨苷作为蜂胶与杨树胶鉴别的指标。

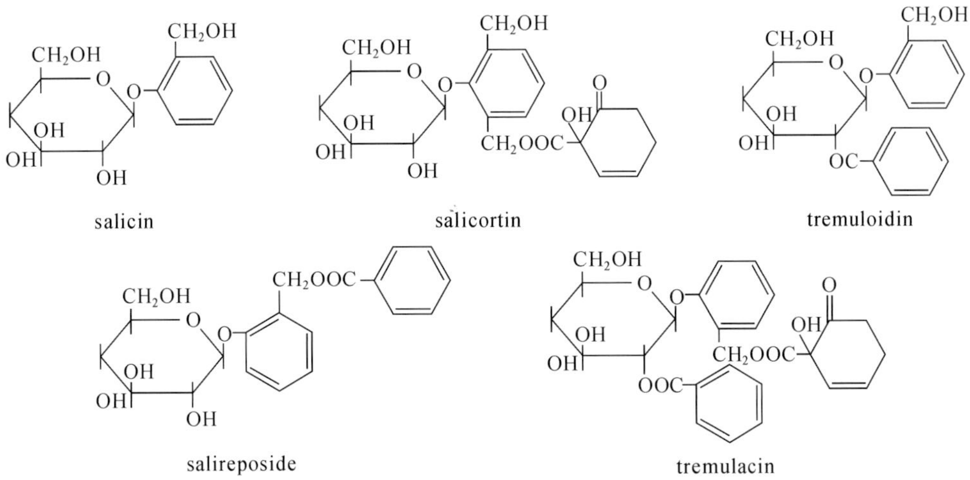

图12.6 典型酚苷结构

Fig.12.6 Structure of typical phenolic glucosides

本实验以水杨苷为参照,建立区分蜂胶与杨树胶的方法。首先,建立杨树芽提取物中水杨苷的提取和 HPLC 检测方法;然后,对蜂胶采集地周围杨树芽和叶子提取物以及蜂箱内不同采集时间的蜂胶进行水杨苷的检测,并通过 HPLC-UV-ESI-MS 对其进行验证。

1.1 实验材料

1.1.1 化学试剂

甲醇,乙腈:色谱纯,Merck 公司;

水为色谱级:杭州永洁达高纯水机自制;

乙醇:分析纯;

磷酸:分析纯;

水杨苷:色谱纯,Sigma 公司。

1.1.2 仪器

Agilent 1200,自动进样品 G1329A,真空脱气器 G 1322A,四元泵 G1311A,紫外检测器(VWD) G1314B,恒温箱 G1316A。

HPLC-MS

仪器:Agilent 1100 LC/MSD SL,配有 ESI,APCI 源。

液相色谱条件:

色谱柱:Sepax HP-C18 (150mm×4.6mm,5μm);流动相:乙腈:0.25%甲酸=5:95;柱温:30℃;检测波长:213nm;进样量:100μL;流速:1mL/min。

质谱条件:

干燥气流量:12.0L/min;雾化器压力:50 psig;干燥气温度:350℃;毛细管电压:正离子:4000 V;负离子:3500 V;裂解电压:100 V;扫描范围:100~1000amu。

1.1.3 实验原料的收集

蜂胶、杨树顶芽和叶子样本采自天津蓟县。选取两个蜂场,分别先将蜂箱内的蜂胶全部刮掉,每 15 箱一组,共 5 组,分别于第 5d(A 组)、10d(B 组)、15d(C 组)、20d(D 组)、30d(E 组)采集蜂胶。蜂场周围仅有北京杨(*Populus×beijingensis*)和加拿大白杨(*Populus canadensis*)分布,采集这两种杨树的顶芽和叶子。

11 个杨树胶样本从市场上购买。

1.2 方法与结果

1.2.1 提取条件的优化

采用正交试验优化北京杨顶芽中水杨苷的提取条件。包含四个因素:(A)乙醇浓度;(B)超声提取时间;(C)料液比;(D)提取次数。正交设计 L9 (3^4)见表 12.2。2g 北京杨顶芽在表 12.2 的条件下超声提取,抽滤,合并滤液蒸干溶剂,然后用甲醇定容到 50mL。过 0.45μm 滤膜,用 HPLC 分析。

杨树芽中水杨苷的不同提取方法结果见表 12.2。各因素对水杨苷提取率的影响次序为 B>C>D>A,通过比较选择最有效的提取方法为:2g 北京杨顶芽用 75%的乙醇超声提取 30min,料液比为 1:15,反复提取 3 次,挥去多余乙醇,定容至 50mL,过 0.45μm 滤膜,HPLC 法测定。

表 12.2　杨树芽中水杨苷提取条件优化的正交实验设计及结果

Table 12.2　Results and analysis of orthogonal design for the optimization of extraction conditions

试验号	A 乙醇浓度/%	B 超声时间/min	C 料液比	D 提取次数	提取率 /%
1	50	15	1∶5	2	0.1437
2	50	30	1∶10	3	0.1975
3	50	45	1∶15	4	0.1765
4	75	15	1∶10	4	0.1576
5	75	30	1∶15	2	0.2078
6	75	45	1∶5	3	0.1774
7	100	15	1∶15	3	0.1776
8	100	30	1∶5	4	0.1853
9	100	45	1∶10	2	0.1806
均值1	0.173	0.160	0.169	0.177	
均值2	0.181	0.197	0.179	0.184	
均值3	0.181	0.178	0.187	0.173	
极差	0.008	0.037	0.018	0.011	
优水平	A2(3)	B2	C3	D2	

1.2.2　样本制备

根据 1.2.1 的优化条件，蜂胶、杨树芽、杨树叶各 2g，分别用 75% 的乙醇超声提取 30min，料液比为 1∶15，反复提取 3 次。挥去多余乙醇，定容至 50mL，过 0.45μm 滤膜，HPLC 法测定。

由于蜂胶中含有 50% 树脂，杨树胶相当于蜂胶中提取出的树脂部分，因此，取 1g 杨树胶样本，用 75% 的乙醇超声提取 30min，料液比为 1∶15，反复提取 3 次。挥去多余乙醇，定容至 50mL，过 0.45μm 滤膜，HPLC 法测定。

1.2.3　色谱条件

色谱柱：Sepax HP-C18(150mm×4.6mm,5μm)；流动相：乙腈∶0.5% H_3PO_4＝5∶95；柱温：30℃；检测波长：213nm；进样量：5μL；流速：1mL/min。

1.2.4　杨树顶芽、叶子和蜂胶中水杨苷的测定

加拿大杨顶芽、叶子，北京杨顶芽、叶子中均检测到水杨苷(图 12.7)。

在本实验条件下，不同树种、同一树种的芽和叶子的 HPLC 色谱度很相似，两种杨树间化学成分相似。但水杨苷含量依树种不同、组织位置不同而存在差异。加拿大杨顶芽、叶子中水杨苷含量分别明显高于北京杨顶芽、叶子中的含量；同一种杨树的顶芽中水杨苷含量明显高于叶子中的含量(表 12.3)。

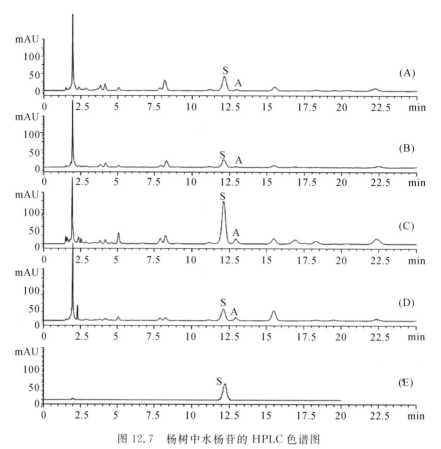

图 12.7 杨树中水杨苷的 HPLC 色谱图

(A)北京杨树芽,(B)北京杨树叶,(C)加拿大杨树芽,(D)加拿大杨树叶,(E)水杨苷标准品。
S:水杨苷;A:未知峰。

Fig. 12.7 HPLC profiles of *Populus* × *beijingensis* buds（A）,*Populus* × *beijingensis* leaves（B）,*Populuscanadensis* buds（C）,*Populuscanadensis* leaves（D）and standard（E）.
S:Salicin;A:Unknown.

表 12.3 杨树芽、叶子中水杨苷含量

Table 12.3 Content of salicin in different *Populus* samples

样本	植物组织	含量/%
北京杨	芽	0.2078
	叶子	0.1203
加拿大杨	芽	0.4835
	叶子	0.1501

对两个蜂场蜂箱内累积不同时间的蜂胶进行检测,均未检测到水杨苷(图 12.8)。不同累积时间采集的蜂胶 HPLC 色谱谱图高度一致,峰形、峰面积均相似。说明不同时间蜜蜂采集的树源一致。

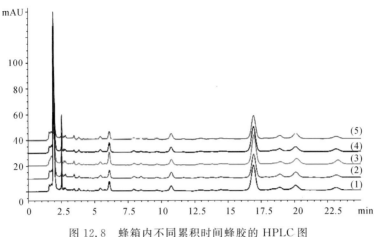

图 12.8 蜂箱内不同累积时间蜂胶的 HPLC 图
(1) A 组;(2) B 组;(3) C 组;(4) D 组;(5) E 组。

Fig. 12.8　HPLC chromatograms of propolis samples for different storing time in hives
(1) Group A;(2) Group B;(3) Group C;(4) Group D;(5) Group E.

1.2.5　蜂胶与杨树胶中水杨苷的对比

蜂胶中检测不到水杨苷,而杨树胶中可检测到水杨苷,而且还有一个峰 A 在杨树胶样本中存在,而在蜂胶中检测不到(图 12.9)。相对于水杨苷的峰面积,峰 A 的峰面积更大。因此,峰 A 是鉴别蜂胶与杨树胶的另一个特征峰。

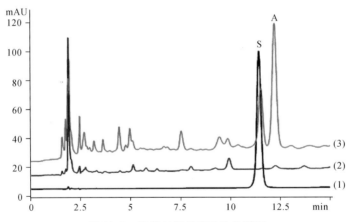

图 12.9　蜂胶与杨树胶对比色谱图
(1)水杨苷标准品;(2)纯蜂胶;(3)杨树胶代表峰;S 为水杨苷;A 为未知峰

Fig. 12.9　Comparative HPLC profile of propolis and poplar three gum
(1) Salicin standard;(2) Pure propolis;(3) Represents peaks of poplar three gum;S is salicin; A is the unknown peak

1.3　蜂胶与杨树胶中水杨苷有无的 HPLC-MS 验证

1.3.1　水杨苷标准品的 HPLC-ESI(±)/MS 图

在全扫描正、负离子两种电离模式下,水杨苷标准品均有信号响应(图 12.10a 和 12.10b)。水杨苷的相对分子质量为 286.28,从图 12.10a 可以看出,m/z 321.1 为负离子状态下水杨苷的准分子离子峰$[M+Cl^-]^+$。从图 12.10b 可以看出，m/z 304.1 为正离子状态下水杨苷的准分子离子峰$[M+NH_4^+]^-$。

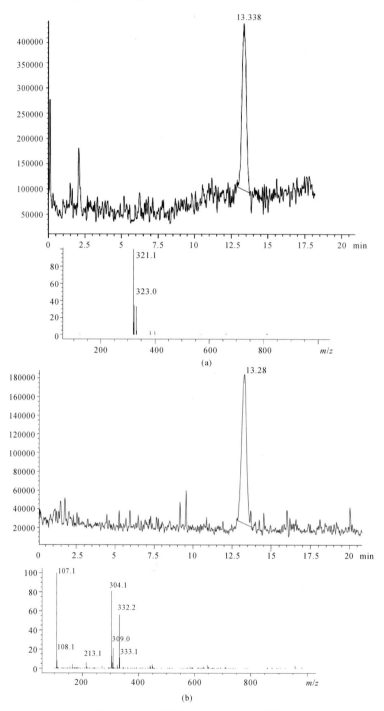

图 12.10　水杨苷 HPLC-ESI/MS 图。
(a) 负离子图；(b) 正离子图。
Fig. 12.10　The HPLC-ESI/MS of salicin.
(a) ESI(－)；(b) ESI(＋).

1.3.2 杨树胶的 HPLC-ESI（±）/MS 图

在全扫描正、负离子两种电离模式下，杨树胶 HPLC-ESI（±）在水杨苷出峰位置均有信号响应（图 12.11a 和 12.11b）。负离子状态下杨树胶中发现 m/z 321.1，与水杨苷标准品的 ESI（－）相同。在正离子模式下，杨树胶中发现 m/z 304.1，与水杨苷标准品的 ESI（+）相同。尽管 A 峰在 HPLC 系统下峰高远远高于水杨苷峰，但在 ESI（±）状态的总离子流图中，未发现 A 峰，因此，A 峰不能通过 ESI（±）进一步确定。我们研究中也尝试了 HPLC-APCI（±）方法探讨 A 峰的结构，也未得到相关信息。有关 A 峰的进一步鉴定有待后续研究。

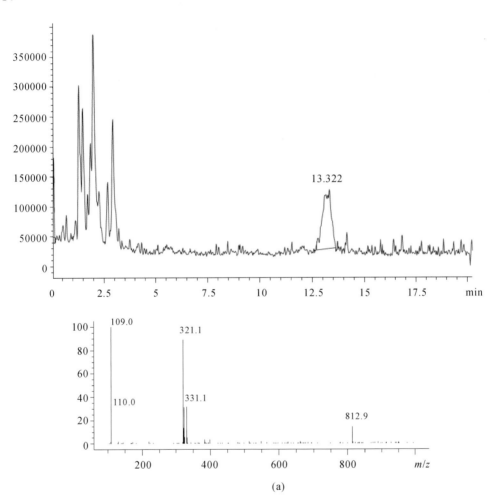

(a)

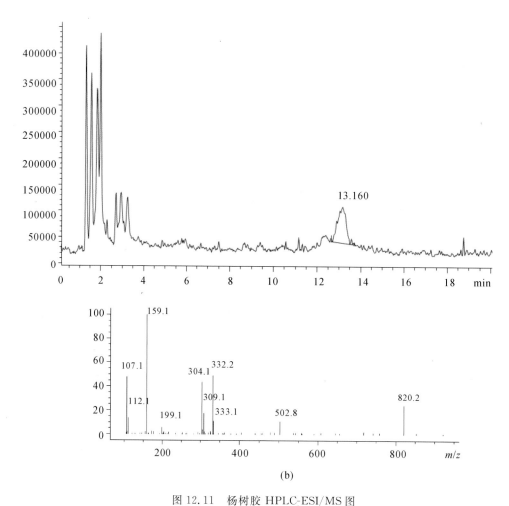

图 12.11 杨树胶 HPLC-ESI/MS 图
(a) 负离子图；(b) 正离子图。
Fig. 12.11 The HPLC-ESI/MS of poplar gum
(a) ESI(−)；(b) ESI(+)。

1.3.3 蜂胶的 HPLC-ESI(±)/MS 图

在全扫描正、负离子两种电离模式下，蜂胶 HPLC-ESI(±)在水杨苷出峰位置均没有信号响应（图 12.12a 和 12.12b）。通过 ESI（±）进一步确证蜂胶中没有水杨苷存在。

1.4 讨论

本研究建立了一种测定杨树芽中水杨苷含量的方法。该方法灵敏，可以作为杨树芽、叶中水杨苷含量的测定方法。由于在杨树胶中检测到水杨苷，而蜂胶中未检测到，因此，该方法可用于蜂胶与杨树胶的鉴别。该方法经 HPLC-ESI（±）/MS 验证，具有可靠性，可用于蜂胶与杨树胶的鉴别。

水杨苷存在于所有杨属植物中。本实验中，北京杨和加拿大杨的顶芽和叶子中均检测到水杨苷，而且在所有的杨树胶样本中也检测到了，但是在所有的蜂胶样本中未检测到水杨苷。这说明，在蜜蜂采集蜂胶的过程中水杨苷可能被降解，而在杨树胶制作过程中水杨苷稳

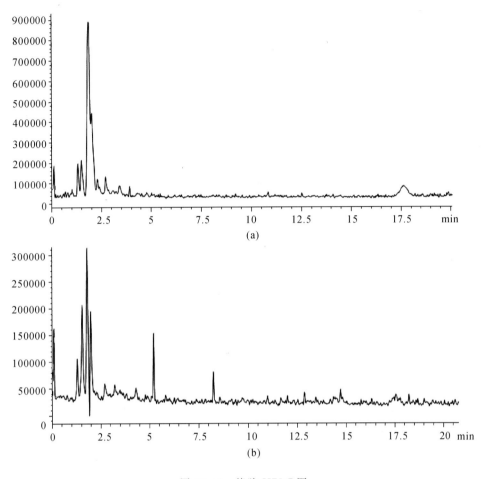

图 12.12　蜂胶 HPLC 图

Fig. 12.12　HPLC profile of Propolis.

定存在。进一步证明以水杨苷为参照鉴别蜂胶与杨树胶的可行性。

除了水杨苷,还有一个峰 A 在杨树胶样本中存在,而在蜂胶中检测不到。相对于水杨苷的峰面积,峰 A 的峰面积更大。因此,峰 A 是鉴别蜂胶与杨树胶的另一个特征峰。而 A 峰在 HPLC-ESI/MS(±)和 HPLC-APCI(±)中均未质子化,A 峰到底是何种物质还需进一步分离鉴定。

2　以邻苯二酚为参照

在杨树芽和杨属型蜂胶化学成分对比研究中,除水杨苷外,在杨树芽提取物的指纹图谱中还有一个未知的特征性成分 A。本研究以 A 物质为研究内容,应用不同有机溶剂、通过 Sephadex LH-20 葡聚糖凝胶色谱以及反相 HPLC 色谱柱等技术对杨树胶中未知物质进行分离纯化,结合液相色谱质谱联用以及核磁共振等现代光谱学分析技术,对 A 物质的化学结构进行鉴定,并探讨了其在蜜蜂加工蜂胶过程中的转化机制。

2.1 未知化合物 A 的鉴别

2.1.1 材料与方法

(1) 试剂

甲醇,乙腈(色谱纯,Merck 公司);超纯水(色谱纯,杭州永洁达高纯水机自制);乙醇(分析纯);石油醚(分析纯);氯仿(分析纯);乙酸乙酯(分析纯);磷酸(分析纯);醋酸:(分析纯);邻苯二酚标准品(色谱纯,Willey 公司);氘代甲醇(青岛一辰实验有限公司);超纯水(色谱纯,杭州永洁达高纯水机自制);三羟甲基氨基甲烷缓冲液(0.1mol/L,pH=8.3);2% H_2O_2(上海镱晨实业有限公司);多酚氧化酶同工酶提取和分离试剂盒(上海镱晨实业有限公司);多酚氧化酶显色剂 A(2g 对苯二胺溶于 18mL 的乙酸中);多酚氧化酶显色剂 B(1.5mL 1%的邻苯二酚溶液、0.3mL 2%的 H_2O_2 置于 60mL 的超纯水中)。

(2) 仪器

冷冻冻干机(Labconco 配有型号为 7960032 冷冻冻干系统);超声波辅助提取仪(SK20GT,上海科岛超声波仪器生产公司);旋转蒸发仪(RE-2000A,上海亚荣生化仪器场);高效液相色谱仪(Agilent 1200,配有真空脱气机 G1322A,自动进样器 G1329A,四元泵 G1311A,紫外检测器(VWD)G1314B,恒温箱 G1316A);液相色谱质谱联用仪(Agilent 6460);核磁共振分析仪(Bruker AVANCE III 600 MHz);低温冷冻离心机(Himac CR22G);分析天平(Denver TB-215D);超纯水机(杭州永洁达净化科技有限公司)。

(3) 杨树芽提取物的制备

杨树芽样、蜜蜂样本均从浙江大学实验蜂场采集。

蜂胶样本采自河南商水。

杨树芽、蜂胶、蜜蜂样品采集后立即于 -18℃ 冷冻储存。

取一定量的杨树芽,加入 100% 的乙醇(料液比 1:10)超声提取 30min,提取 2 次,过滤,合并滤液。向滤液中加入体积比为 1:3 的超纯水,过滤,旋转蒸发去除乙醇,仅留水溶液。

(4) 高效液相色谱条件(分析型液相)

色谱柱:Sepax HP-C18 (150mm×4.6mm,5μm);流动相:乙腈:0.5% H_3PO_4=5:95 (v/v);柱温:30℃;检测波长:213nm;进样量:5μL;流速:1.0mL/min。

(5) 高效液相色谱条件(制备型液相)

色谱柱:ShrmpackODS-C18(20mm×250mm,10μm);流动相:乙腈:水=5:95(v/v);柱温:30℃;检测波长:213nm;进样量:10μL;流速:8.0mL/min。

(6) 未知化合物 A 的粗分离

将杨树芽提取物的水溶液分别与石油醚、氯仿、乙酸乙酯 3 种有机溶剂等体积萃取 3 次,合并乙酸乙酯提取液,低压旋转蒸发除去乙酸乙酯溶液,得到杨树芽的乙酸乙酯提取物。

称取一定质量的杨树芽乙酸乙酯提取物,溶于一定体积的超纯水中,料液比 1:10,将得到的水溶液上样型号为 Sephadex LH-20 羟丙基葡聚糖凝胶树脂柱,上样后的羟丙基葡聚糖凝胶树脂用 10BV(柱床体积)的 100% 乙醇洗脱,流速 3mL/min,将洗脱液分管收集,每管 0.5BV,共收集 20 管,分别用分析型液相色谱仪进行未知峰(出峰时间为 11.5min 左右)检测,将富含有未知峰(出峰时间为 11.5min 左右)的溶液合并,在旋转蒸发仪上 45℃ 蒸干得物质 A 粗提物。

(7) 未知化合物 A 的纯化与鉴定

将得到的物质 A 粗提物溶于超纯水中(料液比 1∶3),采用半制备型高效液相色谱仪对 A 物质进行进一步分离纯化;合并纯化后含有物质 A 的溶液,45℃旋转蒸发,制得物质 A;并利用核磁共振(NMR)、质谱(MS)等技术鉴定其化学结构。然后通过液相色谱法对杨树芽中的未知化合物 A 与标准品进行对比,最终确定未知化合物 A 的化学信息。

(8) 酶提取

分别将冷冻过的蜂胶、杨树胶、杨树芽、蜜蜂头部以及蜜蜂躯干部捣碎,过 30 目筛,混匀。分别取蜂胶、杨树胶、杨树芽、蜜蜂头部以及蜜蜂躯干部粉末 1g 放入 5 个研磨中,加入少量石英砂,然后加入 3mL 三羟甲基氨基甲烷缓冲液(pH=8.3)研磨成糊状。12000r/min 4℃离心 20min,取上清液定容至 5mL。

(9) 聚丙烯胺凝胶电泳

按表配制 12% 的分离胶 10mL,迅速沿玻璃壁灌入玻璃板的细缝至顶端 2cm,然后加入一层超纯水压平凝胶上端,静置 30min。等待凝胶完全聚合后,加入 5% 的浓缩胶至玻璃板顶端,插入梳子,静置 30min。待浓缩胶完全凝固后,拔除梳子,倒入 pH 值为 8.3 的 Tris-HCL 电极缓冲液。分别取蜂胶、杨树芽、杨树胶、蜜蜂头部以及蜜蜂躯干部的粗酶提取液 5μL 上样,每个样品重复 2 次。然后开始电泳,先在 50V 恒压下运行 30min,再在 100V 恒压下运行 2h 后,停止电泳。取下凝胶膜,卸下硅胶框,用镊子撬开短玻璃板,取下凝胶置于多酚氧化酶显色剂 A 中 20min,取出凝胶用超纯水洗去残留的显色剂 A 后,将凝胶置于多酚氧化酶显色剂 B 中 3min。

2.2 结果与分析

2.2.1 未知化合物 A 的质谱检测结果

将分离纯化得到的化合物 A,利用液相色谱-质谱联用技术进行相对分子质量的测定。经过对所得到正负离子质谱图进行分析,选取负离子质谱分析结果为鉴定物质 A 相对分子质量的主要依据。由负离子质谱图(图 12.13)可知,物质 A 的 ESI 质谱信号为:m/z 109 [M-H]-。

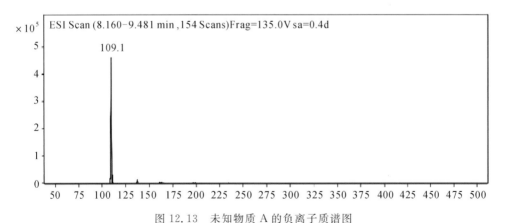

图 12.13 未知物质 A 的负离子质谱图

Fig. 12.13 Negative ions ESI-MS profile of chemical compound A

2.2.2 未知化合物 A 的核磁共振检测结果

由于化合物 A 的极性较大,因此选取氘代甲醇作为溶剂将其溶解,进行核磁共振氢谱以及碳谱测定。检测结果如核磁共振碳谱(图 12.14a 和 12.14b)所示,化合物 A 在苯环区显示出两组相互耦合的氢信号分别为 δH 6.75(2H,m)和 6.65(2H,m),化合物 A 在其碳谱显示出 3 个碳信号:δC 146.3、120.9、116.4(图 12.15)。

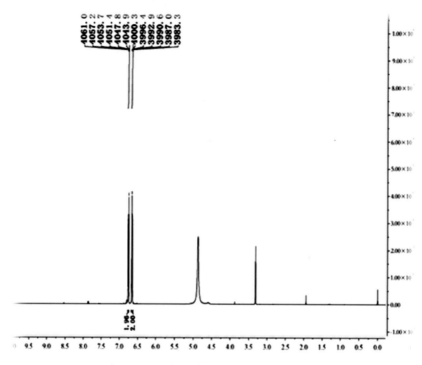

图 12.14　未知化合物 A 的核磁共振氢谱图

Fig. 12.14　1H NMR profile of the unknown chemical compound A (Enlarged profile of Fig. 12.14a)

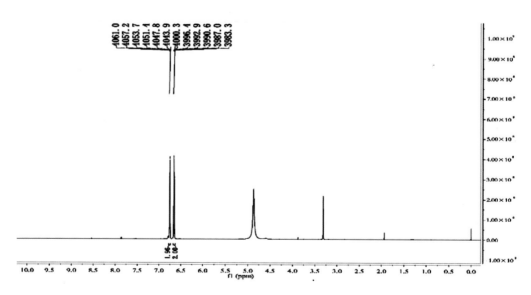

图 12.15　未知化合物 A 的核磁共振碳谱图

Fig. 12.15　13C NMR profile of the unknown compound A

2.2.3 未知化合物 A 化学结构的确定与验证

结合未知化合物 A 的质谱、核磁共振氢谱以及核磁共振碳谱分析结果,我们推断未知化合物 A 的化学名称为邻苯二酚,分子式为 $C_6H_6O_2$,其化学结构式如图 12.16 所示。此外,为了使实验结果更具有真实性,选取不同地理来源的 7 个杨树胶样品,采用高效液相色谱法对其进行了 A 物质的检测,并与邻苯二酚标准品进行比对。实验结果如图 12.17 所示,在 7 个杨树胶样品中均检测到 A 物质,并与邻苯二酚标准品出峰时间完全相同,即未知化合物 A 为邻苯二酚。

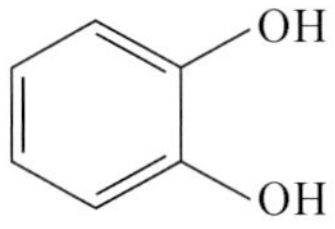

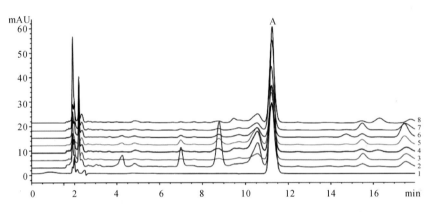

图 12.16　未知化合物 A 的化学结构式

Fig. 12.16　Chemical structure of the unknown compound A

图 12.17　邻苯二酚标准品(1)与杨树胶样本(2—8)的图谱

Fig. 12.17　Profiles of catechol standards (1) and poplar three gum samples (2-8)

2.2.4 邻苯二酚化合物在蜂胶加工过程中的转化机制

邻苯二酚自然状态下为一种无色结晶固体,然而该化合物与可催化氧化邻位二羟基的多酚氧化酶以及氧气同时接触时,无色的邻苯二酚会被氧化成具有红棕色的邻苯二醌化合物。以邻苯二酚作为底物,分别对蜂胶、杨树芽、杨树胶、蜜蜂头部以及蜜蜂躯干部中可以催化氧化邻苯二酚的多酚氧化酶进行检测,发现仅载有杨树芽和蜂胶样品的凝胶条带呈现出红棕色,而载有蜜蜂头部、蜜蜂躯干部以及杨树胶的凝胶条带均没有颜色变化(图 12.18)。说明杨树芽以及蜂胶中含有可以氧化邻苯二酚的多酚氧化酶,而蜜蜂头部、蜜蜂躯干部以及杨树胶中不含有。此外,根据杨树芽和蜂胶中多酚氧化酶在凝胶上不同条带位置可知,二者为具有不同相对分子质量的同工酶,且杨树芽中多酚氧化酶的相对分子质量大于蜂胶中所含多酚氧化酶的相对分子质量。

2.3　讨论

通过对杨树芽中未知化合物 A 分离纯化与鉴定得知,该化合物为邻苯二酚。邻苯二酚也称儿茶酚,是苯的两个邻位氢被取代后形成的化合物,在所研究的杨属植物的芽、叶、花、皮中均有分布。其物理化学性质极其稳定,可在甲醇、酸性、碱性以及中性介质中稳定存在,

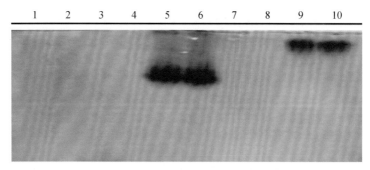

图 12.18 杨树芽、杨树胶、蜂胶、蜜蜂头部以及蜜蜂躯干部蛋白提取液中多酚氧化酶聚丙烯酰胺凝胶电泳检测结果

1—2：蜜蜂头部；3—4：蜜蜂躯干部；5—6：蜂胶；7—8 杨树胶；9—10：杨树芽。

Fig. 12.18　SDS-PAGE results of detecting polyphenol oxidase from protein extracts of poplar buds, poplar tree gum, propolis, bee heads and bee bodies.

1-2, bees' heads; 3-4, bees' bodies; 5-6, propolis; 7-8, poplar tree gum; 9-10, poplar tree buds.

仅在存在多酚氧化酶的条件下，可被氧化为 1,2-苯醌，但其所需的氧化酶应具有较强的特异性，即可以氧化具有邻位二羟基的多酚类化合物。稳定的物理化学特性，为其成为鉴别蜂胶与杨树胶的标志性成分奠定了一定的基础，也为建立有效的蜂胶与杨树胶鉴别方法提供了依据。

本研究通过聚丙烯酰胺凝胶电泳技术，分别对杨树芽、杨树芽提取物（杨树胶）、蜜蜂头部、蜜蜂躯干以及蜂胶中的多酚氧化酶进行了检测，探讨了邻苯二酚在蜜蜂加工蜂胶过程中的转化机制。首次在蜂胶中检测到多酚氧化酶。多酚氧化酶仅存在于蜂胶和杨树芽中，而蜜蜂头部、蜜蜂躯干部以及杨树胶中均未检测到该酶的活性，说明蜂胶中的多酚氧化酶来源于胶源植物，并非自身腺体分泌物。然而在杨树胶中未检测到多酚氧化酶，可能是杨树胶加工过程中加入高浓度乙醇，或者高温高压的提取工艺使得多酚氧化酶失去活性。

杨树芽与蜂胶中多酚氧化酶相对分子质量不同，可能是由于在蜜蜂加工蜂胶的过程中，杨树芽中的多酚氧化酶被蜂蜜体内的蛋白酶或者唾液腺、下颚腺等蜜蜂自身腺体分泌的消化酶所水解，使得蜂胶中多酚氧化酶的相对分子质量小于杨树芽中多酚氧化酶的分子量，该研究还有待进一步证实。

大量研究表明，多酚氧化酶是与白色体、内囊体、前质体以及造粉体结合在一起而存在于植物组织中（Vaughn et al.，1984），天然状态下无活性，仅在植物组织受到损伤后多酚氧化酶被激活，从而使其具有活性。高等植物组织中大部分的酚酸化合物存在于与多酚氧化酶分离的液泡。研究证明，多酚氧化酶并不参与植物组织细胞内酚酸类化合物的生物合成，仅在衰老以及破损的植物组织细胞中发挥其生物活性，催化氧化具有邻位酚羟基的酚酸类化合物生成醌类化合物（Cao et al.，2011）。据此推测，蜜蜂在采集杨树芽制造蜂胶的过程中，将植物组织破坏并激活多酚氧化酶的活性，从而使邻苯二酚与多酚氧化酶接触后氧化成邻苯二醌化合物，使其在蜂胶中消失。综上所述，蜜蜂在加工蜂胶的过程中起着至关重要的作用，不仅加入自身腺体分泌物，而且还能够激活植物组织中的酶从而使胶源植物的化学成分发生变化。

3 以水杨苷和邻苯二酚为参照的蜂胶与杨树胶鉴别方法

本研究以水杨苷和邻苯二酚为指标,对反相高效液相色谱法鉴别蜂胶中是否含有杨树胶的方法进行评价。

3.1 材料与方法

3.1.1 化学试剂

乙腈(色谱纯,Merck 公司);水(色谱级,杭州永洁达高纯水机自制);乙醇(分析纯);磷酸(分析纯);水杨苷(色谱纯,Sigma 公司)。

3.1.2 样本来源

81 个中国蜂胶采自中国 24 个省(市、自治区)。

国外蜂胶样本分布分别为:埃塞俄比亚(1 个)、捷克(1 个)、赞比亚(1 个)、匈牙利(1 个)、马来西亚(1 个)。

3.1.3 仪器

超声波辅助提取仪:SK20GT,上海科岛超声波仪器生产公司。

高效液相色谱仪:Agilent 1200,配有真空脱气机 G1322A,自动进样器 G1329A,四元泵 G1311A,紫外检测器(VWD)G1314B,恒温箱 G1316A。

分析天平:Denver TB-215D。

超纯水机:杭州永洁达净化科技有限公司。

3.1.4 高效液相色谱条件

色谱柱:Sepax HP-C18 (150mm×4.6mm,5μm);流动相:乙腈:0.5% H_3PO_4=5:95(v/v);柱温:30℃;检测波长:213nm;进样量:5μL。流速:1.0mL/min。

3.1.5 样本制备

蜂胶样本 2g,用 85% 的乙醇超声提取 30min,料液比为 1:15,反复提取 3 次,合并提取液,定容至 50mL,过 0.45μm 滤膜,用高效液相色谱法测定。

由于蜂胶中含有 50% 树脂,杨树胶相当于蜂胶中提取出的树脂部分,因此,取 0.5g 杨树胶样本,用 85% 的乙醇超声提取 30min,料液比为 1:15,反复提取 3 次,合并提取液,定容至 50mL,过 0.45μm 滤膜,用高效液相色谱法测定。

3.1.6 水杨苷和邻苯二酚标准溶液的配制

准确称取水杨苷和邻苯二酚对照品各 10.0mg,用甲醇定容至 10mL,配置成浓度为 1mg/mL 的标准溶液。然后用甲醇分别稀释成浓度分别为 0.005、0.001、0.0005、0.00025 以及 0.000125mg/mL 的 6 个不同浓度的对照品溶液。

3.2 实验结果与分析

3.2.1 方法的可行性分析

(1)线性范围考察

以峰面积为纵坐标,以水杨苷和邻苯二酚的浓度为横坐标,建立线性回归方程。研究发现,在该实验条件下,水杨苷和邻苯二酚标准曲线具有良好的线性关系,回归方程分别为 $y=480.82x-0.4425$ 和 $y=18662x+15.549$。

取配置好的水杨苷和邻苯二酚标准溶液,连续稀释所需的浓度,按照3.1.4的色谱条件,分别进入高效液相色谱系统,进行检测分析。准确记录峰面积,以讯噪比(S/N)=3/1为依据计算邻苯二酚的检测限,以讯噪比(S/N)=10/1为依据计算邻苯二酚的定量限。经检测,水杨苷和邻苯二酚的检测限和定量限分别为 1.3μg/mL 和 41.3μg/mL、5.4μg/mL 和 137.6μg/mL。由实验结果可知,该化合物的检测限和定量限均较低,说明该方法可以用来定性和定量检测蜂胶和杨树胶中的邻苯二酚化合物。

(2)稳定性试验

取同一份杨树胶样本提取液,按照3.1.4的色谱条件分别于0、3、6、9、12、24h进样,测得水杨苷和邻苯二酚峰的相对保留时间的RSD分别为0.97%和1.02%,表明该方法具有日内稳定性。取同一份杨树胶样本提取液,分别于1、2、3d内重复测试,测得水杨苷和邻苯二酚峰的相对保留时间的RSD分别为2.94%和3.24%,表明该方法具有日间稳定性。

(3)回收率实验

分别准确称取已知含量的水杨苷和邻苯二酚含的杨树胶样本9份,分为3组,每组按高、中、低三个水平分别加入水杨苷和邻苯二酚对照品溶液,吹干。按照本实验的方法提取并测定每个样品中邻苯二酚化合物的含量。用检测量比上加入量计算得出回收率。由实验结果可知,该化合物的回收率均在90%以上。

表 12.4　水杨苷回收率

Table 12.4　Recovery of salicin ($n=3$)

	理论浓度/(μg/mL)	实测浓度(μg/mL)	RSD(%)	回收率(%)
低浓度	95.16	94.06±0.16	0.17	94.50
中浓度	115.16	112.18±0.23	0.20	92.56
高浓度	155.16	150.02±0.57	0.37	98.14

表 12.5　邻苯二酚回收率

Table 12.5 Recovery of catechol ($n=3$)

	理论浓度(μg/mL)	实际浓度(μg/mL)	RSD(%)	回收率(%)
低浓度	43.96	41.46±0.85	2.05	94.31
中浓度	63.96	61.96±0.80	1.29	96.87
高浓度	103.96	102.19±0.28	0.27	98.30

(4)重复性实验

准确称取同一杨树胶样本6份,按照上述方法提取并进行测定。测得水杨苷和邻苯二酚的保留时间RSD值分别为0.072%和0.28%;峰面积的RSD值分别为1.535和1.09%,表明该方法具有较好的重复性。

3.2.2 不同地理来源蜂胶中水杨苷和邻苯二酚的测定

中国不同地理来源蜂胶样本、国外不同地理来源蜂胶样本的色谱图见图 12.19。所有蜂胶样本在水杨苷和邻苯二酚出峰位置均未检测到色谱峰。

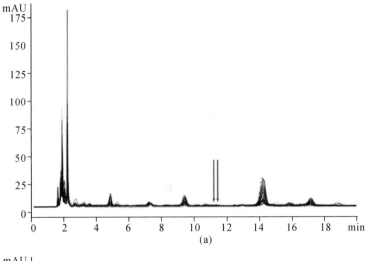

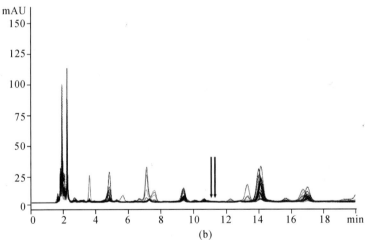

图 12.19　不同地理来源蜂胶中水杨苷的检测

(a-c)中国不同地理来源蜂胶；(d)国外不同来源蜂胶样本

Fig. 12.19　Detection of salicin in the propolis from different geographical sources

(a-c) Propolis from different geographic origins in China；(d) Propolis from different sources abroad

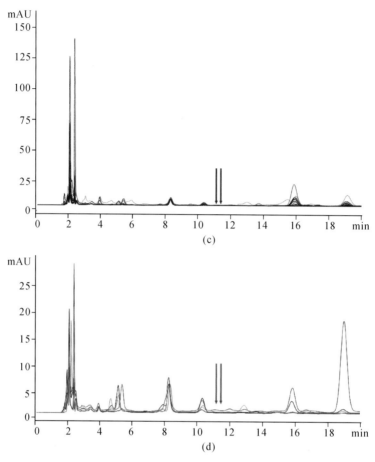

图 12.19 不同地理来源蜂胶中水杨苷的检测(续)

(a-c)中国不同地理来源蜂胶;(d)国外不同来源蜂胶样本

Fig. 12.19 Detection of salicin in the propolis from different geographical sources(Continue)

(a-c) Propolis from different geographic origins in China;(d) Propolis from different sources abroad

3.2.3 杨树胶中水杨苷和邻苯二酚的测定

在 11 个随机购买的杨树胶样本中均检测到水杨苷和邻苯二酚化合物(图 12.20),但是样本间含量存在较大差异,其中水杨苷的含量为 0.0578%～0.1424%,邻苯二酚含量为 0.052～0.132mg/g。

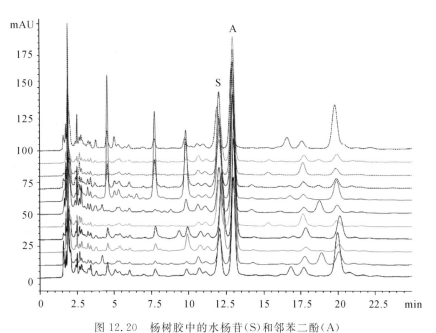

图 12.20 杨树胶中的水杨苷(S)和邻苯二酚(A)

Fig. 12.20 Salicin (S) and catechol (A) in poplar tree gum

小 结

本研究首先建立了水杨苷的测定方法。经检测发现,杨树顶芽和叶子中均检测到水杨苷,而且在所有的杨树胶样本中也检测到,但是所有的蜂胶样本中未检测到。这说明,在蜜蜂采集蜂胶的过程中水杨苷可能被降解,而在杨树胶制作过程中水杨苷稳定存在。除了水杨苷,还有一个峰A在杨树胶样本中存在,而在蜂胶中检测不到。通过对未知化合物A的分离纯化与鉴定得知,该化合物为邻苯二酚。研究发现,蜜蜂在采集杨树芽制造蜂胶的过程中,将植物组织破坏并激活多酚氧化酶的活性,从而使邻苯二酚与多酚氧化酶接触后氧化成邻苯二醌化合物,使其在蜂胶中消失。因此,本研究建立了以水杨苷和邻苯二酚为参照的区分蜂胶与杨树胶的HPLC方法,具有操作简便、灵敏度高、稳定性好的优点。

参考文献

Bankova V (2005) Recent trends and important developments in propolis research[J]. *Evidence-Based Complementary and Alternative Medicine*, 2(1):29-32.

Bankova VS, de Castro SL, Marcucci MC (2000) Propolis: recent advances in chemistry and plant origin[J]. *Apidologie*, 31:3-15.

Bonvehí JS, Coll FV (1994) The composition, active components and bacteriostatic activity of propolis in dietetics[J]. *Journal of the American Oil Chemists' Society*, 71(5):529-532.

Burdock GA (1998) Review of the biological properties and toxicity of bee propolis (propolis)[J]. *Food and Chemical Toxicology*, 36(4):347-363.

Cao C,Wang Z,Yan S,Ma L,Yang C (2011) Effects of Lymantriadispar feeding and mechanical wounding on defense-related enzymes in *Populussimonii*×*Populusnigra*[J]. *African Journal of Biotechnology*,10(36):7034-7039.

Clausen TP,Reichardt PB,Bryant JP,Werner RA,Post K,Frisby K (1989) Chemical model for short-term induction in quaking aspen (*Populustremuloides*) foliage against herbivores[J]. *Journal of Chemical Ecology*,15(9):2335-2346.

Ghisalberti EL (1979) Propolis:a review[J]. *Bee World*,60:59-84.

Greenaway W,Scaysbrook T,Whatley FR (1987) The analysis of bud exudate of *Populus*×*euramericana*, and of propolis, by gas chromatography-mass spectrometry[J]. *Proceedings of the Royal Society of London. Series B,Biological Sciences*,232(1268):249-272.

Krol W,Scheller S,Czuba Z,Matsuno T,Zydowicz G,Shani J,Mos M (1996) Inhibition of neutrophils' chemiluminescence by ethanol extract of propolis (EEP) and its phenolic components[J]. *Journal of Ethnopharmacology*,55(1):19-25.

Marcucci MC (1995) Propolis:chemical composition, biological properties and therapeutic activity[J]. *Apidologie*,26(2):83-99.

Pearl IA,Darling SF (1971) Hot water phenolic extractives of the bark and leaves of diploid *Populustremuloides* * 1[J]. *Phytochemistry*,10(2):483-484.

Thieme H (1967) Phenolic glycosides of the genus *Populus*[J]. *Planta Med*,15:35-40.

Usia T,Banskota AH,Tezuka Y,Midorikawa K,Matsushige K,Kadota S (2002) Constituents of Chinese propolis and their antiproliferative activities[J]. *Journa of Natural Products*,65(5):673-676.

Vaughn KC,Duke SO (1984) Function of polyphenol oxidase in higher plants[J]. *Physiologia Plantarum*,60(1):106-112.

曹炜,符军放,索志荣,陈卫军,郑建斌(2007)蜂胶与杨树芽提取物成分的比较研究[J]. 食品与发酵工业,33(7):162-166.

盛文胜,杜青桃,李树岚,喻建辉,汪玲(2008)蜂胶与杨树胶中八种黄酮与肉桂酸含量的测定[J]. 中国蜂业,3(59):7-8.

王小平,林励,白吉庆(2009)HPLC指纹图谱法鉴别蜂胶和树胶[J]. 陕西农业科学,55(3):133-134.

王欣,王强,徐国钧,徐珞珊(1999)杨属植物化学成分和药理作用的研究进展[J]. 天然产物研究与开发,11:65-74.

余兰平,盛文胜(2006)气相色谱-质谱法分析蜂胶和杨树胶的化学成分[J]. 蜜蜂杂志,(6):3-5.

周萍,陈建清,胡福良,胡元强,邵巧云(2009)不同产地蜂胶HPLC指纹图谱测定及真伪判定[J]. 中国蜂业,60(10):5-8.

周萍,章征天,胡福良,余秀珍(2005)蜂胶HPLC指纹图谱真伪鉴别初探[J]. 蜜蜂杂志,(8):5-6.

第十三章 蜂胶质量控制与质量评价方法研究

第一节 多指标成分定量结合指纹图谱定性的中国蜂胶质量评价方法研究

蜂胶化学成分的复杂多样性,使得蜂胶的质量控制及其标准化成了蜂胶研究的一大难题(Bankova,2005)。目前,大多以蜂胶中特征活性成分的含量作为评价蜂胶质量好坏的指标,如咖啡酸苯乙酯(Banskota et al.,2001a)。然而,对于来源于不同植物组织、含有不同活性成分的蜂胶而言,仅凭借一种或几种特征性成分的含量来评价蜂胶质量的高低,从而达到蜂胶的标准化是无法实现的。因此,根据植物来源的不同将蜂胶进行系统地分类,依据其各自的特有成分制定符合不同类型蜂胶的质量评价方法,是建立蜂胶质量控制体系、实现蜂胶标准化的有效途径。

在我国,主要的蜂胶类型以杨属型蜂胶为主。虽然已建立了以水杨苷和邻苯二酚为指标的蜂胶与杨树胶的鉴别方法,但该方法的建立仅仅解决了待测样本中是否含有杨树胶的问题,对于非杨属来源物质冒充蜂胶以及人为加入类黄酮提高蜂胶理化质量指标的掺假蜂胶,仍需建立更为完善的质量评价方法。

本研究以杨属型蜂胶为研究对象,筛选杨属型蜂胶中含量高、生物活性强的特征性成分,运用现代分析技术对蜂胶的化学信息以图形(图像)的方式进行表征并加以描述,利用高效液相色谱技术,建立杨属型蜂胶的特征性指纹图谱,并对特征活性成分进行定量。以期能够为蜂胶的质量评价方法的建立,实现蜂胶的标准化提供一定的理论依据。

1 材料与方法

1.1 材料

不同地理来源的 66 个蜂胶样品均由蜂农从 17 个省的不同地区采集而得。8 个杨树胶样本从市场上随机购得。蜂胶和杨树胶样本均保存于 -20℃的冰箱中。

1.2 试剂与仪器

甲醇,乙腈:色谱纯,Merck 公司
超纯水:色谱纯,杭州永洁达高纯水机自制
乙醇:分析纯
磷酸:分析纯

香草酸、咖啡酸、阿魏酸、异阿魏酸、p-香豆酸、肉桂酸、3,4-二甲氧基肉桂酸、咖啡酸苯乙酯、芦丁、杨梅酮、芹菜素、高良姜素、柯因、松属素、槲皮素、山奈酚、木樨草素、柚皮素,均为色谱纯,Sigma 公司

短叶松素:色谱纯,宁波海曙奥汉生物化工有限公司

3-O-乙酰基短叶松素:色谱纯,宁波海曙奥汉生物化工有限公司

超声波辅助提取仪:SK20GT,上海科岛超声波仪器生产公司

高效液相色谱仪:Agilent 1200,配有真空脱气机 G1322A,自动进样器 G1329A,四元泵 G1311A,紫外检测器(VWD)G1314B,恒温箱 G1316A

分析天平:Denver TB-215D

超纯水机:杭州永洁达净化科技有限公司

1.3 混合标准溶液的配置

按照实验所需浓度,分别称取一定质量的香草酸、咖啡酸、阿魏酸、异阿魏酸、p-香豆酸、肉桂酸、3,4-二甲氧基肉桂酸、咖啡酸苯乙酯、芦丁、杨梅酮、芹菜素、高良姜素、柯因、松属素、槲皮素、山奈酚、木樨草素、柚皮素、短叶松素、3-O-乙酰基短叶松素,溶于甲醇中制得标准品溶液。按照不同比例将不同单体化合物的标准品溶液混合,用甲醇定容配置成混合标准品溶液。用甲醇按照不同比例稀释混合标准品溶液制得不同浓度的工作混合标准品溶液。所有标准品溶液均放置于棕色瓶中,并储存于 4℃的条件下。

2 结果

2.1 提取条件的优化

以松属素的提取率作为评价指标,采用正交试验优化中国蜂胶中有效成分的提取条件。包含四个因素:(A)乙醇浓度;(B)超声提取时间;(C)料液比;(D)提取次数。正交设计 $L9(3^4)$ 如表 13.1 所示。10g 中国蜂胶样品在表 13.1 的条件下超声提取,抽滤,合并滤液蒸干溶剂,然后用甲醇定容到 50mL。过 $0.45\mu m$ 滤膜,用 HPLC 法检测分析。

中国蜂胶中有效成分的不同提取方法结果如表 13.1 所示。各因素对中国蜂胶中有效成分提取率的影响次序依次为 B>D>C>A,通过对实验结果进行比较分析,选取最有效的提取方法为:10g 中国蜂胶用 95% 的乙醇超声提取 45min,料液比为 1∶15,反复提取 4 次。

2.2 样本制备

根据 2.1 的优化条件,分别取不同地理来源的蜂胶样本 10g,粉碎,用 95% 的乙醇溶液超声提取 45min,料液比 1∶15,反复提取 4 次,过滤,合并滤液后,挥去乙醇溶剂,制得蜂胶浸膏。

分别取蜂胶浸膏、杨树胶 0.2g,溶于 25mL 甲醇中,超声提取 15min。过 $0.45\mu m$ 滤膜后,用 HPLC 法检测分析。

2.3 高效液相色谱条件

色谱柱:Sepax HP-C18 (150mm×4.6mm,5μm)

流动相:甲醇:1‰乙酸,梯度洗脱程序如表13.2所示。
柱温:33℃
检测波长:280nm
进样量:5μL
流速:1.0mL/min

表 13.1 中国蜂胶中有效成分提取条件优化的正交实验设计及其结果
Table 13.1 Results and analysis of orthogonal design for the optimization of extraction condition

试验号	A 乙醇浓度/%	B 超声时间/min	C 料液比	D 提取次数	提取率/%
1	75	15	1:5	2	2.15
2	75	30	1:10	3	2.15
3	75	45	1:15	4	2.14
4	85	15	1:10	4	2.22
5	85	30	1:15	2	2.24
6	85	45	1:5	3	2.19
7	95	15	1:15	3	2.10
8	95	30	1:5	2	3.15
9	95	45	1:10	4	2.12
平均值1	2.147	2.157	2.500	2.170	
平均值2	2.217	2.513	2.160	2.147	
平均值3	2.457	2.150	2.160	2.503	
极差	0.31	0.363	0.340	0.356	
优水平	A3	B2	C1	D3	

表 13.2 杨属型蜂胶中主要生物活性成分分析的梯度洗脱程序
Table 13.2 Gradient elution programs for analyzing the activity chemical compositions in poplar type propolis

	0min	30min	65min	70min	85min
B%(甲醇)	15	40	55	62	100

2.4 方法的可行性

选取香草酸、咖啡酸、阿魏酸、异阿魏酸、p-香豆酸、肉桂酸、3,4-二甲氧基肉桂酸、咖啡酸苯乙酯、芦丁、杨梅酮、芹菜素、高良姜素、柯因、松属素、槲皮素、山柰酚、木樨草素、柚皮素、短叶松素以及 3-O-乙酰基短叶松素 20 种单体化合物作为评价中国蜂胶质量的重要指标。由于化合物所具有的极性不同,所以在同一色谱条件下出峰时间存在着较大的差异。为了使该方法能够达到良好的分离效果,对影响其分离效果的主要因素色谱柱、梯度洗脱程序、流动相、柱温以及流速进行了研究,最终选取 2.3 所述的高效液相色谱条件。此外,为了使该方法具有较高的准确性,本研究分别以每一个单体化合物作为研究对象,对该方法进行了反复验证。

2.4.1 线性范围考察

分别取一系列不同浓度的混合标准品溶液,按照 2.3 的色谱条件进行检测分析。分别以每个单体化合物的峰面积为纵坐标,其浓度为横坐标,建立线性回归方程。研究发现,在该实验条件下,每个指标成分的标准曲线都具有良好的线性关系,其回归方程以及线性范围如表 13.3 所示。

2.4.2 检测限(LOD)和定量限(LOQ)

分别取一系列不同浓度的混合标准品溶液,连续稀释至所需的浓度,按照 2.3 的色谱条件,分别进入高效液相色谱系统,进行检测分析。准确记录峰面积,分别以讯噪比(S/N)=3/1 和(S/N)=10/1 为依据计算每个单体化合物的检测限和定量限,实验结果如表 13.3 所示。根据实验结果可知,每个单体化合物的检测限和定量限均较低,说明该方法能够同时用来定性和定量检测香草酸、咖啡酸、阿魏酸、异阿魏酸、p-香豆酸、肉桂酸、3,4-二甲氧基肉桂酸、咖啡酸苯乙酯、芦丁、杨梅酮、芹菜素、高良姜素、白杨素、松属素、槲皮素、山柰酚、木樨草素、柚皮素、短叶松素以及 3-O-乙酰基短叶松素这 20 种单体化合物。

2.4.3 稳定性试验

取同一份中国蜂胶样本提取液,按照 2.3 的色谱条件分别于 0、3、6、9、12、24h 进样,记录蜂胶中所含每一个指标化合物的相对保留时间以及其峰面积,并分别计算其 RSD 值(表 12.4)。根据研究结果可知,标志性成分相对保留时间的 RSD 值范围为 0.12%~0.23%,峰面积的 RSD 值范围为 0.99%~2.15%,表明样本溶液在 24h 内基本稳定性。

取同一份中国蜂胶样本提取液,按照 2.3 的色谱条件分别于 1、2、3d 内重复进样检测,记录蜂胶中所含每一个指标化合物的相对保留时间,并分别计算其 RSD 值(表 13.4),标志性成分相对保留时间的 RSD 值范围为 0.20%~0.32%,峰面积的 RSD 值范围为 0.94%~2.58%,表明样本具有日间稳定性。

表 13.3 标志性化合物的线性方程、线性范围、定量限、检测限以及回收率

Table 13.3 Regression data, linear range, LOD, LOQ and recovery for markers

化合物名称	线性方程[a]	R^2	线性范围/ (μg/mL)	定量限 (LOD)	检测限 (LOQ)	回收率/ %
香草酸	$y=7.36x+3.3435$	0.9995	10~120	3.31	0.94	101.8~103.9
咖啡酸	$y=17.111x+4.8313$	0.9995	6~72	1.35	0.41	98.6~101.4
p-香豆酸	$y=23.054x+86.406$	0.9994	50~600	2.16	0.39	102.7~106.4
阿魏酸	$y=14.766x+4.1992$	0.9994	5~60	4.0	1.2	103.5~106.7
异阿魏酸	$y=20.84x-3.26$	0.9996	10~50	2.12	0.64	99.9~101.1
3,4-二甲氧基肉桂酸	$y=13.73x+11.921$	0.9994	10~120	2.32	0.70	101.7~112.2
芦丁	$y=3.0759x+3.8067$	0.9991	20~120	11.25	3.38	100.4~110.1
肉桂酸	$y=39.96x+11.055$	0.9994	5~60	0.95	0.28	98.8~102.4
杨梅酮	$y=4.8536x+10.804$	0.9981	10~120	9.0	2.7	100.0~108.3
短叶松素	$y=12.289x+23.356$	0.9994	30~360	0.98	0.29	98.3~106.5
柚皮素	$y=7.9146x+4.9733$	0.9992	10~60	5.14	1.54	98.2~104.3
槲皮素	$y=5.604x+0.8267$	0.9992	10~60	8.57	2.57	94.6~103.5
木樨草素	$y=7.6676x+3.1333$	0.9992	12~72	6.35	1.91	94.4~111.8
山奈酚	$y=15.177x-0.5384$	0.9994	10~120	3.79	1.14	101.3~107.1
芹菜素	$y=10.269x+4.4761$	0.9993	10~120	5.14	1.54	91.7~97.8
松属素	$y=14.659x+25.562$	0.9981	50~600	3.69	1.11	93.7~109.4
3-O-乙酰基短叶松素	$y=10.818x+15.734$	0.9994	20~240	4.97	1.49	92.6~112.6
白杨素	$y=19.344x+46.601$	0.9994	40~480	2.88	0.86	89.9~98.3
咖啡酸苯乙酯	$y=8.6529x+1.8164$	0.9992	20~240	6.55	1.96	96.1~105.2
高良姜素	$y=21.131x-8.1069$	0.9980	20~240	3.03	0.91	97.2~111.8

a: "y"代表化合物的峰面积,"x"代表化合物的浓度。

表 13.4　蜂胶中 9 种黄酮类和 7 种酚酸类标志性化合物精确性以及稳定性检测结果

Table 13.4　Precision and repeatability data of 9 flavonoids and 7 phenolic acids in propolis ($n=6$)

化合物名称	精确性实验（RSD,%）				稳定性实验（RSD,%）	
	日内稳定性		日间稳定性			
	保留时间	峰面积	保留时间	峰面积	保留时间	含量
咖啡酸	0.23	1.18	0.32	1.64	0.24	3.68
p-香豆酸	0.16	1.18	0.24	1.60	0.22	1.90
阿魏酸	0.14	1.29	0.22	1.60	0.18	4.18
异阿魏酸	0.13	1.28	0.21	1.82	0.15	2.16
3,4-二甲氧基肉桂酸	0.13	1.25	0.23	1.65	0.10	2.05
肉桂酸	0.12	0.99	0.21	1.70	0.09	1.24
短叶松素	0.13	1.06	0.20	1.50	0.11	2.07
柚皮素	0.15	1.06	0.21	1.48	0.10	1.78
槲皮素	0.18	1.34	0.26	2.17	0.16	1.62
山柰酚	0.17	1.21	0.23	2.34	0.14	1.87
芹菜素	0.23	1.07	0.32	1.33	0.15	1.29
松属素	0.16	1.15	0.23	1.96	0.10	1.97
3-O-乙酰基短叶松素	0.15	1.27	0.22	2.58	0.10	1.29
白杨素	0.14	1.01	0.23	0.94	0.11	3.51
咖啡酸苯乙酯	0.15	2.15	0.20	2.92	0.09	2.03
高良姜素	0.17	1.28	0.23	2.15	0.10	2.16

2.4.4　回收率实验

取同一已知标志性化合物含量的蜂胶样本 9 份，分为 3 组，每组按高、中、低 3 个浓度水平分别加入混合对照品溶液，吹干。按照 2.2、2.3 所述的实验方法对其进行标志成分的提取与检测。将所测得结果分别代入其各自的线性回归方程中，得出每个标志性化合物的含量，比上其加入量计算得出每种指标化合物的回收率（表 13.3）。

2.4.5　重复性实验

准确称取同一个蜂胶样本 6 份，按照上述方法提取并进行测定，分别测得蜂胶中标志性化合物的出峰时间和峰面积，计算其 RSD 值（表 13.4）。尽管蜂胶样本中酚酸类化合物的出峰时间以及峰面积存在着一定的差异，但其 RSD 值较小，分别小于 0.3% 和 4.2%，因此该方法具有较好的可重复性。

2.5　蜂胶以及杨树胶中标志性化合物的检测分析

分别对不同地理来源的 66 个中国蜂胶以及 8 个杨树胶样本按照本研究建立的方法进行检测，通过与标准品谱图进行对比（图 13.1），得知蜂胶以及杨树胶样本中所含有的标志性化合物的种类，并利用所建立的线性回归方程，分别计算蜂胶以及杨树胶所含标志性化合物的含量(表 13.5)。不同地理来源蜂胶以及杨树胶的 HPLC 图谱如图 13.2 所示。在所有的蜂胶样本中,均检测到了咖啡酸、阿魏酸、p-香豆酸、肉桂酸、3,4-二甲氧基肉桂酸、高良姜素、白杨素、松属素、短叶松素以及 3-O-乙酰基短叶松素;与此同时,异阿魏酸以及咖啡酸苯乙酯也广泛存于中国蜂胶中,仅在个别两三个样本中没有检测到这两种化合物。因此,以上 12 种化合物被认为是中国蜂胶指纹图谱中的标志性化合物。然而在杨树胶样本中,仅检测到蜂胶共有成分中的 9 种化合物,分别为异阿魏酸、肉桂酸、3,4-二甲氧基肉桂酸、咖啡酸苯乙酯、高良姜素、白杨素、松属素、短叶松素以及 3-O-乙酰基短叶松素。此外,还在一些蜂胶以及杨树胶样品中检测到了柚皮素、槲皮素、山奈酚以及芹菜素,但其含量差异性较大,在蜂胶以及杨树胶中这 4 种化合物含量范围分别 0.68～21mg/g、0.99～12.08mg/g、0.1～3.87mg/g、0.77～9.55mg/g 以及 0.56～0.82mg/g、1.75～8.30mg/g、0.38～1.25mg/g、0.42～5.44mg/g。

对比蜂胶和杨树胶中所含有的酚酸以及黄酮类化合物可知,二者在黄酮类化合物的组成方面有着极高的相似性,共同成分包括白杨素、高良姜素、松属素、短叶松素以及 3-O-乙酰基短叶松素在内的 5 种黄酮类化合物,其含量分别占蜂胶以及杨树胶所检测黄酮类化合物总量的 88% 和 86%;杨树胶中异阿魏酸、3,4-二甲氧基肉桂酸以及肉桂酸的含量均高于蜂胶中 3 种化合物的含量,而白杨素、高良姜素、松属素、短叶松素以及 3-O-乙酰基短叶松素的含量均低于蜂胶中所含同类化合物的含量。

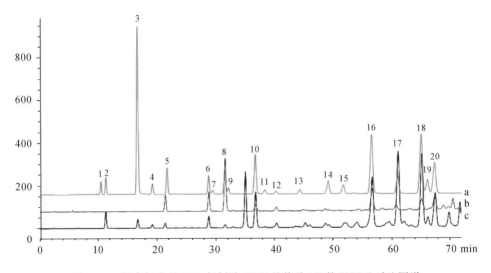

图 13.1 混合标准品(A)、杨树胶(B)以及蜂胶(C)的 HPLC 对比图谱
1.香草酸;2.咖啡酸;3.p-香豆酸;4.阿魏酸;5.异阿魏酸;6.3,4-二甲氧基肉桂酸;7.芦丁;8.肉桂酸;9.杨梅酮;10.短叶松素;11.柚皮素;12.槲皮素;13.木樨草素;14.山奈酚;15.芹菜素;16.槲皮素;17.3-O-乙酰基短叶松素;18.白杨素;19.咖啡酸苯乙酯;20.高良姜素

Fig. 13.1 HPLC chromatograms of the standard solution (A),poplar tree gum (B)and propolis (C)
1. Vanillic;2. Caffeic acid;3. p-Coumaric acid;4. Ferulic acid;5. Isoferulic acid;6. 3,4-Dimethoxycinnamic acid;7. Rutin;8. Cinnamic acid;9. Myricetin;10. Pinobanksin;11. Naringenin;12. Quercetin;13. Luteolin;14. Kaempferol;15. Apigenin;16. Pinocembrin;17. 3-O-acetylpinobanksin;18. Chrysin;19. CAPE;20. Galangin

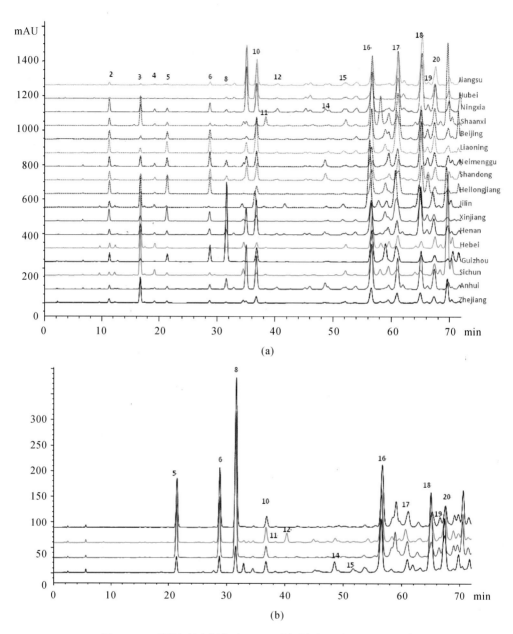

图 13.2 不同地理来源蜂胶(A)以及杨树胶(B)的 HPLC 图谱

Fig. 13.2 HPLC chromatograms of propolis from different geographies (A) and poplar tree gum (B) samples

表 13.5 蜂胶以及杨树胶所含标志性化合物的含量

mg/g

样品地理来源	异阿魏酸	3,4-二甲氧基肉桂酸	肉桂酸	松属素	短叶松素	3-O-乙酰基短叶松素	白杨素	咖啡酸苯乙酯	高良姜素
浙江	2.77±1.47	8.14±6.38	0.42±0.17	18.19±10.31	23.91±21.74	62.21±43.89	37.03±29.85	11.15±7.21	14.92±7.82
安徽	3.74±2.09	11.45±6.29	2.25±1.98	28.0±13.51	43.7±18.85	53.4±20.95	44.82±6.74	15.35±2.43	16.53±5.48
四川	3.35±1.35	10.31±5.84	2.47±3.09	36.63±3.07	62.41±3.06	109.46±26.76	37.23±0.81	12.88±10.28	20.14±4.14
贵州	4.85	16.66	30.49	20.23	48.94	48.02	8.73	2.16	8.90
河北	3.85±2.82	8.56±5.69	1.77±1.87	14.95±8.32	39.02±9.96	50.36±26.11	74.79±19.82	14.95±9.38	14.08±2.22
河南	2.09±0.65	6.13±2.33	2.04±1.65	31.69±16.69	49.17±9.12	57.41±31.49	46.88±22.45	16.60±9.28	16.74±4.71
新疆	2.92±2.09	12.50±5.06	0.22±0.047	18.17±12.96	36.08±7.71	108.79±39.64	42.21±12.58	14.67±11.66	12.33±5.35
吉林	2.11±1.54	3.83±3.55	0.44±0.24	20.30±12.32	68.00±26.79	76.18±23.24	33.92±14.63	9.72±6.54	17.27±5.78
黑龙江	6.59±4.04	9.66±6.13	0.27±0.20	14.79±8.05	46.29±7.14	83.46±48.23	36.37±19.20	26.07±9.04	13.79±4.81
山东	4.35±2.16	9.81±4.65	1.33±0.83	26.39±11.56	36.97±7.14	61.96±19.59	51.28±11.89	23.05±10.02	15.93±3.98
内蒙古	4.13±2.49	7.44±6.62	0.33±0.18	9.58±10.13	58.99±36.67	64.66±55.51	28.86±26.67	9.28±7.29	16.03±6.74
辽宁	5.81±2.90	11.44±5.49	0.36±0.10	21.43±16.04	47.45±11.01	100.95±39.72	38.16±6.51	20.59±7.76	15.38±3.87
北京	7.65±0.22	17.32±0.57	0.82±0.01	13.25±2.56	36.44±3.64	50.74±14.90	54.32±7.83	24.16±2.95	12.84±2.90
陕西	5.21±0.50	14.55±0.01	1.28±1.57	23.87±3.83	27.46±4.16	94.21±69.86	37.78±19.88	21.33±3.68	15.87±2.65
宁夏	2.70±1.73	11.48±5.36	0.42±0.32	38.81±1.74	47.19±17.90	119.95±20.11	56.13±12.88	13.85±2.82	22.10±3.90
湖北	1.99±1.85	3.35±4.40	0.90±0.83	35.44±32.96	37.83±15.59	41.41±34.17	33.52±19.78	8.42±10.48	15.33±5.29
江苏	0.78±0.25	3.30±0.56	0.53±0.25	39.89±6.98	53.50±23.13	64.53±4.62	52.26±7.20	9.95±1.30	22.97±3.01
平均值	4.23±2.31	9.76±4.22	2.73±7.19	24.21±9.47	44.90±11.74	73.39±24.75	39.72±11.35	14.95±6.42	15.95±3.44
杨树胶	6.97±3.63	6.97±3.63	11.73±7.59	6.24±2.23	30.50±3.85	15.47±7.28	19.90±16.23	7.47±2.80	13.08±4.07

3 分析与讨论

本研究利用高效液相色谱技术,建立了一种同时检测蜂胶中多个指标性化合物的分析方法。通过对该方法进行反复验证,并对其进行了线性、精确度、准确度等方法学内容的考察,证明该方法不仅能够准确地对蜂胶中所含有化合物进行定性分析,而且还可以精确地测定其化合物在蜂胶中含量。因此,该方法可广泛地应用于蜂胶的质量评价中。

蜂胶与杨树胶中均不含有芦丁以及杨梅酮;咖啡酸、阿魏酸以及 p-香豆酸仅存在于蜂胶中,而在杨树胶中未检测到;此外,蜂胶与杨树胶中黄酮类化合物的组成极其相似,共同含有柯因、高良姜素、松属素、短叶松素以及 3-O-乙酰基短叶松素。综上所述,选取咖啡酸、阿魏酸以及 p-香豆酸作为评价指标,该方法能够有效地将蜂胶与杨树胶,蜂胶与添加外源性类黄酮类化合物的掺假蜂胶区分开来。与此同时,以白杨素、高良姜素、松属素、短叶松素以及 3-O-乙酰基短叶松素为检测指标,该方法还能够对添加杨属植物以及非杨属来源物质的掺假行为予以区分。

蜂胶与杨树胶的黄酮类化合物组成极其相似,但二者所含的酚酸类化合存在着较大的差异。咖啡酸、阿魏酸以及 p-香豆酸仅存在于蜂胶中,而在杨树胶中未检测到。此外,异阿魏酸、3,4-二甲氧基肉桂酸以及肉桂酸在杨树胶中的含量均高于蜂胶。Truchado 等(2010)通过对成熟蜂蜜和新鲜贮藏蜂蜜的酚酸类化合物进行分析比较,发现了蜜蜂的水解酶在蜂蜜成熟过程对其原始成分进行水解,改变了其原始化学成分。因此,可推断蜂胶中的咖啡酸、阿魏酸以及 p-香豆酸可能是杨树芽中肉桂酸及其衍生物的水解产物。

通过对不同地理来源的蜂胶样品中异阿魏酸、肉桂酸、3,4-二甲氧基肉桂酸、咖啡酸苯乙酯、高良姜素、白杨素、松属素、短叶松素以及 3-O-乙酰基短叶松素的含量进行分析与统计发现,蜂胶中化学物质的含量随其地理来源的不同而存在着一定的差异性。其中高良姜素、白杨素、松属素、短叶松素以及 3-O-乙酰基短叶松素含量相对较高,可作为评价中国蜂胶质量的重要指标。

小 结

本研究利用高效液相色谱技术,建立了一种同时检测蜂胶中多个指标性化合物的分析方法。该方法不仅能够准确地对蜂胶中所含有化合物进行定性分析,而且还可以精确地测定其化合物在蜂胶中的含量。

选取咖啡酸、阿魏酸以及 p-香豆酸为评价指标,该方法能够有效地将蜂胶与杨树胶,蜂胶与添加外源性类黄酮类化合物的掺假蜂胶区分开来。与此同时,以白杨素、高良姜素、松属素、短叶松素以及 3-O-乙酰基短叶松素为检测指标,该方法还能够对添加杨属植物以及非杨属来源物质的掺假行为予以区分。研究发现,蜂胶中化学物质的含量随其地理来源的不同而存在着一定的差异性。其中,高良姜素、柯因、松属素、短叶松素以及 3-O-乙酰基短叶松素含量相对较高,可作为评价中国蜂胶质量的重要指标。

第二节　基于多指标指纹图谱及抗氧化活性的巴西绿蜂胶质量控制研究

胶源植物是影响蜂胶化学成分及生物学活性的重要因素。在巴西东南部及中西部地区,来源于酒神菊属植物(*Baccharis*)的巴西绿蜂胶因其独特的化学成分及强效的保健功效在世界蜂胶市场上备受欢迎。大量研究表明,巴西绿蜂胶显著的抗炎、抗肿瘤、护肝、神经保护及调节血糖血脂等活性都与其抗氧化作用有关。而蜂胶的抗氧化活性是其多种化学成分相互作用的整合结果,这是蜂胶的优势,但同时也给蜂胶的质量控制造成了困难。

本研究采用分光光度法、高效液相色谱/电喷雾-质谱法(HPLC-ESI-MS/MS)以及离线HPLC-DPPH法,对巴西绿蜂胶的化学成分及抗氧化活性进行测定;建立巴西绿蜂胶的化学指纹图谱,确定共有成分峰,同时对特征性化学成分进行定量测定;并筛选评价巴西绿蜂胶的抗氧化活性成分,建立抗氧化活性指纹图谱。

1　巴西绿蜂胶中总黄酮、总酚酸及抗氧化活性测定

蜂胶中的活性成分主要包括黄酮类及酚酸类化合物,其具有抑菌、抗氧化、抗炎、免疫调节及抗肿瘤等多种生物学活性。研究发现许多常见的疾病,包括心血管病、动脉硬化、糖尿病、癌症、白内障、关节炎及老年痴呆等都与体内活跃的自由基有关。因此,蜂胶抗氧化活性的研究一直是其药理活性研究的热点。

本研究采用超声辅助乙醇提取法对巴西绿蜂胶中的活性成分进行提取,并选用四因素三水平的正交试验设计优化其提取工艺;同时结合紫外-可见分光光度法对蜂胶中总黄酮、总酚酸的含量及其DPPH自由基清除活性进行测定,并进一步建立化学成分与抗氧化活性的相关性分析。

1.1　材料与方法

1.1.1　蜂胶样品

22个巴西绿蜂胶(BGP)样本分别由蜂乃宝本铺(南京)保健品有限公司和杭州蜂之语蜂业股份有限公司提供,产地为巴西东南部米纳斯吉纳斯州。所有蜂胶样品均为酒神菊属型巴西绿蜂胶,颜色从黄绿色到绿色不等,条块状固体,具有典型的辛辣味道。所有蜂胶样本均于-18℃下冷冻储存。

1.1.2　仪器及试剂

超纯水机:杭州永洁达净化科技有限公司

多功能高速粉碎机:信强,上海

万分之一分析天平:Denver TB-215D

超声波辅助提取仪:SK20GT,上海科导超声波仪器生产公司

UV-vis2550紫外-可见分光光度计(日本岛津)

全波长酶标仪:Multiskan GO,美国ThermoScientific公司

1,1-二苯基-2-三硝基苯肼试(DPPH):美国Sigma公司

福林酚试剂(Folin-2Ciocalteu)：上海生工科技

槲皮素：色谱纯,美国 Sigma 公司

没食子酸：色谱纯,上海生工科技

甲醇：色谱纯,德国 Merck 公司

超纯水：色谱纯,杭州永洁达高纯水机自制

乙醇：分析纯

乙酸：分析纯

无水碳酸钠：分析纯

九水硝酸铝：分析纯

乙酸钾：分析纯

1.1.3　蜂胶样品处理

准确称取经冷冻粉碎后的巴西绿蜂胶 0.5g(精确到 0.1mg),加入一定量的乙醇-水溶液,超声提取,过滤,合并滤液,置于−18℃冰箱中反复冷冻,过滤,除蜡,将脱蜡后的溶液用 95％的乙醇定容至 50mL,作为蜂胶供试液。同一条件下重复提取 3 次取平均值。

1.1.4　四因素三水平正交实验

分别选取乙醇浓度(75％、85％、95％)、料液比(1∶20、1∶30、1∶40)、超声提取时间(15、30、45min)以及超声提取次数(2、3、4)为影响因素,采用四因素三水平正交实验研究上述 4 个因素对巴西绿蜂胶中多酚类化合物及其抗氧化活性的影响,从而筛选出最佳的提取参数。

1.1.5　总黄酮含量测定

准确吸取 0.2mL 稀释后的待测样品溶液于 10mL 容量瓶中,加 95％乙醇至总体积 3mL,依次加入硝酸铝溶液(100g/L)0.2mL,醋酸钾溶液(9.8g/L) 0.2mL,摇匀,加水至刻度,混合均匀后,室温避光反应 30min,于 415nm 波长处测定其吸光值,重复测定 3 次。以槲皮素(4~24μg/mL)建立标曲,空白对照可用无水乙醇按上述操作处理。

1.1.6　总酚酸含量测定

准确移取 1mL 稀释后的待测样品溶液于 10mL 容量瓶中,加入 1mL Folin-2Ciocalteu 显色剂,混匀,然后加入 5mL 1mol/L 的碳酸钠溶液,用蒸馏水定容至刻度,混合均匀后,室温避光反应 1h,离心,在 760nm 波长处测定上清液吸光值,重复测定 3 次。以没食子酸(1~6μg/mL)建立标曲,空白对照用无水乙醇按上述操作处理。

1.1.7　DPPH 自由基清除活性测定

将 1mg/mL 的样品储备液用无水乙醇稀释至一系列的浓度梯度,蜂胶样品：20~200μg/mL。取上述稀释液 100μL 加入 100μL 新配置的 0.1mg/mL DPPH 工作液,同时以不加 DPPH(以 100μL 无水乙醇代替 DPPH)的供试品溶液各浓度作为对照以消除供试品本身颜色对测试结果的干扰,并设 DPPH 阴性对照(以 100μL 无水乙醇代替供试品),每组设 2 个平行复孔。室温避光反应 30min 后,在 517nm 波长处测定吸光度,使其读数在 0.2~0.8。

$$DPPH 清除率(\%) = (A_0 - A_S)/A_0 \times 100\%$$

A_S：加入样品提取液的 DPPH 溶液的吸光度;

A_0：加入无水乙醇的 DPPH 溶液的吸光度。

以蜂胶醇提液的浓度为横坐标、DPPH自由基清除率为纵坐标建立标准曲线,求线性方程,计算出清除50%DPPH所需供试样品的浓度,即半抑制浓度IC_{50},比较各样品清除DPPH自由基的活性。

1.1.8　数据分析

以SPSS 17.0计算方差估计,求样本平均数及标准差,对数据进行相关性分析及显著性水平分析。

1.2　结果与讨论

1.2.1　四因素三水平正交试验结果与分析

根据正交试验的设计原理,选取乙醇浓度(A)、料液比(B)、提取时间(C)及提取次数(D)4个因素,每个因素的3个不同水平分别以1、2、3表示,以巴西绿蜂胶中总多酚的提取率作为评价指标,优化了巴西绿蜂胶中多酚类化合物的最佳提取工艺。四因素三水平正交试验设计方案及结果见表13.6和表13.7。

表13.6　巴西绿蜂胶中总多酚提取四因素三水平正交表

Table 13.6　Orthogonal table of four factors and three levels for extraction of total phenolics in BGP

水平	因素			
	A 乙醇浓度(%)	B 料液比(g/mL)	C 提取时间(min)	D 提取次数
1	70	1:10	15	2
2	85	1:20	30	3
3	95	1:30	45	4

表13.7　巴西绿蜂胶中总多酚提取四因素三水平正交试验设计方案及结果

Table 13.7　Orthogonal experimental design and results of four factors and three levels for extraction of total phenolics in BGP

试验号	因素				总酚提取率(%)	DPPH清除活性(IC_{50},μg/mL)
	A	B	C	D		
1	1	1	1	1	10.71	103.29
2	1	2	2	2	10.97	102.05
3	1	3	3	3	11.82	102.17
4	2	1	2	3	11.19	101.64
5	2	2	3	1	10.62	104.42
6	2	3	1	2	10.40	104.68
7	3	1	3	2	12.95	101.85
8	3	2	1	3	11.03	102.36
9	3	3	2	1	12.53	101.44

续表

试验号	因素 A	因素 B	因素 C	因素 D	总酚提取率(%)	DPPH 清除活性($IC_{50} \cdot \mu g/mL$)
K1	11.17	11.62	10.71	11.29		
K2	10.74	10.87	11.56	11.44		
K3	12.17	11.58	11.80	11.35		
R	1.43	0.74	1.08	0.15		

从表 13.8 中可直接看出 A 因素对蜂胶多酚类化合物的提取率影响最大;同时结合方差分析的结果发现因素 A 各水平间存在极显著差异($P<0.01$),因素 C 的各水平间存在显著差异($P<0.05$),而因素 B、D 为不显著因子。各因素对蜂胶中多酚类化合物提取率的影响程度大小依次为 A(乙醇浓度)>C(提取时间)>B(料液比)>D(提取次数);最佳的提取条件为 A3(95%乙醇)B3(1:30)C3(45min)D2(3 次)。

表 13.8 巴西绿蜂胶中总多酚提取四因素三水平正交试验方差分析结果

Table 13.8 Results of variance analysis of four factors and three levels for extraction of total phenolics in BGP

方差来源	离差平方和	自由度	均方	F	显著性
A	0.185	2	0.093	0.009	$P<0.01$
B	0.827	2	0.413	0.040	$P<0.05$
C	5.704	2	2.852	0.275	$P>0.05$
D	20.740	2	10.370		
总和	27.456	8			

1.2.2 巴西绿蜂胶中总黄酮、总酚酸含量及 DPPH 自由基清除活性测定

22 个不同批次巴西绿蜂胶醇提物的总黄酮、总酚酸含量及 DPPH 自由基清除活性测定结果见表 13.9。

表 13.9 巴西绿蜂胶中总黄酮、总酚酸含量及 DPPH 自由基清除活性

Table 13.9 Total phenolic, total flavonoid contents and antioxidant activity of BGP ethanol extracts

样品编号	总酚酸含量(mg/g GAE)	总黄酮含量(mg/g QE)	DPPH 清除活性($IC_{50} \cdot \mu g/mL$)
1	148.55 ± 1.06^a	64.99 ± 4.11^a	108.303 ± 7.55^{gh}
2	128.29 ± 4.23^b	67.60 ± 9.94^a	108.44 ± 5.13^{gh}
3	122.47 ± 3.67^c	58.46 ± 6.20^b	116.86 ± 2.92^{fg}
4	122.67 ± 4.78^c	53.74 ± 3.21^{bcd}	128.68 ± 3.66^{de}
5	102.14 ± 5.02^g	46.17 ± 1.34^{def}	154.22 ± 4.70^b
6	87.53 ± 6.26^j	39.48 ± 7.25^{ef}	186.87 ± 8.31^a

续表

样品编号	总酚酸含量 (mg/g GAE)	总黄酮含量 (mg/g QE)	DPPH 清除活性 (IC_{50}, μg/mL)
7	114.71±6.41[de]	50.78±3.54[bcd]	104.04±5.39[h]
8	98.01±1.84[gh]	40.66±5.41[ef]	127.23±5.71[de]
9	115.26±7.23[de]	52.24±3.45[bcd]	93.51±2.90[i]
10	116.89±8.41[d]	45.49±5.78[def]	104.75±6.21[h]
11	111.12±10.20[ef]	49.22±1.27[cd]	121.01±3.69[ef]
12	99.64±4.37[gh]	48.53±4.64[cd]	162.60±7.73[b]
13	116.22±5.38[d]	56.36±2.54[bc]	140.16±6.53[c]
14	107.11±3.48[f]	51.75±4.52[bcd]	140.81±6.63[c]
15	96.13±3.79[i]	38.35±2.10[f]	104.12±5.76[h]
16	95.24±4.16[i]	47.40±1.06[cde]	162.40±11.92[b]
17	97.39±5.86[gh]	49.75±3.42[bcd]	135.85±7.42[cd]
18	99.67±3.51[gh]	39.30±2.57[ef]	134.72±4.59[cd]
19	100.38±4.65[gh]	49.73±3.04[bcd]	157.17±3.49[b]
20	108.52±3.14[f]	40.15±4.13[ef]	113.68±4.73[fgh]
21	109.71±9.04[f]	56.10±5.81[bc]	158.51±9.21[b]
22	110.95±6.20[ef]	50.49±3.14[bcd]	190.27±6.24[a]

注:数据表示为平均值±标准差($n=3$),不同上标字母代表组内差异显著($P<0.05$)。GAE,没食子酸等价值;QE,槲皮素等价值。

由表13.9可知,不同批次巴西绿蜂胶中总黄酮及总酚酸含量均存在较大差异。醇提物中总酚酸的平均含量为109.48mg/g±4.16mg/g,明显高于总黄酮的平均含量(49.85mg/g±2.64mg/g),这与Kumazawa等(2004)的研究结果相一致,表明酚酸类物质是巴西绿蜂胶醇提物中主要的化学成分。不同批次样品间总酚酸含量差异显著,其中样品1中含量最高(148.55mg/g),样品6中含量最低(87.53mg/g)。而总黄酮含量以样品2中含量最高(67.60mg/g),样品15中含量最低(38.35mg/g)。不同批次巴西绿蜂胶样品化学成分的差异主要源于地理位置、采胶季节等因素的影响。

同时,抗氧化活性测定结果显示所有绿蜂胶样均具有良好的DPPH自由基清除活性,IC_{50}值在93.51~190.27μg/mL(IC_{50}值越低,抗氧化活性越强)。已有研究表明,蜂胶的自由基清除活性及还原力与蜂胶中多酚类化合物的含量有着直接的联系(Kawashima et al.,2012)。因此,进一步对总黄酮、总酚酸含量及DPPH自由基清除活性进行相关性分析发现,IC_{50}与总酚酸含量呈现显著的负相关($R^2=-0.506$,$P<0.01$),而与总黄酮含量无显著相关性($R^2=-0.185$,$P>0.05$)。这表明酚酸类化合物对巴西绿蜂胶抗氧化活性的发挥起着更为重要的作用。Bankova等(2005)研究指出巴西绿蜂胶显著的生物学活性主要归因于其丰富的酚酸类化合物,尤其是异戊烯基肉桂酸类化合物。

2 巴西绿蜂胶 HPLC 指纹图谱研究

受外界胶源植物、地理位置、采胶季节、储存条件及蜂种自身的影响,蜂胶化学成分复杂而多变,这对蜂胶的标准化及质量控制来说是一个巨大的挑战。不同植物来源的蜂胶具有各自典型的化学成分,如杨树型蜂胶的 B 环无取代基的黄酮类化合物,酒神菊属型蜂胶的异戊烯基肉桂酸类化合物,血桐属型蜂胶的香叶基黄酮类化合物,克鲁西属型蜂胶的异戊二烯基苯甲酮类化合物。

近年来,兼具整体性和模糊性为一体的指纹图谱分析技术已广泛应用于鉴别中药真伪、评价中药质量一致性和稳定性。指纹图谱包括波谱指纹图谱(UV、IR、MS、NMR 等)和色谱指纹图谱(TLC、GC 和 HPLC)以及两者的联用图谱。其中,液相色谱-质谱联用技术(HPLC-MS^n)是应用最为广泛的分析方法之一。它利用液相色谱的高效分离能力对复杂体系的化学成分进行分离,再以质谱为检测器在线提供不同化合物的分子量信息。同时,电喷雾质谱联用技术(ESI-MS^n)的出现,为研究热不稳定和极性较大的化合物提供了更为简单快速的分析方法,特别是多级质谱串联技术还为分析物提供了更精确的结构信息。

本研究采用反相高效液相色谱法(RP-HPLC)对 22 个不同批次的巴西绿蜂胶进行了化学指纹图谱的构建,筛选出共有特征成分峰;同时结合电喷雾质谱联用技术(ESI-MS^n)进一步定性鉴别了巴西绿蜂胶中的特征性化学成分。根据 HPLC 提供的色谱扫描行为、特征性紫外吸收光谱以及 MS 提供的分子量信息与已有文献数据及标准品的分子量进行对比,快速准确地确定了巴西绿蜂胶中稳定存在的共有成分峰,作为其特有的指标成分。本章建立了一种快速、简便、准确的分析方法,为巴西绿蜂胶的化学标准化及质量控制提供科学的理论依据。

2.1 材料与方法

2.1.1 蜂胶样品

巴西绿蜂胶样品同本章 1.1.1。

2.1.2 仪器及试剂

多功能高速粉碎机:信强,上海;

万分之一分析天平:Denver TB-215D;

超声波辅助提取仪:SK20GT,上海科导超声波仪器生产公司;

高效液相色谱仪:安捷伦 1200,配有四元流动泵 G1311A,在线真空脱气机 G1322A,自动进样器 G1329A,恒温箱 G1316A,紫外检测器(VWD)G1314B 以及化学工作站;

质谱检测仪:安捷伦 6430 串联四极杆质谱仪 QQQ MS;

甲醇:色谱纯,德国 Merck 公司;

超纯水:色谱纯,杭州永洁达高纯水机自制;

乙醇:分析纯;

乙酸:分析纯;

咖啡酸、绿原酸、p-香豆酸、3,5-二咖啡酰奎宁酸、4,5-二咖啡酰奎宁酸、3,4,5-三咖啡酰奎宁酸、山奈酚及山奈素对照品(色谱纯)购于美国 Sigma 公司;阿替匹林 C 标准品(纯度≥

98%)购于日本 Wako 公司。

2.1.3 蜂胶样品处理

准确称取 0.5g(精确至 0.1mg)经粉碎混匀的巴西绿蜂胶样,用 95%乙醇超声提取 45min,料液比为 1∶30,反复提取 3 次,合并滤液,用 95%的乙醇定容至 50mL,取上清液,经 0.45μm 滤膜过滤后加入进样小瓶,供液相色谱测定。

2.1.4 液相色谱分析条件

色谱柱为 Sepax HP-C18(150mm×4.6nm,5μm);流动相:甲醇-0.1%乙酸;流速: 1.0mL/min;波长:280nm;柱温:33℃;进样量:5μL。

梯度洗脱程序为:0~10min,15%~25%甲醇;10~25min,25%~40%甲醇;25~ 55min,40%~60%甲醇;55~75min,60%~75%甲醇;75~90min,75%~85%甲醇。

2.1.5 HPLC-ESI-MS/MS 检测

采用电喷雾电离结合安捷伦 6430 串联四极杆质谱仪 QQQ MS(Agilent Corp,USA) 对蜂胶中化学成分进行定性鉴别。干燥氮气流量:9.0L/min;干燥器温度:350℃;雾化器压力:35Pa;毛细管电压:4000 V;碰撞电压:135V;扫描范围 m/z=100~1000Da。

2.2 实验结果与分析讨论

2.2.1 反相高效液相色谱法的建立与优化

影响色谱分离的因素主要包括流动相组成及其洗脱程序、流速、柱温、检测波长以及色谱柱类型等。反向色谱的色谱柱一般选用极性较弱的十八烷基-硅胶(C18)作为固定相填料,而流动相一般选用极性较强的甲醇、乙腈等有机相和水相缓冲液组成的混合体系。由于乙腈的洗脱能力强,容易造成蜂胶中化学成分出峰时间相对集中,色谱峰分离度欠佳,所以采用甲醇-水为流动相。此外,蜂胶中一些多酚类化合物容易解离导致色谱峰拖尾。因此,我们在水相体系中加入了 1.0%、0.5% 以及 0.1%(v/v)的乙酸缓冲溶液,结果发现上述乙酸溶液均能有效改善色谱峰的拖尾现象,且三者之间无明显的差异。同时,鉴于色谱柱的耐酸性,我们最终选定以甲醇-0.1%乙酸作为流动相。此外,由于蜂胶成分复杂,为了实现各成分间最大的分离效率,我们采用梯度洗脱程序进行色谱峰的分离优化,最终确定的色谱条件见本章 2.1.4。

2.2.2 HPLC 指纹图谱构建

将制备好的巴西绿蜂胶醇提液按本章 2.1.4 中的色谱条件进样检测(图 13.3)。在该色谱条件下,色谱基线平稳,色谱峰形对称,各成分均得到了较好的分离。同时对 22 个不同批次的巴西绿蜂胶进样检测发现,不同批次绿蜂胶化学组成大致相同,但不同色谱成分峰的峰面积存在差异。分别将各蜂胶样的色谱图导入中药色谱指纹图谱相似度评价系统(国家药典委员会,2004A 版)进行分析,选取参照峰,根据时间窗宽度设置对色谱峰进行匹配,确定不同批次巴西绿蜂胶的共有成分峰;同时采用中位数法生成标准化的对照指纹谱图,并进行图谱的差异性分析和整体的相似度评价(图 13.4 和表 13.10)。

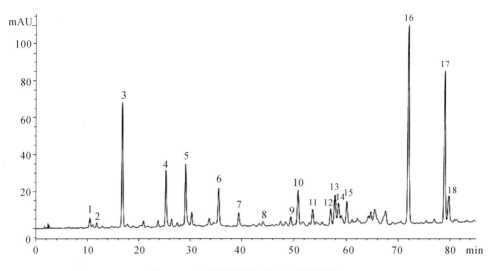

图 13.3 巴西绿蜂胶 HPLC 检测图谱

Fig. 13.3 HPLC chromatogram of BGP sample

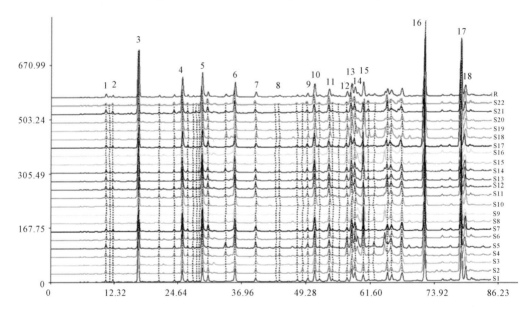

图 13.4 22 个批次巴西绿蜂胶 HPLC 指纹图谱及其对照图谱

Fig. 13.4 HPLC chromatograms of 22 Brazilian green propolis samples and the reference chromatogram

表 13.10 不同批次巴西绿蜂胶相似度评价结果

Table 13.10 Results of similarity values of 22 batches of Brazilian green propolis

样品编号	对照指纹图谱	样品编号	对照指纹图谱
1	0.98	12	0.965
2	0.992	13	0.972
3	0.929	14	0.997

续表

样品编号	对照指纹图谱	样品编号	对照指纹图谱
4	0.966	15	0.948
5	0.941	16	0.946
6	0.955	17	0.996
7	0.996	18	0.941
8	0.943	19	0.996
9	0.952	20	0.99
10	0.971	21	0.955
11	0.971	22	0.97

由图 13.5 可知,22 个不同批次巴西绿蜂胶样的共有色谱成分峰有 18 个,其中 3、16、17 的色谱峰峰高较高,其他色谱峰相对较低,推测这 3 种化合物为巴西绿蜂胶的特征性指标成分。相似度分析结果显示不同绿蜂胶样的相似度较高,均大于 0.9,表明各蜂胶样中化学组成基本一致,而在总体成分的相对含量上存在差异。

同时采用本文建立的 HPLC 方法分别对杨树型蜂胶、桉树型蜂胶、桦树型蜂胶以及血桐属型蜂胶进行化学指纹图谱检测(图 13.5)。巴西绿蜂胶与其他类型蜂胶的色谱图差异极显著,除色谱成分峰 2、3、8、13 存在于杨树型、桉树型及桦树型蜂胶外,其余 14 种色谱峰仅存在于巴西绿蜂胶中,为巴西绿蜂胶的特征性指标成分。这进一步表明胶源植物对蜂胶化学成分的影响具有决定性作用。

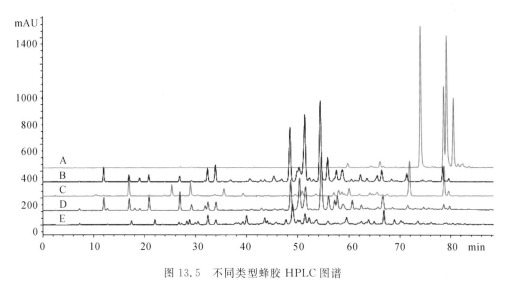

图 13.5 不同类型蜂胶 HPLC 图谱

A. 中国台湾绿蜂胶;B. 俄罗斯桦树型蜂胶;C. 巴西绿蜂胶;D. 中国杨树型蜂胶;E. 澳大利亚桉树型蜂胶

Fig. 13.5 HPLC chromatograms of different types of propolis

A. Taiwanese green propolis; B. Russian betula-type propolis; C. Brazilian green propolis; D. Chinese poplar-type propolis; E. Australian propolis

2.2.3 巴西绿蜂胶中特征性成分峰的 ESI-MS/MS 鉴定

采用液相色谱-电喷雾-质谱法对巴西绿蜂胶中共有成分峰进行定性分析时,为获得全面的质谱信息,先同时采用正、负离子全扫描,得到 HPLC 紫外色谱图和正、负总离子流图(TIC),见图 13.6 和图 13.7。正、负离子检测的信息量相当,在这两种模式下分别出现了大部分分子离子峰[M+H]$^+$ 和[M-H]$^-$ 以及一些碎片离子。考虑到蜂胶中的多酚类化合物在负离子模式下更容易出现去质子的准分子离子峰,故二级质谱定性分析在负离子模式下进行。

利用质谱工作站提供的萃取离子功能,分别对已有文献报道的巴西绿蜂胶中化学成分进行离子萃取,在负离子模式条件下得到了 m/z 为 354(Rt=10.717min),180(Rt=12.027min),164(Rt=16.985min),516(Rt=25.501min),516(Rt=29.354min),302(Rt=35.640min),678(Rt=39.676min),286(Rt=43.522min),232(Rt=50.903min),300(Rt=57.963min),330(Rt=58.612min),300(Rt=71.934min),364(Rt=78.971min),298(Rt=79.802min)的萃取离子,经与现有标准品进行对比,最终鉴定出色谱峰 1、2、3、4、5、7、8、13 及 16 对应的化合物分别为绿原酸、咖啡酸、p-香豆酸、3,5-二咖啡酰奎宁酸、4,5-二咖啡酰奎宁酸、3,4,5-三咖啡酰奎宁酸、山柰酚、山柰素及阿替匹林 C。进一步结合二级质谱离子碎片并与已有文献资料相对比,初步鉴定出色谱峰 6、10、14、17、18 分别为 4′-甲氧基香树素、3-异戊烯基-4-羟基肉桂酸、桦木酚(Betuletol)、3-异戊烯基-4-二氢肉桂酰肉桂酸以及 2,2-二甲基-8-异戊烯基色烯-6-丙烯酸。巴西绿蜂胶中共有成分峰的定性鉴别结果见表 4.2。

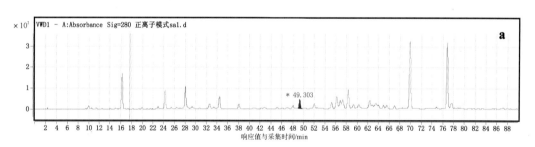

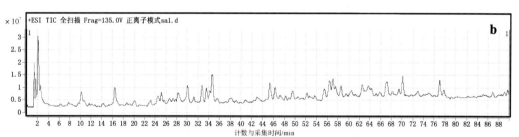

图 13.6 正离子模式巴西绿蜂胶的紫外色谱图(a)和 ESI(+)/MS 总离子流图(b)

Fig. 13.6 UV chromatogram (a) and total ion chromatogram (b) of BGP sample

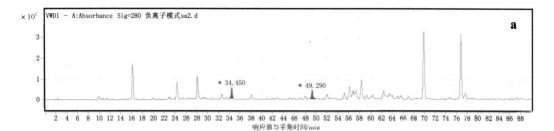
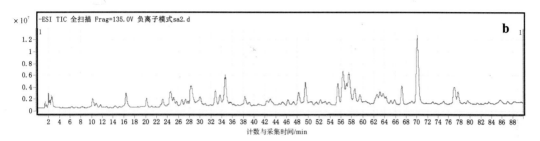

图 13.7 负离子模式巴西绿蜂胶的紫外色谱图(a)和 ESI(-)/MS 总离子流图(b)

Fig. 13.7 UV chromatogram (a) and total ion chromatogram (b) of BGP sample

表 13.11 巴西绿蜂胶 HPLC-ESI-MS^n 光谱及色谱数据

Table 13.11 HPLC-ESI-MS^n spectral and chromatographic data of BGP sample

峰编号	化合物	UVλmax/nm	[M-H]$^-$	MS^2
1	绿原酸[a]	325	353.1	179135191
2	咖啡酸[a]	292322	179.1	135
3	p-香豆酸[a]	310	163.1	119
4	异绿原酸 A[a]	325	515.1	179135191
5	异绿原酸 C[a]	325	515.1	179135191
6	香橙素-4'-甲基醚[b]	290	301.1	213172
7	3,4,5-三咖啡酰奎宁酸[a]	325	677.1	179135191
8	山奈酚[a]	265364	285.1	285257151
10	3-异戊烯基-4-羟基肉桂酸[b]	311	231.2	187133
13	山奈素[a]	265364	299.1	284151
14	桦木酚[b]	290	329.1	314
16	阿替匹林 C[a]	311	299.2	255163151107
17	3-异戊烯基-4-二氢肉桂酰肉桂酸[b]	279	363.2	319187149131
18	2,2-二甲基-8-异戊烯基苯并吡喃-6-丙烯酸[b]	310	297.2	253149

注:具有上标小写字母 a 的物质是通过与已有标准品进行对比后鉴定出的,而具有上标小写字母 b 的物质是通过与已有文献资料对比后鉴定出来的。

3 巴西绿蜂胶中多酚类化学成分定量分析研究

巴西绿蜂胶显著的生理药理活性主要取决于其化学成分的组成及含量。其中,多酚类化合物是蜂胶中主要的活性成分。近年来,对于巴西绿蜂胶中多酚类化合物的研究取得了很大的进展。da Silva 等(2006)研究发现,黄酮类及酚酸类化合物是巴西绿蜂胶发挥抑菌及抗氧化活性的主要成分。Aga 等(1994)通过柱色谱及制备液相色谱分离技术,并结合场解析质谱及核磁共振鉴别技术,从巴西绿蜂胶中分离鉴定出了 3 种强效的抑菌活性成分——阿替匹林 C、3-异戊烯基-4-二氢肉桂酰肉桂酸以及 2,2-二甲基-6-丙烯酸-二氢苯并吡喃,Tazawa 等(1998,1999)采用柱色谱分离方法并结合质谱、红外及核磁共振等多种鉴别技术对巴西绿蜂胶的化学成分展开了一系列的研究,共分离鉴定出了 24 种多酚类化合物,包括 4 种黄酮类、1 种异戊烯基酚酸类、4 种二萜酸类、1 种木脂素类、2 种 p-香豆酸酯类以及 5 种肉桂酸类衍生物。其中,包括二氢山柰酚(香橙素)、6-甲氧基山柰酚、4-羟基-3-异戊烯基苯甲酸、plicatin B、茵陈蒿酸 A(capillartemisin A)以及 7 种新的 p-香豆酸类衍生物首次从巴西绿蜂胶中分离得到。Marcucci 等(2000)对巴西不同地区蜂胶水提物及醇提物的化学组分进行了比较性研究,发现所有蜂胶样中均含有咖啡酸、p-香豆酸、阿魏酸、山柰酚、短叶松素、3-异戊烯基-4-羟基肉桂酸、阿替匹林 C、2,2-二甲基-6-丙烯酸-二氢-1-苯并吡喃及其衍生物。

de Sousa 等(2007)利用 HPLC-PAD 测定了巴西绿蜂胶水醇提取物中肉桂酸、p-香豆酸、阿魏酸、香树素-4′-甲氧基、异樱花素、3-异戊烯基-4-羟基肉桂酸、3-异戊烯基-4-二氢肉桂酰肉桂酸、阿替匹林 C 以及 2,2-二甲基-6-丙烯酸-二氢-1-苯并吡喃酸的含量,认为 p-香豆酸、阿替匹林 C 以及 Baccharin 是巴西绿蜂胶醇提物中主要的多酚类化合物。Rocha 等(2013)通过 HPLC-UV 比较性地研究了巴西绿蜂胶水提物及醇提物的化学成分差异,结果显示两者均含有咖啡酸、p-香豆酸、反式肉桂酸、香橙素及阿替匹林 C,然而在含量上醇提物中反式肉桂酸、香橙素及阿替匹林 C 更具优势。Midorikawa 等(2001)采用 LC-MS 测定了巴西不同来源的绿蜂胶水提物及甲醇提取物的化学指纹图谱,根据检测结果确定了水提物中共有的咖啡酰奎宁酸及其衍生物、p-香豆酸、咖啡酸、阿魏酸、香草醛及松柏醛类强极性多酚类化合物的存在,而甲醇提取物中检测到弱极性物质包括 5 种黄酮类化合物桦木酚、山柰酚、山柰素、ermanin、香橙素-4′-甲氧基,3 种二萜酸类化合物 agathic acid、agathalic acid 以及 cupressic acid,异戊烯基肉桂酸类化合物阿替匹林 C、4-羟基-3-异戊烯基肉桂酸、3-异戊烯基-4-二氢肉桂酰肉桂酸、3-羟基-2,2-二甲基-8-异戊烯基色烷-6-丙烯酸以及 2,2-二甲基色烯-6-羧酸。

本研究结合巴西绿蜂胶中多酚类化合物的 HPLC-ESI-MS^n 定性鉴别并从中筛选出已有标准品的指标成分绿原酸、咖啡酸、p-香豆酸、3,5-二咖啡酰奎宁酸、4,5-二咖啡酰奎宁酸、3,4,5-三咖啡酰奎宁酸、山柰酚、山柰素及阿替匹林 C。以这几种成分为指标,采用 RP-HPLC 方法建立巴西绿蜂胶多指标定量测定。

3.1 实验材料与方法

3.1.1 蜂胶样品

巴西绿蜂胶样品同本章 1.1.1。

3.1.2 仪器及试剂

多功能高速粉碎机：信强，上海

万分之一分析天平：Denver TB-215D

超声波辅助提取仪：SK20GT，上海科导超声波仪器生产公司

高效液相色谱仪：安捷伦1200，配有四元流动泵G1311A，在线真空脱气机G1322A，自动进样器G1329A，恒温箱G1316A，紫外检测器（VWD）G1314B。

甲醇：色谱纯，德国Merck公司

超纯水：色谱纯，杭州永洁达高纯水机自制

乙醇：分析纯

乙酸：分析纯

咖啡酸、绿原酸、p-香豆酸、3,5-二咖啡酰奎宁酸、4,5-二咖啡酰奎宁酸、3,4,5-三咖啡酰奎宁酸、山奈酚及山奈素对照品（色谱纯）购于美国Sigma公司，阿替匹林C标准品（纯度≥98%）购于日本Wako公司。

3.1.3 蜂胶样品处理

巴西绿蜂胶毛胶样品处理同本章2.1.3。

3.1.4 液相色谱条件

色谱柱为Sepax HP-C18（150mm×4.6mm，5μm）；流动相：甲醇-0.1%乙酸；流速：1.0mL/min；波长：280nm；柱温：33℃；进样量：5μL。优化后的梯度洗脱程序为：0～10min，15%～25%甲醇；10～25min，25%～40%甲醇；25～55min，40%～60%甲醇；55～75min，60%～75%甲醇；75～90min，75%～85%甲醇。

3.1.5 混合标准品溶液的配置

标准品储备液的配置：准确称取一定量的绿原酸、咖啡酸、p-香豆酸、3,5-二咖啡酰奎宁酸、4,5-二咖啡酰奎宁酸、3,4,5-三咖啡酰奎宁酸、山奈酚、山奈素及阿替匹林C，用色谱级甲醇分别溶解稀释至1mg/mL，作为标准储备液，于-4℃下保存备用。

混合标准品溶液：分别移取适量的上述标准品储备液至10mL容量瓶，用色谱级甲醇定容至刻度，作为混合标准品母液，于-4℃保存。

3.1.6 方法学验证

3.1.6.1 线性关系考察

将混合标准品母液依次稀释至原来的0.8、0.6、0.4、0.2、0.1倍，得到6个不同浓度梯度的混合标准溶液，经0.45μm微孔滤膜过滤后进样检测，每个浓度平行测定3次，以峰面积为纵坐标，以对应标准品的浓度为横坐标绘制标准曲线，计算线性方程及相关系数。

3.1.6.2 检测限及定量限

将混合标准品进行连续稀释，过0.45μm微孔滤膜进样检测，分别以信噪比S/N为3/1和10/1来确定各标准品的检测限（LOD）和定量限（LOQ）。

3.1.6.3 精密度考察

包括日内精密度和日间精密度。

日内精密度：将适宜浓度的混合标准品每隔2h进样检测，共进样6次，记录各标准品的保留时间及峰面积，计算各自的RSD值。

日间精密度:将适宜浓度的混合标准品连续6d进样检测,每天连续进样3次,记录各标准品的保留时间及峰面积,计算各自的RSD值。

3.1.6.4 重复性及稳定性考察

重复性测定:准确称取同一批巴西绿蜂胶样品6份,按照本章2.1.3的方法平行制备6份蜂胶醇提液,进样检测,记录各标准品的保留时间及峰面积,并计算RSD值。

稳定性测定:分别于第1、2、3、4、5、6d将制备好的同一份巴西绿蜂胶试样进样检测,记录各标准品的保留时间及峰面积,并计算RSD值。

3.1.6.5 加样回收率测定

准确称取已知各标准品含量的同一批巴西绿蜂胶样品9份,每3份一组,分别加入相当于低、中、高剂量的混合标准溶液,按2.1.3中的方法操作,进样检测,记录峰面积,计算加样回收率及RSD值。

3.1.6.6 不同蜂胶中多酚类化合物含量的检测

在上述色谱条件下分别测得混合标准品及22个不同批次绿蜂胶样品的HPLC图谱,记录各指标成分对应的峰面积/峰高并计算其含量。

3.2 实验结果及分析讨论

3.2.1 方法学考察结果

方法学考察结果分别见表13.12、13.13和13.14。线性关系考察结果显示所有对照标准品在上述色谱条件下均呈现出良好的线性关系。同时,所有标准品的检测限及定量限均达到1μg/mL的级别,表明该方法可以用来定性及定量检测巴西绿蜂胶中这9种多酚类化合物。精密度、稳定性及重复性考察的实验结果显示,各对照标准品的色谱峰面积RSD均在3%以内,而保留时间的RSD均在1%以内。此外,各对照品的平均回收率均在90%以上,RSD在0.39%~2.86%。上述方法学考察结果表明该方法准确性好,可行性高。

表13.12 线性关系、检测限及定量限考察结果

Table 13.12 Results of regression data, LODs and LOQs for 9 standards

化合物	标准曲线	R^2	线性范围/(mg/mL)	检测限/(μg/mL)	定量限/(μg/mL)
绿原酸	$y=5858.2x-14.044$	0.9982	0.02~0.1	1.5	5
咖啡酸	$y=15019x-6.9476$	0.9998	0.02~0.1	0.15	0.5
p-香豆酸	$y=23746x+29.4$	0.9993	0.04~0.2	0.12	0.4
3,5-二咖啡酰奎宁酸	$y=10965x-179.93$	0.9961	0.04~0.2	3	10
4,5-二咖啡酰奎宁酸	$y=8352.3x-617.76$	0.9998	0.1~0.3	3	10
3,4,5-三咖啡酰奎宁酸	$y=10267x-46.368$	0.9995	0.1~0.3	2.4	8
山奈酚	$y=13380x-57.578$	0.9996	0.02~0.1	0.11	0.35
山奈素	$y=8463.8x-60.752$	0.9994	0.04~0.2	0.55	1.84
阿替匹林C	$y=10411x-40.379$	0.9993	0.1~0.5	0.3	1

表 13.13 精密度考察结果

Table 13.13 Results of precision data for 9 standards

化合物	日内精密度（RSD％,$n=6$）		日间精密度（RSD％,$n=6$）	
	保留时间	峰面积	保留时间	峰面积
绿原酸	0.21	1.24	0.44	0.92
咖啡酸	0.49	1.05	0.29	1.07
p-香豆酸	0.35	0.53	0.10	0.44
3,5-二咖啡酰奎宁酸	0.13	0.86	0.21	0.82
4,5-二咖啡酰奎宁酸	0.30	1.56	0.10	0.81
3,4,5-三咖啡酰奎宁酸	0.20	0.95	0.32	1.02
山柰酚	0.41	1.28	0.20	2.32
山柰素	0.09	0.11	0.73	0.53
阿替匹林 C	0.07	2.01	0.52	1.07

表 13.14 重复性、稳定性及加样回收率考察结果

Table 13.14 Results of repeatability, stability and recovery rate for 9 standards

化合物	重复性（RSD％,$n=6$）		稳定性（RSD％,$n=6$）		回收率/%
	保留时间	峰面积	保留时间	峰面积	
绿原酸	0.20	2.53	0.12	0.89	95.2~98.6
咖啡酸	0.80	2.46	0.06	2.21	98.4~100.7
p-香豆酸	0.09	1.26	0.18	2.64	96.7~99.5
3,5-二咖啡酰奎宁酸	0.19	0.53	0.24	1.35	92.8~97.6
4,5-二咖啡酰奎宁酸	0.28	1.00	0.15	0.67	94.7~101.4
3,4,5-三咖啡酰奎宁酸	0.12	0.87	0.07	1.69	96.0~99.2
山柰酚	0.08	2.01	0.31	2.94	98.9~99.7
山柰素	0.28	1.28	0.28	1.75	97.8~100.1
阿替匹林 C	0.43	2.86	0.16	0.92	98.4~99.9

3.2.2 不同批次绿蜂胶样中9种多酚类化合物含量的测定

对不同批次巴西绿蜂胶中主要的黄酮及酚酸类化合物进行检测,通过与混合对照品图谱进行对比,并结合各对照品的线性回归方程分别计算其含量。绿原酸、咖啡酸、p-香豆酸、3,5-二咖啡酰奎宁酸、4,5-二咖啡酰奎宁酸、3,4,5-三咖啡酰奎宁酸、山奈酚、山奈素、阿替匹林C的混合对照溶液与绿蜂胶样的HPLC色谱图见图13.8。在该色谱条件下,各物质的分离效果较好,p-香豆酸、3,5-二咖啡酰奎宁酸、4,5-二咖啡酰奎宁酸、山奈素及阿替匹林C的含量相对较高;其中,以阿替匹林C的含量最为丰富;而绿原酸、咖啡酸及3,4,5-三咖啡酰奎宁酸的含量相对较低,山奈酚的含量最低。

对22个不同批次的绿蜂胶检测结果显示所有蜂胶样具有基本一致的化学组成,均含有绿原酸、咖啡酸、p-香豆酸、3,5-二咖啡酰奎宁酸、4,5-二咖啡酰奎宁酸、3,4,5-三咖啡酰奎宁酸、山奈酚、山奈素及阿替匹林C;然而不同蜂胶样间各成分的含量存在较大差异,不同蜂胶样中各对照品的含量见表13.15。其中含量最高的为阿替匹林C,其次为4,5-二咖啡酰奎宁酸、3,5-二咖啡酰奎宁酸、山奈素及p-香豆酸,其他成分含量较低。

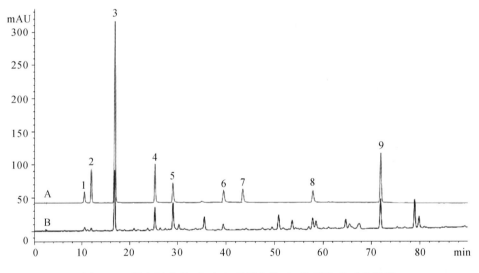

图13.8 混合标准品(A)与巴西绿蜂胶(B)的HPLC对比图谱

1.绿原酸;2.咖啡酸;3.p-香豆酸;4.3,5-二咖啡酰奎宁酸;5.4,5-二咖啡酰奎宁酸;6.3,4,5-三咖啡酰奎宁酸;7.山奈酚;8.山奈素;9.阿替匹林C

Fig. 13.8 HPLC chromatograms of standards (A) and BGP sample S1 (B)

1. Chlorogenic acid; 2. Caffeic acid; 3. p-coumaric acid; 4. 3,5-dicaffeoylquinic acid; 5. 4,5-dicaffeoylquinic acid; 6. 3,4,5-tricaffeoylquinic acid; 7. Kaempferol; 8. Kaempferide; 9. Artepillin C

表 13.15 巴西绿蜂胶中 9 种多酚类化合物的含量

Table 13.15 Content of nine polyphenolic compounds in BGP samples

mg/g

样品编号	绿原酸	咖啡酸	p-香豆酸	3,5-二咖啡酰奎宁酸	4,5-二咖啡酰奎宁酸	3,4,5-三咖啡酰奎宁酸	山柰酚	山柰素	阿替匹林C
1	1.98±0.31	0.75±0.20	10.42±0.54	8.69±1.07	18.20±4.19	2.81±0.35	1.49±0.21	12.73±1.03	26.81±4.58
2	3.78±0.91	0.97±0.11	11.44±3.40	12.54±5.71	25.64±0.86	3.76±0.27	1.66±0.58	12.60±1.29	35.94±0.57
3	3.97±0.58	1.16±0.30	11.22±2.06	9.97±4.07	21.89±3.86	7.16±0.16	1.92±0.57	11.09±0.56	21.66±8.22
4	4.90±2.31	0.76±0.28	10.14±3.10	15.55±2.84	31.99±7.61	4.95±0.61	1.86±0.86	15.97±3.14	60.75±5.22
5	4.57±0.58	0.67±0.26	10.01±0.35	15.12±6.24	30.58±2.16	4.79±0.30	1.74±0.27	18.79±1.02	81.01±3.21
6	4.53±1.15	1.36±0.11	14.13±2.31	15.32±2.10	31.57±5.20	5.46±0.17	2.27±0.26	15.93±4.08	24.80±1.65
7	3.93±0.56	0.99±0.08	10.82±0.95	12.19±0.18	25.61±1.57	3.96±0.32	1.67±0.25	12.54±3.05	36.05±2.10
8	6.07±0.65	0.54±0.04	5.93±0.46	15.36±1.37	30.96±4.02	5.12±0.78	1.25±0.65	13.57±0.85	53.07±8.16
9	3.19±0.57	0.45±0.01	5.74±0.57	10.55±4.03	23.26±2.06	3.34±0.54	0.94±0.51	9.45±0.54	41.14±4.21
10	4.51±0.86	1.05±0.01	7.45±2.06	11.71±1.55	24.74±0.94	4.09±0.28	1.51±0.54	11.19±0.85	40.20±2.17
11	4.71±0.59	1.11±0.06	10.75±4.02	16.33±0.76	33.40±4.90	5.50±0.36	2.19±0.26	14.78±0.54	26.50±8.06
12	4.16±0.57	1.29±0.55	13.86±1.60	13.94±0.63	28.91±8.02	5.12±0.19	2.18±0.68	14.45±4.22	27.61±4.26
13	4.76±0.25	0.80±0.03	10.65±0.89	14.00±4.26	28.71±8.21	4.29±0.55	1.47±0.57	14.02±4.33	65.07±9.22
14	3.24±0.62	0.80±0.41	9.53±0.55	10.36±1.01	23.18±5.03	3.34±0.87	1.80±0.55	13.37±0.51	40.83±2.56
15	4.48±0.42	0.70±0.10	9.82±0.25	14.69±6.21	30.18±7.01	4.89±0.47	1.73±0.24	17.09±1.56	74.22±0.85
16	3.21±0.63	3.35±0.56	13.17±1.85	18.45±6.25	24.31±4.21	7.73±0.10	4.11±0.29	20.10±6.23	35.81±6.21
17	3.36±0.90	0.91±0.02	10.44±0.82	10.93±3.01	22.58±4.21	3.42±0.38	1.59±0.15	12.68±1.56	39.25±1.00
18	4.72±1.02	0.69±0.08	9.55±0.58	14.44±6.21	29.46±9.22	4.74±0.62	1.36±0.21	17.57±9.42	72.76±5.54
19	4.23±0.56	0.86±0.20	9.94±1.07	13.04±3.01	26.76±4.22	4.38±0.74	1.81±0.80	14.28±3.89	42.06±3.55
20	4.56±0.92	0.72±0.21	10.16±1.57	13.42±5.21	27.11±3.47	4.28±0.39	1.28±0.84	12.87±3.01	59.74±8.26
21	6.36±0.76	1.34±0.21	11.61±3.00	14.68±5.01	31.15±4.19	6.58±0.91	2.38±3.01	13.03±089	26.12±2.90
22	3.58±0.51	1.11±0.20	13.07±2.00	12.87±1.53	25.35±1.06	7.55±0.54	2.11±0.43	14.54±3.21	32.81±0.84

注: 数据表示为平均值±标准差($n=3$)。

4 巴西绿蜂胶中抗氧化活性成分的筛选

自由基是指一类含有不成对电子的原子、分子和基团。自由基的存在非常广泛,其氧化反应可引起细胞衰老和各种疾病。研究表明,蜂胶具有较高的抗氧化活性,不仅能够有效地清除多种自由基,而且还能进一步保护机体免受自由基诱导的氧化损伤。同时,大量研究证实,巴西绿蜂胶强效的抗炎、护肝、抗肿瘤、神经保护以及调节血糖血脂等功效均与其抗氧化作用有着直接的联系。

蜂胶中含有多种抗氧化活性成分。研究发现,黄酮及酚酸类化合物是蜂胶发挥抗氧化活性的主要物质基础。Hayashi 等(1999)采用生物自显影导向分离技术从巴西绿蜂胶中分离鉴定出了多种抗氧化活性物质,包括阿替匹林 C、山柰素、3,4-二羟基-5-异戊烯基肉桂酸、桦木酚、山柰酚、4-羟基-3-异戊烯基肉桂酸及 3-异戊烯基-4-二氢肉桂酰肉桂酸。da Silva 等(2006)对巴西 49 个蜂胶样的化学成分及抗氧化活性进行分析测定,发现抗氧化活性与总黄酮及总酚酸含量均存在着显著的相关性。然而,由于蜂胶本身化学成分的多样性和可变性以及不同组分间的相互作用,很难确定其中有效的抗氧化物质基础以及单个抗氧化活性成分的效价值。近年来,国内外研究者将化学指纹图谱分析与抗氧化活性评价相结合,通过离线或在线的活性检测方法,对包括蓝莓、茶叶、薄荷、淫羊藿及黄秋葵等多种水果、蔬菜及中药复杂体系的谱-效关系进行了研究,从而筛选评价其抗氧化物质基础。

本研究采用离线 HPLC-DPPH 法对巴西绿蜂胶中抗氧化活性成分进行了研究,将 HPLC 分离检测得到的化学指纹图谱与 DPPH 自由基清除活性相结合,筛选出巴西绿蜂胶中真正发挥抗氧化活性的物质基础,为巴西绿蜂胶的质量控制及抗氧化作用机制的研究提供了科学依据。

4.1 材料与方法

4.1.1 蜂胶样品

巴西绿蜂胶样品同本章 1.1.1。

4.1.2 仪器及试剂

多功能高速粉碎机:信强,上海

万分之一分析天平:Denver TB-215D

超声波辅助提取仪:SK20GT,上海科导超声波仪器生产公司

高效液相色谱仪:安捷伦 1200,配有四元流动泵 G1311A,在线真空脱气机 G1322A,自动进样器 G1329A,恒温箱 G1316A,紫外检测器(VWD)G1314B。

甲醇:色谱纯,德国 Merck 公司

超纯水:色谱纯,杭州永洁达高纯水机自制

乙醇:分析纯

乙酸:分析纯

咖啡酸、绿原酸、p-香豆酸、3,5-二咖啡酰奎宁酸、4,5-二咖啡酰奎宁酸、3,4,5-三咖啡酰奎宁酸、山柰酚及山柰素对照品(色谱纯)购于美国 Sigma 公司,阿替匹林 C 标准品(纯度≥98%)购于日本 Wako 公司。

4.1.3 蜂胶样品处理

准确称取 0.5g(精确至 0.1mg)经粉碎混匀的巴西绿蜂胶样,用 95%乙醇超声提取 45min,料液比为 1:30,反复提取 3 次,合并滤液,经冷冻除蜡后用 95%的乙醇定容至 50mL,得到终浓度为 10mg/mL 的蜂胶醇提液。

4.1.4 液相色谱条件

色谱柱为 Sepax HP-C18(150mm×4.6mm,5μm);流动相:甲醇-0.1%乙酸;流速:1.0mL/min;波长:280nm;柱温:33℃;进样量:5μL。优化后的梯度洗脱程序为:0～10min,15%～25%甲醇;10～25min,25%～40%甲醇;25～55min,40%～60%甲醇;55～75min,60%～75%甲醇;75～90min,75%～85%甲醇。

4.1.5 离线 HPLC-DPPH 分析

参考 Toshiya 等(2003)的方法。准确移取 10mg/mL 的蜂胶醇提液 500μL 并加入新鲜配置的 3mg/mL DPPH 乙醇溶液 500μL,室温条件下避光反应 1h,经 0.45μm 滤膜过滤后作为实验组;同时用等体积的无水乙醇替代 DPPH 按上述操作处理作为空白对照样。分别进供试样和空白对照各 5μL,用于色谱分析,单个样品重复上述操作 3 次。峰面积下降率(%)

$$=\frac{\text{峰面积}_{\text{反应前}}-\text{峰面积}_{\text{反应后}}}{\text{峰面积}_{\text{反应前}}}\times100\%。$$

4.2 结果与分析

4.2.1 离线 HPLC-DPPH 法的建立与优化

(1)DPPH 浓度的影响

蜂胶样(10mg/mL)与不同浓度的 DPPH 溶液反应后峰面积下降率的测定结果见图 13.9。随着 DPPH 浓度的增大,阿替匹林 C 的峰面积下降率也随之增大;当 DPPH 浓度达到 3mg/mL 时,阿替匹林 C 的峰面积下降率达到最大,并且不再随着 DPPH 浓度的增大而增大。同时,考虑到反应后体系中 DPPH 的色谱吸收峰峰高与蜂胶中化学成分的吸收峰峰高的比例,本实验采用 3mg/mL 的 DPPH 反应工作液。

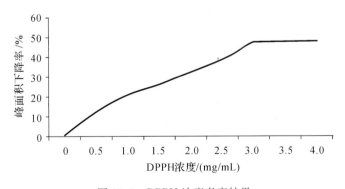

图 13.9 DPPH 浓度考察结果

Fig. 13.9 Result of the impact of DPPH concentration

(2)反应时间的影响

对蜂胶中抗氧化成分的峰面积下降率在不同反应时间的测定结果见图 13.10。蜂胶中抗氧化成分的峰面积下降率在 15～30min 内快速增加,30～45min 内继续保持快速增加趋势，45～60min内缓慢增大并趋于平衡；随后不再随着反应时间的延长而增加。这表明蜂胶与DPPH自由基在60min内已达到了充分的反应，因此确定最佳的反应时间为60min。

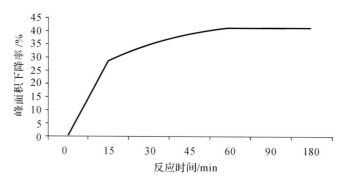

图13.10　反应时间考察结果

Fig. 13.10　Result of the impact of reaction time

因此，基于上述单指标反应条件的优化结果，最终选取10mg/mL的巴西绿蜂胶醇提液按1∶1的反应比例，与3.0mg/mL DPPH自由基于室温避光条件下反应60min。

4.2.2　巴西绿蜂胶的抗氧化成分的HPLC筛选

DPPH在乙醇溶液中以稳定的DPPH自由基形态存在，当加入蜂胶醇提液时，蜂胶中具有抗氧化活性的化学成分因参与DPPH自由基的清除反应而使峰面积下降，而不具有抗氧化活性的化学成分在反应后峰面积无变化。同时，根据峰面积下降的程度可以评价抗氧化成分的活性强弱。

巴西绿蜂胶与DPPH自由基反应前后的色谱图见图13.12。反应后蜂胶中的成分峰1（绿原酸）、2（咖啡酸）、4（3,5-二咖啡酰奎宁酸）、5（4,5-二咖啡酰奎宁酸）、7（3,4,5-三咖啡酰奎宁酸）、8（山奈酚）、12（山奈素）、13（桦木酚）及16（阿替匹林C）的峰面积发生了显著的下降，而其他成分峰在反应后无明显的变化。进一步对各成分峰的峰面积下降率进行计算，各抗氧化成分的抗氧化活性存在显著差异，结果见表13.16。

在黄酮类化合物中，山奈酚、山奈素、桦木酚及香橙素-4'-甲基醚的峰面积下降率均有所不同。其中，山奈酚的峰面积下降率最大，达到100%；山奈素及桦木酚的色谱峰面积下降率平均为70%；而香橙素-4'-甲基醚的峰面积几乎无变化。表明这4种黄酮类化合物的抗氧化活性存在着显著的差异。分析各物质的化学结构发现山奈酚、山奈素及桦木酚的母核结构均具有C环C2-C3之间的双键以及C3上的羟基结构，这种结构有利于与B环形成稳定的共轭结构，从而进一步稳定与DPPH自由基反应后的结构（Catherine，1996）。而香橙素-4'-甲氧基属于二氢黄酮醇，既不具有C环上C2-C3双键，也不含有B环的邻二羟基结构，因此未参与DPPH自由基清除反应。

巴西绿蜂胶中酚酸类化合物的峰面积下降率也均不相同，以咖啡酰奎宁酸类化合物的峰面积下降程度最为显著。其中，个别蜂胶样中3,4,5-三咖啡酰奎宁酸的峰面积下降率达到100%；而p-香豆酸、3-异戊烯基-4-羟基肉桂酸及3-异戊烯基-4-二氢肉桂酰肉桂酸的峰面积几乎无变化。分析上述酚酸类化合物的化学结构发现：苯环上酚羟基的数目、取代基的类型及位置对p-香豆酸及其衍生物的DPPH自由基清除活性有着至关重要的作用。而咖啡酰基的数目和位置是决定咖啡酰奎宁酸类化合物抗氧化活性的主要因素。

此外,在图 13.11 中,反应后出现了新的色谱峰,还有个别成分峰的峰面积在反应后有所增加。分析可能的原因在于蜂胶中某些成分在与 DPPH 自由基反应后生成了新的物质,抑或是反应后的复杂体系在该点的吸光度与蜂胶中原有成分的吸光度发生了叠加。

综上所述,巴西绿蜂胶的 DPPH 自由基清除活性与其含有的多种抗氧化活性成分有关。巴西绿蜂胶中含有多个抗氧化活性成分,包括山奈酚、山奈素、桦木酚、咖啡酸及其衍生物以及绿原酸及其衍生物。

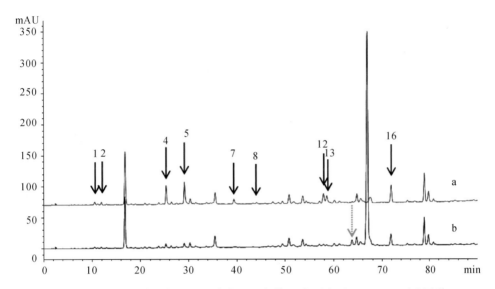

图 13.11　巴西绿蜂胶与 DPPH 自由基反应前(a)与反应后(b)HPLC 对比图谱

1.绿原酸;2.咖啡酸;4.3,5-二咖啡酰奎宁酸;5.4,5-二咖啡酰奎宁酸;7.3,4,5-三咖啡酰奎宁酸;8.山奈酚;12.山奈素;13.桦木酚;16.阿替匹林 C

Fig. 13.11　HPLC chromatograms of BGP sample S1 before (a) and after (b) reacted with DPPH radical
1. Chlorogenic acid;2. Caffeic acid;4. 3,5-dicaffeoylquinic acid;5. 4,5-dicaffeoylquinic acid;7. 3,4,5-tricaffeoylquinic acid;8. Kaempferol;12. Kaempferide;13. Betuletol;16. Artepillin C

表13.16 巴西绿蜂胶中共有成分的峰面积下降率

Table 13.16 Decrease rate of peak areas for common compounds in BGP samples

Decrease rate	S1	S2	S3	S4	S5	S6	S7	S8	S9	S10	S11
P1	49.82±6.99	51.64±4.37	60.69±3.62	37.01±8.02	51.30±2.36	51.14±5.32	55.07±7.02	57.34±3.25	52.74±2.02	54.21±4.37	41.06±4.32
P2	45.65±4.25	48.12±1.54	52.96±2.04	43.31±3.41	45.84±0.98	44.96±2.51	47.32±3.08	48.41±5.04	43.73±1.29	47.17±0.56	51.55±2.57
P3	4.46±1.84	5.45±2.09	4.31±3.04	4.82±0.98	3.43±1.20	3.55±1.03	4.22±0.57	2.40±0.62	2.75±0.75	4.26±1.61	4.75±0.84
P4	76.57±1.38	77.90±1.73	78.44±2.48	72.15±0.95	73.30±1.54	72.26±7.36	74.70±1.34	74.27±1.84	71.52±4.61	71.04±3.84	74.62±2.83
P5	77.08±5.30	77.82±3.73	81.24±6.71	73.23±1.05	73.84±5.21	72.38±5.03	75.54±2.92	75.11±0.62	71.72±1.78	71.86±2.28	75.46±1.62
P6	−1.03±0.46	−2.14±0.21	−0.78±0.16	−1.54±0.10	−3.18±0.16	−5.16±0.37	−6.29±0.23	−4.87±0.72	−5.46±1.02	−4.36±0.63	−3.81±0.36
P7	88.25±3.92	80.70±1.29	88.80±2.01	82.87±0.94	100.00±2.78	83.34±2.76	81.92±3.72	84.95±5.53	83.07±4.25	83.34±1.24	85.37±3.27
P8	85.02±2.73	59.80±5.26	75.68±0.83	73.67±2.19	100.00±1.28	100.00±6.38	100.00±3.28	100.00±6.38	74.64±6.75	79.15±4.39	100.00±2.29
P9	−40.80±3.02	−32.90±6.26	−37.90±7.29	−36.87±2.01	−14.97±0.67	−58.49±3.2	−15.43±1.29	−50.82±3.6	−29.33±7.20	−47.25±1.38	−77.21±0.93
P10	−5.95±2.10	−7.03±0.92	−6.54±0.63	−3.30±0.53	−1.42±0.49	−4.28±1.71	−7.04±1.20	−3.46±0.84	−4.21±0.75	−10.27±2.10	−3.28±0.38
P11	−2.30±0.47	−4.01±1.20	−3.05±0.46	−1.36±0.75	−1.97±0.58	−6.24±0.84	−1.34±0.35	−3.60±2.37	−3.27±0.53	−10.75±1.85	−2.68±0.73
P12	3.31±0.37	2.04±0.89	2.09±1.01	3.39±0.35	2.63±0.78	2.02±0.40	6.72±0.94	4.36±1.30	3.07±0.64	5.12±1.06	6.23±0.47
P13	76.85±2.19	79.56±0.58	84.60±1.23	78.68±2.10	74.73±0.37	61.69±5.39	73.27±1.28	68.45±4.83	61.35±3.29	55.79±3.20	71.64±2.39
P14	81.85±2.38	84.46±1.29	86.51±0.37	78.63±2.75	76.70±4.63	65.22±2.19	80.45±0.69	77.02±7.36	66.26±4.90	62.08±2.38	75.73±5.37
P15	12.35±2.75	11.51±3.27	8.91±1.93	12.96±0.96	9.06±1.31	2.50±0.72	9.83±0.96	3.72±0.53	4.00±0.71	7.67±1.23	2.63±0.79
P16	47.53±5.37	33.43±0.68	44.69±2.38	48.32±1.25	35.67±2.32	27.39±2.79	31.95±0.93	37.96±6.39	28.39±4.32	24.35±1.28	41.61±7.31
P17	1.00±0.32	0.74±0.28	1.10±0.81	1.62±0.03	1.02±0.63	1.58±0.67	1.28±0.25	0.65±0.23	1.84±0.43	1.90±0.21	1.17±0.39
P18	1.27±0.65	1.04±0.18	0.26±0.06	1.81±0.63	1.19±0.35	0.84±0.20	1.19±0.82	2.30±0.73	2.42±0.28	0.88±0.47	2.95±0.53

续表

Decrease rate	S12	S13	S14	S15	S16	S17	S18	S19	S20	S21	S22
P1	53.56±1.06	52.63±3.52	53.92±0.85	47.20±3.05	57.61±2.37	49.61±4.32	54.43±2.54	53.83±2.49	51.86±3.64	54.87±3.64	50.84±2.57
P2	46.96±1.25	45.58±3.07	47.03±4.30	41.33±1.05	50.68±3.53	43.68±1.37	48.45±3.21	48.37±1.07	46.29±2.25	52.30±3.89	48.38±1.57
P3	4.41±1.65	3.77±0.53	6.81±0.38	3.42±1.95	4.29±0.67	4.30±2.01	3.29±1.46	5.22±2.06	2.40±0.61	2.93±0.94	5.37±1.62
P4	74.33±5.69	69.95±7.05	70.05±2.54	67.86±3.58	76.99±6.20	70.41±8.04	72.99±1.48	72.19±4.52	70.39±3.46	70.32±6.25	68.39±2.59
P5	75.06±0.78	71.19±1.25	70.04±5.09	68.96±1.92	76.77±3.21	71.35±4.85	73.77±2.89	72.72±3.27	69.74±0.63	71.55±6.02	68.94±2.75
P6	−2.65±0.16	−5.58±0.35	−0.41±0.10	−2.50±0.38	−1.44±0.38	−0.32±0.09	−0.28±0.06	−2.10±0.71	−0.28±0.12	−3.05±0.15	−1.93±0.72
P7	79.10±0.83	84.94±1.28	79.51±2.18	74.96±3.82	91.81±4.38	86.10±5.63	89.81±2.73	90.27±2.19	87.44±0.66	90.16±3.89	86.48±5.42
P8	61.27±3.95	100.00±3.01	86.28±4.38	70.30±1.28	100.00±8.02	100.00±6.23	100.00±4.28	100.00±2.98	100.00±5.05	100.00±3.29	100.00±6.63
P9	−68.89±3.26	−41.70±0.63	−23.24±6.37	−66.36±2.29	−45.55±3.21	−42.31±1.02	−47.83±2.17	−46.83±5.10	−39.76±4.73	−40.64±2.17	−52.06±3.63
P10	−7.95±2.37	−3.78±0.93	−2.38±0.58	−0.05±0.02	−4.05±0.36	−3.86±0.85	−2.02±0.81	−3.53±1.92	−5.84±0.73	−5.30±1.02	−0.53±0.20
P11	−10.04±0.70	−0.87±0.10	−6.90±0.96	−5.21±0.73	−6.94±1.29	−5.99±2.10	−5.22±1.30	−5.32±1.28	−3.28±1.23	−3.69±1.02	−3.28±0.81
P12	5.49±1.0.68	9.41±1.28	7.17±0.36	4.11±2.17	4.35±0.68	1.73±0.57	0.29±0.15	1.07±0.21	0.27±0.11	0.53±0.47	0.66±0.18
P13	74.73±2.08	66.98±4.84	65.75±6.29	70.77±3.93	62.50±6.29	66.98±1.29	58.93±2.83	60.45±0.83	58.83±1.92	60.27±4.39	56.66±1.27
P14	78.95±2.83	71.96±2.53	71.45±6.50	71.56±1.28	70.22±3.20	68.93±2.67	64.37±0.97	65.88±2.83	63.24±3.82	68.01±1.23	61.34±6.38
P15	7.46±0.80	2.03±0.62	8.17±0.46	3.09±1.24	0.53±0.38	1.04±0.53	1.28±0.42	1.56±0.86	2.30±0.99	3.201±0.56	6.92±1.290
P16	49.59±2.39	34.09±2.30	31.59±1.26	37.08±3.22	30.14±4.39	29.73±7.20	31.29±5.37	32.64±2.35	28.38±1.27	29.57±3.40	34.83±2.70
P17	0.22±0.18	1.14±0.73	3.77±0.51	1.79±0.95	1.37±0.42	2.02±0.60	0.27±0.09	0.42±0.18	0.77±0.29	3.66±0.85	2.84±0.37
P18	0.95±0.23	1.37±0.84	0.25±0.49	2.79±0.36	1.23±0.52	1.87±0.31	0.35±0.19	2.82±0.84	0.37±0.13	2.33±0.36	0.38±0.10

小 结

本研究在对巴西绿蜂胶中总黄酮、总酚酸含量测定及 DPPH 自由基清除活性进行评价的基础上，采用反向高效液相色谱法及电喷雾-质谱联用技术建立了巴西绿蜂胶的 HPLC 指纹图谱,确定了巴西绿蜂胶的共有特征性化学成分。同时采用离线 HPLC-DPPH 法筛选评价了巴西绿蜂胶中的抗氧化活性成分,并结合多指标定量测定建立了巴西绿蜂胶的抗氧化活性指纹图谱。其研究结果如下:

(1) 巴西绿蜂胶中酚酸类化合物含量丰富,具有较强的 DPPH 自由基清除活性。

(2) 建立了巴西绿蜂胶的化学指纹图谱,确定了 18 个共有成分峰。其中绿原酸、3,5-二咖啡酰奎宁酸、4,5-二咖啡酰奎宁酸、3,4,5-三咖啡酰奎宁酸、香橙素-4′-甲基醚、3-异戊烯基-4-羟基肉桂酸、桦木酚、阿替匹林 C、3-异戊烯基-4-二氢肉桂酰肉桂酸以及 2,2-二甲基-8-异戊烯基色烯-6-丙烯酸等 14 种多酚类化合物为巴西绿蜂胶的特征性指标成分。

(3) 对巴西绿蜂胶中标志性成分进行定量分析发现阿替匹林 C 的含量最为丰富,绿原酸及其衍生物 3,5-二咖啡酰奎宁酸、4,5-二咖啡酰奎宁酸及 3,4,5-三咖啡酰奎宁酸的含量相对较高,而山奈酚及咖啡酸的含量最低。

(4) 通过离线 HPLC-DPPH 法筛选出了巴西绿蜂胶中的抗氧化活性成分,研究发现绿原酸、咖啡酸、3,5-二咖啡酰奎宁酸、4,5-二咖啡酰奎宁酸、3,4,5-三咖啡酰奎宁酸、山奈酚、山奈素、桦木酚及阿替匹林 C 是巴西绿蜂胶发挥抗氧化活性的主要物质基础。

参考文献

Aga H, Shibuya T, Sugimoto T, Kurimoto M, Nakajima S (1994) Isolation and identification of antimicrobial compounds in Brazilian propolis[J]. *Biosci Biotech Biochem*, 58:945-946.

Bankova V (2005) Chemical diversity of propolis and the problem of standardization[J]. *Journal of Ethnopharmacology*, 100(1):114-117.

Banskota AH, Tezuka Y, Kadota S (2001) Recent progress in pharmacological research of propolis[J]. *Phytother Research*, 15(7):561-571.

Hayashi K, Komura S, Isaji N, Ohishi N, Yagi K (1999) Isolation of antioxidative compounds from Brazilian propolis: 3,4-dihydroxy-5-prenylcinnamic acid, a novel potent antioxidant[J]. *Chem Pharm Bull*, 47(11):1521.

Kawashima T, Manda S, Uto Y, Ohkubo K, Hori H, Matsumoto K-i, Fukuhara K, Ikota N, Fukuzumi S, Ozawa T (2012) Kinetics and mechanism for the scavenging reaction of the 2, 2-diphenyl-1-picrylhydrazyl radical by synthetic artepillin canalogues [J]. *Bulletin of the Chemical Society of Japan*, 85(8):877-883.

Kumazawa S, Goto H, Hamasaka T. 2004. A new prenylated flavonoid from propolis collected in Okinawa, Japan[J]. *Biosci Biotechnol Biochem*, 68:260-262.

Marcucci MC, Ferreres F, Custódio AR, Ferreira M, Bankova VS, García-Viguera C, Bretz WA (2000) Evaluation of phenolic compounds in Brazilian propolis from different geographic regions[J]. *Z Naturforsch C*, 55(1/2):76-81.

Midorikawa K, Banskota AH, Tezuka Y, Nagaoka T, Matsushige K, Message D, Huertas

AAG, Kadota S (2001) Liquid chromatography-mass spectrometry analysis of propolis [J]. *Phytocheml Anal*, 12(6): 366-373.

Tazawa S, Warashina T, Noro T, Miyase T (1998) Studies on the constituents of Brazilian propolis[J]. *Chem Pharm Bull*, 46: 1477-1479.

Tazawa S, Warashina T, Noro T (1999) Studies on the constituents of Brazilian propolis. II [J]. *Chem Pharm Bull*, 47(10): 1388-1392.

Truchado P, Tourn E, Gallez LM, Moreno DA, Ferreres F, Tomás-Barberán FA (2010) Identification of botanical biomarkers in Argentinean Diplotaxis honeys: flavonoids and glucosinolates[J]. *Journal of Agriculture and Food Chemistry*, 58(24): 12678-12685.

第四篇

蜂胶综合开发利用研究

第十四章　蜂胶中挥发性成分的研究

蜂胶的化学成分复杂,生物学活性广泛。蜂胶中的黄酮类、萜类、酚类等化学成分的存在与协同作用对蜂胶生物学活性的发挥起到积极作用。

近年来,国内外对蜂胶的研究主要集中在非挥发性成分方面,尤其是蜂胶中的黄酮类化合物,其含量的高低通常用以评价蜂胶品质及生物学活性高低。蜂胶的特殊香味被认为有利于蜂胶在蜂巢中发挥防腐、抑菌、驱虫等功效,但目前对于蜂胶中的芳香挥发性成分研究尚少。

目前关于蜂胶挥发性成分的研究还存在以下问题:

(1)挥发性成分的提取效率偏低,尚未报道简单、有效的挥发性成分提取方法;

(2)对不同来源的蜂胶挥发性成分的化学组成及生物学活性尚缺乏有效的比较及评价;

(3)生物学活性方面仅见数篇有关挥发性成分抑菌活性及体外抗氧化活性的报道,尚缺乏体内实验的探讨。

基于以上问题,本研究通过探索蜂胶挥发性成分的较佳制备方法,得到蜂胶挥发性成分,以此为基础,分析比较巴西蜂胶与中国蜂胶中挥发性成分化学组成的差异并对两者体外抑菌及抗氧化活性进行研究,再从整体动物水平评价蜂胶挥发性成分对氧化应激及类焦虑样行为的药理学活性,旨在丰富蜂胶的生物学活性,促进蜂胶产品的多元化及综合利用。

第一节　蜂胶中挥发性成分的提取工艺研究

蜂胶所含挥发性成分主要为在蜂胶中约占10%的芳香挥发油。由于挥发油具有常温下易挥发,能随水蒸气蒸馏且不溶于水的特点,目前用于获得挥发性成分的提取方法主要为水蒸气蒸馏法。蜂胶具有低温硬脆、高温黏稠的质地特异性,因此在常规的水蒸气蒸馏提取过程中,蜂胶在介质水中的分散性差,加热后为熔融态,与蒸汽或是热水的接触面小,难以得到满意的提取效率。曾晞等(2004)研究表明,蜂胶挥发性成分采用水蒸气蒸馏法提取的最高提取率仅为1.32%。目前对蜂胶挥发性成分的提取大部分为实验室少量提取,主要用于成分分析。

在挥发性成分提取方法中,除水蒸气蒸馏提取外,还包括有机溶剂浸提法、超临界流体萃取技术、同时蒸馏-萃取法、固相微萃取法、酶辅助提取法、超声提取法、微波萃取法及半仿生提取法等多种手段(Chen et al.,2001;Gabriele et al.,2009;Pourmortazavi et al.,2007;Schaneberg et al.,2002)。

挥发性成分提取的新技术如微波辅助提取技术、超声波辅助提取技术等备受关注。微波辅助提取技术是近年来新发展起来的一种萃取技术,具有选择性高,快速高效,节能,节省溶剂的优点,特别适合于处理热敏性组分或从天然物质中提取有效成分(Farhat et al.,2010;Phutdhawong et al.,2007),已广泛用于黄酮类、皂普类、多糖、挥发油等活性成分的萃取,亦有研究人员将其应用于蜂胶挥发油的萃取(赵强等,2007b)。

国内外对蜂胶综合开发和深加工研究主要集中在蜂胶的非挥发性成分上如黄酮类化合物、酚类化合物、咖啡酸酯等(Gomez-Caravaca et al.,2006;Popova et al.,2004b;Viuda-Martos et al.,2008)。目前的蜂胶产品主要为蜂胶醇提物制备的胶囊、片剂或酊剂,而关于蜂胶挥发性成分(油)研究较少,但国内外的研究表明,蜂胶挥发性成分中含有丰富的萜类化合物(α-蒎烯、β-芳樟醇、桉叶醇)、苯乙酮、大黄酚等,具有抑菌活性(Bankova et al.,1999b;Melliou et al.,2007;王小平 et al.,2009)。

因此,我们对蜂胶中挥发性成分的提取工艺进行了研究。比较了水蒸气蒸馏法、溶剂常温浸泡法、超声波辅助提取法及微波辅助提取法 4 种提取方式方法对蜂胶挥发性成分的提取工艺、提取得率及挥发性成分性状的影响,并对超声辅助提取工艺及微波辅助提取工艺进行了优化,为蜂胶挥发性成分的高效提取及蜂胶产品的深加工奠定基础。

1 蜂胶挥发性成分的不同提取工艺

1.1 材料与方法

1.1.1 材料与试剂

巴西蜂胶由浙江蜂之语蜂业集团有限公司提供,系该公司从巴西进口的绿色蜂胶;中国蜂胶由杭州天厨蜜源保健品有限公司提供,其产地为山东,胶源植物为杨树;乙醇、乙酸乙酯、石油醚(60~90℃)、乙醚、氯仿、无水硫酸钠均为分析纯。

1.1.2 仪器与设备

旋转蒸发仪 RE-52AA,上海亚荣生化仪器厂;循环水式多用真空泵 SHB-III,郑州长城科工贸有限公司;NJL07-3 型实验专用微波炉,南京杰全微波设备有限公司;KQ-300DE 型数控超声提取器,昆山市超声仪器有限公司;数显智能控温磁力搅拌器,巩义市予华仪器有限责任公司;PB602 电子秤;抽滤装置、真空泵、冷柜、粉碎机、分离筛、分液漏斗等。

1.1.3 材料预处理

蜂胶原料经预选分级,去除杂质,然后放入冷藏柜,置于-18℃冷冻 4h 以上,取出后迅速放入粉碎机粉碎,过 20 目筛备用。

1.1.4 水蒸气蒸馏提取

称量过筛蜂胶原料 50g 于水蒸气蒸馏器中,加入 500mL 超纯水(适时补足),参照《中华人民共和国药典》(2005 年版)方法常压蒸馏,提取至无油状物滴出,提取时间分别选择 3h 及 6h。馏分冷凝后乙醚接收,分液后上面的油层用无水硫酸钠干燥,脱溶即得蜂胶挥发性成分水蒸气蒸馏提取物,同样条件实验重复 3 次,4℃冰箱保存。

1.1.5 常温有机溶剂提取

准确称量过筛蜂胶原料 10g,置于 250mL 圆底烧瓶中,加入 100mL 的溶剂(石油醚、乙醚),常温 28℃下搅拌提取 12h,提取 2 次,过滤,然后用相应溶剂洗涤残渣,合并滤液减压浓缩至 100mL 左右,置于-18℃冰柜中反复冷冻过滤,达到无蜡质析出为止。滤液低于 50℃旋转蒸发脱溶,得到蜂胶挥发性成分,同样条件实验重复 3 次。

1.1.6 超声波辅助提取

准确称取过筛蜂胶原料10g,置于250mL圆底烧瓶中,加入100mL的石油醚,将烧瓶放在超声清洗器的槽中,预设温度28℃,时间设定为30min,进行超声萃取。提取2次,过滤,石油醚洗涤残渣,合并滤液减压浓缩至100mL左右,置于－18℃冰柜中反复冷冻过滤,达到无蜡质析出为止。滤液低于50℃旋转蒸发脱溶,得到蜂胶挥发性成分,4℃冰箱保存。

1.1.7 微波辅助提取

以过筛蜂胶为原料,以石油醚为溶剂,按料液比1∶10投料,装入萃取罐中,设定萃取温度为30℃,萃取时间为30min,进行实验。萃取完毕,取出萃取罐,用风机冷却至室温,将溶液减压抽滤;再将不溶物转入罐中,加入新鲜的石油醚,按同样的条件重新提取一次。然后用石油醚洗涤残渣,滤液合并于一个250mL烧瓶中,在－18℃下反复冷冻过滤多次达到无蜡质析出为止,用旋转蒸发仪以低于50℃蒸去溶剂,得到蜂胶挥发性成分,4℃冰箱保存。

1.1.8 提取率的计算与统计分析

提取率(%)＝蜂胶挥发性成分质量(g)/蜂胶原料质量(g)×100%

提取率以均值表示,两两比较采用t检验,显著性差异水平为$P<0.05$。

1.2 结果与分析

1.2.1 水蒸气蒸馏法对蜂胶挥发性成分提取率的影响

以水蒸气蒸馏法对中国蜂胶和巴西蜂胶中的挥发性成分进行了提取,结果见表14.1。如表14.1所示,水蒸气蒸馏法提取3h得到中国蜂胶中挥发性成分提取率为0.74%,继续提取,至6h提取率为0.79%,两者无显著性差异,随提取时间延长,无油状物滴出。所得中国蜂胶挥发性成分为浅黄色油状物,具有浓郁的蜂胶味。巴西蜂胶挥发性成分水蒸气蒸馏提取率3h为0.87%,6h为0.95%,两时间点间提取率亦无显著性差异。所得巴西蜂胶挥发性成分为浅黄色油状物,具有浓郁的蜂胶味。

表14.1 水蒸气蒸馏法对中国蜂胶和巴西蜂胶中挥发性成分提取率的影响

Table 4.1 Yields of essential oil of Brazilian and Chinese propolis by hydrodistillation

样品来源	生药质量/g	3h提取率/%	6h提取率/%	性状
中国蜂胶	50.07±0.05	0.74±0.09	0.79±0.04	浅黄色、油状
巴西蜂胶	50.04±0.02	0.87±0.15	0.95±0.08	浅黄色、油状

水蒸气蒸馏法提取得到巴西蜂胶中挥发性成分略高于中国蜂胶,两种样品来源蜂胶挥发性成分均呈油状且具有蜂胶特有的香气。提取过程中提取液始终保持微沸状态,蜂胶在高温下变软并聚集结块,可能影响其内部挥发性成分的溢出,降低提取效率。

1.2.2 常温有机溶剂提取法对蜂胶挥发性成分提取率的影响

在常温下,以乙醚及石油醚作为溶剂,提取中国蜂胶和巴西蜂胶中的挥发性成分,结果如表14.2所示。料液比1∶10,常温28℃下搅拌提取12h,以乙醚为溶剂进行提取,得到中国蜂胶中挥发性成分占比为2.35%,提取物呈油状、棕黄色,具有蜂胶特有的气味及刺激性气味;以石油醚为溶剂,提取得到中国蜂胶中挥发性成分占比为2.51%,提取物色泽金黄,呈油状,具有浓郁的蜂胶味。以乙醚为溶剂,提取得到巴西蜂胶中挥发性成分占比为2.99%,

提取物呈深黄色、油状液体,具有蜂胶所特有的气味并混有刺激性气味;以石油醚为溶剂,提取得到巴西蜂胶中挥发性成分占比为 3.06%,提取物为黄色油状液体,有浓郁的蜂胶味。

表 14.2 常温有机溶剂提取法对中国蜂胶和巴西蜂胶中挥发性成分提取率的影响
Table 14.2 Yields of essential oil of Brazilian and Chinese propolis by organic solvents extraction

样品来源	提取时间/h	提取温度/℃	料液比(v/v)	提取溶剂	提取率/%	性状
中国蜂胶	12	28	1:10	乙醚	2.01±0.14	棕黄色、油状
				石油醚	2.51±0.18*	金黄色、油状
巴西蜂胶	12	28	1:10	乙醚	2.99±0.18	深黄色、油状
				石油醚	3.06±0.25△	黄色、油状

* 中国蜂胶挥发性成分石油醚提取 vs 乙醚提取,$P<0.05$
△ 巴西蜂胶挥发性成分石油醚提取 vs 乙醚提取,$P<0.05$

对于中国蜂胶和巴西蜂胶中的挥发性成分,石油醚提取效率显著高于乙醚。同一样品的乙醚提取物色泽较深,提取物常含有刺激性气味。考虑到挥发性成分提取中对溶剂要求具备无色无味、常温下不易大量挥发、选择性强、不溶于水、化学性质为惰性等特点(罗金岳 et al.,2005),考虑两种溶剂的性质及提取效率,在微波辅助及超声辅助提取中选择石油醚作为提取溶剂。

1.2.3 超声波及微波辅助提取对蜂胶挥发性成分提取率的影响

以石油醚为溶剂,采用超声波及微波辅助法对中国蜂胶和巴西蜂胶中的挥发性成分进行提取,结果见表 14.3。如表 14.3 所示,超声波辅助提取得到中国蜂胶中挥发性成分提取率为 6.87%,提出物为金黄色油状物,有浓郁的蜂胶味;巴西蜂胶挥发性成分提取率为 7.11%,提出物为黄色油状物,有浓郁的蜂胶味。微波辅助提取得到中国蜂胶中挥发性成分提取率为 8.27%,提出物亦为金黄色油状物,巴西蜂胶挥发性成分提取率为 8.98%,提出物为黄色油状物。

表 14.3 超声波及微波辅助提取对中国蜂胶和巴西蜂胶中挥发性成分提取率的影响
Table 14.3 Yields of essential oil of Brazilian and Chinese propolis by ultrasonic and microwave assisted extraction

提取方法	提取时间/min	提取温度/℃	提取功率	料液比(v/v)	样品来源	提取率/%	性状
超声波辅助	30	28	100%	1:10	中国蜂胶	6.87±0.28	金黄色、油状
					巴西蜂胶	7.11±0.22	黄色、油状
微波辅助	30	30	2.45GHz	1:10	中国蜂胶	8.80±0.75**	金黄色、油状
					巴西蜂胶	9.32±0.73△△	黄色、油状

** 中国蜂胶挥发性成分超声辅助提取 vs 微波辅助提取,$P<0.01$
△△ 巴西蜂胶挥发性成分超声辅助提取 vs 微波辅助提取,$P<0.01$

采用超声波辅助提取与微波辅助提取得到中国蜂胶和巴西蜂胶中的挥发性成分性状相似,但在设定条件下,微波辅助法提取效率显著高于超声波辅助提取法。较常温溶剂提取法及水蒸气蒸馏提取法,微波辅助及超声波辅助提取显著提高了提取效率。为探索蜂胶中挥发性成分的较佳提取工艺,接下来对超声辅助提取及微波辅助提取工艺条件进行优化。

2 均匀设计法优化超声波辅助提取蜂胶中的挥发性成分

通过比较蜂胶中挥发性成分的水蒸气蒸馏法提取、常温有机溶剂法提取及微波与超声波辅助法提取所得的产品得率及性质,发现超声辅助及微波辅助提取可有效提高蜂胶中挥发性成分的得率,提取所得蜂胶挥发性成分性状较佳。

鉴于超声辅助提取技术在挥发性成分提取当中具有短时、高效等优势(Chizzola et al., 2008;Wang et al., 2009),本节利用超声提取技术对蜂胶中挥发性成分进行提取,优化了超声提取制备蜂胶挥发性成分的工艺参数,对蜂胶挥发性成分的进一步开发提供参考。

2.1 实验材料

2.1.1 材料

巴西蜂胶由浙江蜂之语蜂业集团有限公司提供,系该公司从巴西进口的绿色蜂胶;中国蜂胶由杭州天厨蜜源保健品有限公司提供,其产地为山东,胶源植物为杨树。将蜂胶在-18℃冷冻数日,取出后快速粉碎,过 20 目筛。

2.1.2 主要试剂

石油醚、无水乙醇、正己烷等均为国产分析纯试剂。

2.1.3 主要仪器与设备

旋转蒸发仪 RE-52AA,上海亚荣生化仪器厂;循环水式多用真空泵 SHB-Ⅲ,郑州长城科工贸有限公司;KQ-300DE 型数控超声提取器,昆山市超声仪器有限公司。

2.2 实验方法

2.2.1 蜂胶中挥发性成分的制备

2.2.1.1 蜂胶挥发性成分的提取

蜂胶在-20℃冷冻数日,粉碎,过 20 目筛备用。准确称取 20g 蜂胶粉末,置于 250mL 三角瓶中,在设定的条件下,以石油醚为溶剂进行提取,过滤,然后用石油醚洗涤残渣,合并滤液减压浓缩到 100mL 左右,置于-18℃冰柜中反复冷冻、过滤,去除析出蜡质。将脱蜡处理的滤液于 40℃旋转蒸发脱溶,得到蜂胶挥发性成分粗提物。将蜂胶挥发性成分粗提物与无水乙醇按质量比 1:15 充分混匀,于-20℃反复冷析过滤,达到无蜡质析出为止。采用旋转蒸发仪去除滤液中的溶剂得到蜂胶挥发性成分。同一条件重复 3 次取平均值。

2.2.1.2 提取工艺设计

以中国蜂胶为原料,用均匀设计法优化超声波辅助萃取蜂胶中挥发性成分的萃取工艺,以石油醚为溶剂,选用对提取率有影响的 3 个因素,即提取温度 $X1(30\sim60℃)$、石油醚加入的体积 $X2(30\sim150mL)$(按蜂胶每次加入 10g 计算)、提取时间 $X3(10\sim70min)$,每个因素各取 7 个水平(见表 14.4),按均匀设计表 $U\ 7^*(7^4)$ 确定实验方案。

表 14.4 均匀设计水平、因素编码表

Table 14.4 Factors and levels of uniform design experiment

水平	因素		
	X_1(温度/℃)	X_2(料液比)	X_3(时间/h)
1	30	1∶3	10
2	35	1∶5	20
3	40	1∶7	30
4	45	1∶9	40
5	50	1∶11	50
6	55	1∶13	60
7	60	1∶15	70

2.2.2 蜂胶挥发性成分提取率的计算

提取率(%)＝蜂胶挥发性成分质量(g)/蜂胶原料质量(g)×100%

2.3 结果与分析

2.3.1 超声波辅助提取蜂胶中挥发性成分的工艺优化

对超声波辅助提取蜂胶中挥发性成分的工艺参数：提取温度 X_1(30~60℃)，石油醚加入的体积 X_2(30~150mL)(按蜂胶每次加入 10g 计算)，提取时间 X_3(10~70min)，按表 14.4 均匀设计水平、因素编码表进行优化，各次试验结果见表 14.5。

表 14.5 均匀设计用表和实验结果

Table 14.5 Results of supersonic wave-assisted extraction with uniform design

实验号	X_1(温度/℃)	X_2(料液比)	X_3(时间/h)	提取率(%)
1	40	1∶11	70	7.7
2	55	1∶5	60	6.8
3	30	1∶15	50	8.6
4	45	1∶9	40	7.8
5	60	1∶3	30	6.6
6	35	1∶13	20	7.4
7	50	1∶7	10	6.5

实验结果以 DPS 软件数据处理系统进行二次多项式逐步回归分析，并对该模型进行显著性检验，结果见表 14.6。

得到如下的模型方程：

$$Y = 4.99 + 0.06 \times X_3 - 0.0009 \times X_3 \times X_3 + 0.0022 \times X_1 \times X_2 - 0.0023 \times X_2 \times X_3$$

模型的复相关系数 $R=0.9984$，F 值=153.93，$P=0.0065(<0.05)$，剩余标准差 $S=0.075$，决定系数 $R^2=0.99676$，调整后相关系数 $Ra=0.9951$，可知模型与数据吻合，说明此方程能很好地拟合超声辅助萃取蜂胶挥发性成分的过程。

表 14.6 二次多项式逐步回归法处理的数据结果

Table 14.6 Results of quadratic polynomial regression

因素	偏相关	t 检验值	p 值
$X3$	0.9829	7.5523	0.0048
$X3 \cdot X3$	−0.9907	10.2703	0.0020
$X1 \cdot X2$	0.9383	3.8361	0.0312
$X1 \cdot X3$	0.9814	7.2217	0.0055

从表 14.6 各变量 P 值的大小显著性检验中可知,对提取率的影响的大小为:$X3 \cdot X3 > X3 > X1 \cdot X3 > X1 \cdot X2$。其中 $X3$,$X1 \cdot X2$,$X1 \cdot X3$ 与 Y 呈正相关,$X1 \cdot X3$ 与 Y 呈负相关,说明影响提取率的三因素并非越大提取效果越好,如时间延长提取效果反而不会提高,这是由于超声提取法是利用超声波增大物质分子运动频率和速度,增加溶剂穿透力,提高药物溶出速度和溶出次数,从而缩短提取时间。在短时间内萃取物溶出较多,所以提取率上升较快,但当溶解度达到饱和时,挥发油分子不再被溶解,提取率也就不再有明显提高。且随着超声提取时间的延长,挥发性成分向外溢出的可能性增加。同时从回归的结果可知,因素之间存在交互作用,采用均匀设计法可显著减少实验次数,是一种很好的寻优方案。

2.3.2 验证实验

根据回归分析,再用非线性规划方法进行优化,其最优组合为:$X1=59℃$,$X2=15mL$(按蜂胶 1g 的比例),$X3=62min$,优化条件下最高提取率为 9.54%,比 7 组实验数据都高,说明此优化结果对实验指导有一定的价值。根据均匀设计所确定的最佳工艺条件,取粉碎过筛的中国蜂胶和巴西蜂胶样品,进行蜂胶中挥发性成分的超声辅助提取,实验 3 次,取平均值,得中国蜂胶中挥发性成分提取率为 8.70%±0.77%,巴西蜂胶中挥发性成分提取率为 8.75%±0.76%。

3 正交设计法优化微波辅助提取蜂胶中挥发性成分

微波萃取是利用微波能来提高萃取率的一种新技术,能够选择性地加热组分,从而使得被萃取物质从基体或体系中分离,进入到介电常数较小、微波吸收能力相对差的萃取剂中(Flamini et al., 2007)。鉴于微波萃取技术在挥发性成分提取当中具有设备简单、短时、高效等优势,本节利用微波提取技术对蜂胶挥发性成分进行提取,优化了微波提取制备蜂胶挥发性成分的工艺参数,为深度研究和开发高附加值蜂胶制品提供依据。

3.1 实验材料

3.1.1 材料

巴西蜂胶由浙江蜂之语蜂业集团有限公司提供,系该公司从巴西进口的绿色蜂胶;中国蜂胶由杭州天厨蜜源保健品有限公司提供,其产地为山东,胶源植物为杨树。将蜂胶在 −18℃冷冻数日,取出后快速粉碎,过 20 目筛。

3.1.2 主要试剂

石油醚、无水乙醇、正己烷等均为国产分析纯试剂。

3.1.3 主要仪器与设备

旋转蒸发仪 RE-52AA,上海亚荣生化仪器厂;循环水式多用真空泵 SHB-III,郑州长城科工贸有限公司;NJL07-3型实验专用微波炉,南京杰全微波设备有限公司。

3.2 实验方法

3.2.1 蜂胶挥发性成分的提取

蜂胶在−20℃冷冻数日,粉碎,过20目筛备用。准确称取20g蜂胶粉末,置于微波提取器中,在设定的条件下,以石油醚为溶剂进行提取,过滤,然后用石油醚洗涤残渣,合并滤液减压浓缩至100mL左右,置于−18℃冰柜中反复冷冻、过滤去除析出蜡质。将脱蜡处理的滤液于40℃旋转蒸发脱溶,得到蜂胶挥发性成分粗提物。将蜂胶挥发性成分粗提物与无水乙醇按质量比1∶15充分混匀,于−20℃反复冷析过滤,达到无蜡质析出为止。采用旋转蒸发仪去除滤液中的溶剂得到蜂胶挥发性成分。同一条件重复3次取平均值。

3.2.2 提取工艺设计

首先选择微波处理时间(A)、微波处理温度(B)、料液比(C)、提取次数(D)这四个影响提取的因素做单因素实验,然后根据实验结果选择影响较大的因素和合适的水平做正交实验。

3.3 结果与分析

3.3.1 单因素实验

3.3.1.1 微波处理时间对蜂胶挥发性成分得率的影响

蜂胶原料的总质量约为20g,以石油醚为溶剂,在温度为30℃,在料液比为1∶10,提取1次的条件下,选择微波处理时间分别为5min、10min、15min、20min、25min和30min。

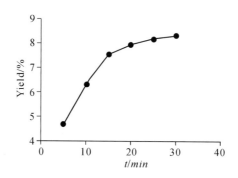

图14.1 不同微波萃取时间对蜂胶中挥发性成分得率的影响

Fig 14.1 Effect of microwave assisted extration time on yield of propolis essential oil

图14.1为微波萃取时间对蜂胶挥发性成分得率的影响曲线,可以看出在萃取时间5~30min内,随着萃取时间的延长挥发性成分得率相应增加,且增加的趋势变缓。一般而言,随着萃取时间的增加,得率也相应增加,但当大部分的油被萃取时,随着萃取时间的增加,得率增加甚少。再者随着微波辐射时间的延长,萃取物进一步破裂,溶解的杂质也会相应增多,因此,微波萃取时间不宜过长。从提高效率,节省能源综合考虑,较佳萃取时间范围为15~25min。

3.3.1.2 微波处理温度对蜂胶挥发性成分得率的影响

蜂胶原料的总质量约为20g,以石油醚为溶剂,在微波处理时间为20min,在料液比为1∶10,提取1次的条件下,选择微波处理温度分别为25℃、30℃、35℃、40℃和45℃。

图14.2为微波萃取温度对蜂胶挥发性成分得率的影响曲线,可以看出在萃取温度25~

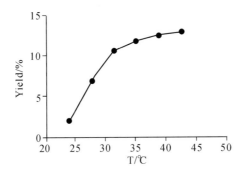

图 14.2 不同微波萃取温度对蜂胶中挥发性成分得率的影响

Fig 14.2 Effect of microwave assisted extration temperature on yield of propolis essential oil

35℃之间,随着萃取温度的增加,蜂胶挥发性成分得率显著增加($P<0.05$),而当萃取温度超过 35℃后,得率则基本保持稳定而稍有降低的趋势。这是因为微波萃取温度对微波萃取过程的影响有两方面:一方面,随着萃取温度的升高,高频电磁波穿透萃取介质到达物料内部,使物料内部的温度迅速上升,从而使物料组织结构破裂,其内的有效成分自由流出,并在较低的温度下溶解于萃取介质中,在一定的温度范围内,增加有效成分提取得率;另一方面,随着萃取物在萃取介质中的饱和,温度的升高不利于热敏性物质的萃取,也增加能源消耗。因此,本研究中选择的较佳工艺萃取温度范围为 30~40℃。

3.3.1.3 料液比对蜂胶挥发性成分得率的影响

蜂胶原料的总质量约为 20g,以石油醚为溶剂,在微波处理时间为 20min,微波处理温度为 35℃,提取 1 次的条件下,选择料液比为 1:5、1:10、1:15、1:20 和 1:25。

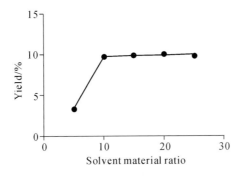

图 14.3 不同料液比对蜂胶中挥发性成分得率的影响

Fig 14.3 Effect of solvent material ratio on yield of propolis essential oil

图 14.3 为不同料液比对蜂胶挥发性成分得率的影响曲线,可以看出随着料液比的增加,蜂胶挥发性成分得率显著增加($P<0.05$),而当料液比超过 1:10 后,得率则基本保持稳定。可见,有效成分在提取溶剂中溶解未达饱和的情况下,增加溶剂量对提高挥发性成分得率没有帮助。从经济效益考虑,本研究中选取的较佳料液比范围为 1:10~1:20。

3.3.1.4 提取次数对蜂胶挥发性成分得率的影响

蜂胶原料的总质量约为 20g,以石油醚为溶剂,在微波处理时间为 20min,微波处理温度为 35℃,料液比为 1:10 的条件下,分别提取 1、2、3、4、5 次。

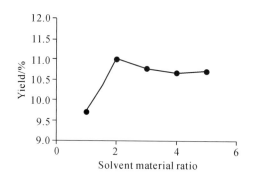

图 14.4 不同提取次数对蜂胶中挥发性成分得率的影响

Fig 14.4 Effect of extraction number on yield of propolis essential oil

图 14.4 为不同提取次数比对蜂胶挥发性成分得率的影响曲线,可以看出随着提取次数的增加,蜂胶挥发性成分得率显著增加($P<0.05$),而当料液比超过 2 次后,得率则基本保持稳定而稍有降低的趋势。这可能是因为增加提取次数有利于挥发性成分的有效提取,但提取次数过多会导致在提取操作过程中挥发性成分损失的增加。从操作方便性考虑,本研究中选取的较佳提取次数为 2~4 次。

3.3.2 正交试验

根据单因素实验的结果,拟定 4 因素 3 水平,按 L9(3^4)正交实验设计表(见表 14.7)做蜂胶挥发性成分提取的正交实验,结果见表 14.8。

表 14.7 正交实验因素水平表

Table 14.7 Factors and levels of orthogonal experiment

水平	A 微波处理时间/min	B 微波处理温度/℃	C 料液比	D 提取次数
1	15	30	1∶10	2
2	20	35	1∶15	3
3	25	40	1∶20	4

表 14.8 正交试验结果与分析

Table 14.8 Results and analysis of orthogonal experiment

试验号	A 微波处理时间/min	B 微波处理温度/℃	C 料液比	D 提取次数	蜂胶挥发性成分得率/%
1	1	1	1	1	9.85
2	1	2	2	2	10.35
3	1	3	3	3	8.42
4	2	1	2	3	10.22
5	2	2	3	1	11.13
6	2	3	1	2	8.26

续表

试验号	A 微波处理时间/min	B 微波处理温度/℃	C 料液比	D 提取次数	蜂胶挥发性成分得率/%
7	3	1	3	2	10.10
8	3	2	1	3	10.96
9	3	3	2	1	8.18
k_1	9.540	10.057	9.690	9.720	
k_2	9.870	10.813	9.583	9.570	
k_3	9.747	8.287	9.883	9.867	
极差	0.330	2.526	0.300	0.297	
优水平	A2	B2	C3	D3	

表 14.9　正交试验方差分析表

Table 14.9　Variance analysis of orthogonal experiment

因素	偏差平方和	自由度	F 比	F 临界值	显著性
A	0.167	2	1.265	19.000	$P>0.05$
B	10.089	2	76.432	19.000	$P<0.05$
C	0.139	2	1.053	19.000	$P>0.05$
D(误差)	0.132	2	1.000	19.000	$P>0.05$

$F_{0.05}(2,2)=19.000$

蜂胶挥发性成分微波辅助萃取正交试验结果与分析见表 14.8、表 14.9。对正交试验数据进行极差和方差分析,结果表明,在试验条件范围内,各因素对蜂胶挥发性成分得率的影响程度依次为 B>A>C>D,因素 B 对蜂胶挥发性成分得率有显著影响($P<0.05$),但因素 A、C、D 对蜂胶挥发性成分的萃取影响不显著。各因素的优组合为 A2B2C3D3,即微波处理时间为 20min,微波处理温度为 35℃,料液比 1∶20,提取 3 次。综合考虑各因素对蜂胶挥发性成分得率的影响,认为 B 因素即微波处理温度为对蜂胶挥发性成分得率影响最为重要,A、C、D 因素对挥发性成分的获得影响不大,所以从节省时间、试剂,提高效率角度考虑,选择 A1、C1、D1,因此选择最佳提取工艺为 A1B2C1D1,即在微波处理温度为 35℃时处理 15min,料液比为 1∶10,提取 2 次。

3.3.3　验证实验结果

精确称取巴西蜂胶和中国蜂胶各 3 份,每份 20g,重复最佳实验条件 A1B2C1D1,即微波处理时间 15min、微波处理温度 35℃、料液比为 1∶10,提取 2 次,进行实验。结果获得巴西蜂胶挥发性成分平均值为 11.21%,此结果与正交表中最好的实验结果基本一致,故本研究确定的工艺较为合理。同样工艺条件提取得到中国蜂胶挥发性成分平均值为 9.73%,两地蜂胶挥发性成分在含量、色泽及气味上均存在差异。

小 结

本研究对蜂胶中的挥发性成分采用 4 种不同的方法进行提取,考察其提取率和提取物性状,通过比较研究得出如下结论:

(1)由于蜂胶的质地特异性,水蒸气蒸馏提取使其变黏聚结,其中的挥发性成分难以充分溢出,提取时间长,得率低;

(2)常温有机溶剂提取蜂胶中的挥发性成分,溶剂对挥发性物质的提取率有影响,采用石油醚提取较乙醇提取得率高,提取物性状好;

(3)微波及超声波辅助萃取可提高蜂胶中挥发性成分的提取率,提取物性状较好,微波辅助提取较超声波辅助萃取在设定条件下得率高;

(4)利用均匀设计法对超声波辅助萃取工艺条件进行优化,得到最优组合为:提取温度 59℃、料液比 1∶15、提取时间 62min,在该条件下进行验证实验得到中国蜂胶和巴西蜂胶中的挥发性成分提取率分别为 8.70% 和 8.75%;

(5)利用正交设计对微波辅助萃取工艺条件进行优化,其萃取较佳工艺条件为微波处理温度为 35℃,微波处理 15min,料液比为 1∶10,提取 2 次,在该条件下进行验证实验得到中国蜂胶和巴西蜂胶中的挥发性成分提取率分别为 9.73% 和 11.21%。

第二节 巴西蜂胶与中国蜂胶中挥发性成分的分析与比较

近年来的研究表明,蜂胶中存在的黄酮类、萜类、酚类等化学成分(Marcucci,1995;Sahinler et al.,2005)在蜂胶的护肝、抗肿瘤、抑菌、抗炎、抗氧化等多种生物学及药理学活性上(Banskota et al.,2001;Burdock,1998;Castaldo et al.,2002)发挥了积极作用。根据蜜蜂所采集的胶源植物不同,各地蜂胶在色泽、香气特征及化学成分上存在很大的差别,特别是来自温带地区的蜂胶和热带地区的蜂胶化学成分相差很大,但有多篇文献报道来自温带及热带地区的蜂胶在抗菌、抗真菌、清除自由基及驱虫方面具有相似的活性(Bankova et al.,2000;Banskota et al.,2000;Kujumgiev et al.,1999;Popova et al.,2004b;Salomao et al.,2004)。目前研究较多的是蜂胶中的非挥发性成分,尤其是其中的黄酮类化合物,通常被作为蜂胶品质及生物活性高低的一大指标进行评价(Popova et al.,2004a;Viuda-Martos et al.,2008)。而温带地区蜂胶及热带地区蜂胶在非挥发性成分的组成上相差甚远,尤其是热带地区的巴西蜂胶,黄酮类化合物含量较低(Daugsch et al.,2008;Park et al.,2002;Silva et al.,2008)。因此探讨温带地区及热带地区蜂胶中挥发性成分的含量及组成,对蜂胶生物学活性的全面评价及不同地区蜂胶化学成分标准的制定具有重要意义。

本研究对利用 GC-MS 技术对来自温带的中国华北地区蜂胶和来自热带的南美洲巴西蜂胶的挥发性成分进行了分析,进一步比较其在含量及化学组成上的差异,以期为蜂胶品质鉴定、蜂胶生物学活性的深入研究及蜂胶产品的开发利用提供依据。

1 材料与方法

1.1 材料与试剂

巴西蜂胶由浙江蜂之语蜂业集团有限公司提供,系该公司从巴西进口的绿色蜂胶;中国蜂胶由杭州天厨蜜源保健品有限公司提供,其产地为山东,胶源植物为杨树;石油醚、无水乙醇、正己烷等均为分析纯试剂。

1.2 主要仪器与设备

旋转蒸发仪 RE-52AA,上海亚荣生化仪器厂;循环水式多用真空泵 SHB-Ⅲ,郑州长城科工贸有限公司;NJL07-3 型实验专用微波炉,南京杰全微波设备有限公司;GC-MS 联用仪 6890N/5975B,安捷伦科技有限公司;HP-5MS 毛细管柱(30m×I.D. 0.25mm×Film 0.25μm)。

1.3 实验方法

1.3.1 蜂胶挥发性成分的提取

蜂胶在−20℃冷冻数日,粉碎,过 20 目筛备用。准确称取 20g 蜂胶粉末,置于微波提取器中,在以石油醚为溶剂,微波处理温度为 35℃,微波处理 15min,料液比为 1∶10,提取 2 次的条件下进行提取,提取液过滤,然用石油醚洗涤残渣,合并滤液减压浓缩至 100mL 左右,置于−18℃冰箱中反复冷冻、过滤去除析出蜡质。将脱蜡处理的滤液于 40℃旋转蒸发脱溶,得到蜂胶挥发性成分粗提物。将蜂胶挥发性成分粗提物与无水乙醇按质量比 1∶15 充分混匀,于−20℃反复冷析过滤,达到无蜡质析出为止。采用旋转蒸发仪去除滤液中的溶剂得到蜂胶挥发性成分,计算得率。同一条件重复 3 次,取平均值。

1.3.2 GC-MS 分析

样品溶于正己烷中过 0.45μm 滤膜后进行 GC-MS 分析。GC 条件:色谱柱为 HP-5MS 毛细管柱(30m×I.D. 0.25mm×Film 0.25μm);载气:氦气,1mL/min;进样口温度 250℃;分流进样量:1.0μL;分流比 20∶1;柱温升温程序:初始 80℃,保持 5min,以 5℃/min 速度升温至 185℃,保持 5min,Post:270℃,2min。

MS 条件:电离方式为 EI;离子源温度 230℃;四极杆温度 150℃;接口温度 280℃;电子能量 70eV;质量扫描范围为 30~500amu。

2 结果与分析

蜂胶挥发性成分微波辅助提取得率及性状见表 14.10。精确吸取供试品溶液 1μL,注入气相色谱仪中,对供试品溶液进行 GC-MS 分析鉴定,在确定各组分质谱数据和扫描峰号后,经 NIST/WILEY 检索数据库进行检索并根据文献进行谱图解析,采用峰面积归一化法计算解析组分的百分含量。中国蜂胶和巴西蜂胶挥发性成分中分别分离鉴定出 47 和 50 个化合物。

表 14.10 中国蜂胶与巴西蜂胶挥发性成分得率

Table 14.10 The yields of propolis volatiles from Brazilian and Chinese propolis

产地	药材质量/g	挥发性成分质量/g	挥发性成分得率/%	色泽	气味
中国	20.0011±0.0008	1.9469±0.0368	9.73±0.18	棕黄色	浓郁的芳香味
巴西	20.0019±0.0005	2.2422±0.0149	11.21±0.07**	金黄色	浓郁的芳香味,微辛辣

t 检验:* $P<0.05$,** $P<0.01$;$n=3$。

由表 14.10 可见,由微波辅助提取技术得到的蜂胶挥发性成分,中国蜂胶挥发性成分得率为 9.73%,巴西蜂胶挥发性成分得率为 11.21%,两者在挥发性成分含量上存在显著性差异。巴西蜂胶与中国蜂胶挥发性成分在色泽与气味上也不同。

表 14.11 中国蜂胶及巴西蜂胶中挥发性成分化学组成及其相对含量

Table 14.11 Chemical composition and their relative content of volatile constituents in Brazilian and Chinese propolis

序号	保留时间/min	化合物	分子式	相对百分含量/% 巴西	相对百分含量/% 中国
1	6.183	Benzyl Alcohol 苯甲醇	C_7H_8O	0.53	2.09
2	8.582	Phenylethyl Alcohol 苯乙醇	$C_8H_{10}O$	1.25	2.05
3	10.402	Benzenecarboxylic acid 苯甲酸	$C_7H_6O_2$	3.60	1.41
4	12.126	Benzoin ethylether 安息香乙醚	$C_{16}H_{16}O_2$	—	0.14
5	12.925	Phenethyl acetate 乙酸苯乙酯	$C_{10}H_{12}O_2$	0.28	—
6	12.931	2-Phenylethylpropinate 丙酸苯乙酯	$C_{11}H_{14}O_2$	—	0.94
7	13.341	trans-Cinnamaldehyde 反式肉桂醛	C_9H_8O	0.20	0.09
8	13.584	4-Ethyl-2-methoxyphenol 4-乙基愈创木酚	$C_9H_{12}O_2$	—	0.10
9	13.692	4-Methoxymandelic acid 对甲氧基苦杏仁酸	$C_9H_{10}O_4$	—	0.05
10	14.335	Cinnamyl alcohol 肉桂醇	$C_9H_{10}O$	1.71	0.66
11	15.010	4-Hydroxy-3-methoxystyrene 对乙烯基愈创木酚	$C_9H_{10}O_2$	—	0.10
12	15.167	3-Phenylpropionic acid 氢化肉桂酸	$C_9H_{10}O_2$	0.33	2.07
13	15.534	Ethyl 3-phenylpropionate 氢化肉桂酸乙酯	$C_{11}H_{14}O_2$	0.33	0.25
14	15.648	α-Longipinene 长叶蒎烯	$C_{15}H_{24}$	1.25	—
15	16.209	Longipinene 环长叶烯	$C_{15}H_{24}$	1.44	—
16	16.334	α-Cubebene 荜澄茄苦素	$C_{15}H_{24}$	0.42	—
17	16.480	Methyl cinnamate 肉桂酸甲酯	$C_{10}H_{10}O_2$	—	0.20
18	16.733	Sativen 蒜头素	$C_{15}H_{24}$	0.90	—
19	16.793	Phenethyl isobutyrate 异丁酸苯乙酯	$C_{12}H_{16}O_2$	—	0.13
20	16.890	Vanillin 香兰素	$C_8H_8O_3$	—	0.11
21	16.987	Alloaromadendrene 别香橙烯	$C_{15}H_{24}$	0.25	—
22	17.036	Methyl eugenol 甲基丁香酚	$C_{11}H_{14}O_2$	0.14	—
23	17.155	Alloaromadenerene 香树烯	$C_{15}H_{24}$	—	0.05
24	17.166	Longipinene 长叶烯	$C_{15}H_{24}$	24.89	—
25	17.333	α-Cedrene α-柏木烯	$C_{15}H_{24}$	—	0.57

续表

序号	保留时间/min	化合物	分子式	相对百分含量/% 巴西	相对百分含量/% 中国
26	17.506	β-Caryophyllene β-丁子香烯	$C_{15}H_{24}$	5.27	—
27	17.549	β-Cedrene β-柏木烯	$C_{15}H_{24}$	—	0.42
28	17.706	Cinnamic acid 肉桂酸	$C_9H_8O_2$	2.05	19.30
29	18.003	Calarene 白菖烯	$C_{14}H_{22}$	0.23	—
30	18.073	Cinnamyl acetate 醋酸肉桂酯	$C_{11}H_{12}O_2$	0.30	—
31	18.273	Humulen 葎草烯	$C_{15}H_{24}$	0.32	—
32	18.386	α-Caryophyllene α-丁子香烯	$C_{15}H_{24}$	1.42	—
33	18.532	Valencene 瓦伦烯（朱栾倍半萜）	$C_{15}H_{24}$	—	0.72
34	18.576	Aromadendrene 香橙烯	$C_{15}H_{24}$	0.40	—
35	18.932	γ-Selinene 芹子烯	$C_{15}H_{24}$	—	3.32
36	18.997	Himachalene 雪松烯	$C_{15}H_{24}$	1.21	0.99
37	19.078	α-Curcumene 姜黄烯	$C_{15}H_{22}$	0.66	2.21
38	19.434	δ-Guaiene 愈创木烯	$C_{15}H_{24}$	0.34	2.16
39	19.553	α-Muurolene 衣兰油烯	$C_{15}H_{24}$	0.79	1.15
40	19.721	B-Bisabolene (甜)没药烯	$C_{15}H_{24}$	0.24	—
41	19.894	γ-Cadinene 杜松烯	$C_{15}H_{24}$	1.12	—
42	19.899	4,9-Cadinadiene	$C_{15}H_{24}$	—	1.31
43	20.110	δ-Cadinene δ-杜松烯	$C_{15}H_{24}$	2.00	—
44	20.110	Calamenene 去氢白菖(蒲)烯	$C_{15}H_{22}$	—	2.64
45	20.207	α-Copaene α-蒎烯	$C_{15}H_{24}$	—	0.21
46	20.337	α-Longipinene 长叶蒎烯	$C_{15}H_{24}$	0.15	—
47	20.596	α-Calacorene 白菖考烯	$C_{15}H_{24}$	0.18	0.83
48	20.731	9-Cedranone		0.14	—
49	20.736	α-Cedrene oxide 雪松烯氧化物		—	0.44
50	21.006	Nerolidol 橙花叔醇		0.24	—
51	21.033	Lauric acid 月桂酸	$C_{12}H_{24}O_2$	—	0.78
52	21.428	Spathulenol 匙叶桉油烯醇		0.10	—
53	21.482	Estragole 茴香脑	$C_{10}H_{12}O$	—	0.07
54	21.563	Caryophyllene Oxide 丁香烯氧化物	$C_{15}H_{24}O$	1.29	—
55	21.719	Ethyl laurate 乙基月桂酸	$C_{14}H_{28}O_2$	0.11	—

续表

序号	保留时间/min	化合物	分子式	相对百分含量/% 巴西	相对百分含量/% 中国
56	21.887	Guaiol 愈创(木)醇	$C_{15}H_{26}O$	5.06	2.16
57	22.017	Cedrol 雪松醇(柏木脑)	$C_{15}H_{26}O$	1.00	2.22
58	22.276	Longifolene-(V4) 长叶烯	$C_{15}H_{24}$	—	0.11
59	22.297	Eremophilene 旱麦草烯	$C_{15}H_{24}$	0.20	—
60	22.346	3-Phenylpropionic acid 氢化肉桂酸	$C_9H_{10}O_2$	—	0.31
61	22.427	Caryophyllene-(I1) 丁香烯	$C_{15}H_{24}$	0.11	0.04
62	22.676	β-Maaliene 马阿里烯	$C_{15}H_{24}$	4.00	—
63	22.692	γ-Eudesmol γ-桉叶(油)醇	$C_{15}H_{26}O$	—	6.75
64	22.876	δ-Cadinene δ-杜松烯	$C_{15}H_{24}$	—	1.87
65	22.881	T-cadinol 杜松醇		2.04	—
66	23.000	Thujopsene-I 3 罗汉柏烯	$C_{15}H_{24}$	0.40	—
67	23.113	β-Eudesmol β-桉叶醇	$C_{15}H_{26}O$	6.08	11.34
68	23.183	α-Eudesmol α-桉叶醇	$C_{15}H_{26}O$	6.94	8.29
69	23.480	Bulnesol 异愈创木醇	$C_{15}H_{26}O$	2.92	1.80
70	23.556	β.-Humulene 葎草烯	$C_{14}H_{22}$	0.78	—
71	23.670	Cadalene 4-异丙-1,6-二甲萘	$C_{15}H_{18}$	-	0.24
72	23.810	α-Bisabolol (甜)没药醇	$C_{15}H_{26}O$	1.35	0.71
73	24.091	Heptadecane 十七烷	$C_{17}H_{36}$	0.75	0.59
74	24.388	Cinnamonitrile 肉桂腈	C_9H_7N	—	1.61
75	24.404	3,4-Dihydro-2-naphthoic acid 3,4-二氢-2-萘甲酸	$C_{11}H_8O_2$	3.94	—
76	24.696	3-甲基茚-2-羧酸	$C_{11}H_{10}O_2$	—	0.20
		Total		91.65	85.8

由表 14.11 可见,从巴西和中国两地产蜂胶的挥发性成分中共鉴定出 76 种化学物质,其中从巴西产蜂胶的挥发性成分中鉴定出 50 种化合物,主要为萜烯类化合物及其含氧衍生物、芳香族化合物等,其中萜烯类化合物占其总挥发性组分的 49.64%。相对含量较高的 5 种化合物为长叶稀(24.89%)、α-桉叶醇(6.94%)、β-桉叶醇(6.08%)、β-丁子香烯(5.27%)及愈创木醇(5.06%);从中国产蜂胶的挥发性成分中鉴定出 47 种化合物,其中萜烯类化合物占其挥发性组分的 18.6%,而芳香酸及酯类化合物占其总挥发性组分的 25.98%,相对含量较高的化合物为肉桂酸(19.3%)、β-桉叶醇(11.34%)、α-桉叶醇(8.29%)、γ-桉叶醇(6.75%)。巴西和中国两地产蜂胶的挥发性成分中的共有组分有 21 种,分别占巴西蜂胶和

中国蜂胶总挥发性组分的 37.15% 和 62.41%。共有组分中反应植物来源的萜烯类物质较少,只有姜黄烯等 4 种组分,且含量很低,绝大多数共有组分为具有定香、抑菌、驱虫活性的芳香酸及萜类含氧衍生物,如苯甲酸、肉桂酸、愈创(木)醇、桉叶醇等。

3 讨论

蜂胶的化学成分由蜜蜂采集的植物来源决定,在北温带地区,以杨树型蜂胶为代表,主要含有酚类、类黄酮糖苷配基、芳香酸及其酯类化合物。在南美洲热带地区,蜂胶化学成分相差很大,以巴西蜂胶为代表,随着地区及季节的不同,常有新成分被鉴定出来,主要成分包括黄酮类、香豆酸、木酯体、萜类物质等(Bankova,2005)。目前对北温带(包括中国北方)不同地区所产蜂胶中的挥发性成分组成及其生物学活性已有一些报道,在蜂胶挥发性成分中所占比例较高的化合物包括:α-蒎烯、桧烯、δ-杜松烯、α-衣兰油烯(希腊);丁香烯、α-姜黄烯、α-布黎烯、δ 杜松烯、愈创木醇、β-桉叶油醇和异愈创木醇(北京);3-甲基-2-丁烯-1-醇、苯乙醇、α-没药醇、薁、菲、芳姜黄酮、十七烷(内蒙古);大黄酚(山东)等(Melliou et al.,2007;付宇新 et al.,2009;王小平 et al.,2009)。报道显示蜂胶挥发性成分中含量较高的绝大部分为萜类化合物及其含氧衍生物,生物活性则体现为抑菌活性。本研究发现中国产蜂胶中除了种类繁多,含量占总挥发性成分约 18.6% 的萜类化合物外,单体化合物所占比例最高的为肉桂酸(19.3%),这与本研究蜂胶主要来源于中国华北地区,胶源植物主要为杨树有关。肉桂酸的存在对蜂胶挥发性成分发挥抑菌效果具有积极作用。此外,在萜类含氧衍生物中含量最高的为 β-桉叶醇(11.34%),这与 Petri 等(1988)所报道的温带蜂胶的特征一致。

巴西蜂胶植物来源包括有异叶南洋杉、大克罗西木、小克罗西木和酒神菊类,所含黄酮类物质较低而萜烯类及 p-香豆酸衍生物含量较高。本研究显示巴西蜂胶挥发性成分中萜烯类成分在挥发性成分中所占比例达 49.64%,远高于中国蜂胶。在关于巴西蜂胶挥发性成分的报道中,Atungulu 等(2007)利用同时蒸馏-萃取法得到巴西蜂胶挥发性物质,鉴定出 30 余个化合物,其中丙烯、苯乙酮、β-芳樟醇是主要成分。Kusumoto 等(2001)鉴定出巴西蜂胶挥发性成分中的两种新物质 2,2-二甲基-8-异戊烯-6-乙烯色原烯及 2,6-二异戊烯-4-乙烯苯酚。本研究鉴定得到巴西蜂胶与中国蜂胶挥发性成分中共有的化合物有 21 种,主要为芳香酸及萜类含氧衍生物,而共有的萜烯类成分仅有姜黄烯、愈创木烯、依兰油烯、雪松烯 4 种,且含量相差很大。以上结果均说明巴西、中国两地蜂胶挥发性成分化学组成差异较大,但均含有相对含量较高的具有抑菌、驱虫等生物活性的化学组分。可见,蜜蜂似乎能因地制宜,以不同的原料制造出功能相似的蜂胶,发挥蜂胶在蜂巢中光滑蜂巢内壁、防腐、抑菌、驱虫的作用。这也为不同地区的蜂胶在胶源植物不同、化学成分差异很大的情况下却具有相似的生物学活性提供了一定的参考。

小 结

本研究利用微波辅助萃取技术得到巴西蜂胶挥发性成分得率为 11.21%,中国蜂胶挥发性成分得率为 9.73%。GC-MS 分析鉴定出巴西蜂胶挥发性组分 50 个,主要是倍半萜类及其含氧衍生物,长叶稀、桉叶醇、愈创木醇等。中国蜂胶挥发性组分 47 个,主要是芳香酸及萜类衍生物,肉桂酸、桉叶醇等。两地蜂胶挥发性成分中含有 21 个具有定香、抑菌、驱虫活性的共有组分。结果说明中国蜂胶和巴西蜂胶挥发性成分在含量及化学组分上存在一定差异,但共有大量生物学活性物质。

第三节　蜂胶中挥发性成分的生物学活性

1　蜂胶中挥发性成分的抑菌与抗氧化活性

有关各类挥发油抑菌、抗氧化活性的报道很多，植物提取所得挥发性成分中含有多种低分子的抑菌物质和抗氧化成分，可作为天然保鲜剂，有效地抑菌及防止果蔬类腐败变质（Ho et al.，2011；Lazarevic et al.，2011；Mkaddem et al.，2010；Tenore et al.，2011；Zeng et al.，2011）。这些低分子的抗菌物质，如肉桂酸、阿魏酸、咖啡酸、桂醛、香茅醇、百里酚、丁香酚等，可抵抗大量的细菌、霉菌和酵母菌。挥发油中往往含有大量的萜烯类化合物及它们的含氧衍生物如醇、醛、酮、酸、酚、醚、酯、内酯等。萜烯类化合物作为一种常见的生物学活性组分，亦具有抗氧化活性，对不同来源的自由基有一定的清除作用（Joulain，2001）。因此，挥发油（精油）除了作为香精香料外，自古以来被作为防腐保鲜剂而广泛地应用（Lacoste et al.，1996）。

蜂胶众多的生理活性中，抑菌和抗氧化活性一直是研究者们最热衷的研究焦点，其抗氧化活性也是蜂胶诸如抗肿瘤、抗炎等药理学活性的基础。最为有趣的是，虽然蜂胶由于产地、胶源植物的不同，在化学成分上有很大的差异，但很多研究发现，化学成分不同的蜂胶似乎能发挥相似的抗氧化及抑菌活性（Bankova et al.，2000；Banskota et al.，2000；Kujumgiev et al.，1999；Popova et al.，2004b；Salomao et al.，2004）。说明蜜蜂能因地制宜，利用不同的原料生产出活性相似的蜂胶，从而发挥蜂胶在蜂群中的作用。

对蜂胶挥发性成分抑菌及抗氧化活性的关注亦是源于蜂胶在蜂群中的作用。研究发现养蜂场周围空气中微生物的数量要显著低于其他地区（Ghisalberti，1979），说明是蜂巢中的挥发性物质发挥了效应。另有研究报道蜂胶中的挥发性成分可防止大米储存过程中的脂质过氧化（Atungulu et al.，2007；Atungulu Griffiths et al.，2008），说明蜂胶的挥发性成分有一定的抗氧化活性。

目前对蜂胶抑菌及抗氧化活性的评价往往利用蜂胶的非挥发性部分，如水提取物或各种有机溶剂提取物，对蜂胶中挥发性成分的相关评价较少。由于蜂胶的非挥发性成分中主要的活性物质为酚酸类及黄酮类化合物，还有很多生物学活性组分由于加工原因被舍弃了，而蜂胶生物活性的发挥往往是其中多种生物学活性组分共同作用的结果，因此，对蜂胶中挥发性成分的生物学活性评价对蜂胶活性的全面评价有重要意义。

本研究比较了巴西蜂胶与中国蜂胶中挥发性成分及醇提物的体外抑菌活性及抗氧化活性，为不同产地及不同蜂胶组成部分在抗氧化及抑菌活性上的异同提供数据支持。

1.1　蜂胶中挥发性成分的抑菌活性

1.1.1　材料与方法

1.1.1.1　材料与试剂

中国蜂胶由杭州天厨蜜源保健品有限公司提供，其产地为山东，胶源植物为杨树；巴西蜂胶由浙江蜂之语蜂业集团有限公司提供，为巴西绿蜂胶；蜂胶挥发性成分为中国蜂胶微波辅助法提取所得;蜂胶乙醇提取物为蜂胶在室温下与80%乙醇以料液比1∶15搅拌提取24h,过滤后所得滤液经冷冻干燥后所得;牛肉膏、蛋白胨、琼脂、蔗糖,均为BR;TW80(吐温80)、氯化钠、氢氧化钠、盐酸,均为AR;红四氮唑(TTC,北京化学试剂公司);新鲜土豆、精密pH试纸、无菌去离子水等。

1.1.1.2 主要仪器及用具

SW-CJ-1F超净工作台,深圳市标雅科技有限公司;YXQ-LS-50G高压蒸汽灭菌锅,北京泽祥恒达科技发展公司;DHP-9052电热恒温培养箱,苏州江东精密仪器有限公司;THZ-D台式恒温振荡器,太仓市实验设备厂。打孔器、电炉、移液枪、玻璃涂布棒、移液管、培养皿、试管、酒精灯、接种环、滤纸、镊子、记号笔等。

1.1.1.3 供试菌种

细菌:大肠埃希氏杆菌(*Escherichia coli*,ATCC 25922)、金黄色葡萄球菌(*Staphylococcus aureus*,ATCC25923)、枯草芽孢杆菌(*Bacillus subtilis*,ATCC6633)、绿脓杆菌(*Pseudomonas aeruginosa*,ATCC 27853);

酵母菌:啤酒酵母(*Saccharomyces cerevisiae*,CCTCC AY92042);

霉菌:黑曲霉(*Aspergillus niger*,CCTCC AF91004)、桔青霉(*Penicillium citrinum*,CCTCCAF93094)。

供试菌种由浙江经贸职业技术学院生物与食品综合实训基地提供。

1.1.1.4 抑菌活性的测定

滤纸片的制备

将厚度为1.5mm的滤纸用打孔器打成直径为6.0mm的圆片,并将滤纸片装在空培养皿内,高压蒸汽灭菌备用。

培养基的制备

牛肉膏蛋白胨培养基:牛肉膏3g,蛋白胨10g,氯化钠5g,琼脂(固体)20g,加水至1000mL,微沸至固形物全部溶解,用0.1mol/L氢氧化钠溶液调pH到7.0~7.2,12L灭菌20min。用于细菌培养。

PDA培养基:去皮马铃薯200g,切成块煮沸30min,然后用纱布过滤,再加葡萄糖20g,琼脂20g,溶化后加水至1000mL,12L灭菌30min。用于真菌培养。

菌悬液的制备

真菌孢子悬液的制备:将保存菌种接入马铃薯固体培养基的试管斜面活化,取一支生长4d的试管斜面,用50mL的0.9%的无菌生理盐水分几次倒入试管中,其间用无菌的玻棒在斜面上轻轻搅动,将孢子刮下来,并用两层纱布过滤,装入三角瓶中,用血球板计数法和平皿计数法测定其生长浊度,将孢子悬液稀释到10^5~10^6cfu/mL。

细菌菌悬液的制备:将保存菌种接入液体牛肉膏蛋白胨培养基中,摇床30℃培养48h后,用血球板计数法和平皿计数法测定其生长浊度,将菌悬液稀释到10^5~10^6cfu/mL,备用。

样品的制备

准确称取蜂胶挥发性成分及蜂胶乙醇提取物,溶于吐温80中制成20mg/mL的溶液,巴氏消毒后待用。

纸片扩散法

取 0.1ML 已选取的合适菌液浓度的菌悬液,分别加入到牛肉膏蛋白胨培养基或 PDA 培养基的固体平板上,涂布均匀。取不同浓度的各种提取物、对照样品各 15μL 转移至已灭菌的圆形滤纸片(D 6mm)。用镊子夹取滤纸片贴于平板表面,并轻轻按压使充分接触。平板中间贴上浸有空白溶剂的滤纸片作为对照。每个菌种设 3 个平行组,贴好滤纸片的含菌平板倒置放入 4℃ 冰箱静置 2h,再置于培养箱培养(细菌 37℃ 培养 24h,真菌 28℃ 培养 48h),取出测量抑菌圈的直径。实验结果重复 3 次取平均值。

最小抑菌浓度(MIC)法

在塑料离心管里用牛肉膏蛋白胨或 PDA 液体培养基将各样品对倍稀释成所需的质量浓度梯度,起始质量浓度(最大质量浓度)设定为 20mg/mL。然后在 96 孔板中分别加入 50μL 稀释好的样品及 50μL 稀释好的菌液。每个质量浓度梯度设 4 个复孔;同时设链霉素、两性霉素 B 阳性对照组、阴性对照组(只加菌液和培养基)和空白对照组(只加培养基),37℃ 培养 24h。培养结束后,每个孔中加入 5μL 质量分数为 0.5% 的 TTC,37℃ 继续培养 1~3h,肉眼观察细菌的生长,培养孔呈现红色为有细菌生长,结果判断以不显色的组其样品质量浓度为该样品对测试菌的最低抑菌浓度(MIC)。

1.1.2 结果与分析

1.1.2.1 蜂胶挥发性成分及乙醇提取物对受试微生物抑菌圈的影响

中国蜂胶及巴西蜂胶中挥发性成分及乙醇提取物对受试微生物的抑菌活性见表 14.12。由表 14.12 可见,巴西蜂胶及中国蜂胶中的挥发性成分对各受试菌种的生长均有抑制作用,其中巴西蜂胶中的挥发性成分对酵母菌、霉菌及革兰阳性细菌均有很强的抑制活性,对革兰阴性细菌抑制作用相对较弱,各受试菌种的抑菌活性大小为酿酒酵母＞黑曲霉＞金黄色葡萄球菌＞桔青霉＞枯草芽孢杆菌＞绿脓假单胞菌＞大肠杆菌;中国蜂胶中的挥发性成分亦对各受试菌种有抑制作用,其中对革兰阳性细菌抑制活性最强,酵母菌及霉菌次之,革兰阴性细菌最弱。巴西蜂胶及中国蜂胶的乙醇提取物对各受试微生物的抑菌活性较相应的挥发性成分弱,其中巴西蜂胶的乙醇提取物对绿脓假单胞菌及大肠杆菌无抑制作用;中国蜂胶的乙醇提取物对大肠杆菌无抑制作用。

表 14.12 蜂胶中挥发性成分及乙醇提取物的抑菌活性

Table 14.12 Antimicrobial activity of PEO and EEP   mm

受试菌种	挥发性成分		乙醇提取物		空白对照
	巴西蜂胶	中国蜂胶	巴西蜂胶	中国蜂胶	
Staphylococcus aureus	14.28±0.42	15.41±0.50	9.80±0.21	9.07±0.28	na^e
Bacillus subtilis	12.60±0.32	14.34±0.39	9.10±0.22	8.01±0.29	na^e
Pseudomonas aeruginosa	10.19±0.41	8.69±0.36	na^e	6.70±0.43	na^e
Escherichia coli	8.70±0.34	7.38±0.40	na^e	na^e	na^e
Saccharomyces cerevisiae	18.25±0.43	11.49±0.35	10.73±0.23	9.11±0.23	na^e
Penicillium citrinum	13.71±0.34	10.20±0.32	8.22±0.34	6.87±0.36	na^e
Aspergillus niger	17.34±0.38	11.87±0.41	9.01±0.27	6.77±0.46	na^e

注:样品浓度 20mg/mL;

na^e:无活性。

1.1.2.2 蜂胶挥发性成分及乙醇提取物对受试微生物的最小抑菌浓度

中国蜂胶与巴西蜂胶中挥发性成分及乙醇提取物对受试微生物的最小抑菌浓度见表14.13。如表14.13所示,巴西蜂胶中的挥发性成分对受试微生物均有抑菌作用,最小抑菌浓度范围在39.06~625μg/mL,其中对酿酒酵母及黑曲霉的抑菌浓度最低,为39.06μg/mL,对大肠杆菌的抑菌浓度最高,为625μg/mL,与前述纸片扩散法结果相一致,表现为最低抑菌浓度越低,抑菌圈越大;中国蜂胶中挥发性成分对受试微生物的最小抑菌浓度范围在39.06~625μg/mL,其中对金黄色葡萄球菌的最小抑菌浓度最低,为39.06μg/mL,对大肠杆菌及绿脓假单胞菌抑菌浓度最高,为625μg/mL。巴西蜂胶与中国蜂胶乙醇提取物对大肠杆菌均无抑制作用,与纸片扩散法结论一致,巴西蜂胶乙醇提取物对受试微生物的最小抑菌浓度范围在156.25~625μg/mL,抑菌活性较巴西蜂胶中的挥发性成分弱;中国蜂胶乙醇提取物对受试微生物的最低抑菌浓度范围在625~2500μg/mL,较其挥发性成分,抑菌活性亦较弱。链霉素作为抑制细菌生长的阳性对照品,其最小抑菌浓度范围在6.25~12.5μg/mL。两性霉素B作为抑制真菌生长的阳性对照品,其最小抑菌浓度为0.78μg/mL。

表14.13 蜂胶中挥发性成分及乙醇提取物的最小抑菌浓度

Table 14.13 Minimum inhibitory concentration of PEO and EEP(μg/mL)

受试菌种	挥发性成分		乙醇提取物		阳性对照	
	巴西蜂胶	中国蜂胶	巴西蜂胶	中国蜂胶	链霉素	两性霉素B
Staphylococcus aureus	78.13	39.06	625.00	625	12.5	nt[f]
Bacillus subtilis	78.13	78.13	625.00	625	12.5	nt[f]
Pseudomonas aeruginosa	156.25	625	na[e]	2500	1.56	nt[f]
Escherichia coli	625.00	625	na[e]	na[e]	6.25	nt[f]
Saccharomyces cerevisiae	39.06	156.25	156.25	625	nt[f]	0.78
Penicillium citrinum	78.13	312.5	625.00	1250	nt[f]	0.78
Aspergillus niger	39.06	156.25	625.00	2500	nt[f]	0.78

[e] na,无活性.
[f] nt,未测.

1.1.3 讨论

从巴西蜂胶及中国蜂胶的挥发性成分和乙醇提取物对7种受试菌种的抑菌圈大小及最小抑菌浓度来看,两地蜂胶的挥发性组分及其乙醇提取物均有一定的抑菌作用。其中挥发性成分的抑菌活性较同地区蜂胶乙醇提取物的抑菌活性要强;不同地区蜂胶挥发性成分对受试的7种微生物抑菌活性相似,均对革兰阳性细菌、酵母菌及霉菌有较强的抑制作用,对革兰阴性细菌则抑菌活性相对较弱。目前对蜂胶抑菌活性的报道显示蜂胶提取物对革兰阳性细菌及真菌的生长有显著的抑制活性(Sforcin et al.,2000;Stepanovic et al.,2003;Uzel et al.,2005),较敏感的微生物包括:蜡样芽孢杆菌、枯草芽孢杆菌、金黄色葡萄球菌、表皮葡萄球菌及白色念珠菌等;对革兰阴性细菌则抑制效果较弱(Erkmen et al.,2008;Scazzocchio et al.,2006)。我们的实验结果与这些报道有一致性。

研究发现黄酮类化合物具有抑菌活性,其机制可能为通过干扰微生物细胞膜的功能及相关酶活性从而抑制微生物的生长。蜂胶乙醇提取物中含有的主要生物学活性成分为黄酮类化合物,其抑菌机制可能与黄酮类化合物的作用相似(Viuda-Martos et al.,2008)。

蜂胶挥发性成分的抑菌作用也与其成分有关。巴西蜂胶挥发性成分中大量含有的萜烯类化合物桉叶醇、愈创木醇等被认为是很多天然植物发挥抑菌活性的主要成分(Perez et al.,1999;Yayli et al.,2005);中国蜂胶挥发性成分中的重要组分肉桂酸,是一种具有强抑菌活性的物质,它可通过抑制 FtsZ 的聚集从而中断细胞分裂,最终抑制革兰阳性细菌的增殖(Rastogi et al.,2008)。另外,蜂胶挥发性成分中大量含量较低的物质如去氢白菖(蒲)烯、芹子烯等,对蜂胶挥发性成分抑菌活性的发挥也起到了协同增强作用(Shunying et al.,2005)。

1.2 蜂胶中挥发性成分的抗氧化活性

1.2.1 材料与方法

1.2.1.1 材料与试剂

中国蜂胶由杭州天厨蜜源有限公司提供,其产地为山东,胶源植物为杨树;巴西蜂胶由浙江蜂之语蜂业集团有限公司提供,为巴西绿蜂胶;蜂胶挥发性成分为中国蜂胶微波辅助法提取所得;蜂胶乙醇提取物为毛胶在室温下与 80% 乙醇以料液比 1∶15 搅拌提取 24h,过滤后所得滤液经冷冻干燥后所得。

2,2-二苯基-1-苦肼基自由基(DPPH),SIGMA 公司,分析纯;2,6-二叔丁基对甲酚(BHT),SUPELCO 公司,分析纯;抗坏血酸(Vc),国药集团化学试剂有限公司,分析纯;磷酸氢二钠荧光素(Fluorescein,FL),2,2-azobis(2-methyl-propionamidine) dihydrochloride (AAPH),6-hydroxy-2,5,7,8-tetramethylchroman-2-carboxylic acid (Trolox),Sigma-Aldrich 公司。

1.2.1.2 主要仪器及用具

RT-02 型二两装高速中药粉碎机,永康市屹立工具厂;R 系列旋转蒸发器,上海申生科技有限公司;SHB-InA 型循环水式多用真空泵,河南太康教材仪器厂;KQ-500E 型超声波清洗仪,昆山市超声仪器有限公司;W201 型恒温水浴锅,上海申生科技有限公司;TB-Z15D 型电子分析天平,DENVER INSTRUMENT 公司,$d=0.01mg(60g)\sim 0.1mg(210g)$;TGL-16G 冷冻离心机,上海安亭科学仪器厂;Thermo Scientific Varioskan Flash 光谱扫描多功能读数仪,美国 Thermo 公司。

1.2.1.3 体外抗氧化能力测定

DPPH 自由基清除能力的测定

DPPH 用无水乙醇配成 2×10^{-4} mol/L 的溶液,置于 4℃ 避光保存。样品和阳性药均用无水乙醇配成 10mg/mL 的原液,测试前用无水乙醇稀释成所需浓度。将不同浓度的供试品溶液 $100\mu L$ 和 2×10^{-4} MDPPH 溶液 $100\mu L$ 加入 96 孔板各孔中,同时以不加 DPPH(以 $100\mu L$ 无水乙醇代替 DPPH)的供试品溶液各浓度作为对照以消除供试品本身颜色对测试结果的干扰,并设 DPPH 阴性对照(以 $100\mu L$ 无水乙醇代替供试品),每组平行设 4 个复孔。将 96 孔板放入酶标仪中,震荡 1min,并于此条件下保存(室温、避光),30min 后测试其在 517nm 波长处的吸光度 OD 值,按如下公式计算供试品的自由基清除率。

$$自由基清除率=[OD_{DPPH.control}-(OD_{sample}-OD_{sample.control})]/OD_{DPPH.control}\times 100\%$$

其中:$OD_{DPPH, control}$为DPPH阴性对照组OD值的平均值;

OD_{sample}为样品组OD值的平均值;

$OD_{sample, control}$为样品乙醇对照组OD值的平均值。

ORAC测定

ORAC(oxygen radical absorbance capacity,抗氧化能力指数)测定:荧光素钠在485nm光激发下,发射527nm荧光,可以被AAPH释放的过氧自由基氧化而使荧光特性消失。当抗氧化剂存在时,可与荧光素钠竞争氧化剂,减缓其荧光消退的速度。根据这一特性,可用来测定样品氧自由基清除活性。实验结果表达为荧光衰退曲线下积分面积,扣除无抗氧化剂的空白曲线下面积,得出抗氧化剂的保护面积(AUC)(Davalos et al.,2004)。确定样品的ORAC值需先建立Trolox标准品的AUC与浓度回归曲线,将被测样品的AUC值代入回归曲线中即得到样品的ORAC值(以μmol/L Trolox eq/g表示)。

测定过程如下:根据样品溶解性,配制质量分数为0.07%的随意甲基化-β-环糊精(RMCD)的丙酮-水(1:1体积比)溶液作为样品溶剂(Miraliakbari et al.,2008)。将样品和阳性药均用溶剂配成10mg/mL的原液,测试前用溶剂稀释成所需浓度。将全黑96孔板每个微孔中加入待测样品溶液20μL,再加入磷酸钾缓冲液20μL和AAPH140μL(终浓度12.8mmol/l),最后添加荧光素钠20μL至终浓度为63nmol/L,立即启动反应并迅速将酶标板置于预温37℃的光谱扫描多功能读数仪中开始测定。采用动力学方式,每2min测定一个点,至荧光强度衰减为零为止。

计算公式为:$AUC = 1 + \sum_{t0=60min}^{t1=60min} Ai/A0$,$A$为荧光强度。

H_2O_2诱导的兔红细胞氧化溶血的保护作用

红细胞悬液的制备:取实验兔血,肝素抗凝,2000r/min离心10min,除去血浆,红细胞用0.9%生理盐水洗涤3次,配制成体积分数为2%的红细胞悬液。分别取兔红细胞悬浮液0.5mL,除正常对照组和模型对照组加入等量的生理盐水外,各实验组加不同浓度的样品0.5ML,混匀,37℃孵育10min后,除正常对照组加入等量的生理盐水外,其他各组均加入H_2O_2 0.5mL(终浓度为2%)启动反应,37℃水浴1h。2500r/min离心10min,取上清液于415nm波长处测定A值,假设模型对照组溶血度为100%。计算各组的溶血度及药物对溶血的抑制率。

溶血度=给药组A_{415}/模型对照组A_{415}×100%;

抑制率=(模型对照组A_{415}-给药组A_{415})/(模型对照组A_{415}-正常组对照组A_{415})×100%。

1.2.2 结果与分析

1.2.2.1 蜂胶挥发性成分及乙醇提取物清除DPPH自由基的能力

中国蜂胶与巴西蜂胶中挥发性成分、乙醇提取物及两种抗氧化剂对DPPH自由基的清除能力见图14.5—14.7。图14.5显示了水溶性抗氧化剂抗坏血酸及脂溶性抗氧化剂BHT对DPPH清除作用的浓度-抑制率曲线,结果显示在2~1000μg/mL浓度范围内,抗坏血酸及BHT对DPPH自由基的清除作用随浓度的增加而增强,其IC_{50}值分别为4.67μg/mL和13.07μg/mL(见图14.7)。巴西蜂胶与中国蜂胶中的挥发性成分及乙醇提取物亦对DPPH自由基有清除作用,其浓度-抑制率曲线如图14.6所示,图14.6显示在0.02~10mg/mL浓度范围内,巴西及中国蜂胶中的挥发性成分及醇提物对 DPPH 自由基的清除率随浓度的增加而增强;图 14.7 显示各样品对 DPPH 自由基清除率的 IC_{50} 值,中国蜂胶醇提物在 4 种样品中 IC_{50} 值最低,为 97.28μg/mL,巴西蜂胶醇提物次之,为 137.17μg/ML,中国蜂胶挥发性成分 IC_{50} 值为 178.30μg/mL,巴西蜂胶中挥发性成分对 DPPH 自由基清除率 IC_{50} 值最大,为 309.5μg/mL。IC_{50} 值越小则抗自由基活性越强。

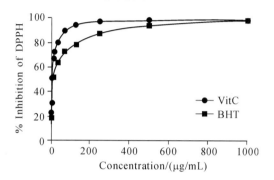

图 14.5 抗坏血酸和 BHT 清除 DPPH 自由基的能力
Fig14.5 Scavenging capacity of Ascorbic acid and BHT against DPPH

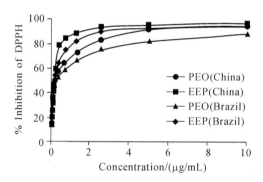

图 14.6 巴西、中国蜂胶中挥发性成分及醇提物清除 DPPH 自由基的能力
Fig14.6 Scavenging capacity of PEO and EEP of China and Brazil against DPPH

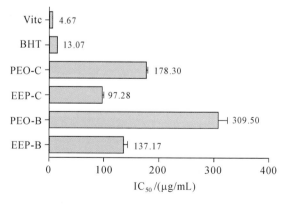

图 14.7 DPPH 自由基清除率-IC_{50}
Fig14.7 Seavenging capacity against DPPH-IC_{50}

PEO-C:中国蜂胶挥发性成分;PEO-B:巴西蜂胶挥发性成分;EEP-C:中国蜂胶乙醇提取物;EEP-B:巴西蜂胶乙醇提取物。

1.2.2.2 蜂胶挥发性成分及乙醇提取物的抗氧化能力指数

Trolox 清除 AAPH 诱发自由基活性的荧光衰退曲线如图 14.8 所示。图 14.8 显示 Trolox 浓度越高,对 AAPH 的清除效率越高,荧光衰退曲线下面积越大。以 AUC 值为纵坐标,以 Trolox 浓度为横坐标作图并进行回归分析,得回归方程为 $y=2.0924x+8.0325$,$R^2=0.9928$。将各样品与阳性对照物的 AUC 值代入回归方程后得到各样品的 ORAC 值,结果见图 14.9。如图 14.9 所示,阳性对照品抗坏血酸 ORAC 值最高,达 1.8μmol/L Trolox eq/g,巴西蜂胶醇提物次之,为 1.58μmol/L Trolox eq/g,中国蜂胶醇提物为 1.2μmol/L Trolox eq/g,BHT、中国蜂胶及巴西蜂胶中挥发性成分的 ORAC 值近似,分别为 0.86、0.82 和 0.88μmol/L Trolox eq/g。

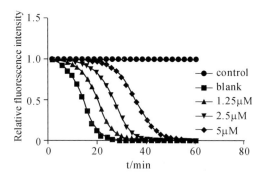

图 14.8 Trolox 清除 AAPH 诱发自由基活性的荧光衰退曲线

Fig14.8 Seavenging activity of Trolox against fluoreseence decay induced with AAPH

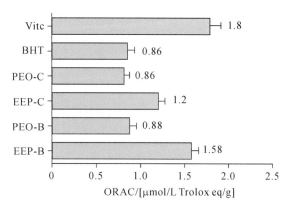

图 14.9 巴西蜂胶和中国蜂胶中挥发性成分及醇提物 ORAC 值(μmol/L Trolox eq/g)

Fig14.9 ORAC value of PEO and EEP from China and Brazil

PEO-C:中国蜂胶挥发性成分;PEO-B:巴西蜂胶挥发性成分;EEP-C:中国蜂胶乙醇提取物;EEP-B:巴西蜂胶乙醇提取物。

1.2.2.3 蜂胶挥发性成分及乙醇提取物对 H_2O_2 诱导的兔红细胞氧化溶血的影响

中国蜂胶与巴西蜂胶中挥发性成分、乙醇提取物及两种抗氧化剂对 H_2O_2 诱导的兔红细胞溶血均有抑制作用,其 IC_{50} 值见图 14.10。图 14.10 显示,抗坏血酸对红细胞溶血的抑制率最强,IC_{50} 值为 0.385mg/mL,其后依次为中国蜂胶醇提物、BHT、巴西蜂胶醇提物、中国蜂胶中的挥发性成分,最后为巴西蜂胶中的挥发性成分,其 IC_{50} 值分别为 0.732、0.763、

0.838、0.871、0.879mg/mL。

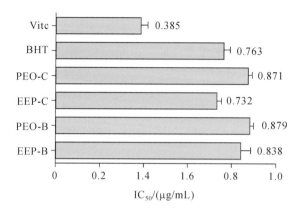

图 14.10 巴西蜂胶与中国蜂胶挥发性成分及醇提物对 H_2O_2 诱导红细胞氧化溶血的保护作用
Fig14.10 Protective effect of propolis on H_2O_2-induced red cell oxidation hemolysis-IC_{50}
PEO-C:中国蜂胶挥发性成分；PEO-B:巴西蜂胶挥发性成分；EEP-C:中国蜂胶乙醇提取物；EEP-B:巴西蜂胶乙醇提取物。

1.2.3 讨论

我们对中国蜂胶与巴西蜂胶中的挥发性成分及乙醇提取物的抗氧化活性通过不同机制的化学模拟体系及细胞体系进行了评价。在化学模拟体系中，应用了基于单电子转移(single electron transfer，SET)模型的 DPPH 实验及基于氢原子转移(hydrogen atom transfer，HAT)模型的 ORAC 实验进行评价；在细胞体系中则观察了对 H_2O_2 诱导的红细胞氧化溶血的保护作用。以此来多方位评价蜂胶挥发性成分和醇提物的体外抗氧化活性。

关于蜂胶抗氧化活性的报道很多(Ahn et al.，2007；Choi et al.，2006；Mohammadzadeh et al.，2007b)，醇提物是蜂胶最为常见的一种产品形式，研究发现蜂胶清除 DPPH 自由基的能力与蜂胶中所含的总酚含量呈正相关(Gregoris et al.，2008)。DPPH 实验是一种广泛应用于评价植物及化学物质抗氧化性的方法。本实验结果显示巴西蜂胶与中国蜂胶中的挥发性成分及乙醇提取物对 DPPH 自由基均有一定的清除能力，通过比较其 IC_{50} 值，强弱依次为：中国蜂胶醇提物＞巴西蜂胶醇提物＞中国蜂胶挥发性成分＞巴西蜂胶挥发性成分。蜂胶挥发性成分的 DPPH 自由基清除活性为首次报道，但基于其一定的活性，可认为蜂胶中的挥发性组分对蜂胶抗氧化活性有一定贡献。

ORAC 法是用于评价化合物体外抗氧化活性的一种较为优越的方法(Miraliakbari et al.，2008)，它可以提供稳定可控的自由基，这些自由基与生命现象中的自由基具有高度的一致性；这种方法采用荧光定量，灵敏度高，是目前唯一可同时测量在特定时间抑制百分率和在特定抑制百分率处的抑制时间的方法。实验结果显示，在所测定的样品中，抗坏血酸的 ORAC 值最高，达 1.8μmol/L Trolox eq/g，各蜂胶样品的 ORAC 值依次排序为巴西蜂胶醇提物＞中国蜂胶醇提物＞巴西蜂胶挥发性成分＞中国蜂胶挥发性成分；其中中国蜂胶与巴西蜂胶中的挥发性成分及巴西蜂胶醇提物 ORAC 值均高于抗氧化剂 BHT。对蜂胶挥发性成分及醇提物 ORAC 值的测定为首次，由于 ORAC 实验被广泛应用于食品及营养补充剂抗氧化活性的评价，因此，对蜂胶挥发性成分及醇提物 ORAC 值的测定可为蜂胶的抗氧化活性提供进一步的依据。

红细胞是评价氧化应激中常用的细胞体系,一些含有酚酸、黄酮、萜类及各类蛋白的天然产物被证明具有抑制由 H_2O_2 诱导的红细胞氧化溶血作用。研究发现,一些黄酮类化合物可通过定位于油相及水相界面间从而起到减少细胞膜氧化损伤的作用(Suwalsky et al.,2007)。本实验结果发现巴西蜂胶与中国蜂胶中的挥发性成分及醇提物对 H_2O_2 诱导红细胞氧化溶血均有保护作用,其作用保护强弱通过 IC_{50} 反映,依次为:中国蜂胶醇提物>巴西蜂胶醇提物>中国蜂胶中的挥发性成分>巴西蜂胶中的挥发性成分。蜂胶醇提物中所含的大量黄酮类化合物被认为对保护 H_2O_2 诱导红细胞氧化溶血起主要作用。

对同一样品,许多体外抗氧化能力测定方法的结果有时不具有一致性。这主要是由于来自于抗氧化物质、自由基以及二者之间的作用机制非常复杂。因此没有任何一种单一方法可以精确地反映所有的自由基或混合物中所有的抗氧化剂的抗氧化能力。在实验研究中,往往需要综合对比各种抗氧化能力测定方法的结果,才能全面和准确地了解样品的抗氧化能力。

小 结

本研究对利用微波辅助法提取得到的巴西蜂胶及中国蜂胶中的挥发性成分及蜂胶乙醇提取的体外抑菌活性及抗氧化活性进行了研究。结果显示来源于两地的蜂胶中的挥发性成分及乙醇提取物均具有一定的抑菌及体外抗氧化活性。蜂胶挥发性成分的抑菌活性强于相应的蜂胶乙醇提取物。本实验中,蜂胶挥发性成分对金黄色葡萄球菌、黑曲霉、啤酒酵母的抑制强度最高;对革兰阴性细菌抑制活性较弱;中国蜂胶与巴西蜂胶的醇提物对大肠杆菌均无抑制作用。蜂胶挥发性成分在本研究所采用的 3 种体外抗氧化评价体系中均显示一定的抗氧化活性,但活性较蜂胶醇提物弱;DPPH 自由基清除实验显示巴西蜂胶、中国蜂胶中的挥发性成分及乙醇提取物对 DPPH 自由基抑制的 IC_{50} 值分别为 309.5、178.3、137.17、97.28μg/mL;ORAC 实验显示巴西蜂胶、中国蜂胶中的挥发性成分及乙醇提取物的 ORAC 值分别为 0.88、0.82、1.58 和 1.2μmol/L Trolox eq/g;H_2O_2 诱导的兔红细胞氧化溶血实验显示巴西蜂胶、中国蜂胶中的挥发性成分及乙醇提取物对红细胞溶血抑制的 IC_{50} 值分别为 0.879、0.871、0.838 和 0.732mg/mL。

2　蜂胶挥发性成分对束缚应激小鼠氧化应激状态的改善作用

自由基是机体正常的代谢产物,参与许多生理和病理过程。正常状态下,机体处于稳态平衡,即代谢产生的自由基受到体内抗氧化体系的调节,从而处于稳定状态。当受到内源性和外源性刺激,机体代谢异常产生过量自由基从而超过体内抗氧化体系的还原能力时,机体处于氧化应激状态,结果导致机体损伤。研究发现自由基的大量堆积和细胞内钙超载,往往是一些损伤因子引起细胞损伤的共同途径(Trueba et al.,2004)。

蜂胶具有较强的抗氧化活性,对不同地区、不同植物来源的蜂胶进行生物学活性研究均发现有一定的抗氧化活性(Ali et al.,2010;Gregoris et al.,2011;Gregoris et al.,2010;Gulcin et al.,2010;Miguel et al.,2010)。对于蜂胶中的挥发性成分,其抗氧化活性研究较少。我们的前期研究发现,蜂胶中的挥发性物质在体外实验中表现出一定的抗氧化活性。但是,体外环境建立的众多方法,几乎都存在着无可回避的内在局限。无论是理化环境,还

是生物源性等体外方法在评价食物中生物活性组分的时候,均缺少生理因素的参与,如生物体内功效组分的消化、吸收、代谢、转化以及排泄等,使得基于体外评价方法的研究结论缺乏科学性和说服力。建立和完善体内抗氧化能力评价技术,用于针对功能性食品功效组分的抗氧化能力评价,探寻对人体具有保护作用的生物活性物质,现实意义十分明显。因此本研究重点探讨蜂胶中挥发性成分对机体氧化应激状态的改善作用,以期为蜂胶中挥发性成分的抗氧化性能提供进一步的证据。

束缚应激模型是经典的心理应激模型之一,也是目前应用最普遍的一种造成应激的方法,其中氧化应激是束缚负荷的可能机制之一。蜂胶挥发性成分具有体外抗氧化活性,我们通过建立小鼠束缚负荷诱发小鼠氧化应激模型考察了蜂胶挥发性成分对束缚小鼠机体氧化应激状态的影响,为蜂胶挥发性成分在医药制品和保健食品行业的应用提供一定的实验依据。

2.1 材料与方法

2.1.1 实验动物

实验动物为 ICR 小鼠,雄性,体重 18~22g,清洁级,由中国科学院上海实验动物中心/上海斯莱克实验动物公司提供,生产许可证:SCXK(沪)2007-0005。

小鼠饲养于屏蔽环境,温度 23±2℃,日温差±1℃,湿度 50%~70%,噪音<50dB,光照:150~200Lx,12h 明暗交替。自来水过滤灭菌,置于高压灭菌的饮水瓶中自由饮用,8:00~8:30am 喂食。实验室使用许可证:SYXK(浙) 2003-0003。

2.1.2 材料与试剂

巴西蜂胶由浙江蜂之语蜂业集团公司提供,系该公司从巴西进口的绿色蜂胶;超氧化物歧化酶(SOD)、丙二醛(MDA)、谷胱甘肽(GSH)、一氧化氮(NO)、黄嘌呤氧化酶(XOD)、过氧化氢酶(CAT)测定试剂盒,购自南京建成生物工程研究所;门冬氨酸氨基转移酶(AST)、丙氨酸氨基转移酶(AST)测定试剂盒购自上海申能-德赛诊断技术有限公司。

2.1.3 主要仪器与设备

7020 型全自动生化分析仪,日本日立公司;725 型－86℃低温冰箱,美国 FORMA;DL-5-B 型离心机,上海安亭科学仪器厂;TGL-16G 冷冻离心机,上海安亭科学仪器厂;DRP-9052 型电热恒温箱,上海森信实验仪器公司。

2.1.4 实验方法

2.1.4.1 动物模型的建立

18h 束缚负荷后诱发小鼠应激模型。

2.1.4.2 分组及给药

7 周龄清洁级昆明种小鼠,雄性,体重 18~22g。将小鼠随机分成正常对照组、模型组、阳性对照组(饲喂维生素 C,200mg/kg)及蜂胶挥发性成分低、中、高剂量组(50mg/kg、100mg/kg、200mg/kg),共 6 组,每组 10 只。各给药组连续灌胃 15d,每天 1 次,正常对照和模型组均灌胃同体积溶剂。第 14d 束缚模型组和蜂胶挥发性成分各组以及阳性对照药组小鼠灌胃 30min 后进行束缚应激实验,使用改造为通风良好的 50mL 尖底聚丙烯塑料离心管,小鼠给予一次性束缚负荷 18h(14:00~8:00),实验期间禁食禁水。

2.1.4.3 指标测定

小鼠全血置于肝素处理过的离心管中,4℃下以 5000r/min 离心 5min 后得到血浆,测定 ALT、AST、MDA、SOD;小鼠肝组织的 5% 生理盐水匀浆液在 4℃,以 10000r/min 离心 10min,测定 SOD、MDA、GSH、XOD、NO、CAT。

2.1.5 统计学分析

所有数值均以均值表示。统计分析使用 SPSS 11.5 统计分析软件完成。先采用单因素方差分析(One Way ANOVA)进行组间比较,如有差异进一步采用 Dunnett's 检验。显著性差异水平为 $P<0.05$。

2.2 结果

2.2.1 蜂胶中挥发性成分对束缚小鼠血液氧化应激标志物的影响

蜂胶中挥发性成分对束缚应激小鼠血清 MDA、SOD 的影响见表 14.14。如表 14.14 所示,经过 18h 一次性束缚应激后 ICR 小鼠血清氧化应激状态标志物发生显著变化,其中 MDA 水平显著升高($P<0.05$),SOD 活性显著下降($P<0.01$)。用蜂胶中的挥发性成分及阳性药物抗坏血酸对动物进行预防性给药可显著改善束缚应激小鼠的血清氧化应激标志物水平,其中,蜂胶挥发性成分各剂量组均可降低束缚应激小鼠血清 MDA 含量($P<0.05$),中、高剂量组可显著提高束缚应激小鼠血清 SOD 活性($P<0.05$,$P<0.01$);抗坏血酸可显著降低束缚应激小鼠血清 MDA 含量($P<0.01$),提高血清 SOD 活性($P<0.01$)。

表 14.14 蜂胶挥发性成分对束缚应激小鼠血清 MDA、SOD 的影响

Table 14.14 Effects of propolis essential oil on serum MDA and SOD of restraint-stressed mice

Group	Dosage(administration)	n	MDA/(nmol/mL)	SOD/(U/mL)
Normal control	0.1mL/10g vehicle ($p.o.$)	10	4.76±1.09*	135.57±15.51**
Model control	0.1mL/10g vehicle ($p.o.$)	10	7.05±1.99	109.51±12.33
Low dose PEO	50mg/kg propolisessential oil($p.o.$)	10	5.33±1.12*	119.38±16.96
Middle dose PEO	100mg/kg propolisessential oil ($p.o.$)	10	5.16±1.06*	127.72±15.71*
High dose PEO	200mg/kg propolisessential oil ($p.o.$)	10	5.14±0.64*	133.85±15.39**
Positive control	200mg/kg VitC ($p.o.$)	10	2.37±0.37**	129.80±16.27**

与模型对照组比:* $P<0.05$,** $P<0.01$。

2.2.2 蜂胶中挥发性成分对束缚应激小鼠肝功能的影响

蜂胶中挥发性成分对束缚应激小鼠肝功能的影响见表 14.15。如表 14.15 所示,18h 一次性束缚使小鼠 ALT、AST 水平显著升高($P<0.01$)。预防性给予蜂胶挥发性成分对束缚应激小鼠血清 ALT 的升高有抑制作用,其中蜂胶挥发性成分高剂量组可显著降低束缚小鼠血清 ALT 的水平;蜂胶挥发性成分对束缚应激小鼠 AST 的水平无影响($P>0.05$);抗坏血酸可显著降低束缚应激小鼠血清 ALT、AST 水平($P<0.05$,$P<0.01$)。

表 14.15　蜂胶挥发性成分对束缚应激小鼠血清 ALT、AST 的影响

Table 14.15　Effects of propolis essential oil on serum ALT and AST of restraint-stressed mice

Group	Dosage(administration)	n	ALT/(IU/L)	AST/(IU/L)
Normal control	0.1mL/10g vehicle (p.o.)	10	31.00±6.28**	123.00±22.11**
Model control	0.1mL/10g vehicle (p.o.)	10	63.00±8.53	279.00±22.10
Low dose PEO	50mg/kg propolis essential oil(p.o.)	10	63.00±11.6	261.80±34.14
Middle dose PEO	100mg/kg propolis essential oil (p.o.)	10	59.83±8.06	259.20±23.64
High dose PEO	200mg/kg propolis essential oil (p.o.)	10	50.50±6.24*	253.67±26.58
Positive control	200mg/kg VitC (p.o.)	10	49.57±9.05*	212.25±28.25**

与模型对照组比：* $P<0.05$，** $P<0.01$。

2.2.3　蜂胶中挥发性成分对束缚应激小鼠肝脏氧化防御体系的影响

蜂胶挥发性成分对束缚应激小鼠肝组织中的抗氧化酶系及抗氧化活性物质的影响见表14.16。如表14.16所示，束缚应激至小鼠肝脏组织中抗氧化酶系 SOD 及 CAT 活性显著降低（$P<0.01$），肝组织中抗氧化活性物质 GSH 含量亦显著降低（$P<0.01$）。预防性给予蜂胶挥发性成分及抗坏血酸对束缚应激小鼠肝组织中抗氧化酶及活性物质有改善作用，蜂胶挥发性成分中、高剂量组可显著提高束缚应激小鼠肝组织中 SOD 活性及 GSH 的含量（$P<0.01$）；蜂胶挥发性成分高剂量组可显著提高肝组织中 CAT 的活性（$P<0.05$）；抗坏血酸可显著提高束缚应激小鼠肝组织中 SOD、CAT 活性（$P<0.01$），并可显著提高小鼠肝组织中 GSH 的含量（$P<0.05$）。

表 14.16　蜂胶挥发性成分对束缚应激小鼠肝组织中 SOD、CAT、GSH 的影响

Table 14.16　Effects of propolis essential oil on SOD, CAT and GSH in liver tissue of restraint-stressed mice

Group	Dosage (administration)	n	SOD (U/mg prot)	CAT (U/mg prot)	GSH (mg/g prot)
Normal control	0.1mL/10g vehicle (p.o.)	10	49.11±8.18**	49.15±11.01**	8.59±1.28**
Model control	0.1mL/10g vehicle (p.o.)	10	31.17±7.46	32.08±6.21	5.30±1.13
Low dose PEO	50mg/kg propolis essential oil (p.o.)	10	37.77±9.79	35.40±10.25	5.81±1.38
Middle dose PEO	100mg/kg propolis essential oil (p.o.)	10	47.37±7.29**	36.82±10.15	7.25±1.38**
High dose PEO	200mg/kg propolis essential oil (p.o.)	10	50.37±8.59**	39.23±7.93*	8.04±1.31**
Positive control	200mg/kg VitC (p.o.)	10	41.66±4.73**	44.29±5.76**	6.57±1.33*

与模型对照组比：* $P<0.05$，** $P<0.01$。

2.2.4　蜂胶中挥发性成分对束缚应激小鼠肝脏氧化损伤的影响

蜂胶中挥发性成分对束缚应激小鼠肝脏氧化损伤的影响见表14.17。如表14.17所示，束缚应激致小鼠肝脏氧化损伤，肝组织中 MDA、NO 含量及 XOD 活性均显著增强（$P<$

0.01)。蜂胶挥发性成分及阳性药物抗坏血酸对束缚应激至小鼠肝损伤有一定保护作用。蜂胶挥发性成分高剂量组可显著降低束缚应激小鼠肝组织中 MDA 的含量($P<0.01$);降低肝组织中 XOD 活性($P<0.05$);蜂胶挥发性成分中、高剂量组可显著降低小鼠肝组织中 NO 的含量($P<0.01$,$P<0.05$);抗坏血酸可显著降低束缚应激小鼠肝组织中 MDA 及 NO 的含量($P<0.05$,$P<0.01$),但对肝组织中 XOD 活性无显著影响($P>0.05$)。

表 14.17 蜂胶挥发性成分对束缚应激小鼠肝组织中 MDA、XOD、NO 的影响

Table 14.17 Effects of propolis essential oil on MDA,XOD and NO in liver tissue of restraint-stressed mice

Group	Dosage (administration)	n	MDA/ (nmol/mg prot)	XOD/ (U/mg prot)	NO/ (nmol/mg prot)
Normal control	0.1mL/10g vehicle (p.o.)	10	3.94±1.52**	18.56±4.40**	0.65±0.11**
Model control	0.1mL/10g vehicle (p.o.)	10	10.10±1.36	30.40±5.32	1.35±0.14
Low dose PEO	50mg/kg propolis essential oil (p.o.)	10	9.62±1.62	31.89±6.86	1.17±0.24
Middle dose PEO	100mg/kg propolis essential oil (p.o.)	10	9.46±1.12	28.54±5.44	1.08±0.18**
High dose PEO	200mg/kg propolis essential oil (p.o.)	10	8.24±1.32**	25.55±3.57*	1.17±0.20*
Positive control	200mg/kg VitC (p.o.)	10	8.61±1.71*	29.21±2.96	0.92±0.10**

与模型对照组比:* $P<0.05$,** $P<0.01$。

2.3 讨论

应激反应是机体对应激原负荷产生的在心理和生理上的不协调症状,是来自各种外界刺激因素(stressor)作用时所引起的一种临床症候群。束缚负荷使动物产生综合的生理及病理性应激反应,其中氧化应激反应被认为是束缚负荷诱发组织损伤的可能机制之一,因此利用束缚应激模型在多靶点整体动物水平上观察药物对机体氧化应激状态的改善作用被国内外广泛应用(Akpinar et al.,2008;Akpinar et al.,2007;Popovic et al.,2009;Saggu et al.,2008;Yoo et al.,2011)。

我们采用束缚方法制备小鼠应激模型,夜晚一般为小鼠活动活跃的时间段,因此束缚时间选择从第一天下午开始至第二天早晨(2:00pm~8:00am),实验动物在束缚应激期间禁水、禁食(Bao et al.,2010)。结果发现,经过 18h 束缚后小鼠血清中 SOD 活性显著下降而MDA 含量显著上升,说明束缚后小鼠处于氧化应激状态。预防性给予小鼠 14d 的蜂胶挥发性成分对小鼠束缚后血清中 SOD 活性剂 MDA 水平起到了显著的调节作用,说明蜂胶挥发性成分对由束缚所致的氧化应激有改善作用。

肝脏是人体最重要的代谢器官之一,是人体应激机制的调节中心,应激反应是否对机体造成损害及损害程度在很大程度上取决于肝功能的正常与否。有研究发现,束缚应激可诱发小鼠急性肝损伤(Chen et al.,2009a;Kaida et al.,2010;Ohashi et al.,2008),束缚后小鼠肝脏的脂质过氧化产物明显增加,总抗氧化能力、维生素 C 及谷胱甘肽的含量明显下降,认为氧化应激是束缚致肝损伤的主要原因之一(Bao et al.,2010;Bao et al.,2008b)。本研究发现,蜂胶挥发性成分对束缚应激所致的小鼠肝损伤具有保护作用,高剂量蜂胶挥发性成分

可显著降低 ALT 活性,部分恢复束缚应激小鼠的肝功能损伤;其作用途径主要是通过有效提高肝脏氧化防御体系的能力和对肝脏氧化损伤的预防或修复作用。

CAT 和 SOD 是重要的抗氧化酶,经常被用来作为揭示 ROS 的形成的生物标志物。蜂胶挥发性成分中、高剂量组显著增加束缚应激小鼠肝脏 SOD 的活性,高剂量组显著增加了 CAT 的活性,说明适当浓度的蜂胶挥发性成分可使肝脏的解毒能力增强,从而调节束缚所致的氧化应激损伤。GSH 是体内重要的内源性抗氧化剂,具有消除自由基和非酶电子亲和物,维持机体蛋白巯基的还原状态,防止蛋白变性,保护机体免受氧化损伤作用。蜂胶挥发性成分的中、高剂量组可使肝脏内 GSH 水平升高,对肝脏中内源性抗氧化活性物质的生成有积极作用。

MDA 是脂质过氧化产物,其水平可间接反应机体细胞受自由基攻击的程度,肝脏中 MDA 的含量可以灵敏地反映机体氧化损伤的状态。一般认为组织损伤时体内或局部产生的活性物质可诱发血管内皮细胞释放 NO,NO 在体内的生物活性与其浓度有关,任何来源的过量 NO 均可通过超氧自由基反应促进细胞膜脂质过氧化,且其增加程度与组织损害呈正相关。本研究发现蜂胶挥发性成分高剂量组可显著降低肝组织中 MDA 及 NO 的含量,对肝脏的脂质过氧化改善作用。蜂胶挥发性成分高剂量组亦显著降低了束缚应激小鼠肝组织中 XOD 的活性,XOD 是检测氧化应激引起组织损伤的最重要的指标之一,它可以催化次黄嘌呤氧化生成黄嘌呤,从而产生大量的自由基,导致细胞结构的破坏和功能障碍,XOD 活性的降低可能是蜂胶挥发性成分改善肝脏脂质过氧化的机制之一。

小　结

本研究通过束缚致小鼠产生氧化应激状态,利用此动物模型对蜂胶挥发性成分能否对机体氧化应激状态起到一定的调节作用进行了研究。结果发现,蜂胶挥发性成分对束缚所致的机体氧化应激状态具有改善作用,可显著降低血清中 MDA 的含量,提高 SOD 的活性;蜂胶挥发性成分还对束缚所致的肝损伤有改善作用,可显著提高肝组织中抗氧化酶系(SOD、CAT)的活性,增加肝组织中内源性抗氧化物质 GSH 的分泌,改善肝组织的脂质过氧化,降低肝组织中 MDA、NO 含量,降低 XOD 活性。

3　蜂胶中挥发性成分对束缚应激小鼠类焦虑样行为的影响及作用途径

焦虑症是以广泛和持续焦虑或反复发作的惊恐不安为主要特征的神经症性障碍。包括广泛性焦虑(GAD)和惊恐障碍(PD)。研究表明,GAD 在普通人群中的发病率达 2%～5%,某些特定人群如老年人、冠心病患者、临考学生发病率可高达 10% 以上(Stein,2004)。随着现代社会工作、人际关系、经济压力等诸多因素作用的加剧,发病率还在上升。虽然一些心理学治疗手段被证明有一定疗效,药物治疗仍然是目前应用最广泛、最有效的手段,尤其是对于重症患者而言(Baldwin et al.,2011)。临床应用的抗焦虑药如安定、丁螺环酮、氟西汀等由于有不同程度的副作用,迫切需要研发疗效高、毒副作用小的抗焦虑新药(Bradley et al.,2007)。

芳香疗法通过植物香气来治疗疾病或是作为人体保健品,已有悠久的历史。国内外许多研究证实,芳香疗法对于神经系统的功效尤为显著,精油可以使人放松精神,减轻忧虑、不安和焦虑的情绪,对于消除疲劳和缓解失眠也有特定效果,同时还可以抑制精神过度兴奋有助于集中注意力、增强记忆力等(Setzer,2009;Tsang et al.,2010)。薰衣草、玫瑰、柠檬及佛手等植物的精油在芳香疗法领域研究较为深入,被证实其中所含的各类化学组分对中枢神经系统有一定的影响(Bradley et al.,2007;Verma et al.,2010)。挥发性成分是蜂胶中的重要生物学活性组分之一。研究发现,蜂胶挥发性成分非常复杂,主要由萜类化合物、脂肪族化合物及芳香族化合物等构成,其化学成分与蜂胶产地及胶原植物有关。对蜂胶挥发性成分的一些基础研究发现,蜂胶挥发性成分具有较强的抑菌活性,且具有抑制脂质过氧化活性。蜂胶挥发性成分对神经系统是否有影响尚无文献报道。

研究发现蜂胶具有神经保护作用,巴西绿蜂胶可降低氧糖剥夺所致视网膜神经节细胞的损伤(Nakajima et al.,2009),并可保护由于缺血引起的神经损伤(Shimazawa et al.,2005);中国蜂胶及其生物学活性组分柯因可通过抑制线粒体凋亡保护神经细胞损伤(Izuta et al.,2008)。Chen等(2008)发现中国蜂胶对神经系统具有保护作用,并可改善东莨菪碱导致的小鼠记忆和学习能力下降。考虑蜂胶对神经系统功能的影响,以及蜂胶挥发性成分所含萜类成分在芳香疗法中的广泛应用,我们初步探索了蜂胶挥发性成分对束缚导致心理应激小鼠焦虑样行为的影响及作用途径。

3.1 材料与方法

3.1.1 实验动物

实验动物为 ICR 小鼠,雄性,体重 18～22g,清洁级,由中国科学院上海实验动物中心/上海斯莱克实验动物公司提供,生产许可证:SCXK(沪)2007-0005。

小鼠饲养于屏蔽环境,温度 23 ± 2℃,日温差 ±1℃,湿度 50%～70%,噪音<50dB,光照:150～200Lx,12h 明暗交替。自来水过滤灭菌,置于高压灭菌的饮水瓶中自由饮用,8:00～8:30am 喂食。实验室使用许可证:SYXK(浙) 2003-0003。

3.1.2 材料与试剂

巴西蜂胶由浙江蜂之语蜂业集团有限公司提供,系该公司从巴西进口的绿色蜂胶;石油醚、无水乙醇、正己烷等均为分析纯试剂;皮质醇(CORT)酶联免疫吸附测定(ELISA)试剂盒,促肾上腺皮质激素(ACTH)酶联免疫吸附测定(ELISA)试剂盒,购自 KapidBio Lab Calabasas(California)美国;超氧化物歧化酶(SOD)、丙二醛(MDA)、谷胱甘肽(GSH)试剂盒,购自南京建成生物工程研究所。

3.1.3 主要仪器与设备

小鼠高架十字迷路,深圳市瑞沃德生命科技有限公司;多功能小鼠自主活动仪,山东省医学科学院设备站;Thermo scientific varioskan flash 光谱多功能读数仪,美国 Thermo 公司。

3.1.4 实验方法

3.1.4.1 小鼠心理应激模型的建立

小鼠拘束装置参考文献方法(Bao et al.,2008a),使用改造为通风良好的 50mL 尖底聚丙烯塑料离心管。给予一次性拘束 18h,制备小鼠心理应激模型。拘束期间小鼠不能进食饮水,与实验组同时间设置不拘束的禁食禁水正常对照组。

3.1.4.2 动物分组及给药

将小鼠按体重随机分成正常对照组、模型组、阳性对照组及蜂胶挥发性成分低、中、高剂量组共6组,每组10只。蜂胶挥发性成分低、中、高剂量组每天灌胃蜂胶挥发性成分50、100、200mg/kg BW,阳性对照组每天灌胃地西泮2mg/kg BW,模型对照组及正常对照组每天灌胃空白溶剂3%吐温80,0.1mL/10g BW。蜂胶挥发性成分及地西泮用3%吐温80配制成相应浓度溶液,按每天0.1mL/10g BW灌胃,为期14d。第14天拘束模型组和蜂胶挥发性成分各组以及阳性对照药组小鼠灌胃30min后进行拘束应激实验,使用改造为通风良好的50mL尖底聚丙烯塑料离心管,小鼠给予一次性拘束负荷18h(14∶00～8∶00),实验期间禁食禁水。

3.1.4.3 动物行为学测定

自主活动实验:解除拘束后,放入小鼠活动光电计数仪活动箱,适应环境10min后,观察5min内小鼠的活动次数。中间用湿布擦拭活动仪,清除粪便用干布擦净后再进行下一只小鼠的测试。

高架十字迷宫实验:小鼠高架十字迷宫包括两个30cm×5cm(长×宽)的开臂和两个30cm×5cm×15cm(长×宽×高)的闭臂,闭臂上部敞开,中央有一个5cm×5cm的开阔部,每种类型的两臂呈相对位置,迷宫距地面45cm。迷路由黑色铁皮构成,为防止小鼠跌落,开臂的周边有一高出0.25cm的边。迷宫上方安置照明系统及摄像头,连接电脑图像分析软件。自主活动实验结束10min后,将小鼠置于迷宫中央,头朝闭臂,观察者距离迷宫中心至少1m。分别记录试验期(通常为5min)内小鼠进入开臂和闭臂的次数和在两臂滞留时间;探头次数(以小鼠头部伸出开臂边缘为准,依据小鼠后段身体所在位置为开臂或闭臂区分为非防卫性和防卫性探头两类);身体伸展行为的次数(小鼠身体后半部不移动,前端向前伸展后回复初始位置,并不连续前行的行为,依据小鼠后段身体所在位置为开臂或闭臂区分为非防卫性和防卫性伸展行为两类)。计算小鼠进入开臂次数和在开臂滞留时间分别占总次数(进入开臂和闭臂次数之和)和占总时间(在开臂与闭臂滞留时间之和)的百分比;防卫性探头次数占总探头次数的百分比;防卫性伸展行为次数占总伸展行为次数的百分比。

3.1.4.4 指标测定

小鼠全血置于肝素处理过的离心管中,4℃以3000r/min离心15min后得到血浆,用ELISA试剂盒测定ACTH和CORT含量。

小鼠脑组织的10%生理盐水匀浆液在4℃,以10000r/min离心10min,测定SOD、MDA和GSH。

3.1.5 统计学分析

所有数值均以均值表示。统计分析使用SPSS 11.5统计分析软件完成。先采用单因素方差分析(One Way ANOVA)进行组间比较,如有差异进一步采用Dunnett's检验。显著性差异水平为$P<0.05$。

3.2 结果

3.2.1 蜂胶挥发性成分对束缚应激小鼠自主活动实验的影响

结果见图14.11。如图14.1所示,束缚应激模型组较正常组小鼠自主活动次数无显著变化($P>0.05$)。蜂胶挥发性成分各剂量组对束缚应激小鼠自主活动次数无显著影响($P>0.05$)。阳性对照药物地西泮显著降低束缚应激小鼠自主活动次数($P<0.05$),显示镇静作用。

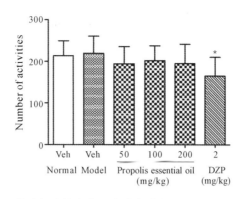

图 14.11 蜂胶挥发性成分对束缚应激小鼠自主活动行为的影响

Fig14.11 Effects of propolis essential oil on spontaneous locomotor activity test in restraint-stressed mice($n=10$).

3.2.2 蜂胶挥发性成分对束缚应激小鼠高架十字迷宫实验的影响

结果见图 14.12 和表 14.18。如图 14.12 和表 14.18 所示,束缚应激模型组较正常组小鼠在高架十字迷宫中开臂停留时间显著减少($P<0.01$),开臂停留时间百分率显著降低($P<0.01$),同时进入开臂次数百分比显著降低($P<0.05$),防卫性探头百分率和防卫性伸展百分率显著增加($P<0.01$,$P<0.05$),显示焦虑状态。束缚应激对小鼠入臂总次数及进入闭臂次数无显著影响($P>0.05$),说明束缚应激对小鼠活动性影响不大。蜂胶挥发性成分对束缚应激小鼠在高架十字迷宫中的焦虑行为有抑制作用:与模型对照组相比,蜂胶挥发性成分低、中剂量组可显著增加小鼠在十字迷路开臂滞留时间的百分率及进入开臂次数百分率($P<0.05$,$P<0.01$),蜂胶挥发性成分各剂量组均可显著减少小鼠防卫性探头次数的百分率($P<0.05$,$P<0.01$),蜂胶挥发性成分中、高剂量组可显著减少小鼠防卫性伸展次数的百分率($P<0.05$,$P<0.01$)。阳性对照药物地西泮可显著增加束缚应激小鼠的开臂滞留时间的百分率及进入开臂次数百分率,降低小鼠防卫性探头百分率及防卫性伸展百分率($P<0.01$)。阳性药物及蜂胶挥发性成分均对小鼠进入闭臂的次数无影响($P>0.05$)。

表 14.18 蜂胶挥发性成分对束缚应激小鼠 EMP 行为学的影响

Table 14.18 Effects of propolis essential oil and diazepam on elevated plus-maze behaviour in restraint-stressed mice

Group	Dosage (administration)	n	Total arm entries	Open arm entries	Closed arm entries	Open arm time(s)
Normal control	0.1mL/10g vehicle (p.o.)	10	13.9±3.8	5.4±1.2	8.5±3.4	153.5±7.4**
Model control	0.1mL/10g vehicle (p.o.)	10	16.1±1.9	4.8±1.3	11.3±2.5	93.6±15.3
Low dose PEO	50mg/kg propolis essential oil (p.o.)	10	16.6±2.0	6.7±1.3**	9.9±1.8	110.0±11.9*
Middle dose PEO	100mg/kg propolis essential oil (p.o.)	10	17.4±2.4	8.1±1.7**	9.3±1.7	136.7±16.3**

续表

Group	Dosage (administration)	n	Total arm entries	Open arm entries	Closed arm entries	Open arm time(s)
High dose PEO	200mg/kg propolis essential oil (p.o.)	10	18.1±2.0*	5.3±1.2	12.7±1.7	102.8±16.7
Positive control	2mg/kg DZP (p.o.)	10	19.9±2.3**	9.8±2.0**	10.1±1.4	141.0±22.4**

与模型对照组比：* $P<0.05$，** $P<0.01$。1

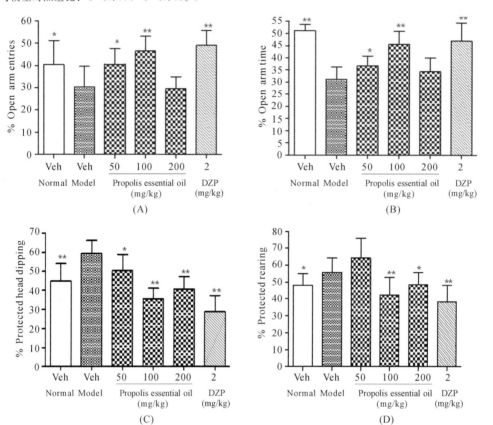

图14.12 蜂胶挥发性成分对束缚应激小鼠EMP行为学的影响
Fig14.12 Effects of propolis essential oil on elevated plus-maze test in restraint-stressed mice($n=10$)

3.2.3 蜂胶挥发性成分对束缚应激小鼠HPA轴基础水平的影响

结果见表14.19。如表14.19所示，束缚应激模型组ACTH及CORT水平较正常组显著升高($P<0.01$，$P<0.05$)；蜂胶挥发性成分各剂量组均对束缚应激小鼠ACTH及CORT水平升高起抑制作用($P<0.05$，$P<0.01$)；阳性对照药物地西泮可显著降低束缚应激小鼠CORT水平($P<0.01$)，对小鼠ACTH水平无显著影响($P>0.05$)。

表14.19 蜂胶挥发性成分对束缚应激小鼠ACTH、CORT的影响
Table 14.19 Effects of propolis essential oil and diazepam on ACTH and CORT in restraint-stressed mice

Group	Dosage (administration)	n	ACTH (pg/mL)	CORT (ng/mL)
Normal control	0.1ML/10g vehicle (p.o.)	10	230.8±33.8**	324.3±45.7*
Model control	0.1ML/10g vehicle (p.o.)	10	322.5±29.9	405.8±46.7

续表

Group	Dosage (administration)	n	ACTH (pg/mL)	CORT (ng/mL)
Low dose PEO	50mg/kg propolis essential oil (*p.o.*)	10	271.4±39.3*	298.8±55.0**
Middle dose PEO	100mg/kg propolis essential oil (*p.o.*)	10	255.1±45.0**	295.9±72.8**
High dose PEO	200mg/kg propolis essential oil (*p.o.*)	10	267.3±46.3*	327.6±56.7*
Positive control	2mg/kg DZP (*p.o.*)	10	300.1±43.5	321.5±41.0**

与模型对照组比：* $P<0.05$，** $P<0.01$。

3.2.4 蜂胶挥发性成分对束缚应激小鼠脑组织脂质过氧化和抗氧化标志物的影响

结果见表14.20。如表14.20所示，束缚应激模型组小鼠脑组织中MDA含量与正常对照组比较显著升高（$P<0.01$），SOD活性及GSH含量较正常对照组显著降低（$P<0.01$）；蜂胶挥发性成分各剂量组均显著降低了束缚应激小鼠脑组织中的MDA含量（$P<0.05$），高剂量组显著提高小鼠脑组织中SOD的活性（$P<0.05$），蜂胶挥发性成分对小鼠脑组织中GSH含量无显著影响（$P>0.05$）；阳性对照药物地西泮可显著降低MDA含量（$P<0.05$），但对SOD的活性无显著影响（$P>0.05$），还显著降低了脑组织中GSH的含量（$P<0.05$）。

表 14.20 蜂胶挥发性成分对束缚应激小鼠脑组织 MDA、SOD、GSH 的影响
Table 14.20 Effects of propolis essential oil and diazepam on MDA, SOD and GSH in brain tissue of restraint-stressed mice

Group	Dosage (administration)	n	MDA (nmol/mg prot)	SOD (U/mg prot)	GSH (mg/g prot)
Normal control	0.1mL/10g vehicle (*p.o.*)	10	4.6±0.4**	87.3±9.5**	9.8±1.0**
Model control	0.1mL/10g vehicle (*p.o.*)	10	5.7±0.6	70.2±11.3	7.4±0.9
Low dose PEO	50mg/kg propolis essential oil (*p.o.*)	10	5.0±0.3*	75.7±10.8	6.9±1.3
Middle dose PEO	100mg/kg propolis essential oil (*p.o.*)	10	4.7±0.7*	78.2±8.6	7.3±1.4
High dose PEO	200mg/kg propolis essential oil (*p.o.*)	10	4.9±0.3*	81.6±9.1*	7.8±1.1
Positive control	2mg/kg DZP (*p.o.*)	10	4.9±0.5*	79.3±11.8	5.9±1.1*

与模型对照组比：* $P<0.05$，** $P<0.01$。

3.3 讨论

行为限制（束缚制动）是制作动物心理应激模型的经典方法之一。动物在束缚应激后，出现明显的行为学变化，包括焦虑、好斗、易怒、学习记忆能力下降等（O'Mahony et al.，2011）。本研究采用一次性束缚小鼠使之心理和躯体疲劳，模拟人类的心理应激状态，建立心理应激致小鼠焦虑模型。研究结果发现，束缚后小鼠在高架十字迷宫实验中在开臂停留时间显著减少，进入开臂次数占总入臂次数的百分比亦显著降低，显示焦虑行为，同时在防卫性探头百分率和防卫性伸展百分率上显著增加，表明受到限制活动导致动物出现紧张、焦

虑等心理应激反应,与人类由于受社会环境和生活压力而产生心理不良应激反应类似(Klenerova et al.,2010;Park et al.,2010)。

高架十字迷宫模型是目前实验室中应用最广泛的检测鼠类动物焦虑相关行为的模型,它以动物对新环境的探究特性和对空旷新环境的回避特性的冲突为基础,产生出一系列的行为指标用以判断动物的焦虑状态(Carobrez et al.,2005)。高架十字迷宫的传统行为指标包括动物进入开臂的次数占总进入次数的百分率以及进入开臂时间占总时间的百分率,抗焦虑药可增加这两个指标。研究表明,在高架十字迷宫实验中结合一系列其他细微的行为学指标,包括探头、站立、伸展、闭臂折返、到达开臂末端等,可提高高架十字迷宫模型的效用,其中防卫性探头和防卫性伸展动作的百分率两个指标被证明是与动物的焦虑状态高度相关(Carobrez et al.,2005;Korte et al.,2003;Yamaguchi et al.,2005)。实验结果显示蜂胶挥发性成分低、中剂量及地西泮作用组小鼠显著增加了开臂停留时间百分比及进入开臂次数百分比,蜂胶挥发性成分高剂量及地西泮对小鼠在高架十字迷宫中的防卫性探头及防卫性伸展有抑制作用,显示蜂胶挥发性成分对小鼠的焦虑样行为有改善作用。另有研究表明,动物的入臂总次数与活动性和焦虑都相关,而进入闭臂的次数只与活动性相关。本研究结果发现,蜂胶挥发性成分及地西泮在抗焦虑剂量下不改变动物进入闭臂的次数,且只增加进入开臂的次数也印证了这一点。

动物的自主活动情况反映其中枢神经系统的功能状态,是评价中枢神经系统兴奋状态的一项重要指标,兴奋时活动次数增加,抑制时活动次数减少(Castaneda et al.,2005)。本实验结果表明地西泮对小鼠的自主活动有抑制作用,显示催眠镇静效果,而蜂胶挥发性成分对束缚应激小鼠自主活动无显著影响,表明蜂胶挥发性成分在抗焦虑剂量下无镇静副作用。

焦虑症发生的病理机制十分复杂,其发病原因和病理环节可能与机体的神经-内分泌-免疫网络(NIM)系统调节紊乱有关。在NIM网络研究中,HPA轴备受关注,研究表明:当机体长期处于应激状态时,HPA轴功能持续亢进,使得交感神经兴奋,神经内分泌可分泌CRH,促使ACTH的产生,此激素作用于肾上腺皮质,使之分泌大量的糖皮质激素,从而适应应激的需要。而HPA轴功能的持续亢进,将严重影响人体的身心健康,从而使机体出现焦虑、抑郁等症状(Kallen et al.,2008;Verma et al.,2010),临床研究亦显示焦虑症患者血浆的CRH、ACTH和CORT浓度结果均较正常对照组高,而经抗焦虑治疗后,随着焦虑症状缓解含量出现下降(Ceccarelli et al.,2004;Saiyudthong et al.,2010),提示HPA轴在焦虑症发病中的作用。本实验研究发现蜂胶挥发性成分对束缚应激小鼠血清CORT及ACTH水平的升高起到了有效的抑制作用,阳性药物地西泮可显著降低束缚小鼠血清CORT水平,对ACTH水平则无显著影响。以上研究结果显示蜂胶挥发性成分抑制束缚应激小鼠焦虑样行为和减少束缚应激小鼠CORT、ACTH水平是一致的。蜂胶挥发性成分这种中枢性影响与抑制焦虑行为是否有关系还有待于进一步研究。

自由基蓄积产生的氧化损伤是焦虑、抑郁、学习能力、记忆力下降等多种神经变性疾病的共同过程,是神经细胞损伤的重要原因之一(Kumar et al.,2010)。临床研究证实焦虑症患者机体脂质过氧化增强,自由基含量明显增加(Matsushita et al.,2010;Ozcan et al.,2004)。本研究发现蜂胶挥发性成分对束缚应激小鼠脑组织脂质过氧化有显著抑制作用,可降低MDA含量,蜂胶挥发性成分高剂量组还可提高脑组织中SOD的活性,显示蜂胶挥发性成分对脑组织氧化应激具有保护作用。

蜂胶挥发性成分中含有的大量萜类化合物,如 α-桉叶醇(6.94%)、β-桉叶醇(6.08%)、蒎烯(1.25%)等,在迷迭香、柠檬草等芳香植物中亦广泛存在,对这些芳香植物挥发性成分的分析及药理学活性研究,支持挥发性成分中萜类成分及具体生物学活性物质对中枢神经系统的保护作用(Bagetta et al.,2010;Blanco et al.,2009;Linck et al.,2010;Satou et al.,2010)。蜂胶挥发性成分对束缚应激小鼠的类焦虑样行为具有改善作用,与降低血清 CORT、ACTH 水平,拮抗 HPA 轴功能亢进及减少脑组织氧化损伤有关。鉴于蜂胶挥发性成分化学组成的复杂性,其发挥抗焦虑活性不仅仅只依赖于一种活性成分,而是多种活性成分协同作用的结果。

小 结

蜂胶挥发性成分可显著改善束缚应激小鼠的类焦虑样行为,其作用途径可能与蜂胶挥发性成分对 HPA 轴功能亢进的拮抗作用及对脑组织氧化损伤的保护作用有关。通过对初步的抗焦虑作用及其相关药效学的研究,结果表明蜂胶挥发性成分具有明显的抗焦虑作用的同时其副作用较小、安全性高,具有一定的应用开发前景。

参考文献

Ahn MR,Kumazawa S,Usui Y,Nakamura J,Matsuka M,Zhu F,Nakayama T(2007). Antioxidant activity and constituents of propolis collected in various areas of China [J]. *Food Chemistry*,101(4):1383-1392.

Akpinar D,Yargicoglu P,Derin N,Aslan M,Agar A(2007). Effect of aminoguanidine on visual evoked potentials (VEPs),antioxidant status and lipid peroxidation in rats exposed to chronic restraint stress[J]. *Brain Res*,1186:87-94.

Akpinar D,Yargicoglu P,Derin N,Aliciguzel Y,& Agar A.(2008). The effect of lipoic acid on antioxidant status and lipid peroxidation in rats exposed to chronic restraint stress [J]. *Physiol Res*,57(6):893-901.

Ali FH,Kassem GM,& Atta-Alla OA(2010). Propolis as a natural decontaminant and antioxidant in fresh oriental sausage[J]. *Vet Ital*,46(2):167-172.

Atungulu G,Miura M,Atungulu E,Satou Y,Suzuki K(2007). Activity of gaseous phase steam distilled propolis extracts on peroxidation and hydrolysis of rice lipids[J]. *Journal of Food Engineering*,80(3):850-858.

Atungulu Griffiths G,Toshitaka U,Fumihiko T,Daisuke H(2008). Effect of vapors from fractionated samples of propolis on microbial and oxidation damage of rice during storage[J]. *Journal of Food Engineering*,88(3):341-352.

Bagetta G,Morrone L A,Rombola L,Amantea D,Russo R,Berliocchi L,Sakurada S,Sakurada T,Rotiroti D,Corasaniti MT(2010). Neuropharmacology of the essential oil of bergamot[J]. *Fitoterapia*,81(6):453-461.

Baldwin DS,Waldman S,Allgulander C(2011). Evidence-based pharmacological treatment of generalized anxiety disorder[J]. *Int J Neuropsychopharmacol*:1-14.

Bankova V,Christov R,Popov S,Marcucci MC,Tsvetkova I,Kujumgiev A(1999).

Antibacterial activity of essential oils from Brazilian propolis[J]. *Fitoterapia*, 70(2): 190-193.

Bankova VS, de Castro SL, Marcucci MC(2000). Propolis: recent advances in chemistry and plant origin[J]. *Apidologie*, 31(1): 3-16.

Bankova V(2005). Chemical diversity of propolis and the problem of standardization[J]. *J Ethnopharmacol*, 100(1-2): 114-117.

Banskota AH, Tezuka Y, Kadota S(2001). Recent progress in pharmacological research of propolis[J]. *Phytother Res*, 15(7): 561-571.

Banskota AH, Tezuka Y, Adnyana IK, Midorikawa K, Matsushige K, Message D, Huertas AA, Kadota S(2000). Cytotoxic, hepatoprotective and free radical scavenging effects of propolis from Brazil, Peru, the Netherlands and China[J]. *J Ethnopharmacol*, 72(1-2): 239-246.

Bao L, Yao XS, Tsi D, Yau CC, Chia CS, Nagai H, Kurihara H(2008). Protective effects of bilberry (*Vaccinium myrtillus* L.) extract on $KBrO_3$-induced kidney damage in mice [J]. *J Agric Food Chem*, 56(2): 420-425.

Bao L, Abe K, Tsang P, Xu JK, Yao XS, Liu HW, Kurihara H(2010). Bilberry extract protect restraint stress-induced liver damage through attenuating mitochondrial dysfunction [J]. *Fitoterapia*, 81(8): 1094-1101.

Bao L, Yao XS, Yau CC, Tsi D, Chia CS, Nagai H, Kurihara H(2008). Protective effects of bilberry (*Vaccinium myrtillus* L.) extract on restraint stress-induced liver damage in mice[J]. *J Agric Food Chem*, 56(17): 7803-7807.

Blanco MM, Costa CA, Freire AO, Santos JG, Jr, Costa M(2009). Neurobehavioral effect of essential oil of *Cymbopogon* citratus in mice[J]. *Phytomedicine*, 16(2-3): 265-270.

Bradley BF, Starkey NJ, Brown SL, Lea RW(2007). The effects of prolonged rose odor inhalation in two animal models of anxiety[J]. *Physiol Behav*, 92(5): 931-938.

Burdock GA(1998). Review of the biological properties and toxicity of bee propolis (propolis) [J]. *Food Chem Toxicol*, 36(4): 347-363.

Carobrez AP, Bertoglio LJ(2005). Ethological and temporal analyses of anxiety-like behavior: the elevated plus-maze model 20 years on[J]. *Neurosci Biobehav Rev*, 29(8): 1193-1205.

Castaldo S, Capasso F(2002). Propolis, an old remedy used in modern medicine[J]. *Fitoterapia*, 73(Suppl 1): S1-6.

Castaneda TR, Jurgens H, Wiedmer P, Pfluger P, Diano S, Horvath T L, Tang-Christensen M, Tschop M H(2005). Obesity and the neuroendocrine control of energy homeostasis: the role of spontaneous locomotor activity[J]. *J Nutr*, 135(5): 1314-1319.

Ceccarelli I, Lariviere WR, Fiorenzani P, Sacerdote P, Aloisi AM(2004). Effects of long-term exposure of lemon essential oil odor on behavioral, hormonal and neuronal parameters in male and female rats[J]. *Brain Research*, 1001(1-2): 78-86.

Chen F, He F, Li J, Luo J, Wu Z, Yu H, Wang L, Lin J(2001). Analysis of essential oil from *Artemisia annul* L. by extraction of different methods[J]. *Zhong Yao Cai*, 24(3):

176-178.

Chen J, Long Y, Han M, Wang T, Chen Q, Wang R (2008). Water-soluble derivative of propolis mitigates scopolamine-induced learning and memory impairment in mice[J]. *Pharmacology Biochemistry and Behavior*, 90(3): 441-446.

Chen M, Wang Y, Zhao Y, Wang L, Gong J, Wu L, Gao X, Yang Z, Qian L (2009). Dynamic proteomic and metabonomic analysis reveal dysfunction and subclinical injury in rat liver during restraint stress[J]. *Biochim Biophys Acta*, 1794(12): 1751-1765.

Chizzola R, Michitsch H, Franz C (2008). Antioxidative properties of Thymus vulgaris leaves: comparison of different extracts and essential oil chemotypes[J]. *J Agric Food Chem*, 56(16): 6897-6904.

Choi YM, Noh DO, Cho SY, Suh HJ, Kim KM, Kim JM (2006). Antioxidant and antimicrobial activities of propolis from several regions of Korea[J]. *LWT-Food Science and Technology*, 39(7): 756-761.

Daugsch A, Moraes C S, Fort P, Park YK (2008). Brazilian red propolis-chemical composition and botanical origin[J]. *Evid Based Complement Alternat Med*, 5(4): 435-441.

Davalos A, Gomez-Cordoves C, Bartolome B (2004). Extending applicability of the oxygen radical absorbance capacity (ORAC-fluorescein) assay[J]. *J Agric Food Chem*, 52(1): 48-54.

Erkmen O, Ozcan MM (2008). Antimicrobial effects of Turkish propolis, pollen, and laurel on spoilage and pathogenic food-related microorganisms[J]. *J Med Food*, 11(3): 587-592.

Farhat A, Fabiano-Tixier AS, Visinoni F, Romdhane M, Chemat F (2010). A surprising method for green extraction of essential oil from dry spices: microwave dry-diffusion and gravity[J]. *J Chromatogr A*, 1217(47): 7345-7350.

Flamini G, Tebano M, Cioni PL, Ceccarini L, Ricci AS, Longo I (2007). Comparison between the conventional method of extraction of essential oil of *Laurus nobilis* L. and a novel method which uses microwaves applied in situ, without resorting to an oven[J]. *J Chromatogr A*, 1143(1-2): 36-40.

Gabriele B, Fazio A, Dugo P, Costa R, Mondello L (2009). Essential oil composition of *Citrus medica* L. Cv. *Diamante* (*Diamante citron*) determined after using different extraction methods[J]. *J Sep Sci*, 32(1): 99-108.

Ghisalberti E (1979). Propolis: a review. Bee world, 60, 59-84.

Gomez-Caravaca AM, Gomez-Romero M, Arraez-Roman D, Segura-Carretero A, Fernandez-Gutierrez A (2006). Advances in the analysis of phenolic compounds in products derived from bees[J]. *J Pharm Biomed Anal*, 41(4): 1220-1234.

Gregoris E, Stevanato R (2008). Correlations between polyphenolic composition and antioxidant activity of Venetian propolis[J]. *Food Chem Toxicol*, 48(1): 76-82.

Gregoris E, Stevanato R (2010). Correlations between polyphenolic composition and antioxidant activity of Venetian propolis[J]. *Food Chem Toxicol*, 48(1): 76-82.

Gregoris E, Fabris S, Bertelle M, Grassato L, Stevanato R (2011). Propolis as potential cosmeceutical sunscreen agent for its combined photoprotective and antioxidant properties [J]. *Int J Pharm*, 405(1-2):97-101.

Gulcin I, Bursal E, Sehitoglu MH, Bilsel M, Goren A C(2010). Polyphenol contents and antioxidant activity of lyophilized aqueous extract of propolis from Erzurum, Turkey [J]. *Food Chem Toxicol*, 48(8-9):2227-2238.

Ho CL, Tseng YH, Wang EI, Liao PC, Chou JC, Lin CN, Su YC(2011). Composition, antioxidant and antimicrobial activities of the seed essential oil of Calocedrus formosana from Taiwan[J]. *Nat Prod Commun*, 6(1):133-136.

Izuta H, Shimazawa M, Tazawa S, Araki Y, Mishima S, Hara H(2008). Protective effects of Chinese propolis and its component, chrysin, against neuronal cell death via inhibition of mitochondrial apoptosis pathway in SH-SY5Y cells[J]. *J Agric Food Chem*, 56(19):8944-8953.

Joulain D, Hochmuth H(2001). *Terpenoids and related constituents of essential oils*[M]. Hamburg: Library of Mass Finder.

Kaida S, Ohta Y, Imai Y, Kawanishi M(2010). Protective effect of L-ascorbic acid against oxidative damage in the liver of rats with water-immersion restraint stress[J]. *Redox Rep*, 15(1):11-19.

Kallen VL, Tulen JH, Utens EM, Treffers PD, De Jong FH, Ferdinand RF (2008). Associations between HPA axis functioning and level of anxiety in children and adolescents with an anxiety disorder[J]. *Depress Anxiety*, 25(2):131-141.

Khalil ML(2006). Biological activity of bee propolis in health and disease[J]. *Asian Pac J Cancer Prev*, 7(1):22-31.

Klenerova V, Krejci I, Sida P, Hlinak Z, & Hynie S (2010). Oxytocin and carbetocin ameliorating effects on restraint stress-induced short-and long-term behavioral changes in rats[J]. *Neuro Endocrinol Lett*, 31(5), 622-630.

Korte SM, De Boer SF(2003). A robust animal model of state anxiety: fear-potentiated behaviour in the elevated plus-maze[J]. *Eur J Pharmacol*, 463(1-3):163-175.

Kujumgiev A, Tsvetkova I, Serkedjieva Y, Bankova V, Christov R, Popov S (1999). Antibacterial, antifungal and antiviral activity of propolis of different geographic origin [J]. *J Ethnopharmacol*, 64(3):235-240.

Kumar A, Garg R, Prakash AK (2010). Effect of St. John's Wort (Hypericum perforatum) treatment on restraint stress-induced behavioral and biochemical alteration in mice[J]. *BMC Complement Altern Med*, 10:18.

Kusumoto T, Miyamoto TR, Doi S, Sugimoto H, Yamada H(2001). Isolation and structures of two new compounds from the essential oil of Brazilian propolis[J]. *Chem Pharm Bull (Tokyo)*, 49(9):1207-1209.

Lacoste E, Chaumont JP, Mandin D, Plumel MM, Matos F J(1996). Antiseptic properties of essential oil of Lippia sidoides Cham. Application to the cutaneous microflora[J]. *Ann Pharm Fr*, 54(5):228-230.

Lazarevic JS, Ethordevic AS, Zlatkovic BK, Radulovic NS, Palic RM (2011). Chemical composition and antioxidant and antimicrobial activities of essential oil of *Allium sphaerocephalon* L. subsp. *sphaerocephalon* (*Liliaceae*) inflorescences[J]. *J Sci Food Agric*, 91(2):322-329.

Linck VM, da Silva AL, Figueiro M, Caramao EB, Moreno PR, Elisabetsky E(2010). Effects of inhaled Linalool in anxiety, social interaction and aggressive behavior in mice[J]. *Phytomedicine*, 17(8-9):679-683.

Marcucci MC(1995). Propolis: chemical composition, biological properties and therapeutic activity[J]. *Apidologie*, 26(2):83-99.

Matsushita M, Kumano-Go T, Suganuma N, Adachi H, Yamamura S, Morishima H, Shigedo Y, Mikami A, Takeda M, Sugita Y (2010). Anxiety, neuroticism and oxidative stress: cross-sectional study in non-smoking college students[J]. *Psychiatry Clin Neurosci*, 64(4):435-441.

Melliou E, Stratis E, & Chinou I(2007). Volatile constituents of propolis from various regions of Greece-Antimicrobial activity[J]. *Food Chemistry*, 103(2):375-380.

Miguel MG, Nunes S, Dandlen SA, Cavaco AM, Antunes MD(2010). Phenols and antioxidant activity of hydro-alcoholic extracts of propolis from Algarve, South of Portugal[J]. *Food Chem Toxicol*, 48(12):3418-3423.

Miraliakbari H, Shahidi F(2008). Antioxidant activity of minor components of tree nut oils[J]. *Food Chemistry*, 111(2):421-427.

Mkaddem MG, Romdhane M, Ibrahim H, Ennajar M, Lebrihi A, Mathieu F, Bouajila J (2010). Essential oil of *Thymus capitatus* Hoff. et Link. from Matmata, Tunisia: gas chromatography-mass spectrometry analysis and antimicrobial and antioxidant activities[J]. *J Med Food*, 13(6):1500-1504.

Mohammadzadeh S, Sharriatpanahi M, Hamedi M, Amanzadeh Y, Sadat Ebrahimi SE, Ostad SN(2007). Antioxidant power of Iranian propolis extract[J]. *Food Chemistry*, 103(3):729-733.

Nakajima Y, Shimazawa M, Mishima S, Hara H(2009). Neuroprotective effects of Brazilian green propolis and its main constituents against oxygen-glucose deprivation stress, with a gene-expression analysis[J]. *Phytother Res*, 23(10):1431-1438.

Ohashi T, Matsui T, Chujo M, Nagao M(2008). Restraint stress up-regulates expression of zinc transporter Zip14 mRNA in mouse liver[J]. *Cytotechnology*, 57(2):181-185.

O'Mahony CM, Clarke G, Gibney S, Dinan TG, Cryan JF (2011). Strain differences in the neurochemical response to chronic restraint stress in the rat: relevance to depression[J]. *Pharmacol Biochem Behav*, 97(4):690-699.

Ozcan ME, Gulec M, Ozerol E, Polat R, Akyol O(2004). Antioxidant enzyme activities and oxidative stress in affective disorders[J]. *Int Clin Psychopharmacol*, 19(2):89-95.

Park YK, Alencar SM, Aguiar CL (2002). Botanical origin and chemical composition of Brazilian propolis[J]. *J Agric Food Chem*, 50(9):2502-2506.

Park HJ, Shim HS, Kim H, Kim KS, Lee H, Hahm DH, Shim I(2010). Effects of Glycyrrhizae Radix on Repeated Restraint Stress-induced Neurochemical and Behavioral Responses[J]. *Korean J Physiol Pharmacol*, 14(6):371-376.

Perez C, Agnese AM, Cabrera JL(1999). The essential oil of *Senecio graveolens* (Compositae): chemical composition and antimicrobial activity tests[J]. *J Ethnopharmacol*, 66(1):91-96.

Petri G, Lembercovics E, Folalvari M(1988). *Examination of differences between propolis (bee glue) produced from different floral environments*[M]. Amsterdam: Elsevier.

Phutdhawong W, Kawaree R, Sanjaiya S, Sengpracha W, Buddhasukh D(2007). Microwave-assisted isolation of essential oil of *Cinnamomum* iners Reinw. ex Bl.: comparison with conventional hydrodistillation[J]. *Molecules*, 12(4):868-877.

Popova M, Bankova V, Butovska D, Petkov V, Nikolova-Damyanova B, Sabatini AG, Marcazzan GL, Bogdanov S(2004a). Validated methods for the quantification of biologically active constituents of poplar-type propolis[J]. *Phytochem Anal*, 15(4):235-240.

Popova M, Bankova V, Naydensky Ch TI, Kujumgiev A(2004b). Comparative study of the biological activity of propolis from different geographic origin: a statistical approach[J]. *Macedonian Pharm Bull*, 50:9-14.

Popovic M, Janicijevic-Hudomal S, Kaurinovic B, Rasic J, Trivic S, & Vojnovic M(2009). Antioxidant effects of some drugs on immobilization stress combinedwith cold restraint stress[J]. *Molecules*, 14(11):4505-4516.

Pourmortazavi SM, Hajimirsadeghi SS(2007). Supercritical fluid extraction in plant essential and volatile oil analysis[J]. *J Chromatogr A*, 1163(1-2):2-24.

Rastogi N, Domadia P, Shetty S, Dasgupta D(2008). Screening of natural phenolic compounds for potential to inhibit bacterial cell division protein FtsZ[J]. *Indian J Exp Biol*, 46(11):783-787.

Saggu S, Kumar R(2008). Effect of seabuckthorn leaf extracts on circulating energy fuels, lipid peroxidation and antioxidant parameters in rats during exposure to cold, hypoxia and restraint (C-H-R) stress and post stress recovery[J]. *Phytomedicine*, 15(6-7):437-446.

Sahinler N, Kaftanoglu O(2005). Natural product propolis: chemical composition[J]. *Nat Prod Res*, 19(2):183-188.

Saiyudthong S, Marsden CA(2010). Acute effects of bergamot oil on anxiety-related behaviour and corticosterone level in rats[J]. *Phytother Res*, doi:10.1002/ptr.3325.

Salomao K, Dantas AP, Borba CM, Campos LC, Machado DG, Aquino Neto FR, de Castro SL(2004). Chemical composition and microbicidal activity of extracts from Brazilian and Bulgarian propolis[J]. *Lett Appl Microbiol*, 38(2):87-92.

Satou T, Murakami S, Matsuura M, Hayashi S, Koike K(2010). Anxiolytic effect and tissue distribution of inhaled Alpinia zerumbet essential oil in mice[J]. *Nat Prod Commun*, 5(1):143-146.

Scazzocchio F, D'Auria FD, Alessandrini D, Pantanella F(2006). Multifactorial aspects of antimicrobial activity of propolis[J]. *Microbiol Res*, 161(4):327-333.

Schaneberg BT, Khan IA(2002). Comparison of extraction methods for marker compounds in the essential oil of lemon grass by GC[J]. *J Agric Food Chem*, 50(6):1345-1349.

Setzer WN(2009). Essential oils and anxiolytic aromatherapy[J]. *Nat Prod Commun*, 4(9):1305-1316.

Sforcin JM, Fernandes A, Jr, Lopes CA, Bankova V, Funari SR(2000). Seasonal effect on Brazilian propolis antibacterial activity[J]. *J Ethnopharmacol*, 73(1-2):243-249.

Sforcin JM, Bankova V(2010). Propolis: is there a potential for the development of new drugs? [J]. *J Ethnopharmacol*, doi:10.1177/0960327110387456.

Shimazawa M, Chikamatsu S, Morimoto N, Mishima S, Nagai H, Hara H(2005). Neuroprotection by Brazilian green propolis against *in vitro* and *in vivo* ischemic neuronal damage[J]. *Evid Based Complement Alternat Med*, 2(2):201-207.

Shunying Z, Yang Y, Huaidong Y, Yue Y, Guolin Z(2005). Chemical composition and antimicrobial activity of the essential oils of Chrysanthemum indicum[J]. *J Ethnopharmacol*, 96(1-2):151-158.

Silva BB, Rosalen PL, Cury JA, Ikegaki M, Souza VC, Esteves A, Alencar SM(2008). Chemical composition and botanical origin of red propolis, a new type of Brazilian propolis[J]. *Evid Based Complement Alternat Med*, 5(3):313-316.

Stein M B(2004). Public health perspectives on generalized anxiety disorder[J]. *J Clin Psychiatry*, 65(Suppl 13):3-7.

Stepanovic S, Antic N, Dakic I, Svabic-Vlahovic M(2003). *In vitro* antimicrobial activity of propolis and synergism between propolis and antimicrobial drugs[J]. *Microbiol Res*, 158(4):353-357.

Suwalsky M, Orellana P, Avello M, Villena F(2007). Protective effect of Ugni molinae Turcz against oxidative damage of human erythrocytes[J]. *Food and Chemical Toxicology*, 45(1):130-135.

Tenore GC, Ciampaglia R, Arnold NA, Piozzi F, Napolitano F, Rigano D, Senatore F(2011). Antimicrobial and antioxidant properties of the essential oil of Salvia lanigera from Cyprus[J]. *Food Chem Toxicol*, 49(1):238-243.

Trueba GP, Sanchez GM, Giuliani A(2004). Oxygen free radical and antioxidant defense mechanism in cancer[J]. *Front Biosci*, 9:2029-2044.

Tsang HW, Ho TY(2010). A systematic review on the anxiolytic effects of aromatherapy on rodents under experimentally induced anxiety models[J]. *Rev Neurosci*, 21(2):141-152.

Uzel A, Sorkun K, Oncag O, Cogulu D, Gencay O, Salih B(2005). Chemical compositions and antimicrobial activities of four different Anatolian propolis samples[J]. *Microbiol Res*, 160(2):189-195.

Verma P, Hellemans KG, Choi FY, Yu W, Weinberg J(2010). Circadian phase and sex effects on depressive/anxiety-like behaviors and HPA axis responses to acute stress

[J]. *Physiol Behav*,99(3):276-285.

Viuda-Martos M,Ruiz-Navajas Y,Fernandez-Lopez J,Perez-Alvarez JA(2008). Functional properties of honey,propolis,and royal jelly[J]. *J Food Sci*,73(9):R117-124.

Wang L,Wang Z,Zhang H,Li X(2009). Ultrasonic nebulization extraction coupled with headspace single drop microextraction and gas chromatography-mass spectrometry for analysis of the essential oil in *Cuminum cyminum* L[J]. *Anal Chim Acta*,647(1):72-77.

Yamaguchi T,Togashi H,Matsumoto M,Yoshioka M(2005). Evaluation of anxiety-related behavior in elevated plus-maze test and its applications[J]. *Nippon Yakurigaku Zasshi*,126(2),99-106.

Yayli N,Yasar A,Gulec C,Usta A,Kolayli S,Coskuncelebi K,Karaoglu S(2005). Composition and antimicrobial activity of essential oils from Centaurea sessilis and Centaurea armena[J]. *Phytochemistry*,66(14),1741-1745.

Yoo KY,Lee CH,Park JH,Hwang IK,Park OK,Kwon SH,Choi JH,Kim DJ,Kwon YG,Kim YM,Won MH(2011). Antioxidant enzymes are differently changed in experimental ischemic hippocampal CA1 region following repeated restraint stress[J]. *J Neurol Sci*,302(1-2):33-42.

Zeng WC,Zhu RX,Jia LR,Gao H,Zheng Y,Sun Q(2011). Chemical composition, antimicrobial and antioxidant activities of essential oil from Gnaphlium affine[J]. *Food Chem Toxicol*.

曾晞,卢玉振,牟兰(2004).GC-MS法分析比较贵州不同产地蜂胶挥发油化学成分[J].生命科学仪器,4(2):28-29.

付宇新,徐元君,陈滨(2009).气相色谱/质谱法分析内蒙古蜂胶挥发性成分[J].分析化学,37(5):745-748.

罗金岳,安鑫南(2005).植物精油和天然色素加工工艺[M].北京:化学工业出版社.

吕泽田(2009).2008年蜂胶市场基本估计与2009年发展方向[J].中国蜂业,60(4):17-19.

王小平,林励,潘建国(2009).不同产地蜂胶挥发油成分的GC-MS比较分析[J].药物分析杂志,29(1):86-90.

赵强,张彬,周武(2007).微波辅助萃取GC/MS联用分析蜂胶挥发油[J].精细化工,24(12):1192-1203.

第十五章 蜂胶残渣中挥发性成分的研究

蜂胶工业化生产是将蜂胶原胶粉碎后,浸泡于高浓度乙醇中,以获取蜂胶中的功效组分,进而加工成蜂胶酊剂、片剂及胶囊等。研究发现,利用乙醇提取加工蜂胶,得到的主要为蜂胶中的非挥发性成分,而作为蜂胶中生物学活性重要组分之一的挥发性成分则在提纯浓缩过程中大量损失。此外,蜂胶工业化生产过程中产生了大量乙醇提取后所剩余的蜂胶残渣,这些蜂胶残渣仍具有浓烈的蜂胶特有香味,但大多数蜂胶加工企业将这些残渣废弃,未合理地开发利用,造成资源的严重浪费。因此,亟须对蜂胶残渣的化学组成、有效成分提取工艺和生物学活性进行系统研究,为蜂胶残渣的再利用提供依据。

本研究通过探索巴西蜂胶工业残渣中挥发性成分的提取方法,挥发性成分组成的分析,并对巴西蜂胶残渣挥发性成分、巴西原胶挥发性成分及巴西蜂胶乙醇提取物的体外抗氧化及抑菌活性的比较研究,探明巴西绿蜂胶工业残渣的利用价值和利用途径,以提高蜂胶资源的利用效率,减少环境污染,推动蜂胶产业的可持续发展。

第一节 响应面法优化微波辅助提取蜂胶残渣中的挥发性成分

1 实验材料

1.1 材料

巴西蜂胶残渣为巴西绿蜂胶经乙醇浸提过滤后剩余残渣再经干燥后所得,由蜂乃宝本铺(南京)保健品有限公司提供。

1.2 主要试剂

石油醚、无水乙醇、正己烷等,均为国产分析纯试剂。

1.3 主要仪器与设备

旋转蒸发器 RE-2000A,上海亚荣生化仪器厂;循环水式多用真空泵 SHB-IIIA,上海豫康教仪器设备有限公司;BQ-MW3 型实验室微波萃取仪,杭州博泉生物科技有限公司。

2 实验方法

2.1 样品处理

巴西绿蜂胶残渣晒干,放入 -18°C 冷冻数日后,粉碎成粉末,过 20 目筛备用。

2.2 蜂胶残渣中挥发性成分的提取

准确称取 100g 蜂胶残渣粉末,置于萃取罐中,在设定的条件下,以石油醚为溶剂进行提取。过滤,石油醚洗涤残渣,合并滤液,45℃下旋转蒸发浓缩至 100mL 左右,置于 −18℃ 冰箱中反复冷冻过滤除蜡。将脱蜡后的溶液于 45℃下旋转蒸发脱溶,得到蜂胶残渣挥发性成分粗提物。将粗提物与无水乙醇按质量体积比 1g∶15mL 充分混匀,于 −18℃ 反复冷冻除蜡,直至无蜡质析出。采用旋转蒸发仪对除蜡后的溶液脱溶,最后得到蜂胶残渣中的挥发性成分。同一条件下重复 3 次取平均值。

2.3 单因素试验

分别选取萃取温度、萃取时间、液固比、提取次数为单独研究因素,在试验条件:微波功率 700W,萃取温度 30℃,萃取时间 30min,液固比 10∶1,提取 1 次不变的情况下,进行单因素试验,研究萃取温度、时间、液固比和提取次数对蜂胶残渣中挥发性成分提取率的影响。

2.4 响应面试验设计

在单因素试验基础上,以萃取温度、萃取时间、液固比、提取次数 4 个因素为试验因素,以挥发性成分提取率为响应指标,采用 Box-Behnken 响应面分析法,利用 Design-Expert 8.0.6.1 软件研究各因素间的交互作用及对挥发性成分提取率的影响,并优化设计微波辅助提取蜂胶残渣中挥发性成分的工艺参数。

2.5 蜂胶残渣中挥发性成分提取率的计算

$$提取率 = \frac{挥发性成分质量}{蜂胶残渣质量} \times 100\%$$

3 结果与讨论

3.1 单因素试验结果与分析

3.1.1 萃取温度对挥发性成分提取率的影响

在萃取时间 30min,液固比 10∶1,提取 1 次的条件下,考查不同萃取温度对巴西蜂胶残渣挥发性成分提取率的影响,如图 15.1 所示。

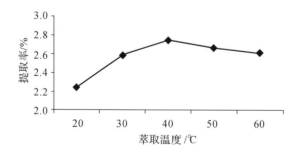

图 15.1 不同萃取温度对蜂胶残渣中挥发性成分提取率的影响

Fig.15.1 Effect of extraction temperature on yield of propolis residue's essential oil

由图 15.1 可知,当萃取温度在 20～40℃时,随着萃取温度的升高,巴西蜂胶残渣挥发性成分提取率逐渐增加,而当萃取温度超过 40℃,提取率有下降趋势。其原因可能是随着温度的上升,分子运动速度加快,渗透和扩散作用加强,使挥发性成分易于溶解在溶剂中,因此,在一定温度范围内,挥发性成分提取率升高。当温度高于一定范围后,提取率开始下降,可能原因是萃取物在溶剂中达到饱和状态,并且溶剂和萃取物挥发增加,造成提取率随着温度的升高而下降。同时萃取温度升高,能耗增加。因此,本研究中选择的较佳工艺萃取温度范围为30~50℃。

3.1.2 液固比对挥发性成分提取率的影响

在萃取温度30℃,萃取时间30min,提取1次的条件下,考查不同液固比对巴西蜂胶残渣挥发性成分提取率的影响,如图15.2所示。

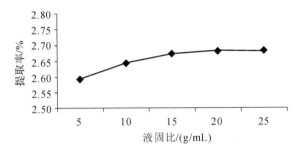

图15.2 不同液固比对蜂胶残渣中挥发性成分提取率的影响

Fig.15.2 Effect of solvent material ratio on yield of propolis residue's essential oil

由图15.2可知,随着液固比的增加,蜂胶残渣挥发性成分的得率相应升高,当液固比大于10g/mL时,挥发性成分提取率曲线逐渐趋于平缓。这可能是因为溶剂用量的增加使得与蜂胶残渣的接触面积增大,从而提高了有效成分的扩散,提取率增加;然而蜂胶残渣中挥发性成分的含量是一定的,当继续增加溶剂的用量时,挥发性成分的提取率增加缓慢。从经济效益考虑,本研究中选取的较佳液固比范围为5~15g/mL。

3.1.3 萃取时间对挥发性成分提取率的影响

在萃取温度30℃,液固比10∶1,提取1次的条件下,考查不同萃取时间对巴西蜂胶残渣挥发性成分提取率的影响,如图15.3所示。

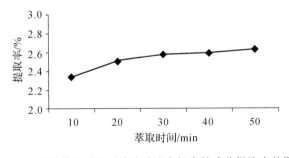

图15.3 不同萃取时间对蜂胶残渣中挥发性成分提取率的影响

Fig.15.3 Effect of extraction time on yield of propolis residue's essential oil

由图15.3可知,在萃取时间10~50min内,随着萃取时间的延长,挥发性成分提取率相应增加。当萃取时间在10~30min之间时,挥发性成分提取率增加较快;当萃取时间在30~50min之间时,挥发性成分增加缓慢。从提高效率和节省能源考虑,较佳萃取时间范围为

10～30min。

3.1.4 提取次数对挥发性成分提取率的影响

在萃取温度30℃,萃取时间30min,液固比10∶1的条件下,考查不同提取次数对巴西蜂胶残渣挥发性成分提取率的影响,如图15.4所示。

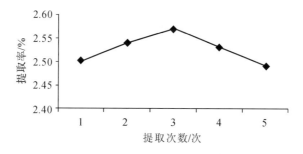

图15.4 不同提取次数对蜂胶残渣中挥发性成分提取率的影响

Fig.15.4 Effect of extraction number on yield of propolis residue's essential oil

由图15.4可知,当提取次数小于3次时,蜂胶残渣中挥发性成分的得率显著增加,而当提取次数超过3次后,提取率呈现降低趋势。可能原因是当提取次数超过3次后,提取溶液总体积增加,导致提取后的冷冻除蜡与蒸发除溶剂过程的时间延长与步骤增加,因此部分挥发性成分损失,另一方面微波辅助提取法容易受机器等因素的影响而导致该方法不稳定,由此可见微波辅助提取法适合定性分析而非定量分析。综合考虑,选取提取次数为1～3次。

3.2 蜂胶残渣挥发性成分提取工艺的响应面分析优化结果与分析

3.2.1 Box-Benhnken设计及响应面分析

根据Box-Benhnken的试验设计原理,选取萃取温度、液固比、萃取时间和提取次数4个因素,对每个因素的低、中、高水平分别以－1、0、1进行编码,以残渣中挥发性成分的提取率作为响应值(Y),对蜂胶残渣挥发性成分的提取工艺进行响应面优化。试验因素与水平设计见表15.1,响应面试验设计方案及结果见表15.2。

表15.1 微波辅助提取蜂胶残渣中挥发性成分工艺优化响应面试验因素水平表

Table 15.1 Factors and their coded levels tested in response surface analysis

因素	水平		
	－1	0	1
萃取温度/℃	40	50	
液固比/mL·g	5	10	15
萃取时间/min	10	20	30
提取次数/次	1	2	3

表 15.2 挥发性成分提取工艺响应面试验设计方案及结果

Table 15.2 Experimental design and results for response surface analysis

Run	萃取温度(A)	液固比(B)	萃取时间(C)	提取次数(D)	提取率/%
1	−1	−1	0	0	2.36
2	0	0	0	0	2.86
3	0	0	0	0	2.88
4	0	0	1	−1	2.72
5	−1	0	1	0	2.52
6	0	1	0	1	2.72
7	1	0	0	1	2.64
8	0	−1	−1	0	2.54
9	0	1	0	−1	2.74
10	0	0	−1	1	2.70
11	0	−1	1	0	2.64
12	1	0	0	−1	2.66
13	0	0	0	0	2.88
14	1	−1	0	0	2.58
15	0	1	1	0	2.68
16	−1	0	0	−1	2.50
17	0	0	1	1	2.70
18	0	0	−1	−1	2.64
19	1	1	0	0	2.56
20	0	0	0	0	2.86
21	−1	0	0	1	2.50
22	0	−1	0	1	2.70
23	−1	1	0	0	2.54
24	0	0	0	0	2.90
25	1	0	−1	0	2.56
26	−1	0	−1	0	2.50
27	1	0	1	0	2.64
28	0	−1	0	1	2.64
29	0	1	−1	0	2.58

运用 Design-Expert 8.0.6.1 软件对表 15.2 中的实验数据进行多元回归拟合、方差分析和显著性检测,得到以响应值挥发性成分提取率 Y 对各条件编码值的二次多项式回归模

型公式为：

$$Y = 2.876 + 0.06A + 0.03B + 0.032C + 0.005D - 0.05AB + 0.015AC - 0.005AD - 0.02BD - 0.02CD - 0.23A^2 - 0.13B^2 - 0.12C^2 - 0.062D^2$$

对所建立的响应模型进行方差分析和回归模型系数显著性检验，结果见表15.3。由表15.3可知，经 F 检验显示回归模型具有极显著性（$P<0.0001$）。回归系数 $R^2 = 0.9764$，表明所建立的模型相关度好，其校正决定系数 $R_{Adj}^2 = 0.9529$，表明此模型能解释95.29%效应值变化，试验误差较小。失拟项 $P=0.1041>0.05$，表明模型的失拟度不显著，该回归模型预测值与实际值能较好地吻合，能够很好地预测残渣中挥发性成分的提取率。此外，由表15.3可知 A、B、C 对挥发性成分提取率影响极显著，D 不显著，以及在两个因素的交互作用中，AB 极显著，其他不显著，A^2、B^2、C^2、D^2 极显著。

根据表15.3中 F 值的大小可以判断各因素对挥发性成分提取率影响的强弱。F 值越大，影响作用越强。因此各因素对挥发性成分提取率的影响程度大小依次为 A（萃取温度）>C（萃取时间）>B（液固比）>D（提取次数）。

表 15.3 响应面试验方差分析结果

Table 15.3 Variance analysis for the developed regression model

方差来源	平方和	自由度	均方和	F 值	P 值
模型	0.4900	14	0.0350	41.4552	<0.0001**
A（萃取温度）	0.0432	1	0.0432	51.1675	<0.0001**
B（液固比）	0.0108	1	0.0108	12.7919	0.0030**
C（萃取时间）	0.0120	1	0.0120	14.2527	0.0020**
D（提取次数）	0.0003	1	0.0003	0.3553	0.5606
AB	0.0100	1	0.0100	11.8443	0.0040**
AC	0.0009	1	0.0009	1.0660	0.3194
AD	0.0001	1	0.0001	0.1184	0.7358
BC	0.0000	1	0.0000	0.0000	1.0000
BD	0.0016	1	0.0016	1.8951	0.1902
CD	0.0016	1	0.0016	1.8951	0.1902
A^2	0.3274	1	0.3274	387.7907	<0.0001**
B^2	0.1176	1	0.1176	139.3286	<0.0001**
C^2	0.0890	1	0.0890	105.4698	<0.0001**
D^2	0.0251	1	0.0251	29.6917	<0.0001**
残差	0.0118	14	0.0008		
失拟项	0.0107	10	0.0011	3.8214	0.1041
纯误差	0.0011	4	0.0003		
总和	0.5018	28			

注：** $P<0.01$，极显著；* $P<0.05$，显著。

3.2.2 响应曲面与等高线分析

采用 Design-Expert 8.0.6.1 软件拟合绘制响应面图和等高线图。根据回归分析方程在考察的区域内绘制的响应曲面图及等高线图可直观反映各因素和响应值以及因素间的两两交互作用,如图 15.5 所示。图 15.5-1 中曲面陡峭,等高线呈椭圆形,表明萃取温度与液固比的交互作用显著。相比较而言,其他因素间的交互作用较小。响应曲面与等高线分析结果与 1.3.2.1 方差分析结果一致。

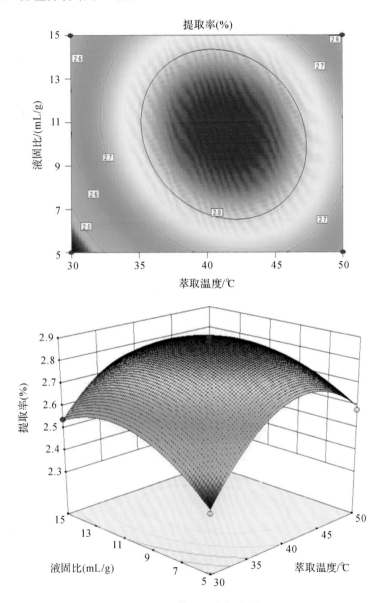

图 15.5-1 萃取温度与液固比

Fig.15.5-1 Extraction temperature and solvent material ratio

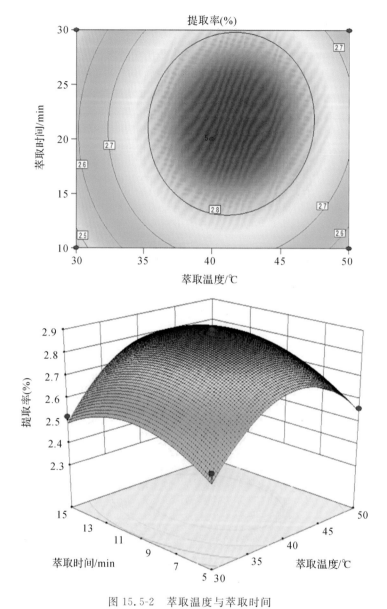

图 15.5-2　萃取温度与萃取时间

Fig. 15.5-2 Extraction temperature and extraction time

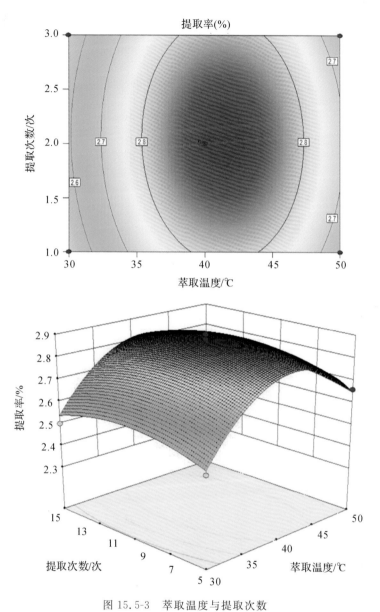

图 15.5-3　萃取温度与提取次数

Fig. 15.5-3　Extraction temperature and extraction number

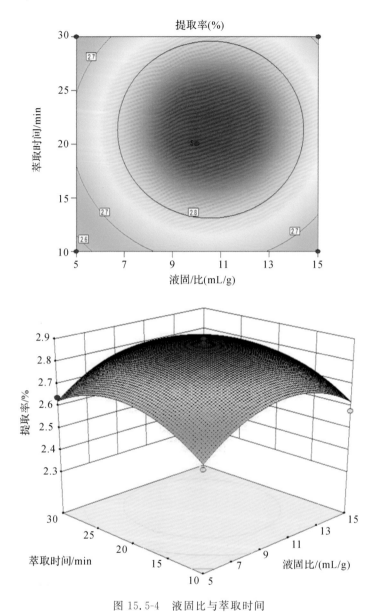

图 15.5-4 液固比与萃取时间

Fig. 15.5-4 Solvent-material ratio and extraction time

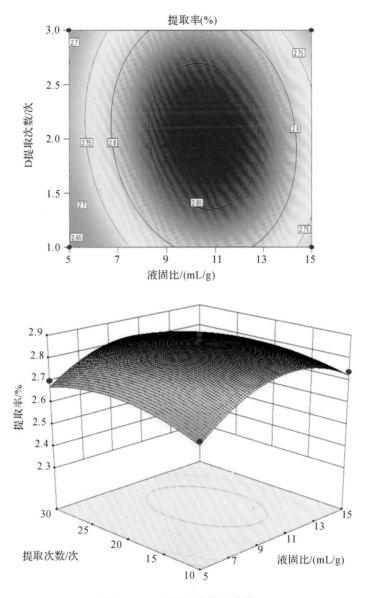

图 15.5-5 液固比与提取次数

Fig. 15.5-5 Solvent-material ratio and extraction number

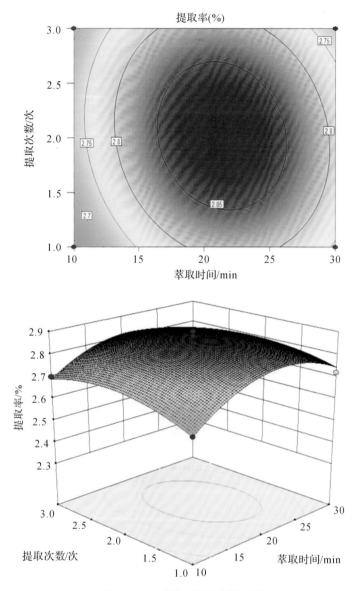

图 15.5-6 萃取时间与提取次数

Fig.15.5-6 Extraction time and extraction number

图 15.5 因素间的两两交互作用的响应曲面及等高线图

Fig.15.5 Response surface and contour polts for the effects of four process parameters on extraction rate

3.3 验证实验

经响应面回归分析得到的最佳提取工艺条件为萃取温度 41.28℃,液固比 10.44∶1 (mL/g),萃取时间 21.44min,提取次数 2 次,在此条件下挥发性成分的提取率理论值为 2.88%。

结合实际操作,把最优条件调整为萃取温度 40℃,液固比 10∶1(mL/g),萃取时间 20min,提取次数 2 次,进行实验,重复 3 次取平均值。在优化条件下获得巴西蜂胶残渣挥发性成分的得率为 2.79%,与理论值较为吻合,相对误差为 3.13%。

小　结

本研究在单因素试验的基础上,采用响应面分析法对微波辅助提取巴西蜂胶残渣中的挥发性成分提取工艺进行优化,并得到回归方程。回归分析结果表明,各因素与响应值之间符合二次关系。各因素对挥发性提取率的影响程度大小依次为萃取温度＞萃取时间＞液固比＞提取次数。其中萃取温度与液固比之间的交互作用对挥发性成分提取率的影响最显著。通过回归模型进行参数优化,结合实际操作的方便,确定挥发性成分的最佳提取工艺为萃取温度 40℃,液固比 10∶1(mL/g),萃取时间 20min,提取次数 2 次。在此条件下,挥发性成分的实际提取率为 2.79%,与理论值 2.88% 较为吻合,说明所建立的模型能较好地预测微波辅助提取的实际提取率。因此,采用响应面法优化巴西蜂胶残渣中挥发性成分的提取工艺是可行的。本研究为蜂胶残渣中挥发性成分的提取提供了一定的参考。

第二节　巴西蜂胶残渣挥发性成分与巴西原胶挥发性成分的分析比较

1　实验材料

1.1　材料

巴西蜂胶原胶由蜂乃宝本铺(南京)保健品有限公司提供;巴西蜂胶残渣为巴西绿蜂胶经乙醇浸提过滤后剩余残渣干燥后所得,由蜂乃宝本铺(南京)保健品有限公司提供。

1.2　主要试剂

石油醚、无水乙醇、正己烷等,均为国产分析纯试剂。

1.3　主要仪器与设备

旋转蒸发器 RE-2000A,上海亚荣生化仪器厂;循环水式多用真空泵 SHB-ⅢA,上海豫康教仪器设备有限公司;BQ-MW3 型实验室微波萃取仪,杭州博泉生物科技有限公司;EQ-500E 型超声波清洗器,昆山市超声仪器有限公司;GC-MS 联用仪 6890N/5975B,安捷伦科技有限公司;HP-5MS 毛细管柱(30m×I.D.0.25mm×Film 0.25μm)。

2　实验方法

2.1　挥发性成分的提取

蜂胶残渣中挥发性成分的提取采用微波辅助有机溶剂提取法。蜂胶残渣放入 −18℃ 冷冻数日后,粉碎成粉末,过 20 目筛备用。准确称取 100g 样品,置于萃取罐中,萃取温度 40℃,液固比 10∶1,萃取时间 20min,提取次数 2 次的条件下,以石油醚为溶剂进行提取,过滤,石油醚洗涤残渣,合并滤液,45℃ 下旋转蒸发浓缩至 100mL 左右,于 −18℃ 中反复冷冻过滤除蜡。将脱蜡后的溶液于 45℃ 下旋转蒸发脱溶,得到残渣挥发性成分粗提物。将粗提物与无水乙醇按质量体积比 1g：15mL 充分混匀,于−18℃反复冷冻除蜡,直至无蜡质析出。采用旋转蒸发仪对除蜡后的溶液脱溶,最后得到蜂胶残渣挥发性成分。同一条件下重复 3 次取平均值。

蜂胶原胶挥发性成分的提取同残渣挥发性成分的提取。

2.2　GC-MS 分析

参照李雅晶等（2011）方法。将样品溶于正己烷,过 0.45μm 滤膜后进行 GC-MS 检测。

GC 条件：色谱柱为 HP-5MS 毛细管柱（30m×I.D.0.25mm×Film 0.25μm）；载气为氦气,1mL/min；进样口温度 250℃；分流进样量为 1.0μL；分流比 20∶1；柱温升温程序为初始 80℃,保持 5min,以 5℃/min 速度升温至 185℃,保持 5min。

MS 条件：电离方式为 EI；离子源温度 230℃；四极杆温度 150℃；接口温度 280℃；电子能量 70eV；质量扫描范围为 30～500amu。

3　结果与讨论

3.1　巴西蜂胶残渣和巴西蜂胶原胶中挥发性成分的得率及性状比较

采用微波辅助石油醚提取法对巴西蜂胶残渣和巴西蜂胶原胶中的挥发性成分进行提取,挥发性成分的得率及色泽、性状、气味情况见表 15.4。由表 15.4 可知,巴西蜂胶残渣及蜂胶原胶的挥发性得率分别为 2.79% 和 5.04%,这可以说明在蜂胶通过乙醇浸提法制备蜂胶酊这一加工过程中,一定量的挥发性物质会溶于蜂胶酊。对比巴西蜂胶残渣与蜂胶原胶中提取得到的挥发性成分,两者在气味上均具有浓郁的蜂胶特有芳香味,其中蜂胶原胶中提取得到的挥发性成分香味略带辛辣。色泽上也存在差异,巴西蜂胶残渣的挥发性成分为深绿色,蜂胶原胶的挥发性成分为金黄色。

表 15.4　巴西蜂胶残渣及蜂胶原胶中挥发性成分得率（$\bar{x}\pm s, n=3$）

Table 15.4　Yield of volatile components in raw material and residues（$\bar{x}\pm s, n=3$）

样品	挥发性成分得率/%	色泽	性状	气味
巴西蜂胶残渣	2.79±0.07	深绿色	油状	浓郁的芳香味
巴西蜂胶原胶	5.04±0.18	金黄色	油状	浓郁的芳香味,微辛辣

3.2　巴西蜂胶残渣和巴西蜂胶原胶中挥发性成分的化学组成

采用 GC-MS 对微波辅助石油醚提取巴西蜂胶残渣与巴西蜂胶原胶的挥发性成分进行分析,总离子流色谱图如图 15.6 和 15.7 所示。从两张图中可以明显看出,挥发性成分的出峰时间主要集中在 16～30min,图谱峰都比较密,挥发性化合物数量整体较多,两者的挥发性成分化学组成复杂,在化合物组成及峰度上存在差异。

在确定各组分质谱数据和扫描峰号后,对 NIST/WILEY 检索数据库进行检索,采用峰面积归一化法计算解析组分的百分含量,并根据文献进行谱图解析。经分析得到巴西蜂胶残渣中挥发性成分的主要物质如表 15.5 所示。由表 15.5 可知,采用 GC-MS 技术对巴西蜂胶残渣挥发性成分与原胶挥发性成分进行分析,鉴定出巴西蜂胶残渣挥发性组分 42 种,主要是萜烯类及其衍生物,单体类化合物所占比例较高的为苯丙酸乙酯（24.462%）、苯丙酸

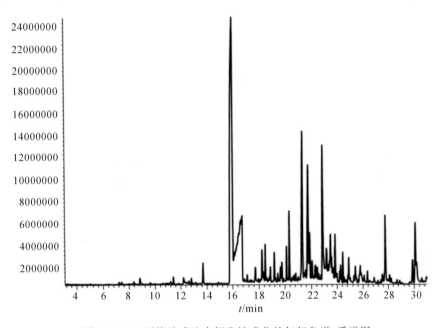

图 15.6 巴西蜂胶残渣中挥发性成分的气相色谱-质谱图

Fig. 15.6 GC-MS spectrum of volatile components in raw material

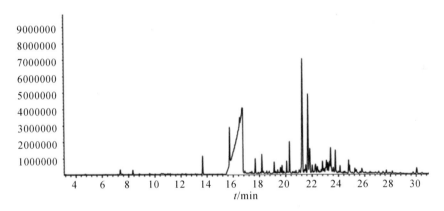

图 15.7 巴西蜂胶原胶中挥发性成分的气相色谱-质谱图

Fig. 15.7 GC-MS spectrum of volatile components in residues

(14.459%)、橙花叔醇(6.185%)和斯巴醇(4.546%)等,占总挥发性成分的 79.283%;巴西原胶挥发性组分 28 种,如苯丙酸(46.528%)、橙花叔醇(10.616%)、斯巴醇(6.441%)和苯丙酸甲酯(5.092%)等,占总挥发性组分含量的 86.073%。本研究获得的巴西蜂胶残渣挥发性组分多于巴西原胶,可能原因是原胶挥发性成分浓度较高,导致部分相对含量低的组分未被解析。

表 15.5　巴西蜂胶残渣与蜂胶原胶中的主要挥发性物质
Table 15.5　Main volatile substances in the raw material and residues

序号	保留时间/min	物质名称	化学式	相对含量/% 蜂胶残渣	相对含量/% 蜂胶原胶
1	7.212	α-甲基-苯甲醇 alpha-methyl-Benzenemethanol	$C_8H_{10}O$	0.072	—
2	7.406	苯乙酮 Acetophenone	C_8H_8O	0.083	0.370
3	8.351	芳樟醇 3,7-dimethyl-1,6-Octadien-3-ol	$C_{10}H_{18}O$	0.072	0.298
4	8.828	苯乙醇 Phenylethyl Alcohol	$C_8H_{10}O$	0.239	—
5	9.627	苄腈 Benzyl nitrile	C_7H_5N	0.049	—
6	11.197	(S)-4-三甲基-3-环己烯-1-甲醇 (S)-4-trimethyl-3-Cyclohexene-1-methanol	$C_{10}H_{18}O$	0.099	—
7	11.412	十二烷 Dodecane	$C_{12}H_{26}$	0.172	—
8	12.204	2,3-二氢苯并呋喃 2,3-dihydro-Benzofuran	C_8H_8O	0.298	—
9	12.803	苯乙酸乙酯 Benzeneacetic acid ethyl ester	$C_{10}H_{12}O_2$	0.130	—
10	13.149	对甲氧基苯甲醛 4-methoxy-Benzaldehyde	$C_8H_8O_2$	0.058	—
11	13.707	苯丙酸甲酯 Benzenepropanoic acid, methyl ester	$C_{10}H_{12}O_2$	0.553	5.092
12	15.993	3-苯丙酸乙酯 Benzenepropanoic acid, ethyl ester	$C_{11}H_{14}O_2$	24.462	—
13	16.717	苯丙酸 Benzenepropanoic acid	$C_9H_{10}O_2$	14.459	46.528
14	17.106	十四烷 n-Tetradecane	$C_{14}H_{30}$	0.249	—
15	17.377	长叶稀 decahydro-4,8,8-trimethyl-9-methylene-1,4-Methanoazulene	$C_{15}H_{24}$	0.373	—
16	17.417	1a,2,3,4,4a,5,6,7b-八氢-1,1,4,7-四甲基-1H-环丙萘 1a,2,3,4,4a,5,6,7b-octahydro-1,1,4,7-tetramethyl-1H-Cyclopropa[a]naphthalene	$C_{15}H_{24}$	—	0.117
17	17.730	石竹烯 Caryophyllene	$C_{15}H_{24}$	0.619	0.948

续表

序号	保留时间/min	物质名称	化学式	相对含量/% 蜂胶残渣	相对含量/% 蜂胶原胶
18	17.962	1a,2,3,5,6,7,7a,7b-八氢-1,1,7,7a-四甲基-1H-环丙萘 1a,2,3,5,6,7,7a,7b-octahydro-1,1,7,7a-tetramethyl-1H-Cyclopropa[a]naphthalene	$C_{15}H_{24}$	0.152	—
19	18.240	香橙烯 decahydro-1,1,7-trimethyl-4-methylene-1H-Cycloprop[e]azulene	$C_{15}H_{24}$	1.413	1.190
20	18.487	茴香酸乙酯 4-methoxy-Benzoic acid ethyl ester	$C_{10}H_{12}O_3$	1.106	—
21	18.559	α-石竹烯 α-Caryophyllene	$C_{15}H_{24}$	—	0.172
22	18.873	(E)-苯丙烯酸乙酯 (E)-3-phenyl-2-Propenoic acid ethyl ester	$C_{11}H_{12}O_2$	0.683	—
23	19.180	7-甲基-4-亚甲基-1-(1-甲基乙烯基)-1,2,3,4,4a,5,6,8a-八氢化萘 1,2,3,4,4a,5,6,8a-octahydro-7-methyl-4-methylene-1-(1-methylethyl)-Naphthalene	$C_{15}H_{24}$	0.959	0.643
24	19.394	十氢-4a-甲基-1-亚甲基-7-(1-甲基乙烯基)-萘 decahydro-4a-methyl-1-methylene-7-(1-methylethenyl)-naphthalene	$C_{15}H_{24}$	—	0.205
25	19.447	4a,8-二甲基-2-(1-甲基乙烯基)-1,2,3,4,4a,5,6,8a-八氢化萘 1,2,3,4,4a,5,6,8a-octahydro-4a,8-dimethyl-2-(1-methylethenyl)-Naphthalene	$C_{15}H_{24}$	0.335	0.503
26	19.654	十氢-1,1,3a-三甲基-7-亚甲基-1H-环丙萘 decahydro-1,1,3a-trimethyl-7-methylene-1H-Cyclopropa[a]naphthalene	$C_{15}H_{24}$	0.519	—
27	19.720	4,7-二甲基-1-(1-甲基乙烯基)-1,2,4a,5,6,8a-六氢化萘 1,2,4a,5,6,8a-hexahydro-4,7-dimethyl-1-(1-methylethyl)-Naphthalene	$C_{15}H_{24}$	—	0.521
28	19.888	9,10-二氢化异长叶烯 9,10-dehydro-Isolongifolene	$C_{15}H_{22}$	0.188	0.144
29	20.124	4,7-二甲基-1-(1-甲基乙烯基)-1,2,4a,5,6,8a-六氢化萘 1,2,4a,5,6,8a-hexahydro-4,7-dimethyl-1-(1-methylethyl)-Naphthalene	$C_{15}H_{24}$	0.929	0.750
30	20.344	4,7-二甲基-1-(1-甲基乙烯基)-1,2,3,5,6,8a-六氢化萘 1,2,3,5,6,8a-hexahydro-4,7-dimethyl-1-(1-methylethyl)-Naphthalene	$C_{15}H_{24}$	1.769	1.951

续表

序号	保留时间/min	物质名称	化学式	相对含量/% 蜂胶残渣	相对含量/% 蜂胶原胶
31	20.559	1,6-二甲基-4-(1-甲基乙烯基)-1,2,3,4,4a,7-六氢化萘 1,2,3,4,4a,7-hexahydro-1,6-dimethyl-4-(1-methylethyl)-Naphthalene	$C_{15}H_{24}$	0.203	0.157
32	20.822	二去氢菖蒲烯 Calacorene	$C_{15}H_{20}$	0.378	—
33	21.341	橙花叔醇 3,7,11-trimethyl-1,6,10-Dodecatrien-3-ol	$C_{15}H_{26}O$	6.185	10.616
34	21.771	(-)-斯巴醇 (-)-Spathulenol	$C_{15}H_{24}O$	4.546	6.441
35	21.878	(-)-蓝桉醇 (-)-Globulol	$C_{15}H_{26}O$	2.136	2.531
36	22.069	1,2,3,3a,4,5,6,7-八氢-1,4-二甲基-7-(1-甲基乙烯基)-甘菊环 1,2,3,3a,4,5,6,7-octahydro-1,4-dimethyl-7-(1-methylethenyl)-Azulene	$C_{15}H_{24}$	0.825	0.759
37	22.312	喇叭茶醇 Ledol	$C_{15}H_{26}O$	0.749	1.094
38	22.450	3,4-二甲基-3-环己烯-1-甲醛 3,4-dimethyl-3-Cyclohexen-1-carboxaldehyde	$C_9H_{14}O$	0.646	—
39	22.509	1,6-二甲基-4-(1-甲基乙烯基)-1,2,3,4,4a,7-六氢化萘 1,2,3,4,4a,7-hexahydro-1,6-dimethyl-4-(1-methylethyl)-Naphthalene	$C_{15}H_{24}$	—	0.246
40	22.569	1a,2,3,4,4a,5,6,7b-八氢化-1,1,4,7-四甲基-1H-环丙烯并[e]奥 1a,2,3,4,4a,5,6,7b-octahydro-1,1,4,7-tetramethy-1H-Cycloprop[e]azulene	$C_{15}H_{24}$	0.361	0.465
41	23.175	τ-杜松醇 tau.-Cadinol	$C_{15}H_{26}O$	1.950	1.224
42	23.489	α-杜松醇 alpha.-Cadinol	$C_{15}H_{26}O$	3.670	1.915
43	24.835	7R,8R-8-羟基-4-异丙基-7-甲基双环[5.3.1]十一碳-1-烯 7R,8R-8-Hydroxy-4-isopropylidene-7-methylbicyclo[5.3.1]undec-1-ene	$C_{15}H_{24}O$	—	0.562
44	24.906	6-异丙烯基-4,8a-二甲基-1,2,3,5,6,7,8,8a-八氢化萘-2-酚 6-Isopropenyl-4,8a-dimethyl-1,2,3,5,6,7,8,8a-octahydro-naphthalen-2-ol	$C_{15}H_{24}O$	1.439	—

续表

序号	保留时间/min	物质名称	化学式	相对含量/% 蜂胶残渣	相对含量/% 蜂胶原胶
45	25.782	苯甲酸苄酯 Benzyl Benzoate	$C_{14}H_{12}O_2$	—	0.488
46	25.798	肉豆蔻酸 Tetradecanoic acid	$C_{14}H_{28}O_2$	1.655	0.143
47	26.357	十四酸乙酯 Tetradecanoic acid ethyl ester	$C_{16}H_{32}O_2$	0.409	—
48	29.894	棕榈酸甲酯 Hexadecanoic acid methyl ester	$C_{17}H_{34}O_2$	0.834	—
49	30.113	1-甲基-4-硝基-2-(3-硝基苯亚甲基氨基)-苯 1-methyl-4-nitro-2-(3-nitrobenzylidenamino)-Benzene	$C_{14}H_{11}N_3O_4$	3.257	—
		总含量		79.283	86.073

分析比较后发现,蜂胶残渣挥发性成分与蜂胶原胶挥发性成分中含有一定量的相同物质,但是各物质的含量差异较大。两者共有的挥发性组分有苯乙酮、芳樟醇、苯丙酸甲酯、苯丙酸、石竹烯、香橙烯、9,10-二氢化异长叶烯、橙花叔醇、(-)-斯巴醇、(-)-蓝桉醇、喇叭茶醇、杜松醇和肉豆蔻酸等21种成分,占蜂胶残渣与蜂胶原胶总挥发性组分含量相对较高,分别为43.659%和83.762%。共有组分可能是蜂胶特有香味的主要来源。

巴西蜂胶植物来源主要是酒神菊属,所含黄酮类物质较少而萜烯类物质含量较高。本研究结果中,巴西蜂胶残渣挥发性成分中萜烯类物质及其含氧衍生物达22种,占总含量的29.770%;巴西蜂胶原胶挥发性成分中萜烯类物质及其含氧衍生物达23种,占总含量的33.452%。两者共有的萜烯类物质有芳樟醇、石竹烯、香橙烯、橙花叔醇和斯巴醇等17种。巴西蜂胶残渣挥发性成分与巴西原胶挥发性成分具有较多相同的萜烯类化合物,但在含量上差异较大。

对蜂胶残渣与蜂胶原胶挥发性成分中的主要物质分析比较后发现,巴西蜂胶原胶挥发性成分中的主要物质为酚酸类化合物、芳香族化合物及大量的萜烯类化合物,而不存在烷烃类化合物,巴西原胶挥发性成分中特有的组分为3-苯丙酸乙酯、茴香酸乙酯、6-异丙烯基-4,8a-二甲基-1,2,3,5,6,7,8,8a-八氢化萘-2-酚、二去氢菖蒲烯等;巴西蜂胶残渣挥发性成分中主要物质为大量的萜烯类化合物及其含氧衍生物和芳香族化合物、少量的酚酸类化合物、有机酸酯、烷烃类化合物、醇类化合物,并且有少量的酯类化合物,其特有的组分为苯甲酸苄酯、α-石竹烯、7R,8R-8-羟基-4-异丙基-7-甲基双环[5.3.1]十一碳-1-烯等。巴西蜂胶残渣挥发性成分中含有较多的酯类化合物,而原胶挥发性成分中却很少,可能原因是巴西蜂胶经乙醇提取,蜂胶原胶挥发性成分中的一些酸类与乙醇起酯化反应生成了酯类化合物,如蜂胶原胶挥发性成分中的苯丙酸易与醇起酯化反应生成3-苯丙酸乙酯。

在关于巴西蜂胶挥发性成分的报道中,Kusumoto等(1987)将巴西蜂胶水蒸气蒸馏5h,蒸馏液经乙醚提取,无水硫酸钠干燥后得到蜂胶挥发性成分,得率为0.34%,得到7种已知成分:2,2-二甲基-6-乙烯色原烯、苯乙酮、2-异戊烯-4-乙烯苯酚、3,4-二甲氧基-苯乙烯、3,4-

二甲氧基-烯丙基苯、4-羟基-3,5-二异戊烯-苯甲醛、匙叶桉油烯醇。并鉴定出 2 种新成分:2,2-二甲基-8-异戊烯-6-乙烯色原烯和 2,6-二异戊烯-4-乙烯苯酚。Atungulu 等(2007)利用同时蒸馏萃取法对巴西蜂胶进行提取,得到了蜂胶的挥发性物质,得率为 0.09%,用 GC-MS 法对其进行分析,鉴定出 30 余个化合物,主要成分为丙烯、苯乙酮、β-芳樟醇。对比以上研究结果,本研究采用微波辅助提取法对巴西蜂胶进行提取,原胶挥发性成分的得率为 5.04%,远远高于以上研究,说明微波辅助提取法在提取率上占有优势,同时可以发现萜烯类化合物是巴西蜂胶挥发性成分中重要的组分。

黄帅等(2013)采用超声波辅助萃取对巴西绿蜂胶原胶、乙醇提取液及醇提后蜂胶渣中的挥发性成分进行提取,GC-MS 分析鉴定出蜂胶原胶挥发性成分中主要活性物质 16 种,主要是芳香酸及烯萜类化合物;蜂胶乙醇提取液及蜂胶渣挥发性成分中主要活性物质分别为 16 种和 14 种,主要为烯萜类化合物和有机酸酯类。蜂胶渣中鉴定出的挥发性成分中主要物质包括萜烯类和有机酸酯类化合物,如 3-苯丙酸乙酯(16.752%)、反式-橙花叔醇(9.798%)、1,2-苯二甲酸二(2-甲基丙基)酯(8.388%)、(+)-d-杜松烯(4.390%)、(-)-a-芹子烯(2.949%)等。与本研究结果中共有的成分仅有 3-苯丙酸乙酯、石竹烯、香橙烯、橙花叔醇 4 种,并且在含量上差异较大,很大的原因是提取方法的不同,还有可能是本研究中所采用的残渣来自巴西蜂胶产业化生产后,而黄帅等人的研究中所采用的为实验室醇提的巴西蜂胶残渣。

萜烯类化合物和芳香族化合物是蜂胶挥发性成分中的重要组成物质,具有多样的生物活性:抑菌、抗炎、降血脂、降血压等(Jorge et al.,2006)。如本研究中巴西蜂胶残渣挥发性成分含有的芳樟醇、苯乙醇、松油醇、长叶烯、石竹烯、香橙烯和橙花叔醇等是市场上最为常见的香精香料,具有抑菌、驱虫等功效;石竹烯具有平喘止咳的功效(吴寿金等,2002);3-苯丙酸乙酯和苯丙酸可用作医药中间体。这些可以为蜂胶残渣挥发性成分的生物学活性研究提供基础,并且为提高蜂胶残渣的利用价值作参考。

小 结

本研究利用微波辅助石油醚萃取技术得到巴西蜂胶残渣挥发性成分,通过对巴西蜂胶原胶及蜂胶残渣中挥发性成分的分析与比较发现,它们的提取率、色泽、性质上存在一定差异,说明在蜂胶工业生产加工过程中,一定量的挥发性物质溶于蜂胶酊中。在利用蜂胶渣进行挥发性成分开发过程中可适当改善蜂胶醇提工艺,减少挥发性物质的损失,以保存更多低沸点易挥发的萜烯类化合物及萜类含氧衍生物,从而保证挥发性成分的性状。

利用 GC-MS 技术分析鉴定出巴西蜂胶残渣挥发性组分 42 个,占总挥发性成分的 79.283%,单体类化合物所占比例较高的为苯丙酸乙酯、苯丙酸、橙花叔醇和斯巴醇等;巴西原胶挥发性组分 28 种,如苯丙酸、橙花叔醇、斯巴醇和苯丙酸甲酯等,占总挥发性组分含量的 86.073%。两者共有苯乙酮、芳樟醇、苯丙酸甲酯、苯丙酸、石竹烯和香橙烯等 21 种组分,其中 17 种为萜烯类化合物。结果说明巴西蜂胶残渣与蜂胶原胶挥发性成分在含量及化学组分上存在一定差异,但共有大量的生物学活性物质。

第三节　巴西蜂胶残渣挥发性成分的生物学活性

1 巴西蜂胶残渣挥发性成分的抗氧化活性

1.1 实验材料

1.1.1 材料

巴西蜂胶、巴西蜂胶残渣：由蜂乃宝本铺(南京)保健品有限公司提供。

1.1.2 主要试剂

石油醚、无水乙醇、正己烷、过硫酸钾等，均为国产分析纯试剂；DPPH(2,2-二苯基-1-苦肼自由基)、Vc(抗坏血酸)、VE(生育酚)、ABTS[2,2-联氮-二(3-乙基苯并噻唑-6-磺酸)二铵盐]均为Sigma分析纯。

1.1.3 主要仪器与设备

旋转蒸发器RE-2000A，上海亚荣生化仪器厂；循环水式多用真空泵SHB-IIIA，上海豫康教仪器设备有限公司；BQ-MW3型实验室微波萃取仪，杭州博泉生物科技有限公司；酶标仪SpectraMax M5，美国Molecular Devices公司。

1.2 实验方法

1.2.1 样品的制备

巴西蜂胶残渣挥发性成分的制备采用微波辅助石油醚提取法。在萃取温度40℃，液固比10∶1(mL/g)，萃取时间20min，提取次数2次的条件下对巴西蜂胶残渣进行提取，过滤后合并滤液，45℃下旋转蒸发浓缩至100mL左右，于−18℃中反复冷冻过滤除蜡。将脱蜡后的溶液于45℃下旋转蒸发脱溶，得到蜂胶残渣挥发性成分粗提物。将粗提物与无水乙醇按质量体积比1g∶15mL充分混匀，于−18℃反复冷冻除蜡，直至无蜡质析出。采用旋转蒸发仪对除蜡后的溶液脱溶，最后得到蜂胶残渣中的挥发性成分。

巴西蜂胶挥发性成分的制备同巴西蜂胶残渣挥发性成分的制备。

蜂胶乙醇提取物的制备：在85%乙醇，液固比15∶1，超声15min，提取2次的条件下对巴西蜂胶进行提取，过滤，合并滤液，于−18℃中反复冷冻过滤除蜡，将脱蜡后的溶液在45℃下旋转蒸发浓缩，将浓缩液倒入平板中并放入50℃烘箱中烘干，置于−18℃冰箱中冷冻保存。

分别精密称取待测挥发性成分、乙醇提取物和蜂胶乙醇提取物样品及Vc、Ve阳性对照品约25mg，用乙醇定容至25mL即为1mg/mL的样品储备溶液和对照品储备液。所有样品及对照品测试前用无水乙醇稀释成所需浓度。

1.2.2 DPPH自由基清除能力的测定

精密称取DPPH标准品约25mg，用无水乙醇定容至25mL，即得DPPH标准工作液(1mg/mL)，低温(4℃)、避光保存备用。使用时稀释10倍(100μg/mL)。

将不同浓度的样品和对照品120μL和DPPH溶液120μL加入到96孔板各孔中，同时以不加 DPPH（以 120μL 无水乙醇代替 DPPH）的供试品溶液各浓度作为对照以消除供试品本身颜色对测试结果的干扰，并设 DPPH 阴性对照（以 120μL 无水乙醇代替供试品），每组平行设 2 个复孔。室温避光反应 30min 后，利用酶标仪在 517nm 波长处测定吸光度。DPPH 自由基的清除率计算公式如下：

$$DPPH 清除率(\%) = (OD_0 - OD_s)/OD_0 \times 100\%$$

式中：OD_s 为加入样品的 DPPH 溶液的吸光度值；

OD_0 为加入无水乙醇的 DPPH 溶液的吸光度值。

以样品溶液的浓度为横坐标，DPPH 自由基清除率为纵坐标作图，求线性方程，并计算清除 50% DPPH 所需各样品的浓度，即半抑制浓度 IC_{50}，比较样品清除 DPPH 自由基活性。

1.2.3 ABTS 自由基清除能力的测定

ABTS 粉末用蒸馏水配制成 7×10^{-3} mol/L 的溶液，过硫酸钾粉末溶于水配置成 0.14mol/L 的溶液，将 7.5mL ABTS 溶液和 132μL 过硫酸钾水溶液室温下避光反应 16h 形成 ABTS 自由基储备液。使用前用乙醇稀释成工作液，使溶液在 732nm 波长下吸光度为 0.7 左右。在 96 孔板中加入 100μL ABTS 自由基储备液，再加入 50μL 不同浓度的各样品，黑暗放置 10min，然后在 734nm 波长下测吸光度，平行设一个复孔。以乙醇代替样品为对照，以蒸馏水代替 ABTS 为阴性对照。ABTS 自由基的清除率计算公式如下：

$$ABTS 自由基清除率 = [1 - (A_i - A_{i0})/A_0] \times 100\%$$

式中：A_i 为加入样品的 ABTS 溶液的吸光度值；

A_0 为加入无水乙醇的 ABTS 溶液的吸光度值；

A_{i0} 为加入蒸馏水和样品的吸光度值。

以样品溶液的浓度为横坐标、ABTS 自由基清除率为纵坐标作图，求线性方程，并计算 IC_{50} 值，比较样品清除 ABTS 自由基活性。

1.3 结果与讨论

1.3.1 蜂胶残渣挥发性成分与蜂胶原胶挥发性成分清除 DPPH 自由基的能力

DPPH 实验是一种广泛应用于评价植物及化学物质抗氧化活性的方法。本实验中，巴西蜂胶残渣挥发性成分、蜂胶原胶挥发性成分、巴西蜂胶乙醇提取物及两种阳性对照物的不同浓度对 DPPH 自由基的清除能力见图 15.8。结果显示，自由基的清除率与各样品的浓度

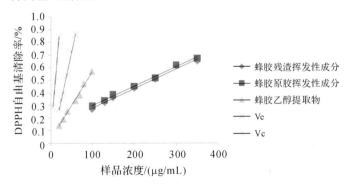

图 15.8 不同浓度样品对 DPPH 自由基清除能力

Fig. 15.8 Scavenging capacity of samples with different concentrations against DPPH

呈线性关系,各样品对 DPPH 自由基的清除率随浓度的增加而增强,呈现浓度依赖性。比较图中各曲线的斜率,斜率越大说明抗氧化能力越强。抗氧化能力从大到小依次为 Vc>VE>蜂胶乙醇提取物>蜂胶原胶挥发性成分>蜂胶残渣挥发性成分。其中蜂胶原胶挥发性成分与蜂胶残渣挥发性成分的抗氧化能力不存在明显差异。

由图 15.8 可知,DPPH 自由基清除率与样品浓度呈现的线性关系较好,因此,可以通过 50% DPPH 自由基清除率(IC_{50})来比较样品清除 DPPH 自由基活性。IC_{50} 值越小,则抗自由基活性越强。各样品的 IC_{50} 如图 15.9 所示,从小到大依次为 Vc<VE<蜂胶乙醇提取物<蜂胶原胶挥发性成分<蜂胶残渣挥发性成分。其中巴西蜂胶残渣挥发性成分与巴西蜂胶原胶挥发性成分的 IC_{50} 分别为 237.19μg/mL 和 236.53μg/mL,无明显差异。总的来说,蜂胶残渣挥发性成分与蜂胶原胶挥发性成分的清除 DPPH 自由基的能力差不多,蜂胶乙醇提取物的清除 DPPH 自由基的能力要强于挥发性成分,可能原因是黄酮类化合物是蜂胶乙醇提取物中最主要的成分,挥发性成分中含有较多的萜类物质,而黄酮类化合物具有较强的抗氧化效果,因此蜂胶乙醇提取物对 DPPH 自由基的清除能力要强于蜂胶挥发性成分。

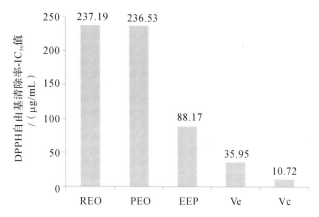

图 15.9　DPPH 自由基清除率-IC_{50} 值/(μg/mL)

Fig.15.9　Scavenging capacity against DPPH-IC_{50}/(μg/ML)

注:REO 为巴西蜂胶残渣挥发性成分,PEO 为巴西原胶挥发性成分,EEP 为巴西蜂胶乙醇提取物。

1.3.2　蜂胶残渣挥发性成分与蜂胶原胶挥发性成分清除 ABTS 自由基的能力

ABTS 阳离子自由基是稳定的有色人工自由基,在特定波长下有吸收。其与抗氧化剂作用后颜色会变浅,特定波长下的吸收变弱。图 15.10 显示了巴西蜂胶残渣与蜂胶原胶挥发性成分、巴西蜂胶乙醇提取物和两种阳性对照物不同浓度与 ABTS 自由基的抑制率的关系。结果表明,随着各样品浓度的增加,ABTS 自由基的清除率也增高。图 15.10 显示了各样品对 ABTS 自由基清除率的 IC_{50} 值,IC_{50} 值从小到大依次为 Vc<VE<蜂胶乙醇提取物<蜂胶原胶挥发性成分<蜂胶残渣挥发性成分。

在 ABTS 自由基清除能力方面,巴西蜂胶原胶挥发性成分的效果略好于巴西蜂胶残渣挥发性成分,可能的原因是巴西蜂胶残渣挥发性成分与巴西原胶挥发性成分共有较多相同的物质,如苯乙酮、芳樟醇、苯丙酸甲酯、苯丙酸、石竹烯和香橙烯等,但是它们也含有较多不同的组分,并且各物质的含量差异较大,因此使得巴西蜂胶原胶挥发性成分清除 ABTS 自由基的效果略好于巴西蜂胶残渣挥发性成分。蜂胶乙醇提取物的清除 ABTS 自由基的能力要强于挥发性成分，这与 DPPH 自由基清除能力的测定结果相同。

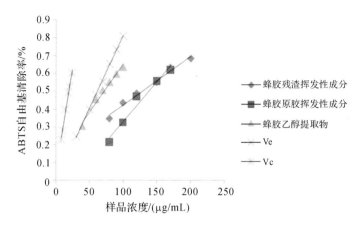

图 15.10　不同浓度样品对 ABTS 自由基清除能力

Fig. 15.10　Scavenging capacity of samples with different concentrations against ABTS

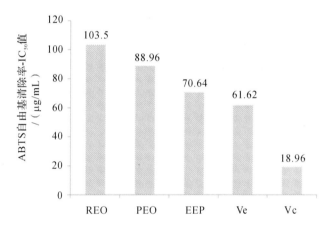

图 15.11　ABTS 自由基清除率-IC_{50} 值/(μg/ML)

Fig. 15.11　Scavenging capacity against ABTS-IC_{50}/(μg/ML)

注：REO 为巴西蜂胶残渣挥发性成分，PEO 为巴西原胶挥发性成分，EEP 为巴西蜂胶乙醇提取物。

小　结

通过对巴西蜂胶残渣中的挥发性成分、蜂胶原胶中的挥发性成分和蜂胶乙醇提取物清除 DPPH 自由基和 ABTS 自由基的能力比较，结果表明，三种样品均具有一定的抗氧化活性，可以推测蜂胶中的挥发性成分对蜂胶抗氧化活性有一定的贡献。

DPPH 自由基清除实验表明巴西蜂胶乙醇提取物对自由基的清除能力最强，蜂胶残渣挥发性成分和蜂胶原胶挥发性成分对自由基的清除能力次之，两者不存在明显差异。三者对 ABTS 自由基清除能力强弱依次为巴西蜂胶乙醇提取物＞蜂胶原胶挥发性成分＞蜂胶残渣挥发性成分，与 DPPH 自由基清除实验结果略微有些差异。推测其原因，可能是蜂胶残渣挥发性成分与蜂胶原胶挥发性成分在种类和含量上存在差异，导致对不同体系的自由基清除率存在差异。因此,在研究中,需要综合对比多种抗氧化能力测定方法的结果,才能全面了解样品的抗氧化能力。

2 巴西蜂胶残渣挥发性成分的抑菌活性

2.1 材料与方法

2.1.1 材料

巴西蜂胶由蜂乃宝本铺(南京)保健品有限公司提供。

巴西蜂胶残渣为巴西绿蜂胶经乙醇浸提过滤后剩余残渣干燥后所得,由蜂乃宝本铺(南京)保健品有限公司提供。

2.1.2 试剂

牛肉膏、蛋白胨、琼脂、氯霉素:生工生物工程(上海)生物有限公司提供;氯化钠、NaOH、无水乙醇,由国药集团化学试剂有限公司提供;pH试纸。

2.1.3 仪器

SW-CJ-IFD单人单面净化工作台,上海一恒科技有限公司;高压灭菌锅 MJ-54A,施都凯仪器设备(上海)有限公司;智能恒温恒湿培养箱,宁波海曙赛福实验仪器厂;立体生物显微镜 CX31,奥林巴斯。

2.1.4 供试菌种

金黄色葡萄球菌(*Staphylococcus aureus*)、大肠杆菌(*Escherichia coli*)由浙江大学动物医学实验室提供。

2.2 研究方法

2.2.1 样品的制备

巴西蜂胶残渣挥发性成分的制备采用微波辅助石油醚提取法。在萃取温度 40℃,液固比 10∶1(mL/g),萃取时间 20min,提取次数 2 次的条件下对巴西蜂胶残渣进行提取,过滤后合并滤液,45℃下旋转蒸发浓缩至 100mL 左右,于-18℃中反复冷冻过滤除蜡。将脱蜡后的溶液于 45℃下旋转蒸发脱溶,得到蜂胶残渣挥发性成分粗提物。将粗提物与无水乙醇按质量体积比 1g∶15mL 充分混匀,于-18℃反复冷冻除蜡,直至无蜡质析出。采用旋转蒸发仪对除蜡后的溶液脱溶,最后得到蜂胶残渣中的挥发性成分。

巴西蜂胶挥发性成分的制备同巴西蜂胶残渣挥发性成分的制备。

蜂胶乙醇提取物的制备:在 85%乙醇,液固比 1∶15,超声 15min,提取 2 次的条件下对巴西蜂胶进行提取,过滤,合并滤液,于-18℃中反复冷冻过滤除蜡,将脱蜡后的溶液在 45℃下旋转蒸发浓缩,将浓缩液倒入平板中并放入 50℃烘箱中烘干,置于-18℃冰箱中冷冻保存。

以上样品均溶于无水乙醇中制成 20mg/mL 的溶液,氯霉素溶于无水乙醇中制成 10mg/mL 的溶液。使用前以 0.22μm 无菌有机滤膜过滤除菌,转移至无菌 EP 管中保存。

2.2.2 培养基的配置

牛肉膏蛋白胨培养基(固体):称取牛肉膏 3g,蛋白胨 10g 和氯化钠 5g,以 1000mL 蒸馏水溶解,加热至微沸,加入琼脂粉 20g。待完全溶解后,用 1N 的 NaOH 调节 pH 为 7.0~7.2。分装后于 121℃下灭菌 20min。

待已灭菌的培养基冷却至50℃左右,在超净工作台内将培养基倒入直径为9cm的已灭菌的培养皿,厚度约为4mm,轻轻摇晃使培养基分布均匀且表面平整,制成平板。

牛肉膏蛋白胨培养基(液体):不加琼脂即可。分装后于121℃下灭菌20min。

2.2.3 菌种的活化及菌悬液的制备

菌种的活化:将供试菌种接入牛肉膏蛋白胨培养基中进行斜面培养,每种菌种重复接多支,与37℃恒温培养箱中培养20h。取出后置于4℃冰箱中冷藏备用。

菌种的分离纯化:用接种环取少许活化后的细菌采用平板划线法进行菌种的分离纯化,37℃下恒温培养箱中培养24h。

菌悬液的制备:用接种环挑取分离开的菌落接入液体牛肉膏蛋白胨培养基中,37℃培养48h。

2.2.4 细菌计数

将菌悬液稀释10倍,采用血球板计数法,在油镜下观察,选取血球计数板中的5个中方格计数。计数公式如下:

$$每毫升细菌个数 = 50000 \times 菌液稀释倍数 \times 5 个中方格中的细菌个数$$

最后用培养基将菌液稀释,制成含菌数10^7个/mL的菌悬液备用。

2.2.5 滤纸片的制备

将厚度为1.5mm的滤纸用打孔器打成直径为5.0mm的圆片,并把滤纸片装在空培养皿中,锡箔纸包好,于121℃下灭菌20min,取出放入烘箱烘干。

2.2.6 样品对细菌的抑制作用

2.2.6.1 纸片扩散法

采用纸片扩散法研究不同样品对大肠杆菌和金黄色葡萄球菌的抑制作用。选择溶剂(无水乙醇)为阴性对照,氯霉素作为阳性对照。

取0.1mL菌悬液均匀涂布于牛肉膏蛋白胨培养基表面。取各样品、对照样品各10μL转移至已灭菌的圆形滤片上,超净工作台内自然晾干。用镊子夹取滤纸片贴于平板表面,并轻轻按压使充分接触。每个菌种设3个平行组。贴好滤纸片的平板倒置放入培养箱,37℃培养24h。用游标卡尺测量滤纸片周围抑菌圈直径。每个实验重复3次,取其平均值进行分析。

2.2.6.2 最小抑菌浓度法(MIC)

在无菌EP管内用无水乙醇对各样品进行7次二倍稀释,样品起始质量浓度为20mg/mL,氯霉素起始质量浓度为10mg/mL。然后取一块无菌的96孔板分别加入100μL不同浓度的样品及100μL稀释好的菌液。每个浓度设一个复孔。同时设氯霉素阳性对照,阴性对照只加菌液和培养基,空白对照只加培养基。37℃下培养24h。肉眼观察细菌的生长。有细菌的生长,培养孔呈现浑浊,无细菌生长则培养孔澄清。结果判断以最后一组澄清孔其样品质量浓度为该样品对测试菌的最小抑菌浓度。每个实验重复3次。

分别取100μL澄清培养孔及其前后各孔中的溶液均匀涂布于牛肉膏蛋白胨培养基的平皿中,37℃条件下培养24h,以进一步验证肉眼观察的结果。

2.3 结果与讨论

2.3.1 蜂胶残渣与原胶挥发性成分及乙醇提取物对细菌的抑制作用

巴西蜂胶残渣与蜂胶原胶挥发性成分及巴西蜂胶乙醇提取物对大肠杆菌和金黄色葡萄

球菌的抑制作用见表15.6和图15.12、15.13。图15.12中,各样品对大肠杆菌没有形成明显的抑菌圈;图15.13中,各样品对金黄色葡萄球菌具有比较清晰的抑菌圈边缘。由表15.6可知,当样品浓度为20mg/ML时,巴西蜂胶残渣与蜂胶原胶挥发性成分及巴西蜂胶乙醇提取物对大肠杆菌的抑制效果不明显,对金黄色葡萄球菌有较强的抑制作用。相关文献报道,蜂胶对革兰阳性菌的抑制作用优于革兰阴性菌(Koru et al.,2007;Grange and Davey,1990)。大肠杆菌为典型的革兰阴性菌,金黄色葡萄球菌为革兰阳性菌,本研究结果与文献所得结果基本一致。

各样品对金黄色葡萄球菌具有一定的抑制作用,阳性对照物氯霉素对金黄色葡萄球菌有强效的抑制作用,空白对照无水乙醇无抑菌圈形成,说明溶剂对本实验结果无影响。在相同浓度下,各样品的抑菌活性从高到低依次为:巴西蜂胶原胶挥发性成分＞巴西蜂胶残渣挥发性成分＞巴西蜂胶乙醇提取物。

表15.6 蜂胶残渣与原胶挥发性成分及乙醇提取物的抑菌圈直径 mm

Table 15.6 Inhibition zone of propolis essential oil, propolis residue's essential oil and ethanol extract of propolis mm

受试菌种	巴西蜂胶残渣挥发性成分	巴西蜂胶原胶挥发性成分	巴西蜂胶乙醇提取物	阳性对照（氯霉素）	空白对照（无水乙醇）
大肠杆菌	无	无	无	23.52±1.76	无
金黄色葡萄球菌	7.41±0.62	8.28±0.17	5.99±0.62	25.22±1.31	无

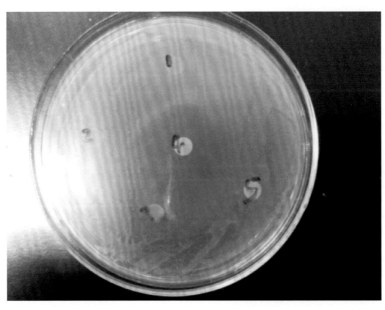

图 15.12 各样品对大肠杆菌的抑制作用

Fig 15.12 Inhibition of samples against *Escherichia coli*

(图中1:巴西蜂胶乙醇提取物;2:巴西蜂胶残渣挥发性成分;3:巴西蜂胶原胶挥发性成分;4:氯霉素;5:无水乙醇)

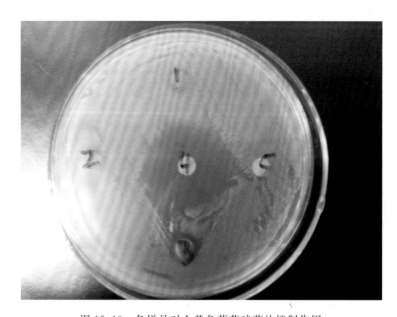

图 15.13 各样品对金黄色葡萄球菌的抑制作用

Fig. 15.13 Inhibition of samples against Staphylococcus aureus

(图中 1：巴西蜂胶乙醇提取物；2：巴西蜂胶残渣挥发性成分；3：巴西蜂胶原胶挥发性成分；4：氯霉素；5：无水乙醇)

2.3.2 蜂胶残渣与原胶挥发性成分及乙醇提取物对细菌的最小抑菌浓度

巴西蜂胶残渣与蜂胶原胶挥发性成分及巴西蜂胶乙醇提取物对大肠杆菌和金黄色葡萄球菌的最小抑菌浓度见表 15.7。各样品对金黄色葡萄球菌的抑制作用呈现浓度依赖效应，浓度越高，抑制效果越强。如表 15.7 所示，阳性对照氯霉素的抑菌效果最强，最小抑菌浓度为 0.078mg/mL，其次为巴西蜂胶原胶与残渣挥发性成分，最小抑菌浓度分别为 1.25mg/mL 和 0.625mg/mL，巴西蜂胶乙醇提取物的抑菌效果最差，最小抑菌浓度为 5.00mg/mL。

由表 15.7 可知，巴西蜂胶残渣与蜂胶原胶挥发性成分及巴西蜂胶乙醇提取物对大肠杆菌的抑制效果不明显，对金黄色葡萄球菌有较强的抑制作用，与纸片扩散法结果相一致，各样品的抑菌活性从高到低依次为：巴西蜂胶原胶挥发性成分＞巴西蜂胶残渣挥发性成分＞巴西蜂胶乙醇提取物。研究发现萜烯类化合物具有很强的抑菌作用，而巴西蜂胶残渣与原胶挥发性成分中含有大量的此类物质，如芳樟醇、石竹烯、香橙烯、橙花叔醇和斯巴醇等，因此表现出较强的抑菌效果。黄酮类化合物也具有一定的抑菌活性，巴西乙醇提取物中含有的主要生物学活性成分是黄酮类化合物。总体来说，黄酮类化合物的抑菌作用不如萜类物质，所以巴西蜂胶残渣与原胶挥发性成分的抑菌效果强于蜂胶乙醇提取物。巴西蜂胶原胶挥发性成分中萜烯类化合物的相对含量为 33.452%，巴西蜂胶残渣中萜烯类化合物的相对含量为 29.770%，略少于巴西蜂胶原胶挥发性成分，因此这可能是巴西蜂胶原胶挥发性成分的抑菌作用强于巴西蜂胶残渣挥发性成分的原因。

表 15.7　蜂胶残渣与原胶挥发性成分及乙醇提取物的最小抑菌浓度

Table 15.7　Minimum inhibitory concentration of propolis essential oil, propolis residue's essential oil and ethanol extract of propois　　　　　　　　　　　　　　　　　　　　　　　　　　　　mg/mL

受试菌种	巴西蜂胶残渣挥发性成分	巴西蜂胶原胶挥发性成分	巴西蜂胶乙醇提取物	阳性对照（氯霉素）	空白对照（无水乙醇）
大肠杆菌	无	无	无	0.156	无
金黄色葡萄球菌	1.25	0.625	5.00	0.078	无

分别取澄清培养孔及其前后各孔中的溶液进行固体培养，培养结果表明，各样品澄清孔及前一孔的溶液均无菌落生成，后一孔的溶液有菌落生成，证明澄清孔无细菌生长，其样品质量浓度为该样品对测试菌的最小抑菌浓度。

小　结

本研究结果表明，巴西蜂胶残渣与蜂胶原胶挥发性成分及巴西蜂胶乙醇提取物都具有一定的抑菌活性，各样品的抑菌活性从高到低依次为：巴西蜂胶原胶挥发性成分＞巴西蜂胶残渣挥发性成分＞巴西蜂胶乙醇提取物。三种样品对金黄色葡萄球菌的抑制效果最明显。目前对蜂胶抑菌活性的报道显示蜂胶提取物对革兰阳性菌的生长有显著的抑制活性，对革兰阴性菌的抑制效果较弱。我们的实验结果与这些报道基本一致。

Melliou 等（2007）研究发现蜂胶的挥发性成分具有良好的抑菌作用，其抑菌活性可能与主要成分 α-蒎烯有关。巴西蜂胶残渣与蜂胶原胶挥发性成分中含有大量的萜烯类化合物如愈创木醇、杜松醇等，它们可能是挥发油中产生抑菌效果的主要物质。巴西蜂胶乙醇提取物中含有的主要活性物质为黄酮类化合物，研究发现黄酮类化合物也具有一定的抑菌活性。因此，蜂胶中的挥发性成分与黄酮类化合物对蜂胶的抑菌活性起到了协同作用。

参考文献

Ahn MR, Kumazawa S, Usui Y, Nakamura J, Matsuka M, Zhu F, Nakayma T (2007) Antioxidant activity and constituents of propolis collected in various areas of China [J]. *Food Chemistry*, 101(4):1383-1392.

Alencar SM, Oldoni TLC, CastroML, Cabral ISR, Costa-Neto CM, Cury JA, Rosalen PL, Ikegaki M*e*(2007) Chemical composition and biological activity of a new type of Brazilian propolis: red propolis[J]. *Journal of Ethnopharmacology*, 113(2):278-283.

Andreas D, Moraes CS, Patricia F, Park YK (2008) Brazilian red propolis-chemical composition and botanical origin[J]. *Evidence-based Complementary and Alternative Medicine*, 5(4):435-441.

Atungulu G, Miura M, Atungulu E (2007) Activity of gaseous phase steam distilled propolis extracts on peroxidation and hydrolysis of rice lipids[J]. *Journal of Food Engineering*, 80(3):850-858.

Awale S, Shrestha S, Tezuka Y, Ueda J, Matsushige K, Kadota S (2005) Neoflavonoids and related constituents from Nepalese propolis and their nitric oxide production inhibitory activity[J]. *Journal of Natural Products*, 68(6):858-864.

Bankova V, Castro S, Marcucci M (2000) Propolis: recent advances in chemistry and plant origin[J]. *Apidologie*, 31(1):3-15.

Bankova V, Christov R, Kujumgiev A, Marcucci MC, Popov S (1995) Chemical composition and antibacterial activity of Brazilian propolis[J]. *Zeitschrift fuer Naturforschung*, 50 (3-4):167-172.

Bankova V, Christov R, Popov S, Marcucci MC, Tsvetkova I, Kujumgiev A (1999) Antibacterial activity of essential oil from Brazilian propolis[J]. *Fitoterapia*, 70(2):190-193.

Bankova V, Dyulgeroy A, Popov S, Evstatieva L, Kuleva L, Pureb O (1992) Propolis produced in Bulgaria and Mongolia: phenolic compounds and plant origin[J]. *Apidologie*, 23(1):79-85.

Bankova V, Marcucci MC, Simova S, Nikolova N, Kujumgiev A, Popov S (1996) Antibacterial diterpenic acids from Brazilian propolis[J]. *Zeitschrift fuer Naturforschung*, 50C:277-280.

Bankova V, Popova M, Bogdanov S, Sabatini A (2002) Chemical composition of European propolis: expected and unexpected results[J]. *Zeitschrift fur Naturforschung C: A Journal of Biosciences*, 57(5-6):530-533.

Bankova V, Castro S, Marcucci M (2000) Propolis: recent advances in chemistry and plant origin[J]. *Apidologie*, 31(1):3-15.

Bonvehi JS, Coll FV, Jorda RE (1994) The composition, active components and bacteriostatic activity of propolis in dietetics[J]. *Journal of the American Oil Chemists' Society*, 71(5):529-532.

Burdock GA (1998) Review of the biological properties and toxicity of bee propolis (propolis)[J]. *Food and Chemical Toxicology*, 36(4):347-363.

Campo FM, Cuesta RO, Rosado PA, Montes DOPR, Marquez HI, Piccinelli AL, Rastrelli L (2008) GC-MS determination of isoflavonoids in seven red Cuban propolis samples[J]. *Journal of Agriculture and Food Chemistry*, 56(21):9927-9932.

Cetin H, Cilek JE, Oz E, Aydin L, Deveci O, Yanikoglu A (2010) Acaricidal activity of *Satureja thymbra* L. essential oil and its major components, carvacrol and gamma-terpinene against adult *Hyalomma marginatum* (Acari: Ixodidae)[J]. *Vet Parasilol*, 170(3-4):287-290.

Chen CR, Lee YN, Lee MR, Chang CMJ (2009) Supercritical fluids extraction of cinnamic acid derivatives from Brazilian propolis and the effect on growth inhibition of colon cancer cells[J]. *Journal of the Taiwan Institute of Chemical Engineers*, 40(2):130-135.

Chen YJ, Huang AC, Chang HH, Liao HF, Jiang CM, Lai LY, Chan JT, Chen YY, Chiang J (2009) Caffeic acid phenethyl ester, an antioxidant from propolis, protects peripheral blood mononuclear cells of competitive cyclists against hyperthermal stress[J]. *Journal of Food Science*, 74(6):162-167.

Garcia-Viguera C, Ferreres F, Tomas-Barberan FA (1993) Study of Canadian propolis by

GC-MS and HPLC[J]. *Zeitschrift fur Naturforschung C:A Journal of Biosciences*,48(5-6):731-735.

Gencay O,Salih B (2009) GC-MS analysis of propolis samples from 17 different regions of Turkey,four different regions of Brazil and one from Japan[J]. *Mellifera*,9(17):19-28.

Ghisalberti EL (1979) Propolis:a review[J]. *Bee World*,60(2):59-84.

Gómez-Caravaca AM,Gómez-Romero M,Arráez Roman D,Segura Carretero A,Fernandez Gutierrez A (2006) Advances in the analysis of phenolic compounds in products derived from bees[J]. *Journal of Pharmaceutical and Biomedical Analysis*,41(4):1220-1234.

Grange JM,Davey RW (1990) Antibacterial properties of propolis (bee glue)[J]. *Journal of the Royal Society of Medicine*,83(3):159-160.

Greenaway W,Scaysbrok T,Whatley FR (1987) The analysis of bud exudate of *Populus euramericana*,and of propolis,by gas chromatography-mass spectrometry[J]. *Proceedings of the Royal Society*,232(1268):249-272.

Greenaway W,Scaysbrook T,Whatley FR (1990) The composition and plant origins of propolis[J]. *Bee World*,71(3):107-118.

Gregory AG,Toshitaka U,Fumihiko T,Daisuke H (2008) Effect of vapors from fractionated samples of propolis on microbial and oxidation damages of rice during storage[J]. *Journal of Food Engineering*,88(3):341-352.

Hernández IM,Fernandez MC,Cuesta-Bubio O,Piccinelli AL,Rastrelli L (2005) Polyprenylated benzophenone derivatives from Cuban propolis[J]. *Journal of Natural Products*,68(6):931-934.

HirotaR,Roger NN,Nakamura H,Song HS,Sawamura M,Suganuma N (2010) Anti-inflammatory effects of limonene from yuzu (*Citrus junos* Tanaka) essential oil on eosinophils[J]. *Journal of Food Science*,75(3):H87-H92.

Jorge AP,Rolando M,América D,Carlos Z,Enrique S (2006) Volatile constituents of propolis from honey bees and stingless bees from Yucatá[J]. *Journal of Essential Oil Research*,(18):53-56.

Junior MRM,Daugsch A,Moraes CS,Queiroga CL,Pastore GM,Yong KP (2008) Comparison of volatile and polyphenolic compounds in Brazilian green propolis and its botanical origin *Baccharis dracunculifolia*[J]. *Food Science and Technology (campinas)*,28(1):178-181.

Koru O,Toksoy F,Acikel CH,Tunca YM,Baysallar M,Guclu AU,Akca E,Tuylu AO,Sorkun K,Tanyuksel M,Salih B (2007) In vitro antimicrobial activity of propolis samples from different geographical origins against certain oral pathogens[J]. *Anaerobe*,13(3-4):140-145.

Kumazawa S,Yoneda M,Shibata I,Kanaeda J,Hamassaka T,Nakayama T (2003) Direct evidence for the plant origin of Brazilian propolis by the observation of honeybee behavior and phytochemical analysis[J]. *Chemical & Pharmaceutical Bulletin*,51(6):

740-742.

Kumazawa S,Nakamura J,Murase M,Miyagawa M,Ahn MR,Fukumoto S (2008) Plant origin of Okinawan propolis:honeybee behavior observation and phytochemical analysis[J]. *Naturwissenschaften*,95(8):781-786.

Kusumoto T,Miyamoto T,Higuchi R,Doi S,Sugimoto H,Yamada H (2001) Isolation and structures of two new compounds from the essential oil of Brazilian propolis[J]. *Chemical and Pharmaceutical Bulletin*,49(9):1207-1209.

Li F,Awale S,Tezuka Y,Kadota S (2008) Cytotoxic constituents from Brazilian red propolis and their structure-activity relationship[J]. *Bioorganic & Medicial Chemistry*,16(10):5434-5440.

Likens ST,Nickerson GB (1964) Detection of certain hop oil constituents in brewing products[J]. *Proceeding of the American Society of Brewing Chemists*,22(7):5-13.

Marcucci MC (1995) Propolis:chemical composition,biological properties and therapeutic activity[J]. *Apidologie*,26(2):83-99.

Markham KR,Mitchell KA,Wilkins AL,Daldy JA,Lu YR (1996) HPLC and GC-MS identification of the major organic constituents in New Zealand propolis[J]. *Phytochemistry*,42(1):205-211.

Melliou E,Stratis E,Chinou I (2007) Volatile constituents of propolis from various regions of Greece-Antimicrobial activity[J]. *Food Chemistry*,103(2):375-380.

Mulyaningsih S,Sporer F,Zimmermann S,Reichling J,Wink M (2010) Synergistic properties of the terpenoids aromadendrene and1,8-cineole frorn the essential oil of Eucalyptus globulus against antibiotic-susceptible and antibiotic-resistant pathogens[J]. *Phytomedicine*,17(13):1061-1066.

Park YK,Alencar SM,Aguiar CL (2002) Botanical origin and chemical composition of Brazilian propolis[J]. *Journal of Agriculture and Food Chemistry*,50(9):2502-2506.

Park YK,Alencar SM,Scamparini ARP,Aguiar CL (2002) Própolis produzida no sul do Brasil,Argentinae Uruguai:evidências fitoquímicas de sua origem vegetal[J]. *Ciência Rural*,32(6):997-1003.

Pereira AS,Nascimento EA,Neto FRA (2002) Lupeol alkanoates in Brazilian propolis[J]. *Zeitschrift fur Naturforschung C:A Journal of Biosciences*,57C:721-726.

Petrova A,Popova M,Kuzmanova C,Tsvetkova I,Naydenski H,Muli E,Bankova V (2010) New biologically active compounds from Kenyan propolis[J]. *Fitoterapia*,81(6):509-514.

Pietta PG,Gardana C,Pietta AM (2002) Analytical methods for quality control of propolis [J]. *Fitoterapia*,73(S1):S7-S20.

Popova MP,Chinou IB,Marekov IN,Bankova VS (2009) Terpenes with antimicrobial activity from Cretan propolis[J]. *Phytochemisstry*,70(10):1262-1271.

Popova M,Graikou K,Chinou I,Bankova VS (2010) GC-MS profiling of diterpene compounds in Mediterranean propolis from Greece[J]. *Journal of Agriculture and Food Chemistry*,58:

3167-3176.

Popravko S, Sokolov M (1980) Plant sources of propolis[J]. *Pchelovodstvo*, 2:28-29.

Popravko SA, Gurevich AI, Kolosov MN (1969) Flavonoid components of propolis[J]. *Khimiya Prirodnykh Soedinenii*, 5(6):476-482.

Righi AA, Alves TR, Negri G, Marques LM, Breyer H, Salatino A (2011) Brazilian red propolis: unreported substances, antioxidant and antimicrobial activities[J]. *Journal of the Science of Food and Agriculture*, 91(13):2363 2370.

Sahinler N, Kaftanoglu O (2005) Natural product propolis: chemical composition[J]. *Natural Product Research*, 19(2):183-188.

Sairam P, Ghosh S, Jena S, Rao KNV, Banji D (2012) Supercritical fluids extraction(SFE)-an overview[J]. *Asian Journal of Pharmaceutical Sciences*, 2(3):112-120.

Salatino A, Teixeira EW, Negri G, Message D (2005) Origin and chemical variation of Brazilian propolis[J]. *Evidence-Based Complementary and Alternative Medicine*, 2(1):33-38.

Santos PA, Miranda PA, Trugo L, Aquino NF (2003) Distribution of quinic acid derivatives and other phenolic compounds in Brazilian propolis[J]. *Zeitschrift fur Naturforschung*, 58C(7/8):590-593.

Smith RM (1999) Supercritical fluids in separation science-the dreams, the reality and the future[J]. *Journal of Chromatography A*, 856(1-2):83-115.

Souza RM, Souza MC, Patitucci ML, Silva JF (2007) Evaluation of antioxidant and antimicrobial activities and characterization of bioactive components of two Brazilian propolis samples using a pKa-guided fractionation[J]. *Zeitschrift fur Naturforschung C: A Journal of Biosciences*, 62(11-12):801-807.

Stefano C, Francesco C (2002) Propolis, an old remedy used in modern medicine[J]. *Fitoterapia*, 73(S):S1-S6.

Walker P, Crane E (1987) Constituents of propolis[J]. *Apidologie*, 18(4):327-334.

Zhou JH, Li Y, Zhao J, Xue XF, Wu LM, Chen F (2008) Geographical traceability of propolis by high-performance liquid-chromatography fingerprints[J]. *Food Chemistry*, 108(2):749-759.

曾晞,卢玉振,牟兰,张长庚(2004)GC-MS法分析比较贵州不同产地蜂胶挥发油化学成分[J].生命科学仪器,2(2):28-29.

董丽,邢钧,吴采樱(2003)香精香料的分析方法进展[J].分析科学学报,19(2):188.

顾青,张燕萍,钟立人(2001)蜂胶中有效成分提取工艺研究[J].浙江农业学报,13(3):161-164.

郭伽,周立东.北京蜂胶挥发油的化学成分研究[J].中国养蜂,2000,51(1):9.

郭伽,周立东.蜂胶的化学成分研究进展(综述)[J].中国养蜂,2000,51(2):17-18.

贺丽萍,李雅萍.SPME-GC/MS联用技术分析蜂胶的挥发性成分[J].中国蜂业,2008,59(4):36-37.

胡福良,詹耀锋,陈民利,应华忠,朱威(2004)蜂胶对高脂血症大鼠血液和肝脏脂质的影响

[J].浙江大学学报(农业与生命科学版),30(5):510-514.

黄帅,卢媛媛,张翠平,胡福良(2013)巴西绿蜂胶乙醇提取前后挥发性成分的分析比较[J].食品与生物技术学报,32(7):680-685.

黄帅,张翠平,胡福良(2013)2008-2012年蜂胶化学成分研究进展[J].天然产物研究与开发,25:1146-1153.

李雅洁,凌建亚,邓勇(2006)蒙山蜂胶超临界二氧化碳萃取挥发性组分的气相色谱-质谱联用分析[J].时珍国医国药,17(10):1975-1976.

李雅晶,胡福良,陆旋,詹忠根(2011)蜂胶中挥发性成分的微波辅助提取工艺研究及中国蜂胶、巴西蜂胶挥发性成分比较[J].中国食品学报,11(5):93-99.

李雅晶(2011)蜂胶中挥发性成分的提取方法、化学组成及生物学活性[D].杭州:浙江大学博士论文.

李雅萍,贺丽萍,陈玉芬,卢占列,郑尧隆,潘建国,林励(2007)SPME-GC/MS联用技术分析蜂胶中挥发性成分的研究[J].现代食品科技,23(7):78-80.

刘安洲,杜金华,王虎(2009)泰山蜂胶挥发性成分检测[J].食品与发酵工业,35(5):163-165.

刘亚男,胡福良,程艳华,李英华(2008)蜂胶黄酮苷的酶解及酶解产物的抗氧化活性[J].天然产物研究与开发,20(B05):181-184.

唐传核.植物生物活性物质[M].北京:化学工业出版社,2005.

王建新.天然活性化妆品[M].北京:中国轻工业出版社,1997.

王小平,林励,潘建国,刘晓涵,卢占列(2009)不同产地蜂胶挥发油成分的GC-MS比较分析[J].药物分析杂志,29(1):86-90.

王艳萍,汪心想,任保增(2009)超临界流体萃取技术的应用[J].河南化工,6(12):25-27.

吴寿金,赵泰,秦永祺(2002)现代中草药成分化学[M].北京:中国医药科技出版社.

徐响,董捷,丁小宇,杨佳林,孙丽萍(2010)不同方法萃取蜂胶挥发油组成及抑菌作用的研究[J].食品科学,31(3):60-63.

徐响,孙丽萍,董捷(2009)响应面法优化蜂胶超临界二氧化碳萃取工艺的研究[J].食品科学,30(8):86-89.

张翠平,胡福良(2009)蜂胶中的黄酮类化合物[J].天然产物研究与开发,21(6):1084-1090.

张翠平,胡福良(2009)蜂胶中的萜类化合物[J].天然产物研究与开发,24(7):976-984.

张翠平,平舜,黄帅,胡福良(2013)蜂胶的地理来源、植物来源及化学成分的研究[J].中国药学杂志,48(22):1889-1892.

张翠平,王凯,胡福良(2013)蜂胶中的酚酸类化合物[J].中国现代应用药学,30(1):102-105.

张德权,胡晓丹(2005)食品超临界CO_2流体加工技术[M].北京:化学工业出版社.

赵强,张彬,周武(2007)微波辅助萃取GC/MS联用分析蜂胶挥发油[J].精细化工,24(12):1192-1203.

赵强,张彬,李岂凡,周武,郭志芳(2008)蜂胶挥发油抗氧化性能及其成分研究[J].天然产物研究与开发,20:82-86.

朱自强(2000)超临界流体技术—原理和应用[M].北京:化学工业出版社.

第十六章　纳米蜂胶的制备及其降血脂降血糖作用研究

纳米技术是指在纳米尺度范围内对物质和材料进行研究处理的技术，是一门新型科技，是现代科学（混沌物理、量子力学、介观物理、分子生物学）和现代技术（计算机技术、微电子和扫描隧道显微镜技术、核分析技术）结合的产物。纳米材料具有 3 个特征：①纳米尺度的结构单元或特征维度尺寸在纳米数量级；②存在大量的界面或自由表面；③各纳米单元之间存在或差或强或弱的相互作用。纳米材料这些特殊的结构特点决定着它具有特殊的性能。当微粒小于 100nm 时，物质的很多性能将发生质变，从而呈现不同于宏观物质的奇异现象：低熔点、高比热容、高膨胀系数；高反应活性、高扩散率；高强度、高韧性；奇特磁性；极强的吸波性。因此，纳米科技具有巨大的市场应用潜力。

由于蜂胶独特的物理性质，如不溶于水，吸湿性强，低温变脆，高温变黏等给蜂胶精加工带来了难度，使传统的蜂胶产品溶解性差，吸收利用度较低，不能充分发挥其药效作用。因此，对蜂胶进行纳米化处理，可将蜂胶颗粒转变成稳定的纳米粒子，通过纳米材料的小尺寸效应、表面效应和宏观量子隧道效应提高其溶解性和药效率，还能解决蜂胶在机体内的作用靶向问题，大幅度地提高蜂胶的生物学活性。这对蜂胶精深加工来说是一项重大的突破，不仅具有重大的理论价值，更具有巨大的实际应用价值。

我们通过对蜂胶纳米化技术的研究，探索出一套拥有自主知识产权，能显著提高蜂胶生物学效价，提高人体吸收率的蜂胶纳米化处理工艺；并研究了纳米蜂胶对高脂血症 SD 大鼠、实验性 2 型糖尿病大鼠体内糖、脂及蛋白质代谢、胰岛素抵抗、抗氧化损伤及肝肾功能的影响，实验性 2 型糖尿病大鼠代谢紊乱与胰岛素抵抗的影响，以期为开发纳米蜂胶作为高脂血症、糖尿病等疾病的预防和治疗药物提供理论依据和技术支撑。

第一节　纳米蜂胶的制备及体外释放实验

1　纳米蜂胶的制备

纳米粉体的制备方法可以分为物理法和化学法。物理法主要有高能球磨法、溅射法、蒸汽冷凝法等；化学法主要有微乳法、热分解法、化学共沉淀法、化学气相和成法、溶胶－凝胶法、化学气相沉淀法、水热法、液相化学还原法等。

蜂胶是一种混合物，成分复杂，不像普通的药物、化工等产品容易利用合成、热分解等方法制作纳米粒，而且其水溶性差。但蜂胶也有其独特的性质，如低温下变脆，溶于酒精、乙醚和丙酮等。不利于蜂胶纳米化的因素有其吸湿性较强，高温下变软、变黏。因此，考虑到蜂胶的特性和生物利用度等问题，我们采用溶胶凝胶法对蜂胶进行纳米化工艺处理。并通过研究温度、pH值以及分散剂比例对蜂胶颗粒大小及其在水中溶解性的影响，探明了蜂胶纳米化处理的最佳工艺。

1.1 工艺流程1(冷冻干燥)

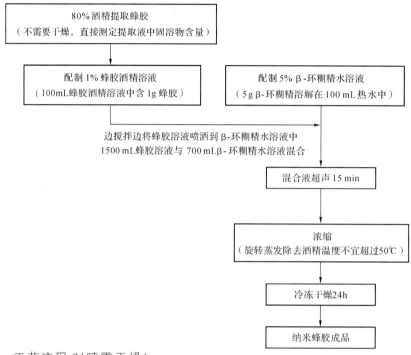

1.2 工艺流程2(喷雾干燥)

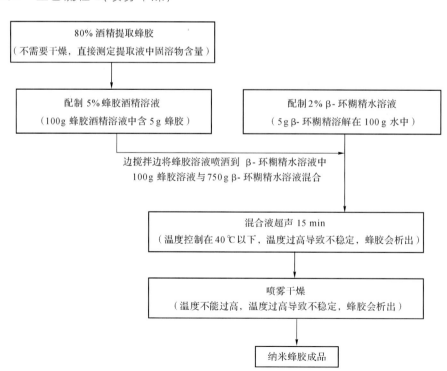

2　纳米蜂胶颗粒及体外释放实验

2.1　实验仪器与材料

2.1.1　实验仪器

微粒粒度与 Zeta 电位测定仪：3000HS，Malvern，UK

高速离心机：3K30，Sigma，Germany

恒温振荡器：HZ—8812S，华利达公司

漩涡振荡仪：XW-80A，上海医科大学仪器厂

721 分光光度计：上海精密科学仪器公司

2.1.2　实验材料

纳米蜂胶：采用溶胶凝胶法，按照本章 1.1 工艺制备所得

药用聚乙二醇(PEG)400：北京市海淀会友精细化工厂

甲醇：上海陆都化学试剂厂

乙醇：杭州大方化学试剂厂

结晶水氯化铝：上海金山区兴塔美兴化工厂

醋酸钾：温州市化学用料厂

2.2　实验方法

2.2.1　纳米蜂胶粒径的测定

取纳米蜂胶适量，以 0.3％泊洛沙姆 188 溶液分散，使纳米蜂胶的终浓度为 0.1mg/mL，漩涡振荡 10min，用微粒粒度与 Zeta 电位测定仪测定粒径。

2.2.2　体外释放实验

称取纳米蜂胶 200mg，投入 20mL 2％ PEG400 介质中，放入 37℃恒温振荡器中，振荡频率为 60 次/min，做蜂胶纳米粒体外释放实验。取样时间点设置为 0h、1h、2h、4h、8h、12h、24h。每次取样 1mL，样品 20000r/min 离心 10min，取上清液按下述的方法测定样品的紫外吸光度。

2.2.3　蜂胶中总黄酮含量的测定

(1)试剂的配制

芦丁标准液的配制：取色谱纯甲醇，配制的浓度为 1.020mg/mL，芦丁标液 10mL，用 70％乙醇定容至 50mL，得 0.204mg/mL 芦丁标准液，备用。

0.1mol/L 氯化铝溶液：称取结晶水氯化铝 2.1445g，加 70％乙醇至 100mL。

1mol/L 醋酸钾溶液：称取醋酸钾 10g，加水至 100mL。

(2)标准曲线的绘制

取 10mL 0.204mg/mL 的芦丁标准液，用 70％乙醇定容至 50mL，得 0.0408mg/mL 芦丁标准液。精密吸取芦丁标准溶液 0、0.5、1、1.5、2、2.5、3、4、5mL，分别置于 15mL 试管中，每管加 0.1mol/L 氯化铝溶液 2mL，再各加 1mol/L 醋酸钾溶液 3mL，最后各加 70％乙醇至总容积为 10mL，漩涡振荡仪混匀，放置 30min，在 420nm 波长下测吸光度。以吸光度为纵坐标，浓度为横坐标作图。

(3) 样品的测定

纯胶用 PEG400 超声至充分分散后稀释至 3mg/mL,20000r/min 离心 10min,取上清液 0.5mL;纳米蜂胶取释放实验各点清液 0.5mL,按标准曲线步骤测定样品吸光度。以释放介质作空白对照,调零。

2.3 结果

2.3.1 纳米蜂胶粒径

经微粒粒度与 Zeta 电位测定仪(3000 HS,Malvern,UK)测得,纳米蜂胶的平均粒径为 220.1nm,其粒度分布见图 16.1。用原子力显微镜测量所得的蜂胶纳米颗粒图见图 16.2。

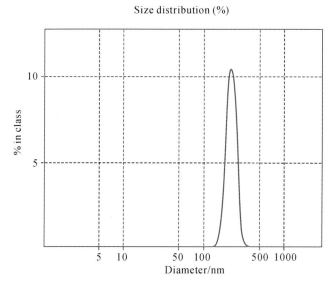

图 16.1 蜂胶纳米粒的粒径分布图

Fig. 16.1 Particle size distribution of propolis nanoparticles

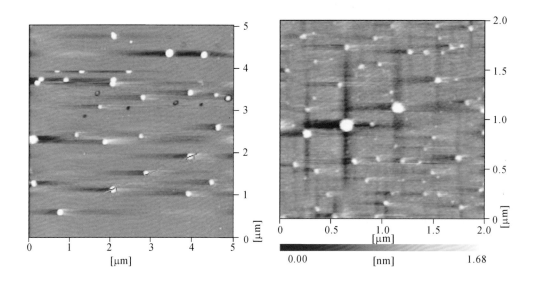

第十六章　纳米蜂胶的制备及其降血脂降血糖作用研究

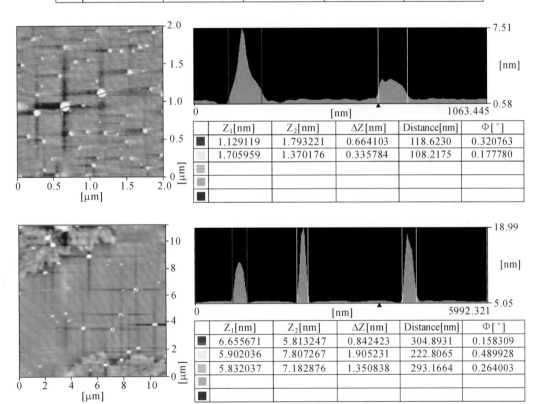

图 16.2　蜂胶纳米颗粒图

Fig. 16.2　Graphs of propolis nanoparticles

2.3.2　蜂胶中黄酮类化合物含量测定

(1)芦丁标准曲线

此法在 1.0~25.0μg/mL 内有良好的线性,最低检测浓度为 0.1μg/mL。

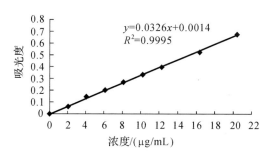

图 16.3　芦丁标准曲线

Fig. 16.3　Standard curve of rutin

(2) 纯胶中总黄酮类的含量

表 16.1　纯胶中总黄酮类的含量

Table 16.1　Contents of total flavonoids in pure propolis

样品	吸光度	浓度/(μg/mL)	含量/%
纯胶 1 (3mg/mL)	0.452	276.45	9.22
纯胶 2 (3mg/mL)	0.445	272.17	9.07
纯胶 3 (3mg/mL)	0.475	290.52	9.68
Mean	0.457	279.71	9.32
S.D	0.016	9.60	0.32

(3) 纳米蜂胶体外释放率

本次实验所用纳米蜂胶包含 30% 纯蜂胶，因此 10mg/mL 纳米蜂胶中总黄酮含量与 3mg/mL 纯蜂胶相同。由图 16.4 可见，纳米蜂胶从纳米粒中缓慢释放，24h 内蜂胶从纳米粒中的释放率为 68%。

蜂胶释放率(%) = (各时间点上清液中总黄酮的含量/纳米蜂胶中总黄酮的含量) × 100%

表 16.2　纳米蜂胶体外释放实验各时间点吸光度

Table 16.2　Absorbance at each time point of nano propolis in vitro release test

| 时间/h | 样品 | | | | |
	纳米蜂胶 1 (10mg/mL)	纳米蜂胶 2 (10mg/mL)	纳米蜂胶 3 (10mg/mL)	Mean	S.D.
0	0.024	0.025	0.024	0.024	0.001
1	0.102	0.097	0.108	0.102	0.005
2	0.163	0.152	0.169	0.161	0.009
4	0.218	0.232	0.201	0.217	0.016
8	0.286	0.308	0.259	0.284	0.025
12	0.312	0.290	0.328	0.310	0.019
24	0.318	0.316	0.312	0.315	0.003

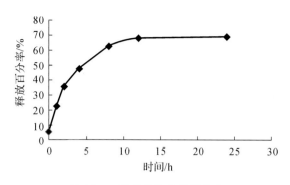

图 16.4 纳米蜂胶体外释放率

Fig. 16.4 In vitro release rate of nano propolis

小 结

根据蜂胶的特性和生物利用度,本研究采用溶胶凝胶法对蜂胶纳米化工艺进行处理,并通过研究温度、pH 值以及分散剂比例对蜂胶颗粒大小及其在水中溶解性的影响,探明了蜂胶纳米化处理的最佳工艺。同时,对纳米蜂胶的体外药剂学性质进行了研究,检验其载药量、包封率和粒径,考察其体外释放度。结果表明,纳米蜂胶包封率达到 98% 以上,平均粒径为 210nm,体外释放 24h 达 68%。因此,所制得的纳米蜂胶的理化性质优良,具有缓释效应。

第二节 纳米蜂胶对高脂血症 SD 大鼠的作用

1 实验材料

1.1 实验动物

大鼠:SD 封闭群,清洁级,雄性,由中国科学院上海实验动物中心供应。动物质量合格证:中科动管第 003 号。

1.2 主要仪器

HH-4 型数控恒温水浴锅:国美电器有限公司

YS2-H 光学显微镜:日本 NiKon 公司显微镜

725 型-86℃ 低温冰箱:美国 FORMA

DL-5-B 型离心机:上海安亭科学仪器厂

全自动生化仪:日立 7020

BT815A 半自动生化仪:上海三科仪器有限公司

WZ-50C2 型单道微量注射泵:浙江大学医学仪器有限公司

OneTouchR Ultra 血糖仪、试纸:强生(上海)医疗器材有限公司

1.3 药物及试剂

纳米蜂胶:采用溶胶凝胶法,按照本章 1.1.1 工艺制备所得,含 30% 蜂胶

洛伐他汀:浙江瑞邦药业有限公司

胆固醇:上海博奥生物科技有限公司

胆盐:杭州微生物食品有限公司

药用聚乙二醇(PEG)400:北京市海淀会友精细化工厂

总胆固醇(TC)试剂盒:宁波慈城生化试剂厂

高密度脂蛋白胆固醇(HDL-C)测定试剂盒:宁波慈城生化试剂厂

低密度脂蛋白胆固醇(LDL-C)测定试剂盒:宁波慈城生化试剂厂

脂蛋白脂酶(LPL):南京建成生物工程研究所

肝脂酶(HL)测定试剂盒:南京建成生物工程研究所

游离胆固醇(FC)测定试剂盒:上海名典生物工程有限公司

2 试验方法

2.1 高脂血症动物模型的建立

2.1.1 高脂血症动物模型的建立及分组

取 84 只 SD 大鼠按空腹血清胆固醇和体重随机分为正常对照组、模型对照组、阳性药物对照组、低剂量纳米蜂胶组、中剂量纳米蜂胶组、高剂量纳米蜂胶组和纯胶组。除正常组喂饲普通饲料外,其余 6 组喂饲高脂饲料(0.5% 胆盐,1% 胆固醇,10% 猪油,10% 蛋黄,78.5% 普通饲料)。在喂饲高脂饲料的同时,各组给予对应剂量纳米蜂胶、纯胶及阳性药物。

2.1.2 剂量设计

蜂胶作为保健品对成年人的推荐剂量为 200mg/d。成年人标准体重以 70 kg 计,则服用蜂胶剂量为 2.86mg/(kg·d)约 3mg/(kg·d)。在保健食品的效果评价试验中,大鼠剂量取人剂量的 10 倍,则大鼠纯胶剂量为 30mg/(kg·d)。

含 30% 纳米蜂胶设置三个剂量组:

低剂量纳米蜂胶组:15mg/(kg·d)

中剂量纳米蜂胶组:30mg/(kg·d)

高剂量纳米蜂胶组:60mg/(kg·d)

2.1.3 实验处理

以每只大鼠 1mL/100g 的量灌胃,其中低剂量纳米蜂胶组灌胃纳米蜂胶 60mg/kg、中剂量纳米蜂胶组灌胃纳米蜂胶 120mg/kg、高剂量纳米蜂胶组灌胃纳米蜂胶 240mg/kg、纯胶组灌胃纯胶 30mg/kg,将纳米蜂胶及纯胶配成相应的浓度。正常组和模型组分别给予等量生理盐水灌胃。阳性药物组给予药物洛伐他汀。连续 6 周,分别于实验 0、2、4、6 周,大鼠禁食 12h,尾静脉取血,制备血清,测 TC、LDL-C、HDL-C,第 6 周时测定禁食 12h 后血清 LCAT、LPL、HL 活力及同一批血清的 HDL-C、HDL_3-C、HDL_2-C。实验第 6 周时,将高脂饲料组改饲普通饲料,1 周后,于晚 11 时至次日凌晨 1 时将大鼠处死,立即取大鼠肝脏,用于测定 HMG C_0A 还原酶比活,并进行肝组织病理学观察。

2.2 观察指标及测定方法

(1)胆固醇(TC)

全自动生化分析仪,于实验 0、2、4、6 周采集空腹血清测定。

(2)低密度脂蛋白胆固醇(LDL-C)

全自动生化分析仪,于实验 0、2、4、6 周采集空腹血清测定。

(3)高密度脂蛋白胆固醇(HDL-C)

采用 HDL-C 及其亚类测定试剂盒,于实验 0、2、4、6 周采集空腹血清测定。

(4)血清卵磷脂胆固醇酰基转移酶(Lecithin Cholesterol Acytransferase,LCAT)

方法:采用改良的 Sperry 法——The self-substrate method(Bartholome et al.,1981)。

操作步骤:将大鼠禁食 12h 后,断尾取血,4℃,3500r/min,离心 15min 制备血清,取两份新鲜血清 0.1mL,其中一份置 56℃水浴,另一份置 37℃水浴,各 1h,立即分别测定 FC 的含量。

计算方法:LCAT 活力=(总 FC-OD 值−反应后剩余的 FC-OD 值)÷标准 FC-OD 值×FC 标准浓度×(总反应液体积÷血浆体积×保温时间)×10^{-3}

$\quad\quad\quad =(\Delta OD \div 0.115) \times (1.29 \times 10^6) \times (0.03 \div 0.1 \times 1) \times 10^{-3}$

单位:nmol FC/(mL·h)

(5)血清脂蛋白脂酶(Lipoprotein,LPL)和肝脂酶(Hepatic Lipase,HL)

方法:参考 Krauss 及 Blache 方法,加以改进(Krauss et al.,1974;Blache et al.,1983;张蓉等,1996)。

样本处理:大鼠禁食 12h 后,尾静脉注射肝素 150 IU/kg 体重,30min 后取血,肝素抗凝,4℃,3500r/min,离心 15min 制备血清。

测定步骤:取两份新鲜血清各 0.1mL,分别加入底物(三酰甘油)0.01mL,pH 8.3,0.1mol/L Tris-HCL 缓冲液,其中一份加入蒸馏水 1.0mL,为总脂酶测定管(LPL+HL),另一份加 NaCl 1.0mL,即为 HL 测定管(LPL 被抑制),混匀,37℃水浴 20min;分别加入提取液 5mL(氯仿:正庚烷:甲醇=49:49:2,v/v,现配),充分混匀 30 s,3500r/min 离心 10min,取下层提取液 2mL,加入铜试剂 2mL,充分混匀 30 s,3500r/min 离心 10min,取上层提取液 2mL,加入显色剂 0.4mL(现配),混匀,静置 10min,550nm 比色。以同样的方法步骤将标准管作对照。该实验的标准品采用南京建成生物有限公司提供的 2 rnEq/L 的棕榈酸标准液。提取液调零。FFA 测定采用南京建成生物工程研究所提供的试剂盒。

表示方法:37℃每毫升血浆每小时在反应体系中所产生的 1 微摩尔的游离脂肪酸 FFA 为 1 个酶活性单位。

计算方法:

总脂酶活性=(测定管 OD 值÷标准管 OD 值)× 标准管 FFA 浓度×(60 分钟÷反应时间)×(1÷反应体积)÷1000=(测定管 OD 值÷0.33)× 500×(60÷20)×(1÷0.1)÷1000

肝脂酶活性:同上

脂蛋白脂酶活性=总脂酶活性−肝脂酶活性

单位:μEqFFA/(mL·h)

(6) 肝脏组织病理学观察

处死大鼠,立即摘取肝脏,用冰生理盐水清洗后,用滤纸吸干;取肝脏左叶一块 3cm×3cm 用 10%福尔马林固定,常规石蜡包埋,切片,HE 染色;镜下观察肝组织病理学变化。

2.3 纳米蜂胶的最大耐受量试验

ICR 小鼠 18~21g(雌雄各 5 只),禁食 4h 后,灌胃 0.8mL/20g(最高给药容量),每隔 2h 一次,共灌胃 3 次(最高频率)。共灌胃纳米蜂胶 24000mg/kg(折合人剂量的 2400 倍)。给药第 1 天每隔 30min 到 1h 观察 1 次,以后每天观察 1 次,连续观察 7d,未见死亡。

2.4 统计方法

实验数据用 SPSS 10.0 软件统计,结果以 $\bar{x}\pm s$ 表示,两两比较采用 t 检验。

3 结 果

3.1 对高脂血症 SD 大鼠体重、肝重、肝指数的影响

高脂饲料喂饲 6 周后,高脂模型组体重、肝脏质量及肝指数均显著高于正常组($P<0.05$,$P<0.01$),各实验组和高脂模型组比较,体重、肝重无显著性差异,纳米蜂胶高剂量组肝指数显著低于模型组($P<0.05$)。结果见表 16.3。

表 16.3 纳米蜂胶对高脂血症 SD 大鼠体重、肝重、肝指数的影响($n=10$, $\bar{x}\pm s$)
Table 16.3 Effects of nanometer propolis on body weight, liver weight, liver index in hyperlipemia SD rats ($n=10$, $\bar{x}\pm s$)

组别	体重/g	肝重/g	肝指数(%)
正常对照组	388.40±38.06	14.75±1.91	3.80±0.35
高脂模型对照组	439.70±45.32*	23.78±3.86**	5.62±0.54**
纳米蜂胶低剂量组	422.10±43.75	22.87±2.66	5.43±0.46
纳米蜂胶中剂量组	429.90±31.50	22.64±3.48	5.25±0.57
纳米蜂胶高剂量组	427.50±48.10	22.09±2.65	5.18±0.36△
纯胶组	429.13±47.37	23.44±3.77	5.45±0.22
洛伐他汀组	433.30±31.98	23.99±1.96	5.55±0.42

注:△ $P<0.05$,△△ $P<0.01$,与模型对照组相比;* $P<0.05$,** $P<0.01$,模型对照组比正常对照组。

3.2 对高脂血症 SD 大鼠血脂水平及动脉硬化指数(AI)的影响

模型组 TC 值显著高于正常组($P<0.01$),说明高脂模型造模成功。纳米蜂胶、纯胶及洛伐他汀对高脂饲料造成的 TC 升高有抑制作用,中、高剂量的纳米蜂胶较模型组 TC 值显著下降($P<0.05$,$P<0.01$)。模型组与正常组相比,HDL-C 显著下降($P<0.01$)而 LDL-C 显著升高($P<0.01$),纳米蜂胶高剂量组 HDL-C、LDL-C 显著低于模型组($P<0.05$)。模型组 AI 较正常组显著升高($P<0.01$),纳米蜂胶、纯胶及洛伐他汀作用可使 AI 较模型组下降,纳米蜂胶低、中、高组 AI 较模型组分别下降了 21.59%、23.48%和 40.04%,其中纳米蜂胶高剂量组 AI 显著低于模型组($P<0.05$)。结果见表 16.4。

表 16.4　纳米蜂胶对高脂血症 SD 大鼠 TC、HDL-C、LDL-C、AI 的影响（$n=10$，$\bar{x}\pm s$）

Table 16.4　Effects of nanometer propolis on serum TC, HDL-C, LDL-C and AI in hyperlipemia SD rats ($n=10$, $\bar{x}\pm s$)

组　别	TC/(mmol/L)	HDL-C/(mmol/L)	LDL-C/(mmol/L)	AI
正常对照组	2.00±0.24	1.38±0.19	0.37±0.12	0.45±0.09
高脂模型对照组	4.03±0.79**	0.71±0.08**	3.27±0.90**	4.77±1.39**
纳米蜂胶低剂量组	3.26±0.88	0.69±0.12	2.58±0.87	3.74±0.83
纳米蜂胶中剂量组	3.01±0.64△	0.65±0.09	2.47±0.65	3.65±0.75
纳米蜂胶高剂量组	2.67±0.76△△	0.62±0.04△	2.05±0.86△	2.86±0.85△
纯胶组	3.14±0.63	0.66±0.13	2.52±0.60	3.84±0.86
洛伐他汀组	3.18±0.80	0.68±0.09	2.53±0.90	3.72±1.03

注：△$P<0.05$，△△$P<0.01$，与模型对照组相比；*$P<0.05$，**$P<0.01$，模型对照组比正常对照组。

3.3　对高脂血症 SD 大鼠脂质代谢相关酶活性的影响

与正常组相比，高脂模型组 LPL、HL 及 LCAT 的活力均显著降低（$P<0.01$）。与高脂模型组相比，纳米蜂胶、纯胶和洛伐他汀可显著提高 LPL 的活性（$P<0.01$），纳米蜂胶和纯胶还可显著提高 HL 的活性（$P<0.05$，$P<0.01$），但洛伐他汀对 HL 活性无显著影响。与模型组相比，纳米蜂胶中、高剂量组、纯胶组及洛伐他汀组 LCAT 活力显著上升（$P<0.05$，$P<0.01$）。纳米蜂胶中剂量组 HL 活性显著高于纯胶组（$P<0.01$），但纳米蜂胶低剂量组 LCAT 活性低于纯胶组（$P<0.05$），结果见表 16.5。

表 16.5　纳米蜂胶对高脂血症 SD 大鼠 LPL、HL、LCAT 活力的影响（$n=10$，$\bar{x}\pm s$）

Table 16.5　Effects of nanometer propolis on plasm LPL, HL, LCAT activities in hyperlipemia SD rats ($n=10$, $\bar{x}\pm s$)

组　别	LPL/ (FFA μmol/(mL·h))	HL/ (FFA μmol/(mL·h))	LCAT/ (nmol FC/(mL·h))
正常对照组	20.04±7.32	12.40±4.77	88.00±24.40
高脂模型对照组	6.19±2.61**	3.47±0.65**	20.00±8.94**
纳米蜂胶低剂量组	10.50±3.14△△	5.59±2.16△	24.29±9.76▲
纳米蜂胶中剂量组	13.67±4.36△△	10.20±3.96△△▲▲	41.67±19.41△
纳米蜂胶高剂量组	11.66±3.99△△	7.02±3.39△	54.29±18.13△△
纯胶组	10.40±3.82△△	4.83±1.41△	44.00±15.17△
洛伐他汀组	13.50±4.82△△	3.30±0.80	73.75±25.04△△

注：△$P<0.05$，△△$P<0.01$，与模型对照组相比；*$P<0.05$，**$P<0.01$，模型对照组比正常对照组；▲$P<0.05$，▲▲$P<0.01$，纳米蜂胶组比普通蜂胶。

3.4　肝组织病理学检验结果

肉眼观察肝脏正常组颜色鲜红，未见异常变化，其余各组都可见肝脏体积不同程度变大,颜色呈浅黄色。光镜检验正常组未见异常,其余各组胞质中均存在数量不等、大小不一的透明空泡,HE染色证明这些空泡主要是中性脂肪滴。其肝脏组织切片见图16.5。

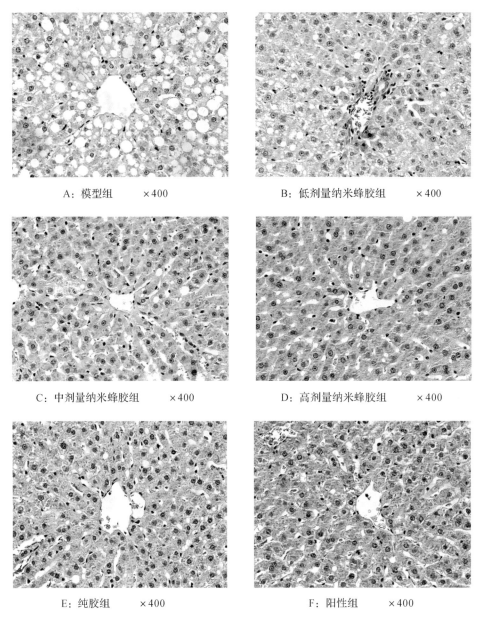

图16.5　肝脏组织切片

A. 模型组:大量肝细胞浊肿,且大泡状脂滴空泡化,肝小叶内细胞索结构紊乱,局部散在中性粒细胞和淋巴细胞为主的炎症细胞浸润。

B. 低剂量纳米蜂胶组:部分肝细胞浊肿,且有大量小泡状脂滴空泡化,门管区微灶性以中性粒细胞和淋巴细胞为主的炎症细胞浸润。

C. 中剂量组纳米蜂胶组:局部肝细胞轻度浊肿,未见明显肝细胞脂滴空泡化,局部区域内有淋巴细胞等炎症细胞散在分布。

D. 高剂量纳米蜂胶组:未见肝细胞明显浊肿,肝小叶结构清晰,肝细胞索状排列,未见明显炎症细胞浸润。

E. 纯胶组:未见肝细胞明显浊肿,肝小叶结构清晰,肝细胞索状排列,仅见局部少量炎症细胞散在分布。

F. 阳性组:未见肝细胞明显浊肿,肝小叶结构清晰,肝细胞索状排列,未见明显炎症细胞浸润。

4 讨 论

本实验所采用的高脂饲料配方中含有 10% 猪油和 10% 蛋黄,另有 1% 胆固醇和 0.5% 胆盐直接加入配方,故认为可造成内源性和外源性相混合、以内源性为主的高胆固醇血症模型(徐叔云,1982)。另外,由于血清甘油三酯(TG)与膳食中单糖和脂肪关系较大,受胆固醇影响较小,本实验喂饲高脂饲料后,只致血清 TC 水平升高而 TG 并不升高。该高胆固醇血症模型着重研究纳米蜂胶对胆固醇代谢的影响。

能量摄入长期超过能量消耗,大多数多余能量以甘油三酯的形式储存在脂肪细胞,从而导致细胞增生和肥大及能量代谢障碍(Djuric et al., 2002)。中医认为肥胖主要由于脾肾气虚又滞食肥甘、脾失健运,水湿内停聚而为痰饮,其内停脏腑、外停筋骨皮肉,从而导致肥胖。本实验研究表明,洛伐他汀、纳米蜂胶、纯胶均可抑制高脂饲料引起的 SD 大鼠体重增长,蜂胶中的有效成分是否参与其中,还有待进一步证实。

血清中 TC、TG、LDL-C 含量升高和 HDL-C 含量降低是诱发动脉硬化(AS)和导致心脑血管病的重要因素,高浓度 LDL-C 血症是 AS 的易患因子,其在血浆中的浓度与动脉硬化的发生率呈明显的负相关。正常人的血脂水平是血总胆固醇(TC)应低于 5.2,甘油三酯(TG)应低于 1.7;如果血清总胆固醇达到或超过 5.2,甘油三酯达到或超过 1.7 则可分别诊断为"高胆固醇血症"。近年来,血脂测定指标中增加了两项重要的内容:一是低密度脂蛋白胆固醇(LDL-C),它可促进冠心病的发生,称为"至动脉粥样硬化性脂蛋白",正常人的含量应不超过 3.12,心脏病患者应控制在 2.6 以下;另外一种为高密度脂蛋白胆固醇(HDL-C),它有利于防止血脂异常引起的冠心病,称为"抗动脉粥样硬化性脂蛋白"。正常人血中含量应超过 1.04,如果低于 0.91,也属于血脂代谢紊乱。根据研究,低密度脂蛋白胆固醇水平每升高 1%,冠心病等动脉粥样硬化性疾病的发病危险会相应地增加 2%;高密度脂蛋白胆固醇水平每降低 1%,冠心病发病危险增加 3%~4%。目前血脂异常与动脉粥样硬化的关系已成定论。本实验研究表明,洛伐他汀、纳米蜂胶、纯胶均可抑制高脂饲料引起的 SD 大鼠血液内 TC、LDL-C 的升高,降低动脉硬化指数(AI),说明对高脂血症大鼠有较好的治疗效果但洛伐他汀、纳米蜂胶、纯胶均对 HDL-C 的影响并不明显。

HL、LPL 和 LCAT 是血浆脂蛋白代谢中的 3 个关键酶,在脂蛋白代谢中发挥着重要的调控作用(吴满平和梅美珍,1991)。

HL 具有两种生物功能:一是表现出脂酶活性,水解各种脂蛋白中的三酰甘油(TG)和磷脂(PL),使各种脂蛋白颗粒的大小和密度发生变化;二是作为配体介导脂蛋白及蛋白多糖和(或)受体的结合与摄取的途径,促进肝细胞摄取高密度脂蛋白(HDL)中的胆固醇或低密度脂蛋白(LDL)残粒,促进肝细胞摄取含载脂蛋白 B 的脂蛋白,从而影响血浆中脂蛋白浓度。所以,肝脂酶的脂解和非脂解功能对于 HDL 的代谢都有重要影响,并可能与冠心病的发病有关。在 HL 方面,从实验结果来看洛伐他汀、纳米蜂胶、纯胶均能提高 HL 活力,中剂量纳米蜂胶、高剂量纳米蜂胶效果最好,其 HL 活力与正常组接近,低剂量纳米蜂胶效果也要好于纯胶和洛伐他汀。蜂胶对 HL 的作用机制尚不清楚。但是有报道蜂胶中含有的姜黄素能提高肝脂酶的活性(沃兴德,2003)。

LPL 是调节脂蛋白代谢的其中一种关键酶,它是由肝外组织分泌的糖蛋白,广泛存在于肝外组织的毛细血管内皮细胞表面。LPL 与硫酸乙酰肝素糖蛋白结合成复合物,附着于内皮细胞表面且十分稳定。注射肝素后，LPL 与肝素结合成复合物，脱离内皮细胞进入血循环，催化 CM 及 VLDL 中的 TG 水解为甘油和脂肪酸。Blades 等（1993）将受检者分为高 TG 血症组及正常 TG 组，结果提示 LPL 活性降低与 HDL-C 水平降低呈平行关系，但这种关系不受 TG 水平是否升高的影响。另一组学者进一步分析 LPL 活性与 HDL 各亚型分布及冠脉病变程度的关系，发现 LPL 活性降低与 HDL 水平降低相关联，LPL 活性越低，冠脉病变越严重。从本次试验结果来看，阳性药物组、中剂量纳米蜂胶、高剂量纳米蜂胶效果非常好，纳米蜂胶在提高 LPL 活力的效果方面要好于纯胶。

LCAT 是由肝脏分泌的一类糖蛋白，存在于血液中，其功能是催化卵磷脂分子二位脂酰基转移到胆固醇第三位轻基，生成 CE，血浆 90％CE 由 LCAT 反应形成。HDL 参与胆固醇的逆向转运，有效地摄取外周多余的胆固醇，抑制细胞对 LDL-C 的摄取，而 HDL 要在 LCAT 的作用下转运至肝脏代谢，从而减少胆固醇在血管壁内沉积。从本次试验结果来看，洛伐他汀、纳米蜂胶、纯胶提高 LCAT，高剂量纳米蜂胶与纯胶相近但纯胶活性略高一点。蜂胶对 LCAT 的作用机制尚不清楚。

血脂紊乱是脂肪肝及肝损伤的一个重要危险因素。研究证明，动脉硬化的形成及肝损伤与脂质过氧化物有关。血液中 TC 和 TG 的含量与血浆和肝脏中的脂质过氧化物呈正相关。血浆中低密度脂蛋白（LDL）在过氧化作用下，形成氧化修饰的低密度脂蛋白（OX-LDL），OX-LDL 目前被认为是引起细胞内脂质聚集和泡沫细胞形成的重要因素。在本研究中肝内脂质代谢加强，肝体积增大，颜色呈浅黄。洛伐他汀及纳米蜂胶、纯胶均能显著减轻脂肪的沉积，使肝重和肝指数明显低于模型组，具有良好的护肝作用。蜂胶的护肝作用可能与其抗氧化性有关。

纳米蜂胶与纯胶相比，在增强 LPL、HL 活力上效果比纯胶要好，可能的原因是：肝是人体的解毒中心，肝对药物有首过效应、降解了部分蜂胶，而纳米粒被载体包埋在里面，具有缓释作用。但是，在其他一些指标上纳米蜂胶的优势没有得到明显体现，这有待今后作进一步的研究。

缓释制剂是通过适当方法，延缓药物在体内的释放、吸收、分布、代谢和排泄过程，以达到延长药物作用的一类制剂。

与一般制剂相比，缓释制剂有 3 个优点：

（1）服用方便。一般制剂常需一日数次服药，而缓释制剂通常只用每日给药 1～2 次。

（2）缓释药较一般制剂作用徐缓，避免了一般制剂频繁给药后，因血药浓度起伏过大而出现有效血药浓度的忽高忽低。

（3）毒副作用较一般制剂小。

本实验以黄酮含量为测定指标，通过体外释放实验表明所用纳米蜂胶具有缓释作用。

小　结

本研究通过纳米蜂胶对高脂血症 SD 大鼠的作用实验，得到了以下结果：

（1）纳米蜂胶和蜂胶有良好的降脂作用。

（2）纳米蜂胶在提高 LPL、HL 活力上，效果要好于蜂胶。

（3）纳米蜂胶和蜂胶对肝脏均具有保护作用。

（4）小鼠最大耐受量实验证明纳米蜂胶是一种低毒物质，不会对机体产生不良影响。

第三节　纳米蜂胶对 2 型糖尿病大鼠代谢紊乱与胰岛素抵抗的影响

1　实验材料

1.1　实验动物

SD 大鼠,雄性,体重 270～370g,清洁级,由中国科学院上海实验动物中心/上海斯莱克实验动物公司提供,实验室使用许可证:SYXK(浙) 2003-0003。

饲养条件:清洁级大鼠实验饲养室,温度 23±2℃,日温差±1℃,湿度 50%～70%,噪音<50dB,光照 150～200lx,12h 明暗交替。自来水过滤灭菌,置于高压灭菌的饮水瓶中自由饮用,8:00～8:30am 喂食。

1.2　主要仪器

HH-4 型数控恒温水浴锅:国美电器有限公司

YS2-H 光学显微镜:日本 NiKon 公司

725 型-86℃低温冰箱:美国 FORMA

DL-5-B 型离心机:上海安亭科学仪器厂

全自动生化仪:日立 7020

BT815A 半自动生化仪:上海三科仪器有限公司

WZ-50C2 型单道微量注射泵:浙江大学医学仪器有限公司

OneTouchR Ultra 血糖仪、试纸:强生(上海)医疗器材有限公司

1.3　主要试剂

纳米蜂胶:采用溶胶凝胶法,按照 1.1.1 工艺制备所得,含 30% 蜂胶

血糖测定试剂盒(GLU):上海申能德赛技术诊断有限公司

果糖胺测定试剂盒(FMN):四川迈克科技有限公司

糖化血红蛋白测定试剂盒(HBA1C):日本和光纯药工业株式会社

胰岛素测定试剂盒(INS):北京科美东雅生物技术有限公司

胆固醇测定试剂盒(TC):上海申能德赛技术诊断有限公司

三酰甘油测定试剂盒(TG):上海申能德赛技术诊断有限公司

高密度脂蛋白测定试剂盒(HDL):上海申能德赛技术诊断有限公司

低密度脂蛋白测定试剂盒(LDL):上海复星长征医学科学有限公司

超氧化物歧化酶测定试剂盒(SOD):南京建成生物工程研究所

丙二醛测定试剂盒(MDA):南京建成生物工程研究所

尿素氮测定试剂盒(BUN):上海申能德赛技术诊断有限公司

肌酐测定试剂盒(CREA):上海申能德赛技术诊断有限公司

尿酸测定试剂盒(UA):上海申能德赛技术诊断有限公司

总蛋白测定试剂盒(TP):上海申能德赛技术诊断有限公司

白蛋白测定试剂盒(ALB):上海申能德赛技术诊断有限公司
谷丙转氨酶测定试剂盒(ALT):上海申能德赛技术诊断有限公司
谷草转氨酶测定试剂盒(AST):上海申能德赛技术诊断有限公司
卡司平:杭州中美华东制药有限公司
链脲佐菌素(STZ):ALEXIS CORPORATION
胰岛素注射液:南京新百药业有限公司
肝素注射液:南京新百药业有限公司

2 实验方法

2.1 2型糖尿病动物模型的建立及处理

2.1.1 2型糖尿病动物模型的建立及分组

取100只体重为270～370g的雄性SD大鼠,喂饲高糖高能饲料,并分4次小剂量注射链脲佐菌素(STZ),以造成实验性2型糖尿病大鼠模型:所有SD大鼠禁食、不禁水16h,尾静脉注射STZ 10mg/kg BW;稳定5d后,第6天所有SD大鼠禁食、不禁水18h,尾静脉注射STZ 5mg/kg BW;再稳定3d后,第4天所有SD大鼠禁食、不禁水18h,尾静脉注射STZ 20mg/kg BW;最后稳定11d后,第12天所有SD大鼠禁食、不禁水16h后,尾静脉注射STZ 10mg/kg BW,稳定1周后称取空腹体重,进行糖耐量与空腹血糖水平测定。所用STZ浓度为1mg/mL,注射容量为0.1mL/100g BW。

选取体重390～470g,糖耐量异常,空腹6h血糖大于7.0mmol/L的大鼠72只,按空腹血糖和体重随机分为6组(每组12只):模型对照组、阳性药对照组(卡司平)、纳米蜂胶低剂量组、纳米蜂胶中剂量组、纳米蜂胶高剂量组和纯胶组。另取正常大鼠12只,作为正常对照组。

2.1.2 实验处理

6组实验组大鼠喂饲高糖高能饲料,正常对照组大鼠喂饲普通饲料。阳性对照组以每天10mg/kg BW灌胃卡司平,纳米蜂胶低、中、高剂量组以50、100、200mg/kg BW灌胃纳米蜂胶,纯胶组以30mg/kg BW灌胃纯胶,模型对照组及正常对照组灌胃2% PEG400。灌胃容积为1mL/100g BW,卡司平、纳米蜂胶及纯胶以2% PEG400配制。

给药10周,每周称重,每2周空腹6h眼眶后静脉丛取血,分离血清,测定血糖(GLU)、果糖胺(FMN)、总胆固醇(TC)、三酰甘油(TG)水平;于实验第6、8、10周测定血清高密度脂蛋白胆固醇(HDL-C)、低密度脂蛋白胆固醇(LDL-C)含量;于实验第4、8周测定空腹血清胰岛素(FINS)水平,计算胰岛素作用指数(IAI);并于实验中期(第5周)及末期(第10周)测定血清中超氧化物歧化酶(SOD)、丙二醛(MDA)水平;实验末期(第10周)测定谷草转氨酶(AST)、谷丙转氨酶(ALT)、尿素氮(BUN)、肌酐(CREA)、尿酸(UA)、总蛋白(TP)、白蛋白(ALB)水平;实验结束,动物禁食12h后用戊巴比妥钠麻醉,腹主动脉取血,测定糖化血红蛋白(HbA_{1c})水平;处死动物,解剖并取部分肝脏、胰腺用10%福尔马林固定,以作病理学检查;取另一侧肾用中性甲醛固定,以做PAS染色,进行病理学检查。

2.2 纳米蜂胶影响 2 型糖尿病大鼠胰岛素敏感性的正糖钳实验

2.2.1 实验动物及分组

另选体重 500～600g、GLU 为 7～11mmol/L 的 2 型糖尿病 SD 大鼠 8 只。分为 4 组（每组 2 只），分别给予 50mg/kg BW、100mg/kg BW 纳米蜂胶、30mg/kg BW 纯胶及 2% PEG400。

2.2.2 手术方法

给予相应剂量纳米蜂胶、纯胶和 2% PEG400 30min 后，腹腔注射 3% 戊巴比妥钠（0.5mL/100g），麻醉后仰位固定于大鼠固定板上，切开颈部正中皮肤。

分离右侧颈外静脉，结扎远心端，用眼科剪作切口，向近心端插入连接三通旋塞的插管（插管中预先注入 80μL/mL 肝素生理盐水溶液）。用眼科镊夹住静脉壁和插管，向前行进 0.5cm，结扎固定插管。

分离左侧颈总动脉，结扎远心端，用小动脉夹夹住近心端，用眼科剪作切口，往近心端反向插入连接 1mL 针筒的插管（插管中预先注入 80μL/mL 肝素生理盐水）。放开小动脉夹，用眼科镊夹住动脉壁和插管，向前行进 0.5cm，结扎固定插管。

颈静脉注射 50μL/mL 肝素 0.1mL/100g 作预防抗凝。

三通旋塞的一个进口连接胰岛素输注微量泵另一个进口连接 20% 葡萄糖输注微量泵，避免管腔中出现气泡。待插管手术和输注装置安装完毕后，开始实验。

2.2.3 实验方法

用 1mL 针筒从右颈动脉取血 0.5mL，用手术钳夹住插管，用 OneTouch^R Ultra 血糖仪和试纸测定基础血糖约需 5μL，放开手术钳，将剩余血再注回颈总动脉，插管头上加 50μL 肝素生理盐水溶液。

胰岛素以速率为 1.0mL/h 恒速输注，10min 再取血测血糖，如血糖低于基础值，则开始输注 20% 葡萄糖，每隔 10min 取血，测定血糖。根据所测血糖值不断调整葡萄糖输注速率，维持在稳定血糖 5mmol/L 上下。达稳态后，每 5min 取血测血糖，记录稳态下 5～6 次葡萄糖输注速率的平均值作为 GIR(glucose infusion rate)，取稳态下 6 次血糖的平均值作为稳态下的血糖(BG)。

2.3 观察指标

2.3.1 血清胰岛素

采用碘$[^{125}I]$-胰岛素(INS)放射免疫分析药盒，于实验第 4、6、8 周分别采集空腹血液，及时分离血清。

2.3.2 胰岛素作用指数

$IAI=ln1/(FBG \times FINS)$

2.3.3 血清 SOD

采用超氧化物歧化酶(SOD)试剂盒，于实验第 5、第 10 周分别采集空腹血液，及时分离血清。

2.3.4 血清 MDA

采用丙二醛(MDA)试剂盒，于实验第 5、第 10 周分别采集空腹血液，及时分离血清。

其余指标均采用日立全自动生化仪结合相应试剂盒进行测定。

2.4 组织病理学观察

处死大鼠,立即摘取肝脏、肾脏和胰腺,用冰生理盐水清洗后滤纸吸干;取肝脏左叶一块 3cm×3cm,胰腺,用 10%甲醛固定,常规石蜡包埋,切片,HE 染色;取一侧肾脏用中性甲醛固定,常规石蜡包埋,切片,PAS 染色;镜下观察肝、肾及胰腺组织病理学变化。

2.5 数据处理

实验数据用 SPSS 11.5 软件统计,结果以 $\bar{x}\pm s$ 表示,两两比较采用 t 检验,多组比较采用方差分析。

3 实验结果

3.1 纳米蜂胶对 2 型糖尿病大鼠体重的影响

实验开始时,各组大鼠体重无明显区别($P>0.05$);给药 10 周内各组大鼠体重稳步增长,给予纳米蜂胶和纯胶的大鼠体重增长和模型对照组基本一致,无显著性差异($P>0.05$),给予阳性药卡司平的大鼠体重增长较快,于给药第 7 周和第 10 周时体重显著高于模型对照组($P<0.05$)。结果见图 16.6。

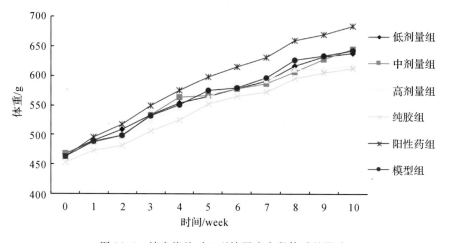

图 16.6 纳米蜂胶对 2 型糖尿病大鼠体重的影响

Fig.16.6 Effects of nanometer propolis on body weight in T2DM Rats

3.2 纳米蜂胶对 2 型糖尿病大鼠糖代谢的影响

3.2.1 纳米蜂胶对 2 型糖尿病大鼠空腹血糖的影响

给药前(实验第 0 周),模型对照组 FBG 显著高于正常对照组($P<0.05$),说明糖尿病模型造模成功;实验期间模型对照组 FBG 呈上升趋势,至实验后期,模型对照组 FBG 分别升高了 17.1%(第 8 周)和 23.1%(第 10 周);纳米蜂胶和阳性药物卡司平对 2 型糖尿病大鼠 FBG 升高有抑制作用,纳米蜂胶低剂量组、阳性药物组在第 6、8、10 周,纳米蜂胶中、高剂量组在第 8、10 周 FBG 显著低于同期模型对照组($P<0.05$);纯胶组在实验期间对糖尿病大鼠 FBG 无显著影响,与同期模型对照组比较无显著差异($P>0.05$)。结果见表 16.6。

第十六章 纳米蜂胶的制备及其降血脂降血糖作用研究

表 16.6 纳米蜂胶对 2 型糖尿病大鼠 FBG 的影响($\bar{x}\pm s, n=12$)

Table 16.6 Effects of nanometer propolis on FBG in T2DM rats ($\bar{x}\pm s, n=12$)

组别	空腹 6h 血糖浓度/(mmol/L)					
	第 0 周	第 2 周	第 4 周	第 6 周	第 8 周	第 10 周
正常对照组	6.07±0.31	6.18±0.21	6.09±0.28	6.09±0.24	6.13±0.24	6.19±0.33
模型对照组	7.35±0.93△△	7.39±0.42△△	7.51±0.83△△	8.23±0.99△△	8.61±0.91△△	9.05±1.32△△
纳米蜂胶低剂量组	7.30±0.81	7.82±0.57	7.67±0.76	7.13±0.50*	7.74±0.55*	7.88±0.48*
纳米蜂胶中剂量组	7.20±1.73	7.79±0.88	7.68±0.95	7.51±0.74	7.59±0.40*	7.51±0.52*
纳米蜂胶高剂量组	7.17±1.20	7.68±0.99	7.63±1.07	7.67±0.44	7.32±0.69*	7.37±1.12*
纯胶组	7.13±1.85	7.44±0.83	7.49±0.81	7.18±0.93	7.98±0.90	8.06±0.87
阳性药(卡司平)组	7.28±1.45	7.68±0.76	6.85±0.60	7.01±0.52*	7.53±0.48*	7.54±0.45*

与模型对照组比:* $P<0.05$, ** $P<0.01$; 与正常对照组比:△ $P<0.05$, △△ $P<0.01$。

3.2.2 纳米蜂胶对 2 型糖尿病大鼠血清果糖胺的影响

模型对照组在实验期间 FMN 逐渐升高, 实验 6 周后, 模型对照组 FMN 显著高于正常对照组($P<0.05, P<0.01$); 实验第 10 周, 阳性药物组 FMN 水平较模型对照组显著下降($P<0.05$); 较之于模型对照组, 纳米蜂胶各剂量组及纯胶组在实验期间对 FMN 水平无显著影响($P>0.05$)。结果见表 16.7。

表 16.7 纳米蜂胶对 2 型糖尿病大鼠 FMN 的影响($\bar{x}\pm s, n=12$)

Table 16.7 Effects of nanometer propolis on FMN in T2DM rats ($\bar{x}\pm s, n=12$)

组别	FMN/(mmol/L)			
	第 2 周	第 6 周	第 8 周	第 10 周
正常对照组	0.73±0.04	0.82±0.02	0.82±0.06	1.09±0.06
模型对照组	0.75±0.03	0.88±0.08△	0.95±0.07△△	1.30±0.11△
纳米蜂胶低剂量组	0.72±0.09	0.82±0.08	0.89±0.07	1.24±0.07
纳米蜂胶中剂量组	0.70±0.06	0.84±0.10	0.88±0.07	1.25±0.06
纳米蜂胶高剂量组	0.71±0.08	0.88±0.06	0.89±0.03	1.23±0.11
纯胶组	0.76±0.07	0.85±0.04	0.94±0.03	1.27±0.06
阳性药(卡司平)组	0.72±0.06	0.82±0.08	0.90±0.05	1.18±0.05*

与模型对照组比:* $P<0.05$, ** $P<0.01$; 与正常对照组比:△ $P<0.05$, △△ $P<0.01$。

3.2.3 纳米蜂胶对 2 型糖尿病大鼠全血糖化血红蛋白的影响

模型对照组 HbA_{1c} 含量显著高于正常对照组($P<0.01$), 纳米蜂胶及阳性药物可降低 2 型糖尿病大鼠 HbA_{1c} 含量, 其中纳米蜂胶低剂量组和阳性药物组与模型对照组比较 HbA_{1c} 含量极显著下降($P<0.01$), 纳米蜂胶中剂量组显著下降($P<0.05$), 纳米蜂胶高剂量组、纯胶组与模型对照组比较无显著差异($P>0.05$)。结果见图 16.7。

3.2.4 纳米蜂胶对 2 型糖尿病大鼠空腹血清胰岛素及胰岛素作用指数的影响

模型对照组与正常对照组比较, FINS 无显著性差异(实验第 4 周, $P>0.05$)或显著升

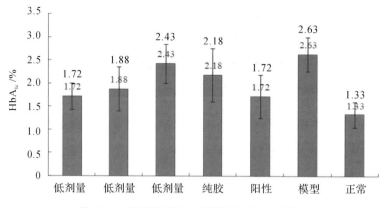

图 16.7 纳米蜂胶对 2 型糖尿病 HbA$_{1c}$ 的影响

Fig. 16.7 Effects of nanometer propolis on HbA$_{1c}$ in T2DM rats

高(实验第 8 周,$P<0.05$),IAI 则显著降低($P<0.05$,$P<0.01$);纳米蜂胶、纯胶及阳性药物可提高 2 型糖尿病大鼠机体 IAI。实验第 8 周,纳米蜂胶低剂量组、阳性药物组 FINS 显著低于模型对照组($P<0.05$),纳米蜂胶中、高剂量组及纯胶组 FINS 较模型组无显著差异($P>0.05$),纳米蜂胶各剂量组、纯胶组及阳性组 IAI 均高于模型组($P<0.05$)。结果见表 16.8。

表 16.8 纳米蜂胶对 2 型糖尿病大鼠 FINS 和 IAI 的影响($\bar{x}±s$,$n=12$)

Table 16.8 Effects of nanometer propolis on FINS and IAI in T2DM rats ($\bar{x}±s$,$n=12$)

组别	FINS/(μIU/mL)		IAI	
	第 4 周	第 8 周	第 4 周	第 8 周
正常对照组	48.00±6.05	39.75±14.70	−5.39±0.19	−5.44±0.41
模型对照组	41.96±8.71	55.85±10.66△	−5.78±0.27△△	−6.17±0.30△△
纳米蜂胶低剂量组	36.96±8.46	43.65±10.42*	−5.65±0.31	−5.79±0.34*
纳米蜂胶中剂量组	37.28±9.02	40.80±16.31	−5.49±0.42	−5.60±0.52*
纳米蜂胶高剂量组	36.27±8.30	44.21±9.13	−5.53±0.43	−5.76±0.24*
纯胶组	36.77±5.89	44.04±8.74	−5.60±0.18	−5.82±0.11*
阳性药(卡司平)组	36.29±9.09	40.55±17.45*	−5.28±0.26	−5.64±0.33*

与模型对照组比:* $P<0.05$,** $P<0.01$;与正常对照组比:△ $P<0.05$,△△ $P<0.01$。

3.3 纳米蜂胶对 2 型糖尿病大鼠脂质代谢的影响

3.3.1 纳米蜂胶对 2 型糖尿病大鼠血清三酰甘油的影响

实验期间模型对照组 TG 水平逐渐升高,实验第 4 周后显著高于正常对照组($P<0.01$,$P<0.05$);纳米蜂胶及阳性药物对糖尿病大鼠 TG 的升高有抑制作用;与模型对照组比较,实验第 4 周,纳米蜂胶中剂量组及阳性药组 TG 水平显著降低($P<0.01$,$P<0.05$);实验第 6 周,纳米蜂胶低、中剂量组及阳性药组 TG 显著降低($P<0.01$,$P<0.05$);实验第 8 周,各剂量纳米蜂胶组、纯胶组及阳性药组 TG 均显著降低($P<0.01$,$P<0.05$);实验第 10 周,纳

米蜂胶中、高剂量组及阳性药组 TG 显著降低($P<0.01,P<0.05$)。结果见表 16.9。

表 16.9　纳米蜂胶对 2 型糖尿病大鼠 TG 的影响($\bar{x}\pm s, n=12$)

Table 16.9　Effects of nanometer propolis on TG in T2DM rats ($\bar{x}\pm s, n=12$)

组别	TG/(mmol/L)				
	第 2 周	第 4 周	第 6 周	第 8 周	第 10 周
正常对照组	1.89±0.34	1.78±0.38	1.80±0.27	1.82±0.39	2.17±0.37
模型对照组	1.59±0.42	2.89±0.77△△	2.98±0.79△△	3.09±0.74△△	3.08±0.74△
纳米蜂胶低剂量组	1.77±0.61	2.27±0.72	2.10±0.44*	2.15±0.63*	2.32±0.59
纳米蜂胶中剂量组	1.79±0.56	2.00±0.35**	1.78±0.54**	1.50±0.50**	1.54±0.59**
纳米蜂胶高剂量组	1.84±0.54	2.43±0.73	2.48±0.59	2.18±0.61*	2.13±0.65*
纯胶组	1.65±0.64	3.26±0.35	2.47±0.74	2.18±0.49*	2.33±0.75
阳性药(卡司平)组	1.60±0.40	2.13±0.49*	1.87±0.76*	1.98±0.69**	1.59±0.46**

与模型对照组比：*$P<0.05$, **$P<0.01$；与正常对照组比：△$P<0.05$, △△$P<0.01$。

3.3.2　纳米蜂胶对 2 型糖尿病大鼠血清总胆固醇的影响

实验期间模型对照组 TC 与正常对照组比较无显著差异($P>0.05$)；第 6 周时纯胶组及阳性组 TC 水平显著低于模型对照组($P<0.05,P<0.01$)；纳米蜂胶各剂量组实验期间 TC 水平较之模型对照组无显著差异($P>0.05$)。结果见表 16.10。

表 16.10　纳米蜂胶对 2 型糖尿病大鼠 TC 的影响($\bar{x}\pm s, n=12$)

Table 16.10　Effects of nanometer propolis on TC in T2DM rats ($\bar{x}\pm s, n=12$)

组别	TC/(mmol/L)				
	第 2 周	第 4 周	第 6 周	第 8 周	第 10 周
正常对照组	1.99±0.16	1.97±0.29	2.02±0.21	2.05±0.27	2.10±0.47
模型对照组	1.89±0.20	2.04±0.30	2.05±0.35	2.28±0.41	2.24±0.33
纳米蜂胶低剂量组	2.02±0.34	2.06±0.32	1.86±0.23	2.01±0.36	2.22±0.46
纳米蜂胶中剂量组	2.04±0.29	2.25±0.39	1.92±0.32	2.20±0.47	2.34±0.50
纳米蜂胶高剂量组	2.10±0.23	2.14±0.22	1.93±0.20	2.03±0.40	2.15±0.37
纯胶组	2.05±0.29	2.12±0.14	1.75±0.21*	1.97±0.21	1.96±0.23
阳性药(卡司平)组	2.04±0.29	1.97±0.33	1.63±0.18**	1.89±0.48	2.00±0.53

与模型对照组比：*$P<0.05$, **$P<0.01$。

3.3.3　纳米蜂胶对 2 型糖尿病大鼠血清高密度脂蛋白胆固醇的影响

实验中后期，模型对照组 HDL-C 水平与正常对照组无显著性差异($P>0.05$)；实验第 6 周及第 8 周阳性药组 HDL-C 显著低于模型对照组($P<0.01$)；纳米蜂胶各剂量组及纯胶组在实验中后期与模型对照组比较 HDL-C 水平无显著差异($P>0.05$)。结果见表 16.11。

表 16.11 纳米蜂胶对 2 型糖尿病大鼠 HDL-C 的影响($\bar{x}\pm s, n=12$)

Table 16.11 Effects of nanometer propolis on HDL-C in T2DM rats ($\bar{x}\pm s, n=12$)

组别	HDL-C/(mmol/L)		
	第 6 周	第 8 周	第 10 周
正常对照组	1.04±0.14	1.03±0.18	1.02±0.21
模型对照组	1.20±0.20	1.18±0.27	1.09±0.29
纳米蜂胶低剂量组	1.26±0.17	1.12±0.15	1.14±0.22
纳米蜂胶中剂量组	1.23±0.19	1.06±0.19	1.12±0.20
纳米蜂胶高剂量组	1.31±0.17	1.11±0.17	1.11±0.18
纯胶组	1.12±0.15	1.14±0.19	1.13±0.09
阳性药(卡司平)组	0.91±0.13**	0.80±0.15**	0.90±0.19

与模型对照组比:* $P<0.05$,** $P<0.01$。

3.3.4 纳米蜂胶对 2 型糖尿病大鼠血清低密度脂蛋白胆固醇的影响

实验第 6 周及第 10 周时,模型对照组 LDL-C 水平显著高于正常对照组($P<0.05$);纳米蜂胶各剂量组、纯胶组和阳性药物组实验中后期 LDL-C 水平与模型对照组相比无显著性差异($P>0.05$)。结果见表 16.12。

表 16.12 纳米蜂胶对 2 型糖尿病大鼠 LDL-C 的影响($\bar{x}\pm s, n=12$)

Table 16.12 Effects of nanometer propolis on LDL-C in T2DM rats ($\bar{x}\pm s, n=12$)

组别	LDL-C/(mmol/L)		
	第 6 周	第 8 周	第 10 周
正常对照组	0.21±0.05	0.24±0.08	0.21±0.08
模型对照组	0.30±0.07△	0.22±0.09	0.32±0.09△
纳米蜂胶低剂量组	0.28±0.06	0.27±0.06	0.27±0.06
纳米蜂胶中剂量组	0.30±0.09	0.28±0.06	0.29±0.09
纳米蜂胶高剂量组	0.25±0.06	0.26±0.06	0.26±0.06
纯胶组	0.25±0.05	0.27±0.07	0.29±0.07
阳性药(卡司平)组	0.28±0.06	0.24±0.06	0.24±0.03

与模型对照组比:* $P<0.05$,** $P<0.01$;与正常对照组比:△ $P<0.05$,△△ $P<0.01$。

3.4 纳米蜂胶对 2 型糖尿病大鼠氧化损伤的影响

与正常对照组比较,模型对照组 SOD 活力显著下降(实验第 10 周,$P<0.01$)而 MDA 含量显著升高($P<0.05$);纳米蜂胶可提高 2 型糖尿病大鼠体内 SOD 活力及降低 MDA 含量。实验第 10 周,纳米蜂胶高剂量组及阳性药组 SOD 活力显著高于模型对照组($P<0.05$);MDA 水平在第 5 周时纳米蜂胶低剂量组显著低于模型组($P<0.05$),其余各组与模型组比较无显著性差异($P>0.05$),第 10 周时纳米蜂胶各剂量组、纯胶组及阳性药组均较模型组显著下降（$P<0.05$，$P<0.01$）。结果见表 16.13。

表 16.13　纳米蜂胶对 2 型糖尿病大鼠血清 SOD、MDA 的影响（$\bar{x}\pm s, n=12$）

Table 16.13　Effects of nanometer propolis on SOD and MDA in T2DM rats（$\bar{x}\pm s, n=12$）

组别	SOD（U/ML）		MDA（nmol/l）	
	第 5 周	第 10 周	第 5 周	第 10 周
正常对照组	307.60±38.28	301.16±14.02	5.97±0.90	7.46±0.82
模型对照组	295.08±33.78	241.50±28.28△△	7.39±0.98△	8.91±1.04△
纳米蜂胶低剂量组	294.33±33.71	260.60±28.26	5.31±0.81*	7.63±0.89*
纳米蜂胶中剂量组	289.85±23.04	288.94±39.52	6.77±0.70	7.49±1.24*
纳米蜂胶高剂量组	286.05±35.61	283.76±22.49*	7.18±1.05	5.78±0.94**
纯胶组	263.40±36.86	276.86±28.99	7.78±0.44	7.30±1.35*
阳性药（卡司平）组	270.21±31.56	284.09±36.11*	7.15±0.77	6.67±0.89**

与模型对照组比：*$P<0.05$，**$P<0.01$；与正常对照组比：△$P<0.05$，△△$P<0.01$。

3.5　纳米蜂胶对 2 型糖尿病大鼠肝肾功能及蛋白质代谢的影响

实验第 10 周，测定血清中各项肝肾功能及蛋白质代谢指标可见：血清 BUN，模型对照组较正常对照组显著降低（$P<0.05$）；纳米蜂胶各剂量组、纯胶组和阳性药组与模型对照组比较无显著差异（$P>0.05$）。血清 CREA、UA、ALT 及 AST，各实验组间未见显著差异（$P>0.05$）。血清 TP 及 ALB，模型对照组均较正常对照组显著下降（$P<0.05$）；与模型对照组比较，纳米蜂胶低剂量组 TP 显著升高（$P<0.05$），纳米蜂胶中、高剂量、纯胶及阳性药组与模型对照组无显著差异（$P>0.05$）。各剂量纳米蜂胶、纯胶及阳性药组 ALB 与模型对照组无显著差异（$P>0.05$）。结果见表 16.14。

表 16.14　纳米蜂胶对 2 型糖尿病大鼠血清 BUN、CREA、UA、ALT、AST、TP、ALB 的影响（$\bar{x}\pm s, n=12$）

Table 16.14　Effects of nanometer propolis on BUN, CREA, UA, ALT, AST, TP, ALB in T2DM rats（$\bar{x}\pm s, n=12$）

组别	实验末期（第 10 周）						
	BUN/(mmol/L)	CREA/(μmol/L)	UA/(μmol/L)	ALT/(IU/L)	AST/(IU/L)	TP/(g/L)	ALB/(g/L)
正常对照组	5.23±0.46	55.00±3.54	73.73±9.51	38.50±5.93	77.60±9.03	69.67±2.18	33.89±0.78
模型对照组	4.40±0.61△	58.10±5.86	67.33±6.04	38.57±6.60	73.50±12.12	66.33±2.07△	31.33±2.16△
纳米蜂胶低剂量组	4.06±0.37	58.55±4.74	69.27±6.78	37.82±6.69	74.44±14.03	69.82±3.49*	33.36±1.57
纳米蜂胶中剂量组	4.20±0.77	58.70±5.85	73.00±7.93	43.00±4.20	77.17±11.25	69.00±5.35	32.00±2.19
纳米蜂胶高剂量组	4.88±0.73	59.91±2.74	67.64±7.97	39.29±9.43	72.89±7.56	66.91±4.04	32.82±2.04
纯胶组	4.55±0.85	58.75±3.96	63.13±10.05	41.33±7.34	66.60±5.03	67.63±4.10	32.38±1.69
阳性药（卡司平）组	4.59±0.80	59.18±3.54	68.64±6.96	37.90±6.01	75.73±14.96	66.18±3.06	30.55±1.92

与模型对照组比：*$P<0.05$，**$P<0.01$；与正常对照组比：△$P<0.05$，△△$P<0.01$。

3.6 纳米蜂胶对2型糖尿病大鼠正糖钳实验下葡萄糖输注速率的影响

在以速率为1.0mL/h恒速输注胰岛素,大鼠稳态血糖值控制在4.8mmol/L左右的情况下,模型对照组稳态下GIR最低,为1.62mL/h;纳米蜂胶低、中剂量组及纯胶组均可显著提高稳态下的GIR($P<0.01$),增强大鼠肌体组织对胰岛素的敏感性。结果见表16.15。

表16.15 纳米蜂胶对2型糖尿病大鼠正糖钳试验稳态下BG和GIR平均值的影响($\bar{x}\pm s, n=6$)

Table 16.15 Effects of nanometer propolis on BG and GIR in T2DM rats ($\bar{x}\pm s, n=6$)

组别	BG/(mmol/L)	GIR-20%GLU/(mL/h)
模型对照组	4.80±0.25	1.62±0.04
纳米蜂胶低剂量组	4.66±0.15	2.44±0.05**
纳米蜂胶中剂量组	4.82±0.29	3.68±0.04**
纯胶组	4.78±0.11	1.98±0.04**

与模型对照组比:* $P<0.05$,** $P<0.01$。

3.7 实验大鼠的病理观察

3.7.1 实验大鼠胰腺的病理观察

如图16.8中,A:正常组大鼠镜下观察胰腺复管状泡腺结构清晰,被膜结构完整,胰小叶内可见染色较浅的胰岛,胰腺泡结构正常,腺泡细胞清晰可见,分布在腺泡周围的胰管结构完整。B:模型组大鼠镜下观察部分胰岛体积减小,胰岛组织与周围界限不清,胰岛β细胞浊肿,部分细胞空泡化。C:低剂量组大鼠胰岛组织与周围界限不清,部分β细胞肿胀或者空泡化,细胞核深染,但未见明显的炎症细胞浸润;D:中剂量组大鼠胰岛组织与周围界限清晰,β细胞轻度肿胀,胰岛组织内未见炎症细胞浸润;E:高剂量组大鼠胰岛组织与周围界限清晰,β细胞轻度肿胀,胰岛组织内未见炎症细胞浸润;F:纯胶组大鼠胰岛组织与周围界限不清,β细胞轻度肿胀,细胞核深染,胰岛组织内未见炎症细胞浸润;G:阳性组大鼠胰岛组织与周围界限清晰,部分β细胞空泡化,胰岛组织内未见炎症细胞浸润。

3.7.2 实验大鼠肝脏的病理观察

如图16.9中,A:正常组大鼠肝组织各小叶结构清晰,小叶中央静脉结构完好,周围肝细胞索呈放射状排列,可见肝血窦和其壁内库普弗细胞。门管区可见小叶间动静脉,肝细胞和肝血窦结构正常。B:模型组大鼠大量肝细胞浊肿,且大泡状脂滴空泡化,肝小叶内细胞索结构紊乱,局部散在中性粒细胞和淋巴细胞为主的炎症细胞浸润;C:低剂量组大鼠部分肝细胞浊肿,且有大量小泡状脂滴空泡化,门管区微灶性以中性粒细胞和淋巴细胞为主的炎症细胞浸润;D:中剂量组大鼠局部肝细胞轻度浊肿,未见明显肝细胞脂滴空泡化,局部区域内有淋巴细胞等炎症细胞散在分布;E:高剂量组大鼠未见肝细胞明显浊肿,肝小叶结构清晰,肝细胞索状排列,未见明显炎症细胞浸润;F:纯胶组大鼠未见肝细胞明显浊肿,肝小叶结构清晰,肝细胞索状排列,仅见局部少量炎症细胞散在分布;G:阳性组大鼠未见肝细胞明显浊肿,肝小叶结构清晰,肝细胞索状排列,未见明显炎症细胞浸润。

3.7.3 实验大鼠肾脏的病理观察

如图16.10中,A:正常组大鼠肾被膜、皮质、髓质结构清晰,皮质部肾小体和肾小管结构

第十六章　纳米蜂胶的制备及其降血脂降血糖作用研究

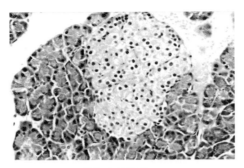

A：正常组大鼠胰腺 HE×400

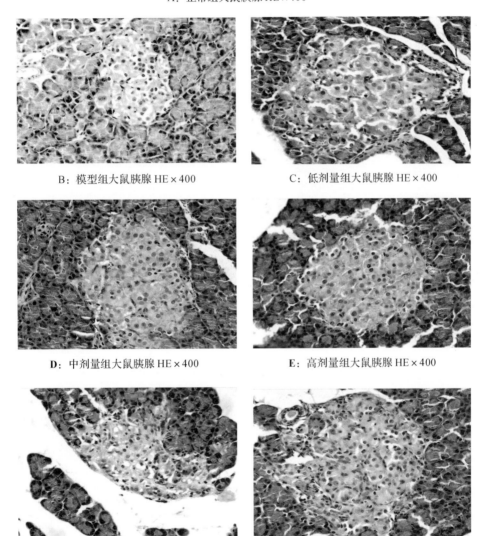

B：模型组大鼠胰腺 HE×400　　　　　　C：低剂量组大鼠胰腺 HE×400

D：中剂量组大鼠胰腺 HE×400　　　　　　E：高剂量组大鼠胰腺 HE×400

F：纯胶组大鼠胰腺 HE×400　　　　　　G：阳性组大鼠胰腺 HE×400

图16.8　各实验组大鼠胰腺典型病理切片

Fig. 16.8　Histological analysis of pancreas in experimental rats

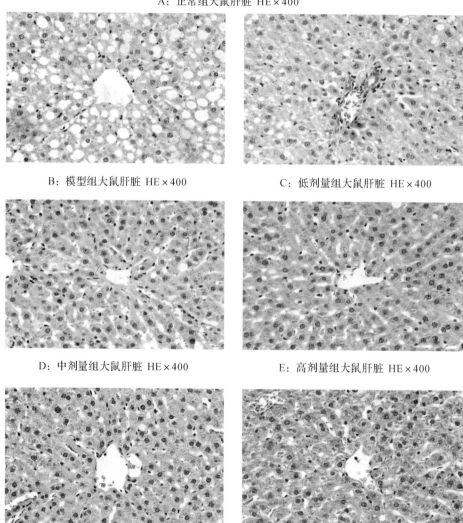

A：正常组大鼠肝脏 HE×400

B：模型组大鼠肝脏 HE×400

C：低剂量组大鼠肝脏 HE×400

D：中剂量组大鼠肝脏 HE×400

E：高剂量组大鼠肝脏 HE×400

F：纯胶组大鼠肝脏 HE×400

G：阳性组大鼠肝脏 HE×400

图16.9　各实验组大鼠肝脏典型病理切片

Fig.16.9　Histological analysis of liver in experimental rats

第十六章 纳米蜂胶的制备及其降血脂降血糖作用研究

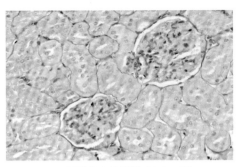
A：正常组大鼠肾脏 PAS×400

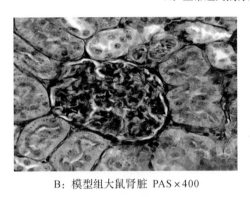
B：模型组大鼠肾脏 PAS×400

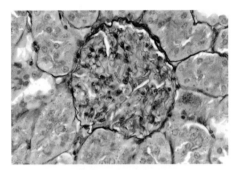
C：低剂量组大鼠肾脏 PAS×400

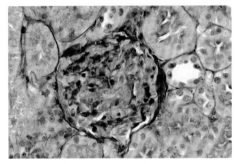
D：中剂量组大鼠肾脏 PAS×400

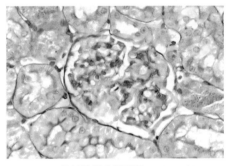
E：高剂量组大鼠肾脏 PAS×400

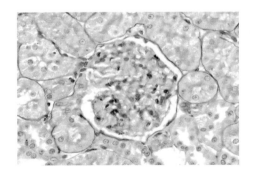
F：纯胶组大鼠肾脏 PAS×400

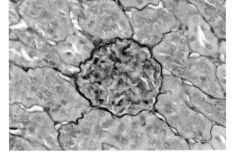
G：阳性组大鼠肾脏 PAS×400

图 16.10　各实验组大鼠肾脏 PAS 染色病理切片

Fig. 16.9　Histological analysis of kidney in experimental rats

完好,未见肾小球增大,无炎症细胞浸润。B:模型组大鼠肾脏,肾小球体积增大,少量染成红染的阳性物质沉积;C:低剂量组大鼠肾脏,肾小球体积增大,少量红染的阳性物质沉积,较模型组明显减少;D:中剂量组大鼠肾脏,肾小球体积增大,少量红染的阳性物质沉积,较模型组明显减少;E:高剂量组大鼠肾脏,肾小球体积增大,极少量红染的阳性物质沉积,较模型组明显减少;F:纯胶组大鼠肾脏,肾小球体积增大,极少量红染的阳性物质沉积,较模型组明显减少;G:阳性组大鼠肾脏,肾小球体积增大,少量红染的阳性物质沉积,较模型组明显减少。

4 讨 论

4.1 2型糖尿病动物模型的建立

长期以来,人们一直在寻找接近于人类2型糖尿病的动物模型,模拟2型糖尿病的发病过程,阐述其发病的分子机制,来寻找预防及治疗2型糖尿病更有效的方法(Morral,2003)。糖尿病动物模型主要分为三类,即遗传性肥胖模型,如C57BL/6J ob/ob小鼠、KKAy小鼠、新西兰肥胖小鼠、WDF/Ta-fa大鼠、Zucker大鼠等(李保春等,2003);化学制剂模型(如链脲霉素、四氧嘧啶等);特殊饲料喂养模型(各种配方饲料)。由于遗传原发性糖尿病动物稀少,繁殖不易,费用高,限制了其大量应用于研究;化学制剂STZ和四氧嘧啶选择性破坏胰岛β细胞,使其分泌胰岛素的能力丧失而造成1型糖尿病;特殊饲料喂养结合小剂量化学制剂可诱导兼具胰岛素抵抗和高血糖的2型糖尿病模型(Reed et al.,2000;Ramadan et al.,2006;Zhang et al.,2003;艾静等,2004)。

本实验通过喂饲高热量饲料,并少量多次尾静脉注射STZ以诱发SD大鼠形成2型糖尿病。造模过程中严密监测其体重的变化,发现通过持续喂饲高热量饲料,动物体重增长迅速,由造模初期平均体重320g增长至成模时的平均体重430g;小剂量注射STZ后监测其血糖的变化,以摸索合适的STZ剂量,使之小到既能保证糖尿病的发生,又能避免胰岛素合成和分泌能力的完全耗竭,不至于成为典型的1型糖尿病模型,通过4次剂量分别为10、5、20、10mg/kg BW的STZ尾静脉注射,得到了血糖显著高于正常SD大鼠的稳定的2型糖尿病大鼠模型。

此模型由高热量饮食导致SD大鼠产生外周胰岛素抵抗,为了克服外周胰岛素抵抗使血糖处于正常范围,胰岛增生,胰岛β细胞代偿性地分泌胰岛素,使胰岛素增高。这时给予一定剂量STZ使胰岛β细胞受损,胰岛素分泌相对减少,此时虽然胰岛素水平仍高于正常,但不能有效代偿外周胰岛素抵抗,因而产生类似糖尿病的表现。小剂量STZ破坏胰岛程度轻,能使机体保持基础胰岛素的分泌(Tobin and Finegood,1993)。因此,将本模型用于纳米蜂胶对2型糖尿病大鼠代谢紊乱及胰岛素抵抗的研究可为纳米蜂胶应用于2型糖尿病的防治提供可靠的生物学资料。

4.2 纳米蜂胶对2糖尿病大鼠糖代谢的影响

高血糖是2型糖尿病最直观的表现,持续性高血糖会殃及全身器官、组织及细胞,是引起糖尿病慢性并发症的主要因素。在糖尿病治疗过程中,控制血糖水平是治疗的首要目标(Bailey,2000)。实验结果显示,模型对照组大鼠随着2型糖尿病病程的延长,血糖水平逐渐升高,至实验中、后期(第6、8、10周)血糖水平与实验初期(第0周)比较分别升高了12.0%、17.1%和23.1%;纳米蜂胶各剂量组、纯胶组及阳性药物组血糖水平未见明显升高,且纳米蜂胶各剂量组、阳性组在第8周及第10周时血糖水平均显著低于同期模型对照组,可见纳米蜂胶及阳性药物卡司平可抑制2型糖尿病大鼠血糖的进一步升高。糖尿病控制及并发症实验(DCCT)及英国糖尿病前瞻性研究(UKPDS)均证实强化血糖控制及降低血糖有利于减少糖尿病慢性并发症的发生(钱容立,1999)。高血糖可能通过多种机制发挥其病理生理作用,其一为提高醛糖还原酶(AR)的活性,活化山梨醇旁路代谢,结果导致细胞内山梨醇堆积,肌醇大量丢失,机体抗氧化酶能力降低,组织蛋白果糖化,最终破坏细胞膜结构与功能的完整性,导致细胞的生理代谢异常(叶山东,2000)。研究证实,蜂胶对AR有抑致作用(金光香等,2005),其抑制活性可能得益于蜂胶中含量丰富的黄酮类化合物(Bent,2002)。可以认为对AR的抑制作用是蜂胶多途径、多靶点、整体调整以降低2型糖尿病大鼠血糖的重要环节之一。

果糖胺(FMN)是由血清蛋白(主要是白蛋白)非酶糖化所形成,能较为稳定地反映采血前2～3周内的血糖控制情况,它对于追踪2型糖尿病病情,观察疗效有一定的参考价值。研究结果显示,随着糖尿病病程的延长,模型组FMN水平逐渐升高,实验第6周开始模型组FMN显著高于正常组;实验期间,纳米蜂胶各剂量组FMN水平虽较同期模型组稍低,但均未呈现显著性差异,阳性药物也仅在第10周出现低于模型组的状况。这可能与2型糖尿病SD大鼠模型血清中白蛋白浓度低,使检测所得FMN结果偏低,不能很好反映血糖实际浓度有关(许曼音等,2003)。

糖化血红蛋白(HbA_{1c})是血红蛋白A组分和葡萄糖分子经非酶促反应而形成的产物,可以反映取血之前机体2个月左右的总体血糖情况。目前,HbA_{1c}已作为糖尿病流行病学研究和疗效考核的有效检测指标在临床中得到了广泛使用。最具权威的两大糖尿病临床研究DCCT和UKPDS均把HbA_{1c}作为糖尿病控制的一个重要评价指标,且都充分肯定了强化治疗在预防血管并发症发生、发展中的重要作用(Davidson,2004;Manley,2003)。HbA_{1c}与平均血糖浓度有很高的相关性,血糖控制的好坏可以直接从HbA_{1c}值看出。本研究测定了实验结束(第10周)时各实验组动物的HbA_{1c}水平,以反应实验期间受试药物对2型糖尿病动物的血糖控制情况。发现模型组HbA_{1c}水平显著高于正常组,纳米蜂胶各剂量组、纯胶组及阳性药物组HbA_{1c}水平均低于模型对照组,其中纳米蜂胶低、中剂量组、阳性药物组有显著性差异。可见,纳米蜂胶及阳性药物对2型糖尿病大鼠血糖的控制起到了较好的效果。

机体的血糖水平是受多种激素相互作用、共同调节的,其中最重要的是由胰岛β细胞分泌的胰岛素(INS)。INS通过抑制肝葡萄糖产生,刺激内脏细胞及周围组织对葡萄糖的摄取等途径调节糖代谢,降低血糖水平。目前普遍认为,胰岛素抵抗和β细胞缺陷是2型糖尿病发病机制的两个主要环节(Szoke and Gerich,2005)。慢性高血糖可通过强化己糖胺途径及激活蛋白激酶C(PKC)等多种机制引发胰岛素抵抗,使内脏及外周组织葡萄糖摄取降低,致使尽管胰岛素水平正常甚或升高都难以维持血糖在正常水平(Petersen and Shulman,2006)。因此,能否改善胰岛素抵抗成为判断药物疗效的一个重要指标。我们以胰岛素作用指数(IAI)来评价胰岛素敏感性(李光伟和Bennett,2005)。研究结果显示,模型组大鼠胰岛素水平并未低于正常大鼠甚至还有升高,但空腹血糖水平却显著高于正常鼠,比较其IAI发现,模型鼠胰岛素敏感性显著低于正常鼠,说明本模型主要因胰岛素抵抗而非胰岛素分泌不足导致血糖升高,对胰岛进行病理观察也证实,2型糖尿病病理改变较轻,β细胞未有显著下降,略有肿胀;纳米蜂胶各剂量组、纯胶组及阳性药物组IAI较模型组比较显著增强,可见纳米蜂胶、纯胶与阳性药物一样可改善2型糖尿病大鼠机体的胰岛素抵抗。说明纳米蜂胶对2型糖尿病动物糖代谢的调节作用与改善胰岛素的敏感性有关。

4.3 纳米蜂胶对2糖尿病大鼠脂质代谢的影响

脂质代谢异常在糖尿病研究领域日益受到重视。在2001年美国糖尿病学会年会上，Banting奖获得者McGarry教授提出：脂代谢障碍为糖尿病及其并发症的原发性病理改变，即2型糖尿病中糖代谢紊乱的根源为脂代谢异常。他甚至将2型糖尿病称为"糖脂病"（Mcgarry，2001）。根据美国国家胆固醇教育方案（NCEP）的资料，2型糖尿病伴有高三酰甘油（TG）血症的发生率高于40%。2型糖尿病病人体内胰岛素对脂肪细胞作用降低，致使TG水解的抑制作用减弱，从而释放大量非酯化脂肪酸（NEFA），增多的NEFA运输至肝脏又可增加肝脏TG的合成。空腹和餐后时，胰岛素对脂肪水解酶，即脂蛋白脂酶（LPL）的作用降低导致对富含TG的脂蛋白清除率降低，这是2型糖尿病发生高TG血症的重要原因（Giovanni et al.，2004；Krentz，2003）。研究发现，纳米蜂胶、纯胶和阳性药物卡司平对2型糖尿病大鼠空腹TG水平的升高有很好的抑制作用，这可能与纳米蜂胶、纯胶及卡司平提高了机体的胰岛素敏感性，从而改善了LPL的活性，加快了TG水解及代谢有关。

研究发现，2型糖尿病患者的高胆固醇（TC）血症与非糖尿病患者无差别，但在同样的TC水平下冠心病的患病率增高。2型糖尿病患者脂质紊乱的主要表现除TG增高外还常伴有高密度脂蛋白胆固醇（HDL-C）减少及小而密的低密度脂蛋白（LDL）片断增加（Watson et al.，2003；Battisti et al.，2003）。HDL-C水平降低与胰岛素抵抗有关，HDL微粒在LPL水解极低密度脂蛋白（VLDL）核心的三酰甘油、载脂蛋白A_1和磷脂后获得新生，胰岛素抵抗时由于LPL活性的降低致使HDL合成降低；另外，在胰岛素抵抗或高胰岛素血症时，HDL微粒变小，密度变大，更易分解，这也使HDL微粒数量减少，HDL水平降低。就2型糖尿病患者而言，由于其肝脏合成VLDL增多、清除减少，因此导致小而密的LDL微粒增多。小而密LDL微粒易于发生氧化和糖化修饰，特别是在2型糖尿病时，从而导致动脉硬化能力增强（Verges，1999；Betteridge，2000）。本实验模型TC、HDL-C水平与正常对照比较未见显著升高，LDL-C水平显著升高，导致模型更易于发生氧化损伤。

对2型糖尿病患者进行血糖控制可以明显降低血清TG水平，在血糖控制良好的情况下HDL-C水平可以升高，并且显示小而密LDL微粒减少（Ginsberg，2005）。本研究显示，纳米蜂胶、纯胶和阳性药物卡司平都可降低2型糖尿病大鼠空腹TG水平，纳米蜂胶对TC、HDL-C及LDL-C水平无影响，阳性药物卡司平使HDL-C水平下降，提示卡司平对脂质调节具有不利影响。可见纳米蜂胶通过有效地降糖作用来改善2型糖尿病的血脂异常，又可通过其改善血脂异常作用，增加胰岛素敏感性，使胰岛素刺激的组织可以有效地对摄取和利用葡萄糖，同时抑制肝糖产生及输出增多，控制高血糖状态，从而形成一个良性循环，起到缓解及控制糖尿病的作用。

4.4 纳米蜂胶对2型糖尿病大鼠氧化损伤的影响

动物实验与临床实验证明，糖尿病情况下存在明显的氧化应激且与糖尿病慢性并发症的发生发展有关。糖尿病患者体内自由基含量升高，主要源于：糖尿病人血糖升高，血糖升高可导致糖化血红蛋白水平上升，单糖（如葡萄糖）和糖基化蛋白（如糖化血红蛋白）可自动氧化而产生自由基；因血中和组织中抗氧化酶如超氧化物歧化酶（SOD）、过氧化氢酶（CAT）

及谷胱甘肽过氧化酶(GSHPx)等活性下降,脂质过氧化物(LPO)增高导致自由基清除系统功能减弱而使自由基增加;另外,碳水化合物(主要为单糖)、脂质和氨基酸等通过代谢反应和非酶反应产生羰基亦可进一步氧化修饰蛋白质等而参与氧化应激的产生。自由基可使 LDL-C 氧化为氧化型的低密度脂蛋白胆固醇(oxLDL-C),oxLDL-C 损伤动脉内皮细胞,引起动脉粥样硬化,进而引起高血压、冠心病、心肌梗死、脑卒中等并发症。自由基还能损伤肾小球微血管引起糖尿病肾病,损伤眼晶体引起白内障,损伤神经引起糖尿病神经病变。这些糖尿病的并发症的发生机制被认为与自由基的增加、体内抗氧化能力的下降密切相关(Wright et al.,2006)。

蜂胶具有很强的抗氧化能力,研究发现在低浓度时蜂胶就能使 SOD 活性显著提高,并清除自由基,减少脂质过氧化物和脂褐素的生成与沉积,保护细胞膜,增强细胞活力,调节器官组织功能(Ichikawa et al.,2002)。

纳米蜂胶在 2 型糖尿病防治中,其抗氧化性能成为其发挥作用不可或缺的一环。研究结果表明,2 型糖尿病大鼠体内 SOD 活力明显降低而丙二醛(MDA)含量明显增高。纳米蜂胶及阳性药物可使血清 SOD 活力增强,提高机体的抗氧化清除自由基性能;另外,纳米蜂胶、纯胶、阳性药物均显著降低了 MDA 含量,对脂质过氧化具有抑制作用。对胰岛进行病理观察发现,纳米蜂胶对胰岛组织具有保护作用,可减轻 β 细胞的肿胀和空泡化,这与蜂胶保护胰岛细胞减少氧化损伤有关。目前认为蜂胶中至少有两类物质起抗氧化作用:一类是黄酮类物质,这类物质主要通过螯合金属离子阻止羟基自由基的形成和清除自由基而起到抗氧化作用,它是蜂胶抗氧化作用的主要成分;另一类是咖啡酸酯类,如 CAPE,这类物质抑制黄嘌呤氧化酶的生成从而减少自由基的产生,同时也具有清除已生成的自由基的作用(Russo et al.,2002)。另外,蜂胶中新分离出的组分 Artepillin C 也证实是一种很强的抗氧化剂(Shimizuk et al.,2004)。纳米蜂胶的抗氧化特性对控制 2 型糖尿病大鼠血糖的升高及调节脂质紊乱从而抑制糖尿病慢性并发症的发生起到了积极作用。

4.5 纳米蜂胶对 2 型糖尿病大鼠蛋白质代谢、肝、肾功能及肝、肾病理的影响

糖尿病不仅引起体内糖、脂质代谢异常,同时也会引起体内蛋白质代谢紊乱。消耗性疾病常引起血清总蛋白(TP)、白蛋白(ALB)的降低(胡福良等,2004)。研究显示,2 型糖尿病大鼠血清 TP、ALB 含量显著低于正常大鼠,而给予纳米蜂胶可使 TP 含量上升,从而降低机体蛋白质消耗,调节体内蛋白质代谢。在 2 型糖尿病时虽然胰岛素抵抗造成肝脏葡萄糖产量增加,周围组织摄取葡萄糖障碍,但脂肪分解和蛋白质消耗对胰岛素作用的反应比葡萄糖代谢更敏感,纳米蜂胶改善了糖尿病大鼠的胰岛素抵抗状况从而间接对蛋白质代谢起到了调控作用(Ginsberg,2005;Hegele and Robert,2003)。

有关糖尿病肝脏损害的研究较少,事实上肝脏亦为长期高血糖累及的靶器官之一,长期高血糖肝脏损害主要表现为脂肪肝、糖原累积性肝肿大、肝硬化、血色素沉淀症及肝炎的发生率增加(Roden and Bernroider,2003;Meyer,2004)。特别是糖尿病性脂肪肝(DFL)已成为非酒精性脂肪肝的主要病因之一。糖尿病状态输入肝脏的脂肪及脂肪酸和肝脏中合成 TG 的速度超过了组织 VLDL 分泌入血的速度,肝脏中 TG 堆积,氧化应激增加,肝脏二次打击,导致脂肪肝发生炎症、坏死和纤维化(Atli et al.,2004;Decristofaro et al.,2003)。研究显示,2 型糖尿病大鼠谷丙转氨酶(ALT)、谷草转氨酶(AST)水平与正常鼠无显著性差异,纳米蜂胶各剂量组、纯胶组及阳性组 ALT、AST 水平与模型组也未产生显著性差异。从肝脏病理观

察发现,模型组大鼠肝细胞浊肿,且大泡状脂滴空泡化,已有脂肪肝产生,局部炎性细胞浸润,但尚未产生肝纤维化,表现为肝功能尚维持正常;纳米蜂胶、纯胶及阳性药物对2型糖尿病大鼠肝脏具有保护作用,肝病理观察可发现肝脏脂肪化程度减轻,炎性细胞浸润减少。

2型糖尿病的高血糖状态很容易引起各种大血管及微血管病变,糖尿病肾病(DN)即是高血糖所致微血管病变的典型糖尿病并发症之一(Agarwal and Rajiv,2006)。目前约有20%~30%的1型和2型糖尿病患者进展为DN。DN的病理改变随病程和DN的不同阶段而不同,DN早期常表现为肾脏的增大、肾小球高滤过和肾小球的肥大,晚期肾小球系膜区细胞外基质增生,肾小管间质纤维化,最终导致肾小球硬化(Hansen and Klavs,1996;Gilbert et al.,1995)。高血糖导致肾脏病变的原因之一可能与糖基化作用有关,长期高血糖症可导致糖化终末产物(AGEs)的形成,AGEs可与近端肾小管上皮细胞中的AGEs受体结合,促进上皮细胞向肌纤维母细胞的转化,导致上述细胞中转移生长因子β(TGF-β)表达增加,TGF-β的过表达可使肾小球肥大以及肾小球、肾小管间质的细胞外基质增多,导致糖尿病性肾病的形成(Oldfield et al.,2001;Vranes et al.,1995;Sharma and Mcgowan,2000)。研究发现,2型糖尿病大鼠血清尿素氮(BUN)、肌酐(CREA)、尿酸(UA)并未高于正常大鼠,肾脏出现轻度形态学变化,观察其肾脏病理发现肾小球肌膜增厚,肾小球肥大,系膜增宽,肾小管间质轻度增多,说明2型糖尿病模型鼠处于早期DN状态,尚无临床表现;纳米蜂胶各剂量组、纯胶组及阳性药物组BUN、CREA、UA与模型组比较无显著差异,观察其肾脏病理发现,纳米蜂胶可减轻肾脏病变,对2型糖尿病引起的肾损伤具有保护作用。

2型糖尿病慢性并发症的许多致病途径是葡萄糖依赖性的,达到最佳血糖控制具有决定性意义(Neyestani et al.,2004;Haukeland et al.,2005)。纳米蜂胶对2型糖尿病血糖的控制,减少AGEs的形成及抗氧化作用对保护2型糖尿病机体的肝肾功能起到了积极作用。

4.6 纳米蜂胶对2型糖尿病大鼠胰岛素抵抗的影响

从理论上说,胰岛素抵抗很普遍,但实践中判定胰岛素抵抗并不容易,因为胰岛素抵抗是指机体胰岛素介导的葡萄糖代谢能力下降,而机体对葡萄糖的代谢不仅受靶组织对胰岛素反应敏感程度(胰岛素抵抗)的影响,而且受机体产生胰岛素量(β细胞分泌功能)的影响。换言之,即胰岛素敏感性不变的情况下,只有那些能"排除"胰岛素缺乏影响的测定葡萄糖代谢能力的方法,才能准确地评估真实的胰岛素敏感性。测定机体胰岛素抵抗的"金标准"是1979年创立的正常血糖胰岛素钳夹技术。同时静脉输入胰岛素和葡萄糖,使体内胰岛素达某种特殊浓度(纠正胰岛素缺乏)。调整葡萄糖输入速度使血葡萄糖水平稳定在4.48~5.04mmol/L,频繁取血测定血糖2h,通过稳态下的葡萄糖输注速率(GIR)可反映机体对葡萄糖的利用率,从而评价机体对外源性胰岛素的敏感性(李秀钧,2001;罗谋伦等,1999)。

本研究为了进一步证明纳米蜂胶对2型糖尿病大鼠胰岛素敏感性的影响,对其进行了正糖钳实验验证。结果表明,与模型组比较,纳米蜂胶及纯胶均可显著提高稳态下葡萄糖的输注速率,纳米蜂胶中剂量组GIR高于低剂量组,存在剂量效应。说明纳米蜂胶及纯胶可增强2型糖尿病大鼠机体对胰岛素的敏感性,这与前述通过IAI评价所得结论一致。纳米蜂胶提高糖尿病大鼠胰岛素敏感性的作用对其控制血糖升高,调节脂质紊乱,抗氧化性能的发挥及对2型糖尿病慢性并发症的预防控制具有重要意义。

小　结

本研究采用动物实验研究了纳米蜂胶对 2 型糖尿病大鼠体内糖、脂质、蛋白质代谢,氧化应激、肝肾功能及胰岛素抵抗的影响。结果表明:

（1）采用持续高能量饲料喂饲,并分次小剂量注射 STZ 的造模方法,能基本模拟 2 型糖尿病发病过程并具备相应特点。

（2）纳米蜂胶能降低 2 型糖尿病大鼠的 FBG 及 HbA_{1c} 水平,提高 IAI 及稳态下 GIR,增强机体胰岛素敏感性。

（3）纳米蜂胶能降低 2 型糖尿病大鼠血清 TG 水平,调节机体脂质代谢能力。

（4）纳米蜂胶对 2 型糖尿病大鼠氧化损伤具有抵抗能力,提高了血清 SOD 活力,并使 MDA 含量降低。

（5）纳米蜂胶对肝、肾、胰腺组织具有保护作用。

参考文献

Agarwal R（2006）Diabetic kidney disease:pathophysiology,implications and management[J]. *Trends in Endocrinology & Metabolism*,17(2):38-39.

Annuzzi G,De Natale C,Iovine C,Patti L,Di Marino L,Coppola S,Rivellese AA（2004）Insulin resistance is independently associated with postprandial alterations of triglyceride-Rich lipoproteins in type 2 diabetes mellitus[J]. *Arteriosclerosis, Thrombosis, and Vascular Biology*,24:2397-2402.

Atli T,Keven K,Avci A,Kutlay S,Turkcapar N,Varli M,Canbolat O（2004）Oxidative stress and antioxidant status in elderly diabetes mellitus and glucose intolerance patients[J]. *Archives of Gerontology and Geriatrics*,39(3):269-275.

Bailey CJ（2000）Potential new treatments for type 2 diabetes[J]. *Trends in Pharmacological Sciences*,21(7):259-265.

Bartholome M,Niedmann D,Wieland H（1981）An optimized methods for measuring LCAT activity,independent of the concentration and quality of the physiological' substrate[J]. *Biochim Biophys Acta*,664(2):327-334.

Battisti WP,Palmisano J,Keane WE（2003）Dyslipidemia in patients with type 2 diabetes:relationships between lipids,kidney disease and cardiovascular disease[J]. *Clinical Chemistry and Laboratory Medicine*,41(9):1174-1181.

Bent HH（2002）The biochemistry and medical significance of the flavonoids[J]. *Pharmacology & Therapeutics*,96(2):67-202.

Betteridge DJ（2000）Diabetic dyslipidaemia[J]. *Diabetes,Obesity and Metabolism*,2(S1):S31-36.

Blache D,Bouthillier D,Davignon J（1983）Simple,reproducible procedure for selective measurement of lipoprotein lipase and hepatic lipase[J]. *Clinical Chemistry*,29(1):154-158.

Bladdes B,Vega GL and Grundy SM（1993）Activities of lipoprotein lipase and hepatic triglyceride lipase in postheparin plasma of patients with low concentration of HDL cholesterol[J]. *Arterioscler Thromb*, 13(8):1227-1235.

Davidson JA (2004) Treatment of the patient with diabetes:importance of maintaining target HbA(1c) levels[J]. *Current Medical Research and Opinion*, 20(12):1919-1927.

Decristofaro R, Rocca B, Vitacolonna E, Vitacolonna, E. ,Falco A, Marchesani P, Ciabattoni G, Davi G (2003) Lipid and protein oxidation contribute to a prothrombotic state in patients with type 2 diabetes mellitus[J]. *Journal of Thrombosis and Haemostasis*, 1(2):250-256.

Djuric Z, Poore KM, Depper JB Uhley VE, Lababidi S, Covington C, Heilbrun LK (2002) Methods to increase fruit and vegetable intake with and without a decrease in fat intake:Comliance and effects on body weight in the nutrition an breast health study [J]. *Nutrition and Cancer*, 43(2):141-151.

Gilbert Re, Cox A, Dziadek M, Cooper ME, Jerums G (1995) Extracellular matrix and its interactions in the diabetic kidney: a molecular biological approach[J]. *Journal of Diabetes and its Complications*, 9(4):252-254.

Ginsberg HN, Zhang YL, Hernandez-Ono A (2005) Regulation of plasma triglycerides in insulin resistance and diabetes[J]. *Archives of Medical Research*, 36(3):232-240.

Hansen KW (1996) Ambulatory blood pressure in insulin-dependent diabetes:the relation to stages of diabetic kidney disease[J]. *Journal of Diabetes and its Complications*, 10(6):331-351.

Haukeland Jw, Konopski Z, Linnestad P, Azimy S, Marit Løberg E, Haaland T, Bjøro K (2005) Abnormal glucose tolerance is a predictor of steatohepatitis and fibrosis in patients with non-alcoholic fatty liver disease[J]. *Scandinavian Journal of Gastroenterology*, 40(12):1469-1477.

Hegele, Robert A (2003) Monogenic forms of insulin resistance: apertures that expose the common metabolic syndrome[J]. *Trends in Endocrinology and Metabolism*, 14, (8):371-377.

Ichikawa H, Satoh K, Tobe T, Yasuda I, Ushio F, Matsumoto K, Ookubo C (2002) Free radical scavenging activity of propolis[J]. *Redox Report*, 7(5):347-350.

Krauss RM, Levy RI, Fredrickson DS (1974) Selective measurement of two lipase activities in postheparin plasma from normal subjects and patients with hyperlipoproteinemia [J]. *Journal of Clinical Investigation*, 54(5):1107-1124.

Krentz AJ (2003) Lipoprotein abnormalities and their consequences for patients with type 2 diabetes[J]. *Diabetes, Obesity and Metabolism*, 5(s1):S19-27.

Manley S (2003) Haemoglobin A1c—a marker for complications of type 2 diabetes:the experience from the UK Prospective Diabetes Study (UKPDS)[J]. *Clinical Chemistry and Laboratory Medicine*, 41(9):1182-1190.

Mcgarry JD (2002) Banting lecture 2001: dysregulation of fatty acid metabolism in the etiology of type 2 diabetes[J]. *Diabetes*, 51(1):7-18.

Meyer C, Woerle HJ, Dostou JM, Welle SL, Gerich JE (2004) Abnormal renal, hepatic, and muscle glucose metabolism following glucose ingestion in type 2 diabetes[J]. *American Journal of Physiology-Endocrinology and Metabolism*, 287(6):1049-1056.

Morral N (2003) Novel targets and therapeutic strategies for type 2 diabetes[J]. *Trends in Endocrinology and Metabolism*, (4):169-175.

Neyestani TR, Alipour-Birgani R, Siassi F, Rajayi M, Djalali M, Mohamadi, M (2004) Glycemic optimization may reduce lipid peroxidation independent of weight and blood lipid changes in Type 2 diabetes mellitus[J]. *Diabetes, Nutrition & Metabolism*, 7(5):275-279.

Oldfield MD, Bach LA, Forbes JM, Nikolic-Paterson D, McRobert A, Thallas V, Cooper ME (2001) Advanced glycation end products cuase epithelial myofibroblast transdifferentiation via receptor for advanced glycation end products(RAGEs)[J]. *The Journal of Clinical Investigation*, 108:1853-1863.

Petersen KF, Shulman GI (2006) Etiology of Insulin Resistance[J]. *The American Journal of Medicine*, 119(5):S10-S16.

RAMADAN W, DEWASMES G, PETITJEAN M, Loos N, Delanaud S, Geloen A, Libert J P (2006) Spontaneous motor activity in fat-fed, streptozotocin-treated rats: a nonobese model of type 2 diabetes[J]. *Physiology & Behavior*, 87(4):765-772.

REED MJ, MESZAROS K, ENTES LJ, Claypool MD, Pinkett JG, Gadbois TM, Reaven GM (2000) A new rat model of type 2 diabetes: the fat-fed, streptozotocin-treated rat[J]. *Metabolism*, 49(11):1390-1396.

Roden M, Bernroider E (2003) Hepatic glucose metabolism in humans—its role in health and disease[J]. *Best Practice & Research Clinical Endocrinology & Metabolism*, 17(3):365-383.

Russo A, Longo R, Vanella A (2002) Antioxidant activity of propolis: role of caffeic acid phenethyl ester and galangin[J]. *Fitoterapia*, 73(s1):S21-29.

Sharma K, Mcgowan TA (2000) TGF-β in diabetic kidney disease: role of novel signaling pathways[J]. *Cytokine & Growth Factor Reviews*, 11(1-2):115-123.

Shimizu K, Ashida H, Matsuura Y, Kanazawa, K (2004) Antioxidative bioavailability of artepillin C in Brazilian propolis[J]. *Archives of Biochemistry and Biophysics*, 424(2):181-188.

Szoke E, Gerich JE (2005) Role of impaired insulin secretion and insulin resistance in the pathogenesis of type 2 diabetes mellitus[J]. *Comprehensive Therapy*, 31(2):106-112.

Tobin BL, Finegood DT (1993) Reduced insulin secretion by repeated low doses of STZ impairs glucose effectiveness but does not induce insulin resistance in dogs[J]. *Diabetes*, 42(3):474-483.

Verges BL(1999) Dyslipidaemia in diabetes mellitus Review of the main lipoprotein abnormalities and their consequences on the development of atherogenesis[J]. *Diabetes & Metabolism*, 25(S3):32-40.

Vranes D,Dilley Rj,Cooper ME (1995) Vascular changes in the diabetic kidney:effects of ACE inhibition[J]. *Journal of Diabetes and its Complications*,9(4):296-300.

Watson KE,Horowitz BN,Matson G (2002) Lipid abnormalities in insulin resistant states [J]. *Reviews in Cardiovascular Medicine*,4(4):228-236.

Wright Ej,Scism-Bacon Jl,Glass LC (2006) Oxidative stress in type 2 diabetes:the role of fasting and postprandial glycaemia[J]. *International Journal of Clinical Practice*,60 (3):308-314.

Zhang F,Ye C,Li G,Ding W,Zhou W,Zhu H,Zhang D (2003) The rat model of type 2 diabetic mellitus and its glycometabolism characters[J]. *Experimental Animals*,52(5):401-407.

艾静,王宁,杜杰,杨梅,刘萍,杜智敏,杨宝峰(2004) Wistar 大鼠 2 型糖尿病动物模型的建立[J]. 中国药理学通报,(11):1309-1312.

胡福良,玄红专,詹耀峰(2004) 蜂胶对糖尿病大鼠肾脏的影响[J]. 养蜂科技,(1):2-3.

金光香,宫海民,段文卓,王红艳(2005) 蜂贝化瘀胶囊对糖尿病大鼠糖脂代谢的调整及微血管并发症的防治[J]. 中国临床康复,9(27):96-98.

李保春,许静,袁伟杰(2003) 2 型糖尿病肾病[M]. 上海:第二军医大学出版社:33-37.

李光伟,Bennett PH(2005) 关于空腹血糖、空腹胰岛素乘积的倒数在流行病学研究中应用的补充说明[J]. 中华糖尿病杂志,13(4):247-249.

李秀钧(2001) 胰岛素抵抗综合征[M]. 北京:人民卫生出版社:44-54.

罗谋伦,郭欲晓,林志彬(1999) 大鼠正糖钳实验方法学[J]. 药学学报,34(4):255-259.

钱容立(1999) 美国糖尿病协会:英国糖尿病前瞻性研究(UKPDS)的意义[J]. 中国糖尿病杂志,7(3):185-188.

沃兴德,崔小强,唐利华(2003) 姜黄素对食饵性高脂血症大鼠血浆脂蛋白代谢相关酶活性的影响[J]. 中国动脉硬化杂志,11(3):223-226.

吴满平,梅美珍(1991) 参与脂蛋白代谢研究的进展[J]. 基础医学和临床,11(2):77-81.

徐叔云(1982) 药理实验方法学[M]. 第一版. 北京:人民卫生出版社:780-781.

许曼音,陆广华,陈名道(2003) 糖尿病学[M]. 上海:上海科学技术出版社:80-81.

叶山东(2000) 糖尿病诊断治疗学[M]. 合肥:安徽科学技术出版社:97-100.

张蓉,刘宇,刘秉文(1996) 血浆脂蛋白及肝脂酶的比色测定法[J]. 华西医科大学学报,27(1):106-110.

缩略词表(Abbreviations)

英文缩写	英文全称	中文全称
ABTS	2,2′-Azino-bis-(3-ethylbenzothiazoline-6-sulfonic acid)diammonium salt	2,2′-连氮基-双-(3-乙基苯并二氢噻唑啉-6-磺酸)二铵盐
AGEs	Advanced glycation end products	糖基化终末产物
ALB	Albumin	白蛋白
ALP	Alkaline phosphatase	碱性磷酸酶
ALT	Alanine aminotransferase	谷丙转氨酶
AMPK	AMP-activated protein kinase	AMP依赖的蛋白激酶
AKT	Protein kinase B	蛋白激酶B
AO	Acridine orange	吖啶橙
Atp1	α subunit of mitochondrial F0F1-ATP synthase	线粒体F0F1-ATP合酶的α亚基
AR	Aldose reductase	醛糖还原酶
Artepillin C	3,5-diprenyl-4-hydroxycinnamic acid	阿替匹林C（3,5-二异戊烯基-4-羟基肉桂酸）
AST	Aspartate aminotransferase	谷草转氨酶
BG	Blood glucose	稳态下的血糖
BP	Brazilian propolis	巴西蜂胶
BGP	Brazillizan green propolis	巴西绿蜂胶
BSA	Bovine serum albunin	牛血清白蛋白
BUN	Blood urea nitrogen	尿素氮
BW	Body weight	体重
CAPE	Caffeic acid phenethyl ester	咖啡酸苯乙酯
CAT	Catalase	过氧化氢酶
CCR	Creatinine clearance rate	肌酐清除率
Con A	Concanavalin A	刀豆蛋白A
CP	Chinese propolis	中国蜂胶
COL1A2	Alpha2 chain of type I collagen	Ⅰ型胶原α2链

续表

英文缩写	英文全称	中文全称
COL3A1	Alpha1 chain of type Ⅲ collagen	Ⅲ型胶原α1链
COX-2	Cyclooxygenase-2	环氧合酶2
CREA	Creatinine	肌酐
CYP450	CytochromeP450	细胞色素P450
DCHF	2′,7′-dichlorodihydrofluorescin	2′,7′-二氯二氢荧光素
DEX	Dexamethasone	地塞米松
D-GalN	D-galactosamine	D-氨基半乳糖
DM	Diabetes mellitus	糖尿病
DMSO	Dimethyl sulphoxide	二甲基亚砜
DN	Diabetic nephropathy	糖尿病肾病
DPPH	1,1-Diphenyl-2-picrylhydrazyl radical	1,1-二苯基-2-三硝基苯肼
DSS	Dextran sodium sulfate	葡聚糖硫酸钠
ECM	Extracellular matrix	细胞外基质
EEBP	Ethanol extract of Brazilian propolis	巴西蜂胶乙醇提取物
EEBGP	Ethanol extracts of *Baccharis* propolis	酒神菊属型蜂胶乙醇提取物
EECP	Ethanol extract of Chinese propolis	中国蜂胶乙醇提取物
EEEP	Ethanolextracts of *Eucalyptus* propolis	桉树型蜂胶乙醇提取物
EEP	Ethanol extract of propolis	蜂胶乙醇提取物
ERK	Extracellular signal-regulated kinase	细胞外信号调节蛋白激酶
FBG	Fasting blood glucose	空腹血糖
FBS	Fetal bovine serum	胎牛血清
FCA	Freund's complete adjuvant	弗氏完全佐剂
FGF	Fibroblast growth factor	纤维母细胞生长因子
FGF-2	Basic fibroblast growth factor	碱性成纤维细胞生长因子
FINS	Fasting insulin	空腹胰岛素
FMN	Fructosamine	果糖胺
FT-NIR	Fourier translation infrared spectroscopy	傅里叶变换近红外光谱法
GCLC	Glutathione-cysteine ligase catalytic subunit	谷胱甘肽-半胱氨酸连接酶催化亚基
GCLM	Glutathione-cysteine ligase modify subunits	谷胱甘肽-半胱氨酸连接酶修饰亚基
GFR	Glomerular filtration rate	肾小球滤过率
GIR	Glucose infusion rate	葡萄糖输注速率
GLU	Glucose	血糖

续表

英文缩写	英文全称	中文全称
GSH-px	Glutathione peroxidase	谷胱甘肽过氧化物酶
GSK-3β	Glycogen synthasc kinase-3β	糖原合酶激酶-3β
GTF	Glucosytransferse	葡糖基转移酶
HAEP	Hydroalcoholic extract of propolis	蜂胶水醇提取物
HbA_{1c}	Glycosylated hemoglobin	糖化血红蛋白
HDL	High density lipoprotein	高密度脂蛋白
HDL-c	High density lipoprotein cholesterol	高密度脂蛋白胆固醇
HL	Hepatic lipase	肝脂酶
HPLC	High-performance liquid chromatography	高效液相色谱
HO-1	Heme oxygenase-1	血红素加氧酶1
HUVECs	Human umbilical vein endothelial cells	人脐静脉血管内皮细胞
IBD	Inflammatory bowel disease	炎症性肠炎
ICAM-1	Intercellular adhesion molecule-1	细胞间黏附因子-1
IEC	Intestinal epithelial cell	肠上皮细胞
IAI	Insulin act index	胰岛素作用指数
IL	Interleukin	白介素
IL-2	Interleukin-2	白介素-2
IL-6	Interleukin-6	白介素-6
IκBα	Nuclear factor of kappa light polypeptide gene enhancer in b-cells inhibitor, alpha	抑制核因子 κB 抑制蛋白 α
iNOS	Induced nitric oxide synthase	诱导性一氧化氮合酶
JNK	Jun N-terminal kinase	氨基末端激酶
JC-1	5,5′6,6′-tetrachloro-1,1′3,3′-tetraethylbenzimidazolylcarbocyanine iodide	5,5′6,6′-四氯-1,1′3,3′-四乙基苯并咪唑羰花青碘化物
LC3	Microtubule-associated protein light chain 3	微管相关蛋白轻链3
LCAT	Lecithin cholesterol acyleransferase	血清卵磷脂胆固醇酰基转移酶
LD_{50}	Median lethal dose	半数致死剂量
LD_{100}	Completely lethal dose	100%致死剂量
LDH	Lactate dehydrogenase	乳酸脱氢酶
LDL	Low density lipoprotein	低密度脂蛋白
LDL-C	Low density lipoprotein cholesterol	低密度脂蛋白胆固醇
LPL	Lipoprotein	血清脂蛋白脂酶

续表

英文缩写	英文全称	中文全称
LPS	Lipopolysaccharide	脂多糖
MAPK	Mitogen-activated protein kinase	蛋白激酶磷酸酶
MBC	Minimum bactericidal concentration	最小杀菌浓度
MCP-1	Monocyte chemoattractant protein-1	单核细胞趋化蛋白-1
MDA	Malonaldehyde	丙二醛
MEP	Methanol extracts of propolis	蜂胶甲醇提取物
MIC	Minimal inhibition concentration,	最小抑菌浓度
MMP-2	Matrix metalloproteinase-2	基质金属蛋白酶2
MS	Mass spectrometric	质谱
MUFA	Monounsaturated fatty acids	单不饱和脂肪酸
NF-κB	Nuclear factor kappa-light-chain-enhancer of activated B cells	核因子活化B细胞κ轻链增强子
NO	Nitric oxide	一氧化氮
NOS	Nitric oxide synthase	一氧化氮合酶
ORAC	Oxygen radical absorbance capacity	抗氧化能力指数
ox-LDL	Oxidized low-density lipoprotein	氧化型低密度脂蛋白
PB	Poplar buds	杨树芽
PBS	Phophate buffer solution	磷酸盐缓冲液
PC-PLC	Phosphatidylcholine-specific phospholipase C	磷脂酰胆碱特异性磷脂酶C
PDA	Photodiode array	光电二极管阵列
PGE2	prostaglandin E_2	前列腺素E_2
PI3K/AKT	Phosphatidylinositol 3-kinase/protein kinase B	磷脂酰肌醇3激酶/蛋白激酶B
PKC	Protein kinase C	蛋白激酶C
PMSF	Phenylmethylsulfonyl fluoride	苯甲基磺酰氟
pNPG	p-nitro-β-D-glucoside	对-硝基β-D-葡萄糖苷
PPARγ	Pemxisome proliferator-aetivated receptor-γ	过氧化物酶体增殖剂激活受体γ
PUFA	Polyunsaturated fatty acids	多不饱和脂肪酸
PVDF	polyvinylidenedifluoride	聚乙烯二氟化物
PVPP	Polyvinyl pyrrolidone	聚乙烯吡咯烷酮
qRT-PCR	Quantitative real-time polymerase chain reaction	实时荧光定量聚合酶链式反应
ROS	Reactive oxygen species	活性氧簇
RP	Reducing power	还原力

续表

英文缩写	英文全称	中文全称
RSD	Relative standard deviation	相对标准偏差
Scr	Serum creatinine	血肌酐
SDS-PAGE	Sodium dodecyl sulfate polyacrylamide gel electrophoresis	十二烷基磺酰钠-聚丙烯酰胺凝胶电泳
SEM	Scanning electron microscopy	扫描电镜
SFA	Saturated fatty acid	饱和脂肪酸
SOD	Superoxide dismutase	超氧化物歧化酶
α-SMA	α-smooth muscle actin	平滑肌肌动蛋白
SPME	Solid phasemicro extraction	固相微萃取
SRB	Sulforhodamine B	磺酰罗丹明B
STZ	Streptozotocin	链佐星
t-BHP	t-butyl hydroperoxide	叔丁基过氧化氢
TER	Transepithelial electrical resistance	跨膜电阻
T1DM	Type 1 diabetes mellitus	1型糖尿病
T2DM	Type 2 diabetes mellitus	2型糖尿病
TC	Total cholesterol	总胆固醇
TCDD	Tetrachlorodibenzo-p-dioxin	四氯二苯-p-二噁英
TDC	Total flavonoid contents	总黄酮含量
TFC	Total phenolic contents	总酚含量
TG	Triglycerides	三酰甘油
TGF-β_1	Transforming growth factor-β_1	转化生长因子β_1
TLRs	Toll like receptors	Toll样受体
TLR2	Toll like receptor 2	Toll样受体2
TLR4	Toll like receptor 4	Toll样受体4
TNF-α	Tumor necrosis factor-α	肿瘤坏死因子-α
tTG	Tissuetransglutaminase	组织型转谷氨酰胺酶
TRAF6	TNF receptor associated factor	肿瘤坏死因子联合受体6
TJ	Tight junction	紧密连接
TP	Total protein	总蛋白
UA	Uric acid	尿酸
UCr	Urine creatinine	尿肌酐
VEC	Vascular endothelial cell	血管内皮细胞
VEGF	Vascular endothelial growth factor	血管内皮生长因子
WEP	Water extract of propolis	蜂胶水提取物

ICS 65.140
B 47

中华人民共和国国家标准

GB/T 24283—2018
代替 GB/T 24283—2009

蜂 胶

Propolis

2018-07-13 发布　　　　　　　　　　　2019-02-01 实施

国家市场监督管理总局
中国国家标准化管理委员会　发布

前　言

本标准按照GB/T 1.1—2009给出的规划起草。

本标准代替GB/T 24283—2009《蜂胶》。

本标准与GB/T 24283—2009相比主要变化如下：
——增加了规范性引用文件(见第2章，2009年版的第2章)；
——修改了蜂胶的定义(见3.1，2009年版的3.1)；
——修改了蜂胶乙醇提取物的定义(见3.2，2009年版的3.2)；
——增加了胶源植物的定义(见3.3)；
——增加了酒神菊属蜂胶的定义(见3.4)；
——修改了蜂胶感官要求的色泽特征指标(见表1，2009年版的表1)；
——修改了蜂胶乙醇提取物感官要求的色泽特征指标(见表2，2009年版的表2)；
——修改了蜂胶乙醇提取物理化要求(见表3，2009年版的表3)；
——修改了真实性要求(见4.3.1，2009年版的4.3)；
——增加了真实性要求的检验方法(见4.3.2)；
——增加了安全卫生要求(见4.4)；
——修改了特殊限制要求(见4.5，2009年版的4.4)；
——修改包装要求(见6.1，2009年版的6.1)；——修改了标志要求(见7.1，2009年版的7.1)；
——增加了酒神菊属蜂胶的特殊要求(见附录A)。

本标准由全国蜂产品标准化工作组(SAC/SWC 2)提出并归口。

本标准起草单位：北京天恩生物工程高新技术研究所、浙江大学动物科学学院、中华人民共和国秦皇岛出入境检验检疫局、农业部蜂产品质量监督检验测试中心(北京)、浙江省缙云县绿纯养蜂专业合作社、杭州蜂之语蜂业股份有限公司、杭州天厨蜜源保健品有限公司、北京百花蜂业科技发展股份有限公司、广州市宝生园股份有限公司、蜂乃宝本铺(南京)保健食品有限公司、江西汪氏蜜蜂园有限公司。

本标准主要起草人：吕泽田、胡福良、张翠平、贾光群、黄京平、胡元强、周萍、王磊、郭利军、许晓宇、刘嘉、汪玲。

本标准所代替标准的历次版本发布情况为：
——GB/T 24283—2009

蜂 胶

1 范围

本标准规定了蜂胶及蜂胶乙醇提取物的术语和定义、要求、试验方法、包装、标志、贮存和运输要求。

本标准适用于蜂胶及蜂胶乙醇提取物的生产、加工、贸易。

2 规范性引用文件

下列文件对于本文件是必不可少的。凡是注日期的引用文件，仅注日期的版本适用于本文件。凡是不注日期的引用文件，其最新版本（包括所有的修改单）适用于本文件。

GB/T 191 包装储运图示标志

GH/T 34782 蜂胶中杨树胶的检测方法 高效液相色谱法

GH/T 1087 蜂胶真实性鉴别 高效液相指纹图谱法

GH/T 1114 蜂胶中阿替匹林C的测业方法 高效液相色谱法

3 术语和定义

下列术语和定义适用于本标准。

3.1 蜂胶 propolis

工蜂采集胶源植物树脂等分泌物与其上颚腺、蜡腺等分泌物混合形成的胶黏性物质。

注：因胶源植物不同,蜂胶可主要分为杨树属、酒神菊属、桉树属、血桐属和地中海型等类型。

3.2 蜂胶乙醇提取物 ethanol extracts of propolis

用乙醇萃取（浸取）蜂胶得到的物质。

3.3 胶源植物 propolis Plants

可被工蜂采集树脂加工成蜂胶的植物。

3.4 酒神菊属蜂胶 baccharis-type propolis

以酒神菊属（*Baccharis*）为胶源植物的蜂胶。

3.5 总黄酮 total flavonoids

黄酮类物质含量的总和。

4 要求

4.1 感官要求

4.1.1 蜂胶的感官要求应符合表1的规定。

表1 蜂胶的感官要求

项目	特征
项目	棕黄色、棕红色、褐色、黄褐色、灰褐色、青绿色、灰黑色等
状态	团块或碎渣状,不透明,约30℃以上随温度升高逐渐变软,且有黏性
气味	有蜂胶所特有的芳香气味,燃烧时有树脂乳香气,无异味
滋味	微苦、略涩,有微麻感和辛辣感

4.1.2 蜂胶乙醇提取物感官要求应符合表2的规定。

表2 蜂胶乙醇提取物的感官要求

项目	特征
结构	断面结构紧密
色泽	棕褐色、深褐色,有光泽
状态	固体状,约30℃以上随温度升高逐渐变软,且有黏性
气味	有蜂胶所特有的芳香味,燃烧时有树脂乳香气,无异味
滋味	微苦、略涩,有微麻感和辛辣感

4.2 理化要求

蜂胶及蜂胶乙醇提取物的理化要求应符合表3的规定。

表3 蜂胶及蜂胶乙醇提取物理化要求

项目		蜂胶		蜂胶乙醇提取物	
		一级品	二级品	一级品	二级品
乙醇提取物含量/(g/100g)	≥	60.0	30.0	98.0	95.0
总黄酮/(g/100g)	≥	15.0	6.0	20.0	17.0
氧化时间/s	≤	22			

4.3 真实性要求

4.3.1 不应加入任何树脂和其他矿物、生物或其提取物质。非蜜蜂采集,人工加工而成的任何树脂胶状物不应称之为"蜂胶"。

4.3.2 按GH/T 34782、GH/T 1087的真实性要求和检测方法检验。

4.4 安全卫生要求

按食品安全标准和法律法规要求规定执行。

4.5 特殊限制要求

4.5.1 不应使用铁纱网或含有污染物质的器具、盖布采集蜂胶。

4.5.2 不应在60℃以上高温加热、室外曝晒。

4.6 酒神菊属蜂胶的特殊要求

酒神菊属蜂胶的感官和理化要求等见附录A。

5 试验方法

5.1 取样方法

从被检样品的不同部位均匀取样,每批取样总量不超过300g。

5.2 感官要求的检验

5.2.1 蜂胶感官要求的检验

5.2.1.1 色泽、状态

在自然光线良好的条件下,观察样品外表色泽。取少许上述样品混匀后,加热至35℃左右,用手揉搓成条,再慢慢向两端拉伸。含胶量越大,黏性越大,拉伸长度越长。

5.2.1.2 气味、滋味

取少许样品,嗅其气味是否有蜂胶特有的明显芳香气味,再点燃,嗅其气味是否异常;口尝其滋味。

5.2.2 蜂胶乙醇提取物感官要求的检验

5.2.2.1 结构

将蜂胶的乙醇提取物样品放在15℃以下2h~3h,用锤砸开,观察其断面。

5.2.2.2 色泽、状态

按5.2.1.1规定的方法检验。

5.2.2.3 气味、滋味

按5.2.1.2规定的方法检验。

5.3 理化要求的检验

5.3.1 样品制备

将被检样品放入10℃以下冰箱中1h后,将其粉碎混合均匀,从中取样100g进行检测。

5.3.2 乙醇提取物含量

5.3.2.1 原理

蜂胶溶于乙醇,用称量乙醇不溶物的重量以减量法获得乙醇提取物重量,计算其占样品重量的百分比。

5.3.2.2 试剂和材料

包括以下试剂和材料:

a) 乙醇:分析纯(≥95%);
b) 定量滤纸 Φ12.5cm。

5.3.2.3 仪器

包括以下仪器：

a) 天平(感量 0.001g)；

b) 250mL 烧杯；

c) 电热鼓风干燥箱；

d) 超声波清洗器；

e) 玻璃漏斗 Φ60mm；

f) 玻璃棒；

g) 250ML 锥形瓶。

5.3.2.4 步骤

称取经过粉碎处理的蜂胶样品 5g，或者蜂胶乙醇提取物 10g（精确至 0.001g），置于 250mL 烧杯中，加适量 95％ 乙醇，放入超声波清洗器中超声，使样品溶解，放冷至室温，此溶液用事先干燥并称至恒重的滤纸过滤，用适量乙醇洗涤烧杯和滤纸 3～5 次。然后将残渣及滤纸在 50℃ 下真空（真空度：－0.096MPa～－0.1MPa)干燥至恒重。在相同条件下作平行实验。

5.3.2.5 计算：

乙醇提取物含量按公式 (1) 计算

$$X_1 = \frac{M_1 - M_2}{M_2} \times 100 \tag{1}$$

式中：X_1——样品中乙醇提取物含量，％；

M_1——样品质量，g；

M_2——残渣质量，g。

平行实验允许误差不超过 1.5％，取 3 次测定的平均值。

5.3.3 总黄酮含量

5.3.3.1 试剂和材料

包括以下试剂和材料：

a) 聚酰胺粉(≥100 目)；

b) 芦丁标准溶液：取 5.0mg 芦丁(≥99％)，加甲醇溶解并定容至 100mL，即得 50μg/mL；

c) 乙醇：分析纯(≥99％)；

d) 甲醇：分析纯(≥95％)。

(e) 甲苯：分析法(≥99％)

5.3.3.2 仪器

包括以下仪器：

a) 紫外可见光分光光度计

b) 层析柱：350mm（长）×15mm（内径）、具活塞、砂芯、抽气嘴、圆底烧瓶，见图 1；

c) 容量瓶：10mL；

d) 移液器：100μL～1000μL；

e) 移液管：1μL～5mL；

f) 玻璃蒸发皿：90mm。

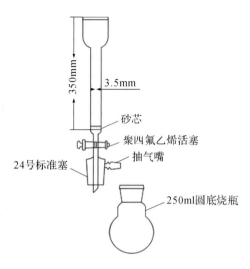

图 1 层析柱示意图

5.3.3.3 步骤

a) 试样处理：称取经过粉碎处理的蜂胶样品 1g（精确到 0.001g），或蜂胶乙醇提取物 0.5g 于容量瓶中，用乙醇定容至 100mL，摇匀后，超声提取 20min，放置，用移液管吸取上清液 1mL 于玻璃蒸发皿中，加入 5mL 乙醇及 1g 聚酰胺粉，用玻璃棒混匀吸附，于 60℃ 水浴上挥去乙醇，然后转入关闭活塞的层析柱中。量取 20mL 苯液，清洗玻璃蒸发皿再将苯液转入层析柱中，分 3 次完成。15min 后开启层析柱活塞，弃去苯液并关闭层析柱活塞，取下圆底烧瓶，将 25mL 容量瓶装于层析柱下方。量取 20mL 甲醇，分 3 次清洗玻璃蒸发皿，再将甲醇转入层析柱中，15min 后开启层析柱活塞将黄酮洗脱于 25mL 容量瓶中，用甲醇定容至 25mL。此液置 1cm 比色皿中于波长 360nm 测定吸收值。同时以芦丁为标准品，用标准曲线法定量。

b) 芦丁标准曲线：分别吸取芦丁标准溶液 0、1.0、2.0、3.0、4.0、5.0mL 于 10mL 容量瓶中，加甲醇至刻度，摇匀，置 1cm 比色皿中于波长 360nm 测定吸收值，绘制标准工作曲线，计算回归方程。

5.3.3.4 计算和结果表示：

按公式(2)计算

$$X_2 = \frac{A \times V_2 \times 100}{V_1 \times M \times 1000} \quad (2)$$

式中：X_2——试样中总黄酮的含量，mg/100g；
　　　A——由标准曲线算得被测液中黄酮量，μg；
　　　M——试样质量，g；
　　　V_1——吸取的上清液体积，mL；
　　　V_2——试样定容总体积，mL；
　　　计算结果保留两位有效数字。

5.3.4 氧化时间

5.3.4.1 原理

通常用高锰酸钾紫红色溶液消退的时间来表示蜂胶中还原性物质的含量。

5.3.4.2 试剂和材料

包括以下试剂和材料：

a) 乙醇：分析纯（95％以上）；

b) 高锰酸钾：（分析纯≥95％）；

c) 0.01mol/L 的高锰酸钾溶液：精确称取 1.580g（±0.005g）高锰酸钾，用水稀释至 1000mL；

d) 硫酸：分析纯（95％～98％）；

e) 20％硫酸溶液：量取 128mL 硫酸，缓缓注入约 700mL 水中，冷却，稀释至 1000mL；

f) 蒸馏水。

5.3.4.3 仪器

包括以下仪器：

a) 天平（感量 0.001g）；

b) 振荡器；

c) 秒表；

d) 250mL 具塞磨口锥形瓶；

e) 50mL、100mL、1000mL 容量瓶；

f) 50mL 锥形瓶；

g) 0.2mL、1.0mL、2.0mL、5.0mL、10.0mL 移液管；

h) 漏斗、定量滤纸；

i) 200μL 微量移液器。

5.3.4.4 步骤

a) 在室温下称取 1g（精确到 0.001g）样品，置于 250 mL 具塞锥形瓶中，加入 25mL 乙醇，盖好瓶塞，于振荡器上低速振荡 1h，然后加入 100mL 蒸馏水，充分摇匀后，过滤，收集滤液。

b) 用移液管吸取 0.5mL 上述滤液放入 50mL 容量瓶中，用蒸馏水稀释至刻度并混匀。

c) 用移液管吸取 10mL 稀释液 b 于 50mL 锥形瓶中，加入 2.0mL 20％硫酸，振荡 1min，然后用 200μL 微量移液器加入 50μL 0.01mol/L 高锰酸钾溶液，在加入高锰酸钾溶液的同时，开动秒表振荡，当溶液的紫红色完全消退时，停止秒表，记录溶液的紫红色完全消退所耗用的时间（以秒计），即是该样品的氧化时间。每个样品平行测定三次，取算术平均值作为该样品的测定值。

6 包装

6.1 应采用符合国家食品安全卫生要求的材料包装。蜂胶乙醇提取物应定量包装。包装场地应符合食品安全卫生要求。包装应严密、牢固。

6.2 应按等级分别包装。

7 标志

7.1 包装上应标明产品名称、等级、净含量、生产日期、保质期和生产者的名称、地址。

7.2 图示标志应符合 GB/T 191 的规定。

8 贮存

8.1 贮存场所应清洁卫生、干燥、阴凉、通风，不应与有毒、有害、有异味、有腐蚀性、有放射性和可能发生污染的物品同场所贮存。

8.2 产品应按品种、规格分别存放。

9 运输

9.1 运输工具应清洁卫生。

9.2 不应与有毒、有害、有异味、易污染的物品混装运输。

9.3 防高温、曝晒、雨淋。

附录 A
（规范性附录）
酒神菊属蜂胶的要求

A1 感官要求

感官要求见表 A1。

表 A1 酒神菊属蜂胶的感官要求

项目	特 征
色泽	黄绿色、绿褐色
状态	条块状，部分条块上有蜂孔，不透明，约 30℃ 以上随温度升高逐渐变软，且有黏性
气味	有明显的酒神菊属植物分泌的树脂、香脂的特殊芳香气味，燃烧时有酒神菊属树脂乳香气，无异味
滋味	微苦涩
杂质	无泥土、蜜蜂肢体等杂物

A2 理化要求

理化要求见表 A2。

表 A2 酒神菊属蜂胶的理化要求

项目		酒神菊属蜂胶
乙醇提取物含量(g/100g)	≥	50.0
总黄酮(g/100g)	≥	4.0
阿替匹林 C(g/100g)	≥	1.4
氧化时间/s	≤	22

A3 真实性要求

按 GH/T 1114 方法，供试样品图谱中含有阿替匹林 C 特征峰，且其含量大于或等于 0.8%，则被检样品判定为酒神菊属蜂胶；如果待测样品中检测不到阿替匹林 C 或其含量小于 0.8%，则该样品判定为非酒神菊属蜂胶样品。

A4 试验方法

A4.1 感官要求按本标准 5.2 方法。

A4.2 理化要求按本标准 5.3 方法。

ICS 65.140
B47

中华人民共和国国家标准

GB/T 34782—2017

蜂胶中杨树胶的检测方法
高效液相色谱法

Method for the Determination of Poplar Tree Gum in Propolis
——High Performance Liquid Chromatography

2017-11-01 发布　　　　　　　　　　　　　　　　2018-05-01 实施

中华人民共和国国家质量监督检验检疫总局
中国国家标准化管理委员会　发布

前　言

本标准按照 GB/T 1.1—2009 给出的规则起草。

本标准由中华供销合作总社提出。

本标准由全国蜂产品标准化工作组(SAC/SWG 2)归口。

本标准起草单位:浙江大学、中国蜂产品协会、农业部蜂产品质量监督检验测试中心(北京)、江山福赐德蜂业科技开发有限公司、厦门思健生物科技有限公司、上海沪郊蜂业联合社有限公司。

本标准主要起草人:胡福良、张翠平、郑火青、吕泽田、李熠、徐水荣、颜鉴翔、孙德官。

引 言

本文件的发布机构提请注意,声明符合本文件时,可能涉及第4~9章与以水杨苷和邻苯二酚为指标鉴别杨树胶的原理及检测方法相关的专利(ZL 201010180675.0 和 ZL 201310721079.2)的使用。

本文件的发布机构对于该专利的真实性、有效性和范围无任何立场。

该专利持有人已向本文件的发布机构保证,他愿意同任何申请人在合理且无歧视的条款和条件下,就专利授权许可进行谈判。该专利持有人的声明已在本文的发布机构备案。相关信息可以通过以下联系方式获得:

专利持有人:浙江大学。

地址:浙江省杭州市西湖区余杭塘路866号。

请注意除上述专利外,本文件的某些内容仍可能涉及专利,本文件的发布机构不承担识别这些专利的责任。

蜂胶中杨树胶的检测方法 高效液相色谱法

1 范围

本标准规定了蜂胶中是否含有杨树胶的高效液相色谱鉴别方法的原理、试剂、材料、仪器、设备和试验方法。

本标准适用于蜂胶及蜂胶乙醇提取物中杨树胶的检测。

2 规范性引用文件

下列文件对于本文件的应用是必不可少的。凡是注日期的引用文件,仅所注日期的版本适用于本文件。凡是不注日期的引用文件,其最新版本(包括所有的修改单)适用于本文件。

GB/T 6682 分析实验室用水规格和试验方法

3 术语和定义

下列术语和定义适用于本标准。

3.1 蜂胶 propolis

工蜂采集树脂等植物分泌物与其上颚腺、蜡腺分泌物等混合形成的胶黏性物质。

3.2 蜂胶乙醇提取物 ethanol extracted propolis

乙醇萃取蜂胶后得到的物质。

3.3 杨树胶 extracted populus buds,leaves or barks

杨属(*Populus*)植物的芽、叶子或树皮等组织经人工熬制加工,再经乙醇提取而成的提取物。

3.4 杨树型蜂胶 poplar type propolis

杨属(*Populus*)为主要植物来源的蜂胶。

3.5 水杨苷 salicin

水杨酸盐类酚苷的基本结构单元。

3.6 邻苯二酚 catechol

苯的两个邻位氢被羟基取代后形成的化合物。

4 原理

水杨苷和邻苯二酚广泛存在于杨属和柳属植物的芽、叶子和树皮中。当这些芽、叶子和树皮经过煎煮之后,水杨苷和邻苯二酚依然存在。而蜂胶中并不存在。因此可以作为蜂胶中是否掺有杨树胶的鉴别指标。

试样经75%乙醇提取后,用配有紫外检测器的反相高效液相色谱仪在213nm处检测水杨苷和邻苯二酚的含量,从而判断蜂胶中是否含有杨树胶。

5 试剂和材料

实验室用水应符合GB/T 6682规定的一级水。除非另有规定,所有试剂均为分析纯。

5.1 甲醇:色谱纯。

5.2 乙腈:色谱纯。

5.3 乙醇。

5.4 磷酸。

5.5 水杨苷标准品:纯度≥98%。

5.6 邻苯二酚标准品:纯度≥99%。

5.7 0.5%磷酸(5.4)溶液:量取5mL磷酸,用水定容至1L。

5.8 标准贮备液:准确称取水杨苷(5.5)和邻苯二酚标准品(5.6)各10mg,分别用甲醇溶液(5.1)配制成1.0mg/mL的标准溶液,于4℃冰箱保存,保持期3个月。

5.9 标准工作液:分别吸取适量的水杨苷和邻苯二酚标准贮备液(5.8),用甲醇溶液(5.1)分别稀释成水杨苷浓度为0.005mg/mL、0.01mg/mL、0.02mg/mL、0.04mg/mL和0.06mg/mL五个浓度,邻苯二酚浓度为0.005mg/mL、0.01mg/mL、0.02mg/mL、0.04mg/mL和0.06mg/mL五个浓度,现用现配。

5.10 滤膜:0.45μm,有机系。

6 仪器和设备

6.1 高效液相色谱仪:配有紫外检测器。

6.2 粉碎机。

6.3 旋转蒸发仪。

6.4 超声波清洗器。

7 样品的取样及预处理

从被检样品的不同部位均匀取样50g,冷冻,粉碎,混匀,作为备检样品。

8 试验方法

8.1 供试样品溶液的制备

称取2g蜂胶试样,用75%乙醇超声波提取30min,料液比为1:15(W/V),反复提取3次,合并滤液,离心过滤后旋转蒸发浓缩,或者称取1g蜂胶乙醇提取物试样,然后用75%乙醇超声溶解定容至50mL,取上清液,0.45μm滤膜过滤,供液相色谱测定。

8.2 色谱条件

色谱柱:Sepax HP-C18,柱长150mm,内径4.6mm,粒度5μm;或性能相当的色谱柱。

流动相流速：1mL/min。

梯度洗脱，洗脱程序参见附录A表A.1。

检测波长：213nm。

柱温：30℃。

进样量：5μL。

8.3 测定

取5μL样品溶液(8.1)、水杨苷和邻苯二酚标准工作液(5.9)分别进样分析。

水杨苷标准品、邻苯二酚标准品、杨树型蜂胶、杨树胶的色谱图分别参见附录B图B.1~B.4。

8.4 结果计算

试样中水杨苷或邻苯二酚含量(X)以毫克每克(mg/g)表示，按式(1)计算：

$$X = \frac{C_s \times V \times A}{m \times A_s} \tag{1}$$

式中：C_s——水杨苷或邻苯二酚标准溶液的浓度，单位为毫克每毫升(mg/mL)；

V——试样最终定容体积，单位为毫升(mL)；

A_s——水杨苷或邻苯二酚标准溶液对应的峰高或峰面积；

A——试样溶液对应的峰高或峰面积；

m——样品质量，单位为克(g)。

将符合重复性要求的两个独立测定值的算术平均值作为测定结果(mg/g)，保留两位有效数字。平行试验相对偏差不得超过2.0%。

8.5 结果判定

本方法所检测的水杨苷和邻苯二酚为杨树胶中含有而在蜂胶中不含有的物质。如果待测样品中检测不到水杨苷，且邻苯二酚检测浓度在蜂胶乙醇提取物中低于0.21mg/g，或在蜂胶原料样品中低于0.12mg/g，则该样品为未检出杨树胶的样品；如果待测样品中检测不到水杨苷，但蜂胶乙醇提取物样品中邻苯二酚检测浓度大于0.21mg/g，或待测原料蜂胶样品中邻苯二酚检测浓度大于0.12mg/g，则该样品判定为含有杨树胶的样品；如果待测样品中检测到水杨苷，则该样品判定为含有杨树胶的样品。

9 检测限

本标准在进样量为5μL时，蜂胶乙醇提取物中水杨苷的最低检测限为0.065mg/g，邻苯二酚的最低检测限为0.063mg/g；蜂胶原料中水杨苷的最低检测限为0.032mg/g，邻苯二酚的最低检测限为0.031mg/g。

附录 A
（资料性附录）

高效液相色谱梯度洗脱程序

高效液相色谱梯度洗脱程序见表 A.1。

表 A.1 高效液相色谱法梯度洗脱程序

时间/min	流速/mL/min	乙腈/%	0.5%磷酸/%
0	1	5	95
20	1	5	95
20.1	1	100	0
30	1	100	0
30.1	1	5	95
38	1	5	95

附录 B
（资料性附录）

水杨苷标准品、邻苯二酚标准品、杨树型蜂胶、杨树胶的色谱图

水杨苷（标准品）色谱图见 B.1。

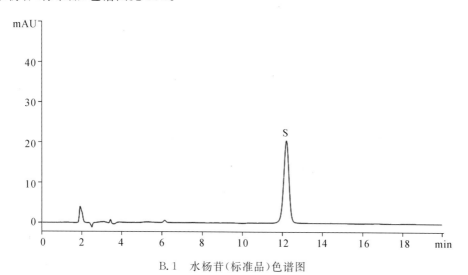

B.1 水杨苷（标准品）色谱图
S：水杨苷

注：标液浓度 0.04mg/mL，出峰时间 12.219min。

邻苯二酚(标准品)色谱图见 B.2。

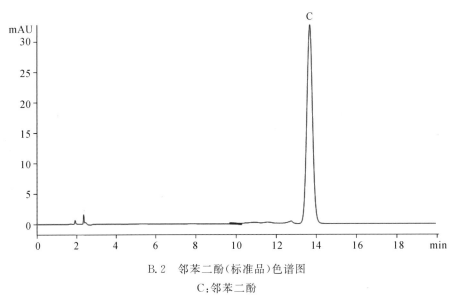

B.2 邻苯二酚(标准品)色谱图
C:邻苯二酚

注:标液浓度 0.03mg/mL,出峰时间 13.851min。

杨树型蜂胶色谱图见 B.3。

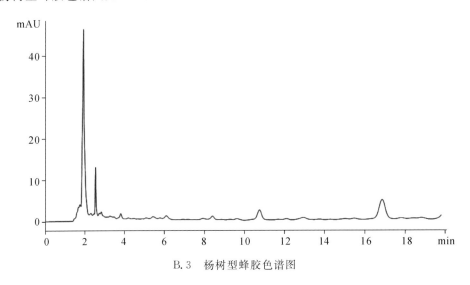

B.3 杨树型蜂胶色谱图

杨树胶色谱图见 B.4。

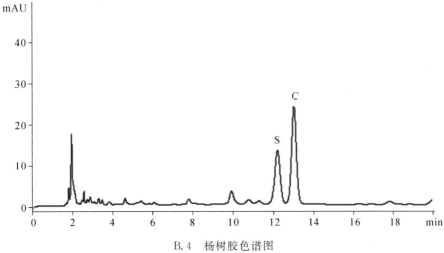

图 B.4　杨树胶色谱图
代表峰:S:水杨苷;C:邻苯二酚

ICS 65.140
B 47
备案号:51234—2015

中华人民共和国供销合作行业标准

GH/T 1114—2015

蜂胶中阿替匹林C的测定方法 高效液相色谱法

Method for the determination of artepillin C in propolis-
High-performance liquid chromatography

2015-08-21 发布　　　　　　　　　　　　2015-10-01 实施

中华全国供销合作总社　发布

前　言

本标准按照 GB/T 1.1—2009 给出的规则起草。
本标准由浙江大学提出。
本标准由中华全国供销合作总社归口。
本标准起草单位:浙江大学、蜂乃宝本铺(南京)保健食品有限公司。
本标准主要起草人:胡福良、张翠平、刘嘉。

引 言

本文件的发布机构提请注意,声明符合本文件时,涉及本文件相关的专利(ZL 201310075055.4)的使用。

本文件的发布机构对于该专利的真实性、有效性和范围无任何立场。

该专利持有人已向本文件的发布机构保证,他愿意同任何申请人在合理且无歧视的条款和条件下,就专利授权许可进行谈判。该专利持有人的声明已在本文件的发布机构备案。相关信息可以通过以下联系方式获得。

专利持有人:浙江大学。

地址:浙江省杭州市西湖区余杭塘路866号。

请注意除上述专利外,本文件的某些内容仍可能涉及专利。本文件的发布机构不承担识别这些专利的责任。

蜂胶中阿替匹林 C 的测定方法高效液相色谱法

1 范围

本标准规定了高效液相色谱法测定蜂胶中阿替匹林 C 的方法。

本标准适用于蜂胶中阿替匹林 C 的定性和定量测定。

本标准的检出限为 0.024mg/g，定量限为 0.080mg/g。

2 规范性引用文件

下列文件对于本文件的应用是必不可少的。凡是注日期的引用文件，仅注日期的版本适用于本文件。凡是不注日期的引用文件，其最新版本（包括所有的修改单）适用于本文件。

GB/T 6682 分析实验室用水规格和试验方法

3 术语和定义

下列术语和定义适用于本文件。

3.1 酒棉菊属型蜂肢 Baccharis-type propolis

巴西绿蜂胶（俗称）

以酒神菊属（Baccharis）为主要植物来源的蜂胶。

3.2 杨属型蜂胶 poplar-type propolis

以杨属（Populus）为主要植物来源的蜂胶。

3.3 阿替匹林 C artepillin C

3,5-二异戊烯基-4-起基肉桂酸，分子结构式见图 1。

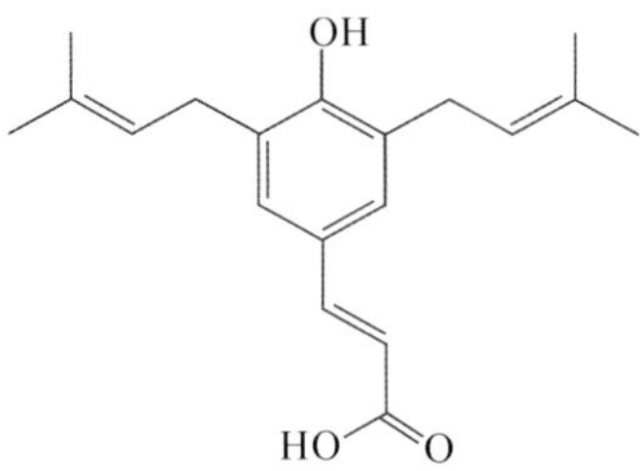

图 1 artepillin C 的分子结构

相对分子质量为 300.40，是一种 2,4,6-三取代苯酚类物质。

阿替匹林 C 存在于 Baccharis、Relhania 和 Flourensia heterolepis 等植物中，特别是在巴西东南部的 Baccharis dracunculifolia DC. 中大量存在，被认为是巴西绿蜂胶的特征性成分。

4 原理

试样经85%乙醇超声提取、离心、过滤、定容,用0.45μm有机系滤膜过滤后得澄清液,用反相高效液相色谱仪经紫外检测器在310nm处检测阿替匹林C,外标法定量,根据保留时间和峰面积进行定性和定量。

5 试剂与材料

除非另有规定,所有试剂均为分析纯,实验用水应符合GB/T 6682中一级水的规定要求。

5.1 甲醇:色谱纯。

5.2 无水乙醇:分析纯。

5.3 乙酸:分析纯。

5.4 1%乙酸溶液(体积分数):量取10mL乙酸(5.3)倒入加有900mL水的容量瓶中,用水定容于1000mL,经0.45μm滤膜过滤。

5.5 阿替匹林C标准品:纯度≥98%,CAS:72044-19-5。

5.6 阿替匹林C标准储备液(1.0mg/mL):准确称取阿替匹林C标准品(5.5)10.0mg于10mL棕色容量瓶中,加甲醇溶液(5.1)使其溶解并定容至刻度,混匀,于4℃冰箱保存,保存期不超过6个月。

5.7 阿替匹林C标准工作液:吸取适量1.0mg/mL的标准储备液(5.6),用甲醇溶液(5.1)分别稀释成0.05mg/mL、0.1mL、0.2mg/mL、0.3mg/mL、0.4mg/mL、0.5mg/mL六个浓度的标准工作液,现用现配。

5.8 滤膜 0.45μm有机系滤膜。

6 仪器与设备

6.1 高效液相色谱仪:配有紫外检测器。

6.2 粉碎机。

6.3 超声波清洗器。

6.4 离心机。

6.5 分析天平:精确至0.0001g。

7 试样制备与保存

从待测蜂胶样品中取出约50g,密封,并做标记,于-18℃冷冻保存。
取出冷冻贮存的试样20~30g,迅速用粉碎机粉碎,过40目筛。

8 操作步骤

8.1 试样溶液的制备

准确称取0.25g蜂胶试样,精确到0.01g,用5mL 85%乙醇超声提取15min(25kHz,功率为50W),然后在离心机上以4000r/min的转速离心15min,取上清液置于50mL容量瓶中,重复提取离心3次,合并上清液,用85%乙醇定容至50mL,经0.45μm滤膜(5.8)过滤,

滤液用于液相色谱测定。

8.2 色谱参考条件

色谱柱:Sepax HP-C18,柱长 150mm,内径 4.6mm,粒皮 5μm;或性能相当的色谱柱。

流动相:以 1%乙酸溶液(5.4)为流动相 A,以甲醇(5.1)为流动相 B,按附录 A 表 A.1 进行梯度洗脱。

流动相流速:1mL/min。

检测波长:310nm。

柱温:30℃。

进样量:5μL。

8.3 测定

8.3.1 定性测定

分别取 5μL 阿替匹林 C 标准工作液(5.7)和试样溶液(8.1),在上述色谱条件(8.2)下进行分析比较,阿替匹林 C 标准物质色谱图和含有阿替匹林 C 的蜂胶样品色谱图参见附录 B,根据保留时间定性。

8.3.2 定量测定

取 5μL 试样溶液(8.1),在上述色谱条件(8.2)下进行分析,根据标准工作液(5.7)绘制的标准工作曲线对样品进行定量。

8.3.3 平行试验

按上述步骤,对同一试样进行平行试验测定。

8.3.4 空白试验

除不称取试样外,均按上述步骤,应不干扰测定。

9 结果计算与表示

试样中阿替匹林 C 含量(X)以克每百克(g/100g)表示,按式(1)计算:

$$X = \frac{C \times V}{m \times 100} \times 100 \tag{1}$$

式中:C——标准工作曲线上计算得到的样液阿替匹林 C 的浓度,单位为毫克每毫升(mg/mL);

V——试样最终定容体积,单位为毫升(mL);

m——样品质量,单位为克(g)。

将符合重复性要求的两个独立测定值的算术平均值作为测定结果(g/100g),保留三位有效数字。

10 精密度

在对同一试样进行两次平行试验获得的两次独立测定结果的绝对差值不应超过算术平均值的 10%。

附录 A
（规范性附录）
高效液相色谱梯度洗脱条件

表 A.1 高效液相色谱法梯度洗脱条件

时间 min	流速 mL/min	甲醇 %	1%乙酸 %
0	1	25	75
30	1	55	45
60	1	80	20
70	1	95	5
80	1	25	75

附录 B
（资料性附录）
阿替匹林 C 标准品、酒神菊属型蜂胶、杨属型蜂胶的色谱图

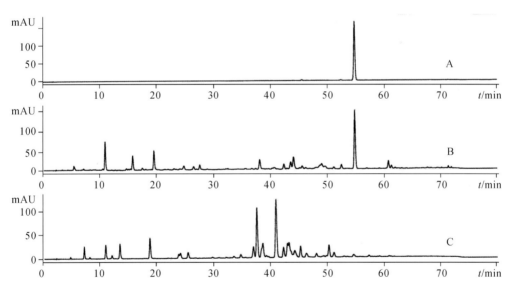

说明：A——阿替匹林 C 标准品；B——酒神菊属型蜂胶；C——杨属型蜂胶。

图 B.1 阿替匹林 C 标准品、酒神菊属型蜂胶、杨属型蜂胶的色谱图

ICS

中华人民共和国供销合作行业标准

××/T ×××××—××××

蜂胶中咖啡酸、p-香豆酸、阿魏酸、短叶松素、松属素、短叶松素 3-乙酸酯、白杨素和高良姜素含量的测定 反相高效液相色谱法

Determination of cafferic acid, p-coumaric acid, ferulic acid, pinobanksin, pinocembrin, 3-acetylpinobanksin, chrysin and galangin in propolis- Reversed-phase high performance liquid chromatography method

（报批稿）

前　言

　　本标准按照 GB/T 1.1—2009《标准化工作导则 第 1 部分:标准的结构和编写》的编写规则。

　　本标准由浙江大学提出。

　　本标准由中华全国供销合作总社归口。

　　本标准起草单位:浙江大学、杭州蜂之语蜂业股份有限公司、杭州天厨蜜源保健品有限公司、缙云县绿纯养蜂专业合作社。

　　本标准主要起草人:胡福良、张翠平、周萍、王磊、胡元强。

蜂胶中咖啡酸、p-香豆酸、阿魏酸、短叶松素、松属素、短叶松素 3-乙酸酯、白杨素和高良姜素含量的测定 反相高效液相色谱法

1 范围

本标准规定了蜂胶中咖啡酸、p-香豆酸、阿魏酸、短叶松素、松属素、短叶松素 3-乙酸酯、白杨素和高良姜素的反相高效液相色谱检测方法。

本标准适用于蜂胶及蜂胶乙醇提取物中咖啡酸、p-香豆酸、阿魏酸、短叶松素、松属素、短叶松素-3-O-乙酸酯、白杨素和高良姜素的测定。

本标准液相色谱的检出限:咖啡酸、p-香豆酸、阿魏酸、短叶松素、松属素、短叶松素 3-乙酸酯、白杨素和高良姜素分别为 0.41、0.39、1.20、0.29、1.11、1.49、0.86、0.91μg/mL。

本标准液相色谱的定量限:咖啡酸、p-香豆酸、阿魏酸、短叶松素、松属素、短叶松素 3-乙酸酯、白杨素和高良姜素分别为 1.35、2.16、4.00、0.98、3.69、4.97、2.88、3.03μg/mL。

2 规范性引用文件

下列文件对于本文件的应用是必不可少的。凡是注日期的引用文件,仅所注日期的版本适用于本文件。凡是不注日期的引用文件,其最新版本(包括所有的修改单)适用于本文件。

GB/T 6682 分析实验室用水规格和试验方法

3 术语和定义

下列术语和定义适用于本标准。

3.1 蜂胶 propolis

工蜂采集树脂等植物分泌物与其上颚腺、蜡腺分泌物等混合形成的胶黏性物质。

4 原理

试样经乙醇提取,浓缩,再溶解于甲醇,用反相高效液相色谱仪紫外检测器在 280nm 处检测,外标法定量。

5 试剂与材料

实验室用水应符合 GB/T 6682 规定的一级水。
除非另有规定,所有试剂均为分析纯。

5.1 甲醇:色谱纯。

5.2 乙醇。

5.3 乙酸,色谱纯。

5.4 流动相:甲醇(5.1)和1%乙酸(5.3)梯度洗脱。

5.5 咖啡酸、p-香豆酸、阿魏酸、短叶松素、松属素、短叶松素3-乙酸酯、白杨素和高良姜素标准品:纯度≥98%

5.6 咖啡酸、p-香豆酸、阿魏酸、短叶松素、松属素、短叶松素3-乙酸酯、白杨素和高良姜素标准储备液:准确称取适量的每种标准物质(5.5),用甲醇溶液(5.1)配成1.0mg/mL的标准储备溶液,于4℃冰箱保存,可保存6个月。

5.7 咖啡酸、p-香豆酸、阿魏酸、短叶松素、松属素、短叶松素3-乙酸酯、白杨素和高良姜素标准工作液:吸取适量1.0mg/mL的标准贮备液(5.6),用甲醇溶液(5.1)分别稀释成适当浓度的标准工作溶液,现用现配。

5.8 滤膜:0.45μm,有机系。

6 仪器和设备

6.1 高效液相色谱仪:配有紫外检测器。

6.2 粉碎机。

6.3 旋转蒸发仪。

6.4 超声波清洗器。

7 试样准备与保存

7.1 试样的准备

取代表性试样适量,冷冻,粉碎,混匀,作为备检试样。

7.2 试样保存

将试样于0℃~4℃保存。

8 操作步骤

8.1 蜂胶样品预处理

准确称取10.0g(精确到0.1g)蜂胶试样,用30mL 95%乙醇超声提取,每次用时15min,共提取3次,合并滤液,定容至100mL(V_1)。取10mL(V_2)减压蒸干,用25mL(V_3)甲醇超声溶解并定容,0.45μm滤膜过滤,供测定。

8.2 色谱条件

色谱柱:Sepax HP-C18,柱长150mm,内径4.6mm,粒度5μm;或性能相当的色谱柱。

流动相流速:1.0mL/min。

梯度洗脱:洗脱程序参见附录A表A.1。

检测波长:280nm。

柱温:33℃。

进样量:5μL。

8.3 测定

取 $5\mu L$ 样品溶液(8.1)、标准工作溶液(5.7)分别进样分析，绘制标准曲线，外标法定量。色谱图参见附录A图A.1。

9 结果计算与表示

9.1 结果计算

试样中咖啡酸、p-香豆酸、阿魏酸、短叶松素、松属素、短叶松素3-乙酸酯、白杨素和高良姜素含量(X)以毫克每克(mg/g)表示，按式(1)计算：

$$X = \frac{C \times V_1 V_3}{m \times V_2} \tag{1}$$

式中：C_S——标准工作曲线上计算得到的样液中化合物的浓度，单位为毫克每毫升(mg/mL)；
V_1——试样溶液定容体积，单位为毫升(mL)；
V_2——从试样溶液中取出的待测体积(mL)；
V_3——待测试样最终定容体积，单位为毫升(mL)；
m——样品质量，单位为克(g)。

9.2 结果表示

将符合重复性要求的两个独立测定值的算术平均值作为测定结果(mg/g)，保留两位有效数字。平行试验相对偏差不得超过2.0%。

附录A
（资料性附录）
高效液相色谱法梯度洗脱程序和色谱图

表A.1 高效液相色谱法梯度洗脱程序

时间 min	流速 mL/min	甲醇 %	1%乙酸 %
0	1	15	85
30	1	40	60
65	1	55	45
70	1	62	38
80	1	100	0
90	1	15	85

图A.1 咖啡酸、p-香豆酸、阿魏酸、短叶松素、松属素、短叶松素3-乙酸酯、白杨素和高良姜素混合标准品和蜂胶的色谱图

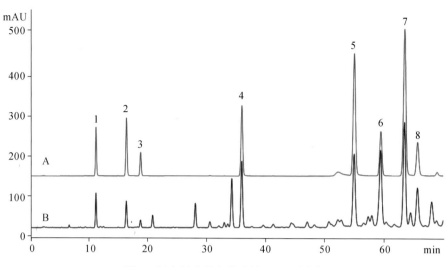

图 1 混合标准品和蜂胶的 HPLC 图谱

A:混合标准品;B:蜂胶

1.咖啡酸、2.p-香豆酸、3.阿魏酸、4.短叶松素、5.松属素、6.短叶松素-3-乙酸酯、7.白杨素、8.高良姜素